Worldwide Guide to Equivalent Nonferrous Metals and Alloys

Third Edition

William C. Mack
Editor

Contributors and Reviewers:

George C. Hsu
Reynolds Metals Company

Paul Kolisnyk
Cominco

Richard F. Lynch, Ph.D.
Lynch & Associates, Inc.

Kurt Moser
Cabot Performance Metals

Lee Ann Olsick
Fry Metals

Rolf Schneider
Siemens Stromberg Carlson

Jack Tosdale
Teledyne Wah Chang Albany

Acquisitions:
Sandra Dunigan

Editorial Coordination:
Fran Cverna

Editorial Assistance:
Jan B. Horesh

Production Coordination:
Randall L. Boring

ASM INTERNATIONAL

The Materials Information Society

Table of Contents

Introduction to the Third Edition

As Editor of the third edition of the *Worldwide Guide to Equivalent Nonferrous Metals and Alloys*, I am proud to present this user-friendly book. We have organized the material to permit users to find similar materials more easily. In addition, we have not eliminated obsolete Designations or Specifications present in th second edition, but rather identified them as "obsolete" as a help to users. The user looking for an obsolete Designation or Specification may be able to locate the composition and find similar materials. Note that there are many standards with historical value that were never included in the first or second editions of the Worldwide Guides. Please inform ASM about any obsolete Designations or Specifications that you feel should be included, and we will try to incorporate them into future editions.

Sections 2 through 12 are, as in the second edition, organized by material group. In this edition, however, alloys within a material group are organized by decreasing concentration of the principal element. This reorganization is not perfect. For each alloy, except those for which the principal element had a specified minimum, we subtracted all of the maximum specified elements from the principal element to obtain a concentration value for the principal element. For concentrations listed with only one value, we choose not to judge if the value was an average, typical or other type of value. Sorting was influenced by the available data, and the user is cautioned to take this into account. For example, an alloy with many restrictions on residual elements will contain less of the principal element than the same alloy in a specification that does not include residual element restrictions.

As was noted in the first edition, the ultimate selection of a given material rests with the reader. Individual requirements relating to product form, heat treatment, cost, and end use may, and often do, vary considerably even for an apparently identical alloy composition. The word "equivalent" as used in the title of this publication, suggests "similar," not necessarily identical. Both ASM and I recommend that a knowledgeable engineer be consulted before any material substitution be considered.

Each line entry in Sections 2 through 12 contains one or more of the following pieces of information:

- Alloy Grade, or common name
- UNS #, the number assigned by the Unified Numbering System jointly published by the Society of Automotive Engineers and the American Society for Testing and Materials
- Country in which the Specification was published
- Specification as an acronym or abbreviation for a standards organization plus the specific document or standard issued by that organization (see also Section 14)
- Designation, typically an alloy identification as it is used in the specification document
- Description including product forms(s), heat treatment, and other descriptive terms (abbreviations are defined in Section 13)
- Chemical composition given in conventional weight percent for each of the chemical elements; in some cases trailing zeroes have unfortunately been dropped

- Tensile properties expressed as MPa for UTS and YS (1 MPa = 0.1450377 ksi), and as percentage for Elongation.

Section 13 lists the abbreviations used in the Description columns of the preceding Sections. Section 14 lists standard organizations by country, along with address, phone, facsimile, WorldWideWeb and E-Mail address if available. Sections 15 and 16 are the alphabetic indices to the alloy entries in Sections 2 through 12.

As certain as I am that you will find this book very helpful in identifying similar alloys, I am equally certain that improvements can be made. The world is changing. ASTM, CEN, and ISO, to name a few standards organizations, are working together to harmonize standards. As always, change is the only constant. Change seems to be constant today, and changes will be needed for tomorrow.

I repeat my request for input concerning obsolete Specifications and any other suggestions. Please send comments to the Center for Materials Data, ASM International, Materials Park, OH 44073 USA (WWW.ASM-INTL.ORG).

William C. Mack

Wadsworth, Ohio
June 1996

Grade UNS #	Country	Specification	Designation	Description	Cr	Cu	Fe	Mg	Mn	Si	Ti	Zn	OE Max	OT Max	Other	UTS	YS	EL
1199	USA					0.006 max	0.006 max	0.006 max	0.002 max	0.006 max	0.002 max	0.006 max	0.002		Al 99.99 min; Ga 0.005 max; V 0.005 max; Zr 0.05			
1199 A91199	Russia	GOST 11069	A99			0.003 max	0.003 max			0.003 max	0.002 max				Al 99.99 min			
1199 A91199	Russia		AV000			0.005 max	0.003 max			0.002 max				0.01	Al 99.99 min			
	Japan	JIS H4170	IN99				0.004								Al 99.99			
	India	IS 739	G1	Wir Ann and CW to hard											Al 99.99	93		
	UK	BS 1470	S1	Plt Sh Strp Ann .2/6 mm diam OBSOLETE Replaced by BS EN 515											Al 99.99	65		30
	UK	BS 1470	S1	Plt Sh Strp SH to 1/2 hard .2/6 mm diam OBSOLETE Replaced by BS EN 515											Al 99.99	80		7
	UK	BS 1470	S1	Plt Sh Strp SH to hard .2/6 mm diam OBSOLETE Replaced by BS EN 515											Al 99.99	100		3
	Austria	ONORM M3426	Al99.98			0.003 max	0.006 max			0.01 max	0.003 max	0.01 max			ET 0.003 max			
	Germany	DIN 1712 pt 3	3.0385/Al99.98R	Obsolete Replaced by DIN EN 10210		0.003 max	0.006 max			0.01 max	0.003 max	0.01 max			ET 0.003 max; Al 99.98 min			
1098	Germany					0.003 max	0.006 max			0.01 max	0.003 max	0.015 max	0.003		Al 99.98 min			
1198	France					0.006 max	0.006 max		0.006 max	0.01 max	0.006 max	0.1 max	0.003		Al 99.98 min; Ga 0.06 max			
1193	USA					0.006 max	0.04 max	0.01 max	0.01 max	0.04 max	0.01 max	0.03 max	0.01		Al 99.93 min; Ga 0.03 max; V 0.05 max			
1188 A91188	Russia		AV1			0.005 max	0.06 max			0.06 max				0.1	Al 99.9 min			
3102	USA					0.1 max	0.7 max		0.05-0.4	0.4 max	0.1 max	0.3 max	0.05	0.15				
1190	France				0.01 max	0.01 max	0.07 max	0.01 max	0.01 max	0.05 max		0.02 max	0.01		0.01 V+Ti; B 0.01 max; Al 99.9 min; B 0.01 max; Ga 0.02 max			
	Austria	ONORM M3426	Al99.9			0.01 max	0.04 max			0.06 max	0.006 max	0.04 max			ET 0.01 max; Ga 0.03 max			
	Germany	DIN 1712 pt 3	3.0305/Al99.9	Obsolete Replaced by DIN EN 10210		0.01 max	0.05 max	0.01 max	0.01 max	0.06 max	0.006 max	0.04 max			ET 0.01 max; Al 99.9 min			
1090	USA					0.02 max	0.07 max	0.01 max	0.01 max	0.07 max	0.01 max	0.03 max	0.01		Al 99.9 min; Ga 0.03 max; V 0.05 max			
	Germany	DIN 1712 pt 1	3.0256/E-AlH	Ingot		0.02 max	0.35 max			0.2 max		0.04 max			ST 0.03 max; ET 0.03 max; ST = Cr + Mn + Ti + V			
	Germany	DIN 1712 pt 3	3.0257/E-Al	Obsolete Replaced by DIN EN 10210		0.02 max	0.4 max	0.05 max		0.25 max		0.05 max			ET 0.03 max; ST 0.03 max; ST = Cr+Mn+Ti+V			
	Austria	ONORM M3426	E-Al			0.05 max	0.4 max			0.25 max		0.07 max			Cr+Mn+Ti 0.03 x each 0.03 x total			

UNS numbers and US grades are provided as a means of cross referencing chemically similar alloys. Exchangability is only possible after independent examination of specifications. Tensile properties are minimum or typical . UTS and YS as Mpa, El as %. See Appendix for list of abbreviations used in Descriptions.

Worldwide Guide to Equivalent Nonferrous Metals and Alloys

Grade UNS #	Country	Specification	Designation	Description	Cr	Cu	Fe	Mg	Mn	Si	Ti	Zn	OE Max	OT Max	Other	UTS	YS	EL
3002	USA					0.15 max	0.1 max	0.05-0.2	0.05-0.25	0.08 max	0.03 max	0.05 max	0.03	0.1	V 0.05 max			
8040	USA					0.2 max			0.05 max			0.2 max	0.03	0.15	1.0 Si+Fe; Zr 0.10-0.30			
	Japan	JIS H4170	IN90				0.03								Al 99.9			
5056A	Europe			Redesignated 5019														
	Europe		AlMg5															
A91188	USA	AWS	ER1188	Weld El Rod		0.005 max	0.06 max	0.01 max	0.01 max	0.06 max	0.01 max	0.03 max	0.01		Al 99.88 min; Be 0.0008 max; Ga 0.03 max; V 0.05 max			
A91188	USA	AWS	R1188	Weld El Rod		0.005 max	0.06 max	0.01 max	0.01 max	0.06 max	0.01 max	0.03 max	0.01		Al 99.88 min; Be 0.0008 max; Ga 0.03 max; V 0.05 max			
1188	USA					0.005 max	0.06 max	0.01 max	0.01 max	0.06 max	0.01 max	0.03 max	0.01		0.0008 max Be for weld; Al 99.88 min; Ga 0.03 max; V 0.05 max			
1285 A91285	Russia		AV2			0.008 max	0.1 max			0.08 max				0.15	Al 99.85 min			
1185	USA					0.01 max		0.02 max	0.02 max		0.02 max	0.03 max	0.01		0.15 Si+Fe; Al 99.85 min; Ga 0.03 max; V 0.05 max			
1385	France				0.01 max	0.02 max	0.12 max	0.02 max	0.01 max	0.05 max		0.03 max	0.01		0.03 V+Ti; Al 99.85 min; B 0.02 max; Ga 0.03 max			
1285	USA					0.02 max	0.08 max	0.01 max	0.01 max	0.08 max	0.02 max	0.03 max	0.01		0.14 Si+Fe; Al 99.85 min; Ga 0.03 max; V 0.05 max			
1085	USA					0.03 max	0.12 max	0.02 max	0.02 max	0.1 max	0.02 max	0.03 max	0.01		Al 99.85 min; Ga 0.03 max; V 0.05 max			
1180	USA					0.01 max	0.09 max	0.02 max	0.02 max	0.09 max	0.02 max	0.03 max	0.02		Al 99.8; Ga 0.03 max; V 0.05 max			
1080 A91080	Czech Republic	CSN 424002	Al99,8			0.02 max	0.15 max			0.15 max	0.03 max	0.05 max			Al 99.8 min			
	India	IS 738	T1A	Wir Ann		0.02	0.15		0.03	0.15		0.06			Al 99.8	79		
	India	IS 738	T1A	Wir Ann and CW to hard 80 mm diam		0.02	0.15		0.03	0.15		0.06			Al 99.8	98		
	India	IS 739	G1A	Wir Ann		0.02	0.15		0.03	0.15		0.06			Al 99.8	79		
	India	IS 739	G1A	Wir Ann and CW to hard		0.02	0.15		0.03	0.15		0.06			Al 99.8	123		
1080 A91080	UK	BS 1470	1080(1A)	Sh Plt Strp OBSOLETE Replaced by BS EN 515		0.02 max	0.15 max		0.03 max	0.15 max		0.06 max			Al 99.8 min			
	UK	BS 1470	S1A	Plt Sh Strp Ann .2/6 mm OBS Replaced by BS EN 515		0.02	0.15		0.03	0.15		0.06			Al 99.8	90		29
	UK	BS 1470	S1A	Plt Sh Strp SH to hard .2/3 mm OBS Replaced by BS EN 515		0.02	0.15		0.03	0.15		0.06			Al 99.8	125		3

UNS numbers and US grades are provided as a means of cross referencing chemically similar alloys. Exchangability is only possible after independent examination of specifications. Tensile properties are minimum or typical . UTS and YS as Mpa, El as %. See Appendix for list of abbreviations used in Descriptions.

Grade UNS #	Country	Specification	Designation	Description	Cr	Cu	Fe	Mg	Mn	Si	Ti	Zn	OE Max	OT Max	Other	UTS	YS	EL
	UK	BS 1475	G1A	Wir Ann 10 mm diam		0.02	0.15		0.03	0.15		0.06			Al 99.8	90		
	UK	BS 1475	G1A	Wir As Mfg 150 mm diam		0.02	0.15		0.03	0.15		0.06			Al 99.8	125		
	Yugoslavia	JUS CC2.100	Al99.8			0.02 max	0.15 max	0.06 max	0.03 max	0.15 max	0.03 max				ST 0.08 max; ET 0.01 max; ST = Zn + Cu			
	Yugoslavia	JUS CC2.100	AlMg0.5			0.02 max	0.15 max	0.2-0.6		0.15 max								
3010	USA				0.4 max	0.03 max	0.2 max		0.2-0.9	0.1 max	0.05 max	0.05 max	0.03	0.1	B 0.05 max; V 0.05 max			
8020	USA					0.005 max	0.1 max		0.005 max	0.1 max		0.005 max	0.03	0.1	Bi 0.10-0.50; Sn 0.10-0.25; V 0.05 max			
1080 A91080	Australia	AS 1734	1080A	Sh Plt, Ann 0.15/0.50mm thk		0.03	0.15	0.02	0.02	0.15	0.02	0.06			Al 99.8			15
1080 A91080	Australia	AS 1734	1080A	Sh Plt, SH to 1/2 hard 0.25/0.30mm Thk		0.03	0.15	0.02	0.02	0.15	0.02	0.06			Al 99.8			1
1080 A91080	Australia	AS 1734	1080A	Sh Plt, Stain hardened to full hard 0.15/0.50mm thk		0.03	0.15	0.02	0.02	0.15	0.02	0.06			Al 99.8			1
1080 A9108C	Australia	AS 1866		Bar Rod, SH		0.03	0.15	0.02	0.02	0.15	0.02	0.06			Al 99.8	53		28
1080A	Europe					0.03 max	0.15 max	0.02 max	0.02 max	0.15 max	0.02 max	0.06 max	0.02		Al 99.8; Ga 0.03 max			
	Europe		Al99.8(A)			0.03 max	0.15 max	0.02 max	0.02 max	0.15 max	0.02 max	0.06 max	0.02		Al 99.8; Ga 0.03 max			
1080 A91080	Finland	SFS 2580	Al99.8	HF		0.03	0.15		0.03	0.15		0.06			Al 99.8			
1080 A91080	France	NF A50411	1080A(A8)	Bar Wir 1/2 hard (drawn bar) 50 mm diam		0.03	0.15 max	0.02	0.02	0.15	0.02	0.06			Al 99.8; Ga 0.03 max	100	70	6
1080 A91080	France	NF A50411	1080A(A8)	Bar Wir Hard (wir) 6 mm diam		0.03	0.15 max	0.02	0.02	0.15	0.02	0.06			Al 99.8; Ga 0.03 max	140	110	4
1080 A91080	France	NF A50411	1080A(A8)	Bar Wir 1/4 hard (tub) 50 mm diam		0.03	0.15 max	0.02	0.02	0.15	0.02	0.06			Al 99.8; Ga 0.03 max	85	60	7
1080 A91080	France	NF A50451	1080A(A8)	Sh Plt Ann .4/3.2 mm diam		0.03	0.15 max	0.02	0.02	0.15	0.02	0.06			Al 99.8	60		38
1080 A91080	France	NF A50451	1080A(A8)	Sh Plt 1/4 hard .4/3.2 mm diam		0.03	0.15 max	0.02	0.02	0.15	0.02	0.06			Al 99.8	80	55	12
1080 A91080	France	NF A50451	1080A(A8)	Sh Plt 1/2 hard .4/1.6 mm diam		0.03	0.15 max	0.02	0.02	0.15	0.02	0.06			Al 99.8	100	70	7
1080 A91080	Germany	DIN 1712 pt 3	3.0285/Al99.8	Obsolete Replaced by DIN EN 10210		0.03 max	0.15 max	0.02 max	0.02 max	0.15 max	0.02 max	0.06 max			ET 0.02 max; Al 99.8 min; Ga 0.03 max			
	India	IS 736	19800	Plt As Mfg 30 mm diam		0.03	0.15		0.03	0.15		0.06			Al 99.8			
	India	IS 736	19800	Plt Ann 34 mm diam		0.03	0.15		0.03	0.15		0.06			Al 99.8			
	India	IS 736	19800	Plt 1/2 hard 8 mm diam		0.03	0.15		0.03	0.15		0.06			Al 99.8			
1080 A91080	International	ISO R827	Al99.8	As Mfg		0.03	0.15		0.03	0.15		0.06			Al 99.8	49		22
1080 A91080	International	ISO TR2136	Al99.8	Sh Plt Ann		0.03	0.15		0.03	0.15		0.06			Al 99.8	90		25
1080 A91080	International	ISO TR2778	Al99.8	Tub Ann OBSOLETE Replace by ISO 6363		0.03	0.15		0.03	0.15		0.06			Al 99.8	50		29
1080 A91080	International	ISO TR2778	Al99.8	Tub SH OBSOLETE Replace by ISO 6363		0.03	0.15		0.03	0.15		0.06			Al 99.8	75		4

UNS numbers and US grades are provided as a means of cross referencing chemically similar alloys. Exchangability is only possible after independent examination of specifications. Tensile properties are minimum or typical . UTS and YS as Mpa, El as %. See Appendix for list of abbreviations used in Descriptions.

Worldwide Guide to Equivalent Nonferrous Metals and Alloys

Grade UNS #	Country	Specification	Designation	Description	Cr	Cu	Fe	Mg	Mn	Si	Ti	Zn	OE Max	OT Max	Other	UTS	YS	EL
1080 A91080	Pan American	COPANT 862	1080			0.03	0.15	0.02	0.02	0.15	0.03	0.03			Al 99.8			
1080	USA					0.03 max	0.15 max	0.02 max	0.02 max	0.15 max	0.03 max	0.03 max	0.02		Al 99.8; Ga 0.03 max; V 0.05 max			
1080 A91080	Japan	JIS H4000	1080	Sh Plt Ann 1.3 mm diam		0.03	0.16	0.02	0.02	0.15	0.03	0.03			Al 99.8	59	20	30
1080 A91080	Japan	JIS H4000	1080	Sh Plt Hard .3 mm diam		0.03	0.16	0.02	0.02	0.15	0.03	0.03			Al 99.8	118		1
1080 A91080	Japan	JIS H4000	1080	Sh Plt HR 12 mm diam		0.03	0.16	0.02	0.02	0.15	0.03	0.03			Al 99.8	69	39	15
8077	USA					0.05 max	0.1-0.4	0.1-0.3		0.1 max		0.05 max	0.03	0.1	B 0.05; Zr 0.02-0.08			
	International	ISO R209 (1971)	Al 99.8			0.03 max	0.15 max		0.03 max	0.15 max		0.06 max			Cu+Si+Fe+Mn+Zn 0.2 max			
1080 A91080	Sweden	MNC 40E	4004	Sh Tub											Al 99.8 min			
1350 A91350	UK		E57S					0.4-0.6							Al 99.8 min			
5457 A95457	Australia	AS 1734	5457	Sh Plt Ann 0.50/0.80 mm thk		0.2	0.1	0.8-1.2	0.15-0.45	0.08		0.05				110		18
5457 A95457	Australia	AS 1734	5457	Sh Plt SH and partially Ann to 5/8 hard 0.50/0/.80 mm thk		0.2	0.1	0.8-1.2	0.15-0.45	0.08		0.05				158		4
5457 A95457	Pan American	COPANT 862	5457			0.2	-0.1	0.8-1.2	0.15-0.45	0.08		0.05			V 0.05			
5457 A95457	USA	ASTM B209	5457	Sh		0.2 max	0.1 max	0.08-1.2	0.15-0.45	0.08 max		0.05 max	0.03	0.1	V 0.05 max			
1075 A91075	Pan American	COPANT 862	1075			0.04	0.2	0.03	0.03	0.2	0.03	0.04			Al 99.75			
1275	Spain					0.05-0.1	0.12 max	0.02 max	0.02 max	0.08 max	0.02 max	0.03 max	0.01		Al 99.75 min; Ga 0.03 max; V 0.03 max			
1175	USA					0.1 max		0.02 max	0.02 max		0.02 max	0.04 max	0.02		0.15 Si+Fe; Al 99.75 min; Ga 0.03 max; V 0.05 max			
8177	USA					0.04 max	0.25-0.45	0.04-0.12		0.1 max		0.05 max	0.03	0.1	B 0.04			
1370	France				0.01 max	0.02 max	0.25 max	0.02 max	0.01 max	0.1 max		0.04 max	0.02	0.1	0.02 V+Ti; Al 99.7 min; B 0.02; Ga 0.03 max			
8004	France					0.03 max	0.15 max	0.02 max	0.02 max	0.15 max	0.3-0.7	0.03 max	0.03	0.15				
1070 A91070	Austria	ONORM M3426	Al99.7			0.03 max	0.25 max	0.03 max	0.03 max	0.2 max	0.03 max	0.07 max			ET 0.03 max			
1070 A91070	Czech Republic	CSN 424003	Al99,7			0.03 max	0.25 max	0.03 max	0.03 max	0.2 max	0.03 max	0.07 max			Al 99.7 min			
1070A	Europe					0.03 max	0.25 max	0.03 max	0.03 max	0.2 max	0.03 max	0.07 max	0.03		Al 99.7 min			
	Europe		Al 99.7			0.03 max	0.25 max	0.03 max	0.03 max	0.2 max	0.03 max	0.07 max	0.03		Al 99.7 min			
1070 A91070	Finland	SFS 2581	Al99.7	HF (shapes)		0.03	0.25		0.03	0.2		0.07			Al 99.7			
1070 A91070	Finland	SFS 2581	Al99.7	Ann (sht)		0.03	0.25		0.03	0.2		0.07			Al 99.7			
1070 A91070	Finland	SFS 2581	Al99.7	Hard (sht) 3 mm diam		0.03	0.25		0.03	0.2		0.07			Al 99.7			
1070 A91070	France	NF A50451	1070A(A7)	Sh Plt Ann .4/3.2 mm diam		0.03	0.25 max	0.03	0.03	0.2	0.03	0.07			Al 99.7	60		38

UNS numbers and US grades are provided as a means of cross referencing chemically similar alloys. Exchangability is only possible after independent examination of specifications. Tensile properties are minimum or typical . UTS and YS as Mpa, El as %. See Appendix for list of abbreviations used in Descriptions.

2-4 Wrought Aluminum

Grade UNS #	Country	Specification	Designation	Description	Cr	Cu	Fe	Mg	Mn	Si	Ti	Zn	OE Max	OT Max	Other	UTS	YS	EL
1070 A91070	France	NF A50451	1070A(A7)	Sh Plt 1/4 hard .4/3.2 mm diam		0.03	0.25 max	0.03	0.03	0.2	0.03	0.07			Al 99.7	80	55	12
1070 A91070	France	NF A50451	1070A(A7)	Sh Plt 1/2 hard .4/1.6 mm diam		0.03	0.25 max	0.03	0.03	0.2	0.03	0.07			Al 99.7	100	70	7
1070 A91070	Germany	DIN 1712 pt 3	3.0275/Al99.7	Obsolete Replaced by DIN EN 10210		0.03 max	0.25 max	0.03 max	0.03 max	0.2 max	0.03 max	0.07 max			ET 0.03 max; Al 99.7 min			
	India	IS 736	19700	Plt As Mfg 30 mm diam		0.03	0.25		0.03	0.2		0.06			Al 99.7	55		30
	India	IS 736	19700	Plt Ann 34 mm diam		0.03	0.25		0.03	0.2		0.06			Al 99.7	90		34
	India	IS 736	19700	Plt 1/2 hard 8 mm diam		0.03	0.25		0.03	0.2		0.06			Al 99.7	85		8
1070 A91070	International	ISO R827	Al99.7	As Mfg		0.03	0.25		0.03	0.2		0.07			Al 99.7	49		22
1070 A91070	International	ISO TR2136	Al99.7	Sh Plt Ann		0.03	0.25		0.03	0.2		0.07			Al 99.7	95		25
1070 A91070	Norway	NS 17015	NS17015/Al99.7	Plt Sh Strp		0.03 max	0.25 max	0.03 max	0.03 max	0.2 max	0.03 max	0.07 max			ET 0.03 max; Al 99.7 max			
	Yugoslavia	JUS CC2.100	Al99.7			0.03 max	0.25 max		0.07 max	0.2 max	0.03 max				ST 0.1 max; ET 0.02 max; ST = Zn+Cu, Si < Fe			
1170	USA				0.03 max	0.03 max		0.02 max	0.03 max		0.03 max	0.04 max	0.03		0.30 Si+Fe; Al 99.7 min; V 0.05 max			
1070 A91070	Australia	AS 1734	1070			0.04 max	0.25 max	0.03 max	0.03 max	0.2 max	0.03 max	0.04 max	0.03		Al 99.7 min; V 0.05 max			
1070 A91070	Denmark	DS 3012	1070	1/2 hard (rolled and drawn products) .5/6 mm diam		0.04	0.25	0.03	0.03	0.2	0.03	0.04			Al 99.7	95	75	6
1070 A91070	Denmark	DS 3012	1070	Hard (rolled and drawn products) .5/6 mm diam		0.04	0.25	0.03	0.03	0.2	0.03	0.04			Al 99.7	130	120	4
1070 A91070	Denmark	DS 3012	1070	As manufactured (extd products)		0.04	0.25	0.03	0.03	0.2	0.03	0.04			Al 99.7	65		18
1070 A91070	Japan	JIS H4000	1070	Sh Plt Ann 1.3 mm diam		0.04	0.25	0.03	0.03	0.2	0.03	0.04			Al 99.7	59		30
1070 A91070	Japan	JIS H4000	1070	Sh Plt Hard .3 mm diam		0.04	0.25	0.03	0.03	0.2	0.03	0.04			Al 99.7	118		1
1070 A91070	Japan	JIS H4000	1070	Sh Plt HR 12 mm diam		0.04	0.25	0.03	0.03	0.2	0.03	0.04			Al 99.7	69	39	15
1070 A91070	Japan	JIS H4040	1070	Bar As Mfg		0.04	0.25	0.03	0.03	0.2	0.03	0.04			Al 99.7	49	20	
1070 A91070	Japan	JIS H4080	1070	Tub As Mfg		0.04	0.25	0.03	0.03	0.2	0.03	0.04			Al 99.7	49	20	
1070 A91070	Japan	JIS H4120	1070	Wir 1/2 hard 25 mm diam		0.04	0.25	0.03	0.03	0.2	0.03	0.04			Al 99.7	76		
1070 A91070	Japan	JIS H4160	1070			0.04	0.25	0.03	0.03	0.2	0.03	0.04			Al 99.7			
1070 A91070	Japan	JIS Z3232	1070	Rod Ann		0.04	0.25	0.03	0.03	0.2	0.03	0.04			Al 99.7; Be 0.001	59		
1070 A91070	Pan American	COPANT 862	1070			0.04	0.25	0.03	0.03	0.2	0.03	0.04			Al 99.7			
1070	USA					0.04 max	0.25 max	0.03 max	0.03 max	0.2 max	0.03 max	0.04 max	0.03		Al 99.7 min; V 0.05 max			
1070 A91070	Belgium	NBN P21-001	1070			0.25 max	0.25 max	0.03 max	0.04 max	0.2 max	0.03 max	0.04 max		0.03	Al 99.7 min; V 0.05 max			
3207A	Norway				0.2 max	0.25 max	0.6 max	0.4 max	0.3-0.8	0.35		0.25 max	0.05	0.15				
5010	USA				0.15 max	0.25 max	0.7 max	0.2-0.6	0.1-0.3	0.4 max	0.1 max	0.3 max	0.05	0.15				
	International	ISO R209 (1971)	Al 99.7			0.03 max	0.25 max		0.03 max	0.20 max		0.07 max			Cu+Si+Fe+Mn+ Zn 0.3 max			
	Austria	ONORM M3430	Al99.85Mg0.5				0.08 max	0.3-0.6		0.06 max	0.02 max	0.05 max			ET 0.02 max; Mg Ti 0.15 max			

UNS numbers and US grades are provided as a means of cross referencing chemically similar alloys. Exchangability is only possible after independent examination of specifications. Tensile properties are minimum or typical . UTS and YS as Mpa, El as %. See Appendix for list of abbreviations used in Descriptions.

Wrought Aluminum 2-5

Worldwide Guide to Equivalent Nonferrous Metals and Alloys

Grade UNS #	Country	Specification	Designation	Description	Cr	Cu	Fe	Mg	Mn	Si	Ti	Zn	OE Max	OT Max	Other	UTS	YS	EL
	Germany	DIN 1725 pt1	3.3307/Al99.8 5Mg0.5				0.08 max	0.3-0.6	0.03 max	0.08 max	0.02 max	0.05 max			ST 0.15 max; ET 0.02 max			
1070 A91070	Sweden	MNC 40E	4005	Sh Tub											Al 99.7 min			
5110	Europe						0.08 max	0.3-0.6	0.03	0.08 max	0.02 max	0.05 max	0.02					
1065	USA					0.05 max	0.3 max	0.03 max	0.03 max	0.25 max	0.03 max	0.05 max	0.03		Al 99.65 min; V 0.05 max			
3007	USA				0.2 max	0.05-0.3	0.7 max	0.6 max	0.3-0.8	0.5 max	0.1 max	0.4 max	0.05	0.15				
5310	Europe						0.008 max	0.35-0.6		0.01 max	0.008 max	0.01 max	0.003		0.008 Fe+Ti			
	Germany	DIN 1725 pt1	3.3309/AlRMg0.5				0.008 max	0.35-0.6		0.01 max	0.008 max	0.01 max			ET 0.003 max; ST 0.02 max; ST=Fe+Ti			
	Germany	DIN 1725 pt1	3.3308/Al99.9 Mg0.5				0.04 max	0.35-0.6	0.03 max	0.06 max	0.1 max	0.04 max			ST 0.1 max; ET 0.01 max			
5210	Europe						0.04 max	0.35-0.6	0.03	0.06 max	0.01 max	0.04 max	0.01					
	Czech Republic	CSN 424406	AlMg0,5R		0.009 max		0.006 max	0.4-0.6	0.006 max	0.01 max								
1060 A91060	Russia	GOST 11069	A6			0.01 max	0.25 max			0.2 max	0.03 max	0.06 max			Al 99.6 min			
1260	USA					0.04 max		0.03 max	0.01 max		0.03 max	0.05 max	0.03		0.40 Si+Fe; 0.0008 max Be welding; Al 99.6 min; V 0.05 max			
1060 A91060	Japan	JIS H4180	1060	Plt 1/2 hard 3/12 mm diam		0.05	0.35	0.03	0.03	0.25	0.03	0.05			Al 99.6	88	59	6
1060 A91060	Japan	JIS H4180	1060	Plt HR 12 mm diam		0.05	0.35	0.03	0.03	0.25	0.03	0.05			Al 99.6	69	39	15
1060 A91060	Japan	JIS H4180	1060	Plt As Mfg (extd plt) 3/30 mm diam		0.05	0.35	0.03	0.03	0.25	0.03	0.05			Al 99.6	59	29	25
1060 A91060	Pan American	COPANT 862	1060			0.05	0.35	0.03	0.03	0.25	0.03	0.05			Al 99.6			
1060 A91060	USA	AMS 4000	1060	Bar Rod Wir Plt Sh Tub Pip Ext		0.05 max	0.35 max	0.03 max	0.03 max	0.25 max	0.03 max	0.05 max		0.03	Al 99.6 min; V 0.05 max			
1060 A91060	USA	ASME SB209	1060	Bar Rod Wir Plt Sh Tub Pip Ext		0.05 max	0.35 max	0.03 max	0.03 max	0.25 max	0.03 max	0.05 max		0.03	Al 99.6 min; V 0.05 max			
1060 A91060	USA	ASME SB210	1060	Bar Rod Wir Plt Sh Tub Pip Ext		0.05 max	0.35 max	0.03 max	0.03 max	0.25 max	0.03 max	0.05 max		0.03	Al 99.6 min; V 0.05 max			
1060 A91060	USA	ASME SB221	1060	Bar Rod Wir Plt Sh Tub Pip Ext		0.05 max	0.35 max	0.03 max	0.03 max	0.25 max	0.03 max	0.05 max		0.03	Al 99.6 min; V 0.05 max			
1060 A91060	USA	ASME SB234	1060	Bar Rod Wir Plt Sh Tub Pip Ext		0.05 max	0.35 max	0.03 max	0.03 max	0.25 max	0.03 max	0.05 max		0.03	Al 99.6 min; V 0.05 max			
1060 A91060	USA	ASME SB241	1060	Bar Rod Wir Plt Sh Tub Pip Ext		0.05 max	0.35 max	0.03 max	0.03 max	0.25 max	0.03 max	0.05 max		0.03	Al 99.6 min; V 0.05 max			
1060 A91060	USA	ASTM B209	1060	Bar Rod Wir Plt Sh Tub Pip Ext		0.05 max	0.35 max	0.03 max	0.03 max	0.25 max	0.03 max	0.05 max		0.03	Al 99.6 min; V 0.05 max			
1060 A91060	USA	ASTM B210	1060	Bar Rod Wir Plt Sh Tub Pip Ext		0.05 max	0.35 max	0.03 max	0.03 max	0.25 max	0.03 max	0.05 max		0.03	Al 99.6 min; V 0.05 max			
1060 A91060	USA	ASTM B211	1060	Bar Rod Wir Plt Sh Tub Pip Ext		0.05 max	0.35 max	0.03 max	0.03 max	0.25 max	0.03 max	0.05 max		0.03	Al 99.6 min; V 0.05 max			
1060 A91060	USA	ASTM B221	1060	Bar Rod Wir Plt Sh Tub Pip Ext		0.05 max	0.35 max	0.03 max	0.03 max	0.25 max	0.03 max	0.05 max		0.03	Al 99.6 min; V 0.05 max			
1060 A91060	USA	ASTM B234	1060	Bar Rod Wir Plt Sh Tub Pip Ext		0.05 max	0.35 max	0.03 max	0.03 max	0.25 max	0.03 max	0.05 max		0.03	Al 99.6 min; V 0.05 max			
1060 A91060	USA	ASTM B241	1060	Bar Rod Wir Plt Sh Tub Pip Ext		0.05 max	0.35 max	0.03 max	0.03 max	0.25 max	0.03 max	0.05 max		0.03	Al 99.6 min; V 0.05 max			
1060 A91060	USA	ASTM B345	1060	Bar Rod Wir Plt Sh Tub Pip Ext		0.05 max	0.35 max	0.03 max	0.03 max	0.25 max	0.03 max	0.05 max		0.03	Al 99.6 min; V 0.05 max			
1060 A91060	USA	ASTM B404	1060	Bar Rod Wir Plt Sh Tub Pip Ext		0.05 max	0.35 max	0.03 max	0.03 max	0.25 max	0.03 max	0.05 max		0.03	Al 99.6 min; V 0.05 max			
1060 A91060	USA	ASTM B483	1060	Bar Rod Wir Plt Sh Tub Pip Ext		0.05 max	0.35 max	0.03 max	0.03 max	0.25 max	0.03 max	0.05 max		0.03	Al 99.6 min; V 0.05 max			

UNS numbers and US grades are provided as a means of cross referencing chemically similar alloys. Exchangability is only possible after independent examination of specifications. Tensile properties are minimum or typical . UTS and YS as Mpa, El as %. See Appendix for list of abbreviations used in Descriptions.

2-6 Wrought Aluminum

Grade UNS #	Country	Specification	Designation	Description	Cr	Cu	Fe	Mg	Mn	Si	Ti	Zn	OE Max	OT Max	Other	UTS	YS	EL
1060 A91060	USA	SAE J454	1060	Bar Rod Wir Plt Sh Tub Pip Ext		0.05 max	0.35 max	0.03 max	0.03 max	0.25 max	0.03 max	0.05 max		0.03	Al 99.6 min; V 0.05 max			
1060	USA					0.05 max	0.35 max	0.03 max	0.03 max	0.25 max	0.03 max	0.05 max	0.03		Al 99.6 min; V 0.05 max			
	USA		Al 99.6			0.05 max	0.35 max	0.03 max	0.03 max	0.25 max	0.03 max	0.05 max	0.03		Al 99.6 min; V 0.05 max			
3207	Germany					0.1 max	0.45 max	0.1 max	0.4-0.8	0.3 max		0.1 max	0.05	0.1				
2036 A92036	USA				0.1 max	2.2 max	0.5 max	0.3-0.6	0.1-0.4	0.5 max	0.15 max	0.25 max		0.15				
8176	USA						0.4-1			0.03-0.15		0.1 max	0.05	0.15	Ga 0.03 max			
8130	USA					0.05-0.15	0.4-1			0.15 max		0.1 max	0.03	0.1	1.0 max Si+Fe			
3107	Spain					0.05-0.15	0.7 max		0.4-0.9	0.6 max	0.1 max	0.2 max	0.05	0.15				
8030	USA					0.15-0.3	0.3-0.8	0.05 max		0.1 max		0.05 max	0.03	0.1	B 0.001-0.01			
1050 A91050	Russia		Al			0.015 max	0.3 max			0.3 max				0.05	Al 99.5 min			
1350A	Germany					0.02 max	0.4 max	0.05 max		0.25 max		0.05 max	0.03		0.03 Cr+Mn+Ti+V; Al 99.5 min			
8276	France				0.01 max	0.035 max	0.5-0.8	0.02 max	0.01 max	0.25 max		0.05 max	0.03	0.1	0.03 V+Ti; B 0.02 max; Ga 0.03 max			
1350 A91350	UK	BS 1474	1350(1E)	Dar Tub		0.04 max									Al 99.5 min			
1350 A91350	Belgium	NBN P21-001	1350		0.01 max	0.05 max	0.04 max		0.01 max	0.1 max		0.05 max			ET 0.03 max; ST 0.02 max; ST=V+Ti; ET = .10 total; Al 99.5 min; B 0.05 min; Be 0.05 max; Ga 0.03 max			
1350 A91350	Germany	DIN	E-A199.5	Wir Strp											Al 99.5 min			
1350 A91350	Norway	NS 17011	NS17011/E-Al99.5	Wir	0.01 max	0.05 max	0.4 max		0.01 max	0.1 max		0.05 max			ET 0.03 max; ST 0.02 max; ET=.10 total; ST=Vi+Ti; Al 99.5 max; B 0.05 max; Be 0.05 max			
1350 A91350	USA	ASTM B230	1350	Wir Strp	0.01 max	0.05 max	0.4 max		0.01 max	0.1 max		0.05 max		0.1	ST 0.02 max; Al 99.5 min; B 0.05 max; Be 0.05 max; Ga 0.03 max			
1350 A91350	USA	ASTM B231	1350	Wir Strp	0.01 max	0.05 max	0.4 max		0.01 max	0.1 max		0.05 max		0.1	ST 0.02 max; Al 99.5 min; B 0.05 max; Be 0.05 max; Ga 0.03 max			
1350 A91350	USA	ASTM B232	1350	Wir Strp	0.01 max	0.05 max	0.4 max		0.01 max	0.1 max		0.05 max		0.1	ST 0.02 max; Al 99.5 min; B 0.05 max; Be 0.05 max; Ga 0.03 max			
1350 A91350	USA	ASTM B233	1350	Wir Strp	0.01 max	0.05 max	0.4 max		0.01 max	0.1 max		0.05 max		0.1	ST 0.02 max; Al 99.5 min; B 0.05 x; Be 0.05 max; Ga 0.03 max			
1350 A91350	USA	ASTM B236	1350	Wir Strp	0.01 max	0.05 max	0.4 max		0.01 max	0.1 max		0.05 max		0.1	ST 0.02 max; Al 99.5 min; B 0.05 x; Be 0.05 x; Ga 0.03 max			

UNS numbers and US grades are provided as a means of cross referencing chemically similar alloys. Exchangability is only possible after independent examination of specifications. Tensile properties are minimum or typical . UTS and YS as Mpa, El as %. See Appendix for list of abbreviations used in Descriptions.

Worldwide Guide to Equivalent Nonferrous Metals and Alloys

Grade UNS #	Country	Specification	Designation	Description	Cr	Cu	Fe	Mg	Mn	Si	Ti	Zn	OE Max	OT Max	Other	UTS	YS	EL
1350 A91350	USA	ASTM B314	1350	Wir Strp OBSOLETE	0.01 max	0.05 max	0.4 max		0.01 max	0.1 max		0.05 max		0.1	ST 0.02 max; Al 99.5 min; B 0.05 max; Be 0.05 max; Ga 0.03 max			
1350 A91350	USA	ASTM B324	1350	Wir Strp	0.01 max	0.05 max	0.4 max		0.01 max	0.1 max		0.05 max		0.1	ST 0.02 max; Al 99.5 min; B 0.05 max; Be 0.05 max; Ga 0.03 max			
1350 A91350	USA	ASTM B400	1350	Wir Strp	0.01 max	0.05 max	0.4 max		0.01 max	0.1 max		0.05 max		0.1	ST 0.02 max; Al 99.5 min; B 0.05 max; Be 0.05 max; Ga 0.03 max			
1350 A91350	USA	ASTM B401	1350	Wir Strp	0.01 max	0.05 max	0.4 max		0.01 max	0.1 max		0.05 max		0.1	ST 0.02 max; Al 99.5 min; B 0.05 max; Be 0.05 max; Ga 0.03 max			
1350 A91350	USA	ASTM B544	1350	Wir Strp OBSOLETE	0.01 max	0.05 max	0.4 max		0.01 max	0.1 max		0.05 max		0.1	ST 0.02 max; Al 99.5 min; B 0.05 max; Be 0.05 max; Ga 0.03 max			
1350 A91350	USA	ASTM B609	1350	Wir Strp	0.01 max	0.05 max	0.4 max		0.01 max	0.1 max		0.05 max		0.1	ST 0.02 max; Al 99.5 min; B 0.05 max; Be 0.05 max; Ga 0.03 max			
1350	USA				0.01 max	0.05 max	0.4 max		0.01 max	0.1 max		0.05 max	0.03	0.1	0.02 V+Ti; Al 99.5 min; B 0.05; Ga 0.03 max			
1350 A91350	Australia	AS 1866	1350		0.01 max	0.05 max	0.4		0.01 max	0.1 max		0.05 max	0.03		V+Ti 0.02 max; OT min 0.1; Al 99.5 min; B 0.05 max; Ga 0.03 max			
1050 A91050	Europe	AECMA prEN2073	Aluminum1050A	Tub H14 .4/2 mm diam		0.05	0.04	0.05	0.05	0.25	0.05	0.07			Al 99.5	100	80	6
1050 A91050	Australia	AS 1734	1050	Sh Plt Ann 0.15/0.50mm Thk		0.05	0.4	0.05	0.05	0.25	0.03	0.05			Al 99.5			
1050 A91050	Australia	AS 1734	1050	Sh Plt SH to 1/2 hard 0.25/0.30mm thk		0.05	0.4	0.05	0.05	0.25	0.03	0.05			Al 99.5	100		2
1050 A91050	Australia	AS 1734	1050	Sh Plt SH to full hard 0.15/0.50mm thk		0.05	0.4	0.05	0.05	0.25	0.03	0.05			Al 99.5	131		1
1050 A91050	Australia	AS 1866	1050	Rod, Bar, SH		0.05	0.4	0.05	0.05	0.25	0.03	0.05			Al 99.5	62		23
1050 A91050	Australia	AS 1867	1050	Tub, Ann		0.05	0.4	0.05	0.05	0.25	0.03	0.05			Al 99.5			30
1050 A91050	Australia	AS 1867	1050	Tub, SH to 1/4 hard		0.05	0.4	0.05	0.05	0.25	0.03	0.05			Al 99.5	82		
1050 A91050	Australia	AS 1867	1050	Tub, SH to full hard		0.05	0.4	0.05	0.05	0.25	0.03	0.05			Al 99.5	131		
1050 A91050	Austria	ONORM M3426	Al99.5			0.05 max	0.4 max	0.05 max	0.05 max	0.25 max	0.05 max	0.07 max			ET 0.03 max			
1056 A91050	Belgium	NBN P21-001	1050			0.05 max	0.4 max	0.05 max	0.05 max	0.25 max	0.03 max	0.05 max		0.03	Al 99.5 min; V 0.05 max			
1050 A91050	Canada	CSA HA.4	0.1050	Plt Sh O 0.006/0.012 in diam		0.05	0.4	0.05	0.05	0.25	0.03	0.05			Al 99.5	59	21	12

UNS numbers and US grades are provided as a means of cross referencing chemically similar alloys. Exchangability is only possible after independent examination of specifications. Tensile properties are minimum or typical . UTS and YS as Mpa, El as %. See Appendix for list of abbreviations used in Descriptions.

2-8 Wrought Aluminum

Grade UNS #	Country	Specification	Designation	Description	Cr	Cu	Fe	Mg	Mn	Si	Ti	Zn	OE Max	OT Max	Other	UTS	YS	EL
1050 A91050	Canada	CSA HA.4	0.1050	Plt Sh H14 .009/.019 in diam		0.05	0.4	0.05	0.05	0.25	0.03	0.05			Al 99.5	103	90	2
1050 A91050	Canada	CSA HA.4	0.1050	Plt Sh H18 .009/.019 in diam		0.05	0.4	0.05	0.05	0.25	0.03	0.05			Al 99.5	131	103	1
1050 A91050	Canada	CSA HA.7	0.1050	Tub Pip Drawn and O 0.1/.5 in diam		0.05	0.4	0.05	0.05	0.25	0.03	0.05			Al 99.5	97		
1050 A91050	Canada	CSA HA.7	0.1050	Tub Pip Drawn and H18 .01/.5 in diam		0.05	0.4	0.05	0.05	0.25	0.03	0.05			Al 99.5	131		
1050 A91050	Canada	CSA HA.7	0.1050	Tub Pip Ext and O		0.05	0.4	0.05	0.05	0.25	0.03	0.05			Al 99.5	97		
1050 A91050	Canada	CSA HA.7.1	0.1050	Tub Pip Drawn and O 0.1/.5 in diam		0.05	0.4	0.05	0.05	0.25	0.03	0.05			Al 99.5	97		
1050 A91050	Canada	CSA HA.7.1	0.1050	Tub Pip Drawn and H14 .01/.5 in diam		0.05	0.4	0.05	0.05	0.25	0.03	0.05			Al 99.5	103		
1050 A91050	Canada	CSA HA.7.1	0.1050	Tub Pip Ext and O		0.05	0.4	0.05	0.05	0.25	0.03	0.05			Al 99.5	97		
1050 A91050	Europe	AECMA prEN2072	1050A	Sht H14 .4/2 mm diam		0.05	0.4	0.05	0.05	0.25	0.05	0.07			Al 99.5	100	80	5
1450	Europe					0.05 max	0.4 max	0.05 max	0.05 max	0.25 max	0.1-0.2	0.07 max	0.03		Al 99.5 min			
1050A	Europe					0.05 max	0.4 max	0.05 max	0.05 max	0.25 max	0.05 max	0.07 max	0.03		Al 99.5 min			
	Europe		Al 99.5			0.05 max	0.4 max	0.05 max	0.05 max	0.25 max	0.05 max	0.07 max	0.03		Al 99.5 min			
1445 A91445	Finland	SFS 2583	E-A199.5	Rod Tub HF (shapes)		0.05	0.4		0.05	0.25		0.03			Al 99.5	70	20	23
1445 A91445	Finland	SFS 2583	E-A199.5	Rod Tub Ann (except shapes)		0.05	0.4		0.05	0.25		0.03			Al 99.5	100		30
1445 A91445	Finland	SFS 2583	E-A199.5	Rod Tub Hard (except shapes)		0.05	0.4		0.05	0.25		0.03			Al 99.5	150		4
1050 A91050	France	NF A50411	1050A(A5)	Bar Wir Tub 1/2 hard (drawn bar) 50 mm diam		0.05	0.4 max	0.05	0.05	0.25	0.05	0.07			Al 99.5	100	75	5
1050 A91050	France	NF A50411	1050A(A5)	Bar Wir Tub 1/2 hard (drawn wir) 12 mm diam		0.05	0.4 max	0.05	0.05	0.25	0.05	0.07			Al 99.5	100	75	5
1050 A91050	France	NF A50411	1050A(A5)	Bar Wir Tub 1/4 hard (tub) 50 mm diam		0.05	0.4 max	0.05	0.05	0.25	0.05	0.07			Al 99.5	85	65	6
1050 A91050	France	NF A50451, A50701	1050A(A5)	Sh Strp Ann 0.4/3.2 mm diam		0.05	0.4 max	0.05	0.05	0.25	0.05	0.07			Al 99.5	65		35
1050 A91050	France	NF A50451, A50701	1050A(A5)	Sh Strp 3/4 hard .4/3.2 mm diam		0.05	0.4 max	0.05	0.05	0.25	0.05	0.07			Al 99.5	120	100	5
1050 A91050	France	NF A50451, A50701	1050A(A5)	Sh Strp Hard .4/3.2 mm diam		0.05	0.4 max	0.05	0.05	0.25	0.05	0.07			Al 99.5	140	125	4
1050 A91050	Germany	DIN 1712 pt 3	3.0255/Al99.5	Obsolete Replaced by DIN EN 10210		0.05 max	0.4 max	0.05 max	0.05 max	0.25 max	0.05 max	0.07 max			ET 0.03 max; Al 99.5 min			
1050 A91050	Japan	JIS H4000	1050	Sh Plt Ann 1.3 mm diam		0.05	0.4	0.05	0.05	0.25	0.03	0.05			Al 99.5	69	20	25
1050 A91050	Japan	JIS H4000	1050	Sh Plt Hard .3 mm diam		0.05	0.4	0.05	0.05	0.25	0.03	0.05			Al 99.5	137		1
1050 A91050	Japan	JIS H4000	1050	Sh Plt Hot rolled 12 mm diam		0.05	0.4	0.05	0.05	0.25	0.03	0.05			Al 99.5	78	49	10
1050 A91050	Japan	JIS H4040	1050	Bar As Mfg		0.05	0.4	0.05	0.05	0.25	0.03	0.05			Al 99.5	59	20	
1050 A91050	Japan	JIS H4080	1050	Tub As Mfg		0.05	0.4	0.05	0.05	0.25	0.03	0.05			Al 99.5	59	20	
1050 A91050	Japan	JIS H4090	1050	Pip Ann 1.3 mm diam		0.05	0.4	0.05	0.05	0.25	0.03	0.05			Al 99.5	69	20	25
1050 A91050	Japan	JIS H4090	1050	Pip 1/2 hard 1.3 mm diam		0.05	0.4	0.05	0.05	0.25	0.03	0.05			Al 99.5	98	69	4

UNS numbers and US grades are provided as a means of cross referencing chemically similar alloys. Exchangability is only possible after independent examination of specifications. Tensile properties are minimum or typical . UTS and YS as Mpa, El as %. See Appendix for list of abbreviations used in Descriptions.

Worldwide Guide to Equivalent Nonferrous Metals and Alloys

Grade UNS #	Country	Specification	Designation	Description	Cr	Cu	Fe	Mg	Mn	Si	Ti	Zn	OE Max	OT Max	Other	UTS	YS	EL
1050 A91050	Japan	JIS H4090	1050	Pip Hard .8 mm diam		0.05	0.4	0.05	0.05	0.25	0.03	0.05			Al 99.5	137		2
1050 A91050	Japan	JIS H4120	1050	Wir 1/2 hard 25 mm diam		0.05	0.4	0.05	0.05	0.25	0.03	0.05			Al 99.5	88		
1050 A91050	Norway	NS 17010	NS17010/Al9 9.5	Plt Sh Strp		0.05 max	0.4 max	0.05 max	0.05 max	0.25 max	0.05 max	0.07 max			ET 0.03 max; Al 99.5 max			
1050 A91050	Pan American	COPANT 862	1050			0.05	0.4	0.03	0.03	0.25	0.05	0.05			Al 99.5			
1050 A91050	USA	ASTM B491	1050	Ext		0.05 max	0.4 max	0.05 max	0.05 max	0.25 max				0.03	Al 99.5 min; V 0.05 max			
1050	USA					0.05 max	0.4 max	0.05 max	0.05 max	0.25 max	0.03 max	0.05 max	0.03		Al 99.5 min; V 0.05 max			
1050 A91050	Czech Republic	CSN 424005	Al99,5			0.05 max	0.4 max			0.3 max	0.05 max	0.07 max			Al 99.5 min			
1050 A91050	Finland	SFS 2582	Al99.5	Sh tub Wir HF (shapes)		0.05	0.4		0.05	0.3		0.1			Al 99.5			
1050 A91050	Finland	SFS 2582	Al99.5	Sh tub Wir Ann (sht, tub, wir)		0.05	0.4		0.05	0.3		0.1			Al 99.5			
1050 A91050	Finland	SFS 2582	Al99.5	Sh tub Wir Hard (tub rod) 3mm diam		0.05	0.4		0.05	0.3		0.1			Al 99.5			
	India	IS 733	19500	Bar Rod		0.05	0.4		0.05	0.3		0.1			Al 99.5	65		23
	India	IS 736	19500	Plt As Mfg 28 mm diam		0.05	0.4		0.05	0.3		0.1			Al 99.5	65		28
	India	IS 736	19500	Plt Ann 30 mm diam		0.05	0.4		0.05	0.3		0.1			Al 99.5	100		30
	India	IS 736	19500	Plt 1/2 hard 7 mm diam		0.05	0.4		0.05	0.3		0.1			Al 99.5	100		7
	India	IS 738	T1B	Wir Ann		0.05	0.4		0.05	0.3		0.1			Al 99.5	93		
	India	IS 738	T1B	Wir Ann and CW to hard 80 mm diam		0.05	0.4		0.05	0.3		0.1			Al 99.5	108		
	India	IS 739	G1B	Wir Ann		0.05	0.4		0.05	0.3		0.1			Al 99.5	93		
	India	IS 739	G1B	Wir Ann and CW to hard		0.05	0.4		0.05	0.3		0.1			Al 99.5	132		
1050 A91050	International	ISO R827	Al99.5	As Mfg		0.05	0.4		0.05	0.3		0.1			Al 99.5	49		22
1050 A91050	International	ISO TR2136	Al99.5	Sh Plt Ann		0.05	0.4		0.05	0.3		0.1			Al 99.5	60		20
1050 A91050	International	ISO TR2136	Al99.5	Sh Plt HB SH		0.05	0.4		0.05	0.3		0.1			Al 99.5	75		60
1050 A91050	International	ISO TR2136	Al99.5	Sh Plt HD SH		0.05	0.4		0.05	0.3		0.1			Al 99.5	95		4
1050 A91050	International	ISO TR2778	Al99.5	Tub Ann OBSOLETE Replace by ISO 6363		0.05	0.4		0.05	0.3		0.1			Al 99.5	60		22
1050 A91050	International	ISO TR2778	Al99.5	Tub SH OBSOLETE Replace by ISO 6363		0.05	0.4		0.05	0.3		0.1			Al 99.5	95		3
1050 A91050	UK	BS 1470	1050A(1B)	Plt Sh Strp OBSOLETE Replaced by BS EN 515		0.05 max	0.4 max		0.05 max	0.3 max		0.1 max			Al 99.5 min			
	UK	BS 1470	S1B	Plt Sh Strp Ann .2/6 mm diam OBSOLETE Replaced by BS EN 515		0.05	0.4		0.05	0.3		0.1			Al 99.5	55		22
	UK	BS 1470	S1B	Plt Sh Strp Ann .2/6 mm diam OBS Replaced by BS EN 515		0.05	0.4		0.05	0.3		0.1			Al 99.5	100		4

UNS numbers and US grades are provided as a means of cross referencing chemically similar alloys. Exchangability is only possible after independent examination of specifications. Tensile properties are minimum or typical . UTS and YS as Mpa, El as %. See Appendix for list of abbreviations used in Descriptions.

Grade UNS #	Country	Specification	Designation	Description	Cr	Cu	Fe	Mg	Mn	Si	Ti	Zn	OE Max	OT Max	Other	UTS	YS	EL
	UK	BS 1470	S1B	Plt Sh Strp SH to hard .2/3 mm diam OBSOLETE Replaced by BS EN 515		0.05	0.4		0.05	0.3		0.1			Al 99.5	135		3
	UK	BS 1471	T1B	Rub Ann 12 mm diam		0.05	0.4		0.05	0.3		0.1			Al 99.5	95		
	UK	BS 1471	T1B	Rub SH to 1/2 hard 12 mm diam		0.05	0.4		0.05	0.3		0.1			Al 99.5	100		
	UK	BS 1471	T1B	Rub SH to hard 12 mm diam		0.05	0.4		0.05	0.3		0.1			Al 99.5	135		
	UK	BS 1472	F1B	As Mfg 150 mm diam Obsolete Replaced by BS EN 586		0.05	0.4		0.05	0.3		0.1			Al 99.5	60		22
	UK	BS 1474	E1B	Bar Tub As Mfg		0.05	0.4		0.05	0.3		0.1			Al 99.5	60		23
	UK	BS 1475	G1B	Wir Ann 10 mm diam		0.05	0.4		0.05	0.3		0.1			Al 99.5	95		
	UK	BS 1475	G1B	Wir As Mfg SH to hard 10 mm diam		0.05	0.4		0.05	0.3		0.1			Al 99.5	135		
	Yugoslavia	JUS CC2.100	Al99.5			0.05 max	0.4 max	0.1 max	0.05 max	0.3 max	0.03 max				ST 0.12 max; ET 0.12 max; ST = Zn+Cu, Si < Fe; ET 0.05 max;			
5005 A95005	Germany	DIN 1725 Part 1	3.3315/AlMg1		0.1 max	0.05 max	0.45 max	0.7-1.1	0.15 max	0.3 max		0.2 max			ET=.15 total			
1050 A91050	Denmark	DS 3012	1050	1/2 hard (rolled and drawn producto) .5/6 mm diam		0.05	0.4	0.05			0.03	0.05			Al 99.5	100	80	6
1050 A91050	Denmark	DS 3012	1050	Hard (rolled and drawn products) .5/6 mm diam		0.05	0.4	0.05			0.03	0.05			Al 99.5	135	125	4
1050 A91050	Denmark	DS 3012	1050	As manufactured (extd products)		0.05	0.4	0.05			0.03	0.05			Al 99.5	55		22
	UK	BS 2898	E1E	Bar Tub As Mfg		0.05									Al 99.5	60		23
	UK	BS 2898	E1E	Bar Tub SH to 1/4 hard		0.05									Al 99.5	85		13
3012 A93012	Argentina				0.2 max	0.1 max	0.7 max	0.1 max	0.5-1.1	0.6 max	0.1 max	0.1 max	0.05	0.15				
5557	USA					0.15 max	0.12 max	0.4-0.8	0.1-0.4	0.1 max				0.1	V 0.05 max			
5005 A95005	Australia	AS 1734	5005	Sh Plt Ann 0.15/0.20 mm thk	0.1	0.2	0.7	0.5-1.1	0.2	0.3		0.25						12
5005 A95005	Australia	AS 1734	5005	Sh Plt SH to 1/2 hard 0.25/0.80 mm thk	0.1	0.2	0.7	0.5-1.1	0.2	0.3		0.25						1
5005 A95005	Australia	AS 1734	5005	Sh Plt SH to full hard 0.15/0.80 mm thk	0.1	0.2	0.7	0.5-1.1	0.2	0.3		0.25						1
5005 A95005	Belgium	NBN P21-001	5005		0.1 max	0.2 max	0.7 max	0.5-1.1	0.2 max	0.3 max		0.25 max			ET 0.05 max; ET = .15 Total			
5005 A95005	France	NF A50411	5005(A-G0.6)	Bar Wir Tub Hard (drawn bar) 10 mm diam	0.1	0.2	0.7	0.5-1.1	0.2	0.3		0.25				185	145	2
5005 A95005	France	NF A50411	5005(A-G0.6)	Bar Wir Tub Hard (wir) 4 mm diam	0.1	0.2	0.7	0.5-1.1	0.2	0.3		0.25				240	200	2
5005 A95005	France	NF A50411	5005(A-G0.6)	Bar Wir Tub 1/2 hard 50 mm	0.1	0.2	0.7	0.5-1.1	0.2	0.3		0.25				145	105	5

UNS numbers and US grades are provided as a means of cross referencing chemically similar alloys. Exchangability is only possible after independent examination of specifications. Tensile properties are minimum or typical . UTS and YS as Mpa, El as %. See Appendix for list of abbreviations used in Descriptions.

Worldwide Guide to Equivalent Nonferrous Metals and Alloys

Grade UNS #	Country	Specification	Designation	Description	Cr	Cu	Fe	Mg	Mn	Si	Ti	Zn	OE Max	OT Max	Other	UTS	YS	EL
5005 A95005	France	NF A50451	5005A(A-G0.6)	Sh Plt Ann .4/3.2 mm diam	0.1	0.2	0.7	0.5-1.1	0.2	0.3		0.25				100		12
5005 A95005	France	NF A50451	5005A(A-G0.6)	Sh Plt 3/4 hard .4/1.6 mm diam	0.1	0.2	0.7	0.5-1.1	0.2	0.3		0.25				160	140	3
5005 A95005	France	NF A50451	5005A(A-G0.6)	Sh Plt Hard .4/1.6 mm diam	0.1	0.2	0.7	0.5-1.1	0.2	0.3		0.25				180	160	3
5005 A95005	Germany	DIN	Al-Mg1	Wir Plt Sh	0.1 max	0.2 max	0.7 max	0.5-1.1	0.2 max	0.3 max		0.25 max	0.05	0.15				
	Norway	NS 17204	NS17204/AIMg1	Bar Plt Sh Strp Tub	0.1 max	0.2 max	0.7 max	0.5-1.1	0.2 max	0.3 max		0.25 max			ET 0.05 max; ET=.15 total			
	Norway	NS 17206	NS17206/AIMg2Mn0.8	Plt Sh Strp	0.1 max	0.2 max	0.7 max	0.5-1.1	0.2 max	0.3 max		0.25 max			ET=.15 total			
5005 A95005	Pan American	COPANT 862	5005		0.1	0.2	0.7	0.5-1.1	0.2	0.3		0.25						
5005 A95005	Sweden	SIS 144106	4106-02	Sh Strp Ann .35/6 mm diam	0.1	0.2	0.7	0.5-1.1	0.2	0.3		0.25			Al 96.9	95	35	25
5005 A95005	Sweden	SIS 144106	4106-14	Sh Strp SH(sh) .35/4 mm diam	0.1	0.2	0.7	0.5-1.1	0.2	0.3		0.25			Al 96.9	145	120	6
5005 A95005	Sweden	SIS 144106	4106-14	Sh Strp SH(strp) .35/2 mm diam	0.1	0.2	0.7	0.5-1.1	0.2	0.3		0.25			Al 96.9	145	120	6
5005 A95005	Sweden	SIS 144106	4106-18	Sh SH .35/1.6 mm diam	0.1	0.2	0.7	0.5-1.1	0.2	0.3		0.25			Al 96.9	205	185	2
5005 A95005	Sweden	SIS 144106	4106-18	Strp Sh SH .35/1.6 mm diam	0.1	0.2	0.7	0.5-1.1	0.2	0.3		0.25			Al 96.9	205	185	2
5005 A95005	Sweden	SIS 144106	4106-24	sh Strp SH ann .35/3 mm diam	0.1	0.2	0.7	0.5-1.1	0.2	0.3		0.25			Al 96.9	145	115	10
5005 A95005	Sweden	SIS 144106	4106-26	Sh Strp SH ann .25/3 mm diam	0.1	0.2	0.7	0.5-1.1	0.2	0.3		0.25			Al 96.9	165	135	8
5005 A95005	Sweden	SIS 144106	4106-40	Sh Strp Ann .35/6 mm diam	0.1	0.2	0.7	0.5-1.1	0.2	0.3		0.25			Al 96.9	95	35	25
5005 A95005	Sweden	SIS 144106	4106-44	Sh Strp SH .35/3 mm diam	0.1	0.2	0.7	0.5-1.1	0.2	0.3		0.25			Al 96.9	145	120	6
5005 A95005	UK	BS N41	5005	Wir Plt Sh	0.1 max	0.2 max	0.7 max	0.5-1.1	0.2 max	0.3 max		0.25 max	0.05	0.15				
5005 A95005	USA	ASTM B209	5005	Wir Plt Sh	0.1 max	0.2 max	0.7 max	0.5-1.1	0.2 max	0.3 max		0.25 max	0.05	0.15				
5005 A95005	USA	ASTM B210	5005	Wir Plt Sh	0.1 max	0.2 max	0.7 max	0.5-1.1	0.2 max	0.3 max		0.25 max	0.05	0.15				
5005 A95005	USA	ASTM B316	5005	Wir Plt Sh	0.1 max	0.2 max	0.7 max	0.5-1.1	0.2 max	0.3 max		0.25 max	0.05	0.15				
5005 A95005	USA	ASTM B396	5005	Wir Plt Sh	0.1 max	0.2 max	0.7 max	0.5-1.1	0.2 max	0.3 max		0.25 max	0.05	0.15				
5005 A95005	USA	ASTM B397	5005	Wir Plt Sh	0.1 max	0.2 max	0.7 max	0.5-1.1	0.2 max	0.3 max		0.25 max	0.05	0.15				
5005 A95005	USA	ASTM B483	5005	Wir Plt Sh	0.1 max	0.2 max	0.7 max	0.5-1.1	0.2 max	0.3 max		0.25 max	0.05	0.15				
5005 A95005	USA	QQ A-403	5005	Wir Plt Sh OBSOLETE	0.1 max	0.2 max	0.7 max	0.5-1.1	0.2 max	0.3 max		0.25 max	0.05	0.15				
5005 A95005	USA	SAE J454	5005	Wir Plt Sh	0.1 max	0.2 max	0.7 max	0.5-1.1	0.2 max	0.3 max		0.25 max	0.05	0.15				
5005 A95005	USA				0.1 max	0.2 max	0.7 max	0.5-1.1	0.2 max	0.3 max		0.25 max	0.05	0.15				
	USA		AlMg1(B)		0.1 max	0.2 max	0.7 max	0.5-1.1	0.2 max	0.3 max		0.25 max	0.05	0.15				
5005 A95005	Denmark	DS 3012	5005	Soft .5/25 mm diam	0.1	0.2	0.7	0.5-1.1	0.2	0.4		0.25				80		24
5005 A95005	Denmark	DS 3012	5005	1/2 hard .5/6 mm diam	0.1	0.2	0.7	0.5-1.1	0.2	0.4		0.25				120	100	5
5005 A95005	Denmark	DS 3012	5005	3/4 hard .5/6 mm diam	0.1	0.2	0.7	0.5-1.1	0.2	0.4		0.25				130	120	4
5005 A95005	Finland	SFS 2586	AlMg1	HF (shapes)	0.1	0.2	0.7	0.5-1.1	0.2	0.4		0.2						
5005 A95005	Finland	SFS 2586	AlMg1	Ann (sht)	0.1	0.2	0.7	0.5-1.1	0.2	0.4		0.2						
5005 A95005	Finland	SFS 2586	AlMg1	Hard (sht) 3/6 mm diam	0.1	0.2	0.7	0.5-1.1	0.2	0.4		0.2						

UNS numbers and US grades are provided as a means of cross referencing chemically similar alloys. Exchangability is only possible after independent examination of specifications. Tensile properties are minimum or typical . UTS and YS as Mpa, El as %. See Appendix for list of abbreviations used in Descriptions.

2-12 Wrought Aluminum

Grade UNS #	Country	Specification	Designation	Description	Cr	Cu	Fe	Mg	Mn	Si	Ti	Zn	OE Max	OT Max	Other	UTS	YS	EL
5005 A95005	International	ISO R827	Al-Mg1	as mfg	0.1	0.2	0.7	0.5-1.1	0.2	0.4		0.2				98		22
5005 A95005	International	ISO TR2136	Al-Mg1	Sh Plt ann	0.1	0.2	0.7	0.5-1.1	0.2	0.4		0.2				95		18
5005 A95005	International	ISO TR2136	Al-Mg1	Sh Plt HB SH	0.1	0.2	0.7	0.5-1.1	0.2	0.4		0.2				115		3
5005 A95005	International	ISO TR2136	Al-Mg1	Sh Plt HD SH	0.1	0.2	0.7	0.5-1.1	0.2	0.4		0.2				140	95	2
5005 A95005	International	ISO TR2778	Al-Mg1	Tub ann OBSOLETE Replace by ISO 6363	0.1	0.2	0.7	0.5-1.1	0.2	0.4		0.2				95		18
5005 A95005	International	ISO TR2778	Al-Mg1	Tub SH OBSOLETE Replace by ISO 6363	0.1	0.2	0.7	0.5-1.1	0.2	0.4		0.2				115		4
5005 A95005	Japan	JIS H4000	5005	sh Plt Ann 1.3 mm diam	0.1	0.2	0.7	0.5-1.1	0.2	0.4		0.25				108	39	20
5005 A95005	Japan	JIS H4000	5005	sh Plt Hard .8 mm diam	0.1	0.2	0.7	0.5-1.1	0.2	0.4		0.25				177		1
5005 A95005	Japan	JIS H4000	5005	sh Plt Hr 12 mm diam	0.1	0.2	0.7	0.5-1.1	0.2	0.4		0.25				118		8
5005 A95005	Yugoslavia	JUS CC2.100	AlMg1		0.1 max	0.2 max	0.7 max	0.5-1.1	0.2 max	0.4 max		0.2 max						
1120	Australia					0.05-0.2			0.05 max	0.05 max	0.03 max	0.05 max	0.03		0.45 Si+Fe; Al 99.5 min			
1150 A91150	Australia	AS 1734	1150	Sh Plt, Ann 0.15/0.50mm thk		0.05-0.2									ST=0.45; ET=0.05; ST=Si+Fe; Al 99.5			15
1150 A91150	Australia	AS 1734	1150	Sh Plt, SH to 1/2 hard 0.25/0.30mm Thk		0.05-0.2									ST=0.45; ET=0.05; ST=Si+Fe; Al 99.5			1
1150 A91150	Australia	AS 1734	1150	Sh Plt, Stain hardened to full hard 0.15/0.50mm thk		0.05-0.2									ST=0.45; ET=0.05; ST=Si+Fe; Al 99.5			1
6463A	Australia					0.25 max	0.15 max	0.3-0.9	0.05 max	0.2-0.6		0.05 max	0.05	0.15				
3105 A93105	Australia	AS 1734	3105		0.2 max	0.3 max	0.7 max	0.2-0.8	0.3-0.8	0.6 max	0.1 max	0.4 max	0.05		OT min 0.15			
3105 A93105	Australia	AS 1734	3105	O .63-1.20 mm thk	0.2 max	0.3 max	0.7 max	0.2-0.8	0.3-0.8	0.6 max	0.1 max	0.4 max	0.05		OT min 0.15	95	35	19
3105 A93105	Australia	AS 1734	3105	H12 .63-1.20 mm thk	0.2 max	0.3 max	0.7 max	0.2-0.8	0.3-0.8	0.6 max	0.1 max	0.4 max	0.05		OT min 0.15	130	105	2
3105 A93105	Australia	AS 1734	3105	H14 .63-1.20 mm thk	0.2 max	0.3 max	0.7 max	0.2-0.8	0.3-0.8	0.6 max	0.1 max	0.4 max	0.05		OT min 0.15	150	125	2
3105 A93105	Australia	AS 1734	3105	H16 .63-1.20 mm thk	0.2 max	0.3 max	0.7 max	0.2-0.8	0.3-0.8	0.6 max	0.1 max	0.4 max	0.05		OT min 0.15	170	145	1
3105 A93105	Australia	AS 1734	3105	H18 .63-1.20 mm thk	0.2 max	0.3 max	0.7 max	0.2-0.8	0.3-0.8	0.6 max	0.1 max	0.4 max	0.05		OT min 0.15	190	165	1
3105 A93105	Australia	AS 1734	3105	H25 .63-1.20 mm thk	0.2 max	0.3 max	0.7 max	0.2-0.8	0.3-0.8	0.6 max	0.1 max	0.4 max	0.05		OT min 0.15	160	130	4
3105A	France				0.2 max	0.3 max	0.7 max	0.2-0.8	0.3-0.8	0.6 max	0.1 max	0.25 max	0.05	0.15				
3105 A93105	Germany	DIN 1725 pt1	3.0505/AlMn0.5Mg0.5		0.2 max	0.3 max	0.7 max	0.2-0.8	0.3-0.8	0.6 max	0.1 max	0.4 max			ET 0.05 max; ET=.15 total			
3105 A93105	Norway	NS 17404	NS17404/AlMn0.5Mg0.5	Plt Sh Strp	0.2 max	0.3 max	0.7 max	0.2-0.8	0.3-0.8	0.6 max	0.1 max	0.4 max			ET 0.05 max; ET=.15 total;			
3105 A93105	Pan American	COPANT 862	3105		0.2	0.3	0.7	0.2-0.8	0.3-0.8	0.6	0.1	0.4						
3105 A93105	UK	BS 1470	3105(N31)	OBSOLETE Replaced by BS EN 485/515/573	0.2 max	0.3 max	0.7 max	0.2-0.8	0.3-0.8	0.6 max	0.1 max	0.4 max			ET 0.05 max; ET=.15 total			
3105 A93105	UK		E4S		0.2 max	0.3 max	0.7 max	0.2-0.8	0.3-0.8	0.6 max	0.1 max	0.4 max			ET 0.05 max; ; ET=.15 total			

UNS numbers and US grades are provided as a means of cross referencing chemically similar alloys. Exchangability is only possible after independent examination of specifications. Tensile properties are minimum or typical . UTS and YS as Mpa, El as %. See Appendix for list of abbreviations used in Descriptions.

Worldwide Guide to Equivalent Nonferrous Metals and Alloys

Grade UNS #	Country	Specification	Designation	Description	Cr	Cu	Fe	Mg	Mn	Si	Ti	Zn	OE Max	OT Max	Other	UTS	YS	EL
3105 A93105	USA	ASTM B209	3105	Sh	0.2 max	0.3 max	0.7 max	0.2-0.8	0.3-0.8	0.6 max	0.1 max	0.4 max	0.05	0.15				
3105	USA				0.2 max	0.3 max	0.7 max	0.2-0.8	0.3-0.8	0.6 max	0.1 max	0.4 max	0.05	0.15				
	USA		AlMn0.5Mg0.5		0.2 max	0.3 max	0.7 max	0.2-0.8	0.3-0.8	0.6 max	0.1 max	0.4 max	0.05	0.15				
3307	USA				0.2 max	0.3 max	0.8 max	0.3 max	0.5-0.9	0.6 max	0.1 max	0.4 max	0.05	0.15				
	International	ISO R209 (1971)	Al 99.5			0.05 max	0.4 max		0.05 max	0.3 max		0.10 max			Cu+Si+Fe+Mn+ Zn 0.5 max			
	UK	BS 1473	R1B	SH to 5/8 hard 12 mm diam			0.4		0.05	0.3		0.1			Al 99.5	110		
1350 A91350	Czech Republic	CSN 424004	Al99,5-E												Al 99.5 min			
1350 A91350	Germany		EAl99.5												Al 99.5 min			
	Germany	DIN 40501Part1	3.0257.10/E-AlF7	Sh Strp CR: Sht 20 mm diam											Al 99.5	70	50	33
	Germany	DIN 40501Part1	3.0257.10/E-AlF7	Sh Strp CR: Strp 3 mm diam											Al 99.5	70	50	33
	Germany	DIN 40501Part1	3.0257.26/E-AlF10	Sh Strp CR:sht 6 mm diam											Al 99.5	100	80	7
	Germany	DIN 40501Part1	3.0257.26/E-AlF10	Sh Strp CR: strp 3 mm diam											Al 99.5	100	80	7
	Germany	DIN 40501Part1	3.0257.30/E-AlF13	Wir As Mfg (rd) 1.5 mm diam											Al 99.5	130	90	2
	Germany	DIN 40501Part1	3.0257.30/E-AlF13	Wir As Mfg (flat) 10 mm diam											Al 99.5	130	90	2
	Germany	DIN 40501Part1	3.0257.30/E-AlF13	Sh Strp CR:sht 4 mm diam											Al 99.5	130	110	3
	Germany	DIN 40501Part1	3.0257.32/E-AlF16	Sh Strp CR 1.5 mm diam											Al 99.5	160	140	2
	Germany	DIN 40501Part2	3.0257.08/E-AlF7	Tub As manufactured 250 mm diam											Al 99.5	70	25	20
	Germany	DIN 40501Part2	3.0257.26/E-AlF10	Tub As Mfg 120 mm diam											Al 99.5	100	70	6
	Germany	DIN 40501Part2	3.320/E-AlMgSi0.5F22	Tub As Mfg 250 mm diam											Al 99.5	215	160	12
	Germany	DIN 40501Part3	3.0257.08/E-AlF6.5	Bar As manufactured (rd,sq,hex) 63 mm diam											Al 99.5	65	25	23
	Germany	DIN 40501Part3	3.0257.08/E-AlF6.5	Bar As manufactured (flat) 200 mm											Al 99.5	65	25	23
	Germany	DIN 40501Part3	3.0257.08/E-Alp	Bar											Al 99.5			
	Germany	DIN 40501Part3	3.0257.09/E-AlF8	Bar As mfg (rd, sq, hex)											Al 99.5	80	50	15
	Germany	DIN 40501Part3	3.0257.09/E-AlF8	Bar As mfg (flat) 120 mm diam											Al 99.5	80	50	15
	Germany	DIN 40501Part3	3.0257.26/E-AlF10	Bar As Mfg (rd, sq, hex) 20 mm diam											Al 99.5	100	70	7
	Germany	DIN 40501Part3	3.0257.26/E-AlF10	Bar As Mfg (flat) 80 mm diam											Al 99.5	100	70	7
	Germany	DIN 40501Part3	3.0257.30/E-AlF13	Bar As Mfg (rd,sq,hex) 10 mm diam											Al 99.5	130	90	5
	Germany	DIN 40501Part3	3.0257.30/E-AlF13	Bar As Mfg (flat) 12 mm diam											Al 99.5	130	90	5
	Germany	DIN 40501Part3	3.320/E-AlMgSi0.5F17	Bar As Mfg (rd,sq,hex)											Al 99.5	170	120	12

UNS numbers and US grades are provided as a means of cross referencing chemically similar alloys. Exchangability is only possible after independent examination of specifications. Tensile properties are minimum or typical . UTS and YS as Mpa, El as %. See Appendix for list of abbreviations used in Descriptions.

Grade UNS #	Country	Specification	Designation	Description	Cr	Cu	Fe	Mg	Mn	Si	Ti	Zn	OE Max	OT Max	Other	UTS	YS	EL
	Germany	DIN 40501Part3	3.320/E-AlMgSi0.5F17	Bar As Mfg (flat) 180 mm diam											Al 99.5	170	120	12
	Germany	DIN 40501Part3	3.320/E-AlMgSi0.5F22	Bar As Mfg (rd,sq,hex)											Al 99.5	215	160	12
	Germany	DIN 40501Part3	3.320/E-AlMgSi0.5F22	Bar As Mfg (flat) 180 mm diam											Al 99.5	215	160	12
	Germany	DIN 40501Part4	3.0257.10/E-AlF7	Wir As manufactured (rd) 3.6 mm diam											Al 99.5	60	60	25
	Germany	DIN 40501Part4	3.0257.10/E-AlF7	Wir As mfg. (flat)											Al 99.5	60	60	25
	Germany	DIN 40501Part4	3.0257.26/E-AlF9	Wir As mfg (rd) 1.5 mm diam											Al 99.5	90	70	3
	Germany	DIN 40501Part4	3.0257.26/E-AlF9	Wir As mfg (flat) 10 mm diam											Al 99.5	90	70	3
	Germany	DIN 40501Part4	3.0257.32/E-AlF17	Wir As Mfg (rd) .2 mm diam											Al 99.5	180		
	Germany	DIN 40501Part4	3.0257.32/E-AlF17	Wir As Mfg (flat)											Al 99.5	180		
	Germany	DIN 40501Part4	3.0257.32/E-AlF17	Wir As Mfg (rd) 3.1 mm diam											Al 99.5	160	130	
	Germany	DIN 40501Part4	3.0257.32/E-AlF17	Wir As Mfg (flat)											Al 99.5	160	130	
	Germany	DIN 40501Part4	3.0257.32/E-AlF17	Wir As Mfg (rd, sq, hex)											Al 99.5	170	130	
	Germany	DIN 40501Part4	3.0257.32/E-AlF17	Wir As Mfg (flat)											Al 99.5	170	130	
1050 A91050	Mexico	DGN W-30	1S-0	Sh Rolled .25/.52 mm diam											Al 99.5			
1050 A91050	Mexico	DGN W-30	1S-H14	Sh Rolled .53/.80 mm diam											Al 99.5			
1050 A91050	Mexico	DGN W-30	1S-H16	Sh Rolled .25/.52 mm diam											Al 99.5			
1050 A91050	Mexico	DGN W-30	1S-H18	Sh Rolled .25/.52 mm diam											Al 99.5			
1050 A91050	Sweden	MNC 40E	4007	Bar Wir Strp Tub Ext											Al 99.5 min			
5005 A95005	Sweden	MNC 40E	4106	Sh						0.8					Al 99.2			
	Sweden	MNC 40E	4008	Bar Wir Strp											Al 99.5			
	Pan American	COPANT 862	X5257			0.1	0.1	0.2-0.6	0.03	0.08		0.03						
1145 A91145	Russia	GOST 1069	AE			0.02 max	0.35 max			0.12 max	0.01 max	0.05 max			Al 99.5 min			
1445	Australia					0.04 max								0.05	0.50 max Si+Fe+Cu; Al 99.45 min			
	France	NF A57703	A5-Y4			0.05	0.05	0.05			0.1	0.1			Al 99.45; Ni 0.05; Sn 0.05			
1145 A91145	Pan American	COPANT 862	1145			0.05	0.05	0.05			0.03	0.05			Al 99.45			
1145 A91145	USA	AMS 4011	1145			0.05 max	0.05 max	0.05 max			0.03 max	0.05 max			ST 0.55 max; Al 99.45 min; V 0.05 max			
1145 A91145	USA	ASTM B373	1145			0.05 max	0.05 max	0.05 max			0.03 max	0.05 max			ST 0.55 max; Al 99.45 min; V 0.05 max			
1145 A91145	USA	QQ A-1876	1145			0.05 max	0.05 max	0.05 max			0.03 max	0.05 max		0.03	ST 0.55 max; Al 99.45 min; V 0.05 max			
1145	USA					0.05 max	0.05 max	0.05 max			0.03 max	0.05 max	0.03		0.55 Si+Fe; Al 99.45 min; V 0.05 max			

UNS numbers and US grades are provided as a means of cross referencing chemically similar alloys. Exchangability is only possible after independent examination of specifications. Tensile properties are minimum or typical . UTS and YS as Mpa, El as %. See Appendix for list of abbreviations used in Descriptions.

Worldwide Guide to Equivalent Nonferrous Metals and Alloys

Grade UNS #	Country	Specification	Designation	Description	Cr	Cu	Fe	Mg	Mn	Si	Ti	Zn	OE Max	OT Max	Other	UTS	YS	EL
1145 A91145	Russia		AT			0.1 max	0.25 max			0.2 max					Al 99.6 min			
1345	USA					0.1 max	0.4 max	0.05 max	0.05 max	0.3 max	0.03 max	0.05 max	0.03		Al 99.45 min; V 0.05 max			
1045	USA					0.1 max	0.45 max	0.05 max	0.05 max	0.3 max	0.03 max	0.05 max	0.03		Al 99.45 min; V 0.05 max			
8005	Italy					0.05 max	0.4-0.8	0.05 max		0.2-0.5		0.05 max	0.05	0.15				
5657 A95657	Pan American	COPANT 862	5657			0.1	0.1	0.6-1	0.03	0.08		0.05			Ga 0.03; V 0.05			
5657 A95657	USA	ASTM B209	5657			0.1 max	0.1 max	0.6-1	0.03 max	0.08 max		0.05 max	0.02	0.05	Ga 0.03 max; V 0.05 max			
5657	USA					0.1 max	0.1 max	0.6-1	0.03 max	0.08 max		0.05 max		0.05	Ga 0.03 max; V 0.05 max			
1040	USA					0.1 max	0.5 max	0.05 max	0.05 max	0.3 max	0.03 max	0.1 max	0.03		Al 99.4 min; V 0.05 max			
5657 A95657	UK	BS 4300	5657(BTR2)			0.15 max	0.1 max	0.7-1.2	0.3 max	0.1 max	0.05 max							
5205	USA				0.1 max	0.03-0.1	0.7 max	0.6-1	0.1 max	0.15 max		0.05 max	0.05	0.15				
1435	USA					0.02 max	0.3-0.5	0.05 max	0.05 max	0.15 max	0.03 max	0.1 max	0.03		Al 99.35 min; V 0.05 max			
6463 A96463	Germany	DIN 1725 Part 1	3.2305/E-AlMgSi			0.02 max	0.1-0.3	0.35-0.6		0.5-0.6		0.15 max			ST 0.03 max; ET 0.03 max; ST=Cr+Mn+Ti+V; ET=.10 total			
6463 A96463	UK	BS 1474	91E			0.04 max	0.5 max	0.4-0.9		0.3-0.7								
6101 A96101	Norway	NS 17301	NS17301/E-AlMgSi0.5			0.05 max	0.1-0.3	0.35-0.6	0.05	0.3-0.6		0.1 max			ET 0.03 max; ET=.15 total			
6101 A96101	UK	BS 2898	E91E	Bar Tub SHT and PT		0.05	0.4	0.4-0.9		0.3-0.7						200	170	8
1235 A91235	Pan American	COPANT 862	1235			0.05		0.05	0.05		0.03	0.1			Al 99.35			
1235	USA					0.05 max		0.05 max	0.05 max		0.06 max	0.1 max	0.03		0.65 Si+Fe; Al 99.35 min; V 0.05 max			
6063 A96063	Finland	SFS 2591	AlMgSi	HF 25 mm diam	0.1	0.058 max	0.5	0.4-0.9		0.3-0.7		0.2				250	80	15
6063 A96063	Finland	SFS 2591	AlMgSi	Aged 25 mm diam	0.1	0.058 max	0.5	0.4-0.9		0.3-0.7		0.2				160	110	10
6063 A96063	Finland	SFS 2591	AlMgSi	ST and aged 25 mm diam	0.1	0.058 max	0.5	0.4-0.9		0.3-0.7		0.2				220	180	10
1035	USA					0.1 max	0.6 max	0.05 max	0.05 max	0.35 max	0.03 max	0.1 max	0.03		Al 99.35 min; V 0.05 max			
6063 A96063	Australia	AS 1865	6063	Wir Rod Bar Strp Ann	0.1	0.1	0.35	0.45-0.9	0.1	0.2-0.6	0.1	0.1				132		
6063 A96063	Australia	AS 1865	6063	Wir Rod Bar Strp SH to full hard 10 mm diam	0.1	0.1	0.35	0.45-0.9	0.1	0.2-0.6	0.1	0.1				186		
6063 A96063	Australia	AS 1866	6063	Rod Bar Ann	0.1	0.1	0.35	0.45-0.9	0.1	0.2-0.6	0.1	0.1				131		16
6063 A96063	Australia	AS 1866	6063	Rod Bar SHT and Nat aged 150.0 mm diam	0.1	0.1	0.35	0.45-0.9	0.1	0.2-0.6	0.1	0.1				131	68	12
6063 A96063	Australia	AS 1866	6063	Rod Bar SHT and Art aged 25.0 mm diam	0.1	0.1	0.35	0.45-0.9	0.1	0.2-0.6	0.1	0.1				206	172	8
6063 A96063	Australia	AS 1867	6063	Tub Ann	0.1	0.1	0.35	0.45-0.9	0.1	0.2-0.6	0.1	0.1				132		20
6063 A96063	Australia	AS 1867	6063	Tub SH to full hard	0.1	0.1	0.35	0.45-0.9	0.1	0.2-0.6	0.1	0.1				179		
6063 A96063	Australia	AS 1867	6063	Tub SHT and Art aged	0.1	0.1	0.35	0.45-0.9	0.1	0.2-0.6	0.1	0.1				200	177	8

UNS numbers and US grades are provided as a means of cross referencing chemically similar alloys. Exchangability is only possible after independent examination of specifications. Tensile properties are minimum or typical . UTS and YS as Mpa, El as %. See Appendix for list of abbreviations used in Descriptions.

Grade UNS #	Country	Specification	Designation	Description	Cr	Cu	Fe	Mg	Mn	Si	Ti	Zn	OE Max	OT Max	Other	UTS	YS	EL
6063 A96063	Canada	CSA GS10	6063	Bar Rod Wir Tub Pip Ext	0.1 max	0.1 max	0.35	0.45-0.9	0.1 max	0.2-0.6	0.1 max	0.1 max			ET 0.05 max; ET 0.15 max;			
6063 A96063	Canada	CSA HA.5	0.6063	Bar Rod Wir Drawn and O	0.1	0.1	0.35	0.45-0.9	0.1	0.2-0.6	0.1	0.1				117		
6063 A96063	Canada	CSA HA.5	0.6063	Bar Rod Wir Drawn and T4 1.5 in diam	0.1	0.1	0.35	0.45-0.9	0.1	0.2-0.6	0.1	0.1				124	55	18
6063 A96063	Canada	CSA HA.5	0.6063	Bar Rod Wir Extd and T62 .125/1 in diam	0.1	0.1	0.35	0.45-0.9	0.1	0.2-0.6	0.1	0.1				207	173	10
6063 A96063	Canada	CSA HA.7	0.6063	Tub Pip Drawn and O	0.1	0.1	0.35	0.45-0.9	0.1	0.2-0.6	0.1	0.1				131		
6063 A96063	Canada	CSA HA.7	0.6063	Tub Pip Drawn and T4 .015/.049 in diam	0.1	0.1	0.35	0.45-0.9	0.1	0.2-0.6	0.1	0.1				152	69	16
6063 A96063	Canada	CSA HA.7	0.6063	Tub Pip Extd and O	0.1	0.1	0.35	0.45-0.9	0.1	0.2-0.6	0.1	0.1				131		18
6063 A96063	Canada	CSA HA.7.1	0.6063	Tub Drawn and O	0.1	0.1	0.35	0.45-0.9	0.1	0.2-0.6	0.1	0.1				131		
6063 A96063	Canada	CSA HA.7.1	0.6063	Tub Drawn and T4 .015/.049 in diam	0.1	0.1	0.35	0.45-0.9	0.1	0.2-0.6	0.1	0.1				152	69	16
6063 A96063	Canada	CSA HA.7.1	0.6063	Tub Extd and O	0.1	0.1	0.35	0.45-0.9	0.1	0.2-0.6	0.1	0.1				131		18
6063 A96063	Denmark	DS 3012	6063	SA 25 mm diam	0.1	0.1	0.35	0.45-0.9	0.1	0.2-0.6	0.1	0.1				145	105	8
6063 A96063	Denmark	DS 3012	6063	Art aged 12 mm diam	0.1	0.1	0.35	0.45-0.9	0.1	0.2-0.6	0.1	0.1				185	145	8
6063 A96063	France	NF A-GS	6063	Bar Rod Wir Tub Pip Ext	0.1 max	0.1 max	0.35 max	0.45-0.9	0.1 max	0.2-0.6	0.1 max	0.1 max			ET 0.05 max; ET 0.15 max			
6063 A96063	Germany	DIN	AlMgSi0.5	Bar Rod Wir Tub Pip Ext	0.1 max	0.1 max	0.35 max	0.45-0.9	0.1 max	0.2-0.6	0.1 max	0.1 max			ET 0.05 max; ET 0.15 max			
A96063	Germany	DIN	3.3206	Bar Rod Wir Tub Pip Ext	0.1 max	0.1 max	0.35 max	0.45-0.9	0.1 max	0.2-0.6	0.1 max	0.1 max			ET 0.05 max; ET 0.15 max			
6063 A96063	Japan	JIS H4040	6063	Bar As Mfg	0.1	0.1	0.35	0.45-0.9	0.1	0.2-0.6	0.1	0.1				118	49	12
6063 A96063	Japan	JIS H4040	6063	Bar Quick cooled and Art Aged	0.1	0.1	0.35	0.45-0.9	0.1	0.2-0.6	0.1	0.1				118	49	8
6063 A96063	Japan	JIS H4040	6063	Bar SHT and Art AH	0.1	0.1	0.35	0.45-0.9	0.1	0.2-0.6	0.1	0.1				206	167	8
6063 A96063	Japan	JIS H4080	6063	Tub As Mfg	0.1	0.1	0.35	0.45-0.9	0.1	0.2-0.6	0.1	0.1				118	49	12
6063 A96063	Japan	JIS H4080	6063	Tub Quick cooled and Art aged	0.1	0.1	0.35	0.45-0.9	0.1	0.2-0.6	0.1	0.1				147	108	8
6063 A96063	Japan	JIS H4080	6063	Tub SHT and Art AH	0.1	0.1	0.35	0.45-0.9	0.1	0.2-0.6	0.1	0.1				206	167	8
6063 A96063	Japan	JIS H4100	6063	As Mfg	0.1	0.1	0.35	0.45-0.9	0.1	0.2-0.6	0.1	0.1				118	49	12
6063 A96063	Japan	JIS H4100	6063	Quick cooled and Art aged	0.1	0.1	0.35	0.45-0.9	0.1	0.2-0.6	0.1	0.1				147	108	8
6063 A96063	Japan	JIS H4100	6063	SHT and Art aged	0.1	0.1	0.35	0.45-0.9	0.1	0.2-0.6	0.1	0.1				206	167	8
6063 A96063	Japan	JIS H4180	6063	Plt SHT and Art AH 3/16 mm diam	0.1	0.1	0.35	0.45-0.9	0.1	0.2-0.6	0.1	0.1				206	167	8
6063 A96063	Norway	NS 17310	NS17301/AIMg0.5Si	Bar Tub	0.1 max	0.1 max	0.35 max	0.45-0.9	0.1 max	0.2-0.6	0.1 max	0.1 max			ET 0.05 max; ET=.15 total;			
6063 A96063	Pan American	COPANT 862	6063		0.1	0.1	0.35	0.45-0.9	0.1	0.2-0.6	0.1	0.1	0.05	0.15				
6063 A96063	Sweden	SIS 144104	4104-04	Bar Tub Nat aged 10/200 mm diam	0.1	0.1	0.35	0.45-0.9	0.1	0.2-0.6	0.1	0.1			Al 98.9	130	70	15
6063 A96063	Sweden	SIS 144104	4104-06	Bar Tub Art aged 10/200 mm diam	0.1	0.1	0.35	0.45-0.9	0.1	0.2-0.6	0.1	0.1			Al 98.9	210	170	12

UNS numbers and US grades are provided as a means of cross referencing chemically similar alloys. Exchangability is only possible after independent examination of specifications. Tensile properties are minimum or typical . UTS and YS as Mpa, El as %. See Appendix for list of abbreviations used in Descriptions.

Worldwide Guide to Equivalent Nonferrous Metals and Alloys

Grade UNS #	Country	Specification	Designation	Description	Cr	Cu	Fe	Mg	Mn	Si	Ti	Zn	OE Max	OT Max	Other	UTS	YS	EL
6063 A96063	UK	BS H19	6063	Bar Rod Wir Tub Pip Ext	0.1 max	0.1 max	0.35 max	0.45-0.9	0.1 max	0.2-0.6	0.1 max	0.1 max			ET 0.05 max; ET 0.15 max			
6063 A96063	USA	AMS 4156	6063	Bar Rod Wir Tub Pip Ext	0.1 max	0.1 max	0.35 max	0.45-0.9	0.1 max	0.2-0.6	0.1 max	0.1 max	0.05	0.15	ET 0.05 max; ET 0.15 max			
6063 A96063	USA	ASME SB221	6063	Bar Rod Wir Tub Pip Ext	0.1 max	0.1 max	0.35 max	0.45-0.9	0.1 max	0.2-0.6	0.1 max	0.1 max	0.05	0.15	ET 0.05 max; ET 0.15 max			
6063 A96063	USA	ASME SB241	6063	Bar Rod Wir Tub Pip Ext	0.1 max	0.1 max	0.35 max	0.45-0.9	0.1 max	0.2-0.6	0.1 max	0.1 max	0.05	0.15	ET 0.05 max; ET 0.15 max			
6063 A96063	USA	ASTM B210	6063	Bar Rod Wir Tub Pip Ext	0.1 max	0.1 max	0.35 max	0.45-0.9	0.1 max	0.2-0.6	0.1 max	0.1 max	0.05	0.15	ET 0.05 max; ET 0.15 max			
6063 A96063	USA	ASTM B221	6063	Bar Rod Wir Tub Pip Ext	0.1 max	0.1 max	0.35 max	0.45-0.9	0.1 max	0.2-0.6	0.1 max	0.1 max	0.05	0.15	ET 0.05 max; ET 0.15 max			
6063 A96063	USA	ASTM B241	6063	Bar Rod Wir Tub Pip Ext	0.1 max	0.1 max	0.35 max	0.45-0.9	0.1 max	0.2-0.6	0.1 max	0.1 max	0.05	0.15	ET 0.05 max; ET 0.15 max			
6063 A96063	USA	ASTM B345	6063	Bar Rod Wir Tub Pip Ext	0.1 max	0.1 max	0.35 max	0.45-0.9	0.1 max	0.2-0.6	0.1 max	0.1 max	0.05	0.15	ET 0.05 max; ET 0.15 max			
6063 A96063	USA	ASTM B429	6063	Bar Rod Wir Tub Pip Ext	0.1 max	0.1 max	0.35 max	0.45-0.9	0.1 max	0.2-0.6	0.1 max	0.1 max	0.05	0.15	ET 0.05 max; ET 0.15 max			
6063 A96063	USA	ASTM B483	6063	Bar Rod Wir Tub Pip Ext	0.1 max	0.1 max	0.35 max	0.45-0.9	0.1 max	0.2-0.6	0.1 max	0.1 max	0.05	0.15	ET 0.05 max; ET 0.15 max			
6063 A96063	USA	ASTM B491	6063	Bar Rod Wir Tub Pip Ext	0.1 max	0.1 max	0.35 max	0.45-0.9	0.1 max	0.2-0.6	0.1 max	0.1 max	0.05	0.15	ET 0.05 max; ET 0.15 max			
6063 A96063	USA	MIL P-25995	6063	Bar Rod Wir Tub Pip Ext	0.1 max	0.1 max	0.35 max	0.45-0.9	0.1 max	0.2-0.6	0.1 max	0.1 max	0.05	0.15	ET 0.05 max; ET 0.15 max			
6063 A96063	USA	QQ A-200/9	6063	Bar Rod Wir Tub Pip Ext OBSOLETE	0.1 max	0.1 max	0.35 max	0.45-0.9	0.1 max	0.2-0.6	0.1 max	0.1 max	0.05	0.15	ET 0.05 max; ET 0.15 max			
6063	USA				0.1 max	0.1 max	0.35 max	0.45-0.9	0.1 max	0.2-0.6	0.1 max	0.1 max	0.05	0.15				
	USA		AlMg0.7Si		0.1 max	0.1 max	0.35 max	0.45-0.9	0.1 max	0.2-0.6	0.1 max	0.1 max	0.05	0.15				
6063 A96063	Czech Republic	CSN 424401	AlMgSi		0.1 max	0.1 max	0.05-0.4	0.4-0.9	0.1 max	0.3-0.7	0.05-0.2	0.2 max						
6063 A96063	UK	BS 1471	HT9	Tub Ann SHT and nat aged 10 mm diam	0.1	0.1	0.4	0.4-0.9	0.1	0.3-0.7	0.2	0.2				155	100	15
6063 A96063	UK	BS 1471	HT9	Tub SHT an PT 10 mm diam	0.1	0.1	0.4	0.4-0.9	0.1	0.3-0.7	0.2	0.2				200	180	8
6063 A96063	UK	BS 1472	HF9	SHT nat Aged 150/200 mm diam Obsolete Replaced by BS EN 586	0.1	0.1	0.4	0.4-0.9	0.1	0.3-0.7	0.2	0.2				125	85	13
6063 A96063	UK	BS 1472	HF9	SHT PT 150/200 mm diam Obsolete Replaced by BS EN 586	0.1	0.1	0.4	0.4-0.9	0.1	0.3-0.7	0.2	0.2				150	130	6
6063 A96063	UK	BS 1474	HE9	Bar Tub Ann 200 mm diam	0.1	0.1	0.4	0.4-0.9	0.1	0.3-0.7	0.2	0.2				140		13
6063 A96063	UK	BS 1474	HE9	Bar Tub AS Mfg 200 mm diam	0.1	0.1	0.4	0.4-0.9	0.1	0.3-0.7	0.2	0.2				100		12
6063 A96063	UK	BS 1474	HE9	Bar Tub SHT and Nat aged 150/200 mm diam	0.1	0.1	0.4	0.4-0.9	0.1	0.3-0.7	0.2	0.2				120	70	16
6063 A96063	UK	BS 1475	HG9	Wir As Mfg SHT and nat aged 10 mm diam	0.1	0.1	0.4	0.4-0.9	0.1	0.3-0.7	0.2	0.2				140		
6063 A96063	UK	BS 1475	HG9	Wir SHT and PT 10 mm diam	0.1	0.1	0.4	0.4-0.9	0.1	0.3-0.7	0.2	0.2				185		
A96101	Germany	DIN	3.3207		0.03 max	0.1 max	0.5 max	0.35-0.8	0.03 max	0.3-0.7		0.1 max	0.03	0.1	Be 0.06 max			
6063 A96063	International	ISO R827	Al-MgSi	PT 1 in; 25 mm diam	0.1	0.1	0.5	0.4-0.9	0.3	0.3-0.7		0.2				147	103	8
6063 A96063	International	ISO R827	Al-MgSi	ST and PT .5 in; 12 mm diam	0.1	0.1	0.5	0.4-0.9	0.3	0.3-0.7		0.2				201	147	8

UNS numbers and US grades are provided as a means of cross referencing chemically similar alloys. Exchangability is only possible after independent examination of specifications. Tensile properties are minimum or typical . UTS and YS as Mpa, El as %. See Appendix for list of abbreviations used in Descriptions.

2-18 Wrought Aluminum

Grade UNS #	Country	Specification	Designation	Description	Cr	Cu	Fe	Mg	Mn	Si	Ti	Zn	OE Max	OT Max	Other	UTS	YS	EL
6063 A96063	International	ISO TR2778	Al-MgSi	Tub SHT and nat aged OBSOLETE Replace by ISO 6363	0.1	0.1	0.5	0.4-0.9	0.3	0.3-0.7		0.2				140	70	14
6063 A96063	International	ISO TR2778	Al-MgSi	Tub ST and PT OBSOLETE Replace by ISO 6363	0.1	0.1	0.5	0.4-0.9	0.3	0.3-0.7		0.2				215	180	8
6101 A96101	Japan	JIS H4180	6101	Plt SHT and Art AH 7 mm diam	0.03	0.1	0.5	0.35-0.8	0.03	0.3-0.7		0.1			B 0.06 min; Be 0.06 max	196	167	10
6101 A96101	Pan American	COPANT 862	6101		0.03	0.1	0.5	0.35-0.8	0.03	0.3-0.7		0.1	0.05	0.15	B 0.06 min; Be 0.06 max			
6063 A96063	Russia	GOST 4784	1310			0.1 max	0.5 max	0.4-0.9	0.1 max	0.3-0.7	0.15 max	0.2 max						
6063 A96063	Russia	GOST 4784	AD31			0.1 max	0.5 max	0.4-0.9	0.1 max	0.3-0.7	0.15 max	0.2 max						
6101 A96101	UK	BS 2898	E91E		0.03 max	0.1 max	0.5 max	0.35-0.8	0.03 max	0.3-0.7		0.1 max	0.03	0.1	Be 0.06 max			
6101 A96101	USA	ASTM B317	6101		0.03 max	0.1 max	0.5 max	0.35-0.8	0.03 max	0.3-0.7		0.1 max	0.05	0.15	B 0.06 max; Be 0.06 max			
6101	USA				0.03 max	0.1 max	0.5 max	0.35-0.8	0.03 max	0.3-0.7		0.1 max		0.1	B 0.06			
6101 A96101	Australia	AS 1866	6101		0.03 max	0.1 max	0.5 max	0.35-0.8	0.03 max	0.3-0.7		0.1 max	0.03		OT min 0.1; B 0.06 max			
6463 A96463	UK		BTRE6			0.2 max	0.15 max	0.4-0.8	0.05 max	0.2-0.5		0.05 max						
6463 A96463	Pan American	COPANT 862	6463			0.2	0.15	0.45-0.9	0.05	0.2-0.6		0.05						
6463 A96463	USA	ASTM B221	6463	Bar Rod Wir Tub Ext Shapes		0.2 max	0.15 max	0.45-0.9	0.05 max	0.2-0.6		0.05 max	0.05	0.15				
6160	USA				0.05 max	0.2 max	0.15 max	0.35-0.6	0.05 max	0.3-0.6		0.05 max	0.05	0.15				
6463	USA					0.2 max	0.15 max	0.45-0.9	0.05 max	0.2-0.6		0.05 max	0.05	0.15				
1135	USA					0.05-0.2	0.05 max		0.04 max		0.03 max	0.1 max	0.03		0.60 Si+Fe; Al 99.35 min; V 0.05 max			
6463 A96463	Australia	AS 1866	6463A	Rod Bar Cooled and Nat aged 12.0 mm diam		0.25	0.15	0.3-0.9	0.05	0.2-0.6						117	62	12
6463 A96463	Australia	AS 1866	6463A	Rod Bar Cooled and Art aged 12.0 mm diam		0.25	0.15	0.3-0.9	0.05	0.2-0.6						151	110	8
6463 A96463	Australia	AS 1866	6463A	Rod Bar SHT and Art aged 3.0 mm diam		0.25	0.15	0.3-0.9	0.05	0.2-0.6						206	172	8
8010	USA				0.2 max	0.1-0.3	0.35-0.7	0.1-0.5	0.1-0.8	0.4 max	0.1 max	0.4 max	0.05	0.15				
6101 A96101	Finland	SFS 2592	E-AlMgSi	Rod Tub ST and aged (shapes)	0.05	0.5	0.3	0.4-0.8	0.1	0.3-0.7		0.2				220	180	10
6101 A96101	Finland	SFS 2592	E-AlMgSi	Rod Tub ST,CF and aged (rod,tub,wir) 6 mm diam	0.05	0.5	0.3	0.4-0.8	0.1	0.3-0.7		0.2				220	180	8
6063 A96063	Sweden	MNC 40E	4104	Bar Tub Ext				0.7		0.4					Al 98.9			
6063 A96063	Finland	SFS 2430	AlMgSi	Wir CF and HT 1.5/3.5 mm diam				0.5		0.5					Al 98	315		3
8017	USA					0.1-0.2	0.55-0.8	0.01-0.05		0.1 max		0.05 max	0.03	0.1	0.003 Li; B 0.04			
8076	USA					0.04 max	0.6-0.9	0.08-0.22		0.1 max		0.05 max	0.03	0.1	B 0.04			
6763	USA					0.04-0.16	0.08 max	0.45-0.9	0.03 max	0.2-0.6		0.03 max	0.05	0.1	V 0.05 max			
	Austria	ONORM M3430	AlMg1		0.1 max	0.05 max	0.45 max	0.7-1.1	0.15 max	0.3 max		0.2 max			ET 0.05 max; ET=.15			

UNS numbers and US grades are provided as a means of cross referencing chemically similar alloys. Exchangability is only possible after independent examination of specifications. Tensile properties are minimum or typical . UTS and YS as Mpa, El as %. See Appendix for list of abbreviations used in Descriptions.

Worldwide Guide to Equivalent Nonferrous Metals and Alloys

Grade UNS #	Country	Specification	Designation	Description	Cr	Cu	Fe	Mg	Mn	Si	Ti	Zn	OE Max	OT Max	Other	UTS	YS	EL
5005A	Germany				0.1 max	0.05 max	0.45 max	0.7-1.1	0.15 max	0.3 max		0.2 max	0.05	0.15				
6101A	UK					0.05 max	0.4 max	0.4-0.9		0.3-0.7				0.1				
	Austria	ONORM M3426	Al99.3			0.05 max	0.05 max	0.05 max	0.05 max		0.05 max	0.1 max			ET 0.05 max; ST 0.07 max; ST=Fe+Si;			
	Yugoslavia	JUS CC2.100	Al99.3			0.08 max	0.6 max			0.25 max	0.04 max				ST 0.15 max; ET 0.03 max; ST = Zn + Cu, Si < Fe			
3103A	Norway				0.1 max	0.1 max	0.7 max	0.3 max	0.7-1.4	0.5 max	0.1 max	0.2 max	0.05	0.15	0.10 Zr+Ti			
8111	USA				0.05 max	0.1 max	0.4-1	0.05 max	0.1 max	0.3-1.1	0.08 max	0.1 max	0.05	0.15				
1230 A91230	Australia	AS 1734	1230	Sh		0.1			0.05		0.03	0.1			ST=0.7; ST=Si+Fe; Al 99.3			
1230 A91230	Canada	CSA HA.4	0.1230	Plt Sh O .010/.032 in diam		0.1			0.05			0.1			Al 99.3	207	97	12
1230 A91230	Canada	CSA HA.4	0.1230	Plt Sh T3 (sheet) .010/.020 in diam		0.1			0.05			0.1			Al 99.3	407	269	12
1230 A91230	Canada	CSA HA.4	0.1230	Plt Sh T42 .010/.02 in diam		0.1			0.05			0.1			Al 99.3		393	234
1230 A91230	Japan	JIS H4160	1N30H			0.1 max		0.05 max	0.05 max			0.05 max			ST 0.7 max; ST=Si+Fe; Al 99.3 min			
	Japan	JIS H4160	IN30			0.1		0.05	0.05			0.05			Al 99.3			
1230	USA					0.1 max		0.05 max	0.05 max		0.03 max	0.1 max	0.03		0.70 Si+Fe; Al 99.3 min; V 0.05 max			
	Austria	ONORM M3430	Al99.85Mg0.5 Si0.5			0.2 max	0.08 max	0.35-0.7	0.03 max	0.35-0.7	0.02 max	0.05 max			ET 0.02 max; ET=.15 total			
3015	USA				0.1 max	0.3 max	0.8 max	0.2-0.7	0.5-0.9	0.6	0.1 max	0.25 max	0.05	0.15				
5305	Europe						0.08 max	0.7-1.1	0.03 max	0.08 max	0.02 max	0.05 max	0.02					
	Germany	DIN 1725 pt1	3.3317/Al99.8 5Mg1				0.08 max	0.7-1.1	0.03 max	0.08 max	0.02 max	0.05 max	0.02	0.15	ST=.15 max			
8079	USA					0.05 max	0.7-1.3			0.05-0.3		0.1 max	0.05	0.15				
6101B	Germany					0.05 max	0.1-0.3	0.35-0.6	0.05 max	0.3-0.6		0.1 max		0.1				
6060 A96060	Austria	ONORM M3430	AlMgSi0.5		0.05 max	0.1 max	0.1-0.3	0.35-0.6	0.1 max	0.3-0.6	0.1 max	0.15 max			ET 0.05 max; ET=.15;			
6060 A96060	Canada	CSA HA.5	0.606	Bar Rod T51 Extd to .126 in diam	0.05 max	0.1 max	0.1-0.3	0.35-0.6	0.1 max	0.3-0.6	0.1 max	0.15 max			ET 0.05 max; ET=.15 total;	150	110	8
6060	Europe				0.05 max	0.1 max	0.1-0.3	0.35-0.6	0.1 max	0.3-0.6	0.1 max	0.15 max	0.05	0.15				
	Europe		AlMgSi		0.05 max	0.1 max	0.1-0.3	0.35-0.6	0.1 max	0.3-0.6	0.1 max	0.15 max	0.05	0.15				
6060 A96060	France	NF A50-701	6060(AGS)		0.05 max	0.1 max	0.1-0.3	0.35-0.6	0.1 max	0.3-0.6	0.1 max	0.15 max			ET 0.05 max; ET=,15			
6060 A96060	France	NF A50411	6060(A-GS)	Bar Wir TubAnn (drawn bar) 25 mm diam	0.05	0.1	0.1-0.3	0.35-0.6	0.1	0.3-0.6	0.15	0.15				180	130	10
6060 A96060	France	NF A50411	6060(A-GS)	Bar Wir Tub Aged (wir) 2/12 mm diam	0.05	0.1	0.1-0.3	0.35-0.6	0.1	0.3-0.6	0.15	0.15				140	80	16
6060 A96060	France	NF A50411	6060(A-GS)	Bar Wir TubQ/T (tub) 150 mm	0.05	0.1	0.1-0.3	0.35-0.6	0.1	0.3-0.6	0.15	0.15				220	170	10

UNS numbers and US grades are provided as a means of cross referencing chemically similar alloys. Exchangability is only possible after independent examination of specifications. Tensile properties are minimum or typical . UTS and YS as Mpa, El as %. See Appendix for list of abbreviations used in Descriptions.

2-20 Wrought Aluminum

Grade UNS #	Country	Specification	Designation	Description	Cr	Cu	Fe	Mg	Mn	Si	Ti	Zn	OE Max	OT Max	Other	UTS	YS	EL
6060 A96060	Germany	DIN 1725 pt1	3.3206/AlMgSi0.5		0.05 max	0.1 max	0.1-0.3	0.35-0.6	0.1 max	0.3-0.6	0.1 max	0.15 max			ET 0.05 max; ET=.15 total			
6060 A96060	Norway	NS 17302	NS17302/AlMgSi	Bar Tub	0.05 max	0.1 max	0.1-0.3	0.35-0.6	0.1 max	0.3-0.6	0.1 max	0.15 max			ET 0.05 max; ET=.15 total			
6060 A96060	Sweden	SIS 144103	4103-00	Bar Tub	0.05	0.1	0.1-0.3	0.35-0.6	0.1	0.3-0.6	0.1	0.15			Al 98.9			
6060 A96060	Sweden	SIS 144103	4103-00	Bar Tub HW 20/200 mm diam	0.05	0.1	0.1-0.3	0.35-0.6	0.1	0.3-0.6	0.1	0.15			Al 98.9	110	50	15
6060 A96060	Sweden	SIS 144103	4103-04	Bar Tub	0.05	0.1	0.1-0.3	0.35-0.6	0.1	0.3-0.6	0.1	0.15			Al 98.9			
6060 A96060	Sweden	SIS 144103	4103-04	Bar Tub Nat aged 10/200 mm diam	0.05	0.1	0.1-0.3	0.35-0.6	0.1	0.3-0.6	0.1	0.15			Al 98.9	120	60	12
6060 A96060	Sweden	SIS 144103	4103-06	Bar Tub	0.05	0.1	0.1-0.3	0.35-0.6	0.1	0.3-0.6	0.1	0.15			Al 98.9			
6060 A96060	Sweden	SIS 144103	4103-06	Bar Tub Art aged 10/200 mm diam	0.05	0.1	0.1-0.3	0.35-0.6	0.1	0.3-0.6	0.1	0.15			Al 98.9	190	150	10
	Germany	DIN 1725 pt1	3.2301/Al99.85MgSi			0.05-0.2	0.08 max	0.35-0.7	0.03 max	0.35-0.7	0.02 max	0.05 max			ST 0.15 max; ET 0.02 max;			
6401	Europe					0.05-0.2	0.4 max	0.35-0.7	0.03 max	0.35-0.7	0.01 max	0.04 max	0.01	0.1				
6106	France				0.2 max	0.25 max	0.35 max	0.4-0.8	0.05-0.2	0.3-0.6		0.1 max		0.1				
6951	USA					0.15-0.4	0.8 max	0.4-0.8	0.1 max	0.2-0.5		0.2 max	0.05	0.15				
6060 A96060	Sweden	MNC 40E	4103	Bar Tub Ext			0.2	0.5		0.4					Al 98.9			
8016	Norway					0.1 max	0.7-1.4	0.1	0.1-0.3	0.2		0.1 max	0.05	0.15				
7072 A97072	USA	ASTM B209	7072	Plt Sh Tub		0.1 max	0.7 max	0.1 max	0.1 max			0.8-1.3	0.05	0.15				
7072 A97072	Germany	DIN 1725 pt1	3.4415/AlZn1			0.1 max		0.1 max	0.1 max			0.8-1.3			ST 0.7 max; ET 0.05 max; ST=Si+Fe; ET=.15 Total;			
7072	USA					0.1 max		0.1 max	0.1 max			0.8-1.3	0.05	0.15	0.7 Si+Fe			
6306	USA					0.05-0.16	0.1 max	0.45-0.9	0.1-0.4	0.2-0.6	0.05 max	0.05 max	0.05	0.15				
6014	Switzerland				0.2 max	0.25 max	0.5 max	0.4-0.8	0.05-0.2	0.3-0.6	0.1 max	0.1 max	0.05	0.15	V 0.05-0.2			
1120	Australia				0.01 max	0.05-0.35	0.4 max	0.2 max	0.01 max	0.1 max		0.05 max	0.03	0.1	0.02 V+Ti; Al 99.2 min; B 0.05; Ga 0.03 max			
	Germany	DIN 1725 pt1	3.3319/AlRMg1			0.8-1.1	0.008 max			0.01 max		0.01 max			ET 0.003 max; ST 0.02 max; ST=Fe+Ti			
5605	Europe						0.008 max	0.8-1.1		0.01 max	0.008 max	0.01 max	0.003		0.008 Fe+Ti			
5505	Europe						0.04 max	0.8-1.1	0.03 max	0.06 max	0.01 max	0.04 max	0.01					
	Germany	DIN 1725 pt1	3.3318/Al99.9Mg1				0.04 max	0.8-1.1	0.03 max	0.06 max	0.01 max	0.04 max			ST 0.1 max; ET 0.01 max			
2219 A92219	Russia		1201						0.8-1.3									
6763 A96763	Pan American	COPANT 862	6763			0.04-0.16	0.08	0.45-0.9	0.03	0.2-0.6		0.03						
6005B	Netherlands				0.1 max	0.1 max	0.3 max	0.4-0.8	0.1 max	0.45-0.8	0.1 max	0.1 max	0.1	0.15				
3011	USA				0.4 max	0.05-0.2	0.7 max		0.8-1.2	0.4 max	0.1 max	0.1 max	0.05	0.15	B 0.1 max; Zr 0.10-0.30			
6006	USA				0.1 max	0.15-0.3	0.35 max	0.45-0.9	0.05-0.2	0.2-0.6	0.1 max	0.1 max	0.1	0.15				
5557 A95557	Australia	AS 1734	5557	Sh Ann 0.50/0.80 mm		0.15	0.12	0.4-0.8	0.1-0.4	0.1						89		17

UNS numbers and US grades are provided as a means of cross referencing chemically similar alloys. Exchangability is only possible after independent examination of specifications. Tensile properties are minimum or typical . UTS and YS as Mpa, El as %. See Appendix for list of abbreviations used in Descriptions.

Worldwide Guide to Equivalent Nonferrous Metals and Alloys

Grade UNS #	Country	Specification	Designation	Description	Cr	Cu	Fe	Mg	Mn	Si	Ti	Zn	OE Max	OT Max	Other	UTS	YS	EL
5557 A95557	Australia	AS 1734	5557	Sh SH and partially Ann to 5/8 hard 0.50/0.80 mm thk		0.15	0.12	0.4-0.8	0.1-0.4	0.1						137		4
1110	France				0.01 max	0.04 max	0.8 max	0.25 max	0.01 max	0.3 max		0.03			0.03 V+Ti; Al 99.1 min; B 0.02			
8016	USA					0.1 max	0.8-1.4	0.1	0.1-0.4	0.3 max		0.1 max	0.05	0.15				
	Austria	ONORM M3430	AlMn		0.1 max	0.1 max	0.7 max	0.3 max	0.9-1.5	0.5 max	0.1 max	0.2 max			ET 0.05 max; ST 0.1 max; ST=Ti+Zn; ET=.15 total			
3103	Europe				0.1 max	0.1 max	0.7 max	0.3 max	0.9-1.5	0.5 max		0.2 max	0.05	0.15	0.10 Zr+Ti			
	Europe		AlMn1		0.1 max	0.1 max	0.7 max	0.3 max	0.9-1.5	0.5 max		0.2 max	0.05	0.15	0.10 Zr+Ti			
3103 A93103	Norway	NS 17405	NS17405/AlMn1	Plt Sh Strp	0.1 max	0.1 max	0.7 max	0.3 max	0.9-1.5	0.5 max		0.2 max			ET 0.05 max; ST 0.1 max; ET=.15 total, ST=Ti+Zr			
	Austria	ONORM M3430	AlFe1Si1			0.1 max	0.5-1	0.1 max	0.1 max	0.4-0.8	0.05 max	0.1 max			ET 0.06 max; . ET=.25 total;			
	Germany	DIN 1725 pt1	3.0915/AlFeSi			0.1 max	0.5-1		0.1 max	0.4-0.8	0.05 max	0.1 max			ET 0.06 max; ET=.25 total;			
8011A	Germany				0.1 max	0.1 max	0.5-1	0.1 max	0.1 max	0.4-0.8	0.05 max	0.1 max	0.05	0.15				
	Norway	NS 17413	NS17413/AlFeSi	Plt Sh Strp		0.1 max	0.5-1		0.1 max	0.4-0.8	0.05 max	0.1 max			ET 0.05 max; ET=.15 total;			
6005A	France				0.3 max	0.3 max	0.35 max	0.4-0.7	0.5 max	0.5-0.9	0.1 max	0.2 max	0.1	0.15	0.12-0.50 Mn+Cr			
	France		AlSiMg(A)		0.3 max	0.3 max	0.35 max	0.4-0.7	0.5 max	0.5-0.9	0.1 max	0.2 max	0.1	0.15	0.12-0.50 Mn+Cr			
7072 A97072	Russia		ATsP1				0.3 max		0.02 max5	0.3 max	0.15 max	0.9-1.3						
	Austria	ONORM M3430	E-AlMgSi			0.02 max	0.1-0.3	0.35-0.6		0.5-0.6		0.15 max			ET 0.03 max; ST 0.03 max; ST=Cr+Mn+Ti+V; ET=.10 total;			
8211	Netherlands				0.15 max	0.1 max	0.5-1	0.1 max	0.05-0.2	0.4-0.8	0.05 max	0.1 max	0.06	0.15				
	South Africa	SABS 712	Al-Mg1		0.05-0.15	0.1	0.5-1.1	0.1	0.1			0.1						
5457	USA					0.2 max	0.1 max	0.8-1.2	0.15-0.45	0.08 max		0.05 max		0.1	V 0.05 max			
5357	USA					0.2 max	0.17 max	0.8-1.2	0.15-0.45	0.12 max		0.05 max		0.15				
6863	France				0.05 max	0.05-0.2	0.15 max	0.5-0.8	0.05 max	0.4-0.6	0.1 max	0.1 max	0.05	0.15				
6008	Switzerland				0.3 max	0.3 max	0.35 max	0.4-0.7	0.3 max	0.5-0.9	0.1 max	0.2 max	0.1	0.15	V 0.05-0.2			
1100 A91100	Russia		A2			0.02 max	0.5 max			0.5 max				1	Al 99 min			
3203	Australia					0.05 max	0.7 max		1-1.5	0.6 max		0.1 max	0.05	0.15	0.0008 max Be for welding			
3203 A93203	Australia	AS 1734	3203	Sh Plt, Ann 0.15/0.20 mm thk		0.05	0.7		1-1.5	0.6		0.1						14
3203 A93203	Australia	AS 1734	3203	Sh Plt SH to 1/2 hard 0.25/0.30 mm thk		0.05	0.7		1-1.5	0.6		0.1						1
3203 A93203	Australia	AS 1734	3203	Sh Plt SH to full hard 0.15/0.50 mm thk		0.05	0.7		1-1.5	0.6		0.1						1

UNS numbers and US grades are provided as a means of cross referencing chemically similar alloys. Exchangability is only possible after independent examination of specifications. Tensile properties are minimum or typical . UTS and YS as Mpa, El as %. See Appendix for list of abbreviations used in Descriptions.

Grade UNS #	Country	Specification	Designation	Description	Cr	Cu	Fe	Mg	Mn	Si	Ti	Zn	OE Max	OT Max	Other	UTS	YS	EL
3203 A93203	Australia	AS 1865	3203	Wir Rod Bar Strp, Ann		0.05	0.7		1-1.5	0.6		0.1						23
3203 A93203	Australia	AS 1865	3203	Wir Rod Bar Strp, SH to 1/4 hard 10 mm diam		0.05	0.7		1-1.5	0.6		0.1						
3203 A93203	Australia	AS 1865	3203	Wir Rod Bar Strp SH to full hard 10 mm diam		0.05	0.7		1-1.5	0.6		0.1						
1200 A91200	Denmark	DS 3012	1200	1/2 hard (rolled and drawn products) .5/6 mm diam		0.05	0.01		0.05			0.1			Al 99	110	90	5
1200 A91200	Denmark	DS 3012	1200	Hard (rolled and drawn products) .5/6 mm diam		0.05	0.01		0.05			0.1			Al 99	140	130	3
1200 A91200	Denmark	DS 3012	1200	As manufactured (extd products)		0.05	0.01		0.05			0.1			Al 99	65		18
1200 A91200	Australia	AS 1734	1200	Sh Plt, Ann 0.15/0.50mm thk		0.05			0.05		0.05	0.1			Si+Fe=1.0 max; Al 99			15
1200 A91200	Australia	AS 1734	1200	Sh Plt, SH to 1/2 hard 0.25/0.30mm Thk		0.05			0.05		0.05	0.1			Si+Fe=1.0 max; Al 99			1
1200 A91200	Australia	AS 1734	1200	Sh Plt, Stain hardened to full hard 0.15/0.50mm thk		0.05			0.05		0.05	0.1			Si+Fe=1.0 max; Al 99			1
1200 A91200	Australia	AS 1865	1200	Wir Rod Bar Strp, Ann		0.05			0.05		0.05	0.1			Si+Fe=1.0 max; Al 99			23
1200 A91200	Australia	AS 1865	1200	Wir Rod Bar Strp, Strain haredened to 1/4 hard 10mm diam		0.05			0.05		0.05	0.1			Si+Fe=1.0 max; Al 99			
1200 A91200	Australia	AS 1865	1200	Wir Rod Bar Strp, Strain haredened to full hard 10 mm diam		0.05			0.05		0.05	0.1			Si+Fe=1.0 max; Al 99			
1200 A91200	Australia	AS 1866	1200	Rod Bar, SH		0.05			0.05		0.05	0.1			Si+Fe=1.0 max; Al 99	75	20	18
1200 A91200	Australia	AS 1867	1200	Tub, Ann		0.05			0.05		0.05	0.1			Si+Fe=1.0 max; Al 99	100		30
1200 A91200	Australia	AS 1867	1200	Tub, SH to 1/4 hard		0.05			0.05		0.05	0.1			Si+Fe=1.0 max; Al 99	96		
1200 A91200	Australia	AS 1867	1200	Tub, SH to full hard		0.05			0.05		0.05	0.1			Si+Fe=1.0 max; Al 99	151		
1200 A91200	Austria	ONORM M3426	Al99			0.05 max			0.05 max		0.05 max	0.1 max			ST 1 max; ST=Si+Fe; ET=.15 total			
1200 A91200	Belgium	NBN P21-001	1200			0.05 max			0.05 max		0.05 max	0.1 max			ET 0.05 max; ST 1 max; ST = Si+Fe; ET = .15 Total; Al 99 min			
1200 A91200	Canada	CSA HA.6	0.1200	Rod Wir O .051/1 in diam		0.05			0.05		0.05	0.1			Al 99	107		
1200 A91200	Canada	CSA HA.6	0.1200	Rod Wir H14 .062/1 in diam		0.05			0.05		0.05	0.1			Al 99	157		
1200 A91200	France	NF A50411	1200(A4)	Bar Wir Tub 1/2 hard (drawn bar) 50 mm diam		0.05			0.05		0.05	0.1			Al 99	110	95	4
1200 A91200	France	NF A50411	1200(A4)	Bar Wir Tub Drawn and ann (wir) 10 mm diam		0.05			0.05		0.05	0.1			Al 99	75	110	30

UNS numbers and US grades are provided as a means of cross referencing chemically similar alloys. Exchangability is only possible after independent examination of specifications. Tensile properties are minimum or typical . UTS and YS as Mpa, El as %. See Appendix for list of abbreviations used in Descriptions.

Grade UNS #	Country	Specification	Designation	Description	Cr	Cu	Fe	Mg	Mn	Si	Ti	Zn	OE Max	OT Max	Other	UTS	YS	EL
1200 A91200	France	NF A50411	1200(A4)	Bar Wir Tub 1/4 hard (tub) in diam		0.05			0.05		0.05	0.1			Al 99	95	75	5
1200 A91200	France	NF A50451	1200(A4)	Sh Plt Ann .4/3.2 mm diam		0.05			0.05		0.05	0.1			Al 99	75		30
1200 A91200	France	NF A50451	1200(A4)	Sh Plt 3/4 hard .4/3.2 mm diam		0.05			0.05		0.05	0.1			Al 99	130	110	4
1200 A91200	France	NF A50451	1200(A4)	Sh Plt Hard .4/3.2 mm diam		0.05			0.05		0.05	0.1			Al 99	150	130	3
1200 A91200	Germany	DIN 1712 pt 3	3.0205/Al99.0	Obsolete Replaced by DIN EN 10210		0.05 max	0.05 max		0.05 max		0.05 max	0.1 max			ST 1 max; ET 0.05 max; ST = Si+Fe; ET = .15 max; Al 99 min			
1200 A91200	Japan	JIS H4000	1200	Sh Plt		0.05			0.05			0.1			Al 99			
1200 A91200	Japan	JIS H4040	1200	Bar As Mfg		0.05			0.05			0.1			Al 99	78	20	
1200 A91200	Japan	JIS H4080	1200	Tub As Mfg		0.05			0.05			0.1			Al 99	78	20	25
1200 A91200	Japan	JIS H4090	1200	Pip Ann 1.3 mm diam		0.05			0.05			0.1			Al 99	78	29	25
1200 A91200	Japan	JIS H4090	1200	Pip 1/2 hard 1.3 mm diam		0.05			0.05			0.1			Al 99	118	98	4
1200 A91200	Japan	JIS H4090	1200	Pip Hard .8 mm diam		0.05			0.05			0.1			Al 99	157		2
1200 A91200	Japan	JIS H4100	1200	As Mfg		0.05			0.05			0.1			Al 99	78	20	
1200 A91200	Japan	JIS H4120	1200	Wir 1/2 hard 25 mm diam		0.05			0.05			0.1			Al 99	108		
1200 A91200	Japan	JIS H4140	1200	die As Mfg 100 mm diam		0.05			0.05			0.1			Al 99	78	29	25
1200 A91200	Japan	JIS Z3232	1200	Rod Ann		0.05			0.05			0.1			Al 99; Be 0.001	78		
1200 A91200	Norway	NS 17005	NS17005/Al99.0	Plt Sh Strp		0.05 max			0.05 max		0.05 max	0.1 max		0.15	ET 0.05 max; ST 1 max; ST=Si+Fe, Al 99 max			
1200 A91200	Pan American	COPANT 862	1200			0.05			0.05		0.05	0.1			Al 99			
1200	USA					0.05 max			0.05 max		0.05 max	0.1 max	0.05	0.15	1.00 Si+Fe; Al 99 min			
	USA		Al 99.0			0.05 max			0.05 max		0.05 max	0.1 max	0.05	0.15	1.00 Si+Fe; Al 99 min			
	India	IS 733	19000	Bar Rod As Mfg	0.058 max	0.7	0.2			0.5		0.1			Al 99	65		18
	India	IS 736	19000	Plt As Mfg 28 mm diam	0.058 max	0.7	0.2			0.5		0.1			Al 99	70		28
	India	IS 736	19000	Plt Ann 28 mm diam	0.058 max	0.7	0.2			0.5		0.1			Al 99	70		28
	India	IS 736	19000	Plt 1/2 hard 7 mm diam	0.058 max	0.7	0.2			0.5		0.1			Al 99	110		7
1200 A91200	Czech Republic	CSN 424009	Al99		0.1 max	0.6 max				0.5 max	0.15 max	0.1 max			Al 99 min			
	India	IS 738	T1C	Wir Ann	0.1	0.7			0.1	0.5					Al 99	98		
	India	IS 738	T1C	Wir Ann and CW to hard 80 mm diam	0.1	0.7			0.1	0.5					Al 99	118		
	India	IS 739	G1C	Wir Ann	0.1	0.7			0.1	0.5					Al 99	98		
	India	IS 739	G1C	Wir Ann and CW to hard	0.1	0.7			0.1	0.5					Al 99	137		
	India	IS 5902	19000	Wir Bar As Drawn	0.1	0.7	0.1		0.1	0.5	0.1	0.1			Al 99; Ni 0.1; Pb 0.05; Sn 0.05 max	108		
	India	IS 5909	19000	Sh Strp Ann 30 mm diam	0.1	0.7	0.1		0.1	0.5	0.15	0.1			Al 99; Ni 0.1; Sn 0.05 max	79		30

UNS numbers and US grades are provided as a means of cross referencing chemically similar alloys. Exchangability is only possible after independent examination of specifications. Tensile properties are minimum or typical. UTS and YS as Mpa, El as %. See Appendix for list of abbreviations used in Descriptions.

2-24 Wrought Aluminum

Grade UNS #	Country	Specification	Designation	Description	Cr	Cu	Fe	Mg	Mn	Si	Ti	Zn	OE Max	OT Max	Other	UTS	YS	EL
	India	IS 5909	19000	Sh Strp Half hard 7 mm diam		0.1	0.7	0.1	0.1	0.5	0.15	0.1			Al 99; Ni 0.1; Sn 0.05 max	108		7
	India	IS 5909	19000	Sh Strp hard 3 mm diam		0.1	0.7	0.1	0.1	0.5	0.15	0.1			Al 99; Ni 0.1; Sn 0.05 max	137		3
1200 A91200	UK		L116			0.1 max	0.7 max		0.1 max	0.5 max		0.1 max			ET 0.05 max; Al 99 min			
	UK	BS 1470	S1C	Plt Sh Strp Ann .2/6 mm diam OBSOLETE Replaced by BS EN 515		0.1	0.7		0.1	0.5		0.1			Al 99	70		20
	UK	BS 1470	S1C	Plt Sh Strp SH to hard .2/3 mm diam OBSOLETE Replaced by BS EN 515		0.1	0.7		0.1	0.5		0.1			Al 99	140		2
	UK	BS 1471	T1C	Tub Ann 12 mm diam		0.1	0.7		0.1	0.5		0.1			Al 99	105		
	UK	BS 1471	T1C	Tub SH to 1/2 hard 12 mm diam		0.1	0.7		0.1	0.5		0.1			Al 99	110		
	UK	BS 1471	T1C	Tub SH to hard 12 mm diam		0.1	0.7		0.1	0.5		0.1			Al 99	140		
	UK	BS 1474	E1C	Bar Tub As Mfg		0.1	0.7		0.1	0.5		0.1			Al 99	65		18
1200 A91200	Finland	SFS 2584	Al99.0	Rod Tub HF (shapes)		0.1	0.8		0.1	0.5		0.1			Al 99			
1200 A91200	Finland	SFS 2584	Al99.0	Rod Tub Ann (except shapes)		0.1	0.8		0.1	0.5		0.1			Al 99			
1200 A91200	Finland	SFS 2584	Al99.0	Rod Tub 1/2 hard (except shapes) 6mm diam		0.1	0.8		0.1	0.5		0.1			Al 99			
	International	ISO R827	Al99.0	As Mfg		0.1	0.8		0.1	0.5		0.1			Al 99	59		18
	International	ISO TR2136	Al99.0	Sh Plt Ann		0.1	0.8		0.1	0.5		0.1			Al 99	70		20
	International	ISO TR2136	Al99.0	Sh Plt HB SH		0.1	0.8		0.1	0.5		0.1			Al 99	90		4
	International	ISO TR2136	Al99.0	Sh Plt HD SH		0.1	0.8		0.1	0.5		0.1			Al 99	105		3
	International	ISO TR2778	Al99.0	Tub ann OBSOLETE Replace by ISO 6363		0.1	0.8		0.1	0.5		0.1			Al 99	70		20
	International	ISO TR2778	Al99.0	Tub SH OBSOLETE Replace by ISO 6363		0.1	0.8		0.1	0.5		0.1			Al 99	105		3
	Yugoslavia	JUS CC2.100	Al99.0			0.1 max	0.8 max	0.1 max	0.1 max	0.5 max	0.04 max				ST 0.15 max; ET 0.04 max; ST = Zn + Cu, Si < Fe; ET=1.0 total;			
6004	USA					0.1 max	0.1-0.3	0.4-0.7	0.2-0.6	0.3-0.6		0.05 max		0.15				
6005 A96005	Pan American	COPANT 862	6005		0.1	0.1	0.35	0.4-0.6	0.1	0.6-0.9	0.1	0.1						
6005 A96005	USA	ASTM B221	6005	Bar Rod Wir Tub Ext Shapes	0.1 max	0.1 max	0.35 max	0.4-0.6	0.1 max	0.6-0.9	0.1 max	0.1 max	0.05	0.15				
6005	USA				0.1 max	0.1 max	0.35 max	0.4-0.6	0.1 max	0.6-0.9	0.1	0.1 max	0.1	0.15				
	USA		AlSiMg		0.1 max	0.1 max	0.35 max	0.4-0.6	0.1 max	0.6-0.9	0.1	0.1 max	0.1	0.15				
1200A	Norway				0.1 max	0.1 max			0.3 max	0.3 max		0.1 max	0.05	0.15	1.00 Si+Fe; Al 99 min			
1100 A91100	International	ISO R827	Al99.0Cu	As mfg		0.05-0.2	0.8		0.1	0.5		0.1			Al 99	59		18

UNS numbers and US grades are provided as a means of cross referencing chemically similar alloys. Exchangability is only possible after independent examination of specifications. Tensile properties are minimum or typical . UTS and YS as Mpa, El as %. See Appendix for list of abbreviations used in Descriptions.

Worldwide Guide to Equivalent Nonferrous Metals and Alloys

Grade UNS #	Country	Specification	Designation	Description	Cr	Cu	Fe	Mg	Mn	Si	Ti	Zn	OE Max	OT Max	Other	UTS	YS	EL
1100 A91100	International	ISO TR2136	Al99.0Cu	Sh Plt Ann		0.05-0.2	0.8		0.1	0.5		0.1			Al 99	75		20
1100 A91100	International	ISO TR2136	Al99.0Cu	Sh Plt HB SH		0.05-0.2	0.8		0.1	0.5		0.1			Al 99	95		4
1100 A91100	International	ISO TR2136	Al99.0Cu	Sh Plt HD SH		0.05-0.2	0.8		0.1	0.5		0.1			Al 99	110		3
1100 A91100	International	ISO TR2778	Al99.0Cu	Tub Ann OBSOLETE Replace by ISO 6363		0.05-0.2	0.8		0.1	0.5		0.1			Al 99	70		20
1100 A91100	International	ISO TR2778	Al99.0Cu	Tub SH OBSOLETE Replace by ISO 6363		0.05-0.2	0.8		0.1	0.5		0.1			Al 99	110		3
3103 A93103	Czech Republic	CSN 424432	AlMn1			0.2 max	0.7 max	0.3 max	1-1.6	0.6 max		0.1 max			Fe+Si:0.70			
1100 A91100	Japan	JIS H4100	1100	As Mfg		0.05-0.2	1		0.05			0.1			Al 99	78	20	
1100 A91100	USA	AMS 4003	1100	Bar Rod Wir Plt Sh Tub Frg Ext		0.05-0.2			0.05 max			0.1 max			ST 0.15; ST=Si+Fe 0.95 max; Al 99 min			
1100 A91100	Australia	AS 1734	1100	Sh, Ann		0.05-0.2			0.05			0.1			0.95 Si+Fe; Al 99	107		15
1100 A91100	Australia	AS 1734	1100	Sh, SH to 1/2 hard 0.25/0.30mm thk		0.05-0.2			0.05			0.1			0.95 Si+Fe; Al 99	110		1
1100 A91100	Australia	AS 1734	1100	Sh, SH to full hard 0.15/0.50mm thk		0.05-0.2			0.05			0.1			0.95 Si+Fe; Al 99	151		1
1100 A91100	Australia	AS 1866	1100	Bar Rod, SH		0.05-0.2			0.05			0.1			Al 99	75	20	18
1100 A91100	Belgium	NBN P21-001	1100			0.05-0.2			0.05 max			0.1 max			ET 0.05 max; ST 0.95 max; ET = .15 total; ST = Si+Fe; Al 99 min			
1100 A91100	Canada	CSA HA.4	0.1100	Plt Sh O .006/.019 in diam		0.05-0.2			0.05			0.1			Al 99	76	24	15
1100 A91100	Canada	CSA HA.4	0.1100	Plt Sh H12 .032/.05 in diam		0.05-0.2			0.05			0.1			Al 99	97	76	6
1100 A91100	Canada	CSA HA.4	0.1100	Plt Sh H14 .009/.012 in diam		0.05-0.2			0.05			0.1			Al 99	110	97	1
1100 A91100	Canada	CSA HA.5	0.1100	Bar Rod Wir Drawn and O		0.05-0.2			0.05			0.1			Al 99	76	21	25
1100 A91100	Canada	CSA HA.5	0.1100	Bar Rod Wir Drawn and H14 .374 in diam		0.05-0.2			0.05			0.1			Al 99	110		
1100 A91100	Canada	CSA HA.5	0.1100	Bar Rod Wir Extruded and O		0.05-0.2			0.05			0.1			Al 99	76	21	25
1100 A91100	Canada	CSA HA.6	0.1100	Rod, Wir		0.05-0.2			0.05			0.1			Al 99; Be 0.001			
1100 A91100	Canada	CSA HA.7	0.1100	Tub Pip Drawn and O .014/.5 in diam		0.05-0.2			0.05			0.1			Al 99.9	107		
1100 A91100	Canada	CSA HA.7	0.1100	Tub Pip Drawn and H18 .014/.5 in diam		0.05-0.2			0.05			0.1			Al 99.9	152		
1100 A91100	Canada	CSA HA.7	0.1100	Tub Pip Ext and O		0.05-0.2			0.05			0.1			Al 99.9	76	21	25
1100 A91100	Canada	CSA HA.7.1	0.1100	Tub Pip Drawn and O 0.14/.5 in diam		0.05-0.2			0.05			0.1			Al 99.9	107		
1100 A91100	Canada	CSA HA.7.1	0.1100	Tub Pip Drawn and H18 .014/.5 in diam		0.05-0.2			0.05			0.1			Al 99.9	152		

UNS numbers and US grades are provided as a means of cross referencing chemically similar alloys. Exchangability is only possible after independent examination of specifications. Tensile properties are minimum or typical . UTS and YS as Mpa, El as %. See Appendix for list of abbreviations used in Descriptions.

2-26 Wrought Aluminum

Grade UNS #	Country	Specification	Designation	Description	Cr	Cu	Fe	Mg	Mn	Si	Ti	Zn	OE Max	OT Max	Other	UTS	YS	EL
1100 A91100	Canada	CSA HA.7.1	0.1100	Tub Pip Ext and O		0.05-0.2			0.05			0.1			Al 99.9	76	21	25
1100 A91100	France	NF A50411	1100(A45)	Bar Wir 1/2 hard (drawn bar) 50 mm diam		0.05-0.2			0.05			0.1			Al 99	110	95	4
1100 A91100	France	NF A50411	1100(A45)	Bar Wir Drawn and ann (wir) 10 mm diam		0.05-0.2			0.05			0.1			Al 99	75	110	30
1100 A91100	France	NF A50411	1100(A45)	Bar Wir 1/4 hard (tub) 50 mm diam		0.05-0.2			0.05			0.1			Al 99	95	75	5
1100 A91100	France	NF A50451	1100(A45)	Sh Plt Ann .4/3.2 mm diam		0.05-0.2			0.05			0.1			Al 99	75		30
1100 A91100	France	NF A50451	1100(A45)	Sh Plt 3/4 hard .4/3.2 mm diam		0.05-0.2			0.05			0.1			Al 99	130	110	4
1100 A91100	France	NF A50451	1100(A45)	Sh Plt Hard .4/3.2 mm diam		0.05-0.2			0.05			0.1			Al 99	150	130	3
1100 A91100	Japan	JIS H4000	1100	Sh Plt Hard .3 mm diam		0.05-0.2			0.05			0.1			Al 99	157		1
1100 A91100	Japan	JIS H4000	1100	Sh Plt HR 12 mm diam		0.05-0.2			0.05			0.1			Al 99	88	49	9
1100 A91100	Japan	JIS H4000	1100	Sh Plt Ann 1.3 mm diam		0.05-0.2			0.05			0.1			Al 99	78	29	25
1100 A91100	Japan	JIS H4040	1100	Bar As Mfg		0.05-0.2			0.05			0.1			Al 99	78	20	
1100 A91100	Japan	JIS H4080	1100	Tub As Mfg		0.05-0.2			0.05			0.1			Al 99	78	20	25
1100 A91100	Japan	JIS H4090	1100	Pip Ann 1.3 mm diam		0.05-0.2			0.05			0.1			Al 99	78	29	25
1100 A91100	Japan	JIS H4090	1100	Pip 1/2 hard 1.3 mm diam		0.05-0.2			0.05			0.1			Al 99	118	98	4
1100 A91100	Japan	JIS H4090	1100	Pip Hard .8 mm diam		0.05-0.2			0.05			0.1			Al 99	157		2
1100 A91100	Japan	JIS H4120	1100	Wir 1/2 hard 25 mm diam		0.05-0.2			0.05			0.1			Al 99	108		
1100 A91100	Japan	JIS Z3232	1100	Rod Ann		0.05-0.2			0.05			0.1			Al 99; Be 0.001	78		
1100 A91100	Norway	NS 17006	NS17006/Al9 9.0Cu	Plt Sh Strp		0.05-0.2			0.05 max			0.1 max			ET 0.05 max; ST 0.95 max; ET=.15 total; ST=Si+Fe; Al 99 max			
1100 A91100	Pan American	COPANT 862	1100			0.05-0.2			0.05			0.1			Al 99			
1100 A91100	USA	AMS 4001	1100	Bar Rod Wir Plt Sh Tub Frg Ext		0.05-0.2			0.05 max			0.1 max			ST 0.15; ST=Si+Fe 0.95 max; Al 99 min			
1100 A91100	USA	AMS 4062	1100	Bar Rod Wir Plt Sh Tub Frg Ext		0.05-0.2			0.05 max			0.1 max			ST 0.15; ST=Si+Fe 0.95 max; Al 99 min			
1100 A91100	USA	AMS 4102	1100	Bar Rod Wir Plt Sh Tub Frg Ext		0.05-0.2			0.05 max			0.1 max			ST 0.15; ST=Si+Fe 0.95 max; Al 99 min			
1100 A91100	USA	AMS 4180	1100	Bar Rod Wir Plt Sh Tub Frg Ext		0.05-0.2			0.05 max			0.1 max			ST 0.15; ST=Si+Fe 0.95 max; Al 99 min			
1100 A91100	USA	ASME SB209	1100	Bar Rod Wir Plt Sh Tub Frg Ext		0.05-0.2			0.05 max			0.1 max			ST 0.15; ST=Si+Fe 0.95 max; Al 99 min			
1100 A91100	USA	ASME SB221	1100	Bar Rod Wir Plt Sh Tub Frg Ext		0.05-0.2			0.05 max			0.1 max			ST 0.15; ST=Si+Fe 0.95 max; Al 99 min			
1100 A91100	USA	ASME SB241	1100	Bar Rod Wir Plt Sh Tub Frg Ext		0.05-0.2			0.05 max			0.1 max			ST 0.15; ST=Si+Fe 0.95 max; Al 99 min			
1100 A91100	USA	ASTM B209	1100	Bar Rod Wir Plt Sh Tub Frg Ext		0.05-0.2			0.05 max			0.1 max			ST 0.15; ST=Si+Fe 0.95 max; Al 99 min			

UNS numbers and US grades are provided as a means of cross referencing chemically similar alloys. Exchangability is only possible after independent examination of specifications. Tensile properties are minimum or typical . UTS and YS as Mpa, El as %. See Appendix for list of abbreviations used in Descriptions.

Worldwide Guide to Equivalent Nonferrous Metals and Alloys

Grade UNS #	Country	Specification	Designation	Description	Cr	Cu	Fe	Mg	Mn	Si	Ti	Zn	OE Max	OT Max	Other	UTS	YS	EL
1100 A91100	USA	ASTM B210	1100	Bar Rod Wir Plt Sh Tub Frg Ext		0.05-0.2			0.05 max			0.1 max			ST 0.15; ST=Si+Fe 0.95 max; Al 99 min			
1100 A91100	USA	ASTM B211	1100	Bar Rod Wir Plt Sh Tub Frg Ext		0.05-0.2			0.05 max			0.1 max			ST 0.15; ST=Si+Fe 0.95 max; Al 99 min			
1100 A91100	USA	ASTM B241	1100	Bar Rod Wir Plt Sh Tub Frg Ext		0.05-0.2			0.05 max			0.1 max			ST 0.15; ST=Si+Fe 0.95 max; Al 99 min			
1100 A91100	USA	ASTM B247	1100	Bar Rod Wir Plt Sh Tub Frg Ext		0.05-0.2			0.05 max			0.1 max			ST 0.15; ST=Si+Fe 0.95 max; Al 99 min			
1100 A91100	USA	ASTM B313	1100	Bar Rod Wir Plt Sh Tub Frg Ext		0.05-0.2			0.05 max			0.1 max			ST 0.15; ST=Si+Fe 0.95 max; Al 99 min			
1100 A91100	USA	ASTM B316	1100	Bar Rod Wir Plt Sh Tub Frg Ext		0.05-0.2			0.05 max			0.1 max			ST 0.15; ST=Si+Fe 0.95 max; Al 99 min			
1100 A91100	USA	ASTM B483	1100	Bar Rod Wir Plt Sh Tub Frg Ext		0.05-0.2			0.05 max			0.1 max			ST 0.15; ST=Si+Fe 0.95 max; Al 99 min			
1100 A91100	USA	ASTM B491	1100	Bar Rod Wir Plt Sh Tub Frg Ext		0.05-0.2			0.05 max			0.1 max			ST 0.15; ST=Si+Fe 0.95 max; Al 99 min			
1100 A91100	USA	ASTM B547	1100	Bar Rod Wir Plt Sh Tub Frg Ext		0.05-0.2			0.05 max			0.1 max			ST 0.15; ST=Si+Fe 0.95 max; Al 99 min			
1100 A91100	USA	QQ A-250/1	1100	Bar Rod Wir Plt Sh Tub Frg Ext		0.05-0.2			0.05 max			0.1 max			ST 0.15; ST=Si+Fe 0.95 max; Al 99 min			
A91100	USA	AWS A 5.10-92	ER1100	Weld El Rod		0.05-0.2			0.05 max			0.1 max	0.05	0.15	Si+Fe<=0.95%; Al 99 min; Be 0.0008 max			
A91100	USA	AWS A 5.10-92	R1100	Weld El Rod		0.05-0.2			0.05 max			0.1 max	0.05	0.15	Si+Fe<=0.95%; Al 99 min; Be 0.0008 max			
A91100	USA	AWS A5.3-91	E1100			0.05-0.2			0.05 max			0.1 max	0.05	0.15	Si + Fe <=0.95%; Al 99 min; Be 0.0008 max			
1100	USA					0.05-0.2			0.05 max			0.1 max	0.05	0.15	0.95 Si+Fe; 0.0008 max Be for welding; Al 99 min			
	USA		Al 99.0 Cu			0.05-0.2			0.05 max			0.1 max	0.05	0.15	0.95 Si+Fe; 0.0008 max Be weld; Al 99 min			
3016	USA				0.1 max	0.3 max	0.8 max	0.5-0.8	0.5-0.9	0.6 max	0.1 max	0.25 max	0.05	0.15				
6005 A96005	France	NF A50411	6005A(A-SG0.5)	Bar Wir Tub Q/T (drawn bar)	0.3	0.3	0.35	0.4-0.7	0.5	0.5-0.9	0.1	0.2				265	235	8
6005 A96005	France	NF A50411	6005A(A-SG0.5)	Bar Wir Tub Q/T (wir) 2/25 mm diam	0.3	0.3	0.35	0.4-0.7	0.5	0.5-0.9	0.1	0.2				265	235	8
6005 A96005	France	NF A50411	6005A(A-SG0.5)	Bar Wir Tub Q/T (tub) 150 mm diam	0.3	0.3	0.35	0.4-0.7	0.5	0.5-0.9	0.1	0.2				265	235	8
6005 A96005	Norway	NS 17303	NS17303/AlSiMg	Bar Tub	0.3 max	0.3 max	0.35 max	0.4-0.7	0.5 max	0.5-0.9	0.1 max	0.2 max			ET 0.05 max; ST 0.12-0.5; ET=.15 total, ST=Mn+Cr;			
	International	ISO R209 (1971)	Al 99.0			0.10 max	0.8 max		0.1 max	0.5 max		0.1 max			Cu+Si+Fe+Mn+Zn 1.0 max			
	Sweden	MNC 40E	4102	Wir				0.5		0.5								
6005 A96005	Sweden	MNC 40E	4107					0.5		0.8					Al 98.7			

UNS numbers and US grades are provided as a means of cross referencing chemically similar alloys. Exchangability is only possible after independent examination of specifications. Tensile properties are minimum or typical . UTS and YS as Mpa, El as %. See Appendix for list of abbreviations used in Descriptions.

2-28 Wrought Aluminum

Grade UNS #	Country	Specification	Designation	Description	Cr	Cu	Fe	Mg	Mn	Si	Ti	Zn	OE Max	OT Max	Other	UTS	YS	EL
1100 A91100	Japan	JIS H4140	1100	Die As Mfg 100 mm diam								0.1			Al 99	78	29	25
1100 A91100	Mexico	DGN W-30	2S-0	Sh Rolled .25/.52 mm diam											Al 99			
1100 A91100	Mexico	DGN W-30	2S-H12	Sh Rolled .53/.80 mm diam											Al 99			
1100 A91100	Mexico	DGN W-30	2S-H14	Sh Rolled .53/.80 mm diam											Al 99			
1100 A91100	Mexico	DGN W-30	2S-H16	Sh Rolled .25/.52 mm diam											Al 99			
1100 A91100	Mexico	DGN W-30	2S-H18	Sh Rolled .25/.52 mm diam											Al 99			
1200 A91200	Sweden	MNC 40E	4010	Sh											Al 99 min			
3003 A93003	Japan	JIS H4000	3203	Sh Plt Ann 1.3 mm diam		0.05	0.7		1-1.5	0.6		0.1				98	39	23
3003 A93003	Japan	JIS H4000	3203	Sh Plt Hard 1.3 mm diam		0.05	0.7		1-1.5	0.6		0.1				186	167	3
3003 A93003	Japan	JIS H4000	3203	Sh Plt HR 12 mm diam		0.05	0.7		1-1.5	0.6		0.1				118	67	8
3003 A93003	Japan	JIS H4040	3203	Bar As Mfg (bar)		0.05	0.7		1-1.5	0.6		0.1				98	39	
3003 A93003	Japan	JIS H4040	3203	Bar 1/2 hard 10 mm diam		0.05	0.7		1-1.5	0.6		0.1				137		
3003 A93003	Japan	JIS H4040	3203	Bar Hard 10 mm diam		0.05	0.7		1-1.5	0.6		0.1				186		
3003 A93003	Japan	JIS H4080	3203	tub As Mfg		0.05	0.7		1-1.5	0.6		0.1				98	39	
3003 A93003	Japan	JIS H4090	3003	Pip Ann		0.05	0.7		1-1.5	0.6		0.1						
3003 A93003	Japan	JIS H4090	3003	Pip		0.05	0.7		1-1.5	0.6		0.1						
3003 A93003	Japan	JIS H4090	3003	Pip		0.05	0.7		1-1.5	0.6		0.1						
6063A	UK				0.05 max	0.1 max	0.15-0.35	0.6-0.9	0.15 max	0.3-0.6	0.1 max	0.15 max	0.05	0.15				
	UK		AlMg0.7Si(A)		0.05 max	0.1 max	0.15-0.35	0.6-0.9	0.15 max	0.3-0.6	0.1 max	0.15 max	0.05	0.15				
3003 A93003	India	IS 736	31000	Plt As Mfg 23 mm diam	0.2	0.1	0.7	0.1	0.8-1.5	0.6	0.2	0.2				95		23
3003 A93003	India	IS 736	31000	Plt ANN 22 mm diam	0.2	0.1	0.7	0.1	0.8-1.5	0.6	0.2	0.2				90		22
3003 A93003	India	IS 736	31000	Plt 1/2 hard 5 mm diam	0.2	0.1	0.7	0.1	0.8-1.5	0.6	0.2	0.2				130		5
3003 A93003	India	IS 5909	31000	Sh Strp ANN 30 mm diam	0.05	0.1	0.7	0.1	1-1.5	0.6	0.15	0.1			Ni 0.2; Pb 0.05; Sn 0.05 max	93		30
3003 A93003	India	IS 5909	31000	Sh Strp 1/2 Hard 7 mm diam	0.05	0.1	0.7	0.1	1-1.5	0.6	0.15	0.1			Ni 0.2; Pb 0.05; Sn 0.05 max	137		7
3003 A93003	India	IS 5909	31000	Sh Strp hard 3 mm diam	0.05	0.1	0.7	0.1	1-1.5	0.6	0.15	0.1			Ni 0.2; Pb 0.05; Sn 0.05 max	177		3
3003 A93003	UK	BS 1470	3103(N3)	OBSOLETE Replaced by BS EN 485, 515, 573	0.2 max	0.1 max	0.7		1-1.5	0.6 max		0.2 max						
6105	USA				0.1 max	0.1 max	0.35 max	0.45-0.8	0.15 max	0.6-1	0.1 max	0.1 max	0.1	0.15				
3003 A93003	Australia	AS 1734	3003	Sh Plt, Ann 0.30/0.80 mm thk		0.05-0.2	0.7		1-1.5	0.6		0.1		0.15				20
3003 A93003	Australia	AS 1734	3003	Sh Plt, SH to 1/2 hard 0.25/0.30 mm thk		0.05-0.2	0.7		1-1.5	0.6		0.1		0.15				1
3003 A93003	Australia	AS 1734	3003	Sh Plt, SH to full hard 0.15/0.50 mm thk		0.05-0.2	0.7		1-1.5	0.6		0.1		0.15				1
	Austria	ONORM M3430	AlMnCu			0.05-0.2	0.7 max		1-1.5	0.6 max		0.1 max			ET 0.05 max;			

UNS numbers and US grades are provided as a means of cross referencing chemically similar alloys. Exchangability is only possible after independent examination of specifications. Tensile properties are minimum or typical. UTS and YS as Mpa, El as %. See Appendix for list of abbreviations used in Descriptions.

Worldwide Guide to Equivalent Nonferrous Metals and Alloys

Grade UNS #	Country	Specification	Designation	Description	Cr	Cu	Fe	Mg	Mn	Si	Ti	Zn	OE Max	OT Max	Other	UTS	YS	EL
3003 A93003	Belgium	NBN P21-001	3003			0.05-0.2	0.7 max		1-1.5	0.6 max		0.1 max			ET 0.05 max; ET = .15 Total			
3003 A93003	Canada	CSA HA.4	0.3003	Sh Plt O		0.05-0.2	0.7		1-1.5	0.6		0.1				97	34	18
3003 A93003	Canada	CSA HA.4	0.3003	Sh Plt H14		0.05-0.2	0.7		1-1.5	0.6		0.1				138	117	1
3003 A93003	Canada	CSA HA.4	0.3003	Sh Plt H18		0.05-0.2	0.7		1-1.5	0.6		0.1					165	1
3003 A93003	Canada	CSA HA.7	0.3003	Tub Pip Drawn and O .025/.049 in diam		0.05-0.2	0.7		1-1.5	0.6		0.1				97	34	30
3003 A93003	Canada	CSA HA.7	0.3003	Tub Pip Drawn and H18 .024 in diam		0.05-0.2	0.7		1-1.5	0.6		0.1				186	165	2
3003 A93003	Canada	CSA HA.7	0.3003	Tub Pip Extd and O		0.05-0.2	0.7		1-1.5	0.6		0.1				97	34	25
3003 A93003	Canada	CSA HA.7.1	0.3003	Tub Pip Drawn and O .025/.049 in diam		0.05-0.2	0.7		1-1.5	0.6		0.1				97	34	30
3003 A93003	Canada	CSA HA.7.1	0.3003	Tub Pip Drawn and H18 .024 in diam		0.05-0.2	0.7		1-1.5	0.6		0.1				186	165	2
3003 A93003	Canada	CSA HA.7.1	0.3003	Tub Pip Extd and O		0.05-0.2	0.7		1-1.5	0.6		0.1				97	34	
3003 A93003	France	NF A50-701	AM1			0.05-0.2	0.7 max		1-1.5	0.6 max		0.1 max			ET 0.05 max; ET=.15;			
3003 A93003	France	NF A50411	3003(A-M1)	Bar Tub HArd (drawn bar) 10 mm diam		0.05-0.2	0.7		1-1.5	0.6		0.1				185	145	2
3003 A93003	France	NF A50411	3003(A-M1)	Bar Tub Hard (wir) 6 mm diam		0.05-0.2	0.7		1-1.5	0.6		0.1				180	140	2
3003 A93003	France	NF A50411	3003(A-M1)	Bar Tub Drawn or extd (tub) 150 mm diam		0.05-0.2	0.7		1-1.5	0.6		0.1				95		12
3003 A93003	France	NF A50451	3003(A-M1)	Sh Plt Ann .4/3.2 mm diam		0.05-0.2	0.7		1-1.5	0.6		0.1				100		25
3003 A93003	France	NF A50451	3003(A-M1)	Sh Plt 3/4 hard .4/1.6 mm diam		0.05-0.2	0.7		1-1.5	0.6		0.1				170	150	3
3003 A93003	France	NF A50451	3003(A-M1)	Sh Plt Hard .4/1.6 mm diam		0.05-0.2	0.7		1-1.5	0.6		0.1				190	170	3
3003 A93003	Germany	DIN 1725 pt1	3.0517/AlMnCu			0.05-0.2	0.7 max		1-1.5	0.6 max		0.1 max			ET 0.15 max; ET=.25 total;			
3003 A93003	International	ISO TR2136	Al-Mn1Cu	Sh Plt Ann		0.05-0.2	0.7		1-1.5	0.6		0.2				95		20
3003 A93003	International	ISO TR2136	Al-Mn1Cu	Sh Plt HB SH		0.05-0.2	0.7		1-1.5	0.6		0.2				115		4
3003 A93003	International	ISO TR2136	Al-Mn1Cu	Sh Plt HD SH		0.05-0.2	0.7		1-1.5	0.6		0.2				140		3
3003 A93003	International	ISO TR2778	Al-Mn1Cu	Tub Ann OBSOLETE Replace by ISO 6363		0.05-0.2	0.7		1-1.5	0.6		0.2				95		20
3003 A93003	International	ISO TR2778	Al-Mn1Cu	Tub SH 50 mm diam OBSOLETE Replace by ISO 6363		0.05-0.2	0.7		1-1.5	0.6		0.2				140		3
3003 A93003	Japan	JIS H4000	3003	Sh Plt Ann 1.3 mm diam		0.05-0.2	0.7		1-1.5	0.6		0.1				96	39	23
3003 A93003	Japan	JIS H4000	3003	Sh Plt Hard 1.3 mm diam		0.05-0.2	0.7		1-1.5	0.6		0.1				186	167	3
3003 A93003	Japan	JIS H4000	3003	Sh Plt HR 12 mm diam		0.05-0.2	0.7		1-1.5	0.6		0.1				118	67	8
3003 A93003	Japan	JIS H4040	3003	Bar As Mfg (bar)		0.05-0.2	0.7		1-1.5	0.6		0.1				98	39	

UNS numbers and US grades are provided as a means of cross referencing chemically similar alloys. Exchangability is only possible after independent examination of specifications. Tensile properties are minimum or typical . UTS and YS as Mpa, El as %. See Appendix for list of abbreviations used in Descriptions.

2-30 Wrought Aluminum

Grade UNS #	Country	Specification	Designation	Description	Cr	Cu	Fe	Mg	Mn	Si	Ti	Zn	OE Max	OT Max	Other	UTS	YS	EL
3003 A93003	Japan	JIS H4040	3003	Bar 1/2 hard 10 mm diam		0.05-0.2	0.7		1-1.5	0.6		0.1				137		
3003 A93003	Japan	JIS H4040	3003	Bar Hard 10 mm diam		0.05-0.2	0.7		1-1.5	0.6		0.1				186		
3003 A93003	Japan	JIS H4080	3003	Tub As Mfg		0.05-0.2	0.7		1-1.5	0.6		0.1				98	39	
3003 A93003	Japan	JIS H4090	3003	Pip Ann 1.3 mm diam		0.05-0.2	0.7		1-1.5	0.6		0.1				98	39	23
3003 A93003	Japan	JIS H4090	3003	Pip 1/2 hard 1.3 mm diam		0.05-0.2	0.7		1-1.5	0.6		0.1				137	118	4
3003 A93003	Japan	JIS H4090	3003	Pip Hard 1.3 mm diam		0.05-0.2	0.7		1-1.5	0.6		0.1				186	167	3
3003 A93003	Japan	JIS H4100	3003	As Mfg		0.05-0.2	0.7		1-1.5	0.6		0.1				98	39	
3003 A93003	Japan	JIS H4160	3003			0.05-0.2	0.7		1-1.5	0.6		0.1						
3003 A93003	Norway	NS 17406	NS17406/AIMn1Cu	Plt Sh Strp		0.05-0.2	0.7 max		1-1.5	0.6 max		0.1 max			ET 0.05 max; ET=.15 total			
3003 A93003	Pan American	COPANT 862	3003			0.05-0.2	0.7		1-1.5	0.6		0.1						
3003 A93003	Russia	GOST 4784	1400			0.2 max	0.7 max	0.05 max	1-1.6	0.6 max		0.1 max						
3003 A93003	Russia		AMts			0.2 max	0.7 max	0.05 max	1-1.6	0.6 max		0.1 max			ET 0.1 max;			
A93003	USA	AWS	E3003			0.05-0.2	0.7 max		1-1.5	0.6 max		0.1 max	0.05	0.15	Be 0.0008 max			
3003	USA					0.05-0.2	0.7 max		1-1.5	0.6 max		0.1 max	0.05	0.15				
3303	USA					0.05-0.2	0.7 max		1-1.5	0.6 max		0.3 max	0.05	0.15				
	USA		AlMn1Cu			0.05-0.2	0.7 max		1-1.5	0.6 max		0.1 max	0.05	0.15				
6205	USA				0.05-0.15	0.2 max	0.7 max	0.4-0.6	0.05-0.15	0.6-0.9	0.15 max	0.25 max	0.15	0.15	Zr 0.05-0.15			
3006	USA				0.2 max	0.1-0.3	0.7 max	0.3-0.6	0.5-0.8	0.5 max	0.1 max	0.15-0.4	0.05	0.15				
6151 A96151	Canada	CSA HA.8	0.6151	Frg T6 (die frg) 4 in diam	0.15-0.35	0.35	1	0.45-0.8	0.2	0.6-1.2	0.15	0.25				303	255	10
6151 A96151	Canada	CSA SG11P	6151	Frg	0.15-0.35	0.35 max	1	0.45-0.8	0.2 max	0.6-1.2	0.15 max	0.25 max	0.05	0.15				
6151 A96151	Japan	JIS H4140	6151	Die SHT and art AH 100 mm diam	0.15-0.35	0.35	1	0.45-0.8	0.2	0.6-1.2	0.15	0.25				304	255	14
6151 A96151	USA	AMS 4125	6151	Frg	0.15-0.35	0.35 max	1 max	0.45-0.8	0.2 max	0.6-1.2	0.15 max	0.25 max	0.05	0.15				
6151 A96151	USA	MIL A-22771	6151	Frg	0.15-0.35	0.35 max	1 max	0.45-0.8	0.2 max	0.6-1.2	0.15 max	0.25 max	0.05	0.15				
6151 A96151	USA	QQ A-367	6151	Frg	0.15-0.35	0.35 max	1 max	0.45-0.8	0.2 max	0.6-1.2	0.15 max	0.25 max	0.05	0.15				
6151 A96151	USA	SAE J454	6151	Frg	0.15-0.35	0.35 max	1 max	0.45-0.8	0.2 max	0.6-1.2	0.15 max	0.25 max	0.05	0.15				
6151	USA				0.15-0.35	0.35 max	1 max	0.45-0.8	0.2 max	0.6-1.2	0.15 max	0.25 max	0.05	0.15				
3003 A93003	Denmark	DS 3012	3003	1/2 hard .5/6 mm diam		0.2-0.5	0.7		1-1.5	0.6		0.1				120	100	5
3003 A93003	Denmark	DS 3012	3003	3/4 hard .5/6 mm diam		0.2-0.5	0.7		1-1.5	0.6		0.1				130	120	4
3003 A93003	Denmark	DS 3012	3003	Hard 1.5 mm diam		0.2-0.5	0.7		1-1.5	0.6		0.1				155	140	3
6151 A96151	Russia	GOST 4784	AV			0.2-0.6	0.5 max	0.45-0.9		0.5-1.2		0.2 max			ST 0.15-0.35; ET 0.1 max; min, ST=Mn or Cr, ET=.80 total			
3003 A93003	Russia		AMtsM				0.61 max		1.37 max	0.21 max	0.09 max							
3003 A93003	Mexico	DGN W-30	3S-0	Sh Rolled .25/.52 mm diam					1.2						Al 97.8			

UNS numbers and US grades are provided as a means of cross referencing chemically similar alloys. Exchangability is only possible after independent examination of specifications. Tensile properties are minimum or typical . UTS and YS as Mpa, El as %. See Appendix for list of abbreviations used in Descriptions.

Worldwide Guide to Equivalent Nonferrous Metals and Alloys

Grade UNS #	Country	Specification	Designation	Description	Cr	Cu	Fe	Mg	Mn	Si	Ti	Zn	OE Max	OT Max	Other	UTS	YS	EL
3003 A93003	Mexico	DGN W-30	3S-H12	Sh Rolled .53/.80 mm diam					1.2						Al 97.8			
3003 A93003	Mexico	DGN W-30	3S-H14	Sh Rolled .53/.80 mm diam					1.2						Al 97.8			
3003 A93003	Mexico	DGN W-30	3S-H16	Sh Rolled .25/.52 mm diam					1.3						Al 97.7			
3003 A93003	Mexico	DGN W-30	3S-H18	Sh Rolled .25/.52 mm diam					1.2						Al 97.8			
6201A	Australia					0.04 max	0.5 max	0.6-0.9		0.5-0.7				0.1	B 0.06			
6201 A96201	Australia	AS 1866	6201A			0.04 max	0.5 max	0.6-0.9		0.5-0.7		0.03			OT min 0.1; B 0.06 max			
	Austria	ONORM M3430	AlMg1.5		0.1 max	0.05 max	0.45 max	1.1-1.7	0.15 max	0.4 max		0.2 max			ET 0.05 max; ET=.15 total;			
	Germany	DIN 1725 pt1	3.3316/AlMg1.5		0.1 max	0.05 max	0.45 max	1.1-1.7	0.15 max	0.4 max		0.2 max			ET 0.05 max; ET=.15 total;			
5050B	Germany				0.1 max	0.05 max	0.45 max	1.1-1.7	0.15 max	0.4 max		0.2 max	0.05	0.15				
	Sweden	SIS 144104	4104-00	Bar Tub Hw 10/200 mm diam	0.1	0.1	0.35	0.45-0.9	0.1	0.2-0.6	0.1	0.1			Al 98.9	120	60	18
6201 A96201	Pan American	COPANT 862	6201		0.03	0.1	0.5	0.6-0.9	0.03	0.5-0.9	0.03	0.1			V 0.06			
6201	USA				0.03 max	0.1 max	0.5 max	0.6-0.7	0.03 max	0.5-0.9		0.1 max		0.1	B 0.06			
6301	USA				0.1 max	0.1 max	0.7 max	0.6-0.9	0.15 max	0.5-0.9	0.15 max	0.25 max	0.15	0.15				
8011 A98011	Australia	AS 1734	8011	O .3-.80 mm thk	0.05 max	0.1 max	0.6-1	0.05 max	0.1 max	0.5-0.9	0.08 max	0.1 max	0.05		OT min 0.15	75		20
8011 A98011	Australia	AS 1734	8011	H12 .80-1.30 mm thk	0.05 max	0.1 max	0.6-1	0.05 max	0.1 max	0.5-0.9	0.08 max	0.1 max	0.05		OT min 0.15	100		6
8011 A98011	Australia	AS 1734	8011	H14 .80-1.30 mm thk	0.05 max	0.1 max	0.6-1	0.05 max	0.1 max	0.5-0.9	0.08 max	0.1 max	0.05		OT min 0.15	115		4
8011 A98011	Australia	AS 1734	8011	H16 .80-1.30 mm thk	0.05 max	0.1 max	0.6-1	0.05 max	0.1 max	0.5-0.9	0.08 max	0.1 max	0.05		OT min 0.15	138		3
8011 A98011	Australia	AS 1734	8011	H18 .50-.80 mm thk	0.05 max	0.1 max	0.6-1	0.05 max	0.1 max	0.5-0.9	0.08 max	0.1 max	0.05		OT min 0.15	160		2
8011 A98011	Australia	AS 1734	8011		0.05 max	0.1 max	0.6-1	0.05 max	0.1 max	0.5-0.9	0.08 max	0.1 max	0.05		OT min 0.15			
8011	USA				0.05 max	0.1 max	0.6-1	0.05 max	0.2 max	0.5-0.9	0.08 max	0.1 max	0.05	0.15				
6201 A96201	USA	ASTM B398	6201	Rod Wir	0.03 max	0.1 max	0.5 max	0.6-0.9	0.03 max	0.5-0.95		0.1 max	0.03	0.1	B 0.06 max; Be 0.06 max			
6201 A96201	USA	ASTM B399	6201	Rod Wir	0.03 max	0.1 max	0.5 max	0.6-0.9	0.03 max	0.5-0.95		0.1 max	0.03	0.1	B 0.06 max; Be 0.06 max			
6201 A96201	USA	SAE J454	6201	Rod Wir	0.03 max	0.1 max	0.5 max	0.6-0.9	0.03 max	0.5-0.95		0.1 max	0.03	0.1	B 0.06 max; Be 0.06 max			
6053	USA				0.15-0.35	0.1 max	0.35 max	1.1-1.4				0.1 max	0.05	0.15	45-65% of Mg			
7072 A97072	Australia	AS 1734	7072	Sh		0.1		0.1	0.1			0.8-1.3		0.15	ST 0.7; ST=Si+Fe			
7075 A97075	Canada	CSA HA.4	.7075Alclad	Plt Sh O .015/.032 in diam		0.1		0.1	0.1			0.8-1.3				248	138	10
7075 A97075	Canada	CSA HA.4	.7075Alclad	Plt Sh T6 .012/.020 in diam		0.1		0.1	0.1			0.8-1.3				483	414	7
7075 A97075	Canada	CSA HA.4	.7075Alclad	Plt Sh T62 (sht) .021/.039 in diam		0.1		0.1	0.1			0.8-1.3				483	414	7
7072 A97072	Pan American	COPANT 862	7072			0.1		0.1	0.1			0.8-1.3						
5050 A95050	Australia	AS 1734	5050A	Sh Plt Ann 0.20/0.50 mm	0.1	0.2	0.7	1.1-1.8	0.3	0.4		0.25					41	16
5050 A95050	Australia	AS 1734	5050A	Sh Plt SH to 1/2 hard 0.25/0.80 mm thk	0.1	0.2	0.7	1.1-1.8	0.3	0.4		0.25					137	3

UNS numbers and US grades are provided as a means of cross referencing chemically similar alloys. Exchangability is only possible after independent examination of specifications. Tensile properties are minimum or typical . UTS and YS as Mpa, El as %. See Appendix for list of abbreviations used in Descriptions.

2-32 Wrought Aluminum

Grade UNS #	Country	Specification	Designation	Description	Cr	Cu	Fe	Mg	Mn	Si	Ti	Zn	OE Max	OT Max	Other	UTS	YS	EL
5050 A95050	Australia	AS 1734	5050A	Sh Plt SH and stab to full hard 0.20/0.80 mm thk	0.1	0.2	0.7	1.1-1.8	0.3	0.4		0.25						2
5050A	Australia				0.1 max	0.2 max	0.7 max	1.1-1.8	0.3 max	0.4 max		0.25 max	0.05	0.15				
5050 A95050	Belgium	NBN P21-001	5050		0.1 max	0.2 max	0.7 max	1.1-1.8	0.1 max	0.4 max		0.25 max			ET 0.05 max; ; ET = .15 Total			
5050 A95050	France	NF A-G1	5050	Bar Rod Wir Plt Sh Tub	0.1 max	0.2 max	0.7 max	1.1-1.8	0.1 max	0.4 max		0.25 max	0.05	0.15				
5050 A95050	International	ISO R827	Al-Mg1.5	As mfg	0.1	0.2	0.7	1.1-1.8	0.3	0.4		0.2				123		12
5050 A95050	International	ISO TR2136	Al-Mg1.5	Sh Plt ann	0.1	0.2	0.7	1.1-1.8	0.3	0.4		0.2				125		18
5050 A95050	International	ISO TR2136	Al-Mg1.5	Sh Plt HB SH	0.1	0.2	0.7	1.1-1.8	0.3	0.4		0.2				145	110	3
5050 A95050	International	ISO TR2136	Al-Mg1.5	Sh Plt HD SH	0.1	0.2	0.7	1.1-1.8	0.3	0.4		0.2				165	140	2
5050 A95050	International	ISO TR2778	Al-Mg1.5	Tub ann OBSOLETE Replace by ISO 6363	0.1	0.2	0.7	1.1-1.8	0.3	0.4		0.2				125		18
5050 A95050	International	ISO TR2778	Al-Mg1.5	Tub 50 mm diam OBSOLETE Replace by ISO 6363	0.1	0.2	0.7	1.1-1.8	0.3	0.4		0.2				165	140	3
5050 A95050	Pan American	COPANT 862	5050		0.1	0.2	0.7	1.1-1.8	0.1	0.4		0.25						
5050 A95050	UK	BS 2L44	5050	Bar Rod Wir Plt Sh Tub	0.1 max	0.2 max	0.7 max	1.1-1.8	0.1 max	0.4 max		0.25 max	0.05	0.15				
5050 A95050	USA	ASTM B209	5050	Bar Rod Wir Plt Sh Tub	0.1 max	0.2 max	0.7 max	1.1-1.8	0.1 max	0.4 max		0.25 max	0.05	0.15				
5050 A95050	USA	ASTM B210	5050	Bar Rod Wir Plt Sh Tub	0.1 max	0.2 max	0.7 max	1.1-1.8	0.1 max	0.4 max		0.25 max	0.05	0.15				
5050 A95050	USA	ASTM B313	5050	Bar Rod Wir Plt Sh Tub	0.1 max	0.2 max	0.7 max	1.1-1.8	0.1 max	0.4 max		0.25 max	0.05	0.15				
5050 A95050	USA	ASTM B483	5050	Bar Rod Wir Plt Sh Tub	0.1 max	0.2 max	0.7 max	1.1-1.8	0.1 max	0.4 max		0.25 max	0.05	0.15				
5050 A95050	USA	ASTM B547	5050	Bar Rod Wir Plt Sh Tub	0.1 max	0.2 max	0.7 max	1.1-1.8	0.1 max	0.4 max		0.25 max	0.05	0.15				
5050 A95050	USA	SAE J454	5050	Bar Rod Wir Plt Sh Tub	0.1 max	0.2 max	0.7 max	1.1-1.8	0.1 max	0.4 max		0.25 max	0.05	0.15				
5050	USA				0.1 max	0.2 max	0.7 max	1.1-1.8	0.1 max	0.4 max		0.25 max	0.05	0.15				
	USA		AlMg1.5(C)		0.1 max	0.2 max	0.7 max	1.1-1.8	0.1 max	0.4 max		0.25 max	0.05	0.15				
	South Africa	SABS 712	Al-MgSi(64-631)			0.05-0.2	0.15	0.45-0.9	0.05	0.2-0.6		0.2						
6162	USA				0.1 max	0.2 max	0.5 max	0.7-1.1	0.1 max	0.4-0.8	0.1 max	0.25 max	0.05	0.15				
6205 A96205	USA				0.05-0.15	0.2 max	0.7 max	0.4-0.6	0.05-0.15	0.6-0.9	0.15 max	0.25 max	0.05	0.15	Zr 0.05-0.15			
	International	ISO R209 (1971)	Al 99.0 Cu			0.05-0.20	0.8 max		0.1 max	0.5 max					Cu+Si+Fe+Mn+ Zn 1.0 max			
6201 A96201	Australia	AS 1865	6201A				0.5 max	0.6-0.9		0.5-0.7		0.03			OT min 0.1; B 0.06 max			
6206	Switzerland				0.1 max	0.2-0.5	0.35 max	0.45-0.8	0.13-0.3	0.35-0.7	0.1 max	0.2 max	0.1	0.15				
	South Africa	SABS 712	Al-MgSi-EC(61-014)			0.05	0.4	0.4-0.9		0.3-0.7								
6003	USA				0.35 max	0.1 max	0.6 max	0.8-1.5	0.8 max	0.35-1	0.1 max	0.2 max	0.1	0.15				
8021	Japan					0.05 max	1.2-1.7			0.15 max			0.05	0.15				
8021A	UK					0.05 max	1.2-1.7	0.02 max	0.03 max	0.2 max	0.05 max	0.05 max	0.03	0.15				

UNS numbers and US grades are provided as a means of cross referencing chemically similar alloys. Exchangability is only possible after independent examination of specifications. Tensile properties are minimum or typical . UTS and YS as Mpa, El as %. See Appendix for list of abbreviations used in Descriptions.

Worldwide Guide to Equivalent Nonferrous Metals and Alloys

Grade UNS #	Country	Specification	Designation	Description	Cr	Cu	Fe	Mg	Mn	Si	Ti	Zn	OE Max	OT Max	Other	UTS	YS	EL
3008	Germany				0.05 max	0.1 max	0.7 max	0.01 max	1.2-1.8	0.4 max	0.1 max	0.05 max	0.05	0.15	Ni 0.05 max; Zr 0.10-0.50			
5006 A95006	Canada	CSA HA.4	0.5006	Plt Sh H 26 .0472/.0984 in diam	0.1 max	0.1 max	0.8 max	0.8-1.3	0.4-0.8	0.4 max	0.1 max	0.25 max			ET 0.05 max; ET=.15 total;	180		4
5006	USA				0.1 max	0.1 max	0.8 max	0.8-1.3	0.4-0.8	0.4 max	0.1 max	0.25 max	0.05	0.15				
6017	USA				0.1 max	0.05-0.2	0.15-0.3	0.45-0.6	0.1 max	0.55-0.7	0.05 max	0.05 max	0.05	0.15				
6015	Italy				0.1 max	0.1-0.25	0.1-0.3	0.8-1.1	0.1 max	0.2-0.4	0.1 max	0.1 max	0.05	0.15				
3005 A93005	Australia	AS 1734	3005	Sh Plt, Ann 0.15/0.20 mm thk	0.1	0.3	0.7	0.2-0.6	1-1.5	0.6	0.1	0.25						14
3005 A93005	Australia	AS 1734	3005	Sh Plt, SH to 1/2 hard 0.15/0.25 mm thk	0.1	0.3	0.7	0.2-0.6	1-1.5	0.6	0.1	0.25						1
3005 A93005	Australia	AS 1734	3005	Sh Plt, SH to full hard 0.30/0.80 mm thk	0.1	0.3	0.7	0.2-0.6	1-1.5	0.6	0.1	0.25						1
	Austria	ONORM M3430	ALMn1Mg0.5		0.1 max	0.3 max	0.7 max	0.2-0.6	1-1.5	0.6 max	0.1 max	0.25 max			ET 0.05 max; ET=.15			
3005 A93005	Belgium	NBN P21-001	3005		0.1 max	0.3 max	0.7 max	0.2-0.6	1-1.5	0.6 max	0.1 max	0.25 max			ET 0.05 max; ; ET = .15 Total			
3005 A93005	France	NF A50451	3005(A-MG0.5)	Sh Plt Ann .4/1.6 mm diam	0.1	0.3	0.7	0.2-0.6	1-1.5	0.6	0.1	0.25				115		22
3005 A93005	France	NF A50451	3005(A-MG0.5)	Sh Plt 3/4 hard .4/1.6 mm diam	0.1	0.3	0.7	0.2-0.6	1-1.5	0.6	0.1	0.25				200	180	3
3005 A93005	France	NF A50451	3005(A-MG0.5)	Sh Plt Hard .4/1.6 mm diam	0.1	0.3	0.7	0.2-0.6	1-1.5	0.6	0.1	0.25				230	200	3
3005 A93005	France	NF A50501	3005(A-MG0.5)	Tub As manufactured	0.1	0.3	0.7	0.2-0.6	1-1.5	0.6	0.1	0.25				170	130	6
3005 A93005	Germany	DIN 1725 pt1	3.0525/AlMn1Mg0.5		0.1 max	0.3 max	0.7 max	0.2-0.6	1-1.5	0.6 max	0.1 max	0.25 max			ET 0.05 max; ET=.15 total			
3005 A93005	Japan	JIS H4001	3005	Sh Ann at 1/4 hard .8/1.2 mm diam	0.1	0.3	0.7	0.2-0.6	1-1.5	0.6	0.1	0.25				137	118	2
3005 A93005	Japan	JIS H4001	3005	Sh Ann /2 hard .8/1.2 mm diam	0.1	0.3	0.7	0.2-0.6	1-1.5	0.6	0.1	0.25				167	147	2
3005 A93005	Japan	JIS H4001	3005	Sh Ann at 3/4 hard .8/1.2 mm diam	0.1	0.3	0.7	0.2-0.6	1-1.5	0.6	0.1	0.25				196	167	2
3005 A93005	Norway	NS 17407	NS17407/AIMn1Mg0.5	Plt Sh Strp	0.1 max	0.3 max	0.7 max	0.2-0.6	1-1.5	0.6 max	0.1 max	0.25 max			ET 0.05 max; ET=.15 total;			
3005 A93005	Pan American	COPANT 862	3005		0.1	0.3	0.7	0.2-0.6	1-1.5	0.6	0.1	0.25						
3005	USA				0.1 max	0.3 max	0.7 max	0.2-0.6	1-1.5	0.6 max	0.1 max	0.25 max	0.05	0.15				
	USA		AlMn1Mg0.5		0.1 max	0.3 max	0.7 max	0.2-0.6	1-1.5	0.6 max	0.1 max	0.25 max	0.05	0.15				
2010	USA				0.15 max	0.7-1.3	0.5 max	0.4-1	0.1-0.4	0.5 max		0.3 max	0.05	0.15				
	Sweden	MNC 40E	4054	Sh					1.2						Al 98.8			
	Japan	JIS H4000	5N01	Sh Plt Ann .3 mm diam		0.2	0.3	0.2-0.6	0.3	0.2		0.03				88		10
	Japan	JIS H4000	5N01	Sh Plt 3/4 hard .3 mm hard		0.2	0.3	0.2-0.6	0.3	0.2		0.03				147		1
	Japan	JIS H4000	5N01	Sh Plt Hard .3 mm diam		0.2	0.3	0.2-0.6	0.3	0.2		0.03				167		1
6016	Switzerland				0.1 max	0.2 max	0.5 max	0.25-0.6	0.2	1-1.5	0.15 max	0.2 max	0.05	0.15				
6002	Italy				0.05 max	0.1-0.25	0.25 max	0.45-0.7	0.1-0.2	0.6-0.9	0.08 max		0.08	0.15	Zr 0.09-.14			
5150	France					0.1 max	0.1 max	1.3-1.7	0.03 max	0.08 max	0.06 max	0.1 max	0.03	0.1				

UNS numbers and US grades are provided as a means of cross referencing chemically similar alloys. Exchangability is only possible after independent examination of specifications. Tensile properties are minimum or typical . UTS and YS as Mpa, El as %. See Appendix for list of abbreviations used in Descriptions.

2-34 Wrought Aluminum

Grade UNS #	Country	Specification	Designation	Description	Cr	Cu	Fe	Mg	Mn	Si	Ti	Zn	OE Max	OT Max	Other	UTS	YS	EL
	UK	BS 1470	NS8	Plt Sh Strp Ann .2/25 mm diam OBSOLETE Replaced by BS EN 515	0.25	0.1	0.4			0.4	0.2	0.2				275	125	12
	UK	BS 1470	NS8	Plt Sh Strp SH to 1/2 hard .2/6 mm diam OBSOLETE Replaced by BS EN 515	0.25	0.1	0.4			0.4	0.2	0.2				310	235	5
4006	France				0.2 max	0.1 max	0.5-0.8	0.01 max	0.05 max	0.8-1.2		0.05 max	0.05	0.15				
3017	Netherlands				0.15 max	0.25-0.4	0.25-0.45	0.1 max	0.8-1.2	0.25 max	0.05 max	0.1 max	0.05	0.15				
	International	ISO R209 (1971)	Al-Mn 1			0.1 max	0.7 max	0.3 max	0.8-1.5	0.6 max					Ti+Zr+Cr 0.2 max			
	International	ISO R209 (1971)	Al-MgSi		0.10 max	0.10 max	0.5 max	0.4-0.9	0.30 max	0.3-0.7					Ti+Zr 0.2 max			
	International	ISO R209 (1971)	Al-Mg 1		0.1 max	0.20 max	0.7 max	0.5-1.1	0.2 max	0.4 max					Ti+Zr 0.2 max			
	USA		5250			0.1 max	0.1 max	1.3-1.8	0.05-0.15	0.08 max		0.05 max	0.03	0.1	Ga 0.03 max; V 0.05 max			
6081 A96081	Austria	ONORM M3430	AlMgSi0.9		0.1 max	0.1 max	0.5 max	0.6-1	0.45 max	0.7-1.1	0.1 max	0.2 max			ET 0.05 max; ET=.15 total			
6061 A96061	France	NF A50451	6081(A-SGM0.3)	Sh Plt Aged .4/6 mm diam	0.1	0.1	0.5	0.6-1	0.1-0.45	0.7-1.1	0.15	0.2				210	110	18
6061 A96061	France	NF A50451	6081(A-SGM0.3)	Sh Plt Q/T .4/3.2 mm diam	0.1	0.1	0.5	0.6-1	0.1-0.45	0.7-1.1	0.15	0.2				250	150	18
8001	USA					0.15 max	0.45-0.7			0.17 max		0.05 max	0.05	0.15	1.0-1.3 C; 1.2-1.4 Li; 0.20-.70 O; Ni 0.9-1.3			
6103 A96103	Australia	AS 1734	6103		0.35 max	0.2-0.3	0.6 max	0.8-1.5	0.8 max	0.35-1	0.1 max	0.2 max	0.05		OT min 0.15			
6103	Australia				0.35 max	0.2-0.3	0.6 max	0.8-1.5	0.8 max	0.35-1	0.1 max	0.2 max	0.1	0.15				
6061 A96061	Australia	AS 1734	6061	Sh Plt Ann 0.25/0.50 mm Thk	0.04-0.35	0.15-0.4	0.7	0.8-1.2	0.15	0.4-0.8	0.15	0.25				152	83	14
6061 A96061	Australia	AS 1734	6061	Sh Plt SHT and Nat aged 0.25/0.50 mm	0.04-0.35	0.15-0.4	0.7	0.8-1.2	0.15	0.4-0.8	0.15	0.25				206	115	14
6061 A96061	Australia	AS 1734	6061	Sh Plt SHT and Art aged 0.25/0.50 mm thk	0.04-0.35	0.15-0.4	0.7	0.8-1.2	0.15	0.4-0.8	0.15	0.25				289	241	8
6061 A96061	Australia	AS 1865	6061	Wir Rod Bar Strp Ann 10 mm diam	0.04-0.35	0.15-0.4	0.7	0.8-1.2	0.15	0.4-0.8	0.15	0.25				152		
6061 A96061	Australia	AS 1865	6061	Wir Rod Bar Strp (H13) 12 mm diam	0.04-0.35	0.15-0.4	0.7	0.8-1.2	0.15	0.4-0.8	0.15	0.25				151		
6061 A96061	Australia	AS 1865	6061	Wir Rod Bar Strp SHT, CW and artidicially aged 6 mm diam	0.04-0.35	0.15-0.4	0.7	0.8-1.2	0.15	0.4-0.8	0.15	0.25				372	324	
6061 A96061	Australia	AS 1866	6061	Bar Rod Ann	0.04-0.35	0.15-0.4	0.7	0.8-1.2	0.15	0.4-0.8	0.15	0.25				152	110	14
6061 A96061	Australia	AS 1866	6061	Bar Rod SHT and Nat aged	0.04-0.35	0.15-0.4	0.7	0.8-1.2	0.15	0.4-0.8	0.15	0.25				179	110	14
6061 A96061	Australia	AS 1866	6061	Bar Rod SHT and Art aged	0.04-0.35	0.15-0.4	0.7	0.8-1.2	0.15	0.4-0.8	0.15	0.25				262	241	8
6061 A96061	Australia	AS 1867	6061	Tub Ann	0.04-0.35	0.15-0.4	0.7	0.8-1.2	0.15	0.4-0.8	0.15	0.25				152	96	15
6061 A96061	Australia	AS 1867	6061	Tub SHT and Nat aged 0.6/1.2 mm wall Thk	0.04-0.35	0.15-0.4	0.7	0.8-1.2	0.15	0.4-0.8	0.15	0.25				216	110	12

UNS numbers and US grades are provided as a means of cross referencing chemically similar alloys. Exchangability is only possible after independent examination of specifications. Tensile properties are minimum or typical . UTS and YS as Mpa, El as %. See Appendix for list of abbreviations used in Descriptions.

Worldwide Guide to Equivalent Nonferrous Metals and Alloys

Grade UNS #	Country	Specification	Designation	Description	Cr	Cu	Fe	Mg	Mn	Si	Ti	Zn	OE Max	OT Max	Other	UTS	YS	EL
6061 A96061	Australia	AS 1867	6061	Tub SHT,CW, and Art aged	0.04-0.35	0.15-0.4	0.7	0.8-1.2	0.15	0.4-0.8	0.15	0.25				324	268	4
6262 A96262	Australia	AS 1865	6262	Wir Rod Bar Strp SHT and Art aged 3/65 mm diam	0.04-0.14	0.15-0.4	0.7	0.8-1.2	0.15	0.4-0.8	0.15	0.25				290	240	9
6262 A96262	Australia	AS 1865	6262	Wir Rod Bar Strp SHT, Art aged, and CW 3/40 mm diam	0.04-0.14	0.15-0.4	0.7	0.8-1.2	0.15	0.4-0.8	0.15	0.25				310		
6262 A96262	Australia	AS 1866	6262	Rod Bar SHT and Art aged	0.04-0.14	0.15-0.4	0.7	0.8-1.2	0.15	0.4-0.8	0.15	0.25				262	241	8
6061 A96061	Canada	CSA HA.4	0.6061	Plt Sh O .016/.020 in	0.04-0.35	0.15-0.4	0.7	0.8-1.2	0.15	0.4-0.8	0.15	0.25				152	83	14
6061 A96061	Canada	CSA HA.4	0.6061	Plt Sh T4 (sht) .016/.020 in	0.04-0.35	0.15-0.4	0.7	0.8-1.2	0.15	0.4-0.8	0.15	0.25				207	16	14
6061 A96061	Canada	CSA HA.5	0.6061	Bar Rod Wir Drawn and O 2.5 in diam	0.04-0.35	0.15-0.4	0.7	0.8-1.2	0.15	0.4-0.8	0.15	0.25				152		16
6061 A96061	Canada	CSA HA.5	0.6061	Bar Rod Wir Drawn and T42 2.50 in diam	0.04-0.35	0.15-0.4	0.7	0.8-1.2	0.15	0.4-0.8	0.15	0.25				207	97	18
6061 A96061	Canada	CSA HA.6	0.6061	Rod Wir H13 .062/1 in diam	0.04-0.35	0.15-0.4	0.7	0.8-1.2	0.15	0.4-0.8	0.15	0.25				152		
6061 A96061	Canada	CSA HA.6	0.6061	Rod Wir T6 .062/1 in diam	0.04-0.35	0.15-0.4	0.7	0.8-1.2	0.15	0.4-0.8	0.15	0.25				290	241	10
6061 A96061	Canada	CSA HA.7	0.6061	Tub Pip Drawn and O	0.04-0.35	0.15-0.4	0.7	0.8-1.2	0.15	0.4-0.8	0.15	0.25				152	97	15
6061 A96061	Canada	CSA HA.7	0.6061	Tub Pip Drawn and T4 .015/.049 in diam	0.04-0.35	0.15-0.4	0.7	0.8-1.2	0.15	0.4-0.8	0.15	0.25				207	110	16
6061 A96061	Canada	CSA HA.7	0.6061	Tub Pip Extd and O	0.04-0.35	0.15-0.4	0.7	0.8-1.2	0.15	0.4-0.8	0.15	0.25				152	110	16
6061 A96061	Canada	CSA HA.7.1	0.6061	Tub Drawn and O	0.04-0.35	0.15-0.4	0.7	0.8-1.2	0.15	0.4-0.8	0.15	0.25				152	97	15
6061 A96061	Canada	CSA HA.7.1	0.6061	Tub Drawn and T4 .015/.049 in	0.04-0.35	0.15-0.4	0.7	0.8-1.2	0.15	0.4-0.8	0.15	0.25				207		16
6061 A96061	Canada	CSA HA.7.1	0.6061	Tub Extruded and T6510 .250 in diam	0.04-0.35	0.15-0.4	0.7	0.8-1.2	0.15	0.4-0.8	0.15	0.25				262	241	10
6061 A96061	Canada	CSA HA.8	0.6061	Frg T6 (die frg) 4 in diam	0.04-0.35	0.15-0.4	0.7	0.8-1.2	0.15	0.4-0.8	0.15	0.25				262	241	7
6061A	Europe				0.04-0.35	0.15-0.4	0.7 max	0.8-1.2	0.15 max	0.4-0.8	0.15 max	0.25 max	0.05	0.15	Pb 0.003 max			
6061 A96061	France	NF A50411	6061	Bar Wir Tub Aged (extd bar)	0.04-0.35	0.15-0.4	0.7	0.8-1.2	0.15	0.4-0.8	0.15	0.25				180	110	15
6061 A96061	France	NF A50411	6061	Bar Wir Tub Aged (wir) 2/12 mm diam	0.04-0.35	0.15-0.4	0.7	0.8-1.2	0.15	0.4-0.8	0.15	0.25				140	80	16
6061 A96061	France	NF A50411	6061	Bar Wir Tub Q/T .4/6 mm diam	0.04-0.35	0.15-0.4	0.7	0.8-1.2	0.15	0.4-0.8	0.15	0.25				290	240	8
6061 A96061	France	NF A50451	6061	Sh Plt Ann .4/3.2 mm diam	0.04-0.35	0.15-0.4	0.7	0.8-1.2	0.15	0.4-0.8	0.15	0.25					80	20
6061 A96061	France	NF A50451	6061	Sh Plt Aged .4/6 mm diam	0.04-0.35	0.15-0.4	0.7	0.8-1.2	0.15	0.4-0.8	0.15	0.25				210	110	
6061 A96061	France	NF A50451	6061	Sh Plt Q/T .4/6 mm diam	0.04-0.35	0.15-0.4	0.7	0.8-1.2	0.15	0.4-0.8	0.15	0.25				290	240	10
6061 A96061	International	ISO R827	Al-Mg1SiCu		0.04-0.35	0.15-0.4	0.7	0.8-1.2	0.15	0.4-0.8		0.25						
6061 A96061	International	ISO R829	Al-Mg1SiCu	Frg ST and PT	0.04-0.35	0.15-0.4	0.7	0.8-1.2	0.15	0.4-0.8		0.25				265	236	8
6061 A96061	International	ISO TR2778	Al-Mg1SiCu	Tub SHT and nat aged OBSOLETE Replace by ISO 6363	0.04-0.35	0.15-0.4	0.7	0.8-1.2	0.15	0.4-0.8		0.25				200	110	14

UNS numbers and US grades are provided as a means of cross referencing chemically similar alloys. Exchangability is only possible after independent examination of specifications. Tensile properties are minimum or typical . UTS and YS as Mpa, El as %. See Appendix for list of abbreviations used in Descriptions.

2-36 Wrought Aluminum

Grade UNS #	Country	Specification	Designation	Description	Cr	Cu	Fe	Mg	Mn	Si	Ti	Zn	OE Max	OT Max	Other	UTS	YS	EL
6061 A96061	International	ISO TR2778	Al-Mg1SiCu	Tub ST and PT OBSOLETE Replace by ISO 6363	0.04-0.35	0.15-0.4	0.7	0.8-1.2	0.15	0.4-0.8		0.25				285	230	8
6061 A96061	Japan	JIS H4000	6061	Sh Plt Strp Ann 1.5 mm diam	0.04-0.35	0.15-0.4	0.7	0.8-1.2	0.15	0.4-0.8	0.15	0.25				147	78	16
6061 A96061	Japan	JIS H4000	6061	Sh Plt Strp SHT and AH 1.5 mm diam	0.04-0.35	0.15-0.4	0.7	0.8-1.2	0.15	0.4-0.8	0.15	0.25				206	106	16
6061 A96061	Japan	JIS H4000	6061	Sh Plt Strp SHT and Art AH 1.5 mm diam	0.04-0.35	0.15-0.4	0.7	0.8-1.2	0.15	0.4-0.8	0.15	0.25				294	245	10
6061 A96061	Japan	JIS H4040	6061	Bar Ann	0.04-0.35	0.15-0.4	0.7	0.8-1.2	0.15	0.4-0.8	0.15	0.25				147	108	16
6061 A96061	Japan	JIS H4040	6061	Bar SHT and AH	0.04-0.35	0.15-0.4	0.7	0.8-1.2	0.15	0.4-0.8	0.15	0.25				177	108	16
6061 A96061	Japan	JIS H4040	6061	Bar SHT and Art AH 7 mm diam	0.04-0.35	0.15-0.4	0.7	0.8-1.2	0.15	0.4-0.8	0.15	0.25				265	245	10
6061 A96061	Japan	JIS H4080	6061	Tub Ann	0.04-0.35	0.15-0.4	0.7	0.8-1.2	0.15	0.4-0.8	0.15	0.25				147	108	16
6061 A96061	Japan	JIS H4080	6061	Tub SHT and AH	0.04-0.35	0.15-0.4	0.7	0.8-1.2	0.15	0.4-0.8	0.15	0.25				177	108	
6061 A96061	Japan	JIS H4080	6061	Tub SHT and Art AH 7 mm diam	0.04-0.35	0.15-0.4	0.7	0.8-1.2	0.15	0.4-0.8	0.15	0.25				265	245	8
6061 A96061	Japan	JIS H4100	6061	Ann	0.04-0.35	0.15-0.4	0.7	0.8-1.2	0.15	0.4-0.8	0.15	0.25				147	108	16
6061 A96061	Japan	JIS H4120	6061	Wir Ann 25 mm diam	0.04-0.35	0.15-0.4	0.7	0.8-1.2	0.15	0.4-0.8	0.15	0.25						
6061 A96061	Japan	JIS H4120	6061	Wir 3/8 hard 25 mm diam	0.04-0.35	0.15-0.4	0.7	0.8-1.2	0.15	0.4-0.8	0.15	0.25						
6061 A96061	Japan	JIS H4120	6061	Wir Ann 25 mm diam	0.04-0.35	0.15-0.4	0.7	0.8-1.2	0.15	0.4-0.8	0.15	0.25				147		
6061 A96061	Japan	JIS H4120	6061	Wir 3/8 hard 25 mm diam	0.04-0.35	0.15-0.4	0.7	0.8-1.2	0.15	0.4-0.8	0.15	0.25				157		
6061 A96061	Japan	JIS H4120	6061	Wir SHT and Art aH 2/25 mm diam	0.04-0.35	0.15-0.4	0.7	0.8-1.2	0.15	0.4-0.8	0.15	0.25				265	245	10
6061 A96061	Japan	JIS H4140	6061	Die SHT and Art AH 100 mm diam	0.04-0.35	0.15-0.4	0.7	0.8-1.2	0.15	0.4-0.8	0.15	0.25				265	245	10
6061 A96061	Japan	JIS H4180	6061	Plt SHT and Art AH 7 mm diam	0.04-0.35	0.15-0.4	0.7	0.8-1.2	0.15	0.4-0.8	0.15	0.25				265	245	8
6061 A96061	Pan American	COPANT 862	6061		0.04-0.35	0.15-0.4	0.7	0.8-1.2	0.15	0.4-0.8	0.15	0.25						
6262 A96262	Pan American	COPANT 862	6262		0.04-0.14	0.15-0.4	0.7	0.8-1.2	0.15	0.4-0.8	0.15	0.25			Bi 0.4-0.7; Pb 0.4-0.7			
6061 A96061	Russia	GOST 4784	1330		0.05-0.35	0.15-0.4	0.7 max	0.8-1.2	0.15 max	0.4-0.8	0.15 max	0.25 max						
6061 A96061	Russia	GOST 4784	AD33		0.05-0.35	0.15-0.4	0.7 max	0.8-1.2	0.15 max	0.4-0.8	0.15 max	0.25 max						
6061 A96061	UK	BS 1473	HB20	SHT ,CW,and PT 12 mm diam	0.04-0.35	0.15-0.4	0.7	0.8-1.2	0.2-0.8	0.4-0.8	0.2	0.2				310	245	
6061 A96061	UK	BS 1474	HE20	Bar Tub SHT and nat aged 150 mm diam	0.04-0.35	0.15-0.4	0.7	0.8-1.2	0.2-0.8	0.4-0.8	0.2	0.2				190	115	4
6061 A96061	UK	BS 1474	HE20	Bar Tub SHT and PT 150 mm diam	0.04-0.35	0.15-0.4	0.7	0.8-1.2	0.2-0.8	0.4-0.8	0.2	0.2				280	240	7
6061 A96061	UK	BS 1475	HG20	Wir SHT,CW, and PT 6/10 mm diam	0.04-0.35	0.15-0.4	0.7	0.8-1.2	0.2-0.8	0.4-0.8	0.2	0.2				355		
6061 A96061	UK		L117		0.04-0.35	0.15-0.4	0.7 max	0.8-1.2	0.2-0.8	0.4-0.8	0.2 max	0.2 max						

UNS numbers and US grades are provided as a means of cross referencing chemically similar alloys. Exchangability is only possible after independent examination of specifications. Tensile properties are minimum or typical . UTS and YS as Mpa, El as %. See Appendix for list of abbreviations used in Descriptions.

Grade UNS #	Country	Specification	Designation	Description	Cr	Cu	Fe	Mg	Mn	Si	Ti	Zn	OE Max	OT Max	Other	UTS	YS	EL
6262 A96262	UK	BS 1471	6061(H20)		0.15-0.35	0.15-0.4	0.7 max	0.8-1.2	0.2-0.8	0.4-0.8	0.2 max	0.2 max						
6061 A96061	USA	AMS 4025	6061	Bar Rod Wir Plt Sh Tub Pip Frg Ext Shapes	0.04-0.35	0.15-0.4	0.7 max	0.8-1.2	0.15 max	0.4-0.8	0.15 max	0.25 max	0.05	0.15				
6061 A96061	USA	AMS 4026	6061	Bar Rod Wir Plt Sh Tub Pip Frg Ext Shapes	0.04-0.35	0.15-0.4	0.7 max	0.8-1.2	0.15 max	0.4-0.8	0.15 max	0.25 max	0.05	0.15				
6061 A96061	USA	AMS 4027	6061	Bar Rod Wir Plt Sh Tub Pip Frg Ext Shapes	0.04-0.35	0.15-0.4	0.7 max	0.8-1.2	0.15 max	0.4-0.8	0.15 max	0.25 max	0.05	0.15				
6061 A96061	USA	AMS 4043	6061	Bar Rod Wir Plt Sh Tub Pip Frg Ext Shapes	0.04-0.35	0.15-0.4	0.7 max	0.8-1.2	0.15 max	0.4-0.8	0.15 max	0.25 max	0.05	0.15				
6061 A96061	USA	AMS 4053	6061	Bar Rod Wir Plt Sh Tub Pip Frg Ext Shapes	0.04-0.35	0.15-0.4	0.7 max	0.8-1.2	0.15 max	0.4-0.8	0.15 max	0.25 max	0.05	0.15				
6061 A96061	USA	AMS 4115	6061	Bar Rod Wir Plt Sh Tub Pip Frg Ext Shapes	0.04-0.35	0.15-0.4	0.7 max	0.8-1.2	0.15 max	0.4-0.8	0.15 max	0.25 max	0.05	0.15				
6061 A96061	USA	AMS 4116	6061	Bar Rod Wir Plt Sh Tub Pip Frg Ext Shapes	0.04-0.35	0.15-0.4	0.7 max	0.8-1.2	0.15 max	0.4-0.8	0.15 max	0.25 max	0.05	0.15				
6061 A96061	USA	AMS 4117	6061	Bar Rod Wir Plt Sh Tub Pip Frg Ext Shapes	0.04-0.35	0.15-0.4	0.7 max	0.8-1.2	0.15 max	0.4-0.8	0.15 max	0.25 max	0.05	0.15				
6061 A96061	USA	ASTM B209	6061	Bar Rod Wir Plt Sh Tub Pip Frg Ext Shapes	0.04-0.35	0.15-0.4	0.7 max	0.8-1.2	0.15 max	0.4-0.8	0.15 max	0.25 max	0.05	0.15				
6061 A96061	USA	ASTM B211	6061	Bar Rod Wir Plt Sh Tub Pip Frg Ext Shapes	0.04-0.35	0.15-0.4	0.7 max	0.8-1.2	0.15 max	0.4-0.8	0.15 max	0.25 max	0.05	0.15				
6061 A96061	USA	ASTM B308	6061	Structural Shp Rolled Ext T6	0.04-0.35	0.15-0.4	0.7 max	0.8-1.2	0.15 max	0.4-0.8	0.15 max	0.25 max	0.05	0.15		260	240	10
6061 A96061	USA	ASTM B632	6061	Bar Rod Wir Plt Sh Tub Pip Frg Ext Shapes	0.04-0.35	0.15-0.4	0.7 max	0.8-1.2	0.15 max	0.4-0.8	0.15 max	0.25 max	0.05	0.15				
6061 A96061	USA	MIL F-17132	6061	Bar Rod Wir Plt Sh Tub Pip Frg Ext Shapes	0.04-0.35	0.15-0.4	0.7 max	0.8-1.2	0.15 max	0.4-0.8	0.15 max	0.25 max	0.05	0.15				
6061 A96061	USA	QQ A-225/8	6061	Bar Rod Wir Plt Sh Tub Pip Frg Ext Shapes	0.04-0.35	0.15-0.4	0.7 max	0.8-1.2	0.15 max	0.4-0.8	0.15 max	0.25 max	0.05	0.15				
6061 A96061	USA	QQ A-250/11	6061	Bar Rod Wir Plt Sh Tub Pip Frg Ext Shapes	0.04-0.35	0.15-0.4	0.7 max	0.8-1.2	0.15 max	0.4-0.8	0.15 max	0.25 max	0.05	0.15				
6262 A96262	USA	ASTM B210	6262	Bar Rod Wir Tub Ext Shapes	0.04-0.14	0.15-0.4	0.7 max	0.8-1.2	0.15 max	0.4-0.8	0.15 max	0.25 max	0.05	0.15	Bi 0.4-0.7; Pb 0.4-0.7			
6262 A96262	USA	ASTM B211	6262	Bar Rod Wir Tub Ext Shapes	0.04-0.14	0.15-0.4	0.7 max	0.8-1.2	0.15 max	0.4-0.8	0.15 max	0.25 max	0.05	0.15	Bi 0.4-0.7; Pb 0.4-0.7			
6262 A96262	USA	ASTM B221	6262	Bar Rod Wir Tub Ext Shapes	0.04-0.14	0.15-0.4	0.7 max	0.8-1.2	0.15 max	0.4-0.8	0.15 max	0.25 max	0.05	0.15	Bi 0.4-0.7; Pb 0.4-0.7			
6262 A96262	USA	ASTM B483	6262	Bar Rod Wir Tub Ext Shapes	0.04-0.14	0.15-0.4	0.7 max	0.8-1.2	0.15 max	0.4-0.8	0.15 max	0.25 max	0.05	0.15	Bi 0.4-0.7; Pb 0.4-0.7			
6262 A96262	USA	QQ A-225/10	6262	Bar Rod Wir Tub Ext Shapes	0.04-0.14	0.15-0.4	0.7 max	0.8-1.2	0.15 max	0.4-0.8	0.15 max	0.25 max	0.05	0.15	Bi 0.4-0.7; Pb 0.4-0.7			
6262 A96262	USA	SAE J454	6262	Bar Rod Wir Tub Ext Shapes	0.04-0.14	0.15-0.4	0.7 max	0.8-1.2	0.15 max	0.4-0.8	0.15 max	0.25 max	0.05	0.15	Bi 0.4-0.7; Pb 0.4-0.7			
6061	USA				0.04-0.35	0.15-0.4	0.7 max	0.8-1.2	0.15 max	0.4-0.8	0.15 max	0.25 max	0.05	0.15				
6091	USA				0.15 max	0.15-0.4	0.7 max	0.8-1.2	0.15 max	0.4-0.8	0.15 max	0.25 max	0.05	0.15	0.05-0.50 O			
6262	USA				0.04-0.14	0.15-0.4	0.7 max	0.8-1.2	0.15 max	0.4-0.8	0.15 max	0.25 max	0.05	0.15	Bi 0.4-0.7; Pb 0.4-0.7			
	USA		AlMg1SiCu		0.04-0.35	0.15-0.4	0.7 max	0.8-1.2	0.15 max	0.4-0.8	0.15 max	0.25 max	0.05	0.15				

UNS numbers and US grades are provided as a means of cross referencing chemically similar alloys. Exchangability is only possible after independent examination of specifications. Tensile properties are minimum or typical . UTS and YS as Mpa, El as %. See Appendix for list of abbreviations used in Descriptions.

Grade UNS #	Country	Specification	Designation	Description	Cr	Cu	Fe	Mg	Mn	Si	Ti	Zn	OE Max	OT Max	Other	UTS	YS	EL
	USA		AlMg1SiPb		0.04-0.14	0.15-0.4	0.7 max	0.8-1.2	0.15 max	0.4-0.8	0.15 max	0.25 max	0.05	0.15	Bi 0.4-0.7; Pb 0.4-0.7			
6061 A96061	Yugoslavia	JUS CC2.100	AlMg1SiCu		0.15-0.35	0.15-0.4	0.7 max	0.8-1.2	0.15 max	0.4-0.8		0.25 max		0.15	ST 0.2 max; ST = Zr + Ti			
6061 A96061	International	ISO TR2136	Al-Mg1SiCu	Sh Plt Ann	0.04-0.35	0.15-0.4		0.8-1.2	0.15	0.4-0.8		0.25				160	90	16
6061 A96061	International	ISO TR2136	Al-Mg1SiCu	Sh Plt SHT and Nat aged	0.04-0.35	0.15-0.4		0.8-1.2	0.15	0.4-0.8		0.25				200	110	15
6061 A96061	International	ISO TR2136	Al-Mg1SiCu	Sh Plt SHT and art aged	0.04-0.35	0.15-0.4		0.8-1.2	0.15	0.4-0.8		0.25				285	235	8
6070 A96070	USA	ASTM B345	6070	Bar Rod Tub Pip Ext Shapes	0.1 max	0.15-0.4	0.5 max	0.5-1.2	0.4-1	1-1.7	0.15 max	0.25 max	0.05	0.15				
6070 A96070	USA	MIL A-12545	6070	Bar Rod Tub Pip Ext Shapes	0.1 max	0.15-0.4	0.5 max	0.5-1.2	0.4-1	1-1.7	0.15 max	0.25 max	0.05	0.15				
6070 A96070	USA	MIL A-46104	6070	Bar Rod Tub Pip Ext Shapes	0.1 max	0.15-0.4	0.5 max	0.5-1.2	0.4-1	1-1.7	0.15 max	0.25 max	0.05	0.15				
6070 A96070	USA	SAE J454	6070	Bar Rod Tub Pip Ext Shapes	0.1 max	0.15-0.4	0.5 max	0.5-1.2	0.4-1	1-1.7	0.15 max	0.25 max	0.05	0.15				
6009	USA				0.1 max	0.15-0.6	0.5 max	0.4-0.8	0.2-0.8	0.6-1	0.1 max	0.25 max	0.05	0.15				
6066 A96066	UK	BS H11	6066	Bar Rod Wir Tub Frg Ext Shapes	0.4 max	0.7-1.2	0.5 max	0.8-1.4	0.6-1.1	0.9-1.8	0.2 max	0.25 max	0.05	0.15				
6066 A96066	USA	ASTM B221	6066	Bar Rod Wir Tub Frg Ext Shapes	0.4 max	0.7-1.2	0.5 max	0.8-1.4	0.6-1.1	0.9-1.8	0.2 max	0.25 max	0.05	0.15				
6066 A96066	USA	QQ A-200/10	6066	Bar Rod Wir Tub Frg Ext Shp OBSOLETE	0.4 max	0.7-1.2	0.5 max	0.8-1.4	0.6-1.1	0.9-1.8	0.2 max	0.25 max	0.05	0.15				
6066 A96066	USA	QQ A-367	6066	Bar Rod Wir Tub Frg Ext Shapes	0.4 max	0.7-1.2	0.5 max	0.8-1.4	0.6-1.1	0.9-1.8	0.2 max	0.25 max	0.05	0.15				
6066 A96066	USA	SAE J454	6066	Bar Rod Wir Tub Frg Ext Shapes	0.4 max	0.7-1.2	0.5 max	0.8-1.4	0.6-1.1	0.9-1.8	0.2 max	0.25 max	0.05	0.15				
	Germany	DIN 1725 pt1	3.3326/AlMg1.8		0.3 max	0.05 max	0.45	1.4-2.1	0.25 max	0.3 max	0.1 max	0.2			EI 0.05 max; ET=.15 total			
5051A	Germany				0.3 max	0.05 max	0.45 max	1.4-2.1	0.25 max	0.3 max	0.1 max	0.2 max	0.05	0.15				
	South Africa	SABS 712	Al-MgSi(60-630)		0.1	0.1	0.35	0.45-0.9	0.1	0.2-0.6	0.1	0.1						
	India	IS 739	NG3	Wir Ann		0.1	0.7		0.93 max1	0.6						123		
	India	IS 739	NG3	Wir Ann and CW to hard		0.1	0.7		0.93 max1	0.6						172		
6081	France				0.1 max	0.1 max	0.5 max	0.6-1	0.1-0.45	0.7-1.1	0.15 max	0.2 max	0.05	0.15				
6181	Europe				0.1 max	0.1 max	0.45 max	0.6-1	0.15 max	0.8-1.2	0.1 max	0.2 max	0.05	0.15				
	Europe		AlSiMg0.8		0.1 max	0.1 max	0.45 max	0.6-1	0.15 max	0.8-1.2	0.1 max	0.2 max	0.05	0.15				
	Germany	DIN 1725 pt1	3.2316/AlMgSi0.8		0.1 max	0.1 max	0.45 max	0.6-1	0.15 max	0.8-1.2	0.1 max	0.2 max			ET 0.05 max; ET=.15 total;			
8014	USA					0.2 max	1.2-1.6	0.1 max	0.2-0.6	0.3 max	0.1 max	0.1 max	0.05	0.15				
8008	Spain					0.2 max	0.9-1.6		0.5-1	0.6 max	0.1 max	0.1 max	0.05	0.15				
2117 A92117	Czech Republic	CSN 424204	AlCu2Mg			2.2-3	0.5 max	0.2-0.5	0.2 max	0.5 max		0.1 max						
	International	ISO R209 (1971)	Al-Si 1 Mg		0.35 max	0.10 max	0.5 max	0.4-1.4	0.4-1.0	0.6-1.6					Ti+Zr 0.2 max			
	International	ISO R209 (1971)	Al-Mg1SiCu		0.04-0.35	0.15-0.40	0.7 max	0.8-1.2	0.15 max	0.4-0.8					Ti+Zr 0.2 max			
8018	UK					0.3-0.6	0.6-1		0.3 max		0.5-0.9	0.006-0.06		0.05	0.15				
5043	USA				0.05 max	0.05-0.35	0.7 max	0.7-1.3	0.7-1.2	0.4 max	0.1 max	0.25 max	0.05	0.15	Ga 0.05 max; V 0.05 max			
6261 A96261	Canada	CSA HA.5	0.6261	Bar Rod O temper	0.1 max	0.15-0.4	0.4	0.7-1	0.2-0.35	0.4-0.7	0.1 max	0.2 max			ET 0.05 max; ET=.15 total;	130		18
6261 A96261	Canada	CSA HA.5	0.6261	Bar Rod T4 temper	0.1 max	0.15-0.4	0.4	0.7-1	0.2-0.35	0.4-0.7	0.1 max	0.2 max			ET 0.05 max; ET=.15 total;	180	110	16

UNS numbers and US grades are provided as a means of cross referencing chemically similar alloys. Exchangability is only possible after independent examination of specifications. Tensile properties are minimum or typical . UTS and YS as Mpa, EI as %. See Appendix for list of abbreviations used in Descriptions.

Worldwide Guide to Equivalent Nonferrous Metals and Alloys

Grade UNS #	Country	Specification	Designation	Description	Cr	Cu	Fe	Mg	Mn	Si	Ti	Zn	OE Max	OT Max	Other	UTS	YS	EL
6261 A96261	Canada	CSA HA.5	0.6261	Bar Rod T6 temper	0.1 max	0.15-0.4	0.4	0.7-1	0.2-0.35	0.4-0.7	0.1 max	0.2 max			ET 0.05 max; ET=.15 total;	260	240	10
6261 A96261	Pan American	COPANT 862	6261		0.1	0.15-0.4	0.4	0.7-1	0.2-0.35	0.4-0.7	0.1	0.2						
6261	USA				0.1 max	0.15-0.4	0.4 max	0.7-1	0.2-0.35	0.4-0.7	0.1 max	0.2 max	0.05	0.15				
6261 A96261	Australia	AS 1866	6261		0.1 max	0.15-0.4	0.4	0.7-1	0.35 max	0.4-0.7	0.1 max	0.2 max	0.05		OT min 0.15			
6261 A96261	Australia	AS 1866	6261	T4 all thk	0.1 max	0.15-0.4	0.4	0.7-1	0.35 max	0.4-0.7	0.1 max	0.2 max	0.05		OT min 0.15	190	115	14
6261 A96261	Australia	AS 1866	6261	T6 all thk	0.1 max	0.15-0.4	0.4	0.7-1	0.35 max	0.4-0.7	0.1 max	0.2 max	0.05		OT min 0.15	295	255	7
6261 A96261	Australia	AS 1866	6261	T62 all thk	0.1 max	0.15-0.4	0.4	0.7-1	0.35 max	0.4-0.7	0.1 max	0.2 max	0.05		OT min 0.15	295	255	7
2008	USA				0.1 max	0.7-1.1	0.4 max	0.25-0.5	0.3 max	0.5-0.8	0.1 max	0.25 max	0.05	0.15	V 0.05 max			
6351 A96351	Australia	AS 1866	6351	Rod Bar SHT and Nat aged 150.0 mm diam		0.1	0.5	0.4-0.8	0.4-0.8	0.7-1.3	0.2	0.2				185	115	16
6351 A96351	Australia	AS 1866	6351	Rod Bar Cooled and Art aged		0.1	0.5	0.4-0.8	0.4-0.8	0.7-1.3	0.2	0.2				262	241	8
6351 A96351	Australia	AS 1866	6351	Rod Bar SHT and Art aged 150.0 mm diam		0.1	0.5	0.4-0.8	0.4-0.8	0.7-1.3	0.2	0.2				293	255	8
6351 A96351	Canada	CSA HA.5	0.6351	Bar Rod Wir Extd and T4		0.1	0.5	0.4-0.8	0.4-0.8	0.7-1.3	0.2	0.2				221	131	16
6351 A96351	Canada	CSA HA.5	0.6351	Bar Rod Wir Extd and T6 .124 in diam		0.1	0.5	0.4-0.8	0.4-0.8	0.7-1.3	0.2	0.2				290	255	8
6351 A96351	Canada	CSA HA.7	0.6351	Tub Pip drawn and O		0.1	0.5	0.4-0.8	0.4-0.8	0.7-1.3	0.2	0.2				152	97	15
6351 A96351	Canada	CSA HA.7	0.6351	Tub Pip Drawn and T6 .015/.049 in diam		0.1	0.5	0.4-0.8	0.4-0.8	0.7-1.3	0.2	0.2				290	255	8
6351 A96351	Canada	CSA HA.7	0.6351	Tub Pip Extd and T4		0.1	0.5	0.4-0.8	0.4-0.8	0.7-1.3	0.2	0.2				221	131	16
6351 A96351	Canada	CSA HA.7.1	0.6351	Tub Pip Drawn and O		0.1	0.5	0.4-0.8	0.4-0.8	0.7-1.3	0.2	0.2				152	97	15
6351 A96351	Canada	CSA HA.7.1	0.6351	Tub Pip Drawn and T6 .015/.049 in diam		0.1	0.5	0.4-0.8	0.4-0.8	0.7-1.3	0.2	0.2				290		8
6351 A96351	Canada	CSA HA.7.1	0.6351	Tub Pip Extd and T4		0.1	0.5	0.4-0.8	0.4-0.8	0.7-1.3	0.2	0.2				221	131	16
6351 A96351	Denmark	DS 3012	6351	Soft (rolled products) .5/25 mm diam		0.1	0.5	0.4-0.8	0.4-0.8	0.7-1.3	0.2	0.2				90		20
6351 A96351	Denmark	DS 3012	6351	SA (rolled products) .5/25 mm diam		0.1	0.5	0.4-0.8	0.4-0.8	0.7-1.3	0.2	0.2				205	120	15
6351 A96351	Denmark	DS 3012	6351	Art aged (extd products) 20/75 mm diam		0.1	0.5	0.4-0.8	0.4-0.8	0.7-1.3	0.2	0.2				295	255	8
6351 A96351	Norway	NS 17311	NS17311/AlSi1Mg0.5Mn	Bar Tub		0.1 max	0.5 max	0.4-0.8	0.4-0.8	0.7-1.3	0.2 max	0.2 max			ET 0.05 max; ET=.15 total;			
6351 A96351	Pan American	COPANT 862	6351			0.1	0.5	0.4-0.8	0.4-0.8	0.7-1.3	0.2	0.2						
6351 A96351	USA	ASTM B221	6351	Bar Rod Wir Tub Pip Shapes		0.1 max	0.5 max	0.4-0.8	0.4-0.8	0.7-1.3	0.2 max	0.2 max	0.05	0.15				
6351 A96351	USA	ASTM B345	6351	Bar Rod Wir Tub Pip Shapes		0.1 max	0.5 max	0.4-0.8	0.4-0.8	0.7-1.3	0.2 max	0.2 max	0.05	0.15				
6351	USA					0.1 max	0.5 max	0.4-0.8	0.4-0.8	0.7-1.3	0.2 max	0.2 max	0.05	0.15				
	USA		AlSiMg0.5Mn			0.1 max	0.5 max	0.4-0.8	0.4-0.8	0.7-1.3	0.2 max	0.2 max	0.05	0.15				

UNS numbers and US grades are provided as a means of cross referencing chemically similar alloys. Exchangability is only possible after independent examination of specifications. Tensile properties are minimum or typical . UTS and YS as Mpa, El as %. See Appendix for list of abbreviations used in Descriptions.

2-40 Wrought Aluminum

Grade UNS #	Country	Specification	Designation	Description	Cr	Cu	Fe	Mg	Mn	Si	Ti	Zn	OE Max	OT Max	Other	UTS	YS	EL
6351 A96351	Finland	SFS 2593	AlSi1Mg	Sh Rod Tub HF (shapes)	0.35	0.1	0.5	0.4-1.4	0.2-1	0.6-1.6		0.2				150	90	12
6351 A96351	Finland	SFS 2593	AlSi1Mg	Sh Rod Tub Ann (except shapes)	0.35	0.1	0.5	0.4-1.4	0.2-1	0.6-1.6		0.2				160	40	20
6351 A96351	Finland	SFS 2593	AlSi1Mg	Sh Rod Tub ST and aged 6 mm	0.35	0.1	0.5	0.4-1.4	0.2-1	0.6-1.6		0.2				300	240	8
5151	USA				0.1 max	0.15 max	0.35 max	1.5-2.1	0.1 max	0.2 max	0.1 max	0.15 max	0.05	0.15				
8006 A98006	Australia	AS 1734	8006			0.3 max	1.2-2	0.1 max	0.3-1	0.4 max		0.1 max	0.05		OT min 0.15			
8006	USA					0.3 max	1.2-2	0.1 max	0.3-1	0.4 max		0.1 max	0.05	0.15				
3014 A93014	Argentina					0.5 max	1 max	0.1 max	1-1.5	0.6 max	0.1 max	0.5-1	0.05	0.15				
	International	ISO R209 (1971)	Al-Mg 1.5		0.1 max	0.20 max	0.7 max	1.1-1.8	0.3 max	0.4 max					Ti+Zr 0.2 max			
6007	USA				0.05-0.25	0.2 max	0.7 max	0.6-0.9	0.05-0.25	0.9-1.4	0.15 max	0.25 max	0.15	0.15	Zr 0.05-0.20			
6018	Switzerland				0.1 max	0.15-0.4	0.7	0.6-1.2	0.3-0.8	0.5-1.2	0.2 max	0.3 max	0.05	0.15	Bi 0.4-0.7; Pb 0.7-1.2			
5351	USA					0.1 max	0.1 max	1.6-2.2	0.1 max	0.08 max		0.05 max	0.03	0.1	V 0.05 max			
	Germany	DIN 1725 pt1	3.0615/AlMgSiPb		0.3 max	0.1 max	0.5 max	0.6-1.2	0.4-1	0.6-1.4	0.2 max	0.3 max			ST 1-2.5; ET 0.05 max; ST=Bi+Cd+Pb+Sn; ET=.15 total;			
6012	Germany				0.3 max	0.1 max	0.5 max	0.6-1.2	0.4-1	0.6-1.4	0.2 max	0.3 max	0.05	0.15	Bi 0.7; Pb 0.40-2.0			
3004A	Australia				0.1 max	0.25 max	0.7 max	0.8-1.5	0.8-1.5	0.4 max	0.05 max	0.25 max	0.05	0.15	Pb 0.03			
3013 A93013	Argentina					0.5 max	1 max	0.2-0.6	0.9-1.4	0.6 max		0.5-1	0.05	0.15				
6110	USA				0.04-0.25	0.2-0.7	0.8 max	0.5-1.1	0.2-0.7	0.7-1.5	0.15 max	0.3 max	0.05	0.15				
6011	USA				0.3 max	0.4-0.9	1 max	0.6-1.2	0.8	0.6-1.2	0.2 max	1.5 max	0.05	0.15	Ni 0.2 max			
8090	Europe				0.1 max	1-1.6	0.3 max	0.6-1.3	0.1 max	0.2 max	0.1 max	0.25 max	0.05	0.15	2.2-2.7 Li; Zr 0.04-0.16			
	South Africa	SABS 712	Al-MgSi-EC(62-014)			0.05	0.4	0.6-0.9		0.5-0.9					B 0.06 min; Be 0.06 max			
3104	USA					0.05-0.25	0.8 max	0.8-1.3	0.8-1.4	0.6 max	0.1 max	0.25 max	0.05	0.15	Ga 0.05 max; V 0.05 max			
	Japan	JIS Z3263	BA23PC	Sh Ann .8 mm diam		0.15-0.4	0.8	0.4-0.8	0.1	0.2-0.5						147		18
8050	Europe				0.05 max	0.05 max	1.1-1.2	0.05 max	0.45-0.55	0.15-0.3		0.1 max	0.03	0.15				
7031	USA					0.1 max	0.8-1.4	0.1 max	0.1-0.4	0.3 max		0.8-1.8	0.05	0.15				
5051 A95051	UK		2L80		0.25 max	0.1 max	0.5 max	2.4 max	0.17-0.5	0.5 max	0.2 max	0.2 max			ST 0.5 max; ST=Mn+Cr			
	Yugoslavia	JUS CC2.100	AlMg2		0.35 max	0.1 max	0.5 max	1.7-2.4	0.4 max	0.5 max	0.2 max	0.2 max		0.15	ST 0.5 max; ST = Cr+Mn+Ti+Zr;			
6082 A96082	Austria	ONORM M3430	AlMgSi1		0.25 max	0.1 max	0.5 max	0.6-1.2	0.4-1	0.7-1.3	0.1 max	0.2 max			ET 0.05 max; ET=.15 total;			
6082	Europe				0.25 max	0.1 max	0.5 max	0.6-1.2	0.4-1	0.7-1.3	0.1 max	0.2 max	0.05	0.15				
	Europe		AlSi1MgMn		0.25 max	0.1 max	0.5 max	0.6-1.2	0.4-1	0.7-1.3	0.1 max	0.2 max	0.05	0.15				
6082A	France				0.25 max	0.1 max	0.5 max	0.6-1.2	0.4-1	0.7-1.3	0.1 max	0.2 max	0.05	0.15	0.003 max Pb			
6082 A96082	Germany	DIN 1725 pt1	3.2315/AlMgSi1		0.25 max	0.1 max	0.5 max	0.6-1.2	0.4-1	0.7-1.3	0.1 max	0.2 max			ET 0.05 max; ET=.15 total;			
6082 A96082	Norway	NS 17305	NS17305/AlSi1Mg	Plt Sh Strp Tub Frg	0.25 max	0.1 max	0.5 max	0.6-1.2	0.4-1	0.7-1.3	0.1 max	0.2 max			ET 0.05 max; ET=.15 total;			

UNS numbers and US grades are provided as a means of cross referencing chemically similar alloys. Exchangability is only possible after independent examination of specifications. Tensile properties are minimum or typical . UTS and YS as Mpa, El as %. See Appendix for list of abbreviations used in Descriptions.

Worldwide Guide to Equivalent Nonferrous Metals and Alloys

Grade UNS #	Country	Specification	Designation	Description	Cr	Cu	Fe	Mg	Mn	Si	Ti	Zn	OE Max	OT Max	Other	UTS	YS	EL
6082 A96082	Sweden	SIS 144212	4212-00	Bar Rod HW 10/200 mm diam	0.25	0.1	0.5	0.6-1.2	0.4-1	0.7-1.3	0.1	0.2				150	90	12
6082 A96082	Sweden	SIS 144212	4212-02	Plt Sh Strp Bar Tub Ann 5/30 mm diam	0.25	0.1	0.5	0.6-1.2	0.4-1	0.7-1.3	0.1	0.2						22
6082 A96082	Sweden	SIS 144212	4212-04	Plt Sh Strp Bar Tub Frg Nat Aged 5/30 mm	0.25	0.1	0.5	0.6-1.2	0.4-1	0.7-1.3	0.1	0.2				205	115	18
6082 A96082	Sweden	SIS 144212	4212-10	Sh Plt Strp SH .5/5 mm diam	0.25	0.1	0.5	0.6-1.2	0.4-1	0.7-1.3	0.1	0.2			Al 97.4	170	140	10
6082 A96082	Sweden	SIS 144212	4212-06	Plt Sh Strp Bar Tub Frg Art Aged .5/5 mm diam	0.25	0.1		0.6-1.2	0.4-1	0.7-1.3	0.1	0.2			Al 97.4	290	245	8
6082 A96082	Czech Republic	CSN 424400	AlMg1Si1Mn			0.1 max	0.5 max	0.7-1.2	0.4-1	0.7-1.4	0.05 max	0.2 max						
5051 A95051	France	NF A50-701	AG2		0.15 max	0.15 max	0.5 max	1.8-2.5	0.4 max	0.4 max	0.15 max	0.15 max						
4014	Norway					0.2 max	0.7 max	0.3-0.8	0.35 max	1.4-2.2		0.2 max	0.05	0.15				
3204	USA					0.1-0.25	0.7 max	0.8-1.5	0.8-1.5	0.3 max		0.25 max	0.05	0.15				
5051 A95051	Denmark	DS 3012	5051	3/4 hard (rolled and drawn products) .5/6 mm diam	0.1	0.25	0.7	1.7-2.2	0.2	0.4	0.1	0.25				245	215	4
5051 A95051	Denmark	DS 3012	5051	Hard (rolled and drawn products) .5/6 mm diam	0.1	0.25	0.7	1.7-2.2	0.2	0.4	0.1	0.25				265	235	3
5051 A95051	Denmark	DS 3012	5051	As manufactured (extrd products)	0.1	0.25	0.7	1.7-2.2	0.2	0.4	0.1	0.25				145	60	12
5051	USA				0.1 max	0.25 max	0.7 max	1.7-2.2	0.2 max	0.4 max	0.1 max	0.25 max	0.05	0.15				
6111	USA				0.1 max	0.5-0.9	0.4 max	0.5-1	0.1-0.45	0.6-1.1	0.1 max	0.15 max	0.05	0.15				
	International	ISO R209 (1971)	Al-Mg 2		0.35 max	0.10 max	0.5 max	1.7-2.4	0.5 max	0.5 max					Ti+Zr 0.2 max; Mn+Cr 0.5 max			
6082 A96082	Sweden	MNC 40E	4212	Bar Sh Tub Ext				0.9-0.9	0.7	1					Al 97.4			
	International	ISO TR2136	Al-Si1Mg	Sh Plt Ann	0.35		0.5	0.4-1.4		0.6-1.6		0.2				160		16
	International	ISO TR2136	Al-Si1Mg	Sh Plt SHT and nat aged	0.35		0.5	0.4-1.4		0.6-1.6		0.2				200	110	15
	International	ISO TR2136	Al-Si1Mg	Sh Plt SHT and art aged	0.35		0.5	0.4-1.4		0.6-1.6		0.2				285	235	8
6010 A96010	USA				0.1 max	0.15-0.6	0.5 max	0.6-1	0.2-0.8	0.8-1.2	0.1 max	0.25 max	0.05	0.15				
	Czech Republic	CSN 424411	AlMg2R			0.008 max	0.08 max	1.8-2.3		0.05 max								
5451	USA				0.15-0.35	0.1 max	0.4 max	1.8-2.4	0.1 max	0.25 max	0.05 max	0.1 max	0.05	0.15	Ni 0.05 max			
	UK	BS 1474	NE8	Bar Tub Ann 150 mm diam	0.25	0.1	0.4		0.5-1	0.4	0.2	0.2				275	125	3
	UK	BS 1474	NE8	Bar Tub As Mgh 150 mm diam	0.25	0.1	0.4		0.5-1	0.4	0.2	0.2				280		11
5251 A95251	Czech Republic	CSN 424412	AlMg2			0.1 max	0.4 max	1.7-2.6	0.05-0.4	0.5 max	0.2 max	0.2 max			Fe+Si:0.60 Sb:0.25			
3004 A93004	Finland	SFS 2585	AlMn1	Sht Ann		0.1	0.7	0.3	0.8-1.5	0.6		0.2				130	40	24
3004 A93004	Finland	SFS 2585	AlMn1	Sht 3/4 hard 6 mm diam		0.1	0.7	0.3	0.8-1.5	0.6		0.2				160	130	4
3004 A93004	Finland	SFS 2585	AlMn1	Sht Hard 3 mm diam		0.1	0.7	0.3	0.8-1.5	0.6		0.2				190	150	3
	India	IS 739	HG9	Wir St and Nat aged	0.1	0.1	0.6	0.4-0.9	0.3	0.3-0.7		0.1				137		

UNS numbers and US grades are provided as a means of cross referencing chemically similar alloys. Exchangability is only possible after independent examination of specifications. Tensile properties are minimum or typical . UTS and YS as Mpa, El as %. See Appendix for list of abbreviations used in Descriptions.

2-42 Wrought Aluminum

Grade UNS #	Country	Specification	Designation	Description	Cr	Cu	Fe	Mg	Mn	Si	Ti	Zn	OE Max	OT Max	Other	UTS	YS	EL
	India	IS 739	HG9	Wir SHT and PHT	0.1	0.1	0.6	0.4-0.9	0.3	0.3-0.7		0.1				186		
	India	IS 738	HT9	Tub as Mfg	0.1	0.1	0.7	0.4-0.9	0.3	0.3-0.7						157		
	India	IS 738	HT9	Tub ST and Nat aged	0.1	0.1	0.7	0.4-0.9	0.3	0.3-0.7						157	93	15
	India	IS 738	HT9	Tub SHT and PHT	0.1	0.1	0.7	0.4-0.9	0.3	0.3-0.7						201	172	8
3004 A93004	Russia		AMts2			0.1 max	0.7 max	0.8-1.3	0.9-1.4	0.7 max								
6351A	France					0.1 max	0.5 max	0.4-0.8	0.7-0.8	0.7-1.3	0.2 max	0.2 max	0.05	0.15	Pb 0.003 max			
5251 A95251	Australia	AS 1734	5251	Sh Plt Ann 0.20/0,50 mm Thk	0.15	0.15	0.5	1.7-2.4	0.1-0.5	0.4	0.15	0.15				170		15
5251 A95251	Australia	AS 1734	5251	Sh Plt SH and Stab to 1/2 hard 0.25/0.50 mm Thk	0.15	0.15	0.5	1.7-2.4	0.1-0.5	0.4	0.15	0.15				231	179	3
5251 A95251	Australia	AS 1734	5251	Sh Plt SH and Stab to full hard 0.20/0.80 mm	0.15	0.15	0.5	1.7-2.4	0.1-0.5	0.4	0.15	0.15				262		3
5251 A95251	Australia	AS 1865	5251	Wir Rod Bar Strp Ann 10 mm diam	0.15	0.15	0.5	1.7-2.4	0.1-0.5	0.4	0.15	0.15						
5251 A95251	Australia	AS 1865	5251	Wir Rod Bar Strp SH and Stab to 1/4 hard 10 mm diam	0.15	0.15	0.5	1.7-2.4	0.1-0.5	0.4	0.15	0.15						
5251 A95251	Australia	AS 1865	5251	Wir Rod Bar Strp SH and Stab to full hard 10 mm diam	0.15	0.15	0.5	1.7-2.4	0.1-0.5	0.4	0.15	0.15						
5251	Europe				0.15 max	0.15 max	0.5 max	1.7-2.4	0.1-0.5	0.4 max	0.15 max	0.15 max	0.05	0.15				
	Europe		AlMg2		0.15 max	0.15 max	0.5 max	1.7-2.4	0.1-0.5	0.4 max	0.15 max	0.15 max	0.05	0.15				
	Norway	NS 17205	NS17205/AlMg2	Plt Sh Strp	0.15 max	0.15 max	0.5 max	1.7-2.4	0.1-0.5	0.4 max	0.15 max	0.15 max			ET 0.05 max; ET=.15 total;			
5016	USA				0.1 max	0.2 max	0.6 max	1.4-1.9	0.4-0.7	0.25 max	0.05 max	0.15 max	0.05	0.15				
3004 A93004	Australia	AS 1734	3004	Sh Plt, Ann 0.15/0.20 mm		0.25	0.7	0.8-1.3	1-1.5	0.3		0.25		0.15				
3004 A93004	Australia	AS 1734	3004	Sh Plt, SH and stab to full full hard 0.15/0.20 mm thk		0.25	0.7	0.8-1.3	1-1.5	0.3		0.25		0.15				1
3004 A93004	Australia	AS 1734	3004	Sh Plt, SH and stab to full hard 0.15/0.20 mm thk		0.25	0.7	0.8-1.3	1-1.5	0.3		0.25						
3004 A93004	Australia	AS 1734	Alclad3004	Sh Plt		0.25	0.7	0.8-1.3	1-1.5	0.3		0.25						
	Austria	ONORM M3430	AlMn1Mg1			0.25 max	0.7 max	0.8-1.3	1-1.5	0.3 max		0.25 max			ET 0.05 max;			
3004 A93004	Belgium	NBN P21-001	3004			0.25 max	0.7 max	0.8-1.3	1-1.5	0.3 max		0.25 max			ET 0.05 max; ET = .15 Total			
3004 A93004	France	NF A-M1G	3004	Pip		0.25 max	0.7 max	0.8-1.3	1-1.5	0.3 max		0.25 max	0.05	0.15				
3004 A93004	France	NF A50451	3004(A-M1G)	Sh Plt Ann .4/3.2 mm diam		0.25	0.7	0.8-1.3	1-1.5	0.3		0.25				155		18
3004 A93004	France	NF A50451	3004(A-M1G)	Sh Plt 3/4 hard .4/1.6 mm diam		0.25	0.7	0.8-1.3	1-1.5	0.3		0.25				240	200	3
3004 A93004	France	NF A50451	3004(A-M1G)	Sh Plt Hard .4/1.6 mm diam		0.25	0.7	0.8-1.3	1-1.5	0.3		0.25				260	210	3
3004 A93004	France	NF A50501	3004(A-M1G)	Tub As mfg		0.25	0.7	0.8-1.3	1-1.5	0.3		0.25				260	210	3

UNS numbers and US grades are provided as a means of cross referencing chemically similar alloys. Exchangability is only possible after independent examination of specifications. Tensile properties are minimum or typical . UTS and YS as Mpa, El as %. See Appendix for list of abbreviations used in Descriptions.

Grade UNS #	Country	Specification	Designation	Description	Cr	Cu	Fe	Mg	Mn	Si	Ti	Zn	OE Max	OT Max	Other	UTS	YS	EL
3004 A93004	Germany	DIN 1725 pt1	3.0526/AlMn1 Mg1			0.25 max	0.7 max	0.8-1.3	1-1.5	0.3 max		0.25 max			ET 0.05 max; ET=.15 total			
3004 A93004	Germany	DIN	AlMn1Mg1	Pip		0.25 max	0.7 max	0.8-1.3	1-1.5	0.3 max		0.25 max	0.05	0.15				
3004 A93004	Japan	JIS H4001	3004	Sh Stab at 1/4 hard .8/1.2 mm diam		0.25	0.7	0.8-1.3	1-1.5	0.3		0.25				196	147	4
3004 A93004	Japan	JIS H4001	3004	Sh Stab at 1/2 hard .8/1.2 mm diam		0.25	0.7	0.8-1.3	1-1.5	0.3		0.25				226	177	3
3004 A93004	Japan	JIS H4001	3004	Sh Stab at 3/4 hard .8/1.2 mm diam		0.25	0.7	0.8-1.3	1-1.5	0.3		0.25				245	196	3
3004 A93004	Norway	NS 17408	NS17408/AlM n1Mg1	Plt Sh Strp		0.25 max	0.7 max	0.8-1.3	1-1.5	0.3 max		0.25 max			ET 0.05 max; ET=.15 total;			
3004 A93004	Pan American	COPANT 862	3004			0.25	0.7	0.8-1.3	1-1.5	0.3		0.25						
3004 A93004	USA	ASTM B209	3004	Pip		0.25 max	0.7 max	0.8-1.3	1-1.5	0.3 max		0.25 max	0.05	0.15				
3004 A93004	USA	ASTM B221	3004	Pip		0.25 max	0.7 max	0.8-1.3	1-1.5	0.3 max		0.25 max	0.05	0.15				
3004 A93004	USA	ASTM B313	3004	Pip		0.25 max	0.7 max	0.8-1.3	1-1.5	0.3 max		0.25 max	0.05	0.15				
3004 A93004	USA	ASTM B547	3004	Pip		0.25 max	0.7 max	0.8-1.3	1-1.5	0.3 max		0.25 max	0.05	0.15				
3004 A93004	USA	SAE J454	3004	Pip		0.25 max	0.7 max	0.8-1.3	1-1.5	0.3 max		0.25 max	0.05	0.15				
3004	USA					0.25 max	0.7 max	0.8-1.3	1-1.5	0.3 max		0.25 max	0.05	0.15				
	USA		AlMn1Mg1			0.25 max	0.7 max	0.8-1.3	1-1.5	0.3 max		0.25 max	0.05	0.15				
3004 A93004	Australia	AS 1734	3004A	O .50-.80 mm thk	0.1 max	0.25 max	0.7 max	0.8-1.5	0.8-1.5	0.4 max	0.05 max	0.25 max	0.05		OT min 0.15; Pb 0.03 max	151	58	14
5251 A95251	Argentina				0.1 max	0.25 max	0.7 max	1.6-2.2	0.2-0.7	0.5 max	0.1 max	0.25 max	0.05	0.15				
2038	USA				0.2 max	0.8-1.8	0.6 max	0.4-1	0.1-0.4	0.5-1.3	0.15 max	0.5 max	0.05	0.15	Ga 0.05 max; V 0.05 max			
2037	USA				0.1 max	1.4-2.2	0.5 max	0.3-0.8	0.1-0.4	0.5 max	0.15 max	0.25 max	0.05	0.15	V 0.05 max			
	Japan	JIS Z3263	BA21PC	Sh Ann .8 mm diam		0.15-0.4	0.8	0.4-0.8	0.1	0.2-0.5		0.2				137		18
	India	IS 738	HT19	Tub ST and Nat aged 1.60 mm diam	0.1	0.1	0.6	0.4-1.5	0.2	0.6-1.3						216	108	12
	India	IS 738	HT19	Tub SHT and PHT 1.60 mm diam	0.1	0.1	0.6	0.4-1.5	0.2	0.6-1.3						294	230	7
5040	USA				0.1-0.3	0.25 max	0.7 max	1-1.5	0.9-1.4	0.3 max		0.25 max	0.05	0.15				
5449	Belgium				0.3 max	0.3 max	0.7 max	1.6-2.6	0.3-1.1	0.4 max	0.1 max	0.3 max	0.05	0.15				
6092	USA				0.15 max	0.7-1	0.3 max	0.8-1.2	0.15 max	0.4-0.8	0.15 max	0.25 max	0.05	0.15	0.05-0.50 O			
6013	USA				0.1 max	0.6-1.1	0.5 max	0.5-1.2	0.2-0.8	0.6-1	0.1 max	0.25 max	0.05	0.15				
8280	USA					0.7-1.3	0.7 max		0.1 max	1-2	0.1 max	0.05 max	0.05	0.15	Ni 0.2-0.7; Sn 5.5-7.0			
8093	France				0.1 max	1-1.6	0.1 max	0.9-1.6	0.1 max	0.1 max	0.1 max	0.25 max	0.05	0.15	1.9-2.6 Li; Zr 0.04-0.16			
	Austria	ONORM M3426	Al98			0.1 max	1 max			0.8 max	0.1 max	0.1 max			ET 0.1 max;			
6012 A96012	Norway	NS 17312	NS17312/AlM gSiPb	Bar	0.3 max	0.1 max	0.5 max	0.6-1.2	0.4-1	0.6-1.4	0.2 max	0.3 max			ET 0.05 max; ET=.15 total; Bi 0.7 max; Pb 0.4-2			

UNS numbers and US grades are provided as a means of cross referencing chemically similar alloys. Exchangability is only possible after independent examination of specifications. Tensile properties are minimum or typical . UTS and YS as Mpa, El as %. See Appendix for list of abbreviations used in Descriptions.

Grade UNS #	Country	Specification	Designation	Description	Cr	Cu	Fe	Mg	Mn	Si	Ti	Zn	OE Max	OT Max	Other	UTS	YS	EL
	Pan American	COPANT 862	X6080			0.1	0.5		0.2-1	1-1.5	0.1	0.1						
	India	IS 736	40800	Plt As Mfg 28 mm diam		0.2	0.95	0.1	0.1	0.6-0.95	0.2	0.2			Al 98		90	28
	India	IS 736	40800	Plt Ann 30 mm diam		0.2	0.95	0.1	0.1	0.6-0.95	0.2	0.2			Al 98	120	85	30
	India	IS 736	40800	Plt 1/2 hard 7 mm diam		0.2	0.95	0.1	0.1	0.6-0.95	0.2	0.2			Al 98	160	120	7
	USA	ASTM B479	98% min Al	Ann Foil		0.2 max		0.05 max	0.1 max		0.08 max	0.1 max	0.05	0.15	Si+Fe 1.8 max; Al 98 min			
	Germany	DIN 1725 pt1	3.1303/AlCu2 Mg0.5		0.1 max	1.8-2.3	0.7 max	0.2-0.8	0.2 max	0.8 max		0.25 max			ET 0.05 max; ET=.15 total;			
	International	ISO R209 (1971)	Al-Mn 1 Cu			0.05-0.20	0.7 max		1.0-1.5	0.6 max					Ti+Zr+Cr 0.2 max			
6070	USA				0.1 max	0.15-0.4	0.5 max	0.5-1.2	0.4-1	1-1.7	0.15 max	0.25 max	0.05	0.15				
5149	Europe				0.3 max	0.05 max	0.4 max	1.6-2.5	0.5-1.1	0.25 max	0.15 max	0.2 max	0.05	0.15				
5249	Europe				0.3 max	0.05 max	0.4 max	1.6-2.5	0.5-1.1	0.25 max	0.15 max	0.2 max	0.05	0.15	Zr 0.10-0.20			
5049	Europe				0.3 max	0.1 max	0.5 max	1.6-2.5	0.5-1.1	0.4 max	0.1 max	0.2 max	0.05	0.15				
	Yugoslavia	JUS CC2.100	AlMg2Mn		0.3 max	0.1 max	0.5 max	1.6-2.5	0.5-1.1	0.4 max	0.1 max	0.2 max		0.15				
	Austria	ONORM M3430	AlMg2Mn0.5		0.3 max	0.1 max	0.55 max	1.6-2.5	0.5-1.1	0.4 max	0.1 max	0.2 max			ET 0.05 max; ET=.15;			
	Germany	DIN 1725 pt1	3.3527/AlMg2 Mn0.8		0.3 max	0.1 max	0.55 max	1.6-2.5	0.5-1.1	0.4 max	0.1 max	0.2 max			ET 0.05 max; ET=.15 total;			
	India	IS 733	63400	Bar Rod ST and Nat aged 150 mm diam	0.1	0.1	0.6	0.4-0.9	0.3	0.3-0.7	0.2	0.2				140	80	14
	India	IS 733	63400	Bar Rod PHT 3 mm diam	0.1	0.1	0.6	0.4-0.9	0.3	0.3-0.7	0.2	0.2				170	140	7
	India	IS 733	63400	Bar Rod SHT and PHT 150 mm diam	0.1	0.1	0.6	0.4-0.9	0.3	0.3-0.7	0.2	0.2				185	150	7
	India	IS 738	HT30	Tub ST and Nat aged 1.60 mm	0.3	0.1	0.6	0.4-1.4	0.4-1	0.6-1.3						216	108	12
	India	IS 738	HT30	Tub SHT and PHT 1.60 mm	0.3	0.1	0.6	0.4-1.4	0.4-1	0.6-1.3						309	246	7
4015	Norway					0.2 max	0.7 max	0.1-0.5	0.6-1.2	1.4-2.2		0.2 max	0.05	0.15				
	Pan American	COPANT 862	3205		0.2	0.3	0.7			0.6	0.1	0.4						
6113	USA				0.1 max	0.6-1.1	0.3 max	0.8-1.2	0.1-0.6	0.6-1	0.1 max	0.25 max	0.05	0.15	0.05-0.50 O			
8091	UK				0.1 max	1.6-2.2	0.5 max	0.5-1.2	0.1 max	0.3 max	0.1 max	0.25 max	0.05	0.15	2.4-2.8 Li; Zr 0.04-0.16			
6053 A96053	Canada	CSA HA.6	0.6053	Rod Wir H13 .062/1 in diam	0.15-0.35	0.1	0.35	1.1-1.4		0.5-0.91		0.1				131		
6053 A96053	Canada	CSA HA.6	0.6053	Rod Wir T61 .062/1 in diam	0.15-0.35	0.1	0.35	1.1-1.4		0.5-0.91		0.1				207	137	14
3018	France				0.1 max	0.1-0.3	0.15-0.25	0.8-1.4	1.1-1.4	0.3 max	0.1 max	0.25 max	0.05	0.15	Pb 0.01			
5652 A95652	USA	ASTM B209	5652	Plt Sh Tub	0.15-0.35	0.04 max		2.2-2.8	0.01 max			0.1 max	0.05	0.15				
5652 A95652	USA	ASTM B241	5652	Plt Sh Tub	0.15-0.35	0.04 max		2.2-2.8	0.01 max			0.1 max	0.05	0.15				
5652 A95652	USA	SAE J454	5652	Plt Sh Tub	0.15-0.35	0.04 max		2.2-2.8	0.01 max			0.1 max	0.05	0.15				
5652	USA				0.15-0.35	0.04 max		2.2-2.8	0.01 max			0.1 max	0.05	0.15	0.40 Si+Fe			
7472	USA					0.05 max	0.6 max	0.9-1.5	0.05 max	0.25 max		1.3-1.9	0.05	0.15				
5552	USA					0.1 max	0.5 max	2.2-2.8	0.1 max	0.04 max		0.05 max	0.03	0.1	V 0.05 max			

UNS numbers and US grades are provided as a means of cross referencing chemically similar alloys. Exchangability is only possible after independent examination of specifications. Tensile properties are minimum or typical . UTS and YS as Mpa, El as %. See Appendix for list of abbreviations used in Descriptions.

Worldwide Guide to Equivalent Nonferrous Metals and Alloys

Grade UNS #	Country	Specification	Designation	Description	Cr	Cu	Fe	Mg	Mn	Si	Ti	Zn	OE Max	OT Max	Other	UTS	YS	EL
5252 A95252	Australia	AS 1734	5252	Sh Plt SH and partially Ann to 5/8 hard 0.50/0.80 mm thk		0.1	0.1	2.2-2.8	0.1	0.08						213		7
5252 A95252	Australia	AS 1734	5252	Sh Plt SH and partially Ann to 7/8 hard 0.50/0.80 mm thk		0.1	0.1	2.2-2.8	0.1	0.08						241		4
5252 A95252	USA	ASTM B209	5252	Sh		0.1 max	0.1 max	2.2-2.8	0.1 max	0.08 max		0.05 max	0.03	0.1	V 0.05 max			
5252 A95252	USA					0.1 max	0.1 max	2.2-2.8	0.1 max	0.08 max		0.05 max	0.03	0.1	V 0.05 max			
5252 A95252	Pan American	COPANT 862	5252			0.1		2.2-2.8	0.1	0.08		0.05						
5052 A95052	Austria	ONORM M3430	AlMg2.5		0.15-0.35	0.1 max	0.4 max	2.2-2.8	0.1 max	0.25 max		0.1 max			ET 0.05 max; ET=.15 total			
5052 A95052	Belgium	NBN P21-001	5052		0.15-0.35	0.1 max	0.4 max	2.2-2.8	0.1 max	0.25 max		0.1 max			ET 0.05 max; ; ET = .15 Total			
5052 A95052	France	NF A50411	5052	Bar 1/2 hard (drawn bar) 50 mm diam	0.15-0.35	0.1	0.4	2.2-2.8	0.1	0.25		0.1				235	180	4
5052 A95052	France	NF A50411	5052	Bar Hard (tub) 50 mm diam	0.15-0.35	0.1	0.4	2.2-2.8	0.1	0.25		0.1				270	210	2
5052 A95052	Germany	DIN 1725 pt1	3.3523/AlMg2.5		0.15-0.35	0.1 max	0.4 max	2.2-2.8	0.1 max	0.25 max		0.1 max			ET 0.05 max; ET=.15 total;			
5052 A95052	Norway	NS 17210	NS17210/AIMg2.5	Plt Sh Strp	0.15-0.35	0.1 max	0.4 max	2.2-2.8	0.1 max	0.25 max		0.1 max			ET 0.05 max; ET=.15 total			
5052 A95052	Pan American	COPANT 862	5052		0.15-0.35	0.1	0.4	2.2-2.8	0.1	0.25		0.1						
5052 A95052	Sweden	SIS 144120	4120-00	Plt Sh HW 5/25 mm diam	0.35	0.1	0.4	2.2-2.8	0.5	0.25		0.1			Al 95.34	190	75	10
5052 A95052	Sweden	SIS 144120	4120-02	Plt Sh Strp Ann 5/20 mm diam	0.35	0.1	0.4	2.2-2.8	0.5	0.25		0.1			Al 95.34	170	65	20
5052 A95052	Sweden	SIS 144120	4120-14	Sh Plt SH 3/5 mm diam	0.35	0.1	0.4	2.2-2.8	0.5	0.25		0.1			Al 95.34	230	180	8
5052 A95052	Sweden	SIS 144120	4120-18	Sh Strp SH .35/4 mm diam	0.35	0.1	0.4	2.2-2.8	0.5	0.25		0.1			Al 95.34	280	240	3
5052 A95052	Sweden	SIS 144120	4120-24	Sh Strp SH ann .35/3 mm diam	0.35	0.1	0.4	2.2-2.8	0.5	0.25		0.1			Al 95.34	220	170	14
5052 A95052	Sweden	SIS 144120	4120-40	Sh Strp Ann .35/6 mm diam	0.35	0.1	0.4	2.2-2.8	0.5	0.25		0.1			Al 95.34	170	65	20
5052 A95052	Sweden	SIS 144120	4120-44	Sh Strp SH .35/4 mm diam	0.35	0.1	0.4	2.2-2.8	0.5	0.25		0.1			Al 95.34	230	180	5
5052 A95052	USA	AMS 4004	5052	Bar Rod Wir Plt Sh Tub Ext Shapes	0.15-0.35	0.1 max	0.4 max	2.2-2.8	0.1 max	0.25 max		0.1 max	0.05	0.15				
5052 A95052	USA	AMS 4015	5052	Bar Rod Wir Plt Sh Tub Ext Shapes	0.15-0.35	0.1 max	0.4 max	2.2-2.8	0.1 max	0.25 max		0.1 max	0.05	0.15				
5052 A95052	USA	AMS 4016	5052	Bar Rod Wir Plt Sh Tub Ext Shapes	0.15-0.35	0.1 max	0.4 max	2.2-2.8	0.1 max	0.25 max		0.1 max	0.05	0.15				
5052 A95052	USA	AMS 4017	5052	Bar Rod Wir Plt Sh Tub Ext Shapes	0.15-0.35	0.1 max	0.4 max	2.2-2.8	0.1 max	0.25 max		0.1 max	0.05	0.15				
5052 A95052	USA	AMS 4069	5052	Bar Rod Wir Plt Sh Tub Ext Shapes	0.15-0.35	0.1 max	0.4 max	2.2-2.8	0.1 max	0.25 max		0.1 max	0.05	0.15				
5052 A95052	USA	AMS 4070	5052	Bar Rod Wir Plt Sh Tub Ext Shapes	0.15-0.35	0.1 max	0.4 max	2.2-2.8	0.1 max	0.25 max		0.1 max	0.05	0.15				
5052 A95052	USA	AMS 4071	5052	Bar Rod Wir Plt Sh Tub Ext Shapes	0.15-0.35	0.1 max	0.4 max	2.2-2.8	0.1 max	0.25 max		0.1 max	0.05	0.15				

UNS numbers and US grades are provided as a means of cross referencing chemically similar alloys. Exchangability is only possible after independent examination of specifications. Tensile properties are minimum or typical . UTS and YS as Mpa, El as %. See Appendix for list of abbreviations used in Descriptions.

2-46 Wrought Aluminum

Grade UNS #	Country	Specification	Designation	Description	Cr	Cu	Fe	Mg	Mn	Si	Ti	Zn	OE Max	OT Max	Other	UTS	YS	EL
5052 A95052	USA	AMS 4114	5052	Bar Rod Wir Plt Sh Tub Ext Shapes	0.15-0.35	0.1 max	0.4 max	2.2-2.8	0.1 max	0.25 max		0.1 max	0.05	0.15				
5052 A95052	USA	ASTM B209	5052	Bar Rod Wir Plt Sh Tub Ext Shapes	0.15-0.35	0.1 max	0.4 max	2.2-2.8	0.1 max	0.25 max		0.1 max	0.05	0.15				
5052 A95052	USA	ASTM B210	5052	Bar Rod Wir Plt Sh Tub Ext Shapes	0.15-0.35	0.1 max	0.4 max	2.2-2.8	0.1 max	0.25 max		0.1 max	0.05	0.15				
5052 A95052	USA	ASTM B211	5052	Bar Rod Wir Plt Sh Tub Ext Shapes	0.15-0.35	0.1 max	0.4 max	2.2-2.8	0.1 max	0.25 max		0.1 max	0.05	0.15				
5052 A95052	USA	ASTM B221	5052	Bar Rod Wir Plt Sh Tub Ext Shapes	0.15-0.35	0.1 max	0.4 max	2.2-2.8	0.1 max	0.25 max		0.1 max	0.05	0.15				
5052 A95052	USA	ASTM B234	5052	Bar Rod Wir Plt Sh Tub Ext Shapes	0.15-0.35	0.1 max	0.4 max	2.2-2.8	0.1 max	0.25 max		0.1 max	0.05	0.15				
5052 A95052	USA	ASTM B241	5052	Bar Rod Wir Plt Sh Tub Ext Shapes	0.15-0.35	0.1 max	0.4 max	2.2-2.8	0.1 max	0.25 max		0.1 max	0.05	0.15				
5052 A95052	USA	ASTM B313	5052	Bar Rod Wir Plt Sh Tub Ext Shapes	0.15-0.35	0.1 max	0.4 max	2.2-2.8	0.1 max	0.25 max		0.1 max	0.05	0.15				
5052 A95052	USA	ASTM B316	5052	Bar Rod Wir Plt Sh Tub Ext Shapes	0.15-0.35	0.1 max	0.4 max	2.2-2.8	0.1 max	0.25 max		0.1 max	0.05	0.15				
5052 A95052	USA	ASTM B404	5052	Bar Rod Wir Plt Sh Tub Ext Shapes	0.15-0.35	0.1 max	0.4 max	2.2-2.8	0.1 max	0.25 max		0.1 max	0.05	0.15				
5052 A95052	USA	ASTM B483	5052	Bar Rod Wir Plt Sh Tub Ext Shapes	0.15-0.35	0.1 max	0.4 max	2.2-2.8	0.1 max	0.25 max		0.1 max	0.05	0.15				
5052 A95052	USA	SAE J454	5052	Bar Rod Wir Plt Sh Tub Ext Shapes	0.15-0.35	0.1 max	0.4 max	2.2-2.8	0.1 max	0.25 max		0.1 max	0.05	0.15				
5052	USA				0.15-0.35	0.1 max	0.4 max	2.2-2.8	0.1 max	0.25 max		0.1 max	0.05	0.15				
	USA		AlMg2.5		0.15-0.35	0.1 max	0.4 max	2.2-2.8	0.1 max	0.25 max		0.1 max	0.05	0.15				
5052 A95052	Russia		AMg			0.1 max	0.4 max	2-2.8	0.4 max						ST 0.15-0.35; ST=Mn or Cr			
5052 A95052	Russia	GOST 4784	1520			0.1 max	0.5 max	1.8-2.8	0.2-0.6	0.4 max	0.1 max	0.2 max						
5052 A95052	Russia	GOST 4784	AMg2			0.1 max	0.5 max	1.8-2.8	0.2-0.6	0.4 max	0.1 max	0.2 max						
5052 A95052	Finland	SFS 2587	AlMg2.5	HF	0.35	0.1	0.5	2.2-2.8	0.5	0.5		0.2						
5052 A95052	Finland	SFS 2587	AlMg2.5	Ann	0.35	0.1	0.5	2.2-2.8	0.5	0.5		0.2						
5052 A95052	Finland	SFS 2587	AlMg2.5	1/2 hard 3mm diam	0.35	0.1	0.5	2.2-2.8	0.5	0.5		0.2						
5052 A95052	International	ISO R827	Al-Mg2.5	As Mfg	0.35	0.1	0.5	2.2-2.8	0.5	0.5		0.2				172	59	12
5052 A95052	International	ISO TR2136	Al-Mg2.5	Sh Plt Ann	0.35	0.1	0.5	2.2-2.8	0.5	0.5		0.2				170	65	16
5052 A95052	International	ISO TR2136	Al-Mg2.5	Sh Plt HB strain	0.35	0.1	0.5	2.2-2.8	0.5	0.5		0.2				205	145	3
5052 A95052	International	ISO TR2136	Al-Mg2.5	Sh Plt HD SH	0.35	0.1	0.5	2.2-2.8	0.5	0.5		0.2				235	180	2
5052 A95052	International	ISO TR2778	Al-Mg2.5	Tub Ann OBSOLETE Replace by ISO 6363	0.35	0.1	0.5	2.2-2.8	0.5	0.5		0.2				170	65	15
5052 A95052	International	ISO TR2778	Al-Mg2.5	Tub SH 80 mm OBSOLETE Replace by ISO 6363	0.35	0.1	0.5	2.2-2.8	0.5	0.5		0.2				235	180	3

UNS numbers and US grades are provided as a means of cross referencing chemically similar alloys. Exchangability is only possible after independent examination of specifications. Tensile properties are minimum or typical . UTS and YS as Mpa, EL as %. See Appendix for list of abbreviations used in Descriptions.

Worldwide Guide to Equivalent Nonferrous Metals and Alloys

Grade UNS #	Country	Specification	Designation	Description	Cr	Cu	Fe	Mg	Mn	Si	Ti	Zn	OE Max	OT Max	Other	UTS	YS	EL
5052 A95052	UK		2L55		0.25 max	0.1 max	0.5 max	1.7-2.8	0.5 max	0.6 max	0.15 max	0.2 max						
	India	IS 739	HG30	Wir Ann	0.3	0.1	0.6	0.4-1.4	0.4-1	0.6-1.3		0.1				172		
	India	IS 739	HG30	Wir ST and Nat aged	0.3	0.1	0.6	0.4-1.4	0.4-1	0.6-1.3		0.1				201		
	International	ISO R827	Al-Si1Mg	ST and nat aged 4 in; 100 mm diam	0.35	0.1	0.5	0.4-1.4	0.4-1	0.6-1.6		0.2				192	118	15
	International	ISO R827	Al-Si1Mg	ST and PT .75 in; 50 mm diam	0.35	0.1	0.5	0.4-1.4	0.4-1	0.6-1.6		0.2				285	245	8
	International	ISO R829	Al-Si1Mg	Frg ST and nat aged	0.35	0.1	0.5	0.4-1.4	0.4-1	0.6-1.6		0.2				186	118	15
	International	ISO R829	Al-Si1Mg	Frg ST and PT	0.35	0.1	0.5	0.4-1.4	0.4-1	0.6-1.6		0.2				294	245	8
	International	ISO TR2778	Al-Si1Mg	Tub Ann OBSOLETE Replace by ISO 6363	0.35	0.1	0.5	0.4-1.4	0.4-1	0.6-1.6		0.2				160		15
	International	ISO TR2778	Al-Si1Mg	Tub SHT and nat aged OBSOLETE Replace by ISO 6363	0.35	0.1	0.5	0.4-1.4	0.4-1	0.6-1.6		0.2				200	110	12
	International	ISO TR2778	Al-Si1Mg	Tub ST and PT OBSOLETE Replace by ISO 6363	0.35	0.1	0.5	0.4-1.4	0.4-1	0.6-1.6		0.2				290	240	8
3009	Germany				0.05 max	0.1 max	0.7 max	0.1 max	1.2-1.8	1-1.8	0.1 max	0.05 max	0.05	0.15	Ni 0.05 max; Zr 0.10			
5052 A95052	Australia	AS 1734	5052	Sh Plt Ann 0.20/0.30 mm thk	0.15-0.35	0.1		2.2-2.8	0.1			0.1					65	14
5052 A95052	Australia	AS 1734	5052	Sh Plt SH and stab to 1/2 hard 0.25/0.50 mm thk	0.15-0.35	0.1		2.2-2.8	0.1			0.1					179	3
5052 A95052	Australia	AS 1734	5052	Sh Plt SH and stab to full hard 0.15/0.20 mm thk	0.15-0.35	0.1		2.2-2.8	0.1			0.1					220	2
5052 A95052	Canada	CSA HA.4	0.5052	Plt Sh O .013/.019 in diam	0.15-0.35	0.1		2.2-2.8	0.1			0.1				172	66	15
5052 A95052	Canada	CSA HA.4	0.5052	Plt Sh H32 .016/.019 in diam	0.15-0.35	0.1		2.2-2.8	0.1			0.1				214	159	4
5052 A95052	Canada	CSA HA.4	0.5052	Plt Sh H112 .250/.499 in diam	0.15-0.35	0.1		2.2-2.8	0.1			0.1				193	110	7
5052 A95052	Canada	CSA HA.5	0.5052	Bar Rod Wir Drawn and O	0.15-0.35	0.1		2.2-2.8	0.1			0.1				172	66	25
5052 A95052	Canada	CSA HA.5	0.5052	Bar Rod Wir Drawn and H34 .374 in diam	0.15-0.35	0.1		2.2-2.8	0.1			0.1				234	179	
5052 A95052	Canada	CSA HA.5	0.5052	Bar Rod Wir Drawn and H38 .374 in diam	0.15-0.35	0.1		2.2-2.8	0.1			0.1				269		
5052 A95052	Canada	CSA HA.6	0.5052	Rod Wir O .062/1 in diam	0.15-0.35	0.1		2.2-2.8	0.1			0.1				221		
5052 A95052	Canada	CSA HA.6	0.5052	Rod Wir H32 .062/1 in diam	0.15-0.35	0.1		2.2-2.8	0.1			0.1				214		
5052 A95052	Canada	CSA HA.7	0.5052	Tub Pip Drawn and O .010/.45 in diam	0.15-0.35	0.1		2.2-2.8	0.1			0.1				172	69	

UNS numbers and US grades are provided as a means of cross referencing chemically similar alloys. Exchangability is only possible after independent examination of specifications. Tensile properties are minimum or typical . UTS and YS as Mpa, El as %. See Appendix for list of abbreviations used in Descriptions.

2-48 Wrought Aluminum

Grade UNS #	Country	Specification	Designation	Description	Cr	Cu	Fe	Mg	Mn	Si	Ti	Zn	OE Max	OT Max	Other	UTS	YS	EL
5052 A95052	Canada	CSA HA.7	0.5052	Tub Pip Drawn and H38 .010/.450 in diam	0.15-0.35	0.1		2.2-2.8	0.1			0.1				269	214	
5052 A95052	Canada	CSA HA.7	0.5052	Tub Pip Extd and H111	0.15-0.35	0.1		2.2-2.8	0.1			0.1				172	69	18
5052 A95052	Canada	CSA HA.7.1	0.5052	Tub Pip Drawn and O .010/.45 in diam	0.15-0.35	0.1		2.2-2.8	0.1			0.1				172	69	
5052 A95052	Canada	CSA HA.7.1	0.5052	Tub Pip Drawn and H38 .010/.450 in diam	0.15-0.35	0.1		2.2-2.8	0.1			0.1				269	214	
5052 A95052	Canada	CSA HA.7.1	0.5052	Tub Pip Drawn and H34 .010/.450 in diam	0.15-0.35	0.1		2.2-2.8	0.1			0.1				234	179	
5052 A95052	France	NF A50451	5052	Sh Plt Ann .4/6 mm diam	0.15-0.35	0.1		2.2-2.8	0.1			0.1				170	60	10
5052 A95052	France	NF A50451	5052	Sh Plt 3/4 hard .4/3.2 mm diam	0.15-0.35	0.1		2.2-2.8	0.1			0.1				250	200	5
5052 A95052	France	NF A50451	5052	Sh Plt Hard .4/3.2 mm diam	0.15-0.35	0.1		2.2-2.8	0.1			0.1				270	220	4
5052 A95052	Japan	JIS H4000	5052	sh Plt Ann 1.3 mm diam	0.15-0.35	0.1		2.2-2.8	0.1			0.1				177	69	18
5052 A95052	Japan	JIS H4000	5052	sh Plt Hard .8 mm diam	0.15-0.35	0.1		2.2-2.8	0.1			0.1				275	226	4
5052 A95052	Japan	JIS H4000	5052	sh Plt Hr 12 mm diam	0.15-0.35	0.1		2.2-2.8	0.1			0.1				196	108	7
5052 A95052	Japan	JIS H4040	5052	Bar As Mfg (bar)	0.15-0.35	0.1		2.2-2.8	0.1			0.1				177	69	
5052 A95052	Japan	JIS H4040	5052	Bar Ann (bar)	0.15-0.35	0.1		2.2-2.8	0.1			0.1				245	69	
5052 A95052	Japan	JIS H4040	5052	Bar Hard 10 mm diam	0.15-0.35	0.1		2.2-2.8	0.1			0.1				275		
5052 A95052	Japan	JIS H4080	5052	Wought Tub As Mfg	0.15-0.35	0.1		2.2-2.8	0.1			0.1				177	69	
5052 A95052	Japan	JIS H4080	5052	Wought Tub Ann	0.15-0.35	0.1		2.2-2.8	0.1			0.1				245	69	20
5052 A95052	Japan	JIS H4090	5052	Pip Ann 1.3 mm diam	0.15-0.35	0.1		2.2-2.8	0.1							177	69	18
5052 A95052	Japan	JIS H4090	5052	Pip 1/2 hard 1.3 mm diam	0.15-0.35	0.1		2.2-2.8	0.1							235	177	4
5052 A95052	Japan	JIS H4090	5052	Pip Hard 1.3 mm diam	0.15-0.35	0.1		2.2-2.8	0.1							275	226	4
5052 A95052	Japan	JIS H4100	5052	As Mfg	0.15-0.35	0.1		2.2-2.8	0.1							177	69	
5052 A95052	Japan	JIS H4100	5052	Ann	0.15-0.35	0.1		2.2-2.8	0.1							245	69	20
5052 A95052	Japan	JIS H4120	5052	Wir Ann and Stab at 1/4 hard 25 mm diam	0.15-0.35	0.1		2.2-2.8	0.1			0.1				216		
5352	USA				0.1 max	0.1 max		2.2-2.8	0.1 max		0.1 max	0.1 max	0.05	0.15	0.45 Si+Fe			
5052 A95052	Germany	DIN 1725 Part 1	3.3525/AlMg2 Mn0.3		0.15 max	0.15 max	0.4 max	1.7-2.4	0.1-0.5	0.4 max	0.15 max	0.15 max			ET 0.05 max; ET=.15 total;			
	International	ISO R209 (1971)	Al-Mg 2.5		0.35 max	0.10 max	0.5 max	2.2-2.8	0.5 max	0.5 max					Ti+Zr 0.2 max; Mn+Cr 0.5 max			
	International	ISO R209 (1971)	Al-Cu2Mg		0.1 max	2.0-3.0	0.7 max	0.2-0.5	0.2 max	0.8 max					Ti+Zr 0.2 max			
	Austria	ONORM M3430	Al99.85Mg2.5				0.08 max	2.2-2.8		0.06 max	0.02 max	0.05 max			ET 0.02 max; ET=.15 total			
5052 A95052	Sweden	MNC 40E	4120	Sh				2.5-2.5							ST 0.3; ST=Cr+Mn; Al 97.2			
	India	IS 738	HT20	Tub ST and Nat aged 1.60 mm diam	0.15-0.35	0.15-0.4	0.7	0.8-1.2	0.2-0.8	0.4-0.8						216	108	12

UNS numbers and US grades are provided as a means of cross referencing chemically similar alloys. Exchangability is only possible after independent examination of specifications. Tensile properties are minimum or typical . UTS and YS as Mpa, El as %. See Appendix for list of abbreviations used in Descriptions.

Worldwide Guide to Equivalent Nonferrous Metals and Alloys

Grade UNS #	Country	Specification	Designation	Description	Cr	Cu	Fe	Mg	Mn	Si	Ti	Zn	OE Max	OT Max	Other	UTS	YS	EL
	India	IS 738	HT20	Tub SHT and PHT 1.60 mm diam	0.15-0.35	0.15-0.4	0.7	0.8-1.2	0.2-0.8	0.4-0.8						294	230	7
8007	USA					0.1 max	1.2-2	0.1 max	0.3-1	0.4 max		0.8-1.8	0.05	0.15				
5021	Norway				0.15 max	0.15 max	0.5 max	2.2-2.8	0.1-0.5	0.4 max		0.15 max	0.05	0.15				
6056	France				0.25 max	0.5-1.1	0.5 max	0.6-1.2	0.4-1	0.7-1.3		0.1-0.7	0.05	0.15	0.20 max Zr+Ti			
	France	NF A50451	5150(A-85GT)	Sh Plt Ann .4/1.6 mmdiam		0.1	0.1	1.3-1.7	0.03	0.08	0.6	0.1				130		25
	France	NF A50451	5150(A-85GT)	Sh Plt 3/4 hard .4/3.2 mm diam		0.1	0.1	1.3-1.7	0.03	0.08	0.6	0.1				190	170	3
	France	NF A50451	5150(A-85GT)	Sh Plt hard .4/3.2 mm diam		0.1	0.1	1.3-1.7	0.03	0.08	0.6	0.1				210	190	2
4007	France				0.05-0.25	0.2 max	0.4-1	0.2 max	0.8-1.5	1-1.7	0.1 max	0.1 max	0.05	0.15	Ni 0.15-0.7			
2002	France				0.2 max	1.5-2.5	0.3 max	0.5-1	0.2 max	0.35-0.8	0.2 max	0.2 max	0.05	0.15	Zr 0.10-0.25			
	France	NF A50411	6101(A-SG)	Bar Aged (bar)	0.1	0.1	0.45	0.6 1	0.15	0.8 1.2	0.1	0.2				200	100	15
	France	NF A50411	6181(A-SG)	Bar Q/T (bar)	0.1	0.1	0.45	0.6-1	0.15	0.8-1.2	0.1	0.2				280	240	8
	France	NF A50411	6181(A-SG)	Bar Q/T (Shapes)	0.1	0.1	0.45	0.6-1	0.15	0.8-1.2	0.1	0.2				275	200	8
	India	IS 733	64430	Bar Rod As Mfg or Ann	0.25	0.1	0.6	0.4-1.2	0.4-1	0.6-1.3	0.2	0.1				110		12
	India	IS 733	64430	Bar Rod ST and Nat aged 150 mm diam	0.25	0.1	0.6	0.4-1.2	0.4-1	0.6-1.3	0.2	0.1				185	120	14
	India	IS 733	64430	Bar Rod SHT and PHT 5 mm diam	0.25	0.1	0.6	0.4-1.2	0.4-1	0.6-1.3	0.2	0.1				295	255	7
	India	IS 736	64430	Plt SHT and Nat aged 15 mm diam	0.25	0.1	0.6	0.4-1.2	0.4-1	0.6-1.3	0.2	0.1				200	115	15
	India	IS 736	64430	Plt SHT and PHT 8 mm diam	0.25	0.1	0.6	0.4-1.2	0.4-1	0.6-1.3	0.2	0.1				285	240	8
2090	USA				0.05 max	2.4-3	0.12 max	0.25 max	0.05 max	0.1 max	0.15 max	0.1 max	0.05	0.15	1.9-2.6 Li; Zr 0.08-0.15			
2117 A92117	India	IS 5902	22500	Bar Wir Ann and CD	0.1	2-3	0.5	0.2-0.5	0.2	0.7		0.2			Ni 0.05; Pb 0.05; Sn 0.05 max			
2117 A92117	Yugoslavia	JUS CC2.100	AlCu3Mg		0.25 max	2-3	0.7 max	0.2-0.5	0.2 max	0.7 max		0.25 max		0.3				
2117 A92117	Austria	ONORM M3430	AlCu2.5Mg0.5		0.1 max	2.2-3	0.7 max	0.2-0.5	0.2 max	0.8 max		0.25 max			ET 0.05 max; ET=.15;			
2117 A92117	Canada	CSA HA.6	0.2117	Rod Wir H15 .062/1 in diam	0.1	2.2-3	0.7	0.2-0.5	0.2	0.8		0.25				193		
2117 A92117	Canada	CSA HA.6	0.2117	Rod Wir T4 .062/1 in diam	0.1	2.2-3	0.7	0.2-0.5	0.2	0.8		0.25				262	124	16
2117 A92117	France	NF A50451	2117(A-U2G)	Sh Plt Aged .4/3.2 mm diam	0.1	2.2-3	0.7	0.2-0.5	0.2	0.8		0.25				250	150	22
2117 A92117	Germany	DIN 1725 Part 1	3.1305/AlCu2.5Mg0.5		0.1 max	2.2-3	0.7 max	0.2-0.5	0.2 max	0.8 max		0.25 max			ET 0.05 max; ET=.15 total			
2117 A92117	Japan	JIS H4120	2117	Wir Ann 25 mm diam	0.1	2.2-3	0.7	0.2-0.5	0.2	0.8		0.25				177		
2117 A92117	Japan	JIS H4120	2117	Wir 5/8 hard 15 mm diam	0.1	2.2-3	0.7	0.2-0.5	0.2	0.8		0.25				196		
2117 A92117	Japan	JIS H4120	2117	Wir SHT and AH 2/25 mm diam	0.1	2.2-3	0.7	0.2-0.5	0.2	0.8		0.25				255	118	18
2117 A92117	USA				0.1 max	2.2-3	0.7 max	0.2-0.5	0.2 max	0.8 max		0.25 max	0.05	0.15				
	USA		Al Cu2.5Mg		0.1 max	2.2-3	0.7 max	0.2-0.5	0.2 max	0.8 max		0.25 max	0.05	0.15				
2117 A92117	Yugoslavia	JUS CC2.100	AlCu2Mg		0.1 max	2-3	0.7 max	0.2-0.5	0.2 max	0.8					ST 0.2 max; ST = Zr + Ti;			

UNS numbers and US grades are provided as a means of cross referencing chemically similar alloys. Exchangability is only possible after independent examination of specifications. Tensile properties are minimum or typical . UTS and YS as Mpa, El as %. See Appendix for list of abbreviations used in Descriptions.

Grade UNS #	Country	Specification	Designation	Description	Cr	Cu	Fe	Mg	Mn	Si	Ti	Zn	OE Max	OT Max	Other	UTS	YS	EL
	International	ISO R209 (1971)	Al-Mg 3		0.35 max	0.10 max	0.5 max	2.4-3.1	0.4 max	0.5 max					Ti+Zr 0.2 max			
	India	IS 736	51000-A	Plt As Mfg 20 mm diam	0.1	0.2	0.7	0.5-1.1	0.2	0.6		0.25				105		20
	India	IS 736	51000-A	Plt Ann 22 mm diam	0.1	0.2	0.7	0.5-1.1	0.2	0.6		0.25				95		22
	India	IS 736	51000-A	Plt 1/2 hard 4 mm diam	0.1	0.2	0.7	0.5-1.1	0.2	0.6		0.25				140		4
	Japan	JIS Z3263	BA11PC	Sh Ann .3 mm		0.05-0.2	0.7		1-1.5	0.6		0.1				137		15
	Japan	JIS Z3263	BA11PC	Sh 1/4 hard .3 mm diam		0.05-0.2	0.7		1-1.5	0.6		0.1				118		2
	Japan	JIS Z3263	BA11PC	Sh 1/2 hard .3 mm diam		0.05-0.2	0.7		1-1.5	0.6		0.1				137		1
	India	IS 739	HG20	Wir ST and PT and drawn 6.30 mm diam	0.15-0.35	0.15-0.4	0.7	0.8-1.2	0.2-0.8	0.4-0.8		0.2				372		
5349	USA					0.18-0.28	0.7 max	1.7-2.6	0.6-1.2	0.4 max	0.09 max		0.05	0.15				
	South Africa	SABS 712	Al-Mn1			0.1	0.7	0.1	0.8-1.5	0.6		0.2						
	UK	BS 1470	NS3	Plt Sh Strp Ann .2/6 mm diam OBSOLETE Replaced by BS EN 515		0.1	0.7	0.1	0.8-1.5	0.6		0.2				90		20
	UK	BS 1470	NS3	Plt Sh Strp SH to 1/2 hard .2/12.5 mm diam OBSOLETE Replaced by BS EN 515		0.1	0.7	0.1	0.8-1.5	0.6		0.2				140		3
	UK	BS 1470	NS3	Plt Sh Strp SH to hard .2/3 mm diam OBSOLETE Replaced by BS EN 515		0.1	0.7	0.1	0.8-1.5	0.6		0.2				175		2
7013	USA					0.1 max	0.7 max		1-1.5	0.6 max		1.5-2	0.05	0.15				
	France	NF A50451	6181(A-SG)	Sh Plt Aged .4/6 mm diam	0.3	0.1	0.45	0.7-1.1	0.15	0.8-1.2	0.1	0.2				210	110	18
	France	NF A50451	6181(A-SG)	Sh Plt Q/T .4/3.2 mm diam	0.3	0.1	0.45	0.7-1.1	0.15	0.8-1.2	0.1	0.2				250	150	18
	South Africa	SABS 712	Al-SiMgMn			0.1	0.5	0.4-0.8	0.4-0.8	0.7-1.3	0.2	0.2						
	Pan American	COPANT 862	X6082			0.1	0.5	0.7-1.3	0.2-1	0.8-1.5	0.1	0.1						
4016	Norway					0.2 max	0.7 max	0.1 max	0.6-1.2	1.4-2.2		0.5-1.3	0.05	0.15				
	India	IS 733	65032	Bar Rod As Mfg or Ann	0.15-0.35	0.15-0.4	0.7	0.7-1.2	0.2-0.8	0.4-0.8	0.2	0.2				110		12
	India	IS 733	65032	Bar Rod ST and Nat aged 150 mm diam	0.15-0.35	0.15-0.4	0.7	0.7-1.2	0.2-0.8	0.4-0.8	0.2	0.2				185	115	14
	India	IS 733	65032	Bar Rod SHT and PHT 150 mm diam	0.15-0.35	0.15-0.4	0.7	0.7-1.2	0.2-0.8	0.4-0.8	0.2	0.2				280	235	7
	India	IS 736	65032	Plt SHT and Nat aged 15 mm	0.15-0.35	0.15-0.4	0.7	0.7-1.2	0.2-0.8	0.4-0.8	0.2	0.2				200	110	15
	India	IS 736	65032	Plt SHT and PHT 8 mm diam	0.15-0.35	0.15-0.4	0.7	0.7-1.2	0.2-0.8	0.4-0.8	0.2	0.2				280	235	8
2096	USA					2.3-3	0.15 max	0.25-0.8	0.25 max	0.12 max	0.1 max	0.25 max	0.05	0.15	1.3-1.9 Li; Ag 0.25-0.6; Zr 0.04-0.18			
	Norway	NS 17211	NS17211/AIMg3	Plt Sh Strp	0.3 max	0.1 max	0.4 max	2.6-3.6	0.5 max	0.4 max	0.15 max	0.2 max			ET 0.05 max; ST 0.1-0.6; ET=.15 total, ST=Mn+Cr;			

UNS numbers and US grades are provided as a means of cross referencing chemically similar alloys. Exchangability is only possible after independent examination of specifications. Tensile properties are minimum or typical . UTS and YS as Mpa, EI as %. See Appendix for list of abbreviations used in Descriptions.

Worldwide Guide to Equivalent Nonferrous Metals and Alloys

Grade UNS #	Country	Specification	Designation	Description	Cr	Cu	Fe	Mg	Mn	Si	Ti	Zn	OE Max	OT Max	Other	UTS	YS	EL
5754	USA				0.3 max	0.1 max	0.4 max	2.6-3.6	0.5 max	0.4 max	0.15 max	0.2 max	0.15	0.15	0.10-0.6 Mn+Cr			
	USA		AlMg3		0.3 max	0.1 max	0.4 max	2.6-3.6	0.5 max	0.4 max	0.15 max	0.2 max	0.15	0.15	0.10-0.6 Mn+Cr			
	Yugoslavia	JUS CC2.100	AlMg3		0.35 max	0.1 max	0.5 max	2.6-3.5	0.4 max	0.4 max					ST 0.2 max; ST 0.5 max; ST = Ti+Zr, Mn+Cr;			
	France	NF A50411	6082(A-SGM0.7)	Bar Tube Aged (extd bar)	0.25	0.1	0.5	0.6-1.2	0.4-1	0.7-1.3	0.1	0.2				200	120	15
	France	NF A50411	6082(A-SGM0.7)	Bar Tube Aged (tub) 150 mm diam	0.25	0.1	0.5	0.6-1.2	0.4-1	0.7-1.3	0.1	0.2				310	282	15
	France	NF A50411	6082(A-SGM0.7)	Bar Tube Aged (shapes)	0.25	0.1	0.5	0.6-1.2	0.4-1	0.7-1.3	0.1	0.2				200	120	15
	UK	BS 1470	HS30	Plt Sh Strp Ann .2/3 mm diam OBSOLETE Replaced by BS EN 515	0.25	0.1	0.5	0.5-1.2	0.4-1	0.7-1.3	0.2	0.2				155		16
	UK	BS 1470	HS30	Plt Sh Strp SHT and nat aged .2/3 mm diam OBSOLETE Replaced by BS EN 515	0.25	0.1	0.5	0.5-1.2	0.4-1	0.7-1.3	0.2	0.2				200	120	15
	UK	BS 1470	HS30	Plt Sh Strp SHT and PT .2/3 mm diam OBSOLETE Replaced by BS EN 515	0.25	0.1	0.5	0.5-1.2	0.4-1	0.7-1.3	0.2	0.2					255	8
	UK	BS 1471	HT30	Tub SHT and nat aged 6 mm diam	0.25	0.1	0.5	0.5-1.2	0.4-1	0.7-1.3	0.2	0.2				215	115	12
	UK	BS 1471	HT30	Tub SHT and PT 6 mm diam	0.25	0.1	0.5	0.5-1.2	0.4-1	0.7-1.3	0.2	0.2				3310	25	7
	UK	BS 1472	HF30	SHT nat Aged 200 mm diam Obsolete Replaced by BS EN 586	0.25	0.1	0.5	0.5-1.2	0.4-1	0.7-1.3	0.2	0.2				185	120	16
	UK	BS 1472	HF30	SHT PT 150 mm diam Obsolete Replaced by BS EN 586	0.25	0.1	0.5	0.5-1.2	0.4-1	0.7-1.3	0.2	0.2				295	255	8
	UK	BS 1473	HB30	SHT and PT 6 mm diam	0.25	0.1	0.5	0.5-1.2	0.4-1	0.7-1.3	0.2	0.2				295	255	
	UK	BS 1473	HR30	SHT and nat aged 25 mm diam	0.25	0.1	0.5	0.5-1.2	0.4-1	0.7-1.3	0.2	0.2				200		
	UK	BS 1474	HE30	Bar Tub Ann 200 mm diam	0.25	0.1	0.5	0.5-1.2	0.4-1	0.7-1.3	0.2	0.2				170		14
	UK	BS 1474	HE30	Bar Tub As Mfg 200 mm diam	0.25	0.1	0.5	0.5-1.2	0.4-1	0.7-1.3	0.2	0.2				110		12
	UK	BS 1474	HE30	Bar Tub SHT and nat aged 150/200 mm diam	0.25	0.1	0.5	0.5-1.2	0.4-1	0.7-1.3	0.2	0.2				170	100	
6253	USA				0.04-0.35	0.1 max	0.5 max	1-1.5				1.6-2.4	0.05	0.15	45-65% of Mg			
2036	USA				0.1 max	2.2-3	0.5 max	0.3-0.6	0.1-0.4	0.5 max	0.15 max	0.25 max	0.05	0.15				
2197	USA					2.5-3.1	0.1 max	0.25 max	0.1-0.5	0.1 max	0.12 max	0.05 max	0.05	0.15	1.3-1.7 Li; Zr 0.08-0.15			

UNS numbers and US grades are provided as a means of cross referencing chemically similar alloys. Exchangability is only possible after independent examination of specifications. Tensile properties are minimum or typical . UTS and YS as Mpa, El as %. See Appendix for list of abbreviations used in Descriptions.

2-52 Wrought Aluminum

Grade UNS #	Country	Specification	Designation	Description	Cr	Cu	Fe	Mg	Mn	Si	Ti	Zn	OE Max	OT Max	Other	UTS	YS	EL
2097	USA					2.5-3.1	0.15 max	0.35 max	0.1-0.6	0.12 max	0.15 max	0.35 max	0.05	0.15	1.2-1.8 Li; Zr 0.08-0.16			
5754 A95754	Czech Republic	CSN 421413	AlMg3			0.1 max	0.4 max	2.6-4	0.05-0.4	0.5 max	0.2 max	0.2 max			Fe+S 0.60; Sb 0.25			
	UK	BS 1471	HT20	Tub SH to 1/2 hard 6 mm diam	0.04-0.35	0.15-0.4	0.7	0.8-1.2	0.2-0.8	0.4-0.8	0.2	0.2				185	160	5
	UK	BS 1471	HT20	Tub SHT and nat aged 6 mmdiam	0.04-0.35	0.15-0.4	0.7	0.8-1.2	0.2-0.8	0.4-0.8	0.2	0.2				215	115	12
	UK	BS 1471	HT20	Tub SHT and PT 6 mm diam	0.04-0.35	0.15-0.4	0.7	0.8-1.2	0.2-0.8	0.4-0.8	0.2	0.2				295	240	7
5017	USA					0.18-0.28	0.7 max	1.9-2.2	0.6-0.8	0.4 max	0.09 max		0.05	0.15				
	Yugoslavia	JUS CC2.100	AlMg3Mn		0.25 max	0.1 max	0.5 max	2.4-3.4	0.3-1	0.5 max		0.2 max		0.15	ST 0.2 max; ST = Ti + Zn;			
	Pan American	COPANT 862	3103		0.1	0.1	0.7	0.3	0.9-1.5	0.5		0.2						
	Sweden	SIS 144054	4054-02	Sh Strp Ann .35/6 mm diam	0.1	0.1	0.7	0.3	0.9-1.5	0.5		0.2				95	35	30
	Sweden	SIS 144054	4054-02	Sh Strp SH .35/5 in diam	0.1	0.1	0.7	0.3	0.9-1.5	0.5		0.2				120	90	10
	Sweden	SIS 144054	4054-12	Strp SH .35/2 mm diam	0.1	0.1	0.7	0.3	0.9-1.5	0.5		0.2				120	90	10
	Sweden	SIS 144054	4054-14	Sh Strp SH .35/4 mm diam	0.1	0.1	0.7	0.3	0.9-1.5	0.5		0.2				14	115	6
	Sweden	SIS 144054	4054-16	Sh Strp SH .35/3 mm diam	0.1	0.1	0.7	0.3	0.9-1.5	0.5		0.2				165	140	5
	Sweden	SIS 144054	4054-18	Sh Strp SH .35/1.6 mm diam	0.1	0.1	0.7	0.3	0.9-1.5	0.5		0.2				205	185	
	Sweden	SIS 144054	4054-24	Sh Strp SH .35/3 mm diam	0.1	0.1	0.7	0.3	0.9-1.5	0.5		0.2				140	110	12
	Sweden	SIS 144054	4054-26	Sh Strp SH Ann .35/3 mm diam	0.1	0.1	0.7	0.3	0.9-1.5	0.5		0.2				165	135	10
	Sweden	SIS 144054	4054-40	Sh Strp Ann .35/6 mm diam	0.1	0.1	0.7	0.3	0.9-1.5	0.5		0.2				95	35	30
	Sweden	SIS 144054	4054-42	Sh Strp SH .35/3 mm diam	0.1	0.1	0.7	0.3	0.9-1.5	0.5		0.2				120	90	10
	Sweden	SIS 144054	4054-44	Sh Strp SH .35/3mm diam	0.1	0.1	0.7	0.3	0.9-1.5	0.5		0.2				140	115	6
	Sweden	SIS 144055	4055-00	Wir		0.1	0.6	0.3	0.9-1.4	0.6		0.2						
	Sweden	SIS 144055	4055-18	Wir SH		0.1	0.6	0.3	0.9-1.4	0.6		0.2				176		
	International	ISO TR2136	Al-Mn1	Sh Plt Ann		0.1	0.7	0.3	0.8-1.5	0.6		0.2				90		20
	International	ISO TR2136	Al-Mn1	Sh Plt HB SH		0.1	0.7	0.3	0.8-1.5	0.6		0.2				115		5
	International	ISO TR2136	Al-Mn1	Sh Plt HD Str hard		0.1	0.7	0.3	0.8-1.5	0.6		0.2				130		3
	UK	BS 1475	NG3	Wir Ann 10 mm diam	0.2	0.1	0.7	0.1	0.8-1.5	0.6	0.2	0.2				130		
	UK	BS 1475	NG3	Wir SH to hard 10 mm daim	0.2	0.1	0.7	0.1	0.8-1.5	0.6	0.2	0.2				175		
	Finland	SFS 2594	AlSi1MgPb	St and aged	0.35	0.1	0.5	0.6-1.2	0.4-1	0.6-1.4		0.5				280	180	8
	International	ISO R209 (1971)	Al-Mg 3 Mn		0.25 max	0.10 max	0.5 max	2.4-3.4	0.3-1.0	0.5 max					Ti+Zr 0.2 max			
5018	Germany				0.3 max	0.05 max	0.4 max	2.6-3.6	0.2-0.6	0.25 max	0.15 max	0.2 max	0.05	0.15	0.20-0.6 Mn+Cr 0.0008 max Be for welding			
5354	Europe				0.05-0.2	0.05 max	0.4 max	2.4-3	0.5-1	0.25 max	0.15 max	0.25 max	0.15	0.15	Zr 0.10-0.20			
5454	USA				0.05-0.2	0.1 max	0.4 max	2.4-3	0.5-1	0.25 max	0.2 max	0.25 max	0.2	0.15				
	USA		AlMg3Mn		0.05-0.2	0.1 max	0.4 max	2.4-3	0.5-1	0.25 max	0.2 max	0.25 max	0.2	0.15				
5254 A95254	Canada	CSA GR40	5254	Plt Sh Tub	0.15-0.35	0.35 max	0.4 max	3.1-3.9	0.01 max	0.25 max	0.05 max	0.2 max			ET 0.05 max; ET=0.15 max;			

UNS numbers and US grades are provided as a means of cross referencing chemically similar alloys. Exchangability is only possible after independent examination of specifications. Tensile properties are minimum or typical . UTS and YS as Mpa, El as %. See Appendix for list of abbreviations used in Descriptions.

Worldwide Guide to Equivalent Nonferrous Metals and Alloys

Grade UNS #	Country	Specification	Designation	Description	Cr	Cu	Fe	Mg	Mn	Si	Ti	Zn	OE Max	OT Max	Other	UTS	YS	EL
5254 A95254	USA	ASTM B209	5254	Plt Sh Tub	0.15-0.35	0.35 max	0.4 max	3.1-3.9	0.01 max	0.25 max	0.05 max	0.2 max			ET 0.05 max; ET=0.15 max			
5254 A95254	USA	ASTM B241	5254	Plt Sh Tub	0.15-0.35	0.35 max	0.4 max	3.1-3.9	0.01 max	0.25 max	0.05 max	0.2 max			ET 0.05 max; ET=0.15 max			
5254 A95254	USA	SAE J454	5254	Plt Sh Tub	0.15-0.35	0.35 max	0.4 max	3.1-3.9	0.01 max	0.25 max	0.05 max	0.2 max			ET 0.05 max; ET=0.15 max			
2091	France				0.1 max	1.8-2.5	0.3 max	1.1-1.9	0.1 max	0.2 max	0.1 max	0.25 max	0.05	0.15	1.7-2.3 Li; Zr 0.04-0.16			
A95554	USA	AWS	ER5554	Weld El Rod	0.05-0.2	0.1 max	0.4 max	2.4-3	0.5-1	0.25 max	0.05-0.2	0.25 max	0.05	0.15	Be 0.0008 max			
A95554	USA	AWS	R5554	Weld El Rod	0.05-0.2	0.1 max	0.4 max	2.4-3	0.5-1	0.25 max	0.05-0.2	0.25 max	0.05	0.15	Be 0.0008 max			
5554	USA				0.05-0.2	0.1 max	0.4 max	2.4-3	0.5-1	0.25 max	0.05-0.2	0.25 max	0.2	0.15	0.0008 max Be for welding; OE min 0.05			
	USA		AlMg3Mn(A)		0.05-0.2	0.1 max	0.4 max	2.4-3	0.5-1	0.25 max	0.05-0.2	0.25 max	0.2	0.15	0.0008 max Be for welding; OE min 0.05			
6003 A96003	Australia	AS 1734	6003		0.35 max	0.1 max	0.6 max	0.8-1.5	0.8 max	0.35-1	0.1 max	0.2 max	0.05		OT min 0.15			
6003 A96003	Canada	CSA HA.4	0.6003	Plt Sh T3 (flat sht) .029/.039 in	0.35	0.1	0.6	0.8-1.5	0.8	0.35-1	0.1	0.2				234	234	14
6003 A06003	Canada	CSA HA.4	0.6003	Plt Sh T451 (plt) .250/.499 in	0.35	0.1	0.6	0.8-1.5	0.8	0.35-1	0.1	0.2				393		15
6003 A96003	Canada	CSA HA.4	0.6003	Plt Sh T6 (sht) .020/.039 in	0.35	0.1	0.6	0.8-1.5	0.8	0.35-1	0.1	0.2				434	379	7
5554 A95554	Canada	CSA HA.6	0.5554	Rod Wir	0.05-0.2	0.1		2.4-3	0.5-1		0.05-0.2	0.25			Be 0.008			
5554 A95554	Japan	JIS Z3232	5554	Rod Ann	0.05-0.2	0.1		2.4-3	0.5-1		0.05-0.2	0.25			Be 0.001	177		
6066	USA				0.4 max	0.7-1.2	0.5 max	0.8-1.4	0.6-1.1	0.9-1.8	0.2 max	0.25 max	0.05	0.15				
2006 A92006	Argentina					1-2	0.7 max	0.5-1.4	0.6-1	0.8-1.3	0.3 max	0.2 max	0.05	0.15	Ni 0.2 max			
2006 A92006	Czech Republic	CSN 424206	AlCu2SiMn		0.01-0.2	1.8-2.6	0.7 max	0.4-0.8	0.4-0.8	0.7-1.2	0.02-0.1	0.3 max			Fe+Ni 0.70; Ni 0.1 max			
5254	USA				0.15-0.35	0.05 max		3.1-3.9	0.01 max		0.05 max	0.2 max	0.05	0.15	0.45 Si + Fe			
5154 A95154	Belgium	NBN P21-001	5154		0.15-0.35	0.1 max	0.4 max	3.1-3.9	0.1 max	0.25 max	0.2 max	0.2 max			ET 0.05 max; ; ET = .15 Total			
5154 A95154	Canada	CSA GR40	5154	Bar Rod Wir Plt Sh Tub	0.15-0.35	0.1 max	0.4 max	3.1-3.9	0.1 max	0.25 max	0.2 max	0.2 max	0.05	0.15				
5154 A95154	France	NF A-G3C	5154	Bar Rod Wir Plt Sh Tub	0.15-0.35	0.1 max	0.4 max	3.1-3.9	0.1 max	0.25 max	0.2 max	0.2 max	0.05	0.15				
5154 A95154	USA	AMS 4018	5154	Bar Rod Wir Plt Sh Tub	0.15-0.35	0.1 max	0.4 max	3.1-3.9	0.1 max	0.25 max	0.2 max	0.2 max	0.05	0.15				
5154 A95154	USA	AMS 4019	5154	Bar Rod Wir Plt Sh Tub	0.15-0.35	0.1 max	0.4 max	3.1-3.9	0.1 max	0.25 max	0.2 max	0.2 max	0.05	0.15				
5154 A95154	USA	ASTM B209	5154	Bar Rod Wir Plt Sh Tub	0.15-0.35	0.1 max	0.4 max	3.1-3.9	0.1 max	0.25 max	0.2 max	0.2 max	0.05	0.15				
5154 A95154	USA	ASTM B210	5154	Bar Rod Wir Plt Sh Tub	0.15-0.35	0.1 max	0.4 max	3.1-3.9	0.1 max	0.25 max	0.2 max	0.2 max	0.05	0.15				
5154 A95154	USA	ASTM B221	5154	Bar Rod Wir Plt Sh Tub	0.15-0.35	0.1 max	0.4 max	3.1-3.9	0.1 max	0.25 max	0.2 max	0.2 max	0.05	0.15				
5154 A95154	USA	ASTM B313	5154	Bar Rod Wir Plt Sh Tub	0.15-0.35	0.1 max	0.4 max	3.1-3.9	0.1 max	0.25 max	0.2 max	0.2 max	0.05	0.15				
5154 A95154	USA	ASTM B547	5154	Bar Rod Wir Plt Sh Tub	0.15-0.35	0.1 max	0.4 max	3.1-3.9	0.1 max	0.25 max	0.2 max	0.2 max	0.05	0.15				
5154 A95154	USA	SAE J454	5154	Bar Rod Wir Plt Sh Tub	0.15-0.35	0.1 max	0.4 max	3.1-3.9	0.1 max	0.25 max	0.2 max	0.2 max	0.05	0.15				
5154	USA				0.15-0.35	0.1 max	0.4 max	3.1-3.9	0.1 max	0.25 max	0.2 max	0.2 max	0.05	0.15	0.0008 max Be for welding			
	USA		AlMg3.5		0.15-0.35	0.1 max	0.4 max	3.1-3.9	0.1 max	0.25 max	0.2 max	0.2 max	0.05	0.15	0.0008 max Be for welding			

UNS numbers and US grades are provided as a means of cross referencing chemically similar alloys. Exchangability is only possible after independent examination of specifications. Tensile properties are minimum or typical . UTS and YS as Mpa, El as %. See Appendix for list of abbreviations used in Descriptions.

2-54 Wrought Aluminum

Grade UNS #	Country	Specification	Designation	Description	Cr	Cu	Fe	Mg	Mn	Si	Ti	Zn	OE Max	OT Max	Other	UTS	YS	EL
5154 A95154	France	NF A50-701	AG3		0.4 max	0.1 max	0.5 max	2.6-3.8	0.1-0.6	0.4 max		0.2 max			ST 0.2 max; ST=+Ti+Zr			
5154 A95154	France	NF A57-350	AG3		0.4 max	0.1 max	0.5 max	2.6-3.8	0.1-0.6	0.4 max		0.2 max			ST 0.2 max; ST=+Ti+Zr;			
5154 A95154	Australia	AS 1734	5154A	Sh Plt Ann 0.50/0.80 mm thk	0.25	0.1	0.5	3.1-3.9	0.1-0.5	0.5	0.2	0.2				216	75	12
5154 A95154	Australia	AS 1734	5154A	Sh Plt SH and Stab to 1/4 hard 0.50/1.30 mm thk	0.25	0.1	0.5	3.1-3.9	0.1-0.5	0.5	0.2	0.2				248	179	5
5154 A95154	Australia	AS 1734	5154A	Sh Plt SH and Stab to 1/2 hard 0.25/1.30 mm thk	0.25	0.1	0.5	3.1-3.9	0.1-0.5	0.5	0.2	0.2				279		4
5154 A95154	Australia	AS 1866	5154A	Rod Bar SH 50.0 mm diam	0.25	0.1	0.5	3.1-3.9	0.1-0.5	0.5	0.2	0.2				217	100	16
5154 A95154	International	ISO R827	Al-Mg3.5	As Mfg 5 in; 127 mm diam	0.35	0.1	0.5	3.1-3.9	0.6	0.5		0.2				206	79	12
5154 A95154	International	ISO TR2136	Al-Mg3.5	Sh Plt Ann	0.35	0.1	0.5	3.1-3.9	0.6	0.5		0.2				210	75	12
5154 A95154	International	ISO TR2136	Al-Mg3.5	Sh Plt HB SH	0.35	0.1	0.5	3.1-3.9	0.6	0.5		0.2				235	165	5
5154 A95154	International	ISO TR2136	Al-Mg3.5	Sh Plt HD Str hard	0.35	0.1	0.5	3.1-3.9	0.6	0.5		0.2				260	195	4
5154A	UK				0.25 max	0.1 max	0.5 max	3.1-3.9	0.5 max	0.5 max	0.2 max	0.2 max	0.05	0.15	0.10-0.50 Mn+Cr; 0.0008 max Be for welding			
	UK		AlMg3.5(A)		0.25 max	0.1 max	0.5 max	3.1-3.9	0.5 max	0.5 max	0.2 max	0.2 max	0.05	0.15	0.10-0.50 Mn+Cr; 0.0008 max De for welding			
5154 A95154	Japan	JIS H4000	5154	Sh Plt Strp Ann 1.3 mm diam	0.15-0.35	0.1		3.1-3.9	0.1		0.2	0.2				206	78	14
5154 A95154	Japan	JIS H4000	5154	Sh Plt Strp Hard 1.3 mm diam	0.15-0.35	0.1		3.1-3.9	0.1		0.2	0.2				314	245	3
5154 A95154	Japan	JIS H4000	5154	Sh Plt Strp HR 12 mm diam	0.15-0.35	0.1		3.1-3.9	0.1		0.2	0.2				226	127	8
5154 A95154	Japan	JIS H4080	5154	Tub As Mfg	0.15-0.35	0.1		3.1-3.9	0.1		0.2	0.2				206	78	
5154 A95154	Japan	JIS H4080	5154	Tub Ann	0.15-0.35	0.1		3.1-3.9	0.1		0.2	0.2				284	78	
5154 A95154	Japan	JIS Z3232	5154	Rod Ann	0.15-0.35	0.1		3.1-3.9	0.1		0.2	0.2			Be 0.001	206		
5154 A95154	Russia	GOST 4784	1530			0.5 max	0.5 max	3.2-3.8	0.3-0.6	0.5-0.8								
5154 A95154	Russia	GOST 4784	AMg3			0.5 max	0.5 max	3.2-3.8	0.3-0.6	0.5-0.8								
	International	ISO R209 (1971)	Al-Mg 3.5		0.35 max	0.10 max	0.5 max	3.1-3.9	0.6 max	0.5 max					Ti+Zr 0.2 max			
6253 A96253	Australia	AS 1865	6253		0.35 max		0.5 max	1-1.5				1.6-2.4	0.05		Si 45-65% of actual Mg content; OT min 0.15			
5654 A95654	Canada	CSA HA.6	0.5654	Rod Wir	0.15-0.35	0.05		3.1-3.9	0.01		0.05-0.15	0.2			Be 0.008			
A95654	USA	AWS		Weld El Rod	0.15-0.35	0.05 max		3.1-3.9	0.01 max		0.05-0.15	0.2 max	0.05	0.15	Si+Fe<=0.45%; Be 0.0008 max			
A95654	USA	AWS		Weld El Rod	0.15-0.35	0.05 max		3.1-3.9	0.01 max		0.05-0.15	0.2 max	0.05	0.15	Si+Fe<=0.45%; Be 0.0008 max			
5654	USA				0.15-0.35	0.05 max		3.1-3.9	0.01 max		0.05-0.15	0.2 max	0.15	0.15	0.45 Si+Fe; 0.0008 max Be for welding; OE min 0.05			
	France	NF A50411	5251(A-G2M)	Bar Tub Wir Hard (drawn bar) 10 mm	0.15	0.15	0.5	1.7-2.4	0.1-0.5	0.4	0.15	0.15				240	200	2

UNS numbers and US grades are provided as a means of cross referencing chemically similar alloys. Exchangability is only possible after independent examination of specifications. Tensile properties are minimum or typical . UTS and YS as Mpa, El as %. See Appendix for list of abbreviations used in Descriptions.

Worldwide Guide to Equivalent Nonferrous Metals and Alloys

Grade UNS #	Country	Specification	Designation	Description	Cr	Cu	Fe	Mg	Mn	Si	Ti	Zn	OE Max	OT Max	Other	UTS	YS	EL
	France	NF A50411	5251(A-G2M)	Bar Tub Wir Hard (wir) 6 mm diam	0.15	0.15	0.5	1.7-2.4	0.1-0.5	0.4	0.15	0.15				240	200	2
	France	NF A50411	5251(A-G2M)	Bar Tub Wir Hard (tub) 25 mm diam	0.15	0.15	0.5	1.7-2.4	0.1-0.5	0.4	0.15	0.15				240	200	2
	Pan American	COPANT 862	X5951		0.15	0.15	0.5	1.7-2.4	0.1-0.5	0.4	0.15	0.15						
	UK	BS 1472	NF4	As Mfg 150 mm diam Obsolete Replaced by BS EN 586	0.25	0.1		1.7-2.4	0.5	0.5	0.2	0.2				170	60	16
	Austria	ONORM M3430	AlMgSiPbCd1		0.3 max	0.1 max	0.5 max	0.6-1.2	0.4-1	0.6-1.4	0.2 max	0.3 max			ET 0.05 max; ET=0.15 total; Cd 0.2-0.4; Pb 1-1.3			
5854	France				0.15-0.35	0.1 max		3.1-3.9	0.1-0.5		0.2 max	0.2 max	0.2	0.15	0.45 Si + Fe			
5042	USA				0.1 max	0.15 max	0.35 max	3-4	0.2-0.5	0.2	0.1 max	0.25 max	0.05	0.15				
5154B	Italy				0.1 max	0.05 max	0.45 max	3.1-3.9	0.15-0.45	0.35 max	0.15 max	0.15 max	0.05	0.15	Ni 0.01 max			
	UK	BS 1474	NE4	Bar Tub As Mfg 150 mm diam	0.25	0.1	0.05	1.7-2.4	0.5	0.5	0.2	0.2				170	60	14
8112 A98112	Pan American	COPANT 862	8112		0.2	0.15-0.4	1	0.3-0.7	0.2-0.6	0.4-1	0.2	1						
8112	USA				0.2 max	0.4 max	1 max	0.7 max	0.6 max	1 max	0.2 max	1 max	0.05	0.15				
5954	USA				0.1 max	0.1 max	0.4 max	3.3-4.1	0.1 max	0.25 max	0.2 max	0.2 max	0.2	0.15				
	International	ISO TR2778	Al-Mn1	Tub Ann OBSOLETE Replace by ISO 6363		0.1	0.7	0.3	1.5	0.6		0.2				90		20
	International	ISO TR2778	Al-Mn1	Tub SH OBSOLETE Replace by ISO 6363		0.1	0.7	0.3	1.5	0.6		0.2				130		3
	France	AIR 9150-B	A-UZG	Wir CF or Ann 1.6/9.6 mm diam	0.1	2.2-3	0.4	0.2-0.5	0.2	0.2-0.7		0.2				290	170	25
5454 A95454	Canada	CSA GM31N	5454	Bar Rod Wir Plt Sh Tub Ext Shapes	0.05-0.2	0.1 max	0.4 max	2.4-3	0.5-1	0.25 max	0.2 max	0.25 max	0.05	0.15				
5454 A95454	France	NF A-G2.5Mc	5454	Bar Rod Wir Plt Sh Tub Ext Shapes	0.05-0.2	0.1 max	0.4 max	2.4-3	0.5-1	0.25 max	0.2 max	0.25 max	0.05	0.15				
5454 A95454	France	NF A50411	5454	Bar WirAnn (extd bar)	0.05-0.2	0.1	0.4	2.4-3	0.5-1	0.25	0.2	0.25				215	80	16
5454 A95454	France	NF A50411	5454	Bar Wir Ann (wir) 25 mm diam	0.05-0.2	0.1	0.4	2.4-3	0.5-1	0.25	0.2	0.25				215	80	16
5454 A95454	France	NF A50411	5454	Bar Wir Ann (tub) 150 mm diam	0.05-0.2	0.1	0.4	2.4-3	0.5-1	0.25	0.2	0.25				215	80	16
5454 A95454	Germany	DIN 1725 pt1	3.3537/AlMg2.7Mn		0.05-0.2	0.1 max	0.4 max	2.4-3	0.5-1	0.25 max	0.2 max	0.25 max			ET 0.05 max; ET=.15 total;			
5454 A95454	Germany	DIN	Al-Mg2.7-Mn	Bar Rod Wir Plt Sh Tub Ext Shapes	0.05-0.2	0.1 max	0.4 max	2.4-3	0.5-1	0.25 max	0.2 max	0.25 max	0.05	0.15				
5454 A95454	Norway	NS 17212	NS17212/AlMg3Mn	Plt Sh Strp	0.2 max	0.1 max	0.4 max	2.4-3	0.5-1	0.25 max	0.2 max	0.25 max			ET 0.05 max; ET=.15 total			
5454 A95454	UK	BS N51	5454	Bar Rod Wir Plt Sh Tub Ext Shapes	0.05-0.2	0.1 max	0.4 max	2.4-3	0.5-1	0.25 max	0.2 max	0.25 max	0.05	0.15				

UNS numbers and US grades are provided as a means of cross referencing chemically similar alloys. Exchangability is only possible after independent examination of specifications. Tensile properties are minimum or typical . UTS and YS as Mpa, El as %. See Appendix for list of abbreviations used in Descriptions.

2-56 Wrought Aluminum

Grade UNS #	Country	Specification	Designation	Description	Cr	Cu	Fe	Mg	Mn	Si	Ti	Zn	OE Max	OT Max	Other	UTS	YS	EL
5454 A95454	USA	ASTM B209	5454	Bar Rod Wir Plt Sh Tub Ext Shapes	0.05-0.2	0.1 max	0.4 max	2.4-3	0.5-1	0.25 max	0.2 max	0.25 max	0.05	0.15				
5454 A95454	USA	ASTM B221	5454	Bar Rod Wir Plt Sh Tub Ext Shapes	0.05-0.2	0.1 max	0.4 max	2.4-3	0.5-1	0.25 max	0.2 max	0.25 max	0.05	0.15				
5454 A95454	USA	ASTM B234	5454	Bar Rod Wir Plt Sh Tub Ext Shapes	0.05-0.2	0.1 max	0.4 max	2.4-3	0.5-1	0.25 max	0.2 max	0.25 max	0.05	0.15				
5454 A95454	USA	ASTM B241	5454	Bar Rod Wir Plt Sh Tub Ext Shapes	0.05-0.2	0.1 max	0.4 max	2.4-3	0.5-1	0.25 max	0.2 max	0.25 max	0.05	0.15				
5454 A95454	USA	ASTM B404	5454	Bar Rod Wir Plt Sh Tub Ext Shapes	0.05-0.2	0.1 max	0.4 max	2.4-3	0.5-1	0.25 max	0.2 max	0.25 max	0.05	0.15				
5454 A95454	USA	QQ A-250/10	5454	Bar Rod Wir Plt Sh Tub Ext Shapes	0.05-0.2	0.1 max	0.4 max	2.4-3	0.5-1	0.25 max	0.2 max	0.25 max	0.05	0.15				
5454 A95454	USA	SAE J454	5454	Bar Rod Wir Plt Sh Tub Ext Shapes	0.05-0.2	0.1 max	0.4 max	2.4-3	0.5-1	0.25 max	0.2 max	0.25 max	0.05	0.15				
5454 A95454	Russia		AMgA			0.1 max	0.4 max	2-2.8		0.4 max					ST 0.15-0.38; ET 0.1 max; ST=Mn or Cr, ET=.80 total			
5454 A95454	International	ISO R827	Al-Mg3Mn	As Mfg 5 in; 127 mm diam	0.25	0.1	0.5	2.4-3.4	0.3-1	0.5		0.2				206	79	12
5454 A95454	International	ISO TR2136	Al-Mg3Mn	Sh Plt Ann	0.25	0.1	0.5	2.4-3.4	0.3-1	0.5		0.2				200	75	12
5454 A95454	International	ISO TR2136	Al-Mg3Mn	Sh Plt HB SH	0.25	0.1	0.5	2.4-3.4	0.3-1	0.5		0.2				235	165	4
5454 A95454	International	ISO TR2136	Al-Mg3Mn	Sh Plt HD Str hard	0.25	0.1	0.5	2.4-3.4	0.3-1	0.5		0.2				260	195	2
5454 A95454	Australia	AS 1734	5454	Sh Plt Ann 0.50/0.80 mm thk	0.05-0.2	0.1		2.4-3	0.5-1		0.2	0.25			ET 0.05; ET=0.15 total	213	82	12
5454 A95454	Australia	AS 1734	5454	Sh Plt SH and Stab to 1/4 hard 0.05/1.30 mm thk	0.05-0.2	0.1		2.4-3	0.5-1		0.2	0.25			ET 0.05; ET=0.15 total	248	179	5
5454 A95454	Australia	AS 1734	5454	Sh Plt Sh and Stab to 1/2 hard 0.05/1.30 mm diam	0.05-0.2	0.1		2.4-3	0.5-1		0.2	0.25			ET 0.05; ET=0.15 total	268		4
5454 A95454	Canada	CSA HA.4	0.5454	Plt Sh O .20/.031 in diam	0.05-0.2	0.1		2.4-3	0.5-1		0.2	0.25				31	12	12
5454 A95454	Canada	CSA HA.4	0.5454	Plt Sh H32 .020/.050	0.05-0.2	0.1		2.4-3	0.5-1		0.2	0.25				36	26	5
5454 A95454	Canada	CSA HA.4	0.5454	Plt Sh H112 .250/.499 in diam	0.05-0.2	0.1		2.4-3	0.5-1		0.2	0.25				32	18	8
5454 A95454	Denmark	DS 3012	5454	Soft .5/25 mm diam	0.05-0.2	0.1		2.4-3.1	0.5-1		0.2	0.25				175	70	18
5454 A95454	Denmark	DS 3012	5454	1/4 hard .5/25 mm diam	0.05-0.2	0.1		2.4-3.1	0.5-1		0.2	0.25				235	175	7
5454 A95454	Denmark	DS 3012	5454	1/2 hard .5/6 mm diam	0.05-0.2	0.1		2.4-3.1	0.5-1		0.2	0.25				265	215	4
5454 A95454	France	NF A50451	5454	Sh Plt Ann .4/6 mm diam	0.05-0.2	0.1		2.4-3	0.5-1		0.2	0.25				210	80	19
5454 A95454	France	NF A50451	5454	Sh Plt /4 hard .4/25 mm diam	0.05-0.2	0.1		2.4-3	0.5-1		0.2	0.25				250	180	9
5454 A95454	France	NF A50451	5454	Sh Plt 1/2 hard .4/25 mm diam	0.05-0.2	0.1		2.4-3	0.5-1		0.2	0.25				270	200	8
	France	NF A50-101	5954(AG3.5M)		0.03 max	0.03 max	0.25 max	3.2-3.8	0.3-0.5	0.2 max	0.1 max	0.1 max			ET 0.05 max; ET=.15; Ni 0.03 max; Zr 0.05 max			

UNS numbers and US grades are provided as a means of cross referencing chemically similar alloys. Exchangability is only possible after independent examination of specifications. Tensile properties are minimum or typical . UTS and YS as Mpa, El as %. See Appendix for list of abbreviations used in Descriptions.

Worldwide Guide to Equivalent Nonferrous Metals and Alloys

Grade UNS #	Country	Specification	Designation	Description	Cr	Cu	Fe	Mg	Mn	Si	Ti	Zn	OE Max	OT Max	Other	UTS	YS	EL
5013	France				0.03 max	0.03 max	0.25 max	3.2-3.8	0.3-0.5	0.2 max	0.1 max	0.1 max	0.05	0.15	Ni 0.03 max; Pb 0.003 max; Zr 0.05			
	Austria	ONORM M3430	AlMg3		0.3 max	0.1 max	0.4 max	2.6-3.6	0.9-1.5	0.4 max	0.15 max	0.2 max			ET 0.05 max; ST 0.1-0.6; ST=Mn+Cr; ET=.15 total;			
	International	ISO R827	Al-Mg2	As Mfg	0.35	0.1	0.5	1.7-2.4	0.5	0.5		0.2				148	59	12
	International	ISO TR2136	Al-Mg2	Sh Plt Ann	0.35	0.1	0.5	1.7-2.4	0.5	0.5		0.2				150	60	16
	International	ISO TR2136	Al-Mg2	Sh Plt HB SH	0.35	0.1	0.5	1.7-2.4	0.5	0.5		0.2				180	110	3
	International	ISO TR2136	Al-Mg2	Sh Plt HD Str hard	0.35	0.1	0.5	1.7-2.4	0.5	0.5		0.2				200	160	2
	International	ISO TR2778	Al-Mg2	Tub Ann OBSOLETE Replace by ISO 6363	0.35	0.1	0.5	1.7-2.4	0.5	0.5		0.2				145	60	15
	International	ISO TR2778	Al-Mg2	Tub SH 80 mm diam OBSOLETE Replace by ISO 6363	0.35	0.1	0.5	1.7-2.4	0.5	0.5		0.2				180	110	
	South Africa	SABS 712	Al-Mg2		0.25	0.1	0.5	1.7-2.4	0.5	0.5		0.2						
	Yugoslavia	JUS CC2.100	AlMg4		0.35 max	0.1 max	0.5 max	3.5-4.6	0.8 max	0.5 max					ST 0; 0T 0.2-9; ST = Ti+Zr 0.20 max; Mn+Cr 0.15-0.9			
	India	IS 733	64423	Bar Rod As Mfg or Ann		0.5-1	0.8	0.5-1.3	1	0.7-1.3						120		10
	India	IS 733	64423	Bar Rod ST and Nat aged		0.5-1	0.8	0.5-1.3	1	0.7-1.3						265	155	13
	India	IS 733	64423	Bar Rod SHT and PHT 6.3 mm diam		0.5-1	0.8	0.5-1.3	1	0.7-1.3						330	265	7
	International	ISO R209 (1971)	Al-Mg 4		0.35 max	0.10 max	0.5 max	3.5-4.6	0.8 max	0.5 max					Ti+Zr 0.2 max; Mn+Cr 0.15-0.9			
	Pan American	COPANT 862	X5754		0.1	0.1		2.6-3.6	0.1-0.5	0.4	0.15	0.2						
	India	IS 736	51000-B	Plt As Mfg 17 mm diam	0.1	0.2	0.7	1.1-1.8	0.7	0.6		0.25				135		17
	India	IS 736	51000-B	Plt Ann 19 mm diam	0.1	0.2	0.7	1.1-1.8	0.7	0.6		0.25				125		19
	India	IS 736	51000-B	Plt 1/2 hard 4 mm diam	0.1	0.2	0.7	1.1-1.8	0.7	0.6		0.25				170		4
7024 A97024	Argentina		7024		0.05-0.35	0.1 max	0.4 max	0.5-1	0.1-0.6	0.3 max	0.1 max	3-5	0.05	0.15				
	India	IS 738	NT4	Wir Ann	0.25	0.1	0.7	1.7-2.8	0.5	0.6						172		18
	India	IS 738	NT4	Wir Ann and CW to 1/2 hard	0.25	0.1	0.7	1.7-2.8	0.5	0.6						230		5
4013	USA					0.05-0.2	0.35 max	0.05-0.2	0.03 max	3.5-4.5	0.02 max	0.05 max	0.05	0.15	Bi 0.6-1.5; Cd 0.05 max			
	Pan American	COPANT 862	5454		0.05-0.2	0.1	0.4	2.4-3	0.05-1	0.25	0.2	0.25						
5086 A95086	Russia	GOST 4784	1540		0.05-0.25	0.05 max	0.4 max	3.8-4.8	0.5-0.8	0.4 max	0.02-0.1	0.2 max						
5086 A95086	Russia	GOST 4784	AMg4		0.05-0.25	0.05 max	0.4 max	3.8-4.8	0.5-0.8	0.4 max	0.02-0.1	0.2 max						
	Pan American	COPANT 862	X5854			0.1	0.4	2.5-3.7	0.3	0.3		0.1						
5086 A95086	Australia	AS 1734	5086	Sh Plt Ann 0.50/1.30 mm	0.05-0.25	0.1	0.5	3.5-4.5	0.2-0.7	0.4	0.15	0.25				241	96	15
5086 A95086	Australia	AS 1734	5086	Sh Plt SH and Stab to 1/2 hard 0.25/0.50 mm	0.05-0.25	0.1	0.5	3.5-4.5	0.2-0.7	0.4	0.15	0.25				303	234	4

UNS numbers and US grades are provided as a means of cross referencing chemically similar alloys. Exchangability is only possible after independent examination of specifications. Tensile properties are minimum or typical . UTS and YS as Mpa, El as %. See Appendix for list of abbreviations used in Descriptions.

2-58 Wrought Aluminum

Grade UNS #	Country	Specification	Designation	Description	Cr	Cu	Fe	Mg	Mn	Si	Ti	Zn	OE Max	OT Max	Other	UTS	YS	EL
5086 A95086	Australia	AS 1734	5086	Sh Plt SH and Stab to full hard 0.15/0.50 mm thk	0.05-0.25	0.1	0.5	3.5-4.5	0.2-0.7	0.4	0.15	0.25				344		3
5086 A95086	Belgium	NBN P21-001	5086		0.05-0.25	0.1 max	0.5 max	3.5-4.5	0.2-0.7	0.4 max	0.15 max	0.25 max			ET 0.05 max; ; ET = .15 Total			
5086 A95086	Canada	CSA HA.4	0.5086	Sh Plt O .020/.050 in diam	0.05-0.25	0.1	0.5	3.5-4.5	0.2-0.7	0.4	0.15	0.25				241	97	15
5086 A95086	Canada	CSA HA.4	0.5086	Sh Plt H32 .020/.050 in diam	0.05-0.25	0.1	0.5	3.5-4.5	0.2-0.7	0.4	0.15	0.25				276	193	6
5086 A95086	France	NF A-G4MC	5086	Bar Rod Wir Plt Sh Tub Ext Shapes	0.05-0.25	0.1 max	0.5 max	3.5-4.5	0.2-0.7	0.4 max	0.15 max	0.25 max			ET 0.05; ET=0.15 total			
5086 A95086	France	NF A50-411	5086(AG4MC)		0.05-0.25	0.1 max	0.5 max	3.5-4.5	0.2-0.7	0.4 max	0.15 max	0.25 max			ET 0.05 max; ET=.15			
5086 A95086	France	NF A50-701	5086(AG4MC)		0.05-0.25	0.1 max	0.5 max	3.5-4.5	0.2-0.7	0.4 max	0.15 max	0.25 max			ET 0.05 max; ET=.16			
5086 A95086	France	NF A50411	5086(A-G4MC)	Bar Wir Tub 1/4 hard (drawn bar) 25 mm diam	0.05-0.25	0.1	0.5	3.5-4.5	0.2-0.7	0.4	0.15	0.25				270	190	4
5086 A95086	France	NF A50411	5086(A-G4MC)	Bar Wir Tub Hard (wir) 25 mm diam	0.05-0.25	0.1	0.5	3.5-4.5	0.2-0.7	0.4	0.15	0.25				24	95	18
5086 A95086	France	NF A50411	5086(A-G4MC)	Bar Wir Tub Ann (tub) 150 mm diam	0.05-0.25	0.1	0.5	3.5-4.5	0.2-0.7	0.4	0.15	0.25				240	95	18
5086 A95086	France	NF A50451	5086(A-G4MC)	Sh Plt Ann .4/6 mm diam	0.05-0.25	0.1	0.5	3.5-4.5	0.2-0.7	0.4	0.15	0.25				240	100	10
5086 A95086	France	NF A50451	5086(A-G4MC)	Sh Plt 1/4 hard .4/1.6 mm diam	0.05-0.25	0.1	0.5	3.5-4.5	0.2-0.7	0.4	0.15	0.25				280	190	8
5086 A95086	France	NF A50451	5086(A-G4MC)	Sh Plt 1/2 hard .4/1.6 mm diam	0.05-0.25	0.1	0.5	3.5-4.5	0.2-0.7	0.4	0.15	0.25				310	230	7
5086 A95086	Germany	DIN 1725 pt1	3.3545/AlMg4 Mn		0.05-0.25	0.1 max	0.5 max	3.5-4.5	0.2-0.7	0.4 max	0.15 max	0.25 max			ET 0.05 max; ET=.15 total;			
5086 A95086	Germany	DIN	Al-Mg4	Bar Rod Wir Plt Sh Tub Ext Shapes	0.05-0.25	0.1 max	0.5 max	3.5-4.5	0.2-0.7	0.4 max	0.15 max	0.25 max			ET 0.05; ET=0.15 total			
5086 A95086	Norway	NS 17213	NS17213/AlMg4	Plt Sh Strp	0.05-0.2	0.1 max	0.5 max	3.5-4.5	0.2-0.7	0.4 max	0.15 max	0.25 max			ET 0.05 max; ET=.15 total			
5086 A95086	Pan American	COPANT 862	5086		0.05-0.25	0.1	0.5	3.5-4.5	0.2-0.7	0.4	0.15	0.25						
5086 A95086	USA	ASTM B209	5086	Bar Rod Wir Plt Sh Tub Ext Shapes	0.05-0.25	0.1 max	0.5 max	3.5-4.5	0.2-0.7	0.4 max	0.15 max	0.25 max			ET 0.05; ET=0.15 total;			
5086 A95086	USA	ASTM B210	5086	Bar Rod Wir Plt Sh Tub Ext Shapes	0.05-0.25	0.1 max	0.5 max	3.5-4.5	0.2-0.7	0.4 max	0.15 max	0.25 max			ET 0.05; ET=0.15 total			
5086 A95086	USA	ASTM B221	5086	Bar Rod Wir Plt Sh Tub Ext Shapes	0.05-0.25	0.1 max	0.5 max	3.5-4.5	0.2-0.7	0.4 max	0.15 max	0.25 max			ET 0.05; ET=0.15 total			
5086 A95086	USA	ASTM B241	5086	Bar Rod Wir Plt Sh Tub Ext Shapes	0.05-0.25	0.1 max	0.5 max	3.5-4.5	0.2-0.7	0.4 max	0.15 max	0.25 max			ET 0.05; ET=0.15 total			
5086 A95086	USA	ASTM B313	5086	Bar Rod Wir Plt Sh Tub Ext Shapes	0.05-0.25	0.1 max	0.5 max	3.5-4.5	0.2-0.7	0.4 max	0.15 max	0.25 max			ET 0.05; ET=0.15 total			
5086 A95086	USA	ASTM B345	5086	Bar Rod Wir Plt Sh Tub Ext Shapes	0.05-0.25	0.1 max	0.5 max	3.5-4.5	0.2-0.7	0.4 max	0.15 max	0.25 max			ET 0.05; ET=0.15 total			
5086 A95086	USA	ASTM B547	5086	Bar Rod Wir Plt Sh Tub Ext Shapes	0.05-0.25	0.1 max	0.5 max	3.5-4.5	0.2-0.7	0.4 max	0.15 max	0.25 max			ET 0.05; ET=0.15 total			
5086 A95086	USA	QQ A-200/5	5086	Bar Rod Wir Plt Sh Tub Ext Shp OBSOLETE	0.05-0.25	0.1 max	0.5 max	3.5-4.5	0.2-0.7	0.4 max	0.15 max	0.25 max			ET 0.05; ET=0.15 total			

UNS numbers and US grades are provided as a means of cross referencing chemically similar alloys. Exchangability is only possible after independent examination of specifications. Tensile properties are minimum or typical . UTS and YS as Mpa, El as %. See Appendix for list of abbreviations used in Descriptions.

Wrought Aluminum 2-59

Worldwide Guide to Equivalent Nonferrous Metals and Alloys

Grade UNS #	Country	Specification	Designation	Description	Cr	Cu	Fe	Mg	Mn	Si	Ti	Zn	OE Max	OT Max	Other	UTS	YS	EL
5086 A95086	USA	QQ A-250/19	5086	Bar Rod Wir Plt Sh Tub Ext Shapes	0.05-0.25	0.1 max	0.5 max	3.5-4.5	0.2-0.7	0.4 max	0.15 max	0.25 max			ET 0.05; ET=0.15 total			
5086 A95086	USA	QQ A-250/7	5086	Bar Rod Wir Plt Sh Tub Ext Shapes	0.05-0.25	0.1 max	0.5 max	3.5-4.5	0.2-0.7	0.4 max	0.15 max	0.25 max			ET 0.05; ET=0.15 total			
5086 A95086	USA	SAE J454	5086	Bar Rod Wir Plt Sh Tub Ext Shapes	0.05-0.25	0.1 max	0.5 max	3.5-4.5	0.2-0.7	0.4 max	0.15 max	0.25 max			ET 0.05; ET=0.15 total			
5086	USA				0.05-0.25	0.1 max	0.5 max	3.5-4.5	0.2-0.7	0.4 max	0.15 max	0.25 max	0.15	0.15				
	USA		AlMg4		0.05-0.25	0.1 max	0.5 max	3.5-4.5	0.2-0.7	0.4 max	0.15 max	0.25 max	0.15	0.15				
5086 A95086	International	ISO R827	Al-Mg4		0.35	0.1	0.5	3.5-4.6	0.8	0.5		0.2						
5086 A95086	International	ISO R827	Al-Mg4	As Mfg 5 in; 127 mm diam	0.35	0.1	0.5	3.5-4.6	0.8	0.5		0.2						
5086 A95086	International	ISO TR2136	Al-Mg4	Sh Plt Ann	0.35	0.1	0.5	3.5-4.6	0.8	0.5		0.2				240	95	12
5086 A95086	International	ISO TR2136	Al-Mg4	Sh Plt HB SH	0.35	0.1	0.5	3.5-4.6	0.8	0.5		0.2				270	185	5
5086 A95086	International	ISO TR2136	Al-Mg4	Sh Plt HD Str hard	0.35	0.1	0.5	3.5-4.6	0.8	0.5		0.2				305	203	4
5086 A95086	International	ISO TR2778	Al-Mg4	Tub Ann OBSOLETE Replace by ISO 6363	0.35	0.1	0.5	3.5-4.6	0.8	0.5		0.2				240	95	15
5086 A95086	International	ISO TR2778	Al-Mg4	Tub SH 50 mm diam OBSOLETE Replace by ISO 6363	0.35	0.1	0.5	3.5-4.6	0.8	0.5		0.2				260	185	5
	UK	BS 1470	NS4	Plt Sh Strp Ann .2/6 mm diam OBSOLETE Replaced by BS EN 515	0.25	0.1	0.5	1.7-2.4	0.5	0.5	0.2	0.2				160	60	18
	UK	BS 1470	NS4	Plt Sh Strp SH to 3/8 hard .2/6 mm diam OBSOLETE Replaced by BS EN 515	0.25	0.1	0.5	1.7-2.4	0.5	0.5	0.2	0.2				200	130	4
	UK	BS 1471	NT4	Tub Ann 10 mm diam	0.25	0.1	0.5	1.7-2.4	0.5	0.5	0.2	0.2				160	60	18
	UK	BS 1471	NT4	Tub SH to 1/2 hard 10 mm diam	0.25	0.1	0.5	1.7-2.4	0.5	0.5	0.2	0.2				225	175	5
	UK	BS 1475	NG4	Wir	0.25	0.1	0.5	1.7-2.4	0.5	0.5	0.2	0.2				170		
	UK	BS 1475	NG4	Wir	0.25	0.1	0.5	1.7-2.4	0.5	0.5	0.2	0.2				260		
A94643	USA	AWS	ER4643	Weld El Rod		0.1 max	0.8 max	0.1-0.3	0.05 max	3.6-4.6	0.15 max	0.1 max	0.05	0.15	Be 0.0008 max			
A94643	USA	AWS	R4643	Weld El Rod		0.1 max	0.8 max	0.1-0.3	0.05 max	3.6-4.6	0.15 max	0.1 max	0.05	0.15	Be 0.0008 max			
4643	USA					0.1 max	0.8 max	0.1-0.3	0.05 max	3.6-4.6	0.15 max	0.1 max	0.05	0.15	0.0008 max Be for welding			
5086 A95086	France	AIR 9051-A	A-G4MC(5086)	Sh Bar Rolled or ann 120 mm diam		3.5-4.5	0.5	0.5-1	0.3-0.8	0.25-0.8	0.2	0.25				240	100	16
5086 A95086	France	AIR 9051-A	A-G4MC(5086)	Sh Bar Ann (bar) 250 mm diam		3.5-4.5	0.5	0.5-1	0.3-0.8	0.25-0.8	0.2	0.25				240	95	16
5086 A95086	France	AIR 9051-A	A-G4MC(5086)	Sh Bar As cast or HT (bil and plt) 500 mm		3.5-4.5	0.5	0.5-1	0.3-0.8	0.25-0.8	0.2	0.25				240	95	16

UNS numbers and US grades are provided as a means of cross referencing chemically similar alloys. Exchangability is only possible after independent examination of specifications. Tensile properties are minimum or typical . UTS and YS as Mpa, El as %. See Appendix for list of abbreviations used in Descriptions.

2-60 Wrought Aluminum

Grade UNS #	Country	Specification	Designation	Description	Cr	Cu	Fe	Mg	Mn	Si	Ti	Zn	OE Max	OT Max	Other	UTS	YS	EL
2030 A92030	Belgium	NBN P21-001	2030		0.1 max	3.5-4.5	0.7 max	0.5-1.3	1 max	0.8 max	0.2 max	0.5 max			ET 0.05 max; ET = .15 Total; Bi 0.2 max; Ni 0.2 maxPb 0.8-1.5			
2030 A92030	Czech Republic	CSN 424254	AlCu4BiPb			3-5	1 max	0.4-1.5	0.3-1.5	1 max		0.7 max			Pb+Bi 0.50-1.50; Sn+Cd 0.40; Ni 0.3 max			
2030 A92030	Sweden	SIS 144355	4355-04	Bar Art aged 10/120 mm diam		5-6	0.7			0.4		0.3			Bi 0.2-0.6; Pb 0.2-0.6	275	125	14
5091	USA						0.3 max	3.7-4.2		0.2 max				0.15	1.0-1.3 C; 1.2-1.4 Li, 0.20-.7 O			
	India	IS 733	52000	Bar Rod As Mfg 150 mm diam	0.25	0.1	0.5	1.7-2.6	0.5	0.6	0.2	0.2				170		14
	India	IS 739	NG4	Wir Ann	0.25	0.1	0.7	1.7-2.8	0.5	0.6		0.2				172		
	India	IS 739	NG4	Wir Ann and CW to hard	0.25	0.1	0.7	1.7-2.8	0.5	0.6		0.2				265		
7025 A97025	Argentina				0.05-0.35	0.1 max	0.4 max	0.8-1.5	0.1-0.6	0.3 max	0.1 max	3-5	0.05	0.15				
2195	USA					3.7-4.3	0.15 max	0.25-0.8	0.25 max	0.12 max	0.1 max	0.25 max	0.05	0.15	0.8-1.2 Li; Ag 0.25-0.6; Zr 0.08-0.16			
	India	IS 736	52000	Plt As Mfg 12 mm diam	0.25	0.1	0.7	1.7-2.6	0.5	0.6	0.2	0.2				190		12
	India	IS 736	52000	Plt Ann 18 mm diam	0.25	0.1	0.7	1.7-2.6	0.5	0.6	0.2	0.2				175	60	18
	India	IS 736	52000	Plt 1/2 hard 5 mm diam	0.25	0.1	0.7	1.7-2.6	0.5	0.6	0.2	0.2				200	160	5
5082 A95082	Australia	AS 1734	5082		0.15 max	0.15 max	0.35 max	4-5	0.15 max	0.2 max	0.1 max	0.25 max	0.05		OT min 0.15			
5082 A95082	Germany	DIN 1725 pt1	3.3345/AlMg4.5		0.15 max	0.15 max	0.35 max	4-5	0.15 max	0.2 max	0.1 max	0.25 max			ET 0.05 max; ET =.15 total			
5082	USA				0.15 max	0.15 max	0.35 max	4-5	0.15 max	0.2 max	0.1 max	0.25 max	0.1	0.15				
2030	Europe				0.1 max	3.3-4.5	0.7 max	0.5-1.3	0.2-1	0.8 max	0.2 max	0.5 max	0.1	0.3	Bi 0.20; Pb 0.8-1.5			
	Europe		AlCu4PbMg		0.1 max	3.3-4.5	0.7 max	0.5-1.3	0.2-1	0.8 max	0.2 max	0.5 max	0.1	0.3	Bi 0.20; Pb 0.8-1.5			
2012	USA					4-5.5	0.7 max			0.4 max		0.3 max	0.05	0.15	Bi 0.20-0.7; Sn 0.20-0.6			
	International	ISO R209 (1971)	Al-Mg4.5Mn		0.25 max	0.10 max	0.5 max	4.0-4.9	0.3-1.0	0.5 max					Ti+Zr 0.2 max			
	France	NF A50451	5754X(A-G3M)	Sh Plt Ann .4/1.6 mm diam	0.1	0.1	0.5	2.6-3.6	0.1-0.5	0.4	0.15	0.2				190	80	20
	France	NF A50451	5754X(A-G3M)	Sh Plt 1/2 hard .4/3.2 mm diam	0.1	0.1	0.5	2.6-3.6	0.1-0.5	0.4	0.15	0.2				240	160	8
	France	NF A50451	5754X(A-G3M)	Sh Plt 3/4 hard .4/3.2 mm diam	0.1	0.1	0.5	2.6-3.6	0.1-0.5	0.4	0.15	0.2				260	190	7
	International	ISO TR2136	Al-Mg3	Sh Plt Ann	0.35	0.1	0.5	2.4-3.1	0.4	0.5		0.2				180	75	17
	International	ISO TR2136	Al-Mg3	Sh Plt HB SH	0.35	0.1	0.5	2.4-3.1	0.4	0.5		0.2				215	140	3
	International	ISO TR2136	Al-Mg3	Sh Plt HD Str hard	0.35	0.1	0.5	2.4-3.1	0.4	0.5		0.2				245	180	2
	International	ISO TR2778	Al-Mg3	Tub Ann OBSOLETE Replace by ISO 6363	0.35	0.1	0.5	2.4-3.1	0.4	0.5		0.2				180	70	15
	International	ISO TR2778	Al-Mg3	Tub SH 80 mm diam OBSOLETE Replace by ISO 6363	0.35	0.1	0.5	2.4-3.1	0.4	0.5		0.2				215	140	3
2031	UK					1.8-2.8	0.6-1.2	0.6-1.2	0.2 max	0.5-1.3	0.2 max	0.2 max	0.05	0.15	Ni 0.6-1.4			

UNS numbers and US grades are provided as a means of cross referencing chemically similar alloys. Exchangability is only possible after independent examination of specifications. Tensile properties are minimum or typical . UTS and YS as Mpa, El as %. See Appendix for list of abbreviations used in Descriptions.

Worldwide Guide to Equivalent Nonferrous Metals and Alloys

Grade UNS #	Country	Specification	Designation	Description	Cr	Cu	Fe	Mg	Mn	Si	Ti	Zn	OE Max	OT Max	Other	UTS	YS	EL
2095	USA					3.9-4.6	0.15 max	0.25-0.8	0.25 max	0.12 max	0.1 max	0.25 max	0.05	0.15	0.7-1.5 Li; Ag 0.25-0.6; Zr 0.04-0.18			
5182 A95182	Australia	AS 1734	5182		0.1 max	0.15 max	0.35 max	4-5	0.5 max	0.2 max	0.1 max	0.25 max	0.05		OT min 0.15			
5182 A95182	USA				0.1 max	0.15 max	0.35 max	4-5	0.2-0.5	0.2 max	0.1 max	0.25 max	0.05	0.15				
5182	USA				0.1 max	0.15 max	0.35 max	4-5	0.2-0.5	0.2 max	0.1 max	0.25 max	0.1	0.15				
2048 A92048	USA					2.8-3.8	0.2 max	1.2-1.8	0.2-0.6	0.15 max	0.1 max	0.25 max		0.15				
2048	USA					2.8-3.8	0.2 max	1.2-1.8	0.2-0.6	0.15 max	0.1 max	0.25 max	0.05	0.15				
2009	USA					3.2-4.4	0.05 max	1-1.6		0.25 max		0.1 max	0.05	0.15	0.6 max Oxygen			
	Germany	DIN 1725 pt1	3.1645/AlCuMgPb			3.3-4.6	0.8 max	0.4-1.8	0.5-1	0.8 max	0.2 max	0.8 max			ST 1-2.5; ET 0.1 max; ST=Bi+Cd+Pb+Sn; ET=.30 total;			
2007	Germany				0.1 max	3.3-4.6	0.8 max	0.4-1.8	0.5-1	0.8 max	0.2 max	0.8 max	0.1	0.3	Bi 0.20; Ni 0.2 max; Pb 0.8-1.5; Sn 0.20			
	Yugoslavia	JUS CC2.100	AlCu4MgSi			3.5-4.7	0.7 max	0.2-1.2	0.3-1	0.2-0.8					ST 0.3 max; ST = Ti + Zr + Cr;			
	Sweden	SIS 144134	4134-00	Wir	0.05	0.1	0.4	3.1-3.9	0.05	0.4	0.05	0.2						
	Sweden	SIS 144134	4134-18	Wir HW	0.05	0.1	0.4	3.1-3.9	0.05	0.4	0.05	0.2				225		
	Pan American	COPANT 862	2062	OBSOLETE		1-2	0.7	0.5-1.4	0.6-1	0.8-1.3	0.3	0.2			Ni 0.2			
	International	ISO R209 (1971)	Al-Cu4MgSi			3.5-4.7	0.7 max	0.3-1.2	0.3-1.0	0.2-0.8					Ti+Zr+Cr 0.3 max; Ni 0.2 max			
	France	NF A50411	5754(A-G3M)	Bar Tub Wir Hard (drawn bar) 10 mm	0.3	0.1	0.4	2.6-3.6	0.5	0.4	0.15	0.2				280	240	2
	France	NF A50411	5754(A-G3M)	Bar Tub Wir Hard (wir) 6 mm	0.3	0.1	0.4	2.6-3.6	0.5	0.4	0.15	0.2				280	240	2
	France	NF A50411	5754(A-G3M)	Bar Wir Tub Ann (tub) 150 mm	0.3	0.1	0.4	2.6-3.6	0.5	0.4	0.15	0.2				200	80	17
	South Africa	SABS 712	Al-Mg3Mn		0.2	0.1		3-3	1			0.25						
2003	Europe					4-5	0.3 max	0.02 max	0.3-0.8	0.3 max	0.15 max	0.1 max	0.05	0.15	Cd 0.02-0.20; V 0.05-0.2; Zr 0.30-0.50			
5083 A95083	Australia	AS 1734	5083	Sh Plt Ann 1.30/40.0 mm thk	0.05-0.25	0.1	0.4	4-4.9	0.4-1	0.4	0.15	0.25					124	14
5083 A95083	Australia	AS 1734	5083	Sh Plt SH less than 1/8 hard 6/40 mm thk	0.05-0.25	0.1	0.4	4-4.9	0.4-1	0.4	0.15	0.25					172	12
5083 A95083	Australia	AS 1734	5083	Sh Plt SH less than 1/4 hard 6/50 mm thk	0.05-0.25	0.1	0.4	4-4.9	0.4-1	0.4	0.15	0.25						10
5083 A95083	Australia	AS 1866	5083	Rod Bar H111 125.0 mm diam	0.05-0.25	0.1	0.4	4-4.9	0.4-1	0.4	0.15					275	165	10
5083 A95083	Australia	AS 1866	5083	Rod Bar SH (H112) 125.0 mm diam	0.05-0.25	0.1	0.4	4-4.9	0.4-1	0.4	0.15					268	110	10
5083 A95083	Austria	ONORM M3430	AlMg4.5Mn			0.1 max	0.4 max	0.05-0.25	0.4-4.9	0.4 max	0.15 max	0.25 max			ET 0.05 max; ET=.15 total;			
5083 A95083	Belgium	NBN P21-001	5083		0.05-0.25	0.1 max	0.4 max	4-4.9	0.4-1	0.4	0.15 max	0.25 max			ET 0.05 max; ; ET = .15 Total			
5083 A95083	Canada	CSA GM41	5083	Bar Rod Wir Plt Sh Strp Tub Pip Ext	0.05-0.25	0.1 max	0.4 max	4-4.9	0.4-1	0.4 max	0.15 max	0.25 max	0.05	0.15	E 0.05 max; T 0.15 max;			

UNS numbers and US grades are provided as a means of cross referencing chemically similar alloys. Exchangability is only possible after independent examination of specifications. Tensile properties are minimum or typical . UTS and YS as Mpa, El as %. See Appendix for list of abbreviations used in Descriptions.

2-62 Wrought Aluminum

Grade UNS #	Country	Specification	Designation	Description	Cr	Cu	Fe	Mg	Mn	Si	Ti	Zn	OE Max	OT Max	Other	UTS	YS	EL
5083 A95083	Canada	CSA HA.4	0.5083	Plt Sh O .051/1.5 in diam	0.05-0.25	0.1	0.4	4-4.9	0.4-1	0.4	0.15	0.25				124	276	16
5083 A95083	Canada	CSA HA.4	0.5083	Plt Sh H112 .051/1.5 in diam	0.05-0.25	0.1	0.4	4-4.9	0.4-1	0.4	0.15	0.25				124	276	12
5083 A95083	Canada	CSA HA.4	0.5083	Plt Sh H321 .188/1.5 in diam	0.05-0.25	0.1	0.4	4-4.9	0.4-1	0.4	0.15	0.25				303	214	12
5083 A95083	Canada	CSA HA.5	0.5083	Bar Rod Wir Extd and O	0.05-0.25	0.1	0.4	4-4.9	0.4-1	0.4	0.15	0.25				269	110	16
5083 A95083	Canada	CSA HA.5	0.5083	Bar Rod Wir Extd and H111	0.05-0.25	0.1	0.4	4-4.9	0.4-1	0.4	0.15	0.25				276	165	12
5083 A95083	Canada	CSA HA.7	0.5083	Tub Pip Extd and H111	0.05-0.25	0.1	0.4	4-4.9	0.4-1	0.4	0.15	0.25				276	165	12
5083 A95083	France	NF A50-411	AG4.5MC		0.05-0.25	0.1 max	0.4 max	4-4.9	0.4-1	0.4 max	0.5 max	0.25 max			ET 0.05 max; ET=.15;			
5083 A95083	France	NF A50-451	AG.5MC		0.05-0.25	0.1 max	0.4 max	4-4.9	0.4-1	0.4 max	0.5 max	0.25 max			ET 0.05 max; ET=.15;			
5083 A95083	France	NF A50411	5083	Bar Tub 1/4 hard (drawn bar) 25 mm diam	0.05-0.25	0.1	0.4	4-4.9	0.4-1	0.4	0.15	0.25				300	200	4
5083 A95083	France	NF A50411	5083	Bar Tub Ann (tub) 150 mm diam	0.05-0.25	0.1	0.4	4-4.9	0.4-1	0.4	0.15	0.25				270	110	17
5083 A95083	France	NF A50411	5083	Bar Tub Ann (shapes) 110 mm diam	0.05-0.25	0.1	0.4	4-4.9	0.4-1	0.4	0.15	0.25				270	110	18
5083 A95083	France	NF A50451	5083	Sh Plt Ann 12/150 mm diam	0.05-0.25	0.1	0.4	4-4.9	0.4-1	0.4	0.15	0.25				270	120	10
5083 A95083	France	NF A50451	5083	Sh Plt 1/4 hard .4/25 mm diam	0.05-0.25	0.1	0.4	4-4.9	0.4-1	0.4	0.15	0.25				300	210	11
5083 A95083	Germany	DIN 1725 Part 1	3.3547/AlMg4.5Mn		0.05-0.25	0.1 max	0.4 max	4-4.9	0.4-1	0.4 max	0.15 max	0.25 max			ET 0.05 max; ET=.15 total,			
5083 A95083	Japan	JIS H4000	5083	Sh Plt Ann 1.3 mm diam	0.05-0.25	0.1	0.4	4-4.9	0.3-1	0.4	0.15	0.25				275	196	16
5083 A95083	Japan	JIS H4000	5083	Sh Plt Stab at 1/4 hard 1.3 mm diam	0.05-0.25	0.1	0.4	4-4.9	0.3-1	0.4	0.15	0.25				314	304	8
5083 A95083	Japan	JIS H4000	5083	Sh Plt HR 12 mm diam	0.05-0.25	0.1	0.4	4-4.9	0.3-1	0.4	0.15	0.25				275	118	12
5083 A95083	Japan	JIS H4040	5083	Bar As fab 130 mm diam	0.05-0.25	0.1	0.4	4-4.9	0.3-1	0.4	0.15	0.25				275	108	12
5083 A95083	Japan	JIS H4040	5083	Bar Ann 130 mm diam	0.05-0.25	0.1	0.4	4-4.9	0.3-1	0.4	0.15	0.25				353	108	14
5083 A95083	Japan	JIS H4080	5083	Tub As Mfg	0.05-0.25	0.1	0.4	4-4.9	0.3-1	0.4	0.15	0.25				275	108	12
5083 A95083	Japan	JIS H4080	5083	Tub Ann	0.05-0.25	0.1	0.4	4-4.9	0.3-1	0.4	0.15	0.25				353	108	14
5083 A95083	Japan	JIS H4100	5083	As Mfg 130 mm diam	0.05-0.25	0.1	0.4	4-4.9	0.3-1	0.4	0.15	0.25				275	108	12
5083 A95083	Japan	JIS H4100	5083	Ann 130 mm diam	0.05-0.25	0.1	0.4	4-4.9	0.3-1	0.4	0.15	0.25				353	108	14
5083 A95083	Japan	JIS H4140	5083	Die As Mfg 100 mm diam	0.05-0.25	0.1	0.4	4-4.9	0.3-1	0.4	0.15					275	127	16
5083 A95083	Norway	NS 17215	NS17215/AlMg4.5Mn	Plt Sh Strp	0.25 max	0.1 max	0.4 max	4-4.9	0.4-1	0.4 max	0.15 max	0.25 max			ET 0.05 max; ET=.15 total;			
5083 A95083	Sweden	SIS 144140	4140-00	Plt HW 6/30 mm diam	0.05-0.25	0.1	0.4	4-4.9	0.4-1	0.4	0.15	0.25			Al 94.65	275	125	12
5083 A95083	Sweden	SIS 144140	4140-02	Sh Plt Strp Ann .5/5 mm diam	0.15	0.1	0.4	4-4.9	0.4-1	0.4	0.15	0.25			Al 94.65	270	120	17
5083 A95083	Sweden	SIS 144140	4140-12	Plt SH	0.15	0.1	0.4	4-4.9	0.4-1	0.4	0.15	0.25			Al 94.65	310	205	12
5083 A95083	Sweden	SIS 144140	4140-22	Sh Strp SH .5/2 mm diam	0.15	0.1	0.4	4-4.9	0.4-1	0.4	0.15	0.25			Al 94.65	310	205	13
5083 A95083	Sweden	SIS 144140	4140-24	Sh Strp SH .5/6 mm diam	0.15	0.1	0.4	4-4.9	0.4-1	0.4	0.15	0.25			Al 94.65	345	270	6
5083 A95083	UK	BS 1471	NT8	Tub Ann 10 mm diam	0.25	0.1	0.4		4-4.9	0.4	0.2	0.2				275	125	12

UNS numbers and US grades are provided as a means of cross referencing chemically similar alloys. Exchangability is only possible after independent examination of specifications. Tensile properties are minimum or typical . UTS and YS as Mpa, El as %. See Appendix for list of abbreviations used in Descriptions.

Worldwide Guide to Equivalent Nonferrous Metals and Alloys

Grade UNS #	Country	Specification	Designation	Description	Cr	Cu	Fe	Mg	Mn	Si	Ti	Zn	OE Max	OT Max	Other	UTS	YS	EL
5083 A95083	UK	BS 1471	NT8	Tub SH to 1.4 hard 10 mm diam	0.25	0.1	0.4		4-4.9	0.4	0.2	0.2				310	235	5
5083 A95083	UK	BS N8	5083	Bar Rod Wir Plt Sh Strp Tub Pip Ext	0.05-0.25	0.1 max	0.4 max	4-4.9	0.4-1	0.4 max	0.15 max	0.25 max	0.05	0.15				
5083 A95083	USA	AMS 4056	5083	Bar Rod Wir Plt Sh Strp Tub Pip Ext	0.05-0.25	0.1 max	0.4 max	4-4.9	0.4-1	0.4 max	0.15 max	0.25 max	0.05	0.15				
5083 A95083	USA	AMS 4057	5083	Bar Rod Wir Plt Sh Strp Tub Pip Ext OBSOLETE	0.05-0.25	0.1 max	0.4 max	4-4.9	0.4-1	0.4 max	0.15 max	0.25 max	0.05	0.15				
5083 A95083	USA	AMS 4058	5083	Bar Rod Wir Plt Sh Strp Tub Pip Ext OBSOLETE	0.05-0.25	0.1 max	0.4 max	4-4.9	0.4-1	0.4 max	0.15 max	0.25 max	0.05	0.15				
5083 A95083	USA	AMS 4059	5083	Bar Rod Wir Plt Sh Strp Tub Pip Ext OBSOLETE	0.05-0.25	0.1 max	0.4 max	4-4.9	0.4-1	0.4 max	0.15 max	0.25 max	0.05	0.15				
5083 A95083	USA	ASTM B209	5083	Bar Rod Wir Plt Sh Strp Tub Pip Ext	0.05-0.25	0.1 max	0.4 max	4-4.9	0.4-1	0.4 max	0.15 max	0.25 max	0.05	0.15				
5083 A95083	USA	ASTM B210	5083	Bar Rod Wir Plt Sh Strp Tub Pip Ext	0.05-0.25	0.1 max	0.4 max	4-4.9	0.4-1	0.4 max	0.15 max	0.25 max	0.05	0.15				
5083 A95083	USA	ASTM B221	5083	Bar Rod Wir Plt Sh Strp Tub Pip Ext	0.05-0.25	0.1 max	0.4 max	4-4.9	0.4-1	0.4 max	0.15 max	0.25 max	0.05	0.15				
5083 A95083	USA	ASTM B241	5083	Bar Rod Wir Plt Sh Strp Tub Pip Ext	0.05-0.25	0.1 max	0.4 max	4-4.9	0.4-1	0.4 max	0.15 max	0.25 max	0.05	0.15				
5083 A95083	USA	ASTM B247	5083	Bar Rod Wir Plt Sh Strp Tub Pip Ext	0.05-0.25	0.1 max	0.4 max	4-4.9	0.4-1	0.4 max	0.15 max	0.25 max	0.05	0.15				
5083 A95083	USA	ASTM B547	5083	Bar Rod Wir Plt Sh Strp Tub Pip Ext	0.05-0.25	0.1 max	0.4 max	4-4.9	0.4-1	0.4 max	0.15 max	0.25 max	0.05	0.15				
5083 A95083	USA	MIL A-45225	5083	Bar Rod Wir Plt Sh Strp Tub Pip Ext	0.05-0.25	0.1 max	0.4 max	4-4.9	0.4-1	0.4 max	0.15 max	0.25 max	0.05	0.15				
5083 A95083	USA	MIL A-46027	5083	Bar Rod Wir Plt Sh Strp Tub Pip Ext	0.05-0.25	0.1 max	0.4 max	4-4.9	0.4-1	0.4 max	0.15 max	0.25 max	0.05	0.15				
5083 A95083	USA	MIL A-46083	5083	Bar Rod Wir Plt Sh Strp Tub Pip Ext	0.05-0.25	0.1 max	0.4 max	4-4.9	0.4-1	0.4 max	0.15 max	0.25 max	0.05	0.15				
5083 A95083	USA	QQ A-200/4	5083	Bar Rod Wir Plt Sh Strp Tub Pip Ext OBSOLETE	0.05-0.25	0.1 max	0.4 max	4-4.9	0.4-1	0.4 max	0.15 max	0.25 max	0.05	0.15				
5083 A95083	USA	QQ A-250/6	5083	Bar Rod Wir Plt Sh Strp Tub Pip Ext	0.05-0.25	0.1 max	0.4 max	4-4.9	0.4-1	0.4 max	0.15 max	0.25 max	0.05	0.15				
5083 A95083	USA	QQ A-367	5083	Bar Rod Wir Plt Sh Strp Tub Pip Ext	0.05-0.25	0.1 max	0.4 max	4-4.9	0.4-1	0.4 max	0.15 max	0.25 max	0.05	0.15				
5083 A95083	USA	SAE J454	5083	Bar Rod Wir Plt Sh Strp Tub Pip Ext	0.05-0.25	0.1 max	0.4 max	4-4.9	0.4-1	0.4 max	0.15 max	0.25 max	0.05	0.15				
5083	USA				0.05-0.25	0.1 max	0.4 max	4-4.9	0.4-1	0.4 max	0.15 max	0.25 max	0.15	0.15				
	USA		AlMg4.5Mn0.7		0.05-0.25	0.1 max	0.4 max	4-4.9	0.4-1	0.4 max	0.15 max	0.25 max	0.15	0.15				
5083 A95083	India	IS 733	54300	Bar Rod Ann 150 mm diam	0.25	0.1	0.7	4-4.9	0.5-1	0.4	0.2	0.2				265	125	13
5083 A95083	India	IS 733	54300	Bar Rod As Mfg 150 mm diam	0.25	0.1	0.7	4-4.9	0.5-1	0.4	0.2	0.2				275	130	11
5083 A95083	India	IS 736	54300	Plt As Mfg 12 mm diam	0.25	0.1	0.7	4-4.9	0.5-1	0.4	0.2	0.2				280	125	12

UNS numbers and US grades are provided as a means of cross referencing chemically similar alloys. Exchangability is only possible after independent examination of specifications. Tensile properties are minimum or typical . UTS and YS as Mpa, El as %. See Appendix for list of abbreviations used in Descriptions.

2-64 Wrought Aluminum

Grade UNS #	Country	Specification	Designation	Description	Cr	Cu	Fe	Mg	Mn	Si	Ti	Zn	OE Max	OT Max	Other	UTS	YS	EL
5083 A95083	India	IS 736	54300	Plt Ann 16 mm diam	0.25	0.1	0.7	4-4.9	0.5-1	0.4	0.2	0.2				270	115	16
5083 A95083	India	IS 736	54300	Plt 1/2 hard 5 mm diam	0.25	0.1	0.7	4-4.9	0.5-1	0.4	0.2	0.2				355	275	5
	Finland	SFS 2588	AlMg3	Sh Rod Tub HF (shapes) 50 mm diam	0.35	0.1	0.5	2.6-3.4	0.5	0.5		0.2						
	Finland	SFS 2588	AlMg3	Sh Rod Tub Ann (sht)	0.35	0.1	0.5	2.6-3.4	0.5	0.5		0.2						
	Finland	SFS 2588	AlMg3	Sh Rod Tub Hard (rod,tub,wir) 6 mm diam	0.35	0.1	0.5	2.6-3.4	0.5	0.5		0.2						
5083 A95083	International	ISO R827	Al-Mg4.5Mn	As Mfg 5 in; 127 mm diam	0.25	0.1	0.5	4-4.9	0.3-1	0.5		0.2				265	106	12
5083 A95083	International	ISO TR2136	Al-Mg4.5Mn	Sh Plt Ann	0.25	0.1	0.5	4-4.9	0.3-1	0.5		0.2				275	125	12
5083 A95083	International	ISO TR2136	Al-Mg4.5Mn	Sh Plt HB Strain	0.25	0.1	0.5	4-4.9	0.3-1	0.5		0.2				305	315	5
5083 A95083	International	ISO TR2136	Al-Mg4.5Mn	Sh Plt HD SH	0.25	0.1	0.5	4-4.9	0.3-1	0.5		0.2				345	270	4
5083 A95083	International	ISO TR2778	Al-Mg4.5Mn	Tub Ann OBSOLETE Replace by ISO 6363	0.25	0.1	0.5	4-4.9	0.3-1	0.5		0.2				270	110	12
7027	Switzerland					0.1-0.3	0.4 max	0.7-1.1	0.1-0.4	0.25 max	0.1 max	3.5-4.5	0.05	0.15	Zr 0.05-0.30			
5083 A95083	Sweden	MNC 10E	4140	Sh	0.15			4.5-4.5	0.7						Al 94.65			
	Pan American	COPANT 862	X7007		0.05-0.35	0.1	0.4	0.5-1	0.1-0.6	0.3	0.1	3-5						
	Japan	JIS H1120	5N02	Wir As Mfg 25 mm diam	0.5	0.1	0.4	3-4	0.3-1	0.4	0.2	0.1				266		20
5058	Germany				0.1 max	0.1 max	0.5 max	4.5-5.6	0.2 max	0.4 max	0.2 max	0.2 max	0.2	0.15	Pb 1.2-1.8			
	Yugoslavia	JUS CC2.100	AlMg5		0.35 max	0.1 max	0.5 max	4.5-5.6	0.5 max	0.5 max		0.2 max		0.15	ST = Ti+Zr 0.20 max; Mn+Cr 0.10-0.30			
4043 A94043	Finland	SFS 2590	AlSi5	HF		0.1	0.5	0.7	0.3	3.5-5.5		0.2				130	70	15
4043 A94043	Germany	DIN 1732 Part 1	3.2245/S-AlSi5			0.1 max	0.5 max	0.7 max	0.3 max	3.5-5.5	0.2 max	0.2 max			ET 0.05 max; ; ET=.15 total			
4043 A94043	Sweden	SIS 144225	4225-00	Wir		0.1	0.6	0.25	0.5	4.5-6		0.2						
4043 A94043	Sweden	SIS 144225	4225-18	Wir Hard		0.1	0.6	0.25	0.5	4.5-6		0.2				176		
4043 A94043	UK	BS 1475	4043(N21)			0.1 max	0.6 max	0.2 max	0.5 max	4.5-6		0.2 max			Ni 0.2 max; Pb 0.05 max; Sn 0.05 max			
	Germany	DIN 1725 pt1	3.4337/Al99.8 ZnMg		0.1 max	0.2 max	0.1 max	0.7-1.2	0.05 max	0.1 max	0.02 max	3.8-4.6			ST 0.2 max; ET 0.02 max; ST=Fe+Si+Ti+Mn			
4043A	Europe					0.3 max	0.6 max	0.2 max	0.15 max	4.5-6	0.15 max	0.1 max	0.05	0.15	0.0008 max Be for welding			
	Europe		Al Si5(A)			0.3 max	0.6 max	0.2 max	0.15 max	4.5-6	0.15 max	0.1 max	0.05	0.15	0.0008 max Be for welding			
4043 A94043	Canada	CSA HA.6	0.4043	Rod Wir		0.3	0.8	0.05	0.05	4.5-6	0.2	0.1			Be 0.001			
4043 A94043	Canada	CSA S5	4043	Rod Wir		0.3 max	0.8 max		0.05 max	4.5-6	0.2 max	0.1 max	0.05	0.15				
4043 A94043	France	NF A-S5	4043	Rod Wir		0.3 max	0.8 max		0.05 max	4.5-6	0.2 max	0.1 max	0.05	0.15				
4043 A94043	Germany	DIN	AlSi5	Rod Wir		0.3 max	0.8 max		0.05 max	4.5-6	0.2 max	0.1 max	0.05	0.15				
4043 A94043	Japan	JIS Z3232	4043	Rod Q/T		0.3	0.8	0.05	0.05	4.5-6	0.2	0.1				167		

UNS numbers and US grades are provided as a means of cross referencing chemically similar alloys. Exchangability is only possible after independent examination of specifications. Tensile properties are minimum or typical . UTS and YS as Mpa, El as %. See Appendix for list of abbreviations used in Descriptions.

Worldwide Guide to Equivalent Nonferrous Metals and Alloys

Grade UNS #	Country	Specification	Designation	Description	Cr	Cu	Fe	Mg	Mn	Si	Ti	Zn	OE Max	OT Max	Other	UTS	YS	EL
4043 A94043	Pan American	COPANT 862	4043			0.3	0.8	0.05	0.05	4.5-6	0.2	0.1			Be 0.001			
4043 A94043	UK	BS N21	4043	Rod Wir		0.3 max	0.8 max		0.05 max	4.5-6	0.2 max	0.1 max	0.05	0.15				
4043 A94043	USA	AMS 4190	4043	Rod Wir		0.3 max	0.8 max		0.05 max	4.5-6	0.2 max	0.1 max	0.05	0.15				
4043 A94043	USA	MIL E-16053	4043	Rod Wir		0.3 max	0.8 max		0.05 max	4.5-6	0.2 max	0.1 max	0.05	0.15				
4043 A94043	USA	MIL W-6712	4043	Rod Wir		0.3 max	0.8 max		0.05 max	4.5-6	0.2 max	0.1 max	0.05	0.15				
4043 A94043	USA	QQ R-566	4043	Rod Wir		0.3 max	0.8 max		0.05 max	4.5-6	0.2 max	0.1 max	0.05	0.15				
4043 A94043	USA	SAE J454	4043	Rod Wir		0.3 max	0.8 max		0.05 max	4.5-6	0.2 max	0.1 max	0.05	0.15				
A94043	USA	AWS	E4043	Core Wire SMAW El		0.3 max	0.8 max	0.05 max	0.05 max	4.5-6	0.2 max	0.1 max	0.05	0.15	Be 0.0008 max			
A94043	USA	AWS	ER4043	Weld El Rod		0.3 max	0.8 max		0.05 max	4.5-6	0.2 max	0.1 max	0.05	0.15	Be 0.0008 max			
A94043	USA	AWS	R4043	Weld El Rod		0.3 max	0.8 max		0.05 max	4.5-6	0.2 max	0.1 max	0.05	0.15	Be 0.0008 max			
4043	USA					0.3 max	0.8 max	0.05 max	0.05 max	4.5-6	0.2 max	0.1 max	0.05	0.15	0.0008 max Be for welding			
	USA		Al Si5			0.3 max	0.8 max	0.05 max	0.05 max	4.5-6	0.2 max	0.1 max	0.05	0.15	0.0008 max Be for welding			
2017 A92017	Russia	GOST 4784	D1P		3.8-4.5	0.5 max	0.4-0.8	0.4-0.8	0.5			0.1 max			ET 0.1 max; , ET=1.2 total			
2017 A92017	France	AIR 9051-A	A-U4G(2017-F)	Sh Bar Water Q/T (sht) 6/12 mm diam		3.5-4.5	0.5	0.5-1	0.3-0.8	0.25-0.8	0.2	0.25				390	240	13
2017 A92017	France	AIR 9051-A	A-U4G(2017-F)	Sh Bar ST (bar) 250 mm diam		3.5-4.5	0.5	0.5-1	0.3-0.8	0.25-0.8	0.2	0.25				380	230	9
2017 A92017	France	AIR 9051-A	A-U4G(2017-F)	Sh Bar ST (bil and plt) 500 mm		3.5-4.5	0.5	0.5-1	0.3-0.8	0.25-0.8	0.2	0.25				400	240	14
2017 A92017	France	AIR 9150-B	A-U4G	Wir CF or Ann 1.6/9.6 mm diam		3.5-4.5	0.5	0.5-1	0.3-0.8	0.3-0.8		0.25				390	240	14
2017 A92017	Austria	ONORM M3430	AlCuMg1		0.1 max	3.5-4.5	0.7 max	0.4-1	0.4-1	0.2-0.8		0.25 max			ET 0.05 max; ST 0.25 max; ST=Ti+Zr; ET=.15 total;			
2017 A92017	Belgium	NBN P21-001	2017		0.1 max	3.5-4.5	0.7 max	0.4-0.8	0.4-1	0.2-0.8	0.15 max	0.25 max			ET 0.05 max; ST 0.2 max; ST=Zr+Ti; ET =.15 Total			
2017A	Europe				0.1 max	3.5-4.5	0.7 max	0.4-1	0.4-1	0.2-0.8		0.25 max	0.05	0.15	0.25 Zr+Ti			
	Europe		AlCu4MgSi(A)		0.1 max	3.5-4.5	0.7 max	0.4-1	0.4-1	0.2-0.8		0.25 max	0.05	0.15	0.25 Zr+Ti			
2017 A92017	France	NF A50411	2017A(A-U4G)	Bar Wir Tub Aged (extd bar)	0.1	3.5-4.5	0.7	0.4-1	0.4-1	0.2-0.8		0.25				390	255	10
2017 A92017	France	NF A50411	2017A(A-U4G)	Bar Wir Tub Aged (tub) 150 mm diam	0.1	3.5-4.5	0.7	0.4-1	0.4-1	0.2-0.8		0.25				390	240	14
2017 A92017	France	NF A50451	2017A(A-U4G)	Sh Plt Ann .4/1.6 mm diam	0.1	3.5-4.5	0.7	0.4-1	0.4-1	0.2-0.8		0.25					150	13
2017 A92017	France	NF A50451	2017A(A-U4G)	Sh Plt Aged .4/1.6 mm diam	0.1	3.5-4.5	0.7	0.4-1	0.4-1	0.2-0.8		0.25				390	240	15
2017 A92017	France	NF A50506	2017.S(A-U4G)	Shapes CR .8/3.2 mm diam	0.1	3.5-4.5	0.7	0.4-1	0.4-1	0.2-0.8		0.25				390	240	15
2017 A92017	Germany	DIN 1725 pt1	3.1325/AlCuMg1		0.1 max	3.5-4.5	0.7 max	0.4-1	0.4-1	0.2-0.8		0.25 max			ST 0.25 max; ET 0.05 max; ST=Ti+Zr; ET=.15 total;			
2017 A92017	Japan	JIS H4000	2017	Sh Plt Ann 1.5 mm diam	0.1	3.5-4.5	0.7	0.2-0.8	0.4-1	0.8		0.25				216	108	12
2017 A92017	Japan	JIS H4000	2017	Sh Plt SHT and AH 1.5 mm diam	0.1	3.5-4.5	0.7	0.2-0.8	0.4-1	0.8		0.25				353	108	12

UNS numbers and US grades are provided as a means of cross referencing chemically similar alloys. Exchangability is only possible after independent examination of specifications. Tensile properties are minimum or typical . UTS and YS as Mpa, El as %. See Appendix for list of abbreviations used in Descriptions.

2-66 Wrought Aluminum

Grade UNS #	Country	Specification	Designation	Description	Cr	Cu	Fe	Mg	Mn	Si	Ti	Zn	OE Max	OT Max	Other	UTS	YS	EL
2017 A92017	Japan	JIS H4000	2017	Sh Plt SHT and CW 1.5 mm diam	0.1	3.5-4.5	0.7	0.2-0.8	0.4-1	0.8		0.25				373	216	15
2017 A92017	Japan	JIS H4040	2017	Bar Ann	0.1	3.5-4.5	0.7	0.2-0.8	0.4-1	0.8		0.25				245	127	16
2017 A92017	Japan	JIS H4040	2017	Bar SHT and AH	0.1	3.5-4.5	0.7	0.2-0.8	0.4-1	0.8		0.25				343	216	12
2017 A92017	Japan	JIS H4080	2017	Tub Ann	0.1	3.5-4.5	0.7	0.2-0.8	0.4-1	0.8		0.25				245	127	16
2017 A92017	Japan	JIS H4080	2017	Tub SHT and AH	0.1	3.5-4.5	0.7	0.2-0.8	0.4-1	0.8		0.25				343	216	12
2017 A92017	Japan	JIS H4100	2017	Ann	0.1	3.5-4.5	0.7	0.2-0.8	0.4-1	0.8		0.25				245	127	16
2017 A92017	Japan	JIS H4100	2017	SHT and AH	0.1	3.5-4.5	0.7	0.2-0.8	0.4-1	0.8		0.25				343	216	12
2017 A92017	Japan	JIS H4120	2017	Wir Ann 25 mm diam	0.1	3.5-4.5	0.7	0.2-0.8	0.4-1	0.8		0.25				245		
2017 A92017	Japan	JIS H4120	2017	Wir 3/8 hard 25 mm diam	0.1	3.5-4.5	0.7	0.2-0.8	0.4-1	0.8		0.25				206		
2017 A92017	Japan	JIS H4120	2017	Wir SHT and AH 2/25 mm diam	0.1	3.5-4.5	0.7	0.2-0.8	0.4-1	0.8		0.25				373	216	12
2017 A92017	Japan	JIS H4140	2017	Die SHT and AH 100 mm diam	0.1	3.5-4.5	0.7	0.2-0.8	0.4-1	0.8						343	216	14
2017 A92017	Norway	NS 17103	NS17103/AlCu4MgSi	Bar Plt Sh Strp Tub Frg	0.1 max	3.5-4.5	0.7 max	0.4-1	0.4-1	0.2-0.8		0.25 max			ET 0.05 max; ST 0.25 max; ST=Ti+Zr, ET=.15 total;			
2017 A92017	Pan American	COPANT 002	2017		0.1	3.5-4.5	0.7	0.4-0.8	0.4-1	0.2-0.8	0.15	0.25						
2017	USA				0.1 max	3.5-4.5	0.7 max	0.4-0.8	0.4-1	0.2-0.8	0.15 max	0.25 max	0.05	0.15				
	USA		Al Cu4MgSi		0.1 max	3.5-4.5	0.7 max	0.4-0.8	0.4-1	0.2-0.8	0.15 max	0.25 max	0.05	0.15				
2017 A92017	Russia		D17			3.5-4.5	0.8 max	0.4-0.8	0.4-0.8	0.8 max								
2017 A92017	International	ISO R827	Al-Cu4MgSi	ST and nat aged 0.75 in 20 mm diam		3.5-4.7	0.7	0.3-1.2	0.3-1	0.2-0.8		0.5			Ni 0.2	353	226	10
2017 A92017	International	ISO R829	Al-Cu4MgSi	Frg ST and nat aged		3.5-4.7	0.7	0.3-1.2	0.3-1	0.2-0.8		0.5			Ni 0.2	353	206	11
2017 A92017	International	ISO TR2136	Al-Cu4MgSi	Sh Plt SHT and nat aged		3.5-4.7	0.7	0.3-1.2	0.3-1	0.2-0.8		0.5			Ni 0.2	385	240	14
2017 A92017	International	ISO TR2778	Al-Cu4MgSi	Tub SHT and nat aged OBSOLETE Replace by ISO 6363		3.5-4.7	0.7	0.3-1.2	0.3-1	0.2-0.8		0.5			Ni 0.2	385	245	12
2017 A92017	Czech Republic	CSN 424201	AlCu4Mg			3.8-4.8	0.7 max	0.4-0.8	0.4-0.8	0.7 max		0.3 max			Fe+Ni 0.70; Ni 0.1 max			
2017 A92017	Czech Republic	CSN 424251	AlCu4Mg pl			3.8-4.8	0.7 max	0.4-0.8	0.4-0.8	0.7 max		0.3 max			Fe+Ni 0.70; Ni 0.1 max			
2017 A92017	Russia		AK1			3.8-4.8	0.7 max	0.4-0.8	0.4-0.8	0.7 max								
2017 A92017	Russia	GOST 4784	D1T		0.4-0.8	3.8-4.8		0.4-0.8										
	Yugoslavia	JUS CC2.100	AlCu5Mg1		0.1 max	3.9-5	1 max	0.2-1	0.4-1.2	1.2 max	0.15 max	0.25 max			ET 0.05 max; ET = .15 Total;			
2011A	Switzerland					4.5-6	0.5 max			0.4 max		0.3 max	0.05	0.15	Bi 0.20-0.6; Pb 0.20-0.6			
	International	ISO R209 (1971)	Al-Mg 5		0.35 max	0.10 max	0.5 max	4.5-5.6	0.5 max	0.5 max					Ti+Zr 0.2 max; Mn+Cr 0.1-0.5			
5056 A95056	Russia		AMg5V			0.05 max	0.5 max	4.8-5.5	0.3-0.6	0.5 max		0.2 max			ET 0.1 max;ET=1.35 total; V 0.02-0.2			
5056 A95056	Australia	AS 1865	5056	Wir Rod Bar Strp Ann 10 mm diam	0.05-0.2	0.1	0.4	4.5-5.6	0.05-0.2	0.3		0.1						20

UNS numbers and US grades are provided as a means of cross referencing chemically similar alloys. Exchangability is only possible after independent examination of specifications. Tensile properties are minimum or typical . UTS and YS as Mpa, El as %. See Appendix for list of abbreviations used in Descriptions.

Worldwide Guide to Equivalent Nonferrous Metals and Alloys

Grade UNS #	Country	Specification	Designation	Description	Cr	Cu	Fe	Mg	Mn	Si	Ti	Zn	OE Max	OT Max	Other	UTS	YS	EL
5056 A95056	Australia	AS 1865	5056	Wir Rod Bar Strp SH and stab to 1/4 hard 10 mm diam	0.05-0.2	0.1	0.4	4.5-5.6	0.05-0.2	0.3		0.1						
5056 A95056	Australia	AS 1865	5056	Wir Rod Bar Strp SH and stab to full hard 10 mm diam	0.05-0.2	0.1	0.4	4.5-5.6	0.05-0.2	0.3		0.1						
5056 A95056	Belgium	NBN P21-001	5056		0.05-0.2	0.1 max	0.4 max	4.5-5.6	0.05-0.2	0.3 max		0.1 max			ET 0.05 max; ET = .15 Total			
5056 A95056	Canada	CSA HA.6	0.5056	Rod Wir O .062/1 in diam	0.05-0.2	0.1	0.4	4.5-5.6	0.05-0.2	0.3		0.1				262		
5056 A95056	Canada	CSA HA.6	0.5056	Rod Wir H32 .062/1 in diam	0.05-0.2	0.1	0.4	4.5-5.6	0.05-0.2	0.3		0.1				303		
5056 A95056	Germany	DIN	Al-Mg5	Rod Wir	0.2 max	0.1 max	0.4 max	4.5-5.6	0.05-0.2	0.3 max		0.1 max	0.05	0.15				
5056 A95056	Japan	JIS H4040	5056	Bar As Mfg (bar	0.05-0.2	0.1	0.4	4.5-5.6	0.05-0.2	0.3		0.1				245	118	
5056 A95056	Japan	JIS H4040	5056	Bar Ann 6 mm diam	0.05-0.2	0.1	0.4	4.5-5.6	0.05-0.2	0.3		0.1				314	98	20
5056 A95056	Japan	JIS H4080	5056	Tub As Mfg	0.05-0.2	0.1	0.4	4.5-5.6	0.05-0.2	0.3		0.1				245	118	
5056 A95056	Japan	JIS H4120	5056	Wir Ann 25 mm diam	0.5	0.1	0.4	4.5-5.6	0.05-0.2	0.3	0.2	0.1				314		
5056 A95056	Japan	JIS H4120	5056	Wir Stab at 1/4 hard 25 mm diam	0.5	0.1	0.4	4.5-5.6	0.05-0.2	0.3	0.2	0.1				304		
5056 A95056	Japan	JIS H4140	5056	Die As Mfg 100 mm diam	0.05-0.2	0.1	0.4	4.5-5.6	0.05-0.2	0.3		0.1				245	118	18
5056 A95056	Pan American	COPANT 862	5056		0.05-0.2	0.1	0.4	4.5-5.6	0.05-0.2	0.3		0.1						
5056 A95056	UK	BS N6 2L.58	5056	Rod Wir	0.2 max	0.1 max	0.4 max	4.5-5.6	0.05-0.2	0.3 max		0.1 max	0.05	0.15				
5056 A95056	USA	AMS 4005	5056	Rod Wir OBSOLETE	0.2 max	0.1 max	0.4 max	4.5-5.6	0.05-0.2	0.3 max		0.1 max	0.05	0.15				
5056 A95056	USA	AMS 4182	5056	Rod Wir	0.2 max	0.1 max	0.4 max	4.5-5.6	0.05-0.2	0.3 max		0.1 max	0.05	0.15				
5056 A95056	USA	ASTM B211	5056	Rod Wir	0.2 max	0.1 max	0.4 max	4.5-5.6	0.05-0.2	0.3 max		0.1 max	0.05	0.15				
5056 A95056	USA	ASTM B316	5056	Rod Wir	0.2 max	0.1 max	0.4 max	4.5-5.6	0.05-0.2	0.3 max		0.1 max	0.05	0.15				
5056 A95056	USA	MIL A-81596	5056	Rod Wir	0.2 max	0.1 max	0.4 max	4.5-5.6	0.05-0.2	0.3 max		0.1 max	0.05	0.15				
5056 A95056	USA	QQ A-430	5056	Rod Wir	0.2 max	0.1 max	0.4 max	4.5-5.6	0.05-0.2	0.3 max		0.1 max	0.05	0.15				
5056 A95056	USA	SAE J454	5056	Rod Wir	0.2 max	0.1 max	0.4 max	4.5-5.6	0.05-0.2	0.3 max		0.1 max	0.05	0.15				
5056	USA				0.05-0.2	0.1 max	0.4 max	4.5-5.6	0.05-0.2	0.3 max		0.1 max		0.15				
	USA		AlMg5Cr		0.05-0.2	0.1 max	0.4 max	4.5-5.6	0.05-0.2	0.3 max		0.1 max		0.15				
5056 A95056	UK	BS 1473	NB6	SH to 1/2 hard 12 mm diam	0.25	0.1	0.5	4.5-5.5	0.5	0.3	0.2	0.2				310	240	
5056 A95056	UK	BS 1473	NR6	ann or as mfg 25 mm diam	0.25	0.1	0.5	4.5-5.5	0.5	0.3	0.2	0.2				255		
5056 A95056	UK	BS 1475	NG6	Wir Ann 10 mm diam	0.25	0.1	0.5	4.5-5.5	0.5	0.3	0.2	0.2				310		
5056 A95056	UK	BS 1475	NG6	Wir SH to hard 10 mm diam	0.25	0.1	0.5	4.5-5.5	0.5	0.3	0.2	0.2						
5056 A95056	Austria	ONORM M3430	AlMg5		0.2 max	0.1 max	0.5 max	4.5-5.6	0.1-0.6	0.4 max	0.2 max	0.2 max			ET 0.05 max; St 0.1-0.6; ST=Mn+Cr; ET=.15 total;			
5056 A95056	Germany	DIN 1725 Part 1	3.3555/AlMg5		0.2 max	0.1 max	0.5 max	4.5-5.6	0.1-0.6	0.4 max	0.2 max	0.2 max			ST 0.1-0.6; ET 0.05 max; min; ST=Mn+Cr; ET=.15 total			

UNS numbers and US grades are provided as a means of cross referencing chemically similar alloys. Exchangability is only possible after independent examination of specifications. Tensile properties are minimum or typical . UTS and YS as Mpa, El as %. See Appendix for list of abbreviations used in Descriptions.

2-68 Wrought Aluminum

Grade UNS #	Country	Specification	Designation	Description	Cr	Cu	Fe	Mg	Mn	Si	Ti	Zn	OE Max	OT Max	Other	UTS	YS	EL
5056 A95056	Germany	DIN 1725 pt1	3.3555/AlMg5		0.2 max	0.1 max	0.5 max	4.5-5.6	0.1-0.6	0.4 max	0.2 max	0.2 max			ST 0.1-0.6; ET 0.05 max; ST=Mn+Cr;			
5056 A95056	Finland	SFS 2589	AlMg5	HF (shapes) 125 mm diam	0.35	0.1	0.5	4.5-5.5	1	0.5		0.2				270	130	12
5056 A95056	Finland	SFS 2589	AlMg5	Ann (sht) 6 mm diam	0.35	0.1	0.5	4.5-5.5	1	0.5		0.2				300	120	16
5056 A95056	Finland	SFS 2589	AlMg5	1/2 hard (sht) 6 mm diam	0.35	0.1	0.5	4.5-5.5	1	0.5		0.2				300	220	8
5056 A95056	UK		A56S		0.25 max	0.1 max	0.5 max	4.5-5.5	0.15-0.5	0.6 max	0.15 max	0.2 max						
5056 A95056	Czech Republic	CSN 424415	AlMg5		0.3 max	0.1 max	0.4 max	4-6	0.04-0.6	0.7 max	0.2 max	0.2 max			Fe+Si 0.60; Sb 0.25			
5056 A95056	Russia	GOST 4784	AMg5			0.2 max	0.4 max	4.7-5.7	0.2-0.6	0.4 max					ET 0.1 max; ET=1.1 total; Ni 0.2 max			
5056 A95056	Russia	GOST 4784	AMg5P			0.2 max	0.4 max	4.7-5.7	0.2-0.6	0.4 max					ET 0.1 max; ET=1.1 total; Ni 0.2 max			
5056 A95056	Russia		AMg5M			0.2 max	0.4 max	4.7-5.7	0.2-0.6	0.4 max					ET 0.1 max; ET=1.1 total; Ni 0.2 max			
7019A A97019	Argentina				0.05-0.35	0.1 max	0.4 max	1.5-2.5	0.1-0.6	0.3 max	0.1 max	3-5	0.05	0.15				
5019	Europe				0.2 max	0.1 max	0.5 max	4.5-5.6	0.1-0.6	0.4 max	0.2 max	0.2 max	0.05	0.15	0.10-0.6 Mn+Cr			
	India	IS 7793	2285/24850	Chill		3.723 max	0.7	1.2-1.8		0.6	0.2	0.2			Ni 1.7-2.3; Sn 0.05 max	225		
5356	USA				0.05-0.2	0.1 max	0.4 max	4.5-5.5	0.05-0.2	0.25 max	0.06-0.2	0.1 max	0.2	0.15	0.0008 max Be for welding; OE min 0.06			
	USA		AlMg5Cr(A)		0.05-0.2	0.1 max	0.4 max	4.5-5.5	0.05-0.2	0.25 max	0.06-0.2	0.1 max	0.2	0.15	0.0008 max Be for welding; OE min 0.06			
	France	AIR 9050	AG5MC		0.05-0.35	0.1 max	0.4 max	4.6-5.4	0.05-0.7	0.3 max		0.1 max			ET 0.05 max; ST 0.2 max; ET=.15; ST=+Ti+Zr;			
	France	AIR 9302	AG5MC		0.05-0.35	0.1 max	0.4 max	4.6-5.4	0.05-0.7	0.3 max		0.1 max			ET 0.05 max; ST 0.2 max; ET=.15; ST=+Ti+Zr;			
2094	USA					4.4-5.2	0.15 max	0.25-0.8	0.25 max	0.12 max	0.1 max	0.25 max	0.05	0.15	0.7-1.4 Li; Ag 0.25-0.6; Zr 0.04-.18			
5119	Germany				0.3 max	0.05 max	0.4 max	4.5-5.6	0.2-0.6	0.25 max	0.15 max	0.2 max	0.05	0.15	0.20-0.6 Mn+Cr; 0.0008 max Be for welding			
A02060	USA	AWS		Weld El Rod		4.2-5	0.15 max	0.15-0.35	0.2-0.5	0.1 max	0.15-0.3	0.1 max	0.05	0.15	Ni 0.05 max; Sn 0.05 max			
	Pan American	COPANT 862	X7010		0.05-0.35	0.1	0.4	0.8-1.5	0.1-0.6	0.3	0.1	3-5						
5183 A95183	Canada	CSA HA.6	0.5183	Rod Wir	0.05-0.25	0.1	0.4	4.3-5.2	0.5-1	0.4	0.15	0.25			Be 0.008			
5183 A95183	France	NF A81-331	AG4.5MC		0.05-0.25	0.1 max	0.4 max	4.3-5.2	0.5-1	0.4 max	0.15 max				ET 0.05 max; ET=.15;			
5183 A95183	Japan	JIS Z3232	5183	Rod Ann	0.05-0.25	0.1	0.4	4.3-5.2	0.5-1	0.4	0.15	0.25			Be 0.001	275		
A95183	USA	AWS	ER5183	Weld El Rod	0.05-0.25	0.1 max	0.4 max	4.3-5.2	0.5-1	0.4 max	0.15 max	0.25 max	0.05	0.15	Be 0.0008 max			
A95183	USA	AWS	R5183	Weld El Rod	0.05-0.25	0.1 max	0.4 max	4.3-5.2	0.5-1	0.4 max	0.15 max	0.25 max	0.05	0.15	Be 0.0008 max			
5183	USA				0.05-0.25	0.1 max	0.4 max	4.3-5.2	0.5-1	0.4 max	0.15 max	0.25 max	0.15	0.15	0.0008 max Be for welding			
	USA		AlMg4.5Mn0.7(A)		0.05-0.25	0.1 max	0.4 max	4.3-5.2	0.5-1	0.4 max	0.15 max	0.25 max	0.15	0.15	0.0008 max Be for welding			
	India	IS 738	NT5	Wir Ann	0.25	0.1	0.7	2.8-4	0.6	0.6						216		18

UNS numbers and US grades are provided as a means of cross referencing chemically similar alloys. Exchangability is only possible after independent examination of specifications. Tensile properties are minimum or typical. UTS and YS as Mpa, El as %. See Appendix for list of abbreviations used in Descriptions.

Wrought Aluminum 2-69

Worldwide Guide to Equivalent Nonferrous Metals and Alloys

Grade UNS #	Country	Specification	Designation	Description	Cr	Cu	Fe	Mg	Mn	Si	Ti	Zn	OE Max	OT Max	Other	UTS	YS	EL
	India	IS 738	NT5	Wir Ann and CW to 1/2 hard	0.25	0.1	0.7	2.8-4	0.6	0.6						245		5
7051	France				0.05-0.25	0.15 max	0.45 max	1.7-2.5	0.1-0.45	0.35 max	0.15 max	3-4	0.05	0.15				
X2080	USA					3.3-4.1	0.2 max	1.5-2.2	0.25 max	0.1 max		0.1 max	0.05	0.15	0.20-0.50 O; Be 0.005 max; Zr 0.08-0.25			
	Yugoslavia	JUS CC2.100	AlCu4Mg1			3.8-4.9	0.5 max	1-1.8	1.2 max	0.5 max					ST 0.3 max; ST = Ti + Cr + Zr;			
2025	USA				0.1 max	3.9-5	1 max	0.05 max	0.4-1.2	0.5-1.2	0.15 max	0.25 max	0.05	0.15				
	International	ISO R209 (1971)	Al-Cu4SiMg			3.8-5.0	0.7 max	0.2-0.8	0.3-1.2	0.5-1.2					Ti+Zr+Cr 0.3 max; Ni 0.2 max			
2618A	Europe					1.8-2.7	0.9-1.4	1.2-1.8	0.25 max	0.15-0.25	0.2 max	0.15 max	0.05	0.15	0.25 Zr+Ti; Ni 0.8-1.4			
7104 A97104	Pan American	COPANT 862	7104			0.03	0.4	0.5-0.9		0.25	0.1	3.6-4.4						
	South Africa	SABS 712	Al-Mg3.5		0.25	0.1	0.5	3.1-3.9	0.5	0.5		0.2						
	India	IS 733	53000	Bar Rod As Mfg 50 mm diam	0.25	0.1	0.5	2.8-4	0.5	0.6	0.2	0.2				215	100	14
5014	Europe				0.2 max	0.2 max	0.4 max	4-5.5	0.2-0.9	0.4 max	0.2 max	0.7-1.5	0.05	0.15				
	Mexico	NOM W-28	A,C-R,C-L,andCEL	Tub Extd/drawn (A) 3.175 mm	0.5		1	1	1	0.8		0.1			Al 95.05			
5283A	Europe				0.05 max	0.03 max	0.3 max	4.5-5.1	0.5-1	0.3 max	0.03 max	0.1 max	0.03	0.15	Ni 0.03 max; Pb 0.0003 max; Zr 0.05			
	France	NF A50-101	5957(AG5MO.7)		0.05 max	0.03 max	0.3 max	4.5-5.1	0.5-1	0.3 max	0.03 max	0.1 max			ET 0.05 max; ET=.15; Ni 0.03 max; Zr 0.03 x			
5283	France				0.05 max	0.03 max	0.3 max	4.5-5.1	0.5-1	0.3 max	0.03 max	0.1 max	0.03	0.15	0.05 Zr; Ni 0.03 x			
7004 A97004	Canada	CSA HA.5	0.7004	Bar Rod T1 to 1.3780 diam	0.05 max	0.05 max	0.35 max	1-2	0.2-0.7	0.25 max	0.05 max	3.8-4.6			ET 0.05 max; ET=.15 total; Zr 0.1-0.2	325	205	10
7004 A97004	Canada	CSA HA.5	0.7004	Bar Rod T5, to 1.3780 diam	0.05 max	0.05 max	0.35 max	1-2	0.2-0.7	0.25 max	0.05 max	3.8-4.6			ET 0.05 max; ET=.15 total; Zr 0.1-0.2	370	315	10
7004 A97004	Canada	CSA HA.5	0.7004	Bar Rod T51, to 1.3780 diam	0.05 max	0.05 max	0.35 max	1-2	0.2-0.7	0.25 max	0.05 max	3.8-4.6			ET 0.05 max; ET=.15 total; Zr 0.1-0.2	290	255	10
7004 A97004	Pan American	COPANT 862	7004		0.05	0.05	0.35	1-2	0.2-0.7	0.25	0.05	3.8-4.6			Zr 0.1-0.2			
7004	USA				0.05 max	0.05 max	0.35 max	1-2	0.2-0.7	0.25 max	0.05 max	3.8-4.6	0.05	0.15	Zr 0.1-0.2			
	Sweden	SIS 144133	4133-00	Wir	0.35	0.1	0.5	3.1-3.9	0.6-0.6	0.5		0.2						
	Sweden	SIS 144133	4133-18	Wir HW	0.35	0.1	0.5	3.1-3.9	0.6-0.6	0.5		0.2				225		
	India	IS 739	NG5	Wir	0.25	0.1	0.7	2.8-4	0.6	0.6		0.2						
	UK	BS 1472	HF12	SHT nat Aged 150 mm OBSLT Replaced by BS EN 586		1.8-2.8	0.6-1.2	0.6-1.2	0.5	0.5-1.3	0.2	0.2			Ni 0.6-1.4	310	160	13
2014 A92014	Japan	JIS H4140	2014	Die	0.1	3.9-3.9	0.7	0.2-0.8	0.4-1.2	0.5-1.2	0.15	0.25						
2014 A92014	Sweden	MNC 40E	4338	Bar Sh Tub Ext		4.5-4.5		0.5	0.8	0.8					Al 93.4			
2014 A92014	Japan	JIS H4040	2014	Bar	0.1	3.9-4.5	0.7	0.2-0.8	0.4-1.2	0.5-1.2	0.15	0.25						
2007 A92007	Norway	NS 17106	NS17106/AlCu4PbMg	Bar	0.1 max	3.3-4.6	0.8 max	0.4--1.8	0.5-1	0.8 max	0.2 max	0.8 max			ET 0.1 max; ET=.30 total; Bi 0.2 max; Ni 0.2 max; Pb 0.8-1.5; Sn 0.2 max			

UNS numbers and US grades are provided as a means of cross referencing chemically similar alloys. Exchangability is only possible after independent examination of specifications. Tensile properties are minimum or typical . UTS and YS as Mpa, El as %. See Appendix for list of abbreviations used in Descriptions.

Grade UNS #	Country	Specification	Designation	Description	Cr	Cu	Fe	Mg	Mn	Si	Ti	Zn	OE Max	OT Max	Other	UTS	YS	EL
2014 A92014	India	IS 738	HT15	Tub ST and Nat aged 1.60 mm diam		3.8-4.8	0.7	0.2-0.8	0.3-1.2	0.5-0.9						402	279	8
2014 A92014	India	IS 739	HG15	Wir St and Nat aged		3.8-4.8	0.7	0.2-0.8	0.3-1.2	0.5-0.9		0.2			Ni 0.2; Pb 0.05; Sb 0.05; Sn 0.05 max	387		
2014 A92014	India	IS 739	HG15	Wir SHT and PHT		3.8-4.8	0.7	0.2-0.8	0.3-1.2	0.5-0.9		0.2			Ni 0.2; Pb 0.05; Sb 0.05; Sn 0.05 max	432		
2005 A92005	Argentina				0.1 max	3.5-5	0.7 max	0.2-1	1 max	0.8 max	0.2 max	0.5 max	0.05	0.15	Ni 0.2 max			
2014 A92014	Argentina				0.1 max	3.9-5	0.5 max	0.2-0.8	0.4-1.2	0.5-0.9	0.15 max	0.25 max	0.05	0.15	0.20Zr+Ti; Ni 0.1			
2005 A92005	Argentina		Al Cu4SiMg(A)		0.1 max	3.9-5	0.5 max	0.2-0.8	0.4-1.2	0.5-0.9	0.15 max	0.25 max	0.05	0.15	0.20Zr+Ti; Ni 0.1 max			
2014 A92014	UK	BS 1470	HC15	Plt Sh Strp SHT and nat aged .2/12.5 mm diam OBSOLETE Replaced by BS EN 515	0.1	3.9-5	0.7	0.2-0.8	0.4-1.2	0.5-0.9	0.2	0.2				375	230	13
2014 A92014	UK	BS 1470	HC15	Plt Sh Strp SHT and PT .2/3 mm diam OBSOLETE Replaced by BS EN 515	0.1	3.9-5	0.7	0.2-0.8	0.4-1.2	0.5-0.9	0.2	0.2				400	325	7
2014 A92014	UK	BS 1470	HS15	Plt Sh Strp SHT and nat aged .2/25 mm diam OBSOLETE Replaced by BS EN 515	0.1	3.9-5	0.7	0.2-0.8	0.4-1.2	0.5-0.9	0.2	0.2				385	245	13
2014 A92014	UK	BS 1470	HS15	Plt Sh Strp SHT and PT .2/3 mm diam OBSOLETE Replaced by BS EN 515	0.1	3.9-5	0.7	0.2-0.8	0.4-1.2	0.5-0.9	0.2	0.2				430	375	6
2014 A92014	UK	BS 1471	HT15	Tub SHT and nat aged 10 mm diam	0.1	3.9-5	0.7	0.2-0.8	0.4-1.2	0.5-0.9	0.2	0.2				400	290	8
2014 A92014	UK	BS 1471	HT15	Tub SHT and PT 10 mm diam	0.1	3.9-5	0.7	0.2-0.8	0.4-1.2	0.5-0.9	0.2	0.2				450	370	6
2014 A92014	UK	BS 1472	HF15	SHT nat aged 150 mm diam Obsolete Replaced by BS EN 586	0.1	3.9-5	0.7	0.2-0.8	0.4-1.2	0.5-0.9	0.2	0.2				370	215	13
2014 A92014	UK	BS 1472	HF15	SHT PT 150 mm diam Obsolete Replaced by BS EN 586	0.1	3.9-5	0.7	0.2-0.8	0.4-1.2	0.5-0.9	0.2	0.2				450	395	6
2014 A92014	UK	BS 1473	HB15	SHT and PT 12 mm diam	0.1	3.9-5	0.7	0.2-0.8	0.4-1.2	0.5-0.9	0.2	0.2				430	390	
2014 A92014	UK	BS 1473	HR15	SHT and nat aged 12 mm diam	0.1	3.9-5	0.7	0.2-0.8	0.4-1.2	0.5-0.9	0.2	0.2				385		
2014 A92014	UK	BS 1474	HE15	Bar Tub SHT and nat aged 20 mm diam	0.1	3.9-5	0.7	0.2-0.8	0.4-1.2	0.5-0.9	0.2	0.2				370	230	10
2014 A92014	UK	BS 1474	HE15	Bar Tub SHT and PT 20 mm diam	0.1	3.9-5	0.7	0.2-0.8	0.4-1.2	0.5-0.9	0.2	0.2				435	370	6
2014 A92014	UK	BS 1475	HG15	Wir SHT and nat aged 10 mm diam	0.1	3.9-5	0.7	0.2-0.8	0.4-1.2	0.5-0.9	0.2	0.2				385		

UNS numbers and US grades are provided as a means of cross referencing chemically similar alloys. Exchangability is only possible after independent examination of specifications. Tensile properties are minimum or typical. UTS and YS as Mpa, El as %. See Appendix for list of abbreviations used in Descriptions.

Worldwide Guide to Equivalent Nonferrous Metals and Alloys

Grade UNS #	Country	Specification	Designation	Description	Cr	Cu	Fe	Mg	Mn	Si	Ti	Zn	OE Max	OT Max	Other	UTS	YS	EL
2014 A92014	UK	BS 1475	HG15	Wir SHT and PT 10 mm diam	0.1	3.9-5	0.7	0.2-0.8	0.4-1.2	0.5-0.9	0.2	0.2				430		
2014 A92014	UK	BS 1470	2014A(H15)	OBSOLETE Replaced by BS EN 515	0.1 max	3.9-5	0.7 max	0.2-0.8	0.4-1.2	0.5-0.9	0.2 max	0.2 max						
2214	USA				0.1 max	3.9-5	0.3 max	0.2-0.8	0.4-1.2	0.5-1.2	0.15 max	0.25 max	0.05	0.15				
2014 A92014	Australia	AS 1866	2014	Rod Bar, SHT and Nat aged 10.0 mm diam	0.1	3.9-5	0.7	0.2-0.8	0.4-1.2	0.5-1.2	0.15	0.25				372	241	15
2014 A92014	Austria	ONORM M3430	AlCuSiMn		0.1 max	3.9-5	0.7 max	0.2-0.8	0.4-1.2	0.5-1.2	0.15 max	0.25 max			ET 0.05 max; ST 0.2 max; ST=Ti+Zr; ET=.15 total;			
2014 A92014	Belgium	NBN P21-001	2014		0.1 max	3.9-5	0.7 max	0.2-0.8	0.4-1.2	0.5-1.2	0.15 max	0.25 max			ET 0.05 max; ST 0.2 max; ST = Zr + Ti; ; ET = .15 Total;			
2014 A92014	Canada	CSA HA.4	.2014Alclad	Plt Sh T3 (flat Sht) .020/.039 in diam	0.1	3.9-5	0.7	0.2-0.8	0.4-1.2	0.5-1.2	0.15	0.25				379	234	14
2014 A92014	Canada	CSA HA.4	.2014Alclad	Plt Sh T451 (plt) .250/.499 in diam	0.1	3.9-5	0.7	0.2-0.8	0.4-1.2	0.5-1.2	0.15	0.25				393	248	15
2014 A92014	Canada	CSA HA.4	.2014Alclad	Plt Sh T6 (sht) .020/.039 in diam	0.1	3.9-5	0.7	0.2-0.8	0.4-1.2	0.5-1.2	0.15	0.25				434	379	7
2014 A92014	Canada	CSA HA.5	0.2014	Bar Rod Wir Extd and O	0.1	3.9-5	0.7	0.2-0.8	0.4-1.2	0.5-1.2	0.15	0.25				207	124	12
2014 A92014	Canada	CSA HA.5	0.2014	Bar Rod Wir Extd and T4 .124 in diam	0.1	3.9-5	0.7	0.2-0.8	0.4-1.2	0.5-1.2	0.15	0.25				345	241	12
2014 A92014	Canada	CSA HA.5	0.2014	Bar Rod Wir Extd and T6 .125/.499 in diam	0.1	3.9-5	0.7	0.2-0.8	0.4-1.2	0.5-1.2	0.15	0.25				414	365	7
2014 A92014	Canada	CSA HA.8	0.2014	Frg T6 (die frg) 1 in diam	0.1	3.9-5	0.7	0.2-0.8	0.4-1.2	0.5-1.2	0.15	0.25				448	386	6
2014 A92014	Canada	CSA HA.8	0.2014	Frg (Hand frg) 2 in diam	0.1	3.9-5	0.7	0.2-0.8	0.4-1.2	0.5-1.2	0.15	0.25				448	386	8
2014 A92014	Finland	SFS 2595	AlCu4SiMg	HF (shapes)		3.8-5	0.7	0.2-0.8	0.3-1.2	0.5-1.2		0.2			Ni 0.2			
2014 A92014	Finland	SFS 2595	AlCu4SiMg	ST and Nat aged (plt) 12 mm diam		3.8-5	0.7	0.2-0.8	0.3-1.2	0.5-1.2		0.2			Ni 0.2			
2014 A92014	Finland	SFS 2595	AlCu4SiMg	ST and aged 12 mm diam		3.8-5	0.7	0.2-0.8	0.3-1.2	0.5-1.2		0.2			Ni 0.2			
2014 A92014	France	NF A50411	2014(A-U4SG)	Bar Tub aged (extd bar)	0.1	3.9-5	0.7	0.2-0.8	0.4-1.2	0.5-1.2	0.15	0.25				390	255	10
2014 A92014	France	NF A50411	2014(A-U4SG)	Bar Tub Q/T (tub) 150 mm diam	0.1	3.9-5	0.7	0.2-0.8	0.4-1.2	0.5-1.2	0.15	0.25				440	390	8
2014 A92014	France	NF A50451	2014(A-U4SG)	Sh Plt Strp Ann .4/1.6 mm diam	0.1	3.9-5	0.7	0.2-0.8	0.4-1.2	0.5-1.2	0.15	0.25					140	13
2014 A92014	France	NF A50451	2014(A-U4SG)	Sh Plt Strp Aged .4/1.6 mm diam	0.1	3.9-5	0.7	0.2-0.8	0.4-1.2	0.5-1.2	0.15	0.25				390	240	15
2014 A92014	Germany	DIN 1725 pt1	3.1255/AlCuSiMn		0.1 max	3.9-5	0.7 max	0.2-0.8	0.4-1.2	0.5-1.2	0.15 max	0.25 max			ST 0.2 max; ET 0.05 max; ST=Ti+Zr; ET=.15 total			
2014 A92014	India	IS 5902	24435	Wir Bar ANN CD	0.1	3.9-5	0.7	0.2-0.8	0.4-1.2	0.5-1.2	0.15	0.25				383		
2014 A92014	International	ISO R827	Al-Cu4SiMg	ST and nat aged .75in; 20 mm diam		3.8-5	0.7	0.2-0.8	0.3-1.2	0.5-1.2		0.2			Ni 0.2	353	226	11

UNS numbers and US grades are provided as a means of cross referencing chemically similar alloys. Exchangability is only possible after independent examination of specifications. Tensile properties are minimum or typical . UTS and YS as Mpa, El as %. See Appendix for list of abbreviations used in Descriptions.

Grade UNS #	Country	Specification	Designation	Description	Cr	Cu	Fe	Mg	Mn	Si	Ti	Zn	OE Max	OT Max	Other	UTS	YS	EL
2014 A92014	International	ISO R827	Al-Cu4SiMg	ST and PT .38in; 10 mm diam		3.8-5	0.7	0.2-0.8	0.3-1.2	0.5-1.2		0.2			Ni 0.2	412	363	7
2014 A92014	International	ISO R829	Al-Cu4SiMg	Frg ST and nat aged		3.8-5	0.7	0.2-0.8	0.3-1.2	0.5-1.2		0.2			Ni 0.2	353	206	11
2014 A92014	International	ISO R829	Al-Cu4SiMg	Frg ST and PT		3.8-5	0.7	0.2-0.8	0.3-1.2	0.5-1.2		0.2			Ni 0.2	412	363	6
2014 A92014	International	ISO TR2136	Al-Cu4SiMg	Sh Plt SHT and nat aged		3.8-5	0.7	0.2-0.8	0.3-1.2	0.5-1.2		0.2			Ni 0.2	385	240	14
2014 A92014	International	ISO TR2136	Al-Cu4SiMg	Sh Plt SHT and art aged		3.8-5	0.7	0.2-0.8	0.3-1.2	0.5-1.2		0.2			Ni 0.2	430	380	6
2014 A92014	International	ISO TR2778	Al-Cu4SiMg	Tub SHT and nat aged OBSOLETE Replace by ISO 6363		3.8-5	0.7	0.2-0.8	0.3-1.2	0.5-1.2		0.2			Ni 0.2	380	235	10
2014 A92014	International	ISO TR2778	Al-Cu4SiMg	Tub ST and PT OBSOLETE Replace by ISO 6363		3.8-5	0.7	0.2-0.8	0.3-1.2	0.5-1.2		0.2			Ni 0.2	450	380	6
2014 A92014	International	ISO TR2778	Al-Cu4SiMg	Tub OBSOLETE Replace by ISO 6363		3.8-5	0.7	0.2-0.8	0.3-1.2	0.5-1.2		0.2			Ni 0.2			
2014 A92014	Japan	JIS H4000	2014	Sh Plt	0.1	3.9-5	0.7	0.2-0.8	0.4-1.2	0.5-1.2	0.15	0.25						
2014 A92014	Japan	JIS H4080	2014	Tub	0.1	3.9-5	0.7	0.2-0.8	0.4-1.2	0.5-1.2	0.15	0.25						
2014 A92014	Japan	JIS H4100	2014		0.1	3.9-5	0.7	0.2-0.8	0.4-1.2	0.5-1.2	0.15	0.25						
2014 A92014	Norway	NS 17105	NS17105/AlCu4SiMg	Bar Plt Sh Strp Tub Frg	0.1 max	3.9-5	0.7 max	0.2-0.8	0.4-1.2	0.5-1.2	0.15 max	0.25 max			ET 0.05 max; ST 0.2 max; ET=.15 total, 3T=Ti+Zr,			
2014 A92014	Pan American	COPANT 862	2014		0.1	3.9-5	0.7	0.2-0.8	0.4-1.2	0.5-1.2	0.15	0.25						
2014 A92014	Sweden	SIS 144338	4338-00	Bar HW 10/150 mm diam	0.1	3.9-5	0.7	0.2-0.8	0.4-1.2	0.5-1.2	0.15	0.25			Al 90.24	300	200	8
2014 A92014	Sweden	SIS 144338	4338-02	Plt Sh Strp Bar Tub Ann 5/30 mm diam	0.1	3.9-5	0.7	0.2-0.8	0.4-1.2	0.5-1.2	0.15	0.25			Al 90.24	220	140	14
2014 A92014	Sweden	SIS 144338	4338-04	Plt Sh Strp Bar Tub Bat aged 5/30 mm diam	0.1	3.9-5	0.7	0.2-0.8	0.4-1.2	0.5-1.2	0.15	0.25			Al 90.24	385	240	14
2014 A92014	Sweden	SIS 144338	4338-06	Plt Sh Strp Bar Tub Frg Art aged	0.1	3.9-5	0.7	0.2-0.8	0.4-1.2	0.5-1.2	0.15	0.25			Al 90.24	440	380	
2014 A92014	Sweden	SIS 144338	4338-10	Plt Sh Strp SH 5/10 mm diam	0.1	3.9-5	0.7	0.2-0.8	0.4-1.2	0.5-1.2	0.15	0.25			Al 90.24	280	260	3
2014 A92014	USA	MIL A-12545	2014	Plt FrgExt	0.1 max	3.9-5	0.7 max	0.2-0.8	0.4-1.2	0.5-1.2	0.15 max	0.25 max	0.15					
2014 A92014	USA	MIL A-22771	2014	Plt FrgExt	0.1 max	3.9-5	0.7 max	0.2-0.8	0.4-1.2	0.5-1.2	0.15 max	0.25 max	0.15					
2014 A92014	USA	QQ A-200/2	2014	Plt Frg Ext OBSOLETE	0.1 max	3.9-5	0.7 max	0.2-0.8	0.4-1.2	0.5-1.2	0.15 max	0.25 max	0.15					
2014 A92014	USA	QQ A-225/4	2014	Plt FrgExt	0.1 max	3.9-5	0.7 max	0.2-0.8	0.4-1.2	0.5-1.2	0.15 max	0.25 max	0.15					
2014 A92014	USA	QQ A-367	2014	Plt Frg Ext	0.1 max	3.9-5	0.7 max	0.2-0.8	0.4-1.2	0.5-1.2	0.15 max	0.25 max	0.15					
2014 A92014	USA	SAE J454	2014	Plt FrgExt	0.1 max	3.9-5	0.7 max	0.2-0.8	0.4-1.2	0.5-1.2	0.15 max	0.25 max	0.15					
2014 A92014	USA	AMS 4014	2014	Plt FrgExt	0.1 max	3.9-5	0.7 max	0.2-0.8	0.4-1.2	0.5-1.2	0.15 max	0.25 max	0.15					
2014 A92014	USA	AMS 4028	2014	Plt FrgExt	0.1 max	3.9-5	0.7 max	0.2-0.8	0.4-1.2	0.5-1.2	0.15 max	0.25 max	0.15					
2014 A92014	USA	AMS 4029	2014	Plt FrgExt	0.1 max	3.9-5	0.7 max	0.2-0.8	0.4-1.2	0.5-1.2	0.15 max	0.25 max	0.15					
2014 A92014	USA	AMS 4121	2014	Plt FrgExt	0.1 max	3.9-5	0.7 max	0.2-0.8	0.4-1.2	0.5-1.2	0.15 max	0.25 max	0.15					

UNS numbers and US grades are provided as a means of cross referencing chemically similar alloys. Exchangability is only possible after independent examination of specifications. Tensile properties are minimum or typical . UTS and YS as Mpa, El as %. See Appendix for list of abbreviations used in Descriptions.

Worldwide Guide to Equivalent Nonferrous Metals and Alloys

Grade UNS #	Country	Specification	Designation	Description	Cr	Cu	Fe	Mg	Mn	Si	Ti	Zn	OE Max	OT Max	Other	UTS	YS	EL
2014 A92014	USA	AMS 4133	2014	Plt FrgExt	0.1 max	3.9-5	0.7 max	0.2-0.8	0.4-1.2	0.5-1.2	0.15 max	0.25 max		0.15				
2014 A92014	USA	AMS 4134	2014	Plt FrgExt	0.1 max	3.9-5	0.7 max	0.2-0.8	0.4-1.2	0.5-1.2	0.15 max	0.25 max		0.15				
2014 A92014	USA	AMS 4135	2014	Plt FrgExt	0.1 max	3.9-5	0.7 max	0.2-0.8	0.4-1.2	0.5-1.2	0.15 max	0.25 max		0.15				
2014 A92014	USA	AMS 4153	2014	Plt FrgExt	0.1 max	3.9-5	0.7 max	0.2-0.8	0.4-1.2	0.5-1.2	0.15 max	0.25 max		0.15				
2014 A92014	USA	ASTM B209	2014	Plt FrgExt	0.1 max	3.9-5	0.7 max	0.2-0.8	0.4-1.2	0.5-1.2	0.15 max	0.25 max		0.15				
2014 A92014	USA	ASTM B210	2014	Plt FrgExt	0.1 max	3.9-5	0.7 max	0.2-0.8	0.4-1.2	0.5-1.2	0.15 max	0.25 max		0.15				
2014 A92014	USA	ASTM B211	2014	Plt FrgExt	0.1 max	3.9-5	0.7 max	0.2-0.8	0.4-1.2	0.5-1.2	0.15 max	0.25 max		0.15				
2014 A92014	USA	ASTM B221	2014	Plt FrgExt	0.1 max	3.9-5	0.7 max	0.2-0.8	0.4-1.2	0.5-1.2	0.15 max	0.25 max		0.15				
2014 A92014	USA	ASTM B247	2014	Plt FrgExt	0.1 max	3.9-5	0.7 max	0.2-0.8	0.4-1.2	0.5-1.2	0.15 max	0.25 max		0.15				
2014	USA				0.1 max	3.9-5	0.7 max	0.2-0.8	0.4-1.2	0.5-1.2	0.15 max	0.25 max	0.05	0.15				
	USA		Al Cu4SiMg		0.1 max	3.9-5	0.7 max	0.2-0.8	0.4-1.2	0.5-1.2	0.15 max	0.25 max	0.05	0.15				
2014 A92014	Yugoslavia	JUS CC2.100	AlCu4SiMg			3.8-5	0.7 max	0.2-0.8	0.3-1.2	0.5-1.2		0.2 max			ST 0.3 max; ST = Ti + Zr + Cr;			
2014 A92014	Europe	AECMA prEN2124	Al-P16-T651	Sht ST Quen and Art aged 6/10 mm diam	0.1	3.9-5		0.2-0.8	0.4-1.2	0.5-1.2	0.15	0.25				460	410	8
2014 A92014	USA	ASTM B241	2014	Plt FrgExt	0.1 max	3.9-5		0.2-0.8	0.4-1.2	0.5-1.2	0.15 max	0.25 max		0.15				
2011 A92011	Sweden	MNC 40E	4355	Bar Ext		5.5-5.5									Al 93.7; Bi 0.4; Pb 0.4			
2011 A92011	Australia	AS 1865	2011	Wir Rod Bar Strp, SHT and CW 3/40mm diam		5-6	0.7				0.4		0.3		Bi 0.2-0.6; Pb 0.2-0.6	310	262	8
2011 A92011	Australia	AS 1865	2011	Wir Rod Bar Strp, SHT,CW, and Art aged 3/75mm diam		5-6	0.7				0.4		0.3		Bi 0.2-0.6; Pb 0.2-0.6	358	275	8
2011 A92011	Australia	AS 1866	2011	Bar Rod, SHT and Nat aged		5-6	0.7				0.4		0.3		Bi 0.2-0.6; Pb 0.2-0.6	275	124	14
2011 A92011	Australia	AS 1866	2011	Bar Rod, SHT and Art aged 25.0mm diam		5-6	0.7				0.4		0.3		Bi 0.2-0.6; Pb 0.2-0.6	351	220	8
2011 A92011	Austria	ONORM M3430	AlCuBiPb			5-6	0.7 max				0.4 max		0.3 max		ET 0.05 max; ET=.15 total; Bi 0.2-0.6			
2011 A92011	Belgium	NBN P21-001	2011			5-6	0.7 max				0.4 max		0.3 max		ET 0.05 max; ET = .15 Total; Bi 0.2-0.6; Pb 0.2-0.6			
2011 A92011	Canada	CSA CB60	2011	Bar Rod wir		5-6	0.7 max				0.4 max		0.3 max	0.15	Pb 0.2-0.6			
2011 A92011	Canada	CSA HA.5	0.2011	Bar Rod Wir T8 .125/2.5 in diam		5-6	0.7				0.4		0.3			372	276	10
2011 A92011	Canada	CSA HA.5	0.20101	Bar Rod Wir T3 .125/1.5 in diam		5-6	0.7				0.4		0.3				310	262
2011 A92011	Denmark	DS 3012	2011	SA and CW 3/40 mm diam		5-6	0.7				0.4		0.3		Al 93; Bi 0.2-0.6; Pb 0.2-0.6	310	260	10
2011 A92011	Denmark	DS 3012	2011	CW and Art aged 3/75 mm diam		5-6	0.7				0.4		0.3		Al 93; Bi 0.2-0.6; Pb 0.2-0.6	370	275	10
2011 A92011	France	NF A50411	2011(A-U5PbBi)	Bar CF (drawn Bar)		5-6	0.7				0.4		0.3		Bi 0.2-0.6; Pb 0.2-0.6	310	265	9
2011 A92011	France	NF A50411	2011(A-U5PbBi)	Bar CF (tub) 150 mm diam		5-6	0.7				0.4		0.3		Bi 0.2-0.6; Pb 0.2-0.6	295	235	11

UNS numbers and US grades are provided as a means of cross referencing chemically similar alloys. Exchangability is only possible after independent examination of specifications. Tensile properties are minimum or typical . UTS and YS as Mpa, El as %. See Appendix for list of abbreviations used in Descriptions.

2-74 Wrought Aluminum

Grade UNS #	Country	Specification	Designation	Description	Cr	Cu	Fe	Mg	Mn	Si	Ti	Zn	OE Max	OT Max	Other	UTS	YS	EL
2011 A92011	Germany	DIN 1725 pt1	3.1655/AlCuBiPb			5-6	0.7 max			0.4 max		0.3 max			ET 0.05 max; ET=.15 total; Bi 0.2-0.6; Pb 0.2-0.6			
2011 A92011	Japan	JIS H4040	2011	Bar SHT and CW 6 mm diam		5-6	0.7			0.4		0.3			Bi 0.2-0.6; Pb 0.2-0.6	314	265	
2011 A92011	Japan	JIS H4040	2011	Bar SHT, CW, and art AH 6 mm diam		5-6	0.7			0.4		0.3			Bi 0.2-0.6; Pb 0.2-0.6	363	275	
2011 A92011	Norway	NS 17107	NS17107/AlCu6BiPb	Bar		5-6	0.7 max			0.4 max		0.3 max			ET 0.05 max; ET=.15 total; Bi 0.2-0.6; Pb 0.2-0.6			
2011 A92011	Pan American	COPANT 862	2011			5-6	0.7			0.4		0.3			Bi 0.2-0.6			
2011 A92011	Sweden	SIS 144355	4355-06	Bar Nat aged 10/75 mm diam		5-6	0.7			0.4		0.3			Bi 0.2-0.6; Pb 0.2-0.6	350	230	8
2011 A92011	Sweden	SIS 144355	4355-08	Bar SH and art aged 10/50 mm		5-6	0.7			0.4		0.3			Bi 0.2-0.6; Pb 0.2-0.6	370	275	8
2014 A92014	Sweden	SIS 144355	4355-03	Bar SH and Nat aged 10/50 mm		5-6	0.7			0.4		0.3			Bi 0.2-0.6; Pb 0.2-0.6	310	260	10
2011 A92011	UK	BS FC1	2011	Bar Rod wir		5-6	0.7 max			0.4 max		0.3 max			0.15 Pb 0.2-0.6			
2011 A92011	USA	ASTM B210	2011	Bar Rod wir		5-6	0.7 max			0.4 max		0.3 max			0.15 Pb 0.2-0.6			
2011 A92011	USA	ASTM B211	2011	Bar Rod wir		5-6	0.7 max			0.4 max		0.3 max			0.15 Pb 0.2-0.6			
2011 A92011	USA	QQ A-255/3	2011	Bar Rod Wir OBSOLETE		5-6	0.7 max			0.4 max		0.3 max			0.15 Pb 0.2-0.6			
2011 A92011	USA	SAE J454	2011	Bar Rod wir		5-6	0.7 max			0.4 max		0.3 max			0.15 Pb 0.2-0.6			
2011 A92011	USA					5-6	0.7 max			0.4 max		0.3 max	0.05	0.15	Bi 0.20-0.6; Pb 0.20-0.6			
2111	USA					5-6	0.7 max			0.4 max		0.3 max	0.05	0.15	Bi 0.20-0.6; Sn 0.10-0.50			
	USA		AlCu6BiPb			5-6	0.7 max			0.4 max		0.3 max	0.05	0.15	Bi 0.20-0.6; Pb 0.20-0.6			
2011 A92011	Yugoslavia	JUS CC2.100	AlCu5PbBi			5-6	0.7 max			0.4 max		0.3 max			ST 0.1 max; ST = Mn + Mg; Bi 0.2-0.6; Pb 0.2-0.6			
2011 A92011	Yugoslavia	JUS CC2.100	AlCu6Pb			5-6	0.7 max	0.2 max	0.2 max	0.4 max		0.3 max			0.15 Bi 0.2-0.6; Pb 0.2-0.6			
2011 A92011	International	ISO 2779	Al-Cu6BiPb	SHT and nat aged .125in; 3 mm diam OBSOLETE Replaced by ISO 6362		5-6				0.4		0.3			Bi 0.2-0.6; Pb 0.2-0.6	275	125	16
2011 A92011	International	ISO 2779	Al-Cu6BiPb	TD .125in; 3 mm diam OBSOLETE Replaced by ISO 6362		5-6				0.4		0.3			Bi 0.2-0.6; Pb 0.2-0.6	310	260	10
2011 A92011	International	ISO 2779	Al-Cu6BiPb	SHT and art aged .125in; 3 mm diam OBSOLETE Replaced by ISO 6362		5-6				0.4		0.3			Bi 0.2-0.6; Pb 0.2-0.6	370	275	10
	Japan	JIS Z3232	7N11	Rod SHT and AH	0.3	0.1	0.3	3-4.6	0.2-0.7	0.25	0.2	1-3				294		
7020 A97020	Czech Republic	CSN 424441	AlZn4Mg1		0.1-0.25	0.1 max	0.5 max	1-1.4	0.1-0.5	0.5 max	0.01-0.2	4-5						
7020 A97020	Austria	ONORM M3430	AlZn4.5Mg1		0.1-0.35	0.2 max	0.4 max	1-1.4	0.05-0.5	0.35 max		4-5			ET 0.05 max; ET=.15 total; Zr 0.08-0.2			

UNS numbers and US grades are provided as a means of cross referencing chemically similar alloys. Exchangability is only possible after independent examination of specifications. Tensile properties are minimum or typical . UTS and YS as Mpa, El as %. See Appendix for list of abbreviations used in Descriptions.

Wrought Aluminum 2-75

Worldwide Guide to Equivalent Nonferrous Metals and Alloys

Grade UNS #	Country	Specification	Designation	Description	Cr	Cu	Fe	Mg	Mn	Si	Ti	Zn	OE Max	OT Max	Other	UTS	YS	EL
7020	Europe				0.1-0.35	0.2 max	0.4 max	1-1.4	0.05-0.5	0.35 max		4-5	0.05	0.15	0.08-0.25 Zr+Ti; Zr 0.08-0.20			
	Europe		AlZn4.5Mg1		0.1-0.35	0.2 max	0.4 max	1-1.4	0.05-0.5	0.35 max		4-5	0.05	0.15	0.08-0.25 Zr+Ti; Zr 0.08-0.20			
7020 A97020	France	NF A50411	7020(A-Z5G)	Bar Wir Tub Q/T (extd bar)	0.1-0.35	0.2	0.4	1-1.4	0.05-0.5	0.35		4-5			Zr 0.08-0.2	340	275	10
7020 A97020	France	NF A50411	7020(A-Z5G)	Bar Wir Tub Q/T (wir) 2/12 mm diam	0.1-0.35	0.2	0.4	1-1.4	0.05-0.5	0.35		4-5			Zr 0.08-0.2	340	275	10
7020 A97020	France	NF A50411	7020(A-Z5G)	Bar Wir Tub Q/T (tub) 150 mm diam	0.1-0.35	0.2	0.4	1-1.4	0.05-0.5	0.35		4-5			Zr 0.08-0.2	295	235	11
7020 A97020	Norway	NS 17410	NS17410/AlZN4Mg	Bar Plt Sh Strp Tub Frg	0.1-0.35	0.2 max	0.4 max	1-1.4	0.05-0.5	0.35 max		4-5			ET 0.05 max; ST 0.08-0.25; ET=.15 total, ST=Ti+Zr; Zr 0.08-0.2			
7020 A97020	Sweden	MNC 40E	4425	Bar Sh Tub Ext	0.2			1.2	0.3			4.5			ST 0.2; ST=Ti+Z; Al 93.6			
7020 A97020	Sweden	SIS 144425	4425-00	Bar Tub HW 10/200 mm diam	0.2			1.2	0.3			4.5			Al 93.6; Zr 0.15 max	240	140	10
7020 A97020	Sweden	SIS 144425	4425-02	Plt Sh Strp Ann 5/30 mm diam	0.2			12	0.3			4.5			Al 93.6; Zr 0.15 max	220	150	15
7020 A97020	Sweden	SIS 144425	4425-04	Bar Tub nat aged 10/200 mm diam	0.2			1.2	0.3			4.5			Al 93.6; Zr 0.15 max	275	145	12
7020 A97020	Sweden	SIS 144425	4425-06	Plt Sh Strp Bar Tub Art aged 5/15 mm diam	0.2			1.2	0.3			4.5			Al 93.6; Zr 0.15 max	350	275	10
7020 A97020	Sweden	SIS 144425	4425-07	Bar Tub Art aged 1/25 mm diam	0.2			1.2	0.3			4.5			Al 93.6; Zr 0.15 max	350	290	10
7011	USA				0.05-0.2	0.05 max	0.2 max	1-1.6	0.1-0.3	0.15 max	0.05 max	4-5.5	0.05	0.15				
	UK	BS 1470	NS5	Plt Sh Strp Ann .2/6 mm diam OBSOLETE Replaced by BS EN 515	0.25	0.1	0.5	3.1-3.9	0.5	0.5	0.2	0.2				215	85	12
	UK	BS 1470	NS5	Plt Sh Strp SH to 1/4 hard .2/6 mm diam OBSOLETE Replaced by BS EN 515	0.25	0.1	0.5	3.1-3.9	0.5	0.5	0.2	0.2				245	165	5
	UK	BS 1470	NS5	Plt Sh Strp SH to 1/2 hard .2/6 mm diam OBSOLETE Replaced by BS EN 515	0.25	0.1	0.5	3.1-3.9	0.5	0.5	0.2	0.2				275	225	4
	UK	BS 1471	NT5	Tub Ann 10 mm diam	0.25	0.1	0.5	3.1-3.9	0.5	0.5	0.2	0.2				215	85	16
	UK	BS 1471	NT5	Tub SH to 1/2 hard 10 mm diam	0.25	0.1	0.5	3.1-3.9	0.5	0.5	0.2	0.2				245	200	4
	UK	BS 1472	NF5	As Mfg 150 mm diam Obsolete Replaced by BS EN 586	0.25	0.1	0.5	3.1-3.9	0.5	0.5	0.2	0.2				215	100	16
	UK	BS 1473	NR5	Ann or as Mfg 25 mm diam	0.25	0.1	0.5	3.1-3.9	0.5	0.5	0.2	0.2				215		
	UK	BS 1474	NE5	Bar Tub Ann 150 mm diam	0.25	0.1	0.5	3.1-3.9	0.5	0.5	0.2	0.2				215	85	16

UNS numbers and US grades are provided as a means of cross referencing chemically similar alloys. Exchangability is only possible after independent examination of specifications. Tensile properties are minimum or typical . UTS and YS as Mpa, El as %. See Appendix for list of abbreviations used in Descriptions.

2-76 Wrought Aluminum

Grade UNS #	Country	Specification	Designation	Description	Cr	Cu	Fe	Mg	Mn	Si	Ti	Zn	OE Max	OT Max	Other	UTS	YS	EL
	UK	BS 1474	NE5	Bar Tub As Mfg 150 mm diam	0.25	0.1	0.5	3.1-3.9	0.5	0.5	0.2	0.2				215		14
	UK	BS 1475	NG5	Wir	0.25	0.1	0.5	3.1-3.9	0.5	0.5	0.2	0.2				250		
4543	USA				0.05 max	0.1 max	0.5 max	0.1-0.4	0.05 max	5-7	0.1 max	0.1 max	0.05	0.15				
	International	ISO R209 (1971)	Al-Cu4Mg1			3.8-4.9	0.5 max	1.0-1.8	0.3-1.2	0.5 max					Ti+Zr+Cr 0.3 max; Ni 0.2 max			
2618 A92618	Czech Republic	CSN 424218	AlCu2Mg2Ni			1.9-2.5	0.9-1.3	1.4-1.8	0.2 max	0.35 max	0.02-0.1	0.3 max			Ni 1-1.5			
2618 A92618	Russia	GOST 4784	1141			1.9-2.5	1.1-1.5	1.4-1.8	0.2 max	0.35 max	0.02-0.1	0.3 max			Ni 1-1.5			
2618 A92618	Norway	NS 17102	NS17102/AlCu2MgNi			1.9-2.7	0.9-1.3	0.9-1.2	1.3-1.8	0.1-0.25	0.04-0.1	0.1 max			ET 0.05 max; ET=.15 total; Ni 0.9-1.2			
2618 A92618	USA	AMS 4132	2618	Frg		1.9-2.7	0.9-1.3	1.3-1.8		0.1-0.25	0.04-0.1	0.1 max	0.05	0.05	Ni 0.9-1.2			
2618 A92618	USA	ASTM B247	2618	Frg		1.9-2.7	0.9-1.3	1.3-1.8		0.1-0.25	0.04-0.1	0.1 max	0.05	0.05	Ni 0.9-1.2			
2618 A92618	USA	MIL A-22771	2618	Frg		1.9-2.7	0.9-1.3	1.3-1.8		0.1-0.25	0.04-0.1	0.1 max		0.05	Ni 0.9-1.2			
2618 A92618	USA	QQ A-367	2618	Frg		1.9-2.7	0.9-1.3	1.3-1.8		0.1-0.25	0.04-0.1	0.1 max	0.05	0.05	Ni 0.9-1.2			
2618 A92618	USA	SAE J454	2618	Frg		1.9-2.7	0.9-1.3	1.3-1.8		0.1-0.25	0.04-0.1	0.1 max	0.05	0.05	Ni 0.9-1.2			
2618	USA					1.9-2.7	0.9-1.3	1.3-1.8		0.1-0.25	0.04-0.1	0.1 max	0.05	0.15	Ni 0.9-1.2			
7019	UK				0.2 max	0.2 max	0.45 max	1.5-2.5	0.15-0.5	0.35 max	0.15 max	3.5-4.5	0.05	0.15	Ni 0.1 max; Zr 0.10-0.25			
7008	USA				0.12-0.25	0.05 max	0.1 max	0.7-1.4	0.05 max	0.1 max	0.05 max	4.5-5.5	0.05	0.1				
7108	USA					0.05 max	0.1 max	0.7-1.4	0.05 max	0.1 max	0.05 max	4.5-5.5	0.05	0.15	Zr 0.12-0.25			
5087	Europe				0.05-0.25	0.05 max	0.4 max	4.5-5.2	0.7-1.1	0.25 max	0.15 max	0.25 max	0.15	0.15	Zr 0.10-0.20			
5456A	Germany				0.05-0.25	0.05 max	0.4 max	4.5-5.2	0.7-1.1	0.25 max	0.15 max	0.25 max	0.15	0.15	0.0008 max Be for welding			
5356 A95356	Canada	CSA GM50P	5356		0.05-0.2	0.1 max	0.4 max	4.5-5.5	0.05-0.2	0.25 max	0.06-0.2	0.1 max	0.05	0.15				
5356 A95356	USA	MIL E-16053	5356		0.05-0.2	0.1 max	0.4 max	4.5-5.5	0.05-0.2	0.25 max	0.06-0.2	0.1 max	0.05	0.15				
5356 A95356	USA	QQ R-566	5356		0.05-0.2	0.1 max	0.4 max	4.5-5.5	0.05-0.2	0.25 max	0.06-0.2	0.1 max	0.05	0.15				
5456 A95456	USA	ASTM B209	5456	Bar Rod Wir Plt Sh Tub Ext Shapes	0.05-0.2	0.1 max	0.4 max	4.7-5.5	0.5-1	0.25 max	0.2 max	0.25 max	0.05	0.15				
5456 A95456	USA	ASTM B210	5456	Bar Rod Wir Plt Sh Tub Ext Shapes	0.05-0.2	0.1 max	0.4 max	4.7-5.5	0.5-1	0.25 max	0.2 max	0.25 max	0.05	0.15				
5456 A95456	USA	ASTM B221	5456	Bar Rod Wir Plt Sh Tub Ext Shapes	0.05-0.2	0.1 max	0.4 max	4.7-5.5	0.5-1	0.25 max	0.2 max	0.25 max	0.05	0.15				
5456 A95456	USA	ASTM B241	5456	Bar Rod Wir Plt Sh Tub Ext Shapes	0.05-0.2	0.1 max	0.4 max	4.7-5.5	0.5-1	0.25 max	0.2 max	0.25 max	0.05	0.15				
5456 A95456	USA	MIL A-45225	5456	Bar Rod Wir Plt Sh Tub Ext Shapes	0.05-0.2	0.1 max	0.4 max	4.7-5.5	0.5-1	0.25 max	0.2 max	0.25 max	0.05	0.15				
5456 A95456	USA	MIL A-46027	5456	Bar Rod Wir Plt Sh Tub Ext Shapes	0.05-0.2	0.1 max	0.4 max	4.7-5.5	0.5-1	0.25 max	0.2 max	0.25 max	0.05	0.15				
5456 A95456	USA	MIL A-46083	5456	Bar Rod Wir Plt Sh Tub Ext Shapes	0.05-0.2	0.1 max	0.4 max	4.7-5.5	0.5-1	0.25 max	0.2 max	0.25 max	0.05	0.15				
5456 A95456	USA	QQ A-200/7	5456	Bar Rod Wir Plt Sh Tub Ext Shp OBSOLETE	0.05-0.2	0.1 max	0.4 max	4.7-5.5	0.5-1	0.25 max	0.2 max	0.25 max	0.05	0.15				

UNS numbers and US grades are provided as a means of cross referencing chemically similar alloys. Exchangability is only possible after independent examination of specifications. Tensile properties are minimum or typical . UTS and YS as Mpa, El as %. See Appendix for list of abbreviations used in Descriptions.

Worldwide Guide to Equivalent Nonferrous Metals and Alloys

Grade UNS #	Country	Specification	Designation	Description	Cr	Cu	Fe	Mg	Mn	Si	Ti	Zn	OE Max	OT Max	Other	UTS	YS	EL
5456 A95456	USA	QQ A-250/20	5456	Bar Rod Wir Plt Sh Tub Ext Shapes	0.05-0.2	0.1 max	0.4 max	4.7-5.5	0.5-1	0.25 max	0.2 max	0.25 max	0.05	0.15				
5456 A95456	USA	QQ A-250/9	5456	Bar Rod Wir Plt Sh Tub Ext Shapes	0.05-0.2	0.1 max	0.4 max	4.7-5.5	0.5-1	0.25 max	0.2 max	0.25 max	0.05	0.15				
5456 A95456	USA	SAE J454	5456	Bar Rod Wir Plt Sh Tub Ext Shapes	0.05-0.2	0.1 max	0.4 max	4.7-5.5	0.5-1	0.25 max	0.2 max	0.25 max	0.05	0.15				
A95356	USA	AWS	ER5356	Weld El Rod	0.05-0.2	0.1 max	0.4 max	4.5-5.5	0.05-0.2	0.25 max	0.06-0.2	0.1 max	0.05	0.15	Be 0.0008 max			
A95356	USA	AWS	R5356	Weld El Rod	0.05-0.2	0.1 max	0.4 max	4.5-5.5	0.05-0.2	0.25 max	0.06-0.2	0.1 max	0.05	0.15	Be 0.0008 max			
5456	USA				0.05-0.2	0.1 max	0.4 max	4.7-5.5	0.5-1	0.25 max	0.2 max	0.25 max	0.2	0.15				
	USA		AlMg5Mn1		0.05-0.2	0.1 max	0.4 max	4.7-5.5	0.5-1	0.25 max	0.2 max	0.25 max	0.2	0.15				
5356 A95356	Canada	CSA HA.6	0.5356	Rod Wir	0.05-0.2	0.1		4.5-5.5	0.05-0.2		0.06-0.2	0.1			Be 0.008			
5280	France				0.05-0.25	0.1 max		3.5-4.5	0.2-0.7			1.5-2.8		0.15	0.35 Si+Fe, Be 0.0008 max; Zr 0.05-0.25			
5356 A95356	Japan	JIS Z3232	5356	Rod	0.05-0.2	0.1		4.5-5.5	0.05-0.2		0.06-0.2	0.1			Be 0.001			
5456 A95456	UK	BS 1475	NG61	Wir	0.05-0.2	0.1		5-5.5	0.6-1		0.05-0.2	0.2						
	India	IS 738	HT14	Tub ST and Nat aged 1.60 mm diam		3.5-4.7	0.7	0.4-1.2	0.4-1	0.2-0.7						402	279	8
5456 A95456	Russia	GOST 4784	AMg6				0.4 max	5.8-7	0.05-0.75	0.4 max	0.1-0.2	0.2 max						
5456 A95456	Russia		45Mg2					4.8-6.5	0.4-1		0.03-0.1							
5456 A95456	Russia		AMg61N					5.8-6.8	0.2-0.4						Zr 0.1-0.2			
7005 A97005	Russia		1910		4.3 max	0.1 max	0.3 max	1.5 max	0.3 max	0.25 max					Zr 0.17 max			
7005 A97005	Denmark	DS 3012	7005	SA	0.06-0.2	0.1	0.4	1-1.8	0.2-0.7	0.35	0.01-0.06	4-5			Zr 0.08-0.2	275	145	12
7005 A97005	Denmark	DS 3012	7005	Art aged	0.06-0.2	0.1	0.4	1-1.8	0.2-0.7	0.35	0.01-0.06	4-5			Zr 0.08-0.2	335	275	10
7005 A97005	Pan American	COPANT 862	7005		0.06-0.2	0.1	0.4	1-1.8	0.2-0.7	0.35	0.01-0.06	4-5						
7005	USA				0.06-0.2	0.1 max	0.4 max	1-1.8	0.2-0.7	0.35 max	0.01-0.06	4-5	0.05	0.15	Zr 0.08-0.20			
	USA		AlZn4.5Mg1.5 Mn		0.06-0.2	0.1 max	0.4 max	1-1.8	0.2-0.7	0.35 max	0.01-0.06	4-5	0.05	0.15	Zr 0.08-0.20			
7005 A97005	Russia		ATsM			0.25 max	0.3 max	1.8 max	0.6 max	0.25 max		4.5 max			Zr 0.17 max			
7005 A97005	Germany	DIN 1725 Part 1	3.4345/AlZnMgCu0.5		0.10-0.30	0.50-1.0	0.50 max	2.6-3.7	0.10-0.40	0.5 max	0.10 max	4.3-5.2			ST 0.2 max; ET 0.05 max; ST=Ti+Zr; ET=.15 total			
A95556	USA	AWS	ER5556	Weld El Rod	0.05-0.2	0.1 max	0.4 max	4.7-5.5	0.5-1	0.25 max	0.05-0.2	0.25 max	0.05	0.15	Be 0.0008 max			
A95556	USA	AWS	R5556	Weld El Rod	0.05-0.2	0.1 max	0.4 max	4.7-5.5	0.5-1	0.25 max	0.05-0.2	0.25 max	0.05	0.15	Be 0.0008 max			
5556	USA				0.05-0.2	0.1 max	0.4 max	4.7-5.5	0.5-1	0.25 max	0.05-0.2	0.25 max	0.2	0.15	0.0008 max Be for welding; OE min 0.05			
5556 A95556	Canada	CSA HA.6	0.5556	Rod Wir	0.05-0.2	0.1		4.7-5.5	0.5-1		0.05-0.2	0.25			Be 0.008			
5556 A95556	Japan	JIS Z3232	5556	Rod Ann	0.05-0.2	0.1		4.7-5.5	0.5-1		0.05-0.2	0.25				275		
5556 A95556	UK	BS 1475	5556A(N61)			0.1 max		5.1-5.5	0.6-1			0.2 max						

UNS numbers and US grades are provided as a means of cross referencing chemically similar alloys. Exchangability is only possible after independent examination of specifications. Tensile properties are minimum or typical. UTS and YS as Mpa, El as %. See Appendix for list of abbreviations used in Descriptions.

2-78 Wrought Aluminum

Grade UNS #	Country	Specification	Designation	Description	Cr	Cu	Fe	Mg	Mn	Si	Ti	Zn	OE Max	OT Max	Other	UTS	YS	EL
7016	USA					0.45-1	0.12 max	0.8-1.4	0.03 max	0.1 max	0.03 max	4-5	0.03	0.1	V 0.05 max			
2014 A92014	Europe	AECMA prEN2125	Al-P16-T6151	Sht ST Quen and Art aged 6/10 mm diam	0.1	3.9-5		0.2-0.8	0.4-1.2	0.5-1.2		0.25				460	410	8
2024 A92024	Czech Republic	CSN 424203	AlCu4Mg1			3.8-1.9	0.5 max	1.2-1.8	0.3-0.9	0.5 max		0.3 max			Ni 0.1 max			
2324	USA				0.1 max	3.8-4.4	0.12 max	1.2-1.8	0.3-0.9	0.1 max	0.15 max	0.25 max	0.05	0.15				
2424	USA					3.8-4.4	0.12 max	1.2-1.6	0.3-0.6	0.1 max	0.1 max	0.2 max	0.05	0.15				
2224	USA				0.1 max	3.8-4.4	0.15 max	1.2-1.8	0.3-0.9	0.12 max	0.15 max	0.25 max	0.05	0.15				
2024 A92024	Russia	GOST 4724	D16P			3.8-4.5	0.5 max	1.2-1.6	0.3-0.7	0.5 max		0.1 max			ET 0.1 max; ET=1.2 total			
2024 A92024	Europe	AECMA prEN2090	Al-P13Pl-T3	Sht ST Quen and Art aged .4/.8 mm diam	0.1	3.8-4.9		1.2-1.8	0.3-0.9	0.05	0.15	0.25			Ni 0.1	405	270	12
2124 A92124	USA	AMS 4101	2124	Plt	0.1 max	3.8-4.9	0.3 max	1.2-1.8	0.3-0.9	0.2 max	0.15 max	0.25 max		0.005				
2124 A92124	USA	ASTM B209	2124	Plt	0.1 max	3.8-4.9	0.3 max	1.2-1.8	0.3-0.9	0.2 max	0.15 max	0.25 max		0.005				
2124 A92124	USA	QQ A-250/29	2124	Plt	0.1 max	3.8-4.9	0.3 max	1.2-1.8	0.3-0.9	0.2 max	0.15 max	0.25 max		0.005				
2124	USA				0.1 max	3.8-4.9	0.3 max	1.2-1.8	0.3-0.9	0.2 max	0.15 max	0.25 max	0.05	0.15				
2024 A92024	Australia	AS 1734	Alclad2024	sh Plt	0.1	3.8-4.9	0.5	1.2-1.8	0.9	0.5	0.15	0.25						
2024 A92024	Austria	ONORM M3430	AlCuMg2		0.1 max	3.8-4.9	0.5 max	1.2-1.8	0.3-0.9	0.5 max	0.15 max	0.25 max			ET 0.05 max; ST 0.2 max; ST=Ti=7r; ET=.15 total;			
2024 A92024	Belgium	NBN P21-001	2024		0.1 max	3.8-4.9	0.5 max	1.2-1.8	0.3-0.9	0.5 max	0.15 max	0.25 max			ET 0.05 max; ST 0.2 max; ST = Zr + Ti; ET = .15 Total			
2024 A92024	Canada	CSA HA.4	0.2024	Plt Sh O .010/.032 in diam	0.1	3.8-4.9	0.5	1.2-1.8	0.3-0.9	0.5	0.15	0.25				221	97	12
2024 A92024	Canada	CSA HA.4	0.2024	Plt Sh T3 (sheet) .010/.020 in	0.1	3.8-4.9	0.5	1.2-1.8	0.3-0.9	0.5	0.15	0.25				441	290	12
2024 A92024	Canada	CSA HA.4	0.2024	Plt Sh T42 .010/.02 in diam	0.1	3.8-4.9	0.5	1.2-1.8	0.3-0.9	0.5	0.15	0.25				427	262	12
2024 A92024	Canada	CSA HA.4	.2024Alclad	Plt Sh O .010/.032 in diam	0.1	3.8-4.9	0.5	1.2-1.8	0.3-0.9	0.5	0.15	0.25				207	97	12
2024 A92024	Canada	CSA HA.4	.2024Alclad	Plt Sh T3 (sht) .010/.020 in diam	0.1	3.8-4.9	0.5	1.2-1.8	0.3-0.9	0.5	0.15	0.25				407	269	12
2024 A92024	Canada	CSA HA.4	.2024Alclad	Plt Sh T42 .010/.020 in diam	0.1	3.8-4.9	0.5	1.2-1.8	0.3-0.9	0.5	0.15	0.25				393	234	12
2024 A92024	Canada	CSA HA.5	0.2024	Bar Rod Wir Drawn and O 2.5 in diam	0.1	3.8-4.9	0.5	1.2-1.8	0.3-0.9	0.5	0.15	0.25				241		16
2024 A92024	Canada	CSA HA.5	0.2024	Bar Rod Wir Drawn and T4 .499 in diam	0.1	3.8-4.9	0.5	1.2-1.8	0.3-0.9	0.5	0.15	0.25				427	310	10
2024 A92024	Canada	CSA HA.5	0.2024	Bar Rod Wir Extd and T3510 .250/.749 in	0.1	3.8-4.9	0.5	1.2-1.8	0.3-0.9	0.5	0.15	0.25				414	303	12
2024 A92024	Canada	CSA HA.6	0.2024	Rod Wir H13 .062/1 in diam	0.1	3.8-4.9	0.5	1.2-1.8	0.3-0.9	0.5	0.15	0.25				221		
2024 A92024	Canada	CSA HA.7	0.2024	Tub Pip Drawn and O	0.1	3.8-4.9	0.5	1.2-1.8	0.3-0.9	0.5	0.15	0.25				221	103	

UNS numbers and US grades are provided as a means of cross referencing chemically similar alloys. Exchangability is only possible after independent examination of specifications. Tensile properties are minimum or typical . UTS and YS as Mpa, El as %. See Appendix for list of abbreviations used in Descriptions.

Worldwide Guide to Equivalent Nonferrous Metals and Alloys

Grade UNS #	Country	Specification	Designation	Description	Cr	Cu	Fe	Mg	Mn	Si	Ti	Zn	OE Max	OT Max	Other	UTS	YS	EL
2024 A92024	Canada	CSA HA.7	0.2024	Tub Pip Drawn and T3 .015/.024 in diam	0.1	3.8-4.9	0.5	1.2-1.8	0.3-0.9	0.5	0.15	0.25				441	290	19
2024 A92024	Canada	CSA HA.7	0.2024	Tub Pip Extruded and O	0.1	3.8-4.9	0.5	1.2-1.8	0.3-0.9	0.5	0.15	0.25				241	19	12
2024 A92024	Czech Republic	CSN 424253	AlCu4Mg1 pl			3.8-4.9	0.5 max	1.2-1.8	0.3-0.9	0.5 max		0.3 max			Ni 0.1 max			
2024 A92024	France	AIR 9051-A	A-U4G1(2024-F)	Sh Bar rolled (sht) 12 mm diam	0.1	3.8-4.9	0.5	1.2-1.8	0.3-0.9	0.5		0.25				420	310	12
2024 A92024	France	AIR 9051-A	A-U4G1(2024-F)	Sh Bar ST (bar) 250 mm diam	0.1	3.8-4.9	0.5	1.2-1.8	0.3-0.9	0.5		0.25				450	290	9
2024 A92024	France	AIR 9051-A	A-U4G1(2024-F)	Sh Bar Water Q/T (bil and plt) 500 mm diam	0.1	3.8-4.9	0.5	1.2-1.8	0.3-0.9	0.5		0.25				410	260	12
2024 A92024	France	NF A50411	2024(A-U4G1)	Bar Wir Aged (extd bar)	0.1	3.8-4.9	0.5	1.2-1.8	0.3-0.9	0.5	0.15	0.25				450	310	10
2024 A92024	France	NF A50411	2024(A-U4G1)	Bar Wir Aged (wir) 2/6 mm diam	0.1	3.8-4.9	0.5	1.2-1.8	0.3-0.9	0.5	0.15	0.25				430	280	15
2024 A92024	France	NF A50411	2024(A-U4G1)	Bar Wir Aged (shapes)	0.1	3.8-4.9	0.5	1.2-1.8	0.3-0.9	0.5	0.15	0.25				430	310	12
2024 A92024	France	NF A50451	2024(A-U4G1)	Sh Plt Ann .4/1.6 mm diam		3.8-4.9	0.5	1.2-1.8	0.3-0.9	0.5	0.15	0.25					140	13
2024 A92024	France	NF A50451	2024(A-U4G1)	Sh Plt Aged .4/1.6 mm diam		3.8-4.9	0.5	1.2-1.8	0.3-0.9	0.5	0.15	0.25				430	270	14
2024 A92024	France	NF A50506	2024(A-U4G1)	CR .4/.8 mm diam	0.1	3.8-4.9	0.5	1.2-1.8	0.3-0.9	0.5	0.15	0.25				390	250	14
2024 A92024	Germany	DIN 1725 pt1	3.1355/AlCuMg2		0.1 max	3.8-4.9	0.5 max	1.2-1.8	0.3-0.9	0.5 max	0.15 max	0.25 max			ST 0.2 max; ET 0.05 max; ST=Ti+Zr; ET=.15 total;	427		
2024 A92024	India	IS 5902	24530	Wir Bar ANN CD	0.1	3.8-4.9	0.5	1.2-1.8	0.3-0.9	0.5		0.25				427		
2024 A92024	International	ISO R827	Al-Cu4Mg1	ST straightened and nat aged 0.25 in 6 mm diam		3.8-4.9	0.5	1-1.8	0.3-1.2	0.5		0.2			Ni 0.2	392	284	12
2024 A92024	International	ISO R829	Al-Cu4Mg1	Frg ST and Nat aged		3.8-4.9	0.5	1-1.8	0.3-1.2	0.5		0.2			Ni 0.2	392	265	11
2024 A92024	International	ISO TR2778	Al-Cu4Mg1	Tub SHT and Nat aged OBSOLETE Replace by ISO 6363		3.8-4.9	0.5	1-1.8	0.3-1.2	0.5		0.2			Ni 0.2	430	285	10
2024 A92024	Japan	JIS H4000	2024	Sh Plt Ann 1.5 mm diam	0.1	3.8-4.9	0.5	1.2-1.8	0.3-0.9	0.5		0.25				216	98	12
2024 A92024	Japan	JIS H4000	2024	Sh Plt SHT and AH 1.5 mm diam	0.1	3.8-4.9	0.5	1.2-1.8	0.3-0.9	0.5		0.25				431	275	15
2024 A92024	Japan	JIS H4000	2024	Sh Plt SHT CW 1.5 mm diam	0.1	3.8-4.9	0.5	1.2-1.8	0.3-0.9	0.5		0.25				441	294	15
2024 A92024	Japan	JIS H4040	2024	Bar Ann	0.1	3.8-4.9	0.5	1.2-1.8	0.3-0.9	0.5		0.25				245	127	12
2024 A92024	Japan	JIS H4040	2024	Bar SHT and AH 7 mm diam	0.1	3.8-4.9	0.5	1.2-1.8	0.3-0.9	0.5		0.25				392	294	12
2024 A92024	Japan	JIS H4080	2024	Tub Ann	0.1	3.8-4.9	0.5	1.2-1.8	0.3-0.9	0.5		0.25				245	127	12
2024 A92024	Japan	JIS H4080	2024	Tub SHT and AH 7 mm diam	0.1	3.8-4.9	0.5	1.2-1.8	0.3-0.9	0.5		0.25				392	294	10
2024 A92024	Japan	JIS H4100	2024	Ann	0.1	3.8-4.9	0.5	1.2-1.8	0.3-0.9	0.5		0.25				245	127	12
2024 A92024	Japan	JIS H4100	2024	SHT and AH 7 mm diam	0.1	3.8-4.9	0.5	1.2-1.8	0.3-0.9	0.5		0.25				392	294	12
2024 A92024	Japan	JIS H4120	2024	Wir SHT and AH 2/25 mm diam	0.1	3.8-4.9	0.5	1.2-1.8	0.3-0.9	0.5		0.25				265	245	10

UNS numbers and US grades are provided as a means of cross referencing chemically similar alloys. Exchangability is only possible after independent examination of specifications. Tensile properties are minimum or typical . UTS and YS as Mpa, El as %. See Appendix for list of abbreviations used in Descriptions.

2-80 Wrought Aluminum

Grade UNS #	Country	Specification	Designation	Description	Cr	Cu	Fe	Mg	Mn	Si	Ti	Zn	OE Max	OT Max	Other	UTS	YS	EL
2024 A92024	Norway	NS 17104	NS17104/AlCu4Mg1	Bar Plt Sh Strp Tub Frg	0.1 max	3.8-4.9	0.5 max	1.2-1.8	0.3-0.9	0.5 max	0.05 max	0.25 max			ET 0.05 max; ST 0.25 max; ET=.15 total, ST=Ti+Zr;			
2024 A92024	Russia		D16AVTV			3.8-4.9	0.5 max	1.2-1.8	0.3-0.9	0.5 max								
2024 A92024	South Africa	SABS 712	Al-Cu4Mgl		0.1	3.8-4.9	0.5	1.2-1.8	0.3-0.9	0.5		0.25						
2024 A92024	Yugoslavia	JUS CC2.100	AlCu5Mg2		0.25 max	3.8-4.9	0.5 max	1.2-1.8	0.3-0.9	0.5 max	0.15 max	0.25 max			ET 0.05 max; ET = .15 Total;			
2024 A92024	Australia	AS 1734	2024	Sh Plt, Ann 0.25/6.0 mm thk	0.1	3.8-4.9	0.7	1.2-1.8	0.3-0.9	0.5	0.15	0.25					97	12
2024 A92024	Europe	AECMA prEN2091	Al-P13PI-T4	Sht ST Quen and Art aged .4/.8 mm diam	0.1	3.8-4.9		1.2-1.8	0.3-0.9	0.5	0.15	0.25			Ni 0.1	390	235	12
2024	USA				0.1 max	3.8-4.9	0.5 max	1.2-1.8	0.3-0.9	0.50 max	0.15 max	0.25 max	0.05	0.15				
	USA		AlCu4Mg1		0.1 max	3.8-4.9	0.5 max	1.2-1.8	0.3-0.9	0.50 max	0.15 max	0.25 max	0.05	0.15				
2024 A92024	Czech Republic	CSN 424250	AlCu4Mg1Mn		0.2 max	3-5		0.4-2	0.3-1.5	1 max		0.5 max			Fe+Ti 0.80; Ni 0.3 max; Pb 0.1 max			
2020 A92020	Russia		1230			4.9-5.8	0.3 max	0.05 max	0.4-0.8	0.3 max	0.15 max	0.1 max						
2024 A92020	Russia		VAD23			4.9-5.8	0.3 max	0.05 max	0.4-0.8	0.3 max	0.15 max	0.1 max						
	UK		4018		0.2 max	0.2 max	0.45 max	0.7-1.5	0.15-0.5	0.35 max	0.15 max	4.5-5.5	0.05	0.15	Ni 0.1 max; Zr 0.10-0.25			
	UK	BS 1475	NG21	Wir		0.1	0.6		0.90 max4	4.5-6		0.2						
5180	USA				0.1 max	0.1 max		3.5-4.5	0.2-0.7		0.06-0.2	1.7-2.8	0.2	0.15	0.35 Si+Fe; 0.0008 max Be weld; OE min 0.06; Zr .08-0.25			
7108A	Norway				0.04 max	0.05 max	0.3 max	0.7-1.5	0.05 max	0.2 max	0.03 max	4.8-5.8	0.05	0.15	Ga 0.03 max; Zr 0.15-0.25			
	Pan American	COPANT 862	X7052		0.05-0.35	0.1	0.4	1.5-2.5	0.1-0.6	0.3	0.1	3-5						
7003	JAPAN				0.2 max	0.2 max	0.35 max	0.5-1	0.3 max	0.3 max	0.2 max	5-6.5	0.05	0.15	Zr 0.05-0.25			
7116	USA					0.5-1.1	0.3 max	0.8-1.4	0.05 max	0.15 max	0.05 max	4.2-5.2	0.05	0.15	Ga 0.03 max; V 0.05 max			
2001	UK					5.5-6.5	0.2 max	0.5 max	0.1 max	0.2 max	0.05 max	0.1 max	0.05	0.15	Bi 0.20; Pb 1.0-2.0			
2519	USA					5.3-6.4	0.3 max	0.05-0.4	0.1-0.5	0.25 max	0.02-0.1	0.1 max	0.05	0.15	0.40 max Si+Fe; V 0.05-0.15; Zr 0.10-0.25			
	UK	BS 1472	HF16	SHT nat Aged 200 mm diam Obsolete Replaced by BS EN 586		1.8-2.7	0.9-1.4	1.2-1.8	0.2	0.25	0.2	0.2			Ni 0.8-1.4	430	340	5
	South Africa	SABS 712	Al-Cu4MgSi		0.1	3.9-4.9	0.7	0.4-1	0.3-0.8	0.005 max		0.25			ST 0.2; ST=Ti+Zr			
2001	France				0.1 max	5.2-6	0.2 max	0.2-0.45	0.15-0.5	0.2 max	0.2 max	0.1 max	0.05	0.15	Ni 0.05 max; Pb 0.003 max			
5556A	UK				0.05-0.2	0.1 max	0.4 max	5-5.5	0.6-1	0.25 max	0.05-0.2	0.2 max	0.2	0.15	0.0008 max Be for welding; OE min 0.05			
2018	USA				0.1 max	3.5-4.5	1 max	0.45-0.9	0.2 max	0.9 max		0.25 max	0.05	0.15	Ni 1.7-2.3			
	France	NF A50451	7020(A-Z5G)	Sh Plt Aged .4/12 mm diam	0.35	0.2	0.4	0.9-1.5	0.05-0.5	0.35		3.7-5			Zr 0.08-0.2	320	210	14
	France	NF A50451	7020(A-Z5G)	Sh Plt Q/T .4/25 mm diam	0.35	0.2	0.4	0.9-1.5	0.05-0.5	0.35		3.7-5			Zr 0.08-0.2	350	280	10
	France	NF A50501	7020(A-Z5G)	Tub As manufactured	0.35	0.2	0.4	0.9-1.5	0.05-0.5	0.35		3.7-5			Zr 0.08-0.2	340	275	10

UNS numbers and US grades are provided as a means of cross referencing chemically similar alloys. Exchangability is only possible after independent examination of specifications. Tensile properties are minimum or typical . UTS and YS as Mpa, El as %. See Appendix for list of abbreviations used in Descriptions.

Worldwide Guide to Equivalent Nonferrous Metals and Alloys

Grade UNS #	Country	Specification	Designation	Description	Cr	Cu	Fe	Mg	Mn	Si	Ti	Zn	OE Max	OT Max	Other	UTS	YS	EL
	Denmark	DS 3012	5083	Sft (rolled and drawn products) 1.3/25 mm diam	0.05-0.25	0.1	0.4	4-4.9	0.4-1	0.4	0.15	0.25				275	125	16
	Denmark	DS 3012	5083	1/2 hard (rolled and drawn products) 1.3/6 mm diam	0.05-0.25	0.1	0.4	4-4.9	0.4-1	0.4	0.15	0.25				345	270	6
	Denmark	DS 3012	5083	As manufactured (extd products)	0.05-0.25	0.1	0.4	4-4.9	0.4-1	0.4	0.15	0.25				345	270	6
	France	AIR 9051-A	A7-U4SG	Sh Bar Rolled (sht) 12 mm diam	0.1	3.9-5	0.35	0.2-0.8	0.4-1.2	0.5-1.2	0.15	0.25				460	410	8
	France	AIR 9051-A	A7-U4SG	Sh Bar ST (bar) 250 mm diam	0.1	3.9-5	0.35	0.2-0.8	0.4-1.2	0.5-1.2	0.15	0.25				450	380	6
	France	AIR 9051-A	A7-U4SG	Sh BarWater Q/T (bil and plt) 500 mm diam	0.1	3.9-5	0.35	0.2-0.8	0.4-1.2	0.5-1.2	0.15	0.25				450	390	8
	UK	BS 1472	NF8	As Mfg 150 mm diam Obsolete Replaced by BS EN 586	0.25	0.1	0.4	4-4.9	0.5-1	0.4	0.2	0.2				280	130	12
7039	USA				0.15-0.25	0.1 max	0.4 max	2.3-3.3	0.1-0.4	0.3 max	0.1 max	3.5-4.5	0.05	0.15				
	France	NF A50411	7051(A-Z3G2)	Wir Aged (wir) 2/12 mm diam	0.05-0.25	0.15	0.45	1.7-2.5	0.1-0.45	0.35	0.15	3-4				340	220	15
	France	NF A50411	7051(A-Z3G2)	Wir Aged (shapes)	0.05-0.25	0.15	0.45	1.7-2.5	0.1-0.45	0.35	0.15	3-4				320	200	12
A33550	USA	AWS		Weld El Rod		1-1.5	0.2 max	0.4-0.6	0.1 max	4.5-5.5	0.2 max	0.1 max	0.05	0.15				
	Japan	JIS H4140	2N01	Die SHT and Art AH 100 mm diam		1.5-2.5	1.5	1.2-1.8	0.2	0.5-1.3	0.2	0.2			Ni 0.6-1.4	373	294	6
A94009	USA	AWS	ER4009	Weld El Rod		1-1.5	0.2 max	0.45-0.6	0.1 max	4.5-5.5	0.2 max	0.1 max	0.05	0.15	Be 0.0008 max			
A94009	USA	AWS	R4009	Weld El Rod		1-1.5	0.2 max	0.45-0.6	0.1 max	4.5-5.5	0.2 max	0.1 max	0.05	0.15	Be 0.0008 max			
4009	USA					1-1.5	0.2 max	0.45-0.6	0.1 max	4.5-5.5	0.2 max	0.1 max	0.05	0.15	0.0008 max Be for welding			
	South Africa	SABS 712	Al-Cu4SiMg		0.1	3.9-4.9	0.7	0.2-0.8	0.4-1.2	0.5-1.2		0.25			ST 0.2; ; ST=Ti+Zr			
7015	Spain				0.15 max	0.06-0.15	0.3 max	1.3-2.1	0.1 max	0.2 max	0.1 max	4.6-5.2	0.05	0.15	Zr 0.10-0.20			
7030	Norway				0.04 max	0.2-0.4	0.3 max	1-1.5	0.05 max	0.2 max	0.03 max	4.8-5.9	0.05	0.15	Ga 0.03 max; Zr 0.03			
7229	USA					0.5-0.9	0.08 max	1.3-21	0.03 max	0.06 max	0.05 max	4.2-5.2	0.03	0.1	V 0.05 max			
7029	USA					0.5-0.9	0.12 max	1.3-2	0.03 max	0.1 max	0.05 max	4.2-5.2	0.03	0.1	V 0.05 max			
7129	USA				0.1 max	0.5-0.9	0.3 max	1.3-2	0.1 max	0.15 max	0.05 max	4.2-5.2	0.05	0.15	Ga 0.03 max; V 0.05 max			
	France	AIR 9051-A	A-UZGN	Sh Bar Rolled (sht) 40 mm diam		1.8-2.7	1.4	1.2-1.8	0.2	0.25	0.2	0.15			Ni 0.8-1.4	420	360	5
	France	AIR 9051-A	A-UZGN	Sh Bar ST (bar) 250 mm diam		1.8-2.7	1.4	1.2-1.8	0.2	0.25	0.2	0.15			Ni 0.8-1.4	410	340	6
	France	AIR 9051-A	A-UZGN	Sh Bar Water Q/T (bil and plt) 500 mm diam		1.8-2.7	1.4	1.2-1.8	0.2	0.25	0.2	0.15			Ni 0.8-1.4	410	340	6
	India	IS 733	24345	Bar Rod ST and Nat aged 10 mm diam	0.3	3.8-5	0.7	0.2-0.8	0.3-1.2	0.5-1.2	0.3	0.2				375	225	10
	India	IS 733	24345	Bar Rod Sht and Pht 10 mm diam	0.3	3.8-5	0.7	0.2-0.8	0.3-1.2	0.5-1.2	0.3	0.2				430	375	6

UNS numbers and US grades are provided as a means of cross referencing chemically similar alloys. Exchangability is only possible after independent examination of specifications. Tensile properties are minimum or typical . UTS and YS as Mpa, El as %. See Appendix for list of abbreviations used in Descriptions.

Grade UNS #	Country	Specification	Designation	Description	Cr	Cu	Fe	Mg	Mn	Si	Ti	Zn	OE Max	OT Max	Other	UTS	YS	EL
	India	IS 736	24345Alclad	Plt SHT and Nat aged 12 mm diam	0.3	3.8-5	0.7	0.2-0.8	0.3-1.2	0.5-1.2	0.3	0.2				375	225	12
	India	IS 736	24345Alclad	Plt SHT and PHT 7 mm diam	0.3	3.8-5	0.7	0.2-0.8	0.3-1.2	0.5-1.2	0.3	0.2				420	310	7
7017	UK				0.35 max	0.2 max	0.45 max	2-3	0.05-0.5	0.35 max	0.15 max	4-5.2	0.05	0.15	0.15 Mn+Cr; Ni 0.1 max; Zr 0.10-0.25			
2419	USA					5.8-6.8	0.18 max	0.02 max	0.2-0.4	0.15 max	0.02-0.1	0.1 max	0.05	0.15	V 0.05-0.15; Zr 0.10-0.25			
2021	UK					5.8-6.8	0.3 max	0.02 max	0.2-0.4	0.2 max	0.02-0.1	0.1 max	0.05	0.15	Cd 0.05-0.20; Sn 0.03-0.08; V 0.05-0.15; Zr 0.10-0.25			
2219	USA					5.8-6.8	0.3 max	0.02 max	0.2-0.4	0.2 max	0.02-0.1	0.1 max	0.05	0.15	V 0.05-0.15; Zr 0.10-0.25			
	USA		AlCu6Mn			5.8-6.8	0.3 max	0.02 max	0.2-0.4	0.2 max	0.02-0.1	0.1 max	0.05	0.15	V 0.05-0.15; Zr 0.10-0.25			
	Sweden	SIS 144163	4163-00			0.1	0.5	4-6	0.5	0.5-1.5	0.2	0.2			Ni 0.05; Pb 0.05; Sn 0.05 max			
	India	IS 733	43000	Bar Rod As Mfg 10 mm diam		0.1	0.6	0.2	0.5	4.5-6		0.2				110		18
	Japan	JIS H4000	7N01	Sh Plt Strp Ann 3 mm diam	0.3	0.25	0.4	1-2.2	0.2-0.9	0.3	0.2	3.8-5				245	147	12
	Japan	JIS H4000	7N01	Sh Plt Strp SHT and AH 3 mm diam	0.3	0.25	0.4	1-2.2	0.2-0.9	0.3	0.2	3.8-5				314	196	1
	Japan	JIS H4000	7N01	Sh Plt Strp SHT and Art AH 3 mm diam	0.3	0.25	0.4	1-2.2	0.2-0.9	0.3	0.2	3.8-5				333	275	10
	Japan	JIS H4040	7N01	Bar Ann	0.3	0.25	0.4	1-2.2	0.2-0.9	0.3	0.2	3.8-5				245	147	12
	Japan	JIS H4040	7N01	Bar SHT and AH	0.3	0.25	0.4	1-2.2	0.2-0.9	0.3	0.2	3.8-5				314	196	11
	Japan	JIS H4040	7N01	Bar SHT and Art AH	0.3	0.25	0.4	1-2.2	0.2-0.9	0.3	0.2	3.8-5				33	275	10
	Japan	JIS H4080	7N01	Tub Ann 1.6/1.2 mm diam	0.3	0.25	0.4	1-2.2	0.2-0.9	0.3	0.2	3.8-5				245	147	12
	Japan	JIS H4080	7N01	Tub SHT and AH 1.6/1.2 mm diam	0.3	0.25	0.4	1-2.2	0.2-0.9	0.3	0.2	3.8-5				314	196	11
	Japan	JIS H4080	7N01	Tub SHT and Art AH 6 mm	0.3	0.25	0.4	1-2.2	0.2-0.9	0.3	0.2	3.8-5				324	235	10
	Japan	JIS H4100	7N01	Ann	0.3	0.25	0.4	1-2.2	0.2-0.9	0.3	0.2	3.8-5				245	147	12
	Japan	JIS H4100	7N01	SHT and AH	0.3	0.25	0.4	1-2.2	0.2-0.9	0.3	0.2	3.8-5				314	196	11
	Japan	JIS H4100	7N01	Quick Cooled and ARt aged	0.3	0.25	0.4	1-2.2	0.2-0.9	0.3	0.2	3.8-5				324	245	10
2319 A92319	USA	MIL E-16053	2319	Wir		5.8-6.8	0.3 max	0.02 max	0.2-0.4	0.2 max	0.1-0.2	0.1 max	0.05	0.15	V 0.05-0.15			
2319	USA					5.8-6.8	0.3 max	0.02 max	0.2-0.4	0.2 max	0.1-0.2	0.1 max	0.05	0.15	0.0008 max Be for welding; V 0.05-0.15; Zr 0.10-0.25			
	France	AIR 9050	AG7			0.1 max	0.4 max	6-8	0.2-1	0.3 max		0.1 max			ST 0.2 max; ST=Zr+Ti;			
	France	NF A-G7	3350			0.1 max	0.4 max	6-8	0.2-1	0.3 max		0.1 max			ST 0.2 max; ST=+Zr+Ti;			
7021	USA				0.05 max	0.25 max	0.4 max	1.2-1.8	0.1 max	0.25 max	0.1 max	5-6	0.05	0.15	Sr 0.08-0.18			
7277	USA				0.18-0.35	0.8-1.7	0.7 max	1.7-2.3		0.5 max	0.1 max	3.7-4.3	0.05	0.15				
	Pan American	COPANT 862	2130	OBSOLETE		3.5-5	0.8	0.2-1	0.1-0.8	0.1-0.8	0.2	0.3			Bi 1-2			
	South Africa	SABS 903	SABS903	Sh As Mfg 1.12/1.28 mm		0.2	0.7	3-3	1.5	0.6		0.25				175		

UNS numbers and US grades are provided as a means of cross referencing chemically similar alloys. Exchangability is only possible after independent examination of specifications. Tensile properties are minimum or typical . UTS and YS as Mpa, El as %. See Appendix for list of abbreviations used in Descriptions.

Worldwide Guide to Equivalent Nonferrous Metals and Alloys

Grade UNS #	Country	Specification	Designation	Description	Cr	Cu	Fe	Mg	Mn	Si	Ti	Zn	OE Max	OT Max	Other	UTS	YS	EL
7028	Spain				0.2 max	0.1-0.3	0.5 max	1.5-2.3	0.15-0.6	0.36 max5	0.05 max	4.5-5.2	0.05	0.15	0.08-0.25 Zr+Ti			
2025 A92025	Japan	JIS H4140	2025	Die SHT and art AH 100 mm diam	0.1	3.9-5	1	0.05	0.4-1.2	0.5-1.2	0.15	0.25				382	226	16
2319 A92319	USA	QQ R-566	2319	Wir		5.8-6.8	0.3 max	0.02 max	0.2-0.4	0.2 max	0.1-0.2	0.1 max	0.05	0.15	V 0.05-0.15			
A92319	USA	AWS A 5.10-92	ER2319	Weld El Rod		5.8-6.8	0.3 max	0.02 max	0.2-0.4	0.2 max	0.1-0.2	0.1 max	0.05	0.15	Be 0.0008 max; V 0.05-0.15; Zr 0.1-0.25			
A92319	USA	AWS	R2319	Weld El Rod		5.8-6.8	0.3 max	0.02 max	0.2-0.4	0.2 max	0.1-0.2	0.1 max	0.05	0.15	Be 0.0008 max; V 0.05-0.15; Zr 0.1-0.25			
2034	USA				0.05 max	4.2-4.8	0.12 max	1.3-1.9	0.8-1.3	0.1 max	0.15 max	0.2 max	0.05	0.15	Zr 0.08-0.15			
	India	IS 739	NG21	Wir		0.1	0.6	0.25	0.5	4.5-6		0.2			Ni 0.2; Pb 0.05; Sn 0.05 max			
2218 A92218	India	IS 7793	2285/24850	Frg		3.723 max	0.7	1.2-1.8		0.6	0.2	0.2			Ni 1.7-2.3; Pb 0.05; Sn 0.05 max	345		
2218 A92218	France	NF A-U4N	2218	Frg	0.1 max	3.5-4.5	1 max	1.2-1.8	0.2 max	0.9 max		0.25 max	0.05	0.15	Ni 1.7-2.3			
2218 A92218	Japan	JIS H4140	2218	Die SHT and Art AH (high temp.) 100 mm diam	0.1	3.5-4.5	1	1.2-1.8	0.2	0.9		0.25			Ni 1.7-2.3	382	275	10
2218 A92218	USA	AMS 4142	2218	Frg	0.1 max	3.5-4.5	1 max	1.2-1.8	0.2 max	0.9 max		0.25 max	0.05	0.15	Ni 1.7-2.3			
2218 A92218	USA	QQ A-367	2218	Frg	0.1 max	3.5-4.5	1 max	1.2-1.8	0.2 max	0.9 max		0.25 max	0.05	0.15	Ni 1.7-2.3			
2218 A92218	USA	SAE J454	2218	Frg	0.1 max	3.5-4.5	1 max	1.2-1.8	0.2 max	0.9 max5		0.25 max	0.05	0.15	Ni 1.7-2.3			
2218	USA				0.1 max	3.5-4.5	1 max	1.2-1.8	0.2 max	0.9 max		0.25 max	0.05	0.15	Ni 1.7-2.3			
	India	IS 739	NG6	Wir Ann	0.25	0.1	0.7	4.5-5.5	0.4	0.6		0.2				245		
	India	IS 739	NG6	Wir Ann and CW to hard	0.25	0.1	0.7	4.5-5.5	0.4	0.6		0.2				387		
7011 A97011	Finland	SFS 2596	AlZn5Mgl	HF (shapes)	0.1-0.25	0.1	0.5	1-1.4	0.5	0.5		4-5				240	140	10
7011 A97011	Finland	SFS 2596	AlZn5Mgl	Ann (plt)	0.1-0.25	0.1	0.5	1-1.4	0.5	0.5		4-5				150	70	15
7011 A97011	Finland	SFS 2596	AlZn5Mgl	ST and aged 12 mm diam	0.1-0.25	0.1	0.5	1-1.4	0.5	0.5		4-5				330	280	10
7023 A97023	Argentina				0.05-0.35	0.5-1	0.5 max	2-3	0.1-0.6	0.5 max	0.1 max	4-6	0.05	0.15				
	India	IS 733	74530	Bar Rod ST and Nat aged for 30 days 6.3 mm diam	0.2	0.2	0.7	1-1.5	0.2-0.7	0.4	0.2	4-5				255	220	9
	India	IS 736	74530	Plt SHT and Nat aged 8 mm diam	0.2	0.2	0.7	1-1.5	0.2-0.7	0.4	0.2	4-5				265	160	8
	India	IS 736	74530	Plt SHT and PHT 7 mm diam	0.2	0.2	0.7	1-1.5	0.2-0.7	0.4	0.2	4-5				305	255	7
A13560	USA	AWS		Weld El Rod		0.2 max	0.2 max	0.25-0.45	0.1 max	6.5-7.5	0.2 max	0.1 max	0.05	0.15				
7026	Italy					0.6-0.9	0.12 max	1.5-1.9	0.05-0.2	0.08 max	0.05 max	4.6-5.2	0.03	0.1	Zr 0.09-0.14			
	Sweden	SIS 144146	4146-00	Wir	0.35	0.1	0.5	4.5-5.5	1	0.5		0.2						
	Sweden	SIS 144146	4146-18	Wir HW	0.35	0.1	0.5	4.5-5.5	1	0.5		0.2				274		
	India	IS 736	55000	Plt As Mfg	0.25	0.1	0.7	4.5-5.5	0.5	0.6	0.2	0.2				275	125	
	India	IS 736	55000	Plt Ann 16 mm diam	0.25	0.1	0.7	4.5-5.5	0.5	0.6	0.2	0.2				265	100	16
	India	IS 736	55000	Plt 1/2 hard 5 mm diam	0.25	0.1	0.7	4.5-5.5	0.5	0.6	0.2	0.2				355	275	5

UNS numbers and US grades are provided as a means of cross referencing chemically similar alloys. Exchangability is only possible after independent examination of specifications. Tensile properties are minimum or typical . UTS and YS as Mpa, El as %. See Appendix for list of abbreviations used in Descriptions.

2-84 Wrought Aluminum

Grade UNS #	Country	Specification	Designation	Description	Cr	Cu	Fe	Mg	Mn	Si	Ti	Zn	OE Max	OT Max	Other	UTS	YS	EL
A94010	USA	AWS	ER4010	Weld El Rod		0.2 max	0.2 max	0.3-0.45	0.1 max	6.5-7.5	0.2 max	0.1 max	0.05	0.15	Be 0.0008 max			
A94010	USA	AWS	R4010	Weld El Rod		0.2 max	0.2 max	0.3-0.45	0.1 max	6.5-7.5	0.2 max	0.1 max	0.05	0.15	Be 0.0008 max			
4010	USA					0.2 max	0.2 max	0.3-0.45	0.1 max	6.5-7.5	0.2 max	0.1 max	0.05	0.15	0.0008 max Be for welding			
4343	USA					0.25 max	0.8		0.1 max	6.8-8.2		0.2 max	0.05	0.15				
4008	USA					0.05 max	0.09 max	0.3-0.45	0.05 max	6.5-7.5	0.04-0.15	0.05 max	0.05	0.15	0.0008 max Be for welding			
	India	IS 738	NT6	Wir Ann	0.25	0.1	0.7	4.5-5.5	1	0.6						265		18
	India	IS 738	NT6	Wir Ann and CW to 1/2 hard	0.25	0.1	0.7	4.5-5.5	1	0.6						279	216	5
A03570	USA	AWS		Weld El Rod		0.05 max	0.15 max	0.45-0.6	0.03 max	6.5-7.5	0.2 max	0.05 max	0.05	0.15				
	India	IS 5902	55000	Wir Bar Ann	0.25	0.1	0.7	4.5-5.5	0.5	0.6	0.2	0.1			Ni 0.2; Pb 0.05; Sn 0.05 max	265		
A13570	USA	AWS		Weld El Rod		0.2 max	0.2 max	0.4-0.7	0.1 max	6.5-7.5	0.04-0.2	0.1 max	0.05	0.15	Be 0.04-0.07			
A94011	USA	AWS	R4011	Weld El Rod		0.2 max	0.2 max	0.45-0.7	0.1 max	6.5-7.5	0.04-0.2	0.1 max	0.05	0.15				
4011	USA					0.2 max	0.2 max	0.45-0.7	0.1 max	6.5-7.5	0.04-0.2	0.1 max	0.05	0.15	Be 0.04-0.07			
4044 A94044	Russia		AK10			0.1 max			0.1 max	7-10		0.2 max						
7075 A07075	UK		M75S		0.5 max	0.3-1.5	0.5 max	2-3.5	0.3-1	0.5 max	0.3 max	4.5-6.5						
X8019	USA						7.3-9.3		0.05 max	0.2 max	0.05 max	0.05 max	0.05	0.15	3.5-4.5 Ce; 0.20-0.50 O			
7022	Europe				0.1-0.3	0.5-1	0.5 max	2.6-3.7	0.1-0.4	0.5 max		4.3-5.2	0.05	0.15	0.20 Ti+Zr			
7046	USA				0.2 max	0.25 max	0.4 max	1-1.6	0.3 max	0.2 max	0.06 max	6.6-7.6	0.05	0.15	0.10-0.18 Zr			
	France	NF A50411	2030(A-U4Pb)	Bar Aged (bar)	0.1	3.5-4.5	0.5	0.5-1.3	1	0.4	0.2	0.5			Ni 0.2; Pb 0.8-1.5	370	235	7
	France	NF A50411	2030(A-U4Pb)	Bar CF (tub) 150 mm diam	0.1	3.5-4.5	0.5	0.5-1.3	1	0.4	0.2	0.5			Ni 0.2; Pb 0.8-1.5	370	235	7
7146	USA						0.4 max	1-1.6		0.2 max	0.06 max	6.6-7.6	0.05	0.15	Zr 0.10-0.18			
	Pan American	COPANT 862	X7073		0.05-0.35	0.5-1	0.5	2-3	0.1-0.6	0.5	0.1	4-6						
4044	USA					0.25 max	0.8 max		0.1 max	7.8-9.2		0.2 max	0.05	0.15				
8022	USA				0.1 max		6.2-6.8		0.1 max	1.2-1.4	0.1 max	0.25 max	0.05	0.15	0.05-0.20 O; V 0.4-0.8			
7014	UK					0.3-0.7	0.5 max	2.2-3.2	0.3-0.7	0.5 max		5.2-6.2	0.05	0.15	0.20 Ti + Zr; Ni 0.1 max			
2018 A92018	Canada	CSA HA.8	0.2018	Frg T61 (die frg) 4 in diam	0.1	3.5-4.5	1	0.45-0.9	0.2	0.9		0.25			Ni 1.7-2.3	379	276	7
2018 A92018	Japan	JIS H4140	2018	Die SHT and Art AH (high temp) 100 mm diam	0.1	3.5-4.5	1	0.45-0.9	0.2	0.9		0.25			Ni 1.7-2.3	382	275	10
2020 A92018	Russia		AK2			3.5-4.5	0.5-1	0.4-0.8	0.1 max	0.5-1		0.3 max			ET 0.1 max; ET=.60 total; Ni 1.8-2.3			
7075 A97075	UK		C77S		0.08-0.25	1-2.2	0.5 max	2-3	0.3 max	0.5 max	0.3 max	5-7.5			Ni 0.1 max; Pb 0.05; Sn 0.05 max			
4343 A94343	Japan	JIS Z3263	BA22PC	Sh		0.25	0.8		0.1	6.8-8.2		0.2						
4343 A94343	Japan	JIS Z3263	BA4343	Wir		0.25	0.8		0.1	6.8-8.2		0.2						
	Japan	JIS Z3263	BA12PC	Sh Ann .3 mm diam		0.25	0.8		0.1	6.8-8.2		0.2				137		15

UNS numbers and US grades are provided as a means of cross referencing chemically similar alloys. Exchangability is only possible after independent examination of specifications. Tensile properties are minimum or typical . UTS and YS as Mpa. El as %. See Appendix for list of abbreviations used in Descriptions.

Worldwide Guide to Equivalent Nonferrous Metals and Alloys

Grade UNS #	Country	Specification	Designation	Description	Cr	Cu	Fe	Mg	Mn	Si	Ti	Zn	OE Max	OT Max	Other	UTS	YS	EL
	Japan	JIS Z3263	BA12PC	Sh 1/4 hard .3 mm diam		0.25	0.8		0.1	6.8-8.2		0.2				118		2
	Japan	JIS Z3263	BA12PC	Sh 1/2 hard .3 mm diam		0.25	0.8		0.1	6.8-8.2		0.2				137		1
7009	Germany				0.1-0.25	0.6-1.3	0.2 max	2.1-2.9	0.1 max	0.2 max	0.2 max	5.5-6.5	0.05	0.15	Ag 0.24-0.40			
7475 A97475	USA	AMS 4084	7475	Plt Sh Shapes	0.18-0.25	1.2-1.9	0.12 max	1.9-2.6	0.06 max	0.1 max		5.2-6.2	0.05	0.15				
7475 A97475	USA	AMS 4085	7475	Plt Sh Shapes	0.18-0.25	1.2-1.9	0.12 max	1.9-2.6	0.06 max	0.1 max		5.2-6.2	0.05	0.15				
7475 A97475	USA	AMS 4089	7475	Plt Sh Shapes	0.18-0.25	1.2-1.9	0.12 max	1.9-2.6	0.06 max	0.1 max		5.2-6.2	0.05	0.15				
7475 A97475	USA	AMS 4090	7475	Plt Sh Shapes	0.18-0.25	1.2-1.9	0.12 max	1.9-2.6	0.06 max	0.1 max		5.2-6.2	0.05	0.15				
7475	USA				0.18-0.25	1.2-1.9	0.12 max	1.9-2.6	0.06 max	0.1 max	0.06 max	5.2-6.2	0.05	0.15				
	USA		AlZn5.5MgCu (A)		0.18-0.25	1.2-1.9	0.12 max	1.9-2.6	0.06 max	0.1 max	0.06 max	5.2-6.2	0.05	0.15				
7175 A97175	USA	AMS 4109	7175	Frg	0.18-0.28	1.2-2	0.2 max	2.1-2.9	0.1 max	0.15 max	0.1 max	5.1-6.1	0.05	0.15				
7175 A97175	USA	AMS 4148	7175	Frg	0.18-0.28	1.2-2	0.2 max	2.1-2.9	0.1 max	0.15 max	0.1 max	5.1-6.1	0.05	0.15				
7175 AC7175	USA	AMS 4149	7175	Frg	0.18-0.28	1.2-2	0.2 max	2.1-2.9	0.1 max	0.15 max	0.1 max	5.1-6.1	0.05	0.15				
7175 A97175	USA	AMS 4179	7175	Frg	0.18-0.28	1.2-2	0.2 max	2.1-2.9	0.1 max	0.15 max	0.1 max	5.1-6.1	0.05	0.15				
7175	USA				0.18-0.28	1.2-2	0.2 max	2.1-2.9	0.1 max	0.15 max	0.1 max	5.1-6.1	0.05	0.15				
7075 A97075	Austria	ONORM M3429	AlZnMgCu1.5		0.18-0.35	1.2-2	0.5 max	2.1-2.9	0.3 max	0.4 max	0.2 max	5.1-6.1			ET 0.05 max .15 tot; Ti+Zr 0.25 x			
7075 A97075	Germany	DIN 1725 pt1	3.4365/AlZnMgCu1.5		0.18-0.28	1.2-2	0.5 max	2.1-2.9	0.3 max	0.4 max	0.2 max	5.1-6.1			ST 0.25 max; ET 0.05 max; ST=Ti+Zr; ET=.15 total			
7075 A97075	Norway	NS 17411	NS17411/AlZn6MgCu	Bar Plt Sh Strp Tub Frg	0.18-0.28	1.2-2	0.5 max	2.1-2.9	0.3 max	0.4 max	0.2 max	5.1-6.1			ST 0.25 max; ST=Ti+Zr			
7075 A97075	USA				0.18-0.28	1.2-2	0.5 max	2.1-2.9	0.3 max	0.4 max	0.2 max	5.1-6.1	0.05	0.15				
	USA		AlZn5.5MgCu		0.18-0.28	1.2-2	0.5 max	2.1-2.9	0.3 max	0.4 max	0.2 max	5.1-6.1	0.05	0.15				
7075 A97075	Czech Republic	CSN 424222	AlZn6Mg2Cu		0.1-0.25	1.4-2	0.5	1.8-2.8	0.2-0.6	0.5 max		5-7						
7075 A97075	Russia	GOST 4784	V95		0.1-0.25	1.4-2	0.5 max	1.8-2.8	0.2-0.6	0.5 max		5-7			ET 0.1 max;			
7012	Italy				0.04 max	0.8-1.2	0.25 max	1.8-2.2	0.05-0.15	0.15 max	0.02-0.08	5.8-6.5	0.05	0.15	0.10-0.18 Zr			
	International	ISO R209 (1971)	Al-Zn6MgCu		0.10-0.35	1.2-2.0	0.50 max	2.1-2.9	0.30 max	0.40 max					Ti+Zr 0.30 max; Mn+Cr 0.50 x; Ni 0.10 max			
	International	ISO 208	Al-Si7Mg			0.2	0.5	0.2-0.4	0.6	6.5-7.5	0.2	0.3			Ni 0.05; Pb 0.05; Sn 0.05 max			
7109	Germany				0.04-0.08	0.6-1.3	0.15 max	2.2-2.7	0.1 max	0.1 max	0.1 max	5.8-6.5	0.05	0.15	Ag 0.25-0.40; Zr 0.10-0.20			
	India	IS 6754	83428			0.7-1.3	0.7	0.3	0.1	1-2	0.1				Ni 0.2-0.7; Sn 5.5-7			
7076	USA					0.3-1	0.6 max	1.2-2	0.3-0.8	0.4 max	0.2 max	7-8	0.05	0.15				
7091	USA					1.1-1.8	0.15 max	2-3		0.12 max		5.8-7.1	0.05	0.15	0.20-0.6 Co; 0.20-0.50 O			
	International	ISO 2779	Al-Cu4PbMg	SHT and nat aged .125in;3 mm diam OBSOLETE Replaced by ISO 6362	0.1	3.5-5	0.8	0.3-1.8	1	0.8	0.2	0.8			Pb 1.5	370	245	10

UNS numbers and US grades are provided as a means of cross referencing chemically similar alloys. Exchangability is only possible after independent examination of specifications. Tensile properties are minimum or typical . UTS and YS as Mpa, El as %. See Appendix for list of abbreviations used in Descriptions.

2-86 Wrought Aluminum

Grade UNS #	Country	Specification	Designation	Description	Cr	Cu	Fe	Mg	Mn	Si	Ti	Zn	OE Max	OT Max	Other	UTS	YS	EL
4045	USA					0.3 max	0.8 max	0.05 max	0.05 max	9-11	0.2 max	0.1 max	0.05	0.15				
4046	Czech Republic	CSN 424261	AlCu8Fe1Si1			7.5-8.5	1-1.6	0.5 max	0.3 max	0.5-1		0.5 max			Ti+Ni 0.10; Bi+Pb+Sn 0.10			
	Europe					0.03 max	0.5 max	0.2-0.5	0.4 max	9-11	0.15 max	0.1 max	0.05	0.15				
7060	France				0.15-0.25	1.8-2.6	0.2 max	1.3-2.1	0.2 max	0.015 max	0.05 max	6.1-7.5	0.05	0.15	Pb 0.003; Zr 0.05			
7010	UK				0.05 max	1.5-2	0.15 max	2.1-2.6	0.1 max	0.12 max	0.06 max	5.7-6.7	0.05	0.15	Ni 0.05 max; Zr 0.10-0.16			
	UK		AlZn6MgCu		0.05 max	1.5-2	0.15 max	2.1-2.6	0.1 max	0.12 max	0.06 max	5.7-6.7	0.05	0.15	Ni 0.05 max; Zr 0.10-0.16			
7075 A97075	Europe	AECMA prEN2092	7075	Sht ST Quen and Art aged .4/.8 mm diam	0.18-0.28	1.2-2		2.1-2.9	0.3	0.4	0.2	5.1-6.1				485	420	7
7075 A97075	Europe	AECMA prEN2126	7075	Sht T651 ST Quen and Art aged 6/10 mm diam	0.18-0.28	1.2-2		2.1-2.9	0.3	0.4	0.2	5.1-6.1				530	470	8
7075 A97075	France	NF A50506	7075(A-Z5GU)	CR .4/.8 mm diam	0.18-0.35	1.2-2		2.1-2.9	0.3	0.4	0.2	5.1-6.1				480	410	7
7075 A97075	France	AIR 9051-A	A-Z5GU(7075)	Sh Bar Rolled (sht) 25 mm diam	0.18-0.35	1.2-2	0.5	2.1-2.9	0.3	0.4		5.1-6.1				530	450	7
7075 A97075	France	AIR 9051-A	A-Z5GU(7075)	Sh Bar ST (bar) 160 mm diam	0.18-0.35	1.2-2	0.5	2.1-2.9	0.3	0.4		5.1-6.1				540	480	6
7075 A97075	Norway	NS 17412	NS17412/AlZn6Mg2Cu	Bar Frg	0.04 max	2-2.6	0.15 max	1.9-2.6	0.1 max	0.12 max	0.06 max	5.7-6.7			ET 0.05 max; ET=.15 total			
7050	USA				0.04 max	2-2.6	0.15 max	1.9-2.6	0.1 max	0.12 max	0.06 max	5.7-6.7	0.05	0.15	Zr 0.08-0.15			
	USA		AlZn6CuMgZr		0.04 max	2-2.6	0.15 max	1.9-2.6	0.1 max	0.12 max	0.06 max	5.7-6.7	0.05	0.15	Zr 0.08-0.15			
7050 A97050	USA	AMS 4050	7050	Plt Frg Ext	0.04 max	2-2.6	0.15 max	1.9-2.6	0.1 max	0.12 max	0.06 max	5.7-6.7	0.05	0.15	Zr 0.08-0.15			
7050 A97050	USA	AMS 4108	7050	Plt Frg Ext	0.04 max	2-2.6	0.15 max	1.9-2.6	0.1 max	0.12 max	0.06 max	5.7-6.7	0.05	0.15	Zr 0.08-0.15			
7075 A97075	International	ISO R827	Al-Zn6MgCu	ST and PT .25 in; 6 mm diam	0.1-0.35	1.2-2	0.5	2.1-2.9	0.3	0.4		5.1-6.4			Ni 0.1	525	471	7
7075 A97075	International	ISO TR2136	Al-Zn6MgCu	Sh Plt SHT and art aged	0.1-0.35	1.2-2	0.5	2.1-2.9	0.3	0.4		5.1-6.4			Ni 0.1	525	450	7
7075 A97075	International	ISO TR2778	Al-Zn6MgCu	Tub ST and PT OBSOLETE Replace by ISO 6363	0.1-0.35	1.2-2	0.5	2.1-2.9	0.3	0.4		5.1-6.4			Ni 0.1	515	440	7
	South Africa	SABS 712	Al-Cu4MglPb		0.4	3.5-5	1	0.2-1.8	1	1		1			Bi 0.5; Ni 1; Pb 0.5-2			
7075 A97075	Australia	AS 1865	7075	Wir Rod Bar Strp	0.18-0.35	1.2-2	0.5	2.1-2.9	0.3	0.4	0.2	5.1-6.1						
7075 A97075	Canada	CSA HA.4	0.7075	Plt Sh O .015/.049 in diam	0.18-0.35	1.2-2	0.5	2.1-2.9	0.3	0.4	0.2	5.1-6.1				276	145	10
7075 A97075	Canada	CSA HA.4	0.7075	Plt Sh T6 .015/.049 in diam	0.18-0.35	1.2-2	0.5	2.1-2.9	0.3	0.4	0.2	5.1-6.1				524	462	7
7075 A97075	Canada	CSA HA.4	0.7075	Plt Sh T651 .250/.499 in diam	0.18-0.35	1.2-2	0.5	2.1-2.9	0.3	0.4	0.2	5.1-6.1				538	462	9
7075 A97075	Canada	CSA HA.4	.7075Alclad	Plt Sh O .015/.032 in diam	0.18-0.35	1.2-2	0.5	2.1-2.9	0.3	0.4	0.2	5.1-6.1				248	138	10
7075 A97075	Canada	CSA HA.4	.7075Alclad	Plt Sh T6 .012/.020 in diam	0.18-0.35	1.2-2	0.5	2.1-2.9	0.3	0.4	0.2	5.1-6.1				483	414	7
7075 A97075	Canada	CSA HA.4	.7075Alclad	Plt Sh T62 (sht) .021/.039 in diam	0.18-0.35	1.2-2	0.5	2.1-2.9	0.3	0.4	0.2	5.1-6.1				483	414	7

UNS numbers and US grades are provided as a means of cross referencing chemically similar alloys. Exchangability is only possible after independent examination of specifications. Tensile properties are minimum or typical . UTS and YS as Mpa, El as %. See Appendix for list of abbreviations used in Descriptions.

Worldwide Guide to Equivalent Nonferrous Metals and Alloys

Grade UNS #	Country	Specification	Designation	Description	Cr	Cu	Fe	Mg	Mn	Si	Ti	Zn	OE Max	OT Max	Other	UTS	YS	EL
7075 A97075	Canada	CSA HA.5	0.7075	Bar Rod Wir Drawn and O .250/1.5 in diam	0.18-0.35	1.2-2	0.5	2.1-2.9	0.3	0.4	0.2	5.1-6.1				276		10
7075 A97075	Canada	CSA HA.5	0.7075	Bar Rod Wir Extd and T6 .249 in diam	0.18-0.35	1.2-2	0.5	2.1-2.9	0.3	0.4	0.2	5.1-6.1				518	483	7
7075 A97075	Canada	CSA HA.5	0.7075	Bar Rod Wir Extd and T6510 .500/2.999 in diam	0.18-0.35	1.2-2	0.5	2.1-2.9	0.3	0.4	0.2	5.1-6.1				558	496	7
7075 A97075	Canada	CSA HA.7	0.7075	Tub Pip drawn and O .025/.259 in diam	0.18-0.35	1.2-2	0.5	2.1-2.9	0.3	0.4	0.2	5.1-6.1				276	145	8
7075 A97075	Canada	CSA HA.7	0.7075	Tub Pip Drawn and T6 .259 in diam	0.18-0.35	1.2-2	0.5	2.1-2.9	0.3	0.4	0.2	5.1-6.1				531	455	8
7075 A97075	Canada	CSA HA.7	0.7075	Tub Pip Drawn and T62 .260/.500 in diam	0.18-0.35	1.2-2	0.5	2.1-2.9	0.3	0.4	0.2	5.1-6.1				531	455	9
7075 A97075	Canada	CSA HA.8	0.7075	Frg T6 (die frg) 1 in diam	0.18-0.35	1.2-2	0.5	2.1-2.9	0.3	0.4	0.2	5.1-6.1				517	441	7
7075 A97075	Canada	CSA HA.8	0.7075	Frg T6 (hand frg) 2 in diam	0.18-0.35	1.2-2	0.5	2.1-2.9	0.3	0.4	0.2	5.1-6.1				510	434	9
7075 A97075	France	NF A50411	7075(A-Z5GU)	Bar Q/T (extd bar)	0.18-0.28	1.2-2	0.5	2.1-2.9	0.3	0.4	0.2	5.1-6.1						2
7075 A97075	France	NF A50411	7075(A-Z5GU)	Bar Q/T (tub) 150 mm diam	0.18-0.28	1.2-2	0.5	2.1-2.9	0.3	0.4	0.2	5.1-6.1				530	450	8
7075 A97075	France	NF A50451	7075(A-Z5GU)	Sh Plt Ann .4/.8 mm diam	0.18-0.35	1.2-2	0.5	2.1-2.9	0.3	0.4	0.2	5.1-6.1					150	
7075 A97075	France	NF A50451	7075(A-Z5GU)	Sh Plt Q/T .8/3.2 mm diam	0.18-0.35	1.2-2	0.5	2.1-2.9	0.3	0.4	0.2	5.1-6.1				530	420	9
7075 A97075	Japan	JIS H4000	7075	Sh Plt Strp Ann 1.5 mm diam	0.18-0.35	1.2-2	0.5	2.1-2.9	0.3	0.4	0.2	5.1-6.1				275	147	10
7075 A97075	Japan	JIS H4000	7075	Sh Plt Strp SHT and Art AH 1 mm diam	0.18-0.35	1.2-2	0.5	2.1-2.9	0.3	0.4	0.2	5.1-6.1				530	461	7
7075 A97075	Japan	JIS H4040	7075	Bar Ann	0.18-0.35	1.2-2	0.5	2.1-2.9	0.3	0.4	0.2	5.1-6.1				275	167	10
7075 A97075	Japan	JIS H4040	7075	Bar SHT and Art AH 7 mm diam	0.18-0.35	1.2-2	0.5	2.1-2.9	0.3	0.4	0.2	5.1-6.1				539	481	7
7075 A97075	Japan	JIS H4080	7075	Tub Ann	0.18-0.35	1.2-2	0.5	2.1-2.9	0.3	0.4	0.2	5.1-6.1				275	167	10
7075 A97075	Japan	JIS H4080	7075	Wought Tub SHT and Art AH 7 mm diam	0.18-0.35	1.2-2	0.5	2.1-2.9	0.3	0.4	0.2	5.1-6.1				539	481	7
7075 A97075	Japan	JIS H4100	7075	Ann	0.18-0.35	1.2-2	0.5	2.1-2.9	0.3	0.4	0.2	5.1-6.1				275	167	10
7075 A97075	Japan	JIS H4100	7075	SHT and Art AH 7 mm diam	0.18-0.35	1.2-2	0.5	2.1-2.9	0.3	0.4	0.2	5.1-6.1				539	481	7
7075 A97075	Pan American	COPANT 862	7075		0.18-0.35	1.2-2	0.5	2.1-2.9	0.3	0.4	0.2	5.1-6.1						
7150	USA				0.04 max	1.9-2.5	0.15 max	2-2.7	0.1 max	0.12 max	0.06 max	5.9-6.9	0.05	0.15	Zr 0.08-0.15			
7075 A97075	Japan	JIS H4140	7075	Die SHT and Art AH 75 mm diam	0.18-0.35	1.2-2	0.5	2.1-2.9	0.3	0.4	0.21	5.1-6.1				520	451	10
7090	USA					0.6-1.3	0.15 max	2-3		0.12 max		7.3-8.7	0.05	0.15	1.0-1.9 Co; 0.20-0.50 O			
4004	USA					0.25 max	0.8 max	1-2	0.1 max	9-10.5		0.2 max	0.05	0.15				
4104	USA					0.25 max	0.8 max	1-2	0.1 max	9-10.5		0.2 max	0.05	0.15	Bi 0.02-0.20			
7278A	Switzerland				0.05 max	1.3-2.1	0.15 max	2.3-3.2	0.25 max	0.12 max	0.05 max	6.4-7.4	0.05	0.15	0.05-0.25 Zr			

UNS numbers and US grades are provided as a means of cross referencing chemically similar alloys. Exchangability is only possible after independent examination of specifications. Tensile properties are minimum or typical . UTS and YS as Mpa, El as %. See Appendix for list of abbreviations used in Descriptions.

2-88 Wrought Aluminum

Grade UNS #	Country	Specification	Designation	Description	Cr	Cu	Fe	Mg	Mn	Si	Ti	Zn	OE Max	OT Max	Other	UTS	YS	EL
7178 A97178	USA	AMS 4158	7178	Bar Rod Wir Tub Ext Shapes	0.18-0.35	1.6-2.4	0.5 max	2.4-3.1	0.3 max	0.4 max	0.2 max	6.3-7.3	0.05	0.15				
7178 A97178	USA	ASTM B209	7178	Bar Rod Wir Tub Ext Shapes	0.18-0.35	1.6-2.4	0.5 max	2.4-3.1	0.3 max	0.4 max	0.2 max	6.3-7.3	0.05	0.15				
7178 A97178	USA	ASTM B221	7178	Bar Rod Wir Tub Ext Shapes	0.18-0.35	1.6-2.4	0.5 max	2.4-3.1	0.3 max	0.4 max	0.2 max	6.3-7.3	0.05	0.15				
7178 A97178	USA	ASTM B241	7178	Bar Rod Wir Tub Ext Shapes	0.18-0.35	1.6-2.4	0.5 max	2.4-3.1	0.3 max	0.4 max	0.2 max	6.3-7.3	0.05	0.15				
7178 A97178	USA	ASTM B316	7178	Bar Rod Wir Tub Ext Shapes	0.18-0.35	1.6-2.4	0.5 max	2.4-3.1	0.3 max	0.4 max	0.2 max	6.3-7.3	0.05	0.15				
7178 A97178	USA	QQ A-200	7178	Bar Rod Wir Tub Ext Shp OBSLT	0.18-0.35	1.6-2.4	0.5 max	2.4-3.1	0.3 max	0.4 max	0.2 max	6.3-7.3	0.05	0.15				
7178 A97178	USA	QQ A-250	7178	Bar Rod Wir Tub Ext Shapes	0.18-0.35	1.6-2.4	0.5 max	2.4-3.1	0.3 max	0.4 max	0.2 max	6.3-7.3	0.05	0.15				
7178	USA				0.18-0.28	1.6-2.4	0.5 max	2.4-3.1	0.3 max	0.4 max	0.2 max	6.3-7.3	0.05	0.15				
7149	USA				0.1-0.22	1.2-1.9	0.2 max	2-2.9	0.2 max	0.15 max	0.1 max	7.2-8.2	0.05	0.15				
7049 A97049	USA	AMS 4111	7049	Frg Ext	0.1-0.22	1.2-1.9	0.35 max	2-2.9	0.2 max	0.25 max	0.1 max	7.2-8.2	0.05	0.15				
7049 A97049	USA	AMS 4157	7049	Frg Ext	0.1-0.22	1.2-1.9	0.35 max	2-2.9	0.2 max	0.25 max	0.1 max	7.2-8.2	0.05	0.15				
7049 A97049	USA	AMS 4159	7049	Frg Ext	0.1-0.22	1.2-1.9	0.35 max	2-2.9	0.2 max	0.25 max	0.1 max	7.2-8.2	0.05	0.15				
7049 A97049	USA	MIL H-6088	7049	Frg Ext	0.1-0.22	1.2-1.9	0.35 max	2-2.9	0.2 max	0.25 max	0.1 max	7.2-8.2	0.05	0.15				
7049 A97049	USA	QQ A-367	7049	Frg Ext	0.1-0.22	1.2-1.9	0.35 max	2-2.9	0.2 max	0.25 max	0.1 max	7.2-8.2	0.05	0.15				
7049	USA				0.1-0.22	1.2-1.9	0.35 max	2-2.9	0.2 max	0.25 max	0.1 max	7.2-8.2	0.05	0.15				
7049 A97049	France	NF A50411	7049A(A-Z8GU)	Bar Q/T	0.05-0.25	1.2-2.9	0.5	2.1-3.1	0.5	0.4		7.2-8.4				610	530	5
4045 A94045	Japan	JIS Z3263	BA4045	Wir		0.3	0.8	0.05	0.05	9-11	0.2	0.1						
	Japan	JIS Z3263	BA24PC	Sh Ann .8 mm		0.3	0.8	0.05	0.05	9-11	0.2	0.1				147		18
7049A	France				0.05-0.25	1.2-1.9	0.5 max	2.1-3.1	0.5 max	0.4		7.2-8.4	0.05	0.15	0.25 Zr+Ti			
7064	USA				0.06-0.25	1.8-2.4	0.15 max	1.9-2.9		0.12 max		6.8-8	0.05	0.15	Zr 0.10-0.50			
7349	France				0.1-0.22	1.4-2.1	0.15 max	1.8-2.7	0.2 max	0.12 max		7.5-8.7	0.05	0.15	0.25 Zr+Ti			
7449	France					1.4-2.1	0.15 max	1.8-2.7	0.2 max	0.12 max		7.5-8.7	0.05	0.15	0.25 Zr+Ti			
7278	Norway				0.17-0.25	1.6-2.2	0.2 max	2.5-3.2	0.02 max	0.15 max	0.03 max	6.6-7.4	0.03	0.1	Ga 0.03 max; V 0.05 max			
7249	USA				0.12-0.18	1.3-1.9	0.12 max	2-2.4	0.1 max	0.1 max	0.06 max	7.5-8.2	0.05	0.15				
4047 A94047	Germany	DIN 1732 Part 1	3.2585/S-AlSi12			0.1 max	0.7 max	0.1 max	0.5 max	11-13.5	0.2 max	0.1 max			Ni 0.1 max; Pb 0.1 max; Sn 0.05 max			
4047 A94047	Russia	GOST 1583	AK12			0.2 max	0.03 max			11.5-13.5								
4047A	Europe					0.3 max	0.6 max	0.1 max	0.15 max	11-13	0.15 max	0.2 max	0.05	0.15	0.0008 max Be for welding			
	Europe		AlSi12(A)			0.3 max	0.6 max	0.1 max	0.15 max	11-13	0.15 max	0.2 max	0.05	0.15	0.0008 max Be for welding			
4047 A94047	Canada	CSA HA.6	0.4047	Rod Wir		0.3	0.8	0.1	0.15	11-13		0.2			Be 0.001			
4047 A94047	Japan	JIS Z3263	BA4047	Wir		0.3	0.8	0.1	0.15	11-13		0.2						
4047 A94047	Pan American	COPANT 862	4047			0.3	0.8	0.1	0.15	11-13		0.2						
A94047	USA	AWS	ER4047	Weld El Rod		0.3 max	0.8 max		0.15 max	11-13		0.2 max	0.05	0.15	Be 0.0008 max			
A94047	USA	AWS	R4047	Weld El Rod		0.3 max	0.8 max		0.15 max	11-13		0.2 max	0.05	0.15	Be 0.0008 max			

UNS numbers and US grades are provided as a means of cross referencing chemically similar alloys. Exchangability is only possible after independent examination of specifications. Tensile properties are minimum or typical . UTS and YS as Mpa, El as %. See Appendix for list of abbreviations used in Descriptions.

Worldwide Guide to Equivalent Nonferrous Metals and Alloys

Grade UNS #	Country	Specification	Designation	Description	Cr	Cu	Fe	Mg	Mn	Si	Ti	Zn	OE Max	OT Max	Other	UTS	YS	EL
4047	USA					0.3 max	0.8 max	0.1 max	0.15 max	11-13		0.2 max	0.05	0.15	0.0008 max Be for welding			
	USA		Al Si12			0.3 max	0.8 max	0.1 max	0.15 max	11-13		0.2 max	0.05	0.15	0.0008 max Be for welding			
7001	USA				0.18-0.35	1.6-2.6	0.4 max	2.6-3.4	0.2 max	0.35 max	0.2 max	6.8-8	0.05	0.15				
4147	USA					0.25 max	0.8 max	0.1-0.5	0.1 max	11-13		0.2 max	0.05	0.15	0.0008 max Be for welding			
8009	USA				0.1 max		8.4-8.9		0.1 max	1.7-1.9	0.1 max	0.25 max	0.05	0.15	0.30 max O; V 1.1-1.5			
	Czech Republic	CSN 055690	B-AlSi12-590/575	Brazing Filler Metal		0.05 max	0.5 max	11 max	0.3-0.5	11-13.5					Cu+Zn 0.10			
7055	USA				0.04 max	2-2.6	0.15	1.8-2.3	0.05 max	0.1 max	0.06 max	7.6-8.4	0.05	0.15	Zr 0.08-0.25			
7093	USA					1.1-1.9	0.15 max	2-3		0.12 max		8.3-9.7	0.05	0.15	0.05-0.50 O; Ni 0.04-0.16; Zr 0.08-0.20			
	India	IS 733	46000	Bar Rod As Mfg 10 mm diam		0.1	0.6	0.2	0.5	10-13		0.2				150		10
	UK	BS 1475	NG2	Wir		0.1	0.6	0.2	0.5	10-13		0.2						
	Germany	DIN 8512	3.2285/L-AlSi12			0.03	0.4		0.1	11-13.5	0.03	0.07						
	Sweden	SIS 144262	4262-00	Wir		0.1	0.6	0.25	0.5	10-13		0.2						
	Sweden	SIS 144262	4262-18	Wir Hard		0.1	0.6	0.25	0.5	10-13		0.2				196		
4032 A94032	Japan	JIS H4140	4032	Die SHT and Art AH 100 mm diam	0.1	0.5-1.3	1	0.8-1.3		11-13.5		0.2			Ni 0.5-1.3	363	294	5
	India	IS 739	NG2	Wir		0.1	0.6	0.25	0.5	10-13		0.2			Ni 0.2; Pb 0.05; Sn 0.05 max			
4145A	UK					3-5	0.6 max	0.1 max	0.15 max	9-11	0.15 max	0.2 max	0.05	0.15	0.0008 max Be for welding			
A94145	USA	AWS	ER4145	Weld El Rod	0.15 max	3.3-4.7	0.8 max		0.15 max	9.3-10.7		0.2 max	0.05	0.15	Be 0.0008 max			
A94145	USA	AWS	R4145	Weld El Rod	0.15 max	3.3-4.7	0.8 max		0.15 max	9.3-10.7		0.2 max	0.05	0.15	Be 0.0008 max			
4145	USA				0.15 max	3.3-4.7	0.8 max	0.15 max	0.15 max	9.3-10.7		0.2 max	0.05	0.15	0.0008 max Be for welding			
4032 A94032	Czech Republic	CSN 424237	AlSi12Ni1Mg			0.8-1.1		0.8-1.3	0.2 max	11.5-13	0.2 max	0.1 max			Fe+Ti 0.80; Ni 0.8-1.5			
4032 A94032	UK		38S			0.7-1.3	0.6 max	0.8-1.5	0.2 max	10.5-13	0.2 max	0.1 max			Ni 0.7-1.3; Pb 0.05 max; Sn 0.05 max			
4032 A94032	Canada	CSA HA.8	0.4032	Frg T6 (die frg) 4 in diam	0.1	0.5-1.3	1	0.8-1.3		11-13.5		0.25			Ni 0.5-1.3	359	290	3
4032 A94032	Canada	CSA HA.8	0.4032		0.1 max	0.5-1.3	1 max	0.8-1.3		11-13.5		0.25 max			ET 0.05 max; ET=.15 total; Ni 0.5-1.3			
4032	USA				0.1 max	0.5-1.3	1 max	0.8-1.3		11-13.5		0.25 max	0.05	0.15	Ni 0.5-1.3			
4145 A94145	Japan	JIS Z3263	BA4145	Wir	0.15	3.3-4.7	0.8	0.15	0.15	9.3-10.7		0.2						
	India	IS 7793	4658/49582	Frg		0.8-1.5	0.8	0.8-1.3	0.2	11-13	0.2	0.35			Ni 1.5; Pb 0.05; Sn 0.05 max	295		
8081	USA					0.7-1.3	0.7 max		0.1 max	0.7 max	0.1 max	0.05 max	0.05	0.15	Sn 18-22			
	India	IS 6754	89200			0.7-1.3	0.7		0.7	0.7					Sn 17.5-22.5			
	India	IS 7793	4928-A/42285	Frg		0.8-1.5	0.7	0.8-1.3	0.2	17-19	0.2	0.2			Ni 0.8-1.3; Pb 0.05; Sn 0.05 max			
4048	USA				0.07 max	3.3-4.7	0.8 max	0.07 max	0.07 max	9.3-10.7		9.3-10.7	0.05	0.15	0.0008 max Be for welding			

UNS numbers and US grades are provided as a means of cross referencing chemically similar alloys. Exchangability is only possible after independent examination of specifications. Tensile properties are minimum or typical . UTS and YS as Mpa, El as %. See Appendix for list of abbreviations used in Descriptions.

Grade UNS #	Country	Specification	Designation	Description	Cr	Cu	Fe	Mg	Mn	Si	Ti	Zn	OE Max	OT Max	Other	UTS	YS	EL
	Germany	DIN 8512	3.2685/L-AlSiSn							10-12					Al 72			
	Germany	DIN EN 10210		Replaces DIN 1712 pt 3														
	UK	BS EN 586		Replaces BS 1472														
	USA	ASME SB211		See ASTM B211														
	USA	ASME SB308		See ASTM B308														

UNS numbers and US grades are provided as a means of cross referencing chemically similar alloys. Exchangability is only possible after independent examination of specifications. Tensile properties are minimum or typical . UTS and YS as Mpa, El as %. See Appendix for list of abbreviations used in Descriptions.

Wrought Aluminum 2-91

Grade UNS #	Country	Specification	Designation	Description	Cu	Fe	Mg	Mn	Ni	Si	Ti	Zn	OE Max	OT Max	Other	UTS	YS	EL
	Canada	CSA HA.2	0.9999												Al 99.99			
	Germany	DIN 1712 pt 1	3.0400/Al99.9 9R	Ingot	0.003 max	0.005 max				0.006 max	0.002 max	0.005 max			ET 0.001 max; Al 99.99 min			
	Canada	CSA HA.2	0.9995												Al 99.95			
	Canada	CSA HA.2	0.999			0.07				0.07					Al 99.9			
	South Africa	SABS 711	Al99.9	Ingot	0.005 max	0.04	0.01 max	0.01 max		0.04 max	0.005 max	0.015 max		0.005	Al 99.9 min; Ga 0.03 max			
	Austria	ONORM M3426	H Al199,9		0.005 max	0.035 max				0.05 max	0.006 max	0.035 max			ET 0.003 max; Ga 0.03 max			
	Germany	DIN 1712 pt 1	3.0300/Al99.9 H	Ingot	0.005 max	0.04 max				0.05 max	0.005 max	0.003 max			ET 0.01 max; Al 99.9 min			
	South Africa	SABS 711	Al99.85	Ingot	0.01 max	0.07	0.01 max	0.02 max		0.05 max	0.01 max	0.02 max		0.01	Al 99.85; Cr 0.02; Ga 0.03 max			
	Canada	CSA HA.2	0.9985		0.02	0.1				0.1					Al 99.85			
	Sweden	MNC 41E	4020	Ingot											Al 99.8			
	South Africa	SABS 711	Al99.8	Ingot	0.01 max	0.1	0.01 max	0.02 max		0.1 max	0.02 max	0.03 max		0.01	Al 99.8 min; Cr 0.02; Ga 0.03 max			
	Austria	ONORM M3426	H Al199,8		0.01 max	0.15 max				0.15 max	0.02 max	0.04 max			ET 0.01 max; Ga 0.03 max			
	Germany	DIN 1712 pt 1	3.0280/Al99.8 H	Ingot	0.01 max	0.15 max				0.15 max	0.02 max	0.04 max			ET 0.02 max; Al 99.8 min			
	Canada	CSA HA.2	0.998		0.02	0.15				0.15					Al 99.8			
	Finland	SFS 2560	G-Al99.8		0.02	0.15				0.15	0.06				Al 99.8			
	International	ISO R115	Al 99.8	Ingot for RemeltIngot	0.02 max	0.15 max				0.15 max	0.06 max		0.03	0.2	Al 99.8 min			
	Sweden	SIS 144020	4020-00		0.02	0.15				0.15	0.06				Al 99.8			
	South Africa	SABS 711	Al 99.75 E	Ingot for remeltIngot	0.01 max	0.15	0.01 max	0.002 max		0.00 max	0.002 max	0.02 max		0.02	Al 99.75 min; B 0.01 max; Be 0.01 max; Cr 0.002 max; Ga 0.03 max; V 0.002 max			
	Canada	CSA HA.2	0.9975		0.02	0.2				0.2					Al 99.75			
	Australia	AS 1874	BA170	Ingot									0.03	0.1	Fe 1.5xSi min; Mn+Ti+Cr+V=0.025 max; Al 99.7 max			
170.1 A01701	International		170.1	Ingot								0.05 max	0.03	0.1	Mn+Cr+Ti+V 0.025 max; Fe/Si 1.5 min; Al 99.7 min			
	Sweden	MNC 41E	4021	Ingot											Al 99.7			
	South Africa	SABS 711	Al 99.70 E	Ingot for remeltIngot	0.01 max	0.2 max	0.02 max	0.003 max		0.08 max	0.003 max	0.02 max	0.02		Al 99.7 min; B 0.02 max; Cr 0.003 max; V 0.003 max			
	South Africa	SABS 711	Al99.7	Ingot	0.01 max	0.2	0.02 max	0.03 max		0.1 max	0.02 max	0.03 max		0.02	Al 99.7 min			
	Austria	ONORM M3426	H Al199,7		0.01 max	0.25 max				0.2 max	0.03 max	0.05 max			ET 0.01 max; Ga .03 max			
	Germany	DIN 1712 pt 1	3.0270/Al99.7 H	Ingot	0.01 max	0.25 max				0.2 max	0.02 max	0.04 max			ET 0.03 max; Al 99.7 min			
170.1 A01701	Australia	AS 1874	AA185	Ingot	0.02 max	0.1 max				0.1 max			0.03	0.1	Al 99.85 max			
170.1 A01701	Australia	AS 1874	AA180	Ingot	0.02 max	0.15 max				0.15 max			0.03	0.1	Al 99.8 max			
170.1 A01701	Australia	AS 1874	AA175	Ingot	0.02 max	0.2 max				0.2 max			0.03	0.1	Al 99.75 max			
170.1 A01701	Australia	AS 1874	AA170	Ingot	0.02 max	0.25 max				0.2 max			0.03	0.1	Al 99.7 max			
	Canada	CSA HA.2	0.997		0.02	0.25				0.2					Al 99.7			

UNS numbers and US grades are provided as a means of cross referencing chemically similar alloys. Exchangability is only possible after independent examination of specifications. Tensile properties are minimum or typical . UTS and YS as Mpa, El as %. See Appendix for list of abbreviations used in Descriptions.

Worldwide Guide to Equivalent Nonferrous Metals and Alloys

Grade UNS #	Country	Specification	Designation	Description	Cu	Fe	Mg	Mn	Ni	Si	Ti	Zn	OE Max	OT Max	Other	UTS	YS	EL
	Finland	SFS 2561	G-Al99.7		0.02	0.25				0.2		0.06			Al 99.7			
	India	IS 23	99.7Al		0.02	0.25		0.03		0.2		0.03			Al 99.7			
	International	ISO R115	Al 99.7	Ingot for RemeltIngot	0.02 max	0.25 max				0.2 max		0.06 max	0.03	0.3	Al 99.7 min			
	South Africa	SABS 989	Al-99.7A		0.02	0.25				0.2		0.06			Al 99.7			
	South Africa	SABS 991	Al-99.7A		0.02	0.25				0.2		0.06			Al 99.7			
	South Africa	SABS 992	Al-99.7A		0.02	0.25				0.2		0.06			Al 99.7			
	Sweden	SIS 144021	4021-00		0.02	0.25				0.2		0.06			Al 99.7			
	Canada	CSA HA.2	0.9965			0.3				0.25					Al 99.65			
	South Africa	SABS 711	Al 99.65 E	Ingot for remeltIngot	0.01 max	0.25 max	0.02 max	0.003 max		0.1 max	0.003 max		0.02		Al 99.65 min; B 0.02 max; Cr 0.003 max; V 0.003 max			
	South Africa	SABS 711	Al 99.65	Ingot for remeltIngot	0.02 max	0.2 max	0.03 max	0.03 max		0.15 max		0.04 max	0.02		Al 99.65 min; Cr 0.03 max			
160.1 A01601	International		160.1	Ingot		0.25 max				0.1 max		0.05 max	0.03	0.1	Mn+Cr+Ti+V 0.025 max; Fe/Si 2.0 min; Al 99.6 min			
160.1 A01601	Australia	AS 1874	AA160	Ingot		0.3 max				0.1 max			0.02	0.1	Fe 2.0xSi min; Mn+Ti+Cr+V=0.01 max; Al 99.6 max			
	South Africa	SABS 711	Al 99.60 E	Ingot for remeltIngot	0.01 max	0.3 max	0.02 max	0.005 max		0.1 max	0.005 max	0.04 max	0.03		Al 99.6 min; B 0.04 max; Cr 0.005 max; V 0.005 max			
160.1 A01601	South Africa	SABS 711	Al-EC99.6		0.02	0.3				0.15		0.07			Al 99.6			
160.1 A01601	South Africa	SABS 711	Al99.65		0.02	0.25	0.03	0.03		0.2		0.06			Al 99.65; Cr 0.03			
	Sweden	MNC 41E	4022	Ingot											Al 99.5			
	South Africa	SABS 711	Al 99.50 E	Ingot for remeltIngot		0.35 max		0.005 max		0.1 max	0.004 max	0.03 max	0.02		Al 99.5 min; B 0.02 max; Cr 0.004 max; V 0.004 max			
150.1 A01501	Australia	AS 1874	AA150	Ingot		0.4 max				0.3 max			0.03	0.15	Al 99.5 max			
	Canada	CSA HA.2	0.995			0.4				0.3					Al 99.5			
	Germany	DIN 1712 pt 1	3.0250/Al99.5 H	Ingot	0.01 max	0.25 max				0.2 max	0.02 max	0.04 max			ET 0.03 max; Al 99.5 min			
	Austria		H Al199,5		0.02 max	0.4 max				0.25 max	0.03 max	0.05 max			ET 0.02 max; Ga 0.03 max			
	UK	BS 1490	LM0		0.03 max	0.4 max	0.03 max	0.3 max	0.03 max	0.03 max		0.07 max			Al 99.5 min; Pb 0.03 max; Sn 0.03 max			
	Finland	SFS 2562	G-Al99.5		0.03	0.4				0.3		0.07			Al 99.5			
	India	IS 23	99.5Al	Bar	0.03	0.4		0.03		0.3		0.05			Al 99.5			
	International	ISO R115	Al 99.5	Ingot for RemeltIngot	0.03 max	0.4 max				0.3 max		0.07 max	0.03	0.5	Al 99.5 min			
150.1 A01501	South Africa	SABS 711	10500 (Al99.5)		0.03	0.4	0.03	0.03		0.3		0.07			Al 99.5			
	South Africa	SABS 989	Al-99.5A		0.03	0.4				0.3		0.07			Al 99.5			
	South Africa	SABS 991	Al-99.5A		0.03	0.4				0.3		0.07			Al 99.5			
	South Africa	SABS 992	Al-99.5A		0.03	0.4				0.3		0.07			Al 99.5			
150.1 A01501	South Africa	SABS 989	10500 (Al99.5)	Ingot	0.03 max	0.4 max				0.3 max		0.07 max	0.03		Al 99.5 min			
	South Africa	SABS 991	Al99.5	Sand	0.03 max	0.4 max				0.3 max		0.07 max	0.03		Al 99.5 min			

UNS numbers and US grades are provided as a means of cross referencing chemically similar alloys. Exchangability is only possible after independent examination of specifications. Tensile properties are minimum or typical . UTS and YS as Mpa, El as %. See Appendix for list of abbreviations used in Descriptions.

3-2 Cast Aluminum

Grade UNS #	Country	Specification	Designation	Description	Cu	Fe	Mg	Mn	Ni	Si	Ti	Zn	OE Max	OT Max	Other	UTS	YS	EL
	South Africa	SABS 992	Al 99.5	Press Die	0.03 max	0.4 max				0.3 max		0.07 max	0.03		Al 99.5 min			
	Sweden	SIS 144022	4022-00		0.03	0.4				0.3		0.07			Al 99.5			
150.1 A01501	International		150.1	Ingot	0.05 max							0.05 max	0.03	0.1	Mn+Cr+Ti+V 0.025 max; Fe/Si 2.0 min; Al 99.5 min			
	South Africa	SABS 711	Al 99.50	Ingot for remeltIngot	0.02 max	0.3 max	0.03 max	0.03 max		0.15 max	0.02 max	0.05 max	0.03		Al 99.5 min			
	Austria	ONORM M3426	H Al99,3		0.03 max	0.5 max				0.4 max	0.03 max	0.08 max			ET 0.03 max;			
	Japan	JIS H2211	C7AV	Ingot	0.05	0.2		0.03	0.03	0.2	0.2	0.03						
	Sweden	MNC 41E	4024	Ingot											Al 99			
	Canada	CSA HA.2	0.99			0.6				0.5					Al 99			
	South Africa	SABS 711	Al99.0	Ingot	0.005 max	0.04	0.01 max	0.01 max		0.04 max	0.005 max	0.015 max		0.005	Al 99 min; Ga 0.03 max			
	India	IS 23	99Al	Bar	0.03	0.6		0.05		0.5		0.06			Al 99			
	Austria	ONORM M3426	H Al99,0		0.03 max	0.8 max				0.5 max	0.03 max	0.08 max			ET 0.03 max;			
	Finland	SFS 2563	G-Al99.0		0.03	0.8				0.5		0.08			Al 99			
	International	ISO R115	Al 99.0	Ingot for RemeltIngot	0.03 max	0.8 max				0.5 max		0.08 max	0.03	1	Al 99 min			
	Sweden	SIS 144024	4024-00		0.03	0.8				0.5		0.08			Al 99			
100.1 A01001	International		100.1	Ingot	0.1 max	0.6 0.8				0.15 max		0.05 max	0.03	0.1	Mn+Cr+Ti+V 0.025 max; Al 99 min			
100.1 A01001	USA	ASTM B179	100.1		0.1	0.6-0.8				0.15		0.05			Al 99			
	South Africa	SABS 991	Al-Si12(Fe)	F Sand	0.15 max	0.65 max	0.1 max	0.55 max	0.1 max	1.5-13.5	0.2 max	0.15 max	0.05	0.15	Pb 0.1 max	150		4
	South Africa	SABS 991	Al-Si12(Fe)	F Gravity Die	0.15 max	0.65 max	0.1 max	0.55 max	0.1 max	1.5-13.5	0.2 max	0.15 max	0.05	0.15	Pb 0.1 max	170		7
	UK	BS 1490	M27		1.5-2.5	0.4 max	0.03 max	0.03 max	0.03 max	0.3 max		0.07 max			Pb 0.03 max; Sn 0.03 max			
515.2 A05142	USA	QQ A-371F	514.2		0.1	0.3	3.6-4.5	0.1		0.3	0.2	0.1						
	Sweden	SIS 144261	4261-03	Sand As cast		0.6	0.1	0.5	0.1		0.2	0.3			Pb 0.1; Sn 0.05 max		80	4
	Sweden	SIS 144261	4261-06	As cast		0.6	0.1	0.5	0.1		0.2	0.3			Pb 0.1; Sn 0.05 max		90	5
	Austria	ONORM M3426	H Al98		0.1 max	1 max	0.05 max	0.05 max		0.8 max	0.1 max	0.1 max			ET 0.05 max;			
	USA	ASTM B37	980A	Rod	0.2		0.5					0.2			Al 98			
	Austria	ONORM M3429	AlMg3(Cu)	perm Mold Sand	0.3 max	0.6 max	2-4	0.6 max		1.3 max	0.2 max	0.3 max			ET 0.05 max; ET=.15;			
	Austria	ONORM M3429	G AlMg3(Cu)		0.3 max	0.6 max	2-4			1.3 max	0.2 max	0.3 max	0.05		ET=.15 total;			
	Austria	ONORM M3429	GS AlMg3(Cu)	Sand	0.3 max	0.6 max	2-4	0.6 max		1.3 max	0.2 max	0.3 max			ET 0.05 max; ET=.15 total;			
	Germany	DIN 1725Sh3	3.2581/V-AlSi12		0.05	0.5	0.05	0.4		1-13.5	0.15	0.1						
	Germany	DIN 1725Sh3	3.0551/V-AlCr5		0.15	0.45	0.5	0.35	0.1	0.4	0.1	0.15			Cr 4-6; Pb 0.1; Sn 0.1			
	Austria	ONORM M3429	AlMg3		0.03 max	0.4 max	2.5-3.5	0.4 max		0.3 max	0.2 max	0.1 max			ET 0.05 max; ET=0.15 total;			
	Austria	ONORM M3429	AlMg3		0.03 max	0.4 max	2.5-3.5	0.4 max		0.3 max	0.2 max	0.1 max			ET 0.05 max; ET=.15 total;			
	Austria	ONORM M3429	GK AlMg3	Perm Mold	0.05 max	0.5 max	2.5-3.5	0.4 max		0.5 max	0.2 max	0.1 max			ET 0.05 max; ET=.15 total;			
	Austria	ONORM M3429	GS AlMg3	Sand	0.05 max	0.5 max	2.5-3.5	0.4 max		0.5 max	0.2 max	0.1 max			ET 0.05 max; ET=.15 total;			
	Belgium	NBN P21-101	SGAlMg3	Sand	0.1 max	0.5 max	2.5-4	0.6 max	0.05 max	0.5 max	0.2 max	0.2 max			Cr 0.1 max; Pb 0.05 max; Sn 0.05 max			

UNS numbers and US grades are provided as a means of cross referencing chemically similar alloys. Exchangability is only possible after independent examination of specifications. Tensile properties are minimum or typical . UTS and YS as Mpa, El as %. See Appendix for list of abbreviations used in Descriptions.

Worldwide Guide to Equivalent Nonferrous Metals and Alloys

Grade UNS #	Country	Specification	Designation	Description	Cu	Fe	Mg	Mn	Ni	Si	Ti	Zn	OE Max	OT Max	Other	UTS	YS	EL
	France	NF A57-702	AG3T	Perm Mold Sand	0.1 max	0.5 max	2.5-3.5	0.5 max	0.05 max	0.5 max	0.2 max	0.2 max			Pb 0.05 max; Sn 0.05 max			
	International	ISO 3522	Al-Mg3	Sand Cast As mfg	0.1 max	0.5 max	2.5-4.5	0.6 max	0.05 max	0.5 max	0.2 max	0.2 max			Cr 0.1 max; Pb 0.05 max; Sn 0.05 max	150		5
	International	ISO 3522	Al-Mg3	Perm Mold As mfg	0.1 max	0.5 max	2.5-4.5	0.6 max	0.05 max	0.5 max	0.2 max	0.2 max			Cr 0.1 max; Pb 0.05 max; Sn 0.05 max	150		5
	South Africa	SABS 991	Al-Mg3	F Sand	0.1 max	0.55 max	2.5-3.5	0.45 max		0.55 max	0.2 max	0.1 max	0.05	0.15		140		3
	South Africa	SABS 991	Al-Mg3	F Gravity Die	0.1 max	0.55 max	2.5-3.5	0.45 max		0.55 max	0.2 max	0.1 max	0.05	0.15		150		5
514.2 A05142	Canada	CSA HA.3	.GS40		0.1	0.3	3.6-4.5	0.1		0.3-0.7	0.2	0.1						
	South Africa	SABS 989	Al-Mg3	Ingot	0.08 max	0.45 max	2.7-3.5	0.45 max		0.45 max	0.15 max	0.1 max	0.05	0.15				
	Germany	DIN 1725Sh3	3.0821/V-AlB3		0.02	0.3	0.02	0.02	0.02	0.2	0.02	0.03			B 2.5-3.4; Cr 0.02			
435.2 A04352	International		435.2	Ingot	0.05 max	0.4 max	0.05 max	0.05 max		3.3-3.9		0.1 max	0.05	0.2				
	Austria	ONORM M3429	AlMg3Si		0.03 max	0.4 max	2.5-3.5	0.4 max		0.9-1.3	0.2 max	0.1 max			ET 0.05 max; ET=.15;			
	Austria	ONORM M3429	G AlMg3Si		0.03 max	0.4 max	2.5-3.5	0.4 max		0.9-1.3	0.2 max	0.1 max			ET 0.05 max; ET=.15 total			
	Austria	ONORM M3429	GK AlMg3Si	Perm Mold	0.05 max	0.5 max	2.5-3.5	0.4 max		0.9-1.3	0.2 max	0.1 max			ET 0.05 max; ET=.15 total;			
	Austria	ONORM M3429	GS AlMg3Si	Sand	0.05 max	0.5 max	2.5-3.5	0.4 max		0.9-1.3	0.2 max	0.1 max			ET 0.05 max; ET=.15 total;			
514.2 A05142	Canada	CSA HA.9	.GS40	Sand	0.1	0.5	3.5-4.5	0.35		0.3-0.7	0.25	0.15						
	International	ISO 3522	Al-Mg3 Si2	Sand Cast Perm Mold	0.1 max	0.5 max	2.5-4.5	0.6 max	0.05 max	0.9-2.2	0.2 max	0.2 max			Cr 0.4 max; Pb 0.05 max; Sn 0.05 max			
515 A05150	International		515	Die	0.2 max	1.3 max	2.5-4	0.4-0.6		0.5-1		0.1 max	0.05	0.15				
514.2 A05142	USA	ASTM B179	514.2		0.1	0.3	3.6-4.5	0.1		0.3	0.2	0.1						
514 A05140	Japan	JIS H5202	AC7A		0.1	0.4	3.5-5.5	0.6		0.3	0.2	0.1				216		12
514 A05140	International	ISO R164	Al-Mg3		0.1	0.05	2-4.5	0.6	0.05	0.5	0.2	0.2			Cr 0.1; Pb 0.05; Sn 0.05 max			
514 A05140	International	ISO R2147	Al-Mg3	Sand OBSOLETE Replaced by ISO 3522	0.1	0.05	2-4.5	0.6	0.05	0.5	0.2	0.2			Cr 0.1; Pb 0.05; Sn 0.05 max	150		5
514 A05140	Russia	GOST 2685	AL13		0.1 max	0.5 max	4.5-5.5	0.1-0.4	0.2 max	0.8-1.3				0.6				
514 A05140	International		514	Sand	0.15 max	0.5 max	3.5-4.5	0.35 max		0.35 max	0.25 max	0.15 max	0.05	0.15				
514 A05140	USA	ASTM B26	514	Sand As fabricated	0.15	0.5	3.5-4.5	0.35		0.35	0.25	0.15				152	62	6
514 A05140	USA	ASTM B618	514	Invest As fabricated .250 in diam	0.15	0.5	3.5-4.5	0.35		0.35	0.25	0.15				152	62	6
514.2 A05142	International		514.2	Ingot	0.1 max	0.3 max	3.6-4.5	0.1 max		0.3 max	0.2 max	0.1 max	0.05	0.15				
514.1 A05141	Canada	CSA	G4		0.1 max	0.4 max	3.5-4.5	0.3 max		0.3 max	0.2 max	0.1 max			ET 0.05 max; ET=.15 total;			
514.1 A05141	International		514.1	Ingot	0.15 max	0.4 max	3.6-4.5	0.35 max		0.35 max	0.25 max	0.15 max	0.05	0.15				
514.1 A05141	USA	ASTM B179	514.1		0.15	0.4	3.6-4.5	0.35		0.35	0.25	0.15						
514.1 A05141	USA	QQ A-371F	514.1		0.15	0.4	3.6-4.5	0.35		0.35	0.25	0.15						
516 A05160	International		516	Die	0.3 max	0.35-1	2.5-4.5	0.15-0.4	0.25-0.4	0.3-1.5	0.1-0.2	0.2 max	0.05		Pb 0.10 max; Sn 0.1 max			
	South Africa	SABS 711	AlCr2		0.1	0.6				0.3		0.07			Al 96.27; Cr 2.5			

UNS numbers and US grades are provided as a means of cross referencing chemically similar alloys. Exchangability is only possible after independent examination of specifications. Tensile properties are minimum or typical . UTS and YS as Mpa, El as %. See Appendix for list of abbreviations used in Descriptions.

3-4 Cast Aluminum

Grade UNS #	Country	Specification	Designation	Description	Cu	Fe	Mg	Mn	Ni	Si	Ti	Zn	OE Max	OT Max	Other	UTS	YS	EL
516.1 A05161	International		516.1	Ingot	0.3 max	0.35-0.7	2.6-4.5	0.15-0.4	0.25-0.4	0.3-1.5	0.1-0.2	0.2 max	0.05		Pb 0.10 max; Sn 0.1 max			
511 A05110	Czech Republic	CSN 424515	AlMg5Si1Mn	Sand Ingot Alloys	0.05 max	0.5 max	4.4-5.5	0.25-0.6		0.6-1.5	0.15 max	0.1 max			Cr 0.15 max			
511 A05110	Czech Republic	CSN 424515	AlMg5Si1Mn	Permanent Mold Ingot Alloys	0.05 max	0.5 max	4.4-5.5	0.25-0.6		0.6-1.5	0.15 max	0.1 max			Cr 0.15 max			
511 A05110	International		511	Sand	0.15 max	0.5 max	3.5-4.5	0.35 max		0.3-0.7	0.25 max	0.15 max	0.05	0.15				
511.2 A05112	International		511.2	Ingot	0.1 max	0.3 max	3.6-4.5	0.1 max		0.3-0.7	0.2 max	0.1 max	0.05	0.15				
511.1 A05111	International		511.1	Ingot	0.15 max	0.4 max	3.6-4.5	0.35 max		0.3-0.7	0.25 max	0.15 max	0.05	0.15				
	International	ISO 3522	Al-Si5 Mg	Sand Cast SHT PHT	0.1 max	0.6 max	0.5-0.9	0.6 max	0.1 max	3.5-6	0.2 max	0.1 max			Pb 0.1 max; Sn 0.05 max	230		1
	International	ISO 3522	Al-Si5 Mg	Perm Mold As mfg	0.1 max	0.6 max	0.5-0.9	0.6 max	0.1 max	3.5-6	0.2 max	0.1 max			Pb 0.1 max; Sn 0.05 max	160		2
	International	ISO 3522	Al-Si5 Mg	Perm Mold SHT PHT	0.1 max	0.6 max	0.5-0.9	0.6 max	0.1 max	3.5-6	0.2 max	0.1 max			Pb 0.1 max; Sn 0.05 max	240		1
	International	ISO 3522	Al-Cu4 Ti	Sand Cast SHT PHT	4-5	0.25 max	0.05 max	0.1 max	0.1 max	0.25 max	0.05-0.3	0.2 max			Pb 0.05 max; Sn 0.05 max	280		4
	International	ISO 3522	Al-Cu4 Ti	Perm Mold SHT PHT	4-5	0.25 max	0.05 max	0.1 max	0.1 max	0.25 max	0.05-0.3	0.2 max			Pb 0.05 max; Sn 0.05 max	310		9
	Germany	DIN 1725Sh3	3.0831/V-AlB4		0.02	0.3	0.02	0.02	0.02	0.2	0.02	0.03			B 3.5-4.5; Cr 0.02; V 0.02; Zr 0.02			
515.2 A05152	South Africa	SABS 989	Al-Mg3A		0.03	0.4	2.8-3.5	0.5	0.05	1.3	0.2	0.1			Pb 0.05; Sn 0.05 max	140		3
515.2 A05152	South Africa	SABS 989	Al Mg3A	Chill	0.03	0.4	2.8-3.5	0.5	0.05	1.3	0.2	0.1			Pb 0.05; Sn 0.05 max	145		3
515.2 A05152	South Africa	SABS 989	Al-Mg3A	SHT and PHT	0.03	0.4	2.8-3.5	0.5	0.05	1.3	0.2	0.1			Pb 0.05; Sn 0.05 max	250		2
515.2 A05152	South Africa	SABS 990	Al Mg3A	Sand	0.05	0.5	2.5-3.5	0.5	0.05	1.3	0.2	0.1			Pb 0.05; Sn 0.5 max	140		3
515.2 A05152	South Africa	SABS 990	Al-Mg3A	Sand SHT and PHT	0.05	0.5	2.5-3.5	0.5	0.05	1.3	0.2	0.1			Pb 0.05; Sn 0.5 max	205		2
515.2 A05152	International		515.2	Ingot	0.1 max	0.6-1	2.7-4	0.4-0.6		0.5-1		0.05 max	0.05	0.15				
	Germany	DIN 1725Sh2	3.3541.01/G-AlMg3	Sand	0.05	0.5	2.5-3.5	0.4		0.5	0.2	0.1				140	70	3
	Germany	DIN 1725Sh2	3.3541.02/GK-AlMg3		0.05	0.5	2.5-3.5	0.4		0.5	0.2	0.1				150	70	5
	Belgium	NBN P21-101	SGAlCu4MgTi	Sand	4-5	0.4 max	0.15-0.35	1 max	0.05 max	0.05-0.3	0.1-0.3	0.2 max			Pb 0.05 max; Sn 0.05 max			
	USA	QQ A-371F	L514.2		0.1	0.6-1	2.7-4	0.4-0.6		0.5-1		0.05						
	Belgium	NBN P21-101	SGAlCu2NiMg	Sand	1.5-2.5	0.8-1.4	0.6-1	0.1 max	0.8-1.8	0.7-2.3	0.35 max	0.5 max			Cr 0.2 max; Pb 0.05; Sn 0.05 max max			
	International	ISO 3522	Al-Cu4 Mg Ti	Sand Cast SHT Aged	4.2-5	0.35 max	0.15-0.35	0.1 max	0.05 max	0.3 max	0.05-0.35	0.1 max			Pb 0.05 max; Sn 0.05 max	290		4
	International	ISO 3522	Al-Cu4 Mg Ti	Perm Mold SHT Aged	4.2-5	0.35 max	0.15-0.35	0.1 max	0.05 max	0.3 max	0.05-0.35	0.1 max			Pb 0.05 max; Sn 0.05 max	330		8
443.1 A04431	USA	QQ A-371F	443.1			0.6	0.05	0.5		4-6	0.25	0.5			Cr 0.25			
	Belgium	NBN P21-101	SGAlMg6	Sand	0.1 max	0.5 max	4.5-7	0.6 max	0.05 max	0.5 max	0.2 max	0.2 max			Cr 0.5 max; Pb 0.05 x; Sn 0.05 x			
	International	ISO 3522	Al-Mg6	Sand Cast As mfg	0.1 max	0.5 max	4.5-7	0.6 max	0.05 max	0.5 max	0.2 max	0.2 max			Cr 0.5 max; Pb 0.05 x; Sn 0.05 x	160		2
	International	ISO 3522	Al-Mg6	Perm Mold As mfg	0.1 max	0.5 max	4.5-7	0.6 max	0.05 max	0.5 max	0.2 max	0.2 max			Cr 0.5 max; Pb 0.05 max; Sn 0.05 max	170		3
	International	ISO 3522	Al-Mg5 Si1	Perm Mold As mfg	0.1 max	0.5 max	4-6	0.5 max	0.05 max	0.5-1.5	0.2 max	0.2 max			Pb 0.05 max; Sn 0.05 max	170		2
B443.0 A24430	Belgium	NBN P21-101	SGAlSi5Mg	Sand	0.1 max	0.6 max	0.4-0.9	0.6 max	0.1 max	3.5-6	0.02 max	0.1 max			Pb 0.1 max; Sn 0.05 max			
B443.1 A24431	Canada	CSA HA.3	S5		0.1	0.6	0.05	0.1		4.5-6	0.2	0.1						

UNS numbers and US grades are provided as a means of cross referencing chemically similar alloys. Exchangability is only possible after independent examination of specifications. Tensile properties are minimum or typical . UTS and YS as Mpa, El as %. See Appendix for list of abbreviations used in Descriptions.

Worldwide Guide to Equivalent Nonferrous Metals and Alloys

Grade UNS #	Country	Specification	Designation	Description	Cu	Fe	Mg	Mn	Ni	Si	Ti	Zn	OE Max	OT Max	Other	UTS	YS	EL
443.2 A14432	International		443.2	Ingot	0.1 max	0.6 max	0.05 max	0.1 max		4.5-6	0.25	0.1 max	0.05	0.15				
B443.1 A24431	Canada	CSA HA.10	S5		0.1	0.8	0.05	0.1		4.5-6	0.2	0.1						
B443.1 A24431	Canada	CSA HA.9	S5	Sand	0.1	0.8	0.05	0.1		4.5-6	0.2	0.1						
	International	ISO 3522	Al-Si5	Sand Cast As mfg	0.1 max	0.8 max	0.1 max	0.5 max	0.1 max	4.5-6	0.2 max	0.1 max			Pb 0.1 max; Sn 0.1 max	120		2
C443.0 A34430	International	ISO R164	Al-Si5Fe		0.1	1.3	0.1	0.5	0.1	4-6	0.2	0.1			Pb 0.1; Sn 0.1 max			
	International	ISO 3522	Al-Si5 Fe	Pressure Die Cast	0.1 max	1.3 max	0.1 max	0.5 max	0.1 max	4.5-6	0.2 max	0.1 max			Pb 0.1 max; Sn 0.1 max			
B443.1 A24431	International		B443.1	Ingot	0.15 max	0.6 max	0.05 max	0.35 max		4.5-6	0.25 max	0.35 max	0.05	0.15				
B443.0 A24430	International		B443.0	Sand, Perm Mold	0.15 max	0.8 max	0.05 max	0.35 max		4.5-6	0.25 max	0.35 max	0.05	0.15				
705 A07050	International		705	Sand, Perm Mold	0.2 max	0.8 max	1.4-1.8	0.4-0.6		0.2 max	0.25 max	2.7-3.3	0.05	0.15	Cr 0.2-0.4			
705 A07050	USA	ASTM B108	705	Cooled and nat aged or art aged	0.2	0.8	1.4-1.8	0.4-0.6		0.2	0.25	2.7-3.3			Cr 0.2-0.4	255	117	10
705 A07050	USA	ASTM B26	705	Sand Cooled and art aged .250 in diam	0.2	0.8	1.4-1.8	0.4-0.6		0.2	0.25	2.7-3.3			Cr 0.2-0.4	207	117	5
705 A07050	USA	ASTM B618	705	Invest Cooled and nat or art aged .250 in diam	0.2	0.8	1.4-1.8	0.4-0.6		0.2	0.25	2.7-3.3			Cr 0.2-0.4	207	117	5
A443.1 A14431	International		A443.1	Ingot	0.3 max	0.6 max	0.05 max	0.5 max		4.5-6	0.25 max	0.5 max		0.35	Cr 0.25 max			
A443.0 A14430	International		A443.0	Sand	0.3 max	0.8 max	0.05 max	0.5 max		4.5-6	0.2 max	0.5 max		0.35	Cr 0.25 max			
443.1 A04431	International		443.1	Ingot	0.6 max	0.6 max	0.05 max	0.5 max		4.5-6	0.25 max	0.5 max		0.35	Cr 0.25 max			
443.1 A04431	USA	ASTM B179	443.1		0.6	0.6	0.05	0.5		4.5-6	0.25	0.5			Cr 0.25			
443 A04430	International		443	Sand, Perm Mold	0.6 max	0.8 max	0.05 max	0.5 max		4.5-6		0.5 max		0.35	Cr 0.25 max			
443 A04430	USA	ASTM B108	443	As fabricated	0.6	0.8	0.05	0.5		4.5-6	0.25	0.5			Cr 0.25	145	49	2
443 A04430	USA	ASTM B26	443	Sand As fabricated .250 in diam	0.6	0.8	0.05	0.5		4.5-6	0.25	0.5			Cr 0.25	117	48	3
443 A04430	USA	ASTM B618	443	Invest As fabricated .250 in diam	0.6	0.8	0.05	0.5		4.5-6	0.25	0.5			Cr 0.25	117	48	3
C443.1 A34431	International		C443.1	Ingot	0.6 max	1.1 max	0.1 max	0.35 max	0.5 max	4.5-6		0.4 max		0.25	Sn 0.15 max			
C443.0 A34430	International		C443.0	Die	0.6 max	2 max	0.1 max	0.35 max	0.5 max	4.5-6	0.25 max	0.5 max		0.25	Sn 0.15 max			
443 A04430	USA	QQ A-591E	443		0.6	2	0.1	0.35	0.5	4.5-6		0.5			Sn 0.15 max	228	97	9
204 A02040	Austria	ONORM M3429	G AlCu4TiMg		4.2-4.9	0.12 max	0.15-0.3	0.05 max		0.12 max	0.15-0.3	0.07 max			ET 0.03 max; ET=.10 total;			
204 A02040	Austria	ONORM M3429	AlCu4TiMg	Perm Mold Sand	4.2-4.9	0.15 max	0.15-0.3	0.05 max		0.15 max	0.15-0.3	0.07 max			ET 0.03 max; ET=0.10 total;			
204 A02040	Austria	ONORM M3429	GK AlCu4TiMg	Perm Mold	4.2-4.9	0.15 max	0.15-0.3	0.05 max		0.15 max	0.15-0.3	0.07 max			ET 0.03 max; ET=.10 total;			
204 A02040	Austria	ONORM M3429	GS AlCu4TiMg	Sand	4.2-4.9	0.15 max	0.15-0.3	0.05 max		0.15 max	0.15-0.3	0.07 max			ET 0.03 max; ET=.10 total;			
A201.0 A12010	International		A201.0	Sand	4-5	0.1 max	0.15-0.35	0.2-0.4		0.05 max	0.15-0.35		0.03	0.1	Silver 0.40-1.0			
201 A02010	South Africa	SABS 989	21000 (Al-Cu4 Mg Ti)	Ingot	4.2-5	0.3 max	0.2-0.35	0.1 max	0.05 max	0.15 max	0.15-0.25	0.1 max	0.03	0.1	Pb 0.05 max; Sn 0.05 max			
204 A02040	International		204	Sand, Perm Mold	4.2-5	0.35 max	0.15-0.35	0.1 max	0.05 max	0.2 max	0.15-0.3	0.1 max	0.05	0.15	Sn 0.05 max			

UNS numbers and US grades are provided as a means of cross referencing chemically similar alloys. Exchangability is only possible after independent examination of specifications. Tensile properties are minimum or typical. UTS and YS as Mpa, El as %. See Appendix for list of abbreviations used in Descriptions.

Grade UNS #	Country	Specification	Designation	Description	Cu	Fe	Mg	Mn	Ni	Si	Ti	Zn	OE Max	OT Max	Other	UTS	YS	EL
201 A02010	South Africa	SABS 991	21000 (Al-Cu4 Mg Ti)	T4 Sand	4.2-5	0.35 max	0.15-0.35	0.1 max	0.05 max	0.2 max	0.15-0.3	0.1 max	0.03	0.1	Pb 0.05 max; Sn 0.05 max	300		5
201 A02010	South Africa	SABS 991	21000 (Al-Cu4 Mg Ti)	T4 Gravity Die	4.2-5	0.35 max	0.15-0.35	0.1 max	0.05 max	0.2 max	0.15-0.3	0.1 max	0.03	0.1	Pb 0.05 max; Sn 0.05 max	320		8
204 A02040	USA	ASTM B108	204	SHt and nat aged	4.2-5	0.35	0.15-0.35	0.1	0.05	0.2	0.15-0.3	0.1			Sn 0.05 max	331	200	8
204 A02040	USA	ASTM B26	204	Sand SHT and nat aged .250 in diam	4.2-5	0.35	0.15-0.35	0.1	0.05	0.2	0.15-0.3	0.1			Sn 0.05 max	310	193	6
204 A02040	USA	ASTM B618	204	Invest SHT and nat aged .250 in diam	4.2-5	0.35	0.15-0.35	0.1	0.05	0.2	0.15-0.3	0.1			Sn 0.05 max	310	193	6
204 A02040	South Africa	SABS 989	Al-Cu5MgTiA	ST and PHT	4.2-5	0.3	0.2-0.35	0.1	0.05	0.25	0.05-0.3	0.1			Sn 0.05 max	295		2
204 A02040	South Africa	SABS 989	Al-Cu5MgTiA	Chill SHT and PHT	4.2-5	0.3	0.2-0.35	0.1	0.05	0.25	0.05-0.3	0.1			Sn 0.05 max	325		3
204 A02040	South Africa	SABS 991	Al-Cu5MgTiA	SHT PHT	5	0.35	0.15-0.35	0.1	0.05	0.3	0.05-0.3	0.1			Sn 0.05 max	325		3
204 A02040	International	ISO R2147	Al-Cu4MgTi	Sand SHT and nat aged OBSOLETE Replaced by ISO 3522	4-5	0.05	0.15-0.35	0.1	0.05	0.35	0.35	0.2			Pb 0.05; Sn 0.05 max			4
201 A02010	International		201	Sand	4-5.2	0.15 max	0.15-0.55	0.2-0.5		0.1 max	0.15-0.35		0.05	0.1	Silver 0.40-1.0			
201 A02010	USA	ASTM B618	201	Invest SHT and art aged .250 in diam	4-5.2	0.15	0.15-0.55	0.2-0.5		0.1	0.15-0.35				Ag 0.4-1	414	345	5
201 A02010	USA	ASTM B618	201	Invest SHT and stab .250 in diam	4-5.2	0.15	0.15-0.55	0.2-0.5		0.1	0.15-0.35				Ag 0.4-1	414	345	3
201 A02010	USA	ASTM B26	201	Sand SHT and art aged .250 in diam	4-5.2	0.15	0.15-0.55	0.2-0.5		0.1	0.15-0.35					414	345	5
201 A02010	USA	ASTM B26	201	Sand SHT and stab .250 in diam	4-5.2	0.15	0.15-0.55	0.2-0.5		0.1	0.15-0.35					414	345	3
A201.1 A12011	International		A201.1	Ingot	4-5	0.07 max	0.2-0.35	0.2-0.4		0.05 max	0.15-0.35		0.03	0.1	Silver 0.40-1.0			
201.2 A02012	International		201.2	Ingot	4-5.2	0.1 max	0.2-0.55	0.2-0.5		0.1 max	0.15-0.35		0.05	0.1	Silver 0.40-1.0			
201.2 A02012	USA	ASTM B179	201.2		4-5.2	1		0.2-0.5		0.1	0.15-0.35				Ag 0.4-1.2			
705.1 A07051	International		705.1	Ingot	0.2 max	0.6 max	1.5-1.8	0.4-0.6		0.2 max	0.25 max	2.7-3.3	0.05	0.15	Cr 0.2-0.4			
705.1 A07051	USA	QQ A-371F	705.1		0.2	0.6	1.5-1.8	0.4-0.6		0.2	0.25	2.7-3.3			Cr 0.2-0.4			
	Germany	DIN 1725Sh2	3.3241./GK-AlMg3Siwa	Q and AH	0.05	0.5	2.5-3.5	0.4		0.9-1.3	0.2	0.1				220	120	3
	Germany	DIN 1725Sh2	3.3241.01/G-AlMg3Si	Sand	0.05	0.5	2.5-3.5	0.4		0.9-1.3	0.2	0.1				140	80	3
	Germany	DIN 1725Sh2	3.3241.02/GK-AlMg3Si		0.05	0.5	2.5-3.5	0.4		0.9-1.3	0.2	0.1				150	80	4
	Germany	DIN 1725Sh2	3.3241.6/G-AlMg3Si	Sand	0.05	0.5	2.5-3.5	0.4		0.9-1.3	0.2	0.1				200	120	2
204.2 A02042	USA	ASTM B179	204.2		4.2-4.9	0.1-0.2	0.2-0.35	0.05	0.03	0.15	0.15-0.25	0.05			Sn 0.05 max			
204.2 A02042	Germany	DIN 1725Sh2	3.1371.4/G-AlCu4TiMg	Sand Q and AH	4.2-4.9	2	0.15-0.3	0.05		0.18	0.15-0.3	0.07				350	240	3
204.2 A02042	Germany	DIN 1725Sh2	3.1371.4/G-AlCu4TiMg	Sand PH	4.2-4.9	2	0.15-0.3	0.05		0.18	0.15-0.3	0.07				300	220	5
	Austria	ONORM M3429	AlCu4Ti		4.5-5.2	0.12 max	0.03 max	0.05 max		0.12 max	0.15-0.3	0.07 max			ET 0.03 max; ET=0.10 total;			
	Austria	ONORM M3429	G AlCu4Ti		4.5-5.2	0.12 max	0.03 max	0.05 max		0.12 max	0.15-0.3	0.07 max			ET 0.03 max; ET=.10 total;			
	Austria	ONORM M3429	GK AlCu4Ti	Perm Mold	4.5-5.2	0.15 max	0.03 max	0.05 max		0.15 max	0.15-0.3	0.07 max			ET 0.03 max; ET=.10 total;			
	Austria	ONORM M3429	GS AlCu4Ti	Sand	4.5-5.2	0.15 max	0.03 max	0.05 max		0.15 max	0.15-0.3	0.07 max			ET 0.03 max; ET=.10 total;			

UNS numbers and US grades are provided as a means of cross referencing chemically similar alloys. Exchangability is only possible after independent examination of specifications. Tensile properties are minimum or typical . UTS and YS as Mpa, El as %. See Appendix for list of abbreviations used in Descriptions.

Cast Aluminum 3-7

Worldwide Guide to Equivalent Nonferrous Metals and Alloys

Grade UNS #	Country	Specification	Designation	Description	Cu	Fe	Mg	Mn	Ni	Si	Ti	Zn	OE Max	OT Max	Other	UTS	YS	EL
204.2 A02042	International		204.2	Ingot	4.2-7.9	0.1-0.2	0.2-0.35	0.05 max	0.03 max	0.15 max	0.15-0.25	0.05 max	0.05	0.15	Sn 0.05 max			
	Finland	SFS 2571	G-AlMg3	Die 20 mm diam	0.1	0.5	2.5-4	0.6	0.05	0.5	0.2	0.2			Cr 0.1; Pb 0.05; Sn 0.05 max	150	70	5
	USA	QQ A-371F	F514.2		0.1	0.3	3.6-4.5	0.1		0.3-0.7	0.2	0.1						
A206.0 A12060	International		A206.0	Sand, Perm Mold	4.2-5	0.1 max	0.15-0.35	0.2-0.5	0.05 max	0.05 max	0.15-0.3	0.1 max	0.05	0.15	Sn 0.05 max			
206 A02060	International		206	Sand, Perm Mold	4.2-5	0.15 max	0.15-0.35	0.2-0.5	0.05 max	0.1 max	0.15-0.3	0.1 max	0.05	0.15	Sn 0.05 max			
295 A02950	Japan	JIS H5202	AC1A		4-5	0.5	0.3	0.3		1.2	0.25	0.3				157		5
295 A02950	Japan	JIS H5202	AC1A	Q/T	4-5	0.5	0.3	0.3		1.2	0.25	0.3				275		3
295.2 A02952	International		295.2	Ingot	4-5	0.8 max	0.03 max	0.3 max		0.7-1.2	0.2 max	0.3 max	0.05	0.15				
295.2 A02952	USA	QQ A-371F	295.2		4-5	0.8	0.03	0.3	0.3	0.7-1.2	0.2							
295.2 A02952	USA	ASTM B179	295.2		4-5	0.8	0.03	0.3		0.7-1.2		0.3						
295.1 A02951	International		295.1	Ingot	4-5	0.8 max	0.03 max	0.35 max		0.7-1.5	0.25 max	0.35 max	0.05	0.15				
295.1 A02951	USA	QQ A-371F	295.1		4-5	0.8	0.03	0.35	0.35	0.7-1.5	0.25							
295 A02950	International		295	Sand	4-5	1 max	0.03 max	0.35 max		0.7-1.5	0.25 max	0.35 max	0.05	0.15				
295 A02950	Russia		AL7Ch		3-5	1 max	0.3 max	0.5 max	0.3 max	1.5 max		0.45 max						
295 A02950	USA	ASTM B26	295		4-5	1	0.03	0.35		0.7-1.5	0.25	0.35				200	90	6
295 A02950	USA	ASTM B26	295	Sand SHT and art aged .250 in diam	4-5	1	0.03	0.35		0.7-1.5	0.25	0.35				221	138	3
295 A02950	USA	ASTM B26	295	Sand SHT and stab .250 in diam	4-5	1	0.03	0.35		0.7-1.5	0.25	0.35				110	110	3
295 A02950	USA	ASTM B618	295	Invest SHT and nat aged .250 in diam	4-5	1	0.03	0.35		0.7-1.5	0.25	0.35				200	90	6
295 A02950	USA	ASTM B618	295	Invest SHT and art aged .250 in diam	4-5	1	0.03	0.35		0.7-1.5	0.25	0.35				221	138	3
295 A02950	USA	ASTM B618	295	Invest SHT and stab .250 in diam	4-5	1	0.03	0.35		0.7-1.5	0.25	0.35				200	110	3
295 A02950	Germany	DIN 1725 pt 2	3.1841/G-AlCu4Ti		4.5-5.2	0.18 max		0.05 max	0.07 max	0.18 max	0.15-0.3							
295 A02950	Germany	DIN 1725 pt 2	3.1841/Gk-AlCu4Ti		4.5-5.2	0.18 max		0.05 max		0.18 max	0.15-0.3	0.07 max			ET 0.03 max; ET=.10 total,			
224.2 A02242	International		224.2	Ingot	4.5-5.5	0.04 max		0.2-0.5		0.02 max	0.25 max		0.03	0.1	Vanadium 0.05-0.15, zirconium 0.10-0.25			
224 A02240	International		224	Sand, Perm Mold	4.5-5.5	0.1 max		0.2-0.5		0.06 max	0.35 max		0.03	0.1	Vanadium 0.05-0.15, zirconium 0.10-0.25			
	South Africa	SABS 991	Al-Mg4MnA		0.1	0.6	3-6	0.3-0.7	0.1	0.3	0.2	0.1			Pb 0.05; Sn 0.05 max	170		5
A206.2 A12062	International		A206.2	Ingot	4.2-5	0.07 max	0.2-0.35	0.2-0.5	0.03 max	0.05 max	0.15-0.25	0.05 max	0.05	0.15	Sn 0.05 max			
206.2 A02062	International		206.2	Ingot	4.2-5	0.1 max	0.2-0.35	0.2-0.5	0.03 max	0.1 max	0.15-0.25	0.05 max	0.05	0.15	Sn 0.05 max			
	Belgium	NBN P21-101	SGAlZn5MgC	Sand	0.35 max	0.8 max	0.2-0.7	0.4 max	0.05 max	0.3 max	0.1-0.3	4.5-6			Cr 0.15-0.4; Pb 0.05 max; Sn 0.05 max			
	Canada	CSA HA.3	C4		4-5	0.2	0.05	0.1	0.1	0.25	0.05-0.2	0.1						
512 A05120	Sweden	MNC 41E	4163	Die Sand Ingot			5-5			1					Al 94			

UNS numbers and US grades are provided as a means of cross referencing chemically similar alloys. Exchangability is only possible after independent examination of specifications. Tensile properties are minimum or typical. UTS and YS as Mpa, El as %. See Appendix for list of abbreviations used in Descriptions.

3-8 Cast Aluminum

Grade UNS #	Country	Specification	Designation	Description	Cu	Fe	Mg	Mn	Ni	Si	Ti	Zn	OE Max	OT Max	Other	UTS	YS	EL
513 A05130	International		513	Perm Mold	0.1 max	0.4 max	3.5-4.5	0.3 max		0.3 max	0.2 max	1.4-2.2	0.05	0.15				
512 A05120	UK	BS 1490	LM5		0.1 max	0.6 max	3-6	0.3-0.7	0.1 max	0.3 max	0.2 max	0.1 max			Pb 0.05 max; Sn 0.05 max			
513 A05130	Belgium	NBN P21-101	SGAlMg4Zn	Sand	0.1 max	0.5 max	3.5-4.5	0.6 max	0.05 max	0.5 max	0.25 max	0.9-1.5			Pb 0.05 max; Sn 0.05 max			
513 A05130	France	NF A57-702	AG4Z	Perm Mold Sand	0.1 max	0.55 max	3.2-4.5	0.3 max	0.05 max	0.5 max	0.2 max	0.9-1.45			Pb 0.05 max; Sn 0.05 max			
512 A05120	International		512	Sand	0.35 max	0.6 max	3.5-4.5	0.8 max		1.4-2.2	0.25 max	0.35 max	0.05	0.15	Cr 0.25 max			
	International	ISO 3522	Al-Si5 Cu1 Mg	Sand Cast SHT PHT	1-1.5	0.6 max	0.4-0.6	0.5 max	0.3 max	4.5-5.5	0.2 max	0.5 max			Pb 0.1 max; Sn 0.1 max	220		1
	International	ISO 3522	Al-Si5 Cu1 Mg	Perm Mold As mfg	1-1.5	0.6 max	0.4-0.6	0.5 max	0.3 max	4.5-5.5	0.2 max	0.5 max			Pb 0.1 max; Sn 0.1 max	160		
	International	ISO 3522	Al-Si5 Cu1 Mg	Perm Mold SHT PHT	1-1.5	0.6 max	0.4-0.6	0.5 max	0.3 max	4.5-5.5	0.2 max	0.5 max			Pb 0.1 max; Sn 0.1 max	290		
	India	IS 202	A-11	Sand ST PT	4-5	0.25	0.1	0.1	0.1	0.25		0.1			Pb 0.05; Sn 0.05 max	278	186	4
	India	IS 202	A-11	Sand ST PT	4-5	0.25	0.1	0.1	0.1	0.25		0.1			Pb 0.05; Sn 0.05 max	278	186	4
513.2 A05132	International		513.2	Ingot	0.1 max	0.3 max	3.6-4.5	0.1 max		0.3 max	0.2 max	1.4-2.2	0.05	0.15				
	Norway	NS 17552	NS17552-00		0.1	0.5	4-6	0.4	0.05	0.4	0.2	0.2			Al 95; Pb 0.05; Sn 0.05 max			
	Norway	NS 17552	NS17552-01	Sand	0.1	0.5	4-6	0.4	0.05	0.4	0.2	0.2			Al 95; Pb 0.05; Sn 0.05 max	157	88	2
	Norway	NS 17552	NS17552-02		0.1	0.5	4-6	0.4	0.05	0.4	0.2	0.2			Al 95; Pb 0.05; Sn 0.05 max	167	98	2
	International	ISO R164	Al-Mg3Si		0.1	0.5	2-4.5	0.6	0.05	1.3	0.2	0.2			Cr 0.4; Pb 0.05; Sn 0.05 max			
	Belgium	NBN P21-101	AlMg6Si	Sand	0.1 max	0.5 max	4.5-7	0.6 max	0.05 max	0.5-1.5	0.2 max	0.2 max			Cr 0.5 max; Pb 0.05 max; Sn 0.05 max			
512.2 A05122	International		512.2	Ingot	0.1 max	0.3 max	3.6-4.5	0.1 max		1.4-2.2	0.2 max	0.1 max	0.05	0.15				
	USA	ASTM B327	G1C	Bar	2	0.8	0.75-1.1	0.5	0.2	0.7		1			Al 95; Cd 0.01 max; Cr 0.2; Pb 0.02; Sn 0.02 max			
	Sweden	MNC 41E	4337	Sand Ingot	4.5		0.2				0.2				Al 95			
	Germany	DIN 1725Sh2	3.1371./GK-AlCu4TiMg		4.2-9.2	0.2	0.15-0.3	0.05		0.18	0.15-0.3	0.07	0.03	0.1				
B201.0 A22010	International		B201.0	Sand	4.5-5	0.05 max	0.25-0.35	0.2-0.5		0.05 max	0.15-0.35		0.05	0.15	Silver 0.50-1.0			
	Germany	DIN 1725Sh2	3.1841./GK-AlCu4Tita		4.5-5.2	0.18		0.05		0.18	0.15-0.3	0.07						
	Germany	DIN 1725Sh2	3.1841.6/G-AlCu4Ti	Sand Selectively hardened	4.5-5.2	0.18		0.05		0.18	0.15-0.3	0.07				280	180	
	Germany	DIN 1725Sh2	3.1841.6/G-AlCu4Tiwa	Sand Q and AH	4.5-5.2	0.18		0.05		0.18	0.15-0.3	0.07				300	200	
	Germany	DIN 1725Sh2	3.1841/GK-AlCu4Tiwa		4.5-5.2	0.18		0.05		0.18	0.15-0.3	0.07						
	South Africa	SABS 990	Al-Mg4MnA	Sand	0.1	0.6	3-6	0.7	0.1	0.3	0.2	0.1			Pb 0.05; Sn 0.05 max	140		3
	USA	MIL A-21180C	A201.0	Class 1	4.5-5	0.1	0.15-0.35	0.2-0.4		0.05	0.15-0.35					414		3
	USA	MIL A-21180C	A201.0	Class 2	4.5-5	0.1	0.15-0.35	0.2-0.4		0.05	0.15-0.35					414		5
C443.2 A34432	International		C443.2	Ingot	0.1 max	0.7-1.1	0.05 max	0.1 max		4.5-6		0.1 max	0.05	0.15				
	USA	QQ A-371F	F514.1		0.15	0.4	3.6-4.5	0.35		0.3-0.7	0.25	0.15						
	Finland	SFS 2564	G-AlCu4Ti	Die SHT and aged 20 mm diam	4-5	0.35	0.05	0.1	0.1	0.35	0.05-0.35	0.2			Sn 0.05 max	320		4
	International	ISO 3522	Al-Zn5 Mg	Sand Cast Cooled Aged	0.35 max	0.8 max	0.5-0.7	0.4 max	0.05 max	0.3 max	0.1-0.3	4.5-6			Cr 0.15-0.6; Pb 0.05 x; Sn 0.05 x	200		3

UNS numbers and US grades are provided as a means of cross referencing chemically similar alloys. Exchangability is only possible after independent examination of specifications. Tensile properties are minimum or typical . UTS and YS as Mpa, El as %. See Appendix for list of abbreviations used in Descriptions.

Worldwide Guide to Equivalent Nonferrous Metals and Alloys

Grade UNS #	Country	Specification	Designation	Description	Cu	Fe	Mg	Mn	Ni	Si	Ti	Zn	OE Max	OT Max	Other	UTS	YS	EL
	International	ISO 3522	Al-Zn5 Mg	Perm Mold Cooled Aged	0.35 max	0.8 max	0.5-0.7	0.4 max	0.05 max	0.3 max	0.1-0.3	4.5-6			Cr 0.15-0.6; Pb 0.05 max; Sn 0.05 max	210		3
	Japan	JIS H2117	C7AS	Ingot	0.1	0.3	3.6-5.5	0.6	0.1	0.3	0.2	0.1			ST 0.5; ST-Fe+Si;			
	Germany	DIN 1725Sh2	3.3543./GK-AlMg3(Cu)		0.3	0.6	2-4	0.6		1.3	0.2	0.3						
	Germany	DIN 1725Sh2	3.3543.0/G-AlMg3(Cu)	Sand	0.3	0.6	2-4	0.6		1.3	0.2	0.3				140	80	2
	International	ISO R164	Al-Cu4Ti		4-5	0.4	0.05	0.1	0.1	0.35	0.05-0.35	0.2			Pb 0.05; Sn 0.05 max			
	Germany	DIN 1725Sh3	3.0841/V-AlBe5		0.05	0.4	0.05	0.03		0.2	0.02	0.1			B 4.5-6.0; Cr 0.03			
	Austria	ONORM M3429	GK AlSi5Mg	Perm Mold	0.05 max	0.5 max	0.4-0.8	0.4 max		5-6	0.2 max	0.1 max			ET 0.05 max; ET=.15 total;			
	Austria	ONORM M3429	GS AlSi5Mg	Sand	0.05 max	0.5 max	0.4-0.8	0.4 max		5-6	0.2 max	0.1 max			ET 0.05 max; ET=.15 total;			
	Canada	CSA HA.9	C4	Sand T4	4.5-5	0.25	0.05	0.1	0.1	0.25	0.05-0.2	0.1				234		8
	Canada	CSA HA.9	C4	Sand T6	4.5-5	0.25	0.05	0.1	0.1	0.25	0.05-0.2	0.1				276		3
	Canada	CSA HA.3	CG50		4.3-5	0.35	0.05	0.1	0.05	0.3	0.15-0.3	0.1						
	France	AIR 9150-B	A-G5MC	Wir CF or Ann 1.6/9.6 mm diam	0.1	0.4	4.5-5.5	0.05-0.7		0.3		0.1				290	140	18
	Australia	AS 1874	BA701	Ingot	0.15 max	0.5 max	0.5-0.7	0.15 max	0.1 max	0.25 max	0.15-0.25	4.8-5.7	0.05	0.15	Cr 0.4-0.6; Pb 0.05 max; Sn 0.05 max			
	Australia	AS 1874	BA701	T1 sand	0.15 max	0.5 max	0.5-0.7		0.1 max	0.25 max	0.15-0.25	4.8-5.7	0.05	0.15	Cr 0.4-0.6; Pb 0.05 max; Sn 0.05 max	215		4
	Australia	AS 1874	BA701	T5 sand	0.15 max	0.5 max	0.5-0.7		0.1 max	0.25 max	0.15-0.25	4.8-5.7	0.05	0.15	Cr 0.4-0.6; Pb 0.05 max; Sn 0.05 max	215		4
	Austria	ONORM M3429	AlSi5Mg		0.03 max	0.3 max	0.5-0.8	0.4 max		5-6	0.2 max	0.1 max			ET 0.05 max; ET=.15 total;			
	Austria	ONORM M3429	G AlSi5Mg		0.03 max	0.3 max	0.5-0.8	0.4 max		5-6	0.2 max	0.1 max			ET 0.05 max; ET=.15 total;			
A305.2 A13052	International		A305.2	Ingot	1-1.5	0.13 max		0.05 max		4.5-5.5	0.2 max	0.05 max	0.05	0.15				
A305.1 A13051	International		A305.1	Ingot	1-1.5	0.15 max	0.1 max	0.1 max		4.5-5.5	0.2 max	0.1 max	0.05	0.15				
A305.0 A13050	International		A305.0	Sand, Perm Mold	1-1.5	0.2 max	0.1 max	0.1 max		4.5-5.5	0.2 max	0.1 max	0.05	0.15				
305 A03050	International		305	Sand, Perm Mold	1-1.5	0.6 max	0.1 max	0.5 max		4.5-5.5	0.25 max	0.35 max	0.05	0.15	Cr 0.25 max			
A03050	Australia	AS 1874	AA309	T51 Sand	1-1.5	1 max	0.5-0.6	0.05 max		4.5-5.5	0.2 max	0.05 max	0.05	0.15		170		
A03050	Australia	AS 1874	AA309	T6 Sand	1-1.5	1 max	0.5-0.6	0.05 max		4.5-5.5	0.2 max	0.05 max	0.05	0.15		220		2
A03050	Australia	AS 1874	AA309	T62 permanent mould cst	1-1.5	1 max	0.5-0.6	0.05 max		4.5-5.5	0.2 max	0.05 max	0.05	0.15		275		
A13051	Australia	AS 1874	AA311	F1 permanent mould	1-1.5	0.15 max	0.05 max	0.05 max		4-6	0.2 max	0.1 max	0.05	0.2				7
A13051	Australia	AS 1874	AA311	Ingot	1-1.5	0.25 max	0.05 max	0.05 max		4-6	0.2 max	0.1 max	0.05	0.2				
A03050	Australia	AS 1874	AA309	T51 permanent mould	1-1.5	0.25 max	0.5-0.6	0.05 max		4.5-5.5	0.2 max	0.05 max	0.05	0.15		185		
A03050	Australia	AS 1874	AA309	T6 permanent mould	1-1.5	0.25 max	0.5-0.6	0.05 max		4.5-5.5	0.2 max	0.05 max	0.05	0.15		255		1.5
	USA	QQ A-371F	C443.2		0.1	0.7-1.1	0.05	0.1		4.5-6		0.1			Sn 0.15 max			
	USA	ASTM B179	C443.2		0.1	0.7-1.1	0.05	0.1		4.5-6		0.1						

UNS numbers and US grades are provided as a means of cross referencing chemically similar alloys. Exchangability is only possible after independent examination of specifications. Tensile properties are minimum or typical . UTS and YS as Mpa, El as %. See Appendix for list of abbreviations used in Descriptions.

3-10 Cast Aluminum

Grade UNS #	Country	Specification	Designation	Description	Cu	Fe	Mg	Mn	Ni	Si	Ti	Zn	OE Max	OT Max	Other	UTS	YS	EL
	Canada	CSA HA.9	CG50	Sand T4	4.3-5	0.4	0.15-0.35	0.1	0.05	0.3	0.15-0.3	0.1				310	193	7
	Germany	DIN 1725Sh3	3.0862/V-AlZr6		0.02	0.21	0.02	0.02	0.04	0.2	0.02	0.03			Cr 0.02; Zr 5-6.5			
	South Africa	SABS 989	Al-Si5MgA	sand	0.08	0.5	0.5-0.8	0.5	0.1	3.5-6	0.2	0.1			Pb 0.1; Sn 0.05 max	125		2
	South Africa	SABS 989	Al-Si5MgA	chill	0.08	0.5	0.5-0.8	0.5	0.1	3.5-6	0.2	0.1			Pb 0.1; Sn 0.05 max	160		3
	South Africa	SABS 989	Al-Si5MgA	Chill , SHT and PHT	0.08	0.5	0.5-0.8	0.5	0.1	3.5-6	0.2	0.1			Pb 0.1; Sn 0.05 max	275		2
	South Africa	SABS 990	Al-Si5MgA	Sand PHT	0.1	0.6	0.4-0.8	0.5	0.1	3.5-6	0.2	0.1			Pb 0.1; Sn 0.05 max	145		1
	South Africa	SABS 990	Al-Si5MgA	Sand SHT	0.1	0.6	0.4-0.8	0.5	0.1	3.5-6	0.2	0.1			Pb 0.1; Sn 0.05 max	160		3
	South Africa	SABS 991	Al-Si5MgA	PHT	0.1	0.6	0.4-0.8	0.5	0.1	3.5-6	0.2	0.1			Pb 0.1; Sn 0.05 max	185		2
	South Africa	SABS 991	Al-Si5MgA	SHT	0.1	0.6	0.4-0.8	0.5	0.1	3.5-6	0.2	0.1			Pb 0.1; Sn 0.05 max	230		5
	South Africa	SABS 991	Al-Si5MgA	PHT and SHT	0.1	0.6	0.4-0.8	0.5	0.1	3.5-6	0.2	0.1			Pb 0.1; Sn 0.05 max	275		2
305.2 A03052	International		305.2	Ingot	1-1.5	0.14-0.25		0.05 max		4.5-5.5	0.2 max	0.05 max	0.05	0.15				
443.2 A04432	USA	ASTM B179	443.2		0.1	0.6	0.05	0.1		4.5-6	0.2	0.1						
443.2 A04432	USA	QQ A-371F	443.2		0.1	0.6	0.05	0.1		4.5-6	0.2	0.1						
712.2 A07122	International		712.2	Ingot	0.25 max	0.4 max	0.5-0.65	0.1 max		0.15 max	0.15-0.25	5-6.5	0.05	0.2	Cr 0.4-0.6			
712 A07120	International		712	Sand	0.25 max	0.5 max	0.5-0.65	0.1 max		0.3 max	0.15 0.25	5-6.5	0.05	0.2	Cr 0.4-0.6			
712 A07120	Canada	CSA	ZG61		0.3 max	1 max	0.5-0.7	0.3 max		0.25 max	0.1-0.3	5-6			ET 0.05 max; ET=.15 total; Cr 0.4-0.6			
	International	ISO R164	Al-Si5Mg		0.1	0.6	0.4-0.9	0.6	0.1	3.5-6	0.2	0.1			Pb 0.1; Sn 0.05 max			
	International	ISO R2147	Al-Si5Mg	Sand SHT and art aged OBSOLETE Replaced by ISO 3522	0.1	0.6	0.4-0.9	0.6	0.1	3.5-6	0.2	0.1			Pb 0.1; Sn 0.05 max	230		1
	Austria	ONORM M3429	GS AlZn5Mg	Sand	0.05 max	0.3 max	0.8-1	0.1 max		0.25 max		5-5.4			ET 0.03 max; ET=.10 total; Cr 0.05-0.15			
	USA	ASTM B179	B514.2		0.1	0.3	3.6-4.5	0.1		1.4-2.2	0.2	0.1						
	USA	QQ A-371F	B514.2		0.1	0.3	3.6-4.5	0.1		1.4-2.2	0.2	0.1						
	Austria	ONORM M3429	AlMg5SiCu		0.4-0.6	0.3 max	4.5-5.5	0.5 max		0.9-1.5	0.2 max	0.1 max			ET 0.05 max; ET=.15;			
	Austria	ONORM M3429	G AlMg5SiCu		0.4-0.6	0.3 max	4.5-5.5	0.5 max		0.9-1.5	0.2 max	0.1 max			ET 0.05 max; ET=.15 total;			
	Austria	ONORM M3429	GK AlMg5SiCu	Perm Mold	0.4-0.6	0.5 max	4.5-5.5	0.5 max		0.9-1.5	0.2 max	0.1 max			ET 0.05 max; ET=.20 total;			
	Austria	ONORM M3429	GS AlMg5SiCu	Sand	0.4-0.6	0.5 max	4.5-5.5	0.5 max		0.9-1.5	0.2 max	0.1 max			ET 0.05 max; ET=.15 total;			
	Austria	ONORM M3429	AlZn5Mg	Perm Mold Sand	0.05 max	0.3 max	0.8-1	0.1 max		0.25 max	0.03-0.15	5-5.4			ET 0.03 max; ET=.10 total; Cr 0.05-0.15			
	Austria	ONORM M3429	G AlZn5Mg		0.05 max	0.3 max	0.8-1	0.1 max		0.25 max	0.03-0.15	5-5.4			ET 0.03 max; ET=.10 total; Cr 0.05-0.15			
	Austria	ONORM M3429	GK AlZn5Mg	Perm Mold	0.05 max	0.3 max	0.8-1	0.1 max		0.25 max	0.03-0.15	5-5.4			ET 0.05 max; ET=.15 total; Cr 0.05-0.15			
	USA	QQ A-371F	D712.2		0.2	0.6	1.5-1.8	0.4-0.6		0.2	0.25	2.7-3.3			Cr 0.2-0.4			
	International	ISO R164	Al-Zn5Mg		0.35		0.2-0.7	0.4	0.05	0.3	0.1-0.3	4.5-6			Cr 0.15-0.6; Sn 0.05 max			
C355.0 A33550	International		C355.0	Sand, Perm Mold	1-1.5	0.2 max	0.4-0.6	0.1 max		4.5-5.5	0.2 max	0.1 max	0.05	0.15				

UNS numbers and US grades are provided as a means of cross referencing chemically similar alloys. Exchangability is only possible after independent examination of specifications. Tensile properties are minimum or typical . UTS and YS as Mpa, El as %. See Appendix for list of abbreviations used in Descriptions.

Worldwide Guide to Equivalent Nonferrous Metals and Alloys

Grade UNS #	Country	Specification	Designation	Description	Cu	Fe	Mg	Mn	Ni	Si	Ti	Zn	OE Max	OT Max	Other	UTS	YS	EL
355.1 A03551	International		355.1	Ingot	1-1.5	0.5 max	0.456-0.6	0.5 max		4.5-5.5	0.25 max	0.35 max	0.05	0.15	If Fe > 0.45 then Mn > (0.5 Fe); Cr 0.25 max			
355.1 A03551	USA	ASTM B179	355.1		1-1.5	0.5	0.45-0.6	0.5		4.5-5.5	0.25	0.35			Cr 0.25			
355.1 A03551	USA	QQ A-371F	355.1		1-1.5	0.5	0.45-0.6	0.25-0.5		4.5-5.5		0.3			Cr 0.25			
355.1 A03551	Canada	CSA	SC51		1-1.5	0.6 max	0.4-0.6	0.3 max		4.5-5.5	0.2 max	0.1 max			ET 0.05 max; ET=.15 total;			
355 A03550	International		355	Sand, Perm Mold	1-1.5	0.6 max	0.4-0.6	0.5 max		4.5-5.5	0.25 max	0.35 max	0.05	0.15	If Fe > 0.45 then Mn > (0.5 Fe); Cr 0.25 max			
355 A03550	USA	ASTM B108	355	T51	1-1.5	0.6	0.4-0.6	0.5		4.5-5.5	0.25	0.35			Cr 0.25	186		
355 A03550	USA	ASTM B108	355	T62	1-1.5	0.6	0.4-0.6	0.5		4.5-5.5	0.25	0.35			Cr 0.25	290		
355 A03550	USA	ASTM B108	355	SHT and stab	1-1.5	0.6	0.4-0.6	0.5		4.5-5.5	0.25	0.35			Cr 0.25	248		
355 A03550	USA	ASTM B26	355	Sand SHT and art aged .250 in diam	1-1.5	0.6	0.4-0.6	0.5		4.5-5.5	0.25	0.35			Cr 0.25	221	138	2
355 A03550	USA	ASTM B26	355	Sand T51 .250 in diam	1-1.5	0.6	0.4-0.6	0.5		4.5-5.5	0.25	0.35			Cr 0.25	172	124	
355 A03550	USA	ASTM B26	355	Sand T71 .250 in diam	1-1.5	0.6	0.4-0.6	0.5		4.5-5.5	0.25	0.35			Cr 0.25	207	152	
355 A03550	USA	ASTM B618	355	Invest SHT and art aged .250 in diam	1-1.5	0.6	0.4-0.6	0.5		4.5-5.5	0.25	0.35			Cr 0.25	221	138	2
355 A03550	USA	ASTM B618	355	Invest T51 .250 in diam	1-1.5	0.6	0.4-0.6	0.5		4.5-5.5	0.25	0.35			Cr 0.25	172	124	
355 A03550	USA	ASTM B618	355	Invest T71 .250 in diam	1-1.5	0.6	0.4-0.6	0.5		4.5-5.5	0.25	0.35			Cr 0.25	207	152	
355 A03550	USA	QQ A-596d	355	SHT and art aged and T6	1-1.5	0.8	0.5	0.4-0.6		4.5-5.5	0.25	0.35			Cr 0.25	255		1
355 A03550	USA	QQ A-596d	355	SHT and art aged and T62	1-1.5	0.8	0.5	0.4-0.6		4.5-5.5	0.25	0.35			Cr 0.25	290		
355 A03550	USA	QQ A-596d	355	Art aged only and temper T51	1-1.5	0.8	0.5	0.4-0.6		4.5-5.5	0.25	0.35			Cr 0.25	186		
355 A03550	International	ISO R164	Al-Si5Cu1		1-1.5	0.1	0.3-0.6	0.5	0.3	4.5-6	0.2	0.5			Pb 0.2; Sn 0.1			
	USA	QQ A-596d	Ternalloy5(603)	Aged and temper T5	0.2	0.8	1.4-1.8	0.4-0.6			0.2	0.25	2.7-3.8			Cr 0.2-0.4	255		10
C355.1 A33551	International		C355.1	Ingot	1-1.5	0.15 max	0.45-0.6	0.1 max		4.5-5.5	0.2 max	0.1 max	0.05	0.15				
A355.0 A13550	International		A355.0	Sand, Perm Mold	1-1.5	0.09 max	0.45-0.6	0.05 max		4.5-5.5	0.04-0.2	0.05 max	0.05	0.15				
	Norway	NS 17550	NS17550-02	Sand	0.1	0.5	4-6	0.5			0.2	0.2			Sn 0.05; Al 94; Pb 0.05; Sn 0.05			
	USA	ASTM B179	A514.2		0.1	0.3	3.6-4.5	0.1			0.3	0.2	1.4-2.2						
	USA	QQ A-371F	A514.2		0.1	0.3	3.6-4.5	0.1			0.3	0.2	1.4-2.2						
	Norway	NS 17550	NS17550-00		0.1	0.5	4-6	0.5	0.05	0.5-1.5	0.2	0.2			Al 94; Pb 0.05; Sn 0.05 max			
B443.0 A24430	International	ISO R164	Al-Si5		0.1	0.8	0.1	0.5	0.1	4-6	0.2	0.1			Pb 0.1; Sn 0.1 max			
C355.2 A33552	International		C355.2	Ingot	1-1.5	0.13 max	0.5-0.6	0.05 max		4.5-5.5	0.2 max	0.05 max	0.05	0.15				
208.2 A02082	International		208.2	Ingot	3.5-4.5	0.8 max	0.03 max	0.3 max		2.5-3.5	0.2 max	0.2 max		0.3				
208.2 A02082	USA	ASTM B179	208.2		3.5-4.5	0.8	0.03	0.3		2.5-3.5	0.2	0.2						
208.2 A02082	USA	QQ A-371F	208.2		3.5-4.5	0.8	0.03	0.3		2.5-3.5	0.25	0.2						
208.1 A02081	International		208.1	Ingot	3.5-4.5	0.9 max	0.1 max	0.5 max	0.35 max	2.5-3.5	0.25 max	1 max		0.5				
208.1 A02081	USA	ASTM B179	208.1		3.5-4.5	0.9	0.1	0.5	0.35	2.5-3.5	0.25	1						

UNS numbers and US grades are provided as a means of cross referencing chemically similar alloys. Exchangability is only possible after independent examination of specifications. Tensile properties are minimum or typical . UTS and YS as Mpa, El as %. See Appendix for list of abbreviations used in Descriptions.

3-12 Cast Aluminum

Grade UNS #	Country	Specification	Designation	Description	Cu	Fe	Mg	Mn	Ni	Si	Ti	Zn	OE Max	OT Max	Other	UTS	YS	EL
208.1 A02081	USA	QQ A-371F	208.1		3.5-4.5	0.9	0.1	0.5	0.35	2.5-3.5	0.25	1						
208 A02080	International		208	Sand, Perm Mold	3.5-4.5	1.2 max	0.1 max	0.5 max	0.35 max	2.5-3.5	0.25 max	1 max		0.5				
208 A02080	USA	ASTM B108	208	SHT and nat aged	3.5-4.5	1.2	0.1	0.5	0.35	2.5-3.5	0.25	1				228	103	5
208 A02080	USA	ASTM B108	208	SHT and art aged	3.5-4.5	1.2	0.1	0.5	0.35	2.5-3.5	0.25	1				241	152	2
208 A02080	USA	ASTM B108	208	SHT and stab	3.5-4.5	1.2	0.1	0.5	0.35	2.5-3.5	0.25	1				228	110	3
208 A02080	USA	ASTM B26	208	Sand As Fabricated .250 in diam	3.5-4.5	1.2	0.1	0.5	0.35	2.5-3.5	0.25	1				131	83	2
208 A02080	USA	ASTM B618	208	Invest As fabricated .250 in diam	3.5-4.5	1.2	0.1	0.5		2.5-3.5	0.25	1				131	63	2
296.2 A02962	International		296.2	Ingot	4-5	0.8 max	0.03 max	0.3 max		2-3	0.2 max	0.3 max	0.05	0.15				
296.1 A02961	International		296.1	Ingot	4-5	0.9 max	0.05 max	0.35 max	0.35 max	2-3	0.25 max	0.5 max		0.35				
296 A02960	Russia	GOST 2685	AL7		4-5	1 max	0.03 max		0.3 max	1.2-3								
296 A02960	International		296	Perm Mold	4-5	1.2 max	0.05 max	0.35 max	0.35 max	2-3	0.25 max	0.5 max		0.35				
A355.2 A13552	International		A355.2	Ingot	1-1.5	0.06 max	0.5-0.6	0.03 max		4.5-5.5	0.04-0.2	0.03 max	0.03	0.1				
	Finland	SFS 2572	G-AlMg5Si1	Die 20 mm diam	0.1	0.5	4-6	0.5	0.05	0.5-1.5	0.2	0.2			Pb 0.05; Sn 0.05 max	180	110	2
	Sweden	SIS 144163	4163-03	Sand	0.1	0.5	4-6	0.5	0.05	0.5-1.5	0.2	0.2			Pb 0.05; Sn 0.05 max	160	90	2
	Sweden	SIS 144163	4163-06	As cast	0.1	0.5	4-6	0.5	0.05	0.5-1.5	0.2	0.2			Pb 0.05; Sn 0.05 max	170	100	2
	International	ISO R164	Al-Cu4MgTi		4-5	0.4	0.15-0.35	0.1	0.5	0.35	0.35	0.2			Pb 0.05; Sn 0.05 max			
355.2 A03552	International		355.2	Ingot	1-1.5	0.14-0.25	0.5-0.6	0.05 max		4.5-5.5	0.2	0.05 max	0.05	0.15				
355.2 A03552	USA	ASTM B179	355.2		1-1.5	0.14-0.25	0.5-0.6	0.05		4.5-5.5	0.2	0.05						
355.2 A03552	USA	QQ A-371F	355.2		1-1.5	0.14-0.25	0.5-0.6	0.05		4.5-5.5		0.05						
	Japan	JIS H2211	C1AV	Ingot	4-5	0.3	0.3	0.03	0.03	1.2	0.25	0.03						
203 A02030	International		203	Sand	4.5-5.5	0.5 max	0.1 max	0.2-0.3	1.3-1.7	0.3 max	0.15-0.25	0.1 max	0.05	0.2	Sb 0.20-0.30, Co 0.20-0.30, Zr 0.10-0.30 Ti+Zr=0.50 max			
	USA	ASTM B108	A514.0	As fabricated	0.1	0.4	3.5-4.5	0.3		0.3	0.2	1.4-2.2				152	83	3
	USA	QQ A-596d	A214.0	As cast	0.1	0.4	3.5-4.5	0.3		0.3	0.2	1.4-2.2				152		3
707 A07070	International		707	Sand, Perm Mold	0.2 max	0.8 max	1.8-2.4	0.4-0.6		0.2 max	0.25 max	4-4.5	0.05	0.15	Cr 0.2-0.4			
707 A07070	USA	ASTM B108	707	Cooled and nat or art aged	0.2	0.8	1.8-2.4	0.4-0.6		0.2	0.25	4-4.5			Cr 0.2-0.4	290	173	4
707 A07070	USA	ASTM B108	707	SHT and stab	0.2	0.8	1.8-2.4	0.4-0.6		0.2	0.25	4-4.5			Cr 0.2-0.4	310	241	3
707 A07070	USA	ASTM B26	707	Sand Cooled anr art aged .250 in diam	0.2	0.8	1.8-2.4	0.4-0.6		0.2	0.25	4-4.5			Cr 0.2-0.4	228	152	2
707 A07070	USA	ASTM B26	707	Sand SHT and stab .250 in diam	0.2	0.8	1.8-2.4	0.4-0.6		0.2	0.25	4-4.5			Cr 0.2-0.4	255	207	1
707 A07070	USA	ASTM B618	707	Invest Cooled and nat and art aged .250 in diam	0.2	0.8	1.8-2.4	0.4-0.6		0.2	0.25	4-4.5			Cr 0.2-0.4	228	152	2

UNS numbers and US grades are provided as a means of cross referencing chemically similar alloys. Exchangability is only possible after independent examination of specifications. Tensile properties are minimum or typical . UTS and YS as Mpa, El as %. See Appendix for list of abbreviations used in Descriptions.

Worldwide Guide to Equivalent Nonferrous Metals and Alloys

Grade UNS #	Country	Specification	Designation	Description	Cu	Fe	Mg	Mn	Ni	Si	Ti	Zn	OE Max	OT Max	Other	UTS	YS	EL
707 A07070	USA	ASTM B618	707	Invest SHT and stab .250 in diam	0.2	0.8	1.8-2.4	0.4-0.6		0.2	0.25	4-4.5			Cr 0.2-0.4	255	207	1
707 A07070	Russia	GOST 2685	AL24		0.2 max	0.5 max	1.5-2	0.2-0.5		0.3 max	0.1-0.2	3.4-4.5		1	Be 0.1 max			
	International	ISO 3522	Al-Si5 Cu3	Sand Cast As mfg	2-4	0.8 max	0.15 max	0.2-0.6	0.3 max	4-6	0.2 max	0.5 max			Pb 0.1 max; Sn 0.05 max	140		1
	International	ISO 3522	Al-Si5 Cu3	Perm Mold As mfg	2-4	0.8 max	0.15 max	0.2-0.6	0.3 max	4-6	0.2 max	0.5 max			Pb 0.1 max; Sn 0.05 max	150		1
	Germany	DIN 1725Sh2	3.3561.01/G-AlMg5	Sand	0.05	0.5	4.5-5.5	0.4		0.5	0.2	0.1				180	100	3
	Germany	DIN 1725Sh2	3.3561.02/GK-AlMg5		0.05	0.5	4.5-5.5	0.4		0.5	0.2	0.1				180	100	4
	USA	ASTM B179	B443.1		0.15	0.6	0.05	0.35		4.5-6	0.25	0.35						
707.1 A07071	International		707.1	Ingot	0.2 max	0.6 max	1.9-2.4	0.4-0.6		0.2 max	0.25 max	4-4.5	0.05	0.15	Cr 0.2-0.4			
707.1 A07071	USA	ASTM B179	705.1		0.2	0.6	1.5-1.8	0.4-0.6		0.2	0.25	2.7-3.3			Cr 0.2-0.4			
707.1 A07071	USA	ASTM B179	707.1		0.2	0.6	1.9-2.4	0.4-0.6		0.2	0.25	4-4.5			Cr 0.2-0.4			
707.1 A07071	USA	QQ A-371F	707.1		0.2	0.6	1.9-2.4	0.4-0.6		0.2	0.25	4-4.5			Cr 0.2-0.4			
	Japan	JIS H2211	C4DV	Ingot	1-1.5	0.3	0.4-0.6	0.03	0.03	4.5-5.5	0.03	0.03						
	USA	QQ A-371F	C355.2		1-1.5	0.13	0.45-0.6	0.05		4.5-5.5	0.2	0.05						
535 A05350	International		535	Sand	0.05 max	0.15 max	6.2-7.5	0.1-0.25		0.15 max	0.1-0.25		0.05	0.15	Beryllium 0.003-0.007, boron 0.005 max			
535 A05350	USA	ASTM B108	535	As fabricated	0.05	0.15	6.2-7.5	0.1-0.25		0.15	0.1-0.25				B 0.005 max; Be 0.003-0.007	241	124	8
535 A05350	USA	ASTM B618	535	Invest As fabricated .250 in diam	0.05	0.15	6.2-7.5	0.1-0.25		0.15	0.1-0.25				B 0.002 max; Be 0.003-0.007	241	124	9
535 A05350	USA	ASTM B26	535	Sand As fabricated .250 in diam	0.05	0.15	6.2-7.5	0.1-0.25		0.15	0.1-0.25					241	124	9
	International	ISO R164	Al-Si5MgFe		0.1	1.3	0.4-0.9	0.6	0.1	3.5-6	0.2	0.1			Pb 0.1; Sn 0.05 max			
535 A05350	Russia	GOST 2685	AL23		0.15 max	0.2 max	6-7	0.1 max		0.2 max	0.05-0.15	0.1 max		0.6	Be 0.02-0.1			
242 A02420	UK	BS 1490	LM14		3.5-4.5	0.6 max	1.2-1.7	0.6 max	1.8-2.3	0.6 max	0.2 max	0.1 max			Pb 0.05 max; Sn 0.05 max			
242 A02420	UK		218		3.5-4.5	0.6 max	1.2-1.7	0.6 max	1.8-2.3	0.6 max	0.2 max	0.1 max			Pb 0.05 max; Sn 0.05 max			
242 A02420	Czech Republic	CSN 424315	AlCu4Ni2Mg2	Permanent Mold Ingot Alloys	3.75-4.5	0.7 max	1.25-1.75			0.6 max	0.2 max	0.1 max						
242 A02420	Czech Republic	CSN 424315	AlCu4Ni2Mg2	Sand Ingot Alloys	3.75-4.5	0.7 max	1.25-1.75		1.75-2.25	0.6 max	0.2 max	0.1 max						
242 A02420	Japan	JIS H5202	AC5A		3.5-4.5	0.8	1.2-1.8	0.3	1.7-2.3	0.6	0.2	0.1				216		
242 A02420	Japan	JIS H5202	AC5A	Temp	3.5-4.5	0.8	1.2-1.8	0.3	1.7-2.3	0.6	0.2	0.1				196		
242 A02420	Japan	JIS H5202	AC5A	Q/T	3.5-4.5	0.8	1.2-1.8	0.3	1.7-2.3	0.6	0.2	0.1				294		
242 A02420	International	ISO R164	Al-Cu4Ni2Mg2		3.5-4.5	0.05	1.2-1.8	0.6	1.7-2.3	0.7	0.2	0.1			Cr 0.2; Pb 0.05; Sn 0.05 max			
242 A02420	Belgium	NBN P21-101	SGAlCu4Ni2Mg2	Sand	3.5-4.5	0.7 max	1.2-1.8	0.6 max	1.7-2.3	0.7 max	0.2 max	0.5 max			Cr 0.2 max; Pb 0.05 max; Sn 0.05 max			
	International	ISO 3522	Al-Cu4 Ni2 Mg2	Sand Cast Ann	3.5-4.5	0.7 max	1.2-1.8	0.6 max	1.7-2.3	0.7 max	0.2 max	0.1 max			Cr 0.2 max; Pb 0.05 x; Sn 0.05 x	150		
	International	ISO 3522	Al-Cu4 Ni2 Mg2	Sand Cast SHT PHT	3.5-4.5	0.7 max	1.2-1.8	0.6 max	1.7-2.3	0.7 max	0.2 max	0.1 max			Cr 0.2 max; Pb 0.05 x; Sn 0.05 x	220		
	International	ISO 3522	Al-Cu4 Ni2 Mg2	Perm Mold SHT PHT	3.5-4.5	0.7 max	1.2-1.8	0.6 max	1.7-2.3	0.7 max	0.2 max	0.1 max			Cr 0.2 max; Pb 0.05 max; Sn 0.05 max	260		

UNS numbers and US grades are provided as a means of cross referencing chemically similar alloys. Exchangability is only possible after independent examination of specifications. Tensile properties are minimum or typical . UTS and YS as Mpa, El as %. See Appendix for list of abbreviations used in Descriptions.

3-14 Cast Aluminum

Grade UNS #	Country	Specification	Designation	Description	Cu	Fe	Mg	Mn	Ni	Si	Ti	Zn	OE Max	OT Max	Other	UTS	YS	EL
242 A02420	Russia	GOST 2685	AL1		3.75-4.5	0.8 max	1.25-1.75		1.75-2.25	0.7 max		0.3 max		1.5				
242 A02420	International		242	Sand, Perm Mold	3.5-4.5	1 max	1.2-1.8	0.35 max	1.7-2.3	0.7 max	0.25 max	0.35 max	0.05	0.15	Cr 0.25 max			
242 A02420	USA	ASTM B108	242	T571	3.5-4.5	1	1.2-1.8	0.35	1.7-2.3	0.7	0.25	0.35			Cr 0.25	234		
242 A02420	USA	ASTM B108	242	T61	3.5-4.5	1	1.2-1.8	0.35	1.7-2.3	0.7	0.25	0.35			Cr 0.25	276		
242 A02420	USA	ASTM B26	242	Sand T61 .250 in diam	3.5-4.5	1	1.2-1.8	0.35	1.7-2.3	0.7	0.25	0.35			Cr 0.25	221	138	
242 A02420	USA	ASTM B26	242	Sand Ann .250 in diam	3.5-4.5	1	1.2-1.8	0.35	1.7-2.3	0.7	0.25	0.35			Cr 0.25	159		
242 A02420	USA	ASTM B618	242	Invest Ann .250 in diam	3.5-4.5	1	1.2-1.8	0.35	1.7-2.3	0.7	0.25	0.35			Cr 0.25	159		
242 A02420	USA	ASTM B618	242	Invest T61 .250 in diam	3.5-4.5	1	1.2-1.8	0.35	1.7-2.3	0.7	0.25	0.35			Cr 0.25	221	138	
	Canada	CSA HA.3	SC51P		1-1.5	0.13	0.5-0.6	0.05		4.5-5.5	0.2	0.05						
	USA	ASTM B179	C355.2		1-1.5	0.13	0.5-0.6	0.05		4.5-5.5	0.2	0.05						
	Canada	CSA HA.3	SC51N	T5	1-1.5	0.14-0.25	0.5-0.6	0.05		4.5-5.5	0.2	0.05				186		
	USA	ASTM B108	B443.0	As fabricated	0.15	0.8	0.05	0.35		4.5-6	0.25	0.35				145	41	3
	USA	ASTM B26	B443.0	Sand As fabricated .250 in diam	0.15	0.8	0.05	0.35		4.5-6	0.25	0.35				117	41	3
	USA	ASTM B618	B443.0	Invest As fabricated .250 In dlam	0.15	0.8	0.05	0.35		4.5-6	0.25	0.35				117	41	3
	Canada	CSA HA.3	CS42		4-5		0.03	0.1		2-3	0.2		0.05	0.15				
203.2 A02032	International		203.2	Ingot	4.8-5.2	0.35 max	0.1 max	0.2-0.3	1.3-1.7	0.2 max	0.15-0.25	0.1 max	0.05	0.2	Sb 0.20-0.30, Co 0.20-0.30, Zr 0.10-0.30 Ti+Zr=0.50 max			
	USA	ASTM B179	295.1		4-5	0.8	0.03	0.35		0.7-1.5	0.25	0.35						
A444.2 A14442	International		A444.2	Ingot	0.05 max	0.12 max	0.05 max	0.05 max		6.5-7.5	0.2 max	0.05 max	0.05	0.15				
	Australia	AS 1874	AB405	T4 permanent mould	0.1 max	0.15 max	0.05 max	0.1 max	0.1 max	6.5-7.5	0.2 max	0.1 max	0.05	0.15		140		20
A444.1 A14441	International		A444.1	Ingot	0.1 max	0.15 max	0.05 max	0.1 max		6.5-7.5	0.2 max	0.1 max	0.05	0.15				
A444.0 A14440	International		A444.0	Perm Mold	0.1 max	0.2 max	0.05 max	0.1 max		6.5-7.5	0.2 max	0.1 max	0.05	0.15				
	Belgium	NBN P21-101	SGAlSi7	Sand	0.1 max	0.5 max	0.6 max	0.6 max	0.05 max	6.5-7.5	0.2 max	0.3 max			Pb 0.05 max; Sn 0.05 max			
444 A04440	International		444	Sand, Perm Mold	0.25 max	0.6 max	0.1 max	0.35 max		6.5-7.5	0.25 max	0.35 max	0.05	0.15				
	Australia	AS 1874	AA507	Ingot	0.3 max	0.4 max	6-7	0.35 max		0.4-0.8	0.1-0.2	0.05 max	0.05	0.15	Pb 0.05 max; Sn 0.05 max			
	Canada	CSA HA.10	SC51P	T6A	1-1.5	0.2	0.4-0.6	0.1		4.5-5.5	0.2	0.1				276	207	3
	Canada	CSA HA.9	SC51P	Sand T6	1-1.5	0.2	0.4-0.6	0.1		4.5-5.5	0.2	0.1				248	172	3
	Canada	CSA HA.9	SC51P	Sand T6A	1-1.5	0.2	0.4-0.6	0.1		4.5-5.5	0.2	0.1				276	207	3
	USA	ASTM B108	C355.0	T61	1-1.5	0.2	0.4-0.6	0.1		4.5-5.5	0.2	0.1				255	207	1
	USA	ASTM B26	C355.0	Sand SHT and art aged .250 in diam	1-1.5	0.2	0.4-0.6	0.1		4.5-5.5	0.2	0.1				248	172	3
	USA	ASTM B618	C355.0	Invest SHT and art aged .250 in diam	1-1.5	0.2	0.4-0.6	0.1		4.5-5.5	0.2	0.1				248	172	3
	USA	MIL A-21180C	C355.0	Class 1	1-1.5	0.2	0.4-0.6	0.1		4.5-5.5	0.2	0.1				283	241	3
	USA	MIL A-21180C	C355.0	Class 2	1-1.5	0.2	0.4-0.6	0.1		4.5-5.5	0.2	0.1				303	228	3

UNS numbers and US grades are provided as a means of cross referencing chemically similar alloys. Exchangability is only possible after independent examination of specifications. Tensile properties are minimum or typical . UTS and YS as Mpa, El as %. See Appendix for list of abbreviations used in Descriptions.

Worldwide Guide to Equivalent Nonferrous Metals and Alloys

Grade UNS #	Country	Specification	Designation	Description	Cu	Fe	Mg	Mn	Ni	Si	Ti	Zn	OE Max	OT Max	Other	UTS	YS	EL
	USA	MIL A-21180C	C355.0	Class 3	1-1.5	0.2	0.4-0.6	0.1		4.5-5.5	0.2	0.1				345	276	2
	USA	QQ A-596d	C355	SHt and art aged T61	1-1.5	0.2	0.1	0.4-0.6		4.5-5.5	0.2	0.1				276		3
242.2 A02422	USA	ASTM B179	242.2		3.5-4.5	0.6	1.3-1.8	0.1	1.7-2.3	0.06	0.2	0.1						
242.2 A02422	International		242.2	Ingot	3.5-4.5	0.6 max	1.3-1.8 max	0.1 max	1.7-2.3	0.6 max	0.2 max	0.1 max	0.05	0.15				
242.2 A02422	USA	QQ A-371F	242.2		3.5-4.5	0.6	1.3-1.8	0.1	1.7-2.3	0.6	0.2	0.1						
242.2 A02422	Canada	CSA	CN42		3.5-4.5	0.8 max	1.2-1.7 max	0.1 max	1.7-2.3	0.6 max	0.2 max	0.1 max			ET .05 max .15 tot			
242.1 A02421	International		242.1	Ingot	3.5-4.5	0.8 max	1.3-1.8	0.35 max	1.7-2.3	0.7 max	0.25 max	0.35 max	0.05	0.15	Cr 0.25 max			
242.1 A02421	USA	ASTM B179	242.1		3.5-4.5	0.8	1.3-1.8	0.35	1.7-2.3	0.7	0.25	0.35			Cr 0.25			
242.1 A02421	USA	QQ A-371F	242.1		3.5-4.5	0.8	1.3-1.8	0.35	1.7-2.3	0.7	0.25	0.35			Cr 0.25			
	Canada	CSA HA.10	CS42	T4	4-5	0.2	0.03	0.1		2-3	0.2					296	124	15
	Canada	CSA HA.10	CS42	T6	4-5	0.2	0.03	0.1		2-3	0.2					338	193	8
	Canada	CSA HA.3	Zg61N		0.25	0.4	0.5-0.65	0.1		0.15	0.15-0.25	5-6.5			Cr 0.4-0.6			
B535.0 A25350	International		B535.0	Sand	0.1 max	0.15 max	6.5-7.5	0.05 max		0.15 max	0.1-0.25		0.05	0.15				
A535.0 A15350	International		A535.0	Sand	0.1 max	0.2 max	6.5-7.5	0.1-0.25		0.2 max	0.25 max		0.05	0.15	Nb 0.25 max			
444.2 A04442	International		444.2	Ingot	0.1 max	0.13-0.25	0.05 max	0.05 max		6.5-7.5	0.2 max	0.05 max	0.05	0.15				
	India	IS 202	A-25	Sand ST PT	4-5	1	0.03	0.3		1.2		0.1				221	128	3
	India	IS 202	A-25	Sand ST PT	4-5	1	0.03	0.3		1.2		0.1				221	128	3
	Germany	DIN 1725Sh2	3.3261.01/G-AlMg5Si	Sand	0.05	0.5	4.5-5.5	0.4		0.9-1.5	0.2	0.1				160	110	2
	Germany	DIN 1725Sh2	3.3261.02/GK-AlMg5Si		0.05	0.5	4.5-5.5	0.4		0.9-1.5	0.2	0.1				180	110	2
	Germany	DIN 1725Sh2	3.2341./GK-AlSi5Mgka	and PH	0.05	0.5	0.4-0.8	0.4		5-6	0.2	0.1				210	160	2
	Germany	DIN 1725Sh2	3.2341.01/G-AlSi5Mg	Sand	0.05	0.5	0.4-0.8	0.4		5-6	0.2	0.1				140	100	1
	Germany	DIN 1725Sh2	3.2341.02/GK-AlSi5Mg		0.05	0.5	0.4-0.8	0.4		5-6	0.2	0.1				160	120	1
	Germany	DIN 1725Sh2	3.2341.4/G-AlSi5Mgka	Sand Ph	0.05	0.5	0.4-0.8	0.4		5-6	0.2	0.1				180	150	2
	Germany	DIN 1725Sh2	3.2341.6/G-AlSi5Mgwa	Sand Q and AH	0.05	0.5	0.4-0.8	0.4		5-6	0.2	0.1				240	220	1
	Germany	DIN 1725Sh2	3.2341/GK-AlSi5Mgwa	Q and AH	0.05	0.5	0.4-0.8	0.4		5-6	0.2	0.1				260	240	1
	USA	ASTM B179	D712.2		0.25	0.5	0.5-0.65	0.1		0.15	0.15-0.25	5-6.5			Cr 0.4-0.6			
356 A03560	South Africa	SABS 989	42100 (Al-Si7 Mg)	Ingot	0.03 max	0.15 max	0.3-0.45	0.1 max		6.5-7.5	0.1-0.18	0.07 max	0.03	0.1				
B535.2 A23532	International		B535.2	Ingot	0.05 max	0.12 max	6.6-7.5	0.05 max		0.1 max	0.1-0.25		0.05	0.15				
356 A03560	South Africa	SABS 991	42100 (Al-Si7 Mg)	T6 Sand	0.05 max	0.19 max	0.25-0.45	0.1 max		6.5-7.5	0.08-0.25	0.07 max	0.03	0.1		230		2
356 A03560	South Africa	SABS 991	421000 (Al-Si7 Mg)	T6 Gravity Die	0.05 max	0.19 max	0.25-0.45	0.1 max		6.5-7.5	0.08-0.25	0.07 max	0.03	0.1		290		4
772 A07720	International		772	Sand	0.1 max	0.15 max	0.6-0.8	0.1 max		0.15 max	0.1-0.2	6-7	0.05	0.15	Cr 0.06-0.2			
A535.1 A15351	International		A535.1	Ingot	0.1 max	0.15 max	6.6-7.5	0.1-0.25		0.2 max	0.25 max		0.05	0.15	Nb 0.25 max			
	Belgium	NBN P21-101	SGAlSi7Mg	Sand	0.1 max	0.3 max	0.2-0.6	0.3 max	0.05 max	6.5-7.5	0.2 max	0.1 max			Pb 0.05 max; Sn 0.05 max			
356 A03560	UK	BS 1490	LM25		0.1 max	0.5	0.2-0.45	0.3 max	0.1 max	6.5-7.5	0.05-0.2	0.1 max			Pb 0.1 max; Sn 0.2 max			
356 A03560	International	ISO R2147	Al-Si7Mg	Sand SHT and art aged	0.2	0.05	0.2-0.4	0.6	0.05	6.5-7.5	0.2	0.3			Pb 0.05; Sn 0.05 max	230		1

UNS numbers and US grades are provided as a means of cross referencing chemically similar alloys. Exchangability is only possible after independent examination of specifications. Tensile properties are minimum or typical . UTS and YS as Mpa, El as %. See Appendix for list of abbreviations used in Descriptions.

3-16 Cast Aluminum

Grade UNS #	Country	Specification	Designation	Description	Cu	Fe	Mg	Mn	Ni	Si	Ti	Zn	OE Max	OT Max	Other	UTS	YS	EL
				OBSOLETE Replaced by ISO 3522														
	International	ISO 3522	Al-Si7 Mg (Fe)	Sand Cast As mfg	0.2 max	0.5 max	0.2-0.4	0.6 max	0.05 max	6.5-7.5	0.2 max	0.3 max			Pb 0.05 max; Sn 0.05 max	140		2
	International	ISO 3522	Al-Si7 Mg (Fe)	Sand Cast SHT PHT	0.2 max	0.5 max	0.2-0.4	0.6 max	0.05 max	6.5-7.5	0.2 max	0.3 max			Pb 0.05 max; Sn 0.05 max	210		1
	International	ISO 3522	Al-Si7 Mg (Fe)	Perm Mold As mfg	0.2 max	0.5 max	0.2-0.4	0.6 max	0.05 max	6.5-7.5	0.2 max	0.3 max			Pb 0.05 max; Sn 0.05 max	150		3
	International	ISO 3522	Al-Si7 Mg (Fe)	Perm Mold SHT PHT	0.2 max	0.5 max	0.2-0.4	0.6 max	0.05 max	6.5-7.5	0.2 max	0.3 max			Pb 0.05 max; Sn 0.05 max	230		2
356 A03560	Japan	JIS H5202	AC4C		0.2	0.5	0.2-0.4	0.3		6.5-7.5	0.2	0.3				157		3
356 A03560	Japan	JIS H5202	AC4C	Temp	0.2	0.5	0.2-0.4	0.3		6.5-7.5	0.2	0.3				177		3
356 A03560	Japan	JIS H5202	AC4C	Q/T	0.2	0.5	0.2-0.4	0.3		6.5-7.5	0.2	0.3				226		3
356 A03560	Czech Republic	CSN 424332	AlSi7Mg(Fe)	Sand Ingot Alloys	0.2 max	0.6 max	0.2-0.4	0.5 max		6-8	0.2 max	0.3 max			Sn 0.01 max			
356 A03560	Russia	GOST 2685	AL9		0.2 max	0.6 max	0.2-0.4	0.5 max		6-8		0.3 max		1	Sn 0.01 max			
356 A03560	Czech Republic	CSN 424332	AlSi7Mg(Fe)	Permanent Mold Ingot Alloys	0.2 max	0.8 max	0.2-0.4	0.5 max		6-8	0.2 max	0.2 max			Sn 0.01 max			
356 A03560	International		356	Sand, Perm Mold	0.25 max	0.6 max	0.2-0.45	0.35 max		6.5-7.5	0.25 max	0.35 max	0.05	0.15	If iron exceeds 0.45, manganese content shall not be less than one-half iron content			
356 A03560	USA	ASTM B108	356	As fabricated	0.25	0.6	0.2-0.4	0.35		6.5-7.5	0.25	0.35				145		3
356 A03560	USA	ASTM B108	356	SHT and art aged	0.25	0.6	0.2-0.4	0.35		6.5-7.5	0.25	0.35				228	152	3
356 A03560	USA	ASTM B108	356	T71	0.25	0.6	0.2-0.4	0.35		6.5-7.5	0.25	0.35				172		3
356 A03560	USA	ASTM B26	356	Sand As fabricated .250 in diam	0.25	0.6	0.2-0.4	0.35		6.5-7.5	0.25	0.35				131		2
356 A03560	USA	ASTM B26	356	Sand SHT and art aged .250 in diam	0.25	0.6	0.2-0.4	0.35		6.5-7.5	0.25	0.35				207	138	3
356 A03560	USA	ASTM B618	356	Invest As fabricated .250 in diam	0.25	0.6	0.2-0.4	0.35		6.5-7.5	0.25	0.35				131		2
356 A03560	USA	ASTM B618	356	Invest SHT and art aged .250 in diam	0.25	0.6	0.2-0.4	0.35		6.5-7.5	0.25	0.35				207	138	3
356 A03560	USA	ASTM B618	356	Invest SHT and stab .250 in diam	0.25	0.6	0.2-0.4	0.35		6.5-7.5	0.25	0.35				214		
356 A03560	USA	QQ A-596d	356	SHT and art aged T6	0.25	0.6	0.35	0.2-0.4		6.5-7.5	0.25	0.35				228		3
356 A03560	USA	QQ A-596d	356	SHt and stab T7	0.25	0.6	0.35	0.2-0.4		6.5-7.5	0.25	0.35				200		4
356 A03560	USA	QQ A-596d	356	Art aghed only and temper T51	0.25	0.6	0.35	0.2-0.4		6.5-7.5	0.25	0.35				172		
	USA	ASTM B179	A443.1		0.3	0.6	0.05	0.5		4.5-6	0.25	0.5			Cr 0.25			
F356.2	International		F356.2	Ingot	0.1 max	0.12 max	0.17-0.25	0.05 max		6.5-7.5	0.04-0.2	0.05 max	0.05	0.15				
F356.0	International		F356.0	Sand, Perm Mold	0.2 max	0.2 max	0.17-0.25	0.1 max		6.5-7.5	0.04-0.2	0.1 max	0.05	0.15				
	Germany	DIN 1725Sh3	3.0861/V-AlTi5B1		0.02	0.3	0.02	0.02	0.04	0.2	5-6.2	0.03			B 0.9-1.4; Cr 0.02; V 0.2			
A356.0 A13560	Austria	ONORM M3429	AlSi7Mg		0.03 max	0.15 max	0.3-0.4	0.05 max		6.5-7.5	0.15 max	0.07 max			ET 0.03 min; ET=.10 total;			
A356.0 A13560	Austria	ONORM M3429	G AlSi7Mg		0.03 max	0.15 max	0.3-0.4	0.05 max		6.5-7.5	0.15 max	0.07 max			ET 0.03 max; ET=.10 total;			

UNS numbers and US grades are provided as a means of cross referencing chemically similar alloys. Exchangability is only possible after independent examination of specifications. Tensile properties are minimum or typical . UTS and YS as Mpa, El as %. See Appendix for list of abbreviations used in Descriptions.

Worldwide Guide to Equivalent Nonferrous Metals and Alloys

Grade UNS #	Country	Specification	Designation	Description	Cu	Fe	Mg	Mn	Ni	Si	Ti	Zn	OE Max	OT Max	Other	UTS	YS	EL
A356.0 A13560	Austria	ONORM M3429	GK AlSi7Mg	Perm Mold	0.05 max	0.18 max	0.2-0.4	0.05 max		6.5-7.5	0.15 max	0.07 max			ET 0.05 max; ET=.15 total;			
A356.0 A13560	Austria	ONORM M3429	GS AlSi7Mg	Sand	0.05 max	0.18 max	0.2-0.4	0.05 max		6.5-7.5	0.15 max	0.07 max			ET 0.03 max; ET=.10 total;			
	Australia	AS 1874	CC601	Ingot	0.05 max	0.2 max	0.25-0.35	0.05 max		6.5-7.5	0.2 max	0.05 max	0.05	0.15				
772.2 A07722	International		772.2	Ingot	0.1 max	0.1 max	0.65-0.8	0.1 max		0.1 max	0.1-0.2	6-7	0.05	0.15	Cr 0.06-0.2			
	International	ISO 3522	Al-Si7 Mg	Sand Cast SHT PHT	0.1 max	0.2 max	0.25-0.45	0.1 max	0.05 max	6.5-7.5	0.2 max	0.1 max	0.05	0.15		230	180	2
	International	ISO 3522	Al-Si7 Mg	Perm Mold SHT PHT	0.1 max	0.2 max	0.25-0.45	0.1 max	0.05 max	6.5-7.5	0.2 max	0.1 max	0.05	0.15		250	190	5
A356.0 A13560	Czech Republic	CSN 424334	AlSiMgTi	Sand Ingot Alloys	0.1 max	0.3 max	0.25-0.45	0.1 max	0.05 max	6.5-7.5	0.1-0.2	0.1 max						
A356.0 A13560	International		A356.0	Sand, Perm Mold	0.2 max	0.2 max	0.25-0.45	0.1 max		6.5-7.5	0.2 max	0.1 max	0.05	0.15				
356.1 A03561	Denmark	DS 3002	4244		0.2	0.5	0.2-0.4	0.5	0.1	6.5-7.5	0.2	0.3			Pb 0.05; Sn 0.05			
356.1 A03561	Norway	NS 17525	NS17525-00		0.2	0.5	0.2-0.4	0.5	0.1	6.5-7.5	0.2	0.3			Al 93; Pb 0.05; Sn 0.05 max			
356.1 A03561	Norway	NS 17525	NS17525-41	Sand	0.2	0.5	0.2-0.4	0.5	0.1	6.5-7.5	0.2	0.3			Al 93; Pb 0.05; Sn 0.05 max			
356.1 A03561	Norway	NS 17525	NS17525-42	H and art aged	0.2	0.5	0.2-0.4		0.1	6.5-7.5	0.2	0.3			Al 93; Pb 0.05; Sn 0.05 max	216		1
42000	South Africa	SABS 991	Al-Si7 Mg 0.3	F Sand	0.2 max	0.55 max	0.2-0.65	0.35 max	0.15 max	6.5-7.5	0.05-0.25	0.15 max	0.05	0.15	Pb 0.15 max; Sn 0.05 max	140		2
42000	South Africa	SABS 991	Al-Si7 Mg 0.3	T6 Sand	0.2 max	0.55 max	0.2-0.65	0.35 max	0.15 max	6.5-7.5	0.05-0.25	0.15 max	0.05	0.15	Pb 0.15 max; Sn 0.05 max	220		1
42000	South Africa	SABS 991	Al-Si7 Mg 0.3	F Gravity Die	0.2 max	0.55 max	0.2-0.65	0.35 max	0.15 max	6.5-7.5	0.05-0.25	0.15 max	0.05	0.15	Pb 0.15 max; Sn 0.05 max	170		2.5
42000	South Africa	SABS 991	Al-Si7 Mg 0.3	T6 Gravity Die	0.2 max	0.55 max	0.2-0.65	0.35 max	0.15 max	6.5-7.5	0.05-0.25	0.15 max	0.05	0.15	Pb 0.15 max; Sn 0.05 max	260		1
356.1 A03561	International		356.1	Ingot	0.25 max	0.5 max	0.25-0.45	0.35 max		6.5-7.5	0.25 max	0.35 max	0.05	0.15	If iron exceeds 0.45, manganese content shall not be less than one-half iron content			
356.1 A03561	USA	ASTM B179	356.1	SHT and stab .250 in diam	0.25	0.5	0.25-0.4	0.35		6.7-7.5	0.25	0.35				214		
356.1 A03561	USA	QQ A-371F	356.1		1-1.5	0.13	0.45-0.6	0.05		4.5-5.5	0.2	0.05						
A242.0 A12420	International		A242.0	Sand	3.7-4.5	0.8 max	1.2-1.7	0.1 max	1.8-2.3	0.6 max	0.07-0.2	0.1 max	0.05	0.15	Cr 0.15-0.25			
C356.0 A33560	International		C356.0	Sand, Perm Mold	0.05 max	0.07 max	0.25-0.45	0.05 max		6.5-7.5	0.04-0.2	0.05 max	0.05	0.15				
B356.0 A23560	International		B356.0	Sand, Perm Mold	0.05 max	0.09 max	0.25-0.45	0.05 max		6.5-7.5	0.04-0.2	0.05 max	0.05	0.15				
535.2 A05352	International		535.2	Ingot	0.05 max	0.1 max	6.6-7.5	0.1-0.25		0.1 max	0.1-0.25		0.05	0.15	Beryllium 0.003-0.007, boron 0.002 max.			
535.2 A05352	USA	ASTM B179	535.2		0.05	0.1	6.6-7.5	0.1-0.25		0.1	0.1-0.25				B 0.002 max; Be 0.003-0.007			
535.2 A05352	USA	QQ A-371F	535.2		0.05	0.1	6.6-7.5	0.1-0.25		0.1	0.1-0.25				B 0.002 max; Be 0.003-0.007			
	Australia	AS 1874	AA601	Ingot	0.05 max	0.2 max	0.3-0.4	0.05 max		6.5-7.5	0.2 max	0.05 max	0.05					
	Australia	AS 1874	AA601	T1 sand	0.05 max	0.2 max	0.3-0.4	0.05 max		6.5-7.5	0.2 max	0.05 max	0.05	0.15		130		2
	Australia	AS 1874	AA601	T1 permanent mould	0.05 max	0.2 max	0.3-0.4	0.05 max		6.5-7.5	0.2 max	0.05 max	0.05			140		3
	Australia	AS 1874	AA601	T5 sand	0.05 max	0.2 max	0.3-0.4	0.05 max		6.5-7.5	0.2 max	0.05 max	0.05	0.15		155		
	Australia	AS 1874	AA601	T5 permanent mould	0.05 max	0.2 max	0.3-0.4	0.05 max		6.5-7.5	0.2 max	0.05 max	0.05			170		
	Australia	AS 1874	AA601	T6 sand	0.05 max	0.2 max	0.3-0.4	0.05 max		6.5-7.5	0.2 max	0.05 max	0.05	0.15		205		3

UNS numbers and US grades are provided as a means of cross referencing chemically similar alloys. Exchangability is only possible after independent examination of specifications. Tensile properties are minimum or typical . UTS and YS as Mpa, El as %. See Appendix for list of abbreviations used in Descriptions.

3-18 Cast Aluminum

Grade UNS #	Country	Specification	Designation	Description	Cu	Fe	Mg	Mn	Ni	Si	Ti	Zn	OE Max	OT Max	Other	UTS	YS	EL
	Australia	AS 1874	AA601	T6 permanent mould	0.05 max	0.2 max	0.3-0.4	0.05 max		6.5-7.5	0.2 max	0.05 max	0.05			220		5
	Australia	AS 1874	AA601	T61 permanent mould	0.05 max	0.2 max	0.3-0.4	0.05 max		6.5-7.5	0.2 max	0.05 max	0.05	0.15		260		3
	Australia	AS 1874	AC601	Ingot	0.05 max	0.2 max	0.3-0.4	0.05 max		6.5-7.5	0.2 max	0.05 max	0.05	0.15				
	Australia	AS 1874	AC601	T1 sand	0.05 max	0.2 max	0.3-0.4	0.05 max		6.5-7.5	0.2 max	0.05 max	0.05	0.15		130		2
	Australia	AS 1874	AC601	T1 permanent mould	0.05 max	0.2 max	0.3-0.4	0.05 max		6.5-7.5	0.2 max	0.05 max	0.05			140		3
	Australia	AS 1874	AC601	T5 sand	0.05 max	0.2 max	0.3-0.4	0.05 max		6.5-7.5	0.2 max	0.05 max	0.05	0.15		155		
	Australia	AS 1874	AC601	T5 permanent mould	0.05 max	0.2 max	0.3-0.4	0.05 max		6.5-7.5	0.2 max	0.05 max	0.05			170		
	Australia	AS 1874	AC601	T6 sand	0.05 max	0.2 max	0.3-0.4	0.05 max		6.5-7.5	0.2 max	0.05 max	0.05	0.15		205		3
	Australia	AS 1874	AC601	T6 permanent mould	0.05 max	0.2 max	0.3-0.4	0.05 max		6.5-7.5	0.2 max	0.05 max	0.05			220		5
	Australia	AS 1874	AC601	T61 permanent mould	0.05 max	0.2 max	0.3-0.4	0.05 max		6.5-7.5	0.2 max	0.05 max	0.05	0.15		260		3
	Australia	AS 1874	CC601	T1 sand	0.05 max	0.2 max	0.3-0.4	0.05 max		6.5-7.5	0.2 max	0.05 max	0.05	0.15		130		2
	Australia	AS 1874	CC601	T1 permanent mould	0.05 max	0.2 max	0.3-0.4	0.05 max		6.5-7.5	0.2 max	0.05 max	0.05			140		3
	Australia	AS 1874	CC601	T5 sand	0.05 max	0.2 max	0.3-0.4	0.05 max		6.5-7.5	0.2 max	0.05 max	0.05	0.15		155		
	Australia	AS 1874	CC601	T5 permanent mould	0.05 max	0.2 max	0.3-0.4	0.05 max		6.5-7.5	0.2 max	0.05 max	0.05			170		
	Australia	AS 1874	CC601	T6 sand	0.05 max	0.2 max	0.3-0.4	0.05 max		6.5-7.5	0.2 max	0.05 max	0.05	0.15		205		3
	Australia	AS 1874	CC601	T6 permanent mould	0.05 max	0.2 max	0.3-0.4	0.05 max		6.5-7.5	0.2 max	0.05 max	0.05			220		5
	Australia	AS 1874	CC601	T61 permanent mould	0.05 max	0.2 max	0.3-0.4	0.05 max		6.5-7.5	0.2 max	0.05 max	0.05	0.15		260		3
535.2 A05352	France	NF A57-702	AG6	Perm Mold Sand	0.1 max	0.5 max	5-7	0.5 max	0.05 max	0.4 max	0.2 max	0.2 max			Cr 0.05 max; Pb 0.05 max; Sn 0.05 max			
A356.2 A13562	Canada	CSA	SC70P		0.1 max	0.11 max	0.3-0.4	0.05 max		6.5-7.5	0.2 max	0.05 max			ET 0.05 max; ET=.15 total;			
A356.2 A13562	International		A356.2	Ingot	0.1 max	0.12 max	0.3-0.45	0.05 max		6.5-7.5	0.2 max	0.05 max	0.05	0.15				
A356.1 A13561	South Africa	SABS 989	Al-Si7MgA	Sand	0.1	0.45	0.25-0.4	0.3	0.05	6.5-7.5	0.2	0.1			Pb 0.05; Sn 0.05 max	125		2
A356.1 A13561	South Africa	SABS 989	Al-Si7MgA	Sand SHT and PHT	0.1	0.45	0.25-0.4	0.3	0.05	6.5-7.5	0.2	0.1			Pb 0.05; Sn 0.05 max	230		
A356.1 A13561	South Africa	SABS 989	Al-Si7MgA	Chill	0.1	0.45	0.25-0.4	0.3	0.05	6.5-7.5	0.2	0.1			Pb 0.05; Sn 0.05 max	160		3
A356.1 A13561	South Africa	SABS 990	Al-Si7MgA	Sand	0.1	0.5	0.2-0.4	0.3	0.05	6.5-7.5	0.2	0.1			Pb 0.05; Sn 0.05 max	125		2
A356.1 A13561	South Africa	SABS 990	Al-Si7MgA	Sand PHT	0.1	0.5	0.2-0.4	0.3	0.05	6.5-7.5	0.2	0.1			Pb 0.05; Sn 0.05 max	145		1
A356.1 A13561	South Africa	SABS 990	Al-Si7MgA	Sand SHT snd Stab	0.1	0.5	0.2-0.4	0.3	0.05	6.5-7.5	0.2	0.1			Pb 0.05; Sn 0.05 max	160		3
A356.1 A13561	South Africa	SABS 991	Al-Si7MgA		0.1	0.5	0.2-0.4	0.3	0.05	6.5-7.5	0.2	0.1			Pb 0.05; Sn 0.05 max			
	Australia	AS 1874	AB405	Ingot	0.1 max	0.15 max	0.05 max	0.1 max		6.8-7.8	0.2 max	0.1 max	0.05	0.15				
42000	South Africa	SABS 989	Al-Si7 Mg 0.3	Ingot	0.15 max	0.45 max	0.25-0.65	0.35 max	0.15 max	6.5-7.5	0.05-0.2	0.15 max	0.05	0.15	Pb 0.15 max; Sn 0.05 max			

UNS numbers and US grades are provided as a means of cross referencing chemically similar alloys. Exchangability is only possible after independent examination of specifications. Tensile properties are minimum or typical . UTS and YS as Mpa, El as %. See Appendix for list of abbreviations used in Descriptions.

Worldwide Guide to Equivalent Nonferrous Metals and Alloys

Grade UNS #	Country	Specification	Designation	Description	Cu	Fe	Mg	Mn	Ni	Si	Ti	Zn	OE Max	OT Max	Other	UTS	YS	EL
535.2 A05352	France	NF A57-703	AG6	Die	0.2 max	1.3 max	5-8.5	0.6 max	0.1 max	1 max	0.2 max	0.4 max			Be 0.05 max; Sn 0.1 max			
A356.1 A13561	International		A356.1	Ingot	0.2 max	0.15 max	0.3-0.45	0.1 max		6.5-7.5	0.2 max	0.1 max	0.05	0.15				
	Canada	CSA HA.9	ZG61N	Sand T5	0.25	0.5	0.5-0.65	0.1		0.3	0.15-0.25	5-6.5			Cr 0.4-0.6	234	172	4
	USA	ASTM B26	D172.0	Sand Cooled and art aged .250 in diam	0.25	0.5	0.5-0.65	0.1		0.3	0.15-0.25	5-6.5		0.05	Cr 0.4-0.6	224	172	4
	USA	ASTM B618	D712.0	Invest Cooled and nat or art aged .250 in	0.25	0.5	0.5-0.65	0.1		0.3	0.15-0.25	5-6.5			Cr 0.4-0.6	234	172	4
	Australia	AS 1874	DA601	Ingot	0.25 max	0.5 max	0.3-0.5	0.35 max		6.5-7.5	0.25 max	0.35 max	0.05	0.15				
	Australia	AS 1874	DA601	T1 sand	0.25 max	0.5 max	0.3-0.5	0.35 max		6.5-7.5	0.25 max	0.35 max	0.05	0.15		130		2
	Australia	AS 1874	DA601	T1 perm mould	0.25 max	0.5 max	0.3-0.5	0.35 max		6.5-7.5	0.25 max	0.35 max	0.05	0.15		140		3
	Australia	AS 1874	DA601	T6 sand	0.25 max	0.5 max	0.3-0.5	0.35 max		6.5-7.5	0.25 max	0.35 max	0.05	0.15		205		3
	Australia	AS 1874	DA601	T6 permanent mould	0.25 max	0.5 max	0.3-0.5	0.35 max		6.5-7.5	0.25 max	0.35 max	0.05	0.15		225		3
C356.2 A33562	International		C356.2	Ingot	0.03 max	0.04 max	0.3-0.45	0.03 max		6.5-7.5	0.04-0.2	0.03 max	0.03	0.1				
B356.2 A23562	International		B356.2	Ingot	0.03 max	0.06 max	0.3-0.45	0.03 max		6.5-7.5	0.04-0.2	0.03 max	0.03	0.1				
A356.0 A13560	Czech Republic	CSN 424334	AlSiMgTi	Permanent Mold Ingot Alloys	0.1 max	0.3 max	0.25-0.45	0.1 max	0.05 max	6.5-7.5	0.1-0.2	0.1 max						
	Japan	JIS H2117	C1AS	Ingot	4-5	0.4	0.3	0.3	0.1	1.2	0.25	0.3						
A242.2 A12422	International		A242.2	Ingot	3.7-4.5	0.6 max	1.3-1.7	0.1 max	1.8-2.3	0.35 max	0.07-0.2	0.1 max	0.05	0.15	Cr 0.15-0.25			
A242.1 A12421	International		A242.1	Ingot	3.7-4.5	0.6 max	1.3-1.7	0.1 max	1.8-2.3	0.6 max	0.07-0.2	0.1 max	0.05	0.15	Cr 0.15-0.25			
	USA	ASTM B108	A443.0	As fabricated	0.3	0.8	0.05	0.5		4.5-6	0.25	0.5			Cr 0.25	145	49	2
	USA	ASTM B26	A443.0	Sand As fabricated .250 in diam	0.3	0.8	0.05	0.5		4.5-6	0.25	0.5			Cr 0.25	117	48	3
	USA	ASTM B618	A443.0	Invest As fabricate .250 in diam	0.3	0.8	0.05	0.5		4.5-6	0.25	0.5			Cr 0.25	117	48	3
850.1 A08501	International		850.1	Ingot	0.7-1.3	0.5 max	0.1 max	0.1 max	0.7-1.3	0.7 max	0.2 max			0.3	Sn 5.5-7			
850.1 A08501	USA	ASTM B179	850.1		0.7-1.3	0.5	0.1	0.1	0.7-1.3	0.7	0.2				Sn 5.5-7			
850.1 A08501	USA	QQ A-371F	850.1		0.7-1.3	0.5	0.1	0.1	0.7-1.3	0.7	0.2				Sn 5.5-7			
850 A08500	International		850	Sand, Perm Mold	0.7-1.3	0.7 max	0.1 max	0.1 max	0.7-1.3	0.7 max	0.2 max			0.3	Sn 5.5-7			
850 A08500	USA	ASTM B108	850	Cjoled and art aged	0.7-1.3	0.7	0.1	0.1	0.7-1.3	0.7	0.2				Sn 5.5-7	124	8	
850 A08500	USA	ASTM B26	850	Sand Cooled and art aged .250 in diam	0.7-1.3	0.7	0.1	0.1	0.7-1.3	0.7	0.2				Sn 5.5-7	110		5
850 A08500	USA	ASTM B618	850	Invest Cooled and art aged .250 in diam	0.7-1.3	0.7	0.1	0.1	0.7-1.3	0.7	0.2				Sn 5.5-7	110		5
850 A08500	USA	QQ A-596d	850	Aged and temper T5	0.7-1.3	0.7		0.1	0.7-1.3	0.7	0.2				Sn 5.7-7	124		8
356.2 A03562	Canada	CSA	SG70		0.1 max	0.12-0.25	0.3-0.4	0.05 max	0.05 max	6.5-7.5	0.2 max				ET 0.05 max; ET=.15 total;			
356.2 A03562	International		356.2	Ingot	0.1 max	0.13-0.25	0.3-0.45	0.05 max		6.5-7.5	0.2 max	0.05 max	0.05	0.15				
356.2 A03562	USA	ASTM B179	356.2		0.1	0.13-0.25	0.3-0.4	0.05		6.5-7.5	0.2	0.05						

UNS numbers and US grades are provided as a means of cross referencing chemically similar alloys. Exchangability is only possible after independent examination of specifications. Tensile properties are minimum or typical . UTS and YS as Mpa, El as %. See Appendix for list of abbreviations used in Descriptions.

3-20 Cast Aluminum

Grade UNS #	Country	Specification	Designation	Description	Cu	Fe	Mg	Mn	Ni	Si	Ti	Zn	OE Max	OT Max	Other	UTS	YS	EL
356.2 A03562	USA	QQ A-371F	356.2		0.1	0.12-0.25	0.3-0.4	0.05		6.5-7.5	0.2	0.05						
B357.0 A23570	International		B357.0	Sand, Perm Mold	0.05 max	0.09 max	0.4-0.6	0.05 max		6.5-7.5	0.04-0.2	0.05 max	0.05	0.15				
A357.0 A13571	International		A357.0	Sand, Perm Mold	0.12 max	0.2 max	0.4-0.7	0.1 max		6.5-7.5	0.04-0.2	0.1 max	0.05	0.15	Beryllium 0.04-0.07			
357.1 A03571	International		357.1	Ingot	0.05 max	0.12 max	0.45-0.6	0.03 max		6.5-7.5	0.2 max	0.05 max	0.05	0.15				
	Australia	AS 1874	AA603	Ingot	0.05 max	0.15 max	0.45-0.7	0.03 max		6.5-7.5	0.2 max	0.05 max	0.05	0.15				
	Australia	AS 1874	AA603	T6 sand	0.05 max	0.15 max	0.45-0.7	0.03 max		6.5-7.5	0.2 max	0.05 max	0.05	0.15		270		1
	Australia	AS 1874	AA603	T6 permanent mould	0.05 max	0.15 max	0.45-0.7	0.03 max		6.5-7.5	0.2 max	0.05 max	0.05	0.15		290		3
	Australia	AS 1874	AC603	Ingot	0.05 max	0.15 max	0.45-0.7	0.03 max		6.5-7.5	0.2 max	0.05 max	0.05	0.15				
	Australia	AS 1874	AC603	T6 Sand	0.05 max	0.15 max	0.45-0.7	0.03 max		6.5-7.5	0.2 max	0.05 max	0.05	0.15		270		1
	Australia	AS 1874	AC603	T6 permanent mould	0.05 max	0.15 max	0.45-0.7	0.03 max		6.5-7.5	0.2 max	0.05 max	0.05	0.15		290		3
357 A03570	International		357	Sand, Perm Mold	0.05 max	0.15 max	0.45-0.6	0.03 max		6.5-7.5	0.2 max	0.05 max	0.05	0.15				
357 A03570	USA	QQ A-596d	357	SHT and art aged T6	0.05	0.15	0.03	0.45-0.6		6.5-7.5	0.2	0.05				310		3
710 A07100	Norway	NS 17570	NS17570-00		0.2-0.5	0.7	0.6-0.8	0.4	0.05	0.3	0.25	6			Al 93; Cr 0.3-0.6; Pb 0.05; Sn 0.05 max			
710 A07100	Canada	CSA HA.3	ZG61P		0.35-0.65	0.4	0.65-0.8	0.05		0.15	0.2	6-7						
710 A07100	Canada	CSA HA.9	ZG61P	Sand	0.35-0.65	0.5	0.6-0.8	0.05		0.15	0.25	6-7						
710 A07100	International		710	Sand	0.35-0.65	0.5 max	0.6-0.8	0.05 max		0.15 max	0.25 max	6-7	0.05	0.15				
B357.2 A23572	International		B357.2	Ingot	0.03 max	0.06 max	0.45-0.6	0.03 max		6.5-7.5	0.04-0.2	0.03 max	0.03	0.1				
C357.0 A33570	International		C357.0	Sand, Perm Mold	0.05 max	0.09 max	0.45-0.7	0.05 max		6.5-7.5	0.04-0.2	0.05 max	0.05	0.15	Beryllium 0.04-0.07			
A357.2 A13572	International		A357.2	Ingot	0.1 max	0.12 max	0.45-0.7	0.05 max		6.5-7.5	0.04-0.2	0.05 max	0.03	0.1	Beryllium 0.04-0.07			
	Sweden	MNC 41E	4245	Sand Die Ingot		0.2 max	0.37			7					Al 93			
	Sweden	MNC 41E	4244	Die Sand Ingot		0.5 max	0.3			7					Fe 0.5 max; Al 93			
	Sweden	MNC 41E	4438	die Sand Ingot	0.4		0.7				0.2	5.5			Al 93; Cr 0.4			
710.1 A07101	Norway	NS 17570	NS17570-31	Nat aged	0.2-0.5	0.7	0.6-0.8	0.4	0.05	0.3	0.25	6			Al 93; Cr 0.3-0.6; Pb 0.05; Sn 0.05 max	216	167	4
710.1 A07101	International		710.1	Ingot	0.35-0.65	0.4 max	0.65-0.8	0.05 max		0.15 max	0.25 max	6-7	0.05	0.15				
213.1 A02131	International		213.1	Ingot	6-8	0.9 max	0.1 max	0.6 max	0.35 max	1-3	0.25 max	2.5 max		0.5				
213.1 A02131	USA	QQ A-371F	213.1		6-8	0.9	0.1	0.6	0.35	1-3	0.25	2.5-2.5						
213.1 A02131	Canada	CSA	CS52		6.5-8	0.95 max	0.07 max	0.5 max	0.2 max	1-3	0.2 max	0.1 max			ET 0.05 max; ET=.20 total;			
213 A02130	International		213	Sand, Perm Mold	6-8	1.2 max	0.1 max	0.6 max	0.35 max	1-3	0.25 max	2.5 max		0.5				
213 A02130	Russia	GOST 2685	AL27		6-8	1.3 max				3 max	1 max							
213 A02130	Canada	CSA HA.3	CS72		6.5-8	0.95-1.4	0.07	0.5	0.2	1-3	0.2	0.1	0.05	0.2				
213 A02130	Canada	CSA HA.9	CS72	Sand	6.5-8	0.95-1.5	0.3	0.5	0.2	1-3	0.2	0.1	0.05	0.2				

UNS numbers and US grades are provided as a means of cross referencing chemically similar alloys. Exchangability is only possible after independent examination of specifications. Tensile properties are minimum or typical . UTS and YS as Mpa, El as %. See Appendix for list of abbreviations used in Descriptions.

Worldwide Guide to Equivalent Nonferrous Metals and Alloys

Grade UNS #	Country	Specification	Designation	Description	Cu	Fe	Mg	Mn	Ni	Si	Ti	Zn	OE Max	OT Max	Other	UTS	YS	EL
213 A02130	UK	BS 1490	LM1		6-8	1 max	0.15 max	0.6 max	0.5 max	2-4	0.2 max	2-4			ST 1 max; ST−Mn+Ni+Pb+Sn; Pb 0.3 max; Sn 0.2 max			
213 A02130	Czech Republic	CSN 424361	AlCu8FeSi	Permanent Mold Ingot Alloys	7-8.5	0.6-1.6	0.5 max	0.3 max		0.8-1.3		0.3 max			Ti+Ni:0.10 Bi+Pb+Sn:0.10			
213 A02130	Czech Republic	CSN 424361	AlCu8FeSi	Sand Ingot Alloys	7-8.5	1.6	0.5 max	0.3 max		0.8-1.3		0.3 max			Ti+Ni:0.10 Bi+Pb+Sn:0.10			
213 A02130	Russia		AM8		7-9	1 max	0.3 max	0.3 max	0.3 max	1-1.25								
213 A02130	Russia		AL18Ch		7.5-9.5	1-1.7	0.8 max	0.3-0.8	0.5 max	1.5-2.5		0.5 max						
	USA	QQ A-371F	B535.2		0.05		6.6-7.5	0.05		0.1	0.1-0.25							
	USA	ASTM B179	A444.2		0.05		0.05	0.05		6.5-7.5	0.2	0.05						
	USA	QQ A-371F	A444.2		0.05		0.05	0.05		6.5-7.5	0.2	0.05						
42200	South Africa	SABS 991	Al-Si7 Mg 0.6	T6 Sand	0.05 max	0.19 max	0.45-0.7	0.1 max		6.5-7.5	0.08-0.25	0.07 max	0.03	0.1		250		1
42200	South Africa	SABS 991	Al-Si7 Mg 0.6	T6 Gravity Die	0.05 max	0.19 max	0.45-0.7	0.1 max		6.5-7.5	0.08-0.25	0.07 max	0.03	0.1		320		3
	Canada	CSA HA.9	CS42	Sand T4	4.5-5	0.2	0.03	0.1		2-3	0.2		0.05	0.15		248	137	4
C357.2 A33572	International		C357.2	Ingot	0.03 max	0.06 max	0.5-0.7	0.03 max		6.5-7.5	0.04-0.2	0.03 max	0.03	0.1	Beryllium 0.04-0.07			
	Germany	DIN 1725Sh3	3.0851/V-AlTi5		0.15	0.45	0.5	0.35	0.1	0.5	4.5-6	0.15			Cr 0.1; Sn 0.1; V 0.25			
320 A03200	Czech Republic	CSN 424353	AlSi6Cu2	Permanent Mold Ingot Alloys	2-3	1 max	0.45 max	0.5 max	0.3 max	5.5-7.5	0.2 max	0.8 max			Sn 0.1 max			
320 A03200	Czech Republic	CSN 424353	AlSi6Cu2	Sand Ingot Alloys	2-3	1 max	0.45 max	0.5 max	0.3 max	5.5-7.5	0.2 max	0.8 max			Sn 0.1 max			
320 A03200	International		320	Sand, Perm Mold	2-4	1.2 max	0.05-0.6	0.8 max	0.35 max	5-8	0.25 max	3 max		0.5				
42200	South Africa	SABS 989	Al-Si7 Mg 0.6	Ingot	0.03 max	0.15 max	0.5-0.7	0.1 max		6.5-7.5	0.1-0.18	0.07 max	0.03	0.1				
445.2 A04452	International		445.2	Ingot	0.1 max	0.6-1.3		0.1 max		6.5-7.5		0.1 max	0.1	0.2				
45400	South Africa	SABS 989	Al-Si5 Cu3	Ingot	2.6-3.6	0.5 max	0.05 max	0.55 max	0.1 max	4.5-6	0.2 max	0.2 max	0.05	0.15	Pb 0.1 max; Sn 0.05 max			
45400	South Africa	SABS 991	Al-Si5 Cu3	T4 Gravity Die	2.6-3.6	0.6 max	0.05 max	0.55 max	0.1 max	4.5-6	0.25 max	0.2 max	0.05	0.15	Pb 0.1 max; Sn 0.05 max	230		6
320.1 A03201	International		320.1	Ingot	2-4	0.9 max	0.1-0.6	0.8 max	0.35 max	5-8	0.25 max	3 max		0.5				
	International	ISO R208	Al-Cu4S		4-5	1	0.03	0.3	0.05	1.2	0.2	0.3			Pb 0.05; Sn 0.05 max			
D357.0	International		D357.0	Sand		0.2 max	0.55-0.6	0.1 max		6.5-7.5	0.1-0.2		0.05	0.15	Beryllium 0.04-0.07			
	Germany	DIN 1725Sh2	3.2371./GK-AlSi7Mgwa	Q AH	0.05	0.18	0.2-0.4	0.05		6.5-7.5	0.15	0.07				250	200	5
	Germany	DIN 1725Sh2	3.2371.6/G-AlSi7Mgwa	Sand Q and AH	0.05	0.18	0.2-0.4	0.05		6.5-7.5	0.15	0.07				230	190	2
	South Africa	SABS 989	Al-Si5 Cu3 Mn	Ingot	2.5-4	0.7 max	0.4 max	0.2-0.55	0.3 max	4.5-6	0.15 max	0.55 max	0.05	0.25	Pb 0.2 max; Sn 0.1 max			
	South Africa	SABS 991	Al-Si5 Cu3 Mn	F Sand	2.5-4	0.8 max	0.4 max	0.2-0.55	0.3 max	4.5-6	0.2 max	0.55 max	0.05	0.25	Pb 0.2 max; Sn 0.1 max	140		1
	South Africa	SABS 991	Al-Si5 Cu3 Mn	T6 Sand	2.5-4	0.8 max	0.4 max	0.2-0.55	0.3 max	4.5-6	0.2 max	0.55 max	0.05	0.25	Pb 0.2 max; Sn 0.1 max	230		<1
	South Africa	SABS 991	Al-Si5 Cu3 Mn	F Gravity Die	2.5-4	0.8 max	0.4 max	0.2-0.55	0.3 max	4.5-6	0.2 max	0.55 max	0.05	0.25	Pb 0.2 max; Sn 0.1 max	160		1
	South Africa	SABS 991	Al-Si5 Cu3 Mn	T6 Gravity Die	2.5-4	0.8 max	0.4 max	0.2-0.55	0.3 max	4.5-6	0.2 max	0.55 max	0.05	0.25	Pb 0.2 max; Sn 0.1 max	280		<1
	Japan	JIS H2211	C4CV	Ingot	0.05	0.3	0.25-0.4	0.03	0.03	6.5-7.5	0.03	0.03						
	USA	ASTM B108	A444.0		0.1	0.2	0.05	0.1		6.5-7.5	0.2	0.1						
	USA	ASTM B108	B514.0	As fabricated	0.35	0.6	3.5-4.5	0.8		1.4-2.2	0.25	0.35			Cr 0.25	131		2

UNS numbers and US grades are provided as a means of cross referencing chemically similar alloys. Exchangability is only possible after independent examination of specifications. Tensile properties are minimum or typical . UTS and YS as Mpa, El as %. See Appendix for list of abbreviations used in Descriptions.

3-22 Cast Aluminum

Grade UNS #	Country	Specification	Designation	Description	Cu	Fe	Mg	Mn	Ni	Si	Ti	Zn	OE Max	OT Max	Other	UTS	YS	EL
	USA	ASTM B26	B514.0	Sand As fabricated	0.35	0.6	3.5-4.5	0.8		1.4-2.2	0.25	0.35			Cr 0.25	117	69	
	USA	ASTM B618	B514.0	Invest As fabricated	0.35	0.6	3.5-4.5	0.8		1.4-2.2	0.25	0.35			Cr 0.25	117	69	
	South Africa	SABS 991	Al-Si5 Cu3 Mg	T4 Gravity Die	2.6-3.6	0.6 max	0.15-0.45	0.55 max	0.1 max	4.5-6	0.25 max	0.2 max	0.05	0.15	Pb 0.1 max; Sn 0.05 max	270		2.5
	South Africa	SABS 991	Al-Si5 Cu3 Mg	T6 Gravity Die	2.6-3.6	0.6 max	0.15-0.45	0.55 max	0.1 max	4.5-6	0.25 max	0.2 max	0.05	0.15	Pb 0.1 max; Sn 0.05 max	320		<1
243 A02430	International		243	Sand	3.5-4.5	0.4 max	1.8-2.3	0.45 max	1.9-2.3	0.35 max	0.06-0.2	0.05 max	0.05	0.15	Cr 0.2-0.4; V 0.06-0.20			
711 A07110	International		711	Perm Mold	0.35-0.65	0.7-1.4	0.25-0.45	0.05 max		0.3 max	0.2 max	6-7	0.05	0.15				
	South Africa	SABS 989	Al-Si5 Cu3 Mg	Ingot	2.6-3.6	0.5 max	0.2-0.45	0.55 max	0.1 max	4.5-6	0.2 max	0.2 max	0.05	0.15	Pb 0.1 max; Sn 0.05 max			
	Canada	CSA HA.3	SG70P		0.1		0.3-0.4	0.05		6.5-7.5	0.2	0.05						
	South Africa	SABS 711	AlMn5		0.1	0.7		6		0.45		0.07						
	USA	ASTM B179	A356.2		0.1		0.3-0.4	0.05		6.5-7.5	0.2	0.05						
	USA	QQ A-371F	A356.2		0.1		0.3-0.4	0.05		6.5-7.5	0.2	0.05						
	Canada	CSA HA.3	SG70N		0.1	0.12-0.25	0.3-0.4	0.05		6.5-7.5	0.2	0.05						
711.1 A07111	International		711.1	Ingot	0.35-0.65	0.7-1.1	0.3-0.45	0.05 max		0.3 max	0.2 max	6-7	0.05	0.15				
243.1 A02431	International		243.1	Ingot	3.5-4.5	0.3 max	1.9-2.3	0.45 max	1.9-2.3	0.35 max	0.06-0.2	0.05 max	0.05	0.15	Cr 0.2-0.4; V 0.06-0.20			
357.1 A03571	USA	QQ A-371F	357.1		0.05		0.45-0.6	0.03		6.5-7.5	0.2	0.05						
771 A07710	International		771	Sand	0.1 max	0.15 max	0.8-1	0.1 max		0.15 max	0.1-0.2	6.5-7.5	0.05	0.15	Cr 0.06-0.2			
771 A07710	USA	ASTM B26	771	Sand Cooled and art aged .250 in diam	0.1	0.15	0.8-1	0.1		0.15	0.1-0.2	6.5-7.5			Cr 0.06-0.2	290	262	2
771 A07710	USA	ASTM B26	771	Sand T51 .250 in diam	0.1	0.15	0.8-1	0.1		0.15	0.1-0.2	6.5-7.5			Cr 0.06-0.2	221	186	3
771 A07710	USA	ASTM B26	771	Sand SHT and art aged .250 in diam	0.1	0.15	0.8-1	0.1		0.15	0.1-0.2	6.5-7.5			Cr 0.06-0.2	290	241	5
771 A07710	USA	ASTM B618	771	Invest Cooled and art aged .250 in diam	0.1	0.15	0.8-1	0.1		0.15	0.1-0.2	6.5-7.5	0.05	0.15	ET 0.05; Cr 0.06-0.2	290	262	2
771 A07710	USA	ASTM B618	771	Invest SHT and art aged .250 in diam	0.1	0.15	0.8-1	0.1		0.15	0.1-0.2	6.5-7.5	0.05	0.15	ET 0.05; Cr 0.06-0.2	290	241	5
771 A07710	USA	ASTM B618	771	Invest T71 .250 in diam	0.1	0.15	0.8-1	0.1		0.15	0.1-0.2	6.5-7.5	0.05	0.15	ET 0.05; Cr 0.06-0.2	331	310	2
	USA	QQ A-371F	A535.1		0.1	0.15	6.6-7.5	0.1-0.25		0.2	0.25							
	USA	MIL A-21180C	A356.0	Class 2	0.2	0.2	0.2-0.4	0.1		6.5-7.5	0.1	0.1		0.15	ET 0.05	276	207	3
	USA	MIL A-21180C	A356.0	Class 1	0.2	0.2	0.2-0.4	0.1		6.5-7.5	0.1	0.1		0.15	ET 0.05	262	193	5
	USA	MIL A-21180C	A356.0	Class 3	0.2	0.2	0.2-0.4	0.1		6.5-7.5	0.1	0.1		0.15	ET 0.05	310	234	3
	Canada	CSA HA.10	SC51N	T6	1-1.5	0.6	0.4-0.6	0.3		4.5-5.5	0.25	0.35				255	159	1
	Canada	CSA HA.10	SC51N	T6B	1-1.5	0.6	0.4-0.6	0.3		4.5-5.5	0.25	0.35				290		
	Canada	CSA HA.9	SG51N	Sand T6	1-1.5	0.6	0.4-0.6	0.3		4.5-5.5	0.25	0.35				221	138	2
	Canada	CSA HA.9	SG51N	Sand T6A	1-1.5	0.6	0.4-0.6	0.3		4.5-5.5	0.25	0.35				255		
	USA	QQ A-371F	A357.2		0.1		0.45-0.7	0.05		6.5-7.5	0.1-0.2	0.05			Be 0.04-0.07			
771.2 A07712	International		771.2	Ingot	0.1 max	0.1 max	0.85-1	0.1 max		0.1 max	0.1-0.2	6.5-7.5	0.05	0.15	Cr 0.06-0.2			
771.2 A07712	USA	ASTM B179	771.2		0.1	0.1	0.85-1	0.1		0.1	0.1-0.2	6.5-7.5			Cr 0.06-0.2			
771.2 A07712	USA	QQ A-371F	771.2		0.1	0.1	0.85-1	0.1		0.1	0.1-0.2	6.5-7.5			Cr 0.06-0.2			

UNS numbers and US grades are provided as a means of cross referencing chemically similar alloys. Exchangability is only possible after independent examination of specifications. Tensile properties are minimum or typical . UTS and YS as Mpa, El as %. See Appendix for list of abbreviations used in Descriptions.

Worldwide Guide to Equivalent Nonferrous Metals and Alloys

Grade UNS #	Country	Specification	Designation	Description	Cu	Fe	Mg	Mn	Ni	Si	Ti	Zn	OE Max	OT Max	Other	UTS	YS	EL
	Finland	SFS 2573	G-AlZn5Mg	Aged 20 mm diam	0.2-0.5	0.7	0.6-0.8	0.4	0.05	0.3	0.15-0.25	5-6			Pb 0.05; Sn 0.05 max	220	170	4
	Sweden	SIS 144438	4438-00		0.2-0.5	0.7	0.6-0.8	0.4	0.05	0.3	0.15-0.25	5-6			Cr 0.3-0.6; Pb 0.05; Sn 0.05 max			
	Sweden	SIS 144438	4438-04	Sand and nat aged	0.2-0.5	0.7	0.6-0.8	0.4	0.05	0.3	0.15-0.25	5-6			Cr 0.3-0.6; Pb 0.05; Sn 0.05 max		170	4
	Sweden	SIS 144438	4438-07	, nat aged	0.2-0.5	0.7	0.6-0.8	0.4	0.05	0.3	0.15-0.25	5-6			Cr 0.3-0.6; Pb 0.05; Sn 0.05 max		180	4
	USA	ASTM B179	C443.1/S5C		0.6	1	0.1	0.35	0.5	4.5-6		0.4			Sn 0.15 max			
	USA	QQ A-371F	C443.1		0.6	1	0.1	0.35	0.5	4.5-6		0.4			Sn 0.15 max			
	India	IS 202	A-26	Sand	4-5	1	0.05	0.3		2-3		0.1						
	India	IS 202	A-26	Sand	4-5	1	0.05	0.3		2-3		0.1						
518 A05180	Czech Republic	CSN 424519	AlMg10SiCa	Die Ingot Alloys	0.05 max	0.8 max	7-10	0.05 max	1.75-2.25	0.01-2	0.05 max	0.15 max			Ca:0.01-0.15			
518 A05180	Germany	DIN 1725 pt 2	3.3292/GD-AlMg9		0.05 max	1 max	7-10	0.2-0.5		2.5 max	0.15 max	0.1 max			ET 0.05 max; , ET=.15 total			
	Canada	CSA HA.10	SG70P	T6	0.2	0.2	0.2-0.4	0.1		6.5-7.5	0.2	0.1	0.05	0.15		262	179	5
	Canada	CSA HA.9	SG70P	Sand T6	0.2	0.2	0.2-0.4	0.1		6.5-7.5	0.2	0.1				234	165	3
	Canada	CSA HA.9	SG70P	Sand T6A	0.2	0.2	0.2-0.4	0.1		6.5-7.5	0.2	0.1				262	179	5
	USA	ASTM B108	A356.0	T61	0.2	0.2	0.2-0.4	0.1		6.5-7.5	0.2	0.1				193	179	3
	USA	ASTM B26	A356.0	Sand SHT and art aged .250 in diam	0.2	0.2	0.2-0.4	0.1		6.5-7.5	0.2	0.1				234	166	4
	USA	QQ A-596d	A356.0	SHt and art aged T61	0.2	0.2	0.1	0.2-0.4		6.5-7.5	0.2	0.1				255		5
	USA	ASTM B618	A356.0	Invest SHT and art aged	0.2	0.2	0.2-0.4	0.1		6.5-7.6	0.2	0.1				234	106	4
518 A05180	International		518	Die	0.25 max	1.8 max	7.5-8.5	0.35 max	0.15 max	0.35 max		0.15 max		0.25	Sn 0.15 max			
518 A05180	South Africa	SABS 992	Al-Mg8FeA		0.25	1.8	7.5-8.5	0.35	0.15	0.35		0.15			Sn 0.15 max	195	115	1
518 A05180	USA	QQ A-591E	518	Die	0.25	1.8	7.5-8.5	0.35	0.15	0.35		0.15			Sn 0.15 max	310	193	5
	Japan	JIS H2117	C4DS	Ingot	1-1.5	0.5	0.4-0.6	0.5	0.1	4.5-5.5	0.2	0.3						
	Australia	AS 1874	BA303	T1 Sand	3.5-4.5	0.7 max	0.2 max	0.5 max	0.3 max	4-5	0.2 max	0.5 max	0.05	0.2	Cr 0.1 max; Pb 0.15 max; Sn 0.2 max	155		
	Australia	AS 1874	BA303	T1 permanent mould	3.5-4.5	0.7 max	0.2 max	0.5 max	0.3 max	4-5	0.2 max	0.5 max	0.05	0.2	Cr 0.1 max; Pb 0.15 max; Sn 0.2 max	185		
	Australia	AS 1874	BA303	T5 sand	3.5-4.5	0.7 max	0.2 max	0.5 max	0.3 max	4-5	0.2 max	0.5 max	0.05	0.2	Cr 0.1 max; Pb 0.15 max; Sn 0.2 max	170		1
	Australia	AS 1874	BA303	T6 sand	3.5-4.5	0.7 max	0.2 max	0.5 max	0.3 max	4-5	0.2 max	0.5 max	0.05	0.2	Cr 0.1 max; Pb 0.15 max; Sn 0.2 max	210		1.5
	Japan	JIS H2211	C5AV	Ingot	3.5-4.5	0.4	1.3-1.8	0.03	1.7-2.3	0.4	0.2	0.03						
	Austria	ONORM M3429	AlMg7Si		0.05 max	0.3 max	6.5-7.5	0.2-0.5		0.9-1.8	0.2 max	0.1 max			ET 0.05 max; ET=.15;			
	Austria	ONORM M3429	G AlMg7Si		0.05 max	0.3 max	6.5-7.5	0.2-0.5		0.9-1.8	0.2 max	0.1 max			ET 0.05 max; ET=.15 total;			
	Austria	ONORM M3429	GK AlMg7Si	Perm Mold	0.05 max	0.4 max	6.5-7.5	0.2-0.5		0.9-1.8	0.2 max	0.1 max			ET 0.05 max; ET=.15 total;			
	Austria	ONORM M3429	GS AlMg7Si	Sand	0.05 max	0.4 max	6.5-7.5	0.2-0.5		0.9-1.8	0.2 max	0.1 max		0.05	ET=.15 total;			
	Austria	ONORM M3429	GD AlMg7Si	Die	0.05 max	0.8 max	6.5-7.5	0.2-0.5		0.9-1.8	0.2 max	0.1 max			ET 0.05 max; ET=.15 total;			
518.2 A05182	Canada	CSA	G8		0.1 max	0.7 max	7.6-8.5	0.1 max	0.05 max	0.25 max					ET 0.05 max; ET=.10;			

UNS numbers and US grades are provided as a means of cross referencing chemically similar alloys. Exchangability is only possible after independent examination of specifications. Tensile properties are minimum or typical . UTS and YS as Mpa, El as %. See Appendix for list of abbreviations used in Descriptions.

3-24 Cast Aluminum

Grade UNS #	Country	Specification	Designation	Description	Cu	Fe	Mg	Mn	Ni	Si	Ti	Zn	OE Max	OT Max	Other	UTS	YS	EL
518.2 A05182	International		518.2	Ingot	0.1 max	0.7 max	7.6-8.5	0.1 max	0.05 max	0.25 max				0.1	Sn 0.05 max			
518.2 A05182	USA	ASTM B179	518.2		0.1	0.7	7.6-8.5	0.1	0.05	0.25					Sn 0.05 max			
518.2 A05182	USA	QQ A-371F	518.2		0.1	0.7	7.6-8.5	0.1	0.05	0.25					Bi 0.05			
518.1 A05181	France	NF A57703	A-G6-Y4		0.2	0.4	5-8.5	0.6		1		0.1			Sn 0.4 max			
	India	IS 202	A-27	Sand ST PT	0.2	0.5	0.2-0.4	0.1		6.5-7.5		0.1				207	123	3
	India	IS 202	A-27	Sand ST PT	0.2	0.5	0.2-0.4	0.1		6.5-7.5		0.1				207	123	3
518.1 A05181	South Africa	SABS 989	Al-Mg8FeA		0.25	1	8.5-8.5	0.35	0.15	0.35		0.15			Sn 0.15 max			
518.1 A05181	USA	QQ A-371F	518.1		0.25	1	7.6-8.5	0.35	0.15	0.35		0.15			Sn 0.15 max			
518.1 A05181	International		518.1	Ingot	0.25 max	1.1 max	7.6-8.5	0.35 max	0.15 max	0.35 max		0.15 max		0.25	Sn 0.15 max			
713 A07130	USA	ASTM B108	713	Cooled nat or art aged	0.4-1		0.2-0.5	0.6	0.15	0.25	0.25	7-8			Cr 0.35	152	152	4
713 A07130	International		713	Sand, Perm Mold	0.4-1	1.1 max	0.2-0.5	0.6 max	0.15 max	0.25 max	0.25 max	7-8	0.1	0.25	Cr 0.35 max			
713 A07130	USA	ASTM B26	713	Sand Cooled and art aged .250 in diam	0.4-1	1.1	0.2-0.5	0.6	0.15	0.25	0.25	7-8			Cr 0.35	207	152	3
713 A07130	USA	ASTM B618	713	Invest	0.4-1	1.1	0.2-0.5	0.6	0.15	0.25	0.25	7-8			Cr 0.35			
	USA	QQ A-371F	B295.2		4-5	0.8	0.03	0.3		2-3	0.2	0.3						
	USA	ASTM B108	A357.0	T61	0.2	0.2	0.4-0.7	0.1		6.5-7.5	0.1-0.2	0.1			Be 0.04-0.07	283	214	3
	USA	MIL A-21180C	A357.0	Class 1	0.2	0.2	0.4-0.7	0.1		6.5-7.5	0.1-0.2	0.1		0.15	Be 0.04-0.07; ET 0.05	310	241	3
	USA	MIL A-21180C	A357.0	Class 2	0.2	0.2	0.4-0.7	0.1		6.5-7.5	0.1-0.2	0.1		0.15	Be 0.04-0.07; ET 0.05	345	276	
	USA	QQ A-596d	Ternalloy7 (607)	Aged and temper T5	0.2	0.8	1.8-2.4	0.4-0.6			0.2	0.25	4-4.5		Cr 0.2-0.4	290		4
	USA	QQ A-596d	Ternalloy7 (607)	SHT amd stab T7	0.2	0.8	1.8-2.4	0.4-0.6			0.2	0.25	4-4.5		Cr 0.2-0.4	310		3
713.1 A07131	USA	ASTM B179	713.1		0.4-1		0.25-0.5	0.6	0.15	0.25	0.25	7-8			Cr 0.35			
713.1 A07131	International		713.1	Ingot	0.4-1	0.8 max	0.25-0.5	0.6 max	0.15 max	0.25 max	0.25 max	7-8	0.1	0.25	Cr 0.35 max			
713.1 A07131	USA	QQ A-371F	713.1		0.4-1	0.8	0.25-0.5	0.6	0.15	0.25	0.25	7-8			Cr 0.35			
364 A03640	International		364	Die	0.2 max	1.5 max	0.2-0.4	0.1 max	0.15 max	7.5-9.5		0.15 max	0.05	0.15	Be 0.02-0.04; Cr 0.25-0.5; Sn 0.15 max			
	USA	ASTM B179	A712.1		0.35-0.65	0.4	0.65-0.8	0.05		0.15	0.25	6-7						
	USA	QQ A-371F	A712.1		0.35-0.65	0.4	0.65-0.8	0.05		0.15	0.25	6-7						
	USA	ASTM B108	C712.0	Cooled and nat aged	0.35-0.65	0.7	0.25-0.45	0.05		0.3	0.2	6-7				193	124	7
	USA	ASTM B26	A712.0	Sand Cooled and art aged .250 in diam	0.35-0.65	0.5	0.6-0.8	0.05		0.15	0.25	6-7				221	138	2
	USA	ASTM B618	A712.0	Invest Cooled and nat aged .250 in diam	0.35-0.65	0.5	0.6-0.8	0.05		0.15	0.25	6-7				221	138	2
	USA	ASTM B179	C712.1		0.35-0.65	0.7-1.1	0.3-0.45	0.05		0.3	0.2	6-7						
	USA	QQ A-371F	C712.1		0.35-0.65	0.7-1.1	0.3-0.45	0.05		0.3	0.2	6-7						
	Germany	DIN 1725Sh3	3.0941/V-AlFe5		0.15	6	0.4	0.35	0.1	0.4	0.1	0.2			Cr 0.1; Pb 0.15; Sn 0.1			
	Japan	JIS H2211	C2AV	Ingot	3.5-4.5	0.3	0.03	0.03	0.03	4-5	0.03	0.03						
	USA	QQ A-371F	A242.2		3.7-4.5	0.6	1.3-1.7	0.1	1.8-2.3	0.35	0.07-0.2	0.1	0.05		Cr 0.15-0.25			

UNS numbers and US grades are provided as a means of cross referencing chemically similar alloys. Exchangability is only possible after independent examination of specifications. Tensile properties are minimum or typical . UTS and YS as Mpa, El as %. See Appendix for list of abbreviations used in Descriptions.

Worldwide Guide to Equivalent Nonferrous Metals and Alloys

Grade UNS #	Country	Specification	Designation	Description	Cu	Fe	Mg	Mn	Ni	Si	Ti	Zn	OE Max	OT Max	Other	UTS	YS	EL
358 A03580	International		358	Sand, Perm Mold	0.2 max	0.3 max	0.4-0.6	0.2 max		7.6-8.6	0.1-0.2	0.2 max	0.05	0.15	Be 0.10-0.30; Cr 0.2 max			
	India	IS 202	A-4	Sand	2-4	0.8	0.15	0.3-0.7	0.3	4-6		0.5			Pb 0.1; Sn 0.05 max	119	84	2
	India	IS 202	A-4	Sand	2-4	0.8	0.15	0.3-0.7	0.3	4-6		0.5			Pb 0.1; Sn 0.05 max	119	83	2
358.2 A03582	International		358.2	Ingot	0.1 max	0.2 max	0.45-0.6	0.2 max		7.6-8.6	0.12-0.2	0.1 max	0.05	0.15	Be 0.15-0.30; Cr 0.05 max			
	International	ISO 3522	Al-Si6 Cu4	Sand Cast As mfg	3-5	1 max	0.3 max	0.2-0.6	0.3 max	5-7	0.2 max	2 max			Pb 0.2 max; Sn 0.1 max	140		
	International	ISO 3522	Al-Si6 Cu4	Perm Mold As mfg	3-5	1 max	0.3 max	0.2-0.6	0.3 max	5-7	0.2 max	2 max			Pb 0.2 max; Sn 0.1 max	150		1
	Belgium	NBN P21-101	SGAlSi6Cu4	Sand	3-5	1.3 max	0.3 max	0.2-0.6	0.3 max	5-7	0.2 max	2 max			Pb 0.2 max; Sn 0.1 max			
	International	ISO 3522	Al-Si6 Cu4 Fe	Pressure Die Cast	3-5	1.3 max	0.3 max	0.2-0.6	0.3 max	5-7	0.2 max	2 max			Pb 0.2 max; Sn 0.1 max			
	Japan	JIS H2117	C4CS	Ingot	0.2	0.4	0.25-0.4	0.3	0.1	6.5-7.5	0.2	0.3						
	South Africa	SABS 989	Al-Si5Cu3MnA		2-4	0.7	0.15	0.3-0.7	0.3	4-6	0.2	0.5			Pb 0.1; Sn 0.05 max	140		2
	South Africa	SABS 989	Al-Si5Cu3MnA	Chill	2-4	0.7	0.15	0.3-0.7	0.3	4-6	0.2	0.5			Pb 0.1; Sn 0.05 max	155		2
	South Africa	SABS 989	Al-Si5Cu3MnA	Chill, SHT and PHT	2-4	0.7	0.15	0.3-0.7	0.3	4-6	0.2	0.5			Pb 0.1; Sn 0.05 max	275		
	USA	QQ A-371F	A242.1		3.7-4.5	0.6	1.3-1.7	0.1	1.8-2.3	0.6	0.07-0.2	0.1			Cr 0.15-0.25			
	Austria	ONORM M3429	G AlSi6Cu4		3-4.5	0.8 max	0.1-0.3	0.2-0.5	0.3 max	5-6.5	0.15 max	1 max			ET 0.05 max ET=.15 total; Pb 0.3 max; Sn 0.1 max			
	Canada	CSA HA.10	SG70N	T6	0.2	0.5	0.2-0.4	0.35		6.5-7.5	0.25	0.35	0.05	0.15		228	152	3
	Canada	CSA HA.9	SG70N	Sand T6	0.2	0.5	0.2-0.4	0.35		6.5-7.5	0.25	0.35				207	138	3
	Canada	CSA HA.9	SG70N	Sand T5	0.2	0.5	0.2-0.4	0.35		6.5-7.5	0.25	0.35				159		
	South Africa	SABS 990	Al-Si5Cu3MnA	Sand	2-4	0.8	0.15	0.3-0.7	0.3	4-6	0.2	0.5			Pb 0.1; Sn 0.05 max			
	South Africa	SABS 991	Al-Si5Cu3MnA		2-4	0.8	0.15	0.3-0.7	0.3	4-6	0.2	0.5			Pb 0.1; Sn 0.05 max	155		2
	South Africa	SABS 991	Al-Si5Cu3MnA	SHT and PHT	2-4	0.8	0.15	0.3-0.7	0.3	4-6	0.2	0.5			Pb 0.1; Sn 0.05 max	275		
	Sweden	SIS 144231	4231-00	OBSOLETE	2-4.5	0.9	0.15	0.2-0.7	0.3	4-6.5	0.2	0.5			Pb 0.1; Sn 0.05 max			
	Finland	SFS 2569	G-AlSi7Mg	Die ST and aged	0.1	0.5	0.2-0.4	0.5	0.05	6.5-7.5	0.2	0.3			Pb 0.05; Sn 0.05 max	260	220	1
343.1 A03431	International		343.1	Ingot	0.5-0.9	0.9 max	0.1 max	0.5 max		6.7-7.7		1.2-1.9	0.1	0.35	Cr 0.1 max; Sn 0.5 max			
343 A03430	International		343	Die	0.5-0.9	1.2 max	0.1 max	0.5 max		6.7-7.7		1.2-2	0.1	0.35	Cr 0.1 max; Sn 0.5 max			
	Japan	JIS H2117	C5AS	Ingot	3.5-4.5	0.7	1.3-1.8	0.3	1.7-2.3	0.6	0.2	0.1						
	USA	QQ A-371F	B295.1		4-5	0.9	0.05	0.35	0.35	2-3	0.25	0.5						
364.2 A03642	International		364.2	Ingot	0.2 max	0.7-1.1	0.25-0.4	0.1 max	0.15 max	7.5-9.5		0.15 max	0.05	0.15	Be 0.02-0.04; Cr 0.25-0.5; Sn 0.15 max			
364.2 A03642	USA	QQ A-371F	364.2		0.2	0.7-1.1	0.25-0.4	0.1	0.15	7.5-9.5		0.15			Be 0.02-0.04; Cr 0.25-0.5; Sn 0.15 max			
851.1 A08511	International		851.1	Ingot	0.7-1.3	0.5 max	0.1 max	0.1 max	0.3-0.7	2-3	0.2 max			0.3	Sn 5.5-7			
851 A08510	International		851	Sand, Perm Mold	0.7-1.3	0.7 max	.01 max	0.1 max	0.3-0.7	2-3	0.2 max			0.3	Sn 5.5-7			
319 A03190	Russia	GOST 2685	AL6		2-3	1.1-1.4	0.1 max	0.3 max		4.5-6		0.3 max						
A03190	Australia	AS 1874	AA303	T1Sand	2-4	0.8 max	0.15 max	0.7 max	0.3 max	4-5	0.2 max	0.5 max	0.05	0.2	Cr 0.1 x; Pb 0.15 max; Sn 0.15 max	135		1.5
A03190	Australia	AS 1874	AA303	T1 Perm mould	2-4	0.8 max	0.15 max	0.7 max	0.3 max	4-5	0.2 max	0.5 max	0.05	0.2	Cr 0.1 x; Pb 0.15 max; Sn 0.15 max	150		2

UNS numbers and US grades are provided as a means of cross referencing chemically similar alloys. Exchangability is only possible after independent examination of specifications. Tensile properties are minimum or typical . UTS and YS as Mpa, El as %. See Appendix for list of abbreviations used in Descriptions.

3-26 Cast Aluminum

Grade UNS #	Country	Specification	Designation	Description	Cu	Fe	Mg	Mn	Ni	Si	Ti	Zn	OE Max	OT Max	Other	UTS	YS	EL
A03190	Australia	AS 1874	AA303	T6Sand	2-4	0.8 max	0.15 max	0.7 max	0.3 max	4-5	0.2 max	0.5 max	0.05	0.2	Cr 0.1 max; Pb 0.15 max; Sn 0.15 max	225		
A03190	Australia	AS 1874	AA303	T6 Permanent mould	2-4	0.8 max	0.15 max	0.7 max	0.3 max	4-5	0.2 max	0.5 max	0.05	0.2	Cr 0.1 max; Pb 0.15 max; Sn 0.15 max	275		
319.1 A03191	Canada	CSA HA.3	SC53		2-4	0.7	0.15	0.3-0.6	0.3	4-6	0.2	0.2						
319.1 A03191	Canada	CSA HA.10	SC53		2-4	0.8	0.15	0.3-0.7	0.3	4-6	0.2	0.2						
319.1 A03191	Canada	CSA HA.9	SC53	Sand	2-4	0.8	0.15	0.3-0.6	0.3	4-6	0.2	0.2						
319 A03190	UK	BS 1490	LM4		2-4	0.8 max	0.15 max	0.2-0.6	0.3 max	4-6	0.2 max	0.5 max			Pb 0.1 max; Sn 0.1 max			
319.2 A03192	International		319.2	Ingot	3-4	0.6 max	0.1 max	0.1 max	0.1 max	5.5-6.5	0.2 max	0.1 max		0.2				
319.2 A03192	USA	ASTM B179	319.2		3-4	0.6	0.1	0.1	0.1	5.5-6.5	0.2	0.1						
319.2 A03192	USA	QQ A-371F	319.2		3-4	0.6	0.1	0.1	0.1	5.5-6.5	0.2	0.1						
319.1 A03191	International		319.1	Ingot	3-4	0.8 max	0.1 max	0.5 max	0.35 max	5.5-6.5	0.25 max	1 max		0.5				
A319.1 A13191	International		A319.1	Ingot	3-4	0.8 max	0.1 max	0.5 max	0.35 max	5.5-6.5	0.25 max	3 max		0.5				
319.1 A03191	USA	ASTM B179	319.1		3-4	0.8	0.1	0.5	0.35	5.5-6.5	0.25	1						
319.1 A03191	USA	QQ A-371F	319.1		3-4	0.8	0.1	0.5	0.35	5.5-6.5	0.25	1						
319 A03190	International		319	Sand, Perm Mold	3-4	1 max	0.1 max	0.5 max	0.35 max	5.5-6.5	0.25 max	1 max		0.5				
A319.0 A13190	International		A319.0	Sand, Perm Mold	3-4	1 max	0.1 max	0.5 max	0.35 max	5.5-6.5	0.25 max	3 max		0.5				
319 A03190	USA	ASTM B108	319	As fabricated	3-4	1	0.1	0.5	0.35	5.5-6.5	0.25	1				186	97	3
319 A03190	USA	ASTM B26	319	Sand As fabricated .250 in diam	3-4	1	0.1	0.5	0.35	5.5-6.5	0.25	1				159	90	2
319 A03190	USA	ASTM B26	319	Sand SHT and art aged .250 in diam	3-4	1	0.1	0.5	0.35	5.5-6.5	0.25	1				221	138	3
319 A03190	USA	ASTM B618	319	Invest As fabricated .250 in diam	3-4	1	0.1	0.5	0.35	5.5-6.5	0.25	1				159	90	2
319 A03190	USA	ASTM B618	319	Invest SHT and art aged .250 in diam	3-4	1	0.1	0.5	0.35	5.5-6.5	0.25	1				221	138	3
319 A03190	Japan	JIS H5202	AC2B		2-4	1	0.5	0.5	0.3	5-7	0.2	1				157		1
319 A03190	Japan	JIS H5202	AC2B	Q/T	2-4	1	0.5	0.5	0.3	5-7	0.2	1				245		1
319 A03190	International	ISO R164	Al-Si5Cu3		2-4.5	0.05	0.15	0.2-0.7	0.3	4-6.5	0.2	0.5			Pb 0.1; Sn 0.05 max			
319 A03190	International	ISO R164	Al-Si5Cu3Fe		2-4.5	0.2	0.15	0.2-0.7	0.3	4-6.5	0.2	0.5			Pb 0.3; Sn 0.2 max			
319 A03190	Sweden	SIS 144230	4230-00	OBSOLETE	2-4.5	0.95	0.3	0.2-0.6	0.3	5-7	0.2	2			Pb 0.2; Sn 0.1 max			
319 A03190	Sweden	SIS 144230	4230-03	Sand As cast OBSOLETE	2-4.5	0.95	0.3	0.2-0.6	0.3	5-7	0.2	2			Pb 0.2; Sn 0.1 max		110	1
319 A03190	Sweden	SIS 144230	4230-06	As cast OBSOLETE	2-4.5	0.95	0.3	0.2-0.6	0.3	5-7	0.2	2			Pb 0.2; Sn 0.1 max		120	
319 A03190	USA	QQ A-596d	319		3.5-4.5	1	0.1	0.5	0.35	5.5-7	0.25	1				193		1
A319.0 A13190	Czech Republic	CSN 424357	AlSi5Cu4Zn	Permanent Mold Ingot Alloys	3-5	1.2 max	0.5 max	0.6 max	0.3 max	3-6		1-2.5			Sn 0.1 max			
A319.0 A13190	Czech Republic	CSN 424357	AlSi5Cu4Zn	Sand Ingot Alloys	3-5	1.2 max	0.5 max	0.6 max	0.3 max	3-6		1-2.5			Sn 0.1 max			

UNS numbers and US grades are provided as a means of cross referencing chemically similar alloys. Exchangability is only possible after independent examination of specifications. Tensile properties are minimum or typical . UTS and YS as Mpa, El as %. See Appendix for list of abbreviations used in Descriptions.

Worldwide Guide to Equivalent Nonferrous Metals and Alloys

Grade UNS #	Country	Specification	Designation	Description	Cu	Fe	Mg	Mn	Ni	Si	Ti	Zn	OE Max	OT Max	Other	UTS	YS	EL
319 A03190	South Africa	SABS 989	45000 (Al-Si6 Cu4 Mn)	Ingot	3-5	0.9 max	0.55 max	0.55-0.65	0.45 max	5-7	0.2 max	2 max	0.05	0.35	Cr 0.15 max; Pb 0.3 max; Sn 0.15 max			
319 A03190	South Africa	SABS 991	45000 (Al-Si6 Cu4 Mn)	F Sand	3-5	1 max	0.55 max	0.2-0.65	0.45 max	5-7	0.25 max	2 max	0.05	0.35	Cr 0.15 max; Pb 0.3 max; Sn 0.15 max	150		1
319 A03190	South Africa	SABS 991	45000 (Al-Si6 Cu4 Mn)	F Gravity Die	3-5	1 max	0.55 max	0.2-0.65	0.45 max	5-7	0.25 max	2 max	0.05	0.35	Cr 0.15 max; Pb 0.3 max; Sn 0.15 max	170		1
A319.0 A13190	International	ISO R164	Al-Si6Cu4		3-5	1.3	0.3	0.2-0.6	0.3	5-7	0.2	2			Pb 0.2; Sn 0.1 max			
	USA	ASTM B85	C443.0/S5C	Die As cast	0.6	2	0.1	0.35	0.5	4.5-6		0.5				230	100	9
354 A03540	USA	MIL A-21180C	354	Class 1	1.6-2	0.2	0.4-0.6	0.1		8.6-9.4	0.2	0.1				324	248	3
B319.0 A23190	International		B319.0	Sand, Perm Mold	3-4	1.2 max	0.1-0.5	0.8 max	0.5 max	5.5-6.5	0.25 max	1 max		0.5				
	Australia	AS 1874	CA327	Ingot	3-4	0.25 max	0.1-0.18	0.5 max	0.1 max	5.5-6.8	0.15 max	0.1 max	0.05	0.15				
	Australia	AS 1874	CA327	T1 Sand	3-4	0.25 max	0.1-0.18	0.5 max	0.1 max	5.5-6.8	0.15 max	0.1 max	0.05	0.15		155		
	Australia	AS 1874	CA327	T1 permanent mould	3-4	0.25 max	0.1-0.18	0.5 max	0.1 max	5.5-6.8	0.15 max	0.1 max	0.05	0.15		185		
	Australia	AS 1874	CA327	T5 sand	3-4	0.25 max	0.1-0.18	0.5 max	0.1 max	5.5-6.8	0.15 max	0.1 max	0.05	0.15		170		1
	Australia	AS 1874	CA327	T6 sand	3-4	0.25 max	0.1-0.18	0.5 max	0.1 max	5.5-6.8	0.15 max	0.1 max	0.05	0.15		210		1.5
	Australia	AS 1874	CA327	T6 permanent mould	3-4	0.25 max	0.1-0.18	0.5 max	0.1 max	5.5-6.8	0.15 max	0.1 max	0.05	0.15				230
B319.0 A23190	Austria	ONORM M3429	AlSi6Cu4	Perm Mold Sand	3-4.5	0.8 max	0.1-0.3	0.2-0.5	0.3 max	5-6.5	0.15 max	1 max			ET 0.05 max ET=.15 total; Pb 0.3 max; Sn 0.1 max			
B319.0 A23190	Austria	ONORM M3429	GK AlSi6Cu4	Perm Mold	3-4.5	0.8 max	0.1-0.3	0.2-0.5	0.3 max	5-6.5	0.15 max	1 max			ET 0.05 max ET=.15 total; Pb 0.3 max; Sn 0.3 max			
B319.0 A23190	Austria	ONORM M3429	GS AlSi6Cu4	Sand	3-4.5	0.8 max	0.1-0.3	0.2-0.5	0.3 max	5-6.5	0.15 max	1 max			ET 0.05 x ET=.15 total; Pb 0.3 max; Sn 0.3 max			
B319.0 A23190	Austria	ONORM M3429	GD AlSi6Cu4	Die	3-4.5	1 max	0.1-0.3	0.2-0.5	0.3 max	5-6.5	0.15 max	1 max			ET 0.05 x ET=.15 total; Pb 0.3 max; Sn 0.1 max			
B319.1 A23191	International		B319.1	Ingot	3-4	0.9 max	0.15-0.5	0.8 max	0.5 max	5.5-6.5	0.25 max	1 max		0.5				
A23191	Australia	AS 1874	AA339	Ingot	2-4	0.8 max	0.5 max	0.5 max	0.3 max	5-7	0.2 max	1 max	0.05	0.2	Cr 0.1 max; Pb 0.15 max; Sn 0.2 max			
A23191	Australia	AS 1874	AA339	F1 permanent mould	2-4	0.8 max	0.5 max	0.5 max	0.3 max	5-7	0.2 max	1 max	0.05	0.2	Cr 0.1 max; Sn 0.2 max			1
A23191	Australia	AS 1874	AA339	T6 permanent mould	2-4	0.8 max	0.5 max	0.5 max	0.3 max	5-7	0.2 max	1 max	0.05	0.2	Cr 0.1 max; Sn 0.2 max			1
852 A08520	International		852	Sand, Perm Mold	1.7-2.3	0.7 max	0.6-0.9	0.1 max	0.9-1.5	0.4 max	0.2 max			0.3	Sn 5.5-7			
	USA	QQ A-596d	B195		4-5	1.2	0.05	0.35	0.35	2-3	0.25	0.5						
	USA	QQ A-596d	B195	SHT and temper T4	4-5	1.2	0.05	0.35	0.35	2-3	0.25	0.5				228		4
	USA	QQ A-596d	B195	SHT and art aged T6	4-5	1.2	0.05	0.35	0.35	2-3	0.25	0.5				241		2
	Czech Republic	CSN 424384	AlSi10CuMn	Die Ingot Alloys	0.5-1.8	1.5 max	0.6 max	0.3-0.6	0.4 max	8-11		1 max			Sn 0.2 max			
852.1 A08521	International		852.1	Ingot	1.7-2.3	0.5 max	0.7-0.9	0.1 max	0.9-1.5	0.4 max	0.2 max			0.3	Sn 5.5-7			

UNS numbers and US grades are provided as a means of cross referencing chemically similar alloys. Exchangability is only possible after independent examination of specifications. Tensile properties are minimum or typical . UTS and YS as Mpa, El as %. See Appendix for list of abbreviations used in Descriptions.

3-28 Cast Aluminum

Grade UNS #	Country	Specification	Designation	Description	Cu	Fe	Mg	Mn	Ni	Si	Ti	Zn	OE Max	OT Max	Other	UTS	YS	EL
	USA	QQ A-371F	B358.2		0.1	0.2	0.45-0.6	0.1		7.6-8.6	0.12-0.2	0.1		0.05	Be 0.15-0.3; Cr 0.05			
	India	IS 6754	8328		0.7-1.3	0.7	0.3	0.1	0.7-1.3	0.7	0.2				Sn 5.5-7			
328 A03280	Czech Republic	CSN 424384	AlSi10CuMn	Sand Ingot Alloys	0.5-1.6	0.9 max	0.6 max	0.3-0.6	0.4 max	8-11		1 max			Sn 0.2 max			
328 A03280	International		328	Sand	1-2	1 max	0.2-0.6	0.2-0.6	0.25 max	7.5-8.5	0.25 max	1.5 max		0.5	Cr 0.35 max			
328 A03280	USA	ASTM B26	328	Sand As fabricated .250 in diam	1-2	1	0.2-0.6	0.2-0.6	0.25	7.5-8.5	0.25	1.5			Cr 0.35	172	97	1
328 A03280	USA	ASTM B26	328	Sand SHT and art aged .250 in diam	1-2	1	0.2-0.6	0.2-0.6	0.25	7.5-8.5	0.25	1.5			Cr 0.35	234	145	1
328 A03280	USA	ASTM B618	328	Invest As fabricated .250 in diam	1-2	1	0.2-0.6	0.2-0.6	0.25	7.5-8.5	0.25	1.5			Cr 0.35	172	97	1
328 A03280	USA	ASTM B618	328	Invest SHT and art aged .250 in diam	1-2	1	0.2-0.6	0.2-0.6	0.25	7.5-8.5	0.25	1.5			Cr 0.35	234	145	1
328.1 A03281	International		328.1	Ingot	1-2	0.8 max	0.25-0.6	0.2-0.6	0.25 max	7.5-8.5	0.25 max	1.5 max		0.5	Cr 0.35 max			
328.1 A03281	USA	ASTM B179	328.1		1-2	0.8	0.25-0.6	0.2-0.6	0.25	7.5-8.5	0.25	1.5			Cr 0.35			
328.1 A03281	USA	QQ A-371F	328.1		1-2	0.8	0.25-0.6	0.2-0.6	0.25	7.5-8.5	0.25	1.5			Cr 0.35			
A03281	Australia	AS 1874	AA317	F1 sand	1.5-2.5	0.8 max	0.35 max	0.2-0.6	0.35 max	6-8	0.2 max	1 max	0.05	0.2	Cr 0.1 max; Pb 0.25 max; Sn 0.15 max	140		1
A03281	Australia	AS 1874	AA317	Ingot	1.5-2.5	1.3 max	0.35 max	0.2-0.6	0.35 max	6-8	0.2 max	1 max	0.05	0.2	Cr 0.1 max; Pb 0.25 max; Sn 0.15 max			
A03281	Australia	AS 1874	AA317	F1 permanent mould	1.5-2.5	1.3 max	0.35 max	0.2-0.6	0.35 max	6-8	0.2 max	1 max	0.05	0.2	Cr 0.1 max; Pb 0.25 max; Sn 0.15 max	160		2
359 A03590	Germany	DIN 1725 pt 2	3.2373/G-AlSi9Mg		0.05 max	0.18 max	0.2-0.4	0.05 max		9-10	0.15 max	0.07 max			ET 0.03 max; ET=.10 total,			
359 A03590	Germany	DIN 1725 pt 2	3.2373/Gk-AlSi9Mg		0.05 max	0.18 max	0.2-0.4	0.05 max		9-10	0.15 max	0.07 max			ET 0.03 max; ET=.10 total;			
359 A03590	Czech Republic	CSN 424331	AlSi10MgMn	Sand Ingot Alloys	0.1 max	0.6 max	0.2-0.45	0.1-0.4		9-10.5	0.15 max	0.1 max						
359 A03590	International		359	Sand, Perm Mold	0.2 max	0.2 max	0.5-0.7	0.2 max		8.5-9.5	0.2 max	0.1 max	0.05	0.15				
359 A03590	USA	ASTM B108	359	T61	0.2	0.2	0.5-0.7	0.1		8.5-9.5	0.2	0.1				276	207	3
359 A03590	USA	ASTM B108	359	T62	0.2	0.2	0.5-0.7	0.1		8.5-9.5	0.2	0.1				276	207	3
359 A03590	USA	MIL A-21180C	359	Class 1	0.2	0.2	0.5-0.7	0.1		8.5-9.5	0.2	0.1				310	241	4
359 A03590	USA	MIL A-21180C	359	Class 2	0.2	0.2	0.5-0.7	0.1		8.5-9.5	0.2	0.1				324	262	3
359 A03590	Czech Republic	CSN 424337	AlSi10CuMnMg	Permanent Mold Ingot	0.8-1.4	0.8 max	0.15-0.4	0.35-0.5	0.2 max	9-11		0.6 max			Sn 0.1 max			
359 A03590	Czech Republic	CSN 424337	AlSi10CuMnMg	Sand Ingot Alloys	0.8-1.4	0.8 max	0.15-0.4	0.35-0.5	0.2 max	9-11		0.6 max			Sn 0.1 max			
308.2 A03082	International		308.2	Ingot	4-5	0.8 max	0.1 max	0.3 max		5-6	0.2 max	0.5 max		0.5				
308.1 A03081	International		308.1	Ingot	4-5	0.8 max	0.1 max	0.5 max		5-6	0.25 max	1 max		0.5				
308.2 A03082	USA	QQ A-371F	308.2		4-5	0.8	0.1	0.3		5-6	0.2	0.5						
308.1 A03081	USA	QQ A-371F	308.1		4-5	0.8	0.1	0.5		5-6	0.25	1						
308 A03080	International		308	Sand, Perm Mold	4-5	1 max	0.1 max	0.5 max		5-6	0.25 max	1 max		0.5				
308 A03080	Russia		AMK6		4-7	1.3 max	0.4 max			3-6		0.5 max						
308 A03080	Russia		AMK5		5-7.5	1.3 max	0.4 max			4-6		0.5 max						

UNS numbers and US grades are provided as a means of cross referencing chemically similar alloys. Exchangability is only possible after independent examination of specifications. Tensile properties are minimum or typical . UTS and YS as Mpa, El as %. See Appendix for list of abbreviations used in Descriptions.

Grade UNS #	Country	Specification	Designation	Description	Cu	Fe	Mg	Mn	Ni	Si	Ti	Zn	OE Max	OT Max	Other	UTS	YS	EL
359.2 A03592	International		359.2	Ingot	0.1 max	0.12 max	0.55-0.7	0.2 max		8.5-9.5	0.2 max	0.1 max	0.05	0.15				
359.2 A03592	USA	ASTM B179	359.2		0.1		0.55-0.7	0.1		8.5-9.5	0.2	0.1						
359.2 A03592	USA	QQ A-371F	359.2		0.1		0.55-0.7	0.1		8.5-9.5	0.2	0.1						
408.2 A04082	International		408.2	Ingot	0.1 max	0.6-1.3		0.1 max		8.5-9.5		0.1 max	0.1	0.2				
	USA	ASTM B26	B850.0	Sand Cooled and art aged .250 in diam	0.7-1.3	0.7	0.6-0.9	0.1	0.9-1.5	0.4	0.2				Sn 5.5-7	166	124	
	Japan	JIS H2211	C4AV	Ingot	0.05	0.3	0.4-0.8	0.3-0.8	0.03	8-10	0.03	0.03						
	International	ISO 3522	Al-Si10 Mg	Sand Cast As mfg	0.1 max	0.6 max	0.15-0.4	0.6 max	0.05 max	9-11	0.2 max	0.1 max			Pb 0.05 max; Sn 0.05 max	150		2
	International	ISO 3522	Al-Si10 Mg	Sand Cast SHT PHT	0.1 max	0.6 max	0.15-0.4	0.6 max	0.05 max	9-11	0.2 max	0.1 max			Pb 0.05 max; Sn 0.05 max	220		1
	International	ISO 3522	Al-Si10 Mg	Perm Mold As mfg	0.1 max	0.6 max	0.15-0.4	0.6 max	0.05 max	9-11	0.2 max	0.1 max			Pb 0.05 max; Sn 0.05 max	170		3
	International	ISO 3522	Al-Si10 Mg	Perm Mold SHT PHT	0.1 max	0.6 max	0.15-0.4	0.6 max	0.05 max	9-11	0.2 max	0.1 max			Pb 0.05 max; Sn 0.05 max	240		1.5
	Austria	ONORM M3429	GK AlSi9Mg	Permanent Mold	0.05 max	0.18 max	0.2-0.4	0.05 max		9-10	0.15 max	0.07 max			ET 0.03 max; ET=.10 total;			
	Austria	ONORM M3429	GS AlSi9Mg	Sand	0.05 max	0.18 max	0.2-0.4	0.05 max		9-10	0.15 max	0.07 max			ET 0.03 max; ET=.10 total;			
	Austria	ONORM M3429	GK AlSi10Mg	Perm Mold	0.05 max	0.5 max	0.2-0.5	0.4 max		9-11	0.15 max	0.1 max			ET 0.05 max; ET=.15 total;			
	Austria	ONORM M3429	GS AlSi10Mg	Sand	0.05 max	0.5 max	0.2-0.5	0.4 max		9-11	0.15 max	0.1 max			ET 0.05 max; ET=.15 total;			
	Austria	ONORM M3429	GD AlSi10Mg(Cu)	Die	0.1 max	1 max	0.2-0.5	0.4 max		9-11	0.15 max	0.1 max			ET 0.05 max; ET=.15 total;			
	South Africa	SABS 991	Al-Si10 Mg(Cu)	F Sand	0.35 max	0.65 max	0.2-0.45	0.55 max	0.15 max	9-11	0.2 max	0.35 max	0.05	0.15	Pb 0.1 max	160		1
	South Africa	SABS 991	Al-Si10 Mg(Cu)	T6 Sand	0.35 max	0.65 max	0.2-0.45	0.55 max	0.15 max	9-11	0.2 max	0.35 max	0.05	0.15	Pb 0.1 max	220		1
	South Africa	SABS 991	Al-Si10 Mg(Cu)	F Gravity Die	0.35 max	0.65 max	0.2-0.45	0.55 max	0.15 max	9-11	0.2 max	0.35 max	0.05	0.15	Pb 0.1 max	180		1
	South Africa	SABS 991	Al-Si10 Mg(Cu)	T6 Gravity Die	0.35 max	0.65 max	0.2-0.45	0.55 max	0.15 max	9-11	0.2 max	0.35 max	0.05	0.15	Pb 0.1 max	240		1
	South Africa	SABS 991	Al-Si10 Mg	F Sand	0.1 max	0.55 max	0.25-0.45	0.45 max	0.05 max	9-11	0.15 max	0.1 max	0.05	0.15	Pb 0.05 max; Sn 0.05 max	150		2
	South Africa	SABS 991	Al-Si10 Mg	T6 Sand	0.1 max	0.55 max	0.25-0.45	0.45 max	0.05 max	9-11	0.15 max	0.1 max	0.05	0.15	Pb 0.05 max; Sn 0.05 max	220		1
	South Africa	SABS 991	Al-Si10 Mg	F Gravity Die	0.1 max	0.55 max	0.25-0.45	0.45 max	0.05 max	9-11	0.15 max	0.1 max	0.05	0.15	Pb 0.05 max; Sn 0.05 max	180		2.5
	South Africa	SABS 991	Al-Si10 Mg	T6 Gravity Die	0.1 max	0.55 max	0.25-0.45	0.45 max	0.05 max	9-11	0.15 max	0.1 max	0.05	0.15	Pb 0.05 max; Sn 0.05 max	260		1
	South Africa	SABS 989	Al-Si10 Mg(Cu)	Ingot	0.3 max	0.55 max	0.25-0.45	0.55 max	0.15 max	9-11	0.15 max	0.35 max	0.05	0.15	Pb 0.1 max			
	Austria	ONORM M3429	AlSi9Mg		0.03 max	0.15 max	0.3-0.4	0.05 max		9-10	0.15 max	0.07 max			ET 0.03 max; ET=.10 total;			
	Austria	ONORM M3429	G AlSi9Mg		0.03 max	0.15 max	0.3-0.4	0.05 max		9-10	0.15 max	0.7 max			ET 0.03 max; ET=.10 total;			
	Austria	ONORM M3429	AlSi10Mg		0.03 max	0.3 max	0.3-0.4	0.4 max		9-11	0.15 max	0.1 max			ET 0.05 max; ET=.15 total;			
	Austria	ONORM M3429	G AlSi10Mg		0.03 max	0.3 max	0.3-0.5	0.4 max		9-11	0.15 max	0.1 max			ET 0.05 max; ET=.15 total;			
	South Africa	SABS 989	Al-Si10 Mg	Ingot	0.08 max	0.45 max	0.3-0.45	0.45 max	0.05 max	9-11	0.15 max	0.1 max	0.05	0.15	Pb 0.05 max; Sn 0.05 max			
	Czech Republic	CSN 424331	AlSi10MgMn	Permanent Mold Ingot Alloys	0.1 max	0.8 max	0.2-0.45	0.1-0.4		9-10.5	0.15 max	0.1 max						
	Australia	AS 1874	BA327	T1 Sand	3.5-4.1	0.3 max	0.03 max	0.03 max	0.03 max	5.8-6.8	0.2 max	0.03 max	0.05	0.15		155		
	Australia	AS 1874	BA327	T1 permanent mould	3.5-4.1	0.3 max	0.03 max	0.03 max	0.03 max	5.8-6.8	0.2 max	0.03 max	0.05	0.15		185		

UNS numbers and US grades are provided as a means of cross referencing chemically similar alloys. Exchangability is only possible after independent examination of specifications. Tensile properties are minimum or typical. UTS and YS as Mpa, El as %. See Appendix for list of abbreviations used in Descriptions.

Grade UNS #	Country	Specification	Designation	Description	Cu	Fe	Mg	Mn	Ni	Si	Ti	Zn	OE Max	OT Max	Other	UTS	YS	EL
	Australia	AS 1874	BA327	T5 sand	3.5-4.1	0.3 max	0.03 max	0.03 max	0.03 max	5.8-6.8	0.2 max	0.03 max	0.05	0.15		170		1
	Australia	AS 1874	BA327	T6 sand	3.5-4.1	0.3 max	0.03 max	0.03 max	0.03 max	5.8-6.8	0.2 max	0.03 max	0.05	0.15		210		1.5
	Australia	AS 1874	BA327	T6 permanent mould	3.5-4.1	0.3 max	0.03 max	0.03 max	0.03 max	5.8-6.8	0.2 max	0.03 max	0.05	0.15		230		
222 A02220	Canada	CSA	CG100		9.2-10.7	1.5 max	0.15-0.35	0.5 max	0.5 max	2 max	0.25 max	0.8 max		0.35				
222 A02220	International		222	Sand, Perm Mold	9.2-10.7	1.5 max	0.15-0.35	0.5 max	0.5 max	2 max	0.25 max	0.8 max		0.35				
222 A02220	USA	ASTM B108	222	T551	9.2-10.7	1.5	0.15-0.35	0.5	0.5	2	0.25	0.8				207		
222 A02220	USA	ASTM B108	222	T65	9.2-10.7	1.5	0.15-0.35	0.5	0.5	2	0.25	0.8				276		
222 A02220	USA	ASTM B26	222	Sand Ann .250 in diam	9.2-10.7	1.5	0.15-0.35	0.5	0.5	2	0.25	0.8				159		
222 A02220	USA	ASTM B26	222	Sand T61 .250 in diam	9.2-10.7	1.5	0.15-0.35	0.5	0.5	2	0.25	0.8				200		
222 A02220	USA	ASTM B618	222	Invest Ann .250 in diam	9.2-10.7	1.5	0.15-0.35	0.5	0.5	2	0.25	0.8				159		
222 A02220	USA	ASTM B618	222	Invest SHT and art aged .250 in diam	9.2-10.7	1.5	0.15-0.35	0.5	0.5	2	0.25	0.8				207		
222 A02220	Russia		AL12		9-11	1.3 max			0.5 max	1 max	0.5 max							
222 A02220	Russia	GOST 2685	AL28		9-11	1.3 max				1.2 max		1.5 max						
A360.0 A13600	Germany	DIN 1725 pt 2	3.2381/G-AlSi10Mg		0.05 max	0.5 max	0.2-0.5	0.4 max		9-11	0.15 max	0.1 max			ET 0.05 max; , ET=.15 total			
A360.0 A13600	Germany	DIN 1725 pt 2	3.2381/Gk-AlSi10Mg		0.05 max	0.5 max	0.2-0.5	0.4 max		9-11	0.15 max	0.1 max			ET 0.05 max; , ET=.15 total			
A360.0 A13600	International	ISO R2147	Al-Si10Mg	Sand SHT and art aged OBSOLETE Replaced by ISO 3522	0.1	0.7	0.15-0.4	0.6	0.1	9-11	0.15	0.1			Pb 0.05; Sn 0.05 max	220		1
A360.0 A13600	UK	BS 1490	LM9		0.1 max	0.6 max	0.2-0.6	0.3-0.7	0.1 max	10-13	0.2 max	0.1 max			Pb 0.1 max; Sn 0.05 max			
A360.0 A13600	Czech Republic	CSN 424331	AlSi10MgMn	Die Ingot Alloys	0.2 max	0.9 max	0.2-0.5	0.1-0.4		9-10.5	0.15 max	0.3 max						
	Austria	ONORM M3429	GK AlSi10Mg(Cu)	Perm Mold	0.3 max	0.6 max	0.2-0.5	0.2-0.5	0.1 max	9-11	0.15 max	0.3 max			ET 0.05 max; ET=.15 total;			
	Austria	ONORM M3429	GS AlSi10Mg(Cu)	Sand	0.3 max	0.6 max	0.2-0.5	0.2-0.5		9-11	0.15 max	0.3 max			ET 0.05 max; ET=.15 total;			
A360.0 A13600	International		A360.0	Die	0.6 max	1.3	0.4-0.6	0.35 max	0.5 max	9-10		0.5 max		0.25	Sn 0.15 max			
360 A03600	International		360	Die	0.6 max	2 max	0.4-0.6	0.35 max	0.5 max	9-10		0.5 max		0.25	Sn 0.15 max			
360 A03600	Russia	GOST 2685	AL4		0.6 max	2 max	0.4-0.6	0.35 max	0.5 max	9-10		0.5 max		0.25	Sn 0.15 max			
360 A03600	USA	ASTM B85	360.0/SG100B	Die As cast	0.6	2	0.4-0.6	0.35	0.5	9-10		0.5			Sn 0.15 max	300	170	3
360 A03600	USA	QQ A-591E	360	Die	0.6	2	0.4-0.6	0.35	0.5	9-10		0.5			Sn 0.15 max	303	172	3
A360.0 A13600	Russia		AL4Ch		1 max	0.8 max	0.25-0.45	0.2-0.5	0.3 max	8-11		0.45 max			ST 0.2 max; ; ST=Ti+Cr			
	USA	ASTM B179	A850.1		0.7-1.3	0.5	0.1	0.1	0.3-0.7	2-3	0.2				Sn 5.5-7			
	USA	QQ A-371F	A850.1	SHT and stab T7	0.7-1.3	0.5	0.1	0.1	0.3-0.7	2-3	0.2				Sn 5.5-7	228		3
A360.0 A13600	International	ISO R164	Al-Si8Cu3Fe		2.5-4.5	1.3	0.15	0.6	0.3	7-9.5	0.2	1.2			Pb 0.3; Sn 0.2 max			
222.1 A02221	International		222.1	Ingot	9.2-10.7	1.2 max	0.2-0.35	0.5 max	0.5 max	2 max	0.25 max	0.8 max		0.35				
222.1 A02221	USA	ASTM B179	222.1		9.2-10.7	1.2	0.2-0.35	0.5	0.5	2	0.25	0.8						
222.1 A02221	USA	QQ A-371F	222.1		9.2-10.7	1.2	0.2-0.35	0.5	0.5	2	0.25	0.8						

UNS numbers and US grades are provided as a means of cross referencing chemically similar alloys. Exchangability is only possible after independent examination of specifications. Tensile properties are minimum or typical . UTS and YS as Mpa, El as %. See Appendix for list of abbreviations used in Descriptions.

Grade UNS #	Country	Specification	Designation	Description	Cu	Fe	Mg	Mn	Ni	Si	Ti	Zn	OE Max	OT Max	Other	UTS	YS	EL
222.1 A02221	UK	BS 1490	LM12		9-11	1 max	0.2-0.4	0.6 max	0.5 max	2.5 max	0.2 max	0.8 max			Pb 0.1 max; Sn 0.1 max			
A360.2 A13602	Sweden	MNC 41E	4253	Die Sand Ingot			0.3			10					Al 90			
A360.2 A13602	International	ISO R164	Al-Si10Mg		0.058 max	0.7	0.15-0.4		0.1	9-11	0.15	0.1			Pb 0.05; Sn 0.05 max			
A360.2 A13602	International		A360.2	Ingot	0.1 max	0.6 max	0.45-0.6	0.05 max		9-10		0.05 max	0.05	0.15				
A360.2 A13602	Belgium	NBN P21-101	SGAlSi10Mg	Sand	0.1 max	0.7 max	0.15-0.4	0.6 max	0.1 max	9-11	0.15 max	0.1 max			Pb 0.05 max; Sn 0.05 max			
A360.2 A13602	Sweden	SIS 144253	4253-00		0.2	0.5	0.2-0.4	0.5	0.1	9-11	0.2	0.3			Pb 0.05; Sn 0.05 max			
A360.2 A13602	Sweden	SIS 144253	4253-04	Sand St and art aged	0.2	0.5	0.2-0.4	0.5	0.1	9-11	0.2	0.3			Pb 0.05; Sn 0.05 max		200	1
A360.2 A13602	Sweden	SIS 144253	4253-07	ST and art aged	0.2	0.5	0.2-0.4	0.5		9-11	0.2	0.3			Pb 0.05; Sn 0.05 max		220	1
A360.2 A13602	Austria	ONORM M3429	G AlSi10Mg(Cu)		0.3 max	0.6 max	0.3-0.5	0.2-0.5	0.1 max	9-11	0.15 max	0.3 max			ET 0.05 max; ET=.15 total;			
A360.1 A13601	International		A360.1	Ingot	0.6 max	1 max	0.45-0.6	0.35 max	0.5 max	9-10		0.4 max		0.25	Sn 0.15 max			
	Japan	JIS H2211	C4BV	Ingot	2-4	0.3	0.03	0.03	0.03	7-10	0.03	0.03						
	USA	QQ A-371F	A360.2		0.2	1.13	0.25-0.4	0.1	0.15	7.5-9.5		0.15			Cr 0.25-0.5; Sn 0.15 max			
520 A05200	Russia		AL27-2		0.05 max		10-11.5			0.05 max	0.03-0.1		0.2		B 0.01 max; Be 0.05-0.12; Zr 0.03-0.1			
520 A05200	Russia	GOST 2685	AL27-1		0.05 max	0.05 max	9.5-11.5	0.1 max		0.05 max	0.05-0.15	0.05 max	0.6		Be 0.05-0.15; Zr 0.05-0.2			
520 A05200	Russia	GOST 2685	AL8		0.05 max	0.05 max	9.5-11.5	0.1 max		0.05 max	0.07 max	0.1 max			Be 0.14 max; Sn 0.01 max			
520 A05200	Germany	DIN 1725 pt 2	3.3591/G-AlMg10		0.05 max	0.3 max	9-11	0.3 max		0.3 max	0.15 max	0.1 max			ET 0.05 max; ET=.15 total; Pb 0.05 max; Sn 0.05 max			
520 A05200	South Africa	SABS 989	Al-Mg10A	SHT	0.08	0.35	9.6-11	0.1	0.1	0.25	0.2	0.1			Pb 0.05; Sn 0.05 max	275		8
520 A05200	South Africa	SABS 989	Al-Mg10A	SHT Chill	0.08	0.35	9.6-11	0.1	0.1	0.25	0.2	0.1			Pb 0.05; Sn 0.05 max	310		12
520 A05200	UK	BS 1490	LM10		0.1 max	0.35 max	9.5-11	0.1 max	0.1 max	0.25 max	0.2 max	0.1 max			Pb 0.05 max; Sn 0.05 max			
520 A05200	South Africa	SABS 990	Al-Mg10A	Sand SHT	0.1	0.4	9.5-11	0.1	0.1	0.25	0.15	0.1			Pb 0.05; Sn 0.05 max	275		8
520 A05200	South Africa	SABS 991	Al-Mg10A	SHT	0.1	0.4	9.5-11	0.1	0.1	0.25	0.15	0.1			Sn 0.05 max	310		12
520 A05200	International	ISO R164	Al-Mg10		0.1	0.3	9-11	0.3	0.1	0.3	0.15	0.1			Pb 0.05; Sn 0.05 max			
520 A05200	International	ISO R2147	Al-Mg10	Sand SHT OBSOLETE Replaced by ISO 3522	0.1	0.3	9-11	0.3	0.1	0.3	0.15	0.1			Pb 0.05; Sn 0.05 max	270		10
	International	ISO 3522	Al-Mg10	Sand Cast SHT Aged	0.1 max	0.3 max	9.5-11	0.15 max	0.1 max	0.3 max	0.15 max	0.1 max			Be 0.05 max; Pb 0.05 max; Sn 0.05 max	260		8
520 A05200	International		520	Sand	0.25 max	0.3 max	9.5-10.6	0.15 max		0.25 max	0.25 max	0.15 max	0.05	0.15				
520 A05200	USA	ASTM B26	520	Sand SHT and nat aged .250 in diam	0.25	0.3	9.5-10.5	0.15		0.25	0.25	0.15				290	152	
520 A05200	USA	ASTM B618	520	Invest SHT and nat aged .250 in diam	0.25	0.3	9.5-10.6	0.15		0.25	0.25	0.15				290	152	12
	Austria	ONORM M3429	AlSi10Mg(Cu)		0.3 max	0.6 max	0.3-0.5	0.2-0.5	0.1 max	9-11	0.15 max	0.3 max			ET 0.05 max; ET=.15 total;			
	USA	QQ A-596d	A750.0	Aged and temper T5	0.7-1.3	0.7		0.1	0.3-0.7	2-3	0.2				Sn 5.5-7	117		3
	Australia	AS 1874	EA313	Ingot	2-4	1.3 max	0.3 max	0.5 max	0.5 max	7.5-9.5	0.2 max	1 max	0.05	0.2	Cr 0.1 max; Pb 0.25 max; Sn 0.3 max			

UNS numbers and US grades are provided as a means of cross referencing chemically similar alloys. Exchangability is only possible after independent examination of specifications. Tensile properties are minimum or typical . UTS and YS as Mpa, El as %. See Appendix for list of abbreviations used in Descriptions.

Grade UNS #	Country	Specification	Designation	Description	Cu	Fe	Mg	Mn	Ni	Si	Ti	Zn	OE Max	OT Max	Other	UTS	YS	EL
520.2 A05202	France	NF A57-702	AG11	Perm Mold Sand	0.05 max	0.3 max	10.5-12	0.1 max	0.05 max	0.2 max	0.1 max	0.1 max			Be 0.05 max; Cr 0.2 max			
409.2 A04092	International		409.2	Ingot	0.1 max	0.6-1.3		0.1 max		9-10		0.1 max	0.1	0.2				
520.2 A05202	Canada	CSA HA.3	.G10		0.2	0.2	9.6-10.6	0.1		0.15	0.2	0.1						
520.2 A05202	International		520.2	Ingot	0.2 max	0.2 max	9.6-10.6	0.1 max		0.15 max	0.2 max	0.1 max	0.05	0.15				
520.2 A05202	USA	ASTM B179	520.2		0.2	0.2	9.6-10.6	0.1		0.15	0.2	0.1						
520.2 A05202	USA	QQ A-371F	520.2		0.2	0.2	9.6-10.6	0.1		0.15	0.2	0.1						
520.2 A05202	France	NF A57-703	AG10	Die	0.2 max	1.3 max	8.5-11	0.6 max	0.1 max	1 max	0.2 max	0.4 max			Be 0.05 max; Sn 0.1 max			
520.2 A05202	Canada	CSA HA.9	.G10	Sand	0.25	0.3	9.5-10.6	0.15		0.25	0.25	0.15						
	USA	ASTM B108	A850.0	Cooled and art aged	0.7-1.3	0.7	0.1	0.1	0.3-0.7	2-3	0.2				Sn 5.5-7	117		3
	USA	ASTM B108	A850.0	SHT and art aged	0.7-1.3	0.7	0.1	0.1	0.3-0.7	2-3	0.2				Sn 5.5-7	124		8
	USA	ASTM B26	A850.0	Sand Cooled and art aged .250 in diam	0.7-1.3	0.7	0.1	0.1	0.3-0.7	2-3	0.2				Sn 5.5-7	117		3
	USA	ASTM B618	A850.0	Invest Cooled and art aged .250 in diam	0.7-1.3	0.7	0.1	0.1	0.3-0.7	2-3	0.2				Sn 5.5-7	117		3
354 A03540	USA	ASTM B108	354	T62	1.6-2	0.2	0.4-0.6	0.1		8.6-9.4	0.2	0.1				297	228	2
	Germany	DIN 1725Sh2	3.2373./GK AlSi9Mgwa	Q AH	0.05	0.18	0.2-0.4	0.05		9-10	0.15	0.07				260	200	4
	Germany	DIN 1725Sh2	3.2373.6/G-AlSi9Mgwa	Sand Q and AH	0.05	0.18	0.2-0.4	0.05		9-10	0.15	0.07				250	200	2
	South Africa	SABS 992	Al-Si8 Cu3	Press Die	2-3.5	0.8 max 0.55	0.06	0.15 0.65	0.35 max	7.5 0.5	0.25 max	1.2 max	0.05	0.25	Cr 0.15 max, Pb 0.25 max; Sn 0.15 max	240		1
	Japan	JIS H5202	AC2A		3.5-4.5	0.8	0.2	0.5		4-5	0.2	0.5				177		2
	Japan	JIS H5202	AC2A	Q/T	3.5-4.5	0.8	0.2	0.5		4-5	0.2	0.5				275		1
	Japan	JIS H2117	C4AS	Ingot	0.2	0.4	0.4-0.8	0.3-0.8	0.1	8-10	0.2	0.2						
	Japan	JIS H5202	AC4A		0.2	0.5	0.4-0.8	0.3-0.8		8-10	0.2	0.2				177		3
	Japan	JIS H5202	AC4A	Q/T	0.2	0.5	0.4-0.8	0.3-0.8		8-10	0.2	0.2				245		2
	Australia	AS 1874	AB325	Ingot	2-3.5	0.8 max	0.1-0.3	0.2-0.5	0.3 max	7.5-9.5	0.15 max	0.05 max	0.05	0.15	Pb 0.2 max; Sn 0.1 max			
	Australia	AS 1874	AC325	Ingot	2-3.5	0.8 max	0.1-0.3	0.2-0.5	0.3 max	7.5-9.5	0.15 max	0.05 max	0.05	0.15	Pb 0.2 max; Sn 0.1 max			
	USA	ASTM B179	518.1/G8A		0.25	1	7.6-8.5	0.35	0.15	0.35		0.15		0.25	Sn 0.15 max			
	Germany	DIN 1725Sh3	3.0881/V-AlTi10		0.02	0.3	0.02	0.02	0.04	0.2	9-11				Cr 0.02; V 0.3			
	Japan	JIS H2117	C2AS	Ingot	3.5-4.5	0.7	0.2	0.5	0.3	4-5	0.2	0.5						
	Sweden	MNC 41E	4255	Sand Die Ingot						10					Al 90			
	South Africa	SABS 989	Al-Si11	Ingot	0.03 max	0.15 max	0.45 max	0.1 max		10-11.8	0.15 max	0.07 max	0.03	0.1				
	South Africa	SABS 991	Al-Si11	F Sand	0.05 max	0.19 max	0.45 max	0.1 max		10-11.8	0.15 max	0.07 max	0.03	0.1		150		6
	South Africa	SABS 991	Al-Si11	F Gravity Die	0.05 max	0.19 max	0.45 max	0.1 max		10-11.8	0.15 max	0.07 max	0.03	0.1		170		7
	Norway	NS 17520	NS17520-00		0.2	0.5	0.2-0.4	0.5	0.1	9-11	0.2	0.3			Al 90; Pb 0.05; Sn 0.05 max			
	Norway	NS 17520	NS17520-41	Sand H and art aged	0.2	0.5	0.2-0.4	0.5	0.1	9-11	0.2	0.3			Al 90; Pb 0.05; Sn 0.05 max	196		1
	Norway	NS 17520	NS17520-42	H and art aged	0.2	0.5	0.2-0.4	0.5	0.1	9-11	0.2	0.3			Al 90; Pb 0.05; Sn 0.05 max	216		1
	Norway	NS 17520	NS17520-XX		0.2	0.5	0.2-0.4	0.5	0.1	9-11	0.2	0.3			Al 90; Pb 0.05; Sn 0.05 max			

UNS numbers and US grades are provided as a means of cross referencing chemically similar alloys. Exchangability is only possible after independent examination of specifications. Tensile properties are minimum or typical . UTS and YS as Mpa, El as %. See Appendix for list of abbreviations used in Descriptions.

Grade UNS #	Country	Specification	Designation	Description	Cu	Fe	Mg	Mn	Ni	Si	Ti	Zn	OE Max	OT Max	Other	UTS	YS	EL
	Australia	AS 1874	DA401	Ingot	0.6 max	1 max	0.25 max	0.5 max	0.5 max	10-13	0.2 max	0.4 max	0.05	0.25	Cr 0.1 max; Pb 0.15 max; Sn 0.15 max			
	USA	ASTM B179	B850.1		1.7-2.3	0.5	0.7-0.9	0.1	0.9-1.5		0.4	0.2			Sn 5.5-7			
	USA	QQ A-371F	B850.1		1.7-2.3	0.5	0.7-0.9	0.1	0.9-1.5		0.4	0.2			Sn 5.5-7			
	Norway	NS 17535	NS17535-00		2-3	0.7	0.3	0.5	0.3	6-8	0.2	2			Al 90; Pb 0.2; Sn 0.1 max			
	Norway	NS 17535	NS17535-01	Sand	2-3	0.7	0.3	0.5	0.3	6-8	0.2	2			Al 90; Pb 0.2; Sn 0.1 max	147	98	2
	Norway	NS 17535	NS17535-02		2-3	0.7	0.3	0.5	0.3	6-8	0.2	2			Al 90; Pb 0.2; Sn 0.1 max	167	108	2
	Sweden	MNC 41E	4251	Sand Die Ingot	3	0.8 max				8					Al 90			
	Sweden	MNC 41E	4250	Die Ingot	3	1.3 max				8					Al 90			
	Australia	AS 1874	BB325	Ingot	2-3.5	0.8 max	0.2-0.5	0.1-0.3	0.3 max	7.5-8.5	0.15 max	0.2-0.5	0.05	0.15	Pb 0.2 max; Sn 0.1 max			
	International	ISO 3522	Al-Si8 Cu3 Fe	Pressure Die Cast	2.5-4	1.3 max	0.3 max	0.6 max	0.5 max	7.5-9.5	0.2 max	1.2 max			Pb 0.3 max; Sn 0.2 max			
333 A03330	South Africa	SABS 992	46000 (Al-Si9 Cu3 Fe)	Press Die	2-4	1.3 max	0.05-0.55	0.55 max	0.55 max	8-11	0.25 max	1.2 max	0.05	0.25	Cr 0.15 max; Pb 0.35 max; Sn 0.25 max	240		<1
	Germany	DIN 1725Sh3	3.2291/V-AlSi20		0.2	0.45	0.4	0.35	0.2	8-21	0.1	0.2			Cr 0.1; Pb 0.1; Sn 0.1			
361 A03610	International		361	Die	0.5 max	1.1 max	0.4-0.6	0.25 max	0.2-0.3	9.5-10.5	0.2 max	0.5 max	0.05	0.15	Cr 0.2-0.3; Sn 0.1 max			
	USA	ASTM B108	B850.0	Cooled and art aged	1.7-2.3	0.7	0.6-0.9	0.1	0.9-1.5		0.4	0.2			Sn 5.5-7	186		3
	USA	ASTM B618	B850.0	Invest Cooled and art aged .250 in diam	1.7-2.3	0.7	0.6-0.9	0.1	0.9-1.5		0.4	0.2			Sn 5.5-7	166	124	
	USA	QQ A-596d	B750.0	Aged and temper T5	1.7-2.3	0.7	0.6-0.9	0.1	0.9-1.5		0.4	0.2			Sn 5.5-7	186		
	Australia	AS 1874	AA605	Ingot	0.1 max	0.7-1.1	0.45-0.6	0.1 max	0.1 max	9-10	0.2 max	0.1 max	0.05	0.2	Sn 0.1 max			
360.2 A03602	International		360.2	Ingot	0.1 max	0.7-1.1	0.45-0.6	0.1 max	0.1 max	9-10		0.1 max		0.2	Sn 0.1 max			
360.2 A03602	USA	ASTM B179	360.2/SG100C		0.1	0.7-1.1	0.45-0.6	0.1	0.1	9-10		0.1			Sn 0.1 max			
360.2 A03602	USA	QQ A-371F	360.2		0.1	0.7-1.1	0.45-0.6	0.1	0.1	9-10		0.1			Sn 0.1 max			
361.1 A03611	International		361.1	Ingot	0.5 max	0.8 max	0.45-0.6	0.25 max	0.2-0.3	9.5-10.5	0.2 max	0.4 max	0.05	0.15	Cr 0.2-0.3; Sn 0.1 max			
	Australia	AS 1874	CA605	Ingot	0.6 max	0.7-1.1	0.45-0.6	0.35 max	0.5 max	9-10		0.5 max	0.05	0.25	Sn 0.15 max			
363 A03630	International		363	Sand, Perm Mold	2.5-3.5	1.1 max	0.15-0.4		0.25 max	4.5-6	0.2 max	3-4.5		0.3	Mn+Cr 0.8 max; Pb 0.25 max; Fe/Si 2.5 min; Sn 0.25 max			
	Australia	AS 1874	AA607	T1 permanent mould	0.15 max	0.6 max	0.2-0.6	0.05 max		10-13	0.25 max	0.15 max	0.05	0.2		190		3
	Australia	AS 1874	AA607	T5 permanent mould	0.15 max	0.6 max	0.2-0.6	0.05 max		10-13	0.25 max	0.15 max	0.05	0.2		230		2
	Australia	AS 1874	AA607	T6 sand	0.15 max	0.6 max	0.2-0.6	0.05 max		10-13	0.25 max	0.15 max	0.05	0.2		240		
363.1 A03631	International		363.1	Ingot	2.5-3.5	0.8 max	0.2-0.4		0.25 max	4.5-6	0.2 max	3-4.5		0.3	Mn+Cr 0.8 max; Pb 0.25 max; Fe/Si 2.5 min; Sn 0.25 max			
	Austria	ONORM M3429	GD AlSi8Cu3	Die	2.5-3.5	1 max	0.3 max	0.2-0.5	0.3 max	7.5-9.5	0.15 max	1 max			ET 0.05 max ET=.15 total; Pb 0.2 max; Sn 0.1 max			
	Japan	JIS H2212	D10V	Ingot	2-4	0.6	0.03	0.03	0.03	7.5-9.5		0.03			Sn 0.03 max			

UNS numbers and US grades are provided as a means of cross referencing chemically similar alloys. Exchangability is only possible after independent examination of specifications. Tensile properties are minimum or typical . UTS and YS as Mpa, El as %. See Appendix for list of abbreviations used in Descriptions.

Grade UNS #	Country	Specification	Designation	Description	Cu	Fe	Mg	Mn	Ni	Si	Ti	Zn	OE Max	OT Max	Other	UTS	YS	EL
	Germany	DIN 1725Sh2	3.3591.43/G-AlMg10ho	Sand As cast and homogenized	0.05	0.3	9-11	0.3		0.3	0.15	0.1			Pb 0.05; Sn 0.05 max	220	140	
	Austria	ONORM M3429	G AlSi8Cu3		2.5-3.5	0.8 max	0.1-0.3	0.2-0.5	0.3 max	7.5-9.5	0.15 max	1 max			ET 0.05 max ET=.15 total; Pb 0.2 max; Sn 0.1 max			
	Austria	ONORM M3429	GK AlSi8Cu3	Perm Mold	2.5-3.5	0.8 max	0.1-0.3	0.2-0.5	0.3 max	7.5-9.5	0.15 max	1 max			ET 0.05 max ET=.15 total; Pb 0.2 max; Sn 0.1 max			
	Austria	ONORM M3429	AlSi8Cu3	perm Mold Sand	2.5-3.5	0.8 max	0.1-0.3	0.2-0.5	0.3 max	7.5-9.5		1 max		0.15	ET 0.05 max ET=.15 total; Pb 0.2 max; Sn 0.1 max			
	Austria	ONORM M3429	GS AlSi8Cu3	Sand	2.5-3.5	0.8 max	0.1-0.3	0.2-0.5		7.5-9.5	0.15 max	1 max			ET 0.05 max; ET=.15 total;			
	Japan	JIS H2117	C2BS	Ingot	2-4	0.8	0.5	0.5	0.3	5-7	0.2	1						
	Japan	JIS H2211	C7BV	Ingot	0.05	0.2	9.6-11	0.03	0.03	0.2	0.2	0.03						
	Germany	DIN 1725Sh2	3.2381./G-AlSi10Mgwa	Sand	0.05	0.5	0.2-0.5	0.4		9-11	0.15	0.1						
	Germany	DIN 1725Sh2	3.2381.0/GK-AlSi10Mg		0.05	0.5	0.2-0.5	0.4		9-11	0.15	0.1				180	90	2
	Germany	DIN 1725Sh2	3.2381.01/G-AlSi10Mg	Sand	0.05	0.5	0.2-0.5	0.4		9-11	0.15	0.1						
	India	IS 6754	8482		0.7-1.3	0.6	0.75-1.25		1.5-1.8	0.35-0.85					Sn 6.5-7.5			
354 A03540	USA	MIL A-21180C	354	Class 2	1.6-2	0.2	0.4-0.6	0.1		8.6-9.4	0.2	0.1				345	290	2
	India	IS 202	A-10	Sand ST	0.1	0.35	9.5-11		0.1	0.25		0.1			Pb 0.05; Sn 0.05 max	278	152	8
	South Africa	SABS 989	Al-Si12	Ingot	0.03 max	0.4 max		0.35 max		10.5-13.5	0.15 max	0.1 max	0.05	0.15				
	Japan	JIS H2211	C3AV	Ingot	0.05	0.3	0.03	0.03	0.03	10-13	0.03	0.03						
	South Africa	SABS 991	Al-Si12	F Sand	0.05 max	0.55 max		0.35 max		10.5-13.5	0.15 max	0.1 max	0.05	0.15		150		5
	South Africa	SABS 991	Al-Si12	F Gravity Die	0.05 max	0.55 max		0.35 max		10.5-13.5	0.15 max	0.1 max	0.05	0.15		170		6
	South Africa	SABS 989	Al-Si12(Fe)	Ingot	0.1 max	0.55 max	0.1 max	0.55 max	0.1 max	10.5-13.5	0.15 max	0.15 max	0.05	0.15	Pb 0.1 max			
	South Africa	SABS 992	Al-Si 12 Fe	Press Die	0.1 max	1 max		0.55 max		10.5-13.5	0.15 max	0.15 max	0.05	0.25		240		1
	Australia	AS 1874	AA607	Ingot	0.15 max	0.6 max	0.2-0.6	0.3-0.7	0.15 max	10-13	0.25 max	0.15 max	0.05	0.2	Sn 0.05 max			
	Australia	AS 1874	AA607	T5 sand	0.15 max	0.6 max	0.2-0.6	0.3-0.7	0.15 max	10-13	0.25 max	0.15 max	0.05	0.2	Sn 0.05 max	170		1.5
354 A03540	Czech Republic	CSN 424384	AlSi10CuMn	Permanent Mold Ingot	0.5-1.6	1 max	0.6 max	0.3-0.6	0.4 max	8-11		1 max			Sn 0.2 max			
A338.0 A13800	Sweden	MNC 41E	4252	Die Ingot	3	1.3 max				9					Al 88			
A338.0 A13800	Czech Republic	CSN 424339	AlSi8Cu2Mn	Die Ingot Alloys	1.5-3	1.1 max	0.4 max	0.3-0.5	0.5 max	7.5-9.5		0.9 max			Sn 0.2 max			
A13802	Australia	AS 1874	AA325	Ingot	2-3.5	0.3 max	0.1-0.3	0.2-0.5	0.3 max	7.5-9.5	0.15 max	0.1 max	0.05	0.15	Pb 0.2 max; Sn 0.1 max			
380 A13802	South Africa	SABS 989	46200 (Al-Si8 Cu3)	Ingot	2-3.5	0.7 max	0.15-0.55	0.15-0.65	0.35 max	7.5-9.5	0.2	1.2 max	0.05	0.25	Cr 0.15 max; Pb 0.25 max; Sn 0.15 max			
380 A13800	South Africa	SABS 991	46200 (Al-Si8 Cu3)	F Sand	2-3.5	0.8 max	0.05-0.55	0.15-0.65	0.35 max	7.5-9.5	0.25 max	1.2 max	0.05	0.25	Cr 0.15 max; Pb 0.25 max; Sn 0.15 max	150		1
380 A13801	South Africa	SABS 991	46200 (Al-Si8 Cu3)	F Gravity Die	2-3.5	0.8 max	0.05-0.55	0.15-0.65	0.35 max	7.5-9.5	0.25 max	1.2 max	0.05	0.25	Cr 0.15 max; Pb 0.25 max; Sn 0.15 max	170		1
A13802	Australia	AS 1874	AA323	Ingot	3-4	0.7 max	0.3-0.5	0.4-0.6	0.03 max	8-9	0.03 max	0.03 max	0.05	0.15				
A338.0 A13800	Canada	CSA	SC84N		3-4	0.6 max	0.1 max	0.1 max	0.1 max	7.5-9.5		0.1 max			ET 0.05 max; ET=.15 total;			

UNS numbers and US grades are provided as a means of cross referencing chemically similar alloys. Exchangability is only possible after independent examination of specifications. Tensile properties are minimum or typical . UTS and YS as Mpa, El as %. See Appendix for list of abbreviations used in Descriptions.

Worldwide Guide to Equivalent Nonferrous Metals and Alloys

Grade UNS #	Country	Specification	Designation	Description	Cu	Fe	Mg	Mn	Ni	Si	Ti	Zn	OE Max	OT Max	Other	UTS	YS	EL
A380.2 A13802	International		A380.2	Ingot	3-4	0.6 max	0.1 max	0.1 max	0.1 max	7.5-9.5		0.1 max	0.05	0.15				
A380.1 A13801	International		A380.1	Ingot	3-4	1 max	0.1 max	0.5 max	0.5 max	7.5-9.5		2.9 max		0.5	Sn 0.35 max			
B380.1 A23801	International		B380.1	Ingot	3-4	1 max	0.1 max	0.5 max	0.5 max	7.5-9.5		0.9 max		0.5	Sn 0.35 max			
	Australia	AS 1874	CA313	Ingot	3-4	1.1 max	0.3 max	0.5 max	0.5 max	7.5-9.5	0.2 max	3 max	0.05	0.2	Cr 0.1 max; Pb 0.35 max; Sn 0.25 max			
A380.0 A13800	International		A380.0	Die	3-4	1.3 max	0.1 max	0.5 max	0.5 max	7.5-9.5		3 max		0.5	Sn 0.35 max			
B380.0 A23800	International		B380.0	Die	3-4	1.3 max	0.1 max	0.5 max	0.5 max	7.5-9.5		1 max		0.5	Sn 0.35 max			
380 A03800	Japan	JIS H5302	ADC10	Die	2-4	1.3	0.3	0.5	0.5	7.5-9.5		1			Sn 0.3 max			
380 A03800	South Africa	SABS 992	Al-Si8Cu4FeA		3-4	1.3	0.1	0.5	0.5	7.5-9.5	0.2	3			Pb 0.3; Sn 0.2 max	230	110	1
380 A03800	International		380	Die	3-4	2 max	0.1 max	0.5 max	0.5 max	7.5-9.5		3 max		0.5	Sn 0.35 max			
380 A03800	USA	ASTM B85	380.0/SC84.B	Die As cast	3-4	2	0.1	0.5	0.5	7.5-9.5		3			Sn 0.35 max	320	160	3
380 A03800	USA	QQ A-591E	380	Die	3-4	2	0.1	0.5	0.5	7.5-9.5		3			Sn 0.35 max	317	159	3
A338.0 A13800	Sweden	SIS 144252	4252-00		2-4	1.1	0.3	0.5	0.3	7.5-10	0.2	1.2			Pb 0.3; Sn 0.2 max			
A338.0 A13800	France	NF A57703	A-S9U3-Y4	Pressure Die	2.5-4	1.3 max	0.3	0.5	0.5	7.5-10	0.2	1.2			Pb 0.2; Sn 0.2 max	200		1
380 A03800	Norway	NS 17532	NS17532-05		2-4	1.3	0.3	0.5	0.3	7.5-10	0.2	3			Al 86; Pb 0.3; Sn 0.2 max	245	176	1
A338.0 A13800	Sweden	SIS 144252	4252-10	As cast	2-4	1.3	0.3	0.5	0.3	7.5-10	0.2	1.2			Pb 0.3; Sn 0.2 max		180	1
	India	IS 202	A-10	Sand ST	0.1	0.35	9.5-11	0.1	0.1	0.25		0.1			Pb 0.05; Sn 0.05 max	278	152	8
	South Africa	SABS 989	Al-Si12(Cu)	Ingot	0.9 max	0.7 max	0.35 max	0.05-0.55	0.3 max	10.5-13.5	0.15 max	0.55 max	0.05	0.25	Cr 0.1 max; Pb 0.2 max; Sn 0.1 max			
	South Africa	SABS 991	Al-Si12(Cu)	F Sand	1 max	0.8 max	0.35 max	0.05-0.55	0.3 max	10.5-13.5	0.2 max	0.55 max	0.05	0.25	Cr 0.1 max; Pb 0.2 max; Sn 0.1 max	150		1
	South Africa	SABS 991	Al-Si12(Cu)	F Gravity Die	1 max	0.8 max	0.35 max	0.05-0.55	0.3 max	10.5-13.5	0.2 max	0.55 max	0.05	0.25	Cr 0.1 max; Pb 0.2 max; Sn 0.1 max	170		2
	Germany	DIN 1725Sh2	GK-AlSi10Mg(Cu)wa/3.	Q AH	0.05	0.6	0.2-0.5	0.2-0.5	0.1	9-11	0.15	0.3				240	210	1
	Australia	AS 1874	AC609	Ingot	0.05 max	0.15 max	0.1-0.15	0.05 max		10.5-11.5	0.2 max	0.05 max	0.05	0.15				
411.2 A04112	International		411.2	Ingot	0.2 max	0.6-1.3		0.1		10-12		0.1 max	0.1	0.2				
354 A03540	International		354	Perm Mold	1.6-2	0.2 max	0.4-0.6	0.1 max		8.6-9.4	0.2 max	0.1 max	0.05	0.15				
354 A03540	USA	ASTM B108	354	T61	1.6-2	0.2	0.4-0.6	0.1		8.6-9.4	0.2	0.1				297	228	2
	Finland	SFS 2567	G-AlSi10Mg	Die ST and aged 20 mm diam	0.1	0.5	0.2-0.5	0.4	0.1	9-11	0.2	0.1			Pb 0.05; Sn 0.05 max	260	220	1
	Germany	DIN 1725Sh2	3.2381/GK-AlSi10Mgwa	Q AH	0.3	0.5	0.2-0.5	0.4		9-11	0.15	0.1				240	210	1
	USA	QQ A-596d	Tenzaloy (613)	Aged and temper T5	0.4-1	1.8	0.2-0.5	0.6	0.15	0.25	0.25	7-8			Cr 0.35	221		4
354.1 A03541	International		354.1	Ingot	1.6-2	0.15 max	0.45-0.6	0.1 max		8.6-9.4	0.2 max	0.1 max	0.05	0.15				
354.1 A03541	USA	ASTM B179	354.1		1.6-2	0.15	0.45-0.6	0.1		8.6-9.4	0.2	0.1						
354.1 A03541	USA	QQ A-371F	354.1		1.6-2	0.15	0.45-0.6	0.1		8.6-9.4	0.2	0.1						

UNS numbers and US grades are provided as a means of cross referencing chemically similar alloys. Exchangability is only possible after independent examination of specifications. Tensile properties are minimum or typical. UTS and YS as Mpa, El as %. See Appendix for list of abbreviations used in Descriptions.

Grade UNS #	Country	Specification	Designation	Description	Cu	Fe	Mg	Mn	Ni	Si	Ti	Zn	OE Max	OT Max	Other	UTS	YS	EL
	Sweden	SIS 144247	4247-00		0.2	0.8-1.3	0.5	0.5	0.1	8-10	0.2	0.3			Pb 0.1; Sn 0.05 max			
	Sweden	SIS 144247	4247-10	As cast	0.2	0.8-1.3	0.5	0.5	0.1	8-10	0.2	0.3			Pb 0.1; Sn 0.05 max		758	2
	Germany	DIN 1725Sh2	3.2000/G-AlSi10Mg(Cu) wa	Sand Q and AH	0.3	0.6	0.2-0.5	0.2-0.5	0.1	9-11	0.15	0.3				220	180	1
	Germany	DIN 1725Sh2	3.2300/GK-AlSi10Mg(Cu)		0.3	0.6	0.2-0.5	0.2-0.5	0.1	9-11	0.15	0.3				200	100	1
	Germany	DIN 1725Sh2	3.238/G-AlSi10Mg(Cu)	Sand	0.3	0.6	0.2-0.5	0.2-0.5	0.1	9-11	0.15	0.3				180	90	1
	Germany	DIN 1725Sh2	3.2900/GD-AlSi10Mg(Cu)		0.3	0.6	0.2-0.5	0.2-0.5	0.1	9-11	0.15	0.3						
	South Africa	SABS 989	Al-Si12 Fe	Ingot	0.08 max	0.45-0.9		0.55 max		10.5-13.5	0.15 max	0.15 max	0.05	0.25				
	Germany	DIN 1725Sh2	3.2382.0/GD-AlSi10Mg		0.1	1	0.2-0.5	0.4		9-11	0.15	0.1						
413 A04130	Russia	GOST 1521	SIL0							10-13								
A413.0 A14130	Russia		CLM0			0.35 max				10.5-13.5					ST 0.15 max; ; ST=Cu+Zn; Ca 0.1 max			
A413.0 A14130	Russia		CLM2			0.7 max				10.5-13.5					ST 0.2 max; ; ST=Cu+Zn; Ca 0.1 max			
413 A04130	Russia	GOST 1521	SLM3			1.2 max				10.5-13.5					ST 0.2 max; ST=Cu+Zn; Ca 0.2 max			
332 A03320	Germany	DIN 1725 pt 2	3.2371/G-AlSi7Mg		0.03 max	0.18 max	0.2-0.4	0.05 max		6.5-7.5	0.15 max	0.07 max			ET 0.03 max; ET=.10 total,			
332 A03320	Germany	DIN 1725 pt 2	3.2371/Gk-AlSi7Mg		0.03 max	0.18 max	0.2-0.4	0.05 max		6.5-7.5	0.15 max	0.07 max			Fe 0.18 max; ET 0.03 max; ET=.10 total;			
B413.0 A24130	Austria	ONORM M3429	AlSi12		0.03 max	0.3 max	0.05 max	0.4 max		11-13.5	0.15 max	0.1 max			ET 0.05 max; ET=.15 total;			
B413.0 A24130	Austria	ONORM M3429	G AlSi12		0.03 max	0.3 max	0.05 max	0.4 max		11-13.5	0.15 max	0.1 max			ET 0.05 max; ET=.15 total;			
	Germany	DIN 1725Sh2	3.3292.05/GD-AlMg9		0.05	1	7-10	0.2-0.5		2.5	0.15	0.1						
A413.2 A14132	Canada		S12N		0.05 max	0.6 max		0.3 max		11-13		0.05 max			ET 0.05 max; ET=.15 total; Ca 0.01 max			
11B413.011 A2411130	Austria	ONORM M3429	GK AlSi12	Perm Mold	0.05 max	0.5 max	0.05 max	0.4 max		11-13.5	0.15 max	0.1 max			ET 0.05 max; ET=.15 total;			
B413.0 A24130	Austria	ONORM M3429	GS AlSi12	Sand	0.05 max	0.5 max	0.05 max	0.4 max		11-13.5	0.15 max	0.1 max			ET 0.05 max; ET=.15 total;			
A413.1 A14131	International	ISO R2147	Al-Si12	Sand OBSOLETE Replaced by ISO 3522	0.058 max	0.7	0.1		0.1	11-13.5	0.15	0.1			Pb 0.1; Sn 0.05 max	160		4
	Australia	AS 1874	BB401	Ingot	0.1 max	0.4 max	0.05 max	0.1 max	0.05 max	11-13	0.2 max	0.1 max	0.05	0.15				
	Australia	AS 1874	BB401	F1 sand	0.1 max	0.4 max	0.05 max	0.1 max	0.05 max	11-13	0.2 max	0.1 max	0.05	0.15		160		5
	Australia	AS 1874	BB401	F1 permanent mould	0.1 max	0.4 max	0.05 max	0.1 max	0.05 max	11-13	0.2 max	0.1 max	0.05	0.15		190		7
	Australia	AS 1874	CA401	F1 sand	0.1 max	0.4 max	0.05 max	0.1 max	0.05 max	11-13	0.2 max	0.1 max	0.05	0.15		160		5
	Australia	AS 1874	CA401	F1 permanent mould	0.1 max	0.4 max	0.05 max	0.1 max	0.05 max	11-13	0.2 max	0.1 max	0.05	0.15		190		7
	Australia	AS 1874	CC401	F1 sand	0.1 max	0.4 max	0.05 max	0.1 max	0.05 max	11-13	0.2 max	0.1 max	0.05	0.15		160		5
	Australia	AS 1874	CC401	F1 permanent mould	0.1 max	0.4 max	0.05 max	0.1 max	0.05 max	11-13	0.2 max	0.1 max	0.05	0.15		190		7

UNS numbers and US grades are provided as a means of cross referencing chemically similar alloys. Exchangability is only possible after independent examination of specifications. Tensile properties are minimum or typical . UTS and YS as Mpa, El as %. See Appendix for list of abbreviations used in Descriptions.

Grade UNS #	Country	Specification	Designation	Description	Cu	Fe	Mg	Mn	Ni	Si	Ti	Zn	OE Max	OT Max	Other	UTS	YS	EL
	Australia	AS 1874	EA401	F1 sand	0.1 max	0.4 max	0.05 max	0.1 max	0.05 max	11-13	0.2 max	0.1 max	0.05	0.15		160		5
B413.1 A24131	International		B413.1	Ingot	0.1 max	0.4 max	0.05 max	0.35 max	0.05 max	11-13	0.25 max	0.1 max	0.05	0.2				
B413.0 A24130	International		B413.0	Sand, Perm Mold	0.1 max	0.5 max	0.05 max	0.35 max	0.05 max	11-13	0.25 max	0.1 max	0.05	0.2				
B413.0 A24130	Czech Republic	CSN 424330	AlSi12Mn	Permanent Mold Ingot Alloys	0.1 max	0.6 max	0.1 max	0.1-0.4		11-13	0.15 max	0.15 max						
B413.0 A24130	Czech Republic	CSN 424330	AlSi12Mn	Sand Ingot Alloys	0.1 max	0.6 max	0.1 max	0.1-0.4		11-13	0.15 max	0.15 max						
A413.2 A14132	International		A413.2	Ingot	0.1 max	0.6 max	0.05 max	0.05 max	0.05 max	11-13		0.05 max		0.1	Sn 0.05 max			
413 A04130	UK	BS 1490	LM6		0.1 max	0.6 max	0.1 max	0.5 max	0.1 max	10-13	0.2 max	0.1 max			Pb 0.1 max; Sn 0.05 max			
A413.0 A14130	UK		L75		0.1 max	0.6 max	0.2-0.6	0.3-0.7	0.1 max	10-13	0.2 max	0.1 max			Pb 0.1 max; Sn 0.05 max			
A413.0 A14130	Belgium	NBN P21-101	SGAlSi12	Sand	0.1 max	0.7 max	0.1 max	0.5 max	0.1 max	11-13.5	0.15 max	0.1 max			Pb 0.01 max; Sn 0.05 max			
A413.2 A14132	International	ISO R164	Al-Si12		0.1 max	0.7 max	0.1 max	0.5 max	0.1 max	11-13.5	0.15 max	0.1 max			Pb 0.1; Sn 0.05 max			
	International	ISO 3522	Al-Si12	Sand Cast As mfg	0.1 max	0.7 max	0.1 max	0.5 max	0.1 max	11-13.5	0.2 max	0.1 max			Pb 0.1 max; Sn 0.05 max	150		3
B413.0 A24130	Austria	ONORM M3429	GD AlSi12	Die	0.1 max	1 max	0.05 max	0.4 max		11-13.5	0.15 max	0.1 max			ET 0.05 max; ET=.15 total;			
A413.0 A14130	International	ISO R164	Al-Si12Fe		0.1 max	1.3	0.1 max	0.5 max	0.1 max	11-13.5	0.15 max	0.1 max			Pb 0.1; Sn 0.05 max			
	International	ISO 3522	Al-Si12 Fe	Pressure Die Cast	0.1 max	1.3 max	0.1 max	0.5 max	0.1 max	11-13.5	0.2 max	0.1 max			Pb 0.1 max; Sn 0.05 max			
	Australia	AS 1874	EA401	Ingot	0.15 max	0.6 max	0.1 max	0.5 max	0.1 max	11-13	0.2 max	0.15 max	0.05	0.2	Cr 0.1 max; Pb 0.15 max; Sn 0.05 max			
332 A03320	Sweden	SIS 144244	4244-00		0.2	0.5	0.2-0.4	0.5	0.1	6.5-7.5	0.2	0.3			Sn 0.05 max			
332 A03320	Sweden	SIS 144244	4244-04	Sand ST and Art aged	0.2	0.5	0.2-0.4	0.5	0.1	6.5-7.5	0.2	0.3			Sn 0.05 max		200	2
332 A03320	Sweden	SIS 144244	4244-07	ST and Art aged	0.2	0.5	0.2-0.4	0.5	0.1	6.5-7.5	0.2	0.32			Sn 0.05 max		200	1
A413.0 A14130	Czech Republic	CSN 424330	AlSi12Mn	Die Ingot Alloys	0.2 max	0.9 max	0.5 max	0.1-0.4		11-13	0.15 max	0.3 max						
413 A04130	UK	BS 1490	LM20		0.4 max	1 max	0.2 max	0.5 max	0.1 max	10-13	0.2 max	0.2 max			Pb 0.1 max; Sn 0.1 max			
413 A04130	USA	QQ A-591E	413	Die	0.6	2	0.1	0.35	0.5	11-13		0.5			Sn 0.15 max	296	145	3
413 A04130	Norway	NS 17512	NS17512-01	Sand	0.6	0.7	0.3	0.5	0.2	11-13.5	0.2	0.5			Al 88; Pb 0.1; Sn 0.1 max	147	78	2
A413.0 A14130	Norway	NS 17512	NS17512-02		0.6	0.7	0.3	0.5	0.2	11-13.5	0.2	0.5			Al 88; Pb 0.1; Sn 0.1 max	157	88	2
A413.1 A14131	Norway	NS 17512	NS17512-00		0.6	0.7	0.3	0.5	0.2	11-13.5	0.2	0.5			Al 88; Pb 0.1; Sn 0.1 max			
413 A04130	France	NF A57703	A-S12-Y4		0.6	1.3	0.2	0.3	0.5	11-13.5	0.2	0.5			Sn 0.1 max	170		2
A413.0 A14130	Norway	NS 17512	NS17512-05		0.6	1.3	0.3	0.5	0.2	11-13.5	0.2	0.5			Al 88; Pb 0.1; Sn 0.1 max	196	147	2
A413.1 A14131	International	ISO R164	Al-Si12Cu		0.931 max	0.8	0.3			0.2	11-13.5	0.2	0.5			Pb 0.1; Sn 0.1 max			
A413.1 A14131	International		A413.1	Ingot	1 max	1 max	0.1 max	0.35 max	0.5 max	11-13		0.4 max		0.25	Sn 0.15 max			
A413.0 A14130	International		A413.0	Die	1 max	1.3 max	0.1 max	0.35 max	0.5 max	11-13		0.5 max		0.25	Sn 0.15 max			
413 A04130	International		413	Die	1 max	2 max	0.1 max	0.35 max	0.5 max	11-13		0.5 max		0.25	Sn 0.15 max			
413 A04130	USA	ASTM B85	413.0/S12B	Die	1	2	0.1	0.35	0.5	11-13		0.5			Sn 0.15 max	300	140	3
	International	ISO 3522	Al-Si12 Cu	Sand Cast As mfg	1.2 max	0.9 max	0.3 max	0.5 max	0.3 max	11-13.5	0.2 max	0.5 max			Pb 0.2 max; Sn 0.1 max	150		1
A413.0 A14130	International	ISO R164	Al-Si12CuFe		1.2	1.3	0.3	0.5	0.2	11-13.5	0.2	0.5			Pb 0.1; Sn 0.1 max			

UNS numbers and US grades are provided as a means of cross referencing chemically similar alloys. Exchangability is only possible after independent examination of specifications. Tensile properties are minimum or typical . UTS and YS as Mpa, El as %. See Appendix for list of abbreviations used in Descriptions.

Grade UNS #	Country	Specification	Designation	Description	Cu	Fe	Mg	Mn	Ni	Si	Ti	Zn	OE Max	OT Max	Other	UTS	YS	EL
	International	ISO 3522	Al-Si12 Cu Fe	Pressure Die Cast	1.2 max	1.3 max	0.3 max	0.5 max	0.3 max	11-13.5	0.2 max	0.5 max			Pb 0.2 max; Sn 0.1 max			
413 A04130	Czech Republic	CSN 424337	AlSi10CuMn Mg	Die Ingot Alloys	0.8-1.4	1.5 max	0.4	0.35-0.5	0.2 max	9-11		0.6 max			Sn 0.1 max			
	USA	ASTM B108	F332.0	Cooled and art aged		1.2	0.5-1.5	0.5	0.5	8.5-10.5	0.25	1				214		
332 A03320	Japan	JIS H5202	AC8B		2-4	1	0.5-1.5	0.5	0.5-1.5	8.5-10.5	0.2	0.5				177		
332 A03320	Japan	JIS H5202	AC8B	Temp	2-4	1	0.5-1.5	0.5	0.5-1.5	8.5-10.5	0.2	0.5				206		
332 A03320	Japan	JIS H5202	AC8B	Q/T	2-4	1	0.5-1.5	0.5	0.5-1.5	8.5-10.5	0.2	0.5				275		
332 A03320	Japan	JIS H5202	AC8C		2-4	1	0.5-1.5	0.5		8.5-10.5	0.2	0.5				177		
332 A03320	Japan	JIS H5202	AC8C		2-4	1	0.5-1.5	0.5		8.5-10.5	0.2	0.5				206		
332 A03320	Japan	JIS H5202	AC8C	Q/T	2-4	1	0.5-1.5	0.5		8.5-10.5	0.2	0.5				275		
332 A03320	South Africa	SABS 989	46701 (Al-Si10 Cu3 Mg)	Ingot	2-4	1 max	0.7-1.5	0.5 max	0.5-1.5	8.5-10.5		0.5 max	0.05	0.5	Sn 0.2 max			
332 A03320	International		332	Perm Mold	2-4	1.2 max	0.5-1.5	0.5 max	0.5 max	8.5-10.5	0.25 max	1 max		0.5				
332 A03320	South Africa	SABS 991	46700 (Al-Si10 Cu3 Mg)	T6 Gravity Die	2-4	1.2 max	0.5-1.5	0.5 max	0.5-1.5	8.5-10.5		0.5 max	0.05	0.5	Sn 0.2 max	215		
333 A03330	South Africa	SABS 989	46000 (Al-Si9 Cu3 Fe)	Ingot	2-4	0.6-1.1	0.15-0.55	0.55 max	0.55 max	8-11	0.2	1.2 max	0.05	0.25	Cr 0.15 max; Pb 0.35 max; Sn 0.25 max			
	Denmark	DS 3002	4253	Sand	0.2	0.5	0.2-0.4	0.5	0.1	9-11	0.2	0.3			Pb 0.05; Sn 0.05 max	235	196	1
	Denmark	DS 3002	4253	Chill	0.2	0.5	0.2-0.4	0.5	0.1	9-11	0.2	0.3			Pb 0.05; Sn 0.05 max	255	216	1
333 A03330	Czech Republic	CSN 424339	AlSi8Cu2Mn	Permanent Mold Ingot Alloys	2-3	0.8 max	0.3 max	0.3-0.5	0.3 max	7.5-9.5		0.7 max			Sn 0.2 max			
333 A03330	UK	BS 1490	LM24		3-4	1.3 max	0.1 max	0.5 max	0.5 max	7.5-9.5	0.2 max	3 max			Pb 0.3 max; Sn 0.2 max			
333 A03330	International		333	Perm Mold	3-4	1 max	0.5	0.5 max	0.5 max	8-10	0.25 max	1 max		0.5				
A333.0 A13330	International		A333.0	Perm Mold	3-4	1 max	0.5	0.5 max	0.5 max	8-10	0.25 max	3 max		0.5				
333 A03330	Japan	JIS H5202	AC4B		2-4	1	0.5	0.5	0.3	7-10	0.2	1				177		
333 A03330	Japan	JIS H5202	AC4B	Q/T	2-4	1	0.5	0.5	0.3	7-10	0.2	1				245		
333 A03330	USA	ASTM B108	333	As fabricated	3-4	1	0.5	0.5	0.5	8-10	0.25	1				193		
333 A03330	USA	ASTM B108	333	Cooled and art aged	3-4	1	0.5	0.5	0.5	8-10	0.25	1				207		
333 A03330	USA	ASTM B108	333	SHT and art aged	3-4	1	0.5	0.5	0.5	8-10	0.25	1				241		
333 A03330	USA	QQ A-596d	333	As cast	3-4	1	0.5	0.5	0.5	8-10	0.25	1				193		
333 A03330	USA	QQ A-596d	333	Aged and temp T5	3-4	1	0.5	0.5	0.5	8-10	0.25	1				207		
333 A03330	USA	QQ A-596d	333	SHT and art aged T6	3-4	1	0.5	0.5	0.5	8-10	0.25	1				242		
A03321	Australia	AS 1874	AA305	Ingot	2.4-	0.9 max	0.6-1.5	0.5 max	0.5 max	8.5-10.5	0.25 max	1 max	0.05	0.2	Cr 0.1 max; Pb 0.25 max; Sn 0.15 max			
333.1 A03331	International		333.1	Ingot	3-4	0.8 max	0.1-0.5	0.5 max	0.5 max	8-10	0.25 max	1 max		0.5				
A333.1 A13331	International		A333.1	Ingot	3-4	0.8 max	0.1-0.5	0.5 max	0.5 max	8-10	0.25 max	3 max		0.5				
333.1 A03331	USA	ASTM B179	333.1		3-4	0.8	0.1-0.5	0.5	0.5	8-10	0.25	1						
333.1 A03331	USA	QQ A-371F	333.1	Ingot	3-4	0.8	0.1-0.5	0.5	0.5	8-10	0.25	1		0.5				
A03321	Australia	AS 1874	AA337	Ingot	2-4	0.8 max	0.6-1.5	0.5 max	0.5-1.5	8.5-10.5	0.2 max	0.5 max	0.05	0.2	Cr 0.1 max; Pb 0.15 x; Sn 0.2 x			

UNS numbers and US grades are provided as a means of cross referencing chemically similar alloys. Exchangability is only possible after independent examination of specifications. Tensile properties are minimum or typical. UTS and YS as Mpa, El as %. See Appendix for list of abbreviations used in Descriptions.

Worldwide Guide to Equivalent Nonferrous Metals and Alloys

Grade UNS #	Country	Specification	Designation	Description	Cu	Fe	Mg	Mn	Ni	Si	Ti	Zn	OE Max	OT Max	Other	UTS	YS	EL
A03321	Australia	AS 1874	AA305	T5 permanent mould	2-4	0.9 max	0.6-1.5	0.5 max	0.5 max	8.5-10.5	0.25 max	1 max	0.05	0.2	Cr 0.1 max; Pb 0.25 max; Sn 0.15 max	215		
332.1 A03321	International		332.1	Ingot	2-4	0.9 max	0.6-1.5	0.5 max	0.5 max	8.5-10.5	0.25 max	1 max		0.5				
	USA	ASTM B179	F332.1			0.9	0.6-1.5	0.5	0.5	8.5-10.5	0.25	1						
	Sweden	SIS 144255	4255-00		0.2	0.65	0.1	0.5	0.1	9-11	0.2	0.3			Pb 0.1; Sn 0.05 max			
	Sweden	SIS 144255	4255-03	Sand As cast	0.2	0.7	0.1	0.5	0.1	9-11	0.2	0.3			Pb 0.1; Sn 0.05 max		80	4
	Austria	ONORM M3429	AlSi12(Cu)	Perm Mold Sand	1 max	0.8 max	0.3 max	0.2-0.5	0.2 max	11-13.5	0.15 max	0.5 max			ET 0.05 max ET=.15 total; Pb 0.2 max; Sn 0.1 max			
	Austria	ONORM M3429	G AlSi12(Cu)		1 max	0.8 max	0.3 max	0.2-0.5	0.2 max	11-13.5	0.15 max	0.5 max			ET 0.05 max ET=.15 total; Pb 0.2 max; Sn 0.1 max			
	Austria	ONORM M3429	GK AlSi12(Cu)	Perm Mold	1 max	0.8 max	0.3 max	0.2-0.5	0.2 max	11-13.5	0.15 max	0.5 max			ET 0.05 max ET=.15 total; Pb 0.2 max; Sn 0.1 max			
	Austria	ONORM M3429	GS AlSi12(Cu)	Sand	1 max	0.8 max	0.3 max	0.2-0.5	0.2 max	11-13.5	0.15 max	0.5 max			ET 0.05 max;ET=.15 total; Pb 0.2 max; Sn 0.1 max			
	Austria	ONORM M3429	GD AlSi12(Cu)	Die	1 max	1 max	0.3 max	0.2-0.5	0.2 max	11-13.5	0.15 max	0.5 max			ET 0.05 max ET=.15 total; Pb 0.2 max; Sn 0.1 max			
A03802	Australia	AS 1874	AA313	Ingot	3-4	0.15 max	0.1 max	0.1 max	0.1 max	7.5-9.5	0.2 max	0.1 max	0.05	0.15	Sn 0.2 max			
380.2 A03802	International		380.2	Ingot	3-4	0.7-1.1	0.1 max	0.1 max	0.1 max	7.5-9.5		0.1 max		0.2	Sn 0.1 max			
380.2 A03802	USA	ASTM B179	380.2/SC84C	Ingot	3-4	0.7-1.1	0.1	0.1	0.1	7.5-9.5		0.1		0.2	Sn 0.1 max			
380.2 A03802	USA	QQ A-371F	380.2	Ingot	3-4	0.7-1.1	0.1	0.1	0.1	7.5-9.5		0.1		0.2	Sn 0.1 max			
380.2 A03802	South Africa	SABS 989	Al-Si8Cu4FeA	Ingot pressure Die Chill	3-4	1.2	0.1	0.5	0.5	7.5-9.5	0.2	3		0.15	Pb 0.3; Sn 0.2 max	180		1
	Germany	DIN 1725Sh3	3.0571/V-AlMo10		0.2	0.45	0.5	9-11	0.2	0.4	0.1	0.2			Cr 0.1; Pb 0.1; Sn 0.1			
369 A03690	International		369	Die	0.5 max	1.3 max	0.25-0.45	0.35 max	0.05 max	11-12		1 max	0.05	0.15	Cr 0.3-0.4; Sn 0.1 max			
369.1 A03691	International		369.1	Ingot	0.5 max	1 max	0.3-0.45	0.35 max	0.05 max	11-12		0.9 max	0.05	0.15	Cr 0.3-0.4; Sn 0.1 max			
332.2 A03322	International		332.2	Ingot	2-4	0.6 max	0.9-1.3	0.1 max	0.1 max	8.5-10	0.2 max	0.1 max		0.3				
324 A03240	International		324	Perm Mold	4-6	1.2 max	0.4-0.7	0.5 max	0.3 max	7-8	0.2 max	1 max	0.15	0.2				
324.1 A03241	USA	QQ A-371F	324.1		0.4-0.6	0.9	0.45-0.7	0.5	0.3	7-8	0.2	1						
324.2 A03242	International		324.2	Ingot	4-6	0.6 max	0.45-0.7	0.1 max	0.1 max	7-8	0.2 max	0.1 max	0.05	0.15				
324.1 A03241	International		324.1	Ingot	4-6	0.9 max	0.45-0.7	0.5 max	0.3 max	7-8	0.2 max	1 max	0.15	0.2				
	India	IS 202	A-9	Sand ST PT	0.1	0.6	0.2-0.6	0.3-0.7	0.1	10-13		0.1			Pb 0.1; Sn 0.05 max	239	201	
	India	IS 202	A-9	Sand ST PT	0.1	0.6	0.2-0.6	0.3-0.7	0.1	10-13		0.1			Pb 0.1; Sn 0.05 max	239	201	
	Belgium	NBN P21-101	SGAlSi12CuNi	Sand	0.5-1.5	0.8 max	0.5-1.5	0.5 max	0.5-1.5	10-13	0.2 max	0.2 max			Pb 0.05 max; Sn 0.05 max			
383 A03830	South Africa	SABS 989	Al-Si10Cu2FeA	Chill Ingot	0.7-2.5	0.9	0.3	0.5	1	9-11.5	0.2	1.2		0.15	Pb 0.3; Sn 0.2 max	145		

UNS numbers and US grades are provided as a means of cross referencing chemically similar alloys. Exchangability is only possible after independent examination of specifications. Tensile properties are minimum or typical . UTS and YS as Mpa, El as %. See Appendix for list of abbreviations used in Descriptions.

3-40 Cast Aluminum

Grade UNS #	Country	Specification	Designation	Description	Cu	Fe	Mg	Mn	Ni	Si	Ti	Zn	OE Max	OT Max	Other	UTS	YS	EL
A03830	Australia	AS 1874	AA307	Ingot	0.7-2.5	1 max	0.3 max	0.5 max	0.5 max	9-11.5	0.2 max	2 max	0.05	0.2	Cr 0.1 max; Pb 0.35 x; Sn 0.25 x			
383 A03830	South Africa	SABS 992	Al-Si10Cu2FeA	As cast Pressure Die	0.7-2.5	1	0.3	0.5	1	9-11.5	0.2	1.2		0.15	Pb 0.3; Sn 0.2 max	215	70	2
383 A03830	UK	BS 1490	LM2		0.7-2.5	1 max	0.3 max	0.5 max	0.5 max	9-11.5	0.2 max	2 max			Pb 0.3 max; Sn 0.2 max			
46100 A03830	South Africa	SABS 992	Al-Si11 Cu2	Press Die	1.5-2.5	1.1 max	0.3 max	0.55 max	0.45 max	10-12	0.25 max	1.7 max		0.5	Pb 0.25 max; Sn 0.25 max	240		<1
383.1 A03831	International		383.1	Ingot	2-3	1 max	0.1 max	0.5 max	0.3 max	9.5-11.5		2.9 max		0.5	Sn 0.15 max			
383.1 A03831	USA	ASTM B179	383.1/SC102A		2-3	0.6-1	0.1	0.5	0.3	9.5-11.5		2.9			Sn 0.15 max			
383 A03830	International		383	Die	2-3	1.3 max	0.1 max	0.5 max	0.3 max	9.5-11.5		3 max		0.5	Sn 0.15 max			
383 A03830	USA	ASTM B85	383.0/SC102A	Die	2-3	1.3	0.1	0.5	0.3	9.5-11.5		3			Sn 0.15 max	310	150	4
383 A03830	USA	QQ A-591E	383	Die	2-3	1.3	0.1	0.5	0.3	9.5-11.5		3			Sn 0.15 max	310	152	3
383 A03830	Czech Republic	CSN 424352	AlSi11Cu2Mn	Die Ingot Alloys	1.5-3	1.1 max	0.4 max	0.2-0.5	0.3 max	10-12		0.9 max			Sn 0.2 max			
383 A03830	Japan	JIS H5302	ADC12	Die	1.5-3.5	1.3	0.3	0.5	0.5	9.6-12		1			Sn 0.3 max			
	USA	ASTM B179	A380.2	Ingot	3-4	0.6	0.1	0.1	0.1	7.5-9.5		0.1		0.5				
	USA	QQ A-371F	A380.2		3-4	0.6	0.1	0.1	0.1	7.5-9.5		0.1						
383 A03830	South Africa	SABS 989	46500 (Al-Si9 Cu3 Fe (Zn))	Ingot	2-4	0.6-1.2	0.15-0.55	0.55 max	0.55 max	8-11	0.2 max	3 max	0.05	0.25	Cr 0.15 max; Pb 0.35 max; Sn 0.25 max			
383 A03830	South Africa	SABS 992	46500 (Al-Si9 Cu3 Fe (Zn))	Press Die	2-4	1.3 max	0.05-0.55	0.55 max	0.55 max	8-11	0.25 max	3 max	0.05	0.25	Cr 0.15 max; Pb 0.35 max; Sn 0.25 max	240		<1
	India	IS 202	A-6	Sand	0.1	0.6	0.1	0.5	0.1	10-13		0.1			Pb 0.1; Sn 0.05 max	162	54	5
	India	IS 202	A-6	Sand	0.1	0.6	0.1	0.5	0.1	10-13		0.1			Pb 0.1; Sn 0.05 max	162	54	5
	Australia	AS 1874	BA323	Ingot	3.5-4	0.3 max	0.3-0.4	0.3-0.5	0.05 max	7.5-8.5	0.15 max	0.03 max	0.05	0.15				
	South Africa	SABS 989	Al-Si12MgMnA	Sand SHT and PHT	0.08	0.5	0.25-0.6	0.3-0.7	0.1	10-13	0.2	0.1			Pb 0.1; Sn 0.05 max	240		
	South Africa	SABS 989	Al-Si12MgMnA	Chill SHT and PHT	0.08	0.5	0.25-0.6	0.3-0.7	0.1	10-13	0.2	0.1			Pb 0.1; Sn 0.05 max	295		
413.2 A04132	Sweden	SIS 144260	4260-00			0.7	0.3	0.5	0.2	15.33 max	0.2	0.5			Pb 0.1; Sn 0.1 max			
413.2 A04132	Sweden	SIS 144260	4260-03	Sand As cast		0.7	0.3	0.5	0.2	15.33 max	0.2	0.5			Pb 0.1; Sn 0.1 max	150	80	2
A04132	Sweden	SIS 144260	4260-06	As cast		0.7	0.3	0.5		15.33 max	0.2	0.5			Pb 0.1; Sn 0.1 max		90	2
413.2 A04132	South Africa	SABS 989	Al-Si12A	#NAME?	0.03	0.4	0.05	0.5		10-13	0.15	0.1			ET 0.05 max; .15 total	175		6
413.2 A04132	South Africa	SABS 989	Al-Si12A	#NAME?	0.03	0.4	0.05	0.5		10-13	0.15	0.1			ET 0.05 max; ; ET = .15 total	200		7
413.2 A04132	South Africa	SABS 991	Al-Si12A		0.05	0.5	0.05	0.5		10-13	0.15	0.1			ET 0.05 max; ; ET = .15 total	235		7
413.2 A04132	South Africa	SABS 989	Al-Si12B	Sand	0.08	0.6	0.1	0.5	0.1	10-13	0.2	0.1			ET 0.05 max, .15 total; Pb 0.1; Sn 0.05 max	160		5
413.2 A04132	South Africa	SABS 989	Al-Si12B	Chill	0.08	0.6	0.1	0.5	0.1	10-13	0.2	0.1			ET 0.05 max; ET = .15 total; Pb 0.1; Sn 0.05 max	185		7
	South Africa	SABS 991	Al-Si12MgMnA	PHT	0.1	0.6	0.2-0.6	0.3-0.7	0.1	10-13	0.2	0.1			Pb 0.1; Sn 0.05 max	230		2
	South Africa	SABS 991	Al-Si12MgMnA	SHT and PHT	0.1	0.6	0.2-0.6	0.3-0.7	0.1	10-13	0.2	0.1			Pb 0.1; Sn 0.05 max	295		
413.2 A04132	South Africa	SABS 991	Al-Si12B		0.1	0.7	0.1	0.5	0.1	10-13	0.2	0.1			ET 0.05 max; ET = .15 total; Pb 0.1; Sn 0.05 max	185		7
413.2 A04132	International		413.2	Ingot	0.1 max	0.7-1.1	0.07 max	0.1 max	0.1 max	11-13		0.1 max		0.2	Sn 0.1 max			

UNS numbers and US grades are provided as a means of cross referencing chemically similar alloys. Exchangability is only possible after independent examination of specifications. Tensile properties are minimum or typical . UTS and YS as Mpa, El as %. See Appendix for list of abbreviations used in Descriptions.

Worldwide Guide to Equivalent Nonferrous Metals and Alloys

Grade UNS #	Country	Specification	Designation	Description	Cu	Fe	Mg	Mn	Ni	Si	Ti	Zn	OE Max	OT Max	Other	UTS	YS	EL
413.2 A04132	USA	ASTM B179	413.2/S12C		0.1	0.7-1.1	0.07	0.1	0.1	11-13		0.1			Sn 0.1 max			
413.2 A04132	USA	QQ A-371F	413.2		0.1	0.7-1.1	0.07	0.1	0.1	11-13		0.1			Sn 0.1 max			
413.2 A04132	USA	QQ A-596d	B443.0		0.15	0.8	0.05	0.35		4.5-6	0.25	0.35						
	Japan	JIS H5202	AC3A		0.2	0.8	0.1	0.3		10-13		0.3				177		5
413.2 A04132	South Africa	SABS 989	Al-Si12C	Chill	0.38	0.6	0.15	0.5	0.1	10-13	0.2	0.2			ET 0.05 max; ET = .15 total; Pb 0.1; Sn 0.05 maxvv	185		5
413.2 A04132	South Africa	SABS 991	Al-Si12C		0.4	0.7	0.15	0.5	0.1	10-13	0.2	0.2			ET 0.05 max; ET = .15 total; Pb 0.1; Sn 0.05 max	185		5
413.2 A04132	South Africa	SABS 989	Al-Si12FeA	Chill	0.6	1	0.1	0.5	0.5	10-13	0.15	0.5			Sn 0.15 max	195		3
413.2 A04132	South Africa	SABS 992	Al-Si12FeA	Chill	0.6	1.2	0.1	0.5	0.5	10-13	0.15	0.5		0.2	Sn 0.15 max	205	115	1
413.2 A04132	Sweden	MNC 41E	4260	Die Sand Ingot	1 max					12					Al 88			
	Czech Republic	CSN 424352	AlSi6Cu2	Permanent Mold Ingot Alloys	1.5-3	1.1 max	0.4 max	0.2-0.5	0.3 max	10-12		0.9 max			Sn 0.2 max			
	Japan	JIS H2211	C8CV	Ingot	2-4	0.4	0.6-1.5	0.03	0.03	8.5-10.5	0.2	0.03						
	Czech Republic	CSN 424356	AlSi12Cu	Die Ingot Alloys	0.6-0.9	0.9 max	0.3 max	0.2-0.5	0.2 max	11-13	0.15 max	0.5 max			Sn 0.1 max			
	Japan	JIS H2212	D12V	Ingot	1.5-3.5	0.6	0.03	0.03	0.03	9.6-12		0.03			Sn 0.03 max			
	USA	QQ A-371F	A413.2		0.1	0.6	0.03	0.05	0.05	11-13					Sn 0.05 max			
	USA	QQ A-596d	A108	As cast	4-5	1	0.1	0.5		5-6	0.25	1				165		
	USA	ASTM B179	A413.2		0.1	0.6	0.03	0.05	0.05	11-13		0.05			Sn 0.05 max			
	Japan	JIS H2117	C3AS	Ingot	0.2	0.7	0.1	0.3	0.1	10-13	0.2	0.3						
	South Africa	SABS 989	Al-Si11 Cu2	Ingot	1.5-2.5	0.45-1	0.3 max	0.55 max	0.45 max	10-12	0.2 max	1.7 max		0.5	Pb 0.25 max; Sn 0.25 max			
	South Africa	SABS 989	Al-Si6Cu4MnMgA		3-5	0.8	0.15-0.3	0.3-0.6	0.3	5-7	0.2	2			Pb 0.2; Sn 0.1 max			
	Canada	CSA HA.3	S12N		0.05	0.6		0.3		11-13		0.05	0.05	0.15				
	Australia	AS 1874	CA401	Ingot	0.1 max	0.4 max	0.05 max	0.1 max	0.05 max	12-13	0.2 max	0.1 max	0.05	0.15				
	Australia	AS 1874	CB401	Ingot	0.1 max	0.4 max	0.05 max	0.05 max	0.05 max	12-13	0.2 max	0.1 max	0.05	0.15				
	Australia	AS 1874	CC401	Ingot	0.1 max	0.4 max	0.05 max	0.05 max	0.05 max	12-13	0.2 max	0.1 max	0.05	0.2				
	Sweden	MNC 41E	4261	Die Sand Ingot	0.2 max	0.6 max				12					Al 88			
	Sweden	MNC 41E	4263	Die Ingot	0.2 max	1 max				12					Al 88			
	Norway	NS 17530	NS17530-00		2-4	1.1	0.3	0.5	0.3	7.5-10	0.2	1.2			Al 88; Pb 0.3; Sn 0.2 max			
	Canada	CSA HA.9	S12N	Sand	0.1	0.6		0.3		11-13		0.1	0.05	0.15				
383.2 A03832	International		383.2	Ingot	2-3	0.6-1	0.1 max	0.1 max	0.1 max	9.5-11.5		0.1 max		0.2	Sn 0.1 max			
383.2 A03832	USA	ASTM B179	383.2		2-3	0.6-1	0.1	0.1	0.1	9.5-11.5		0.1			Sn 0.1 max			
A03832	Australia	AS 1874	AA335	Ingot	1.5-3.5	0.9 max	0.3 max	0.5 max	0.5 max	9.6-12	0.2 max	1 max	0.05	0.2	Cr 0.1 max; Pb 0.25 max; Sn 0.3 max			
	South Africa	SABS 990	Al-Si6Cu4MnMgA	Sand	3-5	1	0.1-0.3	0.3-0.6	0.3	5-7	0.2	2			Pb 0.2; Sn 0.1 max	155		1
	South Africa	SABS 991	Al-Si6Cu4MnMgA		3-5	1	0.1-0.3	0.3-0.6	0.3	5-7	0.2	2			Pb 0.2; Sn 0.1 max	170		1

UNS numbers and US grades are provided as a means of cross referencing chemically similar alloys. Exchangability is only possible after independent examination of specifications. Tensile properties are minimum or typical . UTS and YS as Mpa, El as %. See Appendix for list of abbreviations used in Descriptions.

3-42 Cast Aluminum

Grade UNS #	Country	Specification	Designation	Description	Cu	Fe	Mg	Mn	Ni	Si	Ti	Zn	OE Max	OT Max	Other	UTS	YS	EL
	South Africa	SABS 989	Al-Si9A	Sand	1.6	0.8	0.3	0.5	0.4	7-11	0.15	1.2			Pb 0.2; Sn 0.2 max	150		1
	South Africa	SABS 989	Al-Si9A	Chill	1.6	0.8	0.3	0.5	0.4	7-11	0.15	1.2			Pb 0.2; Sn 0.2 max	165		1
	Germany	DIN 1725Sh2	3.2151.0/GK-AlSi6Cu4		3-5	1	0.1-0.3	0.3-0.6	0.3	5-7.5	0.15	2			Pb 0.3; Sn 0.1 max	180	120	1
	Germany	DIN 1725Sh2	3.2151.01/G-AlSi6Cu4	Sand	3-5	1	0.1-0.3	0.3-0.6	0.3	5-7.5	0.15	2			Pb 0.3; Sn 0.1 max	160	100	1
	Denmark	DS 3002	4251	Sand	2-3	0.7	0.3	0.5	0.3	6-8	0.2	2			Pb 0.2; Sn 0.1 max	147	98	2
	Denmark	DS 3002	4251	Chill	2-3	0.7	0.3	0.5	0.3	6-8	0.2	2			Pb 0.2; Sn 0.1 max	167	108	2
	Sweden	SIS 144251	4251-00		2-3	0.7	0.3	0.5	0.3	6-8	0.2	2			Pb 0.2; Sn 0.1 max			
	Sweden	SIS 144251	4251-03	Sand As cast	2-3	0.7	0.3	0.5	0.3	6-8	0.2	2			Pb 0.2; Sn 0.1 max		100	2
	Sweden	SIS 144251	4251-06	As cast	2-3	0.7	0.3	0.5	0.3	6-8	0.2	2			Pb 0.2; Sn 0.1 max		110	2
	Germany	DIN 1725Sh2	3.2581.01/G-AlSi12	Sand	0.05	0.5	0.05	0.4		11-13.5	0.15	0.1						
	Germany	DIN 1725Sh2	3.2581.02/GK-AlSi12		0.05	0.5	0.05	0.4		11-13.5	0.15	0.1						
	Germany	DIN 1725Sh2	3.2581.44/G-AlSi12g	Sand	0.05	0.5	0.05	0.4		11-13.5	0.15	0.1						
	Germany	DIN 1725Sh2	3.2581.45/GK-AlSi12g	Ann Q	0.05	0.5	0.05	0.4		11-13.5	0.15	0.1				180	80	6
	Japan	JIS H2211	C8BV	Ingot	2-4	0.4	0.6-1.5	0.03	0.5-1.5	8.5-10.5	0.2	0.03						
	France	NF A57703	A-G10-Y4		0.2	1.3	8.5-11	0.6	0.1	1	0.2	0.4			Sn 0.1 max			
	USA	ASTM B179	A360.1/SG100A-B		0.6	1	0.45-0.6	0.35	0.5	9-10		0.4			Sn 0.15 max			
	USA	QQ A-371F	A360.1		0.6	1	0.45-0.6	0.35	0.5	9-10		0.4						
	Japan	JIS H2117	C4BS	Ingot	2-4	0.8	0.5	0.5	0.3	7-10	0.2	1						
	South Africa	SABS 711	AlMn10		0.1	0.7		11		0.45		0.07						
	France	NF A57703	A-59G-Y4		0.3	1.3	0.15-0.9	0.5	0.5	9-11	0.2	0.5			Sn 0.1 max	180		1
	Germany	DIN 1725Sh2	3.2161.0/GK-AlSi8Cu3		2-3.5	0.8	0.1-0.3	0.2-0.5	0.3	7.5-9.5	0.15	1.2			Pb 0.2; Sn 0.1 max	170	110	1
	Germany	DIN 1725Sh2	3.2161.01/G-AlSi8Cu3	Sand	2-3.5	0.8	0.1-0.3	0.2-0.5	0.3	7.5-9.5	0.15	1.2	0.05	0.15	Pb 0.2; Sn 0.1 max	160	100	1
	Germany	DIN 1725Sh2	3.2152.0/GD-AlSi6Cu4		3-5	1.3	0.1-0.3	0.3-0.6	0.3	5-7.5	0.15	2			Pb 0.3; Sn 0.1 max			
	USA	QQ A-371F	F332.2		2-4	0.6	0.9-1.3	0.1	0.1	8.5-10	0.2	0.1						
	Finland	SFS 2570	AlSi6Cu4	Die 20 mm diam	3-5	1.3	0.3	0.2-0.6	0.3	5-7	0.2	2			Pb 0.2; Sn 0.1 max	180	120	1
	Australia	AS 1874	AA613	Ingot	0.05 max	0.2 max	0.6-0.7	0.15 max		12-13		0.05 max	0.05	0.15				
	USA	ASTM B85	A360.0/SG100A	Die	0.6	1.3	0.4-0.6	0.35	0.5	9-10		0.5			Sn 0.15 max	320	170	4
	USA	QQ A-591E	A360.0		0.6	1.3	0.4-0.6	0.35	0.5	9-10		0.5			Sn 0.15 max	317	165	3
	South Africa	SABS 990	Al-Si9A	Sand As cast	1.6	1	0.3	0.5	0.4	7-11	0.15	1.5			Pb 0.2; Sn 0.2 max	150		1
	South Africa	SABS 991	Al-Si9A		1.6	1	0.3	0.5	0.4	7-11	0.15	1.5			Pb 0.2; Sn 0.2 max	165		1
	Germany	DIN 1725Sh2	3.2162.0/GD-AlSi8Cu3		2-3.5	0.8	0.3	0.2-0.5	0.3	7.5-9.5	0.15	1.2			Pb 0.2; Sn 0.1 max			
	Denmark	DS 3002	4254	Press Die	3.723 max	1.1	0.3		0.3	7.5-10	0.2	3			Pb 0.3; Sn 0.2 max	245	176	1
	Japan	JIS H2118	D10S	Ingot	2-4	0.9	0.3	0.5	0.5	7.5-9.5		1			Sn 0.3 max			
	Finland	SFS 2566	G-AlSi12	Die 20 mm diam	0.1	0.6	0.1	0.5	0.1	11-13.5	0.15	0.1			Pb 0.1; Sn 0.05 max	180	90	5
	Germany	DIN 1725Sh2	3.2582.05/GD-AlSi12		0.1	1	0.05	0.4		11-13.5	0.15	0.1						
	South Africa	SABS 991	Al-Si12 Ni Mg Cu	T6 Sand	0.8-1.5	0.7 max	0.8-1.5	0.35 max	0.7-1.3	10.5-13.5	0.2 max	0.35 max	0.05	0.15		250		

UNS numbers and US grades are provided as a means of cross referencing chemically similar alloys. Exchangability is only possible after independent examination of specifications. Tensile properties are minimum or typical. UTS and YS as Mpa, El as %. See Appendix for list of abbreviations used in Descriptions.

Worldwide Guide to Equivalent Nonferrous Metals and Alloys

Grade UNS #	Country	Specification	Designation	Description	Cu	Fe	Mg	Mn	Ni	Si	Ti	Zn	OE Max	OT Max	Other	UTS	YS	EL
	South Africa	SABS 991	Al-Si12 Ni Mg Cu	T6 Gravity Die	0.8-1.5	0.7 max	0.8-1.5	0.35 max	0.7-1.3	10.5-13.5	0.2 max	0.35 max	0.05	0.15		275		
385.1 A03851	International		385.1	Ingot	2-4	1.1 max	0.3 max	0.5 max	0.5 max	11-13		2.9 max		0.5	Sn 0.3 max			
385 A03850	International		385	Die	2-4	2 max	0.3 max	0.5 max	0.5 max	11-13		3 max		0.5	Sn 0.3 max			
	Denmark	DS 3002	4261	Sand	0.2	0.6	0.1	0.5	0.1	11-13.5	0.2	0.3			Pb 0.1; Sn 0.05 max	167	78	4
	Denmark	DS 3002	4261	Chill	0.2	0.6	0.1	0.5	0.1	11-13.5	0.2	0.3			Pb 0.1; Sn 0.05 max	176	88	5
	Sweden	SIS 144261	4261-00		0.2	0.6	0.1	0.5	0.1	11-13.5	0.2	0.3			Pb 0.1; Sn 0.05 max			
240.1 A02401	International		240.1	Ingot	7-9	0.4 max	5.5-6.5	0.3-0.7	0.3-0.7	0.5 max	0.2 max	0.1 max	0.05	0.15				
240 A02400	International		240	Sand	7-9	0.5 max	5.5-6.5	0.3-0.7	0.3-0.7	0.5 max	0.2 max	0.1 max	0.05	0.15				
	South Africa	SABS 989	Al-Si12 Ni Mg Cu	Ingot	0.5-1.3	0.7 max	1-1.5	0.5 max	0.7-2.5	11-13	0.2 max	0.1 max	0.05	0.15	Pb 0.1 max; Sn 0.1 max			
	Sweden	SIS 144255	4255-06		0.2	0.7	0.1	0.5	0.1	9-11	0.2	0.3			Al 86.74; Pb 0.1; Sn 0.05 max		90	5
	Austria	ONORM M3429	AlSi12CuMg Ni		0.8-1.5	0.5 max	0.8-1.3	0.2 max	0.8-1.3	11-13	0.2 max	0.2 max			ET 0.05 max; ET=.15 total;			
	Austria	ONORM M3429	G AlSi12CuMg Ni		0.8-1.5	0.5 max	0.8-1.3	0.2 max	0.8-1.3	11-13	0.2 max	0.2 max			ET 0.05 max; ET=.15 total;			
	Austria	ONORM M3429	GK AlSi12CuMg Ni	Perm Mold	0.8-1.5	0.7 max	0.8-1.3	0.2 max		11-13	0.8-1.3	0.2 max			ET 0.05 max; ET=.15 total;			
	Sweden	SIS 144254	4254-00		2-4	1.1	0.3	0.5	0.3	7.5-10	0.2	1.2-3			Pb 0.3; Sn 0.2 max			
339 A03390	Czech Republic	CSN 424336	AlSi12NiCuMg	Permanent Mold Ingot Alloys	0.8-1.5	0.6 max	0.8-1.3	0.3 max	0.8-1.3	11-13	0.2 max	0.15 max						
339 A03390	Czech Republic	CSN 424336	AlSi12NiCuMg	Sand Ingot Alloys	0.8-1.5	0.6 max	0.8-1.6	0.3 max	0.8-1.3	11-13	0.2 max	0.15 max						
339 A03390	International		339	Perm Mold	1.5-3	1.2 max	0.5-1.5	0.5 max	0.5-1.5	11-13	0.25 max	1 max		0.5				
A03840	Australia	AS 1874	AA315	Ingot	3-4.5	0.9 max	0.1 max	0.5 max	0.5 max	10.5-12	0.2 max	1 max	0.05	0.2	Cr 0.1 max; Pb 0.25 max; Sn 0.35 max			
384.1 A03841	International		384.1	Ingot	3-4.5	1 max	0.1 max	0.5 max	0.5 max	10.5-12		2.9 max		0.5	Sn 0.35 max			
A384.1 A13841	International		A384.1	Ingot	3-4.5	1 max	0.1 max	0.5 max	0.5 max	10.5-12		0.9 max		0.5	Sn 0.35 max			
384.1 A03841	USA	ASTM B179	384.1/SC114A		3-4.5	1	0.1	0.5	0.5	10.5-12		2.9			Sn 0.35 max			
384.1 A03841	USA	QQ A-371F	384.1		3-4.5	1	0.1	0.5	0.5	10.5-12	0.35	2.9						
384 A03840	International		384	Die	3-4.5	1.3 max	0.1 max	0.5 max	0.5 max	10.5-12		3 max		0.5	Sn 0.35 max			
A384.0 A13840	International		A384.0	Die	3-4.5	1.3 max	0.1 max	0.5 max	0.5 max	10.5-12		1 max		0.5	Sn 0.35 max			
384 A03840	USA	QQ A-591E	384	Die	3-4.5	1.3	0.1	0.5	0.5	10.5-12		3		0.5	Sn 0.35 max	331	165	3
339.1 A03391	International		339.1	Ingot	1.5-3	0.9 max	0.6-1.5	0.5 max	0.5-1.5	11-13	0.25 max	1 max		0.5				
	Finland	SFS 2568	G-AlSi9Cu3Fe	20 mm diam	2-4	1.3	0.3	0.5	0.3	7.5-10	0.2	1.2			Pb 0.3; Sn 0.2 max	250	180	1
	Norway	NS 17530	NS17530-05	(pres die)	2-4	1.3	0.3	0.5	0.3	7.5-10	0.2	1.2			Al 88; Pb 0.3; Sn 0.2 max	245	176	1
	Sweden	SIS 144254	4254-10	As cast	2-4	1.3	0.3	0.5	0.3	7.5-10	0.2	1.2-3			Pb 0.3; Sn 0.2 max		180	1
	Japan	JIS H2117	C8BS	Ingot	2-4	0.8	0.6-1.5	0.5	0.5-1.5	8.5-10.5	0.2	0.5						
	Japan	JIS H2117	C8CS	Ingot	2-4	0.8	0.6-1.5	0.5	0.5	8.5-10.5	0.2	0.5						

UNS numbers and US grades are provided as a means of cross referencing chemically similar alloys. Exchangability is only possible after independent examination of specifications. Tensile properties are minimum or typical . UTS and YS as Mpa, El as %. See Appendix for list of abbreviations used in Descriptions.

3-44 Cast Aluminum

Grade UNS #	Country	Specification	Designation	Description	Cu	Fe	Mg	Mn	Ni	Si	Ti	Zn	OE Max	OT Max	Other	UTS	YS	EL
	Austria	ONORM M3429	GS AlSi12 CuMgNi	Sand	0.8-1.5	0.7 max	0.8-1.3	0.2 max	0.8-1.3	11.3-13	0.2 max	0.2 max			ET 0.05 max; ET=.15 total;			
	South Africa	SABS 989	Al-Si10Cu3MgNi A	Chill and PHT	2-4	1	0.7-1.5	0.5	0.5-1.5	8.5-10.5		0.5	0.05	0.05	Sn 0.2 max	215		
	South Africa	SABS 991	Al-Si10Cu3MgNi A	PHT	2-4	1.2	0.5-1.5	0.5	0.5-1.5	8.5-10.5		0.5	0.05	0.05	Sn 0.2 max	215		
	Sweden	SIS 144231	4231-03	Sand As cast OBSOLETE	2-4.5	0.884	0.15	0.2-0.7	0.3	4-6.5	0.2	0.5			Al 86.09; Sn 0.05 max		110	2
	Sweden	SIS 144231	4231-06	As cast OBSOLETE	2-4.5	0.884	0.15	0.2-0.7	0.3	4-6.5	0.2	0.5			Al 86.09; Sn 0.05 max		120	2
	USA	QQ A-371F	A413.1		0.6	1	0.1	0.35	0.5	11-13		0.4			Sn 0.15 max			
853.2 A08532	International		853.2	Ingot	3-4	0.5 max		0.1 max		5.5-6.5	0.2 max			0.3	Sn 5.5-7			
853 A08530	International		853	Sand, Perm Mold	3-4	0.7 max		0.5 max		5.5-6.5	0.2 max			0.3	Sn 5.5-7			
	Norway	NS 17532	NS17532-00		2-4	1.1	0.3	0.5	0.3	7.5-10	0.2	3			Al 86; Pb 0.3; Sn 0.2 max			
	Denmark	DS 3002	4260	Sand	0.6	0.7	0.3	0.5	0.2	11-13.5	0.2	0.5			Pb 0.1; Sn 0.1 max	147	78	2
	Denmark	DS 3002	4260	Chill	0.6	0.7	0.3	0.5	0.2	11-13.5	0.2	0.5			Pb 0.1; Sn 0.1 max	157	88	2
384.2 A03842	International		384.2	Ingot	3-4.5	0.6-1	0.1 max	0.1 max	0.1 max	10.5-12		0.1 max		0.2	Sn 0.1 max			
384.2 A03842	USA	ASTM B179	384.2		3-4.5	0.6-1	0.1	0.1	0.1	10.5-12		0.1			Sn 0.1 max			
384.2 A03842	USA	QQ A-371F	384.2		3-4.5	0.6-1	0.5-0.65	0.1		10.5-12	0.2	0.1						
	Germany	DIN 1725Sh2	3.2583/GK-AlSi12(Cu)		1	0.8	0.3	0.2-0.5	0.2	11-13.5	0.15	0.5			Sn 0.1 max	180	90	2
336 A03360	Japan	JIS H5202	AC8A		0.8-1.3	0.8	0.7-1.3	0.1	1-2.5	11-13	0.2	0.1				177		
336 A03360	Japan	JIS H5202	AC8A	Temp	0.8-1.3	0.8	0.7-1.3	0.1	1-2.5	11-13	0.2	0.1				216		
336 A03360	Japan	JIS H5202	AC8A	Q/T	0.8-1.3	0.8	0.7-1.3	0.1	1-2.5	11-13	0.2	0.1				275		
336 A03360	South Africa	SABS 991	Al-Si12Ni2MgCu A	Gravity Die SHT and PHT	0.5-1.3	0.8	0.7-1.5	0.5	0.7-2.5	11-13	0.2	1	0.05	0.15	Sn 0.1 max	275		
336 A03360	South Africa	SABS 991	Al-Si12Ni2MgCu A	Gravity Die Ful HT and Stab	0.5-1.3	0.8	0.7-1.5	0.5	0.7-2.5	11-13	0.2	1	0.05	0.15	Sn 0.1 max	200		
336 A03360	Russia	GOST 2685	AL30		0.8-1.5	0.7 max	0.8-1.3	0.2 max	0.8-1.3	11-13	0.2 max	0.2 max		1.2	Pb 0.05 max; Sn 0.01 max			
336 A03360	International		336	Perm Mold	0.5-1.5	1.2 max	0.7-1.3	0.35 max	2-3	11-13	0.25 max	0.35 max	0.05					
336 A03360	USA	QQ A-596d	336	Art aged only and temp T551	0.5-1.5	1.3	0.35	0.7-1.8	2-3	11-13	0.25	0.35				214		
336 A03360	USA	QQ A-596d	336	SHT and art aged	0.5-1.5	1.3	0.35	0.7-1.8	2-3	11-13	0.25	0.35				276		
336 A03360	Russia	GOST 2685	AL25		1.5-3	0.8 max	0.8-1.3	0.3-0.6	0.8-1.3	11-13	0.05-0.2	0.5 max		1.2	Cr 0.2 max; Pb 0.1 max; Sn 0.05 max			
	USA	QQ A-371F	F332.1		2-4	0.9	0.6-1.5	0.5	0.5	8.5-10.5	0.25	1						
	Japan	JIS H2211	C8AV	Ingot	0.8-1.3	0.4	0.8-1.3	0.03	1-2.5	11-13	0.2	0.03						
A03361	Australia	AS 1874	AA319	T5 permanent mould	0.8-1.3	0.7 max	0.8-1.3	0.35 max	1-2.5	11-13	0.2 max	0.25 max	0.05	0.2	Cr 0.1 max; Pb 0.15 max; Sn 0.2 max	215		
A03361	Australia	AS 1874	AA319	Ingot	0.8-1.3	0.8 max	0.8-1.3	0.35 max	1-2.5	11-13	0.2 max	0.25 max	0.05	0.2	Cr 0.1 max; Pb 0.15 max; Sn 0.2 max			
336.1 A03361	South Africa	SABS 989	Al-Si12Ni2MgCu A	Chill, SHT and PHT	0.5-1.3	0.884	1-1.5	0.5	0.7-2.5	11-13	0.2	0.1			Sn 0.1 max	275		

UNS numbers and US grades are provided as a means of cross referencing chemically similar alloys. Exchangability is only possible after independent examination of specifications. Tensile properties are minimum or typical . UTS and YS as Mpa, El as %. See Appendix for list of abbreviations used in Descriptions.

Worldwide Guide to Equivalent Nonferrous Metals and Alloys

Grade UNS #	Country	Specification	Designation	Description	Cu	Fe	Mg	Mn	Ni	Si	Ti	Zn	OE Max	OT Max	Other	UTS	YS	EL
336.1 A03361	International		336.1	Ingot	0.5-1.5	0.9 max	0.8-1.3	0.35 max	2-3	11-13	0.25 max	0.35 max	0.05					
	Japan	JIS H2118	D12S	Ingot	1.5-3.5	0.9	0.3	0.5	0.5	9.6-12		1			Sn 0.3 max			
	USA	QQ A-591E	A413.0	Die	0.6	1.3	0.1	0.35	0.5	11-13		0.5			Sn 0.15 max	290	131	3
	USA	ASTM B179	A413.1/S12A-B		1	1	0.1	0.35	0.5	11-13		0.4			Sn 0.15 max			
	Germany	DIN 1725Sh2	3.2583/G-AlSi12(Cu)	Sand	1	0.8	0.3	0.2-0.5	0.2	11-13.5	0.15	0.5			Pb 0.2; Sn 0.1 max			
336.2 A03362	Canada	CSA	SN122		0.5-1.5	0.6 max	0.9-1.3		2-3	11-13	0.2 max				ET 0.05 max; ET=.10 total;			
336.2 A03362	International		336.2	Ingot	0.5-1.5	0.9 max	0.9-1.3	0.1 max	2-3	11-13	0.2 max	0.1 max	0.05	0.15				
	USA	QQ A-371F	A240.1		7-9	0.4	5.6-6.5	0.3-0.7	0.3-0.7	0.5	0.2	0.1						
	USA	QQ A-596d	F132.0	Art aged only and temper T5	2-4	1.2	0.5-1.5	0.5	0.5	8.5-10.5	0.25	1				214		
	Germany	DIN 1725Sh2	3.2982/GD-AlSi12(Cu)		1	1.3	0.3	0.2-0.5	0.2	11-13.5	0.15	0.5			Sn 0.1 max			
	Japan	JIS H2117	C8AS	Ingot	0.8-1.3	0.7	0.8-1.3	0.1	1-2.5	11-13	0.2	0.1						
	USA	ASTM B85	A413.0/S12A	Die	1	1.3	0.1	0.35	0.5	11-13		0.5			Sn 0.15 max	290	130	4
	Finland	SFS 2565	G-AlSi12Cu	Die 20 mm diam	1.2	0.9		0.5	0.3	11-13.5	0.2	0.5			Pb 0.2; Sn 0.1 max	180	90	2
	South Africa	SABS 989	Al-Si12Cu2MgMnA	Ingot Chill SHT and PHT	1-2	0.7	0.45-1	0.5-0.9	0.05	11-12.5	0.25	1				290		
	South Africa	SABS 991	Al-Si12Cu2MgMnA	Gravity Die PHT	1-2	0.9	0.4-1	0.5-0.9	0.05	11-12.5	0.25	1		0.5		220		
	South Africa	SABS 991	Al-Si12Cu2MgMnA	Gravity Die SHT and PHT	1-2	0.9	0.4-1	0.5-0.9	0.05	11-12.5	0.25	1		0.5		290		
	USA	ASTM B179	A380.1/SC84A-B	Ingot	3-4	1	0.1	0.5	0.5	7.5-9.5		2.9		0.5	Sn 0.35 max			
	USA	QQ A-371F	A380.1	Ingot	3-4	1	0.1	0.5	0.5	7.5-9.5		2.9		0.5	Sn 0.35 max			
	India	IS 7793	4658/49582	Chill	0.8-1.5	0.8	0.8-1.3	0.2	1.5	11-13	0.2	0.35			Pb 0.05; Sn 0.05 max	195		
	Canada	CSA HA.3	SN122		0.5-1.5	0.9	0.9-1.3	0.1	2-3	11-13	0.2	0.1						
	USA	ASTM B179	A332.2		0.5-1.5	0.9	0.9-1.3	0.1	2-3	11-13	0.2	0.1						
	USA	QQ A-371F	A332.2		0.5-1.5	0.9	0.9-1.3	0.1	2-3	11-13	0.2	0.1						
	USA	ASTM B85	A380.0/SC84.A	Die As cast	3-4	1.3	0.1	0.5	0.5	7.5-9.5		3			Sn 0.35 max	320	160	4
	USA	QQ A-591E	A380.0	Die	3-4	1.3	0.1	0.5	0.5	7.5-9.5		3			Sn 0.35 max	324	159	3
238.2 A02382	USA	ASTM B179	238.2		9.5-10.5	1.2	0.2-0.35	0.5	0.5	3.5-4.5	0.2	0.5						
238.2 A02382	USA	QQ A-371F	238.2		9.5-10.5	1.2	0.2-0.35	0.5	0.5	3.5-4.5	0.2	0.5						
	USA	ASTM B179	A332.1		0.5-1.5	0.9	0.8-1.3	0.35	2-3	11-13	0.25	0.35						
	USA	QQ A-371F	A332.1		0.5-1.5	0.9	0.8-1.3	0.35	2-3	11-13	0.25	0.35						
	Canada	CSA HA.10	SN122	Mold T5A	0.5-1	1.2	0.7-1.3	0.35	2-3	11-13	0.25	0.35	0.05	0.15		214		
	Canada	CSA HA.10	SN122	Mold T6A	0.5-1	1.2	0.7-1.3	0.35	2-3	11-13	0.25	0.35	0.05	0.15		276		
	USA	ASTM B108	A322.0	T 551	0.5-1.5	1.2	0.7-1.3	0.35	2-3	11-13	0.25	0.35				214		
	USA	ASTM B108	A322.0	T 65	0.5-1.5	1.2	0.7-1.3	0.35	2-3	11-13	0.25	0.35				276		
238.1 A02381	USA	ASTM B179	238.1		9-11	1.2	0.2-0.35	0.6	1	3.5-4.5	0.25	1.5						
238.1 A02381	USA	QQ A-371F	238.1		9-11	1.2	0.2-0.35	0.6	1	3.5-4.5	0.25	1.5						
238 A02380	USA	ASTM B108	238	as fabricated	9-11	1.5	0.15-0.35	0.6	1	3.5-4.5	0.25	1.5						

UNS numbers and US grades are provided as a means of cross referencing chemically similar alloys. Exchangability is only possible after independent examination of specifications. Tensile properties are minimum or typical . UTS and YS as Mpa, El as %. See Appendix for list of abbreviations used in Descriptions.

3-46 Cast Aluminum

Grade UNS #	Country	Specification	Designation	Description	Cu	Fe	Mg	Mn	Ni	Si	Ti	Zn	OE Max	OT Max	Other	UTS	YS	EL
	Australia	AS 1874	AC329	Ingot	1.9-2.2	0.3 max	0.5-0.7	0.35-0.5	1.7-2.2	13.5-14.3	0.1 max		0.05	0.15				
	Australia	AS 1874	AC331	Ingot	1.9-2.2	0.6-1	0.5-0.7	0.35-0.5	1-1.4	13.8-14.5	0.1 max		0.05	0.15				
	Germany	DIN 1725Sh3	3.2292/VR-AlSi20		0.05	0.03	0.05	0.1	0.05	18-21	0.05	0.1			Cr 0.05			
392.1 A03921	South Africa	SABS 711	AlSi20		0.1	0.7				18-22		0.07						
392.1 A03921	International		392.1	Ingot	0.4-0.8	1.1 max	0.8-1.2	0.2-0.6	0.5 max	18-20	0.2 max	0.4 max	0.15	0.5	Sn 0.3 max			
392 A03920	International		392	Die	0.4-0.8	1.5 max	0.8-1.2	0.2-0.6	0.5 max	18-20	0.2 max	0.5 max	0.15	0.5	Sn 0.3 max			
	South Africa	SABS 991	Al-Si18 Ni Mg Cu	T6 Sand	0.8-1.5	0.7 max	0.8-1.3	0.2 max	0.8-1.3	17-19	0.2 max	0.2 max	0.05	0.15		200		
	South Africa	SABS 991	Al-Si18 Ni Mg Cu	T6 Gravity Die	0.8-1.5	0.7 max	0.8-1.3	0.2 max	0.8-1.3	17-19	0.2 max	0.2 max	0.05	0.15		225		
	South Africa	SABS 989	Al-Si18 Ni Mg Cu	Ingot	0.8-1.4	0.6 max	1-1.3	0.2 max	0.8-1.3	17-19	0.2 max	0.2 max	0.05	0.15				
	South Africa	SABS 991	Al-Si18CuMgNi		0.8-1.5	0.7		0.2	0.8-1.3	17-19	0.2	0.2						
A390.0 A13900	International		A390.0	Sand, Perm Mold	4-5	0.5 max	0.45-0.65	0.1 max		16-18	0.2 max	0.1 max	0.1	0.2				
390 A03900	International		390	Die	4-5	1.3 max	0.45-0.65	0.1 max		16-18	0.2 max	0.1 max	0.1	0.2				
B390.0 A23900	International		B390.0	Die	4-5	1.3 max	0.45-0.65	0.5 max	0.1 max	16-18	0.2 max	1.5 max	0.1	0.2				
A390.1 A13901	International		A390.1	Ingot	4-5	0.4 max	0.5-0.65	0.1 max		16-18	0.2 max	0.1 max	0.1	0.2				
B390.1 A23901	International		B390.1	Ingot	4-5	1 max	0.5-0.65	0.5 max	0.1 max	16-18	0.2 max	1.4 max	0.1	0.2				
	USA	ASTM B327	CG71A	Bar	17-19	0.7	0.65-0.95	0.5	0.2	0.6		1			Sn 0.2 max			
	India	IS 7793	4928-A/49285	Chill	0.8-1.5	0.7	0.8-1.3	0.2	0.8-1.3	17-19	0.2	0.2			Pb 0.05; Sn 0.05 max	175		
	South Africa	SABS 989	Al-Si18CuMgNi A	Chill SHT and PHT	0.8-1.4	0.6	1-1.3	0.2	0.8-1.3	17-19	0.2	0.2	0.05	0.05		225		
390.2 A03902	International		390.2	Ingot	4-5	0.6-1	0.5-0.65	0.1 max		16-18	0.2 max	0.1 max	0.1	0.2				
390.2 A03902	USA	QQ A-371F	390.2		4-5	0.6-1	0.5-0.65	0.1		16-18	0.2	0.1						
	USA	QQ A-371F	A390.1		4-5	0.4	0.5-0.65	0.1		16-18	0.2	0.1						
	Sweden	MNC 41E	4282	Die Ingot		0.3 max	0.5			17					Al 78; Cr 4.5			
	Sweden	MNC 41E	4283	Die Ingot	4.5	1.3 max	0.5			17					Al 78			
393 A03930	International		393	Sand, Perm Mold, Die	0.7-1.1	1.3 max	0.7-1.3	0.1 max	2-2.5	21-23	0.1-0.2	0.1 max	0.05	0.15	Vanadium 0.08-0.15			
393 A03930	Czech Republic	CSN 424386	AlSi20Cu2NiMgMn	Permanent Mold Ingot Alloys	1.5-2	0.6 max	0.75-1.1	0.1-0.4	0.5-1	19-22	0.2 max	0.1 max						
393.2 A03932	International		393.2	Ingot	0.7-1.1	0.8 max	0.8-1.3	0.1 max	2-2.5	21-23	0.1-0.2	0.1 max	0.05	0.15	Vanadium 0.08-0.15			
393.1 A03931	International		393.1	Ingot	0.7-1.1	1 max	0.8-1.3	0.1 max	2-2.5	21-23	0.1-0.2	0.1 max	0.05	0.15	Vanadium 0.08-0.15			
	South Africa	SABS 991	Al-Si24 Ni Mg Cu	T6 Sand	0.8-1.5	0.7 max	0.8-1.3	0.2 max	0.8-1.3	23-26	0.2 max	0.2 max	0.05	0.15	Cr 0.3-0.6	150		
	South Africa	SABS 991	Al-Si24 Ni Mg Cu	T6 Gravity Die	0.8-1.5	0.7 max	0.8-1.3	0.2 max	0.8-1.3	23-26	0.2 max	0.2 max	0.05	0.15	Cr 0.3-0.6	175		
	South Africa	SABS 989	Al-Si24 Ni Mg Cu	Ingot	0.8-1.4	0.6 max	1-1.3	0.2 max	0.8-1.3	23-26	0.2 max	0.2 max	0.05	0.15				
	India	IS 7793	4928-B	Chill	0.8-1.5	0.7	0.8-1.3	0.2	0.8-1.3	23-26	0.2	0.2			Cr 0.3-0.6; Pb 0.05; Sn 0.05 max	165		
	South Africa	SABS 989	Al-Si24CuMgNiCrA	Chill SHT and PHT	0.8-1.4	0.6	1-1.3	0.2	0.8-1.3	23-26	0.2	0.2				175		

UNS numbers and US grades are provided as a means of cross referencing chemically similar alloys. Exchangability is only possible after independent examination of specifications. Tensile properties are minimum or typical . UTS and YS as Mpa, El as %. See Appendix for list of abbreviations used in Descriptions.

Worldwide Guide to Equivalent Nonferrous Metals and Alloys

Grade UNS #	Country	Specification	Designation	Description	Cu	Fe	Mg	Mn	Ni	Si	Ti	Zn	OE Max	OT Max	Other	UTS	YS	EL
	South Africa	SABS 991	Al-Si24CuMgNiCr	SHt and PHT	0.8-1.5	0.7	0.8-1.3	0.2	1.3-1.3	23-26	0.2	0.2				175		
	South Africa	SABS 711	AlCu30		33-33	0.4				0.3		0.07						
	Germany	DIN 1725Sh3	3.1191/V-AlCu50		48-52	0.45	0.3	0.35	0.2	0.4	0.1	0.3			Cr 0.1; Pb 0.2; Sn 0.1			
	USA	ASME SB108		See ASTM B108														
	USA	ASME SB26		See ASTM B26														

UNS numbers and US grades are provided as a means of cross referencing chemically similar alloys. Exchangability is only possible after independent examination of specifications. Tensile properties are minimum or typical . UTS and YS as Mpa, El as %. See Appendix for list of abbreviations used in Descriptions.

3-48 Cast Aluminum

Grade UNS #	Country	Specification	Designation	Comment	Al	Cu	Fe	Mn	Ni	P	Pb	Sn	Zn	OT max	Other	UTS	YS	El
C10100	Australia	AS 1567	101	Wrought Bar Rod Hard 10/12 mm diam		99.99										280		6
C10100	Australia	AS 1567	101	Wrought Bar Rod 1/2 hard 25/50 mm diam		99.99										230		22
C10100	Australia	AS 1567	101	Wrought Bar Rod Soft Ann 12/50 mm diam		99.99										230		45
C10100	Japan	JIS H3510	C1011	Wrought Sh Plt Strp Pip Tub Bar Rod 1/2 hard .3/12 mm diam		99.98					0.001				Te 0.001; S 0.002; O 0.001; Bi 0.001 max	245		15
C10100	Japan	JIS H3510	C1011	Wrought Sh Plt Strp Pip Tub Bar Rod Ann .3/12 mm diam		99.98					0.001				Te 0.001; S 0.002; O 00.01; Bi 0.001 max	196		40
C10100	Japan	JIS H3510	C1011	Wrought Sh Plt Strp Pip Tub Bar Rod HArd .3/10 mm diam		99.98					0.001				Te 0.001; S 0.002; O 0.001; Bi 0.001 max	275		
Oxygen Free Copper C10100	USA	ASTM B133	C10100	Wrought Rod Sh Tub Pip		99.99 min												
Oxygen Free Copper C10100	USA	ASTM B152	C10100	Wrought Rod Sh Tub Pip		99.99 min												
Oxygen Free Copper C10100	USA	ASTM B187	C10100	Wrought Rod Sh Tub Pip		99.99 min												
Oxygen Free Copper C10100	USA	ASTM B188	C10100	Wrought Rod Sh Tub Pip		99.99 min												
Oxygen Free Copper C10100	USA	ASTM B2	C10100	Wrought Rod Sh Tub Pip		99.99 min												
Oxygen Free Copper C10100	USA	ASTM B272	C10100	Wrought Rod Sh Tub Pip		99.99 min												
Oxygen Free Copper C10100	USA	ASTM B280	C10100	Wrought Rod Sh Tub Pip		99.99 min												
Oxygen Free Copper C10100	USA	ASTM B3	C10100	Wrought Rod Sh Tub Pip		99.99 min												
Oxygen Free Copper C10100	USA	ASTM B372	C10100	Wrought Rod Sh Tub Pip		99.99 min												
Oxygen Free Copper C10100	USA	ASTM B42	C10100	Wrought Rod Sh Tub Pip		99.99 min												
Oxygen Free Copper C10100	USA	ASTM B432	C10100	Wrought Rod Sh Tub Pip		99.99 min												

UNS numbers and US grades are provided as a means of cross referencing chemically similar alloys. Exchangability is only possible after independent examination of specifications. Tensile properties are minimum or typical . UTS and YS as Mpa, El as %. See Appendix for list of abbreviations used in Descriptions.

Worldwide Guide to Equivalent Nonferrous Metals and Alloys

Grade UNS #	Country	Specification	Designation	Comment	Al	Cu	Fe	Mn	Ni	P	Pb	Sn	Zn	OT max	Other	UTS	YS	El
Oxygen Free Copper C10100	USA	ASTM B48	C10100	Wrought Rod Sh Tub Pip		99.99 min												
Oxygen Free Copper C10100	USA	ASTM B49	C10100	Wrought Rod Sh Tub Pip		99.99 min												
Oxygen Free Copper C10100	USA	ASTM B68	C10100	Wrought Rod Sh Tub Pip		99.99 min												
Oxygen Free Copper C10100	USA	ASTM B75	C10100	Wrought Rod Sh Tub Pip		99.99 min												
Oxygen Free Copper C10100	USA	ASTM F68	C10100	Wrought Rod Sh Tub Pip		99.99 min												
C10100	USA	CDA	OFE	Oxygen free electronic						0.003 max					Sb 0.0004 max; Te 0.0002 max; As 0.0005 max; Cu+Ag 99.99 min; Max as ppm: Bi 1, Cd 1, Fe 10, Pb 5, Mn 0.5, Hg 1, Ni 10, O 5, Se 3, Ag 25, S 15, Sn 2, Zn 1			
C10100	USA			Oxygen free electronic						0.003 max					Sb 0.0004 max; Te 0.0002 max; As 0.0005 max; Cu+Ag 99.99 min; Max as ppm: Bi 1, Cd 1, Fe 10, Pb 5, Mn 0.5, Hg 1, Ni 10, O 5, Se 3, Ag 25, S 15, Sn 2, Zn 1			
High conductivit Cu C14415	India	IS 6912	High conductivity Cu	Wrought Frg Ann 6 mm diam						0.01					ST 99.9 min; ST = Cu+Ag; Bi 0.002 max			
	USA											0.1-0.15			Cu+Ag+Sn 99.96 min			
	France	NF A53301	Cu/c2	Wrought Bar Shapes 3/4 hard 60 mm diam Obsolete Replaced by NF A51118											ST 99.96 min; ST=Cu+Ag			10
	France	NF A53301	Cu/c2	Wrought Bar Shapes 1/2 hard 70 mm diam Obsolete Replaced by NF A51118											ST 99.96 min; ST=Cu+Ag			15
	France	NF A53301	Cu/c2	Wrought Bar Shapes 1/4 hard all diam Obsolete Replaced by NF A51118											ST 99.96 min; ST=Cu+Ag			20
C10200	Australia	AS 1567	102	Wrought Bar Rod Hard 10/12 mm diam		99.95										280		6

UNS numbers and US grades are provided as a means of cross referencing chemically similar alloys. Exchangability is only possible after independent examination of specifications. Tensile properties are minimum or typical . UTS and YS as Mpa, El as %. See Appendix for list of abbreviations used in Descriptions.

4-2 Wrought Copper

Grade UNS #	Country	Specification	Designation	Comment	Al	Cu	Fe	Mn	Ni	P	Pb	Sn	Zn	OT max	Other	UTS	YS	El
C10200	Australia	AS 1567	102	Wrought Bar Rod 1/2 hard 25/50 mm diam		99.95										230		22
C10200	Australia	AS 1567	102	Wrought Bar Rod Soft Ann 12/50 mm diam		99.95										230		45
C10200	Canada	CSA	Cu-OF(102)			99.92 max												
C10200	Czech Republic	CSN 423000	Cu99,95			99.95 min	0.005 max		0.002 max	0.003 max	0.005 max	0.002 max	0.003 max		Bi 0.002; Sb 0.002; S 0.005; As 0.002 max			
C10200	Czech Republic	CSN 423002	Cu99,95			99.95 min	0.01 max		0.002 max	0.002 max	0.005 max	0.002 max	0.003 max		Bi 0.002; Sb 0.002; S 0.005; O 0.0035; Ag 0.003; As 0.002 max			
C10200	France	NF A53301	Cu/cl	Wrought Bar Shapes 3/4 hard 60 mm diam Obsolete Replaced by NF A51118											ST 99.92 min; ST=Cu+Ag			10
C10200	France	NF A53301	Cu/cl	Wrought Bar Shapes 1/2 hard 70 mm diam Obsolete Replaced by NF A51118											ST 99.92 min; ST=Cu+Ag			15
C10200	France	NF A53301	Cu/cl	Wrought Bar Shapes 1/4 hard all diam Obsolete Replaced by NF A51118											ST 99.92 min; ST=Cu+Ag			20
C10200	International	ISO R1337	Cu-OF	Wrought Plt Sh Strp Rod Bar Tub Wir Frg		99.95									Cu includes Ag			
C10200	Japan	JIS H3100	C1020	Wrought Sh Plt 1/2 hard .3/20 mm diam		99.96										245		15
C10200	Japan	JIS H3100	C1020	Wrought Sh Plt Ann .3/30 mm diam		99.96										196		35
C10200	Japan	JIS H3100	C1020	Wrought Sh Plt Hard .3/10 mm diam		99.96										275		
C10200	Japan	JIS H3140	C1020	Wrought Bar 1/2 hard 2/20 mm diam		99.96										245		15
C10200	Japan	JIS H3140	C1020	Wrought Bar Ann 2/30 mm diam		99.96										196		35
C10200	Japan	JIS H3140	C1020	Wrought Bar Hard 2/20 mm diam		99.96										275		
C10200	Japan	JIS H3250	C1020	Wrought Rod As Mfg 6 mm diam		99.96										196		25
C10200	Japan	JIS H3250	C1020	Wrought Rod Ann 6/25 mm diam		99.96										196		30
C10200	Japan	JIS H3250	C1020	Wrought Rod HArd 6/25 mm diam		99.96										275		
C10200	Japan	JIS H3300	C1020	Wrought Pip Ann 4/100 mm diam		99.96										206		40

UNS numbers and US grades are provided as a means of cross referencing chemically similar alloys. Exchangability is only possible after independent examination of specifications. Tensile properties are minimum or typical. UTS and YS as Mpa, El as %. See Appendix for list of abbreviations used in Descriptions.

Worldwide Guide to Equivalent Nonferrous Metals and Alloys

Grade UNS #	Country	Specification	Designation	Comment	Al	Cu	Fe	Mn	Ni	P	Pb	Sn	Zn	OT max	Other	UTS	YS	El
C10200	Japan	JIS H3300	C1020	Wrought Pip 1/2 hard 4/100 mm diam		99.96										245		
C10200	Japan	JIS H3300	C1020	Wrought Pip Hard 25 mm diam		99.96										314		
C10200	UK	BS 2873	C103	Wrought Wir		99.95					0.01				Bi 0.002 max			
C10200	USA	CDA	OF	Oxygen free; High conductivity ann											Cu+Ag 99.95 min			
C10200	USA			Oxygen free; High conductivity ann											Cu+Ag 99.95 min			
Oxygen-Free Copper C10300	USA	ASTM B111	OFXLP	Wrought Rod Plt Sh Strp Tub Pip Shapes		99.95 min				0.01 max				0.05				
Oxygen-Free Copper C10300	USA	ASTM B12	OFXLP	Wrought Rod Plt Sh Strp Tub Pip Shapes		99.95 min				0.01 max				0.05				
Oxygen-Free Copper C10300	USA	ASTM B133	OFXLP	Wrought Rod Plt Sh Strp Tub Pip Shapes		99.95 min				0.01 max				0.05				
Oxygen-Free Copper C10300	USA	ASTM B152	OFXLP	Wrought Rod Plt Sh Strp Tub Pip Shapes		99.95 min				0.01 max				0.05				
Oxygen-Free Copper C10300	USA	ASTM B187	OFXLP	Wrought Rod Plt Sh Strp Tub Pip Shapes		99.95 min				0.01 max				0.05				
Oxygen-Free Copper C10300	USA	ASTM B188	OFXLP	Wrought Rod Plt Sh Strp Tub Pip Shapes		99.95 min				0.01 max				0.05				
Oxygen-Free Copper C10300	USA	ASTM B251	OFXLP	Wrought Rod Plt Sh Strp Tub Pip Shapes		99.95 min				0.01 max				0.05				
Oxygen-Free Copper C10300	USA	ASTM B272	OFXLP	Wrought Rod Plt Sh Strp Tub Pip Shapes		99.95 min				0.01 max				0.05				
Oxygen-Free Copper C10300	USA	ASTM B280	OFXLP	Wrought Rod Plt Sh Strp Tub Pip Shapes		99.95 min				0.01 max				0.05				
Oxygen-Free Copper C10300	USA	ASTM B302	OFXLP	Wrought Rod Plt Sh Strp Tub Pip Shapes		99.95 min				0.01 max				0.05				
Oxygen-Free Copper C10300	USA	ASTM B306	OFXLP	Wrought Rod Plt Sh Strp Tub Pip Shapes		99.95 min				0.01 max				0.05				
Oxygen-Free Copper	USA	ASTM B359	OFXLP	Wrought Rod Plt Sh Strp Tub Pip Shapes		99.95 min				0.01 max				0.05				
Oxygen-Free Copper	USA	ASTM B372	OFXLP	Wrought Rod Plt Sh Strp Tub Pip Shapes		99.95 min				0.01 max				0.05				
Oxygen-Free Copper	USA	ASTM B395	OFXLP	Wrought Rod Plt Sh Strp Tub Pip Shapes		99.95 min				0.01 max				0.05				

UNS numbers and US grades are provided as a means of cross referencing chemically similar alloys. Exchangability is only possible after independent examination of specifications. Tensile properties are minimum or typical . UTS and YS as Mpa, El as %. See Appendix for list of abbreviations used in Descriptions.

4-4 Wrought Copper

Grade UNS #	Country	Specification	Designation	Comment	Al	Cu	Fe	Mn	Ni	P	Pb	Sn	Zn	OT max	Other	UTS	YS	El
Oxygen-Free Copper C10300	USA	ASTM B42	OFXLP	Wrought Rod Plt Sh Strp Tub Pip Shapes		99.95 min				0.01 max				0.05				
Oxygen-Free Copper C10300	USA	ASTM B447	OFXLP	Wrought Rod Plt Sh Strp Tub Pip Shapes		99.95 min				0.01 max				0.05				
Oxygen-Free Copper C10300	USA	ASTM B68	OFXLP	Wrought Rod Plt Sh Strp Tub Pip Shapes		99.95 min				0.01 max				0.05				
Oxygen-Free Copper C10300	USA	ASTM B75	OFXLP	Wrought Rod Plt Sh Strp Tub Pip Shapes		99.95 min				0.01 max				0.05				
Oxygen-Free Copper C10300	USA	ASTM B88	OFXLP	Wrought Rod Plt Sh Strp Tub Pip Shapes		99.95 min				0.01 max				0.05				
C10300	USA	CDA	OFXLP							0.001-0.005					Cu+Ag+P 99.95 min			
C10300	USA									0.001-0.005					Cu+Ag+P 99.95 min			
C10400	USA	CDA	OFS	Oxygen free with Ag; High conductivity ann											Cu+Ag 99.95 min; Ag 0.027 min			
C10400	USA			Oxygen free with Ag; High conductivity ann											Cu+Ag 99.95 min; Ag 0.027 min			
C10500	USA	CDA	OFS	Oxygen free with Ag; High conductivity ann											Cu+Ag 99.95 min; Ag 0.034 min			
C10500	USA			Oxygen free with Ag; High conductivity ann											Cu+Ag 99.95 min; Ag 0.034 min			
C10700	USA	CDA	OFS	Oxygen free with Ag											Cu+Ag 99.95 min; Ag 0.085 min			
C10700	USA			Oxygen free with Ag											Cu+Ag 99.95 min; Ag 0.085 min			
Oxygen-Free Copper C10800	USA	ASTM B111	C10800	Wrought Plt Sh Strp Tub Pip Shapes		99.95 min				0.01								
Oxygen-Free Copper C10800	USA	ASTM B113	C10800	Wrought Plt Sh Strp Tub Pip Shapes		99.95 min				0.01								
Oxygen-Free Copper C10800	USA	ASTM B12	C10800	Wrought Plt Sh Strp Tub Pip Shapes		99.95 min				0.01								
Oxygen-Free Copper C10800	USA	ASTM B133	C10800	Wrought Plt Sh Strp Tub Pip Shapes		99.95 min				0.01								
Oxygen-Free Copper C10800	USA	ASTM B152	C10800	Wrought Plt Sh Strp Tub Pip Shapes		99.95 min				0.01								
Oxygen-Free Copper	USA	ASTM B187	C10800	Wrought Plt Sh Strp Tub Pip Shapes		99.95 min				0.01								

UNS numbers and US grades are provided as a means of cross referencing chemically similar alloys. Exchangability is only possible after independent examination of specifications. Tensile properties are minimum or typical . UTS and YS as Mpa, El as %. See Appendix for list of abbreviations used in Descriptions.

Grade UNS #	Country	Specification	Designation	Comment	Al	Cu	Fe	Mn	Ni	P	Pb	Sn	Zn	OT max	Other	UTS	YS	El
Oxygen-Free Copper C10800	USA	ASTM B188	C10800	Wrought Plt Sh Strp Tub Pip Shapes		99.95 min				0.01								
Oxygen-Free Copper C10800	USA	ASTM B251	C10800	Wrought Plt Sh Strp Tub Pip Shapes		99.95 min				0.01								
Oxygen-Free Copper C10800	USA	ASTM B280	C10800	Wrought Plt Sh Strp Tub Pip Shapes		99.95 min				0.01								
Oxygen-Free Copper C10800	USA	ASTM B302	C10800	Wrought Plt Sh Strp Tub Pip Shapes		99.95 min				0.01								
Oxygen-Free Copper C10800	USA	ASTM B306	C10800	Wrought Plt Sh Strp Tub Pip Shapes		99.95 min				0.01								
Oxygen-Free Copper C10800	USA	ASTM B357	C10800	Wrought Plt Sh Strp Tub Pip Shapes		99.95 min				0.01								
Oxygen-Free Copper C10800	USA	ASTM B360	C10800	Wrought Plt Sh Strp Tub Pip Shapes		99.95 min				0.01								
Oxygen-Free Copper C10800	USA	ASTM B372	C10800	Wrought Plt Sh Strp Tub Pip Shapes		99.95 min				0.01								
Oxygen-Free Copper C10800	USA	ASTM B395	C10800	Wrought Plt Sh Strp Tub Pip Shapes		99.95 min				0.01								
Oxygen-Free Copper C10800	USA	ASTM B42	C10800	Wrought Plt Sh Strp Tub Pip Shapes		99.95 min				0.01								
Oxygen-Free Copper C10800	USA	ASTM B432	C10800	Wrought Plt Sh Strp Tub Pip Shapes		99.95 min				0.01								
Oxygen-Free Copper C10800	USA	ASTM B68	C10800	Wrought Plt Sh Strp Tub Pip Shapes		99.95 min				0.01								
Oxygen-Free Copper C10800	USA	ASTM B75	C10800	Wrought Plt Sh Strp Tub Pip Shapes		99.95 min				0.01								
Oxygen-Free Copper C10800	USA	ASTM B88	C10800	Wrought Plt Sh Strp Tub Pip Shapes		99.95 min				0.01								
C10800	USA	CDA	OFLP							0.005-0.012					Cu+Ag+P 99.95 min			
C10800	USA									0.005-0.012					Cu+Ag+P 99.95 min			
	Norway	NS 16011	Cu-OF	Wrought Sh Strp Plt Bar Tub Rod Wir		99.95												
	Denmark	DS 3003	5030	Wrought											O 0.06-0.1 bal Zn			
	Mexico	DGN W-23	OF	Wrought Tub CF 3.18/15.92 mm diam		99.92										207		

UNS numbers and US grades are provided as a means of cross referencing chemically similar alloys. Exchangability is only possible after independent examination of specifications. Tensile properties are minimum or typical . UTS and YS as Mpa, El as %. See Appendix for list of abbreviations used in Descriptions.

Grade UNS #	Country	Specification	Designation	Comment	Al	Cu	Fe	Mn	Ni	P	Pb	Sn	Zn	OT max	Other	UTS	YS	El
C10920	USA														Cu+Ag 99.90 min; O 0.02 max			
C10930	USA														Cu+Ag 99.90 min; O 0.02 max; Ag 0.044 min			
C10940	USA														Cu+Ag 99.90 min; O 0.02 max; Ag 0.085 min			
C11000	Australia	AS 1567	110	Wrought Bar Rod Hard 10/12 mm diam		99.9										280		6
C11000	Australia	AS 1567	110	Wrought Bar Rod 1/2 hard 25/50 mm diam		99.9										230		22
C11000	Australia	AS 1567	110	Wrought Bar Rod Soft Ann 12/50 mm diam		99.9										230		45
C11000	Canada	CSA HC.4.1	Cu-ETP(110)	Wrought Sh HR and Ann		99.9										206		
C11000	Canada	CSA HC.4.1	Cu-ETP(110)	Wrought Strp HR and Ann		99.9										206		
C11000	Canada	CSA HC.4.1	Cu-ETP(110)	Wrought Plt HR and ann		99.9										206		
C11000	Canada	CSA HC.4.1	Cu-ETP(110)	Wrought Sh CR		99.9										220		
C11000	Canada	CSA HC.4.1	Cu-ETP(110)	Wrought Strp CR		99.9										220		
C11000	Canada	CSA HC.4.1	Cu-ETP(110)	Wrought Plt CR		99.9										220		
C11000	Canada	CSA HC.4.1	Cu-ETP(110)	Wrought Bar CR		99.9										220		
C11000	Canada	CSA HC.4.1	Cu-ETP(110)	Wrought Sh Full Hard		99.9										296		
C11000	Canada	CSA HC.4.1	Cu-ETP(110)	Wrought Plt Full Hard		99.9										296		
C11000	Canada	CSA HC.4.1	Cu-ETP(110)	Wrought Bar Full Hard		99.9										296		
C11000	Canada	CSA HC.4.1	Cu-ETP(110)	Wrought Strp Spring		99.9										344		
C11000	Canada	CSA HC.4.1	Cu-ETP(110)	Wrought Bar Hr and Ann		99.9										2060		
C11000	Finland	SFS 2908	Cu-ETPandCu-FRHC	Wrought Sh Strp Rod Bar Wir Drawn (shapes)											ST 99.9 min; ST = Cu+Ag; O 0.02-0.1	230	120	10
C11000	Finland	SFS 2908	Cu-ETPandCu-FRHC	Wrought Sh Strp Rod Bar Wir Ann (sh, strp, and bar) 1.5/2.5 mm diam											ST 99.9 min; ST = Cu+Ag; O 0.02-0.1	220	40	40
C11000	Finland	SFS 2908	Cu-ETPandCu-FRHC	Wrought Sh Strp Rod Bar Wir HF (rod,wir,and bar) 5 mm diam											ST 99.9 min; ST = Cu+Ag; O 0.02-0.1	230	60	45
C11000	France	NF A53301	Cu/a2	Wrought Bar shapes 3/4 hard 60 mm diam Obsolete Replaced by NF A51118											ST 99.9 min; ST=Cu+Ag	270		10

UNS numbers and US grades are provided as a means of cross referencing chemically similar alloys. Exchangability is only possible after independent examination of specifications. Tensile properties are minimum or typical . UTS and YS as Mpa, El as %. See Appendix for list of abbreviations used in Descriptions.

Worldwide Guide to Equivalent Nonferrous Metals and Alloys

Grade UNS #	Country	Specification	Designation	Comment	Al	Cu	Fe	Mn	Ni	P	Pb	Sn	Zn	OT max	Other	UTS	YS	El
C11000	France	NF A53301	Cu/a2	Wrought Bar shapes 1/2 hard 70 mm diam Obsolete Replaced by NF A51118											ST 99.9 min; ST=Cu+Ag	250		15
C11000	France	NF A53301	Cu/a2	Wrought Bar shapes 1/4 hard all diam Obsolete Replaced by NF A51118											ST 99.9 min; ST=Cu+Ag	230		20
C11000	France	NF A53301	Cu/al	Wrought Bar 3/4 hard 60 mm diam Obsolete Replaced by NF A51118											ST 99.9 min; ST=Cu+Ag			10
C11000	France	NF A53301	Cu/al	Wrought Bar 1/2 hard 70 mm diam Obsolete Replaced by NF A51118											ST 99.9 min; ST=Cu+Ag			15
C11000	France	NF A53301	Cu/al	Wrought Bar 1/4 hard all diam Obsolete Replaced by NF A51118											ST 99.9 min; ST=Cu+Ag			20
ETP Copper C11000	India	IS 6912	ETPCopper	Wrought Frg Ann 6 mm diam							0.01				99.90 min Cu+Ag total, others; Bi 0.001 max			
C11000	International	ISO R1337	CuETP	Wrought Plt Sh Strp Rod Bar Tub Wir Frg		99.9												
C11000	Japan	JIS H3100	C1100	Wrought Sh Plt 1/2 hard .5/20 mm diam		99.9										245		15
C11000	Japan	JIS H3100	C1100	Wrought Sh Plt Ann .5/30 mm diam		99.9										196		35
C11000	Japan	JIS H3100	C1100	Wrought Sh Plt Hard .5/10 mm diam		99.9										275		
C11000	Japan	JIS H3140	C1100	Wrought Bar 1/2 hard 2/20 mm diam		99.9										245		15
C11000	Japan	JIS H3140	C1100	Wrought Bar Ann 2/30 mm diam		99.9										196		35
C11000	Japan	JIS H3140	C1100	Wrought Bar Hard 2/10 mm diam		99.9										275		
C11000	Japan	JIS H3250	C1100	Wrought Rod As Mfg 6 mm diam		99.9										196		25
C11000	Japan	JIS H3250	C1100	Wrought Rod Ann 6/25 mm diam		99.9										196		30
C11000	Japan	JIS H3250	C1100	Wrought Rod HArd 6/25 mm diam		99.9										275		
C11000	Japan	JIS H3260	C1100	Wrought Rod Ann .5/2 mm diam		99.9										196		15
C11000	Japan	JIS H3260	C1100	Wrought Rod 1/2 hard .5/12 mm diam		99.9										255		

UNS numbers and US grades are provided as a means of cross referencing chemically similar alloys. Exchangability is only possible after independent examination of specifications. Tensile properties are minimum or typical . UTS and YS as Mpa, El as %. See Appendix for list of abbreviations used in Descriptions.

4-8 Wrought Copper

Grade UNS #	Country	Specification	Designation	Comment	Al	Cu	Fe	Mn	Ni	P	Pb	Sn	Zn	OT max	Other	UTS	YS	El
C11000	Japan	JIS H3260	C1100	Wrought Rod Hard .5/10 mm diam		99.9										343		
C11000	Japan	JIS H3300	C1100	Wrought Pip Ann 5/250 mm diam		99.9										206		40
C11000	Japan	JIS H3300	C1100	Wrought Pip 1/2 hard 5/250 mm diam		99.9										245		
C11000	Japan	JIS H3300	C1100	Wrought Pip Hard 5/100 mm diam		99.9										275		
C11000	UK	BS 1036	C101	Wrought							0.01 max			0.03	ST 99 min; ST = Cu+Ag; Bi 0.002 max			
C11000	UK	BS 1037	C102	Wrought							0.01 max			0.03	ST 99 min; ST = Cu+Ag			
ETP C11000	USA	AMS 4500	99.95Cu-.040	Wrought Bar Rod Wir Plt Sh Strp Tub Frg Cast		99.9 min												
ETP C11000	USA	AMS 4701	99.95Cu-.040	Wrought Bar Rod Wir Plt Sh Strp Tub Frg Cast		99.9 min												
ETP C11000	USA	ASME SB11	99.95Cu-.040	Wrought Bar Rod Wir Plt Sh Strp Tub Frg Cast		99.9 min												
ETP C11000	USA	ASME SB12	99.95Cu-.040	Wrought Bar Rod Wir Plt Sh Strp Tub Frg Cast		99.9 min												
ETP C11000	USA	ASTM B101	99.95Cu-.040	Wrought Bar Rod Wir Plt Sh Strp Tub Frg Cast		99.9 min												
ETP C11000	USA	ASTM B152	99.95Cu-.040	Wrought Bar Rod Wir Plt Sh Strp Tub Frg Cast		99.9 min												
ETP C11000	USA	ASTM B212	99.95Cu-.040	Wrought Bar Rod Wir Plt Sh Strp Tub Frg Cast		99.9 min												
ETP C11000	USA	ASTM B248	99.95Cu-.040	Wrought Bar Rod Wir Plt Sh Strp Tub Frg Cast		99.9 min												
ETP C11000	USA	ASTM B370	99.95Cu-.040	Wrought Bar Rod Wir Plt Sh Strp Tub Frg Cast		99.9 min												
ETP C11000	USA	ASTM B451	99.95Cu-.040	Wrought Bar Rod Wir Plt Sh Strp Tub Frg Cast		99.9 min												
C11000	USA	CDA	ETP	Electrolytic tough pitch											Cu+Ag 99.90 min			
ETP C11000	USA	MIL C-12166	99.95Cu-.040	Wrought Bar Rod Wir Plt Sh Strp Tub Frg Cast		99.9 min												
ETP C11000	USA	MIL W-3318	99.95Cu-.040	Wrought Bar Rod Wir Plt Sh Strp Tub Frg Cast		99.9 min												

UNS numbers and US grades are provided as a means of cross referencing chemically similar alloys. Exchangability is only possible after independent examination of specifications. Tensile properties are minimum or typical . UTS and YS as Mpa, El as %. See Appendix for list of abbreviations used in Descriptions.

Grade UNS #	Country	Specification	Designation	Comment	Al	Cu	Fe	Mn	Ni	P	Pb	Sn	Zn	OT max	Other	UTS	YS	El
ETP C11000	USA	MIL W-6712	99.95Cu-.040	Wrought Bar Rod Wir Plt Sh Strp Tub Frg Cast		99.9 min												
ETP C11000	USA	QQ A-673	99.95Cu-.040	Wrought Bar Rod Wir Plt Sh Strp Tub Frg Cast		99.9 min												
ETP C11000	USA	QQ B-825	99.95Cu-.040	Wrought Bar Rod Wir Plt Sh Strp Tub Frg Cast		99.9 min												
ETP C11000	USA	QQ C-502	99.95Cu-.040	Wrought Bar Rod Wir Plt Sh Strp Tub Frg Cast		99.9 min												
ETP C11000	USA	QQ C-576	99.95Cu-.040	Wrought Bar Rod Wir Plt Sh Strp Tub Frg Cast		99.9 min												
ETP C11000	USA	QQ W-343	99.95Cu-.040	Wrought Bar Rod Wir Plt Sh Strp Tub Frg Cast		99.9 min												
ETP C11000	USA	SAE J463	99.95Cu-.040	Wrought Bar Rod Wir Plt Sh Strp Tub Frg Cast		99.9 min												
ETP C11000	USA	WW P-377	99.95Cu-.040	Wrought Bar Rod Wir Plt Sh Strp Tub Frg Cast		99.9 min												
C11000	USA			Electrolytic tough pitch											Cu+Ag 99.90 min			
C11010	USA	CDA	RHC	Remelted high conductivity											Cu+Ag 99.90 min			
C11010	USA			Remelted high conductivity											Cu+Ag 99.90 min			
C11020	USA	CDA	FRHC	Fire refined high conductivity											Cu+Ag 99.90 min			
C11020	USA			Fire refined high conductivity											Cu+Ag 99.90 min			
C11030	USA	CDA	CRTP	Chem refined tough pitch											Cu+Ag 99.90 min			
C11030	USA			Chem refined tough pitch											Cu+Ag 99.90 min			
C11040	USA														Sb 0.0004 max; As 0.0005 max; Te 0.0002 max; Cu+Ag 99.90 min; OT 65 ppm max excluding O			
C11100	USA			Elyctrolytic tough pitch ann res											Cu+Ag 99.90 min			
Tough Pitch Copper C11300	USA	AMS 4701	STP	Wrought Bar Rod Wir Plt Sh Strp Tub Pip Shapes		99- 99.9									O 0.04 max; Ag 0.03 max			
Tough Pitch Copper C11300	USA	ASME SB152	STP	Wrought Bar Rod Wir Plt Sh Strp Tub Pip Shapes		99- 99.9									O 0.04 max; Ag 0.03 max			
Tough Pitch Copper C11300	USA	ASTM B1	STP	Wrought Bar Rod Wir Plt Sh Strp Tub Pip Shapes		99- 99.9									O 0.04 max; Ag 0.03 max			

UNS numbers and US grades are provided as a means of cross referencing chemically similar alloys. Exchangability is only possible after independent examination of specifications. Tensile properties are minimum or typical . UTS and YS as Mpa, El as %. See Appendix for list of abbreviations used in Descriptions.

4-10 Wrought Copper

Grade UNS #	Country	Specification	Designation	Comment	Al	Cu	Fe	Mn	Ni	P	Pb	Sn	Zn	OT max	Other	UTS	YS	El
Tough Pitch Copper C11300	USA	ASTM B152	STP	Wrought Bar Rod Wir Plt Sh Strp Tub Pip Shapes		99-99.9									O 0.04 max; Ag 0.03 max			
Tough Pitch Copper C11300	USA	ASTM B187	STP	Wrought Bar Rod Wir Plt Sh Strp Tub Pip Shapes		99-99.9									O 0.04 max; Ag 0.03 max			
Tough Pitch Copper C11300	USA	ASTM B188	STP	Wrought Bar Rod Wir Plt Sh Strp Tub Pip Shapes		99-99.9									O 0.04 max; Ag 0.03 max			
Tough Pitch Copper C11300	USA	ASTM B189	STP	Wrought Bar Rod Wir Plt Sh Strp Tub Pip Shapes		99-99.9									O 0.04 max; Ag 0.03 max			
Tough Pitch Copper C11300	USA	ASTM B2	STP	Wrought Bar Rod Wir Plt Sh Strp Tub Pip Shapes		99-99.9									O 0.04 max; Ag 0.03 max			
Tough Pitch Copper C11300	USA	ASTM B246	STP	Wrought Bar Rod Wir Plt Sh Strp Tub Pip Shapes		99-99.9									O 0.04 max; Ag 0.03 max			
Tough Pitch Copper C11300	USA	ASTM B272	STP	Wrought Bar Rod Wir Plt Sh Strp Tub Pip Shapes		99-99.9									O 0.04 max; Ag 0.03 max			
Tough Pitch Copper C11300	USA	ASTM B272	STP	Wrought Bar Hod Wir Plt Sh Strp Tub Pip Shapes		99-99.9									O 0.04 max; Ag 0.03 max			
Tough Pitch Copper C11300	USA	ASTM B298	STP	Wrought Bar Hod Wir Plt Sh Strp Tub Pip Shapes		99-99.9									O 0.04 max; Ag 0.03 max			
Tough Pitch Copper C11300	USA	ASTM B3	STP	Wrought Bar Rod Wir Plt Sh Strp Tub Pip Shapes		99-99.9									O 0.04 max; Ag 0.03 max			
Tough Pitch Copper C11300	USA	ASTM B334	STP	Wrought Bar Rod Wir Plt Sh Strp Tub Pip Shapes		99-99.9									O 0.04 max; Ag 0.03 max			
Tough Pitch Copper C11300	USA	ASTM B355	STP	Wrought Bar Rod Wir Plt Sh Strp Tub Pip Shapes		99-99.9									O 0.04 max; Ag 0.03 max			
Tough Pitch Copper C11300	USA	ASTM B48	STP	Wrought Bar Rod Wir Plt Sh Strp Tub Pip Shapes		99-99.9									O 0.04 max; Ag 0.03 max			
Tough Pitch Copper C11300	USA	ASTM B49	STP	Wrought Bar Rod Wir Plt Sh Strp Tub Pip Shapes		99-99.9									O 0.04 max; Ag 0.03 max			
Tough Pitch Copper C11300	USA	ASTM B506	STP	Wrought Bar Rod Wir Plt Sh Strp Tub Pip Shapes		99-99.9									O 0.04 max; Ag 0.03 max			
Tough Pitch Copper C11300	USA	ASTM B8	STP	Wrought Bar Rod Wir Plt Sh Strp Tub Pip Shapes		99-99.9									O 0.04 max; Ag 0.03 max			
C11300	USA	CDA	STP	Tough pitch with Ag											Cu+Ag 99.90 mn; Ag .027 mn			
Tough Pitch Copper	USA	SAE J463	STP	Bar Rod Wir Plt Sh Strp Tub Pip Shapes		99-99.9									O 0.04 max; Ag 0.03 max			

UNS numbers and US grades are provided as a means of cross referencing chemically similar alloys. Exchangability is only possible after independent examination of specifications. Tensile properties are minimum or typical . UTS and YS as Mpa, El as %. See Appendix for list of abbreviations used in Descriptions.

Worldwide Guide to Equivalent Nonferrous Metals and Alloys

Grade UNS #	Country	Specification	Designation	Comment	Al	Cu	Fe	Mn	Ni	P	Pb	Sn	Zn	OT max	Other	UTS	YS	El
C11300	USA			Tough pitch with Ag											Cu+Ag 99.90 min; Ag 0.027 min			
C11400	Canada	CSA HC.4.1	Cu-STP(114)	Wrought Sh HR and Ann		99.9										206		
C11400	Canada	CSA HC.4.1	Cu-STP(114)	Wrought Plt HR and Ann		99.9										206		
C11400	Canada	CSA HC.4.1	Cu-STP(114)	Wrought Strp HR and Ann		99.9										206		
C11400	Canada	CSA HC.4.1	Cu-STP(114)	Wrought Bar CR		99.9										220		
C11400	Canada	CSA HC.4.1	Cu-STP(114)	Wrought Sh CR		99.9										220		
C11400	Canada	CSA HC.4.1	Cu-STP(114)	Wrought Plt CR		99.9										220		
C11400	Canada	CSA HC.4.1	Cu-STP(114)	Wrought Strp CR		99.9										220		
C11400	Canada	CSA HC.4.1	Cu-STP(114)	Wrought Bar Full Hard		99.9										296		
C11400	Canada	CSA HC.4.1	Cu-STP(114)	Wrought Sh Full Hard		99.9										296		
C11400	Canada	CSA HC.4.1	Cu-STP(114)	Wrought Plt Full Hard		99.9										296		
C11400	Canada	CSA HC.4.1	Cu-STP(114)	Wrought Strp Spring		99.9										344		
C11400	Canada	CSA HC.4.1	Cu-STP(114)	Wrought Bar HR and Ann		99.9										2060		
C11400	USA	CDA	STP	Tough pitch with Ag											Cu+Ag 99.90 min; Ag 0.034 min			
C11400	USA			Tough pitch with Ag											Cu+Ag 99.90 min; Ag 0.034 min			
C11500	International	ISO 1634	CuAg0.05	Wrought Plt Sh Strp Ann		99.75									Ag 0.02-0.08			30
C11500	International	ISO 1637	CuAg0.05	Wrought HB-SH 5/20 mm diam		99.75									Ag 0.02-0.08	280		5
C11500	International	ISO 1637	CuAg0.05	Wrought HA 5/40 mm diam		99.75									Ag 0.02-0.08	250		15
C11500	International	ISO 1637	CuAg0.05	Wrought Ann 5 mm diam		99.75									Ag 0.02-0.08			35
C11500	International	ISO 1638	CuAg0.05	Wrought Ann 1/1.5 mm diam		99.75									Ag 0.02-0.08	210		25
C11500	International	ISO 1638	CuAg0.05	Wrought HC 1/5 mm diam		99.75									Ag 0.02-0.08	390		
C11500	International	ISO 1638	CuAg0.05	Wrought HD-SH 1/3 mm diam		99.75									Ag 0.02-0.08	420		
C11500	International	ISO R1336	CuAg0.05	Wrought Strp Rod Bar Wir		99.85									O 0.06-0.1; Ag 0.02-0.08			
C11500	USA	CDA	STP	Tough pitch with Ag											Cu+Ag 99.90 min; Ag 0.054 min			
C11500	USA			Tough pitch with Ag											Cu+Ag 99.90 min; Ag 0.054 min			
C11600	Australia	AS 1567	116	Wrought Bar Rod Hard 10/12 mm diam		99.9									Ag 0.09	280		6
C11600	Australia	AS 1567	116	Wrought Bar Rod 1/2 hard 25/50 mm diam		99.9									Ag 0.09	230		22
C11600	Australia	AS 1567	116	Wrought Bar Rod Soft Ann 12/50 mm diam		99.9									Ag 0.09	230		45

UNS numbers and US grades are provided as a means of cross referencing chemically similar alloys. Exchangability is only possible after independent examination of specifications. Tensile properties are minimum or typical . UTS and YS as Mpa, El as %. See Appendix for list of abbreviations used in Descriptions.

4-12 Wrought Copper

Grade UNS #	Country	Specification	Designation	Comment	Al	Cu	Fe	Mn	Ni	P	Pb	Sn	Zn	OT max	Other	UTS	YS	El
C11600	Finland	SFS 2910	CuAg0.1(OF)	Wrought Sh Strp Bar Rod Wir CF (sh and Strp) .2/.5 mm diam											ST 99.9 min; ST = Cu+Ag; Ag 0.08-0.12	250	180	5
C11600	Finland	SFS 2910	CuAg0.1(OF)	Wrought Sh Strp Bar Rod Wir Drawn (bar) 2.5/5 mm diam											ST 99.9 min; ST = Cu+Ag; Ag 0.08-0.12	340	320	5
C11600	Finland	SFS 2910	CuAg0.1(OF)	Wrought Sh Strp Bar Rod Wir Ann (rod and wir) .2/1 mm diam											ST 99.9 min; ST = Cu+Ag; Ag 0.08-0.12	210	40	20
C11600	International	ISO 1634	CuAg0.1	Wrought Plt Sh Strp SH (HA)		98.81									O 1; Ag 0.05 min	250		16
C11600	International	ISO 1637	CuAg0.1	Wrought HB-SH 5/20 mm diam		99.8									Ag 0.08-0.12	280		5
C11600	International	ISO 1637	CuAg0.1	Wrought HA 5/40 mm diam		99.8									Ag 0.08-0.12	250		15
C11600	International	ISO 1637	CuAg0.1	Wrought Ann 5 mm diam		99.8									Ag 0.08-0.12			35
C11600	International	ISO 1638	CuAg0.1	Wrought Ann 1/1.5 mm diam		99.71									O 0.1; Ag 0.05 min	210		25
C11600	International	ISO 1638	CuAg0.1	Wrought HC 1/5 mm diam		99.71									O 0.1; Ag 0.05 min	390		
C11600	International	ISO 1638	CuAg0.1	Wrought HD-SH l/3 mm diam		99.71									O 0.1; Ag 0.05 min	420		
C11600	International	ISO R1336	CuAg0.1	Wrought Strp Rod Bar Wir		99.85									O 0.06-0.1; Ag 0.08-0.12			
C11600	USA	CDA	STP	Tough pitch with Ag											Cu+Ag 99.90 min; Ag 0.085 min			
C11600	USA			Tough pitch with Ag											Cu+Ag 99.90 min; Ag 0.085 min			
C11700	USA									0.04 max					Cu+Ag+B+P 99.9 min; B 0.004-0.02			
C12000	Canada	CSA	Cu-DLP(120)			99.9 max				0.01 max								
C12000	Finland	SFS 2906	Cu-DLP	Wrought Sh CF .2/.5 mm diam						0.01					ST 99.9 min; ST = Cu+Ag	250	180	5
C12000	Finland	SFS 2906	Cu-DLP	Wrought Sh Ann .2/.5 mm diam						0.01					ST 99.9 min; ST = Cu+Ag	220	140	20
C12000	France	NF A53301	Cu/b	Wrought Bar Shapes 3/4 hard 60 mm diam Obsolete Replaced by NF A51118											ST 99.9 min; ST=Cu+Ag	270		10
C12000	France	NF A53301	Cu/b	Wrought Bar Shapes 1/2 hard 70 mm diam Obsolete Replaced by NF A51118											ST 99.9 min; ST=Cu+Ag	250		15
C12000	France	NF A53301	Cu/b	Wrought Bar Shapes 1/4 hard all diam Obsolete Replaced by NF A51118											ST 99.9 min; ST=Cu+Ag	230		20

UNS numbers and US grades are provided as a means of cross referencing chemically similar alloys. Exchangability is only possible after independent examination of specifications. Tensile properties are minimum or typical . UTS and YS as Mpa, El as %. See Appendix for list of abbreviations used in Descriptions.

Wrought Copper 4-13

Grade UNS #	Country	Specification	Designation	Comment	Al	Cu	Fe	Mn	Ni	P	Pb	Sn	Zn	OT max	Other	UTS	YS	El
C12000	International	ISO 1637	Cu-DLP	Wrought HB-SH 5/20 mm diam						0.01					99.90 min Cu+Ag	280		5
C12000	International	ISO 1637	Cu-DLP	Wrought HA 5/40 mm diam						0.01					99.90 min Cu+Ag	250		15
C12000	International	ISO 1637	Cu-DLP	Wrought Ann 5 mm diam						0.01					99.90 min Cu+Ag			35
C12000	International	ISO R1337	Cu-DLP	Wrought Plt Sh Strp Rod Bar Tub		99.9				0.01								
C12000	Japan	JIS H3100	C1201	Wrought Sh Plt 1/2 hard .3/20 mm diam		99.9				0.02						245		15
C12000	Japan	JIS H3100	C1201	Wrought Sh Plt Ann .3/30 mm diam		99.9				0.02						196		35
C12000	Japan	JIS H3100	C1201	Wrought Sh Plt hard .3/10 mm diam		99.9				0.02						275		
C12000	Japan	JIS H3260	C1201	Wrought Rod Ann .5/2 mm diam		99.9				0.02 max						196		15
C12000	Japan	JIS H3260	C1201	Wrought Rod 1/2 hard .5/12 mm diam		99.9				0.02 max						255		
C12000	Japan	JIS H3260	C1201	Wrought Rod Hard .5/10 mm diam		99.9				0.02 max						343		
C12000	Japan	JIS H3300	C1201	Wrought Pip Ann 4/250 mm diam		99.9				0.02 max						206		40
C12000	Japan	JIS H3300	C1201	Wrought Pip 1/2 hard 4/250 mm diam		99.9				0.02 max						245		
C12000	Japan	JIS H3300	C1201	Wrought Pip Hard 25 mm diam		99.9				0.02 max						314		
C12000	Mexico	DGN W-23	DHP	Tub CF 3.18/15.92 mm		99.9				0.01-0.04						207		40
C12000	Mexico	DGN W-23	DLP	Tub CF 3.18/15.92 mm		99.9				0.01 max						207		40
C12000	USA	CDA	DLP	P deox low resid P						0.004-0.012					Cu+Ag 99.90 min			
C12000	USA			P deox low resid P						0.004-0.012					Cu+Ag 99.90 min			
C12100	USA									0.005-0.012					Cu+Ag 99.90 min; Ag 0.014 min			
C12200	Australia	AS 1567	122	Wrought Rod Bar As Mfg 6 mm diam		99.9				0.02-0.04						230		13
C12200	Australia	AS 1567	122	Wrought Rod Bar Soft Ann 6 mm diam		99.9				0.02-0.04						210		33
C12200	Australia	AS 1572		Wrought Tub		99.9				0.01-0.04								
C12200	Canada	CSA HC.4.1	Cu-DHP(122)	Wrought Sh HR and ann		99.9				0.02-0.04						30		
C12200	Canada	CSA HC.4.1	Cu-DHP(122)	Wrought Strp HR and Ann		99.9				0.02-0.04						206		
C12200	Canada	CSA HC.4.1	Cu-DHP(122)	Wrought Bar HR and Ann		99.9				0.02-0.04						206		
C12200	Canada	CSA HC.4.1	Cu-DHP(122)	Wrought Strp CR		99.9				0.02-0.04						220		
C12200	Canada	CSA HC.4.1	Cu-DHP(122)	Wrought Sh CR		99.9				0.02-0.04						220		
C12200	Canada	CSA HC.4.1	Cu-DHP(122)	Wrought Bar CR		99.9				0.02-0.04						220		

UNS numbers and US grades are provided as a means of cross referencing chemically similar alloys. Exchangability is only possible after independent examination of specifications. Tensile properties are minimum or typical . UTS and YS as Mpa, El as %. See Appendix for list of abbreviations used in Descriptions.

4-14 Wrought Copper

Grade UNS #	Country	Specification	Designation	Comment	Al	Cu	Fe	Mn	Ni	P	Pb	Sn	Zn	OT max	Other	UTS	YS	El
C12200	Canada	CSA HC.4.1	Cu-DHP(122)	Wrought Sh Full Hard		99.9				0.02-0.04						296		
C12200	Canada	CSA HC.4.1	Cu-DHP(122)	Wrought Bar Full Hard		99.9				0.02-0.04						296		
C12200	Canada	CSA HC.4.1	Cu-DHP(122)	Wrought Strp Spring		99.9				0.02-0.04						344		
	Czech Republic	CSN 423003	Cu99,85			99.85 min				0.04 max	0.03 max				O 0.02 Cu incl Ni,Ag; As 0.04 max			
	Czech Republic	CSN 423004	Cu99,75			99.8 min									O 0.04 Cu incl Ni,Ag; As 0.04 max			
C12200	Denmark	DS 3003	5015	Wrought		99.85 min				0.02-0.05					Cu includes Ag			
C12200	Finland	SFS 2907	Cu-DHP	Wrought Tub CF (tube) 5 mm diam						0.02-0.05					ST 99.85 min; ST = Cu+Ag	250	180	20
C12200	Finland	SFS 2907	Cu-DHP	Wrought Tub Ann						0.02-0.05					ST 99.85 min; ST = Cu+Ag	210	40	40
	France	NF A53301	Cu/a3	Wrought Obsolete Replaced by NF A51118		99.75 min								0.1				
C12200	International	ISO 1634	Cu-DHP	Wrought Plt Sh Strp HC						0.01					99.90 min Cu+Ag	340		4
C12200	International	ISO 1635	Cu-DHP	Wrought Tub SH (HB) 100 mm diam						0.01-0.05					99.85 min Cu+Ag	290		10
C12200	International	ISO 1635	Cu-DHP	Wrought Tub HA 200 mm diam						0.01-0.05					99.85 min Cu+Ag	250		20
C12200	International	ISO 1635	Cu-DHP	Wrought Tub Ann						0.01-0.05					99.85 min Cu+Ag			35
C12200	International	ISO 1637	Cu-DHP	Wrought HB-SH 5/20 mm diam						0.01-0.05					99.85 min Cu+Ag	280		5
C12200	International	ISO 1637	Cu-DHP	Wrought HA 5/40 mm diam						0.01-0.05					99.85 min Cu+Ag	250		15
C12200	International	ISO 1637	Cu-DHP	Wrought Ann 5 mm diam						0.01-0.05					99.85 min Cu+Ag			35
C12200	International	ISO R1337	CuDHP	Wrought Plt Sh Strp Rod Bar Tub		99.85												
C12200	Japan	JIS H3100	C1220	Wrought Sh Plt 1/2 hard .3/20 mm diam		99.9				0.02-0.04						245		15
C12200	Japan	JIS H3100	C1220	Wrought Sh Plt Ann .3/30 mm diam		99.9				0.02-0.04						196		35
C12200	Japan	JIS H3100	C1220	Wrought Sh Plt Hard .3/10 mm diam		99.9				0.02-0.04						275		
C12200	Japan	JIS H3100	C1221	Wrought Sh Plt 1/2 hard .3/20 mm diam		99.75				0.04 max						245		15
C12200	Japan	JIS H3100	C1221	Wrought Sh Plt Ann .3/30 mm diam		99.75				0.04 max						196		35
C12200	Japan	JIS H3100	C1221	Wrought Sh Plt Hard .5 mm diam		99.75				0.04 max						275		
C12200	Japan	JIS H3250	C1220	Wrought Rod As Mfg 6 mm diam		99.9				0.02-0.04						196		25
C12200	Japan	JIS H3250	C1220	Wrought Rod Ann 6/25 mm diam		99.9				0.02-0.04						196		30

UNS numbers and US grades are provided as a means of cross referencing chemically similar alloys. Exchangability is only possible after independent examination of specifications. Tensile properties are minimum or typical . UTS and YS as Mpa, El as %. See Appendix for list of abbreviations used in Descriptions.

Grade UNS #	Country	Specification	Designation	Comment	Al	Cu	Fe	Mn	Ni	P	Pb	Sn	Zn	OT max	Other	UTS	YS	El
C12200	Japan	JIS H3250	C1220	Wrought Rod Hard 6/25 mm diam		99.9				0.02-0.04						275		
C12200	Japan	JIS H3250	C1221	Wrought Rod As Mfg 6 mm diam		99.75				0.04 max						196		25
C12200	Japan	JIS H3250	C1221	Wrought Rod Ann 6/25 mm diam		99.75				0.04 max						196		30
C12200	Japan	JIS H3250	C1221	Wrought Rod Hard 6/25 mm diam		99.75				0.04 max						275		
C12200	Japan	JIS H3260	C1220	Wrought Rod Ann .5/2 mmdiam		99.9				0.02-0.04						196		15
C12200	Japan	JIS H3260	C1220	Wrought Rod 1/2 hard .5/12 mm diam		99.9				0.02-0.04						255		
C12200	Japan	JIS H3260	C1220	Wrought Rod Hard .5/10 mm diam		99.9				0.02-0.04						343		
C12200	Japan	JIS H3260	C1221	Wrought Rod Ann .5/2 mmdiam		99.75				0.04 max						196		15
C12200	Japan	JIS H3260	C1221	Wrought Rod 1/2 hard .5/12 mm diam		99.75				0.04 max						255		
C12200	Japan	JIS H3260	C1221	Wrought Rod Hard .5/10 mm diam		99.75				0.04 max						343		
C12200	Japan	JIS H3300	C1220	Wrought Pip Ann 4/250 mm diam		99.9				0.02-0.04						206		40
C12200	Japan	JIS H3300	C1220	Wrought Pip 1/2 hard 4/250 mm diam		99.9				0.02-0.04						245		
C12200	Japan	JIS H3300	C1220	Wrought Pip Hard 25 mm diam		99.9				0.02-0.04						314		
C12200	Japan	JIS H3300	C1221	Wrought Pip Ann 4/250 mm diam		99.75				0.04 max						206		40
C12200	Japan	JIS H3300	C1221	Wrought Pip 1/2 hard 4/250 mm diam		99.75				0.04 max						245		
C12200	Japan	JIS H3300	C1221	Wrought Pip Hard 25 mm diam		99.75				0.04 max						314		
C12200	Mexico	NOM W-17	RR	Tub T 6.35/304.8 mm		99.9				0.02-0.04						207		
C12200	Mexico	NOM W-17	RR	Tub CD 6.35/304.8 mm		99.9				0.02-0.04						207		
C12200	Mexico	NOM W-17	RRR	Wrought Tub T 12.7/203.2 mm diam		99.9				0.02-0.04						207		
C12200	Mexico	NOM W-17	RRR	Wrought Tub CD 12.7/203.2 mm diam		99.9				0.02-0.04						207		
	South Africa	SABS 460	Cu-FRHC	Wrought tub		99.9 min					0.01			0.04	Cu includes Ag; Bi 0.002 max			
	South Africa	SABS 804-807	804	Wrought Bar		99.9 min					0.01			0.03	Bi 0.001 max			
	South Africa	SABS 804-807	805	Wrought Bar		99.9 min	0.01 max				0.01							
	South Africa	SABS 804-807	806	Wrought Bar		99.9 min	0.01	0.02			0.01			0.04	As 0.01			
	South Africa	SABS 804-807	807	Wrought Bar		99.75 min	0.04	0.1			0.02	0.03		0.01	Sb 0.01; Bi 0.01; As 0.01			

UNS numbers and US grades are provided as a means of cross referencing chemically similar alloys. Exchangability is only possible after independent examination of specifications. Tensile properties are minimum or typical . UTS and YS as Mpa, El as %. See Appendix for list of abbreviations used in Descriptions.

Grade UNS #	Country	Specification	Designation	Comment	Al	Cu	Fe	Mn	Ni	P	Pb	Sn	Zn	OT max	Other	UTS	YS	El
C12200	USA	CDA	DHP	P deox high resid P						0.015-0.04					Cu+Ag 99.9 min			
C12200	USA			P deox high resid P						0.015-0.04					Cu+Ag 99.9 min			
C12210	USA									0.015-0.025					Cu+Ag 99.90 min			
C12220	USA									0.04-0.065					Cu+Ag 99.9 min			
C12300	USA									0.015-0.04					Cu+Ag 99.90 min			
C14180	USA				0.01 max					0.075 max	0.02 max				Cu+Ag 99.90 min			
C14181	USA									0.002 max	0.002 max		0.002 max		Cu+Ag 99.90 min; Cd 0.002 max; C 0.005 max			
C14300	USA			Cd Cu deox											Cu+Ag+Cd 99.90 min; Cd 0.05-0.15			
Cadmium Copper C14310	USA		99.8Cu-.2Cd	Inactive designation		99.8-99.9									Cd 0.1-0.3			
C14410	USA						0.05 max			0.005-0.02	0.05 max	0.1-0.2			Cu+Ag+Sn 99.90 min			
C14420	USA											0.04-0.15			Cu+Ag+Te 99.90 min; Te 0.005-0.05			
C14500	Australia	AS 1567	145	Wrought Rod Bar Soft Ann 6 mm diam		99.9				0.02 max					Te 0.3-0.7	210		28
C14500	Australia	AS 1567	145	Wrought Rod Bar As Mfg 6/50 mm diam		99.9				0.01-0.03					Te 0.3-0.7	260		8
C14500	International	ISO 1637	CuTe	Wrought HC 5/40 mm diam		98.99									Te 0.3-0.8	250		10
C14500	International	ISO 1637	CuTe	Wrought Ann 5 mm diam		98.99									Te 0.3-0.8			28
C14500	International	ISO R1336	CuTe	Wrought Rod Bar		98.99								0.2	Te 0.3-0.8			
C14500	UK	BS 2874	C109	Rod As Mfg 6/50 mm diam		99.2								0.2	Te 0.3-0.7	260		8
C14500	UK	BS 2874	C109	Rod Ann 6 mm diam		99.2								0.2	Te 0.3-0.7	216		28
Free-Machining C14500	USA	ASTM B124	99.5Cu-.5Te	Rod Strp Frg Extrusions Shapes						0.01 max					ST 99.9 min; Te 0.4-0.6			
Free-Machining Cu C14500	USA	ASTM B283	99.5Cu-.5Te	Wrought Rod Strp Frg Extrusions Shapes						0.01 max					ST 99.9 min; Te 0.4-0.6			
Free-Machining Cu C14500	USA	ASTM B301	99.5Cu-.5Te	Wrought Rod Strp Frg Extrusions Shapes						0.01 max					ST 99.9 min; Te 0.4-0.6			
C14500	USA			Te bearing						0.004-0.012					Cu+Ag+Te 99.90 min; Te 0.4-0.7			
C14510	USA			Te bearing						0.01-0.03	0.05 max				Cu+Ag+Te 99.90 min; Te 0.3-0.7			
C14520	International	ISO 1637	CuTe(P)	Wrought HA 5/40 mm diam		98.978				0.01 max					0.2 P+others; Te 0.3-0.8	250		10
C14520	International	ISO 1637	CuTe(P)	Wrought Ann 5 mm diam		98.978				0.01 max					0.2 P+others; Te 0.3-0.8			28
C14520	International	ISO R1336	CuTe(P)	Wrought Rod Bar		98.978				0.01 max				0.2	Te 0.3-0.8			

UNS numbers and US grades are provided as a means of cross referencing chemically similar alloys. Exchangability is only possible after independent examination of specifications. Tensile properties are minimum or typical . UTS and YS as Mpa, El as %. See Appendix for list of abbreviations used in Descriptions.

Worldwide Guide to Equivalent Nonferrous Metals and Alloys

Grade UNS #	Country	Specification	Designation	Comment	Al	Cu	Fe	Mn	Ni	P	Pb	Sn	Zn	OT max	Other	UTS	YS	El
C14520	USA	CDA	DPTE	P deox Te bearing						0.004-0.02					Cu+Ag+Te 99.40 min; Te 0.4-0.7			
C14520	USA			P deox Te bearing						0.004-0.02					Cu+Ag+Te 99.40 min; Te 0.4-0.7			
C14530	USA									0.001-0.01	0.003-0.025				Cu+Ag+Te+Se+ Sn 99.90 min; Te+Se 0.003-0.023			
C14700	Australia	AS 1567	147	Wrought Rod Bar As Mfg 6/50 mm diam		99.9									S 0.2-0.5	260		8
C14700	Australia	AS 1567	147	Wrought Rod Bar Soft Ann 6 mm diam		99.9									S 0.2-0.5	210		28
C14700	UK	BS 2874	C111	Wrought Rod As Mfg 6 mm diam		99.19								0.2	S 0.3-0.6	260		8
C14700	UK	BS 2874	C111	Wrought Rod Ann 6 mm diam		99.19								0.2	S 0.3-0.6	216		28
Free-Machining Cu C14700	USA	ASTM B301	99.6Cu-.4S	Wrought Rod Strp										0.1	OT 0.1 max; S 0.2-0.5			
C14700	USA			S bearing; Includes O2 free or deox						0.002-0.005					Cu+Ag 99.90 min; S 0.2-0.5			
	Denmark	DS 3003	5010	Wrought		99.9												
	International	ISO R1337	Cu-FRHC	Wrought Plt Sh Strp Rod Bar Tub Frg		99.9												
	Japan	JIS H3250	C1201	Wrought Rod As Mfg 6 mm diam		99.9										196		25
	Japan	JIS H3250	C1201	Wrought Rod Ann 6/25 mm diam		99.9										196		30
	Japan	JIS H3250	C1201	Wrought Rod Hard 6/25 mm diam		99.9										275		
	Sweden	SIS 145030	5030-24	Wrought RodSH		99.9									Ag in Cu content; Ag 0.08-0.12	270	220	10
	UK	BS 1038	C104	Wrought			0.01 max		0.05 max		0.01 max				Se 0.02 max; ST 99.85 min; ST = Cu+Ag; Te 0.01-0.6; Sb 0.01; O 0.1; Bi 0.001 max; As 0.02 max			
C12900	USA	CDA	FRSTP	Fired refined tough pitch with Ag					0.05 max		0.004 max				Cu+Ag 99.88 min; Sb 0.003 max; Bi 0.003 max; As 0.012 max; Ag 0.054 min			
C12900	USA			Fired refined tough pitch with Ag					0.05 max		0.004 max				Cu+Ag 99.88 min; Sb 0.003 max; Bi 0.003 max; As 0.012 max; Ag 0.054 min			
	International	ISO R1337	Cu-FRTP	Wrought Plt Sh Strp Rod Bar		99.85												
	Norway	NS 16013	Cu-FRTP	Wrought Sh Strp Plt Bar		99.85												

UNS numbers and US grades are provided as a means of cross referencing chemically similar alloys. Exchangability is only possible after independent examination of specifications. Tensile properties are minimum or typical . UTS and YS as Mpa, El as %. See Appendix for list of abbreviations used in Descriptions.

4-18 Wrought Copper

Grade UNS #	Country	Specification	Designation	Comment	Al	Cu	Fe	Mn	Ni	P	Pb	Sn	Zn	OT max	Other	UTS	YS	El
	Norway	NS 16015	Cu-DHP	Wrought Sh Strp Plt Bar Tub		99.85				0.02-0.05								
	Sweden	SIS 145013	5013-10	Wrought Strp		99.85									O 0.02-0.1			
	Switzerland	VSM 11557(Part1)	Cu-DHP	Wrought Tub 1/2 hard		99.85				0.01-0.05						245	147	25
	UK	BS 2873	C106	Wrought Wir		99.85	0.03		0.1	0.01-0.05	0.01	0.01			Te 0.01; Sb 0.01; Bi 0.003 max; As 0.05			
C15100	USA				0.005 max			0.005 max							Cu+Ag 99.82 min; Zr 0.05-0.15			
	International	ISO 1634	CuAg0.05(P)	Wrought Plt Sh Strp SH(HB)		99.805				0.1					Ag 0.02-0.08	280		8
	International	ISO 1637	CuAg0.05(P)	Wrought HB-SH 5/20 mm diam		99.805				0.1					Ag 0.02-0.08	280		5
	International	ISO 1637	CuAg0.05(P)	Wrought HA 5/40 mm diam		99.805				0.1					Ag 0.02-0.08	250		15
	International	ISO 1637	CuAg0.05(P)	Wrought Ann 5 mm diam		99.805				0.1					Ag 0.02-0.08			35
	International	ISO 1638	CuAg0.05(P)	Wrought Ann 1/1.5 mm diam		99.805				0.1					Ag 0.08-0.12			
	International	ISO 1638	CuAg0.05(P)	Wrought HC 1/5 mm diam		99.805				0.1					Ag 0.08-0.12			
	International	ISO 1638	CuAg0.05(P)	Wrought HD-SH 1/3 mm diam		99.805				0.1					Ag 0.08-0.12			
	International	ISO R1336	CuAg0.05(P)	Wrought Strp Rod Bar Wir		99.805				0.01 max					Ag 0.02-0.08			
C15000	USA			Zr Cu											Cu+Ag 99.80 min; Zr 0.10-0.20			
C18050	USA			High copper											Cu+Ag+named el 99.8 min; Te 0.01-0.15; Cr 0.05-0.15			
CopperW6 0189	USA	AWS A5.6-84R	ECu	Arc Weld El	0.1 max		0.2 max	0.1 max			0.02 max			0.5	Cu including Ag,Zn,Sn,Ni,P included in OT; Si 0.1 max			
C15760	USA				0.58-0.62		0.01 max				0.01 max				Cu+Ag 99.77 min; Al as Al2O3; 0.04% O as Cu2O; O 0.52-0.6			
	International	ISO 1634	CuAg0.1(P)	Wrought Plt Sh Strp HC		99.765				0.1					Ag 0.08-0.12	340		4
	International	ISO 1637	CuAg0.1(P)	Wrought HB-SH 5/20 mm		99.765				0.1					Ag 0.08-0.12	280		5
	International	ISO 1637	CuAg0.1(P)	Wrought HA 5/40 mm diam		99.765				0.1					Ag 0.08-0.12	250		15
	International	ISO 1637	CuAg0.1(P)	Wrought Ann 5 mm diam		99.765				0.1					Ag 0.08-0.12			35
	International	ISO 1638	CuAg0.1(P)	Wrought Ann 1/1.5 mm diam		99.765				0.1					Ag 0.08-0.12	210		25
	International	ISO 1638	CuAg0.1(P)	Wrought HC 1/5 mm diam		99.765				0.1					Ag 0.08-0.12	390		
	International	ISO 1638	CuAg0.1(P)	Wrought HD-SH 1/3 mm diam		99.765				0.1					Ag 0.08-0.12	420		
	International	ISO R1336	CuAg0.1(P)	Wrought Strp Rod Bar Wir Ann		99.765				0.01 max					Ag 0.08-0.12			45
C15500	USA									0.04-0.08					Cu +Ag 99.75 min; Mg 0.08-0.13; Ag 0.027-0.1			

UNS numbers and US grades are provided as a means of cross referencing chemically similar alloys. Exchangability is only possible after independent examination of specifications. Tensile properties are minimum or typical . UTS and YS as Mpa, El as %. See Appendix for list of abbreviations used in Descriptions.

Worldwide Guide to Equivalent Nonferrous Metals and Alloys

Grade UNS #	Country	Specification	Designation	Comment	Al	Cu	Fe	Mn	Ni	P	Pb	Sn	Zn	OT max	Other	UTS	YS	El
C18150	USA			High copper											Cu+Ag+named el 99.7 min; Zr 0.05-0.25; Cr 0.5-1.5			
C15715	USA				0.13-0.17		0.01 max				0.01 max				Cu+Ag 99.62 min; Al as Al2O3; 0.04% O as Cu2O; O 0.12-0.2			
C19015	USA			High copper					0.5-2.4	0.02-0.2					Cu+Ag+named el 99.8 min; Si 0.1-0.4; Mg 0.02-0.15			
Phosphors Deoxidized	India	IS 6912	PhosphorusDeoxidized	Wrought Frg Ann 6 mm diam			0.03		0.15	0.02-0.1	0.01	0.01		0.07	ST 99.2 min; ST 0.02; ST = Cu+Ag; ST = Te+Se; Te 0.01; As 0.2-0.5			
	UK	Bs 2901 pt3	C8	Wrought Rod Wir	0.1-0.3	99.4 min	0.03 max		0.1 max	0.02 max	0.01 max				ST 0.25-0.5; ST=Al+Ti; Ti 0.1-0.3; Sb 0.01; As 0.05 max;			
C18040	USA			High copper								0.2-0.3	0.05-0.15		Cu+Ag+named el 99.6 min; Cr 0.25-0.35			
C15720	USA				0.18-0.22		0.01 max				0.01 max				Cu+Ag 99.52 min; Al as Al2O3; 0.04% O as Cu2O; O 0.16-0.2			
	Czech Republic	CSN 423005	Cu99,5		0.05 max	99.5 min	0.1 max				0.1 max	0.2 max			Sb 0.08; Bi 0.01; S 0.05; O 0.10; Se+Te:0.03; As 0.1 max			
	Japan	JIS C2532	GCN15	Wrought Wir Ann .2 mm diam		99 min	0.5	1.5 max	8-12						Ni includes Co; Cu includes Ni+Mn	245		20
	Denmark	DS 3003	5031	Wrought		99.5 min									Cu includes Ag; Ag 0.08-0.12			
	Japan	JIS C2532	GCN30	Wrought Wir Ann .2 mm diam		99 min	0.05	1.5 max	20-25						Cu includes Ni+Mn; Ni includes Co	295		20
	South Africa	SABS 460	Cu-ETP	Wrought Tub		99 min					0.01			0.03	Cu includes Ag; Bi 0.001 max			
C16200	Czech Republic	CSN 423058	CuCd1												Cd 0.90-1.20			
C16200	International	ISO 1637	CuCd1	Wrought HB SH 5/18 mm diam		98.39								0.3	Cd 0.7-1.3	410		8
C16200	International	ISO 1637	CuCd1	Wrought HA 18/30 mm diam		98.39								0.3	Cd 0.7-1.3	350		10
C16200	International	ISO 1638	CuCd1	Wrought HC 1/5 mm diam		98.39								0.3	Cd 0.7-1.3	490		
C16200	International	ISO 1638	CuCd1	Wrought HD-SH 1/3 mm diam		98.39								0.3	Cd 0.7-1.3	490		
C16200	International	ISO R1336	CuCd1	Wrought Rod Bar Wir		98.39								0.3	Cd 0.7-1.3	590		
C16200	UK	BS 2873	C108	Wrought Wir		98.75								0.05	Cd 0.5-1.2			
C16200	UK	BS 2875	C108	Wrought Plt Hard 10/16 mm diam		98.75								0.05	Cd 0.5-1.2	309		13
Cadmium Copper C16200	USA	ASTM B105	99Cu-1Cd	Wrought Rod Wir Plt Sh Strp		98.78-99.3	0.02 max								Cd 0.7-1.2			
Cadmium Copper	USA	ASTM B9	99Cu-1Cd	Wrought Rod Wir Plt Sh Strp		98.78-99.3	0.02 max								Cd 0.7-1.2			

UNS numbers and US grades are provided as a means of cross referencing chemically similar alloys. Exchangability is only possible after independent examination of specifications. Tensile properties are minimum or typical . UTS and YS as Mpa, El as %. See Appendix for list of abbreviations used in Descriptions.

4-20 Wrought Copper

Grade UNS #	Country	Specification	Designation	Comment	Al	Cu	Fe	Mn	Ni	P	Pb	Sn	Zn	OT max	Other	UTS	YS	El
Cadmium Copper C16200	USA	SAE J463	99Cu-1Cd	Wrought Rod Wir Plt Sh Strp		98.78-99.3	0.02 max								Cd 0.7-1.2			
Cadmium copper C16200	USA			High copper			0.02 max								Cu+Ag+named el 99.5 min; Cd 0.7-1.2			
C16210	Australia	AS 1567	162	Wrought Rod Bar Soft Ann 1/2 hard 12/25 mm diam											Cd 0.5-1.2	290		15
C16500	USA			High copper			0.02 max					0.5-0.7			Cu+Ag+named el 99.5 min; Cd 0.6-1			
C17410	USA			High copper	0.2 max		0.2 max								Cu+Ag+named el 99.5 min; Si 0.2 max; Co 0.35-0.6; Be 0.15-0.5			
C18000	USA			High copper			0.15 max		2-3						Cu+Ag+named el 99.5 min; Ni includes Co; Si 0.4-0.8; Cr 0.1-0.6			
C18135	USA			High copper											Cu+Ag+named el 99.5 min; Cr 0.2-0.6; Cd 0.2-0.6			
C18140	USA			High copper											Cu+Ag+named el 99.5 min; Zr 0.05-0.25; Si 0.005-0.05; Cr 0.15-0.45			
C18200	International	ISO 1637	CuCr1	Wrought TL 5/25 mm diam		98.49								0.3	Cr 0.3-1.2	500	440	5
C18200	International	ISO 1637	CuCr1	Wrought TH 5/25 mm diam		98.49								0.3	Cr 0.3-1.2	440	350	10
C18200	International	ISO 1637	CuCr1	Wrought ST and PT 5/80 mm diam		98.49								0.3	Cr 0.3-1.2	370	270	18
C18200	International	ISO 1640	CuCr1	Wrought Frg ST and art aged		98.49								0.3	Cr 0.03-1.2	340		14
C18200	International	ISO R1336	CuCr1	Wrought Plt Sh Rod Bar Tub Wir Frg		98.49								0.3	Cr 0.03-1.2			
Chromium Copper C18200	USA	ASTM F9	99Cu-1Cr	Wrought Rod Plt Sh Tub Frg			0.1 max				0.05 max			0.5	Si 0.1 max; Cr 0.6-1.2			
Chromium copper C18200	USA			High copper			0.1 max				0.05 max				Cu+Ag+named el 99.5 min; Si 0.1 max; Cr 0.6-1.2			
Chromium copper C18400	USA			High copper			0.15 max		0.05				0.7 max		Cu+Ag+named el 99.5 min; Ca 0.005; Li 0.05; Si 0.1 max; Cr 0.4-1.2; AS 0.005 max			
Chromium copper C18500	USA			High copper					0.04		0.015 max				Cu+Ag+named el 99.5 min; Cr 0.4-1; Ag 0.08-0.12			
Free-Machining CuC18700	USA	ASTM B301	99Cu-1Pb	Wrought Rod Plt Sh							0.8-1.5			0.1				
Free-Machining CuC18700	USA	SAE J463	99Cu-1Pb	Wrought Rod Plt Sh							0.8-1.5			0.1				

UNS numbers and US grades are provided as a means of cross referencing chemically similar alloys. Exchangability is only possible after independent examination of specifications. Tensile properties are minimum or typical . UTS and YS as Mpa, El as %. See Appendix for list of abbreviations used in Descriptions.

Worldwide Guide to Equivalent Nonferrous Metals and Alloys

Grade UNS #	Country	Specification	Designation	Comment	Al	Cu	Fe	Mn	Ni	P	Pb	Sn	Zn	OT max	Other	UTS	YS	El
C18700	USA			High copper							0.8-1.5				Cu+Ag+named el 99.5 min			
C18900	USA			High copper	0.01 max			0.1-0.3		0.05	0.02 max	0.6-0.9	0.1 max		Cu+Ag+named el 99.5 min; Si 0.15-0.4			
	India	IS 5743	CuMn30(MasterAlloy)	Wrought Bar		99.5		29-31							Mn included in Cu content; bal Zn			
	Japan	JIS Z3202	YCuCopper	Wrought Rod		99.5					0.03 max							
C50100	USA			Copper tin alloy; Phosphor bronze		0.05 max				0.01-0.05	0.05 max	0.5-0.8			Cu+named el 99.5 min			
C15725	USA				0.23-0.27	0.01 max				0.01 max					Cu+Ag 99.43 min; Al as Al2O3; 0.04% O as Cu2O; O 0.2-0.3			
C14200	Canada	CSA	Cu-DPA(142)			99.4 max				0.02-0.04					As 0.15-0.5			
C14200	Czech Republic	CSN 423009	Cu99,2As			0.1 max		0.2-0.5		0.1 max					Sb 0.08; Bi 0.01; S 0.05; O 0.09; Se+Te 0.03; As 0.1-0.5			
C14200	International	ISO 1634	CuAs(P)	Wrought Plt Sh Strp SH		99.14				0.01-0.05						280		8
C14200	International	ISO 1634	CuAs(P)	Wrought Plt Sh Strp Ann		99.14				0.01-0.05								30
C14200	International	ISO 1634	CuAs(P)	Wrought Plt Sh Strp M		99.14				0.01-0.05						220		35
C14200	International	ISO 1635	CuAs(P)	Wrought Tub SH (HB)		99.14				0.01-0.05					As 0.15-0.5	290		10
C14200	International	ISO 1635	CuAs(P)	Wrought Tub HA		99.14				0.01-0.05					As 0.15-0.5	250		20
C14200	International	ISO 1635	CuAs(P)	Wrought Tub Ann		99.14				0.01-0.05					As 0.15-0.5			35
C14200	International	ISO 1637	CuAs(P)	Wrought HB-SH 5/20 mm diam		99.14				0.01-0.05					As 0.15-0.5	280		5
C14200	International	ISO 1637	CuAs(P)	Wrought HA 5/40 mm diam		99.14				0.01-0.05					As 0.15-0.5	250		15
C14200	International	ISO 1637	CuAs(P)	Wrought Ann 5 mm diam		99.14				0.01-0.05					As 0.15-0.5			35
C14200	International	ISO 1640	CuAs(P)	Wrought Frg As cast		99.14				0.01-0.05					As 0.15-0.5	220		40
C14200	USA	CDA	DPA	P deox arsenical						0.015-0.04					Cu+Ag 99.4 min; As 0.15-0.5			
C14200	USA		DPA	P deox arsenical						0.015-0.04					Cu+Ag 99.4 min; As 0.15-0.5			
	International	ISO 1637	CuS(P0.01)	Wrought HA 5/40 mm diam		99.378				0.01 max					0.1 P+others; S 0.2-0.5	250		10
	International	ISO 1637	CuS(P0.01)	Wrought Ann 5 mm diam		99.378				0.01 max					0.1 P+others; S 0.2-0.5			28
	International	ISO R1336	CuS(P0.01)	Wrought Rod Bar		99.378				0.01 max					S 0.2-0.5			
C15735	USA		99.3Cu-.7Al(2)O(3)	Inactive designation	99.19-99.35		0.01 max				0.01 max				ST 0.65-0.75			
	International	ISO 1637	CuS(P0.03)	Wrought HA 5/40 mm diam		99.34				0.01-0.05					0.1 P+others; S 0.2-0.5	250		10
	International	ISO 1637	CuS(P0.03)	Wrought Ann 5 mm diam		99.34				0.01-0.05					0.1 P+others; S 0.2-0.5			28
	International	ISO R1336	CuS(P0.03)	Wrought Rod Bar		99.34				0.01-0.05			0.1		S 0.2-0.5			
	Japan	JIS H3100	C1401	Wrought Sh Plt Hard .5 mm diam		99.3			0.1-0.2									

UNS numbers and US grades are provided as a means of cross referencing chemically similar alloys. Exchangability is only possible after independent examination of specifications. Tensile properties are minimum or typical . UTS and YS as Mpa, El as %. See Appendix for list of abbreviations used in Descriptions.

4-22 Wrought Copper

Grade UNS #	Country	Specification	Designation	Comment	Al	Cu	Fe	Mn	Ni	P	Pb	Sn	Zn	OT max	Other	UTS	YS	El
C18030	USA			High copper						0.005-0.015		0.08-0.12			Cu+Ag+named el 99.9 min; Cr 0.1-0.2			
C19020	USA			High copper					0.5-3	0.01-0.2		0.3-0.9			Cu+Ag+named el 99.8 min; Mn+Si 0.35			
C19750	USA			High copper			1.2	0.05 max	0.05 max	0.1-0.4	0.05 max	0.05-0.4	0.2 max		Cu+named el 99.8 min; Mg 0.01-0.2; Co 0.05 max			
	UK	BS 2901 Part3	C10	Wrought Rod Wir	0.03 max	98.3 min	4.5-5.5			0.02-0.4	0.02 max							
C65100	Canada	CSA HC.4.7	HC.4.S2(651)	Wrought Bar Sh Strp Plt Ann 70 mm diam		96	0.8 max	0.7				0.05	1.5	0.5	Si 0.8-2	262		
C65100	Canada	CSA HC.4.7	HC.4.S2(651)	Wrought Bar Sh Strp Pl Quarter Hard		96	0.8 max	0.7				0.05	1.5	0.5	Si 0.8-2	331		
C65100	Canada	CSA HC.4.7	HC.4.S2(651)	Wrought Bar Sh Strp Plt Full hard		96	0.8 max	0.7				0.05	1.5	0.5	Si 0.8-2	414		
Low-Silicon Bronze C65100	USA	ASME SB315	98.5Cu-1.5Si	Wrought Bar Rod Wir Plt Sh Tub Shapes			0.8 max	0.7 max				0.05 max	1.5 max		Si 0.8-2			
Low-Silicon Bronze C65100	USA	ASME SB98	98.5Cu-1.5Si	Wrought Bar Rod Wir Plt Sh Tub Shapes			0.8 max	0.7 max				0.05 max	1.5 max		Si 0.8-2			
Low-Silicon Bronze C65100	USA	ASTM B315	98.5Cu-1.5Si	Wrought Bar Rod Wir Plt Sh Tub Shapes			0.8 max	0.7 max				0.05 max	1.5 max		Si 0.8-2			
Low-Silicon Bronze C65100	USA	ASTM B97	98.5Cu-1.5Si	Wrought Bar Rod Wir Plt Sh Tub Shapes			0.8 max	0.7 max				0.05 max	1.5 max		Si 0.8-2			
Low-Silicon Bronze C65100	USA	ASTM B98	98.5Cu-1.5Si	Wrought Bar Rod Wir Plt Sh Tub Shapes			0.8 max	0.7 max				0.05 max	1.5 max		Si 0.8-2			
Low-Silicon Bronze C65100	USA	ASTM B99	98.5Cu-1.5Si	Wrought Bar Rod Wir Plt Sh Tub Shapes			0.8 max	0.7 max				0.05 max	1.5 max		Si 0.8-2			
Low-Silicon Bronze C65100	USA	QQ C-591	98.5Cu-1.5Si	Wrought Bar Rod Wir Plt Sh Tub Shapes			0.8 max	0.7 max				0.05 max	1.5 max		Si 0.8-2			
Low siicon bronze BC65100	USA			Silicon bronze			0.8 max	0.7 max				0.05 max	1.5 max		Cu+named el 99.5 min; Cu incl Ag; Si 0.8-2			
	Switzerland	VSM 11557(Part1)	Cu-DPA	Wrought tub 1/2 hard		99.2				0.01-0.05					As 0.15-0.5	245	147	25
	UK	BS 2875	C105	Wrought Plt As Mfg or ann 10/16 mm diam		99.2			0.15			0.03			0.030 Se+Te; Sb 0.01; O 0.10; Bi 0.01; As 0.3-0.5	279		15
	UK	BS 2875	C105	Wrought Plt As Mfg or ann 10 mm diam		99.2			0.15			0.03			0.030 Se+Te; Sb 0.01; O 0.10; Bi 0.01; As 0.3-0.5	221		35
C19200	USA	ASTM B111	98.97Cu-1Fe-.03P	Wrought Strp Tub		98.7-99.19	0.8-1.2			0.01-0.04								
C19200 C19200	USA	ASTM B359	98.97Cu-1Fe-.03P	Wrought Strp Tub		98.7-99.19	0.8-1.2			0.01-0.04								
C19200 C19200	USA	ASTM B395	98.97Cu-1Fe-.03P	Wrought Strp Tub		98.7-99.19	0.8-1.2			0.01-0.04								

UNS numbers and US grades are provided as a means of cross referencing chemically similar alloys. Exchangability is only possible after independent examination of specifications. Tensile properties are minimum or typical . UTS and YS as Mpa, El as %. See Appendix for list of abbreviations used in Descriptions.

Worldwide Guide to Equivalent Nonferrous Metals and Alloys

Grade UNS #	Country	Specification	Designation	Comment	Al	Cu	Fe	Mn	Ni	P	Pb	Sn	Zn	OT max	Other	UTS	YS	El
C19200 C19200	USA	ASTM B469	98.97Cu-1Fe-.03P	Wrought Strp Tub		98.7-99.19	0.8-1.2			0.01-0.04								
C19200	USA			High copper		98.5 min	0.8-1.2			0.01-0.04			0.2 max		Cu+named el 99.8 min			
C19100	USA			High copper			0.2 max		0.9-1.3	0.15-0.35	0.1 max		0.5 max		Cu+Ag+named el 99.9 min; Te 0.35-0.6			
C19210	USA			High copper			0.05-0.15			0.025-0.04					Cu+named el 99.8 min			
C18070	USA			High copper											Cu+Ag+named el 99.8 min; Cu+Ag 99.0 min; Ti 0.01-0.4; Si 0.02-0.07; Cr 0.15-0.4			
C19000	USA			High copper			0.1 max		0.9-1.3	0.15-0.35	0.05 max		0.8 max		Cu+Ag+named el 99.9 min			
C19010	USA			High copper					0.8-1.8	0.01-0.05					Cu+Ag+named el 99.5 min; Si 0.15-0.35			
C19700	USA			High copper			0.3-1.2	0.05 max	0.05 max	0.1-0.4	0.05 max	0.2 max	0.2 max		Cu+named el 99.8 min; Mg 0.01-0.2; Co 0.05 max			
C50200	USA			Copper tin alloy; Phosphor bronze			0.1 max			0.04 max	0.05 max	1-1.5			Cu+named el 99.5 min			
	Japan	JIS H3100	C2051	Wrought Sh Plt Ma Mfg .35 mm diam		98-99									bal Zn	216		38
C50500	International	ISO 1634	CuSn2	Wrought Plt Sh Strp Ann			0.1		0.3	0.01-0.3	0.05	1-2.5	0.3	0.3				40
C50500	International	ISO 1638	CuSn2	Wrought HC 1/3 mm diam			0.1		0.3	0.01-0.3	0.05	1-2.5	0.3	0.3		360		10
C50500	International	ISO 1638	CuSn2	Wrought Ann 1/5 mm diam			0.1		0.3	0.01-0.3	0.05	1-2.5	0.3	0.3		260		35
C50500	International	ISO 427	CuSn2	Strp Rod Bar Tub Wir			0.1		0.3	0.01-0.3	0.05	1-2.5	0.3	0.3				
Phosphor Bronze C50500	USA	ASTM B105	98.7Cu-1.3Sn	Wrought Wir Strp			0.1 max			0.35 max		1-1.7	0.3 max		ST 99.5 min			
C50500	USA			Copper tin alloy; Phosphor bronze 1.25% E			0.1 max			0.03-0.35	0.05 max	1-1.7	0.3 max		Cu+named el 99.5 min			
C19280	USA			High copper			0.5-1.5			0.005-0.015		0.3-0.7	0.3-0.7		Cu+named el 99.8 min			
	Czech Republic	CSN 423010	E CuSn									1.1						
C19600	USA			High copper			0.9-1.2			0.25-0.35			0.35 max		Cu+named el 99.8 min			
C18100	USA			High copper											Cu+Ag+named el 99.5 min; Cu+Ag 98.7 min; Zr 0.08-0.20; Mg 0.03-0.06; Cr 0.4-1.2			
	International	ISO 1634	CuNi44Mn1	Wrought Plt Sh Strp			0.5	0.5-2					0.2	0.1	Ni+Co 43.0-45.0, Sn+P .02; S 0.05; C 0.1			
	International	ISO 1638	CuNi44Mn1	Wrought Rod Wir			0.5	0.5-2					0.2	0.1	Ni+Co 43.0-45.0; Sn+P .02; S 0.05; C 0.1			
	International	ISO 429	CuNi44Mn1	Wrought Wir Strp			0.5	0.5-2					0.2	0.1	Ni+Co 43.0-45.0; Sn+P .02; S 0.05; C 0.1			

UNS numbers and US grades are provided as a means of cross referencing chemically similar alloys. Exchangability is only possible after independent examination of specifications. Tensile properties are minimum or typical . UTS and YS as Mpa, El as %. See Appendix for list of abbreviations used in Descriptions.

4-24 Wrought Copper

Grade UNS #	Country	Specification	Designation	Comment	Al	Cu	Fe	Mn	Ni	P	Pb	Sn	Zn	OT max	Other	UTS	YS	El
	UK	BS 2901 pt3	C7	Wrought Rod Wir	0.03 max	98.5 min	0.03	0.15-0.35	0.1 max	0.02 max	0.1 max	1 max			Si 0.2-0.35; Bi 0.002 max; Bi 0.002 max			
C19260	USA			High copper		98.5 min	0.4-0.8								Cu+named el 99.5 min; Ti 0.2-0.4; Mg 0.02-0.15			
C50700	USA			Copper tin alloy; Phosphor bronze			0.1 max			0.3 max	0.05 max	1.5-2			Cu+named el 99.5 min			
	UK	BS 2901 pt3	C24	Wrought Rod Wir	0.01 max		0.1 max	1.5-2.5			0.02 max		0.2 max		ST 0.5 max; ST=max. impurity limit other elements			
C17510	USA			High copper	0.2 max		0.1 max		1.4-2.2						Cu+Ag+named el 99.5 min; Si 0.2 max; Co 0.3 max; Be 0.2-0.6			
	Sweden	SIS 145055	5055-25	Wrought SAnn 80 mm diam		98.39								0.3	Cd 0.7-1.3	431	370	3
C50710	USA			Copper tin alloy; Phosphor bronze					0.1-0.4	0.15 max		1.7-2.3			Cu+named el 99.5 min			
C50715	USA			Copper tin alloy; Phosphor bronze			0.05-0.15			0.025-0.04	0.02 max	1.7-2.3			Cu+named el 99.5 min; Cu+Sn+Fe+P 99.5 min			
C18990	USA			High copper						0.005-0.015		1.8-2.2			Cu+Ag+named el 99.9 min; Cr 0.1-0.2			
Copper C18980	USA	AWS A5.27-85	RCu	Rod Oxyfuel Gas Weld	0.01 max		0.5 max			0.15 max	0.02 max	1 max		0.5	Cu includes Ag; Si 0.5 max			
Copper C18980	USA	AWS A5.7-91R	ERCu	Bare Weld Rod El	0.01 max	98 min	0.5 max			0.15 max	0.02 max	1 max		0.5	Cu includes Ag; Si 0.5 max	172		
C18980	USA			High copper			0.5 max			0.15	0.02 max	1 max			Cu+Ag+named el 99.5 min; Cu+Ag 98.0 min; Si 0.5			
C64700	Czech Republic	CSN 423054	CuNi2Si				0.3 max		1.8-2.2		0.1 max	0.1 max	0.1 max		Si 0.5-1			
C64700	International	ISO 1634	CuNi2Si	Wrought Plt Sh Strp TD 200 mm diam		94.59			1.6-2.5					0.5	Si 0.5-0.8			
C64700	International	ISO 1634	CuNi2Si	Wrought Plt Sh Strp TH 200 mm diam		94.59			1.6-2.5					0.5	Si 0.5-.8			
C64700	International	ISO 1637	CuNi2Si	Wrought SHT and nat aged 30 mm diam					1.6-2.5					0.5	Si 0.5-.8			
C64700	International	ISO 1637	CuNi2Si	Wrought SHT and art aged 30 mm diam					1.6-2.5					0.5	Si 0.5-.8			
C64700	International	ISO R1187	CuNi2Si	Wrought Strp Rod Bar Wir					1.6-2.5					0.5	Si 0.05-0.08			
C64700	USA			Silicon bronze			0.1 max				0.1 max		0.5 max		Cu+named el 99.5 min; Ni+Co 1.6-2.2; Cu incl Ag; Si 0.4-0.8			
C64900	USA			Silicon bronze	0.1 max		0.1 max				0.05 max	1.2-1.6	0.2 max		Cu+named el 99.5 min; Ni+Co 0.10 max; Cu incl Ag; Si 0.8-1.2			
C70200	USA			Copper nickel alloy			0.1 max	0.4 max			0.05 max				Cu+named el 99.5 min; Ni+Co 2.0-3.0			

UNS numbers and US grades are provided as a means of cross referencing chemically similar alloys. Exchangability is only possible after independent examination of specifications. Tensile properties are minimum or typical . UTS and YS as Mpa, El as %. See Appendix for list of abbreviations used in Descriptions.

Worldwide Guide to Equivalent Nonferrous Metals and Alloys

Grade UNS #	Country	Specification	Designation	Comment	Al	Cu	Fe	Mn	Ni	P	Pb	Sn	Zn	OT max	Other	UTS	YS	El
	UK	BS 2870	CZ125	Wrought Sh Strp		95-98	0.05				0.02			0.25	bal Zn			
	France	NF A51-109	CuBe1.9	Wrought sh Bar Hard 10 mm diam		97.85									0.25 Ni+Co; Be 1.9	690		2
	France	NF A51-109	CuBe1.9	Wrought sh Bar 1/2 hard 10 mm diam		97.85									0.25 Ni+Co; Be 1.9	580		6
	France	NF A51-109	CuBe1.9	Wrought sh Bar Soft 10 mm diam		97.85									0.25 Ni+Co; Be 1.9	410		35
HSM Copper C19400	USA	ASME SB543	Cu-2.3Fe-.03P-.12Zn	Wrought Plt Sh Strp Tub			2.1-2.6			0.15 max	0.03 max	0.03 max	0.05-0.2	0.15				
HSM Copper C19400	USA	ASTM B465	Cu-2.3Fe-.03P-.12Zn	Wrought Plt Sh Strp Tub			2.1-2.6			0.15 max	0.03 max	0.03 max	0.05-0.2	0.15				
HSM Copper C19400	USA	ASTM B543	Cu-2.3Fe-.03P-.12Zn	Wrought Plt Sh Strp Tub			2.1-2.6			0.15 max	0.03 max	0.03 max	0.05-0.2	0.15				
HSM Copper C19400	USA	ASTM B586	Cu-2.3Fe-.03P-.12Zn	Wrought Plt Sh Strp Tub			2.1-2.6			0.15 max	0.03 max	0.03 max	0.05-0.2	0.15				
C19400	USA			High copper		97 min	2.1-2.6			0.015-0.15	0.03 max		0.05-0.2					
C19450	USA			High copper			1.5-3			0.005-0.05		0.8-2.5			Cu+named el 99.8 min			
C40400	USA			Copper zinc tin alloy; tin brass									2-3		Cu+named el 99.7 min; S 0.35-0.7			
Silicon Bronze W60656	USA	AWS A5.6-84R	ECuSi	Arc Weld El	0.01 max		0.5 max	1.5 max			0.02 max	1.5 max		0.5	Cu including Ag,Zn,Ni,P included in OT; Si 2.4-4			
C19030	USA			High copper			0.1 max		1.5-2	0.01-0.03	0.02 max	1-1.5			Cu+Ag+named el 99.5 min			
C19410	USA			High copper			1.8-2.3			0.015-0.05		0.6-0.9	0.1-0.2		Cu+named el 99.8 min			
C70250	USA			Copper nickel alloy			0.2 max	0.1 max			0.05 max		1 max		Cu+named el 99.5 min; Ni+Co 2.2-4.2; Si 0.25-1.2; Mg 0.05-0.3			
C50900	USA			Copper tin alloy; Phosphor bronze			0.1 max			0.03-0.3	0.05 max	2.5-3.8	0.3 max		Cu+named el 99.5 min			
C17500	International	ISO 1634	CuCo2Be	Wrought Plt Sh Strp TH		95.49									0.5 Ni+Fe; Co 2-2.8; Be 0.4-0.7	780	640	5
C17500	International	ISO 1634	CuCo2Be	Wrought Plt Sh Strp ST and PT		95.49									0.5 Ni+Fe; Co 2-2.8; Be 0.4-0.7	690	490	8
C17500	International	ISO 1634	CuCo2Be	Wrought Plt Sh Strp SHT and nat aged		95.49									0.5 Ni+Fe; Co 2-2.8; Be 0.4-0.7	340	150	25
C17500	International	ISO 1637	CuCo2Be	Wrought SHT and art aged 60 mm diam		98.49									0.5 Ni+Fe; Co 2-2.8; Be 0.4-0.7	700	500	8
C17500	International	ISO 1638	CuCo2Be	Wrought TD 1/3 mm diam		96.49									Co 2-2.8; Be 0.4-0.7	490		3
C17500	International	ISO 1638	CuCo2Be	Wrought SGT and art aged 1/3 mm diam		96.49									Co 2-2.8; Be 0.4-0.7	640		8
C17500	International	ISO 1638	CuCo2Be	Wrought SHT and nat aged 1/3 mm diam		96.49									Co 2-2.8; Be 0.4-0.7	290		25
C17500	International	ISO R1187	CuCo2Be	Wrought Strp Rod Bar Wir		95.49									0.5 Ni+Fe; Co 2-2.8; Be 0.4-0.7			

UNS numbers and US grades are provided as a means of cross referencing chemically similar alloys. Exchangability is only possible after independent examination of specifications. Tensile properties are minimum or typical . UTS and YS as Mpa, El as %. See Appendix for list of abbreviations used in Descriptions.

4-26 Wrought Copper

Grade UNS #	Country	Specification	Designation	Comment	Al	Cu	Fe	Mn	Ni	P	Pb	Sn	Zn	OT max	Other	UTS	YS	El
Low-Beryllium Copper C17500	USA	ASTM B441	97Cu-.5Be-2.5Co	Wrought Bar Rod Wir Strp Tub Frg			0.1 max							0.5	Co 2.4-2.7; Be 0.4-0.7			
Low-Beryllium Copper C17500	USA	ASTM B534	97Cu-.5Be-2.5Co	Wrought Bar Rod Wir Strp Tub Frg			0.1 max							0.5	Co 2.4-2.7; Be 0.4-0.7			
Low-Beryllium Copper C17500	USA	MIL C-46087	97Cu-.5Be-2.5Co	Wrought Bar Rod Wir Strp Tub Frg			0.1 max							0.5	Co 2.4-2.7; Be 0.4-0.7			
Low-Beryllium Copper C17500	USA	MIL C-81021	97Cu-.5Be-2.5Co	Wrought Bar Rod Wir Strp Tub Frg			0.1 max							0.5	Co 2.4-2.7; Be 0.4-0.7			
Low-Beryllium Copper C17500	USA	RWMA Class III	97Cu-.5Be-2.5Co	Wrought Bar Rod Wir Strp Tub Frg			0.1 max							0.5	Co 2.4-2.7; Be 0.4-0.7			
Low-Beryllium Copper C17500	USA	SAE J463	97Cu-.5Be-2.5Co	Wrought Bar Rod Wir Strp Tub Frg			0.1 max							0.5	Co 2.4-2.7; Be 0.4-0.7			
Beryllium copper C17500	USA			High copper	0.2 max		0.1 max								Cu+Ag+named el 99.5 min; Si 0.2 max; Co 2.4-2.7; Be 0.4-0.7			
C19220	USA			High copper			0.3		0.1-0.25	0.03-0.07			0.05-0.1		Cu+named el 99.8 min; B 0.005-0.015			
C65600	USA			Silicon bronze	0.01 max		0.5 max	1.5 max			0.02 max	1.5 max	1.5 max		Cu+named el 99.5 min; Cu incl Ag; Si 2.8-4			
C66100	USA			Silicon bronze			0.25 max	1.5 max			0.2-0.8		1.5 max		Cu+named el 99.5 min; Cu incl Ag; Si 2.8-3.5			
	International	ISO 1634	CuNi1Si	Wrought Plt Sh Strp TD 3 mm diam		97.19			1-1.6					0.5	Si 0.4-0.7	450	340	8
	International	ISO 1634	CuNi1Si	Wrought Plt Sh Strp TH 3 mm diam		97.19			1-1.6					0.5	Si 0.4-.07	640	540	10
	International	ISO 1637	CuNi1Si	Wrought SHT and nat aged 30 mm diam		97.19			1-1.6					0.5	Si 0.4-.07			
	International	ISO 1637	CuNi1Si	Wrought SHT and art aged 30 mm diam		97.19			1-1.6					0.5	Si 0.4-.07			
	International	ISO R1187	CuNi1Si	Wrought Strp Rod Bar Wir		97.19			1-1.6					0.5	Si 0.4-.07			
C63400	USA			Aluminum bronze	2.6-3.2		0.15 max				0.05 max	0.2 max	0.5 max		Cu+named el 99.5 min; Ni+Co 0.15 max; Cu incl Ag; Si 0.25-0.45; As 0.15 max			
C19900	USA			High copper											Cu+named el 99.5 min; Ti 2.9-3.4			
C70100	USA			Copper nickel alloy			0.05 max	0.5 max					0.25 max		Cu+named el 99.5 min; Ni+Co 3.0-4.0			
C65500	Australia	AS 1566	655	Wrought Plt Bar Sh Strp Ann (sht,strp)			0.8 max	0.5-1.3	0.6		0.05		1.5		Si 2.8-3.8	360		40
C65500	Australia	AS 1566	655	Plt Bar Sh Strp SH to 1/4 hard (sht,strp)			0.8 max	0.5-1.3	0.6		0.05		1.5		Si 2.8-3.8	370		

UNS numbers and US grades are provided as a means of cross referencing chemically similar alloys. Exchangability is only possible after independent examination of specifications. Tensile properties are minimum or typical . UTS and YS as Mpa, El as %. See Appendix for list of abbreviations used in Descriptions.

Grade UNS #	Country	Specification	Designation	Comment	Al	Cu	Fe	Mn	Ni	P	Pb	Sn	Zn	OT max	Other	UTS	YS	El
C65500	Australia	AS 1567	655	Wrought Rod Bar As Mfg 6/20 mm diam			0.8 max	0.5-1.3	0.6		0.05		1.5		Si 2.8-3.8	480		15
C65500	Australia	AS 1567	655	Wrought Rod Bar Soft Ann 6/70 mm diam			0.8 max	0.5-1.3	0.6		0.05		1.5		Si 2.8-3.8	370		40
C65500	Australia	AS 1568	655	Wrought Frg As Mfg 6 mm diam			0.8 max	0.5-1.3	0.6		0.05		1.5		Si 2.8-3.8	340		20
C65500	Canada	CSA HC.4.7	HC.4.S3(655)	Wrought Bar Sh Strp Plt Ann 70 mm diam		94.8	1.6 max	1.5	0.6		0.06 max			0.5	Si 2.8-3.8	359		
C65500	Canada	CSA HC.4.7	HC.4.S3(655)	Wrought Bar Sh Strp Plt HR		94.8	1.6 max	1.5	0.6		0.06 max			0.5	Si 2.8-3.8	379		
C65500	Canada	CSA HC.4.7	HC.4.S3(655)	Wrought Bar Sh Strp Plt Quarter Hard		94.8	1.6 max	1.5	0.6		0.06 max			0.5	Si 2.8-3.8	427		
C65500	International	ISO 1634	CuSi3Mn1	Wrought Plt Sh Strp M			0.3	0.7-1.5	0.3		0.03		0.5		1.0 Fe+Ni+Pb+Zn; Si 2.7-3.5	410	120	40
C65500	International	ISO 1637	CuSi3Mn1	Wrought M 5 mm diam			0.3	0.7-1.5	0.3		0.03		0.5		1.0 Fe+Ni+Pb+Zn; Si 2.7-3.5	410	120	30
C65500	International	ISO R1187	CuSi3Mn1	Wrought plt Sh Strp Rod Bar Tub Wir Frg			0.3	0.7-1.5	0.3		0.03		0.5		1.0 Fe+Ni+Pb+Zn; Si 2.7-3.5			
C65500	UK	BS 2870	CS101	Wrought Sh Strp As Mfg 2.7/10 mm diam			0.25	0.75-1.25						0.5	Si 2.75-3.25	259		50
C65500	UK	BS 2872	CS101	Wrought Frg As Mfg or Ann 6 mm diam			0.25	0.75-1.25						0.5	Si 2.75-3.25	339		20
C65500	UK	BS 2873	CS101	Wrought Wir			0.25	0.75-1.25						0.5	Si 2.75-3.25			
C65500	UK	BS 2874	CS101	Wrought Rod As Mfg 6/20 mm diam			0.25	0.75-1.25						0.5	Si 2.75-3.25	539		15
C65500	UK	BS 2874	CS101	Wrought Rod Ann 6/70 mm diam			0.25	0.75-1.25						0.5	Si 2.75-3.25	367		40
C65500	UK	BS 2875	CS101	Wrought Plt As Mfg 10 mm diam			0.25	0.75-1.25						0.5	Si 2.75-3.25	339		31
C65500	UK	BS 2875	CS101	Wrought Plt Ann 10 mm diam			0.25	0.75-1.25						0.5	Si 2.75-3.25	319		40
High-Silicon Bronze C65500	USA	AMS 4615	97Cu-3Si	Wrought Rod Wir Sh Tub			0.8 max	1.5 max	0.6 max		0.5 max		1.5 max		Si 2.8-3.8			
High-Silicon Bronze C65500	USA	AMS 4665	97Cu-3Si	Wrought Rod Wir Sh Tub			0.8 max	1.5 max	0.6 max		0.5 max		1.5 max		Si 2.8-3.8			
High-Silicon Bronze C65500	USA	ASME SB315	97Cu-3Si	Wrought Rod Wir Sh Tub			0.8 max	1.5 max	0.6 max		0.5 max		1.5 max		Si 2.8-3.8			
High-Silicon Bronze C65500	USA	ASME SB96	97Cu-3Si	Wrought Rod Wir Sh Tub			0.8 max	1.5 max	0.6 max		0.5 max		1.5 max		Si 2.8-3.8			
High-Silicon Bronze C65500	USA	ASME SB98	97Cu-3Si	Wrought Rod Wir Sh Tub			0.8 max	1.5 max	0.6 max		0.5 max		1.5 max		Si 2.8-3.8			

UNS numbers and US grades are provided as a means of cross referencing chemically similar alloys. Exchangability is only possible after independent examination of specifications. Tensile properties are minimum or typical. UTS and YS as Mpa, El as %. See Appendix for list of abbreviations used in Descriptions.

Grade UNS #	Country	Specification	Designation	Comment	Al	Cu	Fe	Mn	Ni	P	Pb	Sn	Zn	OT max	Other	UTS	YS	El
High-Silicon Bronze C65500	USA	ASTM B100	97Cu-3Si	Wrought Rod Wir Sh Tub			0.8 max	1.5 max	0.6 max		0.5 max		1.5 max		Si 2.8-3.8			
High-Silicon Bronze C65500	USA	ASTM B124	97Cu-3Si	Wrought Rod Wir Sh Tub			0.8 max	1.5 max	0.6 max		0.5 max		1.5 max		Si 2.8-3.8			
High-Silicon Bronze C65500	USA	ASTM B283	97Cu-3Si	Wrought Rod Wir Sh Tub			0.8 max	1.5 max	0.6 max		0.5 max		1.5 max		Si 2.8-3.8			
High-Silicon Bronze C65500	USA	ASTM B315	97Cu-3Si	Wrought Rod Wir Sh Tub			0.8 max	1.5 max	0.6 max		0.5 max		1.5 max		Si 2.8-3.8			
High-Silicon Bronze C65500	USA	ASTM B96	97Cu-3Si	Wrought Rod Wir Sh Tub			0.8 max	1.5 max	0.6· max		0.5 max		1.5 max		Si 2.8-3.8			
High-Silicon Bronze C65500	USA	ASTM B97	97Cu-3Si	Wrought Rod Wir Sh Tub			0.8 max	1.5 max	0.6 max		0.5 max		1.5 max		Si 2.8-3.8			
High-Silicon Bronze C65500	USA	ASTM B98	97Cu-3Si	Wrought Rod Wir Sh Tub			0.8 max	1.5 max	0.6 max		0.5 max		1.5 max		Si 2.8-3.8			
High-Silicon Bronze C65500	USA	ASTM B99	97Cu-3Si	Wrought Rod Wir Sh Tub			0.8 max	1.5 max	0.6 max		0.5 max		1.5 max		Si 2.8-3.8			
High-Silicon Bronze C65500	USA	MIL T-8231	97Cu-3Si	Wrought Rod Wir Sh Tub			0.8 max	1.5 max	0.6 max		0.5 max		1.5 max		Si 2.8-3.8			
High-Silicon Bronze C65500	USA	QQ C-591	97Cu-3Si	Wrought Rod Wir Sh Tub			0.8 max	1.5 max	0.6 max		0.5 max		1.5 max		Si 2.8-3.8			
High-Silicon Bronze C65500	USA	SAE J463	97Cu-3Si	Wrought Rod Wir Sh Tub			0.8 max	1.5 max	0.6 max		0.5 max		1.5 max		Si 2.8-3.8			
High silicon bronze AC65500	USA			Silicon bronze			0.8 max	0.5-1.3			0.05 max		1.5 max		Cu+named el 99.5 min; Ni+Co 0.6 max; Cu incl Ag; Si 2.8-3.8			
Silicon bronze C65600	USA	AWS A5.27-85	RCuSi-A	Rod Oxyfuel Gas Weld			0.5 max	1.5 max			0.02 max	1 max		0.5	Cu includes Ag; Si 2.8-4			
Silicon bronze C65600	USA	AWS A5.7-91R	ERCuSi-A	Bare Weld Rod El	0.01 max		0.5 max	1.5 max			0.02 max	1 max		0.5	Cu included Ag; Si 2.8-4	345		
C17000	France	NF A51-109	CuBe1.7	Wrought Sh Bar Hard 10 mm diam		99.05									0.25 Ni+Co; Be 1.7	690		2
C17000	France	NF A51-109	CuBe1.7	Wrought Sh Bar 1/2 Hard 10 mm diam		99.05									0.25 Ni+Co; Be 1.7	580		6
C17000	France	NF A51-109	CuBe1.7	Wrought Sh Bar Soft 10 mm diam		99.05									0.25 Ni+Co; Be 1.7	410		35
C17000	International	ISO 1634	CuBe1.7	Wrought Plt Sh Strp St and PT		96.49									0.20-0.60 Ni+Co; Be 1.6-1.8	1180	980	2
C17000	International	ISO 1634	CuBe1.7	Wrought Plt Sh StrpSHT and Nat aged		96.49									0.20-0.60 Ni+Co; Be 1.6-1.8	440	200	35

UNS numbers and US grades are provided as a means of cross referencing chemically similar alloys. Exchangability is only possible after independent examination of specifications. Tensile properties are minimum or typical . UTS and YS as Mpa, El as %. See Appendix for list of abbreviations used in Descriptions.

Grade UNS #	Country	Specification	Designation	Comment	Al	Cu	Fe	Mn	Ni	P	Pb	Sn	Zn	OT max	Other	UTS	YS	El
C17000	International	ISO 1634	CuBe1.7	Wrought Plt Sh Strp TH		96.49									0.20-0.60 Ni+Co; Be 1.6-1.8	1280	1080	
C17000	International	ISO 1634	CuBe2	Wrought Plt Sh Strp SHT and nat aged		96.19									0.20-0.60 Ni+Co; Be 1.6-1.8	490	250	35
C17000	International	ISO 1634	CuBe2	Wrought Plt Sh Strp ST and PT		96.19									0.20-0.60 Ni+Co; Be 1.6-1.8	1280	1080	
C17000	International	ISO 1634	CuBe2	Wrought Plt Sh Strp TH		96.19									0.20-0.60 Ni+Co; Be 1.6-1.8	1370	1180	
C17000	International	ISO 1638	CuBe1.7	Wrought SHT and Nat aged 1/5 mm diam		96.99									0.20-0.60 Ni+Co; Be 1.6-1.8	390		30
C17000	International	ISO 1638	CuBe1.7	Wrought TD 1/3 mm diam		96.99									0.20-0.60 Ni+Co; Be 1.6-1.8	780		
C17000	International	ISO 1638	CuBe1.7	Wrought TH 1/3 mm diam		96.99									0.20-0.60 Ni+Co; Be 1.6-1.8	1230		
C17000	International	ISO R1187	CuBe1.7CoNi	Wrought Plt Sh Strp Rod Bar Tub Wir		96.49									0.20-0.60 Ni+Co; Be 1.6-1.8			
C17000	Japan	JIS H3130	C1700	Wrought Sh Plt Hard .16/1.6 mm diam		97.1									0.20 Ni+Co; Be 1.6-1.8	686		2
C17000	Japan	JIS H3130	C1700	Wrought Sh Plt 1/2 hard .16/1.6 mm diam		97.1									0.20 Ni+Co; Be 1.6-1.8	588		5
C17000	Japan	JIS H3130	C1700	Wrought Sh Plt Ann .16/1.6 mm diam		97.1									0.20 Ni+Co; Be 1.6-1.8	412		35
C17000	UK	BS 2870	CB101	Wrought Sh Strp SHT 10 mm diam		97.19									0.05-0.40 Ni+Co; Be 1.7-1.9	293		40
C17000	UK	BS 2873	CB101	Wrought Wir SHT .5/10 mm diam		97.19									0.05-0.40 Ni+Co; Be 1.7-1.9	382		30
C17000	UK	BS 2873	CB101	Wrought Wir SHT and CW 3 mm diam		97.19									0.05-0.40 Ni+Co; Be 1.7-1.9	755		
C17000	UK	BS 2873	CB101	Wrought Wir SHT and PT .5/10 mm diam		97.19									0.05-0.40 Ni+Co; Be 1.7-1.9	1029		
Beryllium Copper C17000	USA	ASTM B194	98Cu-1.7Be-.3Co	Rod Plt Sh Strp Tub Billets Frg Extrusions											ST 0.2-0.6; Be 1.6-1.79			
Beryllium Copper C17000	USA	ASTM B196	98Cu-1.7Be-.3Co	Rod Plt Sh Strp Tub Billets Frg Extrusions											ST 0.2-0.6; Be 1.6-1.79			
Beryllium Copper C17000	USA	ASTM B570	98Cu-1.7Be-.3Co	Rod Plt Sh Strp Tub Billets Frg Extrusions											ST 0.2-0.6; Be 1.6-1.79			
Beryllium Copper C17000	USA	QQ C-533	98Cu-1.7Be-.3Co	Wrought Rod Plt Sh Strp Tub Billets Frg Extrusions											ST 0.2-0.6; Be 1.6-1.79			
Beryllium Copper C17000	USA	SAE J463	98Cu-1.7Be-.3Co	Wrought Rod Plt Sh Strp Tub Billets Frg Extrusions											ST 0.2-0.6; Be 1.6-1.79			
Beryllium copper C17000	USA			High copper	0.2 max										Cu+Ag+named el 99.5 min; Ni+Co 0.20 min; Ni+Fe+Co 0.6 max; Si 0.2 max; Be 1.6-1.79			

UNS numbers and US grades are provided as a means of cross referencing chemically similar alloys. Exchangability is only possible after independent examination of specifications. Tensile properties are minimum or typical . UTS and YS as Mpa, El as %. See Appendix for list of abbreviations used in Descriptions.

Grade UNS #	Country	Specification	Designation	Comment	Al	Cu	Fe	Mn	Ni	P	Pb	Sn	Zn	OT max	Other	UTS	YS	El
C19520	USA			High copper		96.6 min	0.5-1.5			0.01-0.35		0.50-1.5			Cu+named el 99.8 min			
	France	NF A51111	CuSn4P	Wrought Wir Ann .05/3 mm diam			0.1		0.3	0.1-0.4	0.06 max	3-5.5		0.3		295		45
	India	IS 5743	CuBe4	Wrought Bar		94.5-96.5									Be 3-5			
Copper-silicon	India	IS 6912	Copper-silicon	Wrought Frg As Mfg 6 mm diam			0.25 max	0.75-1.25						0.25	Si 2.75-3.25	345		25
Copper-silicon	India	IS 6912	Copper-silicon	Wrought Frg			0.25 max	0.75-1.25						0.25	Si 2.75-3.25			
	Japan	JIS H3270	C5101	Wrought Rod Hard 13 mm diam		93.7-96.5				0.03-0.35		3-5.5			Sn 3-5.5; 99.5 min Cu+Sn+P	451		10
	UK	BS 2901 pt3	C9	Wrought Rod Wir	0.03 max		0.1 max	0.75-1.25	0.1 max	0.02 max	0.02 max		0.5 max		Si 2.75-3.25; Bi 0.002 max; As 0.05 max			
C51100	Czech Republic	CSN 423013	CuSn4							0.3 max	0.1 max	3-5						
C51100	France	NF A51108	CuSn4P	Wrought Sh Hard 20 mm diam			0.1 max			0.35	0.1	3-5	0.5	0.3		560	500	15
C51100	France	NF A51108	CuSn4P	Wrought Sh 1/2 hard 20 mm diam			0.1 max			0.35	0.1	3-5	0.5	0.3		420	350	30
C51100	France	NF A51108	CuSn4P	Wrought Sh Ann 20 mm diam			0.1 max			0.35	0.1	3-5	0.5	0.3		300	150	50
C51100	Norway	NS 16304	CuSn4			96	0.1 max			0.02-0.4	0.05 max	3-5.5	0.3	0.3				
C51100	UK	BS 2870	PB101	Wrought Sh Strp Ann						0.02-0.4	0.02	3-4.5		0.2				
C51100	UK	BS 2870	PB101	Wrought Sh Strp 1/2 hard 10 mm diam						0.02-0.4	0.02	3-4.5		0.2				
C51100	UK	BS 2870	PB101	Wrought Sh Strp Hard 6 mm diam						0.02-0.4	0.02	3-4.5		0.2				
C51100	UK	BS 2875	PB101	Wrought Plt As Mfg or Ann 10 mm diam						0.02-0.4	0.02	3-4.5		0.2				
C51100	UK	BS 2875	PB101	Wrought Plt Hard 10/16 mm diam						0.02-0.4	0.02	3-4.5		0.2				
C51100 C51100	USA	ASTM B100	95.6Cu-4.2Sn-.2P	Wrought Sh Plt		94.5-96.3	0.1 max			0.03-0.35	0.05 max	3.5-4.9	0.3 max					
C51100 C51100	USA	ASTM B103	95.6Cu-4.2Sn-.2P	Wrought Sh Plt		94.5-96.3	0.1 max			0.03-0.35	0.05 max	3.5-4.9	0.3 max					
C51100	USA			Copper tin alloy; Phosphor bronze			0.1 max			0.03-0.35	0.05 max	3.5-4.9	0.3 max		Cu+named el 99.5 min			
	Japan	JIS H3110	C5101	Wrought Sh Plt Extra Hard .3/5 mm diam		93.65-96.47				0.03-0.35		3-5.5			Sn 3-5.5; 99.5 min Cu+Sn+P	539		4
	Japan	JIS H3110	C5101	Wrought Sh Plt Hard .3/5 mm diam		93.65-96.47				0.03-0.35		3-5.5			Sn 3-5.5; 99.5 min Cu+Sn+P	490		7
	Japan	JIS H3110	C5101	Wrought Sh Plt Ann .3/5 mm diam		93.65-96.47				0.03-0.35		3-5.5			Sn 3-5.5; 99.5 min Cu+Sn+P	294		38
C17200	International	ISO R1187	CuBe2CoNi	Wrought Plt Sh Strp Rod Bar Tub Wir Frg		96.19									0.20-0.60 Ni+Co; Be 1.6-1.8			
C17200	Japan	JIS H3130	C1720	Wrought Sh Plt Hard .16/1.6 mm diam		96.9									0.20 Ni+Co; Be 1.8-2	686		2

UNS numbers and US grades are provided as a means of cross referencing chemically similar alloys. Exchangability is only possible after independent examination of specifications. Tensile properties are minimum or typical . UTS and YS as Mpa, El as %. See Appendix for list of abbreviations used in Descriptions.

Worldwide Guide to Equivalent Nonferrous Metals and Alloys

Grade UNS #	Country	Specification	Designation	Comment	Al	Cu	Fe	Mn	Ni	P	Pb	Sn	Zn	OT max	Other	UTS	YS	El
C17200	Japan	JIS H3130	C1720	Wrought Sh Plt 1/2 hard .16/1.6 mm diam		96.9									0.20 Ni+Co; Be 1.8-2	588		5
C17200	Japan	JIS H3130	C1720	Wrought Sh Plt Ann .16/1.6 mm diam		96.9									0.20 Ni+Co; Be 1.8-2	412		35
C17200	Japan	JIS H3270	C1720	Wrought Rod Bar Ann 6 mm diam		96.9									0.20 Ni+Co; Be 1.8-2	412		
C17200	Japan	JIS H3270	C1720	Wrought Rod Bar Hard 6 mm diam		96.9									0.20 Ni+Co; Be 1.8-2	647		
C17200	Norway	NS 16355	CuBe2CoNi												ST 99.5 min; ST=Cu+Be+Co+ Fe+Ni; Be 1.8-2			
Beryllium Copper C17200	USA	AMS 4530	Alloy 25	Wrought Bar Rod Plt Strp Tub Billets Frg							0.1 max			0.5	ST 0.2-0.6; Be 1.8-2			
Beryllium Copper C17200	USA	AMS 4532	Alloy 25	Wrought Bar Rod Plt Strp Tub Billets Frg							0.1 max			0.5	ST 0.2-0.6; Be 1.8-2			
Beryllium Copper C17200	USA	AMS 4650	Alloy 25	Wrought Bar Rod Plt Strp Tub Billets Frg							0.1 max			0.5	ST 0.2-0.6; Be 1.8-2			
Beryllium Copper C17200	USA	AMS 4725	Alloy 25	Wrought Bar Rod Plt Strp Tub Billets Frg							0.1 max			0.5	ST 0.2-0.6; Be 1.8-2			
Beryllium Copper C17200	USA	ASTM B194	Alloy 25	Wrought Bar Rod Plt Strp Tub Billets Frg							0.1 max			0.5	ST 0.2-0.6; Be 1.8-2			
Beryllium Copper C17200	USA	ASTM B196	Alloy 25	Wrought Bar Rod Plt Strp Tub Billets Frg							0.1 max			0.5	ST 0.2-0.6; Be 1.8-2			
Beryllium Copper C17200	USA	ASTM B197	Alloy 25	Wrought Bar Rod Plt Strp Tub Billets Frg							0.1 max			0.5	ST 0.2-0.6; Be 1.8-2			
Beryllium Copper C17200	USA	ASTM B570	Alloy 25	Wrought Bar Rod Plt Strp Tub Billets Frg							0.1 max			0.5	ST 0.2-0.6; Be 1.8-2			
Beryllium Copper C17200	USA	MIL C-21657	Alloy 25	Wrought Bar Rod Plt Strp Tub Billets Frg							0.1 max			0.5	ST 0.2-0.6; Be 1.8-2			
Beryllium Copper C17200	USA	QQ C-530	Alloy 25	Wrought Bar Rod Plt Strp Tub Billets Frg							0.1 max			0.5	ST 0.2-0.6; Be 1.8-2			
Beryllium Copper	USA	RWMA Class IV	Alloy 25	Bar Rod Plt Strp Tub Frg							0.1 max			0.5	ST 0.2-0.6; Be 1.8-2			
Beryllium Copper C17200	USA	SAE J463	Alloy 25	Wrought Bar Rod Plt Strp Tub Billets Frg							0.1 max			0.5	ST 0.2-0.6; Be 1.8-2			
Beryllium copper C17200	USA			High copper	0.2 max										Cu+Ag+named el 99.5 min; Ni+Co 0.20 min; Ni+Fe+Co 0.6 max; Si 0.2 max; Be 1.8-2			
C17300	USA			High copper	0.2 max						0.2-0.6				Cu+Ag+named el 99.5 min; Ni+Co 0.20 min; Ni+Fe+Co 0.6 max; Si 0.2 max; Be 1.8-2			
C63600	USA			Aluminum bronze	3-4		0.15 max				0.05 max	0.2 max	0.5 max		Cu+named el 99.5 min; Ni+Co 0.15 max; Cu incl Ag; Si 0.7-1.3; As 0.15 max			

UNS numbers and US grades are provided as a means of cross referencing chemically similar alloys. Exchangability is only possible after independent examination of specifications. Tensile properties are minimum or typical . UTS and YS as Mpa, El as %. See Appendix for list of abbreviations used in Descriptions.

4-32 Wrought Copper

Grade UNS #	Country	Specification	Designation	Comment	Al	Cu	Fe	Mn	Ni	P	Pb	Sn	Zn	OT max	Other	UTS	YS	El
	Czech Republic	CSN 423053	CuSi3Mn1				0.3 max	1-1.5	0.1 max	0.1 max	0.05 max	0.3 max	0.5 max		Si 2.8-3.5			
C65400	USA			Silicon bronze							0.05 max	1.2-1.9	0.5 max		Cu+named el 99.5 min; Cu incl Ag; Si 2.7-3.4; Cr 0.01-0.12			
C18090	USA			High copper					0.5-1.2			0.5-1.2			Cu+Ag+named el 99.85 min; Cu+Ag 96.0 min; Ti 0.15-0.8; Cr 0.2-1			
C19500	USA			High copper	0.02 max	96 min	1-2			0.01-0.35	0.02 max	0.1-1	0.2 max		Cu+named el 99.8 min; Co 0.3-1.3			
C21000	Czech Republic	CSN 423200	CuZn4			95-97					0.1 max				Fe+Mn+Al+Sn+ Sb 0.02; bal Zn			
C21000	International	ISO 1634	CuZn5	Wrought Plt Sh Strp HB-SH 0.2/5 mm diam		94-96	0.1				0.05				Cu total includes 0.3 Ni; bal Zn	310		8
C21000	International	ISO 1634	CuZn5	Wrought Plt Sh Strp HA		94-96	0.1				0.05				Cu total includes 0.3 Ni; bal Zn	260		19
C21000	International	ISO 1634	CuZn5	Wrought Plt Sh Strp Ann		94-96	0.1				0.05				Cu total includes 0.3 Ni; bal Zn			33
C21000	International	ISO 1638	CuZn5	Wrought HC 1/5 mm diam		94-96	0.1				0.05				0.3 Ni included in Cu; bal Zn	320		5
C21000	International	ISO 1638	CuZn5	Wrought Ann 1/5 mm diam		94-96	0.1				0.05				0.3 Ni included in Cu; bal Zn	220		30
C21000	International	ISO 4261	CuZn5	Wrought Plt Sh Strp Tub Wir		94-96	0.1				0.05				0.3 Pb+Fe; bal Zn			
C21000	Japan	JIS H3100	C2100	Wrought Sh Plt 1/2 hard .3/20 mm diam		94-96									bal Zn	255		18
C21000	Japan	JIS H3100	C2100	Wrought Sh Plt Ann .3/30 mm diam		94-96									bal Zn	206		33
C21000	Japan	JIS H3100	C2100	Wrought Sh Plt Hard .3/10 mm diam		94-96									bal Zn	284		
C21000	Japan	JIS H3260	C2100	Wrought Rod Ann .5 mm diam		94-96									bal Zn	206		20
C21000	Japan	JIS H3260	C2100	Wrought Rod 1/2 hard .5/10 mm diam		94-96									bal Zn	324		
C21000	Japan	JIS H3260	C2100	Wrought Rod Hard .5/10 mm diam		94-96									bal Zn	412		
C21000	Mexico	DGN W-27	TU-95	Wrought Sh 1/4 hard		94-96			0.2		0.03				bal Zn	274		
C21000	Mexico	DGN W-27	TU-95	Wrought Sh 1/2 hard		94-96			0.2		0.03				bal Zn	294		
C21000	Mexico	DGN W-27	TU-95	Wrought Sh Hard		94-96			0.2		0.03				bal Zn	343		
Gilding Metal C21000	USA	ASTM B134	95Cu-5Zn	Wrought Bar Wir Plt Sh Strp		94-96	0.05 max				0.05 max		3.9-6					
Gilding Metal C21000	USA	ASTM B36	95Cu-5Zn	Wrought Bar Wir Plt Sh Strp		94-96	0.05 max				0.05 max		3.9-6					
Gilding Metal C21000	USA	MIL C-21768	95Cu-5Zn	Wrought Bar Wir Plt Sh Strp		94-96	0.05 max				0.05 max		3.9-6					
Gilding Metal C21000	USA	QQ W-321	95Cu-5Zn	Wrought Bar Wir Plt Sh Strp		94-96	0.05 max				0.05 max		3.9-6					
Gilding Metal C21000	USA	SAE J463	95Cu-5Zn	Wrought Bar Wir Plt Sh Strp		94-96	0.05 max				0.05 max		3.9-6					

UNS numbers and US grades are provided as a means of cross referencing chemically similar alloys. Exchangability is only possible after independent examination of specifications. Tensile properties are minimum or typical . UTS and YS as Mpa, El as %. See Appendix for list of abbreviations used in Descriptions.

Worldwide Guide to Equivalent Nonferrous Metals and Alloys

Grade UNS #	Country	Specification	Designation	Comment	Al	Cu	Fe	Mn	Ni	P	Pb	Sn	Zn	OT max	Other	UTS	YS	El
C21000	USA			Copper zinc alloy; brass; Gilding 95%		94-96	0.05 max				0.03 max				Cu+named el 99.8 min; bal Zn			
Penny Bronze C40500	USA	ASTM B591	95Cu-4Zn-1Sn	Wrought Bar Sh Strp		94-96	0.05 max				0.05 max	0.7-1.3	2.6-5.3					
C40500	USA			Copper zinc tin alloy; tin brass		94-96	0.05 max				0.05 max				Cu+named el 99.7 min; bal Zn; S 0.7-1.3			
C40800 C40800	USA	ASTM B591	95Cu-2Sn-3Zn	Wrought Strp		94-96	0.05 max				0.05 max	1.8-2.2						
	Canada	CSA HC.4.2	HC.4.Z5(210)	Wrought Plt Half Hard		94-96	0.05 max				0.03			0.01	bal Zn	289		
	Canada	CSA HC.4.2	HC.4.Z5(210)	Wrought Plt Full Hard		94-96	0.05 max				0.03			0.01	bal Zn	344		
	Canada	CSA HC.4.2	HC.4.Z5(210)	Wrought Plt Spring		94-96	0.05 max				0.03			0.01	bal Zn	413		
	Canada	CSA HC.4.2	HC.4.Z5(210)	Wrought Sh Half Hard		94-96	0.05 max				0.03			0.1	bal Zn	255		
	Canada	CSA HC.4.2	HC.4.Z5(210)	Wrought Bar Half Hard		94-96	0.05 max				0.03			0.1	bal Zn	289		
	Canada	CSA HC.4.2	HC.4.Z5(210)	Wrought Bar Full Hard		94-96	0.05 max				0.03			0.1	bal Zn	344		
	Canada	CSA HC.4.2	HC.4.Z5(210)	Wrought Sh Full Hard		94-96	0.05 max				0.03			0.1	bal Zn	344		
	Canada	CSA HC.4.2	HC.4.Z5(210)	Wrought Bar Spring		94-96	0.05 max				0.03			0.1	bal Zn	413		
	Canada	CSA HC.4.2	HC.4.Z5(210)	Wrought Sh Spring		94-96	0.05 max				0.03			0.1	bal Zn	413		
	Czech Republic	CSN 423042	CuAl5		4-6		0.4 max	0.5 max	0.5 max		0.03 max	0.1 max	0.5 max					
Aluminum Bronze	USA	ASTM B169	95Cu-5Al	Wrought Bar Rod Plt Sh Strp Frg	4-7	92-96	0.5 max							0.5				
Aluminum Bronze	USA	QQ C-450	95Cu-5Al	Wrought Bar Rod Plt Sh Strp Frg	4-7	92-96	0.5 max							0.5				
Aluminum Bronze	USA	QQ C-465	95Cu-5Al	Wrought Bar Rod Plt Sh Strp Frg	4-7	92-96	0.5 max							0.5				
Aluminum Bronze	USA	QQ C-645	95Cu-5Al	Wrought Bar Rod Plt Sh Strp Frg	4-7	92-96	0.5 max							0.5				
Phosphor Bronze W60518	USA	AWS A5.6-84R	ECuSn-A	Arc Weld El	0.01 max		0.25 max			0.05-0.35	0.02 max	4-6		0.5	Cu including Ag, Zn,Mn,Si,Ni, included in OT			
C51800	Australia	AS 1566	518	Wrought Plt Bar Sh Strp Ann (sht,strp)						0.1-0.35	0.02	4-6				310	350	35
C51800	Australia	AS 1566	518	Wrought Plt Bar Sh Strp SH to 1/2 hard (sht,strp)						0.1-0.35	0.02	4-6				500	220	
C51800	Australia	AS 1566	518	Wrought Plt Bar Sh Strp SH to full hard (sht,strp)						0.1-0.35	0.02	4-6				590		
C51800	Australia	AS 1567	518	Wrought Rod Bar As Mfg 6/20 mm diam						0.1-0.35	0.02	4-6				460	350	12
C51800	Denmark	DS 3003	5428	Wrought		94	0.1 max			0.02-0.4	0.05	5.5-7.5	0.3					
Phosphor bronze C51800	USA	AWS A5.7-91R	ERCuSn-A	Bare Weld Rod El	0.01 max					0.1-0.35	0.02 max	4-6		0.5	Cu includes Ag	240		
C51800	USA			Copper tin alloy; Phosphor bronze	0.01 max					0.1-0.35	0.02 max	4-6			Cu+named el 99.5 min			

UNS numbers and US grades are provided as a means of cross referencing chemically similar alloys. Exchangability is only possible after independent examination of specifications. Tensile properties are minimum or typical . UTS and YS as Mpa, El as %. See Appendix for list of abbreviations used in Descriptions.

4-34 Wrought Copper

Grade UNS #	Country	Specification	Designation	Comment	Al	Cu	Fe	Mn	Ni	P	Pb	Sn	Zn	OT max	Other	UTS	YS	El
C51000	France	NF	CuSn5P							0.02-0.4		4.5-5.5						
C51000	International	ISO 1634	CuSn4	Wrought Plt Sh Strp SH			0.1		0.3	0.1-0.4	0.05	3-5.5	0.3	0.3		640	570	1
C51000	International	ISO 1634	CuSn4	Wrought Plt Sh Strp Ann			0.1		0.3	0.1-0.4	0.05	3-5.5	0.3	0.3				45
C51000	International	ISO 1637	CuSn4	Wrought HC 5/15 mm diam			0.1		0.3	0.1-0.4	0.05	3-5.5	0.3	0.3		510	390	10
C51000	International	ISO 1637	CuSn4	Wrought HB-SH 5/50 mm diam			0.1		0.3	0.1-0.4	0.05	3-5.5	0.3	0.3		490	360	12
C51000	International	ISO 1637	CuSn4	Wrought HA 5/100 mm diam			0.1		0.3	0.1-0.4	0.05	3-5.5	0.3	0.3		380	250	20
C51000	International	ISO 1638	CuSn4	Wrought HC 1/3 mm diam			0.1		0.3	0.1-0.4	0.05	3-5.5	0.3	0.3		490		3
C51000	International	ISO 1638	CuSn4	Wrought Ann 1/5 mm diam			0.1		0.3	0.1-0.4	0.05	3-5.5	0.3	0.3		310		40
C51000	International	ISO 1638	CuSn4	Wrought HE 1/3 mm diam			0.1		0.3	0.1-0.4	0.05	3-5.5	0.3	0.3		690		
C51000	International	ISO 427	CuSn4	Wrought Plt Sh Strp Bar Tub Wir			0.1		0.3	0.01-0.4	0.05	3-5.5	0.3	0.3				
C51000	UK	BS 2870	PB102	Wrought Sh Strp Ann 10 mm diam						0.02-0.4	0.02	4.5-6		0.2				
C51000	UK	BS 2870	PB102	Wrought Sh Strp 1/2 hard 10 mm diam						0.02-0.4	0.02	4.5-6		0.2				
C51000	UK	BS 2870	PB102	Wrought Sh Strp Hard 6 mm diam						0.02-0.4	0.02	4.5-6		0.2				
C51000	UK	BS 2873	PB102	Wrought Wir Ann .5/10 mm diam						0.02-0.4	0.02	4.5-6		0.2				
C51000	UK	BS 2873	PB102	Wrought Wir 1/2 hard .5/10 mm diam						0.02-0.4	0.02	4.5-6		0.2				
C51000	UK	BS 2873	PB102	Wrought Wir Hard .5/10 mm diam						0.02-0.4	0.02	4.5-6		0.2				
C51000	UK	BS 2874	PB102	Wrought Rod As Mfg 6/10 mm diam						0.02-0.4	0.02	4.5-6		0.2				
C51000	UK	BS 2875	PB102	Wrought Plt As Mfg or Ann 10 mm diam						0.02-0.4	0.02	4.5-6		0.2				
C51000	UK	BS 2875	PB102	Wrought Plt Hard 10/16 mm diam						0.02-0.4	0.02	4.5-6		0.2				
Phosphor Bronze, 5%A C51000	USA	AMS 4510	94.8Cu-5Sn-.2P	Wrought Bar Rod Wir Plt Sh Strp Tub Shapes		93.6-95.6	0.1 max			0.03-0.35	0.05 max	4.2-5.8	0.03 max					
Phosphor Bronze, 5%A C51000	USA	AMS 4625	94.8Cu-5Sn-.2P	Wrought Bar Rod Wir Plt Sh Strp Tub Shapes		93.6-95.6	0.1 max			0.03-0.35	0.05 max	4.2-5.8	0.03 max					
Phosphor Bronze, 5%A C51000	USA	AMS 4720	94.8Cu-5Sn-.2P	Wrought Bar Rod Wir Plt Sh Strp Tub Shapes		93.6-95.6	0.1 max			0.03-0.35	0.05 max	4.2-5.8	0.03 max					
Phosphor Bronze, 5%A C51000	USA	ASTM B100	94.8Cu-5Sn-.2P	Wrought Bar Rod Wir Plt Sh Strp Tub Shapes		93.6-95.6	0.1 max			0.03-0.35	0.05 max	4.2-5.8	0.03 max					
Phosphor Bronze, 5%A	USA	ASTM B103	94.8Cu-5Sn-.2P	Bar Rod Wir Plt Sh Strp Tub Shapes		93.6-95.6	0.1 max			0.03-0.35	0.05 max	4.2-5.8	0.03 max					

UNS numbers and US grades are provided as a means of cross referencing chemically similar alloys. Exchangability is only possible after independent examination of specifications. Tensile properties are minimum or typical . UTS and YS as Mpa, El as %. See Appendix for list of abbreviations used in Descriptions.

Worldwide Guide to Equivalent Nonferrous Metals and Alloys

Grade UNS #	Country	Specification	Designation	Comment	Al	Cu	Fe	Mn	Ni	P	Pb	Sn	Zn	OT max	Other	UTS	YS	El
Phosphor Bronze, 5%A C51000	USA	ASTM B139	94.8Cu-5Sn-.2P	Wrought Bar Rod Wir Plt Sh Strp Tub Shapes		93.6-95.6	0.1 max			0.03-0.35	0.05 max	4.2-5.8	0.03 max					
Phosphor Bronze, 5%A C51000	USA	ASTM B159	94.8Cu-5Sn-.2P	Wrought Bar Rod Wir Plt Sh Strp Tub Shapes		93.6-95.6	0.1 max			0.03-0.35	0.05 max	4.2-5.8	0.03 max					
Phosphor Bronze, 5%A C51000	USA	MIL B-13501	94.8Cu-5Sn-.2P	Wrought Bar Rod Wir Plt Sh Strp Tub Shapes		93.6-95.6	0.1 max			0.03-0.35	0.05 max	4.2-5.8	0.03 max					
Phosphor Bronze, 5%A C51000	USA	MIL W-6712	94.8Cu-5Sn-.2P	Wrought Bar Rod Wir Plt Sh Strp Tub Shapes		93.6-95.6	0.1 max			0.03-0.35	0.05 max	4.2-5.8	0.03 max					
Phosphor Bronze, 5%A C51000	USA	QQ B-750	94.8Cu-5Sn-.2P	Wrought Bar Rod Wir Plt Sh Strp Tub Shapes		93.6-95.6	0.1 max			0.03-0.35	0.05 max	4.2-5.8	0.03 max					
Phosphor Bronze, 5%A C51000	USA	QQ W-321	94.8Cu-5Sn-.2P	Wrought Bar Rod Wir Plt Sh Strp Tub Shapes		93.6-95.6	0.1 max			0.03-0.35	0.05 max	4.2-5.8	0.03 max					
Phosphor Bronze, 5%A C51000	USA	SAE J463	94.8Cu-5Sn-.2P	Wrought Bar Rod Wir Plt Sh Strp Tub Shapes		93.6-95.6	0.1 max			0.03-0.35	0.05 max	4.2-5.8	0.03 max					
C51000	USA			Copper tin alloy; Phosphor bronze 5% A			0.1 max			0.03-0.35	0.05 max	4.2-5.8	0.3 max		Cu+named el 99.5 min			
C63800	USA			Aluminum bronze	2.5-3.1		0.2 max	0.1 max	0.2 max		0.05 max		0.8 max		Cu+named el 99.5 min; Cu incl Ag; Si 1.5-2.1; Co 0.25-0.55			
	Japan	JIS H3300	C6561	Wrought Pip Ann 10/60 mm diam		93.5-95.5						0.5-1.5			99.5 min Cu+Si+Sn; Si 2.5-3.5	343		55
	Japan	JIS H3300	C6561	Wrought Pip 1/2 hard 10/60 mm diam		93.5-95.5						0.5-1.5			99.5 min Cu+Si+Sn; Si 2.5-3.5	441		
	Japan	JIS H3300	C6561	Wrought Pip Hard 10/60 mm diam		93.5-95.5						0.5-1.5			99.5 min Cu+Si+Sn; Si 2.5-3.5	539		
	UK	BS 2870	CA101	Wrought Sh Strp As Mfg 2.7/10 mm diam	4.5-5.5						0.02		0.5			238		40
C53400	Canada	CSA	HC.9.TN55P-SC				0.25 max	0.2 max			1 max	4.5-6	2.5 max		ST 84-89; ST=Cu+elements with specified limits, % min; Sb 0.25			
C53400	Japan	JIS H3270	C5341	Wrought Rod Hard 13 mm diam		91.85-95.17				0.03-0.35	0.8-1.5	3.5-5.8			99.5 min Cu+Sn+Pb+P; bal Zn	412		10
C53400	USA			Copper tin lead alloy; Leaded phosphor bronze B-1			0.1 max			0.03-0.35	0.8-1.2		0.3 max		Cu+named el 99.5 min; S 3.5-5.8			
C55180	USA			Copper phosphorus alloy						4.8-5.2					Cu+named el 99.85 min			
C64710	USA			Silicon bronze		95 min		0.1 max					0.2-0.5		Cu+named el 99.5 min; Ni+Co 2.9-3.5; Cu incl Ag; Si 0.5-0.9			

UNS numbers and US grades are provided as a means of cross referencing chemically similar alloys. Exchangability is only possible after independent examination of specifications. Tensile properties are minimum or typical . UTS and YS as Mpa, El as %. See Appendix for list of abbreviations used in Descriptions.

4-36 Wrought Copper

Grade UNS #	Country	Specification	Designation	Comment	Al	Cu	Fe	Mn	Ni	P	Pb	Sn	Zn	OT max	Other	UTS	YS	El
	France	NF A51101	CuZn5	Wrought Sh StressH T1		95							5			250		
C60800	France	NF	CuAl6		5-6.5		0.1 max				0.1 max		0.3 max	0.5				
C60800	International	ISO 1634	CuAl5	Wrought Plt Sh Strp As cast	4-6.5		0.5	0.5 max	0.8 max	0.1		0.5			0.8 Fe+Pb+Zn; As 0.4 max	410		45
C60800	International	ISO 1635	CuAl5	Wrought Tub As cast	4-6.5		0.5	0.5 max	0.8 max	0.1		0.5			0.8 Fe+Pb+Zn; As 0.4 max	410	100	40
C60800	International	ISO 1635	CuAl5	Wrought Tub Ann	4-6.5		0.5	0.5 max	0.8 max	0.1		0.5			0.8 Fe+Pb+Zn; As 0.4 max			45
C60800	International	ISO 428	CuAl5	Wrought Plt Sh Strp Rod Bar Tub Wir	4-6.5		0.5	0.5 max	0.8 max	0.1		0.5			0.8 Fe+Pb+Zn; As 0.4 max			
C60800	Switzerland	VSM 11557	CuAl5As	Wrought Tub 1/2 hard	4.5-5.5	93	0.1		0.2		0.02		0.3	0.5	As 0.2-0.35	343	137	40
C60800	Switzerland	VSM 11557	CuAl5As	Wrought Tub Q	4.5-5.5	93	0.1		0.2		0.02		0.3	0.5	As 0.2-0.35	314	108	45
Aluminum Bronze C60800	USA	ASME SB111	95Cu-5Al	Wrought Tub	5-6.5	92.5-94.8	0.1 max				0.1 max			0.1	As 0.02-0.35			
Aluminum Bronze C60800	USA	ASME SB359	95Cu-5Al	Wrought Tub	5-6.5	92.5-94.8	0.1 max				0.1 max			0.1	As 0.02-0.35			
Aluminum Bronze C60800	USA	ASME SB395	95Cu-5Al	Wrought Tub	5-6.5	92.5-94.8	0.1 max				0.1 max			0.1	As 0.02-0.35			
Aluminum Bronze C60800	USA	ASTM B111	95Cu-5Al	Wrought Tub	5-6.5	92.5-94.8	0.1 max				0.1 max			0.1	As 0.02-0.35			
Aluminum Bronze C60800	USA	ASTM B359	95Cu-5Al	Wrought Tub	5-6.5	92.5-94.8	0.1 max				0.1 max			0.1	As 0.02-0.35			
Aluminum Bronze C60800	USA	ASTM B395	95Cu-5Al	Wrought Tub	5-6.5	92.5-94.8	0.1 max				0.1 max			0.1	As 0.02-0.35			
C60800	USA			Aluminum bronze	5-6.5		0.1 max				0.1 max				Cu+named el 99.5 min; Cu incl Ag; As 0.02-0.35			
C51900	Czech Republic	CSN 423016	CuSn6							0.3 max	0.1 max	5-7						
C51900	France	NF A51108	CuSn6P	Wrought Sh Hard 20 mm diam		0.1 max				0.35	0.1	5-7.5	0.5	0.3		620	560	5
C51900	France	NF A51108	CuSn6P	Wrought Sh 1/2 hard 20 mm diam		0.1 max				0.35	0.1	5-7.5	0.5	0.3		460	350	20
C51900	France	NF A51108	CuSn6P	Wrought Sh Ann 20 mm diam		0.1 max				0.35	0.1	5-7.5	0.5	0.3		330	160	50
C51900	India	IS 7814	II	Wrought Sh Strp Ann (strp)						0.02-0.4	0.02	4.6-5.5		0.2				
C51900	India	IS 7814	II	Wrought Sh Strp 1/2 hard (strp)						0.02-0.4	0.02	4.6-5.5		0.2				
C51900	India	IS 7814	II	Wrought Sh Strp Hard (strp)						0.02-0.4	0.02	4.6-5.5		0.2				
C51900	International	ISO 427	CuSn6	Wrought Plt Sh Strp Bar Tub Wir			0.1		0.3	0.01-0.4	0.05	5.5-7.5	0.3	0.3				
C51900	Japan	JIS H3110	C5191	Wrought Rod Extra hard .3/5 mm diam		92.15-93.97				0.03-0.35		5.5-7			99.5 min Cu+Sn+P	637		5
C51900	Japan	JIS H3110	C5191	Wrought Rod Hard .3/5 mm diam		92.15-93.97				0.03-0.35		5.5-7			99.5 min Cu+Sn+P	588		8

UNS numbers and US grades are provided as a means of cross referencing chemically similar alloys. Exchangability is only possible after independent examination of specifications. Tensile properties are minimum or typical . UTS and YS as Mpa, El as %. See Appendix for list of abbreviations used in Descriptions.

Worldwide Guide to Equivalent Nonferrous Metals and Alloys

Grade UNS #	Country	Specification	Designation	Comment	Al	Cu	Fe	Mn	Ni	P	Pb	Sn	Zn	OT max	Other	UTS	YS	El
C51900	Japan	JIS H3110	C5191	Wrought Rod Ann .3/5 mm diam		92.15-93.97				0.03-0.35		5.5-7			99.5 min Cu+Sn+P	314		42
C51900	Japan	JIS H3270	C5191	Wrought Rod Hard 13 mm diam		92.15-93.97				0.03-0.35		5.5-7			99.5 min Cu+Sn+P	588		10
C51900	Japan	JIS H3270	C5191	Wrought Rod 1/2 hard 13 mm diam		92.15-93.97				0.03-0.35		5.5-7			99.5 min Cu+Sn+P	461		13
C51900	Norway	NS	CuSn6			94	1 max			0.02-0.4	0.05 max	5.5-7.5	0.03 max					
C51900	Sweden	SIS 145428	5428-02	Wrought Plt Sh Strp Wir Ann 5 mm diam		94	0.1			0.02-0.4	0.05	5.5-7.5	0.3	0.3		340	130	55
C51900	Sweden	SIS 145428	5428-04	Wrought Wir Rod SH .2/2.5 mm diam		94	0.1			0.02-0.4	0.05	5.5-7.5	0.3	0.3		470	390	5
C51900	Sweden	SIS 145428	5428-05	Wrought Sh Strp SH .1/.5 mm diam		94	0.1			0.02-0.4	0.05	5.5-7.5	0.3	0.3		470	390	15
C51900	USA			Copper tin alloy; Phosphor bronze			0.1 max			0.03-0.35	0.05 max	5-7	0.3 max		Cu+named el 99.5 min			
C63200	UK	BS 2901 pt3	C26	Wrought Rod Wir	8.5-9.5		3-5	0.6-3.5	4-5.5		0.02 max		0.1 max		Si 0.1 max			
	Japan	JIS Z3231	DCuSnA	Wrought Wir As drawn 3.2/6 mm diam					0.3		0.02	5-7			0.50 Si+Mn+Pb+Al+Fe+Ni+Zn	245		15
	Japan	JIS Z3264	BCuP-1	Wrought Wir		94.49				4.8-5.3				0.2				
C70500	USA		Copper nickel 7%	Copper nickel alloy			0.1 max	0.15 max			0.05 max		0.2 max		Cu+named el 99.5 min; Ni+Co 5.8-7.8			
C61000	India	IS 1545	CuAl7	Wrought Tub Ann	6-7.5									0.5	ST 1-2.5; ST = Fe+Mn+Ni	481		
C61000	India	IS 1545	CuAl7	Wrought Tub As drawn	6-7.5									0.5	ST 1-2.5; ST = Fe+Mn+Ni			
C61000	International	ISO 1634	CuAl8	Wrought Plt Sh Strp As Cast	7-9		0.5	0.5 max	0.8 max		0.1		0.5		0.8 Fe+Pb+Zn	470		35
C61000	International	ISO 1634	CuAl8	Wrought Plt Sh Strp Ann	7-9		0.5	0.5 max	0.8 max		0.1		0.5		0.8 Fe+Pb+Zn			35
C61000	International	ISO 1637	CuAl8	Wrought HC 5/15 mm diam	7-9		0.5	0.5 max	0.8 max		0.1		0.5		0.8 Fe+Pb+Zn	610	440	18
C61000	International	ISO 1637	CuAl8	Wrought HB-SH 5/50 mm diam	7-9		0.5	0.5 max	0.8 max		0.1		0.5		0.8 Fe+Pb+Zn	570	390	20
C61000	International	ISO 1637	CuAl8	Wrought As cast 5 mm diam	7-9		0.5	0.5 max	0.8 max		0.1		0.5		0.8 Fe+Pb+Zn	440	150	40
C61000	International	ISO 428	CuAl8	Wrought Plt Sh Strp Bar Tub Wir Frg	7-9		0.5	0.5 max	0.8 max		0.1		0.5		0.8 Fe+Pb+Zn			
Aluminum bronze C61000	USA	AWS A5.7-91R	ERCuAl-A1	Bare Weld Rod El	6-8.5			0.5 max			0.02 max			0.5	Cu includes Ag; Si 0.1 max	380		
C61000	USA			Aluminum bronze	6-8.5		0.5 max				0.02 max		0.2 max		Cu+named el 99.5 min; Cu incl Ag; Si 0.1 max			
	UK	BS 2871	CA102	Wrought Tub	6-7.5										1.0-2.5 Ni+Fe+Mn			
	UK	BS 2875	CA102	Wrought Plt As Mfg 10/50 mm diam	6-7.5										1.0-2.5 Fe+Ni+Mn	481		31
	UK	BS 2901 pt3	C25	Wrought Rod Wir	0.05 max		0.05 max	0.15-0.4	1-1.7	0.02-0.04	0.01 max	4.5-6			Ti 0.1 max; Si 0.4-0.7; S 0.01 max			
	International	ISO 1634	CuSn4Zn4	Wrought Plt Sh Strp SH(HD)			0.1			0.1 max	0.1	3-5	3-5	0.3		640	520	3

UNS numbers and US grades are provided as a means of cross referencing chemically similar alloys. Exchangability is only possible after independent examination of specifications. Tensile properties are minimum or typical . UTS and YS as Mpa, El as %. See Appendix for list of abbreviations used in Descriptions.

4-38 Wrought Copper

Grade UNS #	Country	Specification	Designation	Comment	Al	Cu	Fe	Mn	Ni	P	Pb	Sn	Zn	OT max	Other	UTS	YS	El
	International	ISO 1634	CuSn4Zn4	Wrought Plt Sh Strp Ann			0.1			0.1 max	0.1	3-5	3-5	0.3				40
C70400	Canada	CSA	HC.7.NF52(704)				1.3-1.7	0.3-0.8	4.8-6.2		0.05 max		1 max		ST 99.5 min; ST= Cu+elements within specified limits			
C70400	France	NF	CuNi5Fe1Mn				1-1.5	0.3-0.8	4.5-6									
C70400	International	ISO 1634	CuNi5Fe1Mn	Wrought Plt Sh Strp Ann			1-1.5	0.3-0.8	4.5-6				0.5		0.05 Sn+Pb; S 0.05; C 0.1			30
C70400	International	ISO 1635	CuNi5Fe1Mn	Wrought Tub Ann			1-1.5	0.3-0.8	4.5-6				0.5		0.5 Sn+Pb; S 0.05; Co 0.5; C 0.1			30
C70400	International	ISO 429	CuNi5Fe1Mn	Wrought Sh Plt Strp Tub			1-1.5	0.3-0.8	4.5-6				0.5		0.05 Sn+Pb; S 0.05; C 0.1			
C70400	UK	BS 2870	CN101	Wrought Sh Strp As Mfg 10 mm diam			1.05-1.35	0.3-0.8	5-6	0.00.03	0.01	0.01		0.3	Si 0.05; Sb 0.01; S 0.05; C 0.1; As 0.05	162		35
	UK	BS 2901 Part3	C11	Wrought Rod Wir	0.03 max	92.3 min				0.02-0.4	0.02 max	5.5-7.5						
C70400	USA		Copper nickel 5%	Copper nickel alloy			1.3-1.7	0.3-0.8			0.05 max		1 max		Cu+named el 99.5 min; Ni+Co 4.8-6.2			
C64730	USA			Silicon bronze		93.5 min		0.1 max				1-1.5	0.2-0.5		Cu+named el 99.5 min; Ni+Co 2.9-3.5; Cu incl Ag; Si 0.5-0.9			
C41000	USA			Copper zinc tin alloy; tin brass		91-93	0.05 max				0.05 max				Cu+named el 99.7 min; bal Zn; S 2-2.8			
C41300	USA			Copper zinc tin alloy; tin brass		89-93	0.05 max				0.1 max				Cu+named el 99.7 min; bal Zn; S 0.7-1.3			
C41500 C41500	USA	ASTM B591	91Cu-7.2Zn-1.8Sn	Wrought Strp		89-93	0.05 max				0.1 max	1.5-2.2						
C41500	USA			Copper zinc tin alloy; tin brass		89-93	0.05 max				0.1 max				Cu+named el 99.7 min; bal Zn; S 1.5-2.2			
C55181	USA			Copper phosphorus alloy						7-7.5					Cu+named el 99.85 min			
C61550	USA			Aluminum bronze	5.5-6.5		0.2 max	1 max			0.05 max	0.05 max	0.8 max		Cu+named el 99.5 min; Ni+Co 1.5-2.5; Cu incl Ag			
Al Bronze W60614	USA	AWS A5.6-84R	ECuAl-A2	Arc Weld El	6.5-9		0.5-5				0.02 max			0.5	Cu including Ag, Zn, Sn, Mn, Ni included in OT; Si 1.5 max			
	Japan	JIS Z3231	DCuSiA	Wrought Wir As drawn 3.2/6 mm diam		93		3		0.3	0.02				0.50 Pb+Al+Ni+Zn; Si 1-2	245		22
C52100	Canada	CSA	HC.4.TJ80(521)				0.1 max			0.03-0.35	0.05 max	7-9	0.2 max		ST 99.5 min; ST=Cu+elements within specified limits			
C52100	Czech Republic	CSN 423015	CuSn8							0.3 max	0.1 max	7-9						
C52100	France	NF A51111	CuSn8P	Wrought Wir			1 max		0.3 max	0.1-0.4	0.05 max	7.5-9	0.3 max	0.3				
C52100	International	ISO 427	CuSn8	Wrought Plt Sh Strp Bar Tub Wir			0.1		0.3	0.01-0.4	0.05	7.5-9	0.3	0.3				
Phosphor Bronze C52100	USA	ASTM B103	92Cu-8Sn	Wrought Bar Rod Wir Plt Sh Shapes		90.5-92.8	0.1 max			0.03-0.35	0.05 max	7-9	0.2 max					
Phosphor Bronze	USA	ASTM B139	92Cu-8Sn	Bar Rod Wir Plt Sh Shapes		90.5-92.8	0.1 max			0.03-0.35	0.05 max	7-9	0.2 max					

UNS numbers and US grades are provided as a means of cross referencing chemically similar alloys. Exchangability is only possible after independent examination of specifications. Tensile properties are minimum or typical . UTS and YS as Mpa, El as %. See Appendix for list of abbreviations used in Descriptions.

Grade UNS #	Country	Specification	Designation	Comment	Al	Cu	Fe	Mn	Ni	P	Pb	Sn	Zn	OT max	Other	UTS	YS	El
Phosphor Bronze C52100	USA	ASTM B159	92Cu-8Sn	Wrought Bar Rod Wir Plt Sh Shapes		90.5-92.8	0.1 max			0.03-0.35	0.05 max	7-9	0.2 max					
Phosphor Bronze C52100	USA	MIL E-23765	92Cu-8Sn	Wrought Bar Rod Wir Plt Sh Shapes		90.5-92.8	0.1 max			0.03-0.35	0.05 max	7-9	0.2 max					
Phosphor Bronze C52100	USA	SAE J463	92Cu-8Sn	Wrought Bar Rod Wir Plt Sh Shapes		90.5-92.8	0.1 max			0.03-0.35	0.05 max	7-9	0.2 max					
C52100	USA			Copper tin alloy; Phosphor bronze 8% C			0.1 max			0.03-0.35	0.05 max	7-9	0.2 max		Cu+named el 99.5 min			
Phosphor bronze W60521	USA	AWS A5.6-84R	ECuSn-C	Arc Weld El	0.01 max		0.25 max			0.05-0.35	0.02 max	7-9		0.5	Cu including Ag,Zn,Mn,Si, Ni, included in OT			
	Japan	JIS C2532	GCN10	Wrought Wir Ann .2 mm diam		90 min	0.5 max	1.5 max	4-7						Co incl in Ni; Cu min includes Ni+Mn			20
C61400	International	ISO 1634	CuAl8Fe3	Wrought Plt Sh Strp As cast	6.5-8.5		1.5-3.5	0.8 max	1 max		0.05		0.5		0.5 Pb+Zn	510	195	28
C61400	International	ISO 1637	CuAl8Fe3	Wrought HA 5/50 mm diam	6.5-8.5		1.5-3.5	0.8 max	1 max		0.05		0.5		0.5 Pb+Zn	540	220	20
C61400	International	ISO 1637	CuAl8Fe3	Wrought HB-SH 5/15 mm diam	6.5-8.5		1.5-3.5	0.8 max	1 max		0.05		0.5		0.5 Pb+Zn	590	250	20
C61400	International	ISO 1637	CuAl8Fe3	Wrought As cast 5 mm diam	6.5-8.5		1.5-3.5	0.8 max	1 max		0.05		0.5		0.5 Pb+Zn	510	200	25
C61400	International	ISO 428	CuAl8Fe3	Wrought Plt Sh Rod Bar	6.5-8.5		1.5-3.5	0.8 max	1 max		0.05		0.5		0.5 Pb+Zn			
C61400	UK	BS 2872	CA106	Wrought Frg As Mfg or Ann 6/10 mm diam	6.5-8		2-3.5	0.5	0.5		0.05	0.1	0.4	0.5	Si 0.15; Mg 0.05	539	245	30
C61400	UK	BS 2874	CA106	Wrought Rod Ann 6 mm diam	6.5-8		2-3.5	0.5	0.5		0.05	0.1	0.4	0.5	Si 0.15; Mg 0.05	466	196	30
C61400	UK	BS 2874	CA106	Wrought Rod As Mfg 6/10 mm diam	6.5-8		2-3.5	0.5	0.5		0.05	0.1	0.4	0.5	Si 0.15; Mg 0.05	539	240	30
C61400	UK	BS 2875	CA106	Wrought Plt As Mfg 10/50 mm diam	6.5-8		2-3.5	0.5	0.5		0.05	0.1	0.4	0.5	Si 0.15; Mg 0.05	481		31
Aluminum Bronze C61400	USA	ASME SB150	91Cu-7Al-2Fe	Wrought Bar Rod Sh Tub Pip	1.5-2.5	88-92.5		1 max		0.02 max	0.01 max		0.2 max	0.5				
Aluminum Bronze C61400	USA	ASME SB169	91Cu-7Al-2Fe	Wrought Bar Rod Sh Tub Pip	1.5-2.5	88-92.5		1 max		0.02 max	0.01 max		0.2 max	0.5				
Aluminum Bronze C61400	USA	ASME SB171	91Cu-7Al-2Fe	Wrought Bar Rod Sh Tub Pip	1.5-2.5	88-92.5		1 max		0.02 max	0.01 max		0.2 max	0.5				
Aluminum Bronze C61400	USA	ASTM B150	91Cu-7Al-2Fe	Wrought Bar Rod Sh Tub Pip	1.5-2.5	88-92.5		1 max		0.02 max	0.01 max		0.2 max	0.5				
Aluminum Bronze C61400	USA	ASTM B169	91Cu-7Al-2Fe	Wrought Bar Rod Sh Tub Pip	1.5-2.5	88-92.5		1 max		0.02 max	0.01 max		0.2 max	0.5				
Aluminum Bronze C61400	USA	ASTM B171	91Cu-7Al-2Fe	Wrought Bar Rod Sh Tub Pip	1.5-2.5	88-92.5		1 max		0.02 max	0.01 max		0.2 max	0.5				
Aluminum Bronze C61400	USA	QQ C-450	91Cu-7Al-2Fe	Wrought Bar Rod Sh Tub Pip	1.5-2.5	88-92.5		1 max		0.02 max	0.01 max		0.2 max	0.5				
Aluminum Bronze C61400	USA	QQ C-465	91Cu-7Al-2Fe	Wrought Bar Rod Sh Tub Pip	1.5-2.5	88-92.5		1 max		0.02 max	0.01 max		0.2 max	0.5				

UNS numbers and US grades are provided as a means of cross referencing chemically similar alloys. Exchangability is only possible after independent examination of specifications. Tensile properties are minimum or typical . UTS and YS as Mpa, El as %. See Appendix for list of abbreviations used in Descriptions.

4-40 Wrought Copper

Grade UNS #	Country	Specification	Designation	Comment	Al	Cu	Fe	Mn	Ni	P	Pb	Sn	Zn	OT max	Other	UTS	YS	El
Aluminum Bronze C61400	USA	SAE J463	91Cu-7Al-2Fe	Wrought Bar Rod Sh Tub Pip	1.5-2.5	88-92.5		1 max		0.02 max	0.01 max		0.2 max	0.5				
C61400	USA			Aluminum bronze	6-8		1.5-3.5	1 max		0.015 max	0.01 max		0.2 max		Cu+named el 99.5min; Cu incl Ag			
	Australia	AS 1567	643	Wrought Rod Bar As Mfg 50 mm diam	5.5-7.5			1 max	0.5						Si 1.5-3	550	260	20
	Japan	JIS Z3264	BCuP-2	Wrought Wir		92.29				6.8-7.5			0.2					
C64200	USA			Aluminum bronze	6.3-7.6		0.3 max	0.1 max			0.05 max	0.2 max	0.5 max		Cu+named el 99.5 min; Ni+Co 0.25 max; Cu incl Ag; Si 1.5-2.2; As 0.15 max			
C64210	USA			Aluminum bronze	6.3-7		0.3 max	0.1 max			0.05 max	0.2 max	0.5 max		Cu+named el 99.5 min; Ni+Co 0.25 max; Cu incl Ag; Si 1.5-2; As 0.15 max			
C41100 C41100	USA	ASTM B105	91Cu-8.5Zn-.5Sn	Wrought Bar Rod Wir Plt Sh Strp		89-93	0.05 max				0.1 max	0.3-0.7						
C41100 C41100	USA	ASTM B508	91Cu-8.5Zn-.5Sn	Wrought Bar Rod Wir Plt Sh Strp		89-93	0.05 max				0.1 max	0.3-0.7						
C41100 C41100	USA	ASTM B591	91Cu-8.5Zn-.5Sn	Wrought Bar Rod Wir Plt Sh Strp		89-93	0.05 max				0.1 max	0.3-0.7						
C41100	USA			Copper zinc tin alloy; tin brass		89-92	0.05 max				0.1 max				Cu+named el 99.7 min; bal Zn; S 0.3-0.7			
	Japan	JIS Z3231	DCuSiB	Wrought Wir As drawn 3.2/6 mm diam		92		3		0.3	0.02				0.50 Pb+Al+Ni+Zn; Si 2.5-4	274		20
	UK	BS 2901 Part3	C12	Wrought Rod Wir	6-7.5	90 min					0.01 max		0.2 max		ST 1-2.5; ST=Fe+Ni+Mn optional; Si 0.01 max			
	UK	BS 2901 pt3	C12Fe	Wrought Rod Wir	6.5-8.5	89 min	2.5-3.5				0.01 max		0.2 max		Si 0.1 max			
Aluminum Bronze C61300	USA	QQ C-450	99Cu-7Al-2.7Fe-.3Sn	Wrought Bar Rod Plt Sh Tub Pip	6-7.5	88.5-91.5	2-3	0.1 max			0.01 max	0.2-0.5	0.05 max	0.05	ST 0.15 max			
C61300	USA			Aluminum bronze; weld apps	6-7.5		2-3	0.2 max		0.015 max	0.01 max	0.2-0.5	0.05 max		Cu+named el 99.8 min; Ni+Co 0.15 max; Zr 0.05 max; Si 0.1 max; Cr 0.05 max; Cd 0.05 max			
C61300	USA			Aluminum bronze	6-7.5		2-3	0.2 max		0.015 max	0.01 max	0.2-0.5	0.1 max		Cu+named el 99.8 min; Ni+Co 0.15 max; Cu incl Ag; Si 0.1 max			
C54400	Canada	CSA	HC.5.TP44ZJ (544)			0.1 max				0.01-0.5	3.5-4.5	3.5-4.5	1.5-4.5		ST 99.5 min; ST= Cu+elements with specified limits, % min			
C54400	France	NF A51108	CuSn4Zn4Pb4	Wrought Sh Ann 20 mm diam		85.9	0.1 max				0.2	3-4.5	3-4.5	3-4.5	0.3		320		2
C54400	France	NF A51108	CuSn4Zn4Pb4	Wrought Sh Hard 20 mm diam		85.9	0.1 max				0.2	3-4.5	3-4.5	3-4.5	0.3		500		3

UNS numbers and US grades are provided as a means of cross referencing chemically similar alloys. Exchangability is only possible after independent examination of specifications. Tensile properties are minimum or typical . UTS and YS as Mpa, El as %. See Appendix for list of abbreviations used in Descriptions.

Grade UNS #	Country	Specification	Designation	Comment	Al	Cu	Fe	Mn	Ni	P	Pb	Sn	Zn	OT max	Other	UTS	YS	El
C54400	France	NF A51108	CuSn4Zn4Pb4	Wrought Sh 1/2 hard 20 mm diam		85.9	0.1 max			0.2	3-4.5	3-4.5	3-4.5	0.3		400		25
C54400	Japan	JIS H3270	C5441	Wrought Rod		85.5-90				0.01-0.5	3.5-4.5	3.5-4.5	1.5-4.5		99.5 min Cu+Sn+Pb+Zn+P			
Bearing Bronze C54400	USA	AMS 4520	88Cu-4Pb-4Sn-4Zn	Wrought Bar Rod Sh Strp			0.1 max			0.03-0.35	0.05 max	3.5-4.5	1.5-4.5		ST 99.5 min			
Bearing Bronze C54400	USA	ASTM B103	88Cu-4Pb-4Sn-4Zn	Wrought Bar Rod Sh Strp			0.1 max			0.03-0.35	0.05 max	3.5-4.5	1.5-4.5		ST 99.5 min			
Bearing Bronze C54400	USA	ASTM B139	88Cu-4Pb-4Sn-4Zn	Wrought Bar Rod Sh Strp			0.1 max			0.03-0.35	0.05 max	3.5-4.5	1.5-4.5		ST 99.5 min			
Bearing Bronze C54400	USA	SAE J460	88Cu-4Pb-4Sn-4Zn	Wrought Bar Rod Sh Strp			0.1 max			0.03-0.35	0.05 max	3.5-4.5	1.5-4.5		ST 99.5 min			
Bearing Bronze C54400	USA	SAE J463	88Cu-4Pb-4Sn-4Zn	Wrought Bar Rod Sh Strp			0.1 max			0.03-0.35	0.05 max	3.5-4.5	1.5-4.5		ST 99.5 min			
C54400	USA			Copper tin lead alloy; Leaded phosphor bronze B-2			0.1 max			0.01-0.5	3.5-4.5		1.5-4.5		Cu+named el 99.5 mln; S 3.5-4.5			
	UK	BS 2901 pt3	C23	Wrought Rod Wir	6-6.4		0.5-0.7	0.5 max	0.1 max		0.01 max	0.1 max	0.4 max		Si 2-2.4; Mg 0.05 max			
C55280	USA			Copper silver phosphorus alloy						6.8-7.2					Cu+named el 99.85 min; Ag 1.8-2.2			
C22000	Australia	AS 1566	220	Wrought Plt Bar Sh Strp SH to hard (sht, Strp)		89-91	0.05				0.05				bal Zn	360		
C22000	Australia	AS 1566	220A	Wrought Plt Bar Sh Strp Ann (sht,strp)		89-91	0.05				0.05				bal Zn	245		40
C22000	Australia	AS 1566	220A	Plt Bar Sh Strp SH to 1/2 hard (sht,strp)		89-91	0.05				0.05				bal Zn	310		
C22000	Australia	AS 1567	220	Wrought Rod Bar As Mfg 6 mm diam		89-91	0.05				0.05				bal Zn	280		24
C22000	Canada	CSA HC.4.2	HC.4.Z10(220)	Wrought Sh Half Hard		89-91	0.05 max				0.05			0.1	bal Zn	324		
C22000	Canada	CSA HC.4.2	HC.4.Z10(220)	Wrought Bar Half Hard		89-91	0.05 max				0.05			0.1	bal Zn	324		
C22000	Canada	CSA HC.4.2	HC.4.Z10(220)	Wrought Strp Half Hard		89-91	0.05 max				0.05			0.1	bal Zn	324		
C22000	Canada	CSA HC.4.2	HC.4.Z10(220)	Wrought Plt Half Hard		89-91	0.05 max				0.05			0.1	bal Zn	324		
C22000	Canada	CSA HC.4.2	HC.4.Z10(220)	Wrought Sh Full Hard		89-91	0.05 max				0.05			0.1	bal Zn	393		
C22000	Canada	CSA HC.4.2	HC.4.Z10(220)	Wrought Bar Full Hard		89-91	0.05 max				0.05			0.1	bal Zn	393		
C22000	Canada	CSA HC.4.2	HC.4.Z10(220)	Wrought Strp Full Hard		89-91	0.05 max				0.05			0.1	bal Zn	393		
C22000	Canada	CSA HC.4.2	HC.4.Z10(220)	Wrought Plt Full Hard		89-91	0.05 max				0.05			0.1	bal Zn	393		
C22000	Canada	CSA HC.4.2	HC.4.Z10(220)	Wrought Sh Spring		89-91	0.05 max				0.05			0.1	bal Zn	475		
C22000	Canada	CSA HC.4.2	HC.4.Z10(220)	Wrought Bar Spring		89-91	0.05 max				0.05			0.1	bal Zn	475		
C22000	Canada	CSA HC.4.2	HC.4.Z10(220)	Wrought Strp Spring		89-91	0.05 max				0.05			0.1	bal Zn	475		
C22000	Canada	CSA HC.4.2	HC.4.Z10(220)	Wrought Plt Spring		89-91	0.05 max				0.05			0.1	bal Zn	475		
C22000	Czech Republic	CSN 423201	CuZn10			88-91					0.1 max				Fe+Mn+Al+Sn+Sb 0.20; bal Zn			

UNS numbers and US grades are provided as a means of cross referencing chemically similar alloys. Exchangability is only possible after independent examination of specifications. Tensile properties are minimum or typical . UTS and YS as Mpa, El as %. See Appendix for list of abbreviations used in Descriptions.

4-42 Wrought Copper

Grade UNS #	Country	Specification	Designation	Comment	Al	Cu	Fe	Mn	Ni	P	Pb	Sn	Zn	OT max	Other	UTS	YS	El
C22000	Finland	SFS 2915	CuZn10	Wrought sh CF .2/.5 mm diam		89-91	0.05 max				0.05 max		10	0.4		290	190	2
C22000	Finland	SFS 2915	CuZn10	Wrought sh Ann .2/.5 mm diam		89-91	0.05 max				0.05 max		10	0.4		260	60	25
C22000	France	NF A51101	CuZn10	Wrought Sh StH T1		90							10			270		
C22000	France	NF A51103	CuZn10	Wrought Tub Hard 80 mm diam		89-91								0.05	O 0.05-0.1; bal Zn	350		7
C22000	France	NF A51103	CuZn10	Wrought Tub 3/4 hard 80 mm diam		89-91								0.05	O 0.05-0.1; bal Zn	300		25
C22000	France	NF A51103	CuZn10	Wrought Tub 1/2 hard 80 mm diam		89-91								0.05	O 0.05-0.1; bal Zn	280		35
C22000	France	NF A51104	CuZn10	Wrought Bar Wir Mill Cond 50 mm diam		89-91	0.1				0.05			0.4	bal Zn	320		20
C22000	International	ISO 1634	CuZn10	Wrought Plt Sh Strp SH (HB) 5/10 mm diam		89-91	0.1				0.05				Cu includes 0.3 Ni; bal Zn	340		10
C22000	International	ISO 1634	CuZn10	Wrought Plt Sh Strp Ann		89-91	0.1				0.05				Cu includes 0.3 Ni; bal Zn			35
C22000	International	ISO 1634	CuZn10	Wrought Plt Sh Strp Specially Ann (OS35) 0.2/5 mm diam		89-91	0.1				0.05				Cu includes 0.3 Ni; bal Zn			35
C22000	International	ISO 1638	CuZn10	Wrought HC 1/5 mm diam		89-91	0.1				0.05				0.3 Ni included in Cu; bal Zn	350		5
C22000	International	ISO 1638	CuZn10	Wrought Ann 1/5 mm diam		89-91	0.1				0.05				0.3 Ni included in Cu; bal Zn	240		30
C22000	International	ISO 4261	CuZn10	Wrought Plt Sh Strp Rod Bar Tub Wir		89-91	0.1				0.05				0.4 Pb+Fe; bal Zn			
C22000	Japan	JIS H3100	C2200	Wrought Sh Plt 1/2 hard .3/20 mm diam		89-91									bal Zn	275		20
C22000	Japan	JIS H3100	C2200	Wrought Sh Plt Ann .3/30 mm diam		89-91									bal Zn	226		35
C22000	Japan	JIS H3100	C2200	Wrought Sh Plt hard .3/10 mm diam		89-91									bal Zn	324		
C22000	Japan	JIS H3260	C2200	Wrought Rod Ann .5 mm diam		89-91									bal Zn	226		20
C22000	Japan	JIS H3260	C2200	Wrought Rod !/2 hard .5/12 mm diam		89-91									bal Zn	343		
C22000	Japan	JIS H3260	C2200	Wrought Rod Hard .5/12 mm diam		89-91									bal Zn	471		
C22000	Japan	JIS H3300	C2200	Wrought Pip 1/2 hard 10/150 mm diam		89-91									bal Zn	275		15
C22000	Japan	JIS H3300	C2200	Wrought Pip Ann 10/150 mm diam		89-91									bal Zn	226		35
C22000	Japan	JIS H3300	C2200	Wrought Pip Hard 10/100 mm diam		89-91									bal Zn	363		
C22000	Mexico	DGN W-27	TU-90	Wrought Sh 1/4 hard		89-92			0.2		0.08				bal Zn	274		
C22000	Mexico	DGN W-27	TU-90	Wrought Sh 1/2 hard		89-92			0.2		0.08				bal Zn	294		

UNS numbers and US grades are provided as a means of cross referencing chemically similar alloys. Exchangability is only possible after independent examination of specifications. Tensile properties are minimum or typical . UTS and YS as Mpa, El as %. See Appendix for list of abbreviations used in Descriptions.

Worldwide Guide to Equivalent Nonferrous Metals and Alloys

Grade UNS #	Country	Specification	Designation	Comment	Al	Cu	Fe	Mn	Ni	P	Pb	Sn	Zn	OT max	Other	UTS	YS	El
C22000	Mexico	DGN W-27	TU-90	Wrought Sh Hard		89-92			0.2		0.08				bal Zn	343		
C22000	Norway	NS 16106	CuZn10	Wrought Sh Strp Plt Bar Tub		90							10					
C22000	UK	BS 2870	CZ101	Wrought Sh Strp Hard 10 mm diam		89-91	0.1 max				0.1			0.4	bal Zn	245		3
C22000	UK	BS 2870	CZ101	Wrought Sh Strp 1/2 hard 3.5 mm diam		89-91	0.1 max				0.1			0.4	bal Zn	217		7
C22000	UK	BS 2870	CZ101	Wrought Sh Strp Ann 10 mm diam		89-91	0.1 max				0.1			0.4	bal Zn	172		35
C22000	UK	BS 2873	CZ101	Wrought Wir		89-91	0.1 max				0.1			0.4	bal Zn			
Commercl Bronze C22000	USA	ASTM B130	90Cu-10Zn	Wrought Bar Wir Plt Sh Strp Tub		89-91	0.05 max				0.05 max		8.9-11					
Commercl Bronze C22000	USA	ASTM B131	90Cu-10Zn	Wrought Bar Wir Plt Sh Strp Tub		89-91	0.05 max				0.05 max		8.9-11					
Commercl Bronze C22000	USA	ASTM B134	90Cu-10Zn	Wrought Bar Wir Plt Sh Strp Tub		89-91	0.05 max				0.05 max		8.9-11					
Commercl Bronze C22000	USA	ASTM B135	90Cu-10Zn	Wrought Bar Wir Plt Sh Strp Tub		89-91	0.05 max				0.05 max		8.9-11					
Commercl Bronze C22000	USA	ASTM B36	90Cu-10Zn	Wrought Bar Wir Plt Sh Strp Tub		89-91	0.05 max				0.05 max		8.9-11					
Commercl Bronze C22000	USA	ASTM B372	90Cu-10Zn	Wrought Bar Wir Plt Sh Strp Tub		89-91	0.05 max				0.05 max		8.9-11					
Commercl Bronze C22000	USA	MIL B-18907	90Cu-10Zn	Wrought Bar Wir Plt Sh Strp Tub		89-91	0.05 max				0.05 max		8.9-11					
Commercl Bronze C22000	USA	MIL B-20292	90Cu-10Zn	Wrought Bar Wir Plt Sh Strp Tub		89-91	0.05 max				0.05 max		8.9-11					
Commercl Bronze C22000	USA	MIL C-21768	90Cu-10Zn	Wrought Bar Wir Plt Sh Strp Tub		89-91	0.05 max				0.05 max		8.9-11					
Commercl Bronze C22000	USA	MIL C-3383	90Cu-10Zn	Wrought Bar Wir Plt Sh Strp Tub		89-91	0.05 max				0.05 max		8.9-11					
Commercl Bronze C22000	USA	MIL T-52069	90Cu-10Zn	Wrought Bar Wir Plt Sh Strp Tub		89-91	0.05 max				0.05 max		8.9-11					
Commercl Bronze C22000	USA	MIL W-6712	90Cu-10Zn	Wrought Bar Wir Plt Sh Strp Tub		89-91	0.05 max				0.05 max		8.9-11					
Commercl Bronze C22000	USA	MIL W-85	90Cu-10Zn	Wrought Bar Wir Plt Sh Strp Tub		89-91	0.05 max				0.05 max		8.9-11					
Commercl Bronze C22000	USA	SAE J463	90Cu-10Zn	Wrought Bar Wir Plt Sh Strp Tub		89-91	0.05 max				0.05 max		8.9-11					
C22000	USA			Copper zinc alloy; brass; Commercial bronze 90%		89-91	0.05 max				0.05 max				Cu+named el 99.8 min; bal Zn			
C42000	USA			Copper zinc tin alloy; tin brass		88-91				0.25 max					Cu+named el 99.7 min; bal Zn; S 1.5-2			
Aluminum bronze C61800	USA	AWS A5.7-91R	ERCuAl-A2	Bare Weld Rod El	8.5-11		1.5 max				0.02 max			0.5	Cu includes Ag; Si 0.1 max	414		

UNS numbers and US grades are provided as a means of cross referencing chemically similar alloys. Exchangability is only possible after independent examination of specifications. Tensile properties are minimum or typical . UTS and YS as Mpa, El as %. See Appendix for list of abbreviations used in Descriptions.

4-44 Wrought Copper

Grade UNS #	Country	Specification	Designation	Comment	Al	Cu	Fe	Mn	Ni	P	Pb	Sn	Zn	OT max	Other	UTS	YS	El
C61800	USA			Aluminum bronze	8.5-11		0.5-1.5				0.02 max		0.02 max		Cu+named el 99.5 min; Cu incl Ag; Si 0.1 max			
C70690	USA			Copper nickel alloy			0.005 max	0.001 max			0.001 max		0.001 max		Cu+named el 99.5 min; Ni+Co 9.0-11.0			
C52400	Canada	CSA	HC.4.TJ100(5 24)				1 max			0.03-0.35	0.05 max	9-11	0.2 max		ST 99.5 min; ST=Copper+elements within specified limits			
C52400	France	NF	CuSn9P				1 max			0.35 max	0.1 max	7.5-10	0.5 max	0.3				
C52400	International	ISO 427	CuSn10	Wrought Plt Sh Strp Rod Bar Tub Wir			0.1		0.3	0.01-0.4	0.05	9-11	0.3	0.3				
Phosphor Bronze C52400	USA	ASTM B103	90Cu-10Sn	Wrought Bar Rod Wir Plt Sh Shapes		88.3-90.7	0.1 max			0.03-0.35	0.05 max	9-11	0.2 max					
Phosphor Bronze C52400	USA	ASTM B139	90Cu-10Sn	Wrought Bar Rod Wir Plt Sh Shapes		88.3-90.7	0.1 max			0.03-0.35	0.05 max	9-11	0.2 max					
Phosphor Bronze C52400	USA	ASTM B159	90Cu-10Sn	Wrought Bar Rod Wir Plt Sh Shapes		88.3-90.7	0.1 max			0.03-0.35	0.05 max	9-11	0.2 max					
Phosphor Bronze C52400	USA	QQ B-750	90Cu-10Sn	Wrought Bar Rod Wir Plt Sh Shapes		88.3-90.7	0.1 max			0.03-0.35	0.05 max	9-11	0.2 max					
C52400	USA			Copper tin alloy; Phosphor bronze 10% D			0.1 max			0.03-0.35	0.05 max	9-11	0.2 max		Cu+named el 99.5 min			
C31200	USA			Copper zinc lead alloy; leaded brass		87.5-90.5	0.1 max	0.25 max			0.7-1.2				Cu+named el 99.6 min; bal Zn			
C31400	USA			Copper zinc lead alloy; leaded brass; Leaded commercial bronze		87.5-90.5	0.1 max	0.7 max			1.3-2.5				Cu+named el 99.6 min; bal Zn			
Leaded-Commercl C31600	USA	ASTM B140	89Cu-8.1Zn-1.9Pb-1Ni	Wrought Bar Rod		87.5-90.5	0.1 max		0.7-1.2	0.04-0.1	1.3-2.5			0.5				
C31600	USA			Copper zinc lead alloy; leaded brass; Leaded commercial bronze (nickel bearing)		87.5-90.5	0.1 max		1.2 max	0.04-0.1	1.3-2.5				Cu+named el 99.6 min; bal Zn			
C61500	USA			Aluminum bronze	7.7-8.3						0.015 max				Cu+named el 99.5 min; Ni+Co 1.8-2.2; Cu incl Ag			
C70700	USA			Copper nickel alloy			0.05 max	0.5 max							Cu+named el 99.5 min; Ni+Co 9.5-10.5			
	Czech Republic	CSN 423044	CuAl9Mn2		8-10		0.5 max	1.5-2.5	0.5 max			0.1 max		1 max	Si 0.1 max			
	India	IS 5743	CuP10	Wrought Bar		88.5-90.5	0.25 max			9-11								
	India	IS 5743	CuSi10	Wrought Bar		88.5-90.5									bal Zn; Si 9-11			
C42500	USA	ASTM B591	88.5Cu-9.5Zn-2Sn	Wrought Bar Sh Strp		87-90	0.05 max				0.35 max	0.05 max	1.5-3					
C42500	USA			Copper zinc tin alloy; tin brass		87-90	0.05 max				0.35 max	0.05 max			Cu+named el 99.7 min; bal Zn; S 1.5-3			

UNS numbers and US grades are provided as a means of cross referencing chemically similar alloys. Exchangability is only possible after independent examination of specifications. Tensile properties are minimum or typical . UTS and YS as Mpa, El as %. See Appendix for list of abbreviations used in Descriptions.

Worldwide Guide to Equivalent Nonferrous Metals and Alloys

Grade UNS #	Country	Specification	Designation	Comment	Al	Cu	Fe	Mn	Ni	P	Pb	Sn	Zn	OT max	Other	UTS	YS	El
C70600	Denmark	DS 3003	5667	Wrought		88	1-1.8	0.5-1	9-11						0.05 Pb+Sn; S 0.05; C 0.1			
C70600	Finland	SFS 2938	CuNi10Fe1Mn	Wrought Tub Ann 5 mm diam		88	1-1.8	0.5-1	9-11				0.5 max		ST 0.05 max; ST = Pb+Sn; S 0.05 max; C 0.1 max	290	100	35
C70600	France	NF	CuNi10Fe1Mn				1-2	0.3-1	9-11				0.5 max	0.1	ST 0.5 max; ST=Pb+Sn; S 0.02 max; C 0.1 max			
C70600	International	ISO 1634	CuNi10Fe1Mn	Wrought Plt Sh Strp As cast					9-11				0.5		0.05 Sn+Pb; S 0.05; C 0.1	330	95	30
C70600	International	ISO 1635	CuNi10Fe1Mn	Wrought Tub Ann			3.74 max	0.3-1	9-11				0.5		0.05 Sn+Pb; S 0.05; C 0.1			30
C70600	International	ISO 429	CuNi10Fe1Mn	Wrought Plt Sh Strp Rod Bar Tub			1-2	0.3-1	9-11				0.5		0.05 Sn+Pb; S 0.05; C 0.1			
C70600	Japan	JIS H3100	C7060	Wrought Sh Plt As Mfg .5/50 mm diam		86-89	1-1.8	0.2-1	9-11		0.05		1		99.5 min Cu+Ni+Fe+Mn	275		30
C70600	Japan	JIS H3300	C7060	Wrought Pip Ann 5/50 mm diam		84.6-88.3	1-1.8	0.2-1	9-11		0.06 max				99.5 min Cu+Ni+Fe+Mn	275		30
C70600	Norway	NS 16410	CuNi10Fe1Mn	Wrought Tub		88	1.5 max	0.7	10									
C70600	Sweden	SIS 145667	5667	Wrought Tub			1-1.8	0.5-1	9-11				0.5 max		ST 0.05 max; ST=Pb+Sn; S 0.05 max; C 0.1 max			
C70600	Switzerland	VSM 11557	CuNi10FeMn	Wrought Tub Q			1-1.7	0.5-1	9-11				0.5		0.05 Pb+Sn; S 0.05; C 0.1	294	98	30
Copper Nickel C70600	USA	ASME SB111	90Cu-10Ni	Wrought Wir Plt Sh Tub Pip			1-1.8	1 max	9-11		0.05 max		1 max	0.5				
Copper Nickel C70600	USA	ASME SB171	90Cu-10Ni	Wrought Wir Plt Sh Tub Pip			1-1.8	1 max	9-11		0.05 max		1 max	0.5				
Copper Nickel C70600	USA	ASME SB359	90Cu-10Ni	Wrought Wir Plt Sh Tub Pip			1-1.8	1 max	9-11		0.05 max		1 max	0.5				
Copper Nickel C70600	USA	ASME SB395	90Cu-10Ni	Wrought Wir Plt Sh Tub Pip			1-1.8	1 max	9-11		0.05 max		1 max	0.5				
Copper Nickel C70600	USA	ASME SB402	90Cu-10Ni	Wrought Wir Plt Sh Tub Pip			1-1.8	1 max	9-11		0.05 max		1 max	0.5				
Copper Nickel C70600	USA	ASME SB466	90Cu-10Ni	Wrought Wir Plt Sh Tub Pip			1-1.8	1 max	9-11		0.05 max		1 max	0.5				
Copper Nickel C70600	USA	ASME SB467	90Cu-10Ni	Wrought Wir Plt Sh Tub Pip			1-1.8	1 max	9-11		0.05 max		1 max	0.5				
Copper Nickel C70600	USA	ASTM B111	90Cu-10Ni	Wrought Wir Plt Sh Tub Pip			1-1.8	1 max	9-11		0.05 max		1 max	0.5				
Copper Nickel C70600	USA	ASTM B122	90Cu-10Ni	Wrought Wir Plt Sh Tub Pip			1-1.8	1 max	9-11		0.05 max		1 max	0.5				
Copper Nickel C70600	USA	ASTM B141	90Cu-10Ni	Wrought Wir Plt Sh Tub Pip			1-1.8	1 max	9-11		0.05 max		1 max	0.5				
Copper Nickel C70600	USA	ASTM B171	90Cu-10Ni	Wrought Wir Plt Sh Tub Pip			1-1.8	1 max	9-11		0.05 max		1 max	0.5				
Copper Nickel C70600	USA	ASTM B359	90Cu-10Ni	Wrought Wir Plt Sh Tub Pip			1-1.8	1 max	9-11		0.05 max		1 max	0.5				
Copper Nickel	USA	ASTM B395	90Cu-10Ni	Wrought Wir Plt Sh Tub Pip			1-1.8	1 max	9-11		0.05 max		1 max	0.5				

UNS numbers and US grades are provided as a means of cross referencing chemically similar alloys. Exchangability is only possible after independent examination of specifications. Tensile properties are minimum or typical . UTS and YS as Mpa, El as %. See Appendix for list of abbreviations used in Descriptions.

Grade UNS #	Country	Specification	Designation	Comment	Al	Cu	Fe	Mn	Ni	P	Pb	Sn	Zn	OT max	Other	UTS	YS	El
Copper Nickel C70600	USA	ASTM B402	90Cu-10Ni	Wrought Wir Plt Sh Tub Pip			1-1.8	1 max	9-11		0.05 max		1 max	0.5				
Copper Nickel C70600	USA	ASTM B432	90Cu-10Ni	Wrought Wir Plt Sh Tub Pip			1-1.8	1 max	9-11		0.05 max		1 max	0.5				
Copper Nickel C70600	USA	ASTM B466	90Cu-10Ni	Wrought Wir Plt Sh Tub Pip			1-1.8	1 max	9-11		0.05 max		1 max	0.5				
Copper Nickel C70600	USA	ASTM B467	90Cu-10Ni	Wrought Wir Plt Sh Tub Pip			1-1.8	1 max	9-11		0.05 max		1 max	0.5				
Copper Nickel C70600	USA	ASTM B543	90Cu-10Ni	Wrought Wir Plt Sh Tub Pip			1-1.8	1 max	9-11		0.05 max		1 max	0.5				
Copper Nickel C70600	USA	ASTM B552	90Cu-10Ni	Wrought Wir Plt Sh Tub Pip			1-1.8	1 max	9-11		0.05 max		1 max	0.5				
Copper Nickel C70600	USA	MIL C-15726	90Cu-10Ni	Wrought Wir Plt Sh Tub Pip			1-1.8	1 max	9-11		0.05 max		1 max	0.5				
C70600	USA		Copper nickel 10%	Copper nickel alloy; weld apps			1-1.8	1 max		0.02 max	0.02 max		0.5 max		Cu+named el 99.5 min; Ni+Co 9.0-11.0; S 0.02 max			
C70600	USA		Copper nickel 10%	Copper nickel alloy			1-1.8	1 max			0.05 max		1 max		Cu+named el 99.5 min; Ni+Co 9.0-11.0			
Aluminum Bronze W60619	USA	AWS A5.6-84R	ECuAl-B	Arc Weld El	7.5-10		2.5-5				0.02 max			0.5	Cu including Ag, Zn, Sn, Mn, Ni included in OT; Si 1.5 max			
9%Aluminum Bronze	India	IS 6912	9%Aluminum Bronze	Wrought Frg As Mfg 6 mm diam	8.8-10			0.5			0.05	0.1	0.4		Si 4 max; Si = Fe+Ni; Si 0.1; Mg 0.05	525	215	25
	Japan	JIS H3100	C6161	Wrought Sh Plt Hard .8/50 mm diam	7-10	83-90	2-4	0.5-2	0.5-2						Fe 2-4; 99.5 min Cu+Al+Fe+Ni+Mn	686		10
	Japan	JIS H3100	C6161	Wrought Sh Plt 1/2 hard .8/50 mm diam	7-10	83-90	2-4	0.5-2	0.5-2						Fe 2-4; 99.5 min Cu+Al+Fe+Ni+Mn	637		25
	Japan	JIS H3100	C6161	Wrought Sh Plt Ann 50 mm diam	7-10	83-90	2-4	0.5-2	0.5-2						Fe 2-4; 99.5 min Cu+Al+Fe+Ni+Mn	490		35
	Japan	JIS H3250	C6161	Rod As Mfg 6 mm diam	7-10	83-90	2-4	0.5-2	0.5-2						99.5 min Cu+Al+Fe+MnNi	588		25
	International	ISO 1637	CuAl9Mn2	Wrought HB-SH 15/50 mm	8-10		1.5 max	1.5-3	0.8 max		0.05		0.5		1.0 Pb+Zn	610	250	15
	International	ISO 1637	CuAl9Mn2	Wrought As cast 5mm diam	8-10		1.5 max	1.5-3	0.8 max		0.05		0.5		1.0 Pb+Zn	490	180	20
	International	ISO 1637	CuAl9Mn2	Wrought HA 5/50 mm diam	8-10		1.5 max	1.5-3	0.8 max		0.05		0.5		1.0 Pb+Zn	510	200	20
	International	ISO 1640	CuAl9Mn2	Wrought Frg As cast	8-10		1.5 max	1.5-3	0.8 max		0.05		0.5		1.0 Pb+Zn	540	200	20
	International	ISO 428	CuAl9Mn2	Wrought Bar Rod Frg	8-10		1.5 max	1.5-3	0.8 max		0.05		0.5		1.0 Pb+Zn			
7%Aluminum Bronze	India	IS 6912	7%Aluminum Bronze	Wrought Frg As Mfg 6/12.5 mm diam	6.5-8		2-3.5	0.5	0.5		0.05	0.1	0.4		Si 0.15; Mg 0.05	540	240	35
C72500	USA			Copper nickel alloy			0.6 max	0.2 max			0.05 max	1.8-2.8	0.5 max		Cu+named el 99.8 min; Ni+Co 8.5-10.5			
C62300	Czech Republic	CSN 423045	CuAl9Fe3		8-10		2-4	0.5 max	0.5 max			0.1 max	1 max		Si 0.1 max			
C62300	International	ISO 1639	CuAl10Fe3	Wrought As cast	8.5-11		2-4	2 max	1 max		0.05		0.5		0.5 Pb+Zn	570	200	15
C62300	UK	BS 2872	CA103	Wrought Frg As Mfg or Ann 6 mm diam	8.8-10			0.5	4		0.05	0.1	0.4	0.5	Ni includes Fe; Si 0.1; Mg 0.05	525	211	20

UNS numbers and US grades are provided as a means of cross referencing chemically similar alloys. Exchangability is only possible after independent examination of specifications. Tensile properties are minimum or typical . UTS and YS as Mpa, El as %. See Appendix for list of abbreviations used in Descriptions.

Worldwide Guide to Equivalent Nonferrous Metals and Alloys

Grade UNS #	Country	Specification	Designation	Comment	Al	Cu	Fe	Mn	Ni	P	Pb	Sn	Zn	OT max	Other	UTS	YS	El
C62300	UK	BS 2874	CA103	Wrought Rod As Mfg 6 mm diam	8.8-10			0.5	4		0.05	0.1	0.4	0.5	Ni includes Fe; Si 0.1; Mg 0.05	525	216	22
	UK	BS 2901 pt3	C13	Wrought Rod Wir	9-11	86 min	0.75-1.5	1 max	1 max		0.01 max		0.2 max		Si 0.1 max			
Aluminum Bronze C62300	USA	ASME SB150	87Cu-10Al-3Fe	Wrought Bar Rod	8.5-11	82.2-89.5	2-4	0.5 max				0.6 max		0.5	ST 1 max; Si 0.25 max			
Aluminum Bronze C62300	USA	ASTM B150	87Cu-10Al-3Fe	Wrought Bar Rod	8.5-11	82.2-89.5	2-4	0.5 max				0.6 max		0.5	ST 1 max; Si 0.25 max			
Aluminum Bronze C62300	USA	ASTM B283	87Cu-10Al-3Fe	Wrought Bar Rod	8.5-11	82.2-89.5	2-4	0.5 max				0.6 max		0.5	ST 1 max; Si 0.25 max			
Aluminum Bronze C62300	USA	MIL B-16166	87Cu-10Al-3Fe	Wrought Bar Rod	8.5-11	82.2-89.5	2-4	0.5 max				0.6 max		0.5	ST 1 max; Si 0.25 max			
Aluminum Bronze C62300	USA	SAE J463	87Cu-10Al-3Fe	Wrought Bar Rod	8.5-11	82.2-89.5	2-4	0.5 max				0.6 max		0.5	ST 1 max; Si 0.25 max			
C62300	USA			Aluminum bronze	8.5-10		2-4	0.5 max				0.6 max			Cu+named el 99.5 min; Ni+Co 1.0 max; Cu incl Ag; Si 0.25 max			
C70800	USA		Copper nickel 11%	Copper nickel alloy			0.1 max	0.15 max			0.05 max		0.2 max		Cu+named el 99.5 min; Ni+Co 10.5-12.5			
	India	IS 5743	CuCr10	Wrought Bar		88.5-89.5	0.05								Cr 9-11			
	Japan	JIS Z3231	DCuAlA	Wrought Wir As drawn 3.2/6 mm diam	7-10		1.5 max	2	0.5		0.02				0.50 Pb+Zn; Si 1	392		15
C55281	USA			Copper silver phosphorus alloy						5.8-6.2					Cu+named el 99.85 min; Ag 4.8-5.2			
	International	ISO 1637	CuSn4Zn4	Wrought As cast 5 mm diam		89.39	0.1			0.1 max	0.1	3-5	3-5	0.3		360		50
	International	ISO 427	CuSn4Zn4	Wrought Strp Rod Bar Tub Wir		89.39	0.1			0.1	0.1	3-5	3-5	0.3				
	France	NF A51108	CuSn5Zn4	Wrought Sh Hard 20 mm diam		89.3	0.1			0.2	0.1	3-5	3-5	0.3		570		5
	France	NF A51108	CuSn5Zn4	Wrought Sh 1/2 hard 20 mm diam		89.3	0.1			0.2	0.1	3-5	3-5	0.3		420	350	15
	France	NF A51108	CuSn5Zn4	Wrought Sh Ann 20 mm diam		89.3	0.1			0.2	0.1	3-5	3-5	0.3		310	150	50
C22600	Mexico	DGN W-27	TU-87	Wrought Sh 1/4 hard		86-89			0.2		0.08				bal Zn	304		
C22600	Mexico	DGN W-27	TU-87	Wrought Sh 1/2 hard		86-89			0.2		0.08				bal Zn	323		
C22600	Mexico	DGN W-27	TU-87	Wrought Sh Hard		86-89			0.2		0.08				bal Zn	412		
C22600	USA			Copper zinc alloy; brass; Jewelry bronze 87.5%		86-89	0.05 max				0.05 max				Cu+named el 99.8 min; bal Zn			
C42200 C42200	USA	ASTM B591	87.5Cu-11.4Zn-1.1Sn	Wrought Bar Sh Strp		86-89	0.05 max			0.04 max	0.05 max	0.8-1.4						
C42200	USA			Copper zinc tin alloy; tin brass		86-89	0.05 max			0.35 max	0.05 max				Cu+named el 99.7 min; bal Zn; S 0.8-1.4			
	Japan	JIS Z3231	DCuAlNi	Wrought Wir As drawn 3.2/6 mm diam	7-10		3.65 max	2	2		0.02				0.50 Pb+Zn; bal Cu	490		13

UNS numbers and US grades are provided as a means of cross referencing chemically similar alloys. Exchangability is only possible after independent examination of specifications. Tensile properties are minimum or typical . UTS and YS as Mpa, El as %. See Appendix for list of abbreviations used in Descriptions.

4-48 Wrought Copper

Grade UNS #	Country	Specification	Designation	Comment	Al	Cu	Fe	Mn	Ni	P	Pb	Sn	Zn	OT max	Other	UTS	YS	El
C55282	USA			Copper silver phosphorus alloy						6.5-7					Cu+named el 99.85 min; Ag 4.8-5.2			
C61900	USA			Aluminum bronze	8.5-10		3-4.5				0.02 max	0.6 max	0.8 max		Cu+named el 99.5 min; Cu incl Ag			
Manganese-nickel aluminum bronze C63380	USA	AWS A5.7-91R	ERCuMnNIAl	Bare Weld Rod El	7-8.5		2-4	11-14			0.02 max			0.5	Cu includes Ag Ni+Co = 1.5-3.0; Si 0.1 max	515		
C63380	USA			Aluminum bronze	7-8.5		2-4	11-14			0.02 max		0.15 max		Cu+named el 99.5 min; Ni+Co 1.5-3.0; Cu incl Ag; Si 0.1 max			
C70610	Australia	AS 1566	706	Wrought Plt Bar Sh Strp Ann (sht,strp)			1-2	0.5-1	10-11	0.01						290		
C70610	India	IS 1545	CuNi10Fe1	Wrought Tub Ann			1-2	0.5-1	10-11	0.01				0.3	S 0.05; C 0.1	372		
C70610	India	IS 1545	CuNi10Fe1	Wrought Tub As drawn			1-2	0.5-1	10-11	0.01				0.3	S 0.05; C 0.1	387		
C70610	South Africa	SABS 460	Cu-Ni10FeMn1	Wrought Tub Ann			1-2	0.5-1	9-11	0.01				0.3	S 0.05; C 0.1			
C70610	South Africa	SABS 460	Cu-Ni10FeMn1	Wrought Tub As drawn			1-2	0.5-1	9-11	0.01				0.3	S 0.05; C 0.1			
C70610	UK	BS 2870	CN102	Wrought Sh Strp As Mfg 10 mm diam			1-2	0.5-1	10-11	0.01				0.3	S 0.05; C 0.1	217		30
C70610	UK	BS 2870	CN102	Wrought Sh Strp Ann 10 mm diam			1-2	0.5-1	10-11	0.01				0.3	S 0.05; C 0.1	197		40
C70610	UK	BS 2871	CN102	Wrought Tub Ann			1-2	0.5-1	10-11	0.01				0.3	S 0.05; C 0.1	300		30
C70610	UK	BS 2871	CN102	Wrought Tub As drawn			1-2	0.5-1	10- .	0.01				0.3	S 0.05; C 0.1	430		
C70610	UK	BS 2875	CN102	Wrought Plt As Mfg 10 mm diam			1-2	0.5-1	10-11	0.01				0.3	S 0.05; C 0.1	279		27
C70610	UK	BS 2875	CN102	Wrought Plt Ann 10 mm diam			1-2	0.5-1	10-11	0.01				0.3	S 0.05; C 0.1	279		36
C70610	USA			Copper nickel alloy			1-2	0.5-1			0.01 max				Cu+named el 99.5 min; Ni+Co 10.0-11.0; S 0.05 max; C 0.05 max			
	India	IS 5743	CuAS10	Wrought Bar		88.5									As 9-11			
C63010	USA			Aluminum bronze	9.7-10.9	78 min	2-3.5	1.5 max				0.2 max	0.3 max		Cu+named el 99.8 min; Ni+Co 4.5-5.5; Cu incl Ag			
Aluminum bronze C62400	USA	AWS A5.7-91R	ERCuAl-A3	Bare Weld Rod El	10-11.5		2-4.5				0.02 max			0.5	Cu includes Ag; Si 0.1 max	450		
C62400	USA			Aluminum bronze	10-11.5		2-4.5	0.3 max				0.2 max			Cu+named el 99.5 min; Cu incl Ag; Si 0.25 max			
	Czech Republic	CSN 423046	CuAl10Fe3Mn1,5		9-11		2-4	1-2	0.5 max	0.01 max		0.1 max	0.5 max		Si 0.1 max			
	Japan	JIS H3250	C6191	Wrought Rod As Mfg 6 mm diam	8-11	81-88	3-5	0.5-2	0.5-2						Fe 3-5; 99.5 min Cu+Al+Fe+Mn+Ni	686		15
	UK	BS 2901 pt3	C16	Wrought Rod Wir	0.03 max		1.5-1.8	0.5-1	10-11	0.01 max	0.01 max				Ti 0.2-0.5; S 0.01 max			
C55283	USA			Copper silver phosphorus alloy						7-7.5					Cu+named el 99.85 min; Ag 5.8-6.2			

UNS numbers and US grades are provided as a means of cross referencing chemically similar alloys. Exchangability is only possible after independent examination of specifications. Tensile properties are minimum or typical . UTS and YS as Mpa, El as %. See Appendix for list of abbreviations used in Descriptions.

Grade UNS #	Country	Specification	Designation	Comment	Al	Cu	Fe	Mn	Ni	P	Pb	Sn	Zn	OT max	Other	UTS	YS	El
C66410	USA			Copper zinc alloy			1.8-2.3				0.015 max	0.05 max	11-12		Cu+named el 99.5 min; Cu incl Ag			
C43000 C43000	USA	ASTM B591	87Cu-10.8Zn-2.2Sn	Wrought Sh Strp		84-87	0.05 max				0.1 max	1.7-2.7						
C43000	USA			Copper zinc tin alloy; tin brass		84-87	0.05 max				0.1 max				Cu+named el 99.7 min; bal Zn; S 1.7-2.7			
C43400 C43400	USA	ASTM B591	85Cu-14.3Zn-.7Sn	Wrought Plt Sh Strp		84-87	0.05 max				0.05 max	0.1-1						
C43400	USA			Copper zinc tin alloy; tin brass		84-87	0.05 max				0.05 max				Cu+named el 99.7 min; bal Zn; S 0.4-1			
	Czech Republic	CSN 423048	CuAl9Ni5Fe1Mn1		8-10		0.5-1.5	0.5-1.5	4-6	0.03 max		0.1 max	0.5 max		Si 0.1 max			
	Japan	JIS H3250	C6241	Wrought Rod As Mfg 6 mm diam	9-12	80-87	3-5	0.5-2	0.5-2						99.5 min Cu+Al+Fe+Mn+Ni	686		10
Nickel-aluminum bronze C63280	USA	AWS A5.7-91R	ERCuNiAl	Bare Weld Rod El	8.5-9.5		3-5	0.6-3.5			0.02 max				0.5	Cu includes Ag Ni+Co = 4-5.5; Si 0.1 max	480		
C63280	USA			Aluminum bronze	8.5-9.5		3-5	0.6-3.5			0.02 max					Cu+named el 99.5 min; Ni+Co 4.0-5.5; Cu incl Ag			
	Japan	JIS Z3264	BCuP-3	Wrought Wir		86.79				5.8-6.7						Ag 4.7-6.3			
C32000	USA			Copper zinc lead alloy; Leaded red brass		83.5-86.5	0.1 max		0.25 max		1.5-2.2				Cu+named el 99.6 min; bal Zn			
Nickel Aluminum Bronze W60632	USA	AWS A5.6-84R	ECuNiAl	Arc Weld El	6-8.5		3-6	0.5-3.5	4-6		0.02 max				0.5	Cu including Ag, Zn, Sn, included in OT; Si 1.5 max			
	Czech Republic	CSN 423056	CuMn13Ni		0.1 max		0.3 max	12-14	1.5-3		0.1 max	0.1 max	0.2 max					
	India	IS 5743	CuP14	Wrought Bar		84.5-86.5	0.2			13-15								
C23000	Australia	AS 1566	230	Wrought Plt Bar Sh Strp Ann (sht,strp)		84-86	0.05				0.05				bal Zn	255		45
C23000	Australia	AS 1566	230	Wrought Plt Bar Sh Strp SH to 1/2 hard (sht,strp)		84-86	0.05				0.05				bal Zn	320		
C23000	Australia	AS 1566	230	Wrought Plt Bar Sh Strp SH to full hard (sht,strp)		84-86	0.05				0.05				bal Zn	370		
C23000	Canada	CSA HC.4.2	HC.4.Z15(230)	Wrought Sh Half Hard		84-86	0.05 max				0.05			0.15	bal Zn	351		
C23000	Canada	CSA HC.4.2	HC.4.Z15(230)	Wrought Plt Half Hard		84-86	0.05 max				0.05			0.15	bal Zn	351		
C23000	Canada	CSA HC.4.2	HC.4.Z15(230)	Wrought Bar Half Hard		84-86	0.05 max				0.05			0.15	bal Zn	351		
C23000	Canada	CSA HC.4.2	HC.4.Z15(230)	Wrought Strp Half Hard		84-86	0.05 max				0.05			0.15	bal Zn	351		
C23000	Canada	CSA HC.4.2	HC.4.Z15(230)	Wrought Sh Full Hard		84-86	0.05 max				0.05			0.15	bal Zn	434		
C23000	Canada	CSA HC.4.2	HC.4.Z15(230)	Wrought Plt Full Hard		84-86	0.05 max				0.05			0.15	bal Zn	434		
C23000	Canada	CSA HC.4.2	HC.4.Z15(230)	Wrought Bar Full Hard		84-86	0.05 max				0.05			0.15	bal Zn	434		
C23000	Canada	CSA HC.4.2	HC.4.Z15(230)	Wrought Strp Full Hard		84-86	0.05 max				0.05			0.15	bal Zn	434		

UNS numbers and US grades are provided as a means of cross referencing chemically similar alloys. Exchangability is only possible after independent examination of specifications. Tensile properties are minimum or typical . UTS and YS as Mpa, El as %. See Appendix for list of abbreviations used in Descriptions.

Grade UNS #	Country	Specification	Designation	Comment	Al	Cu	Fe	Mn	Ni	P	Pb	Sn	Zn	OT max	Other	UTS	YS	El
C23000	Canada	CSA HC.4.2	HC.4.Z15(230)	Wrought Sh Spring		84-86	0.05 max				0.05			0.15	bal Zn	537		
C23000	Canada	CSA HC.4.2	HC.4.Z15(230)	Wrought Plt Spring		84-86	0.05 max				0.05			0.15	bal Zn	537		
C23000	Canada	CSA HC.4.2	HC.4.Z15(230)	Wrought Bar Spring		84-86	0.05 max				0.05			0.15	bal Zn	537		
C23000	Canada	CSA HC.4.2	HC.4.Z15(230)	Wrought Strp Spring		84-86	0.05 max				0.05			0.15	bal Zn	537		
C23000	Czech Republic	CSN 423202	CuZn15			84-86					0.1 max				Fe+Mn+Al+Sn+ Sb 0.20; bal Zn			
C23000	Denmark	DS 3003	5112	Wrought		84-86	0.1 max				0.05		15	0.04				
C23000	Finland	SFS 2916	CuZn15	Wrought Sh Strp Bar Rod Wir Tub Drawn (rod and wir) .2/1 mm diam		84-86	0.1 max				0.05 max		15	0.4		370	290	3
C23000	Finland	SFS 2916	CuZn15	Wrought Sh Strp Bar Rod Wir Tub CF (tub) 5 mm diam		84-86	0.1 max				0.05 max		15	0.4		370	290	10
C23000	Finland	SFS 2916	CuZn15	Wrought Sh Strp Bar Rod Wir Tub Ann (sh and strp) .2/.5 mm diam		84-86	0.1 max				0.05 max		15	0.4		270	70	20
C23000	France	NF A51104	CuZn15	Wrought Bar Wir Mill Cond 50 mm diam		84-86	0.1 max				0.05			0.4	bal Zn	330		25
C23000	International	ISO 1634	CuZn15	Wrought Plt sh Strp SH (HB) 5/10 mm diam		84-86	0.1				0.05				Cu includes 0.3 Ni; bal Zn	360		12
C23000	International	ISO 1634	CuZn15	Wrought Plt sh Strp Ann		84-86	0.1				0.05				Cu includes 0.3 Ni; bal Zn			35
C23000	International	ISO 1634	CuZn15	Wrought Plt sh Strp Specially Ann (OS35) 0.2/5 mm diam		84-86	0.1				0.05				Cu includes 0.3 Ni; bal Zn			35
C23000	International	ISO 1635	CuZn15	Wrought Tub SH (HB)		84-86	0.1				0.05				Cu includes 0.3 Ni; bal Zn	390		17
C23000	International	ISO 1635	CuZn15	Wrought Tub Ann		84-86	0.1				0.05				Cu includes 0.3 Ni; bal Zn			40
C23000	International	ISO 1637	CuZn15	Wrought HA		84-86	0.1				0.05				0.3 Ni included in Cu; bal Zn	310		25
C23000	International	ISO 1637	CuZn15	Wrought Ann 5 mm diam		84-86	0.1				0.05				0.3 Ni included in Cu; bal Zn			40
C23000	International	ISO 1638	CuZn15	Wrought HC 1/5 mm diam		84-86	0.1				0.05				0.3 Ni included in Cu; bal Zn	270		5
C23000	International	ISO 1638	CuZn15	Wrought Ann 1/5 mm diam		84-86	0.1				0.05				0.3 Ni included in Cu; bal Zn	260		30
C23000	International	ISO 4261	CuZn15	Wrought Plt Sh Strp Rod Bar Tub Wir		84-86	0.1				0.05				0.4 Pb+Fe; bal Zn			
C23000	Japan	JIS H3100	C2300	Wrought Sh Plt 1/2 hard .3/20 mm diam		84-86									bal Zn	294		23
C23000	Japan	JIS H3100	C2300	Wrought Sh Plt Ann .3/30 mm diam		84-86									bal Zn	245		40
C23000	Japan	JIS H3100	C2300	Wrought Sh Plt hard .3/10 mm diam		84-86									bal Zn	343		
C23000	Japan	JIS H3260	C2300	Wrought Rod Ann .5 mm diam		84-86									bal Zn	245		20
C23000	Japan	JIS H3260	C2300	Wrought Rod !/2 hard .5/12 mm diam		84-86									bal Zn	373		

UNS numbers and US grades are provided as a means of cross referencing chemically similar alloys. Exchangability is only possible after independent examination of specifications. Tensile properties are minimum or typical . UTS and YS as Mpa, El as %. See Appendix for list of abbreviations used in Descriptions.

Worldwide Guide to Equivalent Nonferrous Metals and Alloys

Grade UNS #	Country	Specification	Designation	Comment	Al	Cu	Fe	Mn	Ni	P	Pb	Sn	Zn	OT max	Other	UTS	YS	El
C23000	Japan	JIS H3260	C2300	Wrought Rod Hard .5/12 mm diam		84-86									bal Zn	520		
C23000	Japan	JIS H3300	C2300	Wrought Pip 1/2 hard 10/150 mm diam		84-86									bal Zn	304		20
C23000	Japan	JIS H3300	C2300	Wrought Pip Ann 10/150 mm diam		84-86									bal Zn	275		35
C23000	Japan	JIS H3300	C2300	Wrought Pip Hard 10/100 mm diam		84-86									bal Zn	392		
C23000	Mexico	DGN W-27	TU-85	Wrought Sh Hard		83-86			0.2		0.08				bal Zn	412		40
C23000	Mexico	DGN W-27	TU-85	Wrought Sh 1/4 hard		83-86			0.2		0.08				bal Zn	304		
C23000	Mexico	DGN W-27	TU-85	Wrought Sh 1/2 hard		83-86			0.2		0.08				bal Zn	323		
C23000	Sweden	SIS 145112	5112-02	Wrought Rod Wir Strp Sh Plt SH 2.5/5 mm diam		84-86	0.1						15	0.4		250	70	35
C23000	Sweden	SIS 145112	5112-04	Wrought Sh Strp Wir Rod Plt SH .2/.5 mm diam		84-86	0.1						15	0.4		310	200	8
C23000	Sweden	SIS 145112	5112-05	Wrought Sh Plt Strp SH 5 mm diam		84-86	0.1						15	0.4		360	280	10
C23000	UK	BS 2870	CZ102	Wrought Sh Strp Hard 10 mm diam		84-86	0.1 max							0.4	bal Zn	259		3
C23000	UK	BS 2870	CZ102	Wrought Sh Strp 1/2 hard 3.5 mm diam		84-86	0.1 max							0.4	bal Zn	228		7
C23000	UK	BS 2870	CZ102	Wrought Sh Strp Ann 10 mm diam		84-86	0.1 max							0.4	bal Zn	172		35
C23000	UK	BS 2873	CZ102	Wrought Wir Ann .5/10 mm		84-86	0.1 max							0.4	bal Zn	286		25
C23000	UK	BS 2873	CZ102	Wrought Wir 1/2 hard .5/10 mm diam		84-86	0.1 max							0.4	bal Zn	432		
C23000	UK	BS 2873	CZ102	Wrought Wir Hard .5/10 mm		84-86	0.1 max							0.4	bal Zn	588		
Red Brass C23000	USA	ASME SB111	85Cu-15Zn	Wrought Wir Sh Tub Pip		84-86	0.05 max				0.06 max							
Red Brass C23000	USA	ASME SB359	85Cu-15Zn	Wrought Wir Sh Tub Pip		84-86	0.05 max				0.06 max							
Red Brass C23000	USA	ASME SB395	85Cu-15Zn	Wrought Wir Sh Tub Pip		84-86	0.05 max				0.06 max							
Red Brass C23000	USA	ASME SB43	85Cu-15Zn	Wrought Wir Sh Tub Pip		84-86	0.05 max				0.06 max							
Red Brass C23000	USA	ASTM B111	85Cu-15Zn	Wrought Wir Sh Tub Pip		84-86	0.05 max				0.06 max							
Red Brass C23000	USA	ASTM B134	85Cu-15Zn	Wrought Wir Sh Tub Pip		84-86	0.05 max				0.06 max							
Red Brass C23000	USA	ASTM B135	85Cu-15Zn	Wrought Wir Sh Tub Pip		84-86	0.05 max				0.06 max							
Red Brass C23000	USA	ASTM B359	85Cu-15Zn	Wrought Wir Sh Tub Pip		84-86	0.05 max				0.06 max							
Red Brass C23000	USA	ASTM B36	85Cu-15Zn	Wrought Wir Sh Tub Pip		84-86	0.05 max				0.06 max							
Red Brass C23000	USA	ASTM B395	85Cu-15Zn	Wrought Wir Sh Tub Pip		84-86	0.05 max				0.06 max							
Red Brass C23000	USA	ASTM B43	85Cu-15Zn	Wrought Wir Sh Tub Pip		84-86	0.05 max				0.06 max							

UNS numbers and US grades are provided as a means of cross referencing chemically similar alloys. Exchangability is only possible after independent examination of specifications. Tensile properties are minimum or typical . UTS and YS as Mpa, El as %. See Appendix for list of abbreviations used in Descriptions.

4-52 Wrought Copper

Grade UNS #	Country	Specification	Designation	Comment	Al	Cu	Fe	Mn	Ni	P	Pb	Sn	Zn	OT max	Other	UTS	YS	El
Red Brass C23000	USA	MIL T-20168	85Cu-15Zn	Wrought Wir Sh Tub Pip		84-86	0.05 max				0.06 max							
Red Brass C23000	USA	QQ B-613	85Cu-15Zn	Wrought Wir Sh Tub Pip		84-86	0.05 max				0.06 max							
Red Brass C23000	USA	QQ B-626	85Cu-15Zn	Wrought Wir Sh Tub Pip		84-86	0.05 max				0.06 max							
Red Brass C23000	USA	QQ W-321	85Cu-15Zn	Wrought Wir Sh Tub Pip		84-86	0.05 max				0.06 max							
Red Brass C23000	USA	SAE J463	85Cu-15Zn	Wrought Wir Sh Tub Pip		84-86	0.05 max				0.06 max							
Red Brass C23000	USA	WW P-351	85Cu-15Zn	Wrought Wir Sh Tub Pip		84-86	0.05 max				0.06 max							
Red Brass C23000	USA	WW T-791	85Cu-15Zn	Wrought Wir Sh Tub Pip		84-86	0.05 max				0.06 max							
C23000	USA			Copper zinc alloy; Red brass 85%		84-86	0.05 max				0.05 max				Cu+named el 99.8 min; bal Zn			
C62200	USA			Aluminum bronze	11-12		3-4.2				0.02 max		0.02 max		Cu+named el 99.5 min; Cu incl Ag; Si 0.1 max			
C72700	USA			Copper nickel alloy			0.5 max	0.05-0.3			0.02 max	5.5-6.5	0.5 max		Cu+named el 99.7 min; Ni+Co 8.5-9.5; Nb 0.1 max; Mg 0.115 max			
C72700	USA			Copper nickel alloy; Hot roll			0.5 max	0.05-0.3			0.005 max	5.5-6.5	0.5 max		Cu+named el 99.7 min; Ni+Co 8.5-9.5; Nb 0.1 max; Mg 0.115 max			
	UK	BS 2870	CN103	Wrought Sh Strp Ann .6/2 mm diam		84-86	0.25	0.05-0.5	14-16		0.01			0.3	S 0.02; C 0.1	197		35
	UK	BS 2871	CZ127	Wrought Tub Ann	0.7-1.2	81-86	0.25	0.1	0.8-1.4		0.05	0.1		0.5	bal Zn			
	UK	BS 2871	CZ127	Wrought Tub As drawn	0.7-1.2	81-86	0.25	0.1	0.8-1.4		0.05	0.1		0.5	bal Zn			
C66400	USA			Copper zinc alloy			1.3-1.7				0.015 max	0.05 max	11-12		Cu+named el 99.5 min; Cu incl Ag; Fe+Co 1.8-2.3; Co 0.3-0.7			
	Japan	JIS Z3264	BCuP-4	Wrought Wir		85.79				6.8-7.7					Ag 4.7-6.3			
C23030	USA			Copper zinc alloy; brass		83.5-85.5	0.05 max				0.05 max				Cu+named el 99.8 min; bal Zn; Si 0.2-0.4			
	India	IS 5743	CuFe15	Wrought Bar		83.5-85.5	14-16											
	India	IS 5743	CuSi15	Wrought Bar		83.5-85.5									Si 14-16			
	France	NF A51108	CuSnZn9	Wrought Sh Hard 20 mm		85.3	0.1			0.2	0.1	2-4	7.5-10	0.3		610	580	4
	France	NF A51108	CuSnZn9	Sh 1/2 hard 20 mm diam		85.3	0.1			0.2	0.1	2-4	7.5-10	0.3		460	400	15
	France	NF A51108	CuSnZn9	Wrought Sh Ann 20 mm		85.3	0.1			0.2	0.1	2-4	7.5-10	0.3		310	150	40
C23000	France	NF A51101	CuZn15	Wrought Sh StressH T1		85							15			300		
C23000	Norway	NS 16108	CuZn15	Wrought Sh Strp Plt Bar Tub Rod Wir		85							15					
C62580	USA			Aluminum bronze	12-13		3-5				0.02 max		0.02 max		Cu+named el 99.5 min; Cu incl Ag; Si 0.04 max			
C63000	Canada	CSA	HC.5.AN105F (630)		9-11		2-4	1.5 max	4-5.5			0.2 max			ST 78-85; ST= Cu+elements within specified limits, % min; Si 0.25 max			

UNS numbers and US grades are provided as a means of cross referencing chemically similar alloys. Exchangability is only possible after independent examination of specifications. Tensile properties are minimum or typical . UTS and YS as Mpa, El as %. See Appendix for list of abbreviations used in Descriptions.

Grade UNS #	Country	Specification	Designation	Comment	Al	Cu	Fe	Mn	Ni	P	Pb	Sn	Zn	OT max	Other	UTS	YS	El
C63000	Czech Republic	CSN 423047	CuAl10Fe4Ni4		9.5-11		3.5-5.5	0.3 max	3.5-5.5	0.01 max		0.1 max	0.3 max		Si 0.1 max			
C63000	UK	BS 2870	C105	Wrought Sh Strp 1/2 hard .6/1.3 mm diam		99.2	0.02		0.15		0.02	0.03		0.5	0.030 Si+Te; Sb 0.01; Bi 0.01; As 0.3-0.5	172		10
C63000	UK	BS 2870	C105	Wrought Sh Strp As Mfg or Ann .6/10 mm diam		99.2	0.02		0.15		0.02	0.03		0.5	0.030 Si+Te; Sb 0.01; Bi 0.01; As 0.3-0.5	148		35
C63000	UK	BS 2870	C105	Wrought Sh Strp Hard .6/2.7 mm diam		99.2	0.02		0.15		0.02	0.03		0.5	0.030 Si+Te; Sb 0.01; Bi 0.01; As 0.3-0.5	217		
C63000	UK	BS 2875	CA105	Wrought Plt As Mfg 10/85 mm diam	8.5-10.5	78-85	1.5-3.5	0.5-2	4-7		0.05	0.1	0.4		Si 0.15; Mg 0.05	618		10
Aluminum Bronze C63000	USA	AMS 4640	82Cu-10Al-5Ni-3Fe	Wrought Bar Rod Tub Frg	9-11	78-85	2-4	1.5 max				0.2 max	0.3 max	0.5	ST 4-5.5; Si 0.25 max			
Aluminum Bronze C63000	USA	ASME SB150	82Cu-10Al-5Ni-3Fe	Wrought Bar Rod Tub Frg	9-11	78-85	2-4	1.5 max				0.2 max	0.3 max	0.5	ST 4-5.5; Si 0.25 max			
Aluminum Bronze C63000	USA	ASME SB171	82Cu-10Al-5Ni-3Fe	Wrought Bar Rod Tub Frg	9-11	78-85	2-4	1.5 max				0.2 max	0.3 max	0.5	ST 4-5.5; Si 0.25 max			
Aluminum Bronze C63000	USA	ASTM B124	82Cu-10Al-5Ni-3Fe	Wrought Bar Rod Tub Frg	9-11	78-85	2-4	1.5 max				0.2 max	0.3 max	0.5	ST 4-5.5; Si 0.25 max			
Aluminum Bronze C63000	USA	ASTM B150	82Cu-10Al-5Ni-3Fe	Wrought Bar Rod Tub Frg	9-11	78-85	2-4	1.5 max				0.2 max	0.3 max	0.5	ST 4-5.5; Si 0.25 max			
Aluminum Bronze C63000	USA	ASTM B171	82Cu-10Al-5Ni-3Fe	Wrought Bar Rod Tub Frg	9-11	78-85	2-4	1.5 max				0.2 max	0.3 max	0.5	ST 4-5.5; Si 0.25 max			
Aluminum Bronze C63000	USA	ASTM B283	82Cu-10Al-5Ni-3Fe	Wrought Bar Rod Tub Frg	9-11	78-85	2-4	1.5 max				0.2 max	0.3 max	0.5	ST 4-5.5; Si 0.25 max			
Aluminum Bronze C63000	USA	MIL B-16166	82Cu-10Al-5Ni-3Fe	Wrought Bar Rod Tub Frg	9-11	78-85	2-4	1.5 max				0.2 max	0.3 max	0.5	ST 4-5.5; Si 0.25 max			
Aluminum Bronze C63000	USA	QQ C-465	82Cu-10Al-5Ni-3Fe	Wrought Bar Rod Tub Frg	9-11	78-85	2-4	1.5 max				0.2 max	0.3 max	0.5	ST 4-5.5; Si 0.25 max			
Aluminum Bronze C63000	USA	SAE J463	82Cu-10Al-5Ni-3Fe	Wrought Bar Rod Tub Frg	9-11	78-85	2-4	1.5 max				0.2 max	0.3 max	0.5	ST 4-5.5; Si 0.25 max			
C63000	USA			Aluminum bronze	9-11		2-4	1.5 max				0.2 max	0.3 max		Cu+named el 99.5 min; Ni+Co 4.0-5.5; Cu incl Ag; Si 0.25 max			
	Japan	JIS H3100	C6280	Wrought Sh Plt As Mfg 50 mm diam	8-11	78-85	1.5-3.5	0.5-2	4-7						99.5 min Cu+Al+Fe+Ni+Mn	618		10
	UK	BS 2901 pt 3	C20	Wrought Rod Wir	8-9.5	80.5-85	1.5-3.5	0.5-2	3.5-5		0.01 max		0.2 max		Si 0.1 max; Bi 0.002 max			
C63200	International	ISO 1637	CuAl10Fe5Ni5	Wrought HA 5/50 mm diam	8.5-11.5		2-6	2 max	4-6		0.05		0.5		0.5 Pb+Zn	740	340	10
C63200	International	ISO 1637	CuAl10Fe5Ni5	Wrought As cast 10 mm diam	8.5-11.5		2-6	2 max	4-6		0.05		0.5		0.5 Pb+Zn	690	290	12
C63200	International	ISO 1640	CuAl10Fe5Ni5	Wrought Frg As cast	8.5-11.5		2-6	2 max	4-6		0.05		0.5		0.5 Pb+Zn	740	290	10
C63200	UK	BS 2872	CA104	Wrought Frg As Mfg or Ann 6/10 mm diam	8.5-11		4-6	0.5	4-6		0.05	0.1	0.4	0.5	Si 0.1; Mg 0.05	696	392	10

UNS numbers and US grades are provided as a means of cross referencing chemically similar alloys. Exchangability is only possible after independent examination of specifications. Tensile properties are minimum or typical. UTS and YS as Mpa, El as %. See Appendix for list of abbreviations used in Descriptions.

Grade UNS #	Country	Specification	Designation	Comment	Al	Cu	Fe	Mn	Ni	P	Pb	Sn	Zn	OT max	Other	UTS	YS	El
C63200	UK	BS 2874	CA104	Wrought Rod As Mfg 6/10 mm diam	8.5-11		4-6	0.5	4-6		0.05	0.1	0.4	0.5	Si 0.1; Mg 0.05	701	402	10
C72200	USA			Copper nickel alloy; weld apps			0.5-1	1 max		0.02 max	0.02 max		0.5 max		Cu+named el 99.8 min; Ni+Co 15.0-18.0; Ti 0.03 max; Si 0.03 max; S 0.02 max; Cr 0.3-0.7			
C72200	USA			Copper nickel alloy			0.5-1	1 max			0.05 max	1 max			Cu+named el 99.8 min; Ni+Co 15.0-18.0; Ti 0.03 x; Si 0.03 max; Cr 0.3-0.7			
C23400	USA			Copper zinc alloy; brass		81-84	0.05 max				0.05 max				Cu+named el 99.8 min; bal Zn			
C62500	USA			Aluminum bronze	12.5-13		3.5-5.5	2 max							Cu+named el 99.5 min; Cu incl Ag			
C62581	USA			Aluminum bronze	13-14		3-5				0.02 max		0.02 max		Cu+named el 99.5 min; Cu incl Ag; Si 0.04 max			
C69100	USA			Copper zinc alloy	0.7-1.2	81-84	0.25 max	0.1			0.05 max	0.1 max			Cu+named el 99.5 min; Cu incl Ag; Ni+Co 0.8-1.4; bal Zn; Si 0.8-1.3			
	Japan	JIS H3100	C6301	Wrought Sh As Mfg 50 mm diam	8.5-10.5	77-84	3.5-6	0.5-2	3.5-6						99.5 min Cu+Al+Fe+Ni+Mn	637		15
C43500	USA			Copper zinc tin alloy; tin brass		79-83	0.05 max				0.1 max				Cu+named el 99.7 min; bal Zn; S 0.6-1.2			
C43600	USA			Copper zinc tin alloy; tin brass		80-83	0.05 max				0.05 max				Cu+named el 99.7 min; bal Zn; S 0.2-0.5			
C62582	USA			Aluminum bronze	14-15		3-5				0.02 max		0.02 max		Cu+named el 99.5 min; Cu incl Ag; Si 0.04 max			
Silicon Red Brass C69400	USA	ASTM B371	81.5Cu-14.5Zn-4Si	Wrought Rod		80-83	0.2 max				0.3 max		16.5 max		Si 3.5-4.5			
Silicon red brass C69400	USA			Copper zinc alloy		80-83	0.2 max				0.3 max				Cu+named el 99.5 min; Cu incl Ag; bal Zn; Si 3.5-4.5			
C69430	USA			Copper zinc alloy		80-83	0.2 max				0.3 max				Cu+named el 99.5 min; Cu incl Ag; bal Zn; Si 3.5-4.5; As 0.03-0.06			
C72800	USA			Copper nickel alloy	0.1 max		0.5 max	0.05-0.3		0.005 max	0.005 max	7.5-8.5	1 max		Cu+named el 99.7 min; Ni+Co 9.5-10.5; B 0.001; Ti 0.01 x; Si 0.05 x; Sb 0.02 x; S 0.0025 x; Nb 0.1-0.3; Mg 0.005-0.15; Bi 0.001 max			
C36200	Canada	CSA	HC.4.ZP391(365)			58-61	0.15 max				0.4-0.9	0.25 max		0.1	bal Zn	402	138	30
C36200	USA			Aluminum bronze	8.7-9.5		3.5-4.3	1.2-2			0.03 max				Cu+named el 99.5 min; Ni+Co 4.0-4.8; Fe<Ni; Cu incl Ag; Si 0.1 max			

UNS numbers and US grades are provided as a means of cross referencing chemically similar alloys. Exchangability is only possible after independent examination of specifications. Tensile properties are minimum or typical . UTS and YS as Mpa, El as %. See Appendix for list of abbreviations used in Descriptions.

Worldwide Guide to Equivalent Nonferrous Metals and Alloys

Grade UNS #	Country	Specification	Designation	Comment	Al	Cu	Fe	Mn	Ni	P	Pb	Sn	Zn	OT max	Other	UTS	YS	El
	Australia	AS 1567	627	Wrought Rod Bar As Mfg 6/10 mm diam	8.5-11		4-6	0.5	4-6		0.05	0.1	0.4		Si 0.1	700	310	10
10%AluminumBronze	India	IS 6912	10%AluminumBronze	Wrought Frg As Mfg 6/12.5 mm diam	8.5-11		4-6	0.5	4-6		0.05	0.1	0.4		Si 0.1; Mg 0.05	695	400	12
C24080	USA			Copper zinc alloy; brass	0.1 max	78-82					0.2 max				Cu+named el 99.8 min; bal Zn			
C24000	Canada	CSA HC.4.2	HC.4.Z20(240)	Wrought Sh Half Hard		78.5-81.5	0.05 max				0.05			0.15	bal Zn	379		
C24000	Canada	CSA HC.4.2	HC.4.Z20(240)	Wrought Strp Half Hard		78.5-81.5	0.05 max				0.05			0.15	bal Zn	379		
C24000	Canada	CSA HC.4.2	HC.4.Z20(240)	Wrought Plt Half Hard		78.5-81.5	0.05 max				0.05			0.15	bal Zn	379		
C24000	Canada	CSA HC.4.2	HC.4.Z20(240)	Wrought Bar Half Hard		78.5-81.5	0.05 max				0.05			0.15	bal Zn	379		
C24000	Canada	CSA HC.4.2	HC.4.Z20(240)	Wrought Sh Full Hard		78.5-81.5	0.05 max				0.05			0.15	bal Zn	468		
C24000	Canada	CSA HC.4.2	HC.4.Z20(240)	Wrought Strp Full Hard		78.5-81.5	0.05 max				0.05			0.15	bal Zn	468		
C24000	Canada	CSA HC.4.2	HC.4.Z20(240)	Wrought Plt Full Hard		78.5-81.5	0.05 max				0.05			0.15	bal Zn	468		
C24000	Canada	CSA HC.4.2	HC.4.Z20(240)	Wrought Bar Full Hard		78.5-81.5	0.05 max				0.05			0.15	bal Zn	468		
C24000	Canada	CSA HC.4.2	HC.4.Z20(240)	Wrought Sh Spring		78.5-81.5	0.05 max				0.05			0.15	bal Zn	586		
C24000	Canada	CSA HC.4.2	HC.4.Z20(240)	Wrought Strp Spring		78.5-81.5	0.05 max				0.05			0.15	bal Zn	586		
C24000	Canada	CSA HC.4.2	HC.4.Z20(240)	Wrought Plt Spring		78.5-81.5	0.05 max				0.05			0.15	bal Zn	586		
C24000	Canada	CSA HC.4.2	HC.4.Z20(240)	Wrought Bar Spring		78.5-81.5	0.05 max				0.05			0.15	bal Zn	586		
C24000	Czech Republic	CSN 423203	CuZn20			78.5-81.5					0.1 max				Fe+Mn+Al+Sn+Sb 0.20; bal Zn			
C24000	Denmark	DS 3003	5114	Wrought		78.5-81.5	0.2 max				0.05		20	0.04				
C24000	Finland	SFS 2917	CuZn20	Wrought Sh Strp CF (sh and strp) .2/.5 mm diam		78.5-81.5	0.1 max				0.05 max		20	0.4		320	200	10
C24000	Finland	SFS 2917	CuZn20	Wrought Sh Strp Ann (sh and strp) .2/.5 mm diam		78.5-81.5	0.1 max				0.05 max		20	0.4		280	80	20
C24000	France	NF A51104	CuZn20	Wrought Bar,Wir Mill Cond 50 mm diam		78.5-81.5	0.1				0.05			0.4	bal Zn	330		25
C24000	France	NF A51101	CuZn20	Wrought Sh StressH T1		80							20			330		
C24000	India	IS 4170	CuZn20	Wrought Rod As Mfg 5 mm diam		79-81	0.05				0.01			0.3	bal Zn	314		25
C24000	India	IS 4170	CuZn20	Wrought Rod Ann		79-81	0.05				0.01			0.3	bal Zn	245		50
C24000	International	ISO 1634	CuZn20	Wrought Plt SH(HB) 5/10 mm diam		78.5-81.5	0.1				0.05				Cu includes 0.3 Ni; bal Zn	380		12
C24000	International	ISO 1634	CuZn20	Wrought Plt Ann		78.5-81.5	0.1				0.05				Cu includes 0.3 Ni; bal Zn			40
C24000	International	ISO 1634	CuZn20	Wrought Plt Specially Ann .2/5 mm diam		78.5-81.5	0.1				0.05				Cu includes 0.3 Ni; bal Zn			40
C24000	International	ISO 1638	CuZn20	Wrought HC 1/5 mm diam		78.5-81.5	0.1				0.05				0.3 Ni included in Cu; bal Zn	390		5
C24000	International	ISO 1638	CuZn20	Wrought Ann 1/5 mm diam		78.5-81.5	0.1				0.05				0.3 Ni included in Cu; bal Zn	260		35

UNS numbers and US grades are provided as a means of cross referencing chemically similar alloys. Exchangability is only possible after independent examination of specifications. Tensile properties are minimum or typical . UTS and YS as Mpa, El as %. See Appendix for list of abbreviations used in Descriptions.

Grade UNS #	Country	Specification	Designation	Comment	Al	Cu	Fe	Mn	Ni	P	Pb	Sn	Zn	OT max	Other	UTS	YS	El
C24000	International	ISO 4261	CuZn20	Wrought Plt Sh Strp Rod Bar Tub Wir		78.5-81.5	0.1				0.05				0.4 Pb+Fe; bal Zn			
C24000	Japan	JIS H3100	C2400	Wrought Sh Plt 1/2 hard .3/20 mm diam		78.5-81.5	0.05				0.05				bal Zn	314		25
C24000	Japan	JIS H3100	C2400	Wrought Sh Plt Ann .3/30 mm diam		78.5-81.5	0.05				0.05				bal Zn	255		44
C24000	Japan	JIS H3100	C2400	Wrought Sh Plt hard .3/10 mm diam		78.5-81.5	0.05				0.05				bal Zn	363		
C24000	Japan	JIS H3260	C2400	Wrought Rod Ann .5 mm diam		78.5-81.5	0.05				0.05				bal Zn	255		20
C24000	Japan	JIS H3260	C2400	Wrought Rod !/2 hard .5/12 mm diam		78.5-81.5	0.05				0.05				bal Zn	373		
C24000	Japan	JIS H3260	C2400	Wrought Rod Hard .5/12 mm diam		78.5-81.5	0.05				0.05				bal Zn	588		
C24000	Norway	NS 16110	CuZn20	Wrought Sh Strp Plt		80							20					
C24000	Sweden	SIS 145114	5114-02	Wrought Plt Sh Strp Ann 5 mm diam		78.5-81.5	0.1				0.05		20	0.4		280	80	45
C24000	Sweden	SIS 145114	5114-04	Wrought Plt Sh Strp SH 5 mm diam		78.5-81.5	0.1				0.05		20	0.4		320	200	30
C24000	Sweden	SIS 145114	5114-05	Wrought Plt Sh Strp SH 5 mm diam		78.5-81.5	0.1				0.05		20	0.4		380	310	2
C24000	UK	BS 2870	CZ103	Wrought Sh Strp Hard 10 mm diam		79-81	0.1 max				0.1			0.4	bal Zn	288		5
C24000	UK	BS 2870	CZ103	Wrought Sh Strp 1/2 hard 3.5 mm diam		79-81	0.1 max				0.1			0.4	bal Zn	238		10
C24000	UK	BS 2870	CZ103	Wrought Sh Strp Ann 10 mm diam		79-81	0.1 max				0.1			0.4	bal Zn	186		40
C24000	UK	BS 2873	CZ103	Wrought Wir Ann .5/10 mm diam		79-81	0.1 max				0.1			0.4	bal Zn	309		30
C24000	UK	BS 2873	CZ103	Wrought Wir 1/2 hard .5/10 mm diam		79-81	0.1 max				0.1			0.4	bal Zn	461		
C24000	UK	BS 2873	CZ103	Wrought Wir Hard .5/10 mm diam		79-81	0.1 max				0.1			0.4	bal Zn	608		
C24000	UK	BS 2874	CZ103	Wrought Rod As Mfg 6 mm diam		79-81	0.1 max				0.1			0.4	bal Zn	309		24
C24000	UK	BS 2874	CZ104	Wrought Rod As Mfg 6 mm diam		79-81					0.1-1			0.6	bal Zn	309		22
Low Brass C24000	USA	ASTM B134	80Cu-20Zn	Wrought Bar Rod Wir Plt Sh Strp Frg Shapes		78.5-81.5	0.05 max				0.05 max		18.4-21.5					
Low Brass C24000	USA	ASTM B36	80Cu-20Zn	Bar Rod Wir Plt Sh Strp Frg Shapes		78.5-81.5	0.05 max				0.05 max		18.4-21.5					
Low Brass C24000	USA	QQ B-613	80Cu-20Zn	Bar Rod Wir Plt Sh Strp Frg Shapes		78.5-81.5	0.05 max				0.05 max		18.4-21.5					
Low Brass C24000	USA	QQ B-626	80Cu-20Zn	Bar Rod Wir Plt Sh Strp Frg Shapes		78.5-81.5	0.05 max				0.05 max		18.4-21.5					

UNS numbers and US grades are provided as a means of cross referencing chemically similar alloys. Exchangability is only possible after independent examination of specifications. Tensile properties are minimum or typical . UTS and YS as Mpa, El as %. See Appendix for list of abbreviations used in Descriptions.

Worldwide Guide to Equivalent Nonferrous Metals and Alloys

Grade UNS #	Country	Specification	Designation	Comment	Al	Cu	Fe	Mn	Ni	P	Pb	Sn	Zn	OT max	Other	UTS	YS	El
Low Brass C24000	USA	QQ B-650	80Cu-20Zn	Wrought Bar Rod Wir Plt Sh Strp Frg Shapes		78.5-81.5	0.05 max				0.05 max		18.4-21.5					
Low Brass C24000	USA	QQ W-321	80Cu-20Zn	Wrought Bar Rod Wir Plt Sh Strp Frg Shapes		78.5-81.5	0.05 max				0.05 max		18.4-21.5					
Low Brass C24000	USA	SAE J463	80Cu-20Zn	Wrought Bar Rod Wir Plt Sh Strp Frg Shapes		78.5-81.5	0.05 max				0.05 max		18.4-21.5					
C24000	USA			Copper zinc alloy; Low brass 80%		78.5-81.5	0.05 max				0.05 max				Cu+named el 99.8 min; bal Zn			
C63020	USA			Aluminum bronze	10.5-11.5	74.5 min	4-5.5	1.5 max			0.03 max	0.25 max	0.3 max		Cu+named el 99.5 min; Ni+Co 4.2-6.0; Cu incl Ag; Cr 0.05 max; Co 0.2 max			
C72420	USA			Copper nickel alloy	1-2		0.7-1.2	3.5-5.5		0.01 max	0.02 max	0.1 max	0.2 max		Cu+named el 99.7 min; Ni+Co 13.5-16.5; Si 0.15 max; S 0.15 max; Mg 0.05 max; Cr 0.5 max; C 0.05 max			
C71000	Canada	CSA	HC.7.NF201(710)				1 max		19-23		0.05 max		1 max		ST 99.5 min; ST= Cu+elements within specified limits			
C71000	France	NF	CuNi20Mn1Fe				0.5-1	0.5-1.5	19-22				0.5 max	0.1	ST 0.05 max; ST=Pb+Sn; S 0.05 max; C 0.1 max			
C71000	International	ISO 1634	CuNi20	Wrought Plt Sh Strp			0.3	0.5 max	19-21				0.2		0.05 Sn+Pb; S 0.05; C 0.1			
C71000	International	ISO 1634	CuNi20Mn1Fe	Wrought Plt Sh Strp M			0.4-1	0.5-1.5					0.5		19.0-22.0 Ni+Co; C 0.1	115	350	35
C71000	International	ISO 429	CuNi20	Wrought Plt Strp Rod Bar			0.3 max	0.5 max	19-21				0.2		0.05 Sn+Pb; S 0.05; C 0.1			
C71000	International	ISO 429	CuNi20Mn1Fe	Wrought Plt Sh Strp Tub Ann			0.4-1	0.5-1.5	19-22				0.5		0.05 Sn+Pb; S 0.05; Co 0.5; C 0.1			35
C71000	Japan	JIS H3300	C7100	Wrought Pip Ann 5/50 mm diam		73.5-78.8	0.5-1	0.2-1	19-23		0.06 max				99.5 min Cu+Ni+Fe+Mn	314		30
C71000	UK	BS 2870	CN104	Wrought Sh Strp Ann .6/2 mm diam		79-81	0.3	0.05-0.5	19-21		0.01			0.1	S 0.02; C 0.1	217		35
C71000	UK	BS 2873	NS108	Wrought Wir		60-65	0.3	0.05-0.5	19-21		0.03			0.5	bal Zn			
80-20 Cupronickel C71000	USA	ASME SB111	80Cu-20Ni	Wrought Wir Plt Sh Tub Pip		73.45-81	1 max	1 max	19-23		0.05 max		1 max	0.5				
80-20 Cupronickel C71000	USA	ASME SB359	80Cu-20Ni	Wrought Wir Plt Sh Tub Pip		73.45-81	1 max	1 max	19-23		0.05 max		1 max	0.5				
80-20 Cupronickel C71000	USA	ASME SB395	80Cu-20Ni	Wrought Wir Plt Sh Tub Pip		73.45-81	1 max	1 max	19-23		0.05 max		1 max	0.5				
80-20 Cupronickel C71000	USA	ASME SB466	80Cu-20Ni	Wrought Wir Plt Sh Tub Pip		73.45-81	1 max	1 max	19-23		0.05 max		1 max	0.5				

UNS numbers and US grades are provided as a means of cross referencing chemically similar alloys. Exchangability is only possible after independent examination of specifications. Tensile properties are minimum or typical. UTS and YS as Mpa, El as %. See Appendix for list of abbreviations used in Descriptions.

4-58 Wrought Copper

Grade UNS #	Country	Specification	Designation	Comment	Al	Cu	Fe	Mn	Ni	P	Pb	Sn	Zn	OT max	Other	UTS	YS	El
80-20 Cupronickel C71000	USA	ASME SB467	80Cu-20Ni	Wrought Wir Plt Sh Tub Pip		73.45-81	1 max	1 max	19-23		0.05 max		1 max	0.5				
80-20 Cupronickel C71000	USA	ASTM B111	80Cu-20Ni	Wrought Wir Plt Sh Tub Pip		73.45-81	1 max	1 max	19-23		0.05 max		1 max	0.5				
80-20 Cupronickel C71000	USA	ASTM B122	80Cu-20Ni	Wrought Wir Plt Sh Tub Pip		73.45-81	1 max	1 max	19-23		0.05 max		1 max	0.5				
80-20 Cupronickel C71000	USA	ASTM B206	80Cu-20Ni	Wrought Wir Plt Sh Tub Pip		73.45-81	1 max	1 max	19-23		0.05 max		1 max	0.5				
80-20 Cupronickel C71000	USA	ASTM B359	80Cu-20Ni	Wrought Wir Plt Sh Tub Pip		73.45-81	1 max	1 max	19-23		0.05 max		1 max	0.5				
80-20 Cupronickel C71000	USA	ASTM B395	80Cu-20Ni	Wrought Wir Plt Sh Tub Pip		73.45-81	1 max	1 max	19-23		0.05 max		1 max	0.5				
80-20 Cupronickel C71000	USA	ASTM B466	80Cu-20Ni	Wrought Wir Plt Sh Tub Pip		73.45-81	1 max	1 max	19-23		0.05 max		1 max	0.5				
80-20 C71000	USA	ASTM B467	80Cu-20Ni	Wrought Wir Plt Sh Tub Pip		73.45-81	1 max	1 max	19-23		0.05 max		1 max	0.5				
80-20 C71000	USA	SAE J463	80Cu-20Ni	Wrought Wir Plt Sh Tub Pip		73.45-81	1 max	1 max	19-23		0.05 max		1 max	0.5				
C71000	USA		Copper nickel 20%	Copper nickel alloy			1 max	1 max			0.05 max		1 max		Cu+named el 99.5 min; Ni+Co 19.0-23.0			
Manganese-nickel aluminum bronze W60633	USA	AWS A5.6-84R	ECuMnNiAl	Arc Weld El	5-7.5		2-6	11-13	1-2.5		0.02 max			0.5	Cu including Ag, Zn, Sn included in OT; Si 1.5 max			
C55284	USA			Copper silver phosphorus alloy						4.8-5.2					Cu+named el 99.85 min; Ag 14.5-15.5			
	India	IS 5743	CuFe20	Wrought Bar		78.5-80.5	19-21											
C69700	USA			Copper zinc alloy		75-80	0.2 max	0.4 max			0.5-1.5				Cu+named el 99.5 min; Cu incl Ag; bal Zn; Si 2.5-3.5			
C69710	USA			Copper zinc alloy		75-80	0.2 max	0.4 max			0.5-1.5				Cu+named el 99.5 min; Cu incl Ag; bal Zn; Si 2.5-3.5; As 0.03-0.06			
C68700	Australia	AS 1572	687	Wrought Tub	1.8-2.3	76-78	0.06				0.07				bal Zn; As 0.02-0.06			
C68700	Denmark	DS 3003	5217	Wrought	1.8-2.3	76-79	0.07				0.07		20	0.3	As 0.02-0.035			
C68700	Finland	SFS 2928	CuZn20A12	Wrought Tub Ann 10 mm diam	1.8-2.3	76-79	0.07 max			0.01 max	0.07 max		20		ST 0.035 max; ST = As+P; As 0.02-0.035	330	80	50
C68700	France	NF	CuZn22Al2		1.8-2.5	76-79	0.06 max				0.07 max			0.3	As 0.02-0.06			
C68700	India	IS 1545	CuZn21Al2As	Wrought Tub Ann	1.8-2.3	76-78	0.06				0.08			0.3	bal Zn; As 0.02-0.06	402		
C68700	India	IS 1545	CuZn21Al2As	Wrought Tub As drawn	1.8-2.3	76-78	0.06				0.08			0.3	bal Zn; As 0.02-0.06	416		
C68700	International	ISO 1634	CuZn20Al2	Wrought Plt Sh Strp HA	1.8-2.5	76-79	0.07			0.02	0.07				0.3 Fe+Pb+P; bal Zn; As 0.02-0.06	390		30
C68700	International	ISO 1634	CuZn20Al2	Wrought Plt Sh Strp M	1.8-2.5	76-79	0.07			0.02	0.07				0.3 Fe+Pb+P; bal Zn; As 0.02-0.06	330		42
C68700	International	ISO 1635	CuZn20Al2	Wrought Tub SH	1.8-2.5	76-79	0.07			0.02	0.07				0.3 Fe+Pb+P; bal Zn; As 0.02-0.06	490		20

UNS numbers and US grades are provided as a means of cross referencing chemically similar alloys. Exchangability is only possible after independent examination of specifications. Tensile properties are minimum or typical . UTS and YS as Mpa, El as %. See Appendix for list of abbreviations used in Descriptions.

Grade UNS #	Country	Specification	Designation	Comment	Al	Cu	Fe	Mn	Ni	P	Pb	Sn	Zn	OT max	Other	UTS	YS	El
C68700	International	ISO 1635	CuZn20Al2	Wrought Tub Ann	1.8-2.5	76-79	0.07			0.02	0.07				0.3 Fe+Pb+P; bal Zn; As 0.02-0.06			45
C68700	International	ISO 426-1	CuZn20Al2	Wrought Plt Tub	1.8-2.5	76-79	0.07			0.02	0.07				0.3 Fe+Pb+P; bal Zn; As 0.02-0.06			
C68700	Japan	JIS H3300	C6870	Wrought Pip Ann 5/250 mm diam	1.8-2.5	76-79	0.06				0.07				bal Zn; As 0.02-0.06	373		40
C68700	Norway	NS 16210	CuZn20Al1	Wrought Tub	2-2	77						1	21-38		As 0.04 max			
C68700	South Africa	SABS 460	Cu-Zn21Al2	Wrought Tub Ann	1.8-2.3	76-78	0.06				0.07			0.3	bal Zn; As 0.02-0.06			
C68700	South Africa	SABS 460	Cu-Zn21Al2	Wrought Tub Specially Ann	1.8-2.3	76-78	0.06				0.07			0.3	bal Zn; As 0.02-0.06			
C68700	South Africa	SABS 460	Cu-Zn21Al2	Wrought Tub As drawn	1.8-2.3	76-78	0.06				0.07			0.3	bal Zn; As 0.02-0.06			
C68700	Sweden	SIS 145217	5217-02	Wrought Tub Ann 10 mm diam	1.8-2.3	76-79	0.07			0.01	0.07		20		0.035 As+P; As 0.02-0.035	330	80	50
C68700	Sweden	SIS 145217	5217-12	Wrought Tub Ann 3 mm diam	1.8-2.3	76-79	0.07			0.01	0.07		20		0.035 As+P; As 0.02-0.035	390	160	45
C68700	Switzerland	VSM 11557	CuZn21Al2	Wrought Tub 1/2 hard	1.8-2.3	76-78	0.06		0.5	0.02	0.07				0.05 P+As; bal Zn; As 0.02-0.045	392	157	40
C68700	Switzerland	VSM 11557	CuZn21Al2	Wrought Tub Q	1.8-2.3	76-78	0.06		0.5	0.02	0.07				0.05 P+As; bal Zn; As 0.02-0.045	333	108	50
C68700	UK	BS 2870	CZ110	Wrought Sh Strp As Mfg 10 mm diam	1.8-2.3	76-78	0.06				0.07			0.3	bal Zn; As 0.02-0.06	238		45
C68700	UK	BS 2870	CZ110	Wrought Sh Strp Ann 10 mm diam	1.8-2.3	76-78	0.06				0.07			0.3	bal Zn; As 0.02-0.06	217		50
C68700	UK	BS 2871	CZ110	Wrought Tub T Ann	1.8-2.3	76-78	0.06				0.07			0.3	bal Zn; As 0.02-0.06	350		35
C68700	UK	BS 2871	CZ110	Wrought Tub Ann	1.8-2.3	76-78	0.06				0.07			0.3	bal Zn; As 0.02-0.06	300		40
C68700	UK	BS 2871	CZ110	Wrought Tub As drawn	1.8-2.3	76-78	0.06				0.07			0.3	bal Zn; As 0.02-0.06	450		
C68700	UK	BS 2871	CZ110	Wrought Tub As drawn	1.8-2.3	76-78	0.06				0.07			0.3	bal Zn; As 0.02-0.06			
C68700	UK	BS 2871	CZ110	Wrought Tub T Ann	1.8-2.3	76-78	0.06				0.07			0.3	bal Zn; As 0.02-0.06			
C68700	UK	BS 2871	CZ110	Wrought Tub Ann	1.8-2.3	76-78	0.06				0.07			0.3	bal Zn; As 0.02-0.06			
C68700	UK	BS 2875	CZ110	Wrought Plt As Mfg 10 mm diam	1.8-2.3	76-78	0.06				0.07			0.3	bal Zn; As 0.02-0.06	279		36
C68700	UK	BS 2875	CZ110	Wrought Plt Ann 10 mm diam	1.8-2.3	76-78	0.06				0.07			0.3	bal Zn; As 0.02-0.06	279		40
Aluminum brass asenical C68700	USA			Copper zinc alloy	1.8-2.5	76-79	0.06 max				0.07 max				Cu+named el 99.5 min; Cu incl Ag; bal Zn; As 0.02-0.06			
	Japan	JIS H3300	C6871	Wrought Pip Ann 5/250 mm diam	1.8-2.5	76-79					0.07				bal Zn; Si 0.2-0.5; As 0.02-0.06	373		40
	Japan	JIS H3300	C6872	Wrought Pip Ann 5/250 mm diam	1.8-2.5	76-79		0.2	0.2-1		0.07				bal Zn; Cr 0.1; As 0.02-0.06; Ag 0.02-0.06	373		40
	Japan	JIS Z3264	BCuP-5	Wrought Wir		78.99-78.99			4.8-5.3					0.2	Ag 14.5-15.5			
	UK	BS 2901 pt3	C22	Wrought Rod Wir	7-8.5		2-4	11-14	1.5-3		0.02 max		0.15 max		Si 0.1 max			

UNS numbers and US grades are provided as a means of cross referencing chemically similar alloys. Exchangability is only possible after independent examination of specifications. Tensile properties are minimum or typical . UTS and YS as Mpa, El as %. See Appendix for list of abbreviations used in Descriptions.

Grade UNS #	Country	Specification	Designation	Comment	Al	Cu	Fe	Mn	Ni	P	Pb	Sn	Zn	OT max	Other	UTS	YS	El
C71100	USA			Copper nickel alloy			0.1 max	0.15 max			0.05 max		0.2 max		Cu+named el 99.5 min; Ni+Co 22.0-24.0			
C72900	USA			Copper nickel alloy			0.5 max	0.3 max			0.02 max	7.5-8.5	0.5 max		Cu+named el 99.7 min; Ni+Co 14.5-15.5; Nb 0.1 max; Mg 0.15 max			
C72900	USA			Copper nickel alloy; Hot roll			0.5 max	0.3 max			0.005 max	7.5-8.5	0.5 max		Cu+named el 99.7 min; Ni+Co 14.5-15.5; Nb 0.1 max; Mg 0.15 max			
C71300	International	ISO 1634	CuNi25	Wrought Plt Sh Strp Ann			0.3	0.5 max	24-26				0.5		0.05 Sn+Pb; S 0.02; C 0.1			35
C71300	International	ISO 429	CuNi25	Wrought Plt Sh Strp Wir			0.3	0.5					0.5		0.05 Sn+Pb; S 0.02; C 0.5			
C71300	UK	BS 2870	CN105	Wrought Sh Strp Ann .6/2 mm diam			0.3	0.05-0.4	24-26				0.2	0.35	S 0.02; C 0.1	238		30
C71300	UK	BS 2873	NS109	Wrought Wir		55-60	0.3	0.05-0.75	24-26		0.03			0.5	bal Zn			
C71300	USA			Copper nickel alloy			0.2 max	1 max			0.05 max		1 max		Cu+named el 99.5 min; Ni+Co 23.5-26.5			
C68800	USA			Copper zinc alloy	3-3.8		0.2 max				0.05 max		21.3-24.1		Cu+named el 99.5 min; Cu incl Ag; Al+Zn 25.1-27.1; Co 0.25-0.55			
C72950	USA			Copper nickel alloy			0.6 max	0.6 max			0.05 max	4.5-5.7			Cu+named el 99.7 min; Ni+Co 20.0-22.0			
C73500	Japan	JIS H3110	C7351	Wrought Sh Plt 1/2 hard .15 mm diam		70-75	0.25	0.5			0.1				16.5-19.5 Ni+Co; bal Zn	392		5
C73500	Japan	JIS H3110	C7351	Wrought Sh Plt Ann .3/5 mm diam		70-75	0.25	0.5			0.1				16.5-19.5 Ni+Co; bal Zn	324		20
C73500	USA			Nickel silver		70.5-73.5	0.25 max	0.5 max			0.1 max				Cu+named el 99.5 min; Ni+Co 16.5-19.5; bal Zn			
C74000	USA			Nickel silver		69-73.5	0.25 max	0.5 max			0.1 max				Cu+named el 99.5 min; Ni+Co 9.0-11.0; bal Zn			
	Canada	CSA HC.4.4	HC.4.NZ1810	Wrought Bar Half Hard		70.5-73.5	0.25 max	0.5	16.5-19.5		0.1			0.5	bal Zn	386		
	Canada	CSA HC.4.4	HC.4.NZ1810	Wrought Plt Quarter Hard		70.5-73.5	0.25 max	0.5	16.5-19.5		0.1			0.5	bal Zn	386		
	Canada	CSA HC.4.4	HC.4.NZ1810	Wrought Plt Half Hard		70.5-73.5	0.25 max	0.5	16.5-19.5		0.1			0.5	bal Zn	434		
	Canada	CSA HC.4.4	HC.4.NZ1810	Wrought Sht Half Hard		70.5-73.5	0.25 max	0.5	16.5-19.5		0.1			0.5	bal Zn	434		
	Canada	CSA HC.4.4	HC.4.NZ1810	Wrought Strp Half Hard		70.5-73.5	0.25 max	0.5	16.5-19.5		0.1			0.5	bal Zn	434		
	Canada	CSA HC.4.4	HC.4.NZ1810	Wrought Bar Full Hard		70.5-73.5	0.25 max	0.5	16.5-19.5		0.1			0.5	bal Zn	503		
	Canada	CSA HC.4.4	HC.4.NZ1810	Wrought Sht Full Hard		70.5-73.5	0.25 max	0.5	16.5-19.5		0.1			0.5	bal Zn	503		
	Canada	CSA HC.4.4	HC.4.NZ1810	Wrought Strp Full Hard		70.5-73.5	0.25 max	0.5	16.5-19.5		0.1			0.5	bal Zn	503		
	Canada	CSA HC.4.4	HC.4.NZ1810	Wrought Bar Extra Hard		70.5-73.5	0.25 max	0.5	16.5-19.5		0.1			0.5	bal Zn	544		
	Canada	CSA HC.4.4	HC.4.NZ1810	Wrought Sht Extra Hard		70.5-73.5	0.25 max	0.5	16.5-19.5		0.1			0.5	bal Zn	544		
	Canada	CSA HC.4.4	HC.4.NZ1810	Wrought Strp Extra Hard		70.5-73.5	0.25 max	0.5	16.5-19.5		0.1			0.5	bal Zn	544		

UNS numbers and US grades are provided as a means of cross referencing chemically similar alloys. Exchangability is only possible after independent examination of specifications. Tensile properties are minimum or typical . UTS and YS as Mpa, El as %. See Appendix for list of abbreviations used in Descriptions.

Worldwide Guide to Equivalent Nonferrous Metals and Alloys

Grade UNS #	Country	Specification	Designation	Comment	Al	Cu	Fe	Mn	Ni	P	Pb	Sn	Zn	OT max	Other	UTS	YS	El
	Canada	CSA HC.4.4	HC.4.ZN2010	Wrought Plt Quarter Hard		69-73.5	0.25 max	0.5	9-11		0.1			0.5	bal Zn	379		
	Canada	CSA HC.4.4	HC.4.ZN2010	Wrought Plt Half Hard		69-73.5	0.25 max	0.5	9-11		0.1			0.5	bal Zn	434		
	Canada	CSA HC.4.4	HC.4.ZN2010	Wrought Sh Half Hard		69-73.5	0.25 max	0.5	9-11		0.1			0.5	bal Zn	434		
	Canada	CSA HC.4.4	HC.4.ZN2010	Wrought Strp Half Hard		69-73.5	0.25 max	0.5	9-11		0.1			0.5	bal Zn	434		
	Canada	CSA HC.4.4	HC.4.ZN2010	Wrought Bar Half Hard		69-73.5	0.25 max	0.5	9-11		0.1			0.5	bal Zn	434		
	Canada	CSA HC.4.4	HC.4.ZN2010	Wrought Sh Full Hard		69-73.5	0.25 max	0.5	9-11		0.1			0.5	bal Zn	503		
	Canada	CSA HC.4.4	HC.4.ZN2010	Wrought Strp Full Hard		69-73.5	0.25 max	0.5	9-11		0.1			0.5	bal Zn	503		
	Canada	CSA HC.4.4	HC.4.ZN2010	Wrought Bar Full Hard		69-73.5	0.25 max	0.5	9-11		0.1			0.5	bal Zn	503		
	Canada	CSA HC.4.4	HC.4.ZN2010	Wrought Sh Extra Hard		69-73.5	0.25 max	0.5	9-11		0.1			0.5	bal Zn	544		
	Canada	CSA HC.4.4	HC.4.ZN2010	Wrought Strp Spring		69-73.5	0.25 max	0.5	9-11		0.1			0.5	bal Zn	544		
	Canada	CSA HC.4.4	HC.4.ZN2010	Wrought Bar Extra Hard		69-73.5	0.25 max	0.5	9-11		0.1			0.5	bal Zn	544		
C44300	Canada	CSA	HC.4.ZT281V (443)			70-73	0.06 max				0.07 max	0.8-1.2		0.1	bal Zn; As 0.02-0.1	414	241	20
C44300	India	IS 1545	CuZn29Sn1A s	Wrought Tub Ann		70-73	0.06				0.08	1-1.5		0.3	bal Zn; As 0.02-0.06	372		
C44300	India	IS 1545	CuZn29Sn1A s	Wrought Tub As drawn		70-73	0.06				0.08	1-1.5		0.3	bal Zn; As 0.02-0.06	387		
C44300	International	ISO 1634	CuZn28Sn1	Wrought Plt Sh Strp HA		70-73	0.07				0.07	0.9-1.3			0.3 Fe+Pb+P; bal Zn	390		30
C44300	International	ISO 1634	CuZn28Sn1	Wrought Plt Sh Strp M		70-73	0.07				0.07	0.9-1.3			0.3 Fe+Pb+P; bal Zn	330		40
C44300	International	ISO 1635	CuZn28Sn1	Wrought Tub SH (HB)		70-73	0.07				0.07	0.9-1.3			0.3 Fe+Pb; bal Zn	420		35
C44300	International	ISO 1635	CuZn28Sn1	Wrought Tub Ann		70-73	0.07				0.07	0.9-1.3			0.3 Fe+Pb; bal Zn			45
C44300	International	ISO 426-1	CuZn28Sn1	Wrought Plt Sh Tub		70-73	0.07				0.07	0.9-1.3			0.3 Fe+Pb; bal Zn			
C44300	Japan	JIS H3300	C4430	Wrought Pip Ann 5/250 mm diam		70-73	0.06				0.07	0.9-1.2			bal Zn; As 0.02-0.06	314		30
C44300	South Africa	SABS 460	Cu-Zn28Sn1	Wrought Tub Ann		70-73	0.06				0.08	1-1.5		0.3	bal Zn; As 0.02-0.06			
C44300	South Africa	SABS 460	Cu-Zn28Sn1	Wrought Tub Specially Ann		70-73	0.06				0.08	1-1.5		0.3	bal Zn; As 0.02-0.06			
C44300	South Africa	SABS 460	Cu-Zn28Sn1	Wrought Tub As drawn		70-73	0.06				0.08	1-1.5		0.3	bal Zn; As 0.02-0.06			
C44300	Sweden	SIS 145220	5220-12	Wrought Tub Ann		70-73	0.07			0.01	0.07	0.9-1.3	28		0.035 As+P	370	150	45
C44300	Switzerland	VSM 11557	CuZn28Sn1	Wrought Tub 1/2 hard		70-72.5	0.06		0.5	0.02	0.07	1-1.3		0.3	bal Zn; As 0.02-0.045	372	47	40
C44300	Switzerland	VSM 11557	CuZn28Sn1	Wrought Tub Q		70-72.5	0.06		0.5	0.02	0.07	1-1.3		0.3	bal Zn; As 0.02-0.045	323	98	50
C44300	UK	BS 2871	CZ111	Wrought Tub		70-73	0.06				0.07	1-1.5		0.3	bal Zn; As 0.02-0.06			
Admiralty Brass C44300	USA	ASME SB111	71Cu-28Zn-1Sn	Wrought Plt Tub		70-73	0.06 max				0.07 max	0.9-1.2			As 0.02-0.1			
Admiralty Brass C44300	USA	ASME SB171	71Cu-28Zn-1Sn	Wrought Plt Tub		70-73	0.06 max				0.07 max	0.9-1.2			As 0.02-0.1			
Admiralty Brass C44300	USA	ASME SB359	71Cu-28Zn-1Sn	Wrought Plt Tub		70-73	0.06 max				0.07 max	0.9-1.2			As 0.02-0.1			
Admiralty Brass C44300	USA	ASME SB395	71Cu-28Zn-1Sn	Wrought Plt Tub		70-73	0.06 max				0.07 max	0.9-1.2			As 0.02-0.1			
Admiralty Brass	USA	ASME SB543	71Cu-28Zn-1Sn	Wrought Plt Tub		70-73	0.06 max				0.07 max	0.9-1.2			As 0.02-0.1			

UNS numbers and US grades are provided as a means of cross referencing chemically similar alloys. Exchangability is only possible after independent examination of specifications. Tensile properties are minimum or typical . UTS and YS as Mpa, El as %. See Appendix for list of abbreviations used in Descriptions.

4-62 Wrought Copper

Grade UNS #	Country	Specification	Designation	Comment	Al	Cu	Fe	Mn	Ni	P	Pb	Sn	Zn	OT max	Other	UTS	YS	El
Admiralty Brass C44300	USA	ASTM B111	71Cu-28Zn-1Sn	Wrought Plt Tub		70-73	0.06 max				0.07 max	0.9-1.2			As 0.02-0.1			
Admiralty Brass C44300	USA	ASTM B171	71Cu-28Zn-1Sn	Wrought Plt Tub		70-73	0.06 max				0.07 max	0.9-1.2			As 0.02-0.1			
Admiralty Brass C44300	USA	ASTM B359	71Cu-28Zn-1Sn	Wrought Plt Tub		70-73	0.06 max				0.07 max	0.9-1.2			As 0.02-0.1			
Admiralty Brass C44300	USA	ASTM B395	71Cu-28Zn-1Sn	Wrought Plt Tub		70-73	0.06 max				0.07 max	0.9-1.2			As 0.02-0.1			
Admiralty Brass C44300	USA	ASTM B543	71Cu-28Zn-1Sn	Wrought Plt Tub		70-73	0.06 max				0.07 max	0.9-1.2			As 0.02-0.1			
C44300	USA			Copper zinc tin alloy; tin brass; Admiralty arsenical		70-73	0.06 max				0.07 max				Cu+named el 99.6 min; bal Zn; S 0.8-1.2; As 0.02-0.06			
C44400	USA			Copper zinc tin alloy; tin brass; Admiralty antimonial		70-73	0.06 max				0.07 max				Cu+named el 99.6 min; bal Zn; Sb 0.02-0.1; S 0.8-1.2			
C44500	USA			Copper zinc tin alloy; tin brass; Admiralty phosphorized		70-73	0.06 max			0.02-0.1	0.07 max				Cu+named el 99.6 min; bal Zn; S 0.8-1.2			
	India	IS 1545	CuZn30As	Wrought Tub Ann		70-73	0.06 max				0.08			0.3	bal Zn	372		
	India	IS 1545	CuZn30As	Wrought Tub As drawn		70-73	0.06 max				0.08			0.3	bal Zn	387		
C32510	Australia	AS 1567	351	Wrought Bar Hod Soft Ann 6 mm diam		69-72					0.3-0.7				bal Zn; As 0.02-0.06	260		38
C71900	USA			Copper nickel alloy			0.5 max	0.2-1		0.02 max	0.015 max		0.5 max		Cu+named el 99.5 min; Ni+Co 28.0-33.0; Zr 0.02-0.35; Ti 0.01-0.2; Si 0.25 max; S 0.015 max; Cr 2.2-3; C 0.04 max			
	Czech Republic	CSN 423239	CuZn28AlSn Mn		0.6-1.5	69-72		0.1-0.5			0.1 max	0.6-1.5			Al+Sn 2.70; bal Zn			
C26000	Australia	AS 1566	260	Wrought Plt Bar Sh Strp Ann (sht,strp)		68.5-71.5	0.07				0.05				bal Zn	290		55
C26000	Australia	AS 1566	260	Wrought Plt Bar Sh Strp SH to 1/2 hard (sht,strp)		68.5-71.5	0.07				0.05				bal Zn	360		
C26000	Australia	AS 1566	260	Wrought Plt Bar Sh Strp SH to full hard (sht,strp)		68.5-71.5	0.07				0.05				bal Zn	420		
C26000	Australia	AS 1567	260	Wrought Bar Rod As Mfg 6 mm diam		68.5-71.5	0.05				0.07				bal Zn	340		28
C26000	Australia	AS 1567	260	Wrought Bar Rod Soft Ann 6 mm diam		68.5-71.5	0.05				0.07				bal Zn	280		45
C26000	Canada	CSA HC.4.2	HC.4.Z30(260)	Wrought Sh Half Hard		68.5-71.5	0.05 max				0.07			0.15	bal Zn	393		
C26000	Canada	CSA HC.4.2	HC.4.Z30(260)	Wrought Strp Half Hard		68.5-71.5	0.05 max				0.07			0.15	bal Zn	393		
C26000	Canada	CSA HC.4.2	HC.4.Z30(260)	Wrought Bar Half Hard		68.5-71.5	0.05 max				0.07			0.15	bal Zn	393		
C26000	Canada	CSA HC.4.2	HC.4.Z30(260)	Wrought Plt Half Hard		68.5-71.5	0.05 max				0.07			0.15	bal Zn	393		

UNS numbers and US grades are provided as a means of cross referencing chemically similar alloys. Exchangability is only possible after independent examination of specifications. Tensile properties are minimum or typical . UTS and YS as Mpa, El as %. See Appendix for list of abbreviations used in Descriptions.

Worldwide Guide to Equivalent Nonferrous Metals and Alloys

Grade UNS #	Country	Specification	Designation	Comment	Al	Cu	Fe	Mn	Ni	P	Pb	Sn	Zn	OT max	Other	UTS	YS	El
C26000	Canada	CSA HC.4.2	HC.4.Z30(260)	Wrought Sh Full Hard		68.5-71.5	0.05 max				0.07			0.15	bal Zn	489		
C26000	Canada	CSA HC.4.2	HC.4.Z30(260)	Wrought Strp Full Hard		68.5-71.5	0.05 max				0.07			0.15	bal Zn	489		
C26000	Canada	CSA HC.4.2	HC.4.Z30(260)	Wrought Bar Full Hard		68.5-71.5	0.05 max				0.07			0.15	bal Zn	489		
C26000	Canada	CSA HC.4.2	HC.4.Z30(260)	Wrought Plt Full Hard		68.5-71.5	0.05 max				0.07			0.15	bal Zn	489		
C26000	Canada	CSA HC.4.2	HC.4.Z30(260)	Wrought Sh Spring		68.5-71.5	0.05 max				0.07			0.15	bal Zn	627		
C26000	Canada	CSA HC.4.2	HC.4.Z30(260)	Wrought Strp Spring		68.5-71.5	0.05 max				0.07			0.15	bal Zn	627		
C26000	Canada	CSA HC.4.2	HC.4.Z30(260)	Wrought Bar Spring		68.5-71.5	0.05 max				0.07			0.15	bal Zn	627		
C26000	Canada	CSA HC.4.2	HC.4.Z30(260)	Wrought Plt Spring		68.5-71.5	0.05 max				0.07			0.15	bal Zn	627		
C26000	Czech Republic	CSN 423210	CuZn30			69-72	0.1 max				0.1 max				bal Zn			
C26000	Finland	SFS 2918	CuZn30	Wrought Sh CF .2/.5 mm diam		68.5-71.5	0.1 max				0.05 max		30	0.4		330	150	15
C26000	Finland	SFS 2918	CuZn30	Wrought Sh Ann .2/.5 mm diam		68.5-71.5	0.1 max				0.05 max		30	0.4		300	90	25
C26000	France	NF A51101	CuZn30	Wrought Sh StressH T1		70							30			330		
C26000	France	NF A51103	CuZn30	Wrought Tub Hard 80 mm diam		68.5-71.5								0.05	O 0.05-0.1; bal Zn	480		14
C26000	France	NF A51103	CuZn30	Wrought Tub 3/4 hard 80 mm diam		68.5-71.5								0.05	O 0.05-0.1; bal Zn	410		30
C26000	France	NF A51103	CuZn30	Wrought Tub 1/2 hard 80 mm diam		68.5-71.5								0.05	O 0.05-0.1; bal Zn	370		40
C26000	France	NF A51104	CuZn30	Wrought Bar Wir Mill Cond 50 mm diam		68.5-71.5	0.1				0.05			0.4	bal Zn	340		30
C26000	India	IS 410	CuZn30	Wrought Plt Sh Strp hard		68.5-71.5	0.05				0.05			0.2	bal Zn	461		3
C26000	India	IS 410	CuZn30	Wrought Plt Sh Strp half hard		68.5-71.5	0.05				0.05			0.2	bal Zn	392		20
C26000	India	IS 410	CuZn30	Wrought Strp As rolled .035 mm diam		68.5-71.5	0.05				0.05			0.2	bal Zn	274		45
C26000	India	IS 410	CuZn30	Wrought Plt Sh Strp Ann		68.5-71.5	0.05				0.05			0.2	bal Zn	274		45
C26000	India	IS 4170	CuZn30	Wrought Rod As Mfg 5 mm diam		68-72	0.03				0.03			0.3	bal Zn	343		25
C26000	India	IS 4170	CuZn30	Wrought Rod Ann		68-72	0.03				0.03			0.3	bal Zn	274		50
C26000	India	IS 4413	CuZn30	Wrought Wir Ann		68-79	0.05				0.03			0.3	bal Zn	309		45
C26000	India	IS 4413	CuZn30	Wrought Wir half hard		68-79	0.05				0.03			0.3	bal Zn	461		
C26000	India	IS 4413	CuZn30	Wrought Wir Hard		68-79	0.05				0.03			0.3	bal Zn	618		
C26000	International	ISO 1634	CuZn30	Wrought Plt Sh Strp SH 5/10 mm diam		68.5-71.5	0.1				0.05				Cu includes 0.3 Ni; bal Zn	410		15
C26000	International	ISO 1634	CuZn30	Wrought Plt Sh Strp Specially Ann (OS25) .2/2 mm diam		68.5-71.5	0.1				0.05				Cu includes 0.3 Ni; bal Zn			35
C26000	International	ISO 1634	CuZn30	Wrought Plt Sh Strp Ann		68.5-71.5	0.1				0.05				Cu includes 0.3 Ni; bal Zn			45
C26000	International	ISO 1635	CuZn30	Wrought Tub SH (HB)		68.5-71.5	0.1				0.05				Cu includes 0.3 Ni; bal Zn	440		20

UNS numbers and US grades are provided as a means of cross referencing chemically similar alloys. Exchangability is only possible after independent examination of specifications. Tensile properties are minimum or typical . UTS and YS as Mpa, El as %. See Appendix for list of abbreviations used in Descriptions.

4-64 Wrought Copper

Grade UNS #	Country	Specification	Designation	Comment	Al	Cu	Fe	Mn	Ni	P	Pb	Sn	Zn	OT max	Other	UTS	YS	El
C26000	International	ISO 1635	CuZn30	Wrought Tub Ann		68.5-71.5	0.1				0.05				Cu includes 0.3 Ni; bal Zn			45
C26000	International	ISO 1638	CuZn30	Wrought HC 1/5 mm diam		68.5-71.5	0.1				0.01				0.3 Ni included in Cu; bal Zn	420		7
C26000	International	ISO 1638	CuZn30	Wrought Ann 1/5 mm diam		68.5-71.5	0.1				0.01				0.3 Ni included in Cu; bal Zn	280		35
C26000	International	ISO 4261	CuZn30	Wrought Plt Sh Strp Rod Bar Tub Wir		68.5-71.5	0.1				0.05				0.4 Pb+Fe; bal Zn			
C26000	Japan	JIS H3100	C2600	Wrought Rod Ann 1 mm diam		68.5-71.5	0.05				0.07				bal Zn	275		40
C26000	Japan	JIS H3100	C2600	Wrought Rod Hard .3/10 mm diam		68.5-71.5	0.05				0.07				bal Zn	412		
C26000	Japan	JIS H3100	C2600	Wrought Rod Extra Hard .3/10 mm diam		68.5-71.5	0.05				0.07				bal Zn	520		
C26000	Japan	JIS H3250	C2600	Wrought Rod As Mfg 6 mm diam		68.5-71.5	0.05				0.07				bal Zn	275		35
C26000	Japan	JIS H3250	C2600	Wrought Rod Ann 6/75 mm diam		68.5-71.5	0.05				0.07				bal Zn	275		45
C26000	Japan	JIS H3250	C2600	Wrought Rod Hard 6/20 mm diam		68.5-71.5	0.05				0.07				bal Zn	412		
C26000	Japan	JIS H3260	C2600	Wrought Rod Ann .5 mm diam		68.5-71.5	0.05				0.07				bal Zn	275		20
C26000	Japan	JIS H3260	C2600	Wrought Rod Hard .5/10 mm		68.5-71.5	0.05				0.07				bal Zn	686		
C26000	Japan	JIS H3260	C2600	Wrought Rod Extra Hard .5/10 mm diam		68.5-71.5	0.05				0.07				bal Zn	785		
C26000	Japan	JIS H3300	C2600	Wrought Pip 1/2 hard 4/250 mm diam		68.5-71.5	0.05				0.07				bal Zn	343		20
C26000	Japan	JIS H3300	C2600	Wrought Pip Ann 4/250 mm		68.5-71.5	0.05				0.07				bal Zn	275		45
C26000	Japan	JIS H3300	C2600	Wrought Pip Hard 4/250 mm diam		68.5-71.5	0.05				0.07				bal Zn	451		
C26000	Japan	JIS H3320	C2600	Wrought Pip 1/2 hard 4/76.2 mm diam		68.5-71.5	0.05				0.07				bal Zn	373		20
C26000	Japan	JIS H3320	C2600	Wrought Pip Ann 4/76.2 mm diam		68.5-71.5	0.05				0.07				bal Zn	275		45
C26000	Japan	JIS H3320	C2600	Wrought Pip hard 4/76.2 mm diam		68.5-71.5	0.05				0.07				bal Zn	451		
C26000	Mexico	DGN W-25	LA-70	Wrought Sh 1/2 hard		70-71	0.03		0.02						bal Zn	380		
C26000	Mexico	DGN W-25	LA-70	Wrought Sh Hard		70-71	0.03		0.02						bal Zn	441		
C26000	Mexico	DGN W-25	LA-70	Wrought Sh Spring hard		70-71	0.03		0.02						bal Zn	552		
C26000	Norway	NS 16115	CuZn30	Wrought Sh Strp Plt		70							30					
C26000	Sweden	SIS 145122	5122-02	Wrought Plt Sh Strp Ann 5 mm diam		68.5-71.5	0.1				0.05		30	0.4		300	90	50
C26000	Sweden	SIS 145122	5122-03	Wrought Plt Strp Sh SH 5 mm diam		68.5-68.5	0.1				0.05		30	0.4		330	150	40
C26000	Sweden	SIS 145122	5122-04	Wrought Plt Sh Strp SH 5 mm diam		68.5-71.5	0.1				0.05		30	0.4		340	220	35

UNS numbers and US grades are provided as a means of cross referencing chemically similar alloys. Exchangability is only possible after independent examination of specifications. Tensile properties are minimum or typical . UTS and YS as Mpa, El as %. See Appendix for list of abbreviations used in Descriptions.

Wrought Copper 4-65

Worldwide Guide to Equivalent Nonferrous Metals and Alloys

Grade UNS #	Country	Specification	Designation	Comment	Al	Cu	Fe	Mn	Ni	P	Pb	Sn	Zn	OT max	Other	UTS	YS	El
C26000	Sweden	SIS 145122	5122-05	Wrought Sh Plt Strp SH .2/.5 mm diam		68.5-71.5	0.1				0.05		30	0.4		410	330	5
C26000	UK	BS 2870	CZ106	Wrought Sh Strp Hard 10 mm diam		68.5-71.5	0.05 max				0.05			0.3	bal Zn			5
C26000	UK	BS 2870	CZ106	Wrought Sh Strp 1/2 hard 3.5 mm diam		68.5-71.5	0.05 max				0.05			0.3	bal Zn	245		20
C26000	UK	BS 2870	CZ106	Wrought Sh Strp Ann 10 mm diam		68.5-71.5	0.05 max				0.05			0.3	bal Zn	197		50
C26000	UK	BS 2873	CZ106	Wrought Wir Ann .5/10 mm diam		68.5-71.5	0.05 max				0.05			0.3	bal Zn	309		45
C26000	UK	BS 2873	CZ106	Wrought Wir 1/2 hard .5/10 mm diam		68.5-71.5	0.05 max				0.05			0.3	bal Zn	461		
C26000	UK	BS 2873	CZ106	Wrought Wir Hard .5/10 mm diam		68.5-71.5	0.05 max				0.05			0.3	bal Zn	608		
C26000	UK	BS 2874	CZ106	Wrought Rod As Mfg 6 mm diam		68.5-71.5	0.05 max				0.05			0.3	bal Zn	339		28
C26000	UK	BS 2874	CZ106	Wrought Rod Ann 6 mm diam		68.5-71.5	0.05 max				0.05			0.3	bal Zn	279		45
C26000	UK	BS 2875	CZ106	Wrought Plt Hard 10/16 mm diam		68.5-71.5	0.05 max				0.05			0.3	bal Zn	358		18
C26000	UK	BS 2875	CZ106	Wrought Plt As Mfg or Ann 10 mm diam		68.5-71.5	0.05 max				0.05			0.3	bal Zn	279		40
Cartridge Brass C26000	USA	AMS 4505	70Cu-30Zn	Wrought Rod Wir Sh Tub		68.5-71.5	0.05 max				0.01 max			0.15				
Cartridge Brass C26000	USA	AMS 4555	70Cu-30Zn	Wrought Rod Wir Sh Tub		68.5-71.5	0.01 max				0.01 max			0.15				
Cartridge Brass C26000	USA	ASTM B129	70Cu-30Zn	Wrought Rod Wir Sh Tub		68.5-71.5	0.01 max				0.01 max			0.15				
Cartridge Brass C26000	USA	ASTM B134	70Cu-30Zn	Wrought Rod Wir Sh Tub		68.5-71.5	0.01 max				0.01 max			0.15				
Cartridge Brass C26000	USA	ASTM B135	70Cu-30Zn	Wrought Rod Wir Sh Tub		68.5-71.5	0.01 max				0.01 max			0.15				
Cartridge Brass C26000	USA	ASTM B19	70Cu-30Zn	Wrought Rod Wir Sh Tub		68.5-71.5	0.01 max				0.01 max			0.15				
Cartridge Brass C26000	USA	ASTM B36	70Cu-30Zn	Wrought Rod Wir Sh Tub		68.5-71.5	0.01 max				0.01 max			0.15				
Cartridge Brass C26000	USA	ASTM B569	70Cu-30Zn	Wrought Rod Wir Sh Tub		68.5-71.5	0.01 max				0.01 max			0.15				
Cartridge Brass C26000	USA	ASTM B587	70Cu-30Zn	Wrought Rod Wir Sh Tub		68.5-71.5	0.01 max				0.01 max			0.15				
Cartridge Brass C26000	USA	MIL C-50	70Cu-30Zn	Wrought Rod Wir Sh Tub		68.5-71.5	0.01 max				0.01 max			0.15				
Cartridge Brass C26000	USA	MIL T-20219	70Cu-30Zn	Wrought Rod Wir Sh Tub		68.5-71.5	0.01 max				0.01 max			0.15				
Cartridge Brass C26000	USA	MIL T-6945	70Cu-30Zn	Wrought Rod Wir Sh Tub		68.5-71.5	0.01 max				0.01 max			0.15				

UNS numbers and US grades are provided as a means of cross referencing chemically similar alloys. Exchangability is only possible after independent examination of specifications. Tensile properties are minimum or typical . UTS and YS as Mpa, El as %. See Appendix for list of abbreviations used in Descriptions.

4-66 Wrought Copper

Grade UNS #	Country	Specification	Designation	Comment	Al	Cu	Fe	Mn	Ni	P	Pb	Sn	Zn	OT max	Other	UTS	YS	El
Cartridge Brass C26000	USA	QQ B-613	70Cu-30Zn	Wrought Rod Wir Sh Tub		68.5-71.5	0.01 max				0.01 max			0.15				
Cartridge Brass C26000	USA	QQ B-626	70Cu-30Zn	Wrought Rod Wir Sh Tub		68.5-71.5	0.01 max				0.01 max			0.15				
Cartridge Brass C26000	USA	SAE J463	70Cu-30Zn	Wrought Rod Wir Sh Tub		68.5-71.5	0.01 max				0.01 max			0.15				
C26000	USA			Copper zinc alloy; Cartridge brass 70%		68.5-71.5	0.05 max				0.07 max				Cu+named el 99.7 min; bal Zn			
C26130	Australia	AS 1567	259	Rod Bar As Mfg 6 mm		69-71	0.05				0.07				bal Zn; As 0.02-0.06	340		28
C26130	Australia	AS 1567	259	Rod Bar Soft Ann 6 mm		69-71	0.05				0.07				bal Zn; As 0.02-0.06	280		45
C26130	Australia	AS 1572	259	Wrought Tub		69-71	0.05				0.05				bal Zn; As 0.02-0.06			
C26130	South Africa	SABS 460	Cu-Zn30As	Wrought Tub Ann		69-71	0.06				0.07			0.3	bal Zn; As 0.02-0.06			
C26130	South Africa	SABS 460	Cu-Zn30As	Wrought Tub Specially Ann		69-71	0.06				0.07			0.3	bal Zn; As 0.02-0.06			
C26130	South Africa	SABS 460	Cu-Zn30As	Wrought Tub As drawn		69-71	0.06				0.07			0.3	bal Zn; As 0.02-0.06			
C26130	Switzerland	VSM 11557	CuZn30As	Wrought Tub 1/2 hard	0.02	69-71	0.05		0.2	0.02	0.05	0.05		0.3	bal Zn; As 0.02-0.045	343	137	40
C26130	UK	BS 2871	CZ126	Wrought Tub Ann		69-71	0.06				0.07			0.3	bal Zn; As 0.02-0.06			
C26130	UK	BS 2871	CZ126	Wrought Tub T Ann		69-71	0.06				0.07			0.3	bal Zn; As 0.02-0.06			
C26130	UK	BS 2871	CZ126	Wrought Tub As drawn		69-71	0.06				0.07			0.3	bal Zn; As 0.02-0.06			
C26130	UK	BS 2875	CZ105	Wrought Plt Hard 10/16 mm diam		70-73	0.06				0.08			0.3	bal Zn; As 0.02-0.06	358		18
C26130	UK	BS 2875	CZ105	Wrought Plt As Mfg or Ann 10 mm diam		70-73	0.06				0.08			0.3	bal Zn; As 0.02-0.06	279		40
C26130	USA			Copper zinc alloy; brass		68.5-71.5	0.05 max				0.05 max				Cu+named el 99.7 min; bal Zn; As 0.02-0.08			
Manganes brass C66700	USA			Copper zinc alloy		68.5-71.5	0.1 max	0.8-1.5			0.07 max				Cu+named el 99.5 min; Cu incl Ag; bal Zn			
	India	IS 407	1/CuZn30As	Wrought Tub Ann		68.5-71.5	0.05				0.07				bal Zn	284		
	India	IS 407	1/CuZn30As	Wrought Tub Half Hard		68.5-71.5	0.05				0.07				bal Zn	372		
	India	IS 407	1/CuZn30As	Wrought Tub Hard		68.5-71.5	0.05				0.07				bal Zn	451		
C71580	UK	BS 2870	CN106	Wrought Sh Strp Ann .6/2 mm diam		69-71	0.3	0.05-0.5	29-31		0.01			0.1	S 0.03; C 0.1	259		30
Copper-nickel C71580	USA	AWS A5.27-85	RCuNi	Rod Oxyfuel Gas Weld		0.4-0.75	1 max			0.02 max	0.02 max			0.5	Cu includes Ag, Ni+Co=29-32; Ti 0.2-0.5; Si 0.25 max; S 0.01 max			
Copper-nickel C71580	USA	AWS A5.7-91R	ERCuNi	Bare Weld Rod El		0.4-0.75	1 max			0.02 max	0.02 max			0.5	Cu includes Ag Ni+Co = 29-32; Ti 0.2-0.5; Si 0.25 max	345		
C71580	USA			Copper nickel alloy	0.05 max		0.5 max	0.3 max		0.03 max	0.05 max		0.05 max		Cu+named el 99.5 min; Ni+Co 29.0-33.0; Si 0.15 max; S 0.024 max; C 0.07 max			

UNS numbers and US grades are provided as a means of cross referencing chemically similar alloys. Exchangability is only possible after independent examination of specifications. Tensile properties are minimum or typical . UTS and YS as Mpa, El as %. See Appendix for list of abbreviations used in Descriptions.

Grade UNS #	Country	Specification	Designation	Comment	Al	Cu	Fe	Mn	Ni	P	Pb	Sn	Zn	OT max	Other	UTS	YS	El
C71590	USA			Copper nickel alloy	0.002 max		0.005 max	0.001 max		0.001 max	0.001 max	0.001 max	0.001 max		Cu+named el 99.5 min; Ni+Co 29.0-33.0; Hg 0.005 max; Ti 0.001 max; Si 0.02 max; Sb 0.001 max; S 0.003 max; C 0.03 max; Bi 0.001 max; As 0.001 min			
	Switzerland	VSM 11557	CuZn30	Wrought Tub 1/2 hard		69-71	0.05		0.2	0.02	0.05			0.3	bal Zn	343	137	40
C71500	Canada	CSA	HC.4.NF301(715)				0.7 max	1 max	29-33		0.05 max		1 max		ST 99.5 min; ST= Cu+elements with specified limits			
C71500	Czech Republic	CSN 423063	CuNi30FeMn				0.4-1	0.7-1.4	29-33		0.1 max				Mg 0.05; Si 0.2 max			
C71500	France	NF	CuNi30Mn1Fe				0.4-0.7	0.5-1.5	29-32			0.5 max		0.1	ST 0.05 max; ST=Pb+Sn; S 0.02 max; C 0.1 max			
C71500	International	ISO 1634	CuNi30Mn1Fe	Wrought Plt Sh Strp Ann			0.4-1		29-32				0.5		0.05 Sn+Pb; S 0.08; C 0.1			
C71500	International	ISO 1635	CuNi30Mn1Fe	Wrought Tub Ann			0.94 max	0.5-1.5	29-32				0.5		0.5 Sn+Pb; S 0.08; C 0.1			35
C71500	International	ISO 1637	CuNi30Mn1Fe	Wrought HB- SH 5/15 mm diam			0.4-1		29-32				0.5		0.05 Sn+Pb; S 0.08; C 0.1	420		20
C71500	International	ISO 1637	CuNi30Mn1Fe	Wrought Ann 5 mm diam			0.4-1		29-32				0.5		0.05 Sn+Pb; S 0.08; C 0.1			40
C71500	International	ISO 429	CuNi30Mn1Fe	Wrought Plt Sh Strp Rod Bar Tub			0.4-1	0.5-1.5	29-32				0.5		0.05 Sn+Pb; S 0.08; C 0.1			
C71500	Japan	JIS H3100	C7150	Wrought Sh Plt As Mfg .5/50 mm diam			0.4-0.7	0.2-1	29-33		0.05		1		99.5 min Cu+Ni+Fe+Mn	343		35
C71500	Japan	JIS H3300	C7150	Wrought Pip Ann 5/50 mm diam		63.75-68.85	0.4-0.7	0.2-1	29-33		0.05		1		99.5 min Cu+Ni+Fe+Mn	363		30
C71500	Norway	NS 16415	CuNi30Mn1Fe	Wrought Tub		67	0.7 max	1	31									
C71500	Sweden	SIS 145682	5682	Wrought Tub			0.4-1	0.5-1.5	30-32				0.5 max		ST 0.05 max; ST=Pb+Sn; S 0.05 max; C 0.1 max			
C71500	Switzerland	VSM 11557	CuNi30FeMn	Wrought Tub 1/2 hard			0.4-1	0.5-1.5	30-32				0.5		0.05 Pb+Sn; S 0.05; C 0.1	363	118	30
C71500	UK	BS 2870	CN107	Wrought Sh Strp Ann .6/2 mm diam			0.4-1	0.5-1.5	30-32		0.01			0.3	S 0.08; C 0.1	259		30
C71500	UK	BS 2870	NS107	Wrought Sh Strp		54-56	0.3	0.05-0.35	17-19		0.03			0.5	bal Zn			
C71500	UK	BS 2871	CN107	Wrought Tub Ann			0.4-1	0.5-1.5	30-32		0.01			0.3	S 0.08; C 0.1	370		30
C71500	UK	BS 2871	CN107	Wrought Tub As drawn			0.4-1	0.5-1.5	30-32		0.01			0.3	S 0.08; C 0.1	500		
C71500	UK	BS 2875	CN107	Wrought Plt As Mfg or Ann 10 mm diam			0.4-1	0.5-1.5	30-32		0.01			0.3	S 0.08; C 0.1	309		27
C71500	USA		Copper nickel 30%	Copper nickel alloy; weld apps			0.4-1	1 max		0.02 max	0.02 max		0.5 max		Cu+named el 99.5 min; Ni+Co 29.0-33.0; S 0.02 max			

UNS numbers and US grades are provided as a means of cross referencing chemically similar alloys. Exchangability is only possible after independent examination of specifications. Tensile properties are minimum or typical . UTS and YS as Mpa, El as %. See Appendix for list of abbreviations used in Descriptions.

Grade UNS #	Country	Specification	Designation	Comment	Al	Cu	Fe	Mn	Ni	P	Pb	Sn	Zn	OT max	Other	UTS	YS	El
C71500	USA		Copper nickel 30%	Copper nickel alloy			0.4-1	1 max			0.05 max	1 max			Cu+named el 99.5 min; Ni+Co 29.0-33.0			
C71581	USA			Copper nickel alloy			0.4-0.7	1 max		0.02 max	0.02 max				Cu+named el 99.5 min; Ni+Co 29.0-32.0; Ti 0.2-0.5; Si 0.25 max; S 0.01 max			
C71700	USA			Copper nickel alloy			0.4-1								Cu+named el 99.5 min; Ni+Co 29.0-33.0; Be 0.3-0.7			
C26200	USA			Copper zinc alloy; brass		67-70	0.05 max				0.07 max				Cu+named el 99.7 min; bal Zn			
	Czech Republic	CSN 423212	CuZn32			67-70	0.2 max				0.1 max	0.1 max			bal Zn			
	France	NF A51106	CuZnClass1	Wrought Bar As Mfg 12 mm diam	5	52-70	3 max	4	5		3	2		1	bal Zn	500	260	5
	France	NF A51106	CuZnClass2	Wrought Bar As Mfg 12 mm diam	5	52-70	3	4	5		3	2		1	bal Zn	600	300	7
	South Africa	SABS 460	Cu-Ni30Mn1	Wrought Tub Ann			0.4-1	0.5-1.5	29-32		0.01			0.3	S 0.08; C 0.1			
	South Africa	SABS 460	Cu-Ni30Mn1	Wrought Tub As drawn			0.4-1	0.5-1.5	29-32		0.01			0.3	S 0.08; C 0.1			
Copper/nic kel (70/30)W6 0715	USA	AWS A5.6-84R	ECuNi	Arc Weld El			0.4-0.75	1-2.5	29-33	0.2 max	0.02 max			0.5	Cu including Ag,Zn,Sn,includ ed n OT; Si 0.5 max			
	Czech Republic	CSN 423064	CuNi30Mn		0.1 max		0.3 max	1.5-3	29-33		0.1 max	0.1 max	0.1 max		Mg 0.05; Si 0.1 max			
	India	IS 1545	CuNi31Mn1F e	Wrought Tub Ann			0.4-1	0.5-1.5	30-32		0.01			0.3	S 0.08; C 0.1	461		
	India	IS 1545	CuNi31Mn1F e	Wrought Tub As drawn			0.4-1	0.5-1.5	30-32		0.01			0.3	S 0.08; C 0.1	481		
	UK	BS 2901 pt3	C18	Wrought Rod Wir	0.03 max		0.4-1	0.5-1.5	30-32	0.01 max	0.01 max				Ti 0.2-0.5; Si 0.1 max; S 0.01 max			
C26800	Canada	CSA HC.4.2	HC.4.Z34(26 8)	Wrought Bar Half Hard		64-68.5	0.05 max				0.15			0.15	bal Zn	379		
C26800	Canada	CSA HC.4.2	HC.4.Z34(26 8)	Wrought Sh Half Hard		64-68.5	0.05 max				0.15			0.15	bal Zn	379		
C26800	Canada	CSA HC.4.2	HC.4.Z34(26 8)	Wrought Strp Half Hard		64-68.5	0.05 max				0.15			0.15	bal Zn	379		
C26800	Canada	CSA HC.4.2	HC.4.Z34(26 8)	Wrought Plt Half Hard		64-68.5	0.05 max				0.15			0.15	bal Zn	379		
C26800	Canada	CSA HC.4.2	HC.4.Z34(26 8)	Wrought Bar Full Hard		64-68.5	0.05 max				0.15			0.15	bal Zn	468		
C26800	Canada	CSA HC.4.2	HC.4.Z34(26 8)	Wrought Sh Full Hard		64-68.5	0.05 max				0.15			0.15	bal Zn	468		
C26800	Canada	CSA HC.4.2	HC.4.Z34(26 8)	Wrought Strp Full Hard		64-68.5	0.05 max				0.15			0.15	bal Zn	468		
C26800	Canada	CSA HC.4.2	HC.4.Z34(26 8)	Wrought Plt Full Hard		64-68.5	0.05 max				0.15			0.15	bal Zn	468		
C26800	Canada	CSA HC.4.2	HC.4.Z34(26 8)	Wrought Bar Spring		64-68.5	0.05 max				0.15			0.15	bal Zn	592		
C26800	Canada	CSA HC.4.2	HC.4.Z34(26 8)	Wrought Sh Spring		64-68.5	0.05 max				0.15			0.15	bal Zn	592		
C26800	Canada	CSA HC.4.2	HC.4.Z34(26 8)	Wrought Strp Spring		64-68.5	0.05 max				0.15			0.15	bal Zn	592		
C26800	Canada	CSA HC.4.2	HC.4.Z34(26 8)	Wrought Plt Spring		64-68.5	0.05 max				0.15			0.15	bal Zn	592		
C26800	France	NF A51101	CuZn33	Wrought Sh StressH T1		67						33				330		
C26800	France	NF A51104	CuZn33	Wrought Bar Wir Mill Cond 50 mm diam		65.5-68.5	0.1 max				0.1			0.4	bal Zn	350		32

UNS numbers and US grades are provided as a means of cross referencing chemically similar alloys. Exchangability is only possible after independent examination of specifications. Tensile properties are minimum or typical . UTS and YS as Mpa, El as %. See Appendix for list of abbreviations used in Descriptions.

Worldwide Guide to Equivalent Nonferrous Metals and Alloys

Grade UNS #	Country	Specification	Designation	Comment	Al	Cu	Fe	Mn	Ni	P	Pb	Sn	Zn	OT max	Other	UTS	YS	El
C26800	India	IS 3168	CuZn33	wrought strp As rolled .035 mm diam		64.5-68.5	0.05				0.1			0.2		274		45
C26800	International	ISO 1634	CuZn33	Wrought Plt Sh Strp SH (HB) 5/10 mm diam		65.5-68.5	0.1				0.01				Cu includes 0.3 Ni; bal Zn	420		15
C26800	International	ISO 1634	CuZn33	Wrought Plt Sh Strp Ann		65.5-68.5	0.1				0.01				Cu includes 0.3 Ni; bal Zn			45
C26800	International	ISO 1634	CuZn33	Wrought Plt Sh Strp Specially Ann .2/5 mm diam		65.5-68.5	0.1				0.01				Cu includes 0.3 Ni; bal Zn			45
C26800	International	ISO 1638	CuZn33	Wrought HC 1/5 mm diam		65.5-68.5	0.1				0.1				0.3 Ni included in Cu; bal Zn	430		7
C26800	International	ISO 1638	CuZn33	Wrought Ann 1/5 mm diam		65.5-68.5	0.1				0.1				0.3 Ni included in Cu; bal Zn	280		35
C26800	International	ISO 4261	CuZn33	Wrought Plt Sh Strp Rod Bar Tub Wir		65.5-68.5	0.1				0.1				0.4 Pb+Fe; bal Zn			
C26800	Japan	JIS H3320	C2680	Wrought Pip 1/2 hard 4/76.2 mm diam		64-68									bal Zn	373		20
C26800	Japan	JIS H3320	C2680	Wrought Pip Ann .3/3 mm diam		64-68									bal Zn	294		40
C26800	Japan	JIS H3320	C2680	Wrought Pip Hard 4/76.2 mm diam		64-68									bal Zn	451		
C26800	UK	BS 2870	C107	Wrought Sh Strp 1/2 hard .6/1.3 mm diam		99.2	0.03		0.15	0.01-0.05	0.01	0.01			0.020 Se+Te; Sb 0.01; Bi 0.003 max; As 0.3-0.5	172		10
C26800	UK	BS 2870	C107	Wrought Sh Strp As Mfg or ann .6/10 mm diam		99.2	0.03		0.15	0.01-0.05	0.01	0.01			0.020 Se+Te; Sb 0.01; Bi 0.003 max; As 0.3-0.5	148		35
C26800	UK	BS 2870	C107	Wrought Sh Strp Hard .6/2.7 mm diam		99.2	0.03		0.15	0.01-0.05	0.01	0.01			0.020 Se+Te; Sb 0.01; Bi 0.003 max; As 0.3-0.5	217		
C26800	UK	BS 2870	CZ107	Wrought Sh Strp Hard 10 mm diam		64-67	0.1 max				0.1			0.4	bal Zn	324		5
C26800	UK	BS 2870	CZ107	Wrought Sh Strp 1/2 hard 3.5/10 mm diam		64-67	0.1 max				0.1			0.4	bal Zn	272		20
C26800	UK	BS 2870	CZ107	Wrought Sh Strp Ann 10 mm diam		64-67	0.1 max				0.1			0.4	bal Zn	197		45
C26800	UK	BS 2873	CZ107	Wrought Wir Ann .5/10 mm diam		64-67	0.1 max				0.1			0.4	bal Zn	319		35
C26800	UK	BS 2873	CZ107	Wrought Wir 1/2 hard .5/10 mm diam		64-67	0.1 max				0.1			0.4	bal Zn	461		
C26800	UK	BS 2873	CZ107	Wrought Wir Hard .5/10 mm diam		64-67	0.1 max				0.1			0.4	bal Zn	608		
C26800	UK	BS 2875	C107	Wrought Plt Hard 10/16 mm diam		99.2	0.03		0.15	0.01-0.05	0.01	0.01			0.020 Se+Te; Te 0.01; Sb 0.01; Bi 0.003 max; As 0.3-0.5	279		15
C26800	UK	BS 2875	C107	Wrought Plt As Mfg or Ann 10 mm diam		99.2	0.03		0.15	0.01-0.05	0.01	0.01			0.020 Se+Te; Te 0.01; Sb 0.01; Bi 0.003 max; As 0.3-0.5	211		35

UNS numbers and US grades are provided as a means of cross referencing chemically similar alloys. Exchangability is only possible after independent examination of specifications. Tensile properties are minimum or typical . UTS and YS as Mpa, El as %. See Appendix for list of abbreviations used in Descriptions.

4-70 Wrought Copper

Grade UNS #	Country	Specification	Designation	Comment	Al	Cu	Fe	Mn	Ni	P	Pb	Sn	Zn	OT max	Other	UTS	YS	El
C26800	USA			Copper zinc alloy; Yellow brass 66%		64-68.5	0.05 max				0.15 max				Cu+named el 99.7 min; bal Zn			
C27000	Japan	JIS H3250	C2700	Wrought Rod As Mfg 6 mm diam		63-67	0.05				0.07				bal Zn	294		30
C27000	Japan	JIS H3250	C2700	Wrought Rod Ann 6/75 mm diam		63-67	0.05				0.07				bal Zn	294		40
C27000	Japan	JIS H3250	C2700	Wrought Rod Hard 6/20 mm diam		63-67	0.05				0.07				bal Zn	412		
C27000	Japan	JIS H3260	C2700	Wrought Rod Ann .5 mm diam		63-67	0.05				0.07				bal Zn	294		20
C27000	Japan	JIS H3260	C2700	Wrought Rod Hard .5/10 mm diam		63-67	0.05				0.07				bal Zn	686		
C27000	Japan	JIS H3260	C2700	Wrought Rod Extra Hard .5/10 mm diam		63-67	0.05				0.07				bal Zn	785		
C27000	Japan	JIS H3300	C2700	Wrought Pip 1/2 hard 4/250 mm diam		63-67	0.05				0.07				bal Zn	373		20
C27000	Japan	JIS H3300	C2700	Wrought Pip Ann 4/250 mm diam		63-67	0.05				0.07				bal Zn	294		40
C27000	Japan	JIS H3300	C2700	Wrought Pip Hard 4/250 mm diam		63-67	0.05				0.07				bal Zn	451		
C27000	Sweden	SIS 145150	5150-07	Wrought Sh Strp Wir Rod SH .1/.5 in diam		62-65.5	0.2				0.3	37	0.5			520	470	1
Yellow Brass C27000	USA	ASTM B134	65Cu-35Zn	Wrought Rod Wir		63-68.5	0.07 max				0.1 max							
Yellow Brass C27000	USA	ASTM B135	65Cu-35Zn	Wrought Rod Wir		63-68.5	0.07 max				0.1 max							
Yellow Brass C27000	USA	ASTM B587	65Cu-35Zn	Wrought Rod Wir		63-68.5	0.07 max				0.1 max							
Yellow Brass C27000	USA	MIL W-6712	65Cu-35Zn	Wrought Rod Wir		63-68.5	0.07 max				0.1 max							
Yellow Brass C27000	USA	QQ B-126	65Cu-35Zn	Wrought Rod Wir		63-68.5	0.07 max				0.1 max							
Yellow Brass C27000	USA	QQ B-613	65Cu-35Zn	Wrought Rod Wir		63-68.5	0.07 max				0.1 max							
Yellow Brass C27000	USA	QQ W-321	65Cu-35Zn	Wrought Rod Wir		63-68.5	0.07 max				0.1 max							
Yellow Brass C27000	USA	SAE J463	65Cu-35Zn	Wrought Rod Wir		63-68.5	0.07 max				0.1 max							
C27000	USA			Copper zinc alloy; Yellow brass 65%		63-68.5	0.07 max				0.1 max				Cu+named el 99.7 min; bal Zn			
C33000	USA			Copper zinc lead alloy; Low leaded brass (tube)		65-68	0.07 max				0.25-0.7				Cu+named el 99.6 min; bal Zn			
Free-Cutting Brass C33200	USA	AMS 4558	66Cu-32.4Zn-1.6Pb	Wrought Tub		65-68	0.07 max				1.3-2		0.5					

UNS numbers and US grades are provided as a means of cross referencing chemically similar alloys. Exchangability is only possible after independent examination of specifications. Tensile properties are minimum or typical . UTS and YS as Mpa, El as %. See Appendix for list of abbreviations used in Descriptions.

Grade UNS #	Country	Specification	Designation	Comment	Al	Cu	Fe	Mn	Ni	P	Pb	Sn	Zn	OT max	Other	UTS	YS	El
Free-Cutting Brass C33200	USA	ASTM B135	66Cu-32.4Zn-1.6Pb	Wrought Tub		65-68	0.07 max				1.3-2			0.5				
Free-Cutting Brass C33200	USA	MIL T-46072	66Cu-32.4Zn-1.6Pb	Wrought Tub		65-68	0.07 max				1.3-2			0.5				
C33200	USA			Copper zinc lead alloy; High leaded brass (tube)		65-68	0.07 max				1.5-2.5				Cu+named el 99.6 min; bal Zn			
C67000	Sweden	SIS 145234	5234-00	Wrought Bar Frg Strp Wir HW 25 mm diam	3.5-5.5	64-68	1-3	2.5-4.5					24			670	370	10
Manganes bronze BC67000	USA			Copper zinc alloy	3-6	63-68	2-4	2.5-5			0.2 max	0.5 max			Cu+named el 99.5 min; Cu incl Ag; bal Zn			
	UK	BS 2872	CZ116	Wrought Frg As Mfg or Ann 6 mm diam	4-5	64-68	0.25-1.2	0.3-2						0.5	bal Zn			
	UK	BS 2874	CZ116	Wrought Rod As Mfg 6/100 mm diam	4-5	64-68	0.25-1.2	0.3-2						0.5	bal Zn			
C71640	UK	BS 2871	CN108	Wrought Tub			1.7-2.3	1.5-2.5	29-32					0.3				
C71640	USA			Copper nickel alloy			1.7-2.3	1.5-2.5			0.01 max				Cu+named el 99.5 min; Ni+Co 29.0-32.0; S 0.03 max; C 0.06 max			
C78200	USA			Nickel silver		63-67	0.35 max	0.5 max			1.5-2.5				Cu+named el 99.5 min; Ni+Co 7.0-9.0; bal Zn			
C79000	USA			Nickel silver		63-67	0.35 max	0.5 max			1.5-2.2				Cu+named el 99.5 min; Ni+Co 11.0-13.0; bal Zn			
	Denmark	DS 3003	5682	Wrought		67	0.4-1	0.5-1.5	30-32					0.5	0.05 Pb+Sn; S 0.05; C 0.1			
	Japan	JIS H3270	C7941	Wrought Rod Hard 6.5 mm diam		61-67	0.25	0.5			0.8-1.8				16.5-19.5 Ni+Co; bal Zn	549		
C33530	Australia	AS 1567	335	Wrought Bar Rod As Mfg 6 mm diam		62.5-66.5	0.1				0.3-0.8				bal Zn	310		25
C74500	Canada	CSA HC.4.4	HC.4.ZN2410 (745)	Wrought Plt Quarter Hard		63.5-68.5	0.25 max	0.5	9-11		0.1			0.5	bal Zn	386		
C74500	Canada	CSA HC.4.4	HC.4.ZN2410 (745)	Wrought Bar Half Hard		63.5-68.5	0.25 max	0.5	9-11		0.1			0.5	bal Zn	461		
C74500	Canada	CSA HC.4.4	HC.4.ZN2410 (745)	Wrought Strp Half Hard		63.5-68.5	0.25 max	0.5	9-11		0.1			0.5	bal Zn	461		
C74500	Canada	CSA HC.4.4	HC.4.ZN2410 (745)	Wrought Plt Half Hard		63.5-68.5	0.25 max	0.5	9-11		0.1			0.5	bal Zn	461		
C74500	Canada	CSA HC.4.4	HC.4.ZN2410 (745)	Wrought Bar Full Hard		63.5-68.5	0.25 max	0.5	9-11		0.1			0.5	bal Zn	551		
C74500	Canada	CSA HC.4.4	HC.4.ZN2410 (745)	Wrought Strp Full Hard		63.5-68.5	0.25 max	0.5	9-11		0.1			0.5	bal Zn	551		
C74500	Canada	CSA HC.4.4	HC.4.ZN2410 (745)	Wrought Bar Spring		63.5-68.5	0.25 max	0.5	9-11		0.1			0.5	bal Zn	655		
C74500	Canada	CSA HC.4.4	HC.4.ZN2410 (745)	Wrought Strp Spring		63.5-68.5	0.25 max	0.5	9-11		0.1			0.5	bal Zn	655		
C74500	Canada	CSA HC.4.4	HC.4.ZN2410 (745)	Wrought Sh Half Hard		63.5-68.5	0.25 max	0.5	9-11		0.1			0.5		461		
C74500	Canada	CSA HC.4.4	HC.4.ZN2410 (745)	Wrought Sh Full Hard		63.5-68.5	0.25 max	0.5	9-11		0.1			0.5		551		
C74500	Canada	CSA HC.4.4	HC.4.ZN2410 (745)	Wrought Sh Spring		63.5-68.5	0.25 max	0.5	9-11		0.1			0.5		655		

UNS numbers and US grades are provided as a means of cross referencing chemically similar alloys. Exchangability is only possible after independent examination of specifications. Tensile properties are minimum or typical . UTS and YS as Mpa, El as %. See Appendix for list of abbreviations used in Descriptions.

Grade UNS #	Country	Specification	Designation	Comment	Al	Cu	Fe	Mn	Ni	P	Pb	Sn	Zn	OT max	Other	UTS	YS	El
C74500	France	NF A51107	CuNi10Zn27	Sh 3/4 hard 10 mm diam		61-65		0.5	8-11		0.05			0.3	bal Zn	550		7
C74500	France	NF A51107	CuNi10Zn27	Sh 1/2 hard 10 mm diam		61-65		0.5	8-11		0.05			0.3	bal Zn	470		12
C74500	France	NF A51107	CuNi10Zn27	Wrought Sh 1/4 hard 10 mm diam		61-65		0.5	8-11		0.05			0.3	bal Zn	420		27
C74500	International	ISO 430	CuNi10Zn27	Wrought Plt Sh Strp SH (HB)		61-65	0.3	0.5 max	9-11		0.05			0.3	bal Zn	540		5
C74500	International	ISO 430	CuNi10Zn27	Wrought Plt Sh Strp HA		61-65	0.3	0.5 max	9-11		0.05			0.3	bal Zn	410		15
C74500	International	ISO 430	CuNi10Zn27	Wrought Plt Sh Strp Specially ann (OS35)		61-65	0.3	0.5 max	9-11		0.05			0.3	bal Zn			38
C74500	Japan	JIS H3110	C7451	Wrought Sh Plt 1/2 hard .15 mm diam		62-68	0.25	0.5	8.5-11.5		0.1				8.5-11.5 Ni+Co; bal Zn	392		5
C74500	Japan	JIS H3110	C7451	Wrought Sh Plt Ann .3/5 mm diam		62-68	0.25	0.5	8.5-11.5		0.1				8.5-11.5 Ni+Co; bal Zn	324		20
C74500	Japan	JIS H3270	C7451	Wrought Rod		62-68	0.25	0.5			0.1				8.5-11.5 Ni+Co; bal Zn			
C74500	UK	BS 2870	NS103	Wrought Sh Strp		60-65	0.25	0.05-0.3	9-11		0.04			0.5	bal Zn			
C74500	UK	BS 2873	NS103	Wrought Wir		60-65	0.25	0.05-0.3	9-11		0.04			0.5	bal Zn			
C74500	USA		Nickel silver 65-10	Nickel silver		63.5-66.5	0.25 max	0.5 max			0.1 max				Cu+named el 99.5 min; Ni+Co 9.0-11.0; bal Zn			
C74500	USA		Nickel silver 65-10	Nickel silver; Rod Wir		63.5-66.5	0.25 max	0.5 max			0.05 max				Cu+named el 99.5 min; Ni+Co 9.0-11.0; bal Zn			
C75200	Canada	CSA HC.4.4	HC.4.NZ1817 (752)	Wrought Plt Quarter Hard		63-66.5	0.25 max	0.5	16.5-19.5		0.1			0.5	bal Zn	399		
C75200	Canada	CSA HC.4.4	HC.4.NZ1817 (752)	Wrought Strp Half Hard		63-66.5	0.25 max	0.5	16.5-19.5		0.1			0.5	bal Zn	455		
C75200	Canada	CSA HC.4.4	HC.4.NZ1817 (752)	Wrought Bar Half Hard		63-66.5	0.25 max	0.5	16.5-19.5		0.1			0.5	bal Zn	455		
C75200	Canada	CSA HC.4.4	HC.4.NZ1817 (752)	Wrought Sh Half Hard		63-66.5	0.25 max	0.5	16.5-19.5		0.1			0.5	bal Zn	455		
C75200	Canada	CSA HC.4.4	HC.4.NZ1817 (752)	Wrought Plt Half Hard		63-66.5	0.25 max	0.5	16.5-19.5		0.1			0.5	bal Zn	455		
C75200	Canada	CSA HC.4.4	HC.4.NZ1817 (752)	Wrought Strp Full Hard		63-66.5	0.25 max	0.5	16.5-19.5		0.1			0.5	bal Zn	517		
C75200	Canada	CSA HC.4.4	HC.4.NZ1817 (752)	Wrought Bar Full Hard		63-66.5	0.25 max	0.5	16.5-19.5		0.1			0.5	bal Zn	517		
C75200	Canada	CSA HC.4.4	HC.4.NZ1817 (752)	Wrought Sh Full Hard		63-66.5	0.25 max	0.5	16.5-19.5		0.1			0.5	bal Zn	517		
C75200	Canada	CSA HC.4.4	HC.4.NZ1817 (752)	Wrought Strp Spring		63-66.5	0.25 max	0.5	16.5-19.5		0.1			0.5	bal Zn	670		
C75200	Canada	CSA HC.4.4	HC.4.NZ1817 (752)	Wrought Bar Spring		63-66.5	0.25 max	0.5	16.5-19.5		0.1			0.5	bal Zn	670		
C75200	Canada	CSA HC.4.4	HC.4.NZ1817 (752)	Wrought Sh Spring		63-66.5	0.25 max	0.5	16.5-19.5		0.1			0.5	bal Zn	670		
C75200	Canada	CSA HC.4.4	HC.4.ZN2718 (752)	Wrought Sh Half Hard		53.5-56.5	0.25 max	0.5	16.5-19.5		0.1			0.5	bal Zn	337		
C75200	Canada	CSA HC.4.4	HC.4.ZN2718 (752)	Wrought Sh Full Hard		53.5-56.5	0.25 max	0.5	16.5-19.5		0.1			0.5	bal Zn	634		
C75200	Canada	CSA HC.4.4	HC.4.ZN2718 (752)	Wrought Sh Spring		53.5-56.5	0.25 max	0.5	16.5-19.5		0.1			0.5	bal Zn	744		
C75200	Canada	CSA HC.4.4	HC.4.ZN2718 (752)	Wrought Plt Quarter Hard		53.5-56.5	0.25 max	0.5			0.1				16.5-19.5 Ni+Co; bal Zn	476		
C75200	Canada	CSA HC.4.4	HC.4.ZN2718 (752)	Wrought Plt Half Hard		53.5-56.5	0.25 max	0.5			0.1				16.5-19.5 Ni+Co; bal Zn	517		
C75200	Canada	CSA HC.4.4	HC.4.ZN2718 (752)	Wrought Bar Plt Full Hard		53.5-56.5	0.25 max	0.5			0.1				16.5-19.5 Ni+Co; bal Zn	635		
C75200	International	ISO 1634	CuNi18Zn20	Wrought Plt Sh Strp SH(HB)		60-64	0.3	0.7 max	17-19		0.03			0.3	bal Zn	560		3

UNS numbers and US grades are provided as a means of cross referencing chemically similar alloys. Exchangability is only possible after independent examination of specifications. Tensile properties are minimum or typical . UTS and YS as Mpa, El as %. See Appendix for list of abbreviations used in Descriptions.

Worldwide Guide to Equivalent Nonferrous Metals and Alloys

Grade UNS #	Country	Specification	Designation	Comment	Al	Cu	Fe	Mn	Ni	P	Pb	Sn	Zn	OT max	Other	UTS	YS	El
C75200	International	ISO 1634	CuNi18Zn20	Wrought Plt Sh Strp HA		60-64	0.3	0.7 max	17-19		0.03			0.3	bal Zn	460		8
C75200	International	ISO 1634	CuNi18Zn20	Wrought Plt Sh Strp Specially ann (OS25) 350 mm diam		60-64	0.3	0.7 max	17-19		0.03			0.3	bal Zn			30
C75200	International	ISO 1637	CuNi18Zn20	Wrought HB-SH 5/15 mm diam		60-64	0.3	0.7 max	17-19		0.03			0.3	bal Zn	540		8
C75200	International	ISO 1637	CuNi18Zn20	Wrought HA 5/50 mm diam		60-64	0.3	0.7 max	17-19		0.03			0.3	bal Zn	470		22
C75200	International	ISO 1638	CuNi18Zn20	Wrought Ann 1/5 mm diam		60-64	0.3	0.7 max	17-19		0.03			0.3	bal Zn	390		35
C75200	International	ISO 1638	CuNi18Zn20	Wrought HD-SH 1/3 mm diam		60-64	0.3	0.7 max	17-19		0.03			0.3	bal Zn	640		
C75200	International	ISO 430	CuNi18Zn20	Wrought Plt Sh Strp Rod Bar Tub Wir		60-64	0.3	0.7 max	17-19		0.03			0.3	bal Zn			
C75200	Japan	JIS H3110	C7521	Wrought Sh Plt Hard .15 mm diam		61-67		0.5							16.5-19.5 Ni+Co; bal Zn	539		3
C75200	Japan	JIS H3110	C7521	Wrought Sh Plt 1/2 hard .15 mm diam		61-67		0.5							16.5-19.5 Ni+Co; bal Zn	441		5
C75200	Japan	JIS H3110	C7521	Wrought Sh Plt Ann .3/5 mm diam		61-67		0.5							16.5-19.5 Ni+Co; bal Zn	353		20
C75200	Japan	JIS H3270	C7521	Wrought Rod 1/2 hard 6.5 mm diam		61-67	0.25	0.5	16.5-19.5		0.1				bal Zn	490		
C75200	Japan	JIS H3270	C7521	Wrought Rod Hard 6.5 mm diam		61-67	0.25	0.5	16.5-19.5		0.1				bal Zn	549		
C75200	Sweden	SIS 145246	5246-02	Wrought Plt Sh Strp Ann 5 mm diam		60-64	0.3	0.7 max	17-19		0.03		20			370	160	40
C75200	USA		Nickel silver 65-18	Nickel silver		63.5-66.5	0.25 max	0.5 max			0.05 max				Cu+named el 99.5 min; Ni+Co 16.5-19.5; bal Zn			
C75400	France	NF A51107	CuNi15Zn22	Wrought Sh 3/4 hard 10 mm diam		61-65		0.5	14-16		0.05			0.3	bal Zn	560		5
C75400	France	NF A51107	CuNi15Zn22	Wrought Sh 1/2 hard 10 mm diam		61-65		0.5	14-16		0.05			0.3	bal Zn	490		9
C75400	France	NF A51107	CuNi15Zn22	Wrought Sh 1/4 hard 10 mm diam		61-65		0.5	14-16		0.05			0.3	bal Zn	440		22
C75400	International	ISO 1634	CuNi15Zn21	Wrought Plt Sh Strp SH(HB)		62-66	0.3	0.5 max	14-16		0.05			0.3	bal Zn	440		18
C75400	International	ISO 1634	CuNi15Zn21	Wrought Plt Sh Strp Ann		62-66	0.3	0.5 max	14-16		0.05			0.3	bal Zn			36
C75400	International	ISO 1637	CuNi15Zn21	Wrought HB-SH 5/15 mm diam		62-66	0.3	0.5 max	14-16		0.05			0.3	bal Zn	440		18
C75400	International	ISO 1637	CuNi15Zn21	Wrought Ann 5 mm diam		62-66	0.3	0.5 max	14-16		0.05			0.3	bal Zn			36
C75400	International	ISO 1638	CuNi15Zn21	Wrought HD-SH 1/3 mm diam		62-66	0.3	0.5 max	14-16		0.05			0.3	bal Zn	590		5
C75400	International	ISO 1638	CuNi15Zn21	Wrought Ann 1/5 mm diam		62-66	0.3	0.5 max	14-16		0.05			0.3	bal Zn	360		35
C75400	International	ISO 430	CuNi15Zn21	Wrought Plt Sh Strp Rod Bar Tub Wir		62-66	0.3	0.5 max	14-16		0.05			0.3	bal Zn			

UNS numbers and US grades are provided as a means of cross referencing chemically similar alloys. Exchangability is only possible after independent examination of specifications. Tensile properties are minimum or typical . UTS and YS as Mpa, El as %. See Appendix for list of abbreviations used in Descriptions.

Grade UNS #	Country	Specification	Designation	Comment	Al	Cu	Fe	Mn	Ni	P	Pb	Sn	Zn	OT max	Other	UTS	YS	El
C75400	Japan	JIS H3110	C7541	Wrought Sh Plt Hard .15 mm diam		59-65	0.25	0.5			0.1				12.5-15.5 Ni+Co; bal Zn	490		3
C75400	Japan	JIS H3110	C7541	Wrought Sh Plt 1/2 hard .15 mm diam		59-65	0.25	0.5			0.1				12.5-15.5 Ni+Co; bal Zn	412		5
C75400	Japan	JIS H3110	C7541	Wrought Sh Plt Ann .3/5 mm diam		59-65	0.25	0.5			0.1				12.5-15.5 Ni+Co; bal Zn	353		20
C75400	Japan	JIS H3270	C7541	Wrought Rod 1/2 hard 6.5 mm diam		59-65	0.25	0.5			0.1				12.5-15.5 Ni+Co; bal Zn	441		
C75400	Japan	JIS H3270	C7541	Wrought Rod Hard 6.5 mm diam		59-65	0.25	0.5			0.1				12.5-15.5 Ni+Co; bal Zn	569		
C75400	UK	BS 2873	NS105	Wrought Wir		60-65	0.3	0.05-0.5	14-16		0.04			0.5	bal Zn			
C75400	USA		Nickel silver 65-15	Nickel silver		63.5-66.5	0.25 max	0.5 max			0.1 max				Cu+named el 99.5 min; Ni+Co 14.0-16.0; bal Zn			
C75700	Canada	CSA HC.5 (757)	HC.5.ZN2312	Wrought		63.5-66	0.25 max	0.5	11-13		0.05			0.5	bal Zn			
C75700	Denmark	DS 3003	5243	Wrought		62-66	0.3 max	0.5	11-13		0.05		24	0.3				
C75700	Finland	SFS 2936	CuNi12Zn24	Wrought Sh Strp CF (wir) .1/2.5 mm diam		62-66	0.3 max	0.3			0.05 max		24		ST 11-13; ST = Ni+Co	740	720	1
C75700	Finland	SFS 2936	CuNi12Zn24	Wrought Sh Strp Ann (sh and strp) .2/.5 mm diam		62-66	0.3 max	0.3			0.05 max		24		ST 11-13; ST = Ni+Co	340	120	20
C75700	France	NF A51107	CuNi12Zn24	Wrought Sh 3/4 hard 10 mm diam		62-66		0.5	11-13		0.05			0.3	bal Zn	550		5
C75700	France	NF A51107	CuNi12Zn24	Wrought Sh 1/2 hard 10 mm diam		62-66		0.5	11-13		0.05			0.3	bal Zn	470		8
C75700	France	NF A51107	CuNi12Zn24	Wrought Sh 1/4 hard 10 mm diam		62-66		0.5	11-13		0.05			0.3	bal Zn	420		20
C75700	International	ISO 1634	CuNi12Zn24	Wrought Plt Sh Strp HA		62-66	0.3	0.5 max	11-13		0.05			0.3	bal Zn	410		20
C75700	International	ISO 1634	CuNi12Zn24	Wrought Plt Sh Strp Specially Ann (OS25)		62-66	0.3	0.5 max	11-13		0.05			0.3	bal Zn			35
C75700	International	ISO 1634	CuNi12Zn24	Wrought Plt Sh Strp Specially Ann (OS35)		62-66	0.3	0.5 max	11-13		0.05			0.3	bal Zn			40
C75700	International	ISO 1635	CuNi12Zn24	Wrought Tub SH (HB)		62-66	0.3	0.5 max	11-13		0.05			0.3	bal Zn	440		30
C75700	International	ISO 1635	CuNi12Zn24	Wrought Tub Ann		62-66	0.3	0.5 max	11-13		0.05			0.3	bal Zn			38
C75700	International	ISO 1637	CuNi12Zn24	Wrought HB-SH 5/15 mm diam		62-66	0.3	0.5 max	11-13		0.05			0.3	bal Zn	540		5
C75700	International	ISO 1637	CuNi12Zn24	Wrought HA 5/50 mm diam		62-66	0.3	0.5 max	11-13		0.05			0.3	bal Zn	440		22
C75700	International	ISO 1638	CuNi12Zn24	Wrought HB-SH 1/3 mm diam		62-66	0.3	0.5 max	11-13		0.05			0.3	bal Zn	490		5
C75700	International	ISO 1638	CuNi12Zn24	Wrought Ann 1/5 mm diam		62-66	0.3	0.5 max	11-13		0.05			0.3	bal Zn	340		38
C75700	International	ISO 430	CuNi12Zn24	Plt Sh Strp Rod Bar Tub Wir		62-66	0.3	0.5 max	11-13		0.05			0.3	bal Zn			
C75700	Norway	NS 16424	CuNi12Zn24	Sh Strp Plt Bar Rod		64		0.3	12				24		bal Zn			

UNS numbers and US grades are provided as a means of cross referencing chemically similar alloys. Exchangability is only possible after independent examination of specifications. Tensile properties are minimum or typical . UTS and YS as Mpa, El as %. See Appendix for list of abbreviations used in Descriptions.

Grade UNS #	Country	Specification	Designation	Comment	Al	Cu	Fe	Mn	Ni	P	Pb	Sn	Zn	OT max	Other	UTS	YS	El
C75700	Sweden	SIS 145243	5243-06	Wrought Sh Strp SH .1/.5 mm diam		62-66	0.3	0.5 max	11-13		0.05		24					
C75700	Sweden	SIS 145243	5243-07	Wrought Wir SH .1/2.5 mm diam		62-66	0.3	0.5 max	11-13		0.05		24			740	720	1
C75700	UK	BS 2870	NS104	Wrought Sh Strp		60-65	0.25	0.05-0.3	11-13		0.04			0.5	bal Zn			
C75700	UK	BS 2873	NS104	Wrought Wir		60-65	0.25	0.05-0.3	11-13		0.04			0.5	bal Zn			
C75700	USA		Nickel silver 65-12	Nickel silver		63.5-66.5	0.25 max	0.5 max			0.05 max				Cu+named el 99.5 min; Ni+Co 11.0-13.0; bal Zn			
C79200	USA			Nickel silver		59-66.5	0.25 max	0.5 max			0.8-1.4				Cu+named el 99.5 min; Ni+Co 11.0-13.0; bal Zn			
	Czech Republic	CSN 423256	CuNi15Zn21		0.1 max	63-66.5	0.2 max	0.2 max	14-17		0.1 max	0.1 max			bal Zn			
C47940	USA			Copper zinc tin alloy; tin brass		63-66	1				1-2				Cu+named el 99.6 min; bal Zn; Ni+Co 0.10-0.50; S 1.2-2			
C74300	USA			Nickel silver		63-66	0.25 max	0.5 max			0.1 max				Cu+named el 99.5 min; Ni+Co 7.0-9.0; bal Zn			
	Mexico	DGN W-25	LA-65	Wrought Sh 1/2 hard	0.2	64-66	0.2	0.2	0.5		0.5	0.3			bal Zn	380		
	Mexico	DGN W-25	LA-65	Wrought Sh Hard	0.2	64-66	0.2	0.2	0.5		0.5	0.3			bal Zn	441		
	Mexico	DGN W-25	LA-65	Wrought Sh Spring hard	0.2	64-66	0.2	0.2	0.5		0.5	0.3			bal Zn	552		
C27200	Australia	AS 1566	272	Wrought Plt Bar Sh Strp		62-65	0.07				0.07				bal Zn	290		45
C27200	Australia	AS 1566	272	Wrought Plt Bar Sh Strp		62-65	0.07				0.07				bal Zn	390		
C27200	Australia	AS 1566	272	Wrought Plt Bar Sh Strp		62-65	0.07				0.07				bal Zn	470		
C27200	Czech Republic	CSN 423213	CuZn37			62-65	0.3 max				0.2 max	0.2 max			bal Zn			
C27200	Denmark	DS 3003	5150	Wrought		62-65.5	0.2 max				0.3		37	0.05				
C27200	Finland	SFS 2919	CuZn37	Wrought Sh Strp Bar Rod Wir Tub Drawn (shapes) 5 mm diam		62-65.5	0.2 max				0.3 max		37	0.5		370	240	5
C27200	Finland	SFS 2919	CuZn37	Wrought Sh Strp Bar Rod Wir Tub CF (rod and wir) 1/2.5 mm diam		62-65.5	0.2 max				0.3 max		37	0.5		360	240	12
C27200	Finland	SFS 2919	CuZn37	Wrought Sh Strp Bar Rod Wir Tub Ann (sh and strp) .2/.5 mm diam		62-65.5	0.2 max				0.3 max		37	0.5		310	100	20
C27200	France	NF A51101	CuZn36	Wrought Sh StressH T1		64							36			330		
C27200	France	NF A51103	CuZn36	Wrought Tub Hard 80 mm diam		62-65.5								0.05	bal Zn	480		12
C27200	France	NF A51103	CuZn36	Wrought Tub 3/4 hard 50 mm diam		62-65.5								0.05	bal Zn	400		25
C27200	France	NF A51103	CuZn36	Wrought Tub 1/2 hard 80 mm diam		62-65.5								0.05	bal Zn	380		35

UNS numbers and US grades are provided as a means of cross referencing chemically similar alloys. Exchangability is only possible after independent examination of specifications. Tensile properties are minimum or typical . UTS and YS as Mpa, El as %. See Appendix for list of abbreviations used in Descriptions.

Grade UNS #	Country	Specification	Designation	Comment	Al	Cu	Fe	Mn	Ni	P	Pb	Sn	Zn	OT max	Other	UTS	YS	El
C27200	Japan	JIS H3100	C2720	Wrought Sh Plt 1/2 hard .3/20 mm diam		62-64	0.07				0.07				bal Zn	353		28
C27200	Japan	JIS H3100	C2720	Wrought Sh Plt Ann 1 mm diam		62-64	0.07				0.07				bal Zn	275		40
C27200	Japan	JIS H3100	C2720	Wrought Sh Plt Hard .3/10 mm diam		62-64	0.07				0.07				bal Zn	412		
C27200	Norway	NS 16120	CuZn37	Wrought Sh Strp plt Bar Tub Rod Wir		63							37					
C27200	South Africa	SABS 460	Cu-Zn37	Wrought Tub Ann		62-65.5					0.02			0.1	bal Zn			
C27200	South Africa	SABS 460	Cu-Zn37	Wrought Tub Specially Ann		62-65.5					0.02			0.1	bal Zn			
C27200	South Africa	SABS 460	Cu-Zn37	Wrought Tub As drawn		62-65.5					0.02			0.1	bal Zn			
C27200	Sweden	SIS 145150	5150-02	Wrought Plt Sh Strp Bar Wir Rod Tub Ann 5 mm diam		62-65.5	0.2				0.3		37	0.5		310	100	45
C27200	Sweden	SIS 145150	5150-03	Wrought Plt Sh Strp Bar Wir Rod SH 5 mm diam		62-65.5	0.2				0.3		37	0.5		340	180	35
C27200	Sweden	SIS 145150	5150-04	Wrought Plt Sh Strp Bar Wir Rod Tub SH 5 mm diam		62-65.5	0.2				0.3		37	0.5		360	240	30
C27200	Sweden	SIS 145150	5150-05	Wrought Plt Sh Strp SH 5 mm diam		62-65.5	0.2				0.3		37	0.5		430	350	10
C27200	Sweden	SIS 145150	5150-10	Wrought Tub Strp Wir SH		62-65.5	0.2				0.3		37	0.5		370	250	15
C27200	Sweden	SIS 145150	5150-11	Wrought Plt Sh Strp SH 5 mm diam		62-65.5	0.2				0.3		37	0.5		400	290	15
C27200	UK	BS 2870	CZ108	Wrought Sh Strp 1/2 hard 3.5 mm diam		62-65					0.3			0.4	bal Zn	272		15
C27200	UK	BS 2870	CZ108	Wrought Sh Strp 1/4 hard 10 mm diam		62-65					0.3			0.4	bal Zn	238		30
C27200	UK	BS 2870	CZ108	Wrought Sh Strp Ann 10 mm diam		62-65					0.3			0.4	bal Zn	197		40
C27200	UK	BS 2871	CZ108	Wrought Tub As drawn		62-65					0.3			0.6	bal Zn	350		35
C27200	UK	BS 2871	CZ108	Wrought Tub T Ann		62-65					0.3			0.6	bal Zn	300		40
C27200	UK	BS 2871	CZ108	Wrought Tub Ann		62-65					0.3			0.6	bal Zn	450		
C27200	UK	BS 2873	CZ108	Wrought Wir Ann .5/10 mm diam		62-65					0.3			0.6	bal Zn	319		35
C27200	UK	BS 2873	CZ108	Wrought Wir 1/2 hard .5/10 mm diam		62-65					0.3			0.6	bal Zn	461		
C27200	UK	BS 2873	CZ108	Wrought Wir Hard .5/10 mm diam		62-65					0.3			0.6	bal Zn	608		
C27200	USA			Copper zinc alloy; brass		62-65	0.07 max				0.07 max				Cu+named el 99.7 min; bal Zn			
Low-Leaded Brass C33500	USA	ASTM B121	65Cu-34.5Zn-.5Pb	Wrought Rod Bar Plt Sh Strp Frg Shapes		62.5-66.5	0.1 max				0.3-0.8			0.5				

UNS numbers and US grades are provided as a means of cross referencing chemically similar alloys. Exchangability is only possible after independent examination of specifications. Tensile properties are minimum or typical . UTS and YS as Mpa, El as %. See Appendix for list of abbreviations used in Descriptions.

Grade UNS #	Country	Specification	Designation	Comment	Al	Cu	Fe	Mn	Ni	P	Pb	Sn	Zn	OT max	Other	UTS	YS	El
Low-Leaded Brass C33500	USA	ASTM B453	65Cu-34.5Zn-.5Pb	Wrought Rod Bar Plt Sh Strp Frg Shapes		62.5-66.5	0.1 max				0.3-0.8			0.5				
Low-Leaded Brass C33500	USA	QQ B-613	65Cu-34.5Zn-.5Pb	Wrought Rod Bar Plt Sh Strp Frg Shapes		62.5-66.5	0.1 max				0.3-0.8			0.5				
Low-Leaded Brass C33500	USA	QQ B-626	65Cu-34.5Zn-5Pb	Wrought Rod Bar Plt Sh Strp Frg Shapes		62.5-66.5	0.1 max				0.3-0.8			0.5				
C33500	USA			Copper zinc lead alloy; Low leaded brass		62-65	0.15 max				0.25-0.7				Cu+named el 99.6 min; bal Zn			
C33500	USA			Copper zinc lead alloy; Low leaded brass; Flat products		62-65	0.1 max				0.25-0.7				Cu+named el 99.6 min; bal Zn			
C34000	Canada	CSA HC.4.3	HC.4.ZP341	Wrought Bar Full Hard		62.5-66.5	0.1 max				0.8-1.4			0.5	bal Zn	68		
C34000	Canada	CSA HC.4.3	HC.4.ZP341	Wrought Plt Quarter Hard		62.5-66.5	0.1 max				0.8-1.4			0.5	bal Zn	337		
C34000	Canada	CSA HC.4.3	HC.4.ZP341	Wrought Sh Half Hard		62.5-66.5	0.1 max				0.8-1.4			0.5	bal Zn	379		
C34000	Canada	CSA HC.4.3	HC.4.ZP341	Wrought Bar Half Hard		62.5-66.5	0.1 max				0.8-1.4			0.5	bal Zn	379		
C34000	Canada	CSA HC.4.3	HC.4.ZP341	Wrought Strp Half Hard		62.5-66.5	0.1 max				0.8-1.4			0.5	bal Zn	379		
C34000	Canada	CSA HC.4.3	HC.4.ZP341	Wrought Plt Half Hard		62.5-66.5	0.1 max				0.8-1.4			0.5	bal Zn	379		
C34000	Canada	CSA HC.4.3	HC.4.ZP341	Wrought Sh Full Hard		62.5-66.5	0.1 max				0.8-1.4			0.5	bal Zn	468		
C34000	Canada	CSA HC.4.3	HC.4.ZP341	Wrought Strp Full Hard		62.5-66.5	0.1 max				0.8-1.4			0.5	bal Zn	468		
C34000	Canada	CSA HC.4.3	HC.4.ZP341	Wrought Sh Spring		62.5-66.5	0.1 max				0.8-1.4			0.5	bal Zn	592		
C34000	Canada	CSA HC.4.3	HC.4.ZP341	Wrought Bar Spring		62.5-66.5	0.1 max				0.8-1.4			0.5	bal Zn	592		
C34000	Canada	CSA HC.4.3	HC.4.ZP341	Wrought Strp Spring		62.5-66.5	0.1 max				0.8-1.4			0.5	bal Zn	592		
C34000	Czech Republic	CSN 423214	CuZn36Pb1			65					1-1.9				Fe+Sn 0.30; bal Zn			
C34000	India	IS 2704	CuZn35Pb1	Wrought Wir 1/4 hard		62-65					0.75-1.5			0.5	bal Zn	323		30
C34000	India	IS 2704	CuZn35Pb1	Wrought Wir Hard		62-65					0.75-1.5			0.5	bal Zn	500		15
C34000	India	IS 2704	CuZn35Pb1	Wrought Wir 1/2 hard		62-65					0.75-1.5			0.5	bal Zn	402		20
C34000	International	ISO 1634	CuZn35Pb2	Wrought Plt Sh Strp SH (HB) .3/5 mm diam		61-64	0.2				1.5-2.5				Cu includes 0.3 Ni; bal Zn	430		10
C34000	International	ISO 1634	CuZn35Pb2	Wrought Plt Sh Strp Ann		61-64	0.2				1.5-2.5				Cu includes 0.3 Ni; bal Zn			35
C34000	International	ISO 1637	CuZn35Pb2	Wrought HA 5/15 mm diam		61-64	0.2				1.5-2.5				Cu includes 0.3 Ni; bal Zm	350		20
C34000	International	ISO 1637	CuZn35Pb2	Wrought M 5 mm diam		61-64	0.2				1.5-2.5				Cu includes 0.3 Ni; bal Zm	360		30
C34000	International	ISO 1638	CuZn35Pb2	Wrought HB-SH 1/3 mm diam		61-64	0.2				1.5-2.5				Cu includes 0.3 Ni; bal Zn	390		10
C34000	International	ISO 1638	CuZn35Pb2	Wrought Ann 1/5 mm diam		61-64	0.2				1.5-2.5				Cu includes 0.3 Ni; bal Zn	340		25
C34000	International	ISO 426-2	CuZn35Pb2	Wrought Plt Sh Strp Rod Bar Tub		61-64	0.2				1.5-2.5			0.3	bal Zn			
C34000	UK	BS 2870	CZ118	Wrought Sh Strp 1/2 hard 6 mm diam		63-66					0.75-1.5			0.3	bal Zn			

UNS numbers and US grades are provided as a means of cross referencing chemically similar alloys. Exchangability is only possible after independent examination of specifications. Tensile properties are minimum or typical . UTS and YS as Mpa, El as %. See Appendix for list of abbreviations used in Descriptions.

4-78 Wrought Copper

Grade UNS #	Country	Specification	Designation	Comment	Al	Cu	Fe	Mn	Ni	P	Pb	Sn	Zn	OT max	Other	UTS	YS	El
C34000	UK	BS 2870	CZ118	Wrought Sh Strp Hard 6 mm diam		63-66					0.75-1.5			0.3	bal Zn			
C34000	UK	BS 2870	CZ118	Wrought Sh Strp Extra Hard 6 mm diam		63-66					0.75-1.5			0.3	bal Zn			
Medium-Leaded Brass C34000	USA	ASTM B121	65Cu-34Zn-1Pb	Wrought Rod Wir Sh		62.5-66.5	0.1 max				0.8-1.4			0.5				
Medium-Leaded Brass C34000	USA	ASTM B453	65Cu-34Zn-1Pb	Wrought Rod Wir Sh		62.5-66.5	0.1 max				0.8-1.4			0.5				
Medium-Leaded Brass C34000	USA	QQ B-613	65Cu-34Zn-1Pb	Wrought Rod Wir Sh		62.5-66.5	0.1 max				0.8-1.4			0.5				
Medium-Leaded Brass C34000	USA	QQ B-626	65Cu-34Zn-1Pb	Wrought Rod Wir Sh		62.5-66.5	0.1 max				0.8-1.4			0.5				
C34000	USA			Copper zinc lead alloy; Medium leaded brass 64.5%		62-65	0.15 max				0.8-1.5				Cu+named el 99.6 min; bal Zn			
C34000	USA			Copper zinc lead alloy; Medium leaded brass 64.5%; Flat products		62-65	0.1 max				0.8-1.5				Cu+named el 99.6 min; bal Zn			
C34200	USA			Copper zinc lead alloy; High leaded brass 64.5%		62-66	0.15 max				1.5-2.5				Cu+named el 99.6 min; bal Zn			
C34200	USA			Copper zinc lead alloy; High leaded brass 64.5%; Flat products		62-65	0.1 max				1.5-2.5				Cu+named el 99.6 min; bal Zn			
C34500	USA			Copper zinc lead alloy; leaded brass		62-65	0.15 max				1.5-2.5				Cu+named el 99.6 min; bal Zn			
C46200	Japan	JIS H3100	C4621	Wrought Sh Plt As Mfg 20 mm diam		61-64	0.1				0.2	0.7-1.5			bal Zn	373		20
C46200	USA			Copper zinc tin alloy; tin brass; Naval brass 63.5%		62-65	0.1 max				0.2 max				Cu+named el 99.6 min; bal Zn; S 0.5-1			
	Australia	AS 1566	757	Wrought Plt Bar Sh Strp		60-65	0.25	0.05-0.5	11-13		0.04				bal Zn			
	Japan	JIS H3100	C6711	Wrought Pip Ann 5/50 mm diam		61-65		0.05-1			0.1-1	0.7-1.5			1.0 Fe+Al+Si; bal Zn			
	India	IS 410	CuZn37	Wrought Plt Sh Strp Ann		61.5-64.5	0.1				0.3			0.55	bal Zn			
	India	IS 410	CuZn37	Wrought Plt Sh Strp 1/2 hard		61.5-64.5	0.1				0.3			0.55	bal Zn			
	India	IS 410	CuZn37	Wrought Plt Sh Strp hard		61.5-64.5	0.1				0.3			0.55	bal Zn			
C27400	Canada	CSA	HC.4.237(274)			61-64	0.05 max				0.1 max			0.5	bal Zn			
C27400	Denmark	DS 3003	5140	Wrought		61-64	0.2 max				0.5-1.5		36	0.3				
C27400	France	NF A51104	CuZn37	Wrought Bar Wir Mill Cond 50 mm diam		62.5-65.5	0.2 max				0.3			0.5	bal Zn	370		30

UNS numbers and US grades are provided as a means of cross referencing chemically similar alloys. Exchangability is only possible after independent examination of specifications. Tensile properties are minimum or typical. UTS and YS as Mpa, El as %. See Appendix for list of abbreviations used in Descriptions.

Grade UNS #	Country	Specification	Designation	Comment	Al	Cu	Fe	Mn	Ni	P	Pb	Sn	Zn	OT max	Other	UTS	YS	El
C27400	India	IS 3168	CuZn37	Wrought Strp As rolled .035 mm diam		61.5-64	0.05				0.15			0.25	bal Zn	274		45
C27400	India	IS 4413	CuZn37	Wrought Wir 1/2 hard		61.5-64					0.3			0.6	bal Zn			
C27400	India	IS 4413	CuZn37	Wrought Wir hard		61.5-64					0.3			0.6	bal Zn			
C27400	India	IS 4413	CuZn37	Wrought Wir Ann		61.5-64					0.3			0.6	bal Zn			
C27400	International	ISO 1634	CuZn37	Wrought Plt Sh Strp Specially Ann (OS25) .2/2 mm diam		62-65.5	0.2				0.3				Cu includes 0.3 Ni; bal Zn			30
C27400	International	ISO 1634	CuZn37	Wrought Plt Sh Strp Ann		62-65.5	0.2				0.3				Cu includes 0.3 Ni; bal Zn			40
C27400	International	ISO 1635	CuZn37	Wrought Tub SH (HB)		62-65.5	0.2				0.3				Cu includes 0.3 Ni; bal Zn	470		13
C27400	International	ISO 1635	CuZn37	Wrought Tub Ann		62-65.5	0.2				0.3				Cu includes 0.3 Ni; bal Zn			40
C27400	International	ISO 1637	CuZn37	Wrought HB-SH 1/3 mm diam		62-65.5	0.2				0.03				0.3 Ni included in Cu; bal Zn	430		15
C27400	International	ISO 1637	CuZn37	Wrought HA		62-65.5	0.2				0.03				0.3 Ni included in Cu; bal Zn	360		35
C27400	International	ISO 1637	CuZn37	Wrought Ann 5 mm diam		62-65.5	0.2				0.03				0.3 Ni included in Cu; bal Zn			40
C27400	International	ISO 1638	CuZn37	Wrought HC 1/3 mm diam		62-65.5	0.2				0.3				0.3 Ni included in Cu; bal Zn	540		2
C27400	International	ISO 1638	CuZn37	Wrought HB-SH 1/3 mm diam		62-65.5	0.2				0.3				0.3 Ni included in Cu; bal Zn	440		4
C27400	International	ISO 1638	CuZn37	Wrought Ann 1/5 mm diam		62-65.5	0.2				0.3				0.3 Ni included in Cu; bal Zn	290		30
C27400	International	ISO 4261	CuZn37	Wrought Plt Sh Strp Rod Bar Tub Wir		62-65.5	0.2				0.3				0.5 Pb+Fe; bal Zn			
C27400	USA			Copper zinc alloy; Yellow brass 63%		61-64	0.05 max				0.1 max				Cu+named el 99.7 min; bal Zn			
C34900 C34900	USA		62Cu-37.5Zn-.3Pb	Inactive designation		61-64	0.1 max				0.1-0.5		0.5					
C35330	USA			Copper zinc lead alloy; leaded brass		60.5-64					1.5-3.5				Cu+named el 99.5 min; bal Zn; As 0.02-0.25			
	Denmark	DS 3003	5246	Wrought		60-64	0.3 max	0.7	17-19		0.03		20	0.3				
	Finland	SFS 2935	CuNi18Zn20	Wrought Sh CF .1/.5 mm		60-64	0.3 max	0.7 max			0.03 max		20		ST 17-19; ST = Ni+Co	610	570	1
	Finland	SFS 2935	CuNi18Zn20	Wrought Sh Ann .2/.5 mm		60-64	0.3 max	0.7 max			0.03 max		20		ST 17-19; ST = Ni+Co	370	160	15
	France	NF	CuNi18Zn20			60-64		0.5 max	17-19		0.05 max			0.3				
	India	IS 6912	NavalBrass	Wrought Frg As Mfg 6 mm diam		61-64	0.1				0.2	1-1.5		0.75	bal Zn			
	Japan	JIS H3250	C4622	Wrought Rod As Mfg 6/50 mm diam		61-64						0.7-1.5			bal Zn	343		20
	Mexico	DGN W-25	LA-63	Wrought Sh 1/2 hard	0.2	63-64	0.2	0.2	0.5		0.5	0.3			bal Zn	380		
	Mexico	DGN W-25	LA-63	Wrought Sh Hard	0.2	63-64	0.2	0.2	0.5		0.5	0.3			bal Zn	441		
	Mexico	DGN W-25	LA-63	Wrought Sh Spring hard	0.2	63-64	0.2	0.2	0.5		0.5	0.3			bal Zn	552		
	South Africa	SABS 460	Cu-Zn36Pb1	Wrought Tub Ann		61-64					0.5-1.5			0.3	bal Zn			
	South Africa	SABS 460	Cu-Zn36Pb1	Wrought Tub As drawn		61-64					0.5-1.5			0.3	bal Zn			

UNS numbers and US grades are provided as a means of cross referencing chemically similar alloys. Exchangability is only possible after independent examination of specifications. Tensile properties are minimum or typical . UTS and YS as Mpa, El as %. See Appendix for list of abbreviations used in Descriptions.

Grade UNS #	Country	Specification	Designation	Comment	Al	Cu	Fe	Mn	Ni	P	Pb	Sn	Zn	OT max	Other	UTS	YS	El
	Australia	AS 1566	464	Wrought Plt Bar Sh Strp		61-63.5	0.2				0.2	1-1.4			bal Zn			
	Australia	AS 1567	464	Wrought Rod Bar As Mfg 6/20 mm diam		61-63.5	0.2				0.2	1-1.4			bal Zn	400		18
	Denmark	DS 3003	5240	Wrought		59.5-63.5	0.2 max				0.2	0.5-1.5	38	0.5				
C28000	Australia	AS 1567	280	Wrought Rod Bar As Mfg 6/50 mm diam		59-63	0.07				0.3				bal Zn	340	160	26
C28000	Canada	CSA	HC.9.ZF391-SC		0.5-1.5	55-60	0.4-2	0.1-1.5	1 max		0.4 max	1			bal Zn			
C28000	Czech Republic	CSN 423220	CuZn40			59-62					0.3 max				bal Zn			
C28000	France	NF A51101	CuZn40	Wrought Sh StressH T1		60							40			360		
C28000	France	NF A51103	CuZn40	Wrought Tub Hard 80 mm diam		59-62								0.05	bal Zn	520		10
C28000	France	NF A51103	CuZn40	Wrought Tub 3/4 hard 80 mm diam		59-62								0.05	bal Zn	470		23
C28000	France	NF A51103	CuZn40	Wrought Tub 1/2 hard 80 mm diam		59-62								0.05	bal Zn	430		33
C28000	France	NF A51104	CuZn40	Wrought Bar,Wir Mill Cond 50 mm diam		59-62	0.2				0.3			0.5	bal Zn	390		20
C28000	India	IS 410	CuZn40	Wrought Plt Sh Strp Ann		58.5-61.5	0.15				0.3			0.75	bal Zn			
C28000	India	IS 410	CuZn40	Wrought Plt Sh Strp 1/2 hard		58.5-61.5	0.15				0.3			0.75	bal Zn			
C28000	India	IS 410	CuZn40	Wrought Plt Sh Strp Hard		58.5-61.5	0.15				0.3			0.75	bal Zn			
LeadFreeBrass C28000	India	IS 6912	Lead Free Brass	Wrought Frg As Mfg 6 mm diam		59-62	0.02				0.1			0.2	bal Zn			
C28000	International	ISO 1634	CuZn40	Wrough Plt Sh Strp SH (HA)		59-62	0.2				0.3				Cu includes 0.3 Ni; bal Zn	390		20
C28000	International	ISO 1634	CuZn40	Wrough Plt Sh Strp ANN		59-62	0.2				0.3				Cu includes 0.3 Ni; bal Zn			35
C28000	International	ISO 1634	CuZn40	Wrough Plt Sh Strp M		59-62	0.2				0.3				Cu includes 0.3 Ni; bal Zn	370		40
C28000	International	ISO 1635	CuZn40	Wrought Tub SH(HB)		59-62	0.2				0.03				Cu includes 0.3 Ni; bal Zn	5410		10
C28000	International	ISO 1635	CuZn40	Wrought Tub Ann		59-62	0.2				0.03				Cu includes 0.3 Ni; bal Zn			25
C28000	International	ISO 1637	CuZn40	Wrought Ann 5 mm diam		59-62	0.2				0.03				0.3 Ni included in Cu; bal Zn			30
C28000	International	ISO 1637	CuZn40	Wrought M 5 mm diam		59-62	0.2				0.03				0.3 Ni included in Cu; bal Zn	370		40
C28000	International	ISO 1639	CuZn40	Wrought M		59-62	0.2				0.03				0.3 Ni included in Cu; bal Zn	370		35
C28000	International	ISO 4261	CuZn40	Wrought Plt Sh Strp Rod Bar Tub Wir		59-62	0.2				0.3				0.5 Pb+Fe; bal Zn			
C28000	Japan	JIS H3100	C2801	Wrought Sh Plt 1/2 hard .3/20 mm diam		59-62									bal Zn	412		15
C28000	Japan	JIS H3100	C2801	Wrought Sh Plt Ann 1 mm diam		59-62									bal Zn	324		35
C28000	Japan	JIS H3100	C2801	Wrought Sh Plt Hard .3/10 mm diam		59-62									bal Zn	471		
C28000	Japan	JIS H3250	C2800	Wrought Rod As Mfg 6 mm diam		59-63	0.07				0.1				bal Zn	314		25

UNS numbers and US grades are provided as a means of cross referencing chemically similar alloys. Exchangability is only possible after independent examination of specifications. Tensile properties are minimum or typical . UTS and YS as Mpa, El as %. See Appendix for list of abbreviations used in Descriptions.

Worldwide Guide to Equivalent Nonferrous Metals and Alloys

Grade UNS #	Country	Specification	Designation	Comment	Al	Cu	Fe	Mn	Ni	P	Pb	Sn	Zn	OT max	Other	UTS	YS	El
C28000	Japan	JIS H3250	C2800	Wrought Rod Ann 6/75 mm diam		59-63	0.07				0.1				bal Zn	314		35
C28000	Japan	JIS H3250	C2800	Wrought Rod Hard 6/20 mm diam		59-63	0.07				0.1				bal Zn	451		
C28000	Japan	JIS H3260	C2800	Wrought Rod Ann .5 mm diam		59-63									bal Zn	314		20
C28000	Japan	JIS H3260	C2800	Wrought Rod 3/4 hard .5/10 mm diam		59-63									bal Zn	539		
C28000	Japan	JIS H3260	C2800	Wrought Rod Hard .5/10 mm diam		59-63									bal Zn	686		
C28000	Japan	JIS H3300	C2800	Wrought Pip 1/2 hard 10/250 mm diam		59-63	0.07				0.1				bal Zn	373		15
C28000	Japan	JIS H3300	C2800	Wrought Pip Ann 10/250 mm diam		59-63	0.07				0.1				bal Zn	314		35
C28000	Japan	JIS H3300	C2800	Wrought Pip Hard 10/250 mm diam		59-63	0.07				0.1				bal Zn	451		
C28000	UK	BS 2872	CZ109	Wrought Frg As Mfg 6 mm diam		59-62					0.1			0.3	bal Zn	309		25
C28000	UK	BS 2874	CZ109	Wrought Rod As Mfg 6/50 mm diam		59-62					0.1			0.3	bal Zn	339		26
Muntz Metal C28000	USA	ASME SB111	60Cu-40Zn	Wrought Rod Plt Tub Frg		59-63	0.07 max				0.3 max							
Muntz Metal C28000	USA	ASTM B111	60Cu-40Zn	Wrought Rod Plt Tub Frg		59-63	0.07 max				0.3 max							
Muntz Metal C28000	USA	ASTM B135	60Cu-40Zn	Wrought Rod Plt Tub Frg		59-63	0.07 max				0.3 max							
Muntz Metal C28000	USA	QQ B-613	60Cu-40Zn	Wrought Rod Plt Tub Frg		59-63	0.07 max				0.3 max							
Muntz Metal C28000	USA	QQ B-626	60Cu-40Zn	Wrought Rod Plt Tub Frg		59-63	0.07 max				0.3 max							
Muntz Metal C28000	USA	WW T-791	60Cu-40Zn	Wrought Rod Plt Tub Frg		59-63	0.07 max				0.3 max							
C28000	USA			Copper zinc alloy; brass; Muntz metal 60%		59-63	0.07 max				0.3 max				Cu+named el 99.7 min; bal Zn			
C35000	Canada	CSA H.4.3	HC.4.ZP342	Wrought Sh Half Hard		60.5-64.5	0.1 max				2-3			0.5	bal Zn	379		
C35000	Finland	SFS 2924	CuZn38Pb1	Wrought Bar Rod CF (rod and wir) 2.5/5 mm diam		60-62	0.2 max				0.5-1.5		38	0.3		370	200	20
C35000	Finland	SFS 2924	CuZn38Pb1	Wrought Bar Rod Drawn (bar) 2.5/10 mm diam		60-62	0.2 max				0.5-1.5		38	0.3		410	240	20
C35000	International	ISO 1634	CuZn36Pb1	Wrought Plt Sh Strp SH (HB) .3/5 mm diam		61-64	0.2				0.5-1.5				Cu includes 0.3 Ni; bal Zn	430		10
C35000	International	ISO 1634	CuZn36Pb1	Wrought Plt Sh Strp Ann .3/10 mm diam		61-64	0.2				0.5-1.5				Cu includes 0.3 Ni; bal Zn			35

UNS numbers and US grades are provided as a means of cross referencing chemically similar alloys. Exchangability is only possible after independent examination of specifications. Tensile properties are minimum or typical . UTS and YS as Mpa, El as %. See Appendix for list of abbreviations used in Descriptions.

4-82 Wrought Copper

Grade UNS #	Country	Specification	Designation	Comment	Al	Cu	Fe	Mn	Ni	P	Pb	Sn	Zn	OT max	Other	UTS	YS	El
C35000	International	ISO 1637	CuZn36Pb1	Wrought HA 5/15 mm diam		61-64	0.2				0.5-1.5				Cu includes 0.3 Ni; bal Zm	450		20
C35000	International	ISO 1637	CuZn36Pb1	Wrought M 5 mmdiam		61-64	0.2				0.5-1.5				Cu includes 0.3 Ni; bal Zn	360		30
C35000	International	ISO 1638	CuZn36Pb1	Wrought HB-Sh 1/3 mm diam		61-64	0.2				0.5-1.5				Cu includes 0.3 Ni; bal Zn	390		10
C35000	International	ISO 1638	CuZn36Pb1	Wrought Ann 1/5 mm diam		61-64	0.2				0.5-1.5				Cu includes 0.3 Ni; bal Zn	340		25
C35000	International	ISO 426-2	CuZn36Pb1	Wrought Plt Sh Strp Rod Bar Tub		61-64	0.2				0.5-1.5			0.3	bal Zn			
C35000	Japan	JIS H3260	C3501	Wrought Rod Ann .5 mm diam		60-64	0.2				0.7-1.7				0.40 Fe+Sn; bal Zn	294		20
C35000	Japan	JIS H3260	C3501	Wrought Rod 1/2 hard .5/12 mm diam		60-64	0.2				0.7-1.7				0.40 Fe+Sn; bal Zn	343		
C35000	Japan	JIS H3260	C3501	Wrought Rod Hard .5/10 mm diam		60-64	0.2				0.7-1.7				0.40 Fe+Sn; bal Zn	422		
Medium Leaded Brass C35000	USA	ASTM B121	62.5Cu-36.4Zn-1.1Pb	Wrought Rod Sh		59-64	0.1 max				0.8-1.4			0.5				
Medium Leaded Brass C35000	USA	ASTM B453	62.5Cu-36.4Zn-1.1Pb	Wrought Rod Sh		59-64	0.1 max				0.8-1.4			0.5				
Medium Leaded Brass C35000	USA	QQ B-613	62.5Cu-36.4Zn-1.1Pb	Wrought Rod Sh		59-64	0.1 max				0.8-1.4			0.5				
Medium Leaded Brass C35000	USA	QQ B-626	62.5Cu-36.4Zn-1.1Pb	Wrought Rod Sh		59-64	0.1 max				0.8-1.4			0.5				
Medium Leaded Brass C35000	USA	SAE J463	62.5Cu-36.4Zn-1.1Pb	Wrought Rod Sh		59-64	0.1 max				0.8-1.4			0.5				
C35000	USA			Copper zinc lead alloy; Medium leaded brass 62%		60-63	0.15 max				0.8-2				Cu+named el 99.6 min; bal Zn; Cu 61.0 min for rod			
C35000	USA			Copper zinc lead alloy; Medium leaded brass 62%; Flat products		60-63	0.1 max				0.8-2				Cu+named el 99.6 min; bal Zn; Cu 61.0 min for rod			
C35300	Australia	AS 1567	353	Wrought Rod Bar As Mfg 6/50 mm diam		60-63	0.1				1.5-2.5				bal Zn	350		22
C35300	Canada	CSA HC.4.3	HC.4.ZP352	Wrought Plt Quarter Hard		59-64.5	0.1 max				1.3-2.3			0.5	bal Zn	337		
C35300	Canada	CSA HC.4.3	HC.4.ZP352	Wrought Strp Half Hard		59-64.5	0.1 max				1.3-2.3			0.5	bal Zn	379		
C35300	Canada	CSA HC.4.3	HC.4.ZP352	Wrought Plt Half Hard		59-64.5	0.1 max				1.3-2.3			0.5	bal Zn	379		
C35300	Canada	CSA HC.4.3	HC.4.ZP352	Wrought Sh Half Hard		59-64.5	0.1 max				1.3-2.3			0.5	bal Zn	379		
C35300	Canada	CSA HC.4.3	HC.4.ZP352	Wrought Bar Half Hard		59-64.5	0.1 max				1.3-2.3			0.5	bal Zn	379		
C35300	Canada	CSA HC.4.3	HC.4.ZP352	Wrought Strp Full Hard		59-64.5	0.1 max				1.3-2.3			0.5	bal Zn	468		
C35300	Canada	CSA HC.4.3	HC.4.ZP352	Wrought Sh Full Hard		59-64.5	0.1 max				1.3-2.3			0.5	bal Zn	468		

UNS numbers and US grades are provided as a means of cross referencing chemically similar alloys. Exchangability is only possible after independent examination of specifications. Tensile properties are minimum or typical . UTS and YS as Mpa, El as %. See Appendix for list of abbreviations used in Descriptions.

Worldwide Guide to Equivalent Nonferrous Metals and Alloys

Grade UNS #	Country	Specification	Designation	Comment	Al	Cu	Fe	Mn	Ni	P	Pb	Sn	Zn	OT max	Other	UTS	YS	El
C35300	Canada	CSA HC.4.3	HC.4.ZP352	Wrought Bar Full Hard		59-64.5	0.1 max				1.3-2.3			0.5	bal Zn	468		
C35300	Canada	CSA HC.4.3	HC.4.ZP352	Wrought Strp Spring		59-64.5	0.1 max				1.3-2.3			0.5	bal Zn	592		
C35300	Canada	CSA HC.4.3	HC.4.ZP352	Wrought Sh Spring		59-64.5	0.1 max				1.3-2.3			0.5	bal Zn	592		
C35300	Canada	CSA HC.4.3	HC.4.ZP352	Wrought Bar Spring		59-64.5	0.1 max				1.3-2.3			0.5	bal Zn	592		
C35300	Canada	CSA HC.4.9	HC.4.ZP372T	Wrought Bar Plt Sh Strp Half Hard .375 in diam		59-62	0.1 max				1.3-2.2	0.5-1		0.1	bal Zn	414	241	20
C35300	Canada	CSA HC.4.9	HC.4.ZP372T	Wrought Bar Plt Sh Strp Soft .375 in diam		59-62	0.1 max				1.3-2.2	0.5-1		0.1	bal Zn	359	138	30
C35300	Canada	CSA HC.4.9	HC.4.ZT381P	Wrought Bar Plt Half Hard .375 in diam		59-62	0.1 max				0.2	0.5-1		0.1	bal Zn	414	241	20
C35300	Canada	CSA HC.4.9	HC.4.ZT381P	Wrought Bar Plt Soft .375 in diam		59-62	0.1 max				0.2	0.5-1		0.1	bal Zn	359	138	30
C35300	Denmark	DS 3003	5167	Wrought		59-61	0.2 max				1.5-2.5		38	0.5	bal Zn			
C35300	UK	BS 2870	CZ119	Wrought Sh Strp 1/2 hard 6 mm diam		61-64					1-2.5			0.3	bal Zn			
C35300	UK	BS 2870	CZ119	Wrought Sh Strp Hard 6 mm diam		61-64					1-2.5			0.3	bal Zn			
C35300	UK	BS 2870	CZ119	Wrought Sh Strp Extra Hard 6 mm diam		61-64					1-2.5			0.3	bal Zn			
C35300	UK	BS 2871	CZ119	Wrought Tub Ann		61-64					1-2.5			0.3	bal Zn			
C35300	UK	BS 2871	CZ119	Wrought Tub As drawn		61-64					1-2.5			0.3	bal Zn			
C35300	UK	BS 2873	CZ119	Wrought Wir		61-64					1-2.5			0.3	bal Zn			
C35300	UK	BS 2874	CZ119	Wrought Rod As Mfg 6/50 mm diam		61-64					1-2.5			0.3	bal Zn			
C35300	USA			Copper zinc lead alloy; High leaded brass 62%		60-63	0.15 max				1.5-2.5				Cu+named el 99.6 min; bal Zn; Cu 61.0 min for rod			
C35300	USA			Copper zinc lead alloy; High leaded brass 62%; Flat products		60-63	0.1 max				1.5-2.5				Cu+named el 99.6 min; bal Zn; Cu 61.0 min for rod			
C35600	Finland	SFS 2922	CuZn36Pb3	Wrought Rod Bar Wir Drawn (rod,bar and wir) 2.5/5 mm diam		60-63	0.35 max				2.5-3.5		36	0.5		400	320	5
C35600	International	ISO 1637	CuZn36Pb3	Wrought HB-SH 5/15 mm daim		60-63	0.35				2.5-3.7				Cu includes 0.3 Ni; bal Zn	410		15
C35600	International	ISO 1637	CuZn36Pb3	Wrought HA 5/75 mm diam		60-63	0.35				2.5-3.7				Cu includes 0.3 Ni; bal Zn	310		20
C35600	International	ISO 1637	CuZn36Pb3	Wrought M 5 mm diam		60-63	0.35				2.5-3.7				Cu includes 0.3 Ni; bal Zn	360		30
C35600	International	ISO 426-2	CuZn36Pb3	Wrought Rod Bar Tub Wir		60-63	0.35				2.5-3.7			0.5	bal Zn			
C35600	Japan	JIS H3100	C3560	Wrought Sh Plt Hard .3/10 mm diam		61-64	0.1				2-3				bal Zn	422		2

UNS numbers and US grades are provided as a means of cross referencing chemically similar alloys. Exchangability is only possible after independent examination of specifications. Tensile properties are minimum or typical . UTS and YS as Mpa, El as %. See Appendix for list of abbreviations used in Descriptions.

4-84 Wrought Copper

Grade UNS #	Country	Specification	Designation	Comment	Al	Cu	Fe	Mn	Ni	P	Pb	Sn	Zn	OT max	Other	UTS	YS	El
C35600	Japan	JIS H3100	C3560	Wrought Sh Plt 1/2 hard .3/10 mm diam		61-64	0.1				2-3				bal Zn	373		10
C35600	Japan	JIS H3100	C3560	Wrought Sh Plt 1/4 hard .3/10 mm diam		61-64	0.1				2-3				bal Zn	343		18
High Leaded Brass C35600	USA	ASTM B121	62Cu-35.5Zn-2.5Pb	Wrought Bar Rod Plt Sh Strp Shapes		59-64.5	0.1 max				2-3			0.5				
High Leaded Brass C35600	USA	ASTM B453	62Cu-35.5Zn-2.5Pb	Wrought Bar Rod Plt Sh Strp Shapes		59-64.5	0.1 max				2-3			0.5				
High Leaded Brass C35600	USA	QQ B-613	62Cu-35.5Zn-2.5Pb	Wrought Bar Rod Plt Sh Strp Shapes		59-64.5	0.1 max				2-3			0.5				
High Leaded Brass C35600	USA	QQ B-626	62Cu-35.5Zn-2.5Pb	Wrought Bar Rod Plt Sh Strp Shapes		59-64.5	0.1 max				2-3			0.5				
C35600	USA			Copper zinc lead alloy; Extra high leaded brass		60-63	0.15 max				2-3				Cu+named el 99.5 min; bal Zn			
C35600	USA			Copper zinc lead alloy; Extra high leaded brass; flat products		60-63	0.1 max				2-3				Cu+named el 99.5 min; bal Zn			
C36000	Australia	AS 1567	360	Wrought Bar Rod		60-63	0.35				2.5-3.7				bal Zn			
C36000	Canada	CSA	HC.5.ZP353(360)			60-63	0.35 max				2.5-3.7			0.5	bal Zn			
C36000	Finland	SFS 2923	CuZn36Pb1	Wrought Rod Wir Tub CF (rod and wir) 1/2.5 mm diam		61-64	0.2 max				0.5-1.5		36	0.3		360	240	10
C36000	Finland	SFS 2923	CuZn36Pb1	Wrought Rod Wir Tub Ann (rod and wir) 1/2.5 mm diam		61-64	0.2 max				0.5-1.5		36	0.3		310	100	25
C36000	Finland	SFS 2923	CuZn36Pb1	Wrought Rod Wir Tub HF (except tube)		61-64	0.2 max				0.5-1.5		36	0.3		360	150	40
C36000	France	NF A51105	CuZn36Pb3	Wrought Bar Wir Rolled,Ext or drawn 7 mm diam		60-62	0.35 max				2.5-3.5			0.5	bal Zn	450		6
C36000	India	IS 319	II	Wrought Bar Rod half hard 10/12 mm diam		60-63	0.35				2.5-3.7			0.15	bal Zn			
C36000	India	IS 319	II	Wrought Bar Rod Hard 10/12 mm diam		60-63	0.35				2.5-3.7			0.15	bal Zn			
C36000	India	IS 319	II	Wrought Bar Rod Ann 10/25 mm diam		60-63	0.35				2.5-3.7			0.15	bal Zn			
C36000	Japan	JIS H3250	C3601	Wrought Rod Ann 6/75 mm diam		59-63	0.3				1.8-3.7				0.50 Fe+Sn; bal Zn	294		25
C36000	Japan	JIS H3250	C3601	Wrought Rod 1/2 hard 6/50 mm diam		59-63	0.3				1.8-3.7				0.50 Fe+Sn; bal Zn	343		
C36000	Japan	JIS H3250	C3601	Wrought Rod Hard 6/20 mm		59-63	0.3				1.8-3.7				0.50 Fe+Sn; bal Zn	451		

UNS numbers and US grades are provided as a means of cross referencing chemically similar alloys. Exchangability is only possible after independent examination of specifications. Tensile properties are minimum or typical . UTS and YS as Mpa, El as %. See Appendix for list of abbreviations used in Descriptions.

Grade UNS #	Country	Specification	Designation	Comment	Al	Cu	Fe	Mn	Ni	P	Pb	Sn	Zn	OT max	Other	UTS	YS	El
C36000	Japan	JIS H3250	C3602	Wrought Rod As Mfg 6/75 mm diam		59-63	0.5				1.8-3.7				1.3 Fe+Sn; bal Zn	314		
C36000	Japan	JIS H3250	C3603	Wrought Rod Ann 6/75 mm diam		57-61	0.35				1.8-3.7				0.50 Fe+Sn; bal Zn	314		20
C36000	Japan	JIS H3250	C3603	Wrought Rod 1/2 hard 6/50 mm diam		57-61	0.35				1.8-3.7				0.50 Fe+Sn; bal Zn	363		
C36000	Japan	JIS H3250	C3603	Wrought Rod Hard 6/20 mm diam		57-61	0.35				1.8-3.7				0.50 Fe+Sn; bal Zn	451		
C36000	Japan	JIS H3250	C3604	Wrought Rod As Mfg 6/75 mm diam		57-61	0.35				1.8-3.7				0.6 Fe+Sn; bal Zn	333		
C36000	Norway	NS 16150	CuZn36Pb1	Wrought Strp Bar Tub Rod		63					1		36					
C36000	UK	BS 2874	CZ124	Wrought Rod As Mfg (round and hex rod) 6/25 mm diam		60-63	0.35				2.5-3.7			0.5	bal Zn			
C36000	UK	BS 2874	CZ124	Wrought Rod 1/2 hard 6/12 mm diam		60-63	0.35				2.5-3.7			0.5	bal Zn			
C36000	UK	BS 2874	CZ124	Wrought Rod Hard 3/5 mm diam		60-63	0.35				2.5-3.7			0.5	bal Zn			
Free-Cutting Brass C36000	USA	AMS 4610	61.5Cu-35.5Zn-3Pb	Wrought Rod		60-63	0.35 max				2.5-3.7			0.5				
Free-Cutting Brass C36000	USA	ASTM B16	61.5Cu-35.5Zn-3Pb	Wrought Rod		60-63	0.35 max				2.5-3.7			0.5				
Free-Cutting Brass C36000	USA	QQ B-613	61.5Cu-35.5Zn-3Pb	Wrought Rod		60-63	0.35 max				2.5-3.7			0.5				
Free-Cutting Brass C36000	USA	QQ B-626	61.5Cu-35.5Zn-3Pb	Wrought Rod		60-63	0.35 max				2.5-3.7			0.5				
Free-Cutting Brass C36000	USA	SAE J463	61.5Cu-35.5Zn-3Pb	Wrought Rod		60-63	0.35 max				2.5-3.7			0.5				
C36000	USA			Copper zinc lead alloy; Free cutting leaded brass		60-63	0.35 max				2.5-3.7				Cu+named el 99.5 min; bal Zn			
C66800	USA			Copper zinc alloy	0.25 max	60-63	0.35 max	2-3.5			0.5 max	0.3 max			Cu+named el 99.5 min; Cu incl Ag; Ni+Co 0.25 max; bal Zn; Si 0.5-1.5			
C67300	USA			Copper zinc alloy	0.25 max	58-63	0.5 max	2-3.5			0.4-3	0.3 max			Cu+named el 99.5 min; Cu incl Ag; Ni+Co 0.25 max; bal Zn; Si 0.5-1.5			
C76000	USA			Nickel silver		60-63	0.25 max	0.5 max			0.1 max				Cu+named el 99.5 min; Ni+Co 6.0-9.0; bal Zn			
	Australia	AS 1566	761	Wrought Plt Bar Sh		59-63	0.25	0.5	9-11		0.1				bal Zn			
	Czech Republic	CSN 423238	CuZn38SnMn As			61-63	0.1 max	0.6-1.2			0.1 max	0.7-1.1			bal Zn; As 0.1-0.15			

UNS numbers and US grades are provided as a means of cross referencing chemically similar alloys. Exchangability is only possible after independent examination of specifications. Tensile properties are minimum or typical . UTS and YS as Mpa, El as %. See Appendix for list of abbreviations used in Descriptions.

Grade UNS #	Country	Specification	Designation	Comment	Al	Cu	Fe	Mn	Ni	P	Pb	Sn	Zn	OT max	Other	UTS	YS	El
	France	NF A51105	CuZn35Pb2	Wrought Bar Wir		61-63	0.2				1.5-2.25			0.3	bal Zn			
	India	IS 407	2/CuZn39	Wrought Tub Ann		59-63	0.07				0.8			0.3	bal Zn; As 0.06	284		
	India	IS 407	2/CuZn39	Wrought Tub Half Hard		59-63	0.07				0.8			0.3	bal Zn; As 0.06	372		
	India	IS 407	2/CuZn39	Wrought Tub Hard		59-63	0.07				0.8			0.3	bal Zn; As 0.06	451		
	International	ISO 1634	CuNi10Zn28Pb1	Wrought Plt Sh SH		59-63	0.3	0.5 max	9-11		1-2			0.5	bal Zn	550		4
	International	ISO 1634	CuNi10Zn28Pb1	Wrought Plt Sh HA		59-63	0.3	0.5 max	9-11		1-2			0.5	bal Zn	410		10
	International	ISO 1634	CuNi10Zn28Pb1	Wrought Plt Sh Ann		59-63	0.3	0.5 max	9-11		1-2			0.5	bal Zn			35
	International	ISO 1637	CuNi10Zn28Pb1	Wrought HB-SH 5/15 mm diam		59-63	0.3	0.5 max	9-11		1-2			0.5	bal Zn	480		8
	International	ISO 1637	CuNi10Zn28Pb1	Wrought HA 5/50 mm diam		59-63	0.3	0.5 max	9-11		1-2			0.5	bal Zn	410		15
	International	ISO 1637	CuNi18Zn19Pb1	Wrought HB-SH 5/15 mm diam		59-63	0.3	0.7 max	17-19		0.5-1.5			0.5	bal Zn	490		10
	International	ISO 1637	CuNi18Zn19Pb1	Wrought HA 5/50 mm diam		59-63	0.3	0.7 max	17-19		0.5-1.5			0.5	bal Zn	430		30
	International	ISO 1639	CuNi18Zn19Pb1	Wrought As cast		59-63	0.3	0.7 max	17-19		0.5-1.5			0.5	bal Zn	450		20
	International	ISO 430	CuNi10Zn28Pb1	Wrought Strp Rod Bar		59-63	0.3	0.5 max	9-11		1-2			0.5	bal Zn			
	International	ISO 430	CuNi18Zn19Pb1	Wrought Rod Strp Wir Bar		59-63	0.3	0.7 max	17 10		0.5 1.5			0.5	bal Zn			
	UK	BS 2874	NS111	Wrought Rod		58-63		0.1-0.5	9-11		1-2			0.5	bal Zn			
	UK	BS 2874	NS112	Wrought Rod		60-63		0.1-0.5	14-16		0.5-1			0.5	bal Zn			
	UK	BS 2874	NS113	Wrought Rod		60-63	0.3	0.1-0.5	17-19		0.4-0.8			0.1	bal Zn			
C37000	Australia	AS 1566	370	Wrought Plt Bar Sh Strp Ann (sht,strp)		59-62	0.15				0.9-1.4				bal Zn	310		25
C37000	Australia	AS 1566	370	Wrought Plt Bar Sh Strp SH to 3/4 hard (sht,strp)		59-62	0.15				0.9-1.4				bal Zn	510		
C37000	Australia	AS 1566	370	Wrought Plt Bar Sh Strp SH to full hard (sht,strp)		59-62	0.15				0.9-1.4				bal Zn	570		
C37000	Canada	CSA	HC.9.ZP361-SC		0.5 max	60-65	0.75 max		1 max		0.8-1.5	0.5-1.5			bal Zn			
C37000	Czech Republic	CSN 423221	CuZn38Pb			59-62	0.2 max				0.5-1				bal Zn			
C37000	Czech Republic	CSN 423222	CuZn39Pb1			57-61	0.5 max	0.2 max			0.8-1.9				bal Zn			
C37000	Norway	NS 16140	CuZn40Pb	Wrought sh Strp Plt		60					0.5		40					
C37000	Sweden	SIS 145165	5165-00	Wrought Bar Tub HW		60-62	0.2				0.5-1.5		38	0.3		370	130	35
C37000	Sweden	SIS 145165	5165-02	Wrought Strp Bar Ann .5/2.5 mm diam		60-62	0.2				0.5-1.5		38	0.3		330	110	20
C37000	Sweden	SIS 145165	5165-04	Wrought Strp Bar Wir Tub SH .5/2.5 mm diam		60-62	0.2				0.5-1.5		38	0.3		410	240	10
C37000	Sweden	SIS 145165	5165-10	Wrought Wir SH 5 mm diam		60-62	0.2				0.5-1.5		38	0.3		390	200	10
Free-cutting Muntz C37000	USA	ASTM B135	60Cu-39Zn-1Pb	Wrought Bar Rod Plt Sh Strp Frg		59-62	0.15 max				0.9-1.4			0.5				

NS numbers and US grades are provided as a means of cross referencing chemically similar alloys. Exchangability is only possible after independent examination of specifications. Tensile properties are minimum or typical . UTS and YS as Mpa, El as %. See Appendix for list of abbreviations used in Descriptions.

Worldwide Guide to Equivalent Nonferrous Metals and Alloys

Grade UNS #	Country	Specification	Designation	Comment	Al	Cu	Fe	Mn	Ni	P	Pb	Sn	Zn	OT max	Other	UTS	YS	El
Free-Cutting Muntz C37000	USA	MIL T-46072	60Cu-39Zn-1Pb	Wrought Bar Rod Plt Sh Strp Frg		59-62	0.15 max				0.9-1.4			0.5				
Free-Cutting Muntz C37000	USA	QQ B-613	60Cu-39Zn-1Pb	Wrought Bar Rod Plt Sh Strp Frg		59-62	0.15 max				0.9-1.4			0.5				
Free-Cutting Muntz C37000	USA	QQ B-626	60Cu-39Zn-1Pb	Wrought Bar Rod Plt Sh Strp Frg		59-62	0.15 max				0.9-1.4			0.5				
C37000	USA			Copper zinc lead alloy; Free cutting muntz metal		59-62	0.15 max				0.8-1.5				Cu+named el 99.6 min; bal Zn			
C37100	Denmark	DS 3003	5165	Wrought		60-62	0.2 max				0.5-1.5		38	0.3				
C37100	Japan	JIS H3100	C3710	Wrought Sh Plt 1/2 hard .3/10 mm diam		58-62					0.6-1.2				bal Zn	422		13
C37100	Japan	JIS H3100	C3710	Wrought Sh Plt 1/4 hard .3/10 mm diam		58-62					0.6-1.2				bal Zn	373		20
C37100	Japan	JIS H3100	C3710	Wrought Sh Plt Hard .3/10 mm diam		58-62					0.6-1.2				bal Zn	471		
C37100	Japan	JIS H3100	C3713	Wrought Sh Plt 1/2 hard .3/10 mm diam		58-62	0.1				1-2				bal Zn	422		10
C37100	Japan	JIS H3100	C3713	Wrought Sh Plt 1/4 hard .3/10 mm diam		58-62	0.1				1-2				bal Zn	373		18
C37100	Japan	JIS H3100	C3713	Wrought Sh Plt Hard .3/10 mm diam		58-62	0.1				1-2				bal Zn	471		
C37100	Japan	JIS H3250	C3712	Wrought Rod As Mfg 6 mm		58-62					0.1-1				0.8 Fe+Sn; bal Zn	314		15
C37100	Sweden	SIS 145173	5173-00	Wrought HW		57-59	0.35				0.4-1.2		41	0.7		410	150	30
C37100	Sweden	SIS 145173	5173-10	Wrought SH		57-59	0.35				0.4-1.2		41	0.7		390	180	10
C37100	USA			Copper zinc lead alloy; leaded brass		58-62	0.15 max				0.6-1.2				Cu+named el 99.6 min; bal Zn			
C46400	Canada	CSA HC.4.9	HC.4.ZT391(464)	Wrought Bar Plt Sh Strp Half Hard .375 in		59-62	0.1 max				0.2	0.5-1		0.1	bal Zn			
C46400	Canada	CSA HC.4.9	HC.4.ZT391(464)	Wrought Bar Plt Sh Strp Soft .375 in diam		59-62	0.1 max				0.2	0.5-1		0.1	bal Zn			
C46400	Czech Republic	CSN 423237	CuZn38Sn1			61-63	0.1 max				0.1 max	0.7-1.1			bal Zn			
C46400	International	ISO 1634	CuZn38Sn1	Wrought Plt Sh Strp HA		59.5-63.5	0.2				0.2	0.5-1.5			0.5 Fe+Pb; bal Zn	390		25
C46400	International	ISO 1634	CuZn38Sn1	Wrought Plt Sh Strp M		59.5-63.5	0.2				0.2	0.5-1.5			0.5 Fe+Pb; bal Zn	370		30
C46400	International	ISO 1637	CuZn38Sn1	Wrought AS cast 5 mm diam		59.5-63.5	0.2				0.2	0.5-1.5			0.5 Fe+Pb+OT; bal Zn	390		35
C46400	International	ISO 426-1	CuZn38Sn1	Wrought Plt Sh Rod Bar Tub Frg		59.5-63.5	0.2				0.2	0.5-1.5			0.5 Fe+Pb; bal Zn			
C46400	Norway	NS 16220	CuZn38Sn1			59.5-63.5	0.2 max				0.2 max	0.5-1.5		0.5	bal Zn			
C46400	UK	BS 2870	CZ112	Wrought Sh Strp Hard 10 mm diam		61-63.5						1-1.4		0.75	bal Zn	288		20

UNS numbers and US grades are provided as a means of cross referencing chemically similar alloys. Exchangability is only possible after independent examination of specifications. Tensile properties are minimum or typical . UTS and YS as Mpa, El as %. See Appendix for list of abbreviations used in Descriptions.

4-88 Wrought Copper

Grade UNS #	Country	Specification	Designation	Comment	Al	Cu	Fe	Mn	Ni	P	Pb	Sn	Zn	OT max	Other	UTS	YS	El
C46400	UK	BS 2870	CZ112	Wrought Sh Strp As Mfg or Ann 10 mm diam		61-63.5						1-1.4		0.75	bal Zn	238		25
C46400	UK	BS 2872	CZ112	Wrought Frg As Mfg or Ann 10 mm diam		61-63.5						1-1.4		0.75	bal Zn	339		15
C46400	UK	BS 2874	CZ112	Wrought Rod As Mfg 6/20 mm diam		61-63.5						1-1.4		0.75	bal Zn	402		18
C46400	UK	BS 2874	CZ112	Wrought Rod Hard 10/12.5 mm diam		61-63.5						1-1.4		0.75	bal Zn	402		18
C46400	UK	BS 2875	CZ112	Wrought Plt As Mfg or Ann 10/25 mm diam		61-63.5						1-1.4		0.75	bal Zn			
Naval Brass C46400	USA	AMS 4611	60Cu-39.2Zn-.8Sn	Wrought Bar Rod Sh Tub Extrusions		59-62	0.1 max				0.2 max	0.5-1	36.7-40.5					
Naval Brass C46400	USA	ASME SB171	60Cu-39.2Zn-.8Sn	Wrought Bar Rod Sh Tub Extrusions		59-62	0.1 max				0.2 max	0.5-1	36.7-40.5					
Naval Brass C46400	USA	ASTM B124	60Cu-39.2Zn-.8Sn	Wrought Bar Rod Sh Tub Extrusions		59-62	0.1 max				0.2 max	0.5-1	36.7-40.5					
Naval Brass C46400	USA	ASTM B171	60Cu-39.2Zn-.8Sn	Wrought Bar Rod Sh Tub Extrusions		59-62	0.1 max				0.2 max	0.5-1	36.7-40.5					
Naval Brass C46400	USA	ASTM B21	60Cu-39.2Zn-.8Sn	Wrought Bar Rod Sh Tub Extrusions		59-62	0.1 max				0.2 max	0.5-1	36.7-40.5					
Naval Brass C46400	USA	ASTM B283	60Cu-39.2Zn-.8Sn	Wrought Bar Rod Sh Tub Extrusions		59-62	0.1 max				0.2 max	0.5-1	36.7-40.5					
Naval Brass C46400	USA	ASTM B432	60Cu-39.2Zn-.8Sn	Wrought Bar Rod Sh Tub Extrusions		59-62	0.1 max				0.2 max	0.5-1	36.7-40.5					
Naval Brass C46400	USA	QQ B-637	60Cu-39.2Zn-.8Sn	Wrought Bar Rod Sh Tub Extrusions		59-62	0.1 max				0.2 max	0.5-1	36.7-40.5					
Naval Brass C46400	USA	QQ B-639	60Cu-39.2Zn-.8Sn	Wrought Bar Rod Sh Tub Extrusions		59-62	0.1 max				0.2 max	0.5-1	36.7-40.5					
Naval Brass C46400	USA	SAE J461	60Cu-39.2Zn-.8Sn	Wrought Bar Rod Sh Tub Extrusions		59-62	0.1 max				0.2 max	0.5-1	36.7-40.5					
Naval Brass C46400	USA	SAE J463	60Cu-39.2Zn-.8Sn	Wrought Bar Rod Sh Tub Extrusions		59-62	0.1 max				0.2 max	0.5-1	36.7-40.5					
C46400	USA			Copper zinc tin alloy; tin brass; Naval brass uninhibited		59-62	0.1 max				0.2 max				Cu+named el 99.6 min; bal Zn; S 0.5-1			
C46500	USA			Copper zinc tin alloy; tin brass; Naval brass arsenical		59-62	0.1 max				0.2 max				Cu+named el 99.6 min; bal Zn; S 0.5-1; As 0.02-0.06			
Leaded Naval Brass C48200	USA	ASTM B124	60.5Cu-38Zn-.8Sn.7Pb	Wrought Bar Rod Wir Plt Frg Shapes		59-62	0.1 max				0.2 max	0.5-1						
Leaded Naval Brass C48200	USA	ASTM B21	60.5Cu-38Zn-.8Sn.7Pb	Wrought Bar Rod Wir Plt Frg Shapes		59-62	0.1 max				0.2 max	0.5-1						

UNS numbers and US grades are provided as a means of cross referencing chemically similar alloys. Exchangability is only possible after independent examination of specifications. Tensile properties are minimum or typical . UTS and YS as Mpa, El as %. See Appendix for list of abbreviations used in Descriptions.

Worldwide Guide to Equivalent Nonferrous Metals and Alloys

Grade UNS #	Country	Specification	Designation	Comment	Al	Cu	Fe	Mn	Ni	P	Pb	Sn	Zn	OT max	Other	UTS	YS	El
Leaded Naval Brass C48200	USA	QQ B-626	60.5Cu-38Zn-.8Sn.7Pb	Wrought Bar Rod Wir Plt Frg Shapes		59-62	0.1 max				0.2 max	0.5-1						
Leaded Naval Brass C48200	USA	QQ B-637	60.5Cu-38Zn-.8Sn.7Pb	Wrought Bar Rod Wir Plt Frg Shapes		59-62	0.1 max				0.2 max	0.5-1						
Leaded Naval Brass C48200	USA	QQ B-639	60.5Cu-38Zn-.8Sn.7Pb	Wrought Bar Rod Wir Plt Frg Shapes		59-62	0.1 max				0.2 max	0.5-1						
C48200	USA			Copper zinc tin alloy; tin brass; Naval brass medium leaded		59-62	0.1 max				0.4-1				Cu+named el 99.6 min; bal Zn; S 0.5-1			
Leaded Naval Brass C48500	USA	ASTM B124	60Cu-37.5Zn1.8Pb.7Sn	Wrought Bar Rod Wir Plt Sh Strp Frg Shapes		59-62	0.1 max				0.2 max	0.5-1						
Leaded Naval Brass C48500	USA	ASTM B21	60Cu-37.5Zn1.8Pb.7Sn	Wrought Bar Rod Wir Plt Sh Strp Frg Shapes		59-62	0.1 max				0.2 max	0.5-1						
Leaded Naval Brass C48500	USA	ASTM B283	60Cu-37.5Zn1.8Pb.7Sn	Wrought Bar Rod Wir Plt Sh Strp Frg Shapes		59-62	0.1 max				0.2 max	0.5-1						
Leaded Naval Brass C48500	USA	QQ B-626	60Cu-37.5Zn1.8Pb.7Sn	Wrought Bar Rod Wir Plt Sh Strp Frg Shapes		59-62	0.1 max				0.2 max	0.5-1						
Leaded Naval Brass C48500	USA	QQ B-637	60Cu-37.5Zn1.8Pb.7Sn	Wrought Bar Rod Wir Plt Sh Strp Frg Shapes		59-62	0.1 max				0.2 max	0.5-1						
Leaded Naval Brass C48500	USA	QQ B-639	60Cu-37.5Zn1.8Pb.7Sn	Wrought Bar Rod Wir Plt Sh Strp Frg Shapes		59-62	0.1 max				0.2 max	0.5-1						
C48500	USA			Copper zinc tin alloy; tin brass; Naval brass high leaded		59-62	0.1 max				1.3-2.2				Cu+named el 99.6 min; bal Zn; S 0.5-1			
C48600	Australia	AS 1567	486	Wrought Bar Rod As Mfg 6/20 mm diam		59-62					1-2.5	0.75-1.5			bal Zn; As 0.02-0.25	380		15
C48600	Australia	AS 1568	486	Wrought Frg As Mfg 6 mm		59-62						0.75-1.5			bal Zn; As 0.02-0.25	320		20
C48600	USA			Copper zinc tin alloy; tin brass		59-62					1-2.5				Cu+named el 99.6 min; bal Zn; S 0.8-1.5; As 0.02-0.25; As 0.02-0.25			
	Denmark	DS 3003	5238	Wrought	0.2-1	58-62	0.5-1.5	1-3			0.2-1	0.5-1.5	35					
	Finland	SFS 2929	CuZn35Mn2A lFe	Wrought Rod Tub Drawn (rod,tube,and shapes) 25/50 mm diam	0.2-1	58-62	0.5-1.5	1-3			0.2-1	0.5-1.5	35	0.5		470	270	15
	Finland	SFS 2929	CuZn35Mn2A lFe	Wrought Rod Tub HF	0.2-1	58-62	0.5-1.5	1-3			0.2-1	0.5-1.5	35	0.5		450	200	35
	Japan	JIS H3100	C6712	Wrought Sh Plt Hard .25/1.5 mm diam		58-62		0.05-1			0.1-1				1.0 Fe+Al+Si; bal Zn			
	Japan	JIS H3250	C4641	Wrought Rod As Mfg 6/50 mm diam		59-62						0.5-1			1.0 Pb+Fe; bal Zn	343		20

UNS numbers and US grades are provided as a means of cross referencing chemically similar alloys. Exchangability is only possible after independent examination of specifications. Tensile properties are minimum or typical . UTS and YS as Mpa, El as %. See Appendix for list of abbreviations used in Descriptions.

4-90 Wrought Copper

Grade UNS #	Country	Specification	Designation	Comment	Al	Cu	Fe	Mn	Ni	P	Pb	Sn	Zn	OT max	Other	UTS	YS	El
	Japan	JIS Z3262	BCuZn-1	Wrought Wir	0.1	58-62					0.05				bal Zn			
	Norway	NS 16420	CuNi18Zn20	Wrought Sh Strp Plt		62		0.4	18				20					
	Sweden	SIS 145238	5238-00	Wrought Bar Tub Frg HW	0.2-1	58-62	0.5-1.5	1-3			0.2-1	0.5-1.5	35			430	160	25
	Sweden	SIS 145238	5238-04	Wrought Wir Bar SH 2.5/5 mm diam	0.2-1	58-62	0.5-1.5	1-3			0.2-1	0.5-1.5	35			530	390	5
C76400	France	NF A51107	CuNi18Zn20	Wrought Sh 3/4 hard 10 mm diam		60-64		0.5	17-19		0.05			0.3	bal Zn	570		5
C76400	France	NF A51107	CuNi18Zn20	Wrought Sh 1/2 hard 10 mm diam		60-64		0.5	17-19		0.05			0.3	bal Zn	510		7
C76400	France	NF A51107	CuNi18Zn20	Wrought Sh 1/4 hard 10 mm diam		60-64		0.5	17-19		0.05			0.3	bal Zn	450		20
C76400	UK	BS 2873	NS106	Wrought Wir		60-65	0.3	0.05-0.5	17-19		0.03			0.5	bal Zn			
C76400	USA			Nickel silver		58.5-61.5	0.25 max	0.5 max			0.05 max				Cu+named el 99.5 min; Ni+Co 16.5-19.5; bal Zn			
C36500	Australia	AS 1567	365	Wrought Rod Bar As Mfg 6/50 mm diaM		58-61	0.15				0.25-0.7	0.25		0.5	bal Zn	340		24
C36500	Australia	AS 1568	365	Wrought Frg As Mfg 6 mm diam		58-61	0.15				0.25-0.7	0.25			bal Zn	310		25
C36500	Finland	SFS 2925	CuZn40Pb	Wrought Sh CF 2.5 mm diam		59-61	0.2 max				0.3-0.8		40	0.5		350	200	20
C36500	Finland	SFS 2925	CuZn40Pb	Wrought Sh HR 2.5 mm diam		59-61	0.2 max				0.3-0.8		40	0.5		340	140	30
C36500	India	IS 4170	CuZn40	Wrought Rod As Mfg 5 mm diam		59-62	0.1				0.75			0.3	bal Zn			
C36500	India	IS 4170	CuZn40	Wrought Rod Ann		59-62	0.1				0.75			0.3	bal Zn			
60/40brass C36500	India	IS 6912	60/40brass	Wrought Frg As Mfg 6 mm diam		59-62	0.1				0.3-0.8			0.3	bal Zn	310		30
C36500	International	ISO 1634	CuZn40Pb	Wrought Plt Sh Strp HA		59-61	0.2				0.3-0.8				Cu includes 0.3 Ni; bal Zn	390		20
C36500	International	ISO 1634	CuZn40Pb	Wrought Plt Sh Strp M		59-61	0.2				0.3-0.8				Cu includes 0.3 Ni; bal Zn	370		40
C36500	International	ISO 1637	CuZn40Pb	Wrought HB-SH 5/15 mm diam		59-61	0.2				0.3-0.8				Cu includes 0.3 Ni; bal Zn	440		15
C36500	International	ISO 1637	CuZn40Pb	Wrought HA 5/50 mm diam		59-61	0.2				0.3-0.8				Cu includes 0.3 Ni; bal Zn	350		25
C36500	International	ISO 1637	CuZn40Pb	Wrought M 5 mm diam		59-61	0.2				0.3-0.8				Cu includes 0.3 Ni; bal Zn	370		35
C36500	International	ISO 1638	CuZn40Pb	Wrought HC 1/5 mm diam		59-61	0.2				0.3-0.8				Cu includes 0.3 Ni; bal Zn	450		5
C36500	International	ISO 1638	CuZn40Pb	Wrought Ann 1/5 mm diam		59-61	0.2				0.3-0.8				Cu includes 0.3 Ni; bal Zn	340		25
C36500	International	ISO 426-2	CuZn40Pb	Wrought Plt Sh Strp Rod Bar Wir		59-61	0.2				0.3-0.8			0.5	bal Zn			
C36500	Sweden	SIS 145163	5163-00	Wrought Plt Sh HW 5 mm		59-61	0.2				0.3-0.8		40	0.5		340	140	30
C36500	Sweden	SIS 145163	5163-04	Wrought Plt Sh SH 2.5/5 mm		59-61	0.2				0.3-0.8		40	0.5		350	200	20
C36500	UK	BS 2870	CZ123	Wrought Sh Strp As Mfg 10 mm diam		59-62					0.3-0.8			0.3	bal Zn			

UNS numbers and US grades are provided as a means of cross referencing chemically similar alloys. Exchangability is only possible after independent examination of specifications. Tensile properties are minimum or typical. UTS and YS as Mpa, El as %. See Appendix for list of abbreviations used in Descriptions.

Grade UNS #	Country	Specification	Designation	Comment	Al	Cu	Fe	Mn	Ni	P	Pb	Sn	Zn	OT max	Other	UTS	YS	El
C36500	UK	BS 2872	CZ123	Wrought Frg As Mfg or Ann 6 mm diam		59-62					0.3-0.8			0.3	bal Zn			
C36500	UK	BS 2874	CZ123	Wrought Rod As Mfg 6/50 mm diam		59-62					0.3-0.8			0.3	bal Zn; Sb 0.02			
C36500	UK	BS 2875	CZ123	Wrought Plt As Mfg 10/25 mm diam		59-62					0.3-0.8			0.3	bal Zn			
Leaded Muntz Metal C36500	USA	ASME SB171	60Cu-39.4Zn-.6Pb	Wrought Plt Sh Tub		58-61	0.15 max				0.4-0.9			0.1				
Leaded Muntz Metal C36500	USA	ASTM B171	60Cu-39.4Zn-.6Pb	Wrought Plt Sh Tub		58-61	0.15 max				0.4-0.9			0.1				
Leaded Muntz Metal C36500	USA	ASTM B432	60Cu-39.4Zn-.6Pb	Wrought Plt Sh Tub		58-61	0.15 max				0.4-0.9			0.1				
C36500	USA			Copper zinc lead alloy; Leaded muntz metal uninhibited		58-61	0.15 max				0.25-0.7	0.25 max			Cu+named el 99.6 min; bal Zn			
C37700	France	NF A51101	CuZn38Pb2	Wrought Bar Wir Rolled,Ext, or drawn 7 mm diam		59-61	0.2				1.5-2.5			0.5	bal Zn	450		6
C37700	International	ISO 1634	CuZn38Pb2	Wrought Plt Sh Strp SH .3/5 mm diam		59-61	0.2				1.5-2.5				Cu includes 0.3 Ni; bal Zn	430		10
C37700	International	ISO 1634	CuZn38Pb2	Wrought Plt Sh Strp Ann		59-61	0.2				1.5-2.5				Cu includes 0.3 Ni; bal Zn			35
C37700	International	ISO 1635	CuZn38Pb2	Wrought Tub		59-61	0.2				1.5-2.5				Cu includes 0.3 Ni; bal Zn			
C37700	International	ISO 1637	CuZn38Pb2	Wrought HA 5/15 mm diam		59-61	0.2				1.5-2.5				Cu includes 0.3 Ni; bal Zn	350		20
C37700	International	ISO 1637	CuZn38Pb2	Wrought M 5 mm diam		59-61	0.2				1.5-2.5				Cu includes 0.3 Ni; bal Zn	360		35
C37700	International	ISO 1638	CuZn38Pb2	Wrought HB-SH 1/3 mm diam		59-61	0.2				1.5-2.5				Cu includes 0.3 Ni; bal Zn	390		7
C37700	International	ISO 1638	CuZn38Pb2	Wrought Ann 1/5 mm diam		59-61	0.2				1.5-2.5				Cu includes 0.3 Ni; bal Zn	340		25
C37700	International	ISO 426-2	CuZn38Pb2	Wrought Plt Sh Strp Rod Bar Tub Wir Frg		59-61	0.2				1.5-2.5			0.5	bal Zn			
C37700	Norway	NS 16135	CuZn39Pb2	Wrought Sh Strp Plt Bar Rod Wir		59					2		39					
C37700	Norway	NS 16145	CuZn38Pb1	Wrought Strp Bar Tub Rod Wir		61					1		38					
C37700	Sweden	SIS 145168	5168-00	Wrought Bar HW 5 mm diam		57.5-59.5	0.35				1.5-2.5		39	0.7		410	150	30
C37700	Sweden	SIS 145168	5168-04	Wrought Bar SH 5/10 mm diam		57.5-59.5	0.35				1.5-2.5		39	0.7		430	290	15
C37700	Sweden	SIS 145168	5168-06	Wrought Plt Sh SH 5/10 mm diam		57.5-59.5	0.35				1.5-2.5		39	0.7		500	430	10
C37700	UK	BS 2870	CZ120	Wrought Sh Strp 1/2 hard 6 mm diam		58-60					1.5-2.5			0.3	bal Zn			

UNS numbers and US grades are provided as a means of cross referencing chemically similar alloys. Exchangability is only possible after independent examination of specifications. Tensile properties are minimum or typical . UTS and YS as Mpa, El as %. See Appendix for list of abbreviations used in Descriptions.

4-92 Wrought Copper

Grade UNS #	Country	Specification	Designation	Comment	Al	Cu	Fe	Mn	Ni	P	Pb	Sn	Zn	OT max	Other	UTS	YS	El
C37700	UK	BS 2870	CZ120	Wrought Sh Strp Hard .03 mm diam		58-60					1.5-2.5			0.3	bal Zn			
C37700	UK	BS 2870	CZ120	Wrought Sh Strp Extra Hard 6 mm diam		58-60					1.5-2.5			0.3	bal Zn			
Forging Brass C37700	USA	AMS 4614	60Cu-38Zn-2Pb	Wrought Bar Rod Frg Shapes		58-62	0.3 max				1.5-2.5			0.5				
Forging Brass C37700	USA	ASME SB283	60Cu-38Zn-2Pb	Wrought Bar Rod Frg Shapes		58-62	0.3 max				1.5-2.5			0.5				
Forging Brass C37700	USA	ASTM B124	60Cu-38Zn-2Pb	Wrought Bar Rod Frg Shapes		58-62	0.3 max				1.5-2.5			0.5				
Forging Brass C37700	USA	ASTM B283	60Cu-38Zn-2Pb	Wrought Bar Rod Frg Shapes		58-62	0.3 max				1.5-2.5			0.5				
Forging Brass C37700	USA	MIL C-13351	60Cu-38Zn-2Pb	Wrought Bar Rod Frg Shapes		58-62	0.3 max				1.5-2.5			0.5				
Forging Brass C37700	USA	QQ B-626	60Cu-38Zn-2Pb	Wrought Bar Rod Frg Shapes		58-62	0.3 max				1.5-2.5			0.5				
Forging Brass C37700	USA	SAE J463	60Cu-38Zn-2Pb	Wrought Bar Rod Frg Shapes		58-62	0.3 max				1.5-2.5			0.5				
C37700	USA			Copper zinc lead alloy; Forging brass		58-61	0.3 max				1.5-2.5				Cu+named el 99.5 min; bal Zn			
Naval brass C47000	USA	AWS A5.27-85	RBCuZn-A	Rod Oxyfuel Gas Weld	0.01 max	57-61					0.05 max	0.25-1		0.5	Cu includes Ag; bal Zn			
C47000	USA			Copper zinc tin alloy; tin brass; Naval brass welding and brazing rod	0.01 max	57-61					0.05 max				Cu+named el 99.6 min; bal Zn; S 0.25-1			
C76200	USA			Nickel silver		57-61	0.25 max	0.5 max			0.1 max				Cu+named el 99.5 min; Ni+Co 11.0-13.5; bal Zn			
	Canada	CSA HC.4.4	HC.4.ZN2912	Wrought Plt Quarter Hard		57-61	0.25 max	0.5			0.1				11.0-13.5 Ni+Co; bal Zn	448		
	Canada	CSA HC.4.4	HC.4.ZN2912	Wrought Plt Half Hard		57-61	0.25 max	0.5			0.1				11.0-13.5 Ni+Co; bal Zn	517		
	Canada	CSA HC.4.4	HC.4.ZN2912	Wrought Bar Sh Strp Plt Full Hard		57-61	0.25 max	0.5			0.1				11.0-13.5 Ni+Co; bal Zn	621		
	Denmark	DS 3003	5163	Wrought		59-61	0.2 max				0.3-0.8		40	0.5	bal Zn			
	Japan	JIS Z3202	YCuZnSn NavalBrass	Wrought Rod	0.02	57-61					0.05	0.5-1.5			bal Zn			
	Japan	JIS Z3262	BCuZn-2	Wrought Wir	0.02	57-61					0.05	0.5-1.5			bal Zn			
	Japan	JIS H3250	C6782	Wrought Rod As Mfg 6/50 mm diam	0.2-2	56-60.5		1-3			0.50				bal Zn	461		20
C37710	Australia	AS 1567	377	Wrought Rod Bar As Mfg 6/80 mm diam		56.5-60	0.3				1-3				bal Zn	380		18
C37710	Australia	AS 1567	377	Wrought Rod Bar Hard 6/40 mm diam		56.5-60	0.3				1-3				bal Zn	460		18
C37710	Australia	AS 1568	377	Wrought Frg As Mfg 6 mm		56.5-60	0.3				1-3				bal Zn	310		20
C37710	India	IS 3488	CuZn42Pb2	Wrought Bar Rod		56.5-60					1-2.5			0.25	bal Zn; Sb 0.02			

UNS numbers and US grades are provided as a means of cross referencing chemically similar alloys. Exchangability is only possible after independent examination of specifications. Tensile properties are minimum or typical . UTS and YS as Mpa, El as %. See Appendix for list of abbreviations used in Descriptions.

Worldwide Guide to Equivalent Nonferrous Metals and Alloys

Grade UNS #	Country	Specification	Designation	Comment	Al	Cu	Fe	Mn	Ni	P	Pb	Sn	Zn	OT max	Other	UTS	YS	El
C37710	India	IS 6912	Leaded brass	Wrought Frg As Mfg 6 mm diam		56.5-60	0.3				1-2.5			0.75	bal Zn; Sb 0.02			
C37710	Japan	JIS H3250	C3771	Wrought Rod As Mfg 6 mm diam		57-61					0.5-2.5				1.0 Fe+Sn; bal Zn	314		15
C37710	UK	BS 2872	CZ122	Wrought Frg As Mfg or Ann 6 mm diam		56.5-60	0.3				1-2.5			0.75	bal Zn			
C37710	UK	BS 2874	CZ122	Wrought Rod As Mfg 6/80 mm diam		56.5-60	0.3				1-2.5			0.7	bal Zn; Sb 0.02			
C37710	UK	BS 2874	CZ122	Wrought Rod Hard 6/40 mm diam		56.5-60	0.3				1-2.5			0.7	bal Zn; Sb 0.02			
C37710	USA			Copper zinc lead alloy; leaded brass		56.5-60	0.3 max				1-3				Cu+named el 99.5 min; bal Zn			
C38000	Czech Republic	CSN 423223	CuZn40Pb2			56.5-60	0.5 max				1-2.5	0.3 max			bal Zn			
C38000	Denmark	DS 3003	5168	Wrought		57.5-59.5	0.35 max				1.5-2.5		39	0.7				
C38000	USA			Copper zinc lead alloy; Low leaded brass Architectural bronze	0.5 max	55-60	0.35 max				1.5-2.5	0.3 max			Cu+named el 99.5 min; bal Zn			
Architectural Bronze C38500	USA	ASTM B455	57Cu-40Zn-3Pb	Wrought Extrusions		55-60	0.3 max				2-3.8			0.5				
C67400	USA			Copper zinc alloy	0.5-2	57-60	0.35 max	2-3.5			0.5 max	0.3 max			Cu+named el 99.5 min; Cu incl Ag; Ni+Co 0.25 max; bal Zn; Si 0.5-1.5			
C67500	International	ISO 1635	CuZn39AlFeMn	Wrought TubM	0.2-1.5	56-61	0.2-1.5	0.2-2	2		1.5	1.2		0.5	bal Zn	500	200	18
C67500	International	ISO 1637	CuZn39AlFeMn	Wrought HA 75/150 mm diam	0.2-1.5	56-61	0.2-1.5	0.2-2	2		1.5	1.2		0.5	bal Zn	500	200	18
C67500	International	ISO 1637	CuZn39AlFeMn	Wrought HB-SH 5/75 mm diam	0.2-1.5	56-61	0.2-1.5	0.2-2	2		1.5	1.2		0.5	bal Zn	540	250	18
C67500	International	ISO 1637	CuZn39AlFeMn	Wrought M 5 mm diam	0.2-1.5	56-61	0.2-1.5	0.2-2	2		1.5	1.2		0.5	bal Zn	570	180	18
C67500	International	ISO 1639	CuZn39AlFeMn	Wrought M	0.2-1.5	56-61	0.2-1.5	0.2-2	2		1.5	1.2		0.5	bal Zn	490	180	15
C67500	International	ISO 1640	CuZn39AlFeMn	Wrought Frg M	0.2-1.5	56-61	0.2-1.5	0.2-2	2		1.5	1.2		0.5	bal Zn	500	200	15
C67500	International	ISO 426-1	CuZn39AlFeMn	Wrought Rod Bar Tub Wir	0.2-1.5	56-61	0.2-1.5	0.2-2	2		1.5	1.2		0.5	bal Zn			
Manganese bronze A C67500	USA			Copper zinc alloy	0.25 max	57-60	0.8-2	0.05-0.5			0.2 max	0.5-1.5			Cu+named el 99.5 min; Cu incl Ag; bal Zn			
C67600	USA			Copper zinc alloy		57-60	0.4-1.3	0.05-0.5			0.5-1	0.5-1.5			Cu+named el 99.5 min; Cu incl Ag; bal Zn			
C67820	Australia	AS 1567	678	Wrought Rod As Mfg 6/80 mm diam	0.3-1.3	56-60	0.5-1.2	0.3-2			0.1	0.3-1			bal Zn	460	240	18
Low-fuming brass(Ni)C68000	USA	AWS A5.27-85	RBCuZn-B	Rod Oxyfuel Gas Weld	0.01 max	56-60	0.25-1.2	0.01-0.5			0.05 max	0.8-1.1		0.5	Cu includes Ag, Ni+Co=20-80; bal Zn; Si 0.04-0.15			
Low fuming bronze C68000	USA			Copper zinc alloy	0.01 max	56-60	0.25-1.25	0.01-0.5			0.05 max	0.75-1.1			Cu+named el 99.5 min; Cu incl Ag; Ni+Co 0.20-0.8; bal Zn; Si 0.04-0.15			

UNS numbers and US grades are provided as a means of cross referencing chemically similar alloys. Exchangability is only possible after independent examination of specifications. Tensile properties are minimum or typical . UTS and YS as Mpa, El as %. See Appendix for list of abbreviations used in Descriptions.

4-94 Wrought Copper

Grade UNS #	Country	Specification	Designation	Comment	Al	Cu	Fe	Mn	Ni	P	Pb	Sn	Zn	OT max	Other	UTS	YS	El
Low-fuming brass C68100	USA	AWS A5.27-85	RBCuZn-C	Rod Oxyfuel Gas Weld	0.01 max	56-60	0.25-1.2	0.01-0.5			0.05 max	0.8-1.1		0.5	Cu includes Ag; bal Zn; Si 0.04-0.15			
Low fuming bronze C68100	USA			Copper zinc alloy	0.01 max	56-60	0.25-1.25	0.01			0.05 max	0.75-1.1			Cu+named el 99.5 min; Cu incl Ag; bal Zn; Si 0.04-0.15			
	Australia	AS 1567	380	Wrought Rod Bar As Mfg 6/80 mm diam	0.1-0.6	55-60	0.3				1.5-3				bal Zn	380		12
	Australia	AS 1567	385	Wrought, Rod Bar As Mfg 6/80 mm diam		56-60					2.5-4.5				bal Zn	380		12
	Australia	AS 1567	686	Wrought Rod Bar As Mfg 6/80 mm diam	0.3-1.5	56-60	0.5-1.2	0.3-2			0.5-1.5	0.2-1			bal Zn	460	240	18
	Czech Republic	CSN 423234	CuZn40Mn			57-60	0.3 max	1-2			0.5 max				bal Zn			
	France	NF A51105	CuZn39Pb2	Wrought Bar Wir Rolled Ext or drawn		58-60	0.35				1.5-2.5			0.5	bal Zn	500		4
	International	ISO 1634	CuZn39Pb2	Wrought Plt Sh Strp SH .3/5 mm diam		57-60	0.35				1.5-2.5				Cu includes 0.3 Ni; bal Zn	50		8
	International	ISO 1637	CuZn39Pb2	Wrought HA		57-60	0.35				1.5-2.5				Cu includes 0.3 Ni; bal Zn	390		15
	International	ISO 1637	CuZn39Pb2	Wrought M 5 mm diam		57-60	0.35				1.5-2.5				Cu includes 0.3 Ni; bal Zn	370		25
	International	ISO 1639	CuZn39Pb2	Wrought HA		57-60	35				1.5-2.5			0.7	bal Zn	390		12
	International	ISO 1639	CuZn39Pb2	Wrought M		57-60	35				1.5-2.5			0.7	bal Zn	370		24
	International	ISO 1640	CuZn39Pb2	Wrought Frg M		57-60	0.35				1.5-2.5				0.3 Ni included in Cu; bal Zn	390		25
	International	ISO 426-2	CuZn39Pb2	Wrought Plt Sh Strp Rod Bar Tub Wir Frg		57-60	0.35				1.5-2.5			0.7	bal Zn			
	Japan	JIS Z3262	BCuZn-3	Wrought Wir	0.01	56-60	0.25-1.25	1	1		0.05	1 min			bal Zn; Si 0.25			
	Finland	SFS 2921	CuZn39Pb2	Wrought Rod Bar Wir Tub Drawn		57.5-59.5	0.35 max				1.5-2.5		39	0.7		390	180	10
	Finland	SFS 2921	CuZn39Pb2	Wrought Rod Bar Wir Tub HF 5 mm diam		57.5-59.5	0.35 max				1.5-2.5		39	0.7		410	150	30
C38500	Denmark	DS 3003	5170	Wrought		57-59	0.35 max				2.5-3.5		39	0.7				
C38500	France	NF A51105	CuZn40Pb3	Wrought Bar Wir Rolled,Ext, or drawn 7 mm diam		57-59	0.35				2.5-3.5			0.7	bal Zn	500		4
C38500	International	ISO 1637	CuZn39Pb3	Wrought HB-SH 5/15 mm diam		56-59	0.35				2.5-3.5				Cu includes 0.3 Ni; bal Zn	440		12
C38500	International	ISO 1637	CuZn39Pb3	Wrought HA 5/75 mm diam		56-59	0.35				2.5-3.5				Cu includes 0.3 Ni; bal Zn	360		18
C38500	International	ISO 1637	CuZn39Pb3	Wrought M 5 mm diam		56-59	0.35				2.5-3.5				Cu includes 0.3 Ni; bal Zn	380		24
C38500	International	ISO 1639	CuZn39Pb3	Wrought M		56-59	0.35				2.5-3.5				0.3 Ni included in Cu; bal Zn	380		20
C38500	International	ISO 426-2	CuZn39Pb3	Wrought Rod Bar Tub Frg		56-59	0.35				2.5-3.5			0.7	bal Zn			
C38500	UK	BS 2874	CZ121	Wrought Rod As Mfg 6/80 mm diam		56-59					2-3.5			0.75	bal Zn; Sb 0.02			
C38500	USA			Leaded brass; Architectural bronze		55-59	0.35 max				2.5-3.5				Cu+named el 99.5 min; bal Zn			

UNS numbers and US grades are provided as a means of cross referencing chemically similar alloys. Exchangability is only possible after independent examination of specifications. Tensile properties are minimum or typical . UTS and YS as Mpa, El as %. See Appendix for list of abbreviations used in Descriptions.

Worldwide Guide to Equivalent Nonferrous Metals and Alloys

Grade UNS #	Country	Specification	Designation	Comment	Al	Cu	Fe	Mn	Ni	P	Pb	Sn	Zn	OT max	Other	UTS	YS	El
C38600	Sweden	SIS 145110	5170-00	Wrought Bar Tub HW		57-59	0.35				2.5-3.5		39	0.7		410	150	30
C38600	Sweden	SIS 145170	5170-04	Wrought Wir Rod Bar SH 2.5/5 mm diam		57-59	0.35				2.5-3.5		39	0.7		480	350	8
C38600	Sweden	SIS 145170	5170-10	Wrought Tub SH		57-59	0.35				2.5-3.5		39	0.7		390	180	10
	Australia	AS 1567	671	Wrought Bar Rod As Mfg 6/80 mm diam	0.1-1	56-59		0.5-1.5			0.5-1.5	0.5-1.5			bal Zn	460	240	18
	Finland	SFS 2921	CuZn39Pb3	Wrought Rod Bar Wir Tub Drawn (tub and shapes)		57-59	0.35 max				2.5-3.5		39	0.7		390	180	10
	Finland	SFS 2921	CuZn39Pb3	Wrought Rod Bar Wir Tub HF 5 mm diam		57-59	0.35 max				2.5-3.5		39	0.7		410	150	30
	France	NF A51101	CuZn39Pb2	Wrought Sh Stress H, T2		59					2		39			400		
	Japan	JIS H3250	C6783	Wrought Rod As Mfg 6/50 mm diam	0.2-2	55-59	1	0.5-2.5			0.5				bal Zn	539		12
	Czech Republic	CSN 423231	CuZn40Mn3A l1		0.5-2	55-58.5	0.5 max	0.5-3.5			0.5 max				bal Zn			
C76700	USA		Nickel silver 56.5-15	Nickel silver		55-58			0.5 max						Cu+named el 99.5 min; Ni+Co 14.0-16.0; bal Zn			
	Denmark	DS 3003	5175	Wrought		54-58	0.5 max				0.2-1.5		43	0.5				
	Denmark	DS 3003	5272	Wrought	0.2-0.8	54-58	0.5 max				0.2-1.5		43	0.5				
	Japan	JIS H3130	C7701	Wrought Sh Plt Hard .15 mm diam		54-58	0.25	0.5	16.5-19.5		0.1				Co included in Ni; bal Zn	628		4
	Japan	JIS H3130	C7701	Wrought Sh Plt 1/2 hard .15 mm diam		54-58	0.25	0.5	16.5-19.5		0.1				Co included in Ni; bal Zn	539		8
	Japan	JIS H3130	C7701	Wrought Sh Plt Extra hard .15 mm diam		54-58	0.25	0.5	16.5-19.5		0.1				Co included in Ni; bal Zn	706		
	Japan	JIS H3270	C7701	Wrought Rod 1/2 hard 6.5 mm diam		54-58	0.25	0.5	16.5-19.5		0.1				bal Zn	520		
	Japan	JIS H3270	C7701	Wrought Rod Hard 6.5 mm diam		54-58	0.25	0.5	16.5-19.5		0.1				bal Zn	618		
	Sweden	SIS 145272	5272-00	Wrought HW	0.2-0.8	54-58	0.5				0.2-1.5		43	0.5		490	160	20
C72150	USA			Copper nickel alloy			0.1 max	0.05 max			0.05 max	0.2 max			Cu+named el 99.5 min; Ni+Co 43.0-46.0; Si 0.5 max; C 0.1 max			
	France	NF A51107	CuNi25Zn20	Wrought Sh 1/2 hard 10 mm diam		53-57		0.5	24-26		0.05			0.3	bal Zn	560		10
	France	NF A51107	CuNi25Zn20	Wrought Sh 1/4 hard 10 mm diam		53-57		0.5	24-26		0.05			0.3	bal Zn	500		20
	France	NF A51107	CuNi25Zn20	Wrought Sh 3/4 hard 10 mm diam		53-57		0.5	24-26		0.05			0.3	bal Zn	640		
	International	ISO 1639	CuZn43Pb2	Wrought M	1 max	54-57	0.5				1-2.5				0.3 Ni included in Cu; bal Zn	440		15
	International	ISO 426-2	CuZn43Pb2	Wrought Frg	1 max	54-57	0.5				1-2.5			0.5	bal Zn			
C77000	France	NF A51107	CuNi18Zn27	Wrought Sh 3/4 hard 10 mm diam		53-57		0.5	17-19		0.05			0.3	bal Zn	620		6

UNS numbers and US grades are provided as a means of cross referencing chemically similar alloys. Exchangability is only possible after independent examination of specifications. Tensile properties are minimum or typical . UTS and YS as Mpa, El as %. See Appendix for list of abbreviations used in Descriptions.

4-96 Wrought Copper

Grade UNS #	Country	Specification	Designation	Comment	Al	Cu	Fe	Mn	Ni	P	Pb	Sn	Zn	OT max	Other	UTS	YS	El
C77000	France	NF A51107	CuNi18Zn27	Wrought Sh 1/2 hard 10 mm diam		53-57		0.5	17-19		0.05			0.3	bal Zn	540		11
C77000	France	NF A51107	CuNi18Zn27	Wrought Sh 1/4 hard 10 mm diam		53-57		0.5	17-19		0.05			0.3	bal Zn	480		22
C77000	International	ISO 1634	CuNi10Zn27	Wrought Plt Sh Strp		61-65	0.3	0.5 max	9-11		0.05			0.3	bal Zn			
C77000	International	ISO 1634	CuNi18Zn27	Wrought Plt Sh Strp SH(HD)		53-56	0.3	0.5 max	17-19		0.05			0.3	bal Zn	690		2
C77000	International	ISO 430	CuNi18Zn27	Wrought Plt Sh Strp Rod Bar Wir		53-56	0.3	0.5 max	17-19		0.05			0.3	bal Zn			
C77000	UK	BS 2873	NS107	Wrought Wir		54-56	0.3	0.05-0.35	17-19		0.03			0.5	bal Zn			
Nickel Silver C77000	USA	ASTM B122	55Cu-27Zn-18Ni	Wrought Bar Rod Wir Plt Sh		53.5-56.5	0.25 max	0.5 max	16.5-19.5		0.1 max		30 max	0.5				
Nickel Silver C77000	USA	ASTM B151	55Cu-27Zn-18Ni	Wrought Bar Rod Wir Plt Sh		53.5-56.5	0.25 max	0.5 max	16.5-19.5		0.1 max		30 max	0.5				
Nickel Silver C77000	USA	ASTM B206	55Cu-27Zn-18Ni	Wrought Bar Rod Wir Plt Sh		53.5-56.5	0.25 max	0.5 max	16.5-19.5		0.1 max		30 max	0.5				
Nickel Silver C77000	USA	QQ C-585	55Cu-27Zn-18Ni	Wrought Bar Rod Wir Plt Sh		53.5-56.5	0.25 max	0.5 max	16.5-19.5		0.1 max		30 max	0.5				
Nickel Silver C77000	USA	QQ C-586	55Cu-27Zn-18Ni	Wrought Bar Rod Wir Plt Sh		53.5-56.5	0.25 max	0.5 max	16.5-19.5		0.1 max		30 max	0.5				
Nickel Silver C77000	USA	QQ W-321	55Cu-27Zn-18Ni	Wrought Bar Rod Wir Plt Sh		53.5-56.5	0.25 max	0.5 max	16.5-19.5		0.1 max		30 max	0.5				
C77000	USA		Nickel silver 55-18	Nickel silver		53.5-56.5	0.25 max	0.5 max			0.05 max				Cu+named el 99.5 min; Ni+Co 16.5-19.5; bal Zn			
	Japan	JIS C2521		Wrought Wir		46.5-56.5		2.5 max							ST 40-50; ST = Ni+Co			
	Australia	AS 1566	770	Wrought Plt Bar Sh Strp		54-56	0.3	0.05-0.35	17-19		0.03				bal Zn			
	Norway	NS 16125	CuZn43Pb2	Wrought		56-56					1		43					
	Czech Republic	CSN 423065	CuNi45Mn		0.05 max		0.3 max	1.5-3	43-46		0.1 max	0.1 max	0.1 max		Mg 0.05			
	Japan	JIS Z3262	BCuZn-4	Wrought Wir		48-55	0.1				0.5				bal Zn			
	Japan	JIS Z3262	BCuZn-5	Wrought Wir		50-53	0.1 max				0.05	3-4.5			bal Zn			
	India	IS 5743	CuAl50(MasterAlloy)	Wrought Bar	49-51	48.5-50.5									bal Zn	461		
Nickel-brass C77300	USA	AWS A5.27-85	RBCuZn-D	Rod Oxyfuel Gas Weld	0.01 max	46-50				0.25 max	0.05 max			0.5	Cu includes Ag, Ni+Co= 9-11; bal Zn; Si 0.04-0.25			
C77300	USA			Nickel silver	0.01 max	46-50				0.25 max	0.05 max				Cu+named el 99.5 min; Ni+Co 9.0-11.0; bal Zn; Si 0.04-0.25			
	Czech Republic	CSN 423226	CuZn45Pb3Mn3Fe		0.25 max	46.5-50	0.5-1.3	2-3.5			0.1 max	2-4	0.5 max		Bi 0.01; Sb 0.05; bal Zn; Si 0.2 max; As 0.05 max			
	Japan	JIS Z3202	YCuZnNi NickelSilver	Wrought Rod	0.02	46-50			9-11	0.25	0.05				bal Zn; Si 0.25			
	Japan	JIS Z3262	BCuZn-6	Wrought Wir	0.02	46-50			9-11	0.25	0.05				bal Zn; Si 0.25			
	India	IS 5743	CuNi51(Master Alloy)	Wrought Bar		47.5-49.5	0.4 max		50-52						bal Zn			

UNS numbers and US grades are provided as a means of cross referencing chemically similar alloys. Exchangability is only possible after independent examination of specifications. Tensile properties are minimum or typical . UTS and YS as Mpa, El as %. See Appendix for list of abbreviations used in Descriptions.

Worldwide Guide to Equivalent Nonferrous Metals and Alloys

Grade UNS #	Country	Specification	Designation	Comment	Al	Cu	Fe	Mn	Ni	P	Pb	Sn	Zn	OT max	Other	UTS	YS	El
	Japan	JIS Z3262	BCuZn-7	Wrought Wir		46-49			10-11						bal Zn; Si 0.15; Ag 0.3-1			
C79800	International	ISO 1637	CuNi10Zn42 Pb2	Wrought HB-SH 5/15 mm diam		44-48	0.5	0.5 max	9-11		1-2.5			0.5	bal Zn	540		8
C79800	International	ISO 1637	CuNi10Zn42 Pb2	Wrought HA 5/50 mm diam		44-48	0.5	0.5 max	9-11		1-2.5			0.5	bal Zn	460		15
C79800	International	ISO 1639	CuNi10Zn42 Pb2	Wrought As cast		44-48	0.5	0.5 max	9-11		1-2.5			0.5	bal Zn	490		15
C79800	International	ISO 1640	CuNi10Zn42 Pb2	Wrought Frg As Cast		44-48	0.5	0.5 max	9-11		1-2.5			0.5		490		15
C79800	International	ISO 430	CuNi10Zn42 Pb2	Wrought Rod Bar Frg		44-48	0.5	0.5 max	9-11		1-2.5			0.5	bal Zn			
C79800	UK	BS 2872	NS101	Wrought Frg As Mfg or Ann 6 mm diam		44-47	0.4	0.2-0.5	9-11		1-2.5			0.3	bal Zn			
C79800	UK	BS 2874	NS101	Wrought Rod As Mfg 6 mm diam		44-47	0.4	0.2-0.5	9-11		1-2.5			0.3	bal Zn			
C79800	USA			Nickel silver		45.5-48.5	0.25 max	0.5-2.5			1.5-2.5				Cu+named el 99.5 min; Ni+Co 9.0-11.0; bal Zn			
C79820	Australia	AS 1567	798	Wrought Rod Bar As Mfg 6 mm diam		46-48		0.5	8-11		2-3.5				bal Zn	460		8
	Australia	AS 1567	796	Wrought Rod Bar As Mfg 6 mm diam		46-48		0.5	8-11		2				bal Zn	460		8
	Australia	AS 1568	796	Wrought Frg As Mfg 6 mm diam		46-48			8-11						bal Zn	460		8
	Denmark	DS 3003	5282	Wrought		44-48	0.5 max	0.5	11-11		1-2.5		42	0.5				
C77400	USA			Nickel silver		43-47					0.2 max				Cu+named el 99.5 min; Ni+Co 9.0-11.0; bal Zn			
Leaded10%Nickel Brass	India	IS 6912	Leaded10%Nickel Brass	Wrought Frg As Mfg 6 mm diam		44-47	0.4	0.2-0.5	9-11		1-2.5				bal Zn			
	Japan	JIS Z3262	BCuZn-0	Wrought Wir		32-36	0.1				0.5				bal Zn			

UNS numbers and US grades are provided as a means of cross referencing chemically similar alloys. Exchangability is only possible after independent examination of specifications. Tensile properties are minimum or typical . UTS and YS as Mpa, El as %. See Appendix for list of abbreviations used in Descriptions.

Grade UNS #	Country	Specification	Designation	Description	Al	Cu	Fe	Ni	P	Pb	Sb	Si	Sn	Zn	Other	UTS	YS	El
C80100	Australia	AS 1565	801A	Sand Cast		99.95									OT 0.05 max	160		23
C80100	USA														Cu+Ag 99.95 min			
C81200	USA								0.045-0.065						Cu+Ag 99.9 min			
C81100	USA														Cu+Ag 99.70 min			
	Czech Republic	CSN 423111	E-Cu										0.2-0.3	0.2-0.4	Cu+Sn+Zn:min 99.5			
C81500	Australia	AS 1565	815B	Sand Cast		98.79									Cr 0.4-1.5	270	170	18
	South Africa	SABS 200	Cu-Cr1	Sand Cast, chill cast, continuously cast		98.79									Cr 0.6-1.2			
Beryllium Copper 70C C81400	USA	RWMA Class II	C81400	Cast		98.5 min									Cr 0.6-1; Be 0.02-0.1			
C81400	USA			High copper alloy		98.5 min									Cu+named el=99.5 min; Cr 0.6-1.0; Be 0.02-0.10			
Beryllium Copper C81800	USA	RWMA Class II	C81800	Cast	0.1 max		0.1 max	0.2 max		0.02 max		0.15 max	0.1 max	0.1 max	Cr 0.1 max; Co 1.4-1.7; Be 0.3-0.55; Ag 0.08-0.12			
C81500	USA			High copper alloy	0.10 max	98.0 min	0.10 max			0.02 max		0.15 max	0.10 max	0.10 max	Cu+named el=99.5 min; Cr 0.40-1.5			
Beryllium Copper C82200	USA	RWMA Class III	C82200	Cast	0.1 max		0.1 max	0.2 max		0.02 max		0.15 max	0.1 max	0.1 max	Cr 0.1 max; Co 0.2 max; Be 0.35-0.8			
C82200	USA			High copper alloy		96.5 min		1.0-2.0							Cu+named el=99.5 min; Be 0.35-0.8			
C99400	USA				0.50-2.0		1.0-3.0	1.0-3.5		0.25 max		0.50-2.0		0.50-5.0	Cu+named el 99.7 min; Mn 0.50 max			
Beryllium Copper C82400	USA	QQ C-390	C82400	Cast	0.15 max		0.2 max	0.2 max		0.02 max		0.15 max	0.1 max	0.1 max	Cr 0.1 max; Co 0.2-1.4; Be 1.65-1.75			
C82400	USA			High copper alloy	0.15 max	96.4 min	0.20 max	0.10 max		0.02 max			0.10 max	0.10 max	Cu+named el=99.5 min; Cr 0.10 max; Co 0.20-0.40; Be 1.65-1.75			
	Czech Republic	CSN 423115	CuSn5		0.01 max		0.1 max	0.2 max	0.1 max	0.2 max			4-6	0.3 max	Bi 0.01; Mn 0.2 max			
Beryllium Copper 20C C82500	USA	AMS 4890	C82500	Cast Sand Invest Centrifugal	0.15 max	95.5 min	0.25 max	0.2 max		0.02 max		0.2-0.35	0.1 max	0.1 max	Co 0.35-0.7; Be 1.9-2.15			
Beryllium Copper 20C C82500	USA	MIL C-11866	C82500	Cast Sand Invest Centrifugal	0.15 max	95.5 min	0.25 max	0.2 max		0.02 max		0.2-0.35	0.1 max	0.1 max	Co 0.35-0.7; Be 1.9-2.15			
Beryllium Copper 20C C82500	USA	MIL C-17324	C82500	Cast Sand Invest Centrifugal	0.15 max	95.5 min	0.25 max	0.2 max		0.02 max		0.2-0.35	0.1 max	0.1 max	Co 0.35-0.7; Be 1.9-2.15			
Beryllium Copper 20C C82500	USA	MIL C-19464	C82500	Cast Sand Invest Centrifugal	0.15 max	95.5 min	0.25 max	0.2 max		0.02 max		0.2-0.35	0.1 max	0.1 max	Co 0.35-0.7; Be 1.9-2.15			
Beryllium Copper 20C C82500	USA	MIL C-22087	C82500	Cast Sand Invest Centrifugal	0.15 max	95.5 min	0.25 max	0.2 max		0.02 max		0.2-0.35	0.1 max	0.1 max	Co 0.35-0.7; Be 1.9-2.15			
Beryllium Copper 20C C82500	USA	QQ C-390	C82500	Cast Sand Invest Centrifugal	0.15 max	95.5 min	0.25 max	0.2 max		0.02 max		0.2-0.35	0.1 max	0.1 max	Co 0.35-0.7; Be 1.9-2.15			
Beryllium Copper 20C C82500	USA	QQ C-390	C82500	Cast Sand Invest Centrifugal	0.15 max	95.5 min	0.25 max	0.2 max		0.02 max		0.2-0.35	0.1 max	0.1 max	Co 0.35-0.7; Be 1.9-2.15			

UNS numbers and US grades are provided as a means of cross referencing chemically similar alloys. Exchangability is only possible after independent examination of specifications. Tensile properties are minimum or typical . UTS and YS as Mpa, El as %. See Appendix for list of abbreviations used in Descriptions.

Worldwide Guide to Equivalent Nonferrous Metals and Alloys

Grade UNS #	Country	Specification	Designation	Description	Al	Cu	Fe	Ni	P	Pb	Sb	Si	Sn	Zn	Other	UTS	YS	El
C82500	USA			High copper alloy	0.15 max	95.5 min	0.25 max	0.20 max		0.02 max		0.20-0.35	0.10 max	0.10 max	Cu+named el=99 5 min; Co+Ni 0.35-0.7; Cr 0.10 max; Be 1.90-2.15			
C82510	USA			High copper alloy	0.15 max	95.5 min	0.25 max	0.20 max		0.02 max		0.20-0.35	0.10 max	0.10 max	Cu+named el=99.5 min; Cr 0.10 max; Co 1.0-1.2; Be 1.90-2.15			
C82600	USA			High copper alloy	0.15 max	95.2 min	0.25 max	0.20 max		0.02 max		0.20-0.35	0.10 max	0.10 max	Cu+named el=99.5 min; Cr 0.10 max; Co 0.35-0.7; Be 2.25-2.45			
C81540	USA			High copper alloy	0.10 max	95.1 min	0.15 max			0.02 max		0.40-0.8	0.10 max	0.10 max	Cu+named el=99.5 min; Cu includes Ag; Ni+Co 2.0-3.0; Cr 0.10-0.6			
Beryllium Copper C82000	USA	MIL C-19464	C82000	Cast	0.1 max		0.1 max	0.2 max		0.02 max		0.15 max	0.1 max		Cr 0.1 max; Co 2.4-2.7; Be 0.45-0.8			
Beryllium Copper C82000	USA	QQ C-390(CA820)	C82000	Cast	0.1 max		0.1 max	0.2 max		0.02 max		0.15 max	0.1 max		Cr 0.1 max; Co 2.4-2.7; Be 0.45-0.8			
C82000	USA			High copper alloy	0.10 max	95.0 min	0.10 max	0.20 max		0.02 max		0.15 max	0.10 max	0.10 max	Cu+named el=99.5 min; Co+Ni 2.4-2.7; Cr 0.10 max; Be 0.45-0.8			
Beryllium Copper C82800	USA	MIL C-19464	C82800	Cast	0.15 max	94.8 min	0.25 max	0.2 max		0.02 max		0.2-0.35	0.1 max	0.1 max	Co 0.35-0.7; Be 2.5-2.75			
Beryllium Copper C82800	USA	QQ C-390	C82800	Cast	0.15 max	94.8 min	0.25 max	0.2 max		0.02 max		0.2-0.35	0.1 max	0.1 max	Co 0.35-0.7; Be 2.5-2.75			
Beryllium Copper C82800	USA	MIL T-16243	C82800	Cast	0.15 max	94.8 min	0.25 max	0.2 max		0.02 max		0.2-0.35	0.1 max	0.1 max	Co 0.35-0.7; Be 2.5-2.75			
C82800	USA			High copper alloy	0.15 max	94.8 min	0.25 max	0.20 max		0.02 max		0.20-0.35	0.10 max	0.10 max	Cu+named el=99.5 min; Co+Ni 0.35-0.7; Cr 0.10 max; Be 2.50-2.75			
C82700	USA			High copper alloy	0.15 max	94.6 min	0.25 max	1.0-1.5		0.02 max		0.15 max	0.10 max	0.10 max	Cu+named el=99.5 min; Cr 0.10 max; Be 2.35-2.55			
C83300	USA			Red brass		92.0-94.0				1.0-2.0			1.0-2.0	2.0-6.0	Cu may be Cu+Ni; Cu+named el=99.3 min			
C87300	USA			Silicon bronze/brass		94.0 min	0.20 max			0.20 max		3.5-4.5		0.25 max	Cu+named el 99.5 min; Mn 0.8-1.5			
C87610	USA			Silicon bronze/brass		90.0 min	0.20 max			0.20 max		3.0-5.0		3.0-5.0	Cu+named el 99.5 min; Mn 0.25 max			
C90200	USA			Tin bronze; continuous cast	0.005 max	91.0-94.0	0.20 max		1.5 max	0.30 max	0.20 max	0.005 max	6.0-8.0	0.50 max	Cu may be Cu+Ni; Cu+named el 99.4 min; Ni+Co 0.50 x; S 0.05 x			

UNS numbers and US grades are provided as a means of cross referencing chemically similar alloys. Exchangability is only possible after independent examination of specifications. Tensile properties are minimum or typical . UTS and YS as Mpa, El as %. See Appendix for list of abbreviations used in Descriptions.

Grade UNS #	Country	Specification	Designation	Description	Al	Cu	Fe	Ni	P	Pb	Sb	Si	Sn	Zn	Other	UTS	YS	El
C90200	USA			Tin bronze	0.005 max	91.0-94.0	0.20 max		0.05 max	0.30 max	0.20 max	0.005 max	6.0-8.0	0.50 max	Cu may be Cu+Ni; Cu+named el 99.4 min; Ni+Co 0.50 x; S 0.05 max			
C95600	USA			Aluminum bronze	6.0-8.0	88.0 min						1.8-3.3			Cu+named el 99.0 min; Ni+Co 0.25 max			
C83400	USA			Red brass; continuous cast	0.005 max	88.0-92.0	0.25 max	1.0 max	1.5 max	0.50 max	0.25 max	0.005 max	0.20 max	8.0-12.0	Cu may be Cu+Ni; Cu+named el=99.3 min; Ni incl Co; S 0.08 max			
C83400	USA			Red brass	0.005 max	88.0-92.0	0.25 max	1.0 max	0.03 max	0.50 max	0.25 max	0.005 max	0.20 max	8.0-12.0	Cu may be Cu+Ni; Cu+named el=99.3 min; Ni incl Co; S 0.08 max			
C99500	USA				0.50-2.0		3.0-5.0	3.5-5.5		0.25 max		0.50-2.0		0.50-2.0	Cu+named el 99.7 min; Mn 0.50 max			
	International	ISO 1338	CuAl9	Permanent mould Cast 12/25 mm diam	8-10.5	88-92	1.2	1		0.3		0.2	0.3	0.5	Cu total includes Ni; Mn 0.5	450		15
	India	IS 1458	II	Cast	0.01		0.3		0.05	1-3			5-7	2-3	OT 0.2 max			
	International	ISO 1458	II	Sand Cast (cast-on)	0.01 max		0.3		0.05	1-3	0.1		5-7	2-3	OT 0.2 max	196		8
	International	ISO 1458	II	Sand Cast (separately cast)	0.01 max		0.3		0.05	1-3	0.1		5-7	2-3	OT 0.2 max	216		12
C93100	Australia	AS 1565	931D	Chill Cast				1	0.3	2-5			6.5-8.5	2	0.50 total others (applies to castings only)	220	130	2
C93100	Australia	AS 1565	931D	Sand Cast				1	0.3	2-5			6.5-8.5	2	0.50 total others (applies to castings only)	190	80	3
C93100	Australia	AS 1565	931D	Centrifugal Cast				1	0.3	2-5			6.5-8.5	2	0.50 total others (applies to castings only)	230	130	4
C93100	Australia	AS 1565	931D	Continuous Cast				1	0.3	2-5			6.5-8.5	2	0.50 total others (applies to castings only)	270	130	5
C93100	International	ISO 1338	CuSn8Pb2	Permanent mould Cast 12/25 mm diam	0.01	82-91	0.2	2.5	0.05	0.5-4	0.25	0.01	6-9	3	Cu, total includes Ni; S 0.1	220	130	2
C93100	International	ISO 1338	CuSn8Pb2	Centrifugally Cast 12/25 mm diam	0.01	82-91	0.2	2.5	0.05	0.5-4	0.25	0.01	6-9	3	Cu, total includes Ni; S 0.1	230	130	4
C93100	International	ISO 1338	CuSn8Pb2	Sand Cast 12/25 mm diam	0.01	82-91	0.2	2.5	0.05	0.5-4	0.25	0.01	6-9	3	Cu, total includes Ni; S 0.1	250	130	16
C93100	USA			High leaded tin bronze; continuous cast	0.005 max		0.25 max		1.5 max	2.0-5.0	0.25 max	0.005 max	6.5-8.5	2.0 max	Cu may be Cu+Ni; Cu+named el 99.0 min; Ni+Co 1.0 x; S 0.05 max			

UNS numbers and US grades are provided as a means of cross referencing chemically similar alloys. Exchangability is only possible after independent examination of specifications. Tensile properties are minimum or typical. UTS and YS as Mpa, El as %. See Appendix for list of abbreviations used in Descriptions.

Worldwide Guide to Equivalent Nonferrous Metals and Alloys

Grade UNS #	Country	Specification	Designation	Description	Al	Cu	Fe	Ni	P	Pb	Sb	Si	Sn	Zn	Other	UTS	YS	El
C93100	USA			High leaded tin bronze	0.005 max		0.25 max		0.30 max	2.0-5.0	0.25 max	0.005 max	6.5-8.5	2.0 max	Cu may be Cu+Ni; Cu+named el 99.0 min; Ni+Co 1.0 max; S 0.05 max			
C83410	Australia	AS 1565	835D	Sand Cast		88-91				0.1			1-2		(castings); bal Zn	190		20
C90250	Australia	AS 1565		Chill Cast		89-91	0.25	0.8	0.05	0.3			9-11	0.5		270	140	5
C90250	Australia	AS 1565		Sand Cast		89-91	0.25	0.8	0.05	0.3			9-11	0.5		230	130	6
C90250	Australia	AS 1565		Centrifugal Cast		89-91	0.25	0.8	0.05	0.3			9-11	0.5		280	140	6
C90250	Australia	AS 1565		Continuous Cast		89-91	0.25	0.8	0.05	0.3			9-11	0.5		310	160	9
C90250	Canada	CSA HC.9	T10-SC	Cast		90 max							10 max					
C90250	Denmark	DS 3001	TinBronze 5443	Sand Cast As cast	0.01	88-91	0.2 max	2	0.2	0.8	0.2	0.01	9-11	0.5	S 0.05; Mn 0.2	270	130	18
C90250	Finland	SFS 2213	CuSn10	Sand Cast As cast	0.01	88-91	0.2 max	2	0.2 max	1	0.2	0.01	9-11	0.5	Mn 0.2	270	130	18
C90250	Germany	DIN	2.0151/GB-CuSn10	Cast	0.01 max	88.5-90.5	0.15 max	1.8 max	0.05 min	0.8 max			9.3-11	0.5 max				
C90250	International	ISO 1338	CuSn10	Sand Cast 12/25 mm diam	0.01	88-91	0.2	2	0.2	1	0.2	0.01	9-11	0.5	S 0.05; Mn 0.2	240	130	12
C90250	South Africa	SABS 200	Cu-Sn10	Chill Cast				0.25	0.15	0.25			9-11	0.05	OT 0.8 max	265		5
C90250	South Africa	SABS 200	Cu-Sn10	Sand Cast				0.25	0.15	0.25			9-11	0.05	OT 0.8 max	230		7
C90250	South Africa	SABS 200	Cu-Sn10	Continuously Cast				0.25	0.15	0.25			9-11	0.05	OT 0.8 max	310		10
C90250	South Africa	SABS 200	CuSn10P	Continuously Cast				0.5	0.4	0.75			9.5-9.5	0.5	OT 0.5 max	325		7
C91600	Japan	JIS H2204	PBCln2	Cast		87-91			0.1 max				9-12		OT 0.5 max			
C92400	Japan	JIS H5113	PBC2	As Cast		87-91				0.05-0.2			9-12			196		5
C95300	USA			Aluminum bronze	9.0-11.0	86.0 min	0.8-1.5								Cu+named el 99.0 min			
	Japan	JIS H5113	PBC2B	As Cast		87-91				0.15-0.5			9-12			294		5
	Czech Republic	CSN 423144	CuAl9Mn2		8-10		1 max	1 max	0.1 max	0.1 max		0.2 max	0.2 max	1 max	Sb 0.05 ; As 0.05 max; Mn 1.5-2.5			
C90700	Czech Republic	CSN 423119	CuSn10		0.01 max		0.2 max	1.5 max	0.4 max	0.5 max			9-11	0.5 max	Bi 0.01; Sb 0.25; S 0.05; Mn 0.2 max			
C90700	Czech Republic	CSN 423120	CuSn10P1		0.01 max		0.2 max	1.5 max	0.4-1	0.5 max			9-11	0.3 max	B :0.01; Sb 0.25; Mn 0.2 max			
C90700	Czech Republic	CSN 423123	CuSn12		0.01 max		0.2 max	1.5 max	0.4 max	0.5 max			11-13	0.3 max	Bi 0.01; Sb 0.2; S 0.05; Mn 0.2 max			
C90700	International	ISO 1338	CuSn10P	Permanent Mould Cast 12/25 mm diam	0.01	87-89.5	0.1	0.1	0.5-1	0.25	0.05	0.02	10-11.5	0.05	S 0.05; Mn 0.05	310	170	2
C90700	International	ISO 1338	CuSn10P	Sand Cast 12/25 mm diam	0.01	87-89.5	0.1	0.1	0.5-1	0.25	0.05	0.02	10-11.5	0.05	S 0.05; Mn 0.05	220	130	3
C90700	International	ISO 1338	CuSn10P	Continuously Cast 12/25 mm diam	0.01	87-89.5	0.1	0.1	0.5-1	0.25	0.05	0.02	10-11.5	0.05	S 0.05; Mn 0.05	360	170	6
C90700	Norway	NS 16510	CuSn10	Sand Cast	0.01 max	88-91	0.2 max	2 max	0.2 max	1 max	0.2 max	0.01 max	9-11	0.5 max	S 0.05 max; Mn 0.2 max			

UNS numbers and US grades are provided as a means of cross referencing chemically similar alloys. Exchangability is only possible after independent examination of specifications. Tensile properties are minimum or typical . UTS and YS as Mpa, El as %. See Appendix for list of abbreviations used in Descriptions.

5-4 Cast Copper

Grade UNS #	Country	Specification	Designation	Description	Al	Cu	Fe	Ni	P	Pb	Sb	Si	Sn	Zn	Other	UTS	YS	El
C90700	Norway	NS 16510	CuSn10	Cast Ingot	0.01 max	88-90.8	0.15 max	2 max	0.05 max	0.8 max	0.2 max	0.01 max	9.3-11	0.5 max	ST 0.01 max; ST=Al+Si; S 0.05 max; Mn 0.2 max			
C90700	Pan American	COPANT 801	C90700	Continuous Bar As Mfg	0.01	88-90	0.15		0.1-0.3	0.5			10-12	0.5	1.0 Pb+Zn+Ni	280	180	10
C90700	South Africa	SABS 200	Cu-Sn10P	Continuously Cast	0.01		0.1	0.1	0.5	0.25		0.02		0.05	OT 0.47 max	355		6
C90700	South Africa	SABS 200	Cu-Sn10P	Chill Cast	0.01		0.1	0.1	0.5	0.25		0.02	10-10	0.05	OT 0.47 max	310		2
C90700	South Africa	SABS 200	Cu-Sn10P	Sand Cast	0.01		0.1	0.1	0.5	0.25		0.02	10-10	0.05	OT 0.47 max	215		3
C90700	Sweden	SIS 145443	5443-03	Sand Cast As Cast	0.01	88-91	0.2	2	0.05	1.5	0.2	0.01	9-11	0.5	S 0.1; Mn 0.2	270	130	18
C90700	Sweden	SIS 145443	5443-15	Cast As Cast	0.01	88-91	0.2	2	0.05	1.5	0.2	0.01	9-11	0.5	S 0.1; Mn 0.2	280	130	15
Tin Bronze C90700	USA	ASTM B30	C90700	Cast Perm Mold Sand Continuous	0.01 max	88-90	0.15 max	1 max	0.05 max	0.3 max		0.01 max	10-12	0.5 max	S 0.05 max			
Tin Bronze C90700	USA	ASTM B505	C90700	Cast Perm Mold Sand Continuous	0.01 max	88-90	0.15 max	1 max	0.05 max	0.3 max		0.01 max	10-12	0.5 max	S 0.05 max			
C90700	USA			Tin bronze; continuous cast	0.005 max	88.0-90.0	0.15 max		1.5 max	0.50 max	0.20 max	0.005 max	10.0-12.0	0.50 max	Cu may be Cu+Ni; Cu+named el 99.4 min; Ni+Co 0.50 x; S 0.05 max			
C90700	USA			Tin bronze	0.005 max	88.0-90.0	0.15 max		0.30 max	0.50 max	0.20 max	0.005 max	10.0-12.0	0.50 max	Cu may be Cu+Ni; Cu+named el 99.4 min; Ni+Co 0.50 x; S 0.05 max			
C92200	Canada	CSA HC.9	TZ64P-SC	Cast		88 max				1.5 max			6 max	4.5 max				
C92200	Pan American	COPANT 801	C92200	Continuous Bar As Mfg 127 mm diam	0.01	86-90	0.25	1	0.05	1-2	0.2	0.01	5.5-6.5	3-5	S 0.05	270	130	18
Navel M Bronze C92200	USA	MIL B-15345	C92200	Cast		86-90	0.25 max	1 max	0.05 max	1-2	0.25 max	0.01 max	5.5-6.5	3-5	S 0.05 max			
Navel M Bronze C92200	USA	MIL B-16541	C92200	Cast		86-90	0.25 max	1 max	0.05 max	1-2	0.25 max	0.01 max	5.5-6.5	3-5	S 0.05 max			
Navel M Bronze C92200	USA	QQ B-225	C92200	Cast		86-90	0.25 max	1 max	0.05 max	1-2	0.25 max	0.01 max	5.5-6.5	3-5	S 0.05 max			
Navel M Bronze C92200	USA	ASTM B271	C92200	Cast		86-90	0.25 max	1 max	0.05 max	1-2	0.25 max	0.01 max	5.5-6.5	3-5	S 0.05 max			
Navel M Bronze C92200	USA	ASTM B30	C92200	Cast		86-90	0.25 max	1 max	0.05 max	1-2	0.25 max	0.01 max	5.5-6.5	3-5	S 0.05 max			
Navel M Bronze C92200	USA	ASTM B505	C92200	Cast		86-90	0.25 max	1 max	0.05 max	1-2	0.25 max	0.01 max	5.5-6.5	3-5	S 0.05 max			
Navel M Bronze C92200	USA	ASTM B584	C92200	Cast		86-90	0.25 max	1 max	0.05 max	1-2	0.25 max	0.01 max	5.5-6.5	3-5	S 0.05 max			
Navel M Bronze	USA	ASTM B61	C92200	Cast		86-90	0.25 max	1 max	0.05 max	1-2	0.25 max	0.01 max	5.5-6.5	3-5	S 0.05 max			
Navel M Bronze	USA	SAE J462	C92200	Cast		86-90	0.25 max	1 max	0.05 max	1-2	0.25 max	0.01 max	5.5-6.5	3-5	S 0.05 max			
C92200	USA			Leaded tin bronze	0.005 max	86.0-90.0	0.25 max		1.5 max	1.0-2.0	0.25 max	0.005 max	5.5-6.5	3.0-5.0	Cu amy be Cu+Ni; Cu+named el 99.3 min; Ni+Co 1.0 x; S 0.05 max			

UNS numbers and US grades are provided as a means of cross referencing chemically similar alloys. Exchangability is only possible after independent examination of specifications. Tensile properties are minimum or typical. UTS and YS as Mpa, El as %. See Appendix for list of abbreviations used in Descriptions.

Worldwide Guide to Equivalent Nonferrous Metals and Alloys

Grade UNS #	Country	Specification	Designation	Description	Al	Cu	Fe	Ni	P	Pb	Sb	Si	Sn	Zn	Other	UTS	YS	El
C92200	USA			Leaded tin bronze	0.005 max	86.0-90.0	0.25 max		0.05 max	1.0-2.0	0.25 max	0.005 max	5.5-6.5	3.0-5.0	Cu amy be Cu+Ni; Cu+named el 99.3 min; Ni+Co 1.0 max; S 0.05 max			
C92410	Australia	AS 1565	922C	Chill Cast	0.01			2		2.5-3.5		0.02	6-8	1.5-3	0.40 Fe+As+Sb; Bi 0.05	250	130	5
C92410	Australia	AS 1565	922C	Continuous Cast	0.01			2		2.5-3.5		0.02	6-8	1.5-3	0.40 Fe+As+Sb; Bi 0.05	300	130	13
C92410	Australia	AS 1565	922C	Sand Cast	0.01			2		2.5-3.5		0.02	6-8	1.5-3	0.40 Fe+As+Sb; Bi 0.05	250	130	16
C92410	Australia	AS 1565	924B	Centrifugal Cast	0.01		0.2 max	2		2.5-3.5	0.25	0.01	6-8	1.5-3	0.40 Fe+As+Sb; Bi 0.05	230	130	6
C92410	South Africa	SABS 200	Cu-Sn7Zn3Pb3	Chill Cast	0.01		0.2 max	2		2.5-3.5		0.01	6-8	1.5-3	Bi 0.05; OT 0.18 max	250		5
C92410	South Africa	SABS 200	Cu-Sn7Zn3Pb3	Continuously Cast	0.01		0.2 max	2		2.5-3.5		0.01	6-8	1.5-3	Bi 0.05; OT 0.18 max	300		13
C92410	South Africa	SABS 200	Cu-Sn7Zn3Pb3	Sand Cast	0.01		0.2 max	2		2.5-3.5		0.01	6-8	1.5-3	Bi 0.05; OT 0.18 max	250		16
C92410	USA			Leaded tin bronze	0.005 max		0.20 max		1.5 max	2.5-3.5	0.25 max	0.005 max	6.0-8.0	1.5-3.0	Cu amy be Cu+Ni; Cu+named el 99.3 min; Ni+Co 0.20 max; Mn 0.05 max			
C92410	USA			Leaded tin bronze	0.005 max		0.20 max			2.5-3.5	0.25 max	0.005 max	6.0-8.0	1.5-3.0	Cu amy be Cu+Ni; Cu+named el 99.3 min; Ni+Co 0.20 max; Mn 0.05 max			
C94700	Canada	CSA HC.9	TN55-SC	Cast		88 max		5 max					5 max	2 max				
C94700	Pan American	COPANT 801	C94700(HT)	Continuous Bar As Mfg 127 mm diam		85-89	0.25	4.5-6	0.05	0.1	0.15	0.01	4.5-6	1-2.5	S 0.05; Mn 0.2	310	140	25
C94700	USA			Nickel tin bronze; Pb 0.01 max HT mech props	0.005 max	85.0-90.0	0.25 max		0.05 max	0.10 max	0.15 max	0.005 max	4.5-6.0	1.0-2.5	Cu+named el 98.7 min; Ni+Co 4.5-6.0; S 0.05 max; Mn 0.20 max			
C90710	Australia	AS 1565	924B	Chill Cast			0.1		0.5-1.2	0.25					S 10-12	230	115	5
C90710	Australia	AS 1565	924B	Continuous Cast			0.1		0.5-1.2	0.25					S 10-12	280	115	12
C90710	Australia	AS 1565	930D	Chill Cast			0.1		0.5-1.2	0.25					S 10-12	310	170	2
C90710	Australia	AS 1565	930D	Sand Cast			0.1		0.5-1.2	0.25					S 10-12	220	130	3
C90710	Australia	AS 1565	930D	Centrifugal Cast			0.1		0.5-1.2	0.25					S 10-12	330	170	4
C90710	Australia	AS 1565	930D	Continuous Cast			0.1		0.5-1.2	0.25					S 10-12	360	170	6
C90710	International	ISO 1338	CuSn11P	Permanent Mould Cast 12/25 mm diam	0.01	86-89.5	0.1	0.2	0.15-1.5	0.5		0.02	10-12	0.5		270		2

UNS numbers and US grades are provided as a means of cross referencing chemically similar alloys. Exchangability is only possible after independent examination of specifications. Tensile properties are minimum or typical . UTS and YS as Mpa, El as %. See Appendix for list of abbreviations used in Descriptions.

5-6 Cast Copper

Grade UNS #	Country	Specification	Designation	Description	Al	Cu	Fe	Ni	P	Pb	Sb	Si	Sn	Zn	Other	UTS	YS	El
C90710	International	ISO 1338	CuSn11P	Sand Cast 12/25 mm diam	0.01	86-89.5	0.1	0.2	0.15-1.5	0.5		0.02	10-12	0.5		220		3
C90710	International	ISO 1338	CuSn11P	Continuously Cast 12/25 mm diam	0.01	86-89.5	0.1	0.2	0.15-1.5	0.5		0.02	10-12	0.5		320		6
C90710	USA			Tin bronze; continuous cast	0.005 max		0.10 max		1.5 max	0.25 max		0.005 max	10.0-12.0	0.05 max	Cu may be Cu+Ni; Cu+named el 99.4 min; Ni+Co 0.10 max			
C90710	USA			Tin bronze	0.005 max		0.10 max		0.50-1.2	0.25 max		0.005 max	10.0-12.0	0.05 max	Cu may be Cu+Ni; Cu+named el 99.4 min; Ni+Co 0.10 max			
C93720	USA			High leaded tin bronze; continuous cast		83.0 min	0.35 max		1.5 max	7.0-9.0	0.50 max		3.5-4.5	4.0 max	Cu+named el 99.0 min; Ni+Co 0.50 max			
C93720	USA			High leaded tin bronze		83.0 min	0.35 max			7.0-9.0	0.50 max		3.5-4.5	4.0 max	Cu+named el 99.0 min; Ni+Co 0.50 max			
C95210	Australia	AS 1565	952C	Sand Cast	8.5-9.5	86 min	2.5-4	1	0.05			0.25	0.1	0.5	Mg 0.05; Mn 1	450	170	20
C95210	Finland	SFS 2211	CuAl10Fe3	Sand Cast As cast	8.5-10.5	83-89.5	2-4	3	0.2			0.2	0.3 max	0.4	Mn 1	500	180	13
C96200	Australia	AS 1565		Sand Cast		84.5-87	1-1.8	9-11				0.3			C 0.1; Mn 1.5	310	170	20
C96200	Germany	DIN 17658	2.0815.01/G-CuNi10	Cast			1-1.8	9-11				0.15-0.25			Nb 0.15-0.35; Mn 1-1.5			
C96200	USA			Copper nickel			1.0-1.8		0.02 max	0.01 max		0.50 max			Cu+named el 99.5 min; Ni+Co 9.0-11.0; Nb 0.50-1.0; C 0.10 max; S 0.02 max; Mn 1.5 max			
	France	NF A53-709	U-A9Fe3Y300	Die Cast As Cast	8.5-10		2-4									600		20
	France	NF A53-709	U-A9Y300	Die Cast As Cast	8.5-10.5		2									500		20
C83450	USA			Leaded red brass; continuous cast	0.005 max	87.0-89.0	0.30 max	0.8-2.0	1.5 max	1.5-3.0	0.25 max	0.005 max	2.0-3.5	5.5-7.5	Cu may be Cu+Ni; Cu+named el=99.3 min; Ni incl Co; S 0.09 max			
C83450	USA			Leaded red brass	0.005 max	87.0-89.0	0.30 max	0.8-2.0	0.03 max	1.5-3.0	0.25 max	0.005 max	2.0-3.5	5.5-7.5	Cu may be Cu+Ni; Cu+named el=99.3 min; Ni incl Co; S 0.09 max			
C87200	Canada	CSA HC.9	SC-SC	Cast	1 max	90 max	1.5 max					3 max	0.5 max	3 max	Mn 1 max			
Silicon Bronze C87200	USA	ASTM B271	C87200	Cast	1.5 max	89 min	2.5 max	1 max		0.4 max			1 max	5 max	Mn 1.5 max			
Silicon Bronze C87200	USA	ASTM B30	C87200	Cast	1.5 max	89 min	2.5 max	1 max		0.4 max			1 max	5 max	Mn 1.5 max			
Silicon Bronze C87200	USA	ASTM B584	C87200	Cast	1.5 max	89 min	2.5 max	1 max		0.4 max			1 max	5 max	Mn 1.5 max			

UNS numbers and US grades are provided as a means of cross referencing chemically similar alloys. Exchangability is only possible after independent examination of specifications. Tensile properties are minimum or typical. UTS and YS as Mpa, El as %. See Appendix for list of abbreviations used in Descriptions.

Worldwide Guide to Equivalent Nonferrous Metals and Alloys

Grade UNS #	Country	Specification	Designation	Description	Al	Cu	Fe	Ni	P	Pb	Sb	Si	Sn	Zn	Other	UTS	YS	El
Silicon Bronze C87200	USA	SAE J462	C87200	Cast	1.5 max	89 min	2.5 max	1 max		0.4 max			1 max	5 max	Mn 1.5 max			
C90300	Canada	CSA HC.9	TZ84-SC	Cast		88 max							8 max	4 max				
C90300	Japan	JIS H2203	BCIn2	Cast		86-90							7-9	3-5	OT 0.1 max			
C90300	Japan	JIS H5111	BC2	Cast		86-90				1 max			7-9	3-5	OT 1 max			
C90300	Pan American	COPANT 801	C90300	Continuous Bar As Mfg	0.01	86-89	0.2	1	0.05	0.3	0.2	0.01	7.5-9	3-5	S 0.05	300	160	18
C90300	USA			Tin bronze; continuous cast	0.005 max	86.0-89.0	0.20 max		1.5 max	0.30 max	0.20 max	0.005 max	7.5-9.0	3.0-5.0	Cu may be Cu+Ni; Cu+named el 99.4 min; Ni+Co 1.0 max; S 0.05 max			
Tin Bronze C90300	USA	ASTM B271	C90300	Cast Sand Invest Centrifugal Continuous	0.01 max	86-89	0.15 max	1 max	0.05 max	0.3 max	0.2 max	0.01 max	7.5-9	3-5	S 0.05 max			
Tin Bronze C90300	USA	ASTM B30	C00300	Cast Sand Invest Centrifugal Continuous	0.01 max	86 80	0.15 max	1 max	0.05 max	0.3 max	0.2 max	0.01 max	7.5 9	3 5	S 0.05 max			
Tin Bronze C90300	USA	ASTM B505	C90300	Cast Sand Invest Centrifugal Continuous	0.01 max	86-89	0.15 max	1 max	0.05 max	0.3 max	0.2 max	0.01 max	7.5-9	3-5	S 0.05 max			
Tin Bronze C90300	USA	ASTM B584	C90300	Cast Sand Invest Centrifugal Continuous	0.01 max	86-89	0.15 max	1 max	0.05 max	0.3 max	0.2 max	0.01 max	7.5-9	3-5	S 0.05 max			
Tin Bronze C90300	USA	MIL C-11866	C90300	Cast Sand Invest Centrifugal Continuous	0.01 max	86-89	0.15 max	1 max	0.05 max	0.3 max	0.2 max	0.01 max	7.5-9	3-5	S 0.05 max			
Tin Bronze C90300	USA	MIL C-15345	C90300	Cast Sand Invest Centrifugal Continuous	0.01 max	86-89	0.15 max	1 max	0.05 max	0.3 max	0.2 max	0.01 max	7.5-9	3-5	S 0.05 max			
Tin Bronze C90300	USA	MIL C-22087	C90300	Cast Sand Invest Centrifugal Continuous	0.01 max	86-89	0.15 max	1 max	0.05 max	0.3 max	0.2 max	0.01 max	7.5-9	3-5	S 0.05 max			
Tin Bronze C90300	USA	MIL C-22229	C90300	Cast Sand Invest Centrifugal Continuous	0.01 max	86-89	0.15 max	1 max	0.05 max	0.3 max	0.2 max	0.01 max	7.5-9	3-5	S 0.05 max			
Tin Bronze C90300	USA	QQ C-390	C90300	Cast Sand Invest Centrifugal Continuous	0.01 max	86-89	0.15 max	1 max	0.05 max	0.3 max	0.2 max	0.01 max	7.5-9	3-5	S 0.05 max			
Tin Bronze C90300	USA	QQ C-525	C90300	Cast Sand Invest Centrifugal Continuous	0.01 max	86-89	0.15 max	1 max	0.05 max	0.3 max	0.2 max	0.01 max	7.5-9	3-5	S 0.05 max			
C90300	USA			Tin bronze	0.005 max	86.0-89.0	0.20 max		0.05 max	0.30 max	0.20 max	0.005 max	7.5-9.0	3.0-5.0	Cu may be Cu+Ni; Cu+named el 99.4 min; Ni+Co 1.0 x; S 0.05 max			
C90500	Canada	CSA HC.9	TZ102-SC	Cast		88 max							10 max	2 max				
C90500	Czech Republic	CSN 423138	CuSn10Zn2		0.02 max		0.3 max		0.2 max	0.5 max			9-11	1-3	Bi:0.025 Sb:0.3			
C90500	Denmark	DS 3001	5458	Sand Cast As cast	0.01	86-89	0.25 max	2	0.05	1.5	0.3	0.01	9-11	1-3	S 0.1; Mn 0.2	260	120	15

UNS numbers and US grades are provided as a means of cross referencing chemically similar alloys. Exchangability is only possible after independent examination of specifications. Tensile properties are minimum or typical . UTS and YS as Mpa, El as %. See Appendix for list of abbreviations used in Descriptions.

5-8 Cast Copper

Grade UNS #	Country	Specification	Designation	Description	Al	Cu	Fe	Ni	P	Pb	Sb	Si	Sn	Zn	Other	UTS	YS	El
C90500	Finland	SFS 2208	CuSn10Zn2	Sand Cast	0.01	86-89	0.25 max	2	0.05	1.5	0.25	0.01	9-11	1-3	S 0.1; Mn 0.2	260	120	15
C90500	Germany	DIN 17656	2.1087/GB-CuSn10Zn	Cast Ingot	0.01 max	86-88.5	0.2 max	1.8 max		1.3 max			9.2-11	1-3				
C90500	International	ISO 1338	CuSn10Zn2	Centrifugally ot continuously cast Cast 12/25 mm diam	0.01	86-89	0.25	2	0.05	1.5	0.3	0.01	9-11	1-3	S 0.1; Mn 0.2	270	170	7
C90500	International	ISO 1338	CuSn10Zn2	Sand Cast 12/25 mm diam	0.01	86-89	0.25	2	0.05	1.5	0.3	0.01	9-11	1-3	S 0.1; Mn 0.2	240	120	12
C90500	Japan	JIS H2203	BCIn3	Cast		86.5-89.5							9-11	1-3	OT 1 max			
C90500	Japan	JIS H5111	BC3	Cast		86.5-89.5				1 max			9-11	1-3	OT 1 max			
C90500	Norway	NS 16512	CuSn10Zn2	Cast Sand Centrifugal Continuous	0.01 max	86-89	0.25 max	2 max	0.05 max	1.5 max	0.3 max	0.01 max	9-11	1-3	S 0.1 max; Mn 0.2 max			
C90500	Pan American	COPANT 801	C90500	Continuous Bar As Mfg	0.01	86-89	0.2	1	0.05	0.3	0.2	0.01	9-11	1-3	S 0.05	300	180	10
C90500	South Africa	SABS 200	Cu-Sn10Zn2	Continuously Cast	0.01		0.15 max	1	0.02	1.5		0.02	9.5-10.5	1.5-2.5	Bi 0.03	295		9
C90500	South Africa	SABS 200	Cu-Sn10Zn2	Chill Cast	0.01		0.15	1	0.02	1.5		0.02	9.5-10.5	1.5-2.5	0.05 As+Sb; Bi 0.03	230		3
C90500	South Africa	SABS 200	Cu-Sn10Zn2	Sand Cast	0.01		0.15	1	0.02	1.5		0.02	9.5-10.5	1.5-2.5	0.05 As+Sb; Bi 0.03	265		13
C90500	Sweden	SIS 145458	5458-03	Sand Cast As Cast	0.01	86-89	0.25	2	0.05	1.5	0.3	0.01	9-11	1-3	S 0.1; Mn 0.2	260	120	15
C90500	Sweden	SIS 145458	5458-15	Cast As Cast	0.01	86-89	0.25	2	0.05	1.5	0.3	0.01	9-11	1-3	S 0.1; Mn 0.2	270	140	7
Gun Metal C90500	USA	AMS 4845	C90500	Cast Sand Centrifugal Continuous	0.01 max	86-89	0.15 max	1 max	0.05 max	0.3 max	0.2 max	0.01 max	9-11	1-3	S 0.05 max			
Gun Metal C90500	USA	ASTM B22	C90500	Cast Sand Centrifugal Continuous	0.01 max	86-89	0.15 max	1 max	0.05 max	0.3 max	0.2 max	0.01 max	9-11	1-3	S 0.05 max			
Gun Metal C90500	USA	ASTM B271	C90500	Cast Sand Centrifugal Continuous	0.01 max	86-89	0.15 max	1 max	0.05 max	0.3 max	0.2 max	0.01 max	9-11	1-3	S 0.05 max			
Gun Metal C90500	USA	ASTM B30	C90500	Cast Sand Centrifugal Continuous	0.01 max	86-89	0.15 max	1 max	0.05 max	0.3 max	0.2 max	0.01 max	9-11	1-3	S 0.05 max			
Gun Metal C90500	USA	ASTM B505	C90500	Cast Sand Centrifugal Continuous	0.01 max	86-89	0.15 max	1 max	0.05 max	0.3 max	0.2 max	0.01 max	9-11	1-3	S 0.05 max			
Gun Metal C90500	USA	ASTM B584	C90500	Cast Sand Centrifugal Continuous	0.01 max	86-89	0.15 max	1 max	0.05 max	0.3 max	0.2 max	0.01 max	9-11	1-3	S 0.05 max			
Gun Metal C90500	USA	QQ C-390	C90500	Cast Sand Centrifugal Continuous	0.01 max	86-89	0.15 max	1 max	0.05 max	0.3 max	0.2 max	0.01 max	9-11	1-3	S 0.05 max			
Gun Metal C90500	USA	SAE J462	C90500	Cast Sand Centrifugal Continuous	0.01 max	86-89	0.15 max	1 max	0.05 max	0.3 max	0.2 max	0.01 max	9-11	1-3	S 0.05 max			
C90500	USA			Tin bronze; continuous cast	0.005 max	86.0-89.0	0.20 max		1.5 max	0.30 max	0.20 max	0.005 max	9.0-11.0	1.0-3.0	Cu may be Cu+Ni; Cu+named el 99.7 min; Ni+Co 1.0 max; S 0.05 max			
C90500	USA			Tin bronze	0.005 max	86.0-89.0	0.20 max		0.05 max	0.30 max	0.20 max	0.005 max	9.0-11.0	1.0-3.0	Cu may be Cu+Ni; Cu+named el 99.7 min; Ni+Co 1.0 x; S 0.05 max			

UNS numbers and US grades are provided as a means of cross referencing chemically similar alloys. Exchangability is only possible after independent examination of specifications. Tensile properties are minimum or typical . UTS and YS as Mpa, El as %. See Appendix for list of abbreviations used in Descriptions.

Grade UNS #	Country	Specification	Designation	Description	Al	Cu	Fe	Ni	P	Pb	Sb	Si	Sn	Zn	Other	UTS	YS	El
C90800	Denmark	DS 3001	TinBronze 5465	Sand Cast As cast	0.01	85-88.5	0.25 max	2	0.4	1	0.2	0.01	11-13	0.5	S 0.05; Mn 0.2	280	160	12
C90800	Finland	SFS 2214	CuSn12	Sand Cast As cast	0.01	85-88.5	0.25 max	2	0.05-0.4	1	0.2	0.01	11-13		S 0.05; Mn 0.2	280	160	12
C90800	Germany	DIN 17656	2.1053/GB-CuSn12	Cast Ingot	0.01 max	86-88	0.15 max	1.8 max	0.05	0.8 max			11.3-13	0.5 max				
C90800	International	ISO 1338	CuSn12	Permanent Mould Cast 12/25 mm diam	0.01	85-88.5	0.25	2	0.05-0.4	1	0.2	0.01	10.5-13	2	S 0.05; Mn 0.2	270	150	5
C90800	International	ISO 1338	CuSn12	Continuously or centrifugally Cast 12/25 mm diam	0.01	85-88.5	0.25	2	0.05-0.4	1	0.2	0.01	10.5-13	2	S 0.05; Mn 0.2	270	150	5
C90800	International	ISO 1338	CuSn12	Sand Cast 12/25 mm diam	0.01	85-88.5	0.25	2	0.05-0.4	1	0.2	0.01	10.5-13	2	S 0.05; Mn 0.2	240	130	7
C90800	Norway	NS 16508	CuSn12	Cast Sand Centrifugal Continuous	0.01 max	85-88.5	0.25 max	2 max	0.05-0.4	1 max	0.2 max	0.01 max	11-13	0.5 max	S 0.05 max; Mn 0.2 max			
C90800	Norway	NS 16508	CuSn12	Cast Ingot	0.01 max	85.5-88.3	0.15 max	1.8 max	0.05 max	0.8 max	0.2 max	0.01 max	11.3-13	0.5 max	ST 0.01 max; ST=Al+Si; S 0.05 max; Mn 0.2 max			
C90800	South Africa	SABS 200	Cu-Sn12P	Chill Cast	0.01		0.15	0.5	0.15	0.5		0.02	11-13	0.3	OT 0.02 max	265		3
C90800	South Africa	SABS 200	Cu-Sn12P	Sand Cast	0.01		0.15	0.5	0.15	0.5		0.02	11-13	0.3	OT 0.02 max	215		5
C90800	South Africa	SABS 200	Cu-Sn12P	Continuously Cast	0.01		0.15	0.5	0.15	0.5		0.02	11-13	0.3	OT 0.02 max	310		5
C90800	Sweden	SIS 145465	5465-03	Sand Cast As Cast	0.01	85-88.5	0.25	2	0.05-0.4	1	0.2	0.01	11-13	0.5	S 0.05; Mn 0.2	280	160	12
C90800	Sweden	SIS 145465	5465-06	Cast As Cast	0.01	85-88.5	0.25		0.05-0.4	1	0.2	0.01	11-13	0.5	S 0.05; Mn 0.2	280	160	12
C90800	Sweden	SIS 145465	5465-15	Cast As Cast	0.01	85-88.5	0.25	2	0.05-0.4	1	0.2	0.01	11-13	0.5	S 0.05; Mn 0.2	300	180	8
C90800	USA			Tin bronze; continuous cast	0.005 max	85.0-89.0	0.15 max		1.5 max	0.25 max	0.20 max	0.005 max	11.0-13.0	0.25 max	Cu may be Cu+Ni; Cu+named el 99.4 min; Ni+Co 0.50 max; S 0.05 max			
C90800	USA			Tin bronze	0.005 max	85.0-89.0	0.15 max		0.30 max	0.25 max	0.20 max	0.005 max	11.0-13.0	0.25 max	Cu may be Cu+Ni; Cu+named el 99.4 min; Ni+Co 0.50 max; S 0.05 max			
C90900	Japan	JIS H2204	PBCIn3	Cast		84-88			0.1 max				12-15		OT 0.5 max			
C90900	Japan	JIS H5113	PBC3B	Cast		84-88			0.15-0.5				12-15		OT 1 max			
C90900	USA			Tin bronze; continuous cast	0.005 max	86.0-89.0	0.15 max		1.5 max	0.25 max	0.20 max	0.005 max	12.0-14.0	0.25 max	Cu may be Cu+Ni; Cu+named el 99.4 min; Ni+Co 0.50 max; S 0.05 max			
C90900	USA			Tin bronze	0.005 max	86.0-89.0	0.15 max		0.05 max	0.25 max	0.20 max	0.005 max	12.0-14.0	0.25 max	Cu may be Cu+Ni; Cu+named el 99.4 min; Ni+Co 0.50 max; S 0.05 max			

UNS numbers and US grades are provided as a means of cross referencing chemically similar alloys. Exchangability is only possible after independent examination of specifications. Tensile properties are minimum or typical . UTS and YS as Mpa, El as %. See Appendix for list of abbreviations used in Descriptions.

Grade UNS #	Country	Specification	Designation	Description	Al	Cu	Fe	Ni	P	Pb	Sb	Si	Sn	Zn	Other	UTS	YS	El
C91600	USA			Tin bronze; continuous cast	0.005 max	86.0-89.0	0.20 max		1.5 max	0.25 max	0.20 max	0.005 max	9.7-10.8	0.25 max	Cu may be Cu+Ni; Cu+named el 99.4 min; Ni+Co 1.2-2.0; S 0.05 max			
C91600	USA			Tin bronze	0.005 max	86.0-89.0	0.20 max		0.30 max	0.25 max	0.20 max	0.005 max	9.7-10.8	0.25 max	Cu may be Cu+Ni; Cu+named el 99.4 min; Ni+Co 1.2-2.0; S 0.05 max			
C92210	USA			Leaded tin bronze	0.005 max	86.0-89.0	0.25 max		1.5 max	1.7-2.5	0.20 max	0.005 max	4.5-5.5	3.0-4.5	Cu amy be Cu+Ni; Cu+named el 99.3 min; Ni+Co 0.7-1.0; S 0.05 max			
C92210	USA			Leaded tin bronze	0.005 max	86.0-89.0	0.25 max		0.03 max	1.7-2.5	0.20 max	0.005 max	4.5-5.5	3.0-4.5	Cu amy be Cu+Ni; Cu+named el 99.3 min; Ni+Co 0.7-1.0; S 0.05 max			
C92300	Canada	CSA HC.9	TN55P-SC	Cast		87 max		5 max		1 max			5 max	2 max				
C92300	Pan American	COPANT 801	C92300	Continuous Bar As Mfg 127 mm diam	0.01	85-89	0.25	1	0.05	0.3-1	0.2	0.01	7.5-9	2.5-5	S 0.05	280	130	16
Leaded Tin Bronze C92300	USA	ASTM B271	C92300	Cast Sand Centrifugal Continuous	0.01 max	85-89	0.25 max	1 max	0.05 max	1-2	0.25 max	0.01 max	7-9	2.5-5	S 0.05 max			
Leaded Tin Bronze C92300	USA	ASTM B30	C92300	Cast Sand Centrifugal Continuous	0.01 max	85-89	0.25 max	1 max	0.05 max	1-2	0.25 max	0.01 max	7-9	2.5-5	S 0.05 max			
Leaded Tin Bronze C92300	USA	ASTM B505	C92300	Cast Sand Centrifugal Continuous	0.01 max	85-89	0.25 max	1 max	0.05 max	1-2	0.25 max	0.01 max	7-9	2.5-5	S 0.05 max			
Leaded Tin Bronze C92300	USA	ASTM B584	C92300	Cast Sand Centrifugal Continuous	0.01 max	85-89	0.25 max	1 max	0.05 max	1-2	0.25 max	0.01 max	7-9	2.5-5				
Leaded Tin Bronze C92300	USA	MIL C-15345	C92300	Cast Sand Centrifugal Continuous	0.01 max	85-89	0.25 max	1 max	0.05 max	1-2	0.25 max	0.01 max	7-9	2.5-5	S 0.05 max			
Leaded Tin Bronze C92300	USA	QQ C-390	C92300	Cast Sand Centrifugal Continuous	0.01 max	85-89	0.25 max	1 max	0.05 max	1-2	0.25 max	0.01 max	7-9	2.5-5	S 0.05 max			
Leaded Tin Bronze C92300	USA	SAE J462	C92300	Cast Sand Centrifugal Continuous	0.01 max	85-89	0.25 max	1 max	0.05 max	1-2	0.25 max	0.01 max	7-9	2.5-5	S 0.05 max			
C92300	USA			Leaded tin bronze	0.005 max	85.0-89.0	0.25 max		1.5 max	0.30-1.0	0.25 max	0.005 max	7.5-9.0	2.5-5.0	Cu amy be Cu+Ni; Cu+named el 99.3 min; Ni+Co 1.0 max; S 0.05 max			
C92300	USA			Leaded tin bronze	0.005 max	85.0-89.0	0.25 max		0.05 max	0.30-1.0	0.25 max	0.005 max	7.5-9.0	2.5-5.0	Cu amy be Cu+Ni; Cu+named el 99.3 min; Ni+Co 1.0 x; S 0.05 max			
C92400	USA			Leaded tin bronze	0.005 max	86.0-89.0	0.25 max		1.5 max	1.0-2.5	0.25 max	0.005 max	9.0-11.0	1.0-3.0	Cu amy be Cu+Ni; Cu+named el 99.3 min; Ni+Co 1.0 x; S 0.05 max			

UNS numbers and US grades are provided as a means of cross referencing chemically similar alloys. Exchangability is only possible after independent nxamination of specifications. Tensile properties are minimum or typical . UTS and YS as Mpa, El as %. See Appendix for list of abbreviations used in Descriptions.

Grade UNS #	Country	Specification	Designation	Description	Al	Cu	Fe	Ni	P	Pb	Sb	Si	Sn	Zn	Other	UTS	YS	El
C92400	USA			Leaded tin bronze	0.005 max	86.0-89.0	0.25 max		0.05 max	1.0-2.5	0.25 max	0.005 max	9.0-11.0	1.0-3.0	Cu amy be Cu+Ni; Cu+named el 99.3 min; Ni+Co 1.0 max; S 0.05 max			
C92700	Canada	CSA HC.9	TP102-SC	Cast		88 max				2 max			10 max					
C92700	Pan American	COPANT 801	C92700	Continuous Bar As Mfg 127 mm diam	0.01	86-89	0.15	1	0.25	1-2.5			9-11	0.7		270	140	8
Leaded Tin Bronze C92700	USA	ASTM B30	C92700	Cast Sand Continuous	0.01 max	86-89	0.15 max	1 max	0.25 max	1-2.5			9-11	0.7 max				
Leaded Tin Bronze C92700	USA	ASTM B505	C92700	Cast Sand Continuous	0.01 max	86-89	0.15 max	1 max	0.25 max	1-2.5			9-11	0.7 max				
Leaded Tin Bronze C92700	USA	SAE J462	C92700	Cast Sand Continuous	0.01 max	86-89	0.15 max	1 max	0.25 max	1-2.5			9-11	0.7 max				
C92700	USA			Leaded tin bronze	0.005 max	86.0-89.0	0.20 max	1.5 max		1.0-2.5	0.25 max	0.005 max	9.0-11.0	0.7 max	Cu amy be Cu+Ni; Cu+named el 99.3 min; Ni+Co 1.0 max; S 0.05 max			
C92700	USA			Leaded tin bronze	0.005 max	86.0-89.0	0.20 max		0.25 max	1.0-2.5	0.25 max	0.005 max	9.0-11.0	0.7 max	Cu amy be Cu+Ni; Cu+named el 99.3 min; Ni+Co 1.0 max; S 0.05 max			
C94800	Pan American	COPANT 801	C94800	Continuous Bar As Mfg 127 mm diam	0.01	84-89	0.25	4.5-6	0.05	0.3-1	0.15	0.01	4.5-6	1-2.5	S 0.05; Mn 0.2	280	140	20
C94800	USA			Nickel tin bronze	0.005 max	84.0-89.0	0.25 max		0.05 max	0.30-1.0	0.15 max	0.005 max	4.5-6.0	1.0-2.5	Cu+named el 98.7 min; Ni+Co 4.5-6.0; S 0.05 max; Mn 0.20 max			
C95200	Canada	CSA HC.9	AF93-SC	Cast	9 max	88 max	3 max											
C95200	Czech Republic	CSN 423145	CuAl9Fe3		8.7-10.7		2-4	1 max	0.1 max	0.1 max		0.2 max	0.2 max	1 max	Sb 0.05; As 0.05 max; Mn 0.5 max			
C95200	France	NF A53-709	U-A9Fe3Y200	Sand Cast As Cast	8.5-10.5		0.2 max	1.5		0.2		0.2	2-4	0.5	3.0 Mn+Ni	500	180	15
C95200	Germany	DIN 17656	2.0941/GB-CuAl10Fe	Cast Ingot	8.7-10.7	83-89	2-3.8	2.7 min		0.1 min		0.1-3.8			Mn 0.8 min			
C95200	International	ISO 1338	CuAl10Fe3	Sand Cast 12/25 mm diam	8.5-11	83-89.5	2-5	3		0.2		0.2	0.3	0.4	Mn 1	500	180	13
C95200	International	ISO 1338	CuAl10Fe3	Permanent mould Cast 12/25 mm diam	8.5-11	83-89.5	2-5	3		0.2		0.2	0.3	0.4	Mn 1	550	200	15
C95200	International	ISO 1338	CuAl10Fe3	Continuously or centrifugally cast 12/25 diam	8.5-11	83-89.5	2-5	3		0.2		0.2	0.3	0.4	Mn 1	550	200	15
C95200	Japan	JIS H2206	AlBCln1	Cast	8-10	85 min	1-4	1 max							OT 0.5 max; Mn 1 max			
C95200	Japan	JIS H5114	AlBC1	Cast	8-10	85 min	1-4	1 max							OT 0.5 max; Mn 1 max			

UNS numbers and US grades are provided as a means of cross referencing chemically similar alloys. Exchangability is only possible after independent examination of specifications. Tensile properties are minimum or typical . UTS and YS as Mpa, El as %. See Appendix for list of abbreviations used in Descriptions.

5-12 Cast Copper

Grade UNS #	Country	Specification	Designation	Description	Al	Cu	Fe	Ni	P	Pb	Sb	Si	Sn	Zn	Other	UTS	YS	El
C95200	Norway	NS 16575	CuAl10Fe3	Cast Sand Centrifugal Continuous	8.5-10.5	83-89.5	1-4	3 max		0.2 max		0.2 max	.0.3 max	0.4 max	Mn 1 max			
C95200	Norway	NS 16575	CuAl10Fe3	Cast Ingot	8.7-10.5	83-89.3	1-3.5	3 max		0.1 max		0.1 max	0.2 max	0.4 max	Mn 1 max			
C95200	Pan American	COPANT 801	C95200	Continuous Bar As Mfg 140 mm diam	8.5-9.5	86	2.5-4									470	180	20
C95200	Sweden	SIS 145710	5710-03	Sand Cast As Cast	8.5-10.5	83-89.5	2-4	3		0.2		0.2	0.3	0.4	Mn 1	500	180	13
C95200	Sweden	SIS 145710	5710-06	As Cast	8.5-10.5	83-89.5	2-4	3		0.2		0.2	0.3	0.4	Mn 1	550	200	15
C95200	Sweden	SIS 145710	5710-15	As Cast	8.5-10.5	83-89.5	2-4	3		0.2		0.2	0.3	0.4	Mn 1	550	200	15
C95200	USA			Aluminum bronze	8.5-9.5	86.0 min	2.5-4.0								Cu+named el 99.0 min			
C95300	Canada	CSA HC.9	AF101-SC	Cast	10 max	89 max	1 max											
C95300	Pan American	COPANT 801	C95300(HT)	Cast Continuous Bar HT 140 mm diam	9-11	86	0.8-1.5									480	100	25
C92310	Australia	AS 1565	903C	Chill Cast	0.01			1		0.3-1.5		0.02	7.5-8.5	3.5-4.5	0.20 Fe+As+Sb (castings); Bi 0.03	220	120	3
C92310	Australia	AS 1565	903C	Continuous Cast	0.01			1		0.3-1.5		0.02	7.5-8.5	3.5-4.5	0.20 Fe+As+Sb; Bi 0.03	280	140	10
C92310	Australia	AS 1565	903C	Sand Cast	0.01			1		0.3-1.5		0.02	7.5-8.5	3.5-4.5	0.20 Fe+As+Sb; Bi 0.03	250	115	15
C92310	USA			Leaded tin bronze	0.005 max				1.5 max	0.30-1.5		0.005 max	7.5-8.0	3.5-4.5	Cu amy be Cu+Ni; Cu+named el 99.3 min; Ni+Co 1.0 max; Mn 0.03 max			
C92310	USA			Leaded tin bronze	0.005 max					0.30-1.5		0.005 max	7.5-8.0	3.5-4.5	Cu amy be Cu+Ni; Cu+named el 99.3 min; Ni+Co 1.0 max; Mn 0.03 max			
C90500	Norway	NS 16512	CuSn10Zn2	Cast Ingot	0.01 max	86-88.5	0.2 max	2 max	0.03 max	1.3 max	0.3 max	0.01 max	9.3-11	1-3	ST 0.01 max; ST=Al+Si; S 0.06 max; Mn 0.2 max			
C90810	Australia	AS 1565	904D	Chill Cast	0.01		0.1 max	0.1	0.15-0.8	0.25		0.02	11-13	0.3		270	170	3
C90810	Australia	AS 1565	904D	Centrifugal Cast	0.01		0.1 max	0.1	0.15-0.8	0.25		0.02	11-13	0.3		270	170	3
C90810	Australia	AS 1565	904D	Sand Cast	0.01		0.1 max	0.1	0.15-0.8	0.25		0.02	11-13	0.3		220	130	5
C90810	Australia	AS 1565	904D	Continuous Cast	0.01		0.1 max	0.1	0.15-0.8	0.25		0.02	11-13	0.3		310	170	5
C90810	USA			Tin bronze; continuous cast	0.005 max		0.15 max		1.5 max	0.25 max	0.20 max	0.005 max	11.0-13.0	0.30 max	Cu may be Cu+Ni; Cu+named el 99.4 min; Ni+Co 0.50 x; S 0.05 max			
C90810	USA			Tin bronze	0.005 max		0.15 max		0.15-0.8	0.25 max	0.20 max	0.005 max	11.0-13.0	0.30 max	Cu may be Cu+Ni; Cu+named el 99.4 min; Ni+Co 0.50 x; S 0.05 max			

UNS numbers and US grades are provided as a means of cross referencing chemically similar alloys. Exchangability is only possible after independent examination of specifications. Tensile properties are minimum or typical . UTS and YS as Mpa, El as %. See Appendix for list of abbreviations used in Descriptions.

Grade UNS #	Country	Specification	Designation	Description	Al	Cu	Fe	Ni	P	Pb	Sb	Si	Sn	Zn	Other	UTS	YS	El
C92600	USA			Leaded tin bronze	0.005 max	86.0-88.5	0.20 max		1.5 max	0.8-1.5	0.25 max	0.005 max	9.3-10.5	1.3-2.5	Cu amy be Cu+Ni; Cu+named el 99.3 min; Ni+Co 0.7 max; S 0.05 max			
C92600	USA			Leaded tin bronze	0.005 max	86.0-88.5	0.20 max		0.03 max	0.8-1.5	0.25 max	0.005 max	9.3-10.5	1.3-2.5	Cu amy be Cu+Ni; Cu+named el 99.3 min; Ni+Co 0.7 max; S 0.05 max			
C92610	Australia	AS 1565	905C	Chill Cast	0.01		0.15 max	1		1.5		0.02	9.5-10.5	1.75-2.75	0.20 Fe+As+Sb;(castings) Bi 0.03	270	130	3
C92610	Australia	AS 1565	905C	Centrifugal Cast	0.01		0.15 max	1		1.5		0.02	9.5-10.5	1.75-2.75	0.20 Fe+As+Sb;(castings) Bi 0.03	250	130	5
C92610	Australia	AS 1565	905C	Continuous Cast	0.01		0.15 max	1		1.5		0.02	9.5-10.5	1.75-2.75	0.20 Fe+As+Sb;(castings) Bi 0.03	300	140	9
C92610	Australia	AS 1565	905C	Sand Cast	0.01		0.15 max	1		1.5		0.02	9.5-10.5	1.75-2.75	0.20 Fe+As+Sb;(castings) Bi 0.03	270	130	13
C92610	Denmark	DS 3001	5456	Sand Cast As cast	0.01	84.5-86.5	0.3 max	2	0.05	1-2	0.35	0.01	9-11	1.5-3.5	S 0.1; Mn 0.2	240	120	12
C92610	USA			Leaded tin bronze	0.005 max		0.15 max	1.5 max		0.30-1.5		0.005 max	9.5-10.5	1.7-2.8	Cu amy be Cu+Ni; Cu+named el 99.3 min; Ni+Co 1.0 max; Mn 0.03 max			
C92610	USA			Leaded tin bronze	0.005 max		0.15 max			0.30-1.5		0.005 max	9.5-10.5	1.7-2.8	Cu amy be Cu+Ni; Cu+named el 99.3 min; Ni+Co 1.0 max; Mn 0.03 max			
C95410	Japan	JIS H5114	AlBC2	Cast	8-10.5	78 min	2.5-5	1-3							OT 0.5 max; Mn 1.5 max			
	South Africa	SABS 200	Cu-Sn8Pb4P	Chill Cast				1	0.3 min	2-5			6.5-8.5	2	OT 0.5 max	215		2
	South Africa	SABS 200	Cu-Sn8Pb4P	Sand Cast				1	0.3 min	2-5			6.5-8.5	2	OT 0.5 max	185		3
	South Africa	SABS 200	Cu-Sn8Pb4P	Continuously Cast				1	0.3 min	2-5			6.5-8.5	2	OT 0.5 max	265		5
C83400	Japan	JIS H2202	YBsCln1	Cast		83-88				0.5 max					ST 1 max; ST = Sn+Al+Fe; bal Zn			
C83500	Germany		2.1095/GB-CuSn6ZnNi			83.5-87	0.2	1.5-2.3	0.03 max	2.8-4			5.8-7	1.8-3				
C83500	USA			Leaded red brass; continuous cast	0.005 max	86.0-88.0	0.25 max	0.50-1.0	1.5 max	3.5-5.5	0.25 max	0.005 max	5.5-6.5	1.0-2.5	Cu may be Cu+Ni; Cu+named el=99.3 min; Ni incl Co; S 0.10 max			
C83500	USA			Leaded red rass	0.005 max	86.0-88.0	0.25 max	0.50-1.0	0.03 max	3.5-5.5	0.25 max	0.005 max	5.5-6.5	1.0-2.5	Cu may be Cu+Ni; Cu+named el=99.3 min; Ni incl Co; S 0.10 max			

UNS numbers and US grades are provided as a means of cross referencing chemically similar alloys. Exchangability is only possible after independent examination of specifications. Tensile properties are minimum or typical . UTS and YS as Mpa, El as %. See Appendix for list of abbreviations used in Descriptions.

Grade UNS #	Country	Specification	Designation	Description	Al	Cu	Fe	Ni	P	Pb	Sb	Si	Sn	Zn	Other	UTS	YS	El
C83700	Australia	AS 1565	837D	Sand Cast		83-88				0.5					(castings); bal Zn; As 0.05-0.2; OT 1 max	170	80	18
C83700	Japan	JIS H5101	YBSC1	As Cast		83-88				0.5					1.0 Sn+Al+Fe; bal Zn	147		25
C92500	Germany		2.1065/GB-CuSn12Pb		0.01 max	84.3-87.3	0.15 max	1.8 max	0.05 max	1.2-2			11.3-13	0.5 max				
C92500	International	ISO 1338	CuSn12Pb2	Centrifugally Cast 12/25 mm diam	0.01	84-87.5	0.2	2	0.05-0.4	1-2.5		0.01	11-13	2	Cu total includes Ni; S 0.05; Mn 0.2	280	150	5
C92500	International	ISO 1338	CuSn12Pb2	Sand Cast 12/25 mm diam	0.01	84-87.5	0.2	2	0.05-0.4	1-2.5		0.01	11-13	2	Cu total includes Ni; S 0.05; Mn 0.2	240	130	7
C92500	International	ISO 1338	CuSn12Pb2	Continuously Cast 12/25 mm diam	0.01	84-87.5	0.2	2	0.05-0.4	1-2.5		0.01	11-13	2	Cu total includes Ni; S 0.05; Mn 0.2	280	150	7
C92500	Pan American	COPANT 801	C92500	Continuous Bar As Mfg 127 mm diam	0.01	85-88	0.3	0.8-1.5	0.2-0.3	1-1.5			10-12	0.5		280	170	10
C92500	USA			Leaded tin bronze	0.005 max	85.0-88.0	0.30 max	1.5 max		1.0-1.5	0.25 max	0.005 max	10.0-12.0	0.50 max	Cu amy be Cu+Ni; Cu+named el 99.3 min; Ni+Co 0.8-1.5; S 0.05 max			
C92500	USA			Leaded tin bronze	0.005 max	85.0-88.0	0.30 max		0.30 max	1.0-1.5	0.25 max	0.005 max	10.0-12.0	0.50 max	Cu amy be Cu+Ni; Cu+named el 99.3 min; Ni+Co 0.8-1.5; S 0.05 max			
C95210	USA			Aluminum bronze	8.5-9.5	86.0 min	2.5-4.0			0.05 max		0.25 max	0.10 max	0.50 max	Cu+named el 99.0 min; Ni+Co 1.0 max; Mg 0.05 max; Mn 1.0 max			
	Czech Republic	CSN 423146	CuAl10Fe3Mn1,5		9-11		2-4	0.5 max				0.1 max	0.1 max	0.5 max	Sb 0.002; As 0.01 max; Mn 1-2			
	Czech Republic	CSN 423148	CuAl10Ni2Mn1		9-10.5		0.5 max	3	0.1 max	0.1 max		0.2 max	0.2 max	0.5 max	Mn 1-2			
	France	NF A53-707	CuSn8	Cast As Cast				1.5		0.5-3			7-9	3	OT 1 max	250		16
	South Africa	SABS 200	Cu-Zn14	Sand Cast		83-88				0.5					bal Zn; As 0.05-0.2; OT 1 max	170		20
	Sweden	SIS 145667	5667-02	Cast Tub Ann 5 mm diam		88	1-1.8	9-11		0.05				0.5	Sn in Pb content; C 0.1; S 0.05; Mn 0.5-1	280	80	30
	Sweden	SIS 145667	5667-03	Cast Tub SH 5 mm diam		88	1-1.8	9-11		0.05				0.5	Sn in Pb content; C 0.1; S 0.05; Mn 0.5-1	340	290	20
	Sweden	SIS 145667	5667-12	Cast Tub Ann 3 mm diam		88	1-1.8	9-11		0.05				0.5	Sn in Pb content; C 0.1; S 0.05; Mn 0.5-1	290	100	35
	South Africa	SABS 200	CuAl10Fe2	Gravity Die Cast	8.5-10.5		1.5-3.5	1		0.05		0.25	0.1	0.5	Mg 0.05; Mn 1	650	250	13
	South Africa	SABS 200	CuAl10Fe2	Sand Cast	8.5-10.5		1.5-3.5	1		0.05		0.25	0.1	0.5	Mg 0.05; Mn 1	500		18
	South Africa	SABS 200	CuAl10Fe2	Gravity Die Cast	8.5-10.5		1.5-3.5	1		0.05		0.25	0.1	0.5	Mg 0.05; Mn 1	540		18
C91700	Germany		2.1063/GB-CuSn12Ni		0.01 max	84-87	0.15 max	1.5-2.4	0.05 max	0.15 max			11.3-15	0.3 max				

UNS numbers and US grades are provided as a means of cross referencing chemically similar alloys. Exchangability is only possible after independent examination of specifications. Tensile properties are minimum or typical . UTS and YS as Mpa, El as %. See Appendix for list of abbreviations used in Descriptions.

Worldwide Guide to Equivalent Nonferrous Metals and Alloys

Grade UNS #	Country	Specification	Designation	Description	Al	Cu	Fe	Ni	P	Pb	Sb	Si	Sn	Zn	Other	UTS	YS	El
C91700	International	ISO 1338	CuSn12Ni2	Centrifugally Cast 12/25 mm diam	0.1	84.5-87.5	0.2	1.5-2.5	0.05-0.4	0.3	0.1	0.1	11-13	0.4	S 0.05; Mn 0.2	300	180	8
C91700	International	ISO 1338	CuSn12Ni2	Continuously Cast 12/25 mm diam	0.1	84.5-87.5	0.2	1.5-2.5	0.05-0.4	0.3	0.1	0.1	11-13	0.4	S 0.05; Mn 0.2	300	180	10
C91700	International	ISO 1338	CuSn12Ni2	Sand Cast 12/25 mm diam	0.1	84.5-87.5	0.2	1.5-2.5	0.05-0.4	0.3	0.1	0.1	11-13	0.4	S 0.05; Mn 0.2	280	160	12
Nickel Gear Bronze C91700	USA	ASTM B427, B30	C91700	Cast Perm Mold Sand Centrifugal		85-87.5								11.3-12.5					
C91700	USA			Tin bronze; continuous cast	0.005 max	84.0-87.0	0.20 max		1.5 max	0.25 max	0.20 max	0.005 max	11.3-12.5	0.25 max	Cu may be Cu+Ni; Cu+named el 99.4 min; Ni+Co 1.2-2.0; S 0.05 max			
C91700	USA			Tin bronze	0.005 max	84.0-87.0	0.20 max		0.30 max	0.25 max	0.20 max	0.005 max	11.3-12.5	0.25 max	Cu may be Cu+Ni; Cu+named el 99.4 min; Ni+Co 1.2-2.0; S 0.05 max			
C92710	Australia	AS 1565	906D	Chill Cast				0.2	0.1	4-6			9-11	1		200	140	3
C92710	Australia	AS 1565	906D	Sand Cast				0.2	0.1	4-6			9-11	1		190	80	5
C92710	Australia	AS 1565	906D	Continuous Cast				0.2	0.1	4-6			9-11	1		280	160	9
C92710	South Africa	SABS 200	Cu-Sn10Pb5	Chill Cast				2	0.1	4-6			9-11	1	OT 0.5 max	215		3
C92710	South Africa	SABS 200	Cu-Sn10Pb5	Sand Cast				2	0.1	4-6			9-11	1	OT 0.5 max	185		5
C92710	South Africa	SABS 200	Cu-Sn10Pb5	Continuously Cast				2	0.1	4-6			9-11	1	OT 0.5 max	280		9
C92710	USA			Leaded tin bronze	0.005 max	78.0 min	0.20 max		1.5 max	4.0-6.0	0.25 max	0.005 max	9.0-11.0	1.0 max	Cu amy be Cu+Ni; Cu+named el 99.3 min; Ni+Co 2.0 x; S 0.05 max			
C92710	USA			Leaded tin bronze	0.005 max	78.0 min	0.20 max		0.10 max	4.0-6.0	0.25 max	0.005 max	9.0-11.0	1.0 max	Cu amy be Cu+Ni; Cu+named el 99.3 min; Ni+Co 2.0 x; S 0.05 max			
C95400	Australia	AS 1565		Sand cast	10-11.5	83 min	3-5	1.5							Mn 0.5	520	210	12
C95400	Canada	CSA HC.9	AF114-SC	Cast	11 max	85 max	4 max											
C95400	USA			Aluminum bronze	10.0-11.5	83.0 min	3.0-5.0								Cu+named el 99.5 min; Ni+Co 1.5 x; Mn 0.50 max			
	Sweden	SIS 145444	5444-03	Sand Cast As Cast	0.01	85-87	0.3	0.83 max		4-4	0.3	0.01	8.3-10	1-2.5	As 0.15; S 0.1; Mn 0.1	250	150	16
	South Africa	SABS 200	Cu-Sn7Zn3Ni5-WP	Continuously Cast				5.25-5.75	0.02	0.1-0.5			6.5-7.5	1-3	0.20 As+Fe+Sb; Bi 0.02; S 0.01; Mn 0.2	430		3
	South Africa	SABS 200	Cu-Sn7Zn3Ni5	Sand Cast	0.01			5.25-5.75	0.02	0.1-0.5			6.5-7.5	1-3	0.20 As+Fe+Sb; Bi 0.02; S 0.01; Mn 0.2	280		16
	South Africa	SABS 200	Cu-Sn7Zn3Ni5	Continuously Cast	0.01			5.25-5.75	0.02	0.1-0.5			6.5-7.5	1-3	0.20 As+Fe+Sb; Bi 0.02; S 0.01; Mn 0.2	340		18

UNS numbers and US grades are provided as a means of cross referencing chemically similar alloys. Exchangability is only possible after independent examination of specifications. Tensile properties are minimum or typical . UTS and YS as Mpa, El as %. See Appendix for list of abbreviations used in Descriptions.

Grade UNS #	Country	Specification	Designation	Description	Al	Cu	Fe	Ni	P	Pb	Sb	Si	Sn	Zn	Other	UTS	YS	El
C83810	Australia	AS 1565	838C	Chill Cast	0.01			2		4-6		0.02	2-3.5	7.5-9.5	Bi 0.1	180	80	2
C83810	Australia	AS 1565	838C	Sand Cast	0.01			2		4-6		0.02	2-3.5	7.5-9.5	Bi 0.1	180	80	11
C83810	South Africa	SABS 200	Cu-Sn3Zn9Pb5	Chill Cast	0.01	79.99		2		4-6		0.02	2-3.5	7-9.5	0.75 As+Fe+Sb; Bi 0.1	180		2
C83810	South Africa	SABS 200	Cu-Sn3Zn9Pb5	Sand Cast	0.01	79.99		2		4-6		0.02	2-3.5	7-9.5	0.75 As+Fe+Sb; Bi 0.1	180		11
C83810	USA			Leaded red brass	0.005 max		0.50 max	2.0 max		4.0-6.0		0.10 max	2.0-3.5	7.5-9.5	Cu may be Cu+Ni; Cu+named el=99.3 min; Fe+Sb+As 0.50 max; Ni incl Co			
	Czech Republic	CSN 423121	CuSn10Pb5		0.01 max		0.2 max	0.5-1.5	0.1 max	4-6			9-11	0.5 max	Sb:0.3			
	Denmark	DS 3001	TinBronze 5475	Sand Cast As cast	0.01	84-86.5	0.2 max	1	0.4	1	0.2	0.01	13-15		S 0.05; Mn 0.2	250	170	5
C83600	Australia	AS 1565	836B	Chill Cast	0.01	84-86		2		4-6		0.08	4-6	4-6	Bi 0.05	200	110	6
C83600	Australia	AS 1565	836B	Sand Cast	0.01	84-86		2		4-6		0.08	4-6	4-6	Bi 0.05	200	100	13
C83600	Australia	AS 1565	836B	Continuously Cast	0.01	84-86		2		4-6		0.08	4-6	4-6	Bi 0.05	270	100	13
C83600	Czech Republic	CSN 423135	CuSn5PbZn5		0.05 max		0.4 max	1.5 max		4-6			4-6	4-6	Sb:0.5			
C83600	Denmark	DS 3001	5204	Sand Cast As cast	0.01	84-86	0.3 max	2	0.05	4-6	0.25	0.01	4-6	4-6	S 0.1; Mn 0.1	230	90	15
C83600	Finland	SFS 2209	CuPb5Sn5Zn5	Sand Cast As cast	0.01	84-86	0.25 max	2	0.05	4-6		0.01	4-6	4-6	S 0.1; Mn 0.1	230	90	15
C83600	Germany		2.1097/GB-CuSn5ZnPb			83.5-85.5	0.2	2.3 max		4-6			4.3-6	4.5-6.5				
C83600	International	ISO 1338	CuPb5Sn5Zn5	Sand Cast and permanent mould12/25 mm diam	0.01	84-86	0.3	2.5	0.05	4-6		0.01	4-6	4-6	Cu total includes Ni; S 0.1	200	90	13
C83600	International	ISO 1338	CuPb5Sn5Zn5	Continuously or centrifugally cast 12/25 diam	0.01	84-86	0.3	2.5	0.05	4-6		0.01	4-6	4-6	Cu total includes Ni; S 0.1	250	100	13
C83600	Japan	JIS H2203	BCln6	Cast		83-87				3-6			4-6	4-6				
C83600	Japan	JIS H5111	BC6	Cast		82-87				4-6			4-6	4-7	OT 2 max			
C83600	Norway	NS 16530	CuSn5Pb5Zn5	Cast Ingot	0.01 max	83.5-85.5	0.25 max	2 max	0.03 max	4-5.7	0.25 max	0.01 max	4.3-6	4.5-6	ST 0.01 max; ST=Al+Si; S 0.06 max; Mn 0.1 max			
C83600	Pan American	COPANT 801	C83600	Continuous Bar As Mfg	0.01	84-86	0.3	1	0.05	4-6	0.25 max	0.01	4-6	4-6	S 0.08	250	130	15
C83600	Sweden	SIS 145204	5204-03	Sand Cast As Cast	0.01	84-86	0.3	2	0.05	4-6	0.25	0.01	4-6	4-6	S 0.1; Mn 0.1	230	90	15
C83600	Sweden	SIS 145204	5204-06	Cast As Cast	0.01	84-86	0.3	2	0.05	4-6	0.25	0.01	4-6	4-6	S 0.1; Mn 0.1	200	90	13
C83600	Sweden	SIS 145204	5204-15	Cast Centrifugal As Cast	0.01	84-86	0.3	2	0.05	4-6	0.25	0.01	4-6	4-6	S 0.1; Mn 0.1	250	100	13
Ounce Metal C83600	USA	AMS 4855	C83600	Cast Continuous	0.01 max	84-86	0.3 max	0.2 max		0.02 max	0.25 max	0.2-0.35	4-6	4-6				
Ounce Metal C83600	USA	ASTM B271	C83600	Cast Continuous	0.01 max	84-86	0.3 max	0.2 max		0.02 max	0.25 max	0.2-0.35	4-6	4-6				
Ounce Metal C83600	USA	ASTM B30	C83600	Cast Continuous	0.01 max	84-86	0.3 max	0.2 max		0.02 max	0.25 max	0.2-0.35	4-6	4-6				

UNS numbers and US grades are provided as a means of cross referencing chemically similar alloys. Exchangability is only possible after independent examination of specifications. Tensile properties are minimum or typical . UTS and YS as Mpa, El as %. See Appendix for list of abbreviations used in Descriptions.

Cast Copper 5-17

Worldwide Guide to Equivalent Nonferrous Metals and Alloys

Grade UNS #	Country	Specification	Designation	Description	Al	Cu	Fe	Ni	P	Pb	Sb	Si	Sn	Zn	Other	UTS	YS	El
Ounce Metal C83600	USA	ASTM B505	C83600	Cast Continuous	0.01 max	84-86	0.3 max	0.2 max		0.02 max	0.25 max	0.2-0.35	4-6	4-6				
Ounce Metal C83600	USA	ASTM B584	C83600	Cast Continuous	0.01 max	84-86	0.3 max	0.2 max		0.02 max	0.25 max	0.2-0.35	4-6	4-6				
Ounce Metal C83600	USA	ASTM B62	C83600	Cast Continuous	0.01 max	84-86	0.3 max	0.2 max		0.02 max	0.25 max	0.2-0.35	4-6	4-6				
Ounce Metal C83600	USA	MIL C-15345	C83600	Cast Continuous	0.01 max	84-86	0.3 max	0.2 max		0.02 max	0.25 max	0.2-0.35	4-6	4-6				
Ounce Metal C83600	USA	QQ C-390	C83600	Cast Continuous	0.01 max	84-86	0.3 max	0.2 max		0.02 max	0.25 max	0.2-0.35	4-6	4-6				
Ounce Metal C83600	USA	SAE J462	C83600	Cast Continuous	0.01 max	84-86	0.3 max	0.2 max		0.02 max	0.25 max	0.2-0.35	4-6	4-6				
C83600	USA			Leaded red brass; continuous cast	0.005 max	84.0-86.0	0.30 max	1.0 max	1.5 max	4.0-6.0	0.25 max	0.005 max	4.0-6.0	4.0-6.0	Cu may be Cu+Ni; Cu+named el=99.3 minNi incl Co; S 0.11 max			
C83600	USA			Leaded red brass	0.005 max	84.0-86.0	0.30 max	1.0 max	0.05 max	4.0-6.0	0.25 max	0.005 max	4.0-6.0	4.0-6.0	Cu may be Cu+Ni; Cu+named el=99.3 minNi incl Co; S 0.11 max			
C91000	Pan American	COPANT 801	C91000	Continuous Bar As Mfg		84-86	0.1	0.7	0.05	0.2			13-15	1.5		210		
C91000	USA			Tin bronze; continuous cast	0.005 max	84.0-86.0	0.10 max		1.5 max	0.20 max	0.20 max	0.005 max	14.0-16.0	1.5 max	Cu may be Cu+Ni; Cu+named el 99.4 min; Ni+Co 0.8 max; S 0.05 max			
C91000	USA			Tin bronze	0.005 max	84.0-86.0	0.10 max		0.05 max	0.20 max	0.20 max	0.005 max	14.0-16.0	1.5 max	Cu may be Cu+Ni; Cu+named el 99.4 min; Ni+Co 0.8 max; S 0.05 max			
C91500	Pan American	COPANT 801	C91500	Continuous Bar As Mfg		82-86	0.15	2.8-4	1.5	2-3.3	0.15		9-11		OT 0.5 max	310	180	8
C92710	Japan	JIS H5115	LBC2	Sand Cast		82-86	0.3	1 max		4-6			9-11		OT 1 max			
C92900	USA			Leaded tin bronze	0.005 max	82.0-86.0	0.20 max		1.5 max	2.0-3.2	0.25 max	0.005 max	9.0-11.0	0.25 max	Cu amy be Cu+Ni; Cu+named el 99.3 min; Ni+Co 2.8-4.0; S 0.05 max			
C92900	USA			Leaded tin bronze	0.005 max	82.0-86.0	0.20 max		0.05 max	2.0-3.2	0.25 max	0.005 max	9.0-11.0	0.25 max	Cu amy be Cu+Ni; Cu+named el 99.3 min; Ni+Co 2.8-4.0; S 0.05 max			
C93500	Australia	AS 1565	946D	Chill Cast		83-86		2	0.1	8-10	0.5	0.02	4.3-6	2		200	80	5
C93500	Australia	AS 1565	946D	Centrifugal Cast		83-86		2	0.1	8-10	0.5	0.02	4.3-6	2		220	80	6
C93500	Australia	AS 1565	946D	Sand Cast		83-86		2	0.1	8-10	0.5	0.02	4.3-6	2		160	60	7

UNS numbers and US grades are provided as a means of cross referencing chemically similar alloys. Exchangability is only possible after independent examination of specifications. Tensile properties are minimum or typical . UTS and YS as Mpa, El as %. See Appendix for list of abbreviations used in Descriptions.

Grade UNS #	Country	Specification	Designation	Description	Al	Cu	Fe	Ni	P	Pb	Sb	Si	Sn	Zn	Other	UTS	YS	El
C93500	Australia	AS 1565	946D	Continuous Cast		83-86		2	0.1	8-10	0.5	0.02	4.3-6	2		230	130	9
C93500	International	ISO 1338	CuPb9Sn5	Centrifugally Cast 12/25 mm diam	0.01	80-87	0.25	2	0.1	8-10	0.5	0.01	4-6	2	Cu total includes Ni; S 0.1; Mn 0.2	220	80	6
C93500	International	ISO 1338	CuPb9Sn5	Sand Cast 12/25 mm diam	0.01	80-87	0.25	2	0.1	8-10	0.5	0.01	4-6	2	Cu total includes Ni; S 0.1; Mn 0.2	160	60	7
C93500	International	ISO 1338	CuPb9Sn5	Continuously Cast 12/25 mm diam	0.01	80-87	0.25	2	0.1	8-10	0.5	0.01	4-6	2	Cu total includes Ni; S 0.1; Mn 0.2	230	130	9
C93500	Pan American	COPANT 801	C93500	Continuous Bar As Mfg 127 mm diam	0.01	83-86	0.2	1	0.05	8-10	0.3	0.01	4.3-6	2	S 0.08	210	110	12
C93500	USA			High leaded tin bronze; continuous cast	0.005 max	83.0-86.0	0.20 max	1.5 max		8.0-10.0	0.30 max	0.005 max	4.3-6.0	2.0 max	Cu may be Cu+Ni; Cu+named el 99.0 min; Ni+Co 1.0 max; S 0.08 max			
C93500	USA			High leaded tin bronze	0.005 max	83.0-86.0	0.20 max		0.05 max	8.0-10.0	0.30 max	0.005 max	4.3-6.0	2.0 max	Cu may be Cu+Ni; Cu+named el 99.0 min; Ni+Co 1.0 x; S 0.08 max			
C95420	USA			Aluminum bronze	10.5-12.0	83.5 min	3.0-4.3								Cu+named el 99.5 min; Ni+Co 0.50 x; Mn 0.50 max			
	Czech Republic	CSN 423137	CuSn8Pb3Zn6		0.02 max		0.4 max	0.5 max	0.1 max	2-4			7-9	5-7	Sb 0.5; Mn 0.2 max			
	Sweden	SIS 145475	5475-03	Sand Cast As Cast	0.01	86	0.2	1	0.4	1	0.2	0.01	13-15	0.5	99.0 Cu+Sn+P; S 0.05; Mn 0.2	200	140	3
C87400	USA			Silicon bronze/brass	0.8 max	79.0 min				1.0 max		2.5-4.0		12.0-16.0	Cu+named el 99.2 min			
C95220	USA			Aluminum bronze	9.5-10.5		2.5-4.0								Cu+named el 99.5 min; Ni+Co 2.5 x; Mn 0.50 max			
C95410	Japan	JIS H2206	AlBCln2	Cast	8-10.5	80 min	2.5-5	1-3							OT 0.5 max; Mn 1.5 max			
C95410	Pan American	COPANT 801	C95400(HT)	Cast Continuous Bar HT 140 mm diam	10-11.5	83	3-5	2.5							Mn 0.5	590	230	12
C95410	USA			Aluminum bronze	10.0-11.5	83.0 min	3.0-5.0								Cu+named el 99.5 min; Ni+Co 1.5-2.5; Mn 0.50 max			
	Denmark	DS 3001	5445	Sand Cast As cast	0.01	83.5-85.5	0.3 max	2	0.05	1.5-3		0.01	8-10	3-5	S 0.1; Mn 0.2	230	110	12
C87500	Canada	CSA HC.9	ZS144-SC	Cast		82 max						4 max		14 max				
C87500	Germany		2.0493/GB-CuZn15Si14			78.5-82	0.05 max			0.6 max		4-4.8	0.2 max		Mn 0.1 max			
C87500	Japan	JIS H5112	SzBC2	Cast		78.5-82.5						4-5		14-16	OT 0.6 max			
C87500	Japan	JIS H5112	SzBC3	Cast		80-84						3.2-4.2	13-15		ST 0.5 max; ST = Mn + Fe; OT 0.5 max			
C87500	USA			Silicon bronze/brass	0.50 max	79.0 min				0.50 max		3.0-5.0		12.0-16.0	Cu+named el 99.5 min			
C91100	Canada	CSA HC.9	T16-SC	Cast		84 max							16 max					

UNS numbers and US grades are provided as a means of cross referencing chemically similar alloys. Exchangability is only possible after independent examination of specifications. Tensile properties are minimum or typical . UTS and YS as Mpa, El as %. See Appendix for list of abbreviations used in Descriptions.

Worldwide Guide to Equivalent Nonferrous Metals and Alloys

Grade UNS #	Country	Specification	Designation	Description	Al	Cu	Fe	Ni	P	Pb	Sb	Si	Sn	Zn	Other	UTS	YS	El
C91100	USA			Tin bronze; continuous cast	0.005 max	82.0-85.0	0.25 max		1.5 max	0.25 max	0.20 max	0.005 max	15.0-17.0	0.25 max	Cu may be Cu+Ni; Cu+named el 99.4 min; Ni+Co 0.50 max; S 0.05 max			
C91100	USA			Tin bronze	0.005 max	82.0-85.0	0.25 max		1.0 max	0.25 max	0.20 max	0.005 max	15.0-17.0	0.25 max	Cu may be Cu+Ni; Cu+named el 99.4 min; Ni+Co 0.50 max; S 0.05 max			
C93200	Australia	AS 1565		Sand Cast		81-85		2	0.1	6-8			6.3-7.5	2-4		210	100	12
C93200	Canada	CSA HC.9	PT77	Cast		83 max				7 max			7 max	3 max				
C93200	Denmark	DS 3001	5426	Sand Cast As cast	0.01	81-85	0.2 max	2	0.1	5-7	0.35	0.01	6-8	2-5	S 0.1	240	100	12
C93200	Finland	SFS 2207	CuSn7Pb6Zn 3	Sand Cast As cast	0.01	82-84	0.2 max	2	0.03	5.3-7	0.25	0.01	6.3-8	2.3-5	S 0.06	240	100	12
C93200	France	NF A53-707	CuSn7Pb6Zn 4	Cast As Cast				1.5		5-7			6-8	2-5	OT 1 max	250		16
C93200	Germany		2.1091/GB-CuSn7ZnPb			81-84.5	0.2 max	1.8 max	0.03 max	5.3-7			6.3-8	3.3-5				
C93200	International	ISO 1338	CuSn7Pb7Zn 3	Sand Cast 12/25 mm diam	0.01	81-85	0.2	2	0.1	5-8	0.35	0.01	6-8	2-5	Cu total includes Ni; S 0.1	210	100	12
C93200	International	ISO 1338	CuSn7Pb7Zn 3	Centrifugally or continuously Cast 12/25 mm diam	0.01	81-85	0.2	2	0.1	5-8	0.35	0.01	6-8	2-5	Cu total includes Ni; S 0.1	260	120	12
C93200	Pan American	COPANT 801	C93200	Continuous Bar As Mfg 127 mm diam	0.01	81-85	0.2	1	0.15	6-8	0.35	0.01	6.3-7.5	2-4	S 0.08	250	140	10
C93200	USA			High leaded tin bronze; continuous cast	0.005 max	81.0-85.0	0.20 max		1.5 max	6.0-8.0	0.35 max	0.005 max	6.3-7.5	1.0-4.0	Cu may be Cu+Ni; Cu+named el 99.0 min; Ni+Co 1.0 max; S 0.08 max			
C93200	USA			High leaded tin bronze	0.005 max	81.0-85.0	0.20 max		0.15 max	6.0-8.0	0.35 max	0.005 max	6.3-7.5	1.0-4.0	Cu may be Cu+Ni; Cu+named el 99.0 min; Ni+Co 1.0 x; S 0.08 max			
C93400	Pan American	COPANT 801	C93400	Continuous Bar As Mfg 127 mm diam		82-85	0.15	1	0.5	7-9	0.5		7-9	0.7		240	140	8
C93400	USA			High leaded tin bronze; continuous cast	0.005 max	82.0-85.0	0.20 max		1.5 max	7.0-9.0	0.50 max	0.005 max	7.0-9.0	0.8 max	Cu may be Cu+Ni; Cu+named el 99.0 min; Ni+Co 1.0 x; S 0.08 max			
C93400	USA			High leaded tin bronze	0.005 max	82.0-85.0	0.20 max		0.50 max	7.0-9.0	0.50 max	0.005 max	7.0-9.0	0.8 max	Cu may be Cu+Ni; Cu+named el 99.0 min; Ni+Co 1.0 x; S 0.08 max			
C95900	USA			Aluminum bronze	12.0-13.5		3.0-5.0								Cu+named el 99.5 min; Ni+Co 0.50 x; Mn 1.5 max			

UNS numbers and US grades are provided as a means of cross referencing chemically similar alloys. Exchangability is only possible after independent examination of specifications. Tensile properties are minimum or typical . UTS and YS as Mpa, El as %. See Appendix for list of abbreviations used in Descriptions.

Grade UNS #	Country	Specification	Designation	Description	Al	Cu	Fe	Ni	P	Pb	Sb	Si	Sn	Zn	Other	UTS	YS	El
	Norway	NS 16525	CuSn7Pb6Zn3	Cast Sand Centrifugal Continuous	0.01 max	81-85	0.2 max	2 max	0.1 max	5-7	0.35 max	0.01 max	6.3-8	2-5	S 0.1 max			
	France	NF A53-709	U-A9N3FeY200	Sand Cast As Cast	8.2-10		1.5-3	1.5-4		0.2		0.2	0.2	0.5	Mn 3	500	180	18
C87800	USA			Silicon bronze/brass	0.15 max	80.0 min	0.15 max		0.1 max	0.15 max	0.05 max	3.8-4.2	0.25 max	12.0-16.0	Cu+named el 99.5 min; Mg 0.01 max; As 0.05 max; S 0.05 max; Mn 0.15 max			
	India	IS 1458	IV	Cast	0.01		0.3		0.05	9-11			6-8	0.5	0.5 Fe+Sb; bal Cu			
C94400	USA			High leaded tin bronze; continuous cast	0.005 max		0.15 max	1.5 max		9.0-12.0	0.8 max	0.005 max	7.0-9.0	0.8 max	Cu+named el 99.0 min; Ni+Co 1.0 max; S 0.08 max			
C94400	USA			High leaded tin bronze	0.005 max		0.15 max	0.50 max		9.0-12.0	0.8 max	0.005 max	7.0-9.0	0.8 max	Cu+named el 99.0 min; Ni+Co 1.0 max; S 0.08 max			
	Norway	NS 16525	CuSn7Pb6Zn3	Cast Ingot	0.01 max	82-84	0.2 max	2 max	0.03 max	5.3-7	0.3 max	0.01 max	6.3-8	2.3-5	ST 0.01 max; ST=Al+Si; S 0.06 max			
C96800	USA			Copper nickel	0.1 max		0.50 max		0.005 max	0.005 max	0.02 max	0.05 max	7.5-8.5	1.0 max	Cu+named el 99.5 min; Ni+Co 9.5-10.5; Ti 0.01 max; Nb 0.10-0.30; Mg 0.005-0.15; Bi 0.001 max; S 0.0025 max; Mn 0.05-0.30			
C83800	Canada	CSA HC.9	ZP66-SC	Cast		83 max				6 max			4 max	7 max				
C83800	Pan American	COPANT 801	C83800	Continuous Bar As Mfg	0.01	82-83.8	0.3	1	0.03	5-7	0.25 max	0.01	3.3-4.2	5-8	S 0.08	210	110	16
Hydraulic Bronze C83800	USA	ASTM B271	C83800	Cast	0.01 max	82-83.8	0.3 max	0.2 max		0.02 max	0.25 max	0.2-0.35	3.3-4.2	5-8				
Hydraulic Bronze C83800	USA	ASTM B30	C83800	Cast	0.01 max	82-83.8	0.3 max	0.2 max		0.02 max	0.25 max	0.2-0.35	3.3-4.2	5-8				
Hydraulic Bronze C83800	USA	ASTM B505	C83800	Cast	0.01 max	82-83.8	0.3 max	0.2 max		0.02 max	0.25 max	0.2-0.35	3.3-4.2	5-8				
Hydraulic Bronze C83800	USA	ASTM B584	C83800	Cast	0.01 max	82-83.8	0.3 max	0.2 max		0.02 max	0.25 max	0.2-0.35	3.3-4.2	5-8				
Hydraulic Bronze C83800	USA	QQ C-390	C83800	Cast	0.01 max	82-83.8	0.3 max	0.2 max		0.02 max	0.25 max	0.2-0.35	3.3-4.2	5-8				
Hydraulic Bronze C83800	USA	SAE J462	C83800	Cast	0.01 max	82-83.8	0.3 max	0.2 max		0.02 max	0.25 max	0.2-0.35	3.3-4.2	5-8				
C83800	USA			Leaded red brass	0.005 max	82.0-83.8	0.30 max	1.0 max	0.03 max	5.0-7.0	0.25 max	0.005 max	3.3-4.2	5.0-8.0	Cu may be Cu+Ni; Cu+named el=99.3 min; Ni incl Co; S 0.12 max			
C95510	USA			Aluminum bronze	9.7-10.9	78.0 min	2.0-3.5						0.20 max	0.30 max	Cu+named el 99.8 min; Ni+Co 4.5-5.5; Mn 1.5 max			

UNS numbers and US grades are provided as a means of cross referencing chemically similar alloys. Exchangability is only possible after independent examination of specifications. Tensile properties are minimum or typical . UTS and YS as Mpa, El as %. See Appendix for list of abbreviations used in Descriptions.

Grade UNS #	Country	Specification	Designation	Description	Al	Cu	Fe	Ni	P	Pb	Sb	Si	Sn	Zn	Other	UTS	YS	El
	International	ISO 1458	V	Sand Cast (cast-on)	0.01 max		0.3		0.05	9-11	0.4		6-8	0.5	.5 Fe+Sb	157		2
	International	ISO 1458	V	Sand Cast (separately cast)	0.01 max		0.3		0.05	9-11	0.4		6-8	0.5	.5 Fe+Sb	177		4
	South Africa	SABS 200	Cu-Sn5Pb10	Chill Cast				2	0.1	8-10	0.5	0.02	4-6	2		200		5
	South Africa	SABS 200	Cu-Sn5Pb10	Sand Cast				2	0.1	8-10	0.5	0.02	4-6	2		155		7
	South Africa	SABS 200	Cu-Sn5Pb10	Continuously Cast				2	0.1	8-10	0.5	0.02	4-6	2		230		9
C95800	Denmark	DS 3001	5716	Sand Cast As cast	8.8-10	77-82	0.2 max	4.5-6.5		0.05		0.1	3.5-5.5	0.2	Cr 0.01; Mn 0.3-2.5	640	250	13
C95800	France	NF A53-709	U-A9N5FeY200	Sand Cast As Cast	8.2-10.5		3-6	4.6-5		0.2			0.2	0.5	Mn 3	630	240	12
C95800	Germany	DIN 17656	2.0976/GB-CuAl10Ni	Cast Ingot	9-10.6	76-80.5	3.5-5.3	4.5-6.3		0.03 max		0.07 max			Mn 2.3 max			
C95800	Japan	JIS H2206	AlBCIn3	Cast	8.5-10.5	78 min	3-6	3-6							OT 0.5 max; Mn 1.5 max			
C95800	Japan	JIS H5114	AlBC3	Cast	8.5-10.5	78 min	3-6	3-6							OT 0.5 max; Mn 1.5 max			
C95800	Norway	NS 16570	CuAl10Fe5Ni5	Cast Ingot	9-10	77-82	3.5-5.3	4.5-6.3		0.03 max		0.07 max		0.2 max	Mg 0.04 max; Cr 0.01 max; OT 0.3 max; Mn 0.3-2.3			
C95800	Norway	NS 16570	CuAl20Fe5Ni5	Cast Sand Centrifugal Continuous	8.8-10	77-82	3.5-5.3	4.5-6.5		0.05 max		0.1 max	0.2 max	0.2 max	Mg 0.05 max; Cr 0.01 max; OT 0.4 max; Mn 0.3-2.5			
C95800	South Africa	SABS 200	CuAl10Ni6Fe4	SandCast	8.8-10		4-5.5	4-5.5		0.05		0.1	0.1	0.5	Mn 1.5	640	250	13
C95800	South Africa	SABS 200	CuAl10Ni6Fe4	Gravity Die Cast	8.8-10		4-5.5	4-5.5		0.05		0.1	0.1	0.5	Mn 1.5	650	250	13
C95800	Sweden	SIS 145716	5716-03	Sand Cast As cast	8.8-10	77-82	3.5-5.5	4.5-6.5		0.05		0.1	0.2	0.2	Mg 0.05; Cr 0.01; Mn 2.5	640	250	13
C95800	Sweden	SIS 145716	5716-06	As Cast	8.8-10	77-82	3.5-5.5	4.5-6.5		0.05		0.1	0.2	0.2	Mg 0.05; Cr 0.01; Mn 2.5	650	250	13
C95800	Sweden	SIS 145716	5716-15	As Cast	8.8-10	77-85	3.5-5.5	4.5-6.5		0.05		0.1	0.2	0.2	Mg 0.05; Cr 0.01; Mn 2.5	670	250	13
C95800	USA			Aluminum bronze	8.5-9.5	79.0 min	3.5-4.5			0.05 max		0.10 max	0.10 max		Cu+named el 99.5 min; Ni+Co 4.0-2.0; Fe<Ni; Mn 0.8-1.5			
C95810	Australia	AS 1565	958C	Sand Cast	8.5-9.5	79 min	3.5-4.5	4-5					0.1	0.5	Mn 0.8-1.5	590	240	18
C95810	USA			Aluminum bronze	8.5-9.5	79.0 min	3.5-4.5			0.10 max		0.10 max		0.50 max	Cu+named el 99.5 min; Ni+Co 4.0-5.0; Fe<Ni; Mg 0.05 max; Mn 0.8-1.5			
C93600	USA			High leaded tin bronze; continuous cast	0.005 max	79.0-83.0	0.20 max		1.5 max	11.0-13.0	0.55 max	0.005 max	6.0-8.0	1.0 max	Cu+named el 99.3 min; Ni+Co 1.0 max; S 0.08 max			
C93600	USA			High leaded tin bronze	0.005 max	79.0-83.0	0.20 max		0.15 max	11.0-13.0	0.55 max	0.005 max	6.0-8.0	1.0 max	Cu+named el 99.3 min; Ni+Co 1.0 max; S 0.08 max			
	France	NF A53707	CuSn12	Cast As Cast				2	0.3	2.5-2.5			10.5-13	2	OT 0.5 max	240		5
	France	NF A53-709	U-AlIFe3Y200	Sand Cast As Cast	10-11.5		2-4	1.5		0.2		0.2	0.2	0.5	Mn 3			5
C84200	Pan American	COPANT 801	C84200	Continuous Bar As Mfg		78-82				2-3			4-6	10-16		230	110	13

UNS numbers and US grades are provided as a means of cross referencing chemically similar alloys. Exchangability is only possible after independent examination of specifications. Tensile properties are minimum or typical . UTS and YS as Mpa, El as %. See Appendix for list of abbreviations used in Descriptions.

Grade UNS #	Country	Specification	Designation	Description	Al	Cu	Fe	Ni	P	Pb	Sb	Si	Sn	Zn	Other	UTS	YS	El
C84200	USA			Leaded semi-red brass; continuous cast	0.005 max	78.0-82.0	0.40 max		1.5 max	2.0-3.0		0.005 max	4.0-6.0	10.0-16.0	Cu may be Cu+Ni; Cu+named el 99.3 min; Ni+Co 0.8 max; S 0.08 max			
C84200	USA			Leaded semi-red brass	0.005 max	78.0-82.0	0.40 max		0.05 max	2.0-3.0		0.005 max	4.0-6.0	10.0-16.0	Cu may be Cu+Ni; Cu+named el 99.3 min; Ni+Co 0.8 x; S 0.08 max			
C84400	Canada	CSA HC.9	ZP87-SC	Cast		81 max				7 max			3 max	9 max				
C84400	Denmark	DS 3001	Tombac 5214	Sand Cast As cast	0.01	79.5-82.5	0.3 max	3.25 max		6-8	0.3	0.01	2.5-3.5	6.5-10	S 0.1; Mn 0.2	190	80	18
C84400	Japan	JIS H2203	BCln1	Cast		79-83				3-7			2-4	8-12	OT 1.5 max			
C84400	Japan	JIS H5111	BC1	Cast		79-83				3-7			2-4	8-12				
C84400	Pan American	COPANT 801	C84400	Continuous Bar As Mfg	0.01	78-82	0.4	1	0.02	6-8	0.25	0.01	2.3-3.5	7-10	S 0.08	210	110	16
Valve Metal C84400	USA	ASTM B271	C84400	Cast	0.01 max	78-82	0.4 max	1 max	0.02 max	6-8	0.25 max	0.01 max	2.3-3.5	7-10	S 0.08 max			
Valve Metal C84400	USA	ASTM B30	C84400	Cast	0.01 max	78-82	0.4 max	1 max	0.02 max	6-8	0.25 max	0.01 max	2.3-3.5	7-10	S 0.08 max			
Valve Metal C84400	USA	ASTM B505	C84400	Cast	0.01 max	78-82	0.4 max	1 max	0.02 max	6-8	0.25 max	0.01 max	2.3-3.5	7-10	S 0.08 max			
Valve Metal C84400	USA	ASTM B584	C84400	Cast	0.01 max	78-82	0.4 max	1 max	0.02 max	6-8	0.25 max	0.01 max	2.3-3.5	7-10	S 0.08 max			
Valve Metal C84400	USA	QQ C-390	C84400	Cast	0.01 max	78-82	0.4 max	1 max	0.02 max	6-8	0.25 max	0.01 max	2.3-3.5	7-10	S 0.08 max			
C84400	USA			Leaded semi-red brass; continuous cast	0.005 max	78.0-82.0	0.40 max		1.5 max	6.0-8.0		0.005 max	2.3-3.5	7.0-10.0	Cu may be Cu+Ni; Cu+named el 99.3 min; Ni+Co 1.0 max; S 0.08 max			
C84400	USA			Leaded semi-red brass	0.005 max	78.0-82.0	0.40 max		0.02 max	6.0-8.0		0.005 max	2.3-3.5	7.0-10.0	Cu may be Cu+Ni; Cu+named el 99.3 min; Ni+Co 1.0 max; S 0.08 max			
C91300	Canada	CSA HC.9	T19-SC	Cast		80 max							20 max					
C91300	Pan American	COPANT 801	C91300	Continuous Bar As Mfg		79-82	0.25		1	0.25			18-20	0.25				
C91300	USA			Tin bronze; continuous cast	0.005 max	79.0-82.0	0.25 max		1.5 max	0.25 max	0.20 max	0.005 max	18.0-20.0	0.25 max	Cu may be Cu+Ni; Cu+named el 99.4 min; Ni+Co 0.50 x; S 0.05 max			
C91300	USA			Tin bronze	0.005 max	79.0-82.0	0.25 max		1.0 max	0.25 max	0.20 max	0.005 max	18.0-20.0	0.25 max	Cu may be Cu+Ni; Cu+named el 99.4 min; Ni+Co 0.50 x; S 0.05 max			
C92800	Pan American	COPANT 801	C92800	Continuous Bar As Mfg 127 mm diam		78-82		0.5		4-6			15-17	0.5				
C92800	USA			Leaded tin bronze Continuous Cast	0.005 max	78.0-82.0	0.20 max		1.5 max	4.0-6.0	0.25 max	0.005 max	15.0-17.0	0.8 max	Cu amy be Cu+Ni; Cu+named el 99.3 min; Ni+Co 0.8 x; S 0.05 max			

UNS numbers and US grades are provided as a means of cross referencing chemically similar alloys. Exchangability is only possible after independent examination of specifications. Tensile properties are minimum or typical . UTS and YS as Mpa, El as %. See Appendix for list of abbreviations used in Descriptions.

Cast Copper 5-23

Worldwide Guide to Equivalent Nonferrous Metals and Alloys

Grade UNS #	Country	Specification	Designation	Description	Al	Cu	Fe	Ni	P	Pb	Sb	Si	Sn	Zn	Other	UTS	YS	El
C92800	USA			Leaded tin bronze	0.005 max	78.0-82.0	0.20 max		0.05 max	4.0-6.0	0.25 max	0.005 max	15.0-17.0	0.8 max	Cu amy be Cu+Ni; Cu+named el 99.3 min; Ni+Co 0.8 max; S 0.05 max			
C92810	USA			Leaded tin bronze Continuous Cast	0.005 max	78.0-82.0	0.50 max	1.5 max		4.0-6.0	0.25 max	0.005 max	12.0-14.0	0.50 max	Cu amy be Cu+Ni; Cu+named el 99.3 min; Ni+Co 0.8-1.2; S 0.05 max			
C92810	USA			Leaded tin bronze	0.005 max	78.0-82.0	0.50 max		0.05 max	4.0-6.0	0.25 max	0.005 max	12.0-14.0	0.50 max	Cu amy be Cu+Ni; Cu+named el 99.3 min; Ni+Co 0.8-1.2; S 0.05 max			
C93700	Australia	AS 1565	937B	Chill Cast	0.01	78-82	0.15	2	0.1	8-11	0.5	0.02	9-11	1		220	140	3
C93700	Australia	AS 1565	937B	Centrifugal Cast	0.01	78-82	0.15	2	0.1	8-11	0.5	0.02	9-11	1		230	140	5
C93700	Australia	AS 1565	937B	Continuous Cast	0.01	78-82	0.15	2	0.1	8-11	0.5	0.02	9-11	1		280	160	6
C93700	Australia	AS 1565	937B	Sand Cast	0.01	78-82	0.15	2	0.1	8-11	0.5	0.02	9-11	1		190	80	5
C93700	Canada	CSA HC.9	PT1010-SC	Cast		80 max				10 max			10 max					
C93700	Czech Republic	CSN 423122	CuSn10Pb10		0.01 max		0.2 max	0.3-1	0.1 max	8.5-11			9-11	0.5 max	Sb:0.3			
C93700	Denmark	DS 3001	5640	Sand Cast As cast	0.01	78-82	0.25 max	2	0.05	8-11	0.5	0.01	9-11	2	S 0.1; Mn 0.2	180	80	7
C93700	Finland	SFS 2215	CuPb10Sn10	Sand Cast As cast	0.01	78-82	0.25 max	3.25 max		8-11	0.5	0.01	9-11	2	S 0.1; Mn 0.2	180	80	7
C93700	France	NF A53707	CuPb10Sn10	Cast As Cast	0.01		0.25 max		0.3	8-11		0.01	9-11	2	OT 1 max	180		7
C93700	Germany	DIN 17656	2.1177/GB-CuPb10Sn	Cast Ingot		78-81	0.15 max	1.3 max	0.03 max	8.5-10	0.4 max		9.3-11	0.8 max				
C93700	Japan	JIS H2207	LBCIn3	Cast			0.15		0.1 max	9.5-11			9.5-11	1 max				
C93700	Japan	JIS H5115	LBC3	Cast		77-81	0.3	1 max		9-11			9-11	1 max	OT 1 max			
C93700	Japan	JIS H5115	LBC3C	Cast Continuous		77-81	0.3	1 max		9-11			9-11	1 max	OT 1 max			
C93700	Norway	NS 16540	CuPb10Sn10	Cast Ingot	0.01 max	78-81	0.15 max	2 max	0.05 max	8.5-10.5	0.5 max	0.01 max	9.3-11	2 max	ST 0.01 max; ST=Al+Si; S 0.1 max; Mn 0.2 max			
C93700	Norway	NS 16540	CuPb10Sn10	Cast Sand Centrifugal Continuous	0.01 max	78-82	0.25 max	0.2 max	0.05 max	8-11	0.5 max	0.01 max	9-11	2 max	S 0.1 max; Mn 0.2 max			
C93700	Pan American	COPANT 801	C93700	Continuous Bar As Mfg 127 mm diam	0.01	78-82	0.15	1	0.15	8-11	0.55	0.01	9-11	0.8	S 0.08	250	140	6
C93700	South Africa	SABS 200	Cu-Sn10Pb10	Chill Cast	0.1		0.15		0.1	8.5-11	0.5	0.02	9-11	1		215		3
C93700	South Africa	SABS 200	Cu-Sn10Pb10	Sand Cast	0.1		0.15		0.1	8.5-11	0.5	0.02	9-11	1		185		5
C93700	South Africa	SABS 200	Cu-Sn10Pb10	Continuously Cast	0.1		0.15		0.1	8.5-11	0.5	0.02	9-11	1		280		6
C93700	Sweden	SIS 145640	5640-03	Sand Cast As Cast	0.01	78-82	0.25	2	0.05	8-11	0.5	0.01	9-11	2	S 0.1; Mn 0.2	180	80	7
C93700	Sweden	SIS 145640	5640-15	Cast As Cast	0.01	78-82	0.25	2	0.05	8-11	0.5	0.01	11-13	2	S 0.1; Mn 0.2	220	110	6
Bearing Bronze C93700	USA	AMS 4842	C93700	Cast Sand Centrifugal Continuous		78-82	0.15 max	1 max	0.25 max	1-2.5	0.5 max		9-11	0.7 max				

UNS numbers and US grades are provided as a means of cross referencing chemically similar alloys. Exchangability is only possible after independent examination of specifications. Tensile properties are minimum or typical . UTS and YS as Mpa, El as %. See Appendix for list of abbreviations used in Descriptions.

Grade UNS #	Country	Specification	Designation	Description	Al	Cu	Fe	Ni	P	Pb	Sb	Si	Sn	Zn	Other	UTS	YS	El
Bearing Bronze C93700	USA	MIL B-13506	C93700	Cast Sand Centrifugal Continuous		78-82	0.15 max	1 max	0.25 max	1-2.5	0.5 max		9-11	0.7 max				
Bearing Bronze C93700	USA	ASTM B22	C93700	Cast Sand Centrifugal Continuous		78-82	0.15 max	1 max	0.25 max	1-2.5	0.5 max		9-11	0.7 max				
Bearing Bronze C93700	USA	ASTM B271	C93700	Cast Sand Centrifugal Continuous		78-82	0.15 max	1 max	0.25 max	1-2.5	0.5 max		9-11	0.7 max				
Bearing Bronze C93700	USA	ASTM B30	C93700	Cast Sand Centrifugal Continuous		78-82	0.15 max	1 max	0.25 max	1-2.5	0.5 max		9-11	0.7 max				
Bearing Bronze C93700	USA	ASTM B505	C93700	Cast Sand Centrifugal Continuous		78-82	0.15 max	1 max	0.25 max	1-2.5	0.5 max		9-11	0.7 max				
Bearing Bronze C93700	USA	ASTM B584	C93700	Cast Sand Centrifugal Continuous		78-82	0.15 max	1 max	0.25 max	1-2.5	0.5 max		9-11	0.7 max				
Bearing Bronze C93700	USA	QQ C-390	C93700	Cast Sand Centrifugal Continuous		78-82	0.15 max	1 max	0.25 max	1-2.5	0.5 max		9-11	0.7 max				
Bearing Bronze C93700	USA	SAE J462	C93700	Cast Sand Centrifugal Continuous		78-82	0.15 max	1 max	0.25 max	1-2.5	0.5 max		9-11	0.7 max				
C93700	USA			High leaded tin bronze; continuous cast	0.005 max	78.0-82.0	0.15 max		1.5 max	8.0-11.0	0.55 max	0.005 max	9.0-11.0	0.8 max	Cu+named el 99.0 min; Ni+Co 1.0 max; S 0.08 max			
C93700	USA			High leaded tin bronze	0.005 max	78.0-82.0	0.15 max			0.15 max	8.0-11.0	0.55 max	0.005 max	9.0-11.0	0.8 max	Cu+named el 99.0 min; Ni+Co 1.0 max; S 0.08 max			
C93700	USA			High leaded tin bronze; steel backed bearing	0.005 max	78.0-82.0	0.35 max			0.15 max	8.0-11.0	0.55 max	0.005 max	9.0-11.0	0.8 max	Cu+named el 99.0 min; Ni+Co 1.0 max; S 0.08 max			
	Finland	SFS 2212	CuAl10Fe5Ni5	Sand Cast As cast	8.8-10	77-82	3.5-5.5	4.5-6.5			0.05		0.1	0.2 max	0.2	Mg 0.05; Cr 0.01; Mn 0.3-2.5	640	250	13
	International	ISO 1338	CuPb10Sn10	Permanent mould Cast 12/25 mm diam	0.01	78-82	0.25	2	0.05	8-11	0.5	0.01	9-11	2	Cu total includes Ni; S 0.1; Mn 0.2	220	140	3
	International	ISO 1338	CuPb10Sn10	Sand Cast 12/25 mm diam	0.01	78-82	0.25	2	0.05	8-11	0.5	0.01	9-11	2	Cu total includes Ni; S 0.1; Mn 0.2	180	80	7
	International	ISO 1338	CuPb10Sn10	Continuously or centrifugally cast 12/25 diam	0.01	78-82	0.25	2	0.05	8-11	0.5	0.01	9-11	2	Cu total includes Ni; S 0.1; Mn 0.2		110	
C84410	USA			Leaded semi-red brass	0.01 max					7.0-9.0		0.20 max	3.0-4.5	7.0-11.0	Cu may be Cu+Ni; Cu+named el 99.3 min; Fe+Sb+As 0.8 max; Ni +Co 1.0 maxBi 0.05 max			
C95520	USA			Aluminum bronze	10.5-11.5	74.5-	4.0-5.5		0.25 max	0.03 max		0.15 max		0.30 max	Cu+named el 99.5 min; Ni+Co 4.2-6.0; Cr 0.05 max; Co 0.20 max; Mn 1.5 max			

UNS numbers and US grades are provided as a means of cross referencing chemically similar alloys. Exchangability is only possible after independent examination of specifications. Tensile properties are minimum or typical . UTS and YS as Mpa, El as %. See Appendix for list of abbreviations used in Descriptions.

Grade UNS #	Country	Specification	Designation	Description	Al	Cu	Fe	Ni	P	Pb	Sb	Si	Sn	Zn	Other	UTS	YS	El
C94900	USA			Nickel tin bronze	0.005 max	79.0-81.0	0.30 max		0.05 max	4.0-6.0	0.25 max	0.005 max	4.0-6.0	4.0-6.0	Cu+named el 99.4 min; Ni+Co 4.0-6.0; S 0.08 max; Mn 0.10 max			
	Czech Republic	CSN 423303	CuZn17Si3		0.1 max	79-81	0.6 max			0.5 max		2.5-4.5	0.3 max		Sb 0.1; bal Zn; Mn 1 max			
C96300	USA			Copper nickel			0.50-1.5	0.02 max		0.01 max		0.50 max			Cu+named el 99.5 min; Ni+Co 18.0-22.0; Nb 0.50-1.5; C 0.15 max; S 0.02 max; Mn 0.25-1.5			
C95500	Canada	CSA HC.9	AF114N-SC	Cast	11 max	80 max	4 max	4 max							Mn 1 max			
C95500	Czech Republic	CSN 423147	CuAl10Fe4Ni4		9-11		3.5-5.5	3.5-5.5	0.1 max	0.1 max		0.2 max	0.2 max	0.5 max	Sb 0.05; As 0.05 max; Mn 0.5 max			
C95500	Germany		2.0981/GB-CuAl11Ni		9.3-11.3	72-77	4.3-5.8	5.3-6.8		0.03 max		0.07 max			Mn 2.3 max			
C95500	International	ISO 1338	CuAl10Fe5Ni5	Sand Cast 12/25 mm diam	8-11	76	3.5-5.5	3.5-6.5		0.1		0.1	0.2	0.5	99.2 min Cu+Fe+Ni+Al+Mn; Mn 3	600	250	10
C95500	International	ISO 1338	CuAl10Fe5Ni5	Continuously or centrifugally cast 12/25 diam	8-11	76	3.5-5.5	3.5-6.5		0.1		0.1	0.2	0.5	99.2 min Cu+Fe+Ni+Al+Mn; Mn 3	680	280	12
C95500	Pan American	COPANT 801	C95500(HT)	Cast Continuous Bar HT 140 mm diam	10-11.5	78	3-5	3-5							Mn 3.5	660	290	10
C95500	USA			Aluminum bronze	10.0-11.5	78.0 min	3.0-5.0								Cu+named el 99.5 min; Ni+Co 3.0-5.5; Mn 3.5 max			
	Czech Republic	CSN 423182	CuPb20		0.05 max		0.6 max	0.3 max	0.1 max	20-26				0.1 max	Sn+Sb 0.6; As 0.05 max			
	South Africa	SABS 200	Cu-Zn18Pb4Sn2	Sand Cast	0.1	70-80	0.75	1		2-5			1-3		OT 0.99 max	170		20
C93900	Pan American	COPANT 801	C93900	Continuous Bar As Mfg 127 mm diam		76.5-79.5	0.4	0.8	1.5	14-18			5-7	1.5		180	110	5
79-6-15 C93900	USA	ASTM B30	C93900	Cast Continuous		76.5-79.5	0.4 max	0.8 max	0.05 max	14-18			5-7	1.5 max				
79-6-15 C93900	USA	ASTM B505	C93900	Cast Continuous		76.5-79.5	0.4 max	0.8 max	0.05 max	14-18			5-7	1.5 max				
C93900	USA			High leaded tin bronze; continuous cast	0.005 max	76.5-79.5	0.40 max		1.5 max	14.0-18.0	0.50 max	0.005 max	5.0-7.0	1.5 max	Cu+named el 98.9 min; Ni+Co 0.8 max; S 0.008 max			
	South Africa	SABS 200	Cu-Sn5Zn5Pb5	Chill Cast	0.01	79.19		2		4-6		0.02	4-6	4-6	0.50 As+Fe+Sb; Bi 0.05	200		6
	South Africa	SABS 200	Cu-Sn5Zn5Pb5	Continuously Cast	0.01	79.19		2		4-6		0.02	4-6	4-6	0.50 As+Fe+Sb; Bi 0.05	265		13
	South Africa	SABS 200	Cu-Sn5Zn5Pb5	Sand Cast	0.01	79.19		2		4-6		0.02	4-6	4-6	0.50 As+Fe+Sb; Bi 0.05	200		13
C84500	USA			Leaded semi-red brass; continuous cast	0.005 max	77.0-79.0	0.40 max		1.5 max	6.0-7.5		0.005 max	2.0-4.0	10.0-14.0	Cu may be Cu+Ni; Cu+named el 99.3 min; Ni+Co 1.0 max; S 0.08 max			

UNS numbers and US grades are provided as a means of cross referencing chemically similar alloys. Exchangability is only possible after independent examination of specifications. Tensile properties are minimum or typical . UTS and YS as Mpa, El as %. See Appendix for list of abbreviations used in Descriptions.

Grade UNS #	Country	Specification	Designation	Description	Al	Cu	Fe	Ni	P	Pb	Sb	Si	Sn	Zn	Other	UTS	YS	El
C84500	USA			Leaded semi-red brass	0.005 max	77.0-79.0	0.40 max		0.02 max	6.0-7.5		0.005 max	2.0-4.0	10.0-14.0	Cu may be Cu+Ni; Cu+named el 99.3 min; Ni+Co 1.0 max; S 0.08 max			
C93800	Canada	CSA HC.9	PT147-SC	Cast		78 max				15 max			7 max					
C93800	Finland	SFS 2216	CuPb15Sn8	Sand Cast As cast	0.01	75-79	0.25 max	2	0.1	13-17	0.5	0.01	7-9	2	S 0.1; Mn 0.2	170	80	5
C93800	Germany	DIN 17656	2.1183/GB-CuPb15Sn	Cast Ingot		75-78	0.15 max	1.8 max	0.03 max	13.5-16.5	0.4 max		7.3-9	2 max				
C93800	India	IS 1458	III	Cast	0.01		0.3		0.05	14-16	0.4		6-8	0.5	0.5 Fe+Sb			
C93800	International	ISO 1458	III	Sand Cast (cast-on)	0.01 max		0.3		0.05	14-16	0.4		6-8	0.5	.5 Fe+Sb	137		2
C93800	International	ISO 1458	III	Sand Cast (separately cast)	0.01 max		0.3		0.05	14-16	0.4		6-8	0.5	.5 Fe+Sb	157		4
C93800	International	ISO 1338	CuPb15Sn8	Sand Cast 12/25 mm diam	0.01	75-79	0.25	2	0.1	13-17	0.5	0.01	7-9	2	Cu total includes Ni; S 0.1; Mn 0.2	170	80	5
C93800	International	ISO 1338	CuPb15Sn8	Continuously or centrifugally cast 12/25 diam	0.01	75-79	0.25	2	0.1	13-17	0.5	0.01	7-9	2	Cu total includes Ni; S 0.1; Mn 0.2	220	100	8
C93800	Japan	JIS H2207	LBCln4	Cast			0.15		0.1 max	14.5-16			7.5-9	1 max	OT 1 max			
C93800	Japan	JIS H5115	LBC4	Sand Cast		74-78	0.3	1 max		14-16			7-9	1 max	OT 1 max			
C93800	Pan American	COPANT 801	C93800	Continuous Bar As Mfg 127 mm diam	0.01	75-79	0.15	1	0.05	13-16	0.8	0.01	6.3-7.5	0.8	S 0.08	180	110	5
Anti-acid Metal C93800	USA	MIL B-16261	C93800	Cast Sand Centrifugal Continuous	0.01 max	75-79	0.15 max	0.7 max	0.05 max	13-16	0.7 max	0.005 max	6.3-7.5	0.7 max	S 0.08 max			
Anti-acid Metal C93800	USA	ASTM B271	C93800	Cast Sand Centrifugal Continuous	0.01 max	75-79	0.15 max	0.7 max	0.05 max	13-16	0.7 max	0.005 max	6.3-7.5	0.7 max	S 0.08 max			
Anti-acid Metal C93800	USA	ASTM B30	C93800	Cast Sand Centrifugal Continuous	0.01 max	75-79	0.15 max	0.7 max	0.05 max	13-16	0.7 max	0.005 max	6.3-7.5	0.7 max	S 0.08 max			
Anti-acid Metal C93800	USA	ASTM B505	C93800	Cast Sand Centrifugal Continuous	0.01 max	75-79	0.15 max	0.7 max	0.05 max	13-16	0.7 max	0.005 max	6.3-7.5	0.7 max	S 0.08 max			
Anti-acid Metal C93800	USA	ASTM B584	C93800	Cast Sand Centrifugal Continuous	0.01 max	75-79	0.15 max	0.7 max	0.05 max	13-16	0.7 max	0.005 max	6.3-7.5	0.7 max	S 0.08 max			
Anti-acid Metal C93800	USA	ASTM B66	C93800	Cast Sand Centrifugal Continuous	0.01 max	75-79	0.15 max	0.7 max	0.05 max	13-16	0.7 max	0.005 max	6.3-7.5	0.7 max	S 0.08 max			
Anti-acid Metal C93800	USA	QQ C-390	C93800	Cast Sand Centrifugal Continuous	0.01 max	75-79	0.15 max	0.7 max	0.05 max	13-16	0.7 max	0.005 max	6.3-7.5	0.7 max	S 0.08 max			
Anti-acid Metal C93800	USA	QQ C-525	C93800	Cast Sand Centrifugal Continuous	0.01 max	75-79	0.15 max	0.7 max	0.05 max	13-16	0.7 max	0.005 max	6.3-7.5	0.7 max	S 0.08 max			
Anti-acid Metal C93800	USA	SAE J462	C93800	Cast Sand Centrifugal Continuous	0.01 max	75-79	0.15 max	0.7 max	0.05 max	13-16	0.7 max	0.005 max	6.3-7.5	0.7 max	S 0.08 max			
Anti-acid Metal C93800	USA	QQ L-225	C93800	Cast Sand Centrifugal Continuous	0.01 max	75-79	0.15 max	0.7 max	0.05 max	13-16	0.7 max	0.005 max	6.3-7.5	0.7 max	S 0.08 max			
C93800	USA			High leaded tin bronze; continuous cast	0.005 max	75.0-79.0	0.15 max		1.5 max	13.0-16.0	0.8 max	0.005 max	6.3-7.5	0.8 max	Cu+named el 99.0 min; Ni+Co 1.0 max; S 0.08 max			

UNS numbers and US grades are provided as a means of cross referencing chemically similar alloys. Exchangability is only possible after independent examination of specifications. Tensile properties are minimum or typical . UTS and YS as Mpa, El as %. See Appendix for list of abbreviations used in Descriptions.

Worldwide Guide to Equivalent Nonferrous Metals and Alloys

Grade UNS #	Country	Specification	Designation	Description	Al	Cu	Fe	Ni	P	Pb	Sb	Si	Sn	Zn	Other	UTS	YS	El
C93800	USA			High leaded tin bronze	0.005 max	75.0-79.0	0.15 max		0.05 max	13.0-16.0	0.8 max	0.005 max	6.3-7.5	0.8 max	Cu+named el 99.0 min; Ni+Co 1.0 max; S 0.08 max			
C93900	USA			High leaded tin bronze	0.005 max	76.5-79.5	0.40 max	1.5 max		14.0-18.0	0.50 max	0.005 max	5.0-7.0	1.5 max	Cu+named el 98.9 min; Ni+Co 0.8 max; S 0.008 max			
C94100	Australia	AS 1565		Sand Cast		65-75		2	0.1	15-22	0.5	0.01	4.5-6.5	3		160	60	5
C94100	Australia	AS 1565		Chill Cast		65-75		2	0.1	15-22	0.5	0.01	4.5-6.5	3		170	80	5
C94100	Australia	AS 1565		Centrifugal Cast		65-75		2	0.1	15-22	0.5	0.01	4.5-6.5	3		190	80	7
C94100	Australia	AS 1565		Continuous Cast		65-75		2	0.1	15-22	0.5	0.01	4.5-6.5	3		190	100	8
C94100	France	NF A53-707	CuPb20Sn5	Cast As Cast	0.01		0.25 max	2.5	0.3	18-23		0.01	4-6	2		150		5
C94100	Germany		2.1189/GB-CuPb20Sn			70-75	0.15 max	2.3 max	0.03 max	18.5-23	0.4 max		3.7-5.5	2 max				
C94100	International	ISO 1338	CuPb20Sn5	Sand Cast 12/25 mm diam	0.01	70-78	0.25	2.5	0.1	18-23	0.75	0.01	4-6	2	Cu total includes Ni; S 0.1; Mn 0.2	150	60	5
C94100	International	ISO 1338	CuPb20Sn5	Continuously Cast 12/25 mm diam	0.01	70-78	0.25	2.5	0.1	18-23	0.75	0.01	4-6	2	Cu total includes Ni; S 0.1; Mn 0.2	180	80	7
C94100	Pan American	COPANT 801	C94100	Continuous Bar As Mfg 127 mm diam		65-75				15-22			4.5-6.5	3		180	120	7
C94100	USA			High leaded tin bronze; continuous cast	0.005 max	72.0-79.0	0.25 max	1.5 max		18.0-22.0	0.8 max	0.005 max	4.5-6.5	1.0 max	Cu+named el 98.7 min: Ni+Co 1.0 max; S 0.25 max			
C94100	USA			High leaded tin bronze	0.005 max	72.0-79.0	0.25 max		0.05 max	18.0-22.0	0.8 max	0.005 max	4.5-6.5	1.0 max	Cu+named el 98.7 min: Ni+Co 1.0 max; S 0.08 max			
C98200	USA			Leaded copper			0.7 max	0.50 max	0.10 max	21.0-27.0	0.50 max		0.6-2.0	0.50 max	Cu+named el 99.5 min			
C95700	Japan	JIS H5114	AlBC4	Cast	6-9	71 min	2.5-5	1-4							OT 0.5 max; Mn 7-15			
C95700	USA			Aluminum bronze	7.0-8.5	71.0 min	2.0-4.0			0.03 max		0.10 max			Cu+named el 99.5 min; Ni+Co 1.5-3.0; Mn 11.0-14.0			
C95710	Australia	AS 1565		Sand Cast	7-8.5	71 min	2-4	1.5-3					1	0.5	Mn 11-14	650	280	18
C95710	Australia	AS 1565		Chill Cast	7-8.5	71 min	2-4	1.5-3					1	0.5	Mn 11-14	670	310	27
C95710	USA			Aluminum bronze	7.0-8.5	71.0 min	2.0-4.0		0.05 max	0.05 max		0.15 max	1.0 max	0.50 max	Cu+named el 99.5 min; Ni+Co 1.5-3.0; Mn 11.0-14.0			
C94500	Japan	JIS H2207	LBCln5	Cast			0.15		0.1 max	16.5-22			6.5-8	1 max	OT 1 max			
C94500	Japan	JIS H5115	LBC5	Sand Cast		70-76	0.5 max	1 max		16-22			6-8	1 max	OT 1 max			
C94500	USA			High leaded tin bronze; continuous cast	0.005 max		0.15 max	1.5 max		16.0-22.0	0.8 max	0.005 max	6.0-8.0	1.2 max	Cu+named el 99.0 min; Ni+Co 1.0 max; S 0.08 max			
C94500	USA			High leaded tin bronze	0.005 max		0.15 max		0.05 max	16.0-22.0	0.8 max	0.005 max	6.0-8.0	1.2 max	Cu+named el 99.0 min; Ni+Co 1.0 x; S 0.08 max			

UNS numbers and US grades are provided as a means of cross referencing chemically similar alloys. Exchangability is only possible after independent examination of specifications. Tensile properties are minimum or typical . UTS and YS as Mpa, El as %. See Appendix for list of abbreviations used in Descriptions.

Grade UNS #	Country	Specification	Designation	Description	Al	Cu	Fe	Ni	P	Pb	Sb	Si	Sn	Zn	Other	UTS	YS	El
	Czech Republic	CSN 423183	CuPb22Sn3		0.05 max		0.6 max	0.3 max	0.1 max	20-26			2-3.5	0.1 max	Sb:0.3 S:0.05			
	Japan	JIS H5115	LBC4C	Cast Continuous		74-78	0.3 max	1 max		14-16			7-9	1 max	OT 1 max			
C84800	Canada	CSA HC.9	ZP156-SC	Cast		76 max				6 max			3 max	15 max				
C84800	Pan American	COPANT 801	C84800	Continuous Bar As Mfg	0.01	75-77	0.4	1	0.02	5.5-7	0.25	0.01	2-3	13-17	S 0.08	210	110	16
Plumbing Goods Brass C84800	USA	ASTM B271	C84800	Cast	0.01 max	75-77	0.4 max	1 max	0.02 max	6-8	0.25 max	0.01 max	2-3	13-17	S 0.08 max			
Plumbing Goods Brass C84800	USA	ASTM B30	C84800	Cast	0.01 max	75-77	0.4 max	1 max	0.02 max	6-8	0.25 max	0.01 max	2-3	13-17	S 0.08 max			
Plumbing Goods Brass C84800	USA	ASTM B505	C84800	Cast	0.01 max	75-77	0.4 max	1 max	0.02 max	6-8	0.25 max	0.01 max	2-3	13-17	S 0.08 max			
Plumbing Goods Brass C84800	USA	ASTM B584	C84800	Cast	0.01 max	75-77	0.4 max	1 max	0.02 max	6-8	0.25 max	0.01 max	2-3	13-17	S 0.08 max			
Plumbing Goods Brass C84800	USA	QQ C-390	C84800	Cast	0.01 max	75-77	0.4 max	1 max	0.02 max	6-8	0.25 max	0.01 max	2-3	13-17	S 0.08 max			
C84800	USA			Leaded semi red brass; continuous cast	0.005 max	76.0 77.0	0.10 max	1.5 max		5.5-7.0		0.005 max	2.0-3.0	10.0 17.0	Cu may be Cu+Ni; Cu+named el 99.3 min; Ni +Co 1.0 max; S 0.08 max			
C84800	USA			Leaded semi-red brass	0.005 max	75.0-77.0	0.40 max		0.02 max	5.5-7.0		0.005 max	2.0-3.0	13.0-17.0	Cu may be Cu+Ni; Cu+named el 99.3 min; Ni +Co 1.0 max; S 0.08 max			
C94330	USA			High leaded tin bronze; continuous cast		68.5-75.5	0.35 max	1.5 max		21.0-25.0	0.50 max		3.0-4.0	3.0 max	Cu+named el 99.0 min; Ni+Co 0.50 max			
C94330	USA			High leaded tin bronze		68.5-75.5	0.35 max			21.0-25.0	0.50 max		3.0-4.0	3.0 max	Cu+named el 99.0 min; Ni+Co 0.50 max			
	South Africa	SABS 200	Cu-Sn9Pb15	Chill Cast				2	0.1	13-17	0.5		8-10	1	OT 0.25 max	200		3
	South Africa	SABS 200	Cu-Sn9Pb15	Sand Cast				2	0.1	13-17	0.5		8-10	1	OT 0.25 max	170		4
	South Africa	SABS 200	Cu-Sn9Pb15	Continuously Cast				2	0.1	13-17	0.5		8-10	1	OT 0.25 max	230		9
C85210	Australia	AS 1565	852C	Sand Cast	0.01	70-75	0.75 max	1		2-5			1-3		bal Zn; As 0.02-0.06	170	80	18
C99300	USA		Incramet 800		10.7-11.5		0.40-1.0	13.5-16.5		0.02 max		0.02 max			Cu+named el 99.7 min; Co 1.0-2.0; S 0.05 max			
	South Africa	SABS 200	Cu-Sn5Pb20	Sand Cast				2	0.1	18-23	0.5	0.01	4-6	1		155		5
	South Africa	SABS 200	Cu-Sn5Pb20	Chill Cast				2	0.1	18-23	0.5	0.01	4-6	1		170		5
	South Africa	SABS 200	Cu-Sn5Pb20	Continuously Cast				2	0.1	18-23	0.5	0.01	4-6	1		185		8
C85200	Canada	CSA HC.9	ZP243-SC	Cast		72 max				3 max			1 max	24 max				
Leaded Yellow Brass	USA	ASTM B271	C85200	Cast	0.01 max	70-74	0.6 max	1 max	0.02 max	1.5-3.8	0.2 max	0.05 max	0.7-2	20-27	S 0.05 max			

UNS numbers and US grades are provided as a means of cross referencing chemically similar alloys. Exchangability is only possible after independent examination of specifications. Tensile properties are minimum or typical . UTS and YS as Mpa, El as %. See Appendix for list of abbreviations used in Descriptions.

Worldwide Guide to Equivalent Nonferrous Metals and Alloys

Grade UNS #	Country	Specification	Designation	Description	Al	Cu	Fe	Ni	P	Pb	Sb	Si	Sn	Zn	Other	UTS	YS	El
Leaded Yellow Brass C85200	USA	ASTM B30	C85200	Cast	0.01 max	70-74	0.6 max	1 max	0.02 max	1.5-3.8	0.2 max	0.05 max	0.7-2	20-27	S 0.05 max			
Leaded Yellow Brass C85200	USA	ASTM B584	C85200	Cast	0.01 max	70-74	0.6 max	1 max	0.02 max	1.5-3.8	0.2 max	0.05 max	0.7-2	20-27	S 0.05 max			
Leaded Yellow Brass C85200	USA	MIL C-15345	C85200	Cast	0.01 max	70-74	0.6 max	1 max	0.02 max	1.5-3.8	0.2 max	0.05 max	0.7-2	20-27	S 0.05 max			
Leaded Yellow Brass C85200	USA	QQ C-390	C85200	Cast	0.01 max	70-74	0.6 max	1 max	0.02 max	1.5-3.8	0.2 max	0.05 max	0.7-2	20-27	S 0.05 max			
Leaded Yellow Brass C85200	USA	SAE J462	C85200	Cast	0.01 max	70-74	0.6 max	1 max	0.02 max	1.5-3.8	0.2 max	0.05 max	0.7-2	20-27	S 0.05 max			
C85200	USA			Leaded yellow brass	0.005 max	70.0-74.0	0.6 max		0.02 max	1.5-3.8	0.20 max	0.05 max	0.7-2.0	20.0-27.0	Cu may be Cu+Ni; Cu+named el 99.1 min; Ni+Co 1.0 max; S 0.05 max			
C98400	USA			Leaded copper			0.7 max	0.50 max	0.10 max	26.0-33.0	0.50 max		0.50 max	0.50 max	Cu+named el 99.5 min; Ag 1.5 max			
C94300	Canada	CSA HC.9	PT245-SC	Cast		70 max				25 max			5 max					
C94300	Pan American	COPANT 801	C94300	Continuous Bar As Mfg 127 mm diam	0.01	68.5-73.5	0.15	1	0.05	22-25	0.8	0.01	4.5-6	0.8	S 0.08	150	110	7
Soft Bronze C94300	USA	MIL B-16261	C94300	Sand Cast		68-73.5	0.15 max	0.7 max	0.05 max	22-25	0.7 max		4.5-6	0.5 max	S 0.08 max			
Soft Bronze C94300	USA	ASTM B271	C94300	Sand Cast		68-73.5	0.15 max	0.7 max	0.05 max	22-25	0.7 max		4.5-6	0.5 max	S 0.08 max			
Soft Bronze C94300	USA	ASTM B30	C94300	Sand Cast		68-73.5	0.15 max	0.7 max	0.05 max	22-25	0.7 max		4.5-6	0.5 max	S 0.08 max			
Soft Bronze C94300	USA	ASTM B505	C94300	Sand Cast		68-73.5	0.15 max	0.7 max	0.05 max	22-25	0.7 max		4.5-6	0.5 max	S 0.08 max			
Soft Bronze C94300	USA	ASTM B566	C94300	Sand Cast		68-73.5	0.15 max	0.7 max	0.05 max	22-25	0.7 max		4.5-6	0.5 max	S 0.08 max			
Soft Bronze C94300	USA	ASTM B584	C94300	Sand Cast		68-73.5	0.15 max	0.7 max	0.05 max	22-25	0.7 max		4.5-6	0.5 max	S 0.08 max			
Soft Bronze C94300	USA	SAE J462	C94300	Sand Cast		68-73.5	0.15 max	0.7 max	0.05 max	22-25	0.7 max		4.5-6	0.5 max	S 0.08 max			
Soft Bronze C94300	USA	QQ L-225	C94300	Sand Cast		68-73.5	0.15 max	0.7 max	0.05 max	22-25	0.7 max		4.5-6	0.5 max	S 0.08 max			
C94300	USA			High leaded tin bronze; continuous cast	0.005 max	68.5-73.5	0.15 max		1.5 max	22.0-25.0	0.8 max	0.005 max	4.5-6.0	0.8 max	Cu+named el 99.0 min; Ni+Co 1.0 max; S 0.25 max			
C94300	USA			High leaded tin bronze	0.005 max	68.5-73.5	0.15 max		0.08 max	22.0-25.0	0.8 max	0.005 max	4.5-6.0	0.8 max	Cu+named el 99.0 min; Ni+Co 1.0 max; S 1.0 max			
C85310	Australia	AS 1565	851D	Sand Cast	0.01	68-73	0.75 max	1		2-5			1.5 min		bal Zn; As 0.02-0.06	170		12
	Czech Republic	CSN 423184	CuPb30Fe		0.05 max		0.4 max	0.3 max	0.1 max	27-33				0.1 max	Sn+Sb 0.4; As 0.05 max			
	Czech Republic	CSN 423188	CuPb20Ag3				0.2 max	0.1 max		27-33			0.3 max	0.1 max	Ag:2.75-3.25			

UNS numbers and US grades are provided as a means of cross referencing chemically similar alloys. Exchangability is only possible after independent examination of specifications. Tensile properties are minimum or typical . UTS and YS as Mpa, El as %. See Appendix for list of abbreviations used in Descriptions.

5-30 Cast Copper

Grade UNS #	Country	Specification	Designation	Description	Al	Cu	Fe	Ni	P	Pb	Sb	Si	Sn	Zn	Other	UTS	YS	El
	South Africa	SABS 200	Cu-Zn26Pb4	Sand Cast	0.01	66-73	0.75	1		2-5				1.5	bal Zn; OT 0.99 max	170		12
	Czech Republic	CSN 423187	CuPb30Ag1,5 Sn				0.2 max	0.1 max	0.1 max	27-33			0.1-0.7	0.1 max	Ag:1.4-1.7			
C94000	USA			High leaded tin bronze; continuous cast	0.005 max	69.0-72.0	0.25 max		1.5 max	14.0-16.0	0.50 max	0.005 max	12.0-14.0	0.50 max	Cu+named el 98.7 min; Nl+Co 0.50-1.0; S 0.25 max			
C94000	USA			High leaded tin bronze	0.005 max	69.0-72.0	0.25 max		0.05 max	14.0-16.0	0.50 max	0.005 max	12.0-14.0	0.50 max	Cu+named el 98.7 min; Nl+Co 0.50-1.0; S 0.08 max			
C94320	USA			High leaded tin bronze; continuous cast			0.35 max	1.5 max		24.0-32.0			4.0-7.0		Cu+named el 99.0 min			
C94320	USA			High leaded tin bronze			0.35 max			24.0-32.0			4.0-7.0		Cu+named el 99.0 min			
C94310	USA			High leaded tin bronze; continuous cast			0.50 max	1.5 max		27.0-34.0	0.50 max		1.5-3.0	0.50 max	Cu+named el 99.0 min; Ni+Co 0.25-1.0			
C94310	USA			High leaded tin bronze			0.50 max		0.05 max	27.0-34.0	0.50 max		1.5-3.0	0.50 max	Cu+named el 99.0 min; Ni+Co 0.25-1.0			
C96400	Australia	AS 1565		Sand Cast		65-69	0.25-1.5	28-32				0.5			Nb 0.5-1.12; Mn 1.5	415	220	20
C96400	Germany	DIN 17658	2.0835.01/G-CuNi30	Cast			0.5-1.5	29-31				0.3-0.7			Nb 0.5-1; Mn 0.6-1.2			
C96400	Pan American	COPANT 801	C96400	Cast Continuous Bar HT 140 mm diam		65-69	0.25-1.5	28-32		0.03		0.5			Nb 0.5-1.11; C 0.2; Mn 1.5	450	250	25
70-30 Copper Nickel C96400	USA	ASTM B30	C96400	Cast Sand Centrifugal Continuous		65-69	0.25-1.5	28-32		0.03 max		0.5 max			Nb 0.5-1.5; C 0.2 max; Mn 1.5 max			
70-30 Copper Nickel C96400	USA	ASTM B369	C96400	Cast Sand Centrifugal Continuous		65-69	0.25-1.5	28-32		0.03 max		0.5 max			Nb 0.5-1.6; C 0.2 max; Mn 1.5 max			
70-30 Copper Nickel	USA	ASTM B505	C96400	Cast Sand Centrifugal Continuous		65-69	0.25-1.5	28-32		0.03 max		0.5 max			Nb 0.5-1.7; C 0.2 max; Mn 1.5 max			
70-30 Copper Nickel	USA	MIL C-15345	C96400	Cast Sand Centrifugal Continuous		65-69	0.25-1.5	28-32		0.03 max		0.5 max			Nb 0.5-1.8; C 0.2 max; Mn 1.5 max			
70-30 Copper Nickel	USA	MIL C-20159	C96400	Cast Sand Centrifugal Continuous		65-69	0.25-1.5	28-32		0.03 max		0.5 max			Nb 0.5-1.9; C 0.2 max; Mn 1.5 max			
70-30 Copper Nickel	USA	QQ C-390	C96400	Cast Sand Centrifugal Continuous		65-69	0.25-1.5	28-32		0.03 max		0.5 max			Nb 0.5-1.10; C 0.2 max; Mn 1.5 max			
C96400	USA			Copper nickel			0.25-1.5			0.01 max		0.50 max			Cu+named el 99.5 min; Ni+Co 28.0-32.0; Nb 0.50-1.5; C 0.15 x; Mn 1.5 max			
C96900	USA			Copper nickel			0.50 max			0.02 max			7.5-8.5	0.50 max	Cu+named el 99.5 min; Ni+Co 14.5-15.5; Nb 0.10 x; Mg 0.15 x; Mn 0.05-0.30			

UNS numbers and US grades are provided as a means of cross referencing chemically similar alloys. Exchangability is only possible after independent examination of specifications. Tensile properties are minimum or typical . UTS and YS as Mpa, El as %. See Appendix for list of abbreviations used in Descriptions.

Worldwide Guide to Equivalent Nonferrous Metals and Alloys

Grade UNS #	Country	Specification	Designation	Description	Al	Cu	Fe	Ni	P	Pb	Sb	Si	Sn	Zn	Other	UTS	YS	El
C85400	Australia	AS 1565	854C	Sand Cast	0.1	65-70	0.75 max	1		1.5-3.8			0.5-1.5	24-32		190	70	11
C85400	Canada	CSA HC.9	ZP313-SC	Cast		67 max				3 max			1 max	29 max				
C85400	Denmark	DS 3001	5144	Sand Cast As cast	0.1	63-67	0.8 max	1	0.05	1-3	0.1	0.05	1.5 max		bal Zn; As 0.1 max; Mn 0.2	180	70	12
C85400	Finland	SFS 2203	CuZn33Pb2	Sand Cast As cast	0.1	63-67	0.8 max	1	0.05	1-3	0.1	0.05	1.5		bal Zn; As 0.1 max; Mn 0.2	180	70	12
C85400	International	ISO 1338	CuZn33Pb2	Sand Cast 12/25 mm diam	0.1	63-67	0.8	1	0.05	1-3		0.05	1.5		Mn 0.2	180	70	12
C85400	Norway	NS 16550	CuZn33Pb2	Sand Cast	0.1 max	63-67	0.8 max	1 max	0.05 max	1-3	0.1 max	0.05 max	1.5 max		As 0.1 max; Mn 0.2 max			
C85400	Norway	NS 16550	CuZn33Pb2	Cast Ingot	0.03 max	63-66	0.7 max	1 max	0.02 max	1-2.8	0.05 max	0.03 max	1.5 max		As 0.08 max; Mn 0.2 max			
C85400	Sweden	SIS 145144	5144-03	Sand Cast As Cast	0.1	63-67	0.8	1	0.05	1-3	0.1	0.05	1.5		bal Zn; As 0.1; Mn 0.2	180	70	12
No.1 Yellow Brass C85400	USA	ASTM B271	C85400	Cast	0.35 max	65-70	0.7 max	1 max	0.02 max	1.5-3.8		0.05 max	0.5-1.5	24-32	S 0.05 max			
No.1 Yellow Brass C85400	USA	ASTM B30	C85400	Cast	0.35 max	65-70	0.7 max	1 max	0.02 max	1.5-3.8		0.05 max	0.5-1.5	24-32	S 0.05 max			
No.1 Yellow Brass C85400	USA	ASTM B584	C85400	Cast	0.35 max	65-70	0.7 max	1 max	0.02 max	1.5-3.8		0.05 max	0.5-1.5	24-32	S 0.05 max			
No.1 Yellow Brass C85400	USA	QQ C-390	C85400	Cast	0.35 max	65-70	0.7 max	1 max	0.02 max	1.5-3.8		0.05 max	0.5-1.5	24-32	S 0.05 max			
No.1 Yellow Brass C85400	USA	SAE J462	C85400	Cast	0.35 max	65-70	0.7 max	1 max	0.02 max	1.5-3.8		0.05 max	0.5-1.5	24-32	S 0.05 max			
C85400	USA			Leaded yellow brass	00.35 max	65.0-70.0	0.7 max			1.5-3.8		0.05 max	0.50-1.5	24.0-32.0	Cu may be Cu+Ni; Cu+named el 98.9 min; Ni+Co 1.0 max			
C98600	Japan	JIS H5403	KJ2	Cast		0.8 max				33-37			1 max		Ag 2 max			
C98600	USA			Leaded copper		60.0-70.0	0.35 max			30.0-40.0			0.50 max		Ag 1.5 max			
	South Africa	SABS 200	CuZn32Pb2	Sand Cast	0.1	63-70	0.75	1		1-3			1.5		bal Zn; OT 0.99 max	185		12
C96600	USA			Copper nickel		0.8-1.1				0.01 max		0.15 max			Cu+named el 99.5 min; Ni+Co 29.0-33.0; Be 0.40-0.7; Mn 1.0 max			
C96700	USA			Copper nickel		0.7-1.0				0.01 max		0.15 max			Cu+named el 99.5 min; Ni+Co 29.0-33.0; Zr 0.10-0.20; Ti 0.10-0.20; Be 1.1-1.2; Mn 0.7 max			
C85800	Canada	CSA HC.9	ZF391P-SC	Cast	0.75 max	61 max	1 max	35.5 max		0.75 max			0.75 max					
C85800	Czech Republic	CSN 423319	CuZn40		0.05 max	58-62	0.5 max		0.1 max	2 max			1 max		Sb+A :0.5; bal Zn; Mn 0.2 max			
C85800	Germany	DIN 17656	2.0342/GB-CuZn37Pb	Cast Ingot	0.4-0.8	59-62	0.4 max	0.8 max		0.7-2.2		0.03 max	0.6 max		bal Zn; Mn 0.1 max			
C85800	Germany		2.0383/GB-CuZn39Pb		0.3-0.7	58-63	0.7 max	0.8 max		1.3-2.5		0.05 max	0.9 max		bal Zn; Mn 0.1 max			

UNS numbers and US grades are provided as a means of cross referencing chemically similar alloys. Exchangability is only possible after independent examination of specifications. Tensile properties are minimum or typical . UTS and YS as Mpa, El as %. See Appendix for list of abbreviations used in Descriptions.

Grade UNS #	Country	Specification	Designation	Description	Al	Cu	Fe	Ni	P	Pb	Sb	Si	Sn	Zn	Other	UTS	YS	El
C85800	Norway	NS 16554	CuZn39Pb2Al	Die Cast	0.3-0.5	58-61	0.5-2	1 max		1.5-2.5		0.1 max	1 max		Mn 0.5 max			
C85800	Sweden	SIS 145253	5253-06	Cast As Cast	0.3-0.5	60-62	0.5	1		1-1.8		0.1	1		bal Zn; Mn 0.5	280	120	15
C85800	Sweden	SIS 145253	5253-10	Cast As Cast	0.3-0.5	58-61	0.5	1		1.5-2.5		0.1	1		bal Zn; Mn 0.5	280	120	5
C85800	USA			Leaded yellow brass	0.55 max	57.0 min	0.50 max		0.01 max	1.5 max	0.05 max	0.25 max	1.5 max	31.0-41.0	Cu may be Cu+Ni; Cu+named el 98.7 min; Ni+Co 0.50 max; As 0.05 max; S 0.05 max; Mn 0.25 max			
C99350	USA				9.5-10.5		1.0 max			0.15 max				7.5-9.5	Cu+named el 99.7 min; Ni+Co 14.5-16.0; Mn 0.25 max			
Manganese Bronze C86100	USA	MIL C-15345	C86100	Cast Sand Invest Centrifugal	4.5-5.5	66-68	2-4	1 max		0.2 max			0.2 max	16.1-25	Mn 2.5-5			
Manganese Bronze C86100	USA	MIL C-22087	C86100	Cast Sand Invest Centrifugal	4.5-5.5	66-68	2-4	1 max		0.2 max			0.2 max	16.1-25	Mn 2.5-5			
Manganese Bronze C86100	USA	MIL C-22229	C86100	Cast Sand Invest Centrifugal	4.5-5.5	66-68	2-4	1 max		0.2 max			0.2 max	16.1-25	Mn 2.5-5			
Manganese Bronze C86100	USA	QQ C-390	C86100	Cast Sand Invest Centrifugal	4.5-5.5	66-68	2-4	1 max		0.2 max			0.2 max	16.1-25	Mn 2.5-5			
Manganese Bronze C86100	USA	QQ C-523	C86100	Cast Sand Invest Centrifugal	4.5-5.5	66-68	2-4	1 max		0.2 max			0.2 max	16.1-25	Mn 2.5-5			
Manganese Bronze C86100	USA	SAE J462	C86100	Cast Sand Invest Centrifugal	4.5-5.5	66-68	2-4	1 max		0.2 max			0.2 max	16.1-25	Mn 2.5-5			
C86100	USA			Leaded Mn bronze (leaded high strength yellow brass)	4.5-5.5	66.0-68.0	2.0-4.0			0.20 max			0.20 max		Cu may be Cu+Ni; Cu+named el 99.0 min; bal Zn; Mn 2.5-5.0			
	France	NF A53-703	U-Z35-Y20	Cast As Cast	0.05	63-68	0.8			1-3			1.5 max		bal Zn; Mn 0.2	180		12
	South Africa	SABS 200	Cu-Zn34Pb2	Gravity Die Cast	0.5	62-68	0.5			1-2.5			1		bal Zn; OT 1 max	240		16
C97600	Canada	CSA HC.9	NZ207-SC	Cast		64 max		20 max		4 max			4 max	8 max				
C97600	USA			Nickel silver	0.005 max	63.0-67.0	1.5 max		0.05 max	3.0-5.0	0.25 max	0.15 max	3.5-4.5	3.0-9.0	Cu+named el 99.7 min; Ni+Co 19.0-21.5; S 0.08 max; Mn 1.0 max			
C97800	USA			Nickel silver	0.005 max	64.0-67.0	1.5 max		0.05 max	1.0-2.5	0.20 max	0.15 max	4.0-5.5	1.0-4.0	Cu+named el 99.6 min; Ni+Co 24.0-27.0; S 0.08 max; Mn 1.0 max			
	Sweden	SIS 145682	5682-02	Cast Tub Ann 3 mm diam		67	0.4-1	30-32		0.05				0.5	Sn in Pb content; C 0.1; S 0.05; Mn 0.5-1.5	380	150	30
	Sweden	SIS 145682	5682-12	Cast Tub Ann 5 mm diam		67	0.4-1	30-32		0.05				0.5	Sn in Pb content; C 0.1; S 0.05; Mn 0.5-1.5	370	140	30

UNS numbers and US grades are provided as a means of cross referencing chemically similar alloys. Exchangability is only possible after independent examination of specifications. Tensile properties are minimum or typical . UTS and YS as Mpa, El as %. See Appendix for list of abbreviations used in Descriptions.

Grade UNS #	Country	Specification	Designation	Description	Al	Cu	Fe	Ni	P	Pb	Sb	Si	Sn	Zn	Other	UTS	YS	El
C99700	USA				0.50-3.0	54.0 min	1.0 max	4.0-6.0		2.0 max				19.0-25.0	Cu+named el 99.7 min; S 1.0 max; Mn 11.0-15.0			
	Canada	CSA HC.9	NT265-SC	Cast		66.5 max		25 max		1.5 max			5 max	2 max				
C86100	Germany	DIN	2.0608/GB-CuZn25A15	Cast	4-6.5	60-66	2.5-3.5	2.7 max					0.05 max		bal Zn; Mn 3-5			
C86200	International	ISO 1338	CuZn26Al4Fe3Mn3	Sand Cast 12/25 mm diam	2.5-5	60-66	1.5-4	3		0.2		0.1	0.2		bal Zn; Mn 1.5-4	600	300	18
C86200	International	ISO 1338	CuZn26Al4Fe3Mn3	Centrifugally or continuously Cast 12/25 mm diam	2.5-5	60-66	1.5-4	3		0.2		0.1	0.2		bal Zn; Mn 1.5-4	600	300	18
C86200	Pan American	COPANT 801	C86200	Continuous Bar As Mfg	3-4.9	60-66	2-4	1		0.2			0.2	22-28	Mn 2.5-5	620	310	18
C86200	USA			Leaded Mn bronze (leaded high strength yellow brass)	3.0-4.9	60.0-66.0	2.0-4.00			0.20 max			0.20 max	22.0-28.0	Cu may be Cu+Ni; Cu+named el 99.0 min; Ni+Co 1.0 max; Mn 2.5-5.0			
C86300	Australia	AS 1565		Sand Cast	5-7.5	60-66	2-4	1		0.2			0.2	22-28	Mn 2.5-5	760	415	12
C86300	Czech Republic	CSN 423311	CuZn35Al5Fe3Mn2		4.5-6	64-68	2-4			1 max			1 max		Sb 0.1; bal Zn; Mn 1.5-2.5			
C86300	International	ISO 1338	CuZn25Al6Fe3Mn3	Sand Cast 12/25 mm diam	4.5-7	60-66	2-4	3		0.2		0.1	0.2		bal Zn; Mn 1.5-4	725	400	10
C86300	International	ISO 1338	CuZn25Al6Fe3Mn3	Centrifugally or continuously Cast 12/25 mm diam	4.5-7	60-66	2-4	3		0.2		0.1	0.2		bal Zn; Mn 1.5-4	740	400	10
C86300	Pan American	COPANT 801	C86300	Continuous Bar As Mfg	5-7.5	60-66	2-4	1		0.2			0.2	22-28	Mn 2.5-5	760	430	14
Manganese Bronze C86300	USA	AMS 4862	C86300	Cast Sand Invest Centrifugal Continuous	3-7.5	60-68	2-4	1 max		0.2 max			0.2 max		Mn 2.5-5			
Manganese Bronze C86300	USA	ASTM B22	C86300	Cast Sand Invest Centrifugal Continuous	3-7.5	60-68	2-4	1 max		0.2 max			0.2 max		Mn 2.5-5			
Manganese Bronze C86300	USA	ASTM B271	C86300	Cast Sand Invest Centrifugal Continuous	3-7.5	60-68	2-4	1 max		0.2 max			0.2 max		Mn 2.5-5			
Manganese Bronze C86300	USA	ASTM B30	C86300	Cast Sand Invest Centrifugal Continuous	3-7.5	60-68	2-4	1 max		0.2 max			0.2 max		Mn 2.5-5			
Manganese Bronze C86300	USA	ASTM B505	C86300	Cast Sand Invest Centrifugal Continuous	3-7.5	60-68	2-4	1 max		0.2 max			0.2 max		Mn 2.5-5			
Manganese Bronze C86300	USA	ASTM B584	C86300	Cast Sand Invest Centrifugal Continuous	3-7.5	60-68	2-4	1 max		0.2 max			0.2 max		Mn 2.5-5			
Manganese Bronze C86300	USA	MIL C-11866	C86300	Cast Sand Invest Centrifugal Continuous	3-7.5	60-68	2-4	1 max		0.2 max			0.2 max		Mn 2.5-5			

UNS numbers and US grades are provided as a means of cross referencing chemically similar alloys. Exchangability is only possible after independent examination of specifications. Tensile properties are minimum or typical . UTS and YS as Mpa, El as %. See Appendix for list of abbreviations used in Descriptions.

Grade UNS #	Country	Specification	Designation	Description	Al	Cu	Fe	Ni	P	Pb	Sb	Si	Sn	Zn	Other	UTS	YS	El
Manganese Bronze C86300	USA	MIL C-15345	C86300	Cast Sand Invest Centrifugal Continuous	3-7.5	60-68	2-4	1 max		0.2 max			0.2 max		Mn 2.5-5			
Manganese Bronze C86300	USA	MIL C-22087	C86300	Cast Sand Invest Centrifugal Continuous	3-7.5	60-68	2-4	1 max		0.2 max			0.2 max		Mn 2.5-5			
Manganese Bronze C86300	USA	MIL C-22229	C86300	Cast Sand Invest Centrifugal Continuous	3-7.5	60-68	2-4	1 max		0.2 max			0.2 max		Mn 2.5-5			
Manganese Bronze C86300	USA	QQ C-390	C86300	Cast Sand Invest Centrifugal Continuous	3-7.5	60-68	2-4	1 max		0.2 max			0.2 max		Mn 2.5-5			
Manganese Bronze C86300	USA	QQ C-523	C86300	Cast Sand Invest Centrifugal Continuous	3-7.5	60-68	2-4	1 max		0.2 max			0.2 max		Mn 2.5-5			
C86300	USA			Leaded Mn bronze (leaded high strength yellow brass)	5.0-7.5	60.0-66.0	2.0-4.0			0.20 max			0.20 max	22.0-28.0	Cu may be Cu+Ni; Cu+named el 99.0 min; Ni+Co 1.0 max; Mn 2.5-5.0			
	Czech Republic	CSN 423313	CuZn36Pb1		0.05 max	62-66	0.5 max		0.1 max	0.5-2			1 max		Cu+Zn 97.0 min; Sb+As 1.0; bal Zn; Mn 0.2 max			
	Denmark	DS 3001	5256	Sand Cast As cast	0.5-2.5	57-65	0.5-2	3		0.5		0.1	1		0.40 As+P+Sb; bal Zn; Mn 0.5-3	500	170	20
C85700	Canada	CSA HC.9	ZP361-SC	Cast	0.15 max	61 max				1 max			1 max	36.9 max				
C85700	Denmark	DS 3001	5241	Cast As cast	0.1-0.8	60-64	0.3 max	0.8		1-2		0.1	0.3		bal Zn; Mn 0.5	280	120	15
C85700	Germany	DIN 17656	2.0291/GB-CuZn33Pb	Cast Ingot	0.03 max	62.5-66	0.7 max	0.8 max		1.3-2.8		0.03 max	1.4 max		bal Zn; Mn 0.1 max			
Leaded Yellow Brass C85700	USA	ASTM B176	C85700	Cast Sand Centrifugal	0.55 max	58-64	0.7 max	1 max	0.02 max	1.5-3.8		0.05 max	0.5-1.5	32-40	S 0.05 max			
Leaded Yellow Brass C85700	USA	ASTM B271	C85700	Cast Sand Centrifugal	0.55 max	58-64	0.7 max	1 max	0.02 max	1.5-3.8		0.05 max	0.5-1.5	32-40	S 0.05 max			
Leaded Yellow Brass C85700	USA	ASTM B30	C85700	Cast Sand Centrifugal	0.55 max	58-64	0.7 max	1 max	0.02 max	1.5-3.8		0.05 max	0.5-1.5	32-40	S 0.05 max			
Leaded Yellow Brass C85700	USA	ASTM B584	C85700	Cast Sand Centrifugal	0.55 max	58-64	0.7 max	1 max	0.02 max	1.5-3.8		0.05 max	0.5-1.5	32-40	S 0.05 max			
Leaded Yellow Brass C85700	USA	MIL C-15345	C85700	Cast Sand Centrifugal	0.55 max	58-64	0.7 max	1 max	0.02 max	1.5-3.8		0.05 max	0.5-1.5	32-40	S 0.05 max			
Leaded Yellow Brass C85700	USA	QQ C-390	C85700	Cast Sand Centrifugal	0.55 max	58-64	0.7 max	1 max	0.02 max	1.5-3.8		0.05 max	0.5-1.5	32-40	S 0.05 max			
Leaded Yellow Brass C85700	USA	SAE J462	C85700	Cast Sand Centrifugal	0.55 max	58-64	0.7 max	1 max	0.02 max	1.5-3.8		0.05 max	0.5-1.5	32-40	S 0.05 max			

UNS numbers and US grades are provided as a means of cross referencing chemically similar alloys. Exchangability is only possible after independent examination of specifications. Tensile properties are minimum or typical . UTS and YS as Mpa, El as %. See Appendix for list of abbreviations used in Descriptions.

Worldwide Guide to Equivalent Nonferrous Metals and Alloys

Grade UNS #	Country	Specification	Designation	Description	Al	Cu	Fe	Ni	P	Pb	Sb	Si	Sn	Zn	Other	UTS	YS	El
C85700	USA			Leaded yellow brass	0.8 max	58.0-64.0	0.7 max			0.8-1.5		0.05 max	0.50-1.5	32.0-40.0	Cu may be Cu+Ni; Cu+named el 98.7 min; Ni+Co 1.0 max			
	France	NF A53-703	U-Z40-Y30	Cast As Cast	1 max	58-64	0.6			2.1 max			0.8		bal Zn	340		8
	South Africa	SABS 200	Cu-Zn36Sn1	Sand Cast	0.01	60-64				0.5			1-1.5		bal ZnOT 0.74 max	245		20
C28000	South Africa	SABS 200	Cu-Zn40	Gravity Die Cast	0.5	59-63				0.25		0.05			bal Zn; OT 0.2 max; Mn 0.5	280		25
C85500	USA			Leaded yellow brass		59.0-63.0	0.20 max			0.20 max			0.20 max		Cu may be Cu+Ni; Cu+named el 98.9 min; bal Zn; Ni+Co 0.20 max; Mn 0.20 max			
C85710	Australia	AS 1565	857B	Chill Cast	0.2-0.8	58-63	0.8 max	1		1-2.5		0.05	1		bal Zn; Mn 0.5	300	90	13
C85710	Denmark	DS 3001	5253	Cast As cast	0.3-0.5	58-61	0.5 max	1		1.5-2.5		0.1	1 max		bal Zn; Mn 0.5	280	120	5
C85710	Finland	SFS 2204	CuZn40Pb	Die Cast As cast	0.3-0.5	58-61	0.5 max	1		1.5-2.5		0.1	1 max		bal Zn; Mn 0.5	280	120	15
C85710	International	ISO 1338	CuZn40Pb	Sand Cast 12/25 mm diam	0.2-0.8	58-63	0.8	1		0.5-2.5		0.05	1		Cu total includes Ni; bal Zn; Mn 0.5	220		15
C85710	International	ISO 1338	CuZn40Pb	Pressure die cast Cast or permanent mould 12/25 mm diam	0.2-0.8	58-63	0.8	1		0.5-2.5		0.05	1		Cu total includes Ni; bal Zn; Mn 0.5	280	120	15
	South Africa	SABS 200	Cu-Zn40	Gravity Die Cast	0.2-0.8	58-63	0.5	1		0.5-2.5			1		bal Zn; OT 0.5 max	295		15
C98800	USA			Leaded copper; Pb and Ag adj for hardness		56.5-62.5	0.35 max		0.02 max	37.5-42.5			0.25 max	0.10 max	Cu Includes Ag; Ag 5.5 max			
C86400	Finland	SFS 2205	CuZn35AlFeMn	Sand Cast As cast	0.5-2.5	57-65	0.5-2	3		0.5		0.1	1		0.40 As+P+Sb; bal Zn; Mn 0.5-3	500	170	20
Manganese Bronze C86400	USA	ASTM B271	C86400	Cast Sand Centrifugal	1.5 max	56-62	2 max	1 max		0.5-1.5			1.5 max		Mn 1.5 max			
Manganese Bronze C86400	USA	ASTM B30	C86400	Cast Sand Centrifugal	1.5 max	56-62	2 max	1 max		0.5-1.5			1.5 max		Mn 1.5 max			
Manganese Bronze C86400	USA	ASTM B584	C86400	Cast Sand Centrifugal	1.5 max	56-62	2 max	1 max		0.5-1.5			1.5 max		Mn 1.5 max			
Manganese Bronze C86400	USA	QQ C-390	C86400	Cast Sand Centrifugal	1.5 max	56-62	2 max	1 max		0.5-1.5			1.5 max		Mn 1.5 max			
Manganese Bronze C86400	USA	QQ C-523	C86400	Cast Sand Centrifugal	1.5 max	56-62	2 max	1 max		0.5-1.5			1.5 max		Mn 1.5 max			
C86400	USA			Leaded Mn bronze (leaded high strength yellow brass)	0.50-1.5	56.0-62.0	0.40-2.0			0.50-1.5			0.50-1.5	34.0-42.0	Cu may be Cu+Ni; Cu+named el 99.0 min; Ni+Co 1.0 max; Mn 0.10-1.0			
C98800	Japan	JIS H5403	KJ1	Cast						38-42			1 max		Ag 2 max			
	Czech Republic	CSN 423321	CuZn38Al		0.05-0.2	58-62	0.6 max		0.1 max	1.5 max		0.3 max	0.5 max		Cu+Zn:min 98.0; Sb 0.1; bal Zn; As 0.1 x; Mn 0.2 max			

UNS numbers and US grades are provided as a means of cross referencing chemically similar alloys. Exchangability is only possible after independent examination of specifications. Tensile properties are minimum or typical . UTS and YS as Mpa, El as %. See Appendix for list of abbreviations used in Descriptions.

5-36 Cast Copper

Grade UNS #	Country	Specification	Designation	Description	Al	Cu	Fe	Ni	P	Pb	Sb	Si	Sn	Zn	Other	UTS	YS	El
C97400	USA			Nickel silver		58.0-61.0	1.5 max			4.5-5.5			2.5-3.5		Cu+named el 99.0 min; Ni+Co 15.5-14.0; bal Zn; Mn 0.50 max			
C99750	USA				0.25-3.0	55.0-61.0	1.0 max	5.0 max						17.0-23.0	Cu+named el 99.7 min; S 0.50-2.5; Mn 17.0-23.0			
	Czech Republic	CSN 423320	CuZn38Fe1Al1Mn1		0.7-1.5	58-61	0.8-1.5		0.1 max	0.4 max			0.2-0.7		Sb 0.1; bal Zn; Mn 0.1-0.6			
	Sweden	SIS 145252	5252-10	Cast As Cast	0.1	58-61	0.5	1		1.5-2.5		0.1-0.3	1		bal Zn; Mn 0.5	280	120	5
C86500	Australia	AS 1565	865C	Sand Cast	0.5-1.5	55-60	0.4-2	1		0.5		0.1	1 max	36-42	Mn 0.1-1.5	450	170	20
C86500	Canada	CSA HC.9	ZF391-SC	Cast	1.25 max	58 max	1.5 max							39.3 max	Mn 0.25 max			
C86500	Germany	DIN 17656	2.0602/GB-CuZn35Al1	Cast Ingot	0.7-1.7	56-64	0.5-1.5	2.7 max		0.7 max			0.6 max		bal Zn; Mn 0.5-2.5			
C86500	International	ISO 1338	CuZn35AlFeMn	Pressure die cast Cast or permanent mould 12/25 mm diam	0.5-2.5	57-65	0.5-2	3		0.5		0.1	1	25.5	0.40 Sb + P + As; Mn 0.1-3	475	200	18
C86500	International	ISO 1338	CuZn35AlFeMn	Continously or centrifugally cast 12/25 mm diamCast	0.5-2.5	57-65	0.5-2	3		0.5		0.1	1	25.5	0.40 Sb + P + As; Mn 0.1-3	475	200	18
C86500	International	ISO 1338	CuZn35AlFeMn	Sand Cast 12/25 mm diam	0.5-2.5	57-65	0.5-2	3		0.5		0.1	1	25.5	0.40 Sb + P + As; Mn 0.1-3	450	170	20
C86500	Japan	JIS H2205	HBsCIn1	Cast	0.5-1.5	55-60	0.5-1.5	1 max		0.4 max		0.1 max	0.5-1.5		bal Zn; Mn 1.5 max			
C86500	Norway	NS 16565	CuZn35AlFeMn	Cast Ingot	0.5-2.5	57-65	0.5-2	3 max		0.5 max		0.1 max	1 max		ST 0.4 max; ST=As+Pb+Sb; Mn 0.5-3			
C86500	Norway	NS 16565	CuZn35AlFeMn	Cast Die Sand Continuous	0.5-2.5	57-65	0.5-2	3 max	0.5 max			0.1 max	1 max		ST 0.4 max; ST=As+P+Sb; Mn 0.5-3			
C86500	Pan American	COPANT 801	C86500	Continuous Bar As Mfg	0.5-1.5	55-60	0.4-2	1		0.4			1	36-42	Mn 0.1-1.5	480	180	25
C86500	Sweden	SIS 145256	5256-03	Sand Cast As Cast	0.5-2.5	57-65	0.5	1		1.5-2.5		0.1	1		bal Zn; Mn 0.5	500	170	20
C86500	Sweden	SIS 145256	5256-06	Cast As Cast	0.5-2.5	57-65	0.5	1		1.5-2.5		0.1	1		bal Zn; Mn 0.5	500	200	18
C86500	Sweden	SIS 145256	5256-10	Cast As Cast	0.5-2.5	57-65	0.5	1		1.5-2.5		0.1	1		bal Zn; Mn 0.5	500	200	18
C86500	Sweden	SIS 145256	5256-15	Cast As Cast	0.5-2.5	57-65	0.5	1		1.5-2.5		0.1	1		bal Zn; Mn 0.5	500	200	18
Manganese Bronze C86500	USA	AMS 4860A	C86500	Cast Sand Invest Centrifugal	0.5-1.5	55-60	0.4-2	1 max		0.4 max			1 max		Mn 1.5 max			
Manganese Bronze C86500	USA	ASTM B271	C86500	Cast Sand Invest Centrifugal	0.5-1.5	55-60	0.4-2	1 max		0.4 max			1 max		Mn 1.5 max			
Manganese Bronze C86500	USA	ASTM B30	C86500	Cast Sand Invest Centrifugal	0.5-1.5	55-60	0.4-2	1 max		0.4 max			1 max		Mn 1.5 max			
Manganese Bronze C86500	USA	ASTM B584	C86500	Cast Sand Invest Centrifugal	0.5-1.5	55-60	0.4-2	1 max		0.4 max			1 max		Mn 1.5 max			
Manganese Bronze C86500	USA	MIL C-15245	C86500	Cast Sand Invest Centrifugal	0.5-1.5	55-60	0.4-2	1 max		0.4 max			1 max		Mn 1.5 max			
Manganese Bronze C86500	USA	MIL C-22087	C86500	Cast Sand Invest Centrifugal	0.5-1.5	55-60	0.4-2	1 max		0.4 max			1 max		Mn 1.5 max			
Manganese Bronze C86500	USA	QQ C-390	C86500	Cast Sand Invest Centrifugal	0.5-1.5	55-60	0.4-2	1 max		0.4 max			1 max		Mn 1.5 max			

UNS numbers and US grades are provided as a means of cross referencing chemically similar alloys. Exchangability is only possible after independent examination of specifications. Tensile properties are minimum or typical . UTS and YS as Mpa, El as %. See Appendix for list of abbreviations used in Descriptions.

Worldwide Guide to Equivalent Nonferrous Metals and Alloys

Grade UNS #	Country	Specification	Designation	Description	Al	Cu	Fe	Ni	P	Pb	Sb	Si	Sn	Zn	Other	UTS	YS	El
Manganese Bronze C86500	USA	MIL C22229	C86500	Cast Sand Invest Centrifugal	0.5-1.5	55-60	0.4-2	1 max		0.4 max			1 max		Mn 1.5 max			
Manganese Bronze C86500	USA	SAE J462	C86500	Cast Sand Invest Centrifugal	0.5-1.5	55-60	0.4-2	1 max		0.4 max			1 max		Mn 1.5 max			
C86500	USA			Leaded Mn bronze (leaded high strength yellow brass)	0.50-1.5	55.0-60.0	0.40-2.0			0.40 max			1.0 max	36.0-42.0	Cu may be Cu+Ni; Cu+named el 99.0 min; Ni+Co 1.0 max; Mn 1.0-1.5			
C86700	Canada	CSA HC.9	ZP392-SC	Cast	1.5 max	57 max	1.5 max			0.5 max				37.5 max	Mn 2 max			
C86700	Czech Republic	CSN 423322	CuZn31MnAl1		0.5-2	57-60	1 max	1 max		1.5 max			1.5 max		Cu+Mn+Al+Zn+Ni:min 98.0; bal Zn; Mn 0.5-2			
C86700	Japan	JIS H2205	HBsCln2	Cast	0.5-2	55-60	0.5-2	1 max		0.4 max		0.1 max	1 max		bal Zn; Mn 3.5 max			
C86700	USA			Leaded Mn bronze (leaded high strength yellow brass)	1.0-3.0	55.0-60.0	1.0-3.0			0.50-1.5			1.5 max	30.0-38.0	Cu may be Cu+Ni; Cu+named el 99.0 min; Ni+Co 1.0 max; Mn 1.0-3.5			
C98820	USA			Leaded copper			0.35 max			40.0-44.0			1.0-5.0					
C99600	USA		Incramute 1		1.0-2.8		0.20 max	0.20 max		0.02 max		0.10 max		0.20 max	Cu+named el 99.7 min; Co 0.20 max; C 0.05 max; S 0.10 max; Mn 39.0-45.0			
	Japan	JIS H5102	HBSC3	As Cast	3-5	60-60	2-4		0.2			0.1	0.5		bal Zn; Mn 2.5-5	637		15
	Japan	JIS H5102	HBSC4	As Cast	5-7.5	60-60	2-4		0.2			0.1	0.2		bal Zn; Mn 2.5-5	755		12
C97300	Canada	CSA HC.9	ZN2012-SC	Cast		57 max		12 max		9 max			2 max	20 max				
C97300	USA			Nickel silver	0.005 max	53.0-58.0	1.5 max		0.05 max	8.0-11.0	0.35 max	0.15 max	1.5-3.0	17.0-25.0	Cu+named el 99.0 min; Ni+Co 11.0-14.0; S 0.08 max; Mn 0.50 max			
C86800	Germany		2.0606/GB-CuZn34Al2		1.5-2.5	55-65	2.5-3.5	2.7 max					0.05 max		bal Zn; Mn 3-5			
C86800	USA			Leaded Mn bronze (leaded high strength yellow brass)	2.0 max	53.5-57.0	1.0-2.5			0.20 max			1.0 max		Cu may be Cu+Ni; Cu+named el 99.0 min; bal Zn; Ni+Co 2.5-4.0; Mn 2.5-4.0			
C98840	USA			Leaded copper			0.35 max			44.0-58.0			1.0-5.0					
	Japan	JIS H5102	HBSC1	As Cast	0.5-1.5	55	0.5-1.5	1		0.4		0.1	1 min		bal Zn; Mn 1.5	431		20
	Japan	JIS H5102	HBSC2	As Cast	0.5-1.5	55	0.5-1.5	1		0.4		0.1	1		bal Zn; Mn 1.5	431		20
	South Africa	SABS 200	Cu-Zn37Fe2(AlMnNi)	Sand Cast	5	55	0.5-2.5	2		0.5		0.1	0.5		bal Zn; OT 0.1 max; Mn 3	585	280	14
	South Africa	SABS 200	Cu-Zn39Fe1(AlMn)	Sand Cast	0.5-2.5	55	0.7-2	1		0.5		0.1	1		bal Zn; OT 0.1 max; Mn 3	470	170	18
	South Africa	SABS 200	Cu-Zn39Fe1 (AlMn)	Gravity Die Cast	0.5-2.5	55	0.7-2	1		0.5		0.1	1		bal Zn; OT 0.1 max; Mn 3	500	210	18

UNS numbers and US grades are provided as a means of cross referencing chemically similar alloys. Exchangability is only possible after independent examination of specifications. Tensile properties are minimum or typical . UTS and YS as Mpa, El as %. See Appendix for list of abbreviations used in Descriptions.

5-38 Cast Copper

Grade UNS #	Country	Specification	Designation	Description	Al	Cu	Fe	Ni	P	Pb	Sb	Si	Sn	Zn	Other	UTS	YS	El
	South Africa	SABS 200	CuZn35Al5Fe2(Mn)(3F)	Sand cast	3-6	55	1.5-3.25	1 max		0.2 max			0.2 max		bal Zn; Mn 4 max	740	400	11
	Czech Republic	CSN 423326	CuZn45Mn4Pb3Fe1		0.25 max	47-51	0.5-1.2		0.1 max	2-4		0.2 max	0.7 max		Bi 0.01; Sb 0.1; bal Zn; As 0.05 max; Mn 2.5-4.5			
	Czech Republic	CSN 423356	CuZn39Ni14Fe2Mn1		1.5 max	42-48	1-3	13-15	0.1 max	0.5 max			0.5 max		Cu+Ni+Fe+Zn+Mn 97.5 min; Sb 0.2; bal Zn; Mn 1-2			

UNS numbers and US grades are provided as a means of cross referencing chemically similar alloys. Exchangability is only possible after independent examination of specifications. Tensile properties are minimum or typical . UTS and YS as Mpa, El as %. See Appendix for list of abbreviations used in Descriptions.

Grade UNS #	Country	Specification	Designation	Description	Ag	Al	As	Bi	Cu	Fe	Pb	Sb	Sn	Zn	Other	UTS	YS	EL
L50001	USA			Zone refined lead							99.9999 min							
L50005	USA			Refined soft lead							99.999 min							
	UK	BS 334	Type B		0.002 max				0.01 max	0.003 max					ST Cd As & S			
	UK	BS 334	Type B3		0.005 max				0.01 max	0.005 max					ST As Cd & S			
L50006	USA	ASTM B29	Low Bi Low Ag Pure Lead	Refined	0.0010 max		0.0005 max	0.0015 max	0.0010 max	0.0002 max	99.995 min	0.0005 max	0.0005 max	0.0005 max	Ni 0.0002 max; Te 0.0001 max			
L50010	Australia	AS 1812	9999	Ing	0.001 max				0.01 max	0.001 max	99.99 min				0.001 max ST, ST=Ni+Co			
L50010	USA			Refined soft lead							99.99 min							
	France	NF A55-105	99.985	Ing	0.001 max				0.01 max	0.001 max	99.99 min				0.30 ST ST=As+Sb+Sn 0.0005 max			
	Germany	DIN 1719	2.3010/Pb99.99		0.001 max				0.01 max	0.001 max	99.99 min							
	Germany	DIN 1719	2.3020/Pb99.985		0.001 max				0.01 max	0.001 max	99.99 min							
	UK	BS 334	Type A		0.002 max				0.01 max	0.003 max	99.99 min				ST=trace elements As Cd & S			
	UK	BS 3909	3	Ing					0.01 max	0.003 max	99.99 min				0.001 ST ST=Ni+Co			
L50020	USA			Refined soft lead							99.985 min							
L50021	USA	ASTM B29	Refined Pure Lead	Refined	0.0025 max		0.0005 max	0.025 max	0.0010 max	0.001 max	99.97 min	0.0005 max	0.0005 max	0.0005 max	Ni 0.0002 max; Te 0.0001 max			
LME L50025	USA			Pure lead							99.97 min							
	France	NF A55-401	99.97						0.01 max		99.97 min				0.00 ST ST= As+Sb+Sn - .0005max			
	France	NF A55-105	99.97	Ing	0.001 max				0.03 max	0.001 max	99.97 min				0.30 ST ST=Cd 3ppm Sb 3ppm Sn 3ppm As 1ppm max; As=Sb=Sn 5ppm max			
	France	NF A55-105	99.97	Ing	0.005 max				0.03 max	0.005 max	99.97 min				0.30ST ST=Sb 25ppm Sn 3ppm As 1ppm max			
	Pan America	COPANT 447	PbSb9X		0.01		0.01 max	0.04 max	0.02	0.01			0.02 max	0.001				
L50035	Australia	AS 1812	9995	Ing	0.002 max				0.05 max	0.002 max	99.95 min				0.002 max ST, ST=Ni+Co			
L50035	USA			Refined soft lead							99.95 min							
	Czech Republic	CSN 423701	Pb99,95	Corrosing Lead	0.0015 max				0.03 max	0.0015 max	99.95 min	0.01 max						
	UK	BS 334	Type B1		0.002 max				0.002 max	0.05-0.07					ST=trace elements As Cd & S			
	UK	BS 334	Type B2		0.002 max				0.01 max	0.05-0.07					ST=trace elements As Cd & S			
L50040	USA			Refined soft lead							99.94 min							
L50042	USA			Corroding lead	0.0015 max				0.050 max	0.0015 max	0.002 max	99.94 min			0.001 max	Cu+Ag 0.0025 max; As+Sb+Sn 0.002 max			
L50045	USA			Common lead	0.005 max				0.050 max	0.0015 max	0.002 max	99.94 min			0.001 max	As+Sb+Sn 0.002 max			
L50049	USA	ASTM B29	Pure Lead	Refined	0.005 max		0.001 max	0.05 max	0.0015 max	0.001 max	99.94 min	0.001 max	0.001 max	0.001 max	Ni 0.001 max; Sb+As+Sn 0.002 max			

UNS numbers and US grades are provided as a means of cross referencing chemically similar alloys. Exchangability is only possible after independent examination of specifications. Tensile properties are minimum or typical . UTS and YS as Mpa, El as %. See Appendix for list of abbreviations used in Descriptions.

Worldwide Guide to Equivalent Nonferrous Metals and Alloys

Grade UNS #	Country	Specification	Designation	Description	Ag	Al	As	Bi	Cu	Fe	Pb	Sb	Sn	Zn	Other	UTS	YS	EL
L51125	USA			Copperized soft lead					0.06		99.9 min							
	Germany	DIN 1719	2.3030/Pb99.94		0.001 max			0.05 max	0.001 max		99.94 min							
L50050	Australia	AS 1812	9990	Ing	0.002 max			0.03 max	0.05-0.07		99.9 min				0.002 max ST, ST=Ni+Co			
Grade A USA L50050	USA										99.90 min				OT 0.10 max			
L50510	USA										99.9				Ba 0.05			
L50710	USA										99.9				Ca 0.008			
L50712	USA			Cable sheath							99.9				Ca 0.025			
L50720	USA										99.9				Ca 0.03			
L50722	USA								0.06		99.9				Ca 0.03			
L50725	USA			Cable sheath					0.06		99.9				Ca 0.035			
L50735	USA			Battery grid							99.9				Ca 0.06			
L50760	USA			Battery grid							99.9				Ca 0.07			
L50770	USA			Battery grid							99.9				Ca 0.10			
L51110	USA			Copperized lead					0.05		99.9							
L51120	USA			Chemical lead	0.002-0.02			0.005 max	0.04-0.08	0.002 max	99.90 min			0.001 max	As+Sb+Sn 0.002 max			
L51121	USA			Copper bearing	0.020 max			0.025 max	0.04-0.08	0.002 max	99.90 min			0.001 max	As+Sb+Sn 0.002 max			
L51121	USA	ASTM B29	Chemical Copper Lead	Refined	0.020 max		0.001 max	0.025 max	0.040-0.080	0.002 max	99.90 min	0.001 max	0.001 max	0.001 max	Ni 0.002 max; Sb+As+Sn 0.002 max			
L51705	USA										99.9				Li 0.01			
L51708	USA										99.9				Li 0.02			
L51710	USA										99.9				Li 0.03			
L51720	USA										99.9				Li 0.06			
L52505	USA										99.9	0.1						
	Czech Republic	CSN 423702	Pb99,9		0.002 max	0.01 max	0.03 max	0.002 max	0.01 max	0.01 max	99.9 min	0.01 max		0.01 max				
	France	NF A55-105	99.9	Ing	0.005 max			0.09 max	0.005 max		99.9 min				0.30 ST=Sb 25ppm Sn 3ppm As 1ppm			
	France	NF A55-401	99.9	Ing	0.0001 max			0.09 max	0.001 max		99.9 min				0.30 ST ST=Sn 0.0003 max As 0.0001 max Sb 0.0025 max			
	Germany	DIN 1719	2.3021/Pb99.9Cu					0.01 max	0.04-0.08		99.9 min		0.001 max		0.001ST, ST=0.0025Ag max			
	Japan	JIS H4311	PbT1	Pip							99.9 min							
	Japan	JIS H4311	PbT1	Ext							99.9 min							
	Japan	JIS H4312	PbTW1	Pip							99.9 min				0.10 max ST ST=Sb+Sn+Cu+Ag+As+Zn+Fe+Bi total impurity limit			
	Japan	JIS H4312	PbTW2	Pip								0.1-0.3			ST=Sn+Cu+Ag+Zn+Fe+Bi Total impurity limit			
Type III USA L50060	USA										99.85 min							

UNS numbers and US grades are provided as a means of cross referencing chemically similar alloys. Exchangability is only possible after independent examination of specifications. Tensile properties are minimum or typical . UTS and YS as Mpa, El as %. See Appendix for list of abbreviations used in Descriptions.

Grade UNS #	Country	Specification	Designation	Description	Ag	Al	As	Bi	Cu	Fe	Pb	Sb	Sn	Zn	Other	UTS	YS	EL
GradeD L51123	USA				0.020 max			0.025 max	0.04-0.08	0.002 max	99.85 min			0.001 max	As+Sb+Sn 0.002 max; Te 0.035-0.060			
Grade D L51124	USA				0.020 max			0.025 max	0.04-0.08	0.002 max	99.82 min		0.016 max	0.001 max	As+Sb 0.002 max; Te 0.035-0.055			
L50101	USA			Cabel sheathing	0.2						99.8							
L52510	USA										99.8		0.1					
L52515	USA			Cable sheath			0.015				99.8	0.2						
	Japan	JIS H4312	PbTW3	Pip								0.15-0.35	0.05-0.15		ST=Cu+Ag+As+ Zn+Fe+Bi total impurity limit			
	Germany	DIN 1719	2.3075/Pb99.75						0.1 max		99.75 min			0.05 max				
Grade AA,C L50065	USA										99.7 min	0.02 max						
L50713	USA			Cable sheath							99.7		0.3		Ca 0.025			
L50736	USA			Battery grid							99.7		0.2		Ca 0.065			
L52525	USA			Cable sheath			0.035				99.7	0.3			Te 0.035			
Alloy 1/2C L54030	USA			Cable sheath							99.7		0.2		Cd 0.075			
L50310	USA			Arsenical lead cable sheath			0.16	0.10			99.8		0.10					
L50775	USA			Battery grid							99.6		0.3		Ca 0.10			
L50795	USA			Battery grid							99.6		0.3		Ca 0.12			
L51740	USA										99.6		0.35		Li 0.02			
L51780	USA										99.6		0.1		Li 0.15; Ca 0.15			
L52530	USA										99.6	0.35						
L52535	USA			Cable sheath			0.03				99.6	0.4						
L55210	USA			Battery alloy							99.6		0.3		Ca 0.06; Sr 0.06			
	Czech Republic	CSN 423433	PbSb0,5		0.004 max		0.01 max	0.06 max	0.005 max	0.01 max		0.4-0.6	0.01 max	0.01 max				
L50110	USA			Electrowinning anode	0.5						99.5							
L50728	USA			Battery grid							99.5		0.5		Ca 0.04			
L50800	USA			Battery grid							99.5		0.3		Ca 0.21			
	France	NF A55-105	99.5	Ing							99.5 min							
	Japan	JIS H4311	PbT2	Pip							99.5 min							
	Japan	JIS H4311	PbT3	Pip							99.5 min							
	UK	BS 3909	1	Ing					0.05 max	0.07	99.5 min	0.1 max	0.1 max	0.01 max				
L50730	USA			Electrowinning anode	0.5						99.4				Ca 0.06			
L50737	USA			Battery grid							99.4		0.5		Ca 0.065			
L50780	USA			Battery grid							99.4		0.5		Ca 0.10			
L52520	USA			Cable sheath							99.4	0.2	0.4					
L52545	USA										99.4	0.6						
Alloy C L54050	USA			Cable sheath							99.4		0.4		Cd 0.15			

UNS numbers and US grades are provided as a means of cross referencing chemically similar alloys. Exchangability is only possible after independent examination of specifications. Tensile properties are minimum or typical . UTS and YS as Mpa, El as %. See Appendix for list of abbreviations used in Descriptions.

Worldwide Guide to Equivalent Nonferrous Metals and Alloys

Grade UNS #	Country	Specification	Designation	Description	Ag	Al	As	Bi	Cu	Fe	Pb	Sb	Sn	Zn	Other	UTS	YS	EL
L51730	Czech Republic USA	CSN 423731	PbSb0,9		0.004 max		0.01 max	0.04 max	0.02 max	0.01 max	99.3	0.6-1.2	0.01 max	0.01 max	Li 0.7			
L52550	USA			Cable sheath					0.04		99.3	0.6			Te0.04			
Alloy B L52555	USA										99.3	0.7						
L50115	Pan America USA	COPANT 448	CabPbSb0.75	Electrowinning anode	0.001 max 0.75		0.001	0.05	0.001 max	0.001	99.25	0.65-0.85	0.01	0.001				
	Germany	DIN 17640	2.3201/R-Pb				0.02-0.05					0.75-1.25						
	Pan America	COPANT 447	PbSb(As)				0.02-0.05					0.75-1.25						
	Pan America	COPANT 447	PbSb(As)				0.02-0.05					0.75-1.25						
L50740	USA			Battery grid							99.2		0.7		Ca 0.065			
L50765	USA			Battery grid							99.2		0.7		Ca 0.07			
L51748	USA										99.2		0.7		Li 0.04			
L52540	USA			Cable sheath							99.2	0.5			Cd 0.25			
L52560	USA			Bullet alloy							99.2	0.75						
L52565	USA			Overhead cable							99.2	0.75						
Alloy B L52570	USA			Cable sheath					0.06		99.1	0.85						
L50120	USA			Electrowinning anode	1.0						99.0							
L50520	USA										99.0		1.0		Ba 0.05			
L50840	USA										99.0				Ca 1.0			
L52595	USA			Cable sheath							99.0	0.95						
L52605	Pan America	COPANT 447	CabPbSb1		0.001 max		0.001	0.05	0.001 max	0.001		0.9-1.1	0.01	0.001				
L52605	South Africa	SABS 250	LEADALLOY B	Bar	0.005			0.05	0.06		98	0.85-0.95	0.01	0.001	OT 0.1 max			
L52605	USA										99.0	1.0						
L55230	USA			Battery alloy		0.03					99		0.8		Sr 0.16			
L50530	USA										98.9		1.0		Ba 0.10			
L50745	USA			Battery grid							98.9		1.0		Ca 0.065			
L50790	USA			Battery grid							98.9		1.0		Ca 0.10			
Frary metal L50540	USA										98.8				Ba 0.4; Ca 0.8			
L51770	USA										98.8		1.0		Li 0.08; Ca 0.03			
L51775	USA										98.8		1.0		Li 0.12; Ca 0.03			
L50810	USA			Bearing metal		0.02					98.7				Li 0.04; Ca 0.7; Na 0.6			
L50750	USA			Battery grid							98.6		1.3		Ca 0.065			
L52615	USA			Die cast			0.1				98.6	1.0	0.3		S 0.003			
	Germany	DIN 17640	2.3202/Pb(Sb)				1.2-1.7					0.2-0.3						
L50521	USA										98.5		1.5		Ba 0.05			
L50541	USA			Frary metal							98.5				Ba 1.0; Ca 0.5			

UNS numbers and US grades are provided as a means of cross referencing chemically similar alloys. Exchangability is only possible after independent examination of specifications. Tensile properties are minimum or typical . UTS and YS as Mpa, El as %. See Appendix for list of abbreviations used in Descriptions.

Grade UNS #	Country	Specification	Designation	Description	Ag	Al	As	Bi	Cu	Fe	Pb	Sb	Sn	Zn	Other	UTS	YS	EL
	Germany	DIN 1719	2.3085/Pb98.5						0.5 max		98.5 min			0.05 max				
	USA	ASTM B32	Sn2	Solder	0.015 max	0.005 max	0.02 max	0.25 max	0.08 max	0.02 max		0.50 max	1.5-2.5	0.005 max	Cd 0.001 max			
L50755	USA			Battery grid							98.4		1.5		Ca 0.065			
L51790	USA										98.3		1.0		Li 0.65; Ca 0.02			
L50121	USA			Solder	1.0						98		1.0					
L50122	USA			Electrowinning anode	1.0		1.0				98							
L50140	USA			Cathodic protection anode	2.0						98.0							
L50522	USA										98.0		2.0		Ba 0.05			
Frary metal L50542	USA										98.0				Ba 1.2; Ca 0.8			
L50850	USA										98				Ca 2.0			
L52618	USA										98.0	1.2			Ga 0.8			
L52625	USA			Shot aloy			0.45				98.0	1.55	0.0005 max					
L52630	USA			Battery			0.3				98	1.6	0.1		Se 0.02			
L52705	USA										98.0	2.0						
L54210	Germany	DIN 1707	2.3402/L-PbSn2	DIN 1707 Obsolete Replaced by DIN EN 29453									1.5-2.5		OT 0.08 max			
L54210	Japan	JIS Z3282	H2A	Bar		0.01			0.05	0.03	97.02	0.3	1.5-2.5	0.01				
L54210	Pan America	COPANT 450	2A			0.01 max	0.02 max	0.25 max	0.08 max	0.02 max	98 min	0.12 max	2 max	0.01 max				
L54210	Pan America	COPANT 450	5B			0.01 max	0.02 max	0.25 max	0.08 max	0.02 max	95 min	0.2-0.5	5 max	0.01 max				
L54210	USA			Solder	0.005 max		0.02 max	0.25 max	0.08 max	0.02 max	98	0.12 max	1.5-2.5	0.005 max				
L54211	Pan America	COPANT 450	2B			0.01 max	0.02 max	0.25 max	0.08 max	0.02 max	98 min	0.2-0.5	2 max	0.01 max				
L54211	USA			Solder	0.005 max		0.02 max	0.25 max	0.08 max	0.02 max	98	0.20-0.50	1.5-2.5	0.005 max				
L55290	USA										98				Sr 2			
	USA	EIA/IPC J-STD-006	Sn02A	Electronic Grade Solder							98.0		2.0					
	USA	ASTM B32	Ag1.5	Solder	1.3-1.7	0.005 max	0.02 max	0.25 max	0.30 max	0.02 max		0.40 max	0.75-1.25	0.005 max	Cd 0.001 max			
L50535	USA										97.9		2.0		Ba 0.10			
L51778	USA										97.8		2.0		Li 0.15; Ca 0.04			
	Czech Republic	CSN 055630	B-PbAg2Cu-325/300	Soft Solder	2-3		0.05 max	0.05 max	0.2-0.3	0.01 max		0.1 max		0.01 max				
Ag2.5 L50151	USA			Solder	2.3-2.7	0.005 max	0.02 max	0.25 max	0.30 max	0.02 max		0.40 max	0.25	0.005 max	Cd 0.001 max; OT 0.03 max			
L51050	Pan America	COPANT 450	2.5S		2.3-2.7	0.01 max	0.02 max	0.25 max	0.08 max	0.02 max	97.5 min	0.4 max		0.01 max				
Ag1.5 L50132	USA			Solder	1.3-1.7	0.005 max	0.02 max	0.25 max	0.30 max	0.02 max		0.40 max	0.75-1.25	0.005 max	Cd 0.001 max; OT 0.08 max			
L50131	Australia	AS 1834	1Sn/1.5Ag		1.3-1.7		0.03 max	0.1	0.05 max	0.02 max		0.2 max	0.8-1.2		OT 0.08 max			
L50131	Pan America	COPANT 450	1.5S		1.3-1.7	0.01 max	0.02 max	0.25 max	0.08 max	0.02 max	97.5 min	0.4 max	1 max	0.01 max				
L50131	USA		1.5S	Solder	1.3-1.7	0.005 max	0.02 max	0.25 max	0.08 max	0.02 max	97.5		1.25	0.005 max				

UNS numbers and US grades are provided as a means of cross referencing chemically similar alloys. Exchangability is only possible after independent examination of specifications. Tensile properties are minimum or typical . UTS and YS as Mpa, El as %. See Appendix for list of abbreviations used in Descriptions.

Worldwide Guide to Equivalent Nonferrous Metals and Alloys

Grade UNS #	Country	Specification	Designation	Description	Ag	Al	As	Bi	Cu	Fe	Pb	Sb	Sn	Zn	Other	UTS	YS	EL
L50150	Germany	DIN 1707	2.3403/L-PbAg3	DIN 1707 Obsolete Replaced by DIN EN 29454	2.1-3										OT 0.2 max			
2.5S L50150	USA			Solder	2.3-2.7	0.005 max	0.02 max	0.25 max	0.08 max	0.02 max	97.5		0.25 max	0.005 max				
L52710	USA			Battery			0.15				97.5	2.0	0.3		S 0.003			
L52725	USA			Bullet							97.5	2.5						
	Czech Republic	CSN 423729	PbSb3		0.03 max		0.01 max	0.05 max	0.025 max	0.01 max		2.5-3.5	0.01 max					
	Germany	DIN 1707	2.3412/L-PbAg25Sn2	DIN 1707 Obsolete Replaced by DIN EN 29457	1.5-2								1-3		OT 0.2 max			
	Germany	DIN 1707	2.3412/L-PbAg2Sn2	DIN 1707 Obsolete Replaced by DIN EN 29458	1.5-2								1-3		OT 0.2 max			
	USA	EIA/IPC J-STD-006	Ag02B	Electronic Grade Solder	2.5						97.5							
	USA	EIA/IPC J-STD-006	Pb97B	Electronic Grade Solder	1.5						97.5		1.00					
	UK	BS 334	Type C		0.01 max		0.01 max	0.01 max	0.01			2.5-11			ST=trace elements Cd & S; Ti 0.01			
L52715	USA			Battery			0.25				97.4	2.25	0.1		Se 0.02			
	USA	ASTM B32	Ag2.5	Solder	2.3-2.7	0.005 max	0.02 max	0.25 max	0.30 max	0.02 max		0.40 max	0.25 max	0.005 max	Cd 0.001 max			
Frary metal L50543	USA										97.2				Ba 2.0; Ca 0.8			
L52720	USA			Battery			0.3				97.2	2.25	0.2		S 0.008; Se 0.02			
L50113	USA			Solder	0.5						97.0		2.5					
L52620	USA			Battery							97.0	1.5			Cd 1.45			
L52805	USA										97	3.0						
	USA	EIA/IPC J-STD-006	Sn03A	Electronic Grade Solder							97.0		3.00					
L52760	USA			Battery			0.18		0.075		96.8	2.75	0.2		S 0.008			
	Germany	DIN 17640	2.3203/PbSbAs				1.2-1.7					2-3.8						
	Pan America	COPANT 447	PbSbAs				1.2-1.7					2-3.8						
L52765	USA			Battery			0.3		0.075		96.6	2.75	0.3		S 0.008			
L52770	USA			Battery			0.15		0.04		96.6	2.9	0.3		S 0.001			
L52775	USA			Battery			0.15		0.05		96.6	2.9	0.3		S 0.004			
	South Africa	SABS 24	S23 (PbSn4Ag)	Pouring Temp 300-307 degrees C	0.45-0.55		0.03 max	0.1 max	0.05 max	0.02 max		0.1 max	3-4		OT 0.08 max			
L52750	USA			Battery			0.4				96.5	2.75	0.3		S 0.005; Ca 0.075			
L52810	USA			Battery			0.15				96.5				S 0.003			
	South Africa	SABS 24	S20 (SnAg4)	Pouring Temp 221 degrees C	3.5-4		0.03 max	0.1 max	0.05 max	0.02 max		0.1 max			In 0.05; bal Sn			
L52755	USA			Battery			0.5				96.4	2.75	0.3		S 0.007; Ca 0.075			

UNS numbers and US grades are provided as a means of cross referencing chemically similar alloys. Exchangability is only possible after independent examination of specifications. Tensile properties are minimum or typical . UTS and YS as Mpa, El as %. See Appendix for list of abbreviations used in Descriptions.

6-6 Lead

Grade UNS #	Country	Specification	Designation	Description	Ag	Al	As	Bi	Cu	Fe	Pb	Sb	Sn	Zn	Other	UTS	YS	EL
L52815	USA			Shot			0.6				96.4	3.0	0.0005 max					
L52905	UK	BS 3909	2	Ing								3.8-4.2			OT 0.25 max			
L52901	Japan	JIS H4302	HPbP4	Plt					0.2 max			3.5-4.5	0.5 max					
L52901	Japan	JIS H4313	HPbT4	Pip					0.2 max			3.5-4.5	0.5 max					
L52901	USA										96	4						
	Czech Republic	CSN 423728	PbSb4,5		0.03 max		0.01 max	0.03 max	0.02 max	0.01 max		5	0.01 max					
	USA	EIA/IPC J-STD-006	Pb96A	Electronic Grade Solder							96.0	2.0	2.00					
	Japan	JIS Z3282	H5B	Bar		0.08 min					91.6 min	1 min	3-7	0.01	0.04 ST max ST=Bi+Zn+Fe+Al+As			
L52840	USA			Battery			0.5		0.12		95.7	3.25	0.4		Ca 0.06			
L50152	USA			Solder	2.5						95.5		2.0					
L52905	USA			Battery			0.15				95.5	4.0	0.3		S 0.003			
L52920	USA										95.5	4.5						
L54322	USA			Solder	0.015 max	0.005 max	0.02 max	0.25 max	0.08 max	0.02 max		0.50 max	4.5-5.5	0.005 max	Cd 0.001 max; OT 0.08 max			
	USA	ASTM B32	Sn5	Solder	0.015 max	0.005 max	0.02 max	0.25 max	0.08 max	0.02 max		0.50 max	4.5-5.5	0.005 max	Cd 0.001 max			
	South Africa	SABS 24	322 (PbAg5)	Pouring Temp 296-301 deg C	4.9-5.1		0.1 max		0.02 max	0.03 max		0.1 max			OT 0.00 max			
Grade B L50080	USA										95.0 min				OT 5.0 max			
L50170	USA			Solder	5.0						95.0							
L50180	Germany	DIN 1707	2.3405/L-PbAg5	DIN 1707 Obsolete Replaced by DIN EN 29455	4.5-6										OT 0.2 max			
Ag5.5 L50180	USA			Solder	5.0-6.0	0.005 max	0.02 max	0.25 max	0.30 max	0.002 max		0.40 max	0.25 max	0.05 max	Cd 0.001 max; OT 0.03 max			
L51511	USA			Solder							95.0				In 5.0			
L52730	USA			Electrotype							95	2.5	2.5					
L52910	USA										95	4	1					
L52922	USA			Anode							95	4.5			S 0.5			
L54320	Australia	AS 1834	5Sn				0.03 max	0.1 max	0.05 max	0.02 max		0.2 max	4-6		OT 0.08 max			
L54320	Japan	JIS Z3282	H5A	Bar		0.01	0.03	0.05	0.05	0.03	93.52	0.3	4-6	0.01				
L54320	Pan America	COPANT 450	5A			0.01 max	0.02 max	0.25 max	0.08 max	0.02 max	95 min	0.12 max	5 max	0.01 max				
L54320	USA			Solder		0.005 max	0.02 max	0.25 max	0.08 max	0.02 max	95	0.12 max	4.5-5.5	0.005 max				
L54321	USA			Solder		0.005 max	0.02 max	0.25 max	0.08 max	0.02 max	95	0.20-0.50	4.5-5.5	0.005 max				
L54370	USA			Plated overlay for bearings								5.0-9.0			OT 3.5 max			
	Czech Republic	CSN 426727	PbSb5,5		0.03 max		0.01 max	0.03 max	0.025 max	0.01 max		5-6	0.01 max					
	USA	EIA/IPC J-STD-006	Pb95B	Electronic Grade Solder	2.0						95.0		3.00					
	USA	EIA/IPC J-STD-006	Sn05A	Electronic Grade Solder							95.0		5.00					
L52940	USA			Battery			0.15		0.05		94.8	4.75	0.3		S 0.004			

UNS numbers and US grades are provided as a means of cross referencing chemically similar alloys. Exchangability is only possible after independent examination of specifications. Tensile properties are minimum or typical . UTS and YS as Mpa, El as %. See Appendix for list of abbreviations used in Descriptions.

Worldwide Guide to Equivalent Nonferrous Metals and Alloys

Grade UNS #	Country	Specification	Designation	Description	Ag	Al	As	Bi	Cu	Fe	Pb	Sb	Sn	Zn	Other	UTS	YS	EL
	USA	ASTM B32	Ag5.5	Solder	5.0-6.0	0.005 max	0.02 max	0.25 max	0.30 max	0.02 max		0.40 max	0.25 max	0.005 max	Cd 0.001 max			
L52930	USA			Battery			0.3		0.05		94.6	4.75	0.3		S 0.007			
Babbitt Metal L53346	USA	ASTM B23	Alloy Number 13	White metal bearing alloy		0.005 max	0.25 max	0.10 max	0.50 max	0.10 max		9.5-10.5	5.5-6.5	0.005 max	Cd 0.05 max			
L53346	USA			Bearing		0.005 max	0.25 max	0.10 max	0.50 max	0.10 max		9.5-10.5	5.5-6.5	0.005 max	Cd 0.05 max			
	USA	EIA/IPC J-STD-006	Ag05B	Electronic Grade Solder	5.5						94.5							
L50880	USA										94.0				Ca 6.0			
L52830	USA			Electrotype							94	3	3					
L53105	Japan	JIS H4302	HPbP6	Plt					0.2 max			5.5-6.5	0.5 max					
L53105	Japan	JIS H4313	HPbT6	Pip					0.2 max			5.5-6.5	0.5 max					
L53105	Pan America	COPANT 447	PbSb5								92.99	5-7						
L53105	USA										94	6.0						
L53110	Germany	DIN 17640	2.3205/PbSb5									5-7						
L53110	USA			Rolled Sh							94	6						
	Czech Republic	CSN 423726	PbSb6,5		0.03 max		0.01 max	0.03 max	0.025 max	0.01 max		6-7	0.01 max					
L53131	USA											6.0-7.0	0.25-0.75					
L53115	USA										93.7	6.0	0.3					
L53120	USA			Electrowinning			0.4				93.693.6	6.0						
L53122	USA			High strength Sh			0.4				93.6	6.0						
L50134	Australia	AS 1834	5Sn/1.5Ag		1.3-1.7		0.03 max	0.1 max	0.05 max	0.02 max		0.2 max	4-6		OT 0.08 max			
5S L50134	USA			Solder	1.5						93.5		5.0					
No. 16 L52860	USA			Bearing alloy (overlay)		0.005	0.05	0.10 max	0.10 max			4.0	3.5-4.7	0.005 max	Cd 0.005 max; OT 0.40 max			
	USA	EIA/IPC J-STD-006	Pb94B	Electronic Grade Solder	1.5						93.5		5.00					
L53135	USA			Battery			-0.3		0.07		93.4	6.0	0.3		S 0.006			
L53125	USA			Creep res Pip Sh			0.65				93.3	6.0						
L52915	USA			Type metal							93	4	3					
L54310	USA			Electrotype curved plate							93	3	4					
	Czech Republic	CSN 423725	PbSb8		0.03 max		0.01 max	0.03 max	0.02 max	0.01 max		7-8	0.01 max					
	South Africa	SABS 695	WM 90	Pouring Temp 340-370 degree C					3 max			7 min	90 max					
L51510	USA			Solder	2.38						92.8				In 4.76			
L53130	Japan	JIS H4313	HPbT8	Pip					0.2 max			7.5-8.5	0.5 max					
L53130	USA						0.6				92.8	6.0	0.6					
L53140	USA			Hard shot			1.2				92.6	6.2	0.0005 max					
L51512	USA			Solder	2.5						92.5				In 5.0			
	Japan	JIS H5601	HPbC8						0.2 max			7.5-8.5	0.5 max		0.10 max ST ST=max impurity limit			

UNS numbers and US grades are provided as a means of cross referencing chemically similar alloys. Exchangability is only possible after independent examination of specifications. Tensile properties are minimum or typical . UTS and YS as Mpa, EI as %. See Appendix for list of abbreviations used in Descriptions.

Grade UNS #	Country	Specification	Designation	Description	Ag	Al	As	Bi	Cu	Fe	Pb	Sb	Sn	Zn	Other	UTS	YS	EL
	South Africa	SABS 24	S12 (PbSn2Sb5)	Pouring Temp 240-285 deg C			0.03 max	0.25 max	0.08 max	0.02 max		5-5.25	2.5-2.75					
	USA	EIA/IPC J-STD-006	In05A	Electronic Grade Solder	2.5						92.5				In 5.00			
	USA	EIA/IPC J-STD-006	Pb93B	Electronic Grade Solder	2.5						92.5		5.00					
L52450	USA			Solder		0.005 max	0.40-0.60	0.25 max	0.08 max	0.02 max		4.90-5.40	2.50-2.75	0.005 max	OT 0.08 max			
L53230	Pan America	COPANT 447	PbSb8								91.49	7.5-8.5						
L53230	USA										92	8						
L54410	Germany	DIN 1707	2.3408/L-PbSn8(Sb)	DIN 1707 Obsolete Replaced by DIN EN 29456								0.12-0.5	7.5-8.5		OT 0.08 max			
L54410	USA			Solder							92	0.3	8					
L54510	Japan	JIS Z3282	H10B	Bar		0.08					86.6	1	8-12		0.35 max ST ST=Bi+Zn+Fe+Al+As			
L54510	USA			Plated overlay for bearings									8.0-12.0		OT 3.5 max			
	USA	EIA/IPC J-STD-006	Sn08A	Electronic Grade Solder							92.0		8.0					
L54280	USA			Solder							91.9	5.1	3.0					
L53220	Japan	JIS H4302	HPbP8	Plt					0.2 max			7.5-8.5	0.5 max					
L53220	USA						0.6				91.8	7.0	0.6					
	Czech Republic	CSN 055612	B-Sn8Pb-305/280	Soft Solder			0.05 max	0.1 max	0.15 max	0.05 max		1.5-2.5	7-9	0.02 max				
	Pan America	COPANT 447	PbSb9				0.01 max	0.02 max	0.01	0.01 max		8.7-9	0.01 max	0.01				
	South Africa	SABS 24	S13 (PbSn5Sb4)	Pouring Temp 240-280 deg C			0.03 max	0.25 max	0.08 max	0.02 max		3.8-4.2	5-5.5					
L53305	Germany	DIN 17640	2.3290/PbSb9									8.7-9						
L53305	Germany	DIN 17640	2.3299/PbSb9X									8.7-9						
L53305	Pan America	COPANT 447	PbSb9		0.005		0.01	0.02 max	0.01	0.01	90.89	8.7-9	0.01	0.001				
L53305	Pan America	COPANT 447	PbSb9X		0.01		0.01	0.04 max	0.02	0.01	90.87	8.7-9	0.02	0.001				
L53305	USA										91	9.0						
L53340	USA	ASTM B102	Y10A	Die cast			0.15 max		0.50 max		79-81	9.25-10.75		0.01 max				
L53340	USA			Die Cast			0.15 max		0.50 max		89-91	9.25-10.75		0.01 max				
	Australia	AS 1834	10Sn				0.03 max	0.1 max	0.05 max	0.02 max		0.2 max	9-11		OT 0.08 max			
	USA	ASTM B32	Sn10A	Solder	0.015 max	0.005 max	0.02 max	0.25 max	0.08 max	0.02 max		0.50 max	9.0-11.0	0.005 max	Cd 0.001 max			
L53235	USA			Hard shot			1.25				90.7	8.0	0.0005 max					
	Czech Republic	CSN 423723	PbSb10	Y10A	0.03 max		0.01 max	0.03 max	0.02 max	0.01 max		9.3-10.7	0.01 max					
L54360	USA			Solder			0.5				90.5	4.0	5.0					
	Japan	JIS H5601	HPbC10						0.2 max			9.5-10.5	0.5 max		0.10 max ST ST=max impurity limit			
L50172	USA			Solder	5.0						90.0				In 5.0			

UNS numbers and US grades are provided as a means of cross referencing chemically similar alloys. Exchangability is only possible after independent examination of specifications. Tensile properties are minimum or typical . UTS and YS as Mpa, El as %. See Appendix for list of abbreviations used in Descriptions.

Worldwide Guide to Equivalent Nonferrous Metals and Alloys

Grade UNS #	Country	Specification	Designation	Description	Ag	Al	As	Bi	Cu	Fe	Pb	Sb	Sn	Zn	Other	UTS	YS	EL
L53020	USA		Lyman's 2	Bullet							90.0	5.0	5.0					
L53238	USA			Hard shot			2				90.0	8	0.0005 max					
L53310	USA										90	9	1					
L54520	Japan	JIS Z3282	H10A	Bar		0.01	0.03	0.05	0.05	0.03	88.52	0.3	9-11	0.01				
L54520	Pan America	COPANT 450	10B			0.01 max	0.02 max	0.25 max	0.08 max	0.02 max	90 min	0.2-0.5	10 max	0.01 max				
L54520	USA			Solder		0.005 max	0.02 max	0.25 max	0.08 max	0.02 max	90	0.20-0.50	10	0.005 max				
	USA	EIA/IPC J-STD-006	Sn10A	Electronic Grade Solder							90.0		10.0					
L54525	USA			Solder	1.7-2.4	0.005 max	0.02 max	0.03 max	0.08 max			0.20 max	9.0-11.0	0.005 max	Cd 0.001 max; OT 0.10 max			
	USA	ASTM B32	Sn10B	Solder	1.7-2.4	0.005 max	0.02 max	0.03 max	0.08 max	0.02 max		0.20 max	9-11.0 max	0.005 max	Cd 0.001 max			
L53210	USA			Type metal							89	7	4					
L53405	USA										89	11						
	Czech Republic	CSN 423730	PbSn6Sb6			0.01 max	0.1 max	0.07 max	0.3 max	0.1 max		5.5-6.5	5.5-6.5	0.01 max	Cd 0.05 max; Ni 0.05 max			
L53260	USA			Spin cast							88.9	8.0	3.1					
	Germany	DIN 1707	2.3412/L-PbSn12Sb	DIN 1707 Obsolete Replaced by DIN EN 29459								0.2-0.7	11.5-12.5		OT 0.08 max			
Babbitt Metal L05120	USA			White metal bearing alloy		0.01 max	0.2 max		0.5 max	0.1 max	83-88	8-10	4-6	0.01 max	OT 0.75 max			
L53265	USA			Type metal							88	8.0	4.0					
	Germany	DIN 17640	2.3212/PbSb8									12-13						
	Pan America	COPANT 447	PbSb12									12-13						
	USA	EIA/IPC J-STD-006	Pb88B	Electronic Grade Solder	2.0						88.0		10.0					
L54540	USA			Solder							87.5	0.45	12					
	Pan America	COPANT 447	PbSb12								86.99	12-13						
	Japan	JIS H2231	K13:2/Class4, No.2	Ing			0.75 max		0.3 max	0.05 max		12-14	1.5-2.5	0.02 max				
L53320	USA			Whitemetal bearing alloy							86	9	5					
L53345	Pan America	COPANT 926	B3			0.01 max	0.25 max	0.1 max	0.5 max	0.1 max		9.5-10.5	5.5-6.5	0.01				
L53345	USA			Bearing		0.005 max	0.25 max	0.10 max	0.50 max			9.0-11.0	5.0-7.0	0.005 max	Cd 0.05 max; OT 0.20 max			
L53420	USA			Lintotype							86	11	3					
L53470	USA			CT metal			0.4				86	12.75	0.75					
L54555	USA	SAE J473a	6B									2.75 max	14-15					
L54555	USA			Solder		0.005 max	0.25 max		0.08 max	0.02 max		2.75 max	14.0-15.0	0.005 max	OT 0.08 max			
	Australia	AS 1834	15Sn				0.03 max	0.1 max	0.05 max	0.02 max		0.2 max	14-16		OT 0.08 max			
	South Africa	SABS 24	S15 (PbSn15)	Pouring Temp 226-289 deg C			0.03 max	0.25 max	0.08 max	0.02 max		0.5 max	14-15					
	USA	SAE J473a	6A									0.4 max	14-15					
L54530	USA			Solder			0.5				85.5	4	10					

UNS numbers and US grades are provided as a means of cross referencing chemically similar alloys. Exchangability is only possible after independent examination of specifications. Tensile properties are minimum or typical . UTS and YS as Mpa, El as %. See Appendix for list of abbreviations used in Descriptions.

Grade UNS #	Country	Specification	Designation	Description	Ag	Al	As	Bi	Cu	Fe	Pb	Sb	Sn	Zn	Other	UTS	YS	EL
	USA	ASTM B32	Sn15	Solder	0.015 max	0.005 max	0.02 max	0.25 max	0.08 max	0.02 max		0.50 max	14.5-16.5	0.005 max	Cd 0.001 max			
L53343	USA			Bearing		0.005 max	0.25 max	0.10 max	0.50 max	0.1 max		9.5-10.5	5.5-6.5	0.005 max	Cd 0.05 max			
L53454	USA			Type metal							85	12	3					
L53505	Japan	JIS H2231	K13:1/Class4, No.1	Ing			0.75 max		0.3 max	0.05 max		12-14	0.5-1.5	0.02 max				
L53505	USA			CT metal			1				85	13	1					
L53550	USA										85	15						
L54560	Pan America	COPANT 450	15B			0.01 max	0.02 max	0.25 max	0.08 max	0.02 max	85 min	0.2-0.5	15 max	0.01 max				
L54560	USA			Solder		0.005 max	0.02 max	0.25 max	0.08 max	0.02 max	85	0.20-0.50	15					
L53620	Japan	JIS H5401	WJ10			0.01	0.75-1.25		0.1-0.5	0.1	81.38	14-15.5	0.8-1.2	0.05				
L53620	USA			Bearing		0.005 max	0.8-1.4	0.10 max	0.6 max	0.10 max		14.5-17.5	0.8-1.2	0.005 max	Cd 0.05 max			
	Japan	JIS H2231	K13:4/Class3, No.3.5	Ing			0.5 max		0.3 max	0.03 max		12-14	3.5-4.5	0.01 max				
L53425	South Africa	SABS 12	LINO			0.01			0.05		83.34	10.5-11.5	3.5-5.5	0.01	0.02 Fe+Ni			
L53425	USA			Linotype							84	11	5					
L53455	Sweden	MNC 74E	7410-00	Ing								11.5-12.5	3.8-4.3					
L53455	USA			Linotype B (Eutectic)							84	12	4					
Babbitt Metal L53620	USA	ASTM B23	Alloy Number 15	White metal bearing alloy		0.005 max	0.8-1.4	0.10 max	0.6 max	0.10 max		14.5-17.5	0.8-1.2	0.005 max	Cd 0.05 max			
	Pan America	COPANT 926	B5			0.01 max	0.8-1.4	0.1 max	0.5 max	0.1 max		14.5-17.5	0.8-1.2	0.01				
L53480	USA			G Babbitt			3.0				83.5	12.75	0.75					
L50940	USA			Eutectic							83.0				Cd 17.0			
L53456	USA			Type meta;							83	12	5					
L53555	USA										83	15	2					
	Japan	JIS H2231	K15:3.5/Class 3,No3.5	Ing			0.75 max		0.3 max	0.05 max		14-16	3-4	0.02 max				
	Sweden	SIS 147410	Pb7410-00			0.001	0.15		0.05	0.02	82.96	11.5-12.5	3.8-4.3	0.001	Ti 0.01			
L54570	USA			Solder							82.5	2.5	15					
L53460	USA			Type metal							82	12	6					
	Sweden	MNC 74E	7416-00	Ing								13.5-14.5	4.8-5.3					
L53565	Pan America	COPANT 926	B4			0.01 max	0.3-0.6	0.1 max	0.5 max	0.1 max		14-16	4.5-5.5	0.01				
L53565	USA			White metal bearing		0.005 max	0.30-0.60	0.10 max	0.50 max	0.10 max		14.0-16.0	4.5-5.5	0.005 max	Cd 0.05 max			
	Japan	JIS H2231	K17:3/Class2, No.3	Ing			0.75 max		0.3 max	0.05 max		16-18	2.5-3.5	0.02 max				
	Czech Republic	CSN 423721	PbSn6Sb14CuAs			0.01 max	0.05 max	0.08 max	0.4-1	0.05 max		13-15	5.5-6.5	0.01 max	Zn+Fe+Al:0.05			
L51530	USA										81.0				In 19.0			
L53558	USA			Type metal							81	15	4					
L53560	USA	ASTM B102	Y155A	Die cast		0.01 max	0.15 max		0.50 max		89-91	14-16	4-6	0.01 max				
L53560	USA			Die Cast		0.01 max	0.15 max		0.50 max		79-81	14-16	4-6	0.01 max				
	Australia	AS 1834	20Sn				0.03 max	0.1 max	0.08 max	0.02 max		0.2 max	19-21		OT 0.08 max			

UNS numbers and US grades are provided as a means of cross referencing chemically similar alloys. Exchangability is only possible after independent examination of specifications. Tensile properties are minimum or typical . UTS and YS as Mpa, El as %. See Appendix for list of abbreviations used in Descriptions.

Worldwide Guide to Equivalent Nonferrous Metals and Alloys

Grade UNS #	Country	Specification	Designation	Description	Ag	Al	As	Bi	Cu	Fe	Pb	Sb	Sn	Zn	Other	UTS	YS	EL
	USA	EIA/IPC J-STD-006	In19A	Electronic Grade Solder							81.0				In 19.0			
	USA	SAE J473a	5A									0.4 max	19-20					
L53510	USA			Stereotype general							80.5	13	6.5					
L54580	USA			Type metal							80.5	4.5	15					
	Japan	JIS H2231	K15:6/Class3, No6	Ing			0.5 max		0.5 max	0.03		14-16	5.5-6.5	0.01 max				
	USA	ASTM B32	Sn20A	Solder	0.015 max	0.005 max	0.02 max	0.25 max	0.08 max	0.02 max		0.50 max	19.5-21.5	0.005 max	Cd 0.001 max			
Babbitt Metal L53565	USA	ASTM B23	Alloy Number 8	White metal bearing alloy		0.005 max	0.30-0.60	0.10 max	0.50 max	0.10 max		14.0-16.0	4.5-5.5	0.005 max	Cd 0.05 max			
	South Africa	SABS 24	S14 (PbSn20Sb)	Pouring Temp 185-270 deg C			0.03 max	0.25 max	0.08 max	0.02 max		0.9-1.2	19-21					
	USA	EIA/IPC J-STD-006	Pb80B	Electronic Grade Solder	1.9						80.1		18.0					
L51532	USA										80.0				In 20.0			
L53530	South Africa	SABS 703	WM6	Bar				0.03	0.05		77.33	13-15	5-7	0.01	OT 0.03 max			
L53530	USA			Stereotype flat							80	14	6					
L54710	USA			Solder							80	0.2 max	20					
L54711	Japan	JIS H3282	H20B	Bar		0.08					76.6	1	18-22		0.35 max ST ST=Bi+Zn+Fe+Al+As			
L54711	Japan	JIS Z3282	H20A	Bar		0.01	0.03	0.05	0.05	0.03	78.52	0.3	19-21	0.01				
L54711	Pan America	COPANT 450	20B			0.01 max	0.02 max	0.25 max	0.08 max	0.02 max	80 min	0.2-0.5	20 max	0.01 max				
Alloy 20B L54711	USA			Solder		0.005 max	0.02 max	0.25 max	0.08 max	0.02 max	80	0.20-0.50	20	0.005 max	Sn 20 desired			
	Sweden	MNC 74E	7417-00	Ing								13.5-14.5	6.5-7.5					
	USA	EIA/IPC J-STD-006	Pb80A	Electronic Grade Solder							80.0	0.20-0.50	20.0					
	USA	EIA/IPC J-STD-006	Sn20A	Electronic Grade Solder							80.0		20.0					
	Sweden	SIS 147416	Pb7416-00			0.001	0.2		0.05	0.02	79.91	13.5-14.5	4.8-5.3	0.001	Ti 0.01			
L54610	Germany	DIN 1707	2.3423/L-PbSn20Sb3	DIN 1707 Obsolete Replaced by DIN EN 29460								1.2-3	19.5-20.5		OT 0.08 max			
L54610	USA	SAE J473a	5B									1.25-1.75	19-20					
L54610	USA			Solder		0.005 max		0.25 max	0.08 max	0.02 max		1.25-1.75	19.0-20.0	0.005 max	OT 0.08 max			
	USA	ASTM B32	Sn20B	Solder	0.015 max	0.005 max	0.02 max	0.25 max	0.08 max	0.02 max		0.8-1.2	19.5-21.5	0.005 max	Cd 0.001 max			
L54712	Pan America	COPANT 450	20C			0.01 max	0.02 max	0.25 max	0.08 max	0.02 max	79 min	0.8-1.2	20 max	0.01 max				
Alloy 20C L54712	USA			Solder		0.005 max	0.02 max	0.25 max	0.08 max	0.02 max	79	0.8-1.2	20	0.005 max				
	Czech Republic	CSN 423741	Pb78Bi16Sb	Low Melting Point Alloy				15-17			77-79							
	Czech Republic	CSN 423742	Pb78Bi12Sb	Low Melting Point Alloy				11-13			77-79							
	USA	EIA/IPC J-STD-006	Pb79A	Electronic Grade Solder							79.0		20.0					
L53570	USA			Monotype							78	15	7					

UNS numbers and US grades are provided as a means of cross referencing chemically similar alloys. Exchangability is only possible after independent examination of specifications. Tensile properties are minimum or typical . UTS and YS as Mpa, EL as %. See Appendix for list of abbreviations used in Descriptions.

Grade UNS #	Country	Specification	Designation	Description	Ag	Al	As	Bi	Cu	Fe	Pb	Sb	Sn	Zn	Other	UTS	YS	EL
	France	NF A56-101	201	Ing		0.01 max	0.1 max	0.05 max	0.5-1	0.05 max	78	15	6	0.05 max	Ti 0.2-0.6; 0.30 max ST ST=max impurity limit			
	Sweden	SIS 147417	Pb7417-00			0.001	0.2		0.05	0.02	77.7	13.5-14.5	6.5-7.5	0.001	Ti 0.01			
L53465	USA								0.05		77.5	12.5	10.0					
L53575	South Africa	SABS 12	STEREO			0.01			0.05		73.84	14-16	5-10	0.01	0.02 Fe+Ni			
L53575	USA			Stereotype							77	15	8					
L54713	USA			Solder							77	1.2	22					
L53585	France	NF A56-101	202	Ing		0.01 max	0.1 max	0.05 max	0.5-1	0.05 max	78	15	10	0.05 max	Ti 0.2-0.6; 0.30 max ST ST=max impuirty limit			
L53585	Pan America	COPANT 926	B2			0.01 max	0.3-0.6	0.1 max	0.5 max	0.1		14-15	9.3-10.7	0.01				
Babbitt Metal L53585	USA	ASTM B23	Alloy Number 7	White metal bearing alloy		0.005 max	0.30-0.60	0.10 max	0.50 max	0.10 max		14.0-16.0	9.3-10.7	0.005 max	Cd 0.05 max			
L53585	USA			Whitemetal bearing		0.005 max	0.30-0.60	0.10 max	0.50 max	0.10 max		14.0-16.0	9.3-10.7	0.005 max	Cd 0.05 max			
	Pan America	COPANT 926	B1				0.02 max		0.4-0.6			13.3-13.7	9.8-10.5					
L54720	Australia	AS 1834	25Sn				0.03 max	0.1 max	0.08 max	0.02 max		0.2 max	24-26		OT 0.08 max			
	USA	SAE J473a	4A									0.4 max	24-25					
	Czech Republic	CSN 423720	PbSn10Sb15 Cu1	Alloy 7		0.05 max	0.1 max		0.5-1.5	0.1 max		14.5-16.5	9.5-10.5	0.05 max				
	USA	ASTM B32	Sn25A	Solder	0.015 max	0.005 max	0.02 max	0.25 max	0.08 max	0.02 max		0.50 max	24.5-26.5	0.005 max	Cd 0.001 max			
L51535	USA										75.0				In 25.0			
L53580	Sweden	MNC 74E	7421-00	Ing								14.5-15.5	9.5-10.5					
L53580	Sweden	MNC 74E	7428-00	Ing								15.5-16.5	9.5-10.5					
L53580	USA			Rules monotype							75	15	10					
L53650	Japan	JIS H2231	K17:8/Class2, No.8	Ing			0.5 max		0.5 max	0.03 max		16-18	7.5-8.5	0.01 max				
L53650	USA			Display monotye							75	17	8					
L54720	Pan America	COPANT 450	25A			0.01 max	0.02 max	0.25 max	0.08 max	0.02 max	75 min	0.25 max	25 max	0.01 max				
L54720	USA			Solder		0.005 max	0.02 max	0.25 max	0.08 max	0.02 max	75	0.25 max	25	0.005 max				
L54721	Pan America	COPANT 450	25B			0.01 max	0.02 max	0.25 max	0.08 max	0.02 max	75 min	0.2-0.5	25 max	0.01 max				
Alloy 25B L54721	USA			Solder		0.005 max	0.02 max	0.25 max	0.08 max	0.02 max	75	0.20-0.50	25	0.005 max	Sn 25 desired			
	South Africa	SABS 24	S11 (PbSn26)	Pouring Temp 183-265 degree C			0.03 max	0.25 max	0.08 max	0.02 max		0.3 max	25-27					
	USA	EIA/IPC J-STD-006	In25A	Electronic Grade Solder							75.0				In 25.0			
	South Africa	SABS 12	MONOA			0.01			0.05		74.8	16.5-17.5	6.5-7.5	0.01	0.02 Fe+Ni			
	USA	SAE J473a	4B									1.2-1.75	24-25					
	USA	ASTM B32	Sn25B	Solder	0.015 max	0.005 max	0.02 max	0.25 max	0.08 max	0.02 max		1.1-1.5	24.5-26.5	0.005 max	Cd 0.001 max			
L53655	USA			Type metal							74	17	9					
L53685	South Africa	SABS 12	MONOB			0.01			0.05		70.8	18.5-19.5	8.5-9.5	0.01	0.02 Fe+Ni			
L54715	USA			Solder							74	3	23					

UNS numbers and US grades are provided as a means of cross referencing chemically similar alloys. Exchangability is only possible after independent examination of specifications. Tensile properties are minimum or typical . UTS and YS as Mpa, El as %. See Appendix for list of abbreviations used in Descriptions.

Worldwide Guide to Equivalent Nonferrous Metals and Alloys

Grade UNS #	Country	Specification	Designation	Description	Ag	Al	As	Bi	Cu	Fe	Pb	Sb	Sn	Zn	Other	UTS	YS	EL
	South Africa	SABS 24	S10 (PbSn26Sb)	Pouring Temp 185-260 deg C			0.03 max	0.25 max	0.08 max	0.02 max		1-1.7	25-27					
	USA	EIA/IPC J-STD-006	Pb74A	Electronic Grade Solder							74.0	1.0	25.0					
L54722	Germany	DIN 1707	2.3425/L-PbSn25Sb	DIN 1707 Obsolete Replaced by DIN EN 29461								0.2-1.5	24.5-25.5		OT 0.08 max			
L54722	Pan America	COPANT 450	25C			0.01 max	0.02 max	0.25 max	0.08 max	0.02 max	73.7 min	1.1-1.5	25 max	0.01 max				
Alloy 25C L54722	USA			Solder	0.005 max		0.02 max	0.25 max	0.08 max	0.02 max	73.7	1.1-1.5	25	0.005 max				
	Sweden	SIS 147421	Pb7421-00			0.001	0.2		0.05	0.02	73.7	14.5-15.5	9.5-10.5	0.001	Ti 0.01			
	South Africa	SABS 12	MONOD			0.01			0.05		72.8	15.5-16.5	9.5-10.5	0.01	0.02 Fe+Ni			
	Sweden	SIS 147428	Pb7428-00			0.001	0.2		0.05	0.02	72.7	15.5-16.5	9.5-10.5	0.001	Ti 0.01			
L53685	USA			Lanston standard case type monotype							72	19	9					
	Japan	JIS H2231	K20:10/Class 1,No.10	Ing			0.5 max		1 max	0.03 max		19-21	9.5-10.5	0.01 max				
	Czech Republic	CSN 055618	B-Sn30Pb-250/185	Soft Solder			0.05 max	0.1 max	0.15 max	0.05 max		2 max	29-31	0.02 max				
	South Africa	SABS 24	S9 (PbSn30)	Pouring Temp 183-255 deg C			0.03 max	0.25 max	0.08 max	0.02 max		0.3 max	29-31					
	USA	SAE J473a	3A									0.5 max	29-30					
L54805	USA			Solder							70.5	1.5	28					
	USA	ASTM B32	Sn30A	Solder	0.015 max	0.005 max	0.02 max	0.25 max	0.08 max	0.02 max		0.50 max	29.5-31.5	0.005 max	Cd 0.001 max			
L54815	USA	SAE J473a	3B									0.75-1.25	29-30					
L54815	USA			Solder		0.005 max		0.25 max	0.08 max	0.02 max		0.75-1.25	29.0-30.0	0.005 max	OT 0.08 max			
L54750	USA			Solder	3						70		27					
L54820	Australia	AS 1834	30Sn				0.03 max	0.1 max	0.08 max	0.02 max		0.2 max	29-31		OT 0.08 max			
L54820	Japan	JIS Z3282	H30A	Bar		0.01	0.03	0.05	0.05	0.03	68.52	0.3	29-31	0.01				
L54820	Pan America	COPANT 450	30A			0.01 max	0.02 max	0.25 max	0.08 max	0.02 max	70 min	0.25 max	30 max	0.01 max				
L54820	USA			Solder	0.005 max		0.02 max	0.25 max	0.08 max	0.02 max	70	0.25 max	30	0.005 max	Sn 30 desired			
L54821	Germany	DIN 1707	2.3430/L-PbSn30(Sb)	DIN 1707 Obsolete Replaced by DIN EN 29462								0.12-0.5	29.5-30.5		OT 0.08 max			
L54821	Pan America	COPANT 450	30B			0.01 max	0.02 max	0.25 max	0.08 max	0.02 max	70 min	0.2-0.5	30 max	0.01 max				
L54821	USA			Solder	0.005 max		0.02 max	0.25 max	0.08 max	0.02 max	70	0.20-0.50	30		Sn 30 desired			
	South Africa	SABS 24	S5 (PbSn30Sb)	Pouring Temp 185-250 deg C			0.03 max	0.25 max	0.08 max	0.02 max		1-1.7	29-31					
	USA	EIA/IPC J-STD-006	In30A	Electronic Grade Solder							70.0				In 30.0			
	USA	EIA/IPC J-STD-006	Pb70A	Electronic Grade Solder							70.0	0.20-0.50	30.0					
	USA	EIA/IPC J-STD-006	Sn30A	Electronic Grade Solder							70.0		30.0					

UNS numbers and US grades are provided as a means of cross referencing chemically similar alloys. Exchangability is only possible after independent examination of specifications. Tensile properties are minimum or typical . UTS and YS as Mpa, El as %. See Appendix for list of abbreviations used in Descriptions.

6-14 Lead

Grade UNS #	Country	Specification	Designation	Description	Ag	Al	As	Bi	Cu	Fe	Pb	Sb	Sn	Zn	Other	UTS	YS	EL
	USA	ASTM B32	Sn30B	Solder	0.015 max	0.005 max	0.02 max	0.25 max	0.08 max	0.02 max		1.4-1.8	29.5-31.5	0.005 max	Cd 0.001 max			
L54822	Australia	AS 1834	30Sn/1.8Sb				0.03 max	0.25 max	0.08 max	0.02 max		1.6-2	29-31		OT 0.08 max			
L54822	Germany	DIN 1707	2.3432/L-PbSn30Sb	DIN 1707 Obsolete Replaced by DIN EN 29463								0.5-1.8	29.5-30.5		OT 0.08 max			
L54822	Pan America	COPANT 450	30C			0.01 max	0.02 max	0.25 max	0.08 max	0.02 max	68.4 min	1.4-1.8	30 max	0.01 max				
L54822	USA			Solder		0.005 max	0.02 max	0.25 max	0.08 max	0.02 max	68.4	1.4-1.8	30	0.005 max	Sn 30 desired			
	USA	EIA/IPC J-STD-006	Pb68A	Electronic Grade Solder							68.4	1.6	30.0					
	South Africa	SABS 702	WM15	Bar					0.25-0.75	0.05	67.88	13-15	14-16	0.01	OT 0.03 max			
	Germany	DIN 1707	2.3433/L-PbSn33(Sb)	DIN 1707 Obsolete Replaced by DIN EN 29464								0.12-0.5	33.5		OT 0.08 max			
L53740	USA			Monotype case type							67	24	9					
L53790	USA			type metal							67	28	5					
L54832	USA			Solder							66.7	1.8	31.5					
	Japan	JIS Z3282	H30B	Bar		0.08					66.6	1	28-32		0.35 max ST ST=Bi+Zn+Fe+Al+As			
	South Africa	SABS 24	S7 (PbSn35)	Pouring Temp 183-245 deg C			0.03 max	0.25 max	0.08 max	0.02 max		0.3 max	34-36					
	USA	SAE J473a	8A									0.4 max	34-35					
L53795	USA			Type metal							65.5	29	5.5-5.5					
	USA	ASTM B32	Sn35A	Solder	0.015 max	0.005 max	0.02 max	0.25 max	0.08 max	0.02 max		0.50 max	34.5-36.5	0.005 max	Cd 0.001 max			
L54833	USA			Solder							65	3	32					
L54850	Australia	AS 1834.1	35Sn				0.03 max	0.1 max	0.08 max	0.02 max		0.2 max	34-36		OT 0.08 max			
L54850	Japan	JIS Z3282	H35A	Bar		0.01	0.03	0.05	0.05	0.03	63.52	0.3	34-36	0.01				
L54850	Pan America	COPANT 450	35A			0.01 max	0.02 max	0.25 max	0.08 max	0.02 max	65 min	0.25 max	35 max	0.01 max				
L54850	USA			Solder		0.005 max	0.02 max	0.25 max	0.08 max	0.02 max	65	0.02 max	35	0.005 max				
L54851	Germany	DIN 1707	2.3435/L-PbSn35(Sb)	DIN 1707 Replaced by DIN EN 29465								0.12-0.5	34.5-35.5		OT 0.08 max			
L54851	Pan America	COPANT 450	35B			0.01 max	0.02 max	0.25 max	0.08 max	0.02 max	65 min	0.2-0.5	35 max	0.01 max				
L54851	USA			Solder		0.005 max	0.02 max	0.25 max	0.08 max	0.02 max	65	0.20-0.50	35	0.005 max				
	USA	EIA/IPC J-STD-006	Pb85A	Electronic Grade Solder							65.0	0.20-0.50	35.0					
	USA	EIA/IPC J-STD-006	Sn35A	Electronic Grade Solder							65.0		35.0					
L53750	South Africa	SABS 12	MONOC			0.01			0.05		62.8	23.5-24.5	11.5-12.5	0.01	0.02 Fe+Ni			
L53750	USA			Monotype case type							64	24	12					
L54827	USA			Type metal							64	6	30					
L54852	Australia	AS 1834.1	35Sn/1.8Sb				0.03 max	0.25 max	0.08 max	0.02 max		1.6-2	34-36		OT 0.08 max			

UNS numbers and US grades are provided as a means of cross referencing chemically similar alloys. Exchangability is only possible after independent examination of specifications. Tensile properties are minimum or typical . UTS and YS as Mpa, El as %. See Appendix for list of abbreviations used in Descriptions.

Worldwide Guide to Equivalent Nonferrous Metals and Alloys

Grade UNS #	Country	Specification	Designation	Description	Ag	Al	As	Bi	Cu	Fe	Pb	Sb	Sn	Zn	Other	UTS	YS	EL
L54852	Germany	DIN 1707	2.3437/L-PbSn35Sb	DIN 1707 Obsolete Replaced by DIN EN 29466								0.5-2	34.5-35.5		OT 0.08 max			
L54852	Pan America	COPANT 450	35C			0.01 max	0.02 max	0.25 max	0.08 max	0.02 max	63.2 min	1.6-2	35 max	0.01 max				
L54852	USA			Solder		0.005 max	0.02 max	0.25 max	0.08 max	0.02 max	63.2	1.6-2.0	35	0.005 max				
	USA	EIA/IPC J-STD-006	Pb63A	Electronic Grade Solder							63.2	1.8	35.0					
	Czech Republic	CSN 423722	Pb60Cu36Ni2,5				0.25 max			0.25 max	57-63	0.5 max	0.5 max		P 0.05; Ni 2.3-2.75; bal Cu			
	Australia	AS 1834.1	34Sn/5Ag		4.8-5.2		0.03 max	0.1 max	0.05 max	0.02 max		0.2 max	33-35		OT 0.08 max			
	Japan	JIS Z3282	H35B	Bar		0.08					61.6	1	33-37		0.35 max ST ST=Bi+Zn+Fe+Al+As			
L54855	USA			Solder	3.0						61.5		35.5					
	Czech Republic	CSN 055620	B-Sn40Pb-225/185	Soft Solder			0.05 max	0.1 max	0.1 max	0.05 max		2 max	39-41	0.02 max				
	South Africa	SABS 24	S4 (PbSn40)	Pouring Temp 183-235 deg C			0.03 max	0.25 max	0.08 max	0.02 max		0.3 max	39-41					
	Australia	AS 1834.1	38Sn/2.3Sb				0.03 max	0.25 max	0.08 max	0.02 max		2.1-2.5	37-39		OT 0.08 max			
	USA	SAE J473a	2A									0.4	39-40					
	Japan	JIS Z3282	H38A	Bar		0.01	0.03	0.05	0.05	0.03	60.52	0.3	37-39	0.01				
L53780	USA			Hard foundry type					1.5		60.5	25	13					
L54905	USA			Solder		0.005 max		0.25 max	0.08 max	0.02 max		1.5-2.00	38.0-38.5	0.005 max	OT 0.08 max			
	USA	ASTM B32	Sn40A	Solder	0.015 max	0.005 max	0.02 max	0.25 max	0.08 max	0.02 max		0.50 max	39.5-41.5	0.005 max	Cd 0.001 max			
L54905	USA	SAE J473a	2B									1.7	38-38.5					
L51540	USA										60.0				In 40.0			
L54915	Australia	AS 1834.1	40Sn				0.03 max	0.1 max	0.08 max	0.02 max		0.2 max	39-41		OT 0.08 max			
L54915	Germany	DIN 1707	2.3442/L-PbSn40	DIN 1707 Obsolete Replaced by DIN EN 29468									39.5-40.5		OT 0.08 max			
L54915	Japan	JIS Z3282	H40A	Bar		0.01	0.03	0.05	0.05	0.03	58.52	0.3	39-41	0.01				
L54915	Japan	JIS Z3282	H40S	Bar		0.01	0.03	0.05	0.03	0.02	58.77	0.1	39-41	0.01				
L54915	Pan America	COPANT 450	40A			0.01 max	0.02 max	0.25 max	0.08 max	0.02 max	60 min	0.12 max	40 max	0.01 max				
L54915	USA			Solder		0.005 max	0.02 max	0.25 max	0.08 max	0.02 max	60	0.12 max	40	0.005 max				
L54916	Germany	DIN 1707	2.3440/L-PbSn40(Sb)	DIN 1707 Obsolete Replaced by DIN EN 29467								0.12-0.5	39.5-40.5		OT 0.08 max			
L54916	Pan America	COPANT 450	40B			0.01 max	0.02 max	0.25 max	0.08 max	0.02 max	60 min	0.2-0.5	40 max	0.01 max				
L54916	USA			Solder		0.005 max	0.02 max	0.25 max	0.08 max	0.02 max	60	0.20-0.50	40	0.005 max				
	USA	EIA/IPC J-STD-006	In40A	Electronic Grade Solder							60.0				In 40.0			
	USA	EIA/IPC J-STD-006	Pb80A	Electronic Grade Solder							60.0	0.20-0.50	40.0					
	USA	EIA/IPC J-STD-006	Sn40A	Electronic Grade Solder							60.0		40.0					

UNS numbers and US grades are provided as a means of cross referencing chemically similar alloys. Exchangability is only possible after independent examination of specifications. Tensile properties are minimum or typical . UTS and YS as Mpa, El as %. See Appendix for list of abbreviations used in Descriptions.

Grade UNS #	Country	Specification	Designation	Description	Ag	Al	As	Bi	Cu	Fe	Pb	Sb	Sn	Zn	Other	UTS	YS	EL
L54727	USA			Bearing alloy					3		59	13	25					
	South Africa	SABS 24	S3 (PbSn40Sn2)	Pouring Temp 185-228 deg C			0.03 max	0.25 max	0.08 max	0.02 max		2-2.4	39-41					
	USA	ASTM B32	Sn40B	Solder	0.015 max	0.005 max	0.02 max	0.25 max	0.08 max	0.02 max		1.8-2.4	39.5-41.5	0.005 max	Cd 0.001 max			
L53710	USA			Hard foundry type					1.5		58.5	20	20					
L51180	USA			Bearing alloy						0.35 max	44.0-58.0		1.0-5.0		bal Cu; OT 0.45 max; OE 0.15 max			
L54918	Australia	AS 1834.1	40Sn/2.4Sb				0.03 max	0.25 max	0.08 max	0.02 max		2.2-2.6	39-41		OT 0.08 max			
L54918	Germany	DIN 1707	2.3442/L-PbSn40Sb	DIN 1707 Obsolete Replaced by DIN EN 29469								0.5-2.4	39.5-40.5		OT 0.08 max			
L54918	Pan America	COPANT 450	40C			0.01 max	0.02 max	0.25 max	0.08 max	0.02 max	58 min	1.8-2.4	40 max	0.01 max				
L54918	USA			Solder		0.005 max	0.02 max	0.25 max	0.08 max	0.02 max	58	1.8-2.4	40	0.005 max				
	USA	EIA/IPC J-STD-006	Pb58A	Electronic Grade Solder							57.8	2.2-2.2	40.0					
	Japan	JIS Z3282	H40B	Bar		0.08					56.6	1	38-42		0.35 max ST ST=Bi+Zn+Fe+Al+As			
	USA	SAE J473a	1A									0.4 max	44-45					
L54930	USA			Fusible					4		55.5		40.5					
L54940	USA	SAE J473a	1B									1 7	43-43.5					
L54940	USA			Solder		0.005 max		0.25 max	0.08 max	0.02 max		1.5-2.00	43.0-43.5	0.005 max	OT 0.08 max			
	USA	ASTM B32	Sn45	Solder	0.015 max	0.005 max	0.025 max	0.25 max	0.08 max	0.02 max		0.50 max	44.5-46.5	0.005 max	Cd 0.001 max			
L54945	USA			Solder	1						55		44					
L54950	Australia	AS 1834.1	45Sn				0.03 max	0.1 max	0.08 max	0.02 max		0.2 max	44-46		OT 0.08 max			
L54950	Japan	JIS Z3282	H45S	Bar		0.01	0.03	0.03 max	0.03	0.02	53.77	0.1	44-46	0.01				
L54950	Pan America	COPANT 450	45A			0.01 max	0.03 max	0.25 max	0.08 max	0.02 max	55 min	0.12 max	45 max	0.01 max				
L54950	USA			Solder		0.005 max	0.03 max	0.25 max	0.08 max	0.02 max	55	0.12 max	45	0.005 max				
L54951	Germany	DIN 1707	2.3445/L-PbSn45(Sb)	DIN 1707 Obsolete Replaced by DIN EN 29470								0.12-0.5	44.5-45.5		OT 0.08 max			
L54951	Japan	JIS Z3282	H45A	Bar		0.01	0.03	0.05	0.05	0.03	53.52	0.3	44-46	0.01				
L54951	Pan America	COPANT 450	45B			0.01 max	0.03 max	0.25 max	0.08 max	0.02 max	55 min	0.2-0.5	45 max	0.01 max				
L54951	USA			Solder		0.005 max	0.03 max	0.25 max	0.08 max	0.02 max	55	0.20-0.50	45	0.005 max				
	USA	ASTM B32	Sn35B	Solder	0.015 max	0.005 max	0.02 max	0.25 max	0.08 max	0.02 max		1.6-2.0	34.5-36.5	0.005 max	Cd 0.001 max			
	South Africa	SABS 24	S16 (PbSn34Sb2)	Pouring Temp 185-215 deg C			0.03 max	0.25 max	0.08 max	0.02 max		2.2-2.7	44-45					
	Australia	AS 1834.1	45Sn/2.7Sb				0.03 max	0.25 max	0.08 max	0.02 max		2.5-2.9	44-46		OT 0.08 max			
L54955	USA			Solder							52.5	2.5	45					
L55005	USA			Solder							52		48					

UNS numbers and US grades are provided as a means of cross referencing chemically similar alloys. Exchangability is only possible after independent examination of specifications. Tensile properties are minimum or typical . UTS and YS as Mpa, El as %. See Appendix for list of abbreviations used in Descriptions.

Worldwide Guide to Equivalent Nonferrous Metals and Alloys

Grade UNS #	Country	Specification	Designation	Description	Ag	Al	As	Bi	Cu	Fe	Pb	Sb	Sn	Zn	Other	UTS	YS	EL
L54755	USA			Fusible				21.5			51.5		27					
	USA	ASTM B32	Sn50	Solder	0.015 max	0.005 max	0.025 max	0.25 max	0.08 max	0.02 max		0.50 max	49.5-51.5	0.005 max	Cd 0.001 max			
	Czech Republic	CSN 055622	B-Sn50Pb-215/185	Soft Solder			0.03 max	0.1 max	0.08 max	0.03 max		0.2 max	49-51		Zn+Al+Cd:0.005 S:0.02			
	South Africa	SABS 24	S6 (Sn50Pb)	Pouring Temp 183-215 deg C			0.03 max	0.1 max	0.05 max	0.02 max		0.5 max	49-51					
	USA	SAE J473a	7A									0.4 max	49-51					
	Japan	JIS Z3282	H45B	Bar		0.08					50.84	1	43-47		0.35 max ST ST=Bi+Zn+Fe+Al+As			
	South Africa	SABS 701	WM30	Bar					0.5-1.5	0.05	50.13	15	29-31	0.01	OT 0.03 max			
L51550	USA										50.0				In 50.0			
L54829	USA			Fusible				20			50		30					
L55030	Australia	AS 1834.1	50Sn				0.03 max	0.1 max	0.08 max	0.02 max		0.2 max	49-51		OT 0.08 max			
L55030	Pan America	COPANT 450	50A			0.01 max	0.03 max	0.25 max	0.08 max	0.02 max	50 min	0.12 max	50 max	0.01 max				
L55030	USA			Solder		0.005 max	0.03 max	0.25 max	0.08 max	0.02 max	50	0.12 max	50	0.005 max				
L55031	Pan America	COPANT 450	50B			0.01 max	0.03 max	0.25 max	0.08 max	0.02 max	50 min	0.2-0.5	50 max	0.01 max				
L55031	USA			Solder		0.005 max	0.03 max	0.25 max	0.08 max	0.02 max	50	0.20-0.50	5-50	0.005 max				
	USA	EIA/IPC J-STD-006	In50A	Electronic Grade Solder							50.0				In 50.0			
	Czech Republic	CSN 423744	Pb48Sn32Bi	Low Melting Point Alloy							47-49		31-33		bal Bi			
	South Africa	SABS 24	S2 (Sn50PbSb3)	Pouring Temp 185-204 deg C			0.03 max	0.1 max	0.05 max	0.02 max		2.5-3	49-51					
L54910	USA			Fusible				12.6			47.5		39.9					
L54925	USA			Bearing alloy					2		46	12	40					
	Czech Republic	CSN 423993	Bi50Pb43Cd	Low Melting Point Alloy				49-51			42-44				bal Cd			
L54810	USA			Fusible				28.5			43		28.5					
L54860	USA			Fusible				21			43		36					
L54935	USA			Fusible				14			43		43					
L54865	USA			Fusible				21			42		37					
	South Africa	SABS 700	WM 40	Pouring Temp 350-400 degrees C					3 max		42 max	15 max	40 max					
L13600	Australia	AS 1834.1	60Sn				0.03 max	0.1 max	0.08 max	0.02 max		0.2 max	59-61		0.01 max ST,ST=Al+Zn+Cd			
	Czech Republic	CSN 055624	B-Sn60Pb-190/185	Soft Solder			0.03 max	0.1 max	0.08 max	0.03 max		0.2 max	59-61		Zn+Al+Cd:0.005 S:0.02			
	South Africa	SABS 24	S8 (Sn60Pb)	Pouring Temp 183-190 deg C			0.03 max	0.1 max	0.05 max	0.02 max		0.5 max	59-61					
	Australia	AS 1834.1	60Sn/1.6Cu				0.03 max	0.01 max	1.4-1.8	0.02 max		0.2 max	59-61					
Grade B L50070	USA			Remelted				0.025 max			39.5 min							
	USA	EIA/IPC J-STD-006	In28A	Electronic Grade Solder							38.5		37.5		In 26.0			
L54830	USA			Fusible				30.8			38.4		30.8					

UNS numbers and US grades are provided as a means of cross referencing chemically similar alloys. Exchangability is only possible after independent examination of specifications. Tensile properties are minimum or typical . UTS and YS as Mpa, El as %. See Appendix for list of abbreviations used in Descriptions.

6-18 Lead

Grade UNS #	Country	Specification	Designation	Description	Ag	Al	As	Bi	Cu	Fe	Pb	Sb	Sn	Zn	Other	UTS	YS	EL
	Australia	AS 1834.1	63Sn/0.35Sb				0.03 max	0.25 max	0.05 max	0.02 max		0.2-0.5	62-64					
Alloy 203	Czech Republic	CSN 423990	Bi50Sn16Pb	Low Melting Point Alloy				49-51					15-17					
	South Africa	SABS 24	S1 (Sn65PbSb)	Pouring Temp 183-186 deg C			0.03 max	0.1 max	0.05 max	0.02 max		1 max	64-66					
L54840	USA			Fusible				32			34		34					
L54835	USA			Fusible				33.3			33.4		33.3					
	Czech Republic	CSN 423987	Bi48Sn14,5Sb9Pb	Low Melting Point Alloy				47-49				8-10	13.5-15.5					
	Czech Republic	CSN 423995	Bi55Pb20Sb5Sn	Low Melting Point Alloy				54-56			29-31	5.5			bal Sn			
Alloy 158	Czech Republic	CSN 423989	Bi50Pb27Sn13Cd	Low Melting Point Alloy				49-51			26-28.3		12-14		bal Cd			
	Czech Republic	CSN 423991	Bi50Sn25Pb	Low Melting Point Alloy				49-51					24-26					
	Czech Republic	CSN 423992	Bi50Pb25Sn12Cd	Low Melting Point Alloy				49-51			24-26		11.5-13.5		bal Cd			
	USA	EIA/IPC J-STD-006	Bi46A	Electronic Grade Solder				46.0			20.0		34.0					

UNS numbers and US grades are provided as a means of cross referencing chemically similar alloys. Exchangability is only possible after independent examination of specifications. Tensile properties are minimum or typical . UTS and YS as Mpa, El as %. See Appendix for list of abbreviations used in Descriptions.

Grade UNS #	Country	Specification	Designation	Comment	Al	Cu	Fe	Mg	Mn	Ni	Si	Zn	Zr	OT max	Other	UTS	YS	EL
	Canada	CSA HG.2	0.9999	Cast Ingot		0.001 max		99.99 max	0.001 max	0.01 max	0.001 max				ET 0.005 max; ST 0.002 max; ST=Fe+Ni+Cu; Sn 0.01 max; Pb 0.001 max			
9998A M19998	Canada	CSA HG.2	0.9998	Cast Ingot		0.001 max		99.98 max	0.002 max	0.001 max	0.003 max				ET 0.006 max; ST 0.003 max; ST=Fe+Ni+Cu, ET=.02 total; Sn 0.01 max; Pb 0.001 max			
	Germany	DIN 17800	3.5003/H-Mg99.8			0.02 max	0.05 max	99.8 min	0.1 max	0.002 max	0.1 max							
9998A M19998	USA	ASTM B275	9998A		0.004 max	0.0005 max	0.002 max	99.98 min	0.002 max	0.0005 max	0.003 max				Cd 0.00005 max; Pb 0.001 max; B 0.00003 max; Ti 0.001 max; OE 0.005 max			
9998A M19998	USA	ASTM B92	9998A	Cast	0.004	0.01	0.002		0.002	0.001	0.003				Ti 0.001; Pb 0.002; OE 0.005 max			
M19998	USA			Remelt	0.004 max	0.0005 max	0.002 max	99.98 min	0.002 max	0.0005 max	0.003 max				B 0.00003 max; Cd 0.00005 max; Ti 0.001 max; Pb 0.001 max; OE 0.005 max			
	Germany	DIN 17800	3.5002/H Mg99.95			0.002 max	0.01 max	100 min	0.01 max	0.001 max	0.01 max							
M19995	International	ISO R207	Mg99.95	Cast Ingot Obsolete Replaced by ISO 8287	0.05	0.002	0.003 max		0.01	0.001	0.01		0.01 min		Sn 0.01 max; Pb 0.05			
M19995	Sweden	SIS 144602	SISMg4602-00	Cast	0.01	0.002	0.003 max	99.95 max	0.01	0.001	0.01		0.01 min		ET 0.01 min; ST 0.005 max; ST=Cu+Fe+Ni; Sn 0.01 max; Pb 0.01			
9995A M19995	USA	ASTM B275	9995A		0.010 max	0.003 max		99.95 min	0.004 max	0.001 max	0.005 max				Cd 0.00005 max; B 0.00003 max; Ti 0.01 max; OE 0.005 max			
M19995	USA			Remelt	0.01 max	0.003 max		99.95 min	0.004 max	0.001 max	0.005 max				B 0.00003 max; Cd 0.00005 max; Ti 0.01 max; OE 0.005 max			
9995A M19995	USA	ASTM B92	9995A	Cast	0.01		0.040 max		0.004 min	0.001	0.01				Ti 0.01; OE 0.01 max			
9995A	Canada	CSA HG.2	0.9995	Cast Ingot		0.002 max		99.95 max	0.01 max	0.001 max	0.01 max				ET 0.01 max; ST 0.005 max; ST=FE+Ni+Cu, ET=.03 total; Sn 0.01 max; Pb 0.003 max			
9990A M19990	USA	ASTM B275	9990A		0.003 max		0.04 max	99.90 min	0.004 max	0.001 max	0.005 max				Cd 0.0001 max; B 0.00007 max; OE 0.01 max			
M19990	USA			Remelt	0.003 max		0.04 max	99.90 min	0.004 max	0.001 max	0.005 max				B 0.00007 max; Cd 0.0001 max; OE 0.01 max			
9990A M19990	USA	ASTM B92	9990A	Cast	0.10		0.04 max	99.9	0.004 min	0.01	0.01							
9990B M19991	USA	ASTM B275	9990B		0.005 max		0.011 max	99.90 min	0.004 max	0.001 max	0.005 max				OE 0.01 max			
M19991	USA			Remelt	0.005 max		0.011 max	99.90 min	0.004 max	0.001 max	0.005 max				OE 0.01 max			

UNS numbers and US grades are provided as a means of cross referencing chemically similar alloys. Exchangability is only possible after independent examination of specifications. Tensile properties are minimum or typical. UTS and YS as Mpa, El as %. See Appendix for list of abbreviations used in Descriptions.

Worldwide Guide to Equivalent Nonferrous Metals and Alloys

Grade UNS #	Country	Specification	Designation	Comment	Al	Cu	Fe	Mg	Mn	Ni	Si	Zn	Zr	OT max	Other	UTS	YS	EL
	Canada	CSA HG.2	0.9990	Cast Ingot		0.01 max		99.9 max	0.01 max	0.001 max	0.01 max				ET 0.05 max; ST 0.01 max; ST=Fe+Ni+Cu, ET=.1 total; Sn 0.01 max; Pb 0.01 max			
EK30A M12300	USA	ASTM B275	EK30A	Cast or Wrought		0.10 max				0.01 max		0.30 max	0.20 min	0.30	Rare earth 2.5-4.0			
M12300	USA			Cast		0.10 max				0.01 max		0.30 max	0.20 min	0.30	Rare earth 2.5-4.0			
9998A M19980	Canada	CSA HG.2	0.9980	Cast Ingot		0.02 max		99.8 min	0.15 max	0.001 max					ET 0.05 max; ET=.1 total; Sn 0.01 max; Pb 0.01 max			
M19980	International	ISO R144	Mg99.8	Cast Ingot	0.2	0.02	0.05 max		0.01		0.05				OE 0.05 max			
9980A M19980	USA	ASTM B275	9980A			0.02 max		99.80 min	0.10 max	0.001 max					Pb 0.01 max; Sn 0.01 max; OE 0.05 max			
M19980	USA			Remelt		0.02 max		99.80 min	0.10 max	0.001 max					Sn 0.01 max; Pb 0.01 max; OE 0.05 max			
9980A M19980	USA	ASTM B92	9980A	Cast		0.02		99.8	0.1	0 max					Sn 0.01 max; Pb 0.01; OE 0.05 max			
M19981	Sweden	SIS 144604	SISMg4604-00	Cast	0.05	0.02	0.05 max	99.8	0.01	0.002	0.05				ET 0.05 max;			
9980B M19981	USA	ASTM B275	9980B			0.021 max		99.80 min	0.10 max	0.005 max					Pb 0.01 max; Sn 0.01 max; OE 0.05 max			
	Sweden	MNC 46E	4604	Cast				99.8										
EK41A M12410	USA	ASTM B275	EK41A	Cast or Wrought		0.10 max				0.01 max		0.30 max	0.40-1.0	0.30	Rare earth 3.0-5.0			
M12410	USA			Cast		0.10 max				0.01 max		0.30 max	0.40-1.0	0.30	Rare earth 3.0-5.0			
K1A M18010	USA	ASTM B275	K1A	Cast or Wrought									0.40-1.0	0.30				
K1A M18010	USA	ASTM B403	K1A	Invest Cast As fab									0.4-1	0.3		152	48	14
K1A M18010	USA	ASTM B80	K1A	Sand Cast As fab									0.4-1	0.3		165	41	14
M18010	USA			Cast									0.40-1.0	0.30				
LS141A M14142	USA	ASTM B275	LS141A	Cast or Wrought	0.05 max	0.05 max	0.005 max		0.15 max	0.005 max	0.50-0.6				Li 12.0-15.0; Na 0.005 max			
LS141A M14142	USA	MIL M-46130	LS141A	Wrought Plt Sh Frg As rolled .020/2 in diam	0.05	0.05	0.01 max		0.15	0.01	0.5-0.6			0.2	Na 0.01 max; Li 12-15	124	90	30
M14142	USA			Wrought	0.05 max	0.05 max	0.005 max		0.15 max	0.005 max	0.50-0.6				Na 0.005 max; Li 12.0-15.0			
	UK	BS 2970	MAG6(Mg-RE3ZnZr)	Cast	0.8-3	0.03 max				0.01 max			1 max					
M1C M15102	USA	ASTM B275	M1C	Cast or Wrought	0.01 max	0.02 max	0.03 max		0.9-1.2	0.001 max				0.30	OE 0.05 max			
M15102	USA			Frg	0.01 max	0.02 max	0.03 max		0.9-1.2	0.001 max				0.30	OE 0.05 max			
ZE10A M16100	USA	ASTM B275	ZE10A	Cast or Wrought								1.0-1.5		0.30	Rare earth 0.12-0.22			
ZE10A M16100	USA	ASTM B90	ZE10A	Wrought Sh Ann .251/.500 in								1-1.5		0.3	Rare earth 0.12-0.22	200	83	12
ZE10A M16100	USA	ASTM B90	ZE10A	Wrought Sh H24 .016/.125 in								1-1.5		0.3	Rare earth 0.12-0.22	248	172	4
M16100	USA			Wrought								1.0-1.5		0.30	Rare earth 0.12-0.22			

UNS numbers and US grades are provided as a means of cross referencing chemically similar alloys. Exchangability is only possible after independent examination of specifications. Tensile properties are minimum or typical . UTS and YS as Mpa, El as %. See Appendix for list of abbreviations used in Descriptions.

7-2 Magnesium

Grade UNS #	Country	Specification	Designation	Comment	Al	Cu	Fe	Mg	Mn	Ni	Si	Zn	Zr	OT max	Other	UTS	YS	EL
	UK	BS 3370	Mg-Mn1.5	Wrought Plt Sh Strp As Mfg .5/6 mm diam	0.05	0.02	0.03 max		1-2	0.01	0.02	0.03 min			Ca 0.02 max	200	70	
	UK	BS 3372	Mg-Mn1.5	Wrought Frg As Mfg	0.05	0.02	0.03 max		1-2	0.01	0.02	0.03 min			Ca 0.02 max	200	105	4
	UK	BS 3373	Mg-Mn1.5	Wrought Bar Tub As Mfg 50/100 mm diam	0.05	0.02	0.03 max		1-2	0.01	0.02	0.03 min			Ca 0.02 max	200	105	4
LA141A M14141	USA	ASTM B275	LA141A	Cast or Wrought	1.0-1.5	0.005 max			0.15 min	0.005 max	0.004 max			0.20	Li 13.0-15.0; Na 0.005 max			
LA141A M14141	USA	ASTM B90	LA141A	Wrought Sh	1-1.5	0.04	0.01 max		0.15	0.01	0.1			0.3	Na 0.01 max; Li 13-15			
LA141A M14141	USA	MIL M-46130	LA141A	Wrought Plt Frg Stab .010/.090 in diam	1-1.5	0.04	0.01 max		0.15	0.01	0.1			0.3	Na 0.01 max; Li 13-15	131	103	10
LA141A M14141	USA	MIL R-6944B	LA141A	Wrought Rod	1-1.5	0.04	0.01 max		0.15	0.01	0.1			0.3	Na 0.01 max; Li 13-15			
M14141	USA			Wrought	1.0-1.5	0.005 max			0.15 min	0.005 max	0.004 max			0.20	Na 0.005 max; Li 13.0-15.0			
	Japan	JIS H4204	MS4	Wrought Extrusions		0.03 max				0.01 max		0.8-1.5	0.4-0.8	0.3				
	Japan	JIS H4202	MT4	Wrought Tub Pip		0.03 max				0.01 max		0.8-1.5	0.4-0.8	0.3				
M1B M15101	USA	ASTM B275	M1B	Cast or Wrought		0.08 max			1.3 min	0.01 max	0.10 max			0.20				
M15101	USA			Cast		0.08 max			1.3 min	0.01 max	0.10 max			0.20				
	UK	BS 3370	Mg-Zn1Zr	Wrought Plt Sh Strp As Mfg 25/50 mm diam	0.02	0.03	0.01 max		0.15	0.01	0.01	0.75-1.5	0.4-0.8					
	UK	BS 3372	Mg-Zn1Zr	Wrought Frg As Mfg	0.02	0.03	0.01 max		0.15	0.01	0.01	0.75-1.5	0.4-0.8			200	125	7
	UK	BS 3373	Mg-Zn1Zr	Wrought Bar Tub As Mfg 10/75 mm diam	0.02	0.03	0.01 max		0.15	0.01	0.01	0.75-1.5	0.4-0.8			260	185	8
AZ10A M11100	USA	ASTM B275	AZ10A	Cast or Wrought	1.0-1.5	0.10 max	0.005 max		0.20 min	0.005 max	0.10 max	0.20-0.6		0.30	Ca 0.04 max			
M11100	USA			Cast	1.0-1.5	0.10 max	0.005 max		0.20 min		0.10 max	0.20-0.6			N 0.005 max			
M15100	USA			Ext Shap Wir		0.05 max			1.2 min	0.01 max	0.10			0.30	Ca 0.15			
M1A M15100	USA	ASTM B107	M1A	Wrought Bar Rod Tub Wir As fab (except tub) 2.5/4.999 in diam		0.05			1.2	0.01	0.1			0.3	Ca 0.3	200		2
M1A M15100	USA	ASTM B107	M1A	Wrought Bar Rod Tub Wir As fab (tub) .028/.750 in diam		0.05			1.2	0.01	0.1			0.3	Ca 0.3	193		2
M19995 M15100	USA	WW T-825B	MIA	Wrought Tub As cast .028/.750 in diam		0.05			1.2	0.01	0.1			0.3	Ca 0.3	193		2
M1A M15100	USA	ASTM B275	M1A	Cast or Wrought		0.05 max			1.2 min	0.01 max	0.10 max			0.30	Ca 0.30 max			
AZ21	Canada	CSA HG.5	AZ21	Wrought Bar Rod Wir Extrusions	1.2-2	0.05 max	0.005 max		0.15 max	0.01 max	0.1 max	0.4-0.8		0.3				
M18330	USA			Cast		0.05-0.10				0.01 max			0.40-1.0	0.30	Rare earth 1.5-3.0; Ag 1.3-1.7			

UNS numbers and US grades are provided as a means of cross referencing chemically similar alloys. Exchangability is only possible after independent examination of specifications. Tensile properties are minimum or typical . UTS and YS as Mpa, EI as %. See Appendix for list of abbreviations used in Descriptions.

Grade UNS #	Country	Specification	Designation	Comment	Al	Cu	Fe	Mg	Mn	Ni	Si	Zn	Zr	OT max	Other	UTS	YS	EL
HM21A M13210	USA	ASTM B275	HM21A	Cast or Wrought					0.15-1.1					0.30	Th 1.5-2.5			
M13210	USA			Wrought Frg					0.45-1.1					0.30	Th 1.5-2.5			
HM21A M13210	USA	MIL M-8917A	HM21A	Wrought Sh T8 .016/.250 in diam					0.45-1.1						OE 0.1 max; Th 1.5-2.5	228	124	6
HM21A M13210	USA	MIL M-8917A	HM21A	Wrought Sh T81 .125/.312 in diam					0.45-1.1						OE 0.1 max; Th 1.5-2.5	234	172	4
HM21A M13210	USA	ASTM B90	HM21A	Wrought Sh T8 .501/3 in diam					0.45-1.1					0.3	Th 1.5-2.5	207	145	6
HM21A M13210	USA	ASTM B91	HM21A	Wrought Frg Cooled and art aged 4 in; 102 mm diam					0.45-1.1					0.3	Th 1.5-2.5	228	172	3
HZ32	Canada	CSA HG.9	HZ32	Sand Cast		0.1 max				0.01 max			1.7-2.5	0.5-1	0.3	ST 0.1 max; ST=.10			
QH21A M18210	USA	ASTM B275	QH21A	Cast or Wrought		0.10 max				0.01 max			0.20 max	0.40-1.0	0.30	Rare earth 0.6-1.5; Th 0.6-1.6; Ag 2.0-3.0			
M18210	USA			Cast		0.10 max							0.20 max	0.40-1.0	0.30	N 0.01 max; Rare earth 0.6-1.5; Ag 2.0-3.0			
QE22 M18220	Canada	CSA HG.9	QE22	Sand Cast		0.1 max				0.01 max				0.4-1	0.3	Didymium 1.8-2.5; Ag 2-3			
M18220	Europe	AECMA prEN 2731	Mg-C51-T6	Sand CastSeparately cast T6 temper, up to 20 mm wall thk	0.03 max	0.01 max		0.15 max	0.01 max	0.01 max	0.2 max	0.4-1			Rare earth 1.8; ET 0.05 max; ST 0.75 min; ST=Nd+Pr; ET=.20 total; Ag 2-3	240	175	2
M18220	Europe	AECMA prEN 2732	Mg-C51-T6	CastSeparately cast T6 temper, up to 20 mm wall thk	0.03 max	0.01 max		0.15 max	0.01 max	0.01 max	0.2 max	0.4-1			Rare earth 1.8; ET 0.05 max; ST 0.75 min; ST=Nd+Pr; ET=.20 total; Ag 2-3	240	175	3
QE22A M18220	USA	ASTM B199	QE22A	Cast SHT and art aged		0.1				0.01				0.4-1	0.3	Rare earth 1.8-2.5; Ag 2-3	241	172	2
QE22A M18220	USA	ASTM B403	QE22A	Invest Cast SHT and art aged		0.1				0.01				0.4-1	0.3	Rare earth 1.8-2.5; Ag 2-3	241	172	2
QE22A M18220	USA	ASTM B80	QE22A	Sand Cast SHT and art aged		0.1				0.01				0.4-1	0.3	Rare earth 1.8-2.5; Ag 2-3	241	172	2
QE22A M18220	USA	ASTM B275	QE22A	Cast or Wrought		0.10 max				0.01 max				0.40-1.0	0.30	Rare earth 1.8-2.5; Ag 2.0-3.0			
M18220	USA			Cast		0.10 max				0.01 max				0.40-1.0	0.30	Rare earth 1.8-2.5; Ag 2.0-3.0			
	Europe	AECMA prEN 2735	Mg-C91-T5	Sand CastSeparately cast T5 temper, up to 20 mm wall thk	0.03 max	0.01 max		0.15 max	0.01 max	0.01 max	2-3	0.4-1			Rare earth 2.5-4; ET 0.05 max; ET=.20 total	140	95	3
M12331	Japan	JIS H5203	MC8	Cast Perm Mold Sand		0.1 max				0.01 max		2-3.1	0.5-1		ST 2.5-4			
M12331	USA	AWS A5.19-92	ER EZ33A	Weld El Rod								2-3.1	0.5-1	0.3	Rare earth 2.5-4.0; Be 0.0008 max			
M12331	USA	AWS A5.19-92	R EZ33A	Weld El Rod								2-3.1	0.5-1	0.3	Rare earth 2.5-4.0; Be 0.0008 max			
M12331	USA			Weld Wir								2.0-3.1	0.45-1.0	0.30	Rare earth 2.5-4.0			

UNS numbers and US grades are provided as a means of cross referencing chemically similar alloys. Exchangability is only possible after independent examination of specifications. Tensile properties are minimum or typical . UTS and YS as Mpa, El as %. See Appendix for list of abbreviations used in Descriptions.

Grade UNS #	Country	Specification	Designation	Comment	Al	Cu	Fe	Mg	Mn	Ni	Si	Zn	Zr	OT max	Other	UTS	YS	EL
ZK21A M16210	USA	ASTM B275	ZK21A	Cast or Wrought		0.10 max				0.01 max		2.0-2.6	0.45-0.8	0.30				
M16210	USA											2.0-2.6	0.45-0.8	0.30				
A3A M10030	USA	ASTM B275	A3A	Cast or Wrought	2.5-3.5	0.005 max	0.005 max		0.005 max	0.001 max		0.10 max		0.30	Rare earth 0.001 max; Cd 0.001 max; Ag 0.001 max			
M10030	USA			Cast	2.5-3.5	0.005 max	0.005 max		0.005 max	0.001 max		0.10 max		0.30	Cd 0.001 max; Rare earth 0.001 max; Ag 0.001 max			
AZ21A M11210	USA	ASTM B275	AZ21A	Cast or Wrought	1.6-2.5	0.05 max	0.005 max		0.15 max	0.002 max	0.05 max	0.8-1.6		0.30	Ca 0.10-0.25			
M11210	USA			Cast	1.6-2.5	0.05 max	0.005 max		0.15 max	0.002 max	0.05 max	0.8-1.6		0.30	Ca 0.10-0.25			
M12330	International	ISO 2119	Mg-RE3Zn2Zr	Cast Obsolete replaced by ISO 3115		0.1				0.01		0.8-3	0.4-1		Rare earth 2.5-4			
M12330	International	ISO 3115	Mg-RE3Zn2Zr	Sand Cast TE		0.1				0.01		0.8-3	0.4-1		Rare earth 2.5-4	140	95	2
EZ33A M12330	USA	ASTM B199	EZ33A	Cast Cooled and art aged		0.1				0.01		2-3.1	0.5-1		Rare earth 2.5-4	138	97	2
EZ33A M12330	USA	ASTM B275	EZ33A	Cast or Wrought		0.10 max				0.01 max		2.0-3.1	0.50-1.0	0.30	Rare earth 2.5-4.0			
EZ33A M12330	USA	ASTM B403	EZ33A	Invest Cast Cooled and art aged		0.1				0.01		2-3.1	0.5-1	0.3	Rare earth 2.5-4	138	97	2
EZ33A M12330	USA	ASTM B80	EZ33A	Sand Cast Cooled and art aged		0.1				0.01		2-3.1	0.5-1	0.3	Rare earth 2.5-4	138	96	2
EZ33A M12330	USA	MIL R-6944B	EZ33A	Wrought Rod								2-3.1		0.3				
M12330	USA			Sand Perm Mold Inv Cast		0.10 max				0.01 max		2.0-3.1	0.50-1.0	0.30	Rare earth 2.5-4.0			
EZ33	Canada	CSA HG.9	EZ33	Sand Cast		0.1 max				0.01 max		2-3.1	0.5-1	0.3	Rare earth 2.5-4			
	UK	BS 3370	Mg-Zn2Mn1	Wrought Plt Sh Strp As Mfg 6/25 mm diam	0.2	0.1	0.06 max		0.6-1.3	0.01	0.1	1.5-2.3				220	120	8
	UK	BS 3370	Mg-Zn2Mn1	Wrought Plt Sh Strp Ann .5/6 mm diam	0.2	0.1	0.06 max		0.6-1.3	0.01	0.1	1.5-2.3				220	120	
	UK	BS 3372	Mg-Zn2Mn1	Wrought Frg As Mfg	0.2	0.1	0.06 max		0.6-1.3	0.01	0.1	1.5-2.3				200	125	9
	UK	BS 3373	Mg-Zn2Mn1	Wrought Bar Tub As Mfg 10/75 mm diam	0.2	0.1	0.06 max		0.6-1.3	0.01	0.1	1.5-2.3				245	160	10
AZ31 M11310	Canada	CSA HG.4	AZ31	Wrought Plt Sh	2.5-3.5	0.05 max	0.01 max		0.15-0.4	0.01 max	0.1 max	0.6-1.4		0.3	Ca 0.04 max			
AZ31 M11310	Canada	CSA HG.5	AZ31	Wrought Bar Rod Wir Extrusions	2.5-3.5	0.05 max	0.005 max		0.15 max			0.6-1.4			ET 0.02 max; ET=.10 total			
AZ31A M11310	USA	ASTM B275	AZ31A	Cast or Wrought	2.5-3.5	0.05 max	0.005 max		0.20 min	0.005 max	0.30 max	0.6-1.4		0.30	Ca 0.30 max			
M11310	USA			Sh Plt	2..5-3.5 min	0.05 max	0.005 max		0.20 min	0.005 max	0.30 max	0.6-1.4		0.30	Ca 0.30 max			
HK31	Canada	CSA HG.9	HK31	Sand Cast		0.1 max				0.01 max		0.3 max	0.4-1	0.3	Th 2.5-4			
	Japan	JIS H4203	MB4	Wrought Bar		0.03 max				0.01 max		2.5-4	0.4-0.8	0.3				
	Japan	JIS H4203	MB5	Wrought Bar		0.03 max				0.01 max		2.5-4	0.4-0.8	0.3				
	Japan	JIS H4204	MS5	Wrought Extrusions		0.03 max				0.01 max		2.5-4	0.4-0.8	0.3				

UNS numbers and US grades are provided as a means of cross referencing chemically similar alloys. Exchangability is only possible after independent examination of specifications. Tensile properties are minimum or typical . UTS and YS as Mpa, El as %. See Appendix for list of abbreviations used in Descriptions.

Grade UNS #	Country	Specification	Designation	Comment	Al	Cu	Fe	Mg	Mn	Ni	Si	Zn	Zr	OT max	Other	UTS	YS	EL
M12350	USA			Cast		0.10 max				0.01 max		0.20 max	0.40-1.0	0.30	Rare earth 0.6-1.5; Th+Di 1.5-2.4; Th 0.6-1.6; Ag 2.0-3.0			
ZK31	Canada	CSA HG.5	ZK31	Wrought Bar Rod Wir Extrusions			0.003 max			0.002 max		2.5-3.5	0.5-1		ET 0.02 max; ET=.10			
ZM21	Canada	CSA HG.5	ZM21	Wrought Bar Rod Wir Extrusions	0.02 max	0.003 max	0.003 max		1-1.5	0.001 max	0.01 max	2-2.8			ET 0.02 max; ET=.10 total			
	UK	BS 3370	Mg-Zn3Zr	Wrought Plt Sh Strp As Mfg 6/50 mm	0.02	0.03	0.01		0.15 max	0.01	0.01	2.5-4	0.4-0.8			250	150	8
	UK	BS 3372	Mg-Zn3Zr	Wrought Frg As Mfg	0.02	0.03	0.01 max		0.15	0.01	0.01	2.5-4	0.4-0.8			270	180	7
	UK	BS 3373	Mg-Zn3Zr	Wrought Bar Tub As Mfg 10/100 mm	0.02	0.03	0.01 max		0.15	0.01	0.01	2.5-4	0.4-0.8			305	225	8
AZ31C M11312	France	NF A65-717	G-A3Z1	Wrought Sh Strp Ann (sht and strp) .5/6 mm diam	2.3-3.5	0.1	0.03 max		0.2	0.01	0.1				Ca 0.04	230	130	12
AZ31C M11312	France	NF A65-717	G-A3Z1	Wrought Sh Strp As Mfg (wir and shapes)	2.3-3.5	0.1	0.03 max		0.2	0.01	0.1				Ca 0.04	240	160	10
M11312	International	ISO R503	25	Cast Ingot Obsolete Replaced by ISO 3116	2.4-3.6	0.01	0.03 max		0.15-0.4	0.01	0.1	0.5-1.5						
M11312	International	ISO 3116	Mg-Al3Zn1 (Alloy 21)	Wrought Plt Sh Strp Ann	2.5-3.5	0.05			0.2		0.1	0.5-1.5				220	105	11
M11312	International	ISO 3116	Mg-Al3Zn1 (Alloy 25)	Wrought Plt Sh Strp ANN	2.4-3.6	0.1	0.03 max		0.15-0.4	0.01	0.1	0.5-1.5				220	105	11
M11312	International	ISO 3116	Mg-Al3Zn1 (Alloy 25)	Wrought Plt Sh Strp SH	2.4-3.6	0.1	0.03 max		0.15-0.4	0.01	0.1	0.5-1.5				250	160	5
M11312	Japan	JIS H4201	MP1	Wrought Plt sh	2.4-3.6	0.1 max	0.01 max		0.15 min		0.01 max	0.5-1.5						
AZ31C M11312	USA	ASTM B107	AZ31C	Wrought Bar Tub Rod Wir	2.4-3.6	0.1			0.15	0.03	0.1	0.5-1.5		0.3				
AZ31C M11312	USA	ASTM B275	AZ31C	Cast or Wrought	2.4-3.6	0.10 max			0.15 min	0.03 max	0.10 max	0.50-1.5		0.30				
AZ31C M11312	USA	ASTM B90	AZ31C	Wrought Sh	2.4-3.6	0.1	0.01 max		0.15	0.03	0.1	0.5-1.5		0.3				
M11312	USA			Wrought Ext Shp	2.4-3.6	0.10 min			0.15 min	0.03 max	0.10 max	0.50-1.5		0.30				
	Japan	JIS H4203	MB1	Wrought Bar	2.5-3.5	0.1 max	0.01 max		0.15 min	0.01 max	0.1 max	0.5-1.5		0.3	Ca 0.04 max			
	Japan	JIS H4204	MS1	Wrought Extrusions	2.5-3.5	0.1 max	0.01 max		0.15 min	0.01 max	0.1 max	0.5-1.5		0.3	Ca 0.04 max			
	Japan	JIS H4202	MT1	Wrought Tub Pip	2.5-3.5	0.1 max	0.01 max		0.15 min	0.01 max	0.1 max	0.5-1.5		0.3	Ca 0.04 max			
M11311	International	ISO R503	21	Cast Ingot Obsolete Replace by ISO 3116	2.5-3.5	0.05	0.01 max		0.2	0.01	0.1	0.5-1.5			Ca 0.04			
M11311	International	ISO 3116	Mg-Al13Zn1 (Alloy 21) M	Wrought Tub	2.5-3.5	0.05	0.01 max		0.2	0.01	0.1	0.5-1.5			Ca 0.04	230	150	6
M11311	International	ISO 3116	Mg-Al13Zn1 (Alloy 25) M	Wrought tub	2.4-3.6	0.1	0.03 max		0.15-0.4	0.01	0.1	0.5-1.5			Ca 0.04	230	150	6
M11311	International	ISO 3116	Mg-Al3Zn1 (Alloy 21) M	Wrought Frg	2.5-3.5	0.05	0.01 max		0.2	0.01	0.1	0.5-1.5			Ca 0.04	240	130	6
M11311	International	ISO 3116	Mg-Al3Zn1 (Alloy 21) M	Wrought Bar	2.5-3.5	0.05	0.01 max		0.2	0.01	0.1	0.5-1.5			Ca 0.04	240	150	6
M11311	International	ISO 3116	Mg-Al3Zn1 (Alloy 25) M	Wrought Bar	2.4-3.6	0.1	0.03 max		0.15-0.4	0.01	0.1	0.5-1.5			Ca 0.04	240	150	6
M11311	International	ISO 3116	Mg-Al3Zn1 (Alloy 25) M	Wrought Frg	2.4-3.6	0.1	0.03 max		0.15-0.4	0.01	0.1	0.5-1.5			Ca 0.04	240	130	6

UNS numbers and US grades are provided as a means of cross referencing chemically similar alloys. Exchangability is only possible after independent examination of specifications. Tensile properties are minimum or typical . UTS and YS as Mpa, El as %. See Appendix for list of abbreviations used in Descriptions.

7-6 Magnesium

Grade UNS #	Country	Specification	Designation	Comment	Al	Cu	Fe	Mg	Mn	Ni	Si	Zn	Zr	OT max	Other	UTS	YS	EL
M11311	UK	BS 3373	Mg-A13Zn1Mn	Wrought Bar Tub As Mfg 10/75 mm diam	2.5-3.5	0.1	0.03 max		0.15-0.4	0.01	0.1	0.6-1.4			Ca 0.04 max	245	160	10
M11311	UK	BS 3370	Mg-Al3Zn1Mn	Wrought Plt Sh Strp As Mfg .5/6 mm diam	2.5-3.5	0.1	0.03 max		0.15-0.4	0.01	0.1	0.6-1.4			Ca 0.04 max	250	160	
M11311	UK	BS 3370	Mg-Al3Zn1Mn	Wrought Plt Sh Strp Ann .5/6 mm diam	2.5-3.5	0.1	0.03 max		0.15-0.4	0.01	0.1	0.6-1.4			Ca 0.04 max	220	120	
AZ31B M11311	USA	ASTM B107	AZ31B	Wrought Bar Tub Rod Wir As fab (except tub) 2.5/4.999 in diam	2.5-3.5	0.05	0.01 max		0.2	0.01	0.1	0.6-1.4		0.3	Ca 0.04	221	138	7
AZ31B M11311	USA	ASTM B107	AZ31B	Wrought Bar Tub Rod Wir As fab (tub) .028/.500 in diam	2.5-3.5	0.05	0.01 max		0.2	0.01	0.1	0.6-1.4		0.3	Ca 0.04	221	110	8
AZ31B M11311	USA	ASTM B90	AZ31B	Wrought Sh Ann .501/2 in diam	2.5-3.5	0.05	0.01 max		0.2	0.01	0.1	0.6-1.3		0.3	Ca 0.04	221		10
AZ31B M11311	USA	ASTM B90	AZ31B	Wrought Sh H24 2.001/3 in diam	2.5-3.5	0.05	0.01 max		0.2	0.01	0.1	0.6-1.3		0.3	Ca 0.04	234	124	8
AZ31B M11311	USA	ASTM B90	AZ31B	Wrought Sh H26 1.501/2 in diam	2.5-3.5	0.05	0.01 max		0.2	0.01	0.1	0.6-1.3		0.3	Ca 0.04	241	145	6
AZ31B M11311	USA	ASTM B91	AZ31B	Wrought Frg As fab	2.5-3.5	0.05	0.01 max		0.2	0.01	0.1	0.6-1.4		0.3	Ca 0.04	234	131	6
M19990 M11311	USA	WW T-825B	AZ31B	Wrought Tub As cast .028/.250 in diam	2.5-3.5	0.05	0.01 max		0.2	0.01	0.1	0.6-1.4		0.3	Ca 0.04	221	110	8
AZ31B M11311	USA	ASTM B275	AZ31B	Cast or Wrought	2.5-3.5	0.05 max	0.005 max		0.20 min	0.005 max	0.10 max	0.6-1.4		0.30	Ca 0.04 max			
M11311	USA			Wrough; Ext Shp Frg	2.5-3.5	0.05 min	0.05 max		0.20 min	0.005 max	0.10 max	0.6-1.4		0.30	Ca 0.04 max			
	UK	BS 2970	MAG4(Mg-Zn4.5Zr)	Cast		0.03 max				0.01 max		3.5-5.5	1 max		ST 0.35 max; ST = Cu + Si + Fe + Ni			
	UK	BS 2970	MAG5(Mg-Zn4REZr)	Cast		0.03 max				0.01 max		3.5-5	1 max		Rare earth 1-1.75			
HM31A M13312	USA	ASTM B275	HM31A	Cast or Wrought					1.2 min					0.30	Th 2.5-3.5			
M13312	USA			Cast					1.2 min					0.30	Th 2.5-3.5			
HZ31A M13312	USA	ASTM B80	HZ31A	Sand Cast Cooled and art aged		0.1				0.01		1.7-2.5	0.5-1	0.3	Rare earth 0.1; Th 2.5-4	186	89	4
HK31A M13310	USA	ASTM B275	HK31A	Cast or Wrought		0.10 max				0.01 max		0.30 max	0.40-1.0	0.30	Th 2.5-4.0			
M13310	USA			Cast		0.10 max				0.01 max		0.30 max	0.40-1.0	0.30	Th 2.5-4.0			
HK31A M13310	USA	ASTM B199	HK31A	Wrought SHT and art aged		0.1				0.01		0.3	0.4-1	0.3	Th 2.5-4	186	90	4
HK31A M13310	USA	ASTM B403	HK31A	Invest Cast SHT and art aged		0.1				0.03		0.3	0.4-1	0.3	Th 2.5-4	186	90	4
HK31A M13310	USA	ASTM B80	HK31A	Wrought Sand SHT and art aged		0.1				0.01		0.3	0.4-1	0.3	Th 2.5-4	186	89	4
HK31A M13310	USA	ASTM B90	HK31A	Wrought sh Ann 1.001/3 in diam		0.1				0.01		0.3	0.4-1	0.3	Th 2.5-4	200	97	12

UNS numbers and US grades are provided as a means of cross referencing chemically similar alloys. Exchangability is only possible after independent examination of specifications. Tensile properties are minimum or typical . UTS and YS as Mpa, El as %. See Appendix for list of abbreviations used in Descriptions.

Worldwide Guide to Equivalent Nonferrous Metals and Alloys

Grade UNS #	Country	Specification	Designation	Comment	Al	Cu	Fe	Mg	Mn	Ni	Si	Zn	Zr	OT max	Other	UTS	YS	EL
HK31A M13310	USA	ASTM B90	HK31A	Wrought sh H24 1.001/3 in diam		0.1				0.01		0.3	0.4-1	0.3	Th 2.5-4	228	172	4
HK32A M13310	USA	MIL M-46062	HK32A	Cast T6		0.1				0.01		0.3	0.5-1	0.3	Th 2.5-4	227	110	6
M16410	Europe	AECMA prEN 2738	Mg-C43-T5	Sand CastSeparately cast T5 temper, up to 20 mm wall thk	0.02 max	0.03 max	0.01 max		0.15 max	0.01 max	0.01 max	3.5-5	0.4-1		Rare earth 0.8-1.7; ET 0.05 max; ET=.20 total	210	145	3
M16410	Europe	AECMA prEN 2739	Mg-C43-T5	Cast Separately cast T5 temper, up to 20 mm wall thk	0.02 max	0.03 max	0.01 max		0.15 max	0.01 max	0.01 max	3.5-5	0.4-1		Rare earth 0.8-1.7; ET 0.05 max; ET=.20 total	200	1400	3
M16410	International	ISO 2119	Mg-Zn4REZr	Cast Obsolete Replaced by ISO 3115		0.1				0.01		3.5-5	0.4-1		Rare earth 0.75-1.75			
M16410	International	ISO 3115	Mg-Zn4REZr	Sand Cast TE		0.1				0.01		3.5-5	0.4-1		Rare earth 0.75-1.75	200	135	2
ZE41A M16410	USA	ASTM B275	ZE41A	Cast or Wrought		0.10 max			0.15 max	0.01 max		3.5-5.0	0.40-1.0	0.30	Rare earth 0.75-1.75			
ZE41A M16410	USA	ASTM B80	ZE41A	Sand Cast Cooled and art aged		0.1			0.15	0.01		3.5-5	0.4-1	0.3	Rare earth 0.75-1.75	200	133	3
M16410	USA			Cast		0.10 max			0.15 max	0.01 max		3.5-5.0	0.40-1.0	0.30	Rare earth 0.75-1.75			
ZE41	Canada	CSA HG.9	ZE41	Sand Cast		0.1 max			0.15 max	0.01 max		3.5-5	0.4-1	0.3	Rare earth 0.75-1.75			
ZK40A M16400	USA	ASTM B107	ZK40A	Wrought Bar Tub Rod Wir Cooled and art aged(except tub)								3.5-4.5	0.5 max	0.3		276	255	4
ZK40A M16400	USA	ASTM B107	ZK40A	Wrought Bar Tub Rod Wir Cooled and art aged (tub) .062/.500 in diam								3.5-4.5	0.5 max	0.3		276	248	4
ZK40A M16400	USA	ASTM B275	ZK40A	Cast or Wrought		0.10 max				0.01 max		3.5-4.5	0.45 mn	0.30				
M16400	USA			Ext Shap		0.10 max				0.01 max		3.5-4.5	0.45 min	0.30				
M16510	International	ISO 2119	Mg-Zn5Zr	Cast Obsolete Replaced by ISO 3115		0.1				0.01		3.5-5.5	0.4-1					
M16510	International	ISO 3115	Mg-Zn5Zr	Sand Cast TE		0.1				0.01		3.5-5.5	0.4-1			235	140	4
M16510	Japan	JIS H5203	MC6	Cast Perm Mold Sand		0.1 max				0.01 max		3.6-5.5	0.5-1					
ZK51A M16510	USA	ASTM B275	ZK51A	Cast or Wrought								3.6-5.5	0.50-1.0	0.30				
ZK51A M16510	USA	ASTM B80	ZK51A	Sand Cast Cooled and art aged		0.1				0.01		3.6-5.5	0.5-1	0.3		234	138	5
ZK51A M16510	USA	MIL M-46062	ZK51A	Cast T6		0.1				0.01		3.6-5.5	0.5-1	0.3		248	145	6
M16510	USA			Cast		0.10 max				0.01 max		3.6-5.5	0.50-1.0	0.30				
AS41A M10410	USA	ASTM B275	AS41A	Cast or Wrought	3.5-5.0	0.06 max			0.20-0.50	0.03 max	0.50-1.5	0.12 max		0.30				
AS41A M10410	USA	ASTM B93	AS41A	Cast	3.7-4.8	0.04			0.22-0.48	0.01	0.6-1.4	0.1		0.3				

UNS numbers and US grades are provided as a means of cross referencing chemically similar alloys. Exchangability is only possible after independent examination of specifications. Tensile properties are minimum or typical . UTS and YS as Mpa, El as %. See Appendix for list of abbreviations used in Descriptions.

7-8 Magnesium

Grade UNS #	Country	Specification	Designation	Comment	Al	Cu	Fe	Mg	Mn	Ni	Si	Zn	Zr	OT max	Other	UTS	YS	EL
AS41A M10410	USA	ASTM B94	AS41A	Cast Die As cast	3.4-5	0.06			0.2-0.5	0.03	0.5-1.5	0.12				210	140	6
M10410	USA			Die Cast	3.5-5.0	0.06 max			0.20-0.50	0.03 max	0.50-1.5	0.12 max		0.30				
	UK	BS 2970	MAG8(Mg-Th3Zn2Zr)	Cast		0.03 max	0.01 max		0.15 max	0.01 max	0.01 max	1.7-2.5	1 max		Rare earth 0.1 max; Th 2.5-4			
M13320	Europe	AECMA prEN 2733	Mg-C81-T5	Sand CastSeparately cast T5 temper, up to 20 mm wall thk		0.03 max	0.01 max		0.15 max	0.01 max	0.01 max	1.7-2.5	0.4-1		Rare earth 0.1 max; ET 0.05 max; ET=.20 total; Th 2.5-4			
M13320	International	ISO 2119	Mg-Th3Zn2Zr	Cast Obsolete Replaced by ISO 3115		0.1				0.01		1.7-2.5	0.4-1		Rare earth 0.1; Th 2.5-4			
M13320	International	ISO 3115	Mg-Th3Zn2Zr	Sand Cast TE		0.1				0.01		1.7-2.5	0.4-1		Rare earth 0.1; Th 2.5-4	185	90	3
HZ32A M13320	USA	ASTM B275	HZ32A	Cast or Wrought		0.10 max				0.01 max		1.7-2.5	0.50-1.0	0.30	Rare earth 0.10 max; Th 2.5-4.0			
M13320	USA			Cast		0.10 max				0.01 max		1.7-2.5	0.50-1.0	0.30	Rare earth 0.10 max; Th 2.5-4.0			
ZK60A M16600	France	NF A65-717	G-Z55Zr	Wrought Wir As Mfg (wir)	0.02		0.01 max		0.15	0.01	0.01	4.8-6.2				300	210	6
ZK60A M16600	France	NF A65-717	G-Z55Zr	Wrought Wir T5 1/10 mm diam	0.02		0.01 max		0.15	0.01	0.01	4.8-6.2				300	250	5
M16600	Japan	JIS H4203	MB6	Wrought Bar		0.03 max				0.01 max		4.8-6.2	0.5-0.8	0.3				
M16600	Japan	JIS H4204	MS6	Wrought Extrusions		0.03 max				0.01 max		4.8-6.2	0.5-0.8	0.3				
M16600	UK	BS 3372	Mg-Zn6Zr	Wrought Frg Cooled and Art aged	0.02	0.03	0.01 max		0.15	0.01	0.01	4.8-6.2	0.5-0.8			280	180	7
M16600	UK	BS 3373	Mg-Zn6Zr	Wrought Bar Tub Cooled and art aged 10/50 mm diam	0.02	0.03	0.01 max		0.15	0.01	0.01	4.8-6.2	0.5-0.8			315	230	8
ZK60A M16600	USA	ASTM B107	ZK60A	Wrought Bar Tub Rod Wir As fab (except tube) 5./39.999 in diam								4.8-6.2	0.5	0.3		296	214	4
ZK60A M16600	USA	ASTM B107	ZK60A	Wrought Bar Tub Rod Wir As fab (tub) .028/.750 in diam								4.8-6.2	0.5	0.3		276	193	5
ZK60A M16600	USA	ASTM B107	ZK60A	Wrought Bar Tub Rod Wir Cooled and art aged (except tube)								4.8-6.2	0.5	0.3		310	248	4
ZK60A M16600	USA	ASTM B275	ZK60A	Cast or Wrought								4.8-6.2	0.45 min	0.30				
ZK60A M16600	USA	ASTM B91	ZK60A	Wrought Die Cooled and art aged 3 in;76 mm diam								4.8-6.2	0.5	0.3		290	179	7
ZK60A M16600	USA	WW T-825B	ZK60A	Wrought Tub SHT and art aged .028/.250 in diam								4.8-6.2	0.5	0.3		317	262	4
M16600	USA			Wrought Frg								4.8-6.2	0.45 min	0.30				

UNS numbers and US grades are provided as a means of cross referencing chemically similar alloys. Exchangability is only possible after independent examination of specifications. Tensile properties are minimum or typical . UTS and YS as Mpa, El as %. See Appendix for list of abbreviations used in Descriptions.

Worldwide Guide to Equivalent Nonferrous Metals and Alloys

Grade UNS #	Country	Specification	Designation	Comment	Al	Cu	Fe	Mg	Mn	Ni	Si	Zn	Zr	OT max	Other	UTS	YS	EL
ZK60 M16601	Canada	CSA HG.5	ZK60	Wrought Bar Rod Wir			0.003 max			0.002 max		4.8-6.2	0.5 max		ET 0.02 max; ET=.10 total			
ZK601B M16601	USA	ASTM B275	ZK601B	Cast or Wrought								4.8-6.8	0.45 min	0.30				
M16601	USA											4.8-6.8	0.45 min	0.30				
M10602	USA			Cast	5.5-6.5	0.010 max	0.005 max		0.25 min	0.002 max	0.10 max	0.22 max			If Fe > 0.005 then Fe:Mn<0.021; OE 0.02 max			
M16630	International	ISO R121	Mg-Al6Zn3	Cast Sand Bar As cast 13 mm diam Obsolete	5-7	0.2	0.05 max		0.1-0.5	0.01	0.3	2-3.5				160	75	3
ZE63A M16630	USA	ASTM B275	ZE63A	Cast or Wrought		0.10 max				0.01 max		5.5-6.0	0.40-1.0	0.30	Rare earth 2.1-3.0			
ZE63A M16630	USA	ASTM B80	ZE63A	Sand Cast SHT art aged		0.1				0.01		5.5-6	0.4-1	0.3	Rare earth 2.1-3	276	186	5
ZE63A M16630	USA	MIL M-46062	ZE63A	Cast T6		0.1 min				0.01		5.5-6	0.4-1	0.3	Rare earth 2-3	290	193	6
M16630	USA			Cast								5.5-6.0	0.40-1.0	0.30	Rare earth 2.1-3.0			
M10603	USA			Ingot	5.7-6.3	0.008 max	0.004 max		0.27 min	0.001 max	0.05 max	0.20 max			OE 0.01 max			
M16610	International	ISO 3115	Mg-Zn6Zr	Sand Cast ST and PT		0.1				0.01		5.5-6.5	0.6-1			275	180	4
M16610	Japan	JIS H5203	MC7	Cast Perm Mold Sand		0.1 max				0.01 max		5.5-6.5	0.6-1					
ZK61A M16610	USA	ASTM B275	ZK61A	Cast or Wrought								5.5-6.5	0.6-1.0	0.30				
ZK61A M16610	USA	ASTM B403	ZK61A	Invest Cast SHT art aged		0.1				0.01		5.5-6.5	0.6-1	0.3		276	172	5
ZK61A M16610	USA	ASTM B80	ZK61A	Sand Cast SHT art aged		0.1				0.01		5.5-6.5	0.6-1	0.3		276	179	
ZK61A M16610	USA	MIL M-46062	ZK61A	T6		0.1				0.01		5.5-6.5	0.6-1	0.3		290	200	6
M16610	USA			Cast								5.5-6.5	0.6-1.0	0.30				
ZK61	Canada	CSA HG.9	ZK61	Sand Cast			0.005 max			0.01 max		5.5-6.5	0.6-1					
	Japan	JIS H4202	MT2	Wrought Tub Pip	5.5-7.2	0.1 max	0.01 max		0.15-0.4	0.01 max	0.1 max	0.5-1.5		0.3				
AZ61	Canada	CSA HG.5	AZ61	Wrought Bar Rod Wir Extrusions	5.8-7.2	0.003 max	0.005 max		0.15 max	0.002 max	0.01 max	0.4-1.5			ET 0.02 max; ET=.10 total			
M11610	International	ISO R503	22	Cast Ingot Obsolete Replace by ISO 3116	5.5-7.2	0.05	0.01 max		0.15	0.01	0.1	0.5-1.5						
M11610	International	ISO R503	26	Cast Ingot SH Obsolete Replace by ISO 3116	5.4-7.3	0.01	0.03 max		0.15-0.4	0.01	0.1	0.5-1.5				250	160	5
M11610	International	ISO 3116	Mg-Al6Zn1 (Alloy 22)	Wrought Frg M	5.5-7.2	0.05	0.01 max		0.15	0.01	0.1	0.5-1.5				270	150	5
M11610	International	ISO 3116	Mg-Al6Zn1 (Alloy 22)	Wrought tub M	5.5-7.2	0.05	0.01 max		0.15	0.01	0.1	0.5-1.5				260	150	6
M11610	International	ISO 3116	Mg-Al6Zn1 (Alloy 22)	Wrought Bar M	5.5-7.2	0.05	0.01 max		0.15	0.01	0.1	0.5-1.5				270	180	6
M11610	International	ISO 3116	Mg-Al6Zn1 (Alloy 26)	Wrought Tub M	5.4-7.3	0.1	0.03 max		0.15-0.4	0.01	0.1	0.5-1.5				260	150	6
M11610	International	ISO 3116	Mg-Al6Zn1 (Alloy 26)	Wrought Frg M	5.4-7.3	0.1	0.03 max		0.15-0.4	0.01	0.1	0.5-1.5				270	150	6
M11610	International	ISO 3116	Mg-Al6Zn1 (Alloy 26)	Wrought Bar as cast	5.4-7.3	0.1	0.03 max		0.15-0.4	0.01	0.1	0.5-1.5				270	180	6
M11610	International	ISO 2119	Mg-Zn6Zr	Cast Obsolete Replaced by ISO 3115		0.1				0.01		5.5-6.5	0.6-1					

UNS numbers and US grades are provided as a means of cross referencing chemically similar alloys. Exchangability is only possible after independent examination of specifications. Tensile properties are minimum or typical . UTS and YS as Mpa, EI as %. See Appendix for list of abbreviations used in Descriptions.

7-10 Magnesium

Grade UNS #	Country	Specification	Designation	Comment	Al	Cu	Fe	Mg	Mn	Ni	Si	Zn	Zr	OT max	Other	UTS	YS	EL
M11610	Japan	JIS H4203	MB2	Wrought Bar	5.5-7.2	0.1 max	0.01 max		0.15-0.4	0.01 max	0.1 max	0.5-1.5		0.3				
M11610	Japan	JIS H4204	MS2	Wrought Extrusions	5.5-7.2	0.1 max	0.01 max		0.15-0.4	0.01 max	0.1 max	0.5-1.5		0.3				
M11610	UK	BS 3373	Mg-A16Zn1Mn	Wrought Bar TubAs Mfg 10/75 mm diam	5.5-6.5	0.1	0.03 max		0.15-0.4	0.01	0.1	0.5-1.5				250	160	6
M11610	UK	BS 3372	Mg-Al6Zn1Mn	Wrought Frg As Mfg	5.5-6.5	0.1	0.03 max		0.15-0.4	0.01	0.1	0.5-1.5				270	160	7
AZ61A M11610	USA	MIL R-6944B	AZ61A	Wrought Rod	5.8-7.2	0.05	0.01 max		0.15	0.01	0.05	0.4-1.5		0.3	Be 0.001 max			
AZ61A M11610	USA	ASTM B107	AZ61A	Wrought Bar Tub Rod Wir As fab (except tube) 2.5/4.999 in diam	5.8-7.2	0.05	0.01 max		0.15	0.01	0.1	0.4-1.5		0.3		276	152	7
AZ61A M11610	USA	ASTM B107	AZ61A	Wrought Bar Tub Rod Wir As fab (tub) .028/.750 in diam	5.8-7.2	0.05	0.01 max		0.15	0.01	0.1	0.4-1.5		0.3		248	110	7
AZ61A M11610	USA	ASTM B275	AZ61A	Cast or Wrought	5.8-7.2	0.05 max	0.005 max		0.15 min	0.005 max	0.10 max	0.40-1.5		0.30				
AZ61A M11610	USA	ASTM B91	AZ61A	Wrought Frg As fab	5.8-7.2	0.05	0.01 max		0.15	0.01	0.1	0.4-1.5		0.3		262	152	6
M19991 M11610	USA	WW T-825B	AZ61A	WRought Tub As cast .028/.750 in diam	5.8-7.2	0.05	0.01 max		0.15	0.01	0.1	0.4-1.5		0.3		248	110	7
ZK61A M11610	USA	MIL M-46062	ZK61A	Cast		0.1				0.01		5.5-6.5	0.6-1	0.3				
M11610	USA			Frg Ext Shp	5.8-7.2	0.05 min	0.005 max		0.15 min	0.005 max	0.10 max	0.40-1.5		0.30				
AZ61A M11611	France	NF A65-717	G-A6Z1	Wrought Wir As Mfg (wir)	5.5-6.5	0.1	0.03 max		0.15-0.4	0.01	0.1	0.5-1.5				280	180	8
M11611	USA	AWS A5.19-92	ER AZ61A	Weld El Rod	5.8-7.2	0.05 max	0.01 max		0.15-0.5	0.005 max	0.05 max	0.4-1.5		0.3	Be 0.0002-0.0008			
M11611	USA	AWS A5.19-92	R AZ61A	Weld El Rod	5.8-7.2	0.05 max	0.01 max		0.15-0.5	0.005 max	0.05 max	0.4-1.5		0.3	Be 0.0002-0.0008			
M11611	USA			Weld Wir	5.8-7.2	0.05 min	0.005 max		0.15 min	0.05 max	0.05 max	0.40-1.5		0.30	Be 0.0002-0.0008			
	UK	BS 2970	MAG9(Mg-Zn5.5Th2Zr)	Cast		0.03 max	0.01 max		0.15 max	0.01 max	0.01 max	5.3-6	1 max		Rare earth 0.2 max; Th 1.5-2.3			
M10600	Norway	NS 17705	17705	Sand Cast T4 temper	6 max			93 max	0.2 max									
M10600	Norway	NS 17705	17705	Cast Die F temper	6 max			93 max	0.2 max									
M10600	Norway	NS 17705	17705	Sand Cast F temper	6 max			93 max	0.2 max									
AM60A M10600	USA	ASTM B275	AM60A	Cast or Wrought	5.5-6.5	0.35 max			0.13 min	0.03 max	0.50 max	0.22 max		0.30				
AM60A M10600	USA	ASTM B93	AM60A	Cast	5.7-6.3	0.25			0.15	0.01	0.2	0.2		0.3				
AM60A M10600	USA	ASTM B94	AM60A	Cast Die As cast	5.5-6.5	0.35			0.13	0.03	0.5	0.2				220	130	8
M10600	USA			Die Mold Cast	5.5-6.5	0.35 max			0.13 min	0.03 max	0.50 max	0.22 max		0.30				
M16620	International	ISO 2119	Mg-Zn6Th2Zr	Cast Obsolete Replaced by ISO 3115		0.1				0.01		5-6.2	0.4-1		Th 1.5-2.3			
M16620	International	ISO 3115	Mg-Zn6Th2Zr	Sand Cast TE		0.1				0.01		5-6	0.4-1		Th 1.5-2.3	240	150	4

UNS numbers and US grades are provided as a means of cross referencing chemically similar alloys. Exchangability is only possible after independent examination of specifications. Tensile properties are minimum or typical . UTS and YS as Mpa, EI as %. See Appendix for list of abbreviations used in Descriptions.

Worldwide Guide to Equivalent Nonferrous Metals and Alloys

Grade UNS #	Country	Specification	Designation	Comment	Al	Cu	Fe	Mg	Mn	Ni	Si	Zn	Zr	OT max	Other	UTS	YS	EL
ZH62A M16620	USA	ASTM B275	ZH62A	Cast or Wrought		0.10 max				0.01 max		5.2-6.2	0.50-1.0	0.30	Th 1.4-2.2			
M16620	USA			Cast		0.10 max				0.01 max		5.2-6.2	0.50-1.0	0.30	Th 1.4-2.2			
ZH62A M16620	USA	ASTM B80	ZH62A	Sand Cast Cooled and art aged		0.1				0.01		5.2-6.2	0.5-1	0.3	Th 1.4-2.2	241	152	5
ZH62A M16620	USA	MIL M-46062	ZH62A	Cast T5		0.1 min				0.01		5.2-6.2	0.5-1	0.3	Th 1.4-2.2	262	159	5
TA54A M18540	USA	ASTM B275	TA54A	Cast or Wrought	3.0-4.0	0.05 max			0.20 min	0.01 max	0.30 max	0.30 max		0.30	Sn 4.0-6.0			
M18540	USA				3.0-4.0	0.05 max			0.20 min	0.01 max	0.30 max	0.30 max		0.30	Sn 4.0-6.0			
LZ145A M14145	USA	MIL M-46130	LZ145A	Wrought Plt Sh Frg Stab .020/2 in diam	0.05	0.05	0.01 max		0.15	0.01	1.5-2	4.5-5		0.2	Na 0.01 max; Li 12-15; Ag 2-3	193	165	20
LZ145A M14145	USA	ASTM B275	LZ145A	Cast or Wrought	0.05 max	0.05 max	0.005 max		0.15 max	0.005 max	1.5-2.0	4.5-5.0		0.20	Li 12.0-15.0; Na 0.005 max; Ag 2.0-3.0			
M14145	USA			Wrought	0.05 max	0.05 max	0.005 max		0.15 max	0.005 max	1.5-2.0	4.5-5.0		0.20	Na 0.005 max; Li 12.0-15.0; Ag 2.0-3.0			
	UK	BS 2970	MAG1(Mg-Al8ZnMn)	Cast	7.5-8.5	0.15 max	0.03 max		0.2-0.4	0.01 max	0.2 max	0.3-1			ST 0.035 max; ST = Cu + Si + Fe + Ni			
	UK	BS 2970	MAG2(Mg-Al8ZnMn)	Cast	7.5-8.5	0.01 max	0.002 max		0.2-0.7	0.01 max	0.01 max	0.3-1						
AZ80	Canada	CSA HG.3	AZ80	Cast Ingot	7.6-8.4	0.01 max	0.01 max		0.15-0.4	0.003 max	0.01 max	0.35-0.7			ET 0.05 max; ET=.15 total			
M16631	USA			Cast		2.4-3.0			0.25-0.75	0.010 max	0.20 max	5.5-6.5		0.30				
AM80A M10800	USA	ASTM B275	AM80A	Cast or Wrought	8.0-9.0	0.08 max			0.18 min	0.01 max	0.20 max	0.20 max		0.30				
M10800	USA			Cast	8.0-9.0	0.08 max			0.18 min	0.01 max	0.20 max	0.20 max		0.30				
	International	ISO R121	Mg-Al8Zn1	Cast Sand Bar As cast 13 mm diam Obsolete	7-9.5	0.35	0.05 max		0.15	0.02	0.5	0.3-2				140	75	
M11630	Japan	JIS H5203	MC1	Cast Perm Mold Sand	5.3-6.7	0.1 max			0.15-0.6	0.01 max	0.3 max	2.5-3.5						
M11630	Japan	JIS H2221	MCIn1	Cast Ingot	5.3-6.7	0.08 max			0.15-0.6	0.01 max	0.2 max	2.5-3.5						
AZ63A M11630	USA	ASTM B275	AZ63A	Cast or Wrought	5.3-6.7	0.25 max			0.15 min	0.01 max	0.30 max	2.5-3.5		0.30				
AZ63A M11630	USA	ASTM B80	AZ63A	Sand Cast As fab	5.3-6.7	0.25			0.15	0.01	0.3	2.5-3.5		0.3		179	76	4
AZ63A M11630	USA	ASTM B80	AZ63A	Sand Cast SHT and nat aged	5.3-6.7	0.25			0.15	0.01	0.3	2.5-3.5		0.3		234	76	7
AZ63A M11630	USA	ASTM B80	AZ63A	Sand Cast Cooled and art aged	5.3-6.7	0.25			0.15	0.01	0.3	2.5-3.5		0.3		179	83	2
AZ63A M11630	USA	ASTM B93	AZ63A	Wrought	5.5-6.5	0.2			0.18	0.01	0.2	2.7-3.3		0.3				
M11630	USA			Cast	5.3-6.7	0.15			0.15 min	0.01 max	0.25	2.5-3.5		0.30				
	UK	BS 2970	MAG7(Mg-Al8.5Zn1Mn)	Cast	7.9-9.2	0.3 max	0.05 max		0.15-0.8	0.02 max	0.3 max	0.3-1.5						
M11916	USA			Cast	8.3-9.7	0.030 min	0.005 max		0.15 min	0.002 max	0.10 max	0.35 max			If Fe > 0.005 then Fe:Mn < 0.032; OE 0.20 max			
AZ80	Canada	CSA HG.9	AZ80	Sand Cast	7.5-8.5	0.02 max	0.02 max		0.4-0.15	0.01 max	0.2 max	0.7-0.3			ET 0.07 max; ET=.30 total			
M17914	Japan	JIS H2221	MCIn2	Cast Ingot	8.1-9.3	0.08 max			0.13-0.5	0.01 max	0.2 max	0.4-1						

UNS numbers and US grades are provided as a means of cross referencing chemically similar alloys. Exchangability is only possible after independent examination of specifications. Tensile properties are minimum or typical. UTS and YS as Mpa, El as %. See Appendix for list of abbreviations used in Descriptions.

Grade UNS #	Country	Specification	Designation	Comment	Al	Cu	Fe	Mg	Mn	Ni	Si	Zn	Zr	OT max	Other	UTS	YS	EL
AM90A M10900	USA	ASTM B275	AM90A	Cast or Wrought	8.5-9.5	0.10 max	0.008 max		0.15 min	0.005 max	0.15 max	0.20 max		0.10	OE 0.02 max; Ag 0.008 max			
M10900	USA			Cast	8.5-9.5	0.10 max	0.008 max		0.15 min	0.005 max	0.15 max	0.20 max		0.10	OE 0.02 max; Ag 0.008 max			
AM90	Canada	CSA HG.10	AM90	Cast Perm Mold	8.5-9.5	0.02 max	0.02 max		0.15-0.4	0.01 max	0.2 max	0.2 max			ET 0.07 max; ET=.30			
AM90	Canada	CSA HG.11	AM90	Die Cast	8.5-9.5	0.02 max	0.02 max		0.15-0.4	0.01 max	0.2 max	0.2 max			ET 0.07 max; ET=.30 total			
AZ90A M11900	USA	ASTM B275	AZ90A	Cast or Wrought	8.5-9.5	0.02 max	0.015 max		0.15 min	0.005 max	0.20 max	0.20 max		0.30	OE 0.07 max			
M11900	USA			Cast	8.5-9.5	0.02 min	0.015 max		0.15 min	0.005 max	0.20 max	0.20 max		0.30	OE 0.07 max			
AZ91 M11914	Canada	CSA HG.9	AZ91	Sand Cast	8.1-9.3	0.02 max	0.02 max		0.4-0.13	0.01 max	0.2 max	0.4-1						
M11914	Japan	JIS H5203	MC2	Cast Perm Mold Sand	8.1-9.3	0.1 max			0.13-0.5	0.01 max	0.3 max	0.4-1						
M11914	Sweden	SIS 144635	SISMg4635-00	Cast	8.3-9.8	0.15	0.03 max		0.12-0.6	0.01	0.2	0.3-0.8			Be 0.001-0.002			
AZ91C M11914	USA	ASTM B199	AZ91C	Cast SHT and nat aged	8.1-9.3	0.1			0.13	0.01	0.3	0.4-1		0.3		234	76	7
AZ91C M11914	USA	ASTM B199	AZ91C	Cast Cooled and art aged	8.1-9.3	0.1			0.13	0.01	0.3	0.4-1		0.3		159	83	2
AZ91C M11914	USA	ASTM B199	AZ91C	Cast SHT and art aged	8.1-9.3	0.1			0.13	0.01	0.3	0.4-1		0.3		234	110	3
AZ91C M11914	USA	ASTM B275	AZ91C	Cast or Wrought	8.1-9.3	0.10 max			0.13 min	0.01 max	0.30 max	0.40-1.0		0.30				
AZ91C M11914	USA	ASTM B403	AZ91C	Invest Cast As fab	8.1-9.3	0.1			0.13	0.01	0.3	0.4-1		0.3		124	69	
AZ91C M11014	USA	ASTM B403	AZ91C	Invest Cast SHT and nat aged	8.1-9.3	0.1			0.13	0.01	0.0	0.4-1		0.3		234	69	7
AZ91C M11914	USA	ASTM B403	AZ91C	Invest Cast Cooled and art aged	8.1-9.3	0.1			0.13	0.01	0.3	0.4-1		0.3		138	76	2
AZ91C M11914	USA	ASTM B80	AZ91C	Sand Cast SHT and nat aged	8.1-9.3	0.1			0.13	0.01	0.3	0.4-1		0.3		234	76	7
AZ91C M11914	USA	ASTM B80	AZ91C	Sand Cast Cooled and art aged	8.1-9.3	0.1			0.13	0.01	0.3	0.4-1		0.3		158	83	2
AZ91C M11914	USA	ASTM B80	AZ91C	Sand Cast SHT and art aged	8.1-9.3	0.1			0.13	0.01	0.3	0.4-1		0.3		234	110	3
AZ91C M11914	USA	ASTM B93	AZ91C	Wrought	8.3-9.2	0.08			0.15	0.01	0.2	0.45-0.9		0.3				
AZ91C M11914	USA	MIL M-46062	AZ91C	Cast T6	8.1-9.3	0.1			0.13	0.01	0.3	0.4-1		0.3		241	124	4
M11914	USA			Cast	8.1-9.3	0.10 min			0.13 min	0.01 max	0.30 max	0.40-1.0		0.30				
AZ91	Canada	CSA HG.10	AZ91	Cast Perm Mold	8.3-9.3	0.02 max	0.02 max		0.15-0.4	0.01 max	0.2 max	0.4-1			ET 0.07 max; ET=.30			
AZ91	Canada	CSA HG.11	AZ91	Die Cast	8.3-9.3	0.02 max	0.02 max		0.15-0.4	0.01 max	0.2 max	0.4-1			ET 0.07 max; ET=.30 total			
M11910	Japan	JIS H5303	1A	Die Cast	8.3-9.7	0.1			0.15	0.03	0.5	0.35-1						
M11910	Japan	JIS H5303	1B	Die Cast	8.3-9.7	0.35			0.15	0.03	0.5	0.35-1						
M11910	Japan	JIS H5303	MDC1A	Die Cast	8.3-9.7	0.1 max			0.15 min	0.03 max	0.5 max	0.35-1						
AZ91A M11910	USA	ASTM B275	AZ91A	Cast or Wrought	8.5-9.5	0.08 max			0.15 min	0.01 max	0.20 max	0.45-0.9		0.30				
AZ91A M11910	USA	ASTM B93	AZ91A	Wrought	8.5-9.5	0.08			0.15	0.01	0.2	0.45-0.9		0.3				
AZ91A M11910	USA	ASTM B94	UNSM11910(AZ91A)	Die Cast As cast	8.3-9.7	0.35			0.13	0.03	0.5	0.35-1				230	160	3
M11910	USA			Cast	8.3-9.7	0.10 min			0.13 min		0.50 max	0.35-1.0		0.30	Mo 0.03 max			
M11919	USA			Sand Cast	8.3-9.2	0.015 min	0.005 max		0.17-0.35	0.001 max	0.20 max	0.45-0.90						

UNS numbers and US grades are provided as a means of cross referencing chemically similar alloys. Exchangability is only possible after independent examination of specifications. Tensile properties are minimum or typical . UTS and YS as Mpa, El as %. See Appendix for list of abbreviations used in Descriptions.

Worldwide Guide to Equivalent Nonferrous Metals and Alloys

Grade UNS #	Country	Specification	Designation	Comment	Al	Cu	Fe	Mg	Mn	Ni	Si	Zn	Zr	OT max	Other	UTS	YS	EL
AZ80 M11800	Canada	CSA HG.5	AZ80	Wrought Bar Rod Wir Extrusions	7.8-9.2	0.02 max	0.01 max		0.12 max	0.003 max	0.01 max	0.2-0.8			ET 0.02 max; ET=.10 total			
AZ80A M11800	France	NF A65-717	G-A8Z	Cast Wir As Mfg	7.5-9.2	0.05	0.01 max		0.1-0.4	0.01	0.1	0.2-1				300	220	8
M11800	International	ISO R503	23	Cast Ingot Obsolete Replace by ISO 3116	7.5-9.2	0.05	0.01 max		0.12	0.01	0.1	0.2-1						
M11800	International	ISO 3116	Mg-Al8Zn	Cast Frg M	7.5-9.2	0.05	0.01 max		0.12	0.01	0.1	0.2-1				290	190	5
M11800	International	ISO 3116	Mg-Al8Zn	Cast Frg TE	7.5-9.2	0.05	0.01 max		0.12	0.01	0.1	0.2-1				290	200	4
M11800	International	ISO R121	Mg-Al8Zn	Cast Sand Bar As Cast 13 mm diam Obsolete	7.5-9	0.2	0.05 max		0.15-0.6	0.01	0.3	0.2-1				140	75	1
M11800	International	ISO R121	Mg-Al8Zn	Cast Sand Bar ST 13 mm diam Obsolete	7.5-9	0.2	0.05 max		0.15-0.6	0.01	0.3	0.2-1				230	75	6
M11800	International	ISO R121	Mg-Al8Zn	Cast Sand Bar Fully HT 13 mm diam Obsolete	7.5-9	0.2	0.05 max		0.15-0 6	0.01	0.3	0.2-1				235	95	2
M11800	International	ISO 3116	Mg-Al8Zn (Alloy 23)	Cast Bar M	7.5-9.2	0.05	0.01 max		0.12	0.01	0.1	0.2-1				290	190	5
M11800	Japan	JIS H4203	MB3	Wrought Bar	7.5-9.2	0.05 max	0.01 max		0.1-0.4	0.01 max	0.1 max	0.2-1		0.3				
M11800	Japan	JIS H4204	MS3	Wrought Extrusions	7.5-9.2	0.05 max	0.01 max		0.1-0.4	0.01 max	0.1 max	0.2-1		0.3				
M11800	Sweden	SIS 144637	SISMg4637-03	Sand Cast As cast	7.5-9	0.2	0.05 max	91	0.15-0.6	0.01	0.3	0.2-1				130	90	1
M11800	Sweden	SIS 144637	SISMg4637-04	Sand Cast ST	7.5-9	0.2	0.05 max	91	0.15-0.6	0.01	0.3	0.2-1				170	90	3
AZ80A M11800	USA	ASTM B107	AZ80A	Cast Bar Tub Rod Wir As fab (except tub) 2.5/4.999 in diam	7.8-9.2	0.05	0.01 max		0.12	0.01	0.1	0.2-0.8		0.3		290	186	4
AZ80A M11800	USA	ASTM B107	AZ80A	Cast Bar Tub Rod Wir Cooled and art aged (except tub) 2.5/4.999 in diam	7.8-9.2	0.05	0.01 max		0.12	0.01	0.1	0.2-0.8		0.3		310	207	
AZ80A M11800	USA	ASTM B275	AZ80A	Cast or Wrought	7.8-9.2	0.05 max	0.005 max		0.12 min	0.005 max	0.10 max	0.20-0.8		0.30				
AZ80A M11800	USA	ASTM B91	AZ80A	Cast Frg As fab	7.8-9.2	0.05	0.01 max		0.12	0.01	0.1	0.2-0.8		0.3		290	179	5
AZ80A M11800	USA	ASTM B91	AZ80A	Cast Frg Cooled and art aged	7.8-9.2	0.05	0.01 max		0.12	0.01	0.1	0.2-0.8		0.3		290	193	2
M11800	USA			Frg Ext	7.8-9.2	0.05 min	0.005 max		0.12 min	0.005 max	0.10 max	0.20-0.8		0.30				
M11810	Norway	NS 17708	17708	Sand Cast F temper	8 max			91 max	0.2 max			0.7 max						
M11810	Norway	NS 17708	17708	Sand Cast T4 temper	8 max			91 max	0.2 max			0.7 max						
M11810	Norway	NS 17708	17708	Die Cast F temper	8 max			91 max	0.2 max			0.7 max						
AZ81A M11810	USA	ASTM B199	AZ81A	Cast SHT and nat aged	7-8.1	0.1			0.13	0.01	0.3	0.4-1		0.3		234	76	7
AZ81A M11810	USA	ASTM B275	AZ81A	Cast or Wrought	7.0-8.1	0.10 max			0.13 min	0.01 max	0.30 max	0.40-1.0		0.30				
AZ81A M11810	USA	ASTM B403	AZ81A	Invest Cast SHT and nat aged	7-8.1	0.1			0.13	0.01	0.3	0.4-1		0.3		234	69	7

UNS numbers and US grades are provided as a means of cross referencing chemically similar alloys. Exchangability is only possible after independent examination of specifications. Tensile properties are minimum or typical . UTS and YS as Mpa, El as %. See Appendix for list of abbreviations used in Descriptions.

7-14 Magnesium

Grade UNS #	Country	Specification	Designation	Comment	Al	Cu	Fe	Mg	Mn	Ni	Si	Zn	Zr	OT max	Other	UTS	YS	EL
AZ81A M11810	USA	ASTM B80	AZ81A	Sand Cast SHT and nat aged	7-8.1	0.1			0.13	0.01	0.3	0.4-1		0.3		234	76	7
AZ81A M11810	USA	ASTM B93	AZ81A	Wrought	7.2-8	0.08			0.15	0.01	0.2	0.5-0.9		0.3				
M11810	USA			Cast	7.0-8.1	0.10 min			0.13 min	0.01 max	0.30 max	0.40-1.0		0.30				
AM90	Canada	CSA HG.3	AM90	Cast Ingot	8.6-9.4	0.01 max	0.01 max		0.4-0.15	0.005 max	0.01 max	0.2 max			ET 0.05 max; ET=.15 total			
	Sweden	MNC 46E	4637	Cast	8-8				0.4			0.6-0.6						
	Sweden	SIS 144637	SISMg4637-00	Cast	7.5-8.5	0.15	0.03 max	91	0.2-0.6	0.01	0.2	0.3-0.8						
	Sweden	SIS 144640	SISMg4640-00	Cast	7-9.2	0.3	0.05 max	91	0.2	0.02	0.3	0.4-1.8						
	Sweden	SIS 144640	SISMg4640-03	Sand Cast As cast	7-9.5	0.35	0.05 max	91	0.15	0.02	0.5	0.3-2				130	90	1
	Sweden	SIS 144640	SISMg4640-04	Sand Cast ST	7-9.5	0.35	0.05 max	91	0.15	0.02	0.5	0.3-2				170	90	3
	Sweden	SIS 144640	SISMg4640-06	Cast as cast	7-9.5	0.35	0.05 max	91	0.15	0.02	0.5	0.3-2				140	100	1
	Sweden	SIS 144640	SISMg4640-07	Cast ST	7-9.5	0.35	0.05 max	91	0.15	0.02	0.5	0.3-2				180	100	3
	Sweden	SIS 141640	SISMg4640-10	Cast As cast	7-9.5	0.35	0.05 max	91	0.15	0.02	0.5	0.3-2				200	140	2
AZ91C M19914	USA	ASTM B199	AZ91C	Cast	8.1-9.3	0.1			0.13	0.01	0.3	0.4-1		0.3				
M11917	USA			Cast Ing	8.5-9.5	0.015 min	0.004 max		0.17 min	0.001 max	0.05 max	0.45-0.9			OE 0.01 max			
AZ91	Canada	CSA HG.3	AZ91	Cast Ingot	8.3-9.3	0.01 max	0.01 max		0.4-0.15 max	0.003 max	0.01 max	0.45-0.9			ET 0.05 max; ET=.15 total			
	Sweden	SIS 144635	SISMg4635-10	Cast As cast	8.3-10.3	0.2	0.05 max		0.15-0.6	0.01	0.3	0.2-1			Be 0.002-0.005	230	160	1
M11912	Japan	JIS H5303	MDC1B	Die Cast	8.3-9.7	0.35 max			0.15 min	0.03 max	0.5 max	0.35-1						
M11912	Japan	JIS H2222	MDCIN 1 B	Die Cast	8.5-9.5	0.25			0.15	0.01	0.3	0.45-0.9						
AZ91B M11912	USA	ASTM B275	AZ91B	Cast or Wrought	8.5-9.5	0.25 max			0.15 min	0.01 max	0.20 max	0.45-0.9		0.30				
AZ91B M11912	USA	ASTM B93	AZ91B	Wrought	8.5-9.5	0.25			0.15	0.01	0.2	0.45-0.9		0.3				
AZ91B M11912	USA	ASTM B94	UNSM11912(AZ91B)	Wrought Die	8.3-9.7	0.35			0.13			0.35-1						
M11912	USA			Cast	8.5-9.5	0.25 min			0.15 min	0.01	0.20 max	0.45-0.9		0.30				
	Japan	JIS H2222	MDCIn 1 A	Die	8.5-9.5	0.08			0.15	0.01	0.2	0.45-0.9						
M10100	Japan	JIS H5203	MC5	Cast Perm Mold Sand	9.3-10.7	0.1 max			0.1-0.5	0.01 max	0.3 max	0.3 max						
M10100	Japan	JIS H2221	MCIn5	Cast Ingot	9.3-10.7	0.08 max			0.1-0.5	0.01 max	0.2 max	0.1 max						
AM100 A M10100	USA	ASTM B199	AM100A	Cast As fab	9.3-10.7	0.1			0.1 min	0.01	0.3	0.3		0.3		138	69	
AM100 A M10100	USA	ASTM B199	AM100A	Cast SHT and nat aged	9.3-10.7	0.1			0.1 min	0.01	0.3	0.3		0.3		234	69	6
AM100 A M10100	USA	ASTM B199	AM100A	Cast SHT and art aged	9.3-10.7	0.1			0.1 min	0.01	0.3	0.3		0.3		234	103	2
AM100 A M10100	USA	ASTM B275	AM100A	Cast or Wrought	9.3-10.7	0.10 max			0.10 min	0.01 max	0.30 max	0.30 max		0.30				
AM100 A M10100	USA	ASTM B403	AM100A	Invest Cast As fab	9.3-10.7	0.1			0.1 min	0.01	0.3	0.3		0.3		138	69	
AM100 A M10100	USA	ASTM B403	AM100A	Invest Cast SHT and nat aged	9.3-10.7	0.1			0.1 miin	0.01	0.3	0.3		0.3	234	69	6

UNS numbers and US grades are provided as a means of cross referencing chemically similar alloys. Exchangability is only possible after independent examination of specifications. Tensile properties are minimum or typical . UTS and YS as Mpa, El as %. See Appendix for list of abbreviations used in Descriptions.

Grade UNS #	Country	Specification	Designation	Comment	Al	Cu	Fe	Mg	Mn	Ni	Si	Zn	Zr	OT max	Other	UTS	YS	EL
M10100	USA	ASTM B403	AM100A	Invest Cast SHT art aged	9.3-10.7	0.1			0.1 min	0.01	0.3	0.3		0.3	234	103	2	
M10100	USA	ASTM B80	AM100A	Sand Cast SHT art aged	9.3-10.7	0.1			0.1	0.01	0.3	0.3		0.3	241	117		
AM100A M10100	USA	ASTM B93	AM100A	Cast	9.4-10.6	0.08			0.13	0.01	0.2	0.2		0.3				
AM100A M10100	USA	MIL M-46062	AM100A	Cast T6	9.3-10.7	0.1			0.1	0.01	0.3	0.3		0.3		262	138	3
M10100	USA			Cast	9.3-10.7	0.10 max			0.1 min	0.01 max	0.30 max	0.30 max		0.30				
	Sweden	MNC 46E	4640	Cast	8-8				0.5			1						
	UK	BS 2970	MAG3(Mg-Al10ZnMn)	Cast	9-10.5	0.15 max	0.03 max		0.2-0.4	0.01 max	0.2 max	0.3-1			ST 0.35 max; ST = Cu + Si + Fe + Ni			
AM100B M10102	USA	ASTM B275	AM100B	Cast or Wrought	9.4-10.6	0.08 max			0.13 min	0.01 max	1.0 max			0.30				
M10102	USA			Die Cast	9.4-10.6	0.08 max			0.13 min	0.01 max	1.0 max			0.30				
M11101	International	ISO R121	Mg-Al9Zn	Cast Sand Bar As cast 13 mm diam Obsolete	8.3-10.3	0.2	0.05 max		0.15-0.6	0.01	0.3	0.2-1				140	75	1
M11101	International	ISO R121	Mg-Al9Zn	Cast Sand Bar Fully HT 13 mm diam Obsolete	8.3-10.3	0.2	0.05 max		0.15-0.6	0.01	0.3	0.2-1				230	5	6
M11101	International	ISO R121	Mg-Al9Zn	Cast Sand Bar ST 13 mm diam Obsolete	8.3-10.3	0.2	0.05 max		0.15-0.6	0.01	0.3	0.2-1				235	110	1
M11101	Norway	NS 17709	17709	Die Cast F temper	9 max			90 max	0.2 max			0.7 max						
M11101	Norway	NS 17709	17709	Cast F temper	9 max			90 max	0.2 max			0.7 max						
M11101	Norway	NS 17709	17709	Cast T4 temper	9 max			90 max	0.2 max			0.7 max						
M11101	Norway	NS 17709	17709	Cast T6 temper	9 max			90 max	0.2 max			0.7 max						
M11101	Norway	NS 17709	17709	Sand Cast F temper	9 max			90 max	0.2 max			0.7 max						
M11101	Norway	NS 17709	17709	Sand Cast T4 temper	9 max			90 max	0.2 max			0.7 max						
M11101	Norway	NS 17709	17709	Sand Cast T6 temper	9 max			90 max	0.2 max			0.7 max						
M11101	USA	AWS A5.19-92	ER AZ101A	Weld El Rod	9.5-10.5	0.05 max	0.01 max		0.15-0.5	0.005 max	0.05 max	0.75-1.3		0.3	Be 0.0002-0.0008			
M11101	USA	AWS A5.19-92	R AZ101A	Weld El Rod	9.5-10.5	0.05 max	0.01 max		0.15-0.5	0.005 max	0.05 max	0.75-1.3		0.3	Be 0.0002-0.0008			
M11101	USA			Cast	9.5-10.5	0.05 max	0.005 max		0.13 min	0.005 max	0.05 max	0.75-1.25		0.30	Be 0.0002-0.0008			
AZ101A M11101	USA	MIL R-6944B	AZ101A	Wrought Rod	9.5-10.5	0.05 max	0.01 max			0.01	0.05			0.3	Be 0.001 max			
AZ101A M11101	USA	ASTM B275	AZ101A	Cast or Wrought	9.5-10.5	0.05 max	0.005 max		0.13 min	0.005 max	0.05 max	0.75-1.25		0.30	Be 0.0002-0.0008			
	Sweden	MNC 46E	4635	Cast	9-9				0.4			0.6-0.6			ST=traces Be			
M11922	USA	AWS A5.19-92	ER AZ92A	Weld El Rod	8.3-9.7	0.05 max	0.01 max		0.15-0.5	0.005 max	0.05 max	1.7-2.3		0.3	Be 0.0002-0.0008			
M11922	USA	AWS A5.19-92	R AZ92A	Weld El Rod	8.3-9.7	0.05 max	0.01 max		0.15-0.5	0.005 max	0.05 max	1.7-2.3		0.3	Be 0.0002-0.0008			
M11922	USA			Weld Wir	8.3-9.7	0.05 max	0.005 max		0.15 min	0.005 max	0.05 max	1.7-2.3		0.30	Be 0.0002-0.0008			
AZ92A M11922	USA	ASTM B93	AZ92A	Wrought	8.5-9.5	0.2			0.13	0.01	0.2	1.7-2.3		0.3				

UNS numbers and US grades are provided as a means of cross referencing chemically similar alloys. Exchangability is only possible after independent examination of specifications. Tensile properties are minimum or typical . UTS and YS as Mpa, El as %. See Appendix for list of abbreviations used in Descriptions.

Grade UNS #	Country	Specification	Designation	Comment	Al	Cu	Fe	Mg	Mn	Ni	Si	Zn	Zr	OT max	Other	UTS	YS	EL
M19001	USA	AWS A5.8-92	BMg-1	Braze Weld Fill	8.3-9.7	0.05 max	0.005 max		0.15-1.5	0.005 max	0.05 max	1.7-2.3		0.30	Be 0.0002-0.0008			
M19001	USA			Braz Fill	8.3-9.7	0.05 max	0.005 max		0.15-1.5	0.005 max	0.05 max	1.7-2.3		0.30	Be 0.0002-0.0008			
M11920	International	ISO R121	Mg-Al9Zn2	Cast Sand Bar As cast 13 mm diam Obsolete	8-10	0.2	0.05 max		0.1-0.5	0.01	0.3	1.5-2.5				140	75	1
M11920	International	ISO R121	Mg-Al9Zn2	Cast Sand Bar ST 13 mm diam Obsolte	8-10	0.2	0.05 max		0.1-0.5	0.01	0.3	1.5-2.5				230	75	5
M11920	International	ISO R121	Mg-Al9Zn2	Cast Sand Bar Fully HT 13 mm diam Obsolete	8-10	0.2	0.05 max		0.1-0.5	0.01	0.3	1.5-2.5				235	110	1
M11920	Japan	JIS H5203	MC3	Cast Perm Mold Sand	8.3-9.7	0.1 max			0.1-0.5	0.01 max	0.3 max	1.6-2.4						
M11920	Japan	JIS H2221	MCIn3	Cast Ingot	8.3-9.7	0.08 max			0.1-0.5	0.01 max	0.2 max	1.6-2.4						
AZ92A M11920	USA	MIL R-6944B	AZ92A	Wrought Rod	8.3-9.7	0.05	0.01 max		0.15	0.01	0.05	1.7-2.3		0.3	Be 0.001 max			
AZ92A M11920	USA	ASTM B199	AZ92A	Cast As fab	8.3-9.7	0.25			0.1	0.01	0.3	1.6-2.4		0.3		159	76	6
AZ92A M11920	USA	ASTM B199	AZ92A	Cast SHT and nat aged	8.3-9.7	0.25			0.1	0.01	0.3	1.6-2.4		0.3		234	76	6
AZ92A M11920	USA	ASTM B199	AZ92A	Cast SHT and art aged	8.3-9.7	0.25			0.1	0.01	0.3	1.6-2.4		0.3		159	63	
AZ92A M11920	USA	ASTM B275	AZ92A	Cast or Wrought	8.3-9.7	0.25 max			0.10 min	0.01 max	0.30 max	1.6-2.4		0.30				
AZ92A M11920	USA	ASTM B403	AZ92A	Invest Cast As fab	8.3-9.7	0.1			0.1	0.01	0.3	1.6-2.4		0.3		138	69	
AZ92A M11920	USA	ASTM B403	AZ92A	Invest Cast SHT and nat aged	8.3-9.7	0.1			0.1	0.01	0.3	1.6-2.4		0.3	204	09	0	
AZ92A M11920	USA	ASTM B403	AZ92A	Invest Cast Cooled and art aged	8.3-9.7	0.1			0.1	0.01	0.3	1.6-2.4		0.3	138	76		
AZ92A M11920	USA	ASTM B80	AZ92A	Sand Cast SHT and nat aged	8.3-9.7	0.25			0.1	0.01	0.3	1.6-2.4		0.3		234	76	6
AZ92A M11920	USA	ASTM B80	AZ92A	Sand Cast Cooled and art aged	8.3-9.7	0.25			0.1	0.01	0.3	1.6-2.4		0.3		158	83	
AZ92A M11920	USA	ASTM B80	AZ92A	Sand Cast SHT and art aged	8.3-9.7	0.25			0.1	0.01	0.3	1.6-2.4		0.3		234	124	1
AZ92A M11920	USA	MIL M-46062	AZ92A	Cast T6	8.3-9.7	0.1			0.1	0.01	0.3	1.6-2.4		0.3		276	172	3
M11920	USA			Cast	8.3-9.7	0.25 min			0.10 min	0.01 max	0.30 max	1.6-2.4		0.30				
AZ92	Canada	CSA HG.9	AZ92	Sand Cast	8.3-9.7	0.02 max	0.02 max		0.4-0.1	0.01 max	0.2	1.6-2.4						
AZ92	Canada	CSA HG.3	AZ92	Cast Ingot	8.5-9.5	0.01 max	0.01 max		0.15-0.4	0.003 max	0.01 max	1.7-2.3			ET 0.05 max; ET=.15 total			
AZ125A M11125	USA	ASTM B275	AZ125A	Cast or Wrought	11.0-13.0							4.5-5.5		0.30	Be 0.0002-0.0008			
M11125	USA			Cast	11.0-13.0							4.5-5.5		0.30	Be 0.0002-0.0008			
	Japan	JIS H2502	50MgNiB	Cast			0.15 max	46-54							ST 99 min; ST = Mg + Ni			
	Japan	JIS H2502	50MgNiA	Cast Ingot			0.05 max	47-53			0.01 min				ST 99.3 min; ST=Mg+Ni; Pb 0.01			
	Japan	JIS H2502	20MgNiB	Cast Ingot			0.02 max	17-23							ST 99 min; ST = Mg + Ni			
	Japan	JIS H2502	20MgNiA	Cast Ingot			0.05 max	18-22			0.01 min				ST 99.3 min; ST=Mg+Ni; Pb 0.01			

UNS numbers and US grades are provided as a means of cross referencing chemically similar alloys. Exchangability is only possible after independent examination of specifications. Tensile properties are minimum or typical . UTS and YS as Mpa, El as %. See Appendix for list of abbreviations used in Descriptions.

Grade UNS #	Country	Specification	Designation	Description	C	Cr	Cu	Fe	Mn	Ni	P	S	Si	Ti	Other	UTS	YS	EL
Nickel 270 N02270	USA				0.02 max	0.001 max		0.005 max	0.001 max	99.97 min		0.001 max	0.001 max	0.001 max	Co 0.001 max; Mg 0.001 max			
N02290	USA				0.01 max	0.001 max	0.02 max	0.015 max	0.001 max			0.008 max	0.001 max		N 0.001 max; Al 0.001 max; Mg 0.001 max; O 0.025 max			
	Germany	DIN 17740	2.4051/Ni99.8 Mg		0.05 max	0.03 max	0.05 max	0.01 max		99.8 min		0.005 max	0.01 max		Mg 0.01-0.04			
	Germany	DIN 17740	2.4050/Ni99.8		0.08 max	0.03 max	0.07 max	0.01 max		99.8 min		0.005 max	0.03 max		Mg 0.04 max			
	Japan	JIS H4502	Class 3	Sh Strp	0.04 max			0.01 max	0.02 max			0.005 max	0.01 max	0.005 max	99.80 min ST ST=Ni+Co; Co 0.05 max; Mg 0.01 max			
	Japan	JIS H4522	Class 3	Tub	0.04 max		0.05 max	0.07 max	0.02 max			0.005 max	0.01 max	0.005 max	99.80 min ST ST=Ni+Co; Mg 0.01 max			
	USA	ASTM B39	ASTM B39	Powd	0.03 max		0.02 max	0.02 max	0.005 max	99.8 min	0.005 max	0.01 max	0.005 max		As 0.005 max; Bi 0.005 max; Co 0.15 max; Pb 0.005 max; Sb 0.005 max; Sn 0.005 max; Zn 0.005 max			
	Germany	DIN 17740	2.4052/Ni99.7 Mg0.05		0.05 max	0.03 max	0.07 max	0.01 max		99.7 min		0.005 max	0.01 max		Mg 0.04-0.06			
	Germany	DIN 17740	2.4053/Ni99.7 Mg0.07		0.05 max	0.03 max	0.07 max	0.01 max		99.7 min		0.005 max	0.01 max		Mg 0.06-0.09			
	Germany	DIN	2.4036/Ni99.7		0.1 max	0.06 max	0.1 max	0.05 max		99.7 min			0.08 max		Pb 0.01 max; Zn 0.01 max			
	Czech Republic	CSN 423405	Ni99 6	Wrought	0.1 max	0.1 max	0.1 max	0.05 max		99.6 min			0.1 max		Ni incl Co; Co 0.15 max; Mg 0.1 max			
	Czech Republic	CSN 423403	Ni99 6E	Wrought	0.1 max	0.1 max	0.1 max	0.05 max		99.6 min	0.001 max	0.003 max	0.1 max		As Cd Bi 0.001; Ag Sb Zn 0.002; Al 0.08 max; Co 0.15 max; Mg 0.1 max; Pb 0.002 max; Sn 0.002 max			
	Germany	DIN 17740	2.4060/Ni99.6		0.08 max		0.1 max	0.2 max	0.3 max	99.6 min		0.005 max	0.1 max		Mg 0.15 max			
	Germany	DIN 17740	2.4061/L CNi99.6		0.02 max		0.1 max	0.2 max	0.35 max	99.6 min		0.005 max	0.1 max		Mg 0.15 max			
	Germany	DIN 17740	2.4062/Ni99.4 Fe		0.05 max		0.05 max	0.2-0.6	0.05 max	99.4 min		0.005 max	0.02 max		Mg 0.05 max			
	Germany	DIN 1702	2.4038/Ni99.4 NiO				0.1 max	0.15 max	0.05 max	99.4 min			0.08 max		0.25-1.0 ST ST=NiO; Pb 0.01 max; Zn 0.01 max			
	Germany	DIN 17740	2.4066/Ni99.2		0.1 max		0.25 max	0.4 max	0.35 max	99.2 min		0.005 max	0.2 max		Mg 0.1 max			
	Japan	JIS H4502	Class 1	Sh Strp	0.04 max		0.05 max	0.05 max	0.15 max			0.005 max	0.25 max		99.20 min ST ST=Ni+Co; Mg 0.001 max			
	Japan	JIS H4522	Class 2	Tub	0.1 max		0.1 max	0.2 max	0.2 max			0.01 max	0.01-0.05		99.20 min ST ST=Ni+Co; Mg 0.1 max			
	Japan	JIS H4502	2B	Sh As agreed 0.5mm diam	0.04		0.05	0.07	0.15			0.005	0.05		99.2 min Ni+Co; Mg 0.03-0.06	490		35
	Japan	JIS H4502	2C	Sh As agreed 0.5mm diam	0.04		0.05	0.07	0.15			0.005	0.05		99.2 min Ni+Co; Mg 0.06-0.09	490		35
	Japan	JIS H4502	2A	Sh As agreed 0.5mm diam	0.04		0.5	0.07	0.15			0.005	0.05		99.2 min Ni+Co; Mg 0.01	490		35
N02205	Japan	JIS H4522	VCNiT 1 A	Tub	0.1		0.1	0.2	0.2			0.01	0.05-0.25		99.2 min Ni+Co; Mg 0.01-0.15			
	Germany	DIN 1702	2.4040/Ni99.0		0.2 max		0.1 max	0.2 max	0.1 max	99.0 min			0.1 max		Pb 0.05 maxZn 0.02 max			

UNS numbers and US grades are provided as a means of cross referencing chemically similar alloys. Exchangability is only possible after independent examination of specifications. Tensile properties are minimum or typical . UTS and YS as Mpa, El as %. See Appendix for list of abbreviations used in Descriptions.

Worldwide Guide to Equivalent Nonferrous Metals and Alloys

Grade UNS #	Country	Specification	Designation	Description	C	Cr	Cu	Fe	Mn	Ni	P	S	Si	Ti	Other	UTS	YS	EL
	Germany	DIN 1702	2.4042/Ni99C Si		0.25-0.35		0.1 max	0.2 max	0.1 max	99.0 min			0.25-0.35		Pb 0.01 max; Zn 0.01 max			
N02200	Germany	DIN 1701	WNi99		0.1 max					99.0 min		0.01 max						
N02200	Japan	JIS H4561	NNCP	Plt Sh Obsolete	0.15 max		0.25 max		0.35 max	99.0 min		0.01 max	0.35 max					
N02200	Japan	JIS H4562	NNCB	Bar Rod	0.15 max		0.25 max	0.4 max	0.35 max	99.0 min		0.01 max	0.35 max					
N02200	UK	BS 3072	NA11	Sh Cold rolled and annealed 0.5/1.5mm diam	0.15		0.25	0.4	0.35			0.01	0.35	0.1	Mg 0.2; 99.0 min Ni+Co	380	105	35
N02200	UK	BS 3072	NA11	Sh Hot rolled	0.15		0.25	0.4	0.35			0.01	0.35	0.1	Mg 0.2; 99.0 min Ni+Co	380	105	35
N02200	UK	BS 3072	NA11	Sh Hot rolled and annealed	0.15		0.25	0.4	0.35			0.01	0.35	0.1	Mg 0.2; 99.0 min Ni+Co	380	105	35
N02200	UK	BS 3073	NA11	Strp	0.15		0.25	0.4	0.35			0.01	0.35	0.1	Mg 0.2; 99.0 min Ni+Co			
N02200	UK	BS 3074	NA11	Tub Cold rolled and annealed 115mm diam	0.15		0.25	0.4	0.35			0.01	0.35	0.1	Mg 0.2; 99.0 min Ni+Co	380	105	40
N02200	UK	BS 3074	NA11	Tub Cold worked andstress relieved 115mm diam	0.15		0.25	0.4	0.35			0.01	0.35	0.1	Mg 0.2; 99.0 min Ni+Co	450	275	15
N02200	UK	BS 3074	NA11	Tub Hot worked and annealed 125mm diam	0.15		0.25	0.4	0.35			0.01	0.35	0.1	Mg 0.2; 99.0 min Ni+Co	380	105	40
N02200	UK	BS 3075	NA11	Wir Cold drawn 3.20mm diam	0.15		0.25	0.4	0.35			0.01	0.35	0.1	Mg 0.2; 99.0 min Ni+Co	540		
N02200	UK	BS 3075	NA11	Wir Cold drawn and annealed 0.45mm diam	0.15		0.25	0.4	0.35			0.01	0.35	0.1	Mg 0.2; 99.0 min Ni+Co	380		20
N02200	USA	ASTM B161	N02200	Pip Annealed 5.0in; 127.0mm diam	0.15		0.25	0.4	0.35			0.01	0.35		99.0 min Ni+Co	380	100	40
N02200	USA	ASTM B161	N02200	Pip Stress relieved	0.15		0.25	0.4	0.35			0.01	0.35		99.0 min Ni+Co	380	100	40
N02200	USA	ASTM B160	N02200	Rod Bar Annealed	0.15		0.25	0.4	0.35			0.01	0.35		99.0 min Ni+Co	380	100	40
N02200	USA	ASTM B160	N02200	Rod Bar Cold drawn (rounds) 1.0in; 25.4mm diam	0.15		0.25	0.4	0.35			0.01	0.35		99.0 min Ni+Co	380	100	40
N02200	USA	ASTM B160	N02200	Rod Bar Hot finished	0.15		0.25	0.4	0.35			0.01	0.35		99.0 min Ni+Co	380	100	40
N02200	USA	ASTM B163	N02200	Tub Annealed	0.15		0.25	0.4	0.35			0.01	0.35		99.0 min Ni+Co	380	100	40
N02200	USA	ASTM B163	N02200	Tub Stress relieved	0.15		0.25	0.4	0.35			0.01	0.35		99.0 min Ni+Co	380	100	40
Nickel 200 N02200	USA				0.15 max		0.25 max	0.40 max	0.35 max	99.0 min		0.010 max	0.35 max					
N02201	France	NF A54-101	Ni-01	Sh 1/2 hard 0.3mm diam	0.12		0.1	0.15	0.25			0.01	0.1		99.5 Ni+Co; Mg 0.05	490		15
N02201	France	NF A54-101		Sh 3/4 hard 0.3mm diam	0.12		0.1	0.15	0.25			0.01	0.1		99.5 Ni+Co; Mg 0.05	490		15
N02201	France	NF A54-101		Sh Hard 0.3mm diam	0.12		0.1	0.15	0.25			0.01	0.1		99.5 Ni+Co; Mg 0.05	490		15
N02201	France	NF A54-101	Ni-02	Sh	0.02		0.1	0.1	0.3			0.01	0.1		99.5 Ni+Co; Mg 0.05			
N02201	Germany	DIN 17740	2.4068/LC-Ni99		0.02 max		0.25 max	0.4 max	0.35 max	99.0 min		0.005 max	0.2 max		Mg 0.15 max			

UNS numbers and US grades are provided as a means of cross referencing chemically similar alloys. Exchangability is only possible after independent examination of specifications. Tensile properties are minimum or typical . UTS and YS as Mpa, El as %. See Appendix for list of abbreviations used in Descriptions.

Grade UNS #	Country	Specification	Designation	Description	C	Cr	Cu	Fe	Mn	Ni	P	S	Si	Ti	Other	UTS	YS	EL
N02201	Japan	JIS H4562	NLCB	Bar Rod Obsolete	0.02 max		0.25 max	0.4 max	0.35 max	99.0 min		0.01 max	0.35 max					
N02201	Japan	JIS H4561	NLCP	Plt Sh	0.02 max		0.25 max	0.4 max	0.35 max	99.0 min		0.01 max	0.35 max					
N02201	UK	BS 3076	NA12	Bar Cold worked and annealed	0.02		0.25	0.4	0.35			0.01	0.35	0.1	Mg 0.2; 99.0 min Ni+Co	340	70	35
N02201	UK	BS 3076	NA12	Bar Hot worked and annealed	0.02		0.25	0.4	0.35			0.01	0.35	0.1	Mg 0.2; 99.0 min Ni+Co	340	70	35
N02201	UK	BS 3072	NA12	Sh Cold rolled and annealed 1.5/4mm diam	0.02		0.25	0.4	0.35			0.01	0.35	0.1	Mg 0.2; 99.0 min Ni+Co	350	85	40
N02201	UK	BS 3072	NA12	Sh Hot rolled	0.02		0.25	0.4	0.35			0.01	0.35	0.1	Mg 0.2; 99.0 min Ni+Co	350	85	30
N02201	UK	BS 3072	NA12	Sh Hot rolled and annealed	0.02		0.25	0.4	0.35			0.01	0.35	0.1	Mg 0.2; 99.0 min Ni+Co	350	85	40
N02201	UK	BS 3073	NA12	Strp Cold rolled material 0.25/0.5mm	0.02		0.25	0.4	0.35			0.01	0.35	0.1	Mg 0.2; 99.0 min Ni+Co	350		30
N02201	UK	BS 3074	NA12	Tub Cold rolled and annealed 115mm diam	0.02		0.25	0.4	0.35			0.01	0.35	0.1	Mg 0.2; 99.0 min Ni+Co	350	85	40
N02201	UK	BS 3074	NA12	Tub Cold worked and stress relieved 115mm diam	0.02		0.25	0.4	0.35			0.01	0.35	0.1	Mg 0.2; 99.0 min Ni+Co	410	205	15
N02201	UK	BS 3074	NA12	Tub Hot worked and annealed 125mm diam	0.02		0.25	0.4	0.35			0.01	0.35	0.1	Mg 0.2; 99.0 min Ni+Co	350	85	40
N02201	UK	BS 3076	NA11	Bar Cold worked 25/55mm	0.15		0.25	0.4	0.35			0.01	0.35	0.1	Mg 0.2; 99.0 min Ni+Co	520	345	14
N02201	UK	BS 3076	NA11	Bar Cold worked and annealed	0.15		0.25	0.4	0.35			0.01	0.35	0.1	Mg 0.2; 99.0 min Ni+Co	380	105	35
N02201	UK	BS 3076	NA11	Bar Hot worked and annealed	0.15		0.25	0.4	0.35			0.01	0.35	0.1	Mg 0.2; 99.0 min Ni+Co	380	105	35
N02201	USA	AMS 5553B	AMS 5553B	Sh Strp	0.02 max		0.25 max	0.4 max	0.35 max			0.01 max	0.35 max		99.00 min ST ST=Ni+Co			
N02201	USA	ASTM B161	N02201	Pip Annealed 5.0in; 127.0mm	0.02		0.25	0.4	0.35			0.01	0.35		99.0 min Ni+Co	380	100	40
N02201	USA	ASTM B161	N02201	Pip Stress relieved	0.02		0.25	0.4	0.35			0.01	0.35		99.0 min Ni+Co	380	100	40
N02201	USA	ASTM B162	N02201	Plt Hot rolled annealed plate	0.02		0.25	0.4	0.35			0.01	0.35		99.0 min Ni+Co	380	100	40
N02201	USA	ASTM B160	N02201	Rod Bar Annealed	0.02		0.25	0.4	0.35			0.01	0.35		99.0 min Ni+Co	380	100	40
N02201	USA	ASTM B160	N02201	Rod Bar Hot finished	0.02		0.25	0.4	0.35			0.01	0.35		99.0 min Ni+Co	380	100	40
N02201	USA	ASTM B162	N02201	Sh Hot rolled annealed sheet	0.02		0.25	0.4	0.35			0.01	0.35		99.0 min Ni+Co	380	100	40
N02201	USA	ASTM B162	N02201	Strp Cold rolled annealed strip	0.02		0.25	0.4	0.35			0.01	0.35		99.0 min Ni+Co	380	100	40
N02201	USA	ASTM B163	N02201	Tub Annealed	0.02		0.25	0.4	0.35			0.01	0.35		99.0 min Ni+Co	380	100	40
N02201	USA	ASTM B163	N02201	Tub Stress relieved	0.02		0.25	0.4	0.35			0.01	0.35		99.0 min Ni+Co	380	100	40
Nickel 201 N02201	USA				0.02 max		0.25 max	0.40 max	0.35 max	99.0 min		0.010 max	0.35 max					

UNS numbers and US grades are provided as a means of cross referencing chemically similar alloys. Exchangability is only possible after independent examination of specifications. Tensile properties are minimum or typical . UTS and YS as Mpa, El as %. See Appendix for list of abbreviations used in Descriptions.

Worldwide Guide to Equivalent Nonferrous Metals and Alloys

Grade UNS #	Country	Specification	Designation	Description	C	Cr	Cu	Fe	Mn	Ni	P	S	Si	Ti	Other	UTS	YS	EL
N02205	Japan	JIS H4501	VNiR	Sh Strp As agreed 0.5mm diam min.	0.1		0.1	0.2	0.3			0.01	0.2		99.00 min ST ST=Ni+Co; Mg 0.1	490		35
N02205	Japan	JIS H4501	VNip	Sh Strip As agreed 0.5mm diam minimum	0.1		0.1	0.2	0.3			0.01	0.2		99.00 min ST ST=Ni+Co; Mg 0.1	490		35
N02205	USA	MIL N-46025B	205	Bar Wir Strp	0.15 max		0.2 max	0.3 max	0.35 max			0.01 max	0.2 max		99.00 min ST ST=Ni+Co			
Nickel 205 N02205	USA				0.15 max		0.15 max	0.20 max	0.35 max	99.0 min		0.008 max	0.15 max	0.01-0.05	Mg 0.01-0.08			
Nickel 220 N02220	USA				0.15 max		0.10 max	0.10 max	0.20 max	99.00 min		0.008 max	0.01-0.05	0.01-0.05	Mg 0.01-0.08			
Nickel 225 N02225	USA				0.15 max		0.10 max	0.10 max	0.20 max	99.00 min		0.008 max	0.15-0.25	0.01-0.05	Mg 0.01-0.08			
Nickel 230 N02230	USA				0.15 max		0.10 max	0.10 max	0.15 max	99.00 min		0.008 max	0.010-0.035	0.005 max	Mg 0.04-0.08			
Nickel 233 N02233	USA				0.15 max		0.10 max	0.10 max	0.30 max	99.00 min		0.008 max	0.10 max	0.005 max	Mg 0.01-0.10			
M220C N03220	USA			Pt Hard Cast	0.30-0.50										Be 1.80-2.30			
N03260	USA			Thoria dispersion strengthened	0.02 max	0.05 max	0.15 max	0.05 max				0.0025 max		0.05 max	ThO2 1.80-2.60; Co 0.20 max			
N03260	USA	AMS 5865	AMS 5865	Sh STrp	0.02 max	0.05 max	0.15 max	0.05 max						0.05 max	ST=Ni+Co ST=S 0.0025max; Th 1.8-2.6			
N03260	USA	AMS 5890	AMS 5890	Bar Frg Ext	0.02 max	0.05 max	0.15 max	0.05 max						0.05 max	ST=Ni+Co ST=S 0.0025max; Th 1.8-2.6			
	Germany	DIN 17741	2.4106/NiMn1		0.1 max			0.5 max	0.3-1.0	98.0 min		0.01 max	0.2 max		Mg 0.15 max			
N99640	USA	AWS A5.8-92	BNi-4	Braze Weld Fill	0.06 max			1.5 max			0.02 max	0.02 max	3.0-4.0	0.05 max	Se 0.005 max; Al 0.05 max; B 1.5-2.2; Co 0.1 max; Zr 0.05 max			
N99640	USA			Braz Fill metal	0.06 max			1.5 max			0.02 max	0.02 max	3.0-4.0	0.05 max	Se 0.005 max; Al 0.05 max; B 1.5-2.2; Co 0.10 max; Zr 0.05 max; OT 0.50 max			
N03360	USA			Pt Hard										0.4-0.6	Be 1.85-2.05			
	Czech Republic	CSN 423519	NiSiCr1 5Mn	Wrought		1.3-2.	0.2 max	0.4 max	1.0 max				1.8-2.5					
	Germany	DIN 17741	2.4108/NiMn1C		0.1 max		0.2 max	0.3 max	1.5-2.5	97.0 min		0.01 max	0.2 max		Mg 0.15 max			
	Germany	DIN 17741	2.4110/NiMn2		0.1 max		0.2 max	0.3 max	1.5-2.5	97.0 min		0.01 max	0.2 max		Mg 0.15 max			
	USA	MIL W-19487A	Grade B	Wir	0.4 max		0.25 max	0.6 max	0.5 max			0.01 max	0.35 max	0.6 max	97 min ST ST=Ni+Co; Mg 0.2-0.5			
Permanickel 300 N03300	USA			Pt Hard	0.40 max		0.25 max	0.60 max	0.50 max	97.0 min		0.01 max	0.35 max	0.20-0.60	Mg 0.20-0.50			
N99630	USA	AWS A5.8-92	BNi-3	Braze Weld Fill	0.06 max			0.5 max			0.02 max	0.02 max	4.0-5.0	0.05 max	Se 0.005 max; Al 0.05 max; B 2.75-3.5; Co 0.1 max; Zr 0.05 max			

UNS numbers and US grades are provided as a means of cross referencing chemically similar alloys. Exchangability is only possible after independent examination of specifications. Tensile properties are minimum or typical . UTS and YS as Mpa, El as %. See Appendix for list of abbreviations used in Descriptions.

Grade UNS #	Country	Specification	Designation	Description	C	Cr	Cu	Fe	Mn	Ni	P	S	Si	Ti	Other	UTS	YS	EL
N99630	USA			Braz Fill metal	0.06 max			0.5 max			0.02 max	0.02 max	4.0-5.0	0.05 max	Se 0.005 max; Al 0.05 max; B 2.75-3.50; Co 0.10 max; Zr 0.05 max; OT 0.50 max			
	Czech Republic	CSN 423515	NiMn2 5	Wrought	0.1 max		0.2 max	0.4 max	2.3-3.3	95.9 min		0.015 max	0.3 max		Ni incl Co Ag; 0.05 Sb; 0.002 Zn; Al 0.05 max; Co 0.5 max; Mg 0.1 max; Pb 0.002 max; Sn 0.005 max			
Alumel N02016	USA			Thermocouple	0.15 max			0.50 max	2.0-3.0				1.6 max		Al 1.8-2.25			
	Japan	JIS H4502	Class 4	Sh Strp	0.04 max	0.05 max	0.07 max	0.15 max				0.005 max	0.06 max	0.005 max	94.50 min ST ST=NiCo; Mg 0.01-0.08; W 3.5-4.5			
	Japan	JIS H4522	Class 4	Tub	0.1 max		0.2 max	0.2 max	0.2 max			0.008 max	0.02-0.06	0.02 max	94.50 min ST ST=Ni+Co; Mg 0.01-0.08; W 3.5-4.5			
	Germany	DIN 17741	2.4116/NiMn5		0.1 max		0.2 max	0.3 max	4.5-5.5	94.0 min			0.2 max		Mg 0.15 max			
	Germany	DIN 17741	2.4122/NiMn3 Al						1.0-3.0	94.0 min			1.0-2.0		Al 1-2			
Nickel 211 N02211	USA			Soln strengthened	0.2 max		0.25 max	0.75 max	4.25-5.25	93.7 min		0.015 max	0.15 max					
	Germany	DIN 1736	2.4155/S-NiTi4							93.0 min				1.0-4.0	OT 0.5 max			
	Germany	DIN 1736	2.4156/S-NiTi3							93.0 min				1.0-4.0	OT 0.5 max			
	USA	MIL W-19487A	Grade A	Wir	0.3 max		0.25 max	0.6 max	0.5 max				1.0 max	0.25-1.0	93.00 min ST ST=Ni+Co; Al 4-4.75; Sa 0.01 max			
	USA	AWS A5.30-79R	IN61	Weld consumable insert grpe.	0.15 max		0.25 max	1.0 max	1.0 max		0.03 max	0.015 max	0.75 max	2.0-3.5	Ni + Co= 93 min; Al 1.5 max; OT 0.5 max			
N02061	UK	BS 2901 Part 5	NA32		0.15 max		0.25 max	1.0 max	1.0 max		0.03 max	0.015 max	0.75 max	2.0-3.5	Al 1.5 max; 93.00 min ST ST=Ni+Co			
N02061	USA	AWS A5.14-89	ERNi-1	Bare Weld El Rod	0.15 max		0.25 max	1.0 max	1.0 max	93.0 min	0.03 max	0.015 max	0.75 max	2.0-3.5	Ni includes Co; Al 1.5 max; OT 0.5 max	380		
FM61 N02061	USA			Weld Fill	0.15 max		0.25 max	1.0 max	1.0 max	93.0 min	0.03 max	0.015 max	0.75 max	2.0-3.5	Al 1.5 max			
N03301	Germany	DIN	2.4128											0.5	Al 4.5			
N03301	Japan	JIS H4562	NDB	Bar Rod	0.3 max		0.25 max	0.6 max	0.5 max	93.0 min		0.01 max	1.0 max	0.25-1.0	Al 4-4.75			
N03301	Japan	JIS H4561	NDP	Plt Sh	0.3 max		0.25 max	0.6 max	0.5 max	93.0 min		0.01 max	1.0 max	0.25-1.0	Al 4-4.75			
Duranickel 301 N03301	USA			Pt Hard	0.30 max		0.25 max	0.60 max	0.50 max	93.0 min		0.01 max	1.00 max	0.25-1.00	Al 4.00-4.75			
	Japan	JIS Z3265	BNi-4	Wir	0.06			1.5					3.0-4.0		92.23 min Ni+Co; B 1-2.2			
	Czech Republic	CSN 423420	Ni92 0	Wrought			2.0 max	3.0 max	3.5 max	92.0 min			0.5 max		Al 0.15 max; Mg 0.1 max			
W82141	USA	AWS A5.11-90	ENi-1	Weld El SMAW	0.1 max		0.25 max	0.75 max	0.75 max	92.0 min	0.03 max	0.02 max	1.25 max	1.0-4.0	Ni includes Co; Al 1 max; OT 0.5 max			
N02100	USA	ASTM A743	CZ-100	Sand As cast annealed	1.		1.25	3.0	1.5		0.03	0.03	2.0			34	125	10
CZ-100 N02100	USA				1.00 max		1.25 max	3.00 max	1.50 max				2.00 max					

UNS numbers and US grades are provided as a means of cross referencing chemically similar alloys. Exchangability is only possible after independent nexamination of specifications. Tensile properties are minimum or typical . UTS and YS as Mpa, El as %. See Appendix for list of abbreviations used in Descriptions.

Worldwide Guide to Equivalent Nonferrous Metals and Alloys

Grade UNS #	Country	Specification	Designation	Description	C	Cr	Cu	Fe	Mn	Ni	P	S	Si	Ti	Other	UTS	YS	EL
N99700	USA	AWS A5.8-92	BNi-6	Braze Weld Fill	0.06 max						10.0-12.0	0.02 max		0.05 max	Se 0.005 max; Al 0.05 max; Co 0.1 max; Zr 0.05 max			
N99700	USA			Braz Fill metal	0.01 max						10.0-12.0	0.02 max		0.05 max	Se 0.005 max; Al 0.05 max; Co 0.10 max; Zr 0.05 max; OT 0.50 max			
	Japan	JIS Z3265	BNi-3	Wir	0.06			1.5					4.0-5.0		89.43 min Ni+Co; B 2.75-3.5			
ERNi-CI N02215	USA			Weld Fill for cast iron	1.0 max		4.0 max	4.0 max	2.5 max			0.03 max	0.75 max		OT 1.0 max			
Colmonoy 5 N99644	USA			Weld Fill metal Hard Surf	0.30-0.60	8.0-14.0		1.25-3.25					3.25		Se 0.005 max; B 2.00-3.00; Co 1.50 max			
N06010	Germany	DIN 17742	2.4870/NiCr10		0.1 max	9.0-11.0	0.1 max	0.5 max	0.5 max	87.0 min			0.6 max					
Chromel N06010	USA			Thermocouple	0.15 max	9.0-11.0		0.50 max	0.10 max				1.60 max		Al 0.20 max			
N99620	USA	AWS A5.8-92	BNi-2	Braze Weld Fill	0.06 max	6.0-8.0		2.5-3.5			0.02 max	0.02 max	4.0-5.0	0.05 max	Se 0.005 max; Al 0.05 max; B 2.75-3.5; Co 0.1 max; Zr 0.05 max			
N99620	USA			Braz Fill metal	0.06 max	6.0-8.0		2.5-3.5			0.02 max	0.02 max	4.0-5.0	0.05 max	Al 0.05 max; B 2.75-3.50; Co 0.10 max; Zr 0.05 max; OT 0.50 max			
Inconel 702 N07702	USA			Pt Hardenable	0.10 max	14.0-17.0	0.5 max	2.0 max	1.0 max			0.01 max	0.7 max	0.25-1.00	Al 2.75-3.75			
N99612	USA			Braz Fill metal	0.06 max	13.5-16.5		1.5 max			0.02 max	0.02 max		0.05 max	Se 0.005 max; Al 0.05 max; B 3.25-4.0; Co 0.10 max; Zr 0.05 max; OT 0.50 max			
N99612	USA	AWS A5.8-92	BNi-9	Braze Weld Fill	0.06 max	13.5-16.5		1.5 max			0.02 max	0.02 max		0.05 max	Se 0.005 max; Al 0.05 max; B 3.25-4; Co 0.1 max; Zr 0.05 max			
	UK	BS 2901 Part 5	NA34		0.26 max	18.0-21.0	0.2 max	0.5 max	1.2 max			0.01 max	0.5 max		bal Ni+Co; Mo 0.03 max			
Colmonoy 6 N99645	USA			Weld Fill metal Hard Surf	0.40-0.80	10.0-16.0		3.00-5.00					3.00-5.00		Se 0.005 max; B 2.00-4.00; Co 1.25 max			
Nimonic 75 N06075	USA			Solid Soln strengthened	0.08-0.15	18.0-21.0	0.50 max	5.00 max	1.00 max				1.00 max	0.20-0.60				
TPM N07002	USA				0.05	16.00			2.30-2.30				0.05	3.10	Al 0.05; Co 0.50			
N26055	USA			Cast	0.05 max	11.0-14.0		2.0 max	1.5 max		0.03 max	0.03 max	0.5 max		Bi 3.0-5.0; Mo 2.0-3.5; Sn 3.0-5.0			
	UK	DTD 703B	DTD 703B	Sh	0.1 max	20.0								0.5 max				
N07080	UK	DTD 5077	DTD 5077			19.5								2.25	Al 1.4			
Nimonic 80A N07080	USA			Pt Hardenable	0.10 max	18.0-21.0	0.2 max	3.0 max	1.0 max		0.045 max	0.015 max	1.00 max	1.8-2.7	Al 1.0-1.8; B 0.008 max; Co 2.0 max			
Inconel 721 N07721	USA			Pt Hardenable	0.07 max	15.0-17.0	0.20 max	8.00 max	2.00-2.50			0.01 max	0.15 max	2.75-3.35	Al 0.10 max			

UNS numbers and US grades are provided as a means of cross referencing chemically similar alloys. Exchangability is only possible after independent examination of specifications. Tensile properties are minimum or typical . UTS and YS as Mpa, El as %. See Appendix for list of abbreviations used in Descriptions.

Grade UNS #	Country	Specification	Designation	Description	C	Cr	Cu	Fe	Mn	Ni	P	S	Si	Ti	Other	UTS	YS	EL
N07751	USSR	GOST B637	KH70MVTIU B		0.12 max	17.5		5.0 max						2.5	Al 1.3; Mo 5; Nb 0.9; W 2.7			
Inconel MA754 N07754	USA			Y Dispersion strengthened	0.05 max	19.0-23.0		2.5 max						0.3-0.6	Y2O3 0.5-0.7; Al 0.2-0.5			
N06075	USSR	GOST B464	KH78T		0.12 max	20.5		6.0 max						0.2				
	Japan	JIS Z3265	BNi-6	Wir	0.15						10.0-20.0				79.84 min Ni+Co			
214 N07214	USA			Pt Hardenable	0.05 max	15.0-17.0		2.0-4.0	0.5 max		0.015 max	0.015 max	0.2 max	0.5 max	Y 0.002-0.040; Al 4.0-5.0; B 0.006 max; Co 2.0 max; Mo 0.5 max; W 0.5 max; Zr 0.05 max			
N10003	USA	AWS A5.14-89	ERNiMo-2	Bare Weld El Rod	0.04-0.08	6.0-8.0	0.5 max	5.0 max	1.0 max		0.015 max	0.01 max	1.0 max		Ni includes Co; Co 0.2 max; Mo 15-18; V 0.5 max; W 0.5 max; OT 0.5 max	690		
Hastelloy N N10003	USA			Solid Soln strengthened	0.04-0.08	6.0-8.0	0.35 max	5.0 max	1.00 max		0.015 max	0.020 max	1.00 max		Al 0.50 max; B 0.010 max; Co 0.20 max; Mo 15.0-18.0; V 0.50 max; W 0.50 max			
	Japan	JIS Z3265	BNi-2	Wir		6.0-8.0		2.0-4.0					3.05 max		78.84 min Ni+Co; B 2.75-3.5			
	Germany	DIN 17745	2.4540/NiFe1 5Mo		0.05 max			14.0-17.0	1.0 max	78.5 min			0.3 max		Mo 3-5			
	Czech Republic	CSN 423483	NiFe16CuCr	Wrought		1.5-2.8	4.4-6.0			0.1-1.0 74.5-78.					bal Fe			
	Czech Republic	CSN 423480	NiFe17CuCr	Wrought		1.5-2.8	4.4-6.0			0.1-1.0 74.5-78.					bal Fe			
N99646	USA			Weld Fill metal Hard Surf	0.50-1.00	12.0-18.0		3.50-5.50					3.50-5.50		Se 0.005 max; B 2.50-4.50; Co 1.00 max			
	Czech Republic	CSN 423487	NiFe15CuCr	Wrought		1.8-2.5	4.8-5.5			0.1-0.6 75.5-77.5					bal Fe			
	USSR	GOST B463	KH77TLUR		0.06 max	20.5		4.0 max						2.5	Al 0.7			
N99710	USA	AWS A5.8-92	BNi-7	Braze Weld Fill	0.06 max	13.0-15.0		0.2 max	0.04 max		9.7-10.5	0.02 max	0.1 max	0.05 max	Se 0.005 max; Al 0.05 max; B 0.01 x; Co 0.1 x; Zr 0.05 max			
N99710	USA			Braz Fill metal	0.08 max	13.0-15.0		0.2 max	0.4 max		9.7-10.5	0.02 max	0.10 max	0.05 max	Se 0.005 x; Al 0.05 x; B 0.01 x; Co 0.10 x; Zr 0.05 x; OT 0.50x			
	USSR	GOST B468	KH80TBIU			16.5		3.0						2.1	Al 0.7; Nb 1.2			
	Germany	DIN 1736	2.4808/S-NiCr20			18.0-21.0				76.0 min					OT 0.5 max			
	UK	DTD 725	DTD 725	Frg	0.1 max	19.0								2.0	Al 1; Co 2 max			
IN-713 N07713	USA			Pt Hardenable	0.08-0.20	12.00-14.00		2.50 max	0.25 max				0.50 max	0.5-1.0	Al 5.5-6.5; B 0.005-0.015; Mo 3.8-5.2; Nb 1.8-2.8; Zr 0.05-0.15			
N08001	UK	BS 2857	A	Strp				24.0		76.0								
N99600	USA	AWS A5.8-92	BNi-1	Braze Weld Fill	0.6-0.9	13.0-15.0		4.0-5.0			0.02 max	0.02 max	4.0-5.0	0.05 max	Se 0.005 max; Al 0.05 max; B 2.75-3.5; Co 0.1 max; Zr 0.05 max			

UNS numbers and US grades are provided as a means of cross referencing chemically similar alloys. Exchangability is only possible after independent examination of specifications. Tensile properties are minimum or typical . UTS and YS as Mpa, El as %. See Appendix for list of abbreviations used in Descriptions.

Grade UNS #	Country	Specification	Designation	Description	C	Cr	Cu	Fe	Mn	Ni	P	S	Si	Ti	Other	UTS	YS	EL
N99600	USA			Braz Fill metal	0.6-0.9	13.0-15.0		4.0-5.0			0.02 max	0.02 max	4.0-5.0	0.05 max	Se 0.005 max; Al 0.05 max; B 2.75-3.50; Co 0.10 max; Zr 0.05 max; OT 0.50 max			
N99610	USA	AWS A5.8-92	BNi-1a	Braze Weld Fill	0.06 max	13.0-15.0		4.0-5.0			0.02 max	0.02 max	4.0-5.0	0.05 max	Se 0.005 max; Al 0.05 max; B 2.75-3.5; Co 0.1 max; Zr 0.05 max			
N99610	USA			Braz Fill metal	0.06	13.0-15.0		4.0-5.0			0.02 max	0.02 max	4.0-5.0	0.05 max	Se 0.005 max; Al 0.05 max; B 2.75-3.50; Co 0.10 max; Zr 0.05 max; OT 0.50 max			
	UK	DTD 736	DTD 736	Frg	0.1 max	19.0		5.0						2.2	Al 1; Co 2 max			
	Germany	DIN 17745	2.4500/NiFe1CuCr		0.05 max	1.5-2.5	4.0-6.0	15.0-18.0	1.0 max	75.5 min			0.3 max					
N10276	Germany	DIN 17745	2.4520/NiFe16CuMo		0.05 max		4.0-6.0	13.0-17.0	1.0 max	75.5 min			0.3 max		Mo 3-5			
	UK	DTD 328	DTD 328	Sh	0.2 max	15.0		10.0	1.0									
	UK	BS 3146	ANC9	Inv	0.8	20.0								2.8	Al 1.4			
N06003	Germany	DIN 17470	2.4869/NiCr8010		0.15 max	19.0-21.0	0.5 max	1.0 max	1.0 max	75.0 min			0.5-2.0					
Nichrome V N06003	USA			Solid Soln strengthened	0.15 max	19-21		1.0 max	2.5 max			0.01 max	0.75-1.6					
N06009	USA			Columbium stabilized	0.15 max	19.0-21.0		1.00 max	2.5 max			0.010 max	0.75-1.60		Nb 0.75-1.50			
N06040	USA	ASTM A743	CY-40	Sand As cast annealed	0.4	14.0-17.0		11.0	1.5		0.03	0.03	3.0			48	195	30
CY-40 N06040	USA			Solid Soln strengthened	0.40 max	14.0-17.0		11.0 max	1.50 max				3.00 max					
N06076	USA			Weld Fill metal	0.08-0.15	19.0-21.0	0.50 max	2.00 max	1.00 max	75.0 min	0.030 max	0.015 max	0.30 max	0.15-0.50	Al 0.40 max; Pb 0.010 max			
	USSR	GOST B471	KHN60V		0.1 max	25.0		4.0 max						0.45	Al 0.5			
Hastelloy H-9M N06920	USA			Solid Soln strengthened	0.03 max	20.5-23.0		17.0-20.0	1.0 max		0.040 max	0.030 max	1.0 max		Co 5.0 max; Mo 8.0-10.0; W 1.0-2.0			
N10665	UK	BS 2901 Part 5	NA44		0.02 max	1.0 max	0.5 max	2.0 max	1.0 max		0.04 max	0.03 max	0.1 max		Co 1 max; Mo 26-30; bal Ni+Co			
N10665	USA	AWS A5.14-89	ERNiMo-7	Bare Weld El Rod	0.02 max	1.0 max	0.5 max	2.0 max	1.0 max		0.04 max	0.03 max	0.1 max		Ni includes Co; Co 1 max; Mo 26-30; W 1 max; OT 0.5 max	760		
Hastelloy B-2 N10665	USA				0.02 max	1.0 max		2.0 max	1.0 max		0.04 max	0.03 max	0.10 max		Co 1.0 max; Mo 26.0-30.0			
	Germany	DIN 17742	2.4872/NiCr20AlSi		0.05 max	19.0-21.0	0.1 max	1.0 max	0.7 max	73.0 min			0.5-2.0		Al 2.5-4			
N99651	USA	AWS A5.8-92	BNi-5a	Braze Weld Fill	0.1 max	18.5-19.5		0.5 max			0.02 max	0.02 max	7.0-7.5	0.05 max	Se 0.005 max; Al 0.05 max; B 1-1.5; Co 0.1 max; Zr 0.05 max			
N99651	USA			Braz Fill metal	0.10 max	18.5-19.5		0.5 max			0.02 max	0.02 max	7.0-7.5	0.05 max	Se 0.005 max; Al 0.05 max; B 1.0-1.5; Co 0.10 max; Zr 0.05 max			

UNS numbers and US grades are provided as a means of cross referencing chemically similar alloys. Exchangability is only possible after independent examination of specifications. Tensile properties are minimum or typical . UTS and YS as Mpa, El as %. See Appendix for list of abbreviations used in Descriptions.

Grade UNS #	Country	Specification	Designation	Description	C	Cr	Cu	Fe	Mn	Ni	P	S	Si	Ti	Other	UTS	YS	EL
W86455	USA	AWS A5.11-90	ENiCrMo-7	Weld El SMAW	0.015 max	14.0-18.0	0.5 max	3.0 max	1.5 max		0.04 max	0.03 max	0.2 max	0.7 max	Ni includes Co; Co 2 max; Mo 14-17; W 0.5 max; OT 0.5 max			
	Germany	DIN 17742	2.4815/LC-NiCr15Fe		0.03 max	14.0-17.0	0.5 max	6.0-10.0	1.0 max	72.0 min	0.03 max	0.01 max	0.5 max	0.3 max				
	Germany	DIN 17742	2.4816/NiCr15Fe		0.08 max	14.0-17.0	0.05 max	6.0-11.0	1.0 max	72.0 min	0.03 max	0.01 max	0.5 max	0.3 max				
	Germany	DIN 17742	2.4951/NiCr20Ti		0.08-0.15	18.0-21.0	0.5 max	0.5 max	1.0 max	72.0 min	0.03 max	0.01 max	1.0 max	0.2-0.6				
N06102	USA	ASTM B518	N06102	Bar Rod	0.08 max	14.0-16.0		5.0-9.0	0.75 max		0.01 max	0.01 max	0.4 max	0.4-0.7	Al 0.3-0.6; B 0.003-0.008; Mg 0.01-0.05; Mo 2.75-3.25; Nb 2.75-3.25; W 2.75-3.25; Zr 0.01-0.05			
N06102	USA	ASTM B519	N06102	Plt Sh Strp	0.08 max	14.0-16.0		5.0-9.0	0.75 max		0.01 max	0.01 max	0.4 max	0.4-0.7	Al 0.3-0.6; B 0.003-0.008; Mg 0.01-0.05; Mo 2.75-3.25; Nb 2.75-3.25; W 2.75-3.25; Zr 0.01-0.05			
N06102	USA	ASTM B445	N06102	Tub Pip	0.08 max	14.0-16.0		5.0-9.0	0.75 max		0.01 max	0.01 max	0.4 max	0.4-0.7	Al 0.3-0.6; B 0.003-0.008; Mg 0.01-0.05; Mo 2.75-3.25; Nb 2.75-3.25; W 2.75-3.25; Zr 0.01-0.05			
IN-102 N06102	USA			Solid Soln strengthened	0.08 max	14.0-16.0		5.0-9.0	0.75 max		0.010 max	0.010 max	0.40 max	0.40-0.70	Al 0.30-0.60; B 0.003-0.008; Mg 0.01-0.05; Mo 2.75-3.25; Nb 2.75-3.25; W 2.75-3.25; Zr 0.01-0.05			
N06455	UK	BS 2901 Part 5	NA45		0.015 max	14.0-18.0	0.5 max	3.0 max	1.0 max		0.04 max	0.03 max	0.08 max	0.7 max	Co 2 max; Mo 14-17; bal Ni+Co			
N06455	USA	AWS A5.14-89	ERNiCrMo-7	Bare Weld El Rod	0.015 max	14.0-18.0	0.5 max	3.0 max	1.0 max		0.04 max	0.03 max	0.08 max	0.7 max	Ni includes Co; Co 2 max; Mo 14-18; W 0.5 max; OT 0.5 max	620		
N06455	USA	ASTM B619	N06455	Pip	0.01 max	14.0-18.0		3.0 max	1.0 max		0.04 max	0.03 max	0.08 max	0.7 max	Mo 14-17; Co 2 max			
N06455	USA	ASTM B575	N06455	Plt Sh Strp	0.01 max	14.0-18.0		3.0 max	1.0 max		0.04 max	0.03 max	0.08 max	0.7 max	Mo 14-17; Co 2 max			
N06455	USA	ASTM B574	N06455	Rod	0.01 max	14.0-18.0		3.0 max	1.0 max		0.04 max	0.03 max	0.08 max	0.7 max	Mo 14-17; Co 2 max			
N06455	USA	ASTM B626	N06455	Tub	0.01 max	14.0-18.0		3.0 max	1.0 max		0.04 max	0.03 max	0.08 max	0.7 max	Mo 14-17; Co 2 max			
N06455	USA	ASTM B622	N06455	Tub Pip	0.01 max	14.0-18.0		3.0 max	1.0 max		0.04 max	0.03 max	0.08 max	0.7 max	Mo 14-17; Co 2 max			
Hastelloy C-4 N06455	USA				0.02 max	14.0-18.0		3.0 max	1.0 max		0.04 max	0.03 max	0.08 max	0.70 max	Co 2.0 max; Mo 14.0-17.0			
N06600	Germany	DIN	2.4640			15.0		8.0										
N06600	USA	ASTM B166	N06600	Bar Rod	0.15 max	14.0-17.0	0.5 max	6.0-10.0	1.0 max			0.01 max	0.5 max					
N06600	USA	ASTM B564	N06600	Frg	0.15 max	14.0-17.0	0.5 max	6.0-10.0	1.0 max			0.01 max	0.5 max					
N06600	USA	ASTM B517	N06600	Pip	0.15 max	14.0-17.0	0.5 max	6.0-10.0	1.0 max			0.01 max	0.5 max					

UNS numbers and US grades are provided as a means of cross referencing chemically similar alloys. Exchangability is only possible after independent examination of specifications. Tensile properties are minimum or typical . UTS and YS as Mpa, El as %. See Appendix for list of abbreviations used in Descriptions.

Grade UNS #	Country	Specification	Designation	Description	C	Cr	Cu	Fe	Mn	Ni	P	S	Si	Ti	Other	UTS	YS	EL
N06600	USA	ASTM B168	N06600	Plt Sh Strp	0.15 max	14.0-17.0	0.5 max	6.0-10.0	1.0 max			0.01 max	0.5 max					
N06600	USA	ASTM B516	N06600	Strp	0.15 max	14.0-17.0	0.5 max	6.0-10.0	1.0 max			0.01 max	0.5 max					
N06600	USA	ASTM B163	N06600	Tub	0.15 max	14.0-17.0	0.5 max	6.0-10.0	1.0 max			0.01 max	0.5 max					
N06600	USA	ASTM B167	N06600	Tub Pip	0.15 max	14.0-17.0	0.5 max	6.0-10.0	1.0 max			0.01 max	0.5 max					
N06600	USA	ASTM B366	N06600	Tub Pip	0.15 max	14.0-17.0	0.5 max	6.0-10.0	1.0 max			0.01 max	0.5 max					
Inconel 600 N06600	USA			Solid Soln strengthened	0.15 max	14.00-17.00	0.50 max	6.00-10.00	1.00 max	72.0 min		0.015 max	0.50 max					
N06602	USA			Solid Soln strengthened	0.02 max	14.0-17.0	0.5 max	6.0-10.0	1.0 max	72.0 min		0.015 max	0.5 max					
N24130	USA			Cast	0.30 max		26.0-33.0	3.50 max	1.50 max		0.03 max	0.03 max	1.0-2.0		Nb 1.0-3.0			
N24135	USA			Cast	0.35 max		26.0-33.0	3.50 max	1.50 max		0.03 max	0.03 max	1.25 max		Nb 0.5 max			
N99624	USA			Braz Fill metal	0.30-0.50	9.0-11.75		2.5-4.0			0.02 max	0.02 max	3.25-4.25	0.05 max	Al 0.05 max; B 2.2-3.1; Co 0.10 max; W 11.50-12.75; Zr 0.05 max; OT 0.50 max			
N99624	USA	AWS A5.8-92	BNi-11	Braze Weld Fill	0.3-0.5	9.0-11.75		2.5-4.0			0.02 max	0.02 max	3.35-4.25	0.05 max	Se 0.005 max; Al 0.05 max; B 2.2-3.1; Co 0.1 max; W 11.5-12.75; Zr 0.05 max			
W80665	USA	AWS A5.11-90	ENiMo-7	Weld El SMAW	0.02 max	1.0 max	0.5 max	2.0 max	1.75 max		0.04 max	0.03 max	0.2 max		Ni includes Co; Co 1 max; Mo 26-30; W 1 max; OT 0.5 max			
N13010	USA			Cast	0.08-0.13	7.50-8.50		0.35 max	0.20 max		0.015 max	0.015 max	0.25 max	0.80-1.20	Al 5.75-6.25; B 0.010-0.020; Bi 0.00005 max; Co 9.50-10.50; Mo 5.75-6.25; Nb 0.10 max; Pb 0.0005 max; Sn 4.00 max; Ta 4.50 min; W 0.10 max; Zr 0.05-0.10			
N99650	USA	AWS A5.8-92	BNi-5	Braze Weld Fill	0.06 max	18.5-19.5					0.02 max	0.02 max	9.75-10.5	0.05 max	Se 0.005 max; Al 0.05 max; B 0.03 max; Co 0.1 max; Zr 0.05 max			
N99650	USA			Braz Fill metal	0.10	18.5-19.5					0.02 max	0.02 max	9.75-10.50	0.05 max	Se 0.005 max; Al 0.05 max; B 0.03 max; Co 0.10 max; Zr 0.05 max; OT 0.50 max			
	Germany	DIN 1736	2.4805/Si-NiCr15FeNb			14.0-17.0		6.0-12.0	1.0-7.0	70.0 min					Nb 1-4; OT 0.5 max			
	USA	AWS A5.30-79R	IN62	Weld consumable insert grpe.	0.08 max	14.0-17.0	0.5 max	6.0-10.0	1.0 max		0.03 max	0.015 max	0.35 max		Ni + Co= 70 min; Co 0.1 max; Nb 1.2-2.7; Ta 0.3 max; OT 0.5 max			
	USSR	GOST B626	KH70IU		0.1 max	29.5		1.0 max										
N04400	Czech Republic	CSN 423431	Ni70Cu30	Wrought	0.2 max		28.0-34.0	2.5 max	1.8 max			0.02 max	0.3 max		Al 0.12 max; Mg 0.1 max			

UNS numbers and US grades are provided as a means of cross referencing chemically similar alloys. Exchangability is only possible after independent examination of specifications. Tensile properties are minimum or typical . UTS and YS as Mpa, EI as %. See Appendix for list of abbreviations used in Descriptions.

Grade UNS #	Country	Specification	Designation	Description	C	Cr	Cu	Fe	Mn	Ni	P	S	Si	Ti	Other	UTS	YS	EL
N04400	Germany	DIN 17743	2.4360/NiCu30Fe		0.15 max		28.0-34.0	1.0-2.5	1.25 max	63.0 min		0.02 max	0.05 max		Al 0.5 max			
N04400	Japan	JIS H4554	NiCuW	Wir	0.03 max		30.0	2.5 max	2.0 max	63.0-70.0		0.02 max	0.5 max					
N04400	Japan	JIS H4553	NiCuB	Bar Rod	0.03 max		30.0	2.5 max	2.0 max	63.0-70.0		0.24 max	0.5 max					
N04400	Japan	JIS H4555	NiCuR	Strp Obsolete	0.03 max		30.0	2.5 max	2.0 max	63.0-70.0		0.24 max	0.5 max					
N04400	UK	BS 3076	NA13	Bar Cold worked and annealed	0.3		28.0-34.0	2.5	2.0			0.02	0.5		63.0 min Ni+Co	480	170	35
N04400	UK	BS 3076	NA13	Bar Cold worked and stress relieved 40mm diam	0.3		28.0-34.0	2.5	2.0			0.02	0.5		63.0 min Ni+Co	600	415	20
N04400	UK	BS 3076	NA13	Bar Hot worked and annealed	0.3		28.0-34.0	2.5	2.0			0.02	0.5		63.0 min Ni+Co	480	170	35
N04400	UK	BS 3072	NA13	Sh Cold rolled and annealed 0.5/4mm diam	0.3		28.0-34.0	2.5	2.0			0.02	0.5		63.0 min Ni+Co	480	195	35
N04400	UK	BS 3072	NA13	Sh Hot rolled	0.3		28.0-34.0	2.5	2.0			0.02	0.5		63.0 min Ni+Co	510	275	25
N04400	UK	BS 3072	NA13	Sh Hot rolled and annealed	0.3		28.0-34.0	2.5	2.0			0.02	0.5		63.0 min Ni+Co	480	195	35
N04400	UK	BS 3073	NA13	Strp Cold rolled to 1/2 hard 0.5/4mm	0.3		28.0-34.0	2.5	2.0			0.02	0.5		63.0 min Ni+Co	480	195	35
N04400	UK	BS 3073	NA13	Strp Cold rolled to 1/4hard 0.5/4mm diam	0.3		28.0-34.0	2.5	2.0			0.02	0.5		63.0 min Ni+Co	480	195	35
N04400	UK	BS 3073	NA13	Strp Cold rolled to hared 0.5/4mm diam	0.3		28.0-34.0	2.5	2.0			0.02	0.5		63.0 min Ni+Co	480	195	35
N04400	UK	BS 3074	NA13	Tub Cold worked and annealed 115mm diam	0.3		28.0-34.0	2.5	2.0			0.02	0.5		63.0 min Ni+Co	480	195	35
N04400	UK	BS 3074	NA13	Tub Cold worked and stress relieved 115mm diam	0.3		28.0-34.0	2.5	2.0			0.02	0.5		63.0 min Ni+Co	590	380	15
N04400	UK	BS 3074	NA13	Tub Hot worked and annealed 125mm diam	0.3		28.0-34.0	2.5	2.0			0.02	0.5		63.0 min Ni+Co	480	195	35
N04400	UK	BS 3075	NA13	Wir Cold drawn 3.20mm diam	0.3		28.0-34.0	2.5	2.0			0.02	0.5		63.0 min Ni+Co	770		
N04400	UK	BS 3075	NA13	Wir Cold drawn and annealed 0.45 mm diam	0.3		28.0-34.0	2.5	2.0			0.02	0.5		63.0 min Ni+Co	480		20
N04400	USA	AMS 4731	AMS 4731	Wir	0.2 max			2.5 max	2.0 max		0.02 max	0.01 max	0.5 max		Al 0.5 max; Co 1 max; 63.00-70.00 ST ST=Ni+Co; Pb 0.01 max; Zn 0.02 max; bal Cu			
N04400	USA	AMS 4544C	AMS 4544C	Plt Sh Strp	0.3 max			2.5 max	2.0 max			0.02 max	0.5 max		Co 1 max; 63.00-70.00 ST ST=Ni+Co; bal Cu			
N04400	USA	AMS 4574B	AMS 4574B	Tub	0.3 max			2.5 max	2.0 max			0.02 max	0.5 max		Co 1 max; 63.00-70.00 ST ST=Ni+Co; bal Cu			

UNS numbers and US grades are provided as a means of cross referencing chemically similar alloys. Exchangability is only possible after independent examination of specifications. Tensile properties are minimum or typical . UTS and YS as Mpa, El as %. See Appendix for list of abbreviations used in Descriptions.

Worldwide Guide to Equivalent Nonferrous Metals and Alloys

Grade UNS #	Country	Specification	Designation	Description	C	Cr	Cu	Fe	Mn	Ni	P	S	Si	Ti	Other	UTS	YS	EL
N04400	USA	AMS 4575B	AMS 4575B	Tub	0.3 max			2.5 max	2.0 max			0.02 max	0.5 max		Co 1 max; 63.00-70.00 ST ST=Ni+Co; bal Cu			
N04400	USA	AMS 4577A	AMS 4577A	Tub	0.3 max			2.5 max	2.0 max			0.02 max	0.5 max		Co 1 max; 63.00-70.00 ST ST=Ni+Co; bal Cu			
N04400	USA	AMS 4675A	AMS 4675A	Bar Frg	0.3 max			2.5 max	2.0 max			0.02 max	0.5 max		Co 1 max; 63.00-70.00 ST ST=Ni+Co; bal Cu			
N04400	USA	AMS 4730D	AMS 4730D	Wir	0.3 max			2.5 max	2.0 max			0.02 max	0.5 max		Co 1 max; 63.00-70.00 ST ST=Ni+Co; bal Cu			
N04400	USA	QQ N-281D	Class B	Bar Rod Wir Plt Sh Strp Frg	0.3 max			2.5 max	2.0 max		0.02 max	0.03-0.06	0.5 max		Al 0.5 max; Sn 0.01 max; 63.00-70.00 ST ST=Ni+Co; Pb 0.01 max; Zn 0.02 max; bal Cu			
N04400	USA	ASTM F96	N04400				28.0-34.0	2.5	2.0		0.02	0.01	0.5		62.0 min Ni+Co; Pb 0.01; Zn 0.02; bal Cu	485	195	35
N04400	USA	ASTM B164	N04400	Rod Bar Annealed rods and bars	0.3		28.0-34.0	2.5				0.02	0.5		63.0 min Ni+Co; bal Cu	483	193	35
N04400	USA	ASTM B164	N04400	Rod Bar Cold drawn rounds 0.50in; 12.70mm diam	0.3		28.0-34.0	2.5				0.02	0.5		63.0 min Ni+Co; bal Cu	483	193	35
N04400	USA	ASTM B164	N04400	Rod Bar Hot finished rounds squares rectangles	0.3		28.0-34.0	2.5				0.02	0.5		63.0 min Ni+Co; bal Cu	483	193	35
N04400	USA	ASTM B163	N04400	Tub Annealed	0.3		28.0-34.0	2.5				0.02	0.5		63.0 min Ni+Co; bal Cu	483	193	35
N04400	USA	ASTM B163	N04400	Tub Stress relieved	0.3		28.0-34.0	2.5				0.02	0.5		63.0 min Ni+Co; bal Cu	483	193	35
N04400	USA	ASTM F467	N04400					2.5	2.0			0.02	0.5		63.0-70.0 Ni+Co; bal Cu			
N04400	USA	ASTM F468	N04400	As manaufactured 0.250/0.750in; 0.635/1.905mm diam	0.3			2.5	2.0			0.02	0.5		63.0-70.0 Ni+Co; bal Cu	550	275	
N04400	USA	ASTM F468	N04400	Hot formed	0.3			2.5	2.0			0.02	0.5		63.0-70.0 Ni+Co; bal Cu	480	205	
N04400	USA	ASTM B127	N04400	Plt Hot rolled annealed plate	0.3			2.5	2.0			0.02	0.5		63.0-70.0 Ni+Co; bal Cu	485	195	35
N04400	USA	ASTM B127	N04400	Sh Hot rolled annealed and pickled sheet	0.3			2.5	2.0			0.02	0.5		63.0-70.0 Ni+Co; bal Cu	485	195	35
N04400	USA	ASTM B127	N04400	Strp Cold rolled annealed strip	0.3			2.5	2.0			0.02	0.5		63.0-70.0 Ni+Co; bal Cu	485	195	35
Monel 400 N04400	USA			Solid Soln strengthened	0.3 max			2.50 max	2.00 max	63.00-70.00		0.024 max	0.50 max		bal Cu			
N04405	Japan	JIS H4551	NCu P-O	Plt Sh Annealed 0.5-50mm diam	0.3			2.5	2.0			0.02	0.5		63.00-70.00 ST ST=Ni+Co	481		35

UNS numbers and US grades are provided as a means of cross referencing chemically similar alloys. Exchangability is only possible after independent examination of specifications. Tensile properties are minimum or typical . UTS and YS as Mpa, El as %. See Appendix for list of abbreviations used in Descriptions.

Grade UNS #	Country	Specification	Designation	Description	C	Cr	Cu	Fe	Mn	Ni	P	S	Si	Ti	Other	UTS	YS	EL
N04405	Japan	JIS H45521	NCuT	Tub Pip Annealed 15-51mm diam	0.3				2.5	2.0			0.24	0.5	63.00-70.00 ST ST=Ni+Co	481		35
N04405	USA	QQ N-281D	Class A	Bar Rod Wir Plt Sh Strp Frg	0.2 max			2.5 max	2.0 max		0.02 max	0.01 max	0.5 max		Al 0.5 max; Pb 0.01 max; Sn 0.01 max; 63.00-70.00 ST ST=Ni+Co; Zn 0.02 max; bal Cu			
N04405	USA	AMS 4674D	AMS 4674D	Bar Frg	0.3 max			2.5 max	2.0 max			0.03 max	0.5 max		Co 1 max; 63.00-70.00 ST ST=Ni+Co; bal Cu			
N04405	USA	ASTM B162	N02200	Plt Hot rolled annealed plate	0.15		0.25	0.4	0.35			0.01	0.35		99.0 min Ni+Co	380	100	40
N04405	USA	ASTM B162	N02200	Sh Hot rolled annealed sheet	0.15		0.25	0.4	0.35			0.01	0.35		99.0 min Ni+Co	380	100	40
N04405	USA	ASTM B162	N02200	Strp Cold rolled annealed strip	0.15		0.25	0.4	0.35			0.01	0.35		99.0 min Ni+Co	380	100	40
N04405	USA	ASTM B165	N04400	Pip Annealed 5.0in; 127.0 mm diam	0.3		28.0-34.0	2.5				0.02	0.5		63.0 min Ni+Co; bal Cu	483	193	35
N04405	USA	ASTM B165	N04400	Pip Stress relieved	0.3		28.0-34.0	2.5				0.02	0.5		63.0 min Ni+Co; bal Cu	483	193	35
N04405	USA	ASTM F467	N04405		0.3			2.5	2.0			0.03-0.06	0.5		63.0-70.0 Ni+Co; bal Cu			
N04405	USA	ASTM B164	N04405	Rod Bar Annealed rods and bars	0.3		28.0-34.0	2.5	2.0			0.03-0.06	0.5		63.0 min Ni+Co; bal Cu	480	170	35
N04405	USA	ASTM B164	N04405	Rod Bar Cold drawn rounds 0.50in; 12.70mm diam	0.3		28.0-34.0	2.5	2.0			0.03-0.06	0.5		63.0 min Ni+Co; bal Cu	585	345	8
N04405	USA	ASTM B164	N04405	Rod Bar Hot finished rounds 3.0in; 76.2mm diam	0.3		28.0-34.0	2.5	2.0			0.03-0.06	0.5		63.0 min Ni+Co; bal Cu	515	240	30
N04405	USA	ASTM F468	N04405	As manufacturred	0.3			2.5	2.5			0.03-0.06	0.5		63.0-70.0 Ni+Co; bal Cu	480	205	
Monel R405 N04405	USA			Solid Soln strengthened	0.30 max			2.5 max	2.0 max	63.0-70.0		0.025-0.060	0.50 max		bal Cu			
N05500	Germany	DIN 17743	2.4375/NiCu30Al				27.0-34.0	0.5-2.		63.0 min				0.3-1.0	Al 2-4			
N05500	Germany	DIN	2.4374				30.0	1.5						0.8	Al 3			
N05500	UK	BS 3076	NA18	Bar Cold worked and precipitation treated	0.25		27.0-33.0	2.0	1.5			0.01	0.5	0.35-0.85	63.00 min Ni+Co; Al 2.3-3.2	970	690	16
N05500	UK	BS 3076	NA18	Bar Cold worked solution and precipitation treated	0.25		27.0-33.0	2.0	1.5			0.01	0.5	0.35-0.85	63.00 min Ni+Co; Al 2.3-3.2	900	285	20
N05500	UK	BS 3076	NA18	Bar Hot worked and precipitation treated	0.25		27.0-33.0	2.0	1.5			0.01	0.5	0.35-0.85	63.00 min Ni+Co; Al 2.3-3.2	830	550	15
N05500	UK	BS 3072	NA18	Sh Cold rolled solution and precipitation treated 0.5/4.0mm	0.25		27.0-33.0	2.0	1.5			0.01	0.5	0.35-0.85	63.00 min Ni+Co; Al 2.3-3.2	900	620	15

UNS numbers and US grades are provided as a means of cross referencing chemically similar alloys. Exchangability is only possible after independent examination of specifications. Tensile properties are minimum or typical . UTS and YS as Mpa, El as %. See Appendix for list of abbreviations used in Descriptions.

Nickel 8-13

Grade UNS #	Country	Specification	Designation	Description	C	Cr	Cu	Fe	Mn	Ni	P	S	Si	Ti	Other	UTS	YS	EL
N05500	UK	BS 3072	NA18	Sh Hot rolled and precipitation treated	0.25		27.0-33.0	2.0	1.5			0.01	0.5	0.35-0.85	63.00 min Ni+Co; Al 2.3-3.2	970	690	15
N05500	UK	BS 3072	NA18	Sh Hot rolled solution and precipitation treated	0.25		27.0-33.0	2.0	1.5			0.01	0.5	0.35-0.85	63.00 min Ni+Co; Al 2.3-3.2	900	620	15
N05500	UK	BS 3073	NA18	Strp Cold rolled solution treated and precipitation treated 0.50/4.0mm diam	0.25		27.0-33.0	2.0	1.5			0.01	0.5	0.35-0.85	63.00 min Ni+Co; Al 2.3-3.2	900	620	15
N05500	UK	BS 3073	NA18	Strp Cold rolled to 1/2 hard and precipitation treated 0.50/4.0 mm diam	0.25		27.0-33.0	2.0	1.5			0.01	0.5	0.35-0.85	63.00 min Ni+Co; Al 2.3-3.2	1000	760	8
N05500	UK	BS 3073	NA18	Strp Cold rolled to hard and precipitation treated 0.50/4.0mm diam	0.25		27.0-33.0	2.0	1.5			0.01	0.5	0.35-0.85	63.00 min Ni+Co; Al 2.3-3.2	1170	900	5
N05500	UK	BS 3074	NA18	Tub Cold worked solution and precipitation treated	0.25		27.0-33.0	2.0	1.5			0.01	0.5	0.35-0.85	63.00 min Ni+Co; Al 2.3-3.2	900	620	15
N05500	UK	BS 3075	NA18	Wir Cold 10 mm diam	0.25		27.0-33.0	2.0	1.5			0.01	0.5	0.35-0.85	63.00 min Ni+Co; Al 2.3-3.2	760		
N05500	UK	BS 3075	NA18	Wir Cold drawn and pricipitation treated 10mm diam	0.25		27.0-33.0	2.0	1.5			0.01	0.5	0.35-0.85	63.00 min Ni+Co; Al 2.3-3.2	1070		
N05500	UK	BS 3075	NA18	Wir Cold drawn and solution treated 10mm diam	0.25		27.0-33.0	2.0	1.5			0.01	0.5	0.35-0.85	63.00 min Ni+Co; Al 2.3-3.2	760		
N05500	UK	BS 3127	4	Tub			30.0			66.0				1.0	Al 3			
N05500	USA	AMS 4676A	AMS 4676A	Bar Frg	0.25 max		2.0 max	1.5 max		0.02 max	0.01 max	1.0 max	0.25-1.0	Al 2-4; Co 1 max; 63.00-70.00 ST ST=Ni+Co; Pb 0.01 max; Zn 0.02 max; bal Cu			
N05500	USA	QQ N-286D	Class A	Bar Rod Wir Sh Strp	0.25 max		2.0 max	1.5 max		0.02 max	0.01 max	0.5 max	0.35-0.85	Al 2.3-3.15; Pb 0.01 max; Sn 0.01 max; 63.0-70.0 Ni+Co; Zn 0.02 max; bal Cu			
N05500	USA	ASTM F467	N05500		0.25			2.0	1.5				0.5	0.35-0.85	Al 2.3-3.15; 63.0-70.0 Ni+Co; bal Cu			
N05500	USA	ASTM F468	N05500	As manufactrued 0.250/0.875in; 6.35/22.2mm	0.25			2.0	1.5			0.01	0.5	0.35-0.85	Al 2.3-3.15; 63.0-70.0 Ni+Co; bal Cu	900	620	

UNS numbers and US grades are provided as a means of cross referencing chemically similar alloys. Exchangability is only possible after independent examination of specifications. Tensile properties are minimum or typical . UTS and YS as Mpa, El as %. See Appendix for list of abbreviations used in Descriptions.

Grade UNS #	Country	Specification	Designation	Description	C	Cr	Cu	Fe	Mn	Ni	P	S	Si	Ti	Other	UTS	YS	EL
Monel K500 N05500	USA			Pt Hard	0.25 max			2.00 max	1.50 max	63.0-70.0		0.01 max	0.50 max	0.35-0.85	Al 2.30-3.15; bal Cu			
N05502	USA	AMS 4677	AMS 4677	Bar Frg														
N05502	USA	QQ N-286D	Class B	Bar Rod Plt	0.1 max			2.0 max	1.5 max		0.02 max	0.01 max	0.5 max	0.5 max	Al 2.5-3.5; Pb 0.01 max; Sn 0.01 max; 63.00-70.00 ST ST=Ni+Co; Zn 0.02 max; bal Cu			
Monel 502 N05502	USA			Pt Hard	0.10 max			2.00 max	1.50 max	63.0-70.0		0.010 max	0.5 max	0.50 max	Al 2.50-3.50; bal Cu			
N05504	USA			Weld Fill Metal	0.25 max			2.00 max	1.50 max	63.0-70.0	0.030 max	0.015 max	1.00 max	0.25-1.00	Al 2.0-4.0; Pb 0.010 max; bal Cu			
Inconel FM62 N06062	USA			Solid Soln strengthened	0.08 max	14.00-17.00	0.50 max	6.00-10.00	1.00 max	70.00 min	0.030 max	0.015 max	0.35 max		Nb 1.50-3.00			
Inconel FM69 N07069	USA			Weld Fill metal	0.08	14.0-17.0	0.50 max	5.0-9.0	1.0 max	70.0 min	0.03 max	0.015 max	0.50 max	2.00-2.75	Nb includes Ta; Al 0.40-1.00; Mo 5.0-7.0; Nb 0.70-1.20			
Inconel 722 N07722	USA			Pt Hardenable	0.08 max	14.0-17.0	0.5 max	5.0-9.0	1.0 max	70.0 min		0.01 max	0.07 max	2.00-2.75	Al 0.4-1.0			
Inconel X750 N07750	USA			Pt Hardenable	0.08 max	14.0-17.0	0.5 max	5.0-9.0	1.0 max	70.0 min		0.01 max	0.50 max	2.25-2.75	Al 0.40-1.0; Nb 0.70-1.20			
Inconel 751 N07751	USA			Pt Hardenable	0.10 max	14.0-17.0	0.50 max	5.00-9.00	1.00 max	70.0 min		0.01 max	0.50 max	2.00-2.60	Al 0.90-1.50; Nb 0.70-1.20			
N10001	France	NF A54-401	Ni-Mo28	Bar Sh	0.05 max	1.0 max		4.0-6.0	1.0 max		0.03 max	0.03 max	1.0 max		Co 2.5 max; Mo 26-30; V 0.2-0.4			
N10001	Germany	DIN	2.4600					5.0							Mo 28			
N10001	USA	AWS A5.14-89	ERNiMo-1	Bare Weld El Rod	0.08 max	1.0 max	0.5 max	4.0-7.0	1.0 max		0.025 max	0.03 max	1.0 max		Ni includes Co; Co 2.5 max; Mo 26-30; V 0.2-0.4; W 1 max; OT 0.5 max	690		
Hastelloy B N10001	USA			Solid Soln strengthened	0.12 max	1.00 max		6.00 max	1.00 max		0.040 max	0.030 max	1.00 max		Co 2.50 max; Mo 26.0-33.0; V 0.60 max			
N10004	USA	AWS A5.14-89	ERNiMo-3	Bare Weld El Rod	0.12 max	4.0-6.0	0.5 max	4.0-7.0	1.0 max		0.04 max	0.03 max	1.0 max		Ni includes Co; Co 2.5 max; Mo 23-26; V 0.6 max; W 1 max; OT 0.5 max	690		
Hastelloy W N10004	USA			Solid Soln strengthened	0.12 max	4.00-6.00		4.00-7.00	1.00 max		0.050 max	0.050 max	1.00 max		Mo 23.00-26.00; V 0.60 max			
N24030	USA			Cast	0.30 max		27.0-33.0	3.50 max	1.50 max		0.03 max	0.03 max	2.7-3.7					
N26455	USA			Cast	0.02 max	15.0-17.5		2.0 max	1.00 max		0.03 max	0.03 max	0.80 max		Mo 15.0-17.5; W 1.0 max			
N30007	USA			Cast	0.07 max	1.0 max		3.00 max	1.00 max		0.040 max	0.030 max	1.00 max		Mo 30.0-33.0			
N30012	USA			Cast	0.12 max	1.00 max		4.0-6.0	1.00 max		0.040 max	0.030 max	1.00 max		Mo 26.0-30.0; V 0.20-0.60			
W80001	USA	AWS A5.11-90	ENiMo-1	Weld El SMAW	0.07 max	1.0 max	0.5 max	4.0-7.0	1.0 max		0.04 max	0.03 max	1.0 max		Ni includes Co; Co 2.5 max; Mo 26-30; V 0.6 max; W 1 max; OT 0.5 max			

UNS numbers and US grades are provided as a means of cross referencing chemically similar alloys. Exchangability is only possible after independent examination of specifications. Tensile properties are minimum or typical . UTS and YS as Mpa, El as %. See Appendix for list of abbreviations used in Descriptions.

Worldwide Guide to Equivalent Nonferrous Metals and Alloys

Grade UNS #	Country	Specification	Designation	Description	C	Cr	Cu	Fe	Mn	Ni	P	S	Si	Ti	Other	UTS	YS	EL
W80004	USA	AWS A5.11-90	ENiMo-3	Weld El SMAW	0.12 max	2.5-5.5	0.5 max	4.0-7.0	1.0 max		0.04 max	0.03 max	1.0 max		Ni includes Co; Co 2.5 max; Mo 23-27; V 0.6 max; W 1 max; OT 0.5 max			
W86040	USA	AWS A5.11-90	ENiCrMo-12	Weld El SMAW	0.03 max	20.5-22.5	0.5 max	5.0 max	2.2 max		0.03 max	0.02 max	0.7 max		Ni includes Co Nb+Ta=1.0-2.8; Mo 8.8-10; OT 0.5 max			
	UK	BS 3146	ANC18A	Inv	0.2		30.0						1.0					
	USA	AWS A5.30-79R	IN60	Weld consumable insert grpe.	0.15 max			2.5 max	4.0 max		0.02 max	0.015 max	1.25 max	1.5-3.0	Ni+Co=62-69; Al 1.25 max; bal Cu; OT 0.5 max			
N04060	UK	BS 2901 Part 5	NA33		0.15 max				3.0-4.0		0.02 max	0.015 max	1.25 max	1.5-3.0	Al 1.25 max; 62.00-69.00 ST; ST=Ni+Co; bal Cu			
N04060	USA	AWS A5.14-89	ERNiCu-7	Bare Weld El Rod	0.15 max			2.5 max	4.0 max	62.0-69.	0.02 max	0.015 max	1.25 max	1.5-3.0	Ni includes Co; Al 1.25 max; bal Cu; OT 0.5 max	480		
Monel FM60 N04060	USA			Weld Fill Metal	0.15 max			2.5 max	4.0 max	62.0-69.0	0.02 max	0.015 max	1.25 max	1.5-3.0	Al 1.25 max; 62.00-69.00 ST; ST=Ni+Co; bal Cu			
N06625	UK	BS 2901 Part 5	NA43		0.1 max	20.0-23.0	0.5 max		0.5 max		0.015 max	0.01-0.4	0.5 max	0.4 max	Co 1 max; ST=Ni+Co 58.0 max; Nb+Ta 3.15-4.15; Mo 8-10			
N06625	USA	AWS A5.14-89	ERNiCrMo-3	Bare Weld El Rod	0.1 max	20.0-23.0	0.5 max	5.0 max	0.5 max	58.0 min	0.02 max	0.015 max	0.5 max	0.4 max	Ni includes Co; Nb+Ta=3.15-4.15; Al 0.4 max; Mo 8-10; OT 0.5 max	590		
N06625	USA	ASTM B446	N06625	Bar Rod	0.1 max	20.0-23.0		5.0 max	0.5 max		0.01 max	0.01 max	0.5 max	0.4 max	Al 0.4 max; Mo 8-10; Nb 3.15-4.15			
N06625	USA	ASTM B443	N06625	Plt Sh Strp	0.1 max	20.0-23.0		5.0 max	0.5 max		0.01 max	0.01 max	0.5 max	0.4 max	Al 0.4 max; Mo 8-10; Nb 3.15-4.15			
N06625	USA	ASTM B366	N06625	Tub Pip	0.1 max	20.0-23.0		5.0 max	0.5 max		0.01 max	0.01 max	0.5 max	0.4 max	Al 0.4 max; Mo 8-10; Nb 3.15-4.15			
N06625	USA	ASTM B444	N06625	Tub Pip	0.1 max	20.0-23.0		5.0 max	0.5 max		0.01 max	0.01 max	0.5 max	0.4 max	Al 0.4 max; Mo 8-10; Nb 3.15-4.15			
Inconel 625 N06625	USA			Solid Soln strengthened	0.10 max	20.0-23.0		5.0 max	0.50 max		0.015 max	0.015 max	0.50 max	0.40 max	Al 0.40 max; Mo 8.0-10.0; Nb 3.15-4.15			
N26625	USA			Cast	0.06 max	20.0-23.0		5.0 max	1.00 max		0.015 max	0.015 max	1.00 max		Mo 8.0-10.0; Nb 3.15-4.50			
N99622	USA	AWS A5.8-92	BNi-10	Braze Weld Fill	0.4-0.55	10.0-13.0		2.5-4.5			0.02 max	0.02 max	3.0-4.0	0.05 max	Se 0.005 max; Al 0.05 x; B 2-3; Co 0.1 x; W 15-17; Zr 0.05 max			
N99622	USA			Braz Fill metal	0.40-0.55	10.0-13.0		2.5-4.5			0.02 max	0.02 max	3.0-4.0	0.05 max	Al 0.05 max; B 2.0-3.0; Co 0.10 x; W 15.0-17.0; Zr 0.05 max; OT 0.50 max			
W84190	USA	AWS A5.11-90	ENiCu-7	Weld El SMAW	0.15 max			2.5 max	4.0 max	62.0-69.	0.02 max	0.015 max	1.5 max	1. max	Al 0.75 max; Ni includes Co; bal Cu; OT 0.5 max			
Hastelloy S N06635	USA				0.02 max	14.5-17.0	0.35 max	3.00 max	0.30-1.00		0.020 max	0.015 max	0.20-0.75		La 0.01-0.10; Al 0.10-0.50; B 0.015 x; Co 2.00 x; Mo 14.0-16.5; W 1.00 max			

UNS numbers and US grades are provided as a means of cross referencing chemically similar alloys. Exchangability is only possible after independent examination of specifications. Tensile properties are minimum or typical. UTS and YS as Mpa, El as %. See Appendix for list of abbreviations used in Descriptions.

Grade UNS #	Country	Specification	Designation	Description	C	Cr	Cu	Fe	Mn	Ni	P	S	Si	Ti	Other	UTS	YS	EL
	UK	BS 3146	ANC12	Inv	0.1 max	21.0								2.5 max	Al 0.8; Co 1; Mo 10			
	USSR	GOST B366	KH70VMTIU		0.08 max	18.5		4.0 max						2.6	Al 1.2; Mo 4.5; W 4.5			
N06006	USA	ASTM B582	N06007	Plt Sh Strp	0.05 max	21.0-23.5	1.5-2.5	18.0-21.0	1.0-2.0		0.04 max	0.03 max	1.0 max		Mo 5.5-7.5; Co 2.5 max; Nb 1.75-2.5; W 1 max			
HX N06006	USA			Solid Soln strengthened	0.35-0.75	15.0-19.0			2.00 max	64.0-68.0	0.04 max	0.04 max	2.50 max		Mo 0.50 max; bal Fe			
Eatonite 3 N06013	USA			Hard Facing	1.80-2.20	28.0-30.0		1.0-8.0	1.0 max		0.030 max	0.030 max	0.80-1.20					
HX-50 N06050	USA			Solid Soln strengthened	0.40-0.60	15.0-19.0			1.50 max	64.0-68.0	0.04 max	0.04 max	0.50-2.00	1.0 max	bal Fe; Al+Ti 1.50 max; Mo 0.50 max; OT 0.50 max			
N99800	USA	AWS A5.8-92	BNi-8	Braze Weld Fill	0.06 max		4.0-5.0		21.5-24.5		0.02 max	0.02 max	6.0-8.0	0.05 max	Se 0.005 max; Al 0.05 max; Co 0.1 max; Zr 0.05 max			
N99800	USA			Braz Fill metal	0.10 max		4.0-5.0		21.5-24.5		0.02 max	0.02 max	6.0-8.0	0.05 max	Se 0.005 max; Al 0.05 max; Co 0.10 max; Zr 0.05 max; OT 0.50 max			
N06008	USA			Solid Soln strengthened	0.15 max	29.0-31.0		1.0 max	0.10 max		0.030 max	0.010 max	0.75-1.60		Al 0.20 max			
	France	NF A54-401	Ni-Mo16Cr15	Bar Sh	0.02 max	14.5-16.5			1.0 max		0.03 max	0.03 max	0.05 max		Co 2.5 max; Mo 15-17; V 0.04 max; W 3-4.5			
	Germany	DIN 1736	2.1806/S NiCr20Nb			18.0-22.0			2.5-3.5	67.0 min					Nb 2-3; OT 0.5 max			
	Germany	DIN 1736	2.4807/S-NiCr15FeMn			13.0-17.0		2.0-9.0	5.0-10.0	67.0 min					Nb 1-3.5; OT 0.5 max			
	Germany	DIN 1736	2.4803/S-NiCr15FeTi			14.0-17.0		6.0-10.0	2.0-3.5	67.0 min				3.5	OT 0.5 max			
	UK	BS 3146	ANC15	Inv	0.1			5.0							Mo 27			
	UK	BS 3146	ANC17	Inv	0.1			5.0							Mo 27			
	UK	DTD 204A	DTD 204A	Bar tub			31.0	2.0		67.0								
	UK	BS 3146	ANC18B	'nv	0.1		30.0						2.7					
	USA	AWS A5.30-79R	IN82	Weld consumable insert grpe.	0.1 max	18.0-22.0	0.5 max	3.0 max	2.5-3.5		0.03 max	0.015 max	0.5 max	0.75 max	Ni + Co= 67 min; Co 0.1 max; Nb 1.7-2.7; Ta 0.3 max; OT 0.5 max			
	USA	AWS A5.14-89	ERNiCr-3	Bare Weld El Rod	0.1 max	18.0-22.0	0.5 max	3.0 max	2.5-3.5	67.0 min	0.03 max	0.015 max	0.5 max	0.75 max	Ni includes Co Nb+Ta = 2-3; Co 0.12 max; Ta 0.03 max; OT 0.5 max	550		
	USA	AWS A5.30-79R	IN6A	Weld consumable insert grpe.	0.08 max	14.0-17.0	0.5 max	8.0 max	2.0-2.75		0.03 max	0.015 max	0.35 max	2.5-3.5	Ni+ Co= 67 min; Co 0.1 max; OT 0.5 max			
N06082	UK	BS 2905 Part 5	NA35	Obsolete Replaced by BS 5194.2	0.1 max	18.0-22.0	0.5 max	3.0 max	2.5-3.5			0.015 max	0.5 max	0.75 max	ST=Ni+Co 67.0 min;Nb+Ta=2.0 min 3.0 max			
N06082	USA	AWS A5.14-89	ERNiCrFe-5	Bare Weld El Rod	0.08 max	14.0-17.0	0.5 max	6.0-10.0	1.0 max	70.0 min	0.03 max	0.015 max	0.35 max		Ni includes Co Nb+Ta= 1.5-3.0; Co 0.12 max; Ta 0.03 max; OT 0.5 max	550		

UNS numbers and US grades are provided as a means of cross referencing chemically similar alloys. Exchangability is only possible after independent examination of specifications. Tensile properties are minimum or typical . UTS and YS as Mpa, El as %. See Appendix for list of abbreviations used in Descriptions.

Grade UNS #	Country	Specification	Designation	Description	C	Cr	Cu	Fe	Mn	Ni	P	S	Si	Ti	Other	UTS	YS	EL
Inconel FM82 N06082	USA			Weld Fill metal	0.10 max	18.0-22.0	0.50 max	3.0 max	2.5-3.5	67.0 min	0.03 max	0.015 max	0.50 max	0.75 max	Nb includes Ta; Nb 2.0-3.0			
N07092	UK	BS 2901 Part 5	NA39		0.08 max	14.0-17.0	0.5 max	8.0 max	2.0-2.7		0.03 max	0.015 max	0.35 max	2.5-3.5	67.00 max ST ST=Ni+Co			
N07092	USA	AWS A5.14-89	ERNiCrFe-6	Bare Weld El Rod	0.08 max	14.0-17.0	0.5 max	8.0 max	2.0-2.7	67.0 min	0.03 max	0.015 max	0.35 max	2.5-3.5	Ni includes Co; OT 0.5 max	550		
Inconel FM92 N07092	USA			Pt Hardenable	0.08 max	14.00-17.00	0.50 max	8.0 max	2.00-2.75	67.00 min	0.030 max	0.015 max	0.35 max	2.50-3.50				
N07626	USA			Pt Hardenable	0.05 max	20.0-23.0	0.50 max	6.00 max	0.50 max		0.020 max	0.015 max	0.50 max	0.60 max	N 0.05 max; Al 0.40-0.80; Co 1.00 max; Mo 8.0-10.0; Nb 4.50-5.50			
IN-100 N13100	USA			Pt Hardenable	0.15-0.20	8.0-11.0		1.0 max	0.20 max			0.015 max	0.20 max	4.50-5.00	Al 5.00-6.00; B 0.01-0.02; Co 13.0-17.0; Mo 2.0-4.0; V 0.70-1.20; Zr 0.03-0.09			
	UK	DTD 196	DTD 196	Bar			32.0		1.5	66.5								
	Germany	DIN 17744	2.4969/NiCr20Co18Ti		0.1 max	18.0-21.0	0.2 max	2.0 max	1.0 max		0.03 max	0.01 max	1.0 max	2.0-3.0	Al 1-2; Co 15-21			
	UK	BS 3146	ANC18C	Inv	0.1		30.0						4.0					
N24025	USA			Cast	0.25 max		27.0-33.0	3.50 max	1.50 max		0.03 max	0.03 max	3.5-4.5					
N30107	USA			Cast	0.07 max	17.0-20.0		3.0 max	1.00 max		0.040 max	0.030 max	1.00 max		Mo 17.0-20.0			
	UK	DTD 477	DTD 477	Tub			30.0		2.0	1.5								
	USSR	GOST B622	KH70		0.07	29.5		5.0										
N07090	UK	DTD 747A	DTD 747A		0.1	20.0		5.0						2.2	Al 1.3; Co 17			
Nimonic 90 N07090	USA			Pt Hardenable	0.13 max	18.0-21.0		3.0 max	1.0 max				1.5 max	1.8-3.0	Al 0.8-2.0; Co 15.0-21.0			
	Germany	DIN 17742	2.4952/NiCr20TiAl		0.1 max	18.0-21.0	0.2 max	3.0 max	1.0 max	65.0 min	0.03 max	0.01 max	1.0 max	1.5-2.7	Al 1-1.8; B 0.01 max; Co 2 max			
	UK	DTD 487	DTD 487	Bar			30.0		2.0	2.0					Al 3			
Inconel FM 52 N06052	USA			Weld Fill metal	0.04 max	28.0-31.5	0.30 max	7.0-11.0	1.0 max		0.020 max	0.015 max	0.50 max		Mo 0.50 max; Nb 0.10 max			
Allcorr N06110	USA			Austenitic Corros Resist	0.15 max	27.0-33.0								1.50 max	Al 1.50 max; Co 12.0 max; Mo 8.00-12.0; Nb 2.00 max; W 4.00 max			
Hastelloy B-3 N10675	USA				0.01 max	1.0-3.0	0.20 max	1.0-3.0	3.0 max	65.0 min	0.030 max	0.010 max	0.10 max	0.20 max	Ni+Mo 94.0-98.0; Al 0.50 max; Co 3.0 max; Mo 27.0-32.0; Nb 0.20 x; Ta 0.20 min; V 0.20 max; W 3.0 x; Zr 0.10 max			
N10276	Germany	DIN	2.4537		0.02 max	15.5									Mo 16; W 3.7			
N10276	Germany	DIN 17742	NiMo16Cr			16.0		6.0							Mo 16; W 4			
N10276	USA	AWS A5.14-89	ERNiCrMo-4	Bare Weld El Rod	0.02 max	14.5-16.5	0.5 max	4.0-7.0	1.0 max		0.04 max	0.03 max	0.08 max		Ni includes Co; Co 2.5 max; Mo 15-17; V 0.35 max; W 3-4.5; OT 0.5 max	590		

UNS numbers and US grades are provided as a means of cross referencing chemically similar alloys. Exchangability is only possible after independent examination of specifications. Tensile properties are minimum or typical . UTS and YS as Mpa, El as %. See Appendix for list of abbreviations used in Descriptions.

Grade UNS #	Country	Specification	Designation	Description	C	Cr	Cu	Fe	Mn	Ni	P	S	Si	Ti	Other	UTS	YS	EL
Hastelloy C276 N10276	USA			Solid Soln strengthened	0.02 max	14.5-16.5		4.0-7.0	1.0 max		0.030 max	0.030 max	0.08 max		Co 2.5 max; Mo 15.0-17.0; V 0.35 max; W 3.0-4.5			
	UK	BS 2901 Part 5	NA36		0.13 max	18.0-21.0	0.2 max	1.5 max	1.0 max				0.01 max	1 max	2.0-3.0	Ag 0.001 max; Al 1-2; B 0.02 max; Co 15-21; Pb 0.002 max; Zr 0.15 max			
Udimet 500 N07500	USA			Pt Hardenable	0.15 max	15.00-20.00	0.15 max	4.00 max	0.75 max		0.015 max	0.015 max	0.75 max	2.50-3.25	Al 2.50-3.25; B 0.003-0.01; Co 13.00-20.00; Mo 3.00-5.00			
N13009	USA			Cast	0.12-0.17	8.00-10.00	0.10 max	1.50 max	0.20 max			0.015 max	0.20 max	1.75-2.25	Al 4.75-5.25; B 0.010-0.020; Bi 0.00005 max; Co 9.00-11.00; Nb 0.75-1.25; Pb 0.0010 max; W 11.5-13.5; Zr 0.03-0.08			
	Germany	DIN 1736	2.4373/S-NiCu30A				27.0-34.0	0.5-2.		63.0 min					Al 2-4; OT 0.5 max			
	UK	DTD 10B	DTD 10B	Sh			35.0	2.3		63.0								
N04020	USA			Solid Soln strengthened	0.04 max		28.0-34.0	2.5 max	2.0 max	63.0 min		0.025 max	0.5 max					
N04020	USA			Weldable	0.35 max		26.0-33.0	2.50 max	1.5 max				2.0 max		Al 0.50 max			
N06022	USA	AWS A5.14-89	ERNiCrMo-10	Bare Weld El Rod	0.015 max	20.0-22.5	0.5 max	2.0-6.0	0.5 max		0.02 max	0.01 max	0.08 max		Ni includes Co; Co 2.5 max; Mo 12.5-14.5; V 0.35 max; W 2.5-4.5; OT 0.5 max	690		
Hastelloy C-22 N06022	USA				0.015 max	20.0-22.5		2.0-6.0	0.50 max		0.02 max	0.02 max	0.08 max		Co 2.5 max; Mo 12.5-14.5; V 0.35 max; W 2.5-3.5			
Alloy 59 N06059	USA				0.010 max	22.0-24.0		1.5 max	0.5 max		0.015 max	0.005 max	0.10 max		Al 1.10 max; Co 0.3 max; Mo 15.0-16.5			
Alloy 230 N06230	USA			Solid Soln strengthened	0.05-0.15	20.0-24.0		3.0 max	0.30-1.00		0.03 max	0.015 max	0.25-0.75		La 0.005-0.05; Al 0.20-0.50; B 0.015 max; Co 5.0 max; Mo 1.0-3.0; W 13.0-15.0			
Haynes 230-W N06231	USA			Weld Fill metal	0.05-0.15	20.0-24.0	0.5 max	3.0 max	0.3-1.0		0.030 max	0.015 max	0.25-0.75		La 0.050 max; Al 0.2-0.5; B 0.003 max; Co 3.0 max; Mo 1.0-3.0; W 13.0-15.0			
Inconel 601 N06601	USA			Solid Soln strengthened	0.1 max	21.0-25.0	1.0 max		1.0 max	58.0-63.0		0.015 max	0.50 max		Al 1.0-7.0; bal Fe			
N07013	USA			Cast	0.07-0.20	12.2-13.0		0.50 max	0.10 max		0.015 max	0.015 max	0.10 max	3.85-4.15	Hf 0.75-1.05; Al+Ti 7.30-7.70; Al 3.20-3.60; B 0.010-0.020; Co 8.50-9.50; Mo 1.70-2.10; Nb 0.10 max; Ta 3.85-4.50; W 3.85-4.50; Zr 0.05-0.14			

UNS numbers and US grades are provided as a means of cross referencing chemically similar alloys. Exchangability is only possible after independent examination of specifications. Tensile properties are minimum or typical . UTS and YS as Mpa, El as %. See Appendix for list of abbreviations used in Descriptions.

Grade UNS #	Country	Specification	Designation	Description	C	Cr	Cu	Fe	Mn	Ni	P	S	Si	Ti	Other	UTS	YS	EL
N07716	USA			Pt Hardenable	0.03 max	19.0-22.0			0.20 max	57.0-63.0	0.015 max	0.010 max	0.20 max	1.0-1.60	Al 0.35 max; Mo 7.00 9.50; Nb 2.75-4.00; bal Fe			
N26022	USA			Cast	0.02 max	20.0-22.5		2.0-6.0	1.00 max		0.025 max	0.025 max	0.80 max		Mo 12.5-14.5; V 0.35 max; W 2.5-3.5			
W80002	USA	AWS A5.11-90	ENiCrMo-5	Weld El SMAW	0.1 max	14.5-16.5	0.5 max	4.0-7.0	1.0 max		0.04 max	0.03 max	1.0 max		Ni includes Co; Co 2.5 max; Mo 15-17; V 0.35 max; W 3-4.5; OT 0.5 max			
W80276	USA	AWS A5.11-90	ENiCrMo-4	Weld El SMAW	0.02 max	14.5-16.5	0.5 max	4.0-7.0	1.0 max		0.04 max	0.03 max	0.2 max		Ni includes Co; Co 2.5 max; Mo 15-17; V 0.35 max; W 3-4.5; OT 0.5 max			
W86022	USA	AWS A5.11-90	ENiCrMo-10	Weld El SMAW	0.02 max	20.0-22.5	0.5 max	2.0-6.0	1.0 max		0.03 max	0.015 max	0.2 max		Ni includes Co; Co 2.5 max; Mo 12.5-14.5; V 0.35 x; W 2.5-3.5; OT 0.5 max			
	UK	DTD 192	DTD 192	Bar Frg			35.0	2.5		62.5								
	UK	DTD 200A	DTD 200A	Bar strp			35.0	2.5		62.5								
N07001	USA	ASTM B637	N07001	Bar Frg	0.03-0.1	18.0-21.0	0.5 max	2.0 max	1.0 max		0.03 max	0.03 max	0.75 max	2.75-3.25	Al 1.2-1.6; B 0.003-0.01; Co 12-15; Mo 3.5-5; Zr 0.02-0.12			
Waspaloy N07001	USA			Pt Hardenable	0.03-0.10	18.00-21.00	0.50 max	2.00 max	1.00 max		0.030 max	0.030 max	0.75 max	2.75-3.25	Al 1.20-1.60; Co 12.00-15.00; Mo 3.50-5.00; Zr 0.02-0.12			
MAR-M Hf mod N13246	USA			Pt Hardenable	0.13-0.17	8.0-10.0	0.10 max	1.0 max	0.20 max			0.015 max	0.20 max	1.25-1.75	Hf 1.5-2.0; Al 5.25-5.75; B 0.01-0.02; Co 9.0-10.0; Mo 2.25-2.75; Ta 1.25-1.75; W 9.0-11.0; Zr 0.03-0.08			
	Germany	DIN 1736	2.4366/S-NiCu30Mn				27.0-35.0	2.0-2.5	1.0-4.0	62.0 min0.					Nb 1-3; OT 0.5 max			
	Germany	DIN 17744	2.4810/NiMo		0.05 max	1.0 max	0.5 max	4.0-7.0	1.0 max	62.0 min0.	0.03 max	0.01 max	0.05 max		Co 1-2.5; Mo 26-30; V 0.06 max			
	Germany	DIN 1736	2.4377/NiCu30MnTi				28.0-34.0	1.0-2.5	3.0-4.0	62.0 min				1.5-3.	OT 0.5 max			
	UK	BS 2901 Part 5	NA37		0.07 max	16.0-20.0	0.2 max	1.0 max	0.5 max			0.01 max	0.5 max	1.5-3.0	Al 1.7-2.5; B 0.005 x; Co 12-16; Mo 6-8; Pb 0.005 x;Zr 0.06x			
N06617	USA	AWS A5.14-89	ERNiCrCoMo-1	Bare Weld El Rod	0.05-0.15	20.0-24.0	0.5 max	3.0 max	0.05-0.15		0.03 max	0.015 max	1.0 max	0.6 max	Ni includes Co; Al 0.8-1.5; Co 10-15; Mo 8-10; OT 0.5 max	690		
Inconel 617 N06617	USA			Solid Soln strengthened	0.05-0.15	20.0-24.0	0.50 max	3.00 max	1.00 max	44.5 min		0.015 max	1.00 max	0.60 max	Al 0.80-1.50; B 0.006 max; Co 1.0-15.0; Mo 8.00-10.0			
HW N08001	USA			Solid Soln strengthened	0.35-0.75	10.0-14.0			2.00 max	58.0-62.0	0.04 max	0.04 max	2.50 max		Mo 0.50 max; bal Fe			
HW-50 N08006	USA			Solid Soln strengthened	0.40-0.60	10.0-14.0			1.50 max	58.0-62.0	0.04 max	0.04 max	0.50-2.00		Mo 0.50 max; bal Fe			
W86117	USA	AWS A5.11-90	ENiCrCoMo-1	Weld El SMAW	0.05-0.15	21.0-26.0	0.5 max	5.0 max	0.3-2.5		0.03 max	0.015 max	0.75 max		Ni includes Co Nb=Ta=1; Co 9-15; Mo 8-10; OT 0.5 max			

UNS numbers and US grades are provided as a means of cross referencing chemically similar alloys. Exchangability is only possible after independent examination of specifications. Tensile properties are minimum or typical . UTS and YS as Mpa, El as %. See Appendix for list of abbreviations used in Descriptions.

Grade UNS #	Country	Specification	Designation	Description	C	Cr	Cu	Fe	Mn	Ni	P	S	Si	Ti	Other	UTS	YS	EL
W86132	USA	AWS A5.11-90	ENiCrFe-1	Weld El SMAW	0.08 max	13.0-17.0	0.5 max	11.0 max	3.5 max	62.0 min	0.03 max	0.015 max	0.75 max		Ni includes Co Nb+Ta= 1.5-4.0; Ta 0.3 max; OT 0.5 max			
W86133	USA	AWS A5.11-90	ENiCrFe-2	Weld El SMAW	0.1 max	13.0-17.0	0.5 max	12.0 max	1.0-3.5	62.0 min	0.03 max	0.02 max	0.75 max		Ni includes Co Nb+Ta= 0.5-3.0; Co 0.12 x; Mo 0.5-2.5; Ta 0.3 x; OT 0.5 max			
	UK	BS 3071	NA1	As cast	0.1-0.3		28.0-32.0	3.0	0.5-1.5			0.05	0.5-1.5		61.57 min Ni+Co; Mg 0.08-0.12; Pb 0.005	386	138	16
	UK	DTD 5075	DTD 5075	Sh.Strp		18.0									Co 14; Mo 7			
	UK	BS 3146	ANC10	Inv	0.1	20.0									Al 1.2; Co 17			
Eatonite 5 N06015	USA			Hard Facing	1.80-2.20	28.0-30.0			1.0-8.0	1.0 max		0.030 max	0.030 max	0.80-1.20		Mo 7.0-9.0			
	UK	BS 3071	NA2	As cast	0.15		28.0-32.0	3.0	0.5-1.5			0.05	2.5-3.		60.22 min Ni+Co; Mg 0.08-0.12; Pb 0.005	386	138	16
	Germany	DIN 1736	2.4800/S-NiMn30						4.0-7.0	60.0 min					Mo 26-30; OT 0.8 max			
	Germany	DIN 1736	2.4802/S-NiMn30						4.0-7.0	60.0 min					Mo 26-30; OT 0.8 max			
	Germany	DIN 1736	2.4831/S-NiCr21Mo9Nb				20.0-23.0		5.0 max	0.05 max	60.0 min		0.01 max	0.5 max		Al 0.4 max; Co 0.01 max; Mo 8-10; Nb 3-4.5			
	UK	BS 3146	ANC5C	Inv		15.0		25.0		60.0								
	UK	BS 3976	NA19	Rod		19.0								2.2	Al 1.3; Co 18			
	UK	DTD 5027	DTD 5027	Sh	0.13 max	20.0								2.2	Al 1.3; Co 17			
	USA	ASTM A494	M-35	Plt annealed(plate)	0.3			1.0	1.25-2.25			0.02	0.75		55-60 Ni+Co; bal Cu	552	379	10
	USA	ASTM A494	M-35	Rod Bar Cold worked(rod bar) 0.5in diam	0.3			1.0	1.25-2.25			0.02	0.75		55-60 Ni+Co; bal Cu	552	379	10
	USA	ASTM A494	M-35	Wir Cold drawn(wire) 0.5in diam	0.3			1.0	1.25-2.25			0.02	0.75		55-60 Ni+Co; bal Cu	552	379	10
	USA	QQ N-288	Composition E		0.3 max		26.0-33.0	3.5 max	1.5 max				1.0-2.0		Al 0.5 max; 60.00 min ST ST=Ni+Co 1.00-3.00ST ST=Nb+Ta			
N04019	UK	BS 3071	NA3				30.0		1.0	65.0			4.0					
N04019	USA	QQ N-288	Composition A		0.35 max		26.0-33.0	2.5 max	1.5 max				2.0 max		Al 0.5 max; 62.00-68.00 ST ST=Ni+Co			
N04019	USA	QQ N-288	Composition B		0.3 max		27.0-33.0	2.5 max	1.5 max				2.7-3.7		Al 0.5 max; 61.00-68.00 ST St=Ni+Co			
N04019	USA	QQ N-288	Composition C		0.2 max		27.0-31.0	2.5 max	1.5 max				3.3-4.3		Al 0.5 max; 60.00 min ST ST=Ni+Co			
N04019	USA	QQ N-288	Composition D		0.25 max		27.0-31.0	2.5 max	1.5 max				3.5-4.5		Al 0.5 max; 60.00 min ST ST=Ni+Co			
N04019	USA			Cast	0.25 max		27.0-31.0	2.50 max	1.50 max	60.0 min		0.015 max	3.50-4.50					
SM2060 Mo N06060	USA				0.03 max	19.0-22.0	1.00 max		1.50 max	54.0-60.0	0.030 max	0.005 max	0.50 max		Al 0.1-0.4; Mo 12.0-14.0; ; Nb 1.25 max; W 1.25 max; bal Fe			

UNS numbers and US grades are provided as a means of cross referencing chemically similar alloys. Exchangability is only possible after independent examination of specifications. Tensile properties are minimum or typical . UTS and YS as Mpa, El as %. See Appendix for list of abbreviations used in Descriptions.

Worldwide Guide to Equivalent Nonferrous Metals and Alloys

Grade UNS #	Country	Specification	Designation	Description	C	Cr	Cu	Fe	Mn	Ni	P	S	Si	Ti	Other	UTS	YS	EL
X-782 N06782	USA			Solid Soln strengthened	2.00	26.00		4.0 max	0.30				0.30		Co 0.50; W 8.75			
AF2-1DA N07012	USA				0.30-0.35	11.5-12.5		1.00 max	0.10 max		0.015 max	0.015 max	0.10 max	2.75-3.25	N 0.005 max; Al 4.20-4.80; B 0.01-0.02; Bi 0.00005 max; Co 9.50-10.50; Mo 2.50-3.50; O 0.010 max; Pb 0.002 max; Ta 1.00-2.00; W 5.50-6.50; Zr 0.05-0.15			
N10002	France	NF A54-401	NiMo16Cr15 C	Bar Sh	0.08 max	14.5-16.5		4.0-7.0	1.0 max		0.04 max	0.03 max	1.0 max		Co 2.5 max; Mo 15-17; V 0.35 max; W 3-4.5			
N10002	Germany	DIN	2.4602			16.0		6.0							Mo 17; W 4			
N10002	UK	BS 3146	ANC16	Inv	0.1	17.0		6.0							Mo 17; W 4.5			
Hastelloy C N10002	USA			Solid Soln strengthened	0.08 max	14.5-16.5		4.0-7.0	1.00 max		0.040 max	0.030 max	1.00 max		Co 2.5 max; Mo 15.0-17.0; V 0.35 max; W 3.0-4.5			
N30002	USA			Cast	0.12 max	15.5-17.5		4.5-7.5	1.00 max		0.040 max	0.030 max	1.00 max		Mo 16.0-18.0; V 0.20-0.40; W 3.75-5.25			
W86134	USA	AWS A5.11-90	ENiCrFe-4	Weld El SMAW	0.2 max	13.0-17.0	0.5 max	12.0 max	1.0-3.5	60.0 min	0.03 max	0.02 max	1.0 max		Ni includes Co Nb+Ta= 1.0-3.5; Mo 1-3.5; OT 0.5 max			
	UK	BS 3146	ANC11	Inv	0.32	21.0									Co 10; Mo 10			
N07048	USA			Age Hardenable	0.015 max	20.0-23.5	1.0-2.2	18.0-21.0	0.8 max		0.02 max	0.01 max	0.1 max	1.5-2.1	Al 0.4-0.9; Co 2.0 max; Nb 0.5 max			
M252 N07252	USA			Pt Hardenable	0.10-0.20	18.00-20.00		5.00 max	0.50 max		0.015 max	0.015 max	0.50 max	2.25-2.75	Al 0.75-1.25; B 0.003-0.1; Co 9.00-11.00; Mo 9.00-10.50			
Inconel 725 N07725	USA			Pt Hardenable	0.03 max	19.0-22.5			0.35 max	55.0-59.0	0.015 max	0.010 max	0.20 max	1.00-1.70	Al 0.35 max; Mo 7.00-9.50; Nb 2.75-4.00; bal Fe			
N13021	USA			Pt Hardenable	0.12-0.17	14.0-15.7	0.2 max	1.0 max	1.0 max			0.015 max	1.0 max	0.9-1.5	Ag 0.0005 max; Al 4.5-4.9; B 0.003-0.010; Bi 0.0001 max; Co 18.0-22.0; Mo 4.5-5.5; Pb 0.0015 max			
W86182	USA	AWS A5.11-90	ENiCrFe-3	Weld El SMAW	0.1 max	13.0-17.0	0.5 max	10.0 max	5.0-9.5	59.0 min	0.03 max	0.015 max	1.0 max	1. max	Ni includes Co Nb+Ta= 1.0-2.5; Co 0.12 max; Ta 0.3 max; OT 0.5 max			
	Germany	DIN 17744	2.4811/NiCr20Mo15		0.03 max	19.0-21.0	0.5 max	2.5 max	0.8 max	58.0 min	0.03 max	0.01 max	0.05 max		Co 1-2.5; Mo 14-17; V 0.35 max			
	USSR	GOST B462	KH75MBTIU		0.1 max	20.5		8.0 max						0.5	Al 0.5; Mo 19.5; Nb 1.1			
Inconel Filler Metal 72 N06072	USA			Filler metal	0.01-0.10	42.0-46.0	0.50 max	0.50 max	0.20 max		0.020 max	0.015 max	0.20 max	0.30-1.00	OT 0.50 max			
Inconel 690 N06690	USA			Solid Soln strengthened	0.05 max	27.0-31.0	0.50 max	7.0-11.0	0.50 max	58.0 min		0.015 max	0.50 max					

UNS numbers and US grades are provided as a means of cross referencing chemically similar alloys. Exchangability is only possible after independent examination of specifications. Tensile properties are minimum or typical . UTS and YS as Mpa, El as %. See Appendix for list of abbreviations used in Descriptions.

Grade UNS #	Country	Specification	Designation	Description	C	Cr	Cu	Fe	Mn	Ni	P	S	Si	Ti	Other	UTS	YS	EL
Promet 31 N07031	USA			Pt Hardenable	0.03-0.06		0.60-1.20		0.20 max	55.0-58.0	0.015 max	0.015 max	0.20 max	2.10-2.60	Al 1.00-1.70; B 0.003-0.007; Mo 1.70-2.30; Nb 0.75-0.95; bal Fe			
Pyromet 31V N07032	USA			Pt Hardenable	0.03-0.06	22.3-22.9				55.0-58.0	0.015 max	0.015 max		2.10-2.40	Al 1.15-1.40; Co 1.00 max; Mo 1.70-2.30; bal Fe			
Astroloy M N13017	USA				0.02-0.06	14.0-16.0	0.10 max	0.50 max	0.15 max		0.015 max	0.015 max	0.20 max	3.35-3.65	N 0.0050 max; Al 3.85-4.15; B 0.020-0.030; Bi 0.00005 max; Co 16.0-18.0; Mo 4.50-5.50; O 0.010 max; Pb 0.0002 max; W 0.05 max; Zr 0.06 max			
N13020	USA				0.03-0.10	14.0-16.0	0.10 max	2.00 max	0.15 max					2.75-3.75	Al 3.75-4.75; B 0.025-0.035; Bi 0.00005 max; Co 17.0-20.0; Mo 4.50-5.50; Zr 0.06 max			
	UK	DTD 5067	DTD 5067			15.0								3.8	Al 4.8; Co 15; Mo 3.5			
N04404	USA	ASTM F96	N04404		0.15			0.5	0.1		0.03	0.01	0.1		Al 0.05; 52.0-57.0 Ni+Co; Pb 0.01; Zn 0.02; bal Cu			
Monel 404 N04404	USA			Solid Soln strengthened	0.15 max			0.50 max	0.10 max	52.0-57.0		0.024 max	0.10 max		Al 0.05 max; bal Cu			
N06004	Germany	DIN 17742	2.4867/NiCr6015		0.15 max	14.0-19.0	0.5 max	19.0-25.0	2.0 max	59.0 min			0.5-2.0					
Nichrome N06004	USA			Solid Soln strengthened	0.15 max	14-18			1.0 max	57 min		0.01 max	0.75-1.6		bal Fe			
NIC 52 N06952	USA				0.03 max	23.0-27.0	0.5-1.0		1.0 max	48.0-56.0	0.03 max	0.003 max		0.6-1.5	Mo 6.0-8.0; bal Fe			
	UK	DTD 5007	DTD 5007		0.2	14.5								1.2	Al 4.7; Co 20; Mo 5			
Rene 41 N07041	USA			Pt Hardenable	0.12 max	18.00-20.00		5.00 max	0.10 max			0.015 max	0.50 max	3.00-3.30	Al 1.40-1.80; B 0.0030-0.010; Co 10.00-12.00; Mo 9.00-10.50			
N07718	USA	AWS A5.14-89	ERNiFeCr-2	Bare Weld El Rod	0.08 max	17.0-21.0	0.3 max		0.35 max	50.0-55.0	0.015 max	0.015 max	0.35 max	0.65-1.15	Al 0.2-0.8; B 0.006 max; Mo 2.8-3.3; Ni includes Co; Nb + Ta= 4.75-5.50; bal Fe; OT 0.5 max	1138		
Inconel 718 N07718	USA			Pt Hardenable	0.08 max	17.0-21.0	0.30 max		0.35 max	50.0-55.0	0.015 max	0.015 max	0.35 max	0.65-1.15	Al 0.20-0.80; B 0.006 max; Co 1.00 x; Mo 2.80-3.30; Nb 4.75-5.50; bal Fe			
W86112	USA	AWS A5.11-90	ENiCrMo-3	Weld El SMAW	0.1 max	20.0-23.0	0.5 max	7.0 max	1.0 max	55.0 min	0.03 max	0.02 max	0.75 max		Ni includes Co Nb+Ta= 3.15-4.15; Co 0.12 max; Mo 5-10; OT 0.5 max			
W86620	USA	AWS A5.11-90	ENiCrMo-6	Weld El SMAW	0.1 max	12.0-17.0	0.5 max	10.0 max	2.0-4.0	55.0 min	0.03 max	0.02 max	1.0 max		Ni includes Co; Nb+Ta 0.5-2.0; Mo 5-9; W 1-2; OT 0.5 max			

UNS numbers and US grades are provided as a means of cross referencing chemically similar alloys. Exchangability is only possible after independent examination of specifications. Tensile properties are minimum or typical . UTS and YS as Mpa, El as %. See Appendix for list of abbreviations used in Descriptions.

Worldwide Guide to Equivalent Nonferrous Metals and Alloys

Grade UNS #	Country	Specification	Designation	Description	C	Cr	Cu	Fe	Mn	Ni	P	S	Si	Ti	Other	UTS	YS	EL
	France	NF A54-301	Fe-Ni54	Bar Rod Wir Plt Sh Strp Frg	0.02			45.5-45.5	0.7	54.5-54.5	0.03 max	0.03	0.2					
	USA	ASTM A743	CW-12M	Sand As cast annealed	0.12	15.5-20.0		7.5	1.0		0.04	0.03	1.5		Mo 16-20; V 0.4; W 5.25	49	320	4
	USSR	GOST	KH60IU		0.1 max	16.5		26.0							Al 3.2			
N06002	UK	BS 2901 Part 5	NA40		0.05-0.15	20.5-23.0	0.5 max	17.0-20.0	1.0 max		0.04 max	0.01 max	1.0 max		B 0.01 max; Co 0.5-2.5; Mo 8-10; Pb 0.001 max; W 0.2-1			
N06002	USA	AWS A5.14-89	ERNiCrMo-2	Bare Weld El Rod	0.05-0.15	20.5-23.0	0.5 max	17.0-20.0	1.0 max		0.04 max	0.03 max	1.0 max		Ni includes Co; Co 0.5-2.5; Mo 8-10; W 0.2-1; OT 0.5 max	590		
N06002	USA	ASTM B619	N06002	Pipe	0.05-0.15	20.5-23.0		17.0-20.0	1.0 max		0.04 max	0.03 max	1.0 max		Co 0.5-2.5; Mo 8-10; W 0.2-1			
N06002	USA	ASTM B435	N06002	Plt Sh Strp	0.05-0.15	20.5-23.0		17.0-20.0	1.0 max		0.04 max	0.03 max	1.0 max		Co 0.5-2.5; Mo 8-10; W 0.2-1			
N06002	USA	ASTM B572	N06002	Rod	0.05-0.15	20.5-23.0		17.0-20.0	1.0 max		0.04 max	0.03 max	1.0 max		Co 0.5-2.5; Mo 8-10; W 0.2-1			
N06002	USA	ASTM B626	N06002	Tub	0.05-0.15	20.5-23.0		17.0-20.0	1.0 max		0.04 max	0.03 max	1.0 max		Co 0.5-2.5; Mo 8-10; W 0.2-1			
N06002	USA	ASTM B366	N06002	Tub Pip	0.05-0.15	20.5-23.0		17.0-20.0	1.0 max		0.04 max	0.03 max	1.0 max		Co 0.5-2.5; Mo 8-10; W 0.2-1			
N06002	USA	ASTM B622	N06002	Tub Pip	0.05-0.15	20.5-23.0		17.0-20.0	1.0 max		0.04 max	0.03 max	1.0 max		Co 0.5-2.5; Mo 8-10; W 0.2-1			
N06002	USA	ASTM A567	N06002		0.05-0.15	20.5-23.0		17.0-20.0	1.0 max		0.04 max	0.03 max	1.0 max		Co 0.5-2.5; Mo 8-10; W 0.2-1			
Hastelloy X N06002	USA			Solid Soln strengthened	0.05-0.15	20.5-23.0		17.0-20.0	1.00 max		0.040 max	0.030 max	1.00 max		Co 0.5-2.5; Mo 8.0-10.0; W 0.20-1.0			
N06985	USA	AWS A5.14-89	ERNiCrMo-9	Bare Weld El Rod	0.015 max	21.0-23.5	1.5-2.5	18.0-21.0	1.0 max		0.04 max	0.03 max	1.0 max		Ni includes Co Nb+Ta=0.50; Co 5 max; Mo 6-8; W 1.5 max; OT 0.5 max	760		
N06985	USA	ASTM B619	N06985	Pip	0.15 max	21.0-23.5	1.5-2.5	18.0-21.0	1.0 max		0.04 max	0.03 max	1.0 max		0.50 max ST ST=Nb+Ta; Co 5 max; Mo 6-8; W 1.5 max			
N06985	USA	ASTM B582	N06985	Plt Sh Strp	0.15 max	21.0-23.5	1.5-2.5	18.0-21.0	1.0 max		0.04 max	0.03 max	1.0 max		0.50 max ST ST=Nb+Ta; Co 5 max; Mo 6-8; W 1.5 max			
N06985	USA	ASTM B581	N06985	Rod	0.15 max	21.0-23.5	1.5-2.5	18.0-21.0	1.0 max		0.04 max	0.03 max	1.0 max		0.50 max ST ST=Nb+Ta; Co 5 max; Mo 6-8; W 1.5 max			
N06985	USA	ASTM B622	N06985	Strp Tub	0.15 max	21.0-23.5	1.5-2.5	18.0-21.0	1.0 max		0.04 max	0.03 max	1.0 max		0.50 max ST ST=Nb+Ta; Co 5 max; Mo 6-8; W 1.5 max			
N06985	USA	ASTM B626	N06985	Tub	0.15 max	21.0-23.5	1.5-2.5	18.0-21.0	1.0 max		0.04 max	0.03 max	1.0 max		0.50 max ST ST=Nb+Ta; Co 5 max; Mo 6-8; W 1.5 max			
Hastelloy G-3 N06985	USA				0.015 max	21.0-23.5	1.5-2.5	18.0-21.0	1.0 max		0.04 max	0.03 max	1.0 max		Nb+Ta 0.50 max; Co 5.0 max; Mo 6.0-8.0; W 1.5 max			
	Germany	DIN 17745	2.4420/NiFe44		0.05 max			43.0-46.0	0.05 max	53.0 min				0.3 max				
	Germany	DIN 17745	2.4472/Ni45		0.05 max			45.0-47.0	0.6 max	53.0 min				0.3 max				
SM2050 N06250	USA				0.02 max0	20.0-23.0	1.00 max	1.0 max		50.0-53.0	0.030 max	0.005 max	0.09 max		Mo 10.0-12.0; W 1.00 max; bal Fe			

UNS numbers and US grades are provided as a means of cross referencing chemically similar alloys. Exchangability is only possible after independent examination of specifications. Tensile properties are minimum or typical . UTS and YS as Mpa, El as %. See Appendix for list of abbreviations used in Descriptions.

Grade UNS #	Country	Specification	Designation	Description	C	Cr	Cu	Fe	Mn	Ni	P	S	Si	Ti	Other	UTS	YS	EL
W86002	USA	AWS A5.11-90	ENiCrMo-2	Weld El SMAW	0.05-0.15	20.5-23.0	0.5 max	17.0-20.0	1.0 max		0.04 max	0.03 max	1.0 max		Ni includes Co; Co 0.5-2.5; Mo 8-10; W 0.2-1; OT 0.5 max			
W86985	USA	AWS A5.11-90	ENiCrMo-9	Weld El SMAW	0.02 max	21.0-23.5	1.5-2.5	18.0-21.0	1.0 max		0.04 max	0.03 max	1.0 max		Ni includes Co Nb+Ta=0.5; Co 5 max; Mo 6-8; W 1.5 max; OT 0.5 max			
N06030	USA	AWS A5.14-89	ERNiCRMo-11	Bare Weld El Rod	0.03 max	28.0-31.5	1.0-2.4	13.0-17.0	0.03 max		0.04 max	0.02 max	0.8 max		Ni includes Co Nb+Ta=0.30-1.50; Co 5 max; Mo 4-6; W 1.5-4; OT 0.5 max	690		
Hastelloy G-30 N06030	USA			Solid Soln strengthened	0.03 max	28.0-31.5	1.0-2.4	13.0-17.0	1.5 max		0.04 max	0.02 max	0.8 max		Co 5.0 max; Mo 4.0-6.0; Nb 0.30-1.50; W 1.5-4.0			
SM2550 N06255	USA				0.03 max	23.0-26.0	1.20 max		1.00 max	47.0-52.0	0.03 max	0.03 max	1.0 max	0.69 max	Mo 6.0-9.0; W 3.0 max; bal Fe			
N06975	USA	AWS A5.14-89	ERNiCrMo-8	Bare Weld El Rod	0.03 max	23.0-26.0	0.7-1.2		1.0 max	47.0-52.0	0.03 max	0.03 max	1.0 max	0.7-1.5	Ni includes Co; Mo 5-7; bal Fe; OT 0.5 max	660		
N06975	USA	ASTM B619	N06975	Pip	0.03 max	23.0-26.0	0.7-1.2		1.0 max	47.0-52.0	0.03 max	0.03 max	1.0 max	0.7-1.5	Mo 5-7; bal Fe			
N06975	USA	ASTM B582	N06975	Plt Sh Strp	0.03 max	23.0-26.0	0.7-1.2		1.0 max	47.0-52.0	0.03 max	0.03 max	1.0 max	0.7-1.5	Mo 5-7; bal Fe			
N06975	USA	ASTM B581	N06975	Rod	0.03 max	23.0-26.0	0.7-1.2		1.0 max	47.0-52.0	0.03 max	0.03 max	1.0 max	0.7-1.5	Mo 5-7; bal Fe			
N06975	USA	ASTM B366	N06975	Tub Pip	0.03 max	23.0-26.0	0.7-1.2		1.0 max	47.0-52.0	0.03 max	0.03 max	1.0 max	0.7-1.5	Mo 5-7; bal Fe			
Hastelloy G-2 N06975	USA				0.03 max	23.0-26.0	0.70-1.20		1.0 max	47.0-52.0	0.03 max	0.03 max	1.0 max	0.70-1.50	Mo 5.0-7.0; bal Fe			
N07263	UK	BS 2901 Part 5	NA38		0.04-0.08	19.0-21.0	0.2 max	0.7 max	0.6 max			0.007 max	0.4 max	1.9-2.4	Ag 0.001 max; Al 0.3-0.6; B 0.005 max; Co 19-21; Mo 5.6-6.1; Pb 0.002 max; 2.40-2.80 ST ST=Al+Ti			
Nimonic 263 N07263	USA			Pt Hardenable	0.04-0.08	19.0-21.0	0.20 max	0.7 max	0.60 max		0.015 max	0.007 max	0.40 max	1.9-2.4	Al+Ti 2.4-2.8; Al 0.3-0.6; Co 19.0-21.0; Mo 5.6-6.1			
W86007	USA	AWS A5.11-90	ENiCrMo-1	Weld El SMAW	0.05 max	21.0-23.5	1.5-2.5	18.0-21.0	1.0-2.0		0.04 max	0.03 max	1.0 max		Ni includes Co; Nb+Ta= 1.75-2.50; Co 2.5 max; Mo 5.5-7.5; W 1 max; OT 0.5 max			
	Czech Republic	CSN 423484	NiFe49	Wrought					0.1-0.6	48.0-51.5					bal Fe			
	Czech Republic	CSN 423482	NiFe48	Wrought					0.1-0.6	50.0-51.5					Nb 0.6 max; bal Fe			
	France	NF A54-301	Fe-Ni51.5	Bar Rod Wir Plt Sh Strp Frg	0.01			48.75	0.4	51.25	0.01	0.01	0.2					
N06007	USA	AWS A5.14-89	ERNiCrMo-1	Bare Weld El Rod	0.05 max	21.0-23.5	1.5-2.5	18.0-21.0	1.0-2.0		0.04 max	0.03 max	1.0 max		Ni includes Co; Nb+Ta=1.75-2.50; Co 2.5 max; Mo 5.5-7.5; W 1 max; OT 0.5 max	590		
N06007	USA	ASTM B619	N06007	Pip	0.05 max	21.0-23.5	1.5-2.5	18.0-21.0	1.0-2.0		0.04 max	0.03 max	1.0 max		Mo 5.5-7.5; Co 2.5 max; Nb 1.75-2.5; W 1 max			

UNS numbers and US grades are provided as a means of cross referencing chemically similar alloys. Exchangability is only possible after independent examination of specifications. Tensile properties are minimum or typical . UTS and YS as Mpa, El as %. See Appendix for list of abbreviations used in Descriptions.

Worldwide Guide to Equivalent Nonferrous Metals and Alloys

Grade UNS #	Country	Specification	Designation	Description	C	Cr	Cu	Fe	Mn	Ni	P	S	Si	Ti	Other	UTS	YS	EL
N06007	USA	ASTM B581	N06007	Rod	0.05 max	21.0-23.5	1.5-2.5	18.0-21.0	1.0-2.0		0.04 max	0.03 max	1.0 max		Mo 5.5-7.5; Co 2.5 max; Nb 1.75-2.5; W 1 max			
N06007	USA	ASTM B366	N06007	Tub Pip	0.05 max	21.0-23.5	1.5-2.5	18.0-21.0	1.0-2.0		0.04 max	0.03 max	1.0 max		Mo 5.5-7.5; Co 2.5 max; Nb 1.75-2.5; W 1 max			
Hastelloy G N06007	USA				0.05 max	21.0-23.5	1.5-2.5	18.0-21.0	1.0-2.0		0.04 max	0.03 max	1.0 max		Co 2.5 max; Mo 5.5-7.5; Nb 1.75-2.5; W 1.0 max			
	Germany	DIN 17745	2.4475/NiFe46		0.05 max			46.0-48.0	0.6 max	51.0 min			0.3 max					
	UK	BS 3146	ANC8	Inv	0.1	20.0								0.4	Al 0.8			
	France	NF A54-301	Fe-Ni50	Bar Rod Wir Plt Sh Strp Frg	0.02			49.5	0.7	50.5	0.03	0.03	0.2					
	France	NF A54-301	Fe-Ni50.5	Bar Rod Wir Plt Sh Strp Frg	0.01			49.5	0.4	50.5	0.01	0.05	0.2					
Alloy 52 N14052	USA				0.05 max	0.25 max			0.60 max	50.5	0.025 max	0.025 max	0.30 max		Al 0.10 max; Co 0.50 max; bal Fe			
	France	NF A54-301	Fe-Ni50Cr1	Bar Rod Wir Plt Sh Strp Frg	0.02	1.0		49.0	0.25	50.0	0.03	0.03	0.2					
	Germany	DIN 1736	2.4801/NiMo16Cr			14.5 min		2.0-7.0		50.0 min					Mo 14-17; OT 0.8 max			
	Germany	DIN 1736	2.4813/NiMo15Cr			14.5 min.		2.0-7.0		50.0 min					Mo 14-17; OT 0.8 max			
	Germany	DIN 17745	2.4478/NiFe47		0.05 max			47.0-49.0	0.6 max	50.0 min			0.3 max					
	Germany	DIN 17745	2.4480/NiFe48Cr		0.05 max	0.7-1.0		47.0-49.0	1.0 max	50.0 min			0.3 max					
Hastelloy G50 N06950	USA			Solid Soln strengthened	0.015 max	19.0-21.0	0.5 max	15.0-20.0	1.0 max	50.0 min	0.04 max	0.015 max	1.0 max		Co 2.5 max; Mo 8.0-10.0; Nb 0.5 max; W 1.0 max			
W86030	USA	AWS A5.11-90	ENiCrMo-11	Weld El SMAW	0.03 max	28.0-31.5	1.0-2.4	13.0-17.0	1.5 max		0.04 max	0.02 max	1.0 max		Ni includes Co Nb+Ta= 0.3-1.5; Co 5 max; Mo 4-6; W 1.5-4; OT 0.5 max			
	France	NF A54-301	Fe-Ni48	Bar Rod Wir Plt Sh Strp Frg	0.02			52.0	0.5	48.0	0.03	0.03	0.2					
	Germany	DIN 17745	1.3921/Ni49		0.05 max	0.7-1.0		49.0-51.0	1.0 max	48.0 min			0.3 max					
D979 N09979	USA			Pt Hardenable	0.08 max	14.00-16.00			0.75 max	42.00-48.00	0.015 max	0.015 max	0.75 max	2.70-3.30	Al 0.75-1.30; B 0.008-0.016; Mo 3.75-4.50; W 3.75-4.50; Zr 0.050 max; bal Fe			
	Germany	DIN 17745	2.4486/NiFe47Cr		0.05 max	5.5-6.5		47.0-49.0	1.0 max	47.0 min			0.3 max					
	UK	DTD 237	DTD 237	Sh Strp			30.0								Zn 23			
K94800	France	NF A54-301	Fe-Ni47Cr5	Bar Rod Wir Plt Sh Strp Frg	0.02	5.0		48.0	0.4	47.0	0.03	0.03	0.2					
RA333 N06333	USA			Solid Soln strengthened	0.08 max	24.00-27.00	0.50 max		2.00 max	44.00-47.00	0.030 max	0.030 max	0.75-1.50		Co 2.50-4.00; Mo 2.50-4.00; Pb 0.025 max; Sn 0.025 max; W 2.50-4.00; bal Fe			
	Germany	DIN 17745	1.3922/Ni48		0.05 max			49.0-53.0	0.5 max	46.0 min			0.3 max		O 0.3 max			

UNS numbers and US grades are provided as a means of cross referencing chemically similar alloys. Exchangability is only possible after independent examination of specifications. Tensile properties are minimum or typical . UTS and YS as Mpa, El as %. See Appendix for list of abbreviations used in Descriptions.

Grade UNS #	Country	Specification	Designation	Description	C	Cr	Cu	Fe	Mn	Ni	P	S	Si	Ti	Other	UTS	YS	EL
	UK	DTD 232	DTD 232	Sh			28.0	1.5	1.5						Zn 23			
N08065	UK	BS 2901 Part 5	NA41		0.05 max	19.5-23.5	1.5-3.0		1.0 max		0.03 max	0.03 max	0.5 max	0.6-1.2	Al 0.2 max; 38.00-46.00 ST ST=Ni+Co;Co 2 max; bal Fe			
Incoloy FM65 N08065	USA			Weld Fill metal	0.05 max	19.5-23.5	1.50-3.0	22.0 min	1.0 max	38.0-46.0	0.03 max	0.03 max	0.50 max	0.60-1.2	Al 0.20 max; Mo 2.5-3.5			
N08221	USA				0.025 max	20.0-22.0	1.50-3.00		1.00 max	36.0-46.0		0.03 max	0.50 max	0.60-1.00	Al 0.20 max; Mo 5.00-6.50; bal Fe			
N0826	USA			Cast	0.05 max	19.5-23.5	1.5-3.0	22.0 min	1.00 max	38.0-46.0	0.03 max	0.03 max	1.00 max		Mo 2.5-3.5; Nb 0.6-1.2			
Incoloy 825 N08825	USA			Solid Soln strengthened	0.05 max	19.5-23.5	1.5-3.0		1.0 max	38.0-46.0	0.03 max	0.03 max	0.5 max	0.6-1.2	Al 0.2 max; Mo 2.5-3.5; bal Fe			
N09925	USA			Pt Hardenable	0.03 max	19.5-23.5	1.50-3.00	22.0 min	1.00 max	38.0-46.0		0.03 max	0.50 max	1.90-2.40	Al 0.10-0.50; Mo 2.50-3.50; Nb 0.50 max			
	UK	DTD 5047	DTD 5047	Sh Strp		16.5		35.0							Mo 3.2			
	UK	BS 2901 Part 5	NA42		0.04-0.08	15.5-17.5	0.5 max		0.2 max			0.01 max	0.05 max	1.1-1.3	Ag 0.001 max; Al 1.1-1.3; B 0.005 max; Co 2 max; Mo 2.8-3.8; Pb 0.002 max; Zr 0.01-0.04; 42.00-45.00 ST ST=Ni+Co; bal Fe			
ERNiFe Mn-Cl N02216	USA			Weld Fill for cast iron	0.50 max		2.5 max		10.0-14.0	35.0-45.0		0.03 max	1.0 max		Al 1.0 max; bal Fe; OT 1.0 max			
Monel 401 N04401	USA			Solid Soln strengthened	0.10 max			0.75 max	2.25 max	40.0-45.0		0.015 max	0.25 max		Co 0.25 max; bal Cu			
Incoloy 901 N09901	USA			Pt Hardenable	0.10 max	11.00-14.00	0.50 max		1.00 max	4.00-45.0		0.030 max	0.60 max	2.35-3.10	Al 0.35 max; B 0.010-0.020; Mo 5.00-7.00; bal Fe			
NIC 42M N08042	USA				0.03 max	20.0-23.0	1.5-3.0		1.0 max	40.0-44.0	0.03 max	0.003 max	0.5 max	0.6-1.2	Mo 5.0-7.0; bal Fe			
Inconel 706 N09706	USA			Pt Hardenable	0.06 max	14.5-17.5	0.30 max		0.35 max	39.0-44.0	0.020 max	0.015 max	0.35 max	1.5-2.0	Al 0.40 max; B 0.006 max; Nb 2.5-3.3; bal Fe			
Ni-Span-C 902 N09902	USA			Pt Hardenable	0.06 max	4.90-5.75			0.80 max	41.0-43.5	0.04 max	0.04 max	1.0 max	2.20-2.75	Al 0.30-0.80; bal Fe			
	UK	DTD 268	DTD 268	Bar Wir			30.0	2.0	2.0						Zn 23			
Incoloy 804 N06804	USA			Solid Soln strengthened	0.10 max	28.0-31.0	0.50 max		1.50 max	39.0-43.0		0.015 max	0.75 max	1.20 max	Al 0.60 max; bal Fe			
K94101	France	NF A54-301	Fe-Ni42	Bar Rod Wir Plt Sh Strp Frg	0.02			58.0	0.7	42.0	0.03	0.03	0.2					
K94760	France	NF A54-301	Fe-Ni42Cr6	Bar Rod Wir Plt Sh Strp Frg	0.02	6.0		52.0	0.4	42.0	0.03	0.03	0.2					
	Germany	DIN 17745	1.3917/Ni42		0.05 max			56.0-58.0	1.0 max	41.0 min			0.3 max					
HU N08004	USA			Solid Soln strengthened	0.35-0.75	17.0-21.0			2.00 max	37.0-41.0	0.04 max	0.04 max	2.50 max		Mo 0.50 max; bal Fe			
HU-50 N08005	USA			Solid Soln strengthened	0.40-0.60	17.0-21.0			1.50 max	37.0-41.0	0.04 max	0.04 max	0.50-2.00		Mo 0.50 max; bal Fe			
N08421	USA				0.025 max	20.0-22.0	1.5-2.0		1.00 max	39.0-41.0		0.03 max	0.5 max	0.6-1.0	Al 0.2 max; Mo 5.0-6.5; bal Fe			

UNS numbers and US grades are provided as a means of cross referencing chemically similar alloys. Exchangability is only possible after independent examination of specifications. Tensile properties are minimum or typical . UTS and YS as Mpa, El as %. See Appendix for list of abbreviations used in Descriptions.

Worldwide Guide to Equivalent Nonferrous Metals and Alloys

Grade UNS #	Country	Specification	Designation	Description	C	Cr	Cu	Fe	Mn	Ni	P	S	Si	Ti	Other	UTS	YS	EL
	Czech Republic	CSN 423481	NiFe63	Wrought					0.1-0.7	36.0-40.0					Nb 0.7 max; bal Fe			
N08004	UK	BS 3076	NA17	Rod		18.0		40.0-40.0	1.2				2.2					
N08024	USA			Nb stabilized	0.03 max	22.5-25.0	0.50-1.50		1.00 max	35.0-40.0	0.035 max	0.035 max	0.50 max		Mo 3.50-5.00; Nb 0.15-0.35; bal Fe			
N08421	UK	BS 3076	NA16	Rod		21.0	2.2	32.0						0.9	Mo 3			
Incoloy 903 N19903	USA			Pt Hardenable	0.06 max	1.00 max	0.50 max		1.00 max	36.0-40.0		0.015 max	0.35 max	1.00-1.25	Al 0.30-1.15; B 0.012 max; Co 13.0-17.0; Nb 2.40-3.50; bal Fe			
N19907	USA			Pt Hardenable	0.06 max	1.0 max	0.5 max		1.0 max	35.0-40.0	0.015 max	0.015 max	0.35 max	1.2-1.8	Al 0.20 max; B 0.012 max; Co 12.0-16.0; Nb 4.3-5.2; bal Fe			
N19909	USA			Pt Hardenable	0.06 max	1.0 max	0.5 max		1.0 max	35.0-40.0	0.015 max	0.015 max	0.25-0.50	1.3-1.8	Al 0.15 max; B 0.012 max; Co 12.0-16.0; Nb 4.3-5.2; bal Fe			
CG27 N09027	USA			Pt Hardenable	0.02-0.08	12.50-14.00			0.25 max	36.50-39.50	0.015 max	0.015 max	0.25 max	2.30-2.70	Al 1.45-1.75; B 0.003-0.015; Mo 5.00-6.00; Nb 0.60-1.10bal Fe			
Eatonite N06005	USA			Solid Soln strengthened	2.40	29.00		6.5 max		39.00-39.00			0.70		Co 10.00; W 15.00			
	Germany	DIN 17744	2.4858/NiCr2 Mo		0.03 max	19.5-23.5	1.5-3.0		1.0 max	38.0 min	0.03 max	0.01 max	0.5 max	0.6-1.2	Al 0.2 max; Mo 2.5-3.5bal Fe			
N08020	USA	ASTM B464	N08020	Pip	0.07 max	19.0-21.0	3.0-4.0		2.0 max	32.0-38.0	0.05 max	0.04 max	1.0 max		1.00 max ST ST=Nb Nb min=5xC; Mo 2-3; bal Fe			
N08020	USA	ASTM B463	N08020	Plt Sh Strp	0.07 max	19.0-21.0	3.0-4.0		2.0 max	32.0-38.0	0.05 max	0.04 max	1.0 max		1.00 max ST ST=Nb Nb min=5xC; Mo 2-3; bal Fe			
N08020	USA	ASTM B468	N08020	Tub	0.07 max	19.0-21.0	3.0-4.0		2.0 max	32.0-38.0	0.05 max	0.04 max	1.0 max		1.00 max ST ST=Nb Nb min=5xC; Mo 2-3; bal Fe			
N08020	USA	ASTM B366	N08020	Tub Pip	0.07 max	19.0-21.0	3.0-4.0		2.0 max	32.0-38.0	0.05 max	0.04 max	1.0 max		1.00 max ST ST=Nb Nb min=5xC; Mo 2-3; bal Fe			
N08020	USA	ASTM B471	N08020	Wir	0.07 max	19.0-21.0	3.0-4.0		2.0 max	32.0-38.0	0.05 max	0.04 max	1.0 max		1.00 max ST ST=Nb Nb min=5xC; Mo 2-3; bal Fe			
N08020	USA	ASTM B462	N08020		0.07 max	19.0-21.0	3.0-4.0		2.0 max	32.0-38.0	0.05 max	0.04 max	1.0 max		1.00 max ST ST=Nb Nb min=5xC; Mo 2-3; bal Fe			
Carpenter 20Cb3 N08020	USA			Solid Soln strengthened	0.07 max	19.00-21.00	3.00-4.00		2.00 max	32.00-38.00	0.045 max	0.035 max	1.00 max		Mo 2.00-3.00; Nb 8xC-1.00; bal Fe			
SM 2035 N08135	USA				0.03 max	20.5-23.5	0.70 max		1.00 max	33.0-38.0	0.03 max	0.03 max	0.75 max		Mo 4.0-5.0; W 0.20-0.80; bal Fe			
Carpenter 20Mo6 N08026	USA			Wrought	0.03 max	22.0-26.0	2.00-4.00		1.00 max	33.0-37.2	0.03 max	0.03 max	0.50 max		N 0.10-0.16; Mo 5.00-6.70; bal Fe			
N08036	USA				0.06 max	22.0-26.0	1.00-3.00		1.00 max	33.0-37.2	0.030 max	0.030 max	0.50 max		Mo 5.00-6.70; N 0.17-0.40; bal Fe			

UNS numbers and US grades are provided as a means of cross referencing chemically similar alloys. Exchangability is only possible after independent examination of specifications. Tensile properties are minimum or typical . UTS and YS as Mpa, El as %. See Appendix for list of abbreviations used in Descriptions.

8-28 Nickel

Grade UNS #	Country	Specification	Designation	Description	C	Cr	Cu	Fe	Mn	Ni	P	S	Si	Ti	Other	UTS	YS	EL
HT N08002	USA			Solid Soln strengthened	0.35-0.75	13.0-17.0			2.00 max	33.0-37.0	0.04 max	0.04 max	2.50 max		Mo 0.50 max; bal Fe			
HT-50C N08008	USA			Solid Soln strengthened	0.40-0.60	13.0-17.0				33.0-37.0					Mo 0.50 max; Nb 0.75-1.25; bal Fe			
HT-30 N08030	USA			Solid Soln strengthened	0.25-0.35	13.0-17.0			2.00 max	33.0-37.0	0.040 max	0.040 max	2.50 max		Mo 0.50 max; bal Fe			
HT-50 N08050	USA			Solid Soln strengthened	0.40-0.60	15.0-19.0			1.50 max	33.0-37.0	0.04 max	0.04 max	0.50-2.00		Mo 0.50 max; bal Fe			
RA 330 N08330	USA			Solid Soln strengthened	0.08 max	17.0-20.0	1.00 max		2.00 max	34.0-37.0	0.03 max	0.03 max	0.75-1.50		Pb 0.005 max; Sn 0.025 max; bal Fe			
ER330 N08331	USA			Weld Fill metal	0.18-0.25	15.0-17.0	0.75 max		1.0-2.5	34.0-37.0	0.03 max	0.03 max	0.30-0.65		Mo 0.75 max; bal Fe			
RA 330TX N08332	USA				0.05-0.10	17.0-20.0	1.00 max		2.00 max	34.0-37.0	0.03 max	0.03 max	0.75-1.50	0.20-0.60	Al 0.10-0.50; Pb 0.005 max; Sn 0.025 max; bal Fe			
RA 330-04 N08334	USA			Solid Soln strengthened	0.18-0.29	17.0-20.0	0.5 max		4.25-6.5	33.0-37.0	0.025 max	0.02 max	0.65-1.3		Mo 0.7 max; bal Fe			
HT-30 N08603	USA			Cast	0.25-0.35	13.0-17.0			2.00 max	33.0-37.0	0.040 max	0.040 max	2.50 max		Mo 0.50 max; bal Fe			
HT N08605	USA			Cast	0.35-0.75	15.0-19.0			2.00 max	33.0-37.0	0.04 max	0.04 max	2.50 max		bal Fe			
HP N08705	USA			Cast	0.35-0.75	24.0-28.0			2.00 max	35.0-37.0	0.04 max	0.04 max	2.50 max		Mo 0.50 max; bal Fe			
N08535	USA				0.03 max	24.0-27.0	1.50 max		1.00 max	29.0-36.5	0.03 max	0.03 max	0.50 max		Mo 2.5-4.0; bal Fe			
K93601	France	NF A54-301	Fe-Ni36	Bar Rod Wir Plt Sh Strp Frg				64.0	0.35	36.0-36.0	0.03	0.03	0.2		Co 0.02			
N08021	USA			Weld Fill metal	0.07 max	19.0-21.0	3.0-4.0		2.5 max	32.0-36.0	0.03 max	0.03 max	0.60 max		Nb+Ta 8xC-1.0; Mo 2.0-3.0			
N08022	USA			Weld Fill metal	0.025 max	19.0-21.0	3.0-4.0		1.5-2.0	32.0-36.0	0.015 max	0.020 max	0.15 max		Nb+Ta 8xC-0.40; Mo 2.0-3.0			
N08321	USA			Solid Soln strengthened	0.035 max	19.0-21.0	3.0-4.0		1.5-2.5	32.0-36.0	0.02 max	0.015 max	0.3 max		Mo 2.0-3.0; Nb 8xC-0.4 max; bal Fe			
Incoloy 800 N08800	USA			Solid Soln strengthened	0.10 max	19.0-23.0	0.75 max		1.5 max	30.0-35.0		0.015 max	1.0 max		Al 0.15-0.60; bal Fe			
Incoloy 802 N08802	USA			Solid Soln strengthened	0.20-0.50	19.0-23.0	0.75 max		1.50 max	30.0-35.0		0.015 max	0.75 max	0.25-1.25	Al 0.15-1.00; bal Fe			
Incoloy 800H N08810	USA				0.05-0.10	19.0-23.0	0.75 max		1.5 max	30.0-35.0		0.015 max	1.0 max	0.15-0.60	Al 0.15-0.60; bal Fe			
Incoloy 800HT N08811	USA			Wrought	0.06-0.10	19.0-23.0	0.75 max	39.5 min	1.5 max	30.0-35.0		0.015 max	1.0 max	0.15-0.60	Al+Ti 0.85-1.20; Al 0.15-0.60			
INCO 032 N08032	USA				0.03 max	20.0-23.0			1.0 max	30.0-34.0	0.03 max	0.005 max	0.05 max		Mo 4.0-5.0; bal Fe			
N08151	USA				0.05-0.15	19.0-21.0			0.50-1.50	31.0-34.0	0.03 max	0.03 max	0.50-1.50		Nb 0.50-1.50;; bal Fe			
Incoloy 801 N08801	USA			Solid Soln strengthened	0.10 max	19.0-22.0	0.5 max		1.5 max	30.0-34.0		0.015 max	1.0 max	0.75-1.5	bal Fe			
Sanicro 28 N08028	USA			Solid Soln strengthened	0.03 max	26.0-28.0	0.6-1.4		2.50 max	29.5-32.5	0.030 max	0.030 max	1.00 max		Mo 3.0-4.0; bal Fe			
	USA	AWS A5.30-79R	IN67	Weld consumable insert grpe.				0.4-0.75	1.0 max		0.02 max	0.01 max	0.15 max	0.2-0.5	Ni + Co = 29-32; bal Cu; OT 0.5 max			
Alloy 31 N08031	USA				0.015 max	26.0-28.0	1.0-1.4		2.0 max	30.0-32.0	0.03 max	0.005 max	0.05 max		N 0.15-0.25; Mo 6.0-7.0; bal Fe			

UNS numbers and US grades are provided as a means of cross referencing chemically similar alloys. Exchangability is only possible after independent examination of specifications. Tensile properties are minimum or typical . UTS and YS as Mpa, El as %. See Appendix for list of abbreviations used in Descriptions.

Worldwide Guide to Equivalent Nonferrous Metals and Alloys

Grade UNS #	Country	Specification	Designation	Description	C	Cr	Cu	Fe	Mn	Ni	P	S	Si	Ti	Other	UTS	YS	EL
CN-7M N08007	USA			Solid Soln strengthened	0.07 max	19.0-22.0	3.00-4.00		1.50 max	27.5-30.5			1.50 max		Mo 2.00-3.00; bal Fe			
Incoloy 926 N09926	USA			Pt Hardenable	0.04 max	14.0-18.0	3.5-5.5	39.0 min	1.5 max	26.0-30.0		0.015 max	0.75 max	1.5-2.3	Al 0.3 max; Mo 2.5-3.5			
	France	NF A54-301	Fe-Ni29Co17	Bar Rod Wir Plt Sh Strp Frg	0.02			54.0	0.25	29.0	0.03	0.03	0.2		Co 17			
N94630	USA			Metal/Ceramic sealing	0.02 max	0.03 max	0.20 max		0.35 max	29	0.006 max	0.006 max	0.15 max	0.01 max	Al 0.01 max; Co 17; Mg 0.01 max; Mo 0.06 max; Zr 0.01 max			
	Germany	DIN 17745	1.3981/NiCo2 918		0.05 max			53.0-55.0	0.05 max	28.0 min			0.3 max		Co 17-18			
N08904	USA				0.02 max0	19.0-23.0	1.00-2.00		2.00 max	23.0-28.0	0.045 max	0.035 max	1.00 max		Mo 4.00-5.00; bal Fe			
	Germany	DIN 17745	1.3982/NiCo2 823		0.05 max			47.0-49.0	0.05 max	27.0 min			0.3 max		Co 22-24			
Haynes 20 Mod N08320	USA				0.05 max	21.0-23.0			2.5 max	25.0-27.0	0.04 max	0.03 max	1.0 max		Mo 4.0-6.0; Ti 4xC min; bal Fe			
N94620	USA			Metal/Ceramic sealing	0.02 max	0.03 max	0.20 max		0.15 max	27	0.006 max	0.006 max	0.15 max	0.01 max	Al 0.01 max; Co 25; Mg 0.01 max; Mo 0.06 max; Zr 0.01 max			
N08065	USA	AWS A5.14-89	ERNiCr-1	Bare Weld El Rod	0.05 max	19.5-22.5	1.5-3.0	22.0 min	1.0 max		0.03 max	0.03 max	0.5 max	0.6-1.2	Ni includes Co; Al 0.2 max; Mo 2.5-3.5; OT 0.5 max	550		
JS 700 N08700	USA				0.04 max	19.0-23.0	0.50 max		2.00 max	24.0-26.0	0.04 max	0.03 max	1.00 max		Mo 4.3-5.0; Nb 8xC-0.50; Pb 0.005 max; Sn 0.035 max; bal Fe			
N08925	USA				0.020 max	19.0-21.0	0.50-1.50		1.00 max	24.0-26.0	0.045 max	0.030 max	0.50 max		Mo 6.00-7.00; N 0.10-0.20; bal Fe			
N08926	USA				0.020 max	19.0-21.0	0.5-1.5		2.00 max	24.0-26.0	0.030 max	0.010 max	0.50 max		N 0.15 max; Mo 6.0-7.0; bal Fe			
Creusot URSB8 N08932	USA				0.020 max	24.0-26.0	1.0-2.0		2.0 max	24.0-26.0	0.025 max	0.010 max	0.50 max		Mo 4.7-5.7; N 0.17-0.25; bal Fe			
AL 6X N08366	USA			Solid Soln strengthened	0.035 max	20.0-22.0			2.00 max	23.5-25.5	0.030 max	0.030 max	1.00 max		Mo 6.00-7.00; bal Fe			
AL 6XN N08367	USA			Wrought or Cast	0.030 max	20.0-22.0			2.00 max	23.50-25.50	0.040 max	0.030 max	1.00 max		Mo 6.00-7.00; N 0.18-0.25; bal Fe			
N08310	USA				0.020 max	24.0-26.0			2.00-4.00	18.0-22.0	0.035 max	0.015 max	0.050 max		Mo 2.00-4.00; N 0.20-0.40; bal Fe			
HL N08604	USA			Cast	0.20-0.60	28.0-32.0			2.00 max	16.0-22.0	0.04 max	0.04 max	2.00 max		Mo 0.50 max; bal Fe			
HL-30 N08613	USA			Cast	0.25-0.35	28.0-32.0			1.50 max	18.0-22.0	0.04 max	0.04 max	0.50-2.00		Mo 0.50 max; bal Fe			
HL-40 N08614	USA			Cast	0.35-0.45	28.0-32.0			1.50 max	18.0-22.0	0.04 max	0.04 max	0.50-2.00		Mo 0.50 max; bal Fe			

UNS numbers and US grades are provided as a means of cross referencing chemically similar alloys. Exchangability is only possible after independent examination of specifications. Tensile properties are minimum or typical . UTS and YS as Mpa, EL as %. See Appendix for list of abbreviations used in Descriptions.

8-30 Nickel

Grade UNS #	Country	Specification	Designation	Description	Al	As	Bi	Cu	Fe	Pb	Sb	Sn	Zn	OT Max	Other	UTS	YS	EL
	UK	BS 3252	T3									99-100			0.90 max ST 0.10 max OT ST=Pb+Sb+Bi +Cu+ As+Fe			
L13002	USA			Pig tin		0.0005 max	0.001 max	0.002 max	0.005 max	0.010 max	0.008	99.98 max	0.001 max		Cd 0.001 max;S 0.002 max; Ni+Co 0.005 max; Cd 0.001 max			
	Germany	DIN 1704	2.3500/Sn99.95	Obsolete Replaced by DIN EN 610		0.01 max	0.002 max	0.01 max	0.01 max	0.02 max	0.02 max	99.95 min			0.002 max ST, ST=Al+Cd+Zn			
	Pan America	COPANT 925	A1	Ing		0.01 max	0.01 max	0.02 max	0.01 max	0.02 max	0.02 max	99.95 min	0.01 max		0.01 max ST ST=Ni+Co; Ag 0.005 max; Cd 0.001 max			
	USA	ASTM B339	Ultra Pure	Pig tin		0.005 max	0.015 max	0.005 max	0.010 max	0.001 max	0.005 max	99.95 min	0.005 max	0.010	Cd 0.001 max; Ni+Co 0.010 max; S 0.010 max; Ag 0.010 max			
L13004	USA			Pig tin		0.01 max	0.01 max	0.02 max	0.01 max	0.02 max	0.02 max	99.95 max	0.005 max		Cd 0.001 max; S 0.01 max; Ni+Co 0.01 max; Cd 0.001 max			
	Germany	DIN 1704	2.3501/Sn99.90	Obsolete Replaced by DIN EN 611		0.03 max	0.01 max	0.03 max	0.01 max	0.04 max	0.03 max	99.9 min			0.002 max ST, ST=Al+Cd+Zn			
	Pan America	COPANT 925	A2			0.01 max	0.03 max	0.02 max	0.01 max	0.03 max	0.02 max	99.9 min	0.05 max		0.01 max ST ST=Ni+Co; Ag 0.005 maxC; Cd 0.06 max			
	UK	BS 3252	I1			0.04 max	0.04 max	0.04 max	0.01 max	0.04 max	0.04 max	99.9 min			0.10 max ST ST=Pb+Sb+Bi +Cu+As+Fe max impurity limit			
	USA	EIA/IPC J-STD-006	Sn99A	Electronic Grade Solder								99.9						
	Pan America	COPANT 925	A3	Ing		0.05 max	0.05 max	0.04 max	0.01 max	0.05 max	0.04 max	99.8 min			0.07 max ST=Cd+Zn+S+ Ag+Ni+Co			
	USA	ASTM B339	Grade A	Pig tin		0.05 max	0.030 max	0.04 max	0.010 max	0.05 max	0.04 max	99.85 min	0.005 max		Cd 0.001 max; Ni+Co 0.01 max; S 0.01 max; Ag 0.01 max			
	USA	ASTM B339	Grade A	Pig tin for tinplate		0.05 max	0.030 max	0.04 max	0.010 max	0.020 max	0.04 max	99.85 min	0.005 max	0.010	Cd 0.001 max; Ni+Co 0.01 max; S 0.01 max; Ag 0.01 max			
L13007	USA			Pig tin	0.01 max	0.04 max	0.015 max	0.03 max	0.015 max	0.05 max	0.04 max	99.80 min	0.001 max		Cd 0.001 max; Ag 0.01 max; S 0.003 max; Ni+Co 0.015 max; Cd 0.001 max			
L13008	USA			Pig tin		0.05 max	0.015 max	0.04 max	0.015 max	0.05 max	0.04 max	99.80 min	0.005 max		Cd 0.001 max; S 0.01 max; Ni+Co 0.01 max; Cd 0.001 max			
L13006	USA			Pig tin		0.05 max						99.80 min						
	Germany	DIN 1704	2.3502/Sn99.75	Obsolete Replaced by DIN EN 612		0.05 max	0.02 max	0.05 max	0.01 max	0.08 max	0.08 max	99.75 min			0.005 max ST, ST=Al+Cd+Zn			

UNS numbers and US grades are provided as a means of cross referencing chemically similar alloys. Exchangability is only possible after independent examination of specifications. Tensile properties are minimum or typical . UTS and YS as Mpa, El as %. See Appendix for list of abbreviations used in Descriptions.

Worldwide Guide to Equivalent Nonferrous Metals and Alloys

Grade UNS #	Country	Specification	Designation	Description	Al	As	Bi	Cu	Fe	Pb	Sb	Sn	Zn	OT Max	Other	UTS	YS	EL
	UK	BS 3252	T2			0.05 max	0.08 max	0.05 max	0.01 max	0.08 max	0.08 max	99.75 min			0.20 max ST 0.05 max OT ST=Pb+Cu+As +Fe+Sb+Bi max impurity limit			
L13010	USA			Pig tin								99.65 min						
	Australia	AS 1834.1	99.3Sn/0.7Cu		0.001 max	0.05 max	0.1 max	0.45-0.85	0.02 max	0.1 max	0.1 max	99.15-99.55	0.01 max		Ag 0.05 max; Cd 0.005 max			
	Australia	AS 1834.1	99.5Sn/4Cu/0.5Ag		0.001 max	0.01 max	0.1 max	3.2-4.8	0.02 max	0.1 max	0.1 max		0.01 max		Ag 0.3-0.7; Cd 0.005 max			
	Germany	DIN 1704	2.3503/Sn99.50	Obsolete Replaced by DIN EN 613		0.08 max	0.03 max	0.05 max	0.01 max	0.4 max	0.1 max	99.5 min			ST=Al+Cd+Zn			
	Pan America	COPANT 925	A4	Ing								99.5 min		0.5				
L13012	USA			Pig tin								99.50 min						
	USA	EIA/IPC J-STD-006	Cu.7C	Electronic Grade Solder				0.7				99.3						
	Germany	DIN 1704	2.3505/Sn99.00	Obsolete Replaced by DIN EN 614		0.1 max	0.03 max	0.1 max	0.05 max	1 max	0.2 max	99 min			0.005 max ST, ST–Al+Cd+Zn			
L13014	Pan America	COPANT 925	A5	Ing								99 min		1				
L13014	USA			Pig tin								99.00 min						
L13963	USA			Pewter		0.05 max		1.0-2.0	0.015 max	0.05 max	1.0-3.0	95-98	0.005 max					
	Czech Republic	CSN 423648	SnSb2,5					0.05 max		0.5 max	1.9-3.1							
L13963	Germany	DIN 17810	2.3702/SnSb2Cu1.5	Obsolete Replaced by DIN EN 611.1				1-2		0.5 max	1-3							
L13965	Australia	AS 1834.1	96.5Sn/3.5Ag		0.005 max	0.03 max	0.1 max	0.08 max	0.02 max		0.2 max		0.01 max	0.08	Ag 3.3-3.7; Cd 0.005 max			
L13965	Pan America	COPANT 450	96.5TS		0.005 max	0.005 max	0.15 max	0.08 max	0.02 max	0.2 max	0.2-0.5	96.5 max	0.01 max		Ag 3.3-3.7			
L13965	USA			Pewter	0.005 max	0.05 max	0.15 max	0.08 max	0.02 max	0.20 max	0.20-0.50	96.5	0.005 max		Ag 3.3-3.7			
L13961	USA			Solder	0.05 max			0.20 max		0.10 max			0.005 max		Ag 3.6-4.4; Cd 0.005 max			
	USA	EIA/IPC J-STD-006	Ag03A	Electronic Grade Solder				0.7			0.5	96.3			Ag 2.5			
	USA	EIA/IPC J-STD-006	Sn96A	Electronic Grade Solder								96.3			Ag 3.7			
L13961	Germany	DIN 1707	2.3620/L-SnAg5	Obsolete Replaced by DIN EN 29454										0.2	Ag 3-5			
L13961	USA	QQ S571d	Sn96			4				0.1 max								
L13960	USA			Solder	0.005 max	0.05 max	0.15 max	0.08 max	0.02 max	0.20 max	0.20-0.50	96	0.005 max		Ag 3.6-4.4			
L13950	Germany	DIN 1707	2.3695/L-SnSb5	Obsolete Replaced by DIN EN 29471							4.5-5.5			0.2				
	USA	EIA/IPC J-STD-006	Cd01A	Electronic Grade Solder								95.0			Ag 4.0; Cd 1.0			
	USA	EIA/IPC J-STD-006	Sn95E	Electronic Grade Solder								95.0			Ag 5.0			
	USA	EIA/IPC J-STD-006	Sb05A	Electronic Grade Solder							4.0-6.0	95.0						
	Australia	AS 1834.1	95Sn/5Ag		0.005 max	0.03 max	0.1 max	0.08 max	0.02 max	0.1 max	0.2 max		0.01 max		Ag 4.8-5.2; Cd 0.005 max			
L13950	Pan America	COPANT 450	95TA		0.005 max	0.05 max	0.15 max	0.08 max	0.04 max	0.2 max	4.5-5.5	95 max	0.01 max					
L13940	USA	QQ S571d	Sb5								5							

UNS numbers and US grades are provided as a means of cross referencing chemically similar alloys. Exchangability is only possible after independent examination of specifications. Tensile properties are minimum or typical . UTS and YS as Mpa, El as %. See Appendix for list of abbreviations used in Descriptions.

9-2 Tin

Grade UNS #	Country	Specification	Designation	Description	Al	As	Bi	Cu	Fe	Pb	Sb	Sn	Zn	OT Max	Other	UTS	YS	EL
L13950	USA			Solder	0.005 max	0.05 max	0.15 max	0.08 max	0.04 max	0.20 max	4.5-5.5	95	0.005 max					
L13912	Germany	DIN 17810	2.3705/SnSb5 Cu1.5					1-2		0.5 max	3.1-7							
L13940	USA			Solder	0.03 max	0.06 max		0.08 max	0.08 max	0.20 max	4.0-6.0	94 min	0.03 max		OT 0.03 max; Cd 0.03 max			
L13911	USA			Pewter cast		0.05 max		0.25-2.0	0.015 max	0.05 max	6-8	90-93	0.005 max					
L13912	USA			Pewter sheet		0.05 max		1.5-3.0	0.015 max	0.05 max	5-7.5	90-93	0.005 max					
L13910	Pan America	COPANT 926	A1		0.005 max	0.1 max	0.08 max	4-5	0.08 max	0.35 max	4-5	90-92	0.01 max		Cd 0.05 max			
	USA	QQ M161/1	M161/1					4			4							
Babbitt metal L13910	USA			White metal bearing alloy	0.005 max	0.10 max	0.08 max	4.0-5.0	0.08 max	0.35 max	4.0-5.0	90.0-92.0	0.005 max		Cd 0.05 max			
	USA	ASTM B102	CY44A	Die cast	0.01 max	0.08 max		4-5	0.08 max	0.35 max	4-5	90-92	0.01 max					
L13913	USA			Die cast	0.01 max	0.08 max		4-5	0.08 max	0.35 max	4-5	90-92	0.01 max					
	Germany	DIN 1707	2.3680/L-Sn90Pb	Obsolete Replaced by DIN EN 29470								90.5		0.08	bal Pb			
	USA	EIA/IPC J-STD-006	Sn90A	Electronic Grade Solder							10.0	90.0						
	France	NF A56-101	101	Ing	0.005 max	0.1 max	0.05 max	3.5	0.05 max	0.35 max	6.5	90	0.01 max		0.25 max ST			
	France	NF A56-101	111	Ing	0.005 max	0.1 max	0.05 max	3.5	0.05 max	0.1 max	6.5	90	0.01 max		0.25 max ST ST=max impurity limit			
L13890	Pan America	COPANT 926	A2		0.005 max	0.1 max	0.08 max	3-4	0.08 max	0.35 max	7-8	88-90	0.01 max		Cd 0.05 max			
Babbitt metal L13890	USA			White metal bearing alloy	0.005 max	0.10 max	0.08 max	3.0-4.0	0.08 max	0.35 max	7.0-8.0	88.0-90.0	0.005 max		Cd 0.05 max			
L13890	UK	BS 3332	1/Tin Base Alloy	Ing				3-3.5		0.35 max	7.25-7.75			0.5	OT includes Pb			
	Czech Republic	CSN 423753	SnSb10Cu3Ni			0.1 max		2.0-4.0	0.1 max	0.5 max	9.0-11.0		0.015 max		Ni 0.4-0.8			
L13870	Pan America	COPANT 926	A4		0.005 max	0.1 max	0.08 max	5-6.5	0.08 max	0.5 max	6-7.5	86-89	0.01 max		Cd 0.05 max			
Babbitt metal L13870	USA			White metal bearing alloy	0.005 max	0.10 max	0.08 max	5.0-6.5	0.08 max	0.50 max	6.0-7.5	86.0-89.0	0.005 max		Cd 0.05 max			
	Pan America	COPANT 926	A3			0.02 max	0.02 max	2.8-3.2	0.01 max		7.4-7.8	87.5-88.9			ST=Cd+Zn+S+ Ag+Ni+Co+In; Cd 0.8-1.2; Ni 0.15-0.25			
	USA	QQ M161/2	M161/2					4			7.5							
L13890	France	NF A56-101	102	Ing	0.001 max	0.1 max	0.05 max	3.5	0.05 max	0.35 max	7.5	88			0.25 max ST ST=max impurity limit; Cd 0.6-1; Ni 0.1-0.4			
L13890	France	NF A56-101	112	Ing	0.005 max	0.1 max	0.05 max	3.5	0.05 max	0.1 max	7.5	88	0.01 max		0.25 max ST ST=max impurity limit; Ni 0.1-0.4			
	UK	BS 3332	2/Tin Base Alloy	Ing				3.5-5		0.35 max	8.5-10			0.5	OT includes Pb			
	Pan America	COPANT 926	A5		0.005 max	0.1 max	0.08 max	7.5-8.5	0.08 max	0.35 max	7.5-8.5	83-85	0.01 max		Cd 0.05 max			
Babbitt metal L13840	USA			White metal bearing alloy	0.005 max	0.10 max	0.08 max	7.5-8.5	0.08 max	0.35 max	7.5-8.5	83.0-85.0	0.005 max		Cd 0.05 max			
	Pan America	COPANT 926	A6			0.02 max	0.02 max	6.1-6.9	0.01 max	1.9-2.1	7.5-8.4	82.5-84.4			Cd 0.05 max			

UNS numbers and US grades are provided as a means of cross referencing chemically similar alloys. Exchangability is only possible after independent examination of specifications. Tensile properties are minimum or typical . UTS and YS as Mpa, EL as %. See Appendix for list of abbreviations used in Descriptions.

Worldwide Guide to Equivalent Nonferrous Metals and Alloys

Grade UNS #	Country	Specification	Designation	Description	Al	As	Bi	Cu	Fe	Pb	Sb	Sn	Zn	OT Max	Other	UTS	YS	EL
	USA	QQ M161/3	M161/3					8		0.3	8							
	USA	ASTM B102	YC135A	Die cast	0.01 max	0.08 max		4-6	0.08 max	0.35 max	12-14	80-84	0.01 max					
L13820	USA			Die cst	0.01 max	0.08 max		4-6	0.08 max	0.35 max	12-14	80-84	0.01 max					
	France	NF A56-101	103	Ing	0.005 max	0.1 max	0.05 max	6	0.05 max	0.35 max	11	83	0.01 max		0.25 max ST ST=max impurity limit			
	France	NF A56-1C1	113	Ing	0.005 max	0.1 max	0.05 max	6	0.05 max	0.1 max	11	83	0.01 max		0.25 max ST ST=max impurity limit			
	USA	QQ M161/4	M161/4					5.5		0.2	13							
	Pan America	COPANT 926	A7			0.02 max	0.02 max	5.1-5.9	0.01 max	2.9-3.1	11.1-11.9	79-81						
	France	NF A56-101	104	Ing	0.005 max	0.1 max	0.05 max	10	0.05 max	0.35 max	10	80	0.01 max		0.25 max ST ST=max impurity limit			
	France	NF A56-101	106	Ing	0.005 max	0.1 max	0.05 max	6	0.05 max	2	12	80	0.01 max		0.25 max ST ST=max impurity limit			
	France	NF A56-101	114	Ing	0.005 max	0.1 max	0.05 max	10	0.05 max	0.1 max	10	80	0.01 max		0.25 max ST ST=max impurity limit			
	Germany	DIN 1707	2.3620/L-SnCd20	Obsolete Replaced by DIN EN 29455										0.2	Cd 19.5-20.5			
	UK	BS 3332	3/Tin Base Alloy	Ing				4-6			3-4	9-11	80 min	0.3				
	France	NF A56-101	105	Ing	0.005 max	0.1 max	0.05 max	9	0.05 max	0.35 max	13	78	0.01 max		0.25 max ST ST=max impurity limit			
	UK	BS 3332	4/Tin Base Alloy	Ing				3-4			11-13	74-76		0.35	bal Pb			
	UK	BS 3332	5/Tin Base Alloy	Ing				2.5-3.5			6-8	74-76						
	France	NF A56-101	107	Ing	0.005 max	0.1 max	0.05 max	4	0.05 max	15	7	74	0.01 max		0.25 max ST ST=max impurity limit			
L13700	Australia	AS 1834	70Sn		0.001 max	0.03 max	0.1 max	0.08 max	0.02 max		0.2 max	69.00-71.00		0.08	Cd 0.005 max			
	Czech Republic	CSN 55635	B-Sn70Zn-320/200	Soft Solder				0.05 max	0.5 max			69.0-71.0			Bi+As 0.10; Cd 0.1 x; bal Zn			
	UK	BS 3332	9/Tin-Zinc Alloy	Ing				1-2				66-71		1	bal Zn			
L13700	Germany	DIN 1707	2.3670/L-Sn70Pb	Obsolete Replaced by DIN EN 29469								69.5-70.5		0.08	bal Pb			
L13700	USA	QQ S571d	Sn70							29	0.45							
	USA	EIA/IPC J-STD-006	Sn70A	Electronic Grade Solder						30.0		70.0						
	USA	EIA/IPC J-STD-006	Pb30A	Electronic Grade Solder						30.0	0.20-0.50	70.0						
L13700	Pan America	COPANT 450	70A		0.005 max	0.03 max	0.25 max	0.08 max	0.02 max	30 min	0.12 max	70 max	0.01 max					
L13701	Pan America	COPANT 450	70B		0.005 max	0.03 max	0.25 max	0.08 max	0.02 max	30 min	0.2-0.5	70 max	0.01 max					
	USA	EIA/IPC J-STD-006	In12A	Electronic Grade Solder						18.018.0		70.0			In 12.0			
L13700	USA			Solder	0.005 max	0.03 max	0.25 max	0.08 max	0.02 max	30	0.20	70	0.005 max					
L13701	USA			Solder	0.005 max	0.03 max	0.25 max	0.08 max	0.02 max	30	0.20-0.50	70	0.005 max					
	France	NF A56-101	301	Ing	0.005 max	0.1 max	0.05 max	1.5	0.05 max			67.5	31		0.25 max ST impurity limit			
	Australia	AS 1834.1	65Sn		0.001 max	0.03 max	0.1 max	0.08 max	0.02 max		0.2 max	64.00-66.00		0.08	0.005 max ST ST=Al+Zn+Cd; Cd 0.005 max			

UNS numbers and US grades are provided as a means of cross referencing chemically similar alloys. Exchangability is only possible after independent examination of specifications. Tensile properties are minimum or typical. UTS and YS as Mpa, El as %. See Appendix for list of abbreviations used in Descriptions.

9-4 Tin

Grade UNS #	Country	Specification	Designation	Description	Al	As	Bi	Cu	Fe	Pb	Sb	Sn	Zn	OT Max	Other	UTS	YS	EL
	USA	ASTM B102	PY1815A	Die cast	0.01 max	0.15 max		1.5-2.5	0.08 max	17-19	14-16	64-66	0.01 max					
L13650	USA			Die cast	0.01 max	0.15 max		1.5-2.5	0.08 max	17-19	14-16	64-66	0.01 max					
L13630	Australia	AS 1834.1	63Sn		0.001 max	0.03 max	0.1 max	0.08 max	0.02 max		0.2 max	62.00-64.00		0.08	0.005 max ST ST=Al+Zn+Cd; Cd 0.005 max			
	Germany	DIN 1707	2.3666/L-Sn63PbAg	Obsolete Replaced by DIN EN 29467								62.5-63.5			Ag 1.3-1.5; bal Pb			
L13630	Germany	DIN 1707	2.3663/L-Sn63Pb	Obsolete Replaced by DIN EN 29465								62.5-63.5		0.002	bal Pb			
	Australia	AS 1834	62Sn/2Ag		0.001 max	0.03 max	0.1 max	0.08 max	0.02 max		0.2 max	61.0-63.0		0.08	Cd 0.005 max			
	USA	EIA/IPC J-STD-006	Sn63A	Electronic Grade Solder						37.0		63.0						
	USA	EIA/IPC J-STD-006	Sn63C	Electronic Grade Solder						37.0		63.0						
	USA	EIA/IPC J-STD-006	Pb37A	Electronic Grade Solder						37.0	0.20-0.50	63.0						
L13631	Pan America	COPANT 450	63B		0.005 max	0.03 max	0.25 max	0.08 max	0.02 max	37 min	0.2-0.5	63 max	0.01 max		Si 0.5 max			
L13630	Pan America	COPANT 450	63A		0.005 max	0.03 max	0.25 max	0.08 max	0.02 max	37 min	0.12 max	63 max	0.01 max					
L13630	USA			Solder	0.005 max	0.03 max	0.25 max	0.08 max	0.02 max	37	0.12 max	63	0.005 max					
L13631	USA			Solder	0.005 max	0.03 max	0.25 max	0.08 max	0.02 max	37	0.20-0.50	63	0.05 max					
L13630	USA	QQ S571d	Sn63							37	0.45							
	USA	EIA/IPC J-STD-006	Pb36A	Electronic Grade Solder						38.0	0.20-0.50	62.0			Ag 2.0			
	USA	EIA/IPC J-STD-006	Pb36B	Electronic Grade Solder						36.0		62.0			Ag 2.0			
	USA	EIA/IPC J-STD-006	Pb36C	Electronic Grade Solder						36.0		62.0			Ag 2.0			
	USA	QQ M161/5	M161/5			.		3		25	10							
	USA	QQ S571d	Sn62							38	0.45							
	Germany	DIN 1707	2.3667/L-Sn60PbAg	Obsolete Replaced by DIN EN 29468								59.5-60.5		0.08	Ag 3-4; bal Pb			
L13600	Germany	DIN 1707	2.3660/L-Sn60Pb	Obsolete Replaced by DIN EN 29462								59.5-60.5		0.08	bal Pb			
	Germany	DIN 1707	2.3661/L-Sn60PbCu	Obsolete Replaced by DIN EN 29463				0.1-0.3				59.5-60.5		0.08	bal Pb			
	Germany	DIN 1707	2.3662/L-Sn60PbCu2	Obsolete Replaced by DIN EN 29464				1.6-2				59.5-60.5		0.08	bal Pb			
L13601	Germany	DIN 1707	2.3665/L-Sn60Pb(Sb)	Obsolete Replaced by DIN EN 29466							0.12-0.5	59.5-60.5		0.08	bal Pb			
	USA	EIA/IPC J-STD-006	Sn 60A	Electronic Grade Solder						40.0		60.0						
	USA	EIA/IPC J-STD-006	Sn60C	Electronic Grade Solder						40.0		60.0						
	USA	EIA/IPC J-STD-006	Pb40A	Electronic Grade Solder						40.0	0.20-0.50	60.0						
	USA	EIA/IPC J-STD-006	Bi02A	Electronic Grade Solder			2.5			38.5		60.0						
	USA	EIA/IPC J-STD-006	Cu02A	Electronic Grade Solder				2.0		38.0		60.0						
L13600	Pan America	COPANT 450	60A		0.005 max	0.03 max	0.25 max	0.08 max	0.02 max	40 min	0.12 max	60 max	0.01 max					

UNS numbers and US grades are provided as a means of cross referencing chemically similar alloys. Exchangability is only possible after independent nexamination of specifications. Tensile properties are minimum or typical . UTS and YS as Mpa, El as %. See Appendix for list of abbreviations used in Descriptions.

Worldwide Guide to Equivalent Nonferrous Metals and Alloys

Grade UNS #	Country	Specification	Designation	Description	Al	As	Bi	Cu	Fe	Pb	Sb	Sn	Zn	OT Max	Other	UTS	YS	EL
L13601	Pan America	COPANT 450	60B		0.005 max	0.03 max	0.25 max	0.08 max	0.02 max	40 min	0.2-0.5	60 max	0.01 max					
	UK	BS 3332	6/Tin Base Alloy	Ing				2-4				9-11	58-60		0.35	bal Pb			
L13600	USA			Solder	0.005 max	0.03 max	0.25 max		0.02 max	40	0.12 max	60	0.005 max		Co 0.08 max			
L13601	USA			Solder	0.005 max	0.03 max	0.25 max	0.08 max	0.02 max	40	0.20-0.50	60	0.005 max					
L13600	USA	QQ S571b	Sn60							40	0.45							
	Czech Republic	CSN 423982	Bi8Sn57Pb	Low Melting Point Alloy			6.5-8.5					56.5-58.5			bal Pb			
	USA	EIA/IPC J-STD-006	In20A	Electronic Grade Solder						26.0		54.0			In 20.0			
	Czech Republic	CSN 55631	B-SnPbCd18-145/145	Soft Solder	0.002 max	0.03 max	0.1 max	0.08 max	0.03 max	31.0-33.0	0.2 max		0.01 max		Cd 17.0-19.0			
	Germany	DIN 1707	2.3618/L-SnPbCd18	Obsolete Replaced by DIN EN 29453						31-32				0.2	Cd 17.5-18.5			
	Czech Republic	CSN 423984	Bi13Sn50Pb	Low Melting Point Alloy			12.0-14.0					49.0-51.0			bal Pb			
	Germany	DIN 1707	2.3657/L-Sn50PbAg	Obsolete Replaced by DIN EN 29461								49.5-50.5		0.08	Ag 34; bal Pb			
	Germany	DIN 1707	2.3620/L-SnIn50	Obsolete Replaced by DIN EN 29456								49.5-50.5		0.2	bal In			
	Germany	DIN 1707	2.3651/L-Sn50PbCu	Obsolete Replaced by DIN EN 29458				1.2-1.6				49.5-50.5		0.08	bal Pb			
	Germany	DIN 1707	2.3653/L-Sn50PbSb	Obsolete Replaced by DIN EN 29459							0.5-3	49.5-50.5		0.08	bal Pb			
L55031	Germany	DIN 1707	2.3655/L-Sn50Pb(Sb)	Obsolete Replaced by DIN EN 29460							0.12-0.5	49.5-50.5		0.08	bal Pb			
L55030	Germany	DIN 1707	2.3650/L-Sn50Pb	Obsolete Replaced by DIN EN 29457								49.5-50.05		0.08	bal Pb			
	USA	EIA/IPC J-STD-006	Cd18A	Electronic Grade Solder						32.0		50.0			Cd 18.0			
	USA	EIA/IPC J-STD-006	Sn50A	Electronic Grade Solder						50.0		50.0						
	USA	EIA/IPC J-STD-006	Sn50C	Electronic Grade Solder						50.0		50.0						
	USA	EIA/IPC J-STD-006	Pb50A	Electronic Grade Solder						50.0	0.20-0.50	50.0						
	USA	EIA/IPC J-STD-006	Cu01A	Electronic Grade Solder				1.5		48.5		50.0						
	USA	QQ S571A	S571A							50								
L55031	USA	QQ S571d	Sn50							50	0.45							
	Australia	AS 1834	47Sn/17Cd		0.005 max	0.03 max	0.1 max	0.08 max	0.02 max		0.2 max	46.0-48.0	0.01 max	0.08	Cd 16.5-17.5; bal Pb			
	USA	EIA/IPC J-STD-006	Bi08A	Electronic Grade Solder			8.0			48.0		46.0						
	USA	EIA/IPC J-STD-006	Sn45A	Electronic Grade Solder						55.0		45.0						
	USA	EIA/IPC J-STD-006	Bi14A	Electronic Grade Solder			14.0			43.0		43.0						
L54821	USA	QQ S571d	Pb70								0.45	31			bal Pb			
	UK	BS 3332	7/Lead Base Alloy	Ing				0.5-1		75 max	12-15	10-13		0.5				
L53560	UK	BS 3332	8/Lead Base Alloy	Ing				0.5 max			14-17	4.5-6.5			bal Pb			

UNS numbers and US grades are provided as a means of cross referencing chemically similar alloys. Exchangability is only possible after independent examination of specifications. Tensile properties are minimum or typical. UTS and YS as Mpa, El as %. See Appendix for list of abbreviations used in Descriptions.

Grade UNS #	Country	Spec.	Designation	Comment	Al	C	Fe	H	N	O	S	V	W	OT x	Other	UTS	YS	ELG
	USA	ASTM B299	SL	Sponge Na reduced Leached	0.05 max	0.02 max	0.05 max	0.05 max	0.015 max	0.10 max				0.05	H2O 0.02 max; Cl 0.20 max; Na 0.19 max; Si 0.04 max			
	USSR	GOST 17746	TG90	Powd		0.02 max	0.06 max		0.02 max	0.04 max					Cl 0.1 max; Ni 0.1 max			
	USA	ASTM B299	EL	Sponge Electrolytic	0.03 max	0.02 max	0.05 max	0.02 max	0.008 max	0.08 max				0.05	H2O 0.02 max; Cl 0.10 max; Mg 0.08 max; Na 0.10 max; Si 0.04 max			
	USA	ASTM B299	MD	Sponge Mg reduced Distilled		0.02 max	0.12 max	0.010 max	0.015 max	0.10 max				0.05	H2O 0.02 max; Cl 0.12 max; Mg 0.08 max; Si 0.04 max			
	Japan	JIS H2151	TS-140S, Class 3S	Powd		0.03 max	0.07 max	0.02 max	0.03 max	0.15 max					Cl 0.2 max; Mn 0.05 max; Na 0.15 max; Ti 99.3 max			
	Japan	JIS H2151	TS-160M, Class 4M	Powd		0.03 max	0.2 max	0.01 max	0.03 max	0.25 max					Cl 0.2 max; Mg 0.08 max; Mn 0.05 max; Ti 99.2 max			
	Japan	JIS H2151	TS-160S, Class 4S	Powd		0.03 max	0.07 max	0.02 max	0.03 max	0.25 max					Cl 0.2 max; Mn 0.05 max; Na 0.15 max; Ti 99.2 max			
	USA	ASTM B299	GP	Sponge Gen Purpose	0.05 max	0.03 max	0.15 max	0.03 max	0.02 max	0.15 max				0.05	H2O 0.02 max; Na or Mg 0.50 max; Cl 0.20 max; Si 0.04 max			
	USSR	GOST 17746	TG100	Powd		0.03 max	0.07 max		0.02 max	0.02 max					Cl 0.1 max; Ni 0.1 max			
	USSR	GOST 17746	TG110	Powd		0.03 max	0.09 max		0.02 max	0.05 max					Cl 0.1 max; Ni 0.1 max			
	USSR	GOST 17746	TG120	Powd		0.04 max	0.11 max		0.03 max	0.06 max					Cl 0.1 max; Ni 0.1 max			
	USSR	GOST 17746	TG130	Powd		0.04 max	0.13 max		0.03 max	0.08 max					Cl 0.1 max; Ni 0.1 max			
	USSR	GOST 17746	TG150	Powd		0.05 max	0.2 max		0.04 max	0.1 max					Cl 0.1 max; Ni 0.1 max			
	USSR	GOST 17746	TGTV	Powd		0.15 max	2 max		0.3 max						Cl 0.3 max; ET 0.1 max			
	Russia	GOST	AT1-1Sv	Weld el		0.08	0.12	0.006	0.04	0.12					Si 0.1			
R50250	China	GB 3620	TA-1			0.05	0.15	0.015	0.03	0.15					Si 0.1			
R50250	Europe	AECMA prEN2525	Ti-P01	Sh Strp		0.08	0.2	0.0125	0.05	0.2				0.6	OE 0.1 max			
R50250	Europe	AECMA prEN3441	Ti-P01	Sh Strp Ann HR		0.08	0.2	0.0125	0.05	0.2				0.6	OE 0.1 max			
R50250	Europe	AECMA prEN3487	Ti-P01	Sh Strp Ann CR		0.08	0.2	0.0125	0.05	0.2				0.6	OE 0.1 max			
R50250	France	AIR 9182	T-35	Sh Strp		0.08	0.12	0.01	0.05						Si 0.04	440	395	30
R50250	Germany	DIN 17850	3.7025	Plt Sh Strp Rod Wir Frg Ann		0.08	0.2	0.013	0.05	0.1						350	180	25
R50250	Germany	DIN 17860	3.7025	Sh Strp		0.08	0.2	0.013	0.05	0.1								
R50250	Germany	DIN 17862	3.7025	Rod		0.08	0.2	0.013	0.05	0.1								
R50250	Germany	DIN 17863	3.7025	Wire		0.08	0.2	0.013	0.05	0.1								
R50250	Germany	DIN 17864	3.7025	Frg		0.08	0.2	0.013	0.05	0.1								
R50250	Germany	DIN 17850	Ti I	Sh Strp Plt Rod Wir Frg Ann		0.08	0.2	0.013	0.05	0.1						350	180	30
R50250	Japan	JIS H4650	TB28C/H Class 1	HW CD Bar			0.2	0.015	0.05	0.15								

UNS numbers and US grades are provided as a means of cross referencing chemically similar alloys. Exchangability is only possible after independent examination of specifications. Tensile properties are minimum or typical . UTS and YS as Mpa, El as %. See Appendix for list of abbreviations used in Descriptions.

Worldwide Guide to Equivalent Nonferrous Metals and Alloys

Grade UNS #	Country	Spec.	Designation	Comment	Al	C	Fe	H	N	O	S	V	W	OT x	Other	UTS	YS	ELG
	Japan	JIS H4650	TB28H, Class 1	Bar			0.2 max	0.02 max	0.05 max	0.15 max								
R50250	Japan	JIS Class 1	Ti Class 1				0.2	0.015	0.05	0.15						345	165	27
	Japan	JIS H4600	TP28C, Class 1	Sh			0.2 max	0.01 max	0.05 max	0.15 max								
	Japan	JIS H4600	TP28H, Class 1	Sh			0.2 max	0.01 max	0.05 max	0.15 max								
R50250	Japan	JIS H4600	TP28H/C Class 1	HR CR Sh			0.2	0.013	0.05	0.15								
	Japan	JIS H4650	TR28C, Class 1	Bar Rod			0.2 max	0.02 max	0.05 max	0.15 max								
	Japan	JIS H4600	TR28C, Class 1	Strp			0.2 max	0.01 max	0.05 max	0.15 max								
	Japan	JIS H4600	TR28H, Class 1	Strp			0.2 max	0.01 max	0.05 max	0.15 max								
R50250	Japan	JIS H4600	TR28H/C Class 1	HR CR Strp.			0.2	0.013	0.05	0.15								
R50250	Japan	JIS H4631	TTH28D Class 1	Smls tube for heat exch			0.2	0.015	0.05	0.15								
R50250	Japan	JIS H4631	TTH28W/WD Class 1	Weld tube for heat exch			0.2	0.015	0.05	0.15								
	Japan	JIS H4630	TTP28D, Class 1	Tub Pip CD 10-80 mm diam OD			0.2 max	0.02 max	0.05 max	0.15 max						275		27
R50250	Japan	JIS H4630	TTP28D/E Class 1	Smls pipe			0.2	0.015	0.05	0.15								
	Japan	JIS H4630	TTP28E, Class 1	Tub Pip HE 10-80 mm diam OD			0.2 max	0.02 max	0.05 max	0.15 max						275		27
	Japan	JIS H4630	TTP28W, Class 1	Tub Pip As weld 10-150 mm OD			0.2 max	0.02 max	0.05 max	0.15 max						275		27
R50250	Japan	JIS H4630	TTP28W/WD Class 1	As-weld/weld & drawn pipe			0.2	0.015	0.05	0.15								
	Japan	JIS H4630	TTP28WD, Class 1	Tub Pip CD 10-150 mm OD			0.2 max	0.02 max	0.05 max	0.15 max						275		27
R50250	Japan	JIS H4670	TW28 Class 1	Wire			0.2	0.015	0.05	0.15								
R50250	Russia	GOST	VT1		0.1		0.3	0.015	0.04	0.15					Si 0.15			
R50250	Russia	GOST 1.90013-71	VT1-00	Sh Plt Strp Foil Rod Frg Ann	0.05		0.2	0.008	0.04	0.1				0.1	Si 0.08	353	245	24
R50250	Russia	GOST	VT1D		0.05		0.3	0.01	0.04	0.15					Si 0.15			
R50250	Russia	GOST	VT1D-1	All forms	0.05		0.3	0.01	0.04	0.15					Si 0.15			
R50250	Russia	GOST	VT1D-2		0.05		0.3	0.01	0.04	0.15					Si 0.15			
R50250	UK	BS 2TA.1	BS 2TA.1	Sh Strp HT			0.2	0.01							Ti 99.8 min	290	200	25
R50250	UK	DTD 5013		Bar Bil			0.2	0.013										
R50250	USA	MIL T-81556A	CP-4	Ext Bar Shap Ann	0.08		0.2	0.015	0.05	0.15				0.3		275	205	22
R50250	USA	MIL T-9046J	CP-4	Sh Strp Plt Ann	0.08		0.2	0.015	0.05	0.15				0.3		240	240	24
R50250	USA	MIL T-46035A	ELI	Ann or HT .25/2.5 in diam	0.1			0.01 max	0.02	0.1 max					Ti 99.8		827	13
R50250	USA	MIL T-46038B	ELI	Rod Bar Ann or HT	0.1			0.01 max	0.02	0.1 max					Ti 99.8		827	13
R50100	USA	AWS A5.16-70	ERTi-1	Weld fill met	0.03	0.1		0.005	0.01	0.1								
R50100	USA	AWS A5.16-90	ERTi-1	Weld El Rod	0.03 max	0.1 max		0.01 max	0.02 max	0.1 max				0.2	OE 0.05 max			
R50120	USA	AWS A5.16-70	ERTi-2	Weld fill met	0.05	0.2		0.008	0.02	0.1								
R50120	USA	AWS A5.16-90	ERTi-2	Weld El Rod	0.03 max	0.2 max		0.01 max	0.02 max	0.1 max				0.2	OE 0.05 max			

UNS numbers and US grades are provided as a means of cross referencing chemically similar alloys. Exchangability is only possible after independent examination of specifications. Tensile properties are minimum or typical. UTS and YS as Mpa, El as %. See Appendix for list of abbreviations used in Descriptions.

10-2 Titanium

Grade UNS #	Country	Spec.	Designation	Comment	Al	C	Fe	H	N	O	S	V	W	OT x	Other	UTS	YS	ELG
R50125	USA	AWS A5.16-70	ERTi-3	Weld fill met		0.05	0.2	0.008	0.02	0.1-0.15								
R50125	USA	AWS A5.16-90	ERTi-3	Weld El Rod		0.03 max	0.2 max	0.01 max	0.02 max	0.1-0.15				0.2	OE 0.05 max			
	USA	ASTM B381	F-1	Frg		0.10 max	0.20 max	0.0150 max	0.03 max	0.18 max				0.4	OE 0.1 max	240	170	24
R50250	USA	ASME SB-381	F-1	Frg Ann		0.1	0.2	0.015	0.03	0.18				0.4	OE 0.1 max			
R50250	USA	MIL T-46035A	LI	Ann or HT .25/2.5 in diam		0.1		0.01 max	0.03	0.12 max					Ti 99.7		827	13
R50250	USA	MIL T-46038B	LI	Rod Bar Ann or HT		0.1		0.01 max	0.03	0.12 max					Ti 99.7		827	13
R50250	USA	MIL T-46035A	NI	Ann or HT .25/2.5 in diam		0.1		0.01 max	0.04	0.18 max					Ti 99.7		827	13
R50250	USA	MIL T-46038B	NI	Rod Bar Ann or HT		0.1		0.01 max	0.04	0.18 max					Ti 99.7		827	13
	USA	ASTM B337	Ti Grade 1	Smls Weld Pip		0.10 max	0.20 max	0.015 max	0.03 max	0.18 max				0.4	OE 0.1 max	240	170	24
	USA	ASTM B338	Ti Grade 1	Smls Weld Tub Exch Conds Ann		0.10 max	0.20 max	0.015 max	0.03 max	0.18 max				0.4	OE 0.1 max	240	170	24
	USA	ASTM B348	Ti Grade 1	Bar Bil (Bar H max 0.100)		0.10 max	0.20 max	0.0125 max	0.03 max	0.18 max				0.4	OE 0.1 max	240	170	24
	USA	ASTM B861	Ti Grade 1	Smls Pip Ann		0.10 max	0.20 max	0.015 max	0.03 max	0.18 max				0.4	OE 0.1 max	240	170	24
	USA	ASTM B863	Ti Grade 1	Wir		0.10 max	0.20 max	0.015 max	0.03 max	0.18 max				0.4	OE 0.1 max	240	170	20
R50250	USA	ASTM B265	Ti Grade 1	Sh Strp Plt Ann		0.10 max	0.20 max	0.015 max	0.03 max	0.18 max				0.4	OE 0.1 max	240	170	24
R50250	USA	ASTM F467-84a	Ti Grade 1	Nut		0.1	0.2	0.0125	0.05	0.18								
R50250	USA	ASTM F467M-84b	Ti Grade 1	Metric Nut		0.1	0.2	0.0125	0.05	0.18								
R50250	USA	ASTM F468-84a	Ti Grade 1	Bolt Screw Stud		0.1	0.2	0.0125	0.05	0.18								
R50250	USA	ASTM F468M-84b	Ti Grade 1	Metric Bolt Screw Stud		0.1	0.2	0.0125	0.05	0.18								
R50250	USA	ASTM F67-88	Ti Grade 1	Surg imp HW CW Frg Ann		0.1	0.2	0.015	0.03	0.18					.	240	170	24
R50250	USA	ASME SB-265	Ti Grade 1	Sh Strp Plt Ann		0.1	0.2	0.015	0.03	0.18				0.4	OE 0.1 max			
	USA	ASTM B363	WPT1	Smls Weld Fittings (Composition as in ASTM B337, B338, B265, B348, B367, B381)														
R50125	USA					0.05	0.2	0.008	0.02	0.1-0.15								
R50120	USA					0.05	0.2	0.008	0.02	0.1								
R50100	USA					0.03 max	0.1 max	0.005 max	0.01 max	0.1 max								
R50250	USA	MIL T-81915A		Invest Cast		0.08	0.2	0.015	0.05	0.2				0.6				
R50250	USA					0.1	0.2	0.015	0.03	0.18					.			
R50130	USA	AWS A5.16-70	ERTi-4	Weld Fill Met		0.05	0.3	0.008	0.02	0.15-0.25				0.4		690	620	11
R50130	USA	AWS A5.16-90	ERTi-4	Weld El Rod		0.03 max	0.3 max	0.01 max	0.02 max	0.25				0.2	OE 0.05 max			
R50130	USA					0.05	0.3	0.008	0.02	0.15-0.25				0.4	OE 0.1 max			
	Russia	GOST	AT1-2Sv	Weld el		0.1	0.3	0.008	0.04	0.15								
R50400	China	GB 3620	TA-2			0.1 max	0.3 max	0.015 max	0.05 max	0.2 max					Si 0.15 max			

UNS numbers and US grades are provided as a means of cross referencing chemically similar alloys. Exchangability is only possible after independent examination of specifications. Tensile properties are minimum or typical . UTS and YS as Mpa, El as %. See Appendix for list of abbreviations used in Descriptions.

Worldwide Guide to Equivalent Nonferrous Metals and Alloys

Grade UNS #	Country	Spec.	Designation	Comment	Al	C	Fe	H	N	O	S	V	W	OT x	Other	UTS	YS	ELG
	Europe	AECMA prEN2518	TI-P02	Bar up to 200 mm		0.08 max	0.2 max		0.06 max	0.25 max					ET 0.1 max	390	290	20
R50400	Europe	AECMA prEN2518	Ti-P02	Sh Strp Bar		0.08	0.2	0.01	0.06	0.25				0.4		965	915	8
R50400	Europe	AECMA prEN2526	Ti-P02	Sh Strp		0.08 max	0.25 max	0.0125 max	0.05 max	0.25 max				0.4	Cu 0.35-1; OE 0.1 max			
	Europe	AECMA prEN2526	Ti-P02	Sh Strp Sht and Strps up to 5 mm		0.08 max	0.2 max	0.01 max	0.06 max	0.25 max					ET 0.1 max	390	290	22
R50400	Europe	AECMA prEN3378	Ti-P02	Wir		0.08 max	0.25 max	0.0125 max	0.05 max	0.25 max				0.4		1140	1070	7
R50400	Europe	AECMA prEN3442	Ti-P02	Sh Strp Ann HR		0.08 max	0.25 max	0.0125 max	0.05 max	0.25 max					Cu 0.7			
R50400	Europe	AECMA prEN3451	Ti-P02	Frg NHT		0.08 max	0.25 max	0.0125 max	0.05 max	0.25 max						1035	950	10
R50400	Europe	AECMA prEN3452	Ti-P02	Frg Ann		0.08 max	0.25 max	0.0125 max	0.05 max	0.25 max				0.4	Cu 0.35-1; OE 0.1 max			
R50400	Europe	AECMA prEN3460	Ti-P02	Bar Ann		0.08 max	0.25 max	0.0125 max	0.05 max	0.25 max				0.4		1140	1070	9
R50400	Europe	AECMA prEN3498	Ti-P02	Sh Strp Ann CR		0.08 max	0.25 max	0.0125 max	0.05 max	0.25 max				0.4	Cu 0.35-1; OE 0.1 max			
R50400	France	AIR 9182	T-35	Sh CR		0.08	0.12	0.015	0.05					0.4	V 5-6; Cu 0.35-1; Si 0.04; Ti 99.69 min	1020	950	7
R50400	France	AIR 9182	T-40	Sh Ann and CF 2 mm		0.08	0.12 max	0.02 max	0.05						Ti 99.7	390	295	28
R50400	France	AIR 9182	T-40	Sh		0.08	0.12	0.015	0.05					0.4	Cu 0.35-1; Si 0.04; Ti 99.69 min	1120	1035	7
R50400	France	AIR 9182	T-50	Sh Ann and CF 2 mm		0.08	0.25 max	0.02 max	0.07						Ti 99.5	490	390	24
R50400	France	AIR 9182	T-60	Sh Ann and CF 2 mm diam		0.08	0.3 max	0.02 max	0.08						Ti 99.6	590	470	20
R50400	Germany	DIN	3.7024	Sh Wir Ann		0.08	0.2	0.0125	0.05	0.2				0.4	Cu 0.35-1;	1035	1050	9
R50400	Germany	DIN	3.7034	Sh Bar Frg Wir Ann		0.08	0.25	0.0125	0.06	0.25				0.4	Cu 0.35-1	1120	1035	
R50400	Germany	DIN 17850	3.7035	Plt Sh Strp Rod Wir Frg Ann		0.08	0.25	0.013	0.06	0.2				0.4	Cu 0.35-1	1035	1050	9
R50400	Germany	DIN 17860	3.7035	Sh Strp		0.08 max	0.25 max	0.013 max	0.06 max	0.2 max				0.4		1070	1000	9
R50400	Germany	DIN 17862	3.7035	Rod		0.08 max	0.25 max	0.013 max	0.06 max	0.2 max				0.3	Cu 0.35-1	1120	1035	8
R50400	Germany	DIN 17863	3.7035	Wir		0.08 max	0.25 max	0.013 max	0.06 max	0.2 max						895	825	10
R50400	Germany	DIN 17864	3.7035	Frg		0.08 max	0.25 max	0.013 max	0.06 max	0.2 max				0.3	Cu 0.35-1	985	915	10
R50400	Germany	DIN 17850	3.7055	Sh Plt Strp Rod Wir Frg Ann	0.1		0.3	0.013	0.06	0.25				0.4	Cu 0.35-1	1120	1025	5
R50400	Germany	DIN 17850	Ti II	Sh Strp Plt Rod Wir Frg Ann	0.08		0.25	0.013	0.06	0.2				0.4	Cu 0.35-1	1000	1015	10
R50400	Germany	DIN 17850	Ti III	Sh Strp Plt Rod Wir Frg Ann	0.1		0.3	0.013	0.06	0.25				0.4	Cu 0.35-1	1015	1035	
R50400	Japan	JIS	Class 2				0.25	0.015	0.05	0.2				0.3	Mn 0.1 max; Cu 0.1 max; OE 0.1 max			
R50400	Japan	JIS H4650	TB 35 C/H Class 2	Bar Rod HW CD			0.25	0.015	0.05	0.2				0.3	Cu 0.35-1	1105	1035	8
	Japan	JIS H4650	TB35C, Class 2	Bar Rod			0.25	0.02 max	0.05	0.15 max								
	Japan	JIS H4650	TB35H, Class 2	Bar Rod			0.25 max	0.02 max	0.05 max	0.15 max								
R50400	Japan	JIS H4600	TP 35 H/C Class 2	Sh HR CR			0.25	0.013	0.05	0.2				0.4	Cu 0.35-1	1200	1100	6

UNS numbers and US grades are provided as a means of cross referencing chemically similar alloys. Exchangability is only possible after independent examination of specifications. Tensile properties are minimum or typical . UTS and YS as Mpa, El as %. See Appendix for list of abbreviations used in Descriptions.

10-4 Titanium

Grade UNS #	Country	Spec.	Designation	Comment	Al	C	Fe	H	N	O	S	V	W	OT x	Other	UTS	YS	ELG
	Japan	JIS H4600	TP35C, Class 2	Sh			0.25 max	0.01 max	0.05 max	0.2 max								
	Japan	JIS H4600	TP35H, Class 2	Sh			0.25 max	0.01 max	0.05 max	0.2 max								
R50400	Japan	JIS H4600	TR 35 H/C Class 2	Strp HR CR			0.25	0.013	0.05	0.2				0.3	Cu 0.35-1	965	895	8
	Japan	JIS H4600	TR35C, Class 2	Strp			0.25 max	0.01 max	0.05 max	0.2 max								
R50400	Japan	JIS H4631	TTH 35 W/WD Class 2	Weld Tub			0.25	0.015	0.05	0.2				0.4	Cu 0.35-1	1035	965	7
R50400	Japan	JIS H4361	TTH 35D Class 2	Smls Tub			0.25	0.015	0.05	0.2				0.3	Cu 0.35-1	1140	1070	6
R50400	Japan	JIS H4630	TTP 35 D/E Class 2	Smls Pip			0.25	0.015	0.05	0.2								
R50400	Japan	JIS H4630	TTP 35 W/WD Class 2	Weld Pip			0.25	0.015	0.05	0.2				0.3	Cu 0.35-1	1000	1015	10
	Japan	JIS H4630	TTP35D, Class 2	Tub Pip CD seamless pipe, 10-80 mm OD			0.25 max	0.02 max	0.05 max	0.2 max						343		23
	Japan	JIS H4630	TTP35E, Class 2	Tub Pip HE seamless pipe, 10-80 mm diam OD			0.2 max	0.02 max	0.05 max	0.15 max						343		23
	Japan	JIS H4630	TTP35W, Class 2	Tub Pip As weld 10-150 mm OD			0.25 max	0.02 max	0.05 max	0.2 max						343		23
	Japan	JIS H4630	TTP35WD, Class 2	Tub Pip CD weld pipe 10-150 mm OD			0.25 max	0.02 max	0.05 max	0.2 max						343		23
R50400	Japan	JIS H4670	TW 35 Class 2	Wir			0.25	0.015	0.05	0.2				0.3	Mn 0.1 max; Cu 0.1 max			
	Japan	JIS H4670	TW35, Class 2	Wir			0.25 max	0.02 max	0.05 max	0.2 max								
R50400	Russia	GOST 1.90000-76	VT1-O	Mult Forms Ann	0.07		0.3	0.01	0.04	0.2				0.4	Si 0.1	895	825	10
R50400	Russia	GOST 1.90060-72	VT1L	Cast	0.15		0.3	0.015	0.05	0.2				0.4	Si 0.15	725	620	13
R50400	Spain	UNE 38-711	L-7001	Sh Plt Strp Bar Wir Ext Ann	0.08		0.2	0.0125	0.05	0.2				0.4		1000	930	10
R50400	Spain	UNE 38-712	L-7002	Sh Plt Strp Bar Wir Ext Ann	0.08		0.25	0.0125	0.05	0.25				0.4	Cu 0.35-1; OE 0.1 max			
R50400	UK	BS 2TA.2	BS 2TA.2	Sh Strp HT			0.2 max	0.01 max							Ti 99.8 min	390	290	22
R50400	UK	BS 2TA.3	BS 2TA.3	Bar HT			0.2 max	0.01 max							Ti 99.8 min	390	290	20
R50400	UK	BS 2TA.4	BS 2TA.4	Frg HT			0.2	0.01						0.3	Mn 0.1 max; Cu 0.1 max; Ti 99.8 min	1125	1055	5
R50400	UK	BS 2TA.5	BS 2TA.5	Frg HT			0.2 max	0.02 max							Ti 99.8 min	390	290	20
R50400	UK	DTD 5073		Tub		0.01 max	0.2 max	0.015 max						0.3		1105	1035	7
R50400	USA	MIL T-81556A	Code CP-3	Ext Bar Shp Ann	0.08		0.3	0.015	0.05	0.2				0.4		920	870	8
R50400	USA	MIL T-9046J	Code CP-3	Sh Strp Plt Ann	0.08		0.3	0.015	0.05	0.2								
	USA	ASTM B381	F-2	Frg	0.10 max		0.30 max	0.0150 max	0.03 max	0.25 max				0.4	OE 0.1 max	345	275	20
R50400	USA	ASME SB-381	F-2	Frg Ann	0.1 max		0.3 max	0.015 max	0.03 max	0.25 max				0.4	Cu 0.35-1; OE 0.1 max			
	USA	ASTM B265	Ti Grade 2	Sh Strp Plt Ann	0.10 max		0.30 max	0.015 max	0.03 max	0.25 max				0.4	OE 0.1 max	345	275	20
	USA	ASTM B337	Ti Grade 2	Smls Weld Pip	0.10 max		0.30 max	0.015 max	0.03 max	0.25 max				0.4	W 0.2; OE 0.1 max	345	275	20

UNS numbers and US grades are provided as a means of cross referencing chemically similar alloys. Exchangability is only possible after independent examination of specifications. Tensile properties are minimum or typical . UTS and YS as Mpa, El as %. See Appendix for list of abbreviations used in Descriptions.

Grade UNS #	Country	Spec.	Designation	Comment	Al	C	Fe	H	N	O	S	V	W	OT x	Other	UTS	YS	ELG
	USA	ASTM B338	Ti Grade 2	Smls Weld Tub Exch Conds Ann		0.10 max	0.30 max	0.015 max	0.03 max	0.25 max				0.4	OE 0.1 max	345	275	20
	USA	ASTM B348	Ti Grade 2	Bar Bil (Bar H max 0.100)		0.10 max	0.30 max	0.0125 max	0.03 max	0.25 max				0.4	OE 0.1 max	345	275	20
	USA	ASTM B861	Ti Grade 2	Smls Pip Ann		0.10 max	0.30 max	0.015 max	0.03 max	0.25 max				0.4	OE 0.1 max	345	275	20
	USA	ASTM B863	Ti Grade 2	Wir		0.10 max	0.30 max	0.015 max	0.03 max	0.25 max				0.4	OE 0.1 max	345	275	18
R50400	USA	ASTM F467-84	Ti Grade 2	Nut		0.1 max	0.3 max	0.0125 max	0.05 max	0.25 max				0.4	Cu 0.35-1; OE 0.1 max			
R50400	USA	ASTM F467M-84a	Ti Grade 2	Nut Met		0.1 max	0.3 max	0.0125 max	0.05 max	0.25 max				0.4		1105	1035	8
R50400	USA	ASTM F468-84	Ti Grade 2	Blt Scr Std		0.1 max	0.3 max	0.0125 max	0.05 max	0.25 max				0.4	Cu 0.35-1; OE 0.1 max			
R50400	USA	ASTM F468M-84b	Ti Grade 2	Blt Scr Std Met		0.1 max	0.3 max	0.0125 max	0.05 max	0.25 max								
R50400	USA	ASTM F67	Ti Grade 2	Surg imp HW CW Frg Ann		0.1	0.3	0.0125 max	0.03	0.25				0.4	.	860	795	8
R50400	USA	ASME SB-265	Ti Grade 2	Sh Strp Plt Ann		0.1 max	0.3 max	0.015 max	0.03 max	0.25 max				0.4		985	915	10
R50400	USA	MIL T-81915	Type I Comp A	Air/chem/marine apps Cast Ann	0.08	0.2	0.015	0.05	0.2					0.4	OE 0.1 max			
	USA	ASTM B363	WPT2	Smls Weld Fittings (Composition as in ASTM B337, B338, B265, B348, B367, B381)														
R50400	USA	AMS 4902E		Sh Strp Plt Ann	0.08	0.3	0.015	0.05	0.2						Cu 0.5			
R50400	USA	AMS 4941C		Weld Tub Ann	0.1	0.2	0.015	0.05	0.25					0.4		1105	1035	
R50400	USA	AMS 4942C		Smls Tube Ann	0.1	0.3	0.015	0.03	0.25					0.3	.	910	735	9
R50400	USA				0.1	0.3	0.015	0.03	0.25							1005	820	9
	USA	ASTM B381	F-13	Frg		0.10 max	0.20 max	0.015 max	0.03 max	0.10 max				0.4	Ru 0.04-0.06; Ni 0.40-0.60; OE 0.1 max	275	170	24
	USA	ASTM B381	F-14	Frg		0.10 max	0.30 max	0.015 max	0.03 max	0.15 max				0.4	Ru 0.04-0.06; Ni 0.40-0.60; OE 0.1 max	410	275	20
	USA	ASTM B381	F-15	Frg		0.10 max	0.30 max	0.015 max	0.05 max	0.25 max				0.4	Ru 0.04-0.06; Ni 0.40-0.60; OE 0.1 max	483	380	18
	USA	ASTM B265	Ti Grade 13	Sh Strp Plt Ann		0.10 max	0.20 max	0.015 max	0.03 max	0.10 max				0.4	Ru 0.04-0.06; Ni 0.40-0.60; OE 0.1 max	275	170	24
	USA	ASTM B338	Ti Grade 13	Smls Weld Tub Exch Conds Ann		0.10 max	0.20 max	0.015 max	0.03 max	0.10 max				0.4	Ru 0.04-0.06; Ni 0.40-0.60; OE 0.1 max	275	170	24
	USA	ASTM B348	Ti Grade 13	Bar Bil (Bar H max 0.100)		0.10 max	0.20 max	0.0125 max	0.03 max	0.10 max				0.4	Ru 0.04-0.06; Ni 0.40-0.60; OE 0.1 max	275	170	24
	USA	ASTM B861	Ti Grade 13	Smls Pip Ann		0.10 max	0.20 max	0.0125 max	0.03 max	0.10 max				0.4	Ru 0.04-0.06; Ni 0.40-0.60; OE 0.1 max	275	170	24
	USA	ASTM B863	Ti Grade 13	Wir		0.10 max	0.20 max	0.0125 max	0.03 max	0.10 max				0.4	Ru 0.04-0.06; Ni 0.40-0.60; OE 0.1 max	275	170	18
	USA	ASTM B265	Ti Grade 14	Sh Strp Plt Ann		0.10 max	0.30 max	0.015 max	0.03 max	0.15 max				0.4	Ru 0.04-0.06; Ni 0.40-0.60; OE 0.1 max	410	275	20
	USA	ASTM B338	Ti Grade 14	Smls Weld Tub Exch Conds Ann		0.10 max	0.30 max	0.015 max	0.03 max	0.15 max				0.4	Ru 0.04-0.06; Ni 0.40-0.60; OE 0.1 max	410	275	20

UNS numbers and US grades are provided as a means of cross referencing chemically similar alloys. Exchangability is only possible after independent examination of specifications. Tensile properties are minimum or typical . UTS and YS as Mpa, El as %. See Appendix for list of abbreviations used in Descriptions.

Grade UNS #	Country	Spec.	Designation	Comment	Al	C	Fe	H	N	O	S	V	W	OT x	Other	UTS	YS	ELG
	USA	ASTM B348	Ti Grade 14	Bar Bil (Bar H max 0.100)		0.10 max	0.30 max	0.0125 max	0.03 max	0.15 max				0.4	Ru 0.04-0.06; Ni 0.40-0.60; OE 0.1 max	410	275	20
	USA	ASTM B861	Ti Grade 14	Smls Pip Ann		0.10 max	0.30 max	0.0125 max	0.03 max	0.15 max				0.4	Ru 0.04-0.06; Ni 0.40-0.60; OE 0.1 max	410	275	20
	USA	ASTM B863	Ti Grade 14	Wir		0.10 max	0.30 max	0.0125 max	0.03 max	0.15 max				0.4	Ru 0.04-0.06; Ni 0.40-0.60; OE 0.1 max	410	275	20
	USA	ASTM B265	Ti Grade 15	Sh Strp Plt Ann		0.10 max	0.30 max	0.015 max	0.05 max	0.25 max				0.4	Ru 0.04-0.06; Ni 0.40-0.60; OE 0.1 max	483	380	18
	USA	ASTM B338	Ti Grade 15	Smls Weld Tub Exch Conds Ann		0.10 max	0.30 max	0.015 max	0.05 max	0.25 max				0.4	Ru 0.04-0.06; Ni 0.40-0.60; OE 0.1 max	483	380	18
	USA	ASTM B348	Ti Grade 15	Bar Bil (Bar H max 0.100)		0.10 max	0.30 max	0.0125 max	0.05 max	0.25 max				0.4	Ru 0.04-0.06; Ni 0.40-0.60; OE 0.1 max	483	380	18
	USA	ASTM B861	Ti Grade 15	Smls Pip Ann		0.10 max	0.30 max	0.0125 max	0.05 max	0.25 max				0.4	Ru 0.04-0.06; Ni 0.40-0.60; OE 0.1 max	483	380	18
	USA	ASTM B863	Ti Grade 15	Wir		0.10 max	0.30 max	0.0125 max	0.05 max	0.25 max				0.4	Ru 0.04-0.06; Ni 0.40-0.60; OE 0.1 max	483	345	15
	USA	ASTM B363	WPT13	Smls Weld Fittings (Composition as in ASTM B337, B338, B265, B348, B381)														
	USA	ASTM B363	WPT14	Smls Weld Fittings (Composition as in ASTM B337, B338, B265, B348, B381)														
	USA	ASTM B363	WPT15	Smls Weld Fittings (Composition as in ASTM B337, B338, B265, B348, B381)														
	USA	ASTM B381	F-16	Frg		0.10 max	0.30 max	0.015 max	0.03 max	0.25 max				0.4	Pd 0.04-0.08; OE 0.1 max	345	275	20
	USA	ASTM B381	F-17	Frg		0.10 max	0.20 max	0.015 max	0.03 max	0.18 max				0.4	Pd 0.04-0.08; OE 0.1 max	240	170	24
	USA	ASTM B265	Ti Grade 16	Sh Strp Plt Ann		0.10 max	0.30 max	0.015 max	0.03 max	0.25 max				0.4	Pd 0.04-0.08; OE 0.1 max	345	275	20
	USA	ASTM B338	Ti Grade 16	Smls Weld Tub Exch Conds Ann		0.10 max	0.30 max	0.015 max	0.03 max	0.25 max				0.4	Pd 0.04-0.08; OE 0.1 max	345	275	20
	USA	ASTM B348	Ti Grade 16	Bar Bil (Bar H max 0.100)		0.10 max	0.30 max	0.0125 max	0.03 max	0.25 max				0.4	Pd 0.04-0.08; OE 0.1 max	345	275	20
	USA	ASTM B861	Ti Grade 16	Smls Pip Ann		0.10 max	0.30 max	0.0125 max	0.03 max	0.25 max				0.4	Pd 0.04-0.08; OE 0.1 max	345	275	20
	USA	ASTM B863	Ti Grade 16	Wir		0.10 max	0.30 max	0.0125 max	0.03 max	0.25 max				0.4	Pd 0.04-0.08; OE 0.1 max	345	275	20
	USA	ASTM B265	Ti Grade 17	Sh Strp Plt Ann		0.10 max	0.20 max	0.015 max	0.03 max	0.18 max				0.4	Pd 0.04-0.08; OE 0.1 max	240	170	24
	USA	ASTM B338	Ti Grade 17	Smls Weld Tub Exch Conds Ann		0.10 max	0.20 max	0.015 max	0.03 max	0.18 max				0.4	Pd 0.04-0.08; OE 0.1 max	240	170	24
	USA	ASTM B348	Ti Grade 17	Bar Bil (Bar H max 0.100)		0.10 max	0.20 max	0.0125 max	0.03 max	0.18 max				0.4	Pd 0.04-0.08; OE 0.1 max	240	170	24
	USA	ASTM B861	Ti Grade 17	Smls Pip Ann		0.10 max	0.20 max	0.0125 max	0.03 max	0.18 max				0.4	Pd 0.04-0.08; OE 0.1 max	240	170	24

UNS numbers and US grades are provided as a means of cross referencing chemically similar alloys. Exchangability is only possible after independent examination of specifications. Tensile properties are minimum or typical . UTS and YS as Mpa, El as %. See Appendix for list of abbreviations used in Descriptions.

Worldwide Guide to Equivalent Nonferrous Metals and Alloys

Grade UNS #	Country	Spec.	Designation	Comment	Al	C	Fe	H	N	O	S	V	W	OT x	Other	UTS	YS	ELG
	USA	ASTM B863	Ti Grade 17	Wir		0.10 max	0.20 max	0.0125 max	0.03 max	0.18 max				0.4	Pd 0.04-0.08; OE 0.1 max	240	170	20
	USA	ASTM B367	Ti-Pd16	Cast		0.10 max	0.30 max	0.0150 max	0.03 max	0.18 max				0.40	Pd 0.04-0.08; OE 0.10 max	345	275	15
	USA	ASTM B367	Ti-Pd17	Cast		0.10 max	0.20 max	0.0150 max	0.03 max	0.25 max				0.40	Pd 0.04-0.08; OE 0.10 max	240	170	20
	USA	ASTM B363	WPT16	Smls Weld Fittings (Composition as in ASTM B337, B338, B265, B348, B381)														
	USA	ASTM B363	WPT17	Smls Weld Fittings (Composition as in ASTM B337, B338, B265, B348, B381)														
	USA	ASTM B381	F-12	Frg		0.08 max	0.30 max	0.015 max	0.03 max	0.25 max				0.4	Ni 0.6-0.9; OE 0.1 max; Mo 0.2-0.4	483	345	18
Ti-0.3Mo-0.8Ni	USA	ASTM B265	Ti Grade 12	Sh Strp Plt Ann		0.08 max	0.30 max	0.015 max	0.03 max	0.25 max	2.5-3.5			0.4	Ni 0.6-0.9; OE 0.1 max; Mo 0.2-0.4	483	345	18
Ti-0.3Mo-0.8Ni	USA	ASTM B337	Ti Grade 12	Smls Weld Pip		0.08 max	0.3 max	0.015 max	0.03 max	0.25 max				0.3	Ni 0.6-0.9; OE 0.05 max; Mo 0.2-0.4	483	345	18
Ti-0.3Mo-0.8Ni	USA	ASTM B338	Ti Grade 12	Smls Weld Tub Exch Conds Ann		0.08 max	0.30 max	0.015 max	0.03 max	0.25 max				0.4	Ni 0.6-0.9; OE 0.1 max; Mo 0.2-0.4	483	345	18
	USA	ASTM B348	Ti Grade 12	Bar Bil (Bar H max 0.100)		0.08 max	0.3 max	0.0125 max	0.03 max	0.25 max				0.4	Ni 0.6-0.9; OE 0.1 max; Mo 0.2-0.4	483	345	18
	USA	ASTM B861	Ti Grade 12	Smls Pip Ann		0.08 max	0.30 max	0.015 max	0.03 max	0.25 max				0.4	Ni 0.6-0.9; OE 0.1 max; Mo 0.2-0.4	483	345	18
	USA	ASTM B863	Ti Grade 12	Wir		0.08 max	0.30 max	0.015 max	0.03 max	0.25 max				0.4	Ni 0.6-0.9; OE 0.1 max; Mo 0.2-0.4	483	345	18
	USA	ASTM B363	WPT12	Smls Weld Fittings (Composition as in ASTM B337, B338, B265, B348, B381)														
R50550	France	AIR 9182	T-50	Sh Ann		0.08	0.25	0.015	0.07						Pd 0.12-0.25; Si 0.04; Ti 99.5 min			
R50550	Germany	DIN 17860	3.7055	Sh Strp		0.1 max	0.3 max	0.013 max	0.06 max	0.25 max				0.4		845	810	10
R50550	Germany	DIN 17862	3.7055	Rod		0.1 max	0.3 max	0.013 max	0.06 max	0.25 max				0.4		1240		6
R50550	Germany	DIN 17863	3.7055	Wir		0.1 max	0.3 max	0.013 max	0.06 max	0.25 max				0.4		725	620	15
R50550	Germany	DIN 17864	3.7055	Frg		0.1 max	0.3 max	0.013 max	0.06 max	0.25 max				0.4		1240	1205	6
R50550	Germany	DIN 17850	3.7065	Plt Sh Strp Rod Wir Frg Ann		0.1	0.35	0.013	0.07	0.3				0.4		1445	965	5
R50550	Germany	DIN 17850	Ti IV	Sh Strp Plt Rod Wir Frg Ann		0.1	0.35	0.013	0.07	0.3								
R50550	Japan	JIS	Class 3				0.3	0.015	0.07	0.3				0.4				
R50550	Japan	JIS H4650	TB 49 C/H Class 3	Bar HW CD			0.3	0.015	0.07	0.3				0.4		860	825	10

UNS numbers and US grades are provided as a means of cross referencing chemically similar alloys. Exchangability is only possible after independent examination of specifications. Tensile properties are minimum or typical. UTS and YS as Mpa, El as %. See Appendix for list of abbreviations used in Descriptions.

10-8 Titanium

Grade UNS #	Country	Spec.	Designation	Comment	Al	C	Fe	H	N	O	S	V	W	OT x	Other	UTS	YS	ELG
	Japan	JIS H4650	TB49C, Class 3	Bar Rod			0.3 max	0.02 max	0.07 max	0.3 max								
	Japan	JIS H4650	TB49H, Class 3	Bar Rod			0.3 max	0.02 max	0.07 max	0.3 max								
R50550	Japan	JIS H4600	TP 49 H/C Class 3	Sh HR CR			0.3	0.013	0.07	0.3				0.4		1170	1105	3
	Japan	JIS H4600	TP49C Class 3	Sh			0.3 max	0.01 max	0.07 max	0.3 max								
	Japan	JIS H4600	TP49H, Class 3	Sh			0.3 max	0.01 max	0.07 max	0.3 max								
R50550	Japan	JIS H4600	TR 49 H/C Class 3	Strp HR CR			0.3	0.013	0.07	0.3				0.4		895	825	10
	Japan	JIS H4600	TR35H, Class 3	Strp			0.25 max	0.01 max	0.05 max	0.2 max								
	Japan	JIS H4600	TR49H, Class 3	Strp			0.3 max	0.01 max	0.07 max	0.3 max								
R50550	Japan	JIS H4631	TTH 49 D Class 3	Smls Tub CD			0.3	0.015	0.07	0.3								
R50550	Japan	JIS H4631	TTH 49 W/WD Class 3	Weld Tub			0.3	0.015	0.07	0.3				0.3		1240	1105	4
R50550	Japan	JIS H4630	TTP 49 D/E Class 3	Smls Pip Hot Ext CD			0.3	0.015	0.07	0.3				0.3		1195	1105	4
R50550	Japan	JIS H4630	TTP 49 W/WD Class 3	Weld Pip			0.3	0.015	0.07	0.3				0.3		1105	1000	6
	Japan	JIS H4630	TTP49D, Class 3	Tub Pip CD seamless pipe, 10-80 mm OD			0.25 max	0.02 max	0.05 max	0.2 max						481		18
	Japan	JIS H4630	TTP49E, Class 3	Tub Pip HE seamless pipe, 10-00 mm OD			0.25 max	0.02 max	0.05 max	0.2 max						481		18
	Japan	JIS H4630	TTP49W, Class 3	Tub Pip As weld 10-150 mm OD			0.3 max	0.02 max	0.07 max	0.3 max						481		18
	Japan	JIS H4630	TTP49WD, Class 3	Tub Pip CD weld pipe 10-150 mm OD			0.3 max	0.02 max	0.07 max	0.3 max						481		18
R50550	Japan	JIS H4670	TW 49 Class 3	Wir			0.3	0.015	0.07	0.3				0.3		965	895	8
	Japan	JIS H4670	TW49, Class 3	Wir			0.3 max	0.02 max	0.07 max	0.3 max								
R50550	UK	BS 2TA.6	BS 2TA.6	Sh Strp HT			0.2	0.01						0.4	Ti 99.8 min	860	825	9
R50550	UK	BS 2TA.7	BS 2TA.7	Bar HT			0.2	0.01						0.4	Ti 99.8 min	1170	1105	4
R50550	UK	BS 2TA.8	BS 2TA.8	Frg			0.2	0.01						0.4	Ti 99.8 min	885	850	9
R50550	UK	BS 2TA.9	BS 2TA.9	Frg HT			0.2	0.015						0.4	Ti 99.8 min	1170	1105	6
R50550	UK	DTD 5023		Sh Strp			0.2 max	0.0125 max						0.4		690	620	10
R50550	UK	DTD 5273		Bar			0.2 max	0.0125 max						0.4		690	620	10
R50550	UK	DTD 5283		Frg			0.2 max	0.0125 max						0.4		690	620	10
R50550	USA	AMS 4951E	AMS 4951	Fill met gas-met W arc weld		0.08	0.2	0.005	0.05	0.18				0.6				
	USA	ASTM B367	C-2	Cast	0.10 max		0.20 max	0.015 max	0.05 max	0.40 max				0.40	OE 0.10 max	345	275	15
	USA	ASTM B367	C-3	Cast	0.10 max		0.25 max	0.015 max	0.05 max	0.40 max				0.40	OE 0.10 max	450	380	12
R50550	USA	MIL T-81556A	Code CP-2	Ext Bar Shp Ann		0.08	0.3	0.015	0.05	0.3					Pd 0.12-0.25			
R50550	USA	MIL T-9046J	Code CP-2	Sh Strp Plt Ann		0.08	0.3	0.015	0.05	0.3					Pd 0.12-0.25			
	USA	ASTM B381	F-3	Frg	0.10 max		0.30 max	0.0150 max	0.05 max	0.35 max				0.4	OE 0.1 max	450	380	18

UNS numbers and US grades are provided as a means of cross referencing chemically similar alloys. Exchangability is only possible after independent examination of specifications. Tensile properties are minimum or typical . UTS and YS as Mpa, El as %. See Appendix for list of abbreviations used in Descriptions.

Worldwide Guide to Equivalent Nonferrous Metals and Alloys

Grade UNS #	Country	Spec.	Designation	Comment	Al	C	Fe	H	N	O	S	V	W	OT x	Other	UTS	YS	ELG
R50550	USA	ASME SB-381	F-3	Frg An		0.1 max	0.3 max	0.015 max	0.05 max	0.35 max								
	USA	ASTM B337	Ti Grade 3	Smls Weld Pip		0.10 max	0.30 max	0.015 max	0.05 max	0.35 max				0.4	OE 0.1 max	450	380	18
	USA	ASTM B338	Ti Grade 3	Smls Weld Tub Exch Conds Ann		0.10 max	0.30 max	0.015 max	0.05 max	0.35 max				0.4	OE 0.1 max	450	380	18
	USA	ASTM B348	Ti Grade 3	Bar Bil (Bar H max 0.100)		0.10 max	0.30 max	0.0125 max	0.05 max	0.35 max				0.4	OE 0.1 max	450	380	18
	USA	ASTM B861	Ti Grade 3	Smls Pip Ann		0.10 max	0.30 max	0.015 max	0.05 max	0.35 max				0.4	OE 0.1 max	450	380	18
	USA	ASTM B863	Ti Grade 3	Wir		0.10 max	0.30 max	0.015 max	0.05 max	0.35 max				0.4	OE 0.1 max	450	380	18
R50550	USA	ASTM B265	Ti Grade 3	Sh Strp Plt Ann		0.10 max	0.30 max	0.015 max	0.05 max	0.35 max				0.4	OE 0.1 max	450	380	18
R50550	USA	ASTM F 67	Ti Grade 3	Surg Imp		0.1	0.3	0.0125 max	0.05	0.35					Pd 0.12-0.25			
R50550	USA	ASME SB-265	Ti Grade 3	Sh Strp Plt Ann		0.1 max	0.3 max	0.015 max	0.05 max	0.35 max				0.4		1205	1140	7
	USA	ASTM B367	Ti-Pd7B	Cast		0.10 max	0.20 max	0.015 max	0.05 max	0.40 max				0.40	Pd 0.12 min; OE 0.10 max	345	275	15
	USA	ASTM B367	Ti-Pd8A	Cast		0.10 max	0.25 max	0.015 max	0.05 max	0.40 max				0.40	Pd 0.12 min; OE 0.10 max	450	380	12
	USA	ASTM B363	WPT3	Smls Weld Fittngs Comp. as in ASTM B337, B338, B265, B348, B367, B381														
R50550	USA	AMS 4900J		Sh Strp Plt Ann		0.08	0.3	0.015	0.05	0.3								
R50550	USA	AMS 4951E		Weld Wir		0.08 max	0.2 max	0.005 max	0.05 max	0.18 max				0.4		1240	1170	8
R50550	USA					0.1	0.3	0.015	0.05	0.35				0.4		1655		5
	USA	ASTM B299	ML	Sponge Mg reduced Leached/Inert Gas Sweep	0.05 max	0.02 max	0.15 max	0.03 max	0.015 max	0.10 max				0.05	H2O 0.02 x; Cl 0.20 x; Mg 0.50 x; Si 0.04x			
	Japan	JIS H2152	TC-1, Class 1	Powd		0.05 max	0.6 max	0.01 max	0.03 max						Cl 0.2 max; Mg 0.1 x; Mn 0.03 max; Ti 99 max			
R50700	China	GB 3620	TA-3			0.1 max	0.4 max	0.015 max	0.05 max	0.3 max					Pd 0.12-0.25; Si 0.15 max			
R52550	Europe	AECMA prEN2519	Ti-P04	Bar up to 200 mm		0.08 max	0.35 max		0.07 max	0.4 max					ET 0.1 max	540	440	15
R50700	Europe	AECMA prEN2519	Ti-P04	Bar Frg Sh Strp		0.08	0.35	0.0125	0.07	0.4					Pd 0.12-0.25			
R50700	Europe	AECMA prEN2520	Ti-P04	Frg		0.08 max	0.2 max	0.0125 max	0.07 max	0.4 max					Pd 0.12-0.25			
R52550	Europe	AECMA prEN2520	Ti-P04	Frg up to 200 mm		0.08 max	0.35 max	0.01 max	0.07 max	0.4 max					ET 0.1 max	540	440	15
R50700	Europe	AECMA prEN2527	Ti-P04	Sh Strp		0.08 max	0.2 max	0.0125 max	0.07 max	0.4 max					Pd 0.12-0.25			
R52550	Europe	AECMA prEN2527	Ti-P04	Sh Strp Sht and Strps up to 5 mm		0.08 max	0.35 max	0.01 max	0.07 max	0.4 max					ET 0.1 max	570	460	15
R50700	Europe	AECMA prEN3443	Ti-P04	Strp Sh Ann CR		0.08 max	0.2 max	0.0125 max	0.07 max	0.4 max					Pd 0.12-0.25			
R50700	Europe	AECMA prEN3453	Ti-P04	Frg NHT		0.08 max	0.2 max	0.0125 max	0.07 max	0.4 max					Pd 0.12-0.25			
R50700	Europe	AECMA prEN3461	Ti-P04	Bar Ann		0.08 max	0.2 max	0.0125 max	0.07 max	0.4 max					Pd 0.12-0.25			
R50700	Europe	AECMA prEN3496	Ti-P04	Frg Ann		0.08 max	0.2 max	0.0125 max	0.07 max	0.4 max					Pd 0.12-0.25			
R50700	Europe	AECMA prEN3499	Ti-P04	Sh Strp Ann CR		0.08 max	0.2 max	0.0125 max	0.07 max	0.4 max					Pd 0.12-0.25			
R50700	France	AIR 9182	T-60	Sh Ann		0.08	0.3	0.015	0.08						Pd 0.12-0.25; Si 0.04; Ti 99.6 min			

UNS numbers and US grades are provided as a means of cross referencing chemically similar alloys. Exchangability is only possible after independent examination of specifications. Tensile properties are minimum or typical . UTS and YS as Mpa, El as %. See Appendix for list of abbreviations used in Descriptions.

10-10 Titanium

Grade UNS #	Country	Spec.	Designation	Comment	Al	C	Fe	H	N	O	S	V	W	OT x	Other	UTS	YS	ELG
R50700	Germany	DIN	3.7064	Sh Rod Bar Frg Ann		0.08	0.35	0.0125	0.07	0.4					Pd 0.12-0.25			
R50700	Germany	DIN 17860	3.7065	Sh Strp		0.1 max	0.35 max	0.013 max	0.07 max	0.3 max					Pd 0.12-0.25			
R50700	Germany	DIN 17862	3.7065	Rod		0.1 max	0.35 max	0.013 max	0.07 max	0.3 max					Pd 0.12-0.25			
R50700	Germany	DIN 17863	3.7065	Wir		0.1 max	0.35 max	0.013 max	0.07 max	0.3 max					Pd 0.12-0.25			
R50700	Germany	DIN 17864	3.7065	Frg		0.1 max	0.35 max	0.013 max	0.07 max	0.3 max					Pd 0.12-0.25			
R50700	Spain	UNE 38-714	L-7004	Mult Forms Ann		0.1	0.4	0.0125	0.07	0.4					Pd 0.12-0.25			
R50700	USA	MIL T-9046H	A(40KSI-YS)	Sh Strp Ann		0.08	0.5 max	0.02 max	0.05	0.2 max					Ti 99.4	345	276	20
R50700	USA	MIL T-9046H	B(70KSI-YS)	Sh Strp Ann		0.08	0.5 max	0.02 max	0.05	0.4 max					Ti 99.2	552	483	15
R50700	USA	MIL T-9046H	C(55KSI-YS)	Sh Strp Ann		0.08	0.5 max	0.02 max	0.05	0.3 max					Ti 99.4	448	379	18
R50700	USA	MIL T-81556A	Code CP-1	Ext Bar Shp Ann		0.08	0.5	0.015	0.05	0.4					Pd 0.15-0.25			
R50700	USA	MIL T-9046J	Code CP-1	Sh Strp Plt Ann		0.08	0.5	0.015	0.05	0.4				0.4	Pd 0.12-0.25	345	275	20
R50700	USA	MIL F-83142	Comp 1	Frg Ann		0.08	0.5	0.0125	0.05	0.4					Pd 0.12-0.25			
	USA	ASTM B381	F-4	Frg		0.10 max	0.50 max	0.0150 max	0.05 max	0.40 max				0.4	OE 0.1 max	550	483	15
R50700	USA	MIL T-9047-G	SP-70	Bar		0.08 max	0.5 max	0.015 max	0.05 max	0.4 max					Pd 0.12-0.25			
	USA	ASTM B348	Ti Grade 4	Bar Bil (Bar H max 0.100)		0.10 max	0.50 max	0.0125 max	0.05 max	0.40 max				0.4	OE 0.1 max	550	483	15
	USA	ASTM B863	Ti Grade 4	Wir		0.10 max	0.50 max	0.015 max	0.05 max	0.40 max				0.4	OE 0.1 max	550	483	15
R50700	USA	ASTM B265	Ti Grade 4	Sh Strp Plt Ann		0.10 max	0.50 max	0.015 max	0.05 max	0.40 max				0.4	OE 0.1 max	550	483	15
R50700	USA	ASTM F467-84	Ti Grade 4	Nut		0.1 max	0.5 max	0.0125 max	0.07 max	0.4 max					Pd 0.12-0.25			
R50700	USA	ASTM F468-84	Ti Grade 4	Blt Scrw Std		0.1 max	0.5 max	0.0125 max	0.07 max	0.4 max					Pd 0.12-0.25			
R50700	USA	ASTM F67	Ti Grade 4	Sh Strp Bar HR CR Ann Frg		0.1	0.5	0.0125 max	0.05	0.4					Pd 0.12-0.25			
R50700	USA	MIL T-9047G	Ti-CP-70	Bar Bil Ann		0.08	0.5	0.0125	0.05	0.4 max			0.01	0.4	Pd 0.12	345	275	15
R50700	USA	AMS 4901L		Sh Strp Plt Ann		0.08	0.5	0.015	0.05	0.4					Pd 0.12-0.25			
R50700	USA	AMS 4921F		Bar Wir Frg Bil Rng Ann		0.08	0.5	0.0125	0.05	0.4					Pd 0.12-0.25			
R50700	USA					0.1	0.5	0.015	0.05	0.4			0.01	0.4	Pd 0.12-0.25	345	275	20
	USA	ASTM B861	Ti Grade 9	Smls Pip Ann		0.05 max	0.25 max	0.015 max	0.02 max	0.15 max		2.0-3.0		0.4	OE 0.1 max	620	483	15
	USA	ASTM B861	Ti Grade 7	Smls Pip Ann	2.5-3.5	0.10 max	0.30 max	0.015 max	0.03 max	0.25 max				0.4	Pd 0.12-0.25; OE 0.1 max	345	275	20
	Europe	AECMA prEN2521	Ti-P11	Bar up to 200 mm		0.08 max	0.2 max		0.05 max	0.2 max					Cu 2-3; ET 0.1 max	540	400	16
	Europe	AECMA prEN2522	Ti-P11	Frg up to 200 mm		0.08 max	0.2 max	0.01 max	0.05 max	0.2 max					Cu 2-3; ET 0.1 max	540	400	16
	Europe	AECMA prEN2523	Ti-P11	Bar up to 75 mm		0.08 max	0.2 max		0.05 max	0.2 max					Cu 2-3; ET 0.1 max	650	530	10
	Europe	AECMA prEN2524	Ti-P11	Frg up to 200 mm		0.08 max	0.2 max	0.01 max	0.05 max	0.2 max					Cu 2-3; ET 0.1 max	650	530	10
	Europe	AECMA prEN2528	Ti-P11	Sh Strp Sht and Strps up to 5 mm		0.08 max	0.2 max	0.01 max	0.05 max	0.2 max					Cu 2-3; ET 0.1 max	540		18
Ti-0.2Pd R52401	USA	AWS A5.16-70	ERTi-0.2Pd	Weld Fill Met		0.05	0.25	0.008	0.02	0.15	2.5-3.5	0.01				900	830	8
R52401	USA	AWS A5.16-90	ERTi-7	Weld El Rod		0.03 max	0.2 max	0.01 max	0.02 max	0.1 max				0.2	OE 0.05 max			

UNS numbers and US grades are provided as a means of cross referencing chemically similar alloys. Exchangability is only possible after independent examination of specifications. Tensile properties are minimum or typical. UTS and YS as Mpa, El as %. See Appendix for list of abbreviations used in Descriptions.

Grade UNS #	Country	Spec.	Designation	Comment	Al	C	Fe	H	N	O	S	V	W	OT x	Other	UTS	YS	ELG
Ti-0.2Pd R52401	USA					0.05	0.25	0.008	0.02	0.15	2.5-3.5		0.01		O+N=0.25	1020	825	10
	Japan	JIS H2152	TC-2, Class 2	Powd		0.1 max	2 max	0.01 max	0.1 max						Cl 0.2 max; Mg 0.5 max; Mn 0.05 max; Si 0.1 max; Ti 97 max			
Ti-0.2Pd R52250	Russia	GOST	4200			0.07	0.18	0.01	0.04	0.12	2.5-3.5		0.01		V. 3.5-4.5; Si 0.1	900	830	8
	USA	ASTM B381	F-11	Frg	0.10 max	0.20 max	0.0150 max	0.03 max	0.18 max					0.4	Pd 0.12-0.25; OE 0.1 max	240	170	24
Ti-0.2Pd	USA	ASTM B265	Ti Grade 11	Sh Strp Plt Ann	0.10 max	0.20 max	0.015 max	0.03 max	0.18 max	2.5-3.5				0.4	Pd 0.12-0.25; OE 0.1 max	240	170	24
Ti-0.2Pd	USA	ASTM B337	Ti Grade 11	Smls Weld Pip	0.10 max	0.20 max	0.015 max	0.03 max	0.18 max					0.4	Pd 0.12-0.25; OE 0.1 max	240	170	24
Ti-0.2Pd	USA	ASTM B338	Ti Grade 11	Smls Weld Tub Exch Conds Ann	0.10 max	0.20 max	0.015 max	0.03 max	0.18 max					0.4	Pd 0.12-0.25; OE 0.1 max	240	170	24
	USA	ASTM B348	Ti Grade 11	Bar Bil (Bar H max 0.100)	0.10 max	0.20 max	0.0125 max	0.03 max	0.18 max					0.4	Pd 0.12-0.25; OE 0.1 max	240	170	24
	USA	ASTM B861	Ti Grade 11	Smls Pip Ann	0.10 max	0.20 max	0.015 max	0.03 max	0.18 max					0.4	Pd 0.12-0.25; OE 0.1 max	240	170	24
	USA	ASTM B863	Ti Grade 11	Wir	0.10 max	0.20 max	0.015 max	0.03 max	0.18 max					0.4	Pd 0.12-0.25; OE 0.1 max	240	170	20
	USA	ASTM B363	WPT11	Smls Weld Fittings (Composition as in ASTM B337, B338, B265, B348, B367, B381)														
Ti-0.2Pd R52250	USA					0.1	0.2	0.015	0.03	0.18	2.5-3.5		0.01			1100	970	8
Ti-0.2Pd R52400	Germany	DIN 17851	3.7225			0.06 max	0.15 max	0.0013 max	0.05 max	0.12 max	2.5-3.5		0.01	0.4	OE 0.1 max			
Ti-0.2Pd R52400	Germany	DIN 17851	3.7235			0.06 max	0.2 max	0.0013 max	0.05 max	0.18 max	2.5-3.5		0.01	0.4	OE 0.1 max			
Ti-0.2Pd R52400	Germany	DIN 17851	3.7255			0.06 max	0.25 max	0.0013 max	0.05 max	0.25 max	2.5-3.5		0.01	0.4	OE 0.1 max			
Ti-0.2Pd R52400	Japan	JIS H 4655	TB28PdC	Rod Bar CD			0.2 max	0.015 max	0.05 max	0.15 max	2.5-3.5		0.01			900	830	8
Ti-0.2Pd R52400	Japan	JIS H 4655	TB28PdH	Rod Bar HW			0.2 max	0.015 max	0.05 max	0.15 max	2.5-3.5		0.01					
Ti-0.2Pd R52400	Japan	JIS H 4655	TB35PdC	Bar Rod CD			0.25 max	0.015 max	0.05 max	0.2 max	2.5-3.5		0.01	0.3		915	845	9
Ti-0.2Pd R52400	Japan	JIS H 4655	TB35PdH	Bar Rod HW			0.25 max	0.015 max	0.05 max	0.2 max	2.5-3.5		0.01	0.4		1035	965	6
Ti-0.2Pd R52400	Japan	JIS H 4655	TB49PdC	Bar Rod CD			0.3 max	0.015 max	0.07 max	0.3 max	2.5-3.5		0.01	0.3		915	845	9
Ti-0.2Pd R52400	Japan	JIS H 4655	TB49PdH	Bar Rod HW			0.3 max	0.015 max	0.07 max	0.3 max	2.5-3.5		0.01	0.3		1070	1000	9
Ti-0.2Pd R52400	Japan	JIS H 4636	TTH28PdD	Smls Pip CD			0.2 max	0.015 max	0.05 max	0.15 max	2.5-3.5		0.01	0.2	Mn 0.1 max; Cu 0.1 max			
Ti-0.2Pd R52400	Japan	JIS H 4636	TTH28PdW	Weld Pip			0.2 max	0.015 max	0.05 max	0.15 max	2.5-3.5		0.01	0.4	.	1050	965	9
Ti-0.2Pd R52400	Japan	JIS H 4636	TTH28PdWD	Weld Pip CD			0.2 max	0.015 max	0.05 max	0.15 max	2.5-3.5		0.01					
Ti-0.2Pd R52400	Japan	JIS H 4636	TTH35PdD	Smls Pip CD			0.25 max	0.015 max	0.05 max	0.2 max	2.5-3.5		0.01	0.4				
Ti-0.2Pd R52400	Japan	JIS H 4636	TTH35PdW	Weld Pip			0.25 max	0.015 max	0.05 max	0.2 max	2.5-3.5		0.01	0.3		880	810	8
Ti-0.2Pd R52400	Japan	JIS H 4636	TTH35PdWD	Weld Pip CD			0.25 max	0.015 max	0.05 max	0.2 max	2.5-3.5		0.01		Ta 0.15-1.5			
Ti-0.2Pd R52400	Japan	JIS H 4636	TTH49PdD	Smls Pip CD			0.3 max	0.015 max	0.07 max	0.3 max	2.5-3.5		0.01	0.4	Ta 0.5-1.5	710	655	10
Ti-0.2Pd R52400	Japan	JIS H 4636	TTH49PdW	Weld Pip			0.3 max	0.015 max	0.07 max	0.3 max	2.5-3.5		0.01	0.4	Ta 0.5-1.5	710	655	10
Ti-0.2Pd R52400	Japan	JIS H 4636	TTH49PdWD	Weld Pip CD			0.3 max	0.015 max	0.07 max	0.3 max	2.5-3.5		0.01		Ta 1			

UNS numbers and US grades are provided as a means of cross referencing chemically similar alloys. Exchangability is only possible after independent examination of specifications. Tensile properties are minimum or typical . UTS and YS as Mpa, El as %. See Appendix for list of abbreviations used in Descriptions.

10-12 Titanium

Grade UNS #	Country	Spec.	Designation	Comment	Al	C	Fe	H	N	O	S	V	W	OT x	Other	UTS	YS	ELG
Ti-0.2Pd) R52400	Japan	JIS H 4635	TTP28PdD	Smls Pip CD			0.2 max	0.015 max	0.05 max	0.15 max	2.5-3.5			0.01				
Ti-0.2Pd R52400	Japan	JIS H 4635	TTP28PdE	Smls Pip HE			0.2 max	0.015 max	0.05 max	0.15 max	2.5-3.5			0.01	0.4	OE 0.1 max			
Ti-0.2Pd R52400	Japan	JIS H 4635	TTP28PdW	Weld Pip			0.2 max	0.015 max	0.05 max	0.15 max	2.5-3.5			0.01					
Ti-0.2Pd R52400	Japan	JIS H 4635	TTP28PdWD	Weld Pip CD			0.2 max	0.015 max	0.05 max	0.15 max	2.5-3.5			0.01	0.2	OE 0.1 max			
Ti-0.2Pd R52400	Japan	JIS H 4635	TTP35PdD	Smls Pip CD			0.25 max	0.015 max	0.05 max	0.2 max	2.5-3.5			0.01			880	810	10
Ti-0.2Pd R52400	Japan	JIS H 4635	TTP35PdE	Smls Pip HE			0.25 max	0.015 max	0.05 max	0.2 max	2.5-3.5			0.01					
Ti-0.2Pd R52400	Japan	JIS H 4635	TTP35PdW	Weld Pip			0.25 max	0.015 max	0.05 max	0.2 max	2.5-3.5			0.01	0.4	.	870	770	9
Ti-0.2Pd R52400	Japan	JIS H 4635	TTP35PdWD	Weld Pip CD			0.25 max	0.015 max	0.05 max	0.2 max	2.5-3.5			0.01	0.4	.	895	825	8
Ti-0.2Pd R52400	Japan	JIS H 4635	TTP49PdD	Smls Pip CD			0.3 max	0.015 max	0.07 max	0.3 max	2.5-3.5			0.01	0.4	.			
Ti-0.2Pd R52400	Japan	JIS H 4635	TTP49PdE	Smls Pip HE			0.3 max	0.015 max	0.07 max	0.3 max	2.5-3.5			0.01	0.4	.	1000	930	6
Ti-0.2Pd R52400	Japan	JIS H 4635	TTP49PdW	Weld Pip			0.3 max	0.015 max	0.07 max	0.3 max	2.5-3.5			0.01	0.4	.	895	825	6
Ti-0.2Pd R52400	Japan	JIS H 4635	TTP49PdWD	Weld Pip CD			0.3 max	0.015 max	0.07 max	0.3 max	2.5-3.5			0.01	0.4	Na 0.15; Cl 0.15	895	825	10
Ti-0.2Pd R52400	Japan	JIS H 4675	TW28Pd	Wir			0.2 max	0.015 max	0.05 max	0.15 max	2.5-3.5			0.01	0.3		895	825	10
Ti-0.2Pd R52400	Japan	JIS H 4675	TW35Pd	Wir			0.25 max	0.015 max	0.05 max	0.2 max	2.5-3.5			0.01	0.3		1000	930	10
Ti 0.2Pd R52400	Japan	JIS H 4675	TW49Pd	Wir			0.3 max	0.015 max	0.07 max	0.25 max	2.5-3.5			0.01	0.3	OE 0.1 max			
Ti-0.2Pd R52400	Spain	UNE 38-715	L-7021	ShPltStrp Bar Wir Ext Ann	0.08		0.25	0.0125	0.05	0.25	2.5-3.5			0.01					
	UK	BS 2TA.21	BS 2TA.21	Sh Strp HT			0.2 max	0.01 max								Cu 2-3; Ti 96.8 max	540	460	15
	UK	BS 2TA.22	BS 2TA.22	Bar HT			0.2 max	0.01 max								Cu 2-3; Ti 96.8 max	540	400	16
	UK	BS 2TA.23	BS 2TA.23	Frg HT			0.2 max	0.01 max								Cu 2-3; Ti 96.8 max	540	400	16
	UK	BS 2TA.24	BS 2TA.24	Frg HT			0.2 max	0.02 max								Cu 2-3; Ti 96.8 max	540	400	16
	UK	BS TA.52	BS TA.52	Sh Strp HT			0.2 max	0.01 max								Cu 2-3; Ti 96.8 max	690	870	10
	UK	BS TA.53	BS TA.53	Bar			0.2 max	0.01 max								Cu 2-3; Ti 96.8 max			
	UK	BS TA.54	BS TA.54	Frg HT			0.2 max	0.01 max								Cu 2-3; Ti 96.8 max	650	525	10
	UK	BS TA.55	BS TA.55	Frg HT			0.2 max	0.02 max								Cu 2-3; Ti 96.8 max	650	525	10
	UK	BS TA.58	BS TA.58	Plt HT			0.2 max	0.01 max								Cu 2-3; Ti 96.8 max	520	420	20
	USA	ASTM B381	F-7	Frg	0.10 max	0.30 max	0.0150 max	0.03 max	0.25 max					0.4	Pd 0.12-0.25; OE 0.1 max	345	275	20
Ti-0.2Pd	USA	ASTM B265	Ti Grade 7	Sh Strp Plt Ann	0.10 max	0.30 max	0.015 max	0.03 max	0.25 max	2.5-3.5				0.4	Pd 0.12-0.25; OE 0.1 max	345	275	20
Ti-0.2Pd	USA	ASTM B337	Ti Grade 7	Smls Weld Pip	0.10 max	0.30 max	0.015 max	0.03 max	0.25 max					0.4	Pd 0.12-0.25; OE 0.1 max	345	275	20
Ti-0.2Pd	USA	ASTM B338	Ti Grade 7	Smls Weld Tub Exch Conds Ann	0.10 max	0.30 max	0.015 max	0.03 max	0.25 max					0.4	Pd 0.12-0.25; OE 0.1 max	345	275	20
Ti-0.2Pd R52400	USA	ASTM F467-84	Ti Grade 7	Nut	0.1 max	0.3 max	0.0125 max	0.05 max	0.25 max	2.5-3.5			0.01	0.4	OE 0.1 max			
Ti-0.2Pd R52400	USA	ASTM F467M-84a	Ti Grade 7	Met Nut	0.1 max	0.3 max	0.0125 max	0.05 max	0.25 max	2.5-3.5			0.01	0.4	OE 0.1 max			
Ti-0.2Pd R52400	USA	ASTM F468-84	Ti Grade 7	Blt Scrw Std	0.1 max	0.3 max	0.0125 max	0.05 max	0.25 max	2.5-3.5			0.01	0.4	OE 0.1 max			
Ti-0.2Pd R52400	USA	ASTM F468M-84b	Ti Grade 7	Met Blt Scrw Std	0.1 max	0.3 max	0.0125 max	0.05 max	0.25 max	2.5-3.5			0.01	0.4	OE 0.1 max			
	USA	ASTM B348	Ti Grade 7	Bar Bil (Bar H max 0.100)	0.10 max	0.30 max	0.0125 max	0.03 max	0.25 max					0.4	Pd 0.12-0.25; OE 0.1 max	345	275	20

UNS numbers and US grades are provided as a means of cross referencing chemically similar alloys. Exchangability is only possible after independent examination of specifications. Tensile properties are minimum or typical . UTS and YS as Mpa, El as %. See Appendix for list of abbreviations used in Descriptions.

Grade UNS #	Country	Spec.	Designation	Comment	Al	C	Fe	H	N	O	S	V	W	OT x	Other	UTS	YS	ELG
	USA	ASTM B863	Ti Grade 7	Wir		0.10 max	0.30 max	0.015 max	0.03 max	0.25 max				0.4	Pd 0.12-0.25; OE 0.1 max	345	275	18
	USA	ASTM B363	WPT7	Smls Weld Fittings (Composition as in ASTM B337, B338, B265, B348, B367, B381)														
Ti-0.2Pd R52400	USA					0.1	0.3	0.015	0.03	0.25	2.5-3.5		0.01		N+O=0.25;			
R53400	USA	AWS A5.16-90	ERTi-12	Weld El Rod		0.03 max	0.3 max	0.01 max	0.02 max	0.25 max				0.2	OE 0.05 max			
TI-0.3Mo-0.8Ni R53400	USA					0.08	0.3	0.015	0.03	0.25	2.5-3.5		0.01	0.4	Ni 0.6-0.9; OE 0.1 max; Mo 0.2-0.4			
	USA	ASTM B348	Ti Grade 9	Bar Bil (Bar H max 0.100)	2.5-3.5	0.10 max	0.25 max	0.0125 max	0.02 max	0.15 max		2.0-3.0		0.4	OE 0.1 max	620	483	15
	USA	ASTM B863	Ti Grade 9	Wir	2.5-3.5	0.05 max	0.25 max	0.013 max	0.02 max	0.15 max		2.0-3.0		0.4	OE 0.1 max	620	483	15
	USA	ASTM B363	WPT9	Smls Weld Fittings (Composition as in ASTM B337, B338, B265, B348, B381)														
	USA	ASTM B381	F-18	Frg	2.5-3.5	0.10 max	0.25 max	0.015 max	0.02 max	0.15 max		2.0-3.0		0.4	Pd 0.04-0.08; OE 0.1 max	620	483	15
	USA	ASTM B265	Ti Grade 18	Sh Strp Plt Ann	2.5-3.5	0.10 max	0.25 max	0.015 max	0.02 max	0.15 max		2.0-3.0		0.4	Pd 0.04-0.08; OE 0.1 max	620	483	15
	USA	ASTM B338	Ti Grade 18	Smls Weld Tub Exch Conds Ann	2.5-3.5	0.10 max	0.30 max	0.015 max	0.02 max	0.15 max		2.0-3.0		0.4	Pd 0.04-0.08; OE 0.1 max	620	483	15
	USA	ASTM B348	Ti Grade 18	Bar Bil (Bar H max 0.100)	2.5-3.5	0.10 max	0.25 max	0.0125 max	0.02 max	0.15 max		2.0-3.0		0.4	Pd 0.04-0.08; OE 0.1 max	620	483	15
	USA	ASTM B861	Ti Grade 18	Smls Pip Ann	2.5-3.5	0.10 max	0.25 max	0.0125 max	0.05 max	0.15 max		2.0-3.0		0.4	Pd 0.04-0.08; OE 0.1 max	620	483	15
	USA	ASTM B863	Ti Grade 18	Wir	2.5-3.5	0.10 max	0.25 max	0.0125 max	0.05 max	0.15 max		2.0-3.0		0.4	Pd 0.04-0.08; OE 0.1 max	620	483	10
	USA	ASTM B367	Ti-Pd18	Cast	2.5-3.5	0.10 max	0.25 max	0.0150 max	0.05 max	0.15 max		2.0-3.0		0.40	Pd 0.04-0.08; OE 0.10 max	620	483	15
	USA	ASTM B363	WPT18	Smls Weld Fittings (Composition as in ASTM B337, B338, B265, B348, B381)														
Ti-3Al-2.5V R56321	USA	AWS A5.16-70	ERTi-3Al-2.5V	Weld Fill Met	2.5-3.5	0.05	0.25	0.008	0.02	0.12		2-3	0.01	0.4	Pd 0.12-0.25;	240	170	24
Ti-3Al-2.5V R56321	USA	AWS A5.16-70	ERTi-3Al-2.5V-1	Weld Fill Met	2.5-3.5	0.04	0.25	0.005	0.01	0.1		2-3	0.01	0.4	Pd 0.12-0.25;	240	240	24
R56321	USA	AWS A5.16-90	ERTi-ELI	Weld El Rod	2.5-3.5	0.03 max	0.2 max	0.01 max	0.01 max	0.1 max		2-3		0.2	OE 0.05 max			
Ti-3Al-2.5V R56321	USA	ASTM B265	Ti Grade 9	Sh Strp Plt Ann	2.5-3.5	0.10 max	0.25 max	0.015 max	0.02 max	0.15 max		2.0-3.0		0.4	OE 0.1 max	620	483	15
Ti-3Al-2.5V R56321	USA	ASTM B337	Ti Grade 9	Smls Weld Pip CW SR	2.5-3.5	0.05 max	0.25 max	0.013 max	0.02 max	0.12 max		2.0-3.0		0.4	OE 0.1 max	860	725	10
Ti-3Al-2.5V R56321	USA	ASTM B337	Ti Grade 9	Smls Weld Pip Ann	2.5-3.5	0.05 max	0.25 max	0.013 max	0.02 max	0.12 max		2.0-3.0		0.4	OE 0.1 max	620	485	15
Ti-3Al-2.5V R56321	USA	ASTM B338	Ti Grade 9	Smls Weld Tub Exch Conds Ann	2.5-3.5	0.10 max	0.25 max	0.015 max	0.02 max	0.15 max		2.0-3.0		0.4	OE 0.1 max	620	483	15

UNS numbers and US grades are provided as a means of cross referencing chemically similar alloys. Exchangability is only possible after independent examination of specifications. Tensile properties are minimum or typical . UTS and YS as Mpa, El as %. See Appendix for list of abbreviations used in Descriptions.

Grade UNS #	Country	Spec.	Designation	Comment	Al	C	Fe	H	N	O	S	V	W	OT x	Other	UTS	YS	ELG
Ti-3Al-2.5V R56321	USA			Weld Fill Wir	2.5-3.5	0.04	0.25	0.005	0.01	0.1		2-3	0.01 max		Pd 0.12-0.25;			
Ti-3Al-2.5V R56320	China		Ti-3Al-2.5V		2.5-3.5	0.08 max	0.3 max	0.015 max	0.05 max	0.12 max		2-3	0.01	0.4	Pd 0.12-0.25; Si 0.15 max	345	360	20
Ti-3Al-2.5V R56320	Europe	AECMA prEN3120	Ti-P69	Tub CW SR	2.5-3.5	0.05 max	0.3 max	0.015 max	0.02 max	0.12 max		2.5-3.5	0.01 max	0.4	Pd 0.12-0.25;	345	360	20
Ti-3Al-2.5V R56320	Russia	GOST	AK2		3					0.35		2.5	0.01			485	345	18
Ti-3Al-2.5V R56320	Russia	GOST	IMP-7	Powd	3		0.3	0.01	0.03	0.16		2	0.01	0.3	Si 0.6	485	345	18
Ti-3Al-2.5V R56320	USA	MIL T-9046J	Code AB-5	Sh Strp Plt Ann	2.5-3.5	0.05	0.3	0.015	0.02	0.12		2-3	0.01	0.4	Pd 0.12-0.25;	240	240	24
R56320	USA	AWS A5.16-90	ERTi-9	Weld El Rod	2.5-3.5	0.03 max	0.25 max	0.01 max	0.02 max	0.12 max		2-3		0.2	OE 0.05 max			
Ti-3Al-2.5V R56320	USA	MIL T-9047G	Ti-3Al-2.5V	Bar Bil Ann	2.5-3.5	0.05	0.3	0.015	0.02	0.12		2-3	0.01	0.3	Pd 0.15-0.3;	471	402	25
Ti-3Al-2.5V R56320	USA	AMS 4945		Smls Tub	2.5-3.5	0.05 max	0.3 max	0.015 max	0.02 max	0.12 max		2-3	0.01 max		Pd 0.12-0.25;	390	275	22
Ti-3Al-2.5V R56320	USA	AMS 4943D		Tub Ann	2.5-3.5	0.05	0.3	0.015	0.02	0.12		2-3	0.01			485	345	18
Ti-3Al-2.5V R56320	USA	AMS 4944D		Smls Tub CW SR	2.5-3.5	0.05 max	0.3 max	0.015 max	0.02 max	0.12 max		2-3	0.01 max		Pd 0.15-0.25;			
Ti-3Al-2.5V R56320	USA	AMS 4944D		Tub CW SR	2.5-3.5	0.05	0.3	0.015	0.02	0.12		2-3	0.01	0.3		485	345	18
Ti-3Al-2.5V R56320	USA				2.5-3.5	0.05	0.25	0.013	0.02	0.12		2-3	0.01 max		Pd 0.12-0.25;			
	USA	ASTM B381	F-9	Frg	2.5-3.5	0.10 max	0.25 max	0.0150 max	0.02 max	0.15 max		2.0-3.0		0.4	OE 0.1 max	620	483	15
	USA	ASTM B367	C-6	Cast	4.00-6.00	0.10 max	0.50 max	0.015 max	0.05 max	0.20 max				0.40	Sn 2.0-3.0; OE 0.10 max	795	725	8
	USA	ASTM B381	F-6	Frg	4.0-6.0	0.10 max	0.50 max	0.0200 max	0.05 max	0.30 max				0.4	Sn 2.0-3.0; OE 0.1 max	828	795	10
	USA	ASTM B348	Ti Grade 6	Bar Bil (Bar H max 0.100)	4.0-6.0	0.10 max	0.50 max	0.0125 max	0.05 max	0.20 max				0.4	Sn 2.0-3.0; OE 0.1 max	828	795	10
	USA	ASTM B863	Ti Grade 6	Wir	4.0-6.0	0.10 max	0.50 max	0.020 max	0.05 max	0.20 max				0.4	Sn 2.0-3.0; OE 0.1 max	828	793	10
Ti-5Al-2.5Sn R54520	China	GB 3620	TA-7		4-6	0.1 max	0.3 max	0.015 max	0.05 max	0.2 max	2-3		0.01	0.4	Si 0.15 max	810	770	10
Ti-5Al-2.5Sn R54520	Germany	DIN	3.7114		4.5-5.5	0.08	0.5	0.015-0.02	0.05	0.2	2-3		0.01	0.4		895	830	
Ti-5Al-2.5Sn R54520	Germany	DIN 17851	3.7115	Plt Sh Strp Ann	4-6	0.08	0.5	0.02	0.05	0.2	2-3		0.01	0.4		915	845	9
Ti-5Al-2.5Sn R54520	Germany	DIN 17851	Ti-5Al-2.5Sn	Sh Strp Plt Rod Wir	4-6	0.08	0.5	0.02	0.05	0.2	2-3		0.01	0.4		915	845	8
Ti-5Al-2.5Sn R54520	Russia	GOST 19807-74	VT5-1	Sh Plt Strp Rod Frg Ann	4-6	0.1	0.3	0.015	0.05	0.15	2-3		0.01	0.4	Si 0.15; Zr 0.3	865	795	10
Ti-5Al-2.5Sn R54520	Russia	GOST	VT5-1KT		4-5.5	0.05	0.2	0.008	0.04	0.12	2-3		0.01	0.4	Si 0.1; Zr 0.2	880	815	10
Ti-5Al-2.5Sn R54520	Spain	UNE 38-716	L-7101	Sh Strp Plt Bar Frg Ext	4.5-5.5	0.15	0.5	0.02	0.07	0.2	2-3		0.01					

UNS numbers and US grades are provided as a means of cross referencing chemically similar alloys. Exchangability is only possible after independent examination of specifications. Tensile properties are minimum or typical . UTS and YS as Mpa, El as %. See Appendix for list of abbreviations used in Descriptions.

Worldwide Guide to Equivalent Nonferrous Metals and Alloys

Grade UNS #	Country	Spec.	Designation	Comment	Al	C	Fe	H	N	O	S	V	W	OT	x	Other	UTS	YS	ELG
Ti-5Al-2.5Sn R54520	UK	BS TA.14	BS TA.14	Sh Obsolete	4-6	0.08 max	0.5 max	0.0125 max				2-3		0.01						
Ti-5Al-2.5Sn R54520	UK	BS TA.15	BS TA.15	Bar Obsolete	4-6	0.08 max	0.5 max	0.0125 max				2-3		0.01						
Ti-5Al-2.5Sn R54520	UK	BS TA.16	BS TA.16	Frg Obsolete	4-6	0.08 max	0.5 max	0.0125 max				2-3		0.01						
Ti-5Al-2.5Sn R54520	UK	BS TA.17	BS TA.17	Frg Obsolete	4-6		0.5 max	0.015 max				2-3		0.01						
Ti-5Al-2.5Sn R54520	USA	MIL T-81915	5Al-2.5Sn	Ann	4.5-5.8	0.08	0.5 max	0.02 max	0.05	0.2 max						Sn 2-3; Ti 90	758	724	10
Ti-5Al-2.5Sn R54520	USA	MIL T-009047F	5Al-5.2Sn	Bar	4.5-5.8	0.08	0.5 max	0.02 max	0.05	0.2 max						Sn 2-3; Ti 90.4			
Ti-5Al-2.5Sn R54520	USA	MIL T-9047E	5Al-5.2Sn	Bar Ann or ST	4.5-5.8	0.08	0.5 max	0.02 max	0.05	0.2 max						Sn 2-3; ti 90.4	793	758	10
Ti-5Al-2.5Sn R54520	USA	MIL T-81556A	Code A-1	Ext Bar Shp Ann	4.5-5.75	0.08	0.5	0.02	0.05	0.2	2-3		0.01	0.3			690	620	10
Ti-5Al-2.5Sn R54520	USA	MIL T-9046J	Code A-1	Sh Strp Plt Ann	4.5-5.75	0.08	0.5	0.02	0.05	0.2	2-3		0.01	0.3			690	655	8
Ti-5Al-2.5Sn R54520	USA	MIL F-83142A	Comp 2	Frg HT	4.5-5.75	0.08	0.5	0.02	0.05	0.2	2-3		0.01	0.3		O+Fe=0.32;	690	655	8
Ti-5Al-2.5Sn R54520	USA	MIL F-83142A	Comp 2	Frg Ann	4.5-5.75	0.08	0.5	0.02	0.05	0.2	2-3		0.01	0.4		O+Fe=0.32;	690	620	10
Ti-5Al-2.5Sn R54520	USA	MIL T-9047G	Ti-5Al-2.5Sn	Bar Bil Ann	4.5-5.75	0.08	0.5	0.02	0.05	0.2	2-3		0.01	0.3			690	620	10
Ti-5Al-2.5Sn R54520	USA	MIL T-81915	Type II Comp A	Cast Ann	4.5-5.75	0.08	0.5	0.02	0.05	0.2	2-3		0.01	0.3			690	655	10
Ti-5Al-2.5Sn R54520	USA	AMS 4910J		Sh Strp Plt Ann	4.5-5.75	0.08	0.5	0.02	0.05	0.2	2-3		0.01	0.4			795	725	8
Ti-5Al-2.5Sn R54520	USA	AMS 4926H		Bar Wir Bil Rng Ann	4-6	0.08	0.5	0.02	0.05	0.2	2-3		0.01	0.4			825	795	10
Ti-5Al-2.5Sn R54520	USA	AMS 4966J		Frg Ann	4-6	0.08	0.5	0.02	0.05	0.2	2-3		0.01	0.4			795	760	12
Ti-5Al-2.5Sn R54520	USA				4-6	0.1	0.5	0.02	0.05	0.2	2-3		0.01	0.4			795	760	10
Ti-5Al-2.5Sn R54523	USA	AWS A5.16-70	ERTi-5Al-2.5Sn-1	ELI Weld Fill Met	4.7-5.6	0.04	0.25	0.005	0.01	0.1	2-3		0.01	0.4			895	825	10
R54523	USA	AWS A5.16-90	ERTi-6ELI	Weld El Rod	4.5-5.8	0.03 max	0.2 max	0.01 max	0.01 max	0.1 max					0.2	Sn 2-3; OE 0.05 max			
Ti-5Al-2.5Sn R54523	USA	ASTM B265	Ti Grade 6	Sh Strp Plt Ann	4.0-6.0	0.10 max	0.50 max	0.020 max	0.05 max	0.20 max	2-3				0.4	Sn 2.0-3.0; OE 0.1 max	828	793	10
Ti-5Al-2.5Sn R54523	USA			ELI Weld Fill Met	4.7-5.6	0.04	0.25	0.005	0.01	0.1	2-3		0.01	0.4			795	760	10
Ti-5Al-2.5Sn R54522	USA	AWS A5.16-70	ERTi-5Al-2.5Sn	Weld Fill Met	4.7-5.6	0.05	0.4	0.008	0.03	0.12	2-3		0.01	0.4			825	795	10
R54522	USA	AWS A5.16-90	ERTi-6	Weld El Rod	4.5-5.8	0.08 max	0.5 max	0.02 max	0.05 max	0.18 max					0.2	Sn 2-3; OE 0.05 max			
Ti-5Al-2.5Sn	USA	AMS 4953D		Weld Fill Wir	4.5-5.75	0.08	0.5	0.015	0.05	0.18	2-3		0.01	0.4			825	795	8

UNS numbers and US grades are provided as a means of cross referencing chemically similar alloys. Exchangability is only possible after independent examination of specifications. Tensile properties are minimum or typical. UTS and YS as Mpa, El as %. See Appendix for list of abbreviations used in Descriptions.

10-16 Titanium

Grade UNS #	Country	Spec.	Designation	Comment	Al	C	Fe	H	N	O	S	V	W	OT x	Other	UTS	YS	ELG
Ti-5Al-2.5Sn R54522	USA			Weld Fill Met	4.7-5.6	0.05	0.4	0.008	0.03	0.12	2-3		0.01	0.4				
R54521	USA	MIL T-009047F	5Al-2.5SnELI	Bar	4.7-5.6	0.05	0.25 max	0.01 max	0.04	0.12 max					Mn 0.1; Sn 2-3; Ti 90.8			
Ti-5Al-2.5Sn ELI R54521	USA	MIL T-9047E	5Al-2.5SnELI	Ann or ST 2 in diam	4.7-5.6	0.05	0.25 max	0.01 max	0.04	0.12 max					Mn 0.1; Sn 2-3; Ti 90.8	689	621	10
R54521	USA	MIL T-9046H	A(5Al-2.5Sn)	SH Strp Plt Ann	4.5-5.8	0.08	0.5 max	0.02 max	0.05	0.2 max					Sn 2-3; Ti 90.4	827	779	10
R54521	USA	MIL T-9046H	B(5Al-2.5SnELI)	Sh Strp Plt Ann	4.5-5.8	0.05	0.25	0.01 max	0.04	0.12 max					Sn 2-3; Ti 90.8	689	655	10
Ti-5Al-2.5Sn R54521	USA	MIL T-81556A	Code A-2	ELI Ext Bar Shp Ann	4.5-5.75	0.05	0.25	0.0125	0.04	0.12	2-3		0.01					
Ti-5Al-2.5Sn R54521	USA	MIL T-9046J	Code A-2	ELI Sh Strp Plt Ann	4.5-5.75	0.05	0.25	0.0125	0.04	0.12	2-3		0.01	0.4		810	770	9
Ti-5Al-2.5Sn R54521	USA	MIL F-83142A	Comp 3	ELI Frg Ann	4.5-5.75	0.05	0.25	0.0125	0.04	0.12	2-3		0.01		Mn 0.1;			
Ti-5Al-2.5Sn R54521	USA	MIL T-9047G	Ti-5Al-2.5Sn ELI	ELI Bar Bil Ann	4.5-5.75	0.05	0.25	0.0125	0.04	0.12	2-3		0.01		2n 2-3;	790	760	7
Ti-5Al-2.5Sn R54521	USA	AMS 4909D		ELI Sh Strp Plt Ann	4.5-5.75	0.05	0.25	0.0125	0.04	0.12	2-3		0.01			790	760	8
Ti-5Al-2.5Sn R54521	USA	AMS 4924D		ELI Bar Frg Rng Ann	4.7-5.6	0.05	0.25	0.0125	0.04	0.12	2-3		0.01	0.3		855	635	10
Ti-5Al-2.5Sn R54521	USA			ELI	5						2.5-2.5		0.01	0.3		895	915	10
	Europe	AECMA prEN2532	Ti-P68	Bar Bars up to 25 mm for frgs	3-5	0.08 max	0.2 max		0.05 max	0.25 max					ET 0.1 max; Sn 1.5-2.5; Si 0.3-0.7; Mo 3-5	1100	960	9
	Europe	AECMA prEN2533	Ti-P68	Bar Bars 25-100 mm for frgs	3-5	0.08 max	0.2 max		0.05 max	0.25 max					ET 0.1 max; Sn 1.5-2.5; Si 0.3-0.7; Mo 3-5	1050	920	9
	Europe	AECMA prEN2534	Ti-P68	Bar Bars 100-150 mm for frgs	3-5	0.08 max	0.2 max		0.05 max	0.25 max					ET 0.1 max; Sn 1.5-2.5; Si 0.3-0.7; Mo 3-5	1000	870	9
Ti-6211 R56210	USA	MIL T-9046J	Code A-3	Sh Strp Plt Ann	5.5-6.5	0.05	0.25	0.0125	0.03	0.1	1.8-2.2		0.01	0.4	Mo 0.5-1			
R56210	USA	AWS A5.16-90	ERTi-15	Weld El Rod	5.5-6.5	0.03 max	0.15 max	0.01 max	0.02 max	0.1 max				0.2	OE 0.05 max			
Ti-6211 R56210	USA	AWS A5.16-70	ERTi-6Al-2Nb-1Ta-1Mo	Weld Fill Met	5.5-6.5	0.04	0.15	0.005	0.01	0.1	1.8-2.2		0.01		Mo 0.5-1.5			
Ti-6211 R56210	USA	MIL T-9046H	G(6Al-2Nb-1Ta-0.8Mo)	Sh Strp Plt Ann (plt)	5.5-6.5	0.05	0.25 max	0.01 max	0.03	0.1 max					Cd 1.5-2.5; Ta 0.5-1.5; Ti 88.1; Mo 0.5-1	710	655	10
Ti-6211 R56210	USA	MIL T-9047G	Ti-6Al-2Nb-1Ta-0.8Mo	Bar Bill Roll/Frg	5.5-6.5	0.05	0.25	0.0125	0.03	0.1	1.8-2.2		0.01	0.3	Mo 0.5-1	550	485	15
Ti-6211 R56210	USA				6						1.8-2.2		0.01	0.4	OE 0.1 max; Mo 0.8-8			
	USA	ASTM B367	C-5	Cast	5.5-6.75 max	0.10 max	0.40 max	0.015 max	0.05 max	0.25 max		3.5-4.5		0.40	OE 0.10 max	895	825	6
	USA	ASTM B381	F-23	Frg	5.5-6.5	0.08 max	0.25 max	0.0125 max	0.05 max	0.13 max		3.5-4.5		0.4	OE 0.1 max	828	759	10
	USA	ASTM B381	F-5	Frg	5.50-6.75	0.10 max	0.30 max	0.0125 max	0.05 max	0.20 max		3.5-4.5		0.4	OE 0.1 max	895	828	10
	USA	ASTM B265	Ti Grade 23	Sh Strp Plt Ann SHT	5.5-6.5	0.08 max	0.25 max	0.0125 max	0.05 max	0.13 max		3.5-4.5		0.4	OE 0.1 max	828	759	10
	USA	ASTM B861	Ti Grade 23	Smls Pip Ann	5.5-6.5	0.08 max	0.25 max	0.0125 max	0.05 max	0.13 max		3.5-4.5		0.4	OE 0.1 max	828	759	10
	USA	ASTM B863	Ti Grade 23	Wir	5.5-6.5	0.08 max	0.25 max	0.0125 max	0.05 max	0.13 max		3.5-4.5		0.4	OE 0.1 max	793	759	10

UNS numbers and US grades are provided as a means of cross referencing chemically similar alloys. Exchangability is only possible after independent examination of specifications. Tensile properties are minimum or typical . UTS and YS as Mpa, El as %. See Appendix for list of abbreviations used in Descriptions.

Worldwide Guide to Equivalent Nonferrous Metals and Alloys

Grade UNS #	Country	Spec.	Designation	Comment	Al	C	Fe	H	N	O	S	V	W	OT	x	Other	UTS	YS	ELG
	USA	ASTM B861	Ti Grade 24	Smls Pip Ann	5.5-6.75	0.10 max	0.40 max	0.015 max	0.05 max	0.20 max		3.5-4.5			0.4	Pd 0.04-0.08; OE 0.1 max	895	828	10
	USA	ASTM B861	Ti Grade 25	Smls Pip Ann	5.5-6.75	0.10 max	0.40 max	0.015 max	0.05 max	0.20 max		3.5-4.5			0.4	Pd 0.04-0.08; Ni 0.3-0.8; OE 0.1 max	895	828	10
	USA	ASTM B861	Ti Grade 5	Smls Pip Ann	5.5-6.75	0.10 max	0.40 max	0.015 max	0.05 max	0.20 max		3.5-4.5			0.4	OE 0.1 max	895	828	10
	USA	ASTM B863	Ti Grade 5	Wir	5.5-6.75	0.10 max	0.40 max	0.015 max	0.05 max	0.20 max		3.5-4.5			0.4	OE 0.1 max	895	828	10
	USA	ASTM B363	WPT23	Smls Weld Fittings (Composition as in ASTM B337, B265, B348, B381)															
	USA	ASTM B381	F-24	Frg	5.5-6.75	0.10 max	0.40 max	0.015 max	0.05 max	0.20 max		3.5-4.5			0.4	Pd 0.04-0.08; OE 0.1 max	895	828	10
	USA	ASTM B381	F-25	Frg	5.5-6.75	0.10 max	0.40 max	0.0125 max	0.05 max	0.20 max		3.5-4.5			0.4	Pd 0.04-0.08; Ni 0.3-0.8; OE 0.1 max	895	828	10
	USA	ASTM B265	Ti Grade 24	Sh Strp Plt Ann	5.5-6.75	0.10 max	0.40 max	0.015 max	0.05 max	0.20 max		3.5-4.5			0.4	Pd 0.04-0.08; OE 0.1 max	895	828	10
	USA	ASTM B348	Ti Grade 24	Bar Bil	5.5-6.75	0.10 max	0.40 max	0.015 max	0.05 max	0.20 max		3.5-4.5			0.4	Pd 0.04-0.08; OE 0.1 max	895	828	10
	USA	ASTM B863	Ti Grade 24	Wir	5.5-6.75	0.10 max	0.40 max	0.015 max	0.05 max	0.20 max		3.5-4.5			0.4	Pd 0.04-0.08; OE 0.1 max	895	828	10
	USA	ASTM B363	WPT24	Smls Weld Fittings (Composition as in ASTM B337, B265, B348, B381)															
R56402	USA	AWS A5.16-90	ERTi-5ELI	Weld El Rod	5.5-6.5	0.03 max	0.15 max	0.01 max	0.01 max	0.1 max		3.5-4.5			0.2	OE 0.05 max			
Ti-6Al-4V ELI R56402	USA	AMS 4956B		ELI Fill Met Wir	5.5-6.75	0.03	0.15	0.005	0.01	0.08	0.1 max	3.5-4.5	0.01				345	360	20
Ti-6Al-4V ELI R56402	USA			Fill Met	5.5-6.75	0.04	0.15	0.005	0.01	0.1	0.1 max	3.5-4.5	0.01	0.3			345	360	20
	USA	ASTM B348	Ti Grade 25	Bar Bil	5.5-6.75	0.10 max	0.40 max	0.0125 max	0.05 max	0.20 max		3.5-4.5			0.4	Pd 0.04-0.08; Ni 0.3-0.8; OE 0.1 max	895	828	10
	USA	ASTM B863	Ti Grade 25	Wir	5.5-6.75	0.10 max	0.40 max	0.015 max	0.05 max	0.20 max		3.5-4.5			0.4	Pd 0.04-0.08; Ni 0.3-0.8; OE 0.1 max	895	728	10
	USA	ASTM B363	WPT25	Smls Weld Fitngs Comp. as in ASTM B337, B265, B348, B381															
Ti-6Al-4V R56400	Europe	AECMA prEN2517	Ti-P63	Sh Strp Plt Bar Ann	5.5-6.75	0.08	0.3	0.01	0.05	0.2	0.1 max	3.5-4.5	0.01				640	390	15
R56400	Europe	AECMA prEN2517	TI-P63	Plt Sh Strp Ann up to 100 mm	5.5-6.8	0.08 max	0.3 max	0.01 max	0.05 max	0.2 max		3.5-4.5				ET 0.1 max	920	870	8
R56400	Europe	AECMA prEN2530	Ti-P63	Bar Ann bars up to 150 mm for frgs	5.5-6.8	0.08 max	0.3 max		0.05 max	0.2 max		3.5-4.5				ET 0.1 max	900	830	10
Ti-6Al-4V R56400	Europe	AECMA prEN2530		Bar Ann	5.5-6.75	0.08 max	0.3 max	0.0125 max	0.05 max	0.2 max	0.1 max	3.5-4.5					390	290	20
Ti-6Al-4V R56400	Europe	AECMA prEN2531		Frg Ann	5.5-6.75	0.08 max	0.3 max	0.0125 max	0.05 max	0.2 max	0.1 max	3.5-4.5							
Ti-6Al-4V R56400	Europe	AECMA prEN3310		Frg NHT	5.5-6.75	0.08 max	0.3 max	0.0125 max	0.05 max	0.2 max	0.1 max	3.5-4.5					410	215	23
Ti-6Al-4V R56400	Europe	AECMA prEN3311		Bar Ann	5.5-6.75	0.08 max	0.3 max	0.0125 max	0.05 max	0.2 max	0.1 max	3.5-4.5							
Ti-6Al-4V R56400	Europe	AECMA prEN3312		Frg Ann	5.5-6.75	0.08 max	0.3 max	0.0125 max	0.05 max	0.2 max	0.1 max	3.5-4.5							

UNS numbers and US grades are provided as a means of cross referencing chemically similar alloys. Exchangability is only possible after independent examination of specifications. Tensile properties are minimum or typical . UTS and YS as Mpa, El as %. See Appendix for list of abbreviations used in Descriptions.

Grade UNS #	Country	Spec.	Designation	Comment	Al	C	Fe	H	N	O	S	V	W	OT x	Other	UTS	YS	ELG
Ti-6Al-4V R56400	Europe	AECMA prEN3313		Frg NHT	5.5-6.75	0.08 max	0.3 max	0.0125 max	0.05 max	0.2 max	0.1 max	3.5-4.5		0.6	OE 0.1 max			
Ti-6Al-4V R56400	Europe	AECMA prEN3314		Bar STA	5.5-6.75	0.08 max	0.3 max	0.0125 max	0.05 max	0.2 max	0.1 max	3.5-4.5						
Ti-6Al-4V R56400	Europe	AECMA prEN3315		Frg STA	5.5-6.75	0.08 max	0.3 max	0.0125 max	0.05 max	0.2 max	0.1 max	3.5-4.5		0.6	OE 0.1 max			
Ti-6Al-4V R56400	Europe	AECMA prEN3352		Inv Cast Ann HIP	5.5-6.75	0.1 max	0.3 max	0.015 max	0.05 max	0.22 max	0.1 max	3.5-4.5						
Ti-6Al-4V R56400	Europe	AECMA prEN3353		Bar Wir STA	5.5-6.75	0.08 max	0.3 max	0.0125 max	0.05 max	0.2 max	0.1 max	3.5-4.5		0.6	OE 0.1 max			
Ti-6Al-4V R56400	Europe	AECMA prEN3354		Sh Ann	5.5-6.75	0.08 max	0.3 max	0.0125 max	0.05 max	0.2 max	0.1 max	3.5-4.5						
Ti-6Al-4V R56400	Europe	AECMA prEN3355		Ext Ann	5.5-6.75	0.08 max	0.3 max	0.0125 max	0.05 max	0.2 max	0.1 max	3.5-4.5		0.6	OE 0.1 max			
Ti-6Al-4V R56400	Europe	AECMA prEN3456		Sh Strp Ann	5.5-6.75	0.08 max	0.3 max	0.0125 max	0.05 max	0.2 max	0.1 max	3.5-4.5				440	395	30
Ti-6Al-4V R56400	Europe	AECMA prEN3457		Frg NHT	5.5-6.75	0.08 max	0.3 max	0.0125 max	0.05 max	0.2 max	0.1 max	3.5-4.5						
Ti-6Al-4V R56400	Europe	AECMA prEN3458		Bar Wir Ann	5.5-6.75	0.08 max	0.3 max	0.0125 max	0.05 max	0.2 max	0.1 max	3.5-4.5				525	320	18
Ti-6Al-4V R56400	Europe	AECMA prEN3464		Plt Ann	5.5-6.75	0.08 max	0.3 max	0.0125 max	0.05 max	0.2 max	0.1 max	3.5-4.5						
Ti-6Al-4V R56400	Europe	AECMA prEN3467		Remelt NHT	5.5-6.75	0.08 max	0.3 max	0.0125 max	0.05 max	0.2 max	0.1 max	3.5-4.5		0.6		355	200	30
Ti-6Al-4V R56400	France	AIR 9183	T-A6V	Bar Rod Frg	5.5-7	0.08	0.25	0.012	0.07	0.2	0.1 max	3.5-4.5	0.01			540	430	16
Ti-6Al-4V R56400	France	AIR 9184	T-A6V	Blt	5.5-7	0.08 max	0.25 max	0.12 max	0.07 max	0.2 max	0.1 max	3.5-4.5	0.01			570	460	15
R56400	France	AIR 9183	T-A6V	Bar As Mfg	5.5-7	0.08	0.25 max		0.07			3.5-4.5			Ti 87.9	880	820	10
Ti-6Al-4V R56400	Germany	DIN	3.7164	Sh Strp Plt Bar Frg Ann	5.5-6.75	0.08	0.3	0.0125 -0.015	0.05	0.2	0.1 max	3.5-4.5	0.01					
Ti-6Al-4V R56400	Germany	DIN 17850	3.7165	Plt Sh Strp Rod Wir Ann	5.5-6.75	0.08	0.3	0.015	0.05	0.2	0.1 max	3.5-4.5	0.01	0.6		240	170	24
Ti-6Al-4V R56400	Germany	DIN 17851	3.7165	Sh Plt Strp Rod Wir Ann	5.5-6.75	0.08	0.3	0.015	0.05	0.2	0.1 max	3.5-4.5	0.01					
Ti-6Al-4V R56400	Germany	DIN	3.7264	Cast Ann	5.5-6.75	0.1	0.3	0.015	0.05	0.2	0.1 max	3.5-4.5	0.01	0.3	W 0.2;	343	294	10
TiAl6V4 R56400	Germany	DIN 17851	3.7615		5.5-6.8	0.08 max	0.3 max	0.02 max	0.05 max	0.2 max		3.5-4.5						
Ti-6Al-4V R56400	Germany	DIN 17860	3.7615	Sh Strp	5.5-6.75	0.2 max	0.3 max	0.015 max	0.05 max		0.1 max	3.5-4.5	0.01	0.6		390	290	22
Ti-6Al-4V R56400	Germany	DIN 17862	3.7615	Rod	5.5-6.75	0.08 max	0.3 max	0.015 max	0.05 max	0.2 max	0.1 max	3.5-4.5	0.01					
Ti-6Al-4V R56400	Germany	DIN 17864	3.7615	Frg	5.5-6.75	0.08 max	0.3 max	0.015 max	0.05 max	0.2 max	0.1 max	3.5-4.5	0.01					
Ti-6Al-4V R56400	Russia	GOST 1.90000-70	VT6	Sh Plt Strp Foil Rod Frg Ann	5.5-7	0.1	0.3	0.015	0.05	0.2	0.1 max	4.2-6	0.01	0.3	Si 0.15	450	465	18
Ti-6Al-4V R56400	Russia	GOST 1.90060-72	VT6L	Cast	5-6.5	0.1	0.3	0.015	0.05	0.15	0.1 max	3.5-4.5	0.01	0.4	Si 0.15; Zr 0.3			
Ti-6Al-4V R56400	Russia	GOST 19807-74	VT6S	Sh Plt Strp Foil Rod Ann	5.3-6.8	0.08	0.25	0.007	0.05	0.02	0.1 max	3.5-4.5	0.01	0.4	Si 0.15; Zr 0.3	450	465	18
Ti-6Al-4V R56400	Spain	UNE 38-723	L-7301	Sh Plt Strp Bar Ex HT	5.5-6.75	0.1	0.3	0.125	0.05	0.2	0.1 max	3.5-4.5	0.01					
Ti-6Al-4V R56400	Spain	UNE 38-723	L-7301	Sh Plt Strp Bar Ex Ann	5.5-6.75	0.1	0.3	0.125	0.05	0.2	0.1 max	3.5-4.5	0.01			500	430	16
Ti-6Al-4V R56400	UK	BS 2TA.10	BS 2TA.10	Sh Strp HT	5.5-6.75		0.3	0.01			0.1 max				Ti 88.2 max	390	290	20
Ti-6Al-4V R56400	UK	BS 2TA.11	BS 2TA.11	Bar HT	5.5-6.8		0.3 max	0.01 max	0.05 max	0.2 max		3.5-4.5			Ti 88.2 max	900	830	8
Ti-6Al-4V R56400	UK	BS 2TA.12	BS 2TA.12	Frg HT	5.5-6.8		0.3 max	0.1 max	0.05 max	0.2 max		3.5-4.5			Ti 88.2 max	900	830	8
Ti-6Al-4V R56400	UK	BS 2TA.13	BS 2TA.13	Frg HT	5.5-6.8		0.3 max	0.02 max		0.2 max		3.5-4.5			Ne 0.05 max; Ti 88.2 max	900	830	8
Ti-6Al-4V R56400	UK	BS 2TA.28	BS 2TA.28	Wir Frg HT Quen	5.5-6.75		0.3	0.01	0.05	0.2	0.1 max	3-5			Ti 88.2 max	525	320	18
Ti-6Al-4V R56400	UK	BS 2TA.28	BS 2TA.28	Frg HT and Q	5.5-6.8		0.3 max	0.01 max	0.05 max	0.2 max		3.5-4.5			Ti 88.2 max	1100	970	8

UNS numbers and US grades are provided as a means of cross referencing chemically similar alloys. Exchangability is only possible after independent examination of specifications. Tensile properties are minimum or typical . UTS and YS as Mpa, El as %. See Appendix for list of abbreviations used in Descriptions.

Worldwide Guide to Equivalent Nonferrous Metals and Alloys

Grade UNS #	Country	Spec.	Designation	Comment	Al	C	Fe	H	N	O	S	V	W	OT x	Other	UTS	YS	ELG
Ti-6Al-4V R56400	UK	BS TA.56	BS TA.56	Plt to 100 mm HT	5.5-6.75		0.3				0.1 max	3.5-4.5			Ti 88.2 max	465	250	26
Ti-6Al-4V R56400	UK	BS TA.56	BS TA.56	Plt HT 5-10 mm diam	5.5-6.8		0.3 max					3.5-4.5			Ti 88.2 max	895	825	10
Ti-6Al-4V R56400	UK	BS TA.59	BS TA.59	Sh Strp	5.5-6.75	0.08 max	0.3 max	0.0125 max			0.1 max	3.5-4.5		0.6		465	290	21
Ti-6Al-4V R56400	UK	DTD 5303		Bar Ann	5.5-6.75	0.2 max	0.3 max	0.0125 max	0.05 max		0.1 max	3.5-4.5						
Ti-6Al-4V R56400	UK	DTD 5323		Frg Ann	5.5-6.75		0.3 max	0.015 max	0.05 max	0.2 max	0.1 max	3.5-4.5						
Ti-6Al-4V R56400	UK	DTD 5363		Cast	5.5-6.75		0.3 max	0.15 max	0.05 max	0.25 max	0.1 max	3.5-4.5						
Ti-6Al-4V R56400	UK	BS 3531 Part 2		Srg Imp	5.5-6.75	0.08 max	0.3 max	0.015 max		0.2 max	0.1 max	3.5-4.5						
Ti-5Al-4V R56400	UK	DTD 5313		Frg Ann	5.5-6.75		0.3 max	0.01 max	0.05 max	0.2 max	0.1 max	3.5-4.5				390	295	27
Ti-6Al-4V R56400	USA	MIL T-9047E	6Al-4V	Bar Ann or ST	5.5-6.8	0.08	0.3 max	0.01 max	0.05	0.2 max		3.5-4.5			Ti 88.1	896	827	10
Ti-6Al-4V R56400	USA	MIL T-9047E	6Al-4V	Bar HT 0.5 in diam	5.5-6.8	0.08	0.3 max	0.01 max	0.05	0.2 max		3.5-4.5			Ti 88.1	1103	1034	10
Ti-6Al-4V R56400	USA	MIL T-9047F	6Al-4V	Bar	6.5-6.8	0.08	0.3 max	0.01 max	0.05	0.2 max		3.5-4.5			Cu 0.1; Mn 0.1; Sn 0.1; Zr 0.1; Ti 87.6; Mo 0.1			
Ti-6Al-4V R56400	USA	MIL T-81915	A(6Al-4V)	Ann	5.5-6.8	0.08	0.3 max	0.02 max	0.05	0.2 max		3.5-4.5			Ti 87.7	862	793	8
Ti-6Al-4V R56400	USA	MIL T-9046H	C(6Al-4V)	Sh Strp Plt Ann (sht,strp)	5.5-6.5	0.08	0.3 max	0.02 max	0.05	0.2 max		3.5-4.5			Ti 88.4	924	869	8
Ti-6Al-4V R56400	USA	MIL T-9046H	C(6Al-4V)	Sh Strp Plt ST (sht,strp)	5.5-6.5	0.08	0.3 max	0.02 max	0.05	0.2 max		3.5-4.5			Ti 88.4		1034	6
Ti-6Al-4V R56400	USA	MIL T-9046H	C(6Al-4V)	Sh Strp Plt Ann (plt)	5.5-6.5	0.08	0.3 max	0.02 max	0.05	0.2 max		3.5-4.5			Ti 88.4	896	827	10
Ti-6Al-4V R56400	USA	MIL T-81556A	Code AB-1	Ext Bar Shp Ann	5.5-6.75	0.08	0.3	0.0125	0.05	0.2	0.1 max	3.5-4.5	0.01					
Ti-6Al-4V R56400	USA	MIL T-81556A	Code AB-1	Ext Bar Shp STA	5.5-6.75	0.08	0.3	0.0125	0.05	0.2	0.1 max	3.5-4.5	0.01					
Ti-6Al-4V R56400	USA	MIL T-9046J	Code AB-1	Sh Strp Plt STA	5.5-6.75	0.08	0.3	0.0125	0.05	0.2	0.1 max	3.5-4.5	0.01					
Ti-6Al-4V R56400	USA	MIL T-9046J	Code AB-1	Sh Strp Plt Ann	5.5-6.75	0.08	0.3	0.0125	0.05	0.2	0.1 max	3.5-4.5	0.01					
Ti-6Al-4V R56400	USA	MIL F-83142A	Comp 6	Frg HT	5.5-6.75	0.08	0.3	0.015	0.05	0.2	0.1 max	3.5-4.5	0.01	0.4	OE 0.1 max			
Ti-6Al-4V R56400	USA	MIL F-83142A	Comp 6	Frg Ann	5.5-6.75	0.08	0.3	0.015	0.05	0.2	0.1 max	3.5-4.5	0.01	0.4	OE 0.1 max			
R56400	USA	AWS A5.16-90	ERTi-5	Weld El Rod	5.5-6.7	0.05 max	0.3 max	0.02 max	0.03 max	0.18 max		3.5-4.5		0.2	OE 0.05 max			
Ti-6Al-4V R56400	USA	AWS A5.16-70	ERTi-6A1-4V	Weld fill met	5.5-6.75	0.05	0.25	0.008	0.02	0.15	0.1 max	3.5-4.5	0.01					
R56400	USA	MIL T-9046H	H(6Al-4V-SPL)	Sh Strp Plt Ann	5.5-6.8	0.08	0.25 max	0.01 max	0.05	0.13 max		3.5-4.5			Ti 88.2	896	827	10
Ti-6Al-4V R56400	USA	MIL T-9047G	MIL-T-9047G	Bar Bil Ann	5.5-6.75	0.08	0.3	0.015	0.05	0.2	0.1 max	3.5-4.5	0.01			540	430	16
Ti-6Al-4V R56400	USA	ASTM B265	Ti Grade 5	Sh Strp Plt Ann	5.5-6.75	0.10 max	0.40 max	0.015 max	0.05 max	0.20 max	0.1 max	3.5-4.5		0.4	OE 0.1 max	895	828	10
Ti-6Al-4V R56400	USA	ASTM B348	Ti Grade 5	Bar Bil (Bar H max 0.100)	5.5-6.75	0.10 max	0.40 max	0.0125 max	0.05 max	0.20 max		3.5-4.5		0.4	OE 0.1 max	895	828	10
Ti-6Al-4V R56400	USA	ASTM F467-84	Ti Grade 5	Blt Scr Std	5.5-6.75	0.1 max	0.4 max	0.0125 max	0.05 max	0.2 max	0.1 max	3.5-4.5	0.01					
Ti-6Al-4V R56400	USA	MIL T-81915	Type III, Comp A	Cast Ann	5.5-6.75	0.08	0.3	0.015	0.05	0.2	0.1 max	3.5-4.5	0.01			640	390	16
Ti-6Al-4V R56400	USA	AMS 4906		Sh Strp	5.5-6.75	0.08 max	0.3 max	0.0125 max	0.05 max	0.2 max	0.1 max	3.5-4.5	0.01 max					
Ti-6Al-4V R56400	USA	AMS 4920		Frg Ann	5.5-6.75	0.1	0.3	0.0125	0.05	0.2	0.1 max	3.5-4.5	0.01			345	275	20
Ti-6Al-4V R56400	USA	AMS 4996		Bill Powd Ann	5.5-6.75	0.1	0.3	0.0125	0.04	0.13-0.19	0.1 max	3.5-4.5			Zr 0.1 max; Mo 0.1 max	290	180	24
Ti-6Al-4V R56400	USA	AMS 4998		Powd	5.5-6.75	0.1	0.3	0.012	0.04	0.13-0.18	0.1 max	3.5-4.5			Nb 1.5-2.5; Zr 0.1 max; Mo 0.1 max	390	290	22

UNS numbers and US grades are provided as a means of cross referencing chemically similar alloys. Exchangability is only possible after independent examination of specifications. Tensile properties are minimum or typical . UTS and YS as Mpa, El as %. See Appendix for list of abbreviations used in Descriptions.

Grade UNS #	Country	Spec.	Designation	Comment	Al	C	Fe	H	N	O	S	V	W	OT x	Other	UTS	YS	ELG
Ti-6Al-4V R56400	USA	AMS 4905A		Plt Beta Ann	5.6-6.3	0.05	0.25	0.0125	0.03	0.12	0.1 max	3.6-4.4	0.01	0.4		345	275	20
Ti-6Al-4V R56400	USA	AMS 4911F		Sh Strp Plt Ann	5.5-6.75	0.08	0.3	0.015	0.05	0.2	0.1 max	3.5-4.5	0.01	0.6	OE 0.1 max			
Ti-6Al-4V R56400	USA	AMS 4928K		Bar Wir Frg Bil Rng Ann	5.5-6.75	0.1	0.3	0.0125	0.05	0.2	0.1 max	3.5-4.5	0.01					
Ti-6Al-4V R56400	USA	AMS 4934A		Ex Rng STA	5.5-6.75	0.1	0.3	0.0125	0.05	0.2	0.1 max	3.5-4.5	0.01	0.3		345	275	15
Ti-6Al-4V R56400	USA	AMS 4935E		Ex Rng Ann	5.5-6.75	0.1	0.3	0.0125	0.05	0.2	0.1 max	3.5-4.5	0.01					
Ti-6Al-4V R56400	USA	AMS 4954D		Fill met gas-met/W-arc weld	5.5-6.75	0.05	0.3	0.015	0.03	0.18	0.1 max	3.5-4.5	0.01	0.3	Nb 2	345	360	20
Ti-6Al-4V R56400	USA	AMS 4965E		Bar Frg Rng STA/Mach Press ves	5.5-6.75	0.08	0.3	0.0125	0.05	0.2	0.1 max	3.5-4.5	0.01	0.6	OE 0.1 max			
Ti-6Al-4V R56400	USA	AMS 4967F		Bar Frg Rng Mach/STA Press ves	5.5-6.75	0.08	0.3	0.0125	0.05	0.2	0.1 max	3.5-4.5	0.01	0.4	OE 0.1 max			
Ti-6Al-4V R56400	USA	AMS 4985A		Cast Ann	5.5-6.75	0.1	0.3	0.015	0.05	0.2	0.1 max	3.5-4.5	0.01		Nb1.5-2.5			
Ti-6Al-4V R56400	USA	AMS 4993A		Powd Sint Nuts	5.5-6.75	0.1	0.3	0.01	0.05	0.3	0.1 max	3.5-4.5		0.3	Nb 1.5-2.5; Si 0.05	430		22
Ti-6Al-4V R56400	USA	MIL A-46077D		Weld armor plt Ann	5.5-6.5	0.04	0.25	0.0125	0.02	0.14	0.1 max	3.5-4.5		0.15		345	360	20
Ti-6Al-4V R56400	USA	ASTM F468-84		Blt Scr Std	5.5-6.75	0.1 max	0.4 max	0.0125 max	0.05 max	0.2 max	0.1 max	3.5-4.5		0.4	OE 0.1 max			
Ti-6Al-4V R56400	USA	MIL T-9047G		Bar Bil STA	5.5-6.75	0.08	0.3	0.015	0.05	0.2	1.8-2.2	3.5-4.5	0.01					
Ti-6Al-4V R56400	USA			Weld Wir	5.5-6.75	0.1	0.4	0.015	0.05	0.2	0.1 max	3.5-4.5		0.4		345	360	20
Ti-6Al-4V ELI R56401	USA	MIL T-9047E	6Al-4V-ELI	Bar Ann or ST 1.50 in diam	5.5-6.8	0.08	0.15 max	0.02 max	0.04	0.13 max		5-6			Ti 86.8	827	758	10
Ti-6Al-4V ELI R56401	USA	MIL T-9047E	6Al-4V-ELI	Bar HT .5 in diam	5.5-6.8	0.08	0.15 max	0.02 max	0.04	0.13 max		5-6			Ti 86.8	1034	965	12
Ti-6Al-4V ELI R56401	USA	MIL T-9047F	6Al-4VELI	Bar	6.5-6.8	0.08	0.15 max	0.01 max	0.05	0.13 max		3.5-4.5			Cu 0.1; Mn 0.1; Sn 0.1; Zr 0.1; Ti 87.8; Mo 0.1			
Ti-6Al-4V ELI R56401	USA	MIL T-81556A	Code AB-2	ELI Ext Bar Ann	5.5-6.5	0.08	0.25	0.0125	0.05	0.13	0.1 max	3.5-4.5	0.01					
Ti-6Al-4V ELI R56401	USA	MIL T-9046J	Code AB-2	ELI Sh Strp Plt Ann	5.5-6.5	0.08	0.25	0.0125	0.05	0.13	0.1 max	3.5-4.5	0.01	0.4	OE 0.1 max			
Ti-6Al-4V ELI R56401	USA	MIL F-83142A	Comp 7	ELI Frg Ann	5.5-6.5	0.08	0.2-0.25	0.0125	0.05	0.13	0.1 max	3.5-4.5	0.01	0.6	OE 0.1 max			
Ti-6Al-4V ELI R56401	USA	MIL F-83142A	Comp 7	ELI Frg HT	5.5-6.5	0.08	0.25	0.0125	0.05	0.13	0.1 max	3.5-4.5	0.01			550	345	18
R56401	USA	MIL T-9046H	D(6Al-4VELI)	Sh Strp Plt Ann	5.5-6.8	0.08	0.25 max	0.01 max	0.05	0.13 max		3.5-4.5			Ti 88.2	896	827	10
R56401	USA	AWS A5.16-70	ERTi-6Al-4V-1	ELI Fill Met Wir Rod	5.5-6.75	0.04	0.15	0.005	0.01	0.1	0.1 max	3.5-4.5	0.01					
Ti-6Al-4V ELI R56401	USA	ASTM B348	Ti Grade 23	Bar Bil (Bar H max 0.100)	5.5-6.5	0.08 max	0.25 max	0.0125 max	0.05 max	0.13 max		3.5-4.5		0.4	OE 0.1 max	828	759	10
Ti-6Al-4V ELI R56401	USA	ASTM B265	Ti Grade 25	Sh Strp Plt Ann	5.5-6.75	0.10 max	0.40 max	0.0125 max	0.05 max	0.20 max		3.5-4.5		0.4	Pd 0.04-0.08; Ni 0.3-0.8; OE 0.1 max	895	828	10
Ti-6Al-4V ELI R56401	USA	AMS 4931		ELI Bar Frg Bil Rng	5.5-6.5	0.08	0.25	0.0125	0.03	0.13	1.8-2.2	3.5-4.5	0.01					
Ti-6Al-4V ELI R56401	USA	AMS 4996		ELI Bil	5.5-6.75	0.1 max	0.3 max	0.0125 max	0.04 max	0.13-0.19	0.1 max	3.5-4.5		0.3			360	20

UNS numbers and US grades are provided as a means of cross referencing chemically similar alloys. Exchangability is only possible after independent examination of specifications. Tensile properties are minimum or typical . UTS and YS as Mpa, El as %. See Appendix for list of abbreviations used in Descriptions.

Worldwide Guide to Equivalent Nonferrous Metals and Alloys

Grade UNS #	Country	Spec.	Designation	Comment	Al	C	Fe	H	N	O	S	V	W	OT x	Other	UTS	YS	ELG
Ti-6Al-4V ELI R56401	USA	AMS 4998		ELI Powd	5.5-6.75	0.1 max	0.3 max	0.0125 max	0.04 max	0.13-0.19	0.1 max							
Ti-6Al-4V ELI R56401	USA	AMS 4905A		ELI Plt	5.6-6.3	0.05 max	0.25 max	0.0125 max	0.03 max	0.12 max	0.1 max	3.6-4.4	0.01 max					
Ti-6Al-4V ELI R56401	USA	AMS 4907D		ELI Sh Strp Plt Ann	5.5-6.5	0.08	0.25	0.0125	0.05	0.13	1.8-2.2	3.5-4.5	0.01					
Ti-6Al-4V ELI R56401	USA	AMS 4930C		ELI Bar Wir Frg Bil Rng Ann	5.5-6.5	0.08	0.25	0.0125	0.05	0.13	0.1 max	3.5-4.5	0.01			345	275	20
Ti-6Al-4V ELI R56401	USA	ASTM F136		ELI Wrought Ann for Surg Imp	5.5-6.5	0.08	0.25	0.012	0.05	0.13	0.1 max	3.5-4.5	0.01	0.4		345	360	20
Ti-6Al-4V ELI R56401	USA	SAE J467		ELI	6.18	0.02	0.22	0.008	0.03	0.1	0.1 max					390	275	20
Ti-6Al-4V ELI R56401	USA	MIL T-9047G		ELI Bar Bil Ann	5.5-6.5	0.08	0.25	0.0125	0.05	0.13	0.1 max	3.5-4.5	0.01	0.6	OE 0.1 max			
Ti-6Al-4V R56401	USA	AMS 4991A		Cast Ann	5.5-6.75	0.1	0.3	0.015	0.05	0.2	0.1 max	3.5-4.5	0.01	0.4	OE 0.1 max			
Ti-6Al-4V R56401	USA				6						1.8-2.2	4	0.01					
R56440	France	AIR 9183	T-A4M	Bar As Mfg	5.5-5	0.08	0.15 max	0.01 max	0.05	0.2 max					Mn 3.5-5; Ti 89.6	930	830	10
Ti-811 R54810	China		Ti-8Al-1Mo-1V		7.5-8.5	0.1 max	0.3 max	0.015 max	0.04 max	0.15 max	2-3	0.8-1.3	0.01 max	0.4	Si 0.15 max; Mo 0.75-1.25			
Ti-811 R54810	Spain	UNE 38-717	L-7102	Sh Strp Plt Bar Ext Ann	7.35-8.35	0.08	0.3	0.015	0.05	0.12	2-3	0.8-1.3	0.01 max		Mo 0.75-1.25			
Ti-811 R54810	USA	MIL T-009047F	8Al-1Mo-1V	Bar	7.4-8.4	0.08	0.3 max	0.01 max	0.05	0.12 max		0.8-1.3			Ti 88.5; Mo 0.75-1.25			
Ti-811 R54810	USA	MIL T-9047E	8Al-1Mo-1V	Bar Ann or ST 2.5 in diam	7.4-8.4	0.08	0.3 max	0.01 max	0.05	0.12 max		0.8-1.3			Ti 88.6; Mo 0.75-1.25	896	827	10
Ti-811 R54810	USA	MIL T-81556A	Code A-4		7.35-8.35	0.08	0.3	0.015	0.05	0.15	2-3	0.8-1.3	0.01	0.4	Mo 0.75-1.25	895	830	10
Ti-811 R54810	USA	MIL T-9046J	Code A-4	Sh Strp Plt Ann	7.35-8.35	0.08	0.3	0.015	0.05	0.15	2-3	0.8-1.3	0.01 max		Mo 0.75-1.25			
Ti-811 R54810	USA	MIL F-83142A	Comp 5	Frg Ann	7.35-8.35	0.08	0.3	0.015	0.05	0.15	2-3	0.8-1.3	0.01 max	0.4	Mo 0.75-1.25	915	845	7
Ti-811 R54810	USA	AWS A5.16-70	ERTi-8Al-1Mo-1V	Weld Fill Met	7.35-8.35	0.05	0.25	0.008	0.03	0.12	2-3	0.8-1.3	0.01	0.3	Mo 0.75-1.25	895	825	10
Ti-811 R54810	USA	MIL T-9046H	F(8Al-1Mo-1V)	Sh Strp Plt Ann (sht,strp)	7.3-8.3	0.08	0.3 max	0.02 max	0.05	0.15 max		0.8-1.3			Ti 88.6; Mo 0.75-1.25	1000	931	10
Ti-811 R54810	USA	MIL T-9046H	F(8Al-1Mo-1V)	Sh Strp Plt Ann (plt) .1875/.25 in diam	7.3-8.3	0.08	0.3 max	0.02 max	0.05	0.15 max		0.8-1.3			Ti 88.6; Mo 0.75-1.25	896	827	10
Ti-811 R54810	USA	SAE J467	Ti-8-1-1		8	0.04 max	0.15 max		0.02 max		2-3	1	0.01 max	0.4	Ni 0.01 max; Si 0.07 max; Mo 1	865	795	10
Ti-811 R54810	USA	MIL T-9047G	Ti-8Al-1Mo-1V	Bar Bil Dup Ann	7.35-8.35	0.08	0.3	0.015	0.05	0.15	2-3	0.8-1.3	0.01	0.4	Mo 0.75-1.25	915	845	8
Ti-811 R54810	USA	AMS 4915C		Sh Strp Plt Ann	7.35-8.35	0.08	0.3	0.015	0.05	0.12	2-3	0.8-1.3	0.01 max	0.4	Mo 0.75-1.25	880	795	8
Ti-811 R54810	USA	AMS 4915F		Sh Strp Plt Ann	7.35-8.35	0.08	0.3	0.015	0.05	0.12	2-3	0.8-1.3	0.01	0.4	Mo 0.75-1.25	915	845	8
Ti-811 R54810	USA	AMS 4916E		Sh Strp Plt Dup Ann	7.35-8.35	0.08	0.3	0.015	0.05	0.12	2-3	0.8-1.3	0.01	0.4	Mo 0.75-1.25	860	795	8
Ti-811 R54810	USA	AMS 4933A		Ext Rng SHT/Stab	7.35-8.35	0.08	0.3	0.015	0.05	0.12	2-3	0.8-1.3	0.01		Mo 0.75-1.25			
Ti-811 R54810	USA	AMS 4955B		Weld Fill Wir	7.35-8.35	0.08	0.3	0.01	0.05	0.12	2-3	0.8-1.3	0.01	0.3	Mo 0.75-1.25	1000	930	9
Ti-811 R54810	USA	AMS 4972C		Bar Wir Rng Bil SHT/Stab	7.35-8.35	0.08	0.3	0.015	0.05	0.12	2-3	0.8-1.3	0.01		Mo 0.75-1.25			
Ti-811 R54810	USA	AMS 4973C		Frg Bil SHT/Stab	7.35-8.35	0.08	0.3	0.015	0.05	0.12	2-3	0.8-1.3	0.01	0.3	Mo 0.75-1.25	1140	1070	9

UNS numbers and US grades are provided as a means of cross referencing chemically similar alloys. Exchangability is only possible after independent examination of specifications. Tensile properties are minimum or typical . UTS and YS as Mpa, El as %. See Appendix for list of abbreviations used in Descriptions.

Grade UNS #	Country	Spec.	Designation	Comment	Al	C	Fe	H	N	O	S	V	W	OT x	Other	UTS	YS	ELG
Ti-811 R54810	USA				8						2-3	1	0.01	0.4	OE 0.1 max; Mo 1			
Ti-7Al-4Mo R56740	USA	MIL T-9047E	7Al-4Mo	Bar Ann or ST 2 in diam	6.5-7.3	0.1	0.3 max	0.01 max	0.05	0.2 max					Ti 87.5; Mo 3.5-4.5	1000	931	10
Ti-7Al-4Mo R56740	USA	MIL T-9047E	7Al-4Mo	Bar HT 1 in diam	6.5-7.3	0.1	0.3 max	0.01 max	0.05	0.2 max					Ti 87.5; Mo 3.5-4.5	1172	1103	8
Ti-7Al-4Mo R56740	USA	MIL T-9047F	7Al-4Mo	Bar	6.5-7.3	0.1	0.3 max	0.01 max	0.05	0.2 max					Ti 86.6; Mo 3.5-4.5			
Ti-7Al-4Mo R56740	USA	MIL F-83142A	Comp 9	Frg Ann	6.5-7.3	0.1	0.3	0.013	0.05	0.2	1.8-2.25		0.01	0.3	Mo 3.5-4.5	550	485	15
Ti-7Al-4Mo R56740	USA	MIL F-83142A	Comp 9	Frg HT	6.5-7.3	0.1	0.3	0.013	0.05	0.2	1.8-2.25		0.01	0.4	Mo 3.5-4.5	550	485	15
Ti-7Al-4Mo R56740	USA	MIL T-9047G	Ti-7Al-4Mo	Bar Bil Ann	6.5-7.3	0.1	0.3	0.013	0.05	0.2	1.8-2.25		0.01	0.3	Mo 3.5-4.5	550	485	15
Ti-7Al-4Mo R56740	USA	MIL T-9047G	Ti-7Al-4Mo	Bar Bil STA	6.5-7.3	0.1	0.3	0.013	0.05	0.2	1.8-2.25		0.01		Mo 3.5-4.5	550	485	15
Ti-7Al-4Mo R56740	USA	AMS 4970E		Frg Bar Wir Bil STA	6.5-7.3	0.1	0.3	0.013	0.05	0.2	1.8-2.25		0.01	0.3	Mo 3.5-4.5	550	640	15
Ti-7Al-4Mo R66710	USA				7						1.8-2.25		0.01	0.4	Mo 4	450	380	12
Ti-662 R56620	China	GB 3620	TC-10		5.5-6.5	0.1 max	0.5	0.015 max	0.04 max	0.2 max	1.5-2.5	5.5-6.5	0.01	0.4	Si 0.15 max	450	465	18
Ti-662 R56620	Europe	AECMA prEN3316	Ti-P64	Sh Strp Ann	5-6	0.05 max	0.35-1	0.0125 max	0.04 max	0.2 max	1.5-2.5	5-6	0.01	0.3		450	465	18
Ti-662 R56620	Europe	AECMA prEN3317	Ti-P64	Plt Ann	5-6	0.05 max	0.35-1	0.0125 max	0.04 max	0.2 max	1.5-2.5	5-6	0.01	0.3		450	380	15
Ti-662 R56620	Europe	AECMA prEN3318	Ti-P64	Frg NHT	5-6	0.05 max	0.35-1	0.0125 max	0.04 max	0.2 max	1.5-2.5	5-6	0.01			450	380	18
Ti-662 R56620	Europe	AECMA prEN3319	Ti-P64	Bar Ann	5-6	0.05 max	0.35-1	0.0125 max	0.04 max	0.2 max	1.5-2.5	5-6	0.01	0.4		450	380	18
Ti-662 R56620	Europe	AECMA prEN3320	Ti-P64	Frg Ann	5-6	0.05 max	0.35-1	0.0125 max	0.04 max	0.2 max	1.5-2.5	5-6	0.01	0.4		450	380	18
Ti-662 R56620	Germany	DIN	3.7174	Sh Strp Plt Bar Frg Ann	5-6	0.05	0.35-1	0.0125 -0.015	0.04	0.2	1.5-2.5	5-6	0.01	0.6		540	440	15
Ti-662 R56620	Germany	DIN	3.7174	Sh Strp Plt Bar Frg STA	5-6	0.05	0.35-1	0.0125 -0.015	0.04	0.2	1.5-2.5	5-6	0.01					
Ti-662 R56620	Spain	UNE 38-725	L-7303	Sh Strp Plt Bar Ext HT	5-6	0.05	0.35-1	0.0125	0.04	0.2	1.5-2.5	5-6	0.01	0.3				
Ti-662 R56620	Spain	UNE 38-725	L-7303	Sh Strp Plt Bar Ext Ann	5-6	0.05	0.35-1	0.0125	0.04	0.2	1.5-2.5		0.01					
Ti-662 R56620	USA	MIL T-9047E	6Al-6V-2Sn	Bar Ann or ST	5-6	0.05	0.35-1	0.02 max	0.04	0.2 max		5-6			Cu 0.35-1; Sn 1.5-2.5; Ti 83.2	1034	965	8
Ti-662 R56620	USA	MIL T-9047E	6Al-6V-2Sn	Bar HT 1 in diam	5-6	0.05	0.35-1	0.02 max	0.04	0.2 max		5-6			Cu 0.35-1; Sn 1.5-2.5; Ti 83.2	1207	1103	6
Ti-662 R56620	USA	MIL T-9047F	6Al-6V-2Sn	Bar	5-6	0.05	0.35-1	0.02 max	0.04	0.2 max		5-6			Cu 0.35-1; Sn 1.5-2.5; Ti 83.2			
Ti-662 R56620	USA	MIL T-81556A	Code AB-3	Ext Bar Shp STA	5-6	0.05	0.35-1	0.015	0.04	0.2	1.5-2.5	5-6	0.01					
Ti-662 R56620	USA	MIL T-81556A	Code AB-3	Ext Bar Shp Ann	5-6	0.05	0.35-1	0.015	0.04	0.2	1.5-2.5	5-6	0.01					
Ti-662 R56620	USA	MIL T-9046J	Code AB-3	Sh Strp Plt STA	5-6	0.05	0.35-1	0.015	0.04	0.2	1.5-2.5	5-6	0.01					
Ti-662 R56620	USA	MIL T-9046J	Code AB-3	Sh Strp Plt ST	5-6	0.05	0.35-1	0.015	0.04	0.2	1.5-2.5	5-6	0.01					
Ti-662 R56620	USA	MIL T-9046J	Code AB-3	Sh Strp Plt Ann	5-6	0.05	0.35-1	0.015	0.04	0.2	1.5-2.5	5-6	0.01					
Ti-662 R56620	USA	MIL F-83142A	Comp 8	Frg Ann	5-6	0.05	0.35-1	0.015	0.04	0.2	1.5-2.5	5-6	0.01	0.6	OE 0.1 max			

UNS numbers and US grades are provided as a means of cross referencing chemically similar alloys. Exchangability is only possible after independent examination of specifications. Tensile properties are minimum or typical . UTS and YS as Mpa, El as %. See Appendix for list of abbreviations used in Descriptions.

Grade UNS #	Country	Spec.	Designation	Comment	Al	C	Fe	H	N	O	S	V	W	OT x	Other	UTS	YS	ELG
Ti-662 R56620	USA	MIL F-83142A	Comp 8	Frg HT	5-6	0.05	0.35-1	0.015	0.04	0.2	1.5-2.5	5-6	0.01					
Ti-662 R56620	USA	MIL T-9046H	E(6Al-6V-2Sn)	Sh Strp Plt Ann (sht,strp)	5-6	0.05	0.35-1	0.02 max	0.05	0.2 max		5-6			Cu 0.35-1; Sn 1.5-2.5; Ti 83.2	1069	1000	10
Ti-662 R56620	USA	MIL T-9046H	E(6Al-6V-2Sn)	Sh Strp Plt ST (sht,strp)	5-6	0.05	0.35-1	0.02 max	0.05	0.2 max		5-6			Cu 0.35-1; Sn 1.5-2.5; Ti 83.2		1103	10
Ti-662 R56620	USA	MIL T-9046H	E(6Al-6V-2Sn)	Sh Strp Plt Ann (plt)	5-6	0.05	0.35-1	0.02 max	0.05	0.2 max		5-6			Cu 0.35-1; Sn 1.5-2.5; Ti 83.2	1034	965	10
Ti-662 R56620	USA	MIL T-9047G	Ti-6Al-6V-2Sn	Bar Bil STA	5-6	0.05	0.35-1	0.015	0.04	0.2	1.5-2.5	5-6	0.01					
Ti-662 R56620	USA	MIL T-9047G	Ti-6Al-6V-2Sn	Bar Bil Ann	5-6	0.05	0.35-1	0.015	0.04	0.2	1.5-2.5	5-6	0.01					
Ti-662 R56620	USA	SAE J467	Ti662		5.5-5.5	0.02 max	0.7		0.02 max		2	5.5	0.01		Ni 0.01 max; Si 0.1 max			
Ti-662 R56620	USA	AMS 4918F		Sh Strp Plt Ann	5-6	0.05	0.35-1	0.015	0.04	0.2	1.5-2.5	5-6	0.01	0.6	OE 0.1 max			
Ti-662 R56620	USA	AMS 4936B		Ext Rng STA	5-6	0.05	0.35-1	0.015	0.04	0.2	1.5-2.5	5-6	0.01	0.6	OE 0.1 max			
Ti-662 R56620	USA	AMS 4936B		Ext Rng Ann	5-6	0.05	0.35-1	0.015	0.04	0.2	1.5-2.5	5-6	0.01	0.6	OE 0.1 max			
Ti-662 R56620	USA	AMS 4936C		Beta Ext Ann Rng Flsh Wld	5-6	0.05 max	0.35-1	0.015 max	0.04 max	0.2 max	1.5-2.5	5-6	0.01 max					
Ti-662 R56620	USA	AMS 4971C		Bar Frg Wir Rng Bil Ann	5-6	0.05	0.35-1	0.015	0.04	0.2	1.5-2.5	5-6	0.01	0.6	OE 0.1 max			
Ti-662 R56620	USA	AMS 4978B		Bar Wir Frg Bil Rng Ann	5-6	0.05	0.35-1	0.015	0.04	0.2	1.5-2.5	5-6	0.01	0.6	OE 0.1 max			
Ti-662 R56620	USA	AMS 4978C		Bar Frg Rng Ann	5-6	0.05 max	0.35-1	0.015 max	0.04 max	0.2 max	1.5-2.5	5-6	0.01 max			565	390	22
Ti-662 R56620	USA	AMS 4979B			5-6	0.05	0.35-1	0.015	0.04	0.2	1.5-2.5	5-6	0.01	0.6	OE 0.1 max			
Ti-662 R56620	USA				5.5-5.5						2	5.5	0.01					
Ti-6242 R54620	Germany	DIN	3.7144		5.5-6.5	0.05	0.25	0.015	0.05	0.15	1.8-2.2		0.01	0.4	Zr 3.6-4.4; Mo 1.8-2.2	620	515	
Ti-6242 R54620	Spain	UNE 38-718	L-7103	Sh Strp Plt HT	5.5-6.5	0.05	0.25	0.015	0.05	0.12	1.8-2.2		0.01	0.4	Zr 3.6-4.4; Mo 1.8-2.2	860	725	8
Ti-6242 R54620	Spain	UNE 38-718	L-7103	Sh Strp Plt Ann	5.5-6.5	0.05	0.25	0.015	0.05	0.12	1.8-2.2		0.01		Zr 3.6-4.4; Mo 1.8-2.2			
Ti-6242 R54620	USA	MIL T-009047F	6Al-2Sn-4Zr-2Mo	Bar		0.05	0.25 max	0.01 max	0.05	0.15 max					Sn 1.75-2.25; Si 0.3; Zr 3.5-4.5; Mo 1.8-2.2			
Ti-6242 R54620	USA	MIL T-9047E	6Al-2Sn-4Zr-2Mo	Bar Ann or ST	5.5-6.5	0.05	0.25 max	0.01 max	0.05	0.15 max					Sn 1.8-2.2; Zr 3.6-4.4; Ti 84.2; Mo 1.8-2.2	896	827	10
Ti-6242 R54620	USA	MIL T-9047E	6Al-2Sn-4Zr-2Mo	Bar HT 1 in diam	5.5-6.5	0.05	0.25 max	0.01 max	0.05	0.15 max					Sn 1.8-2.2; Zr 3.6-4.4; Ti 84.2; Mo 1.8-2.2	1034	951	10
Ti-6242 R54620	USA	MIL T-81915	B(6Al-2Sn-4Zr-2Mo)	Ann	5.5-6.5	0.08	0.35	0.02 max	0.05	0.12 max					Sn 1.5-2.5; Zr 3.6-4.4; Ti 79; Mo 1.5-2.5	862	793	8
Ti-6242 R54620	USA	MIL T-81556A	Code AB-4	Ext Bar Shp Ann	5.5-6.5	0.05	0.25	0.015	0.04	0.15	1.8-2.2		0.01		Si 0.06-0.1; Zr 3.6-4.4; Mo 1.8-2.2			
Ti-6242 R54620	USA	MIL T-81556A	Code AB-4	Ext Bar Shp STA	5.5-6.5	0.05	0.25	0.015	0.04	0.15	1.8-2.2		0.01		Si 0.06-0.1; Zr 3.6-4.4; Mo 1.8-2.2			
Ti-6242 R54620	USA	MIL T-9046J	Code AB-4	Sh Strp Plt TA	5.5-6.5	0.05	0.25	0.015	0.04	0.15	1.8-2.2		0.01	0.4	Zr 3.6-4.4; OE 0.1 max; Mo 1.8-2.2			
Ti-6242 R54620	USA	MIL T-9046J	Code AB-4	Sh Strp Plt DA	5.5-6.5	0.05	0.25	0.015	0.04	0.15	1.8-2.2		0.01	0.4	Zr 3.6-4.4; Mo 1.8-2.2	620	485	15
Ti-6242 R54620	USA	MIL T-9046H	G(6Al-2Sn-4Zr-2Mo)	Sh Strp Plt Ann	5.5-6.5	0.08	0.35	0.02 max	0.05	0.12 max					Sn 1.5-2.5; Zr 3.6-4.4; Ti83.5; Mo 1.5-2.5	931	862	8
Ti-6242 R54620	USA	MIL T-9047G	Ti-6Al-2Sn-4Zr-2Mo	Bar Bil STA	5.5-6.5	0.05	0.25	0.015	0.04	0.15	1.8-2.2		0.01	0.4	Zr 3.6-4.4; OE 0.1 max; Mo 1.8-2.2			

UNS numbers and US grades are provided as a means of cross referencing chemically similar alloys. Exchangability is only possible after independent examination of specifications. Tensile properties are minimum or typical. UTS and YS as Mpa, El as %. See Appendix for list of abbreviations used in Descriptions.

Grade UNS #	Country	Spec.	Designation	Comment	Al	C	Fe	H	N	O	S	V	W	OT x	Other	UTS	YS	ELG
Ti-6242 R54620	USA	MIL T-9047G	Ti-6Al-2Sn-4Zr-2Mo	Bar Bil DA	5.5-6.5	0.05	0.25	0.015	0.04	0.15	1.8-2.2			0.01	0.4 Zr 3.6-4.4; Mo 1.8-2.2	860	725	10
Ti-6242 R54620	USA	MIL T-81915	Type III Comp B	Cast Ann	5.5-6.5	0.08	0.35	0.015	0.05	0.12	1.5-2.5			0.01	0.4 Zr 3.6-4.4; Mo 1.5-2.5	450	465	18
Ti-6242 R54620	USA	AMS 4919C		Sh Strp Plt	5.5-6.5	0.05 max	0.25 max	0.015 max	0.05 max	0.12 max	1.8-2.2			0.01 max	0.4 Si 0.06-0.1; Zr 3.6-4.4; Mo 1.8-2.2	620	515	15
Ti-6242 R54620	USA	AMS 4919G		Sh Strp Plt DA	5.5-6.5	0.05	0.25	0.015	0.05	0.12	1.8-2.2			0.01	0.4 Si 0.1; Zr 3.6-4.4; Mo 1.8-2.2	620	485	15
Ti-6242 R54620	USA	AMS 4975E		Bar Wir Rng Bil STA	5.5-6.5	0.05	0.25	0.0125	0.05	0.15	1.8-2.2			0.01	Si 0.1; Zr 3.6-4.4; Mo 1.8-2.2			
Ti-6242 R54620	USA	AMS 4975F		Bar Rng SHT PHT	5.5-6.5	0.05 max	0.1 max	0.0125 max	0.05 max	0.15 max	1.8-2.2			0.01 max	Si.06-.1; Zr3.6-4.4; Mo 1.8-2.2			
Ti-6242 R54620	USA	AMS 4976C		Frg STA	5.5-6.5	0.05	0.25	0.0125	0.05	0.15	1.8-2.2			0.01	0.4 Si 0.1; Zr 3.6-4.4; Mo 1.8-2.2	620	485	15
Ti-6242 R54620	USA	AMS 4976D		Frg SHT PHT	5.5-6.5	0.05 max	0.1 max	0.0125 max	0.05 max	0.15 max	1.8-2.2			0.01 max	0.4 Si.06-.1; Zr3.6-4.4; Mo 1.8-2.2	620	515	15
Ti-6242 R54620	USA			Weld Fill Met	5.5-6.5	0.04	0.05	0.015	0.15	0.3	1.8-2.2			0.01	Cr 0.25; Zr 3.6-4.4; Mo 1.8-2.2			
Ti-6242 R54620	USA				6						2			0.01	Zr 4; Mo 2			
	UK	BS TA.45	BS TA.45	Bar HT	3-5		0.2 max	0.01 max	0.05 max	0.25 max					Sn 1.5-2.5; Si 0.3-0.7; Ti 86.3 max; Mo 3-5	1100	960	8
	UK	BS TA.46	BS TA.46	Bar HT	3-5		0.2 max	0.01 max	0.05 max	0.25 max					Sn 1.5-2.5; Si 0.3-0.7; Ti 86.3 max; Mo 3-5	1050	920	9
	UK	BS TA.47	BS TA.47	Frg HT 25-100 mm diam	3-5		0.2 max	0.01 max	0.05 max	0.25 max					Sn 1.5-2.5; Si 0.0-0.7; Ti 86.3 max; Mo 3-5	1050	920	9
	UK	BS TA.48	BS TA.48	Frg HT	3-5		0.2 max	0.02 max	0.05 max	0.25 max					Sn 1.5-2.5; Si 0.3-0.7; Ti 86.3 max; Mo 3-5	1050	920	9
	UK	BS TA.49	BS TA.49	Bar Ht	3-5		0.2 max	0.01 max	0.05 max	0.25 max					Sn 1.5-2.5; Si 0.3-0.7; Ti 86.3 max; Mo 3-5	1000	870	9
	UK	BS TA.50	BS TA.50	Frg HT 100-150 mm diam	3-5		0.2 max	0.01 max	0.05 max	0.25 max					Sn 1.5-2.5; Si 0.3-0.7; Ti 86.3 max; Mo 3-5	1000	870	9
	UK	BS TA.51	BS TA.51	Frg HT	3-5		0.2 max	0.02 max		0.25 max					Sn 1.5-2.5; Si 0.3-0.7; Ti 86.3 max; Mo 3-5	1000	870	9
	UK	BS TA.57	BS TA.57	Plt HT 5-10 mm diam	3-5		0.2 max	0.01 max	0.05 max	0.25 max					Sn 1.5-2.5; Si 0.3-0.7; Ti 86.3 max; Mo 3-5	1030	900	9
	UK	BS TA.43	BS TA.43	Frg HT and Q	5.7-6.3		0.2 max	0.01 max	0.05 max	0.25 max					Si 0.1-0.4; Zr 4-6; Ti 86 max; Mo 0.25-0.75	990	850	8
	UK	BS TA.44	BS TA.44	Frg HT	5.7-6.3		0.2 max	0.01 max	0.05 max	0.25 max					Si 0.1-0.4; Zr 4-6; Ti 86 max; Mo 0.25-0.75	990	850	8
	USA	ASTM B337	Ti Grade 10	Smls Weld Pip ST		0.10 max	0.35 max	0.020 max	0.05 max	0.18 max				0.4	Zr 4.50-7.50; OE 0.1 max; Mo 10.0-13.0	690	620	10
Ti-17 R58650	USA	AMS 4995		Bil STA	4.5-5.5	0.05	0.3	0.0125	0.04	0.08-0.13	1.5-2.5			0.01	Cr 3.5-4.5; Zr 1.5-2.5; Mo 3.5-4.5	540	440	16
Ti-17 R58650	USA	AMS 4997		Powd	4.5-5.5	0.05	0.3	0.0125	0.04	0.08-0.12	1.5-2.5			0.01	Cr 3.5-4.5; Zr 1.5-2.5; Mo 3.5-4.5	660	470	18
Ti-17 R58650					4.5-5.5	0.05 max	0.3 max	0.0125 max	0.04 max	0.08-0.13	1.5-2.5			0.01 max	Cr 3.5-4.5; Zr 1.5-2.5; Mo 3.5-4.5	470	460	16
	UK	BS TA.39	BS TA.39	Frg HT	3-5	0.1-0.2	0.2 max	0.01 max	0.05 max	0.25 max					Sn 3-5; Si 0.3-0.7; Ti 83.8 max; Mo 3-5	1250	1095	8
	UK	BS TA.39	BS TA.39	Bar HT	3-5	0.1-0.2	0.2 max	0.01 max	0.05 max	0.25 max					Sn 3-5; Si 0.3-0.7; Ti 83.8 max; Mo 3-5	1250	1095	8

UNS numbers and US grades are provided as a means of cross referencing chemically similar alloys. Exchangability is only possible after independent examination of specifications. Tensile properties are minimum or typical . UTS and YS as Mpa, El as %. See Appendix for list of abbreviations used in Descriptions.

Worldwide Guide to Equivalent Nonferrous Metals and Alloys

Grade UNS #	Country	Spec.	Designation	Comment	Al	C	Fe	H	N	O	S	V	W	OT x	Other	UTS	YS	ELG
	UK	BS TA.40	BS TA.40	Bar	3-5	0.1-0.2	0.2 max	0.01 max	0.05 max	0.25 max					Sn 3-5; Si 0.3-0.7; Ti 83.6 max; Mo 3-5			
	UK	BS TA.41	BS TA.41	Frg HT	3-5	0.1-0.2	0.2 max	0.01 max	0.05 max	0.25 max					Sn 3-5; Si 0.3-0.7; Ti 83.6 max; Mo 3-5	1205	1065	8
	UK	BS TA.42	BS TA.42	Frg HT	3-5	0.1-0.2	0.2 max	0.01 max	0.05 max	0.25 max					Sn 3-5; Si 0.3-0.7; Ti 83.6 max; Mo 3-5	1205	1065	8
8Mo-8V-2Fe-3Al R58820	USA	MIL T-9046H	D(8Mo-8V-2Fe-3Al)	SH Strp Plt ST (sht,strp)		0.05	1.6-2.4	0.02 max	0.05	0.02 max		7.5-8.5			Mo 7.5-8.5	862	827	8
8Mo-8V-2Fe-3Al R58820	USA	MIL T-9046H	D(8Mo-8V-2Fe-3Al)	SH Strp Plt Ann (plt)		0.05	1.6-2.4	0.02 max	0.05	0.02 max		7.5-8.5			Mo 7.5-8.5	862	827	10
	USA	AMS 4984	Ti-10-2-3	Frg STA	2.6-3.4	0.05	1.6-2.2	0.015	0.05	0.13	3.8-5.25	9-11	0.01	0.4		1035	965	7
	USA	AMS 4987	Ti-10-2-3	Frg STOA	2.6-3.4	0.05	1.6-2.2	0.015	0.05	0.13	3.8-5.25	9-11	0.01	0.4		1050	965	6
	USA	AMS 4983A	Ti-10-2-3	Frg STA	2.6-3.4	0.05	1.6-2.2	0.015	0.05	0.13	3.8-5.25	9-11	0.01	0.4		910	850	9
	USA	AMS	Ti-10-2-3	Frg STOA	2.6-3.4	0.05	1.6-2.2	0.015	0.05	0.13	3.8-5.25	9-11	0.01	0.4		895	915	10
Ti-6246 R56260	USA	MIL T-9047F	6Al-2Sn-4Zr-6Mo	Bar	5.5-6	0.04 max	0.15 max	0.01 max	0.04	0.15 max					Sn 1.75-2.25; Zr 3.5-4.5; Ti 80.4; Mo 5.5-6.5			
Ti-6246 R56260	USA	MIL F-83142A	Comp 11	Frg Ann	5.5-6.5	0.04	0.15	0.0125	0.04	0.15	1.8-2.25		0.01	0.3	Zr 3.6-4.4; Mo 5.5-6.5	550	550	15
Ti-6246 R56260	USA	MIL F-83142A	Comp 11	Frg HT	5.5-6.5	0.04	0.15	0.0125	0.04	0.15	1.8-2.25		0.01	0.4	Zr 3.6-4.4; Mo 5.5-6.5	550	485	15
Ti-6246 R56260	USA	MIL T-9047G	Ti-6Al-2Sn-4Zr-6Mo	Bar Bil DA	5.5-6.5	0.04	0.15	0.0125	0.04	0.15	1.8-2.25		0.01	0.3	Zr 3.6-4.4; Mo 5.5-6.5	550	485	15
Ti-6246 R56260	USA	MIL T-9047G	Ti-6Al-2Sn-4Zr-6Mo	Bar Bil STA	5.5-6.5	0.04	0.15	0.0125	0.04	0.15	1.8-2.25		0.01	0.4	Zr 3.6-4.4; Mo 5.5-6.5	550	570	15
Ti-6246 R56260	USA	AMS 4981B		Bar Wir Frg Bil STA	5.5-6.5	0.04	0.15	0.0125	0.04	0.15	1.8-2.25		0.01	0.4	Zr 3.5-4.5; Mo 5.5-6.5	345	275	15
Ti-6246 R56260	USA				6						2		0.01		Zr 4; Mo 6			
	USA	ASTM B381	F-21	Frg ST	2.5-3.5	0.05 max	0.40 max	0.015 max	0.05 max	0.17 max				0.4	Nb 2.2-3.2; Si 0.15-0.25; OE 0.1 max; Mo 14.0-16.0	793	759	15
	USA	ASTM B265	Ti Grade 21	Sh Strp Plt Ann SHT	2.5-3.5	0.05 max	0.40 max	0.015 max	0.05 max	0.17 max				0.4	Nb 2.2-3.2; Si 0.15-0.25; OE 0.1 max; Mo 14.0-16.0	793	759	15
	USA	ASTM B348	Ti Grade 21	Bar Bil ST	2.5-3.5	0.05 max	0.40 max	0.015 max	0.05 max	0.17 max				0.4	Nb 2.2-3.2; Si 0.15-0.25; OE 0.1 max; Mo 14.0-16.0	793	759	15
	USA	ASTM B861	Ti Grade 21	Smls Pip Ann	2.5-3.5	0.05 max	0.40 max	0.015 max	0.05 max	0.17 max				0.4	Nb 2.2-3.2; Si 0.15-0.25; OE 0.1 max; Mo 14.0-16.0	793	759	15
	USA	ASTM B863	Ti Grade 21	Wir ST	2.5-3.5	0.05 max	0.40 max	0.015 max	0.05 max	0.17 max				0.4	Nb 2.2-3.2; Si 0.15-0.25; OE 0.1 max; Mo 14.0-16.0	793	759	10
	USA	ASTM B363	WPT21	Smls Weld Fitngs Comp. as in ASTM B337, B265, B348, B381														
Beta III R58030	Spain	UNE 38-730	L-7702	Sh Str Bar Frg Tub HT		0.1	0.35	0.02	0.05	0.18	3.8-5.25		0.01	0.4	Zr 4.5-7.5; Mo 10-13	895	825	6
Beta III R58030	USA	MIL T-9047E	11.5Mo-6.0Zr-4.5Sn	Bar Ann ST 1.675 in diam		0.1	0.35 max	0.02 max	0.05	0.18 max					Sn 3.75-5.25; Zr 4.5-7.5; Ti 74.6; Mo 10-12	689	671	15

UNS numbers and US grades are provided as a means of cross referencing chemically similar alloys. Exchangability is only possible after independent examination of specifications. Tensile properties are minimum or typical . UTS and YS as Mpa, El as %. See Appendix for list of abbreviations used in Descriptions.

10-26 Titanium

Grade UNS #	Country	Spec.	Designation	Comment	Al	C	Fe	H	N	O	S	V	W	OT x	Other	UTS	YS	ELG
Beta III R58030	USA	MIL T-9047E	11.5Mo-6.0Zr-4.5Sn	Bar HT 1.58 in diam		0.1	0.35 max	0.02 max	0.05	0.18 max					Sn 3.75-5.25; Zr 4.5-7.5; Ti 74.6; Mo 10-12	1241	1207	8
Beta III R58030	USA	MIL T-9047E	11.5Mo-6.0Zr-4.5Sn	Bar HT 2 in diam		0.1	0.35 max	0.02 max	0.05	0.18 max					Sn 3.75-5.25; Zr 4.5-7.5; Ti 74.6; Mo 10-12	1172	1103	4
Beta III R58030	USA	MIL T-9046H	B(11.5Mo-6Zr-4.5Sn)	Sh Strp Plt ST (Sht,Strp)		0.1	0.35 max	0.02 max	0.05	0.18 max					Sn 3.75-5.25; Zr 4.5-7.5; Ti 73.5; Mo 10-13	689	621	12
Beta III R58030	USA	MIL T-9046J	Code B-2	ST		0.1	0.35	0.02	0.05	0.18	3.8-5.25		0.01	0.4	Zr 4.5-7.5; Mo 10-13	910	850	
Beta III R58030	USA	MIL F-83142A	Comp 13	Frg Ann		0.1	0.35	0.02	0.05	0.18	3.8-5.25		0.01	0.3	Zr 4.5-7.5; Mo 10-13	880	810	10
Beta III R58030	USA	MIL F-83142A	Comp 13	Frg HT		0.1	0.35	0.02	0.05	0.18	3.8-5.25		0.01	0.4	Zr 4.5-7.5; Mo 10-13	1015	950	9
Beta III R58030	USA	MIL T-9047G	Ti-4.5Sn-6Zr-11.5Mo	Bar Bil SHT		0.1	0.35	0.02	0.05	0.18	3.8-5.25		0.01	0.4	Zr 4.5-7.5; Mo 10-13	1015	950	10
Beta III R58030	USA	AMS 4980B		Bar Wir STA		0.1	0.35	0.015	0.05	0.18	3.8-5.25		0.01	0.1	Zr 4.5-7.5; Mo 10-13			
Beta III R58030	USA	AMS 4980B		Bar Wir SHT		0.1	0.35	0.015	0.05	0.18	3.8-5.25		0.01	0.4	Zr 4.5-7.5; Mo 10-13	845	775	9
Beta III R58030	USA					0.1	0.35	0.02	0.05	0.18	3.8-5.25		0.01	0.4	Zr 4.5-7.5; OE 0.1 max; Mo 10-13			
	USA	ASTM B381	F-19	Frg ST	3.0-4.0	0.05 max	0.30 max	0.0200 max	0.03 max	0.12 max		7.5-8.5		0.4	Cr 5.5-6.5; Zr 3.5-4.5; OE 0.15 max; Mo 3.5-4.5	793	759	15
	USA	ASTM B381	F-20	Frg ST	3.0-4.0	0.05 max	0.30 max	0.0200 max	0.03 max	0.12 max		7.5-8.5		0.4	Pd 0.04-0.08, Cr 5.5-6.5; Zr 3.5-4.5; OE 0.15 max; Mo 3.5-4.5	793	759	15
H54790	UK	BS TA.18	BS TA.18	Bar HT and Q Obsolete	2-2.5		0.2 max	0.01 max							Si 0.1-0.5; Zr 4-6; Ti 78.1 max; Mo 0.8-1.2	992	866	8
R54790	UK	BS TA.19	BS TA.19	Frg HT and Q Obsolete	2-2.5		0.2 max	0.01 max							Si 0.1-0.5; Zr 4-6; Ti 78.1 max; Mo 0.8-1.2	992	866	8
R54790	UK	BS TA.20	BS TA.20	Frg HT and Q Obsolete	2-2.5		0.2 max	0.02 max							Si 0.1-0.5; Zr 4-6; Ti 78.1 max; Mo 0.8-1.2	992	866	8
R54790	UK	BS TA.25	BS TA.25	Bar HT Obsolete	2-2.5		0.2 max	0.01 max							Si 0.1-0.5; Zr 4-6; Ti 78.1 max; Mo 0.8-1.2	920	786	8
R54790	UK	BS TA.26	BS TA.26	Frg Obsolete	2-2.5		0.2 max	0.01 max							Si 0.1-0.5; Zr 4-6; Ti 78.1 max; Mo 0.8-1.2			
R54790	UK	BS TA.27	BS TA.27	Frg HT Obsolete	2-2.5		0.2 max	0.02 max							Si 0.1-0.5; Zr 4-6; Ti 78.1 max; Mo 0.8-1.2	920	786	8
R54790	USA	MIL T-9047E	11Sn-5Zr-2Al-1Mo	Bar Ann or ST	2-2.5	0.04 max	0.12 max	0.01 max	0.05	0.15 max					Sn 10.5-11.5; Si 0.2; Zr 4-6; Ti 78.2; Mo 0.8-1.2	965	896	10
R54790	USA	MIL T-9047E	11Sn-5Zr-2Al-1Mo	Bar HT 1 in diam	2-2.5	0.04 max	0.12 max	0.01 max	0.05	0.15 max					Sn 10.5-11.5; Si 0.2; Zr 4-6; Ti 78.2; Mo 0.8-1.2	1000	931	12
R54790	USA	MIL T-9047F	11Sn-5Zr-2Al-1Mo	Bar	2-2.5	0.04 max	0.12 max	0.01 max	0.05	0.15 max					Sn 10.5-11.5; Si 0.2-0.3; Zr 4-6; Ti 78.2; Mo 0.8-1.2			

UNS numbers and US grades are provided as a means of cross referencing chemically similar alloys. Exchangability is only possible after independent examination of specifications. Tensile properties are minimum or typical . UTS and YS as Mpa, El as %. See Appendix for list of abbreviations used in Descriptions.

Worldwide Guide to Equivalent Nonferrous Metals and Alloys

Grade UNS #	Country	Spec.	Designation	Comment	Al	C	Fe	H	N	O	S	V	W	OT x	Other	UTS	YS	ELG
	USA	AMS 4914	Ti-15-3	Sh Strp SHT	2.5-3.5	0.05	0.25	0.015	0.05	0.13	2.5-3.5	14-16	0.01	0.3	Cr 2.5-3.5; OE 0.1 max			
	USA	AMS 4914	Ti-15-3	Sh Strp STA	2.5-3.5	0.05	0.25	0.015	0.05	0.13	2.5-3.5	14-16	0.01	0.3	Cr 2.5-3.5; OE 0.1 max			
	USA	ASTM B265	Ti Grade 19	Sh Strp Plt Ann SHT	3.0-4.0	0.05 max	0.30 max	0.0200 max	0.03 max	0.12 max		7.5-8.5		0.4	Cr 5.5-6.5; Zr 3.5-4.5; OE 0.15 max; Mo 3.5-4.5	793	759	15
	USA	ASTM B348	Ti Grade 19	Bar Bil ST	3.0-4.0	0.05 max	0.30 max	0.0200 max	0.03 max	0.12 max		7.5-8.5		0.4	Cr 5.5-6.5; Zr 3.5-4.5; OE 0.15 max; Mo 3.5-4.5	793	759	15
	USA	ASTM B861	Ti Grade 19	Smls Pip Ann	3.0-4.0	0.05 max	0.30 max	0.0200 max	0.03 max	0.12 max		7.5-8.5		0.4	Cr 5.5-6.5; Zr 3.5-4.5; OE 0.15 max; Mo 3.5-4.5	793	759	15
	USA	ASTM B863	Ti Grade 19	Wir ST	3.0-4.0	0.05 max	0.30 max	0.0200 max	0.03 max	0.12 max		7.5-8.5		0.4	Cr 5.5-6.5; Zr 3.5-4.5; OE 0.15 max; Mo 3.5-4.5	793	759	10
	USA	ASTM B348	Ti Grade 20	Bar Bil ST	3.0-4.0	0.05 max	0.30 max	0.0200 max	0.03 max	0.12 max		7.5-8.5		0.4	Pd 0.04-0.08; Cr 5.5-6.5; Zr 3.5-4.5; OE 0.15 max; Mo 3.5-4.5	793	759	15
	USA	ASTM B861	Ti Grade 20	Smls Pip Ann	3.0-4.0	0.05 max	0.30 max	0.0200 max	0.03 max	0.12 max		7.5-8.5		0.4	Pd 0.04-0.08; Cr 5.5-6.5; Zr 3.5-4.5; OE 0.15 max; Mo 3.5-4.5	793	759	15
	USA	ASTM B863	Ti Grade 20	Wir ST	3.0-4.0	0.05 max	0.30 max	0.0200 max	0.03 max	0.12 max		7.5-8.5		0.4	Pd 0.04-0.08; Cr 5.5-6.5; Zr 3.5-4.5; OE 0.15 max; Mo 3.5-4.5	793	759	10
	USA	ASTM B363	WPT19	Smls Weld Fittings (Composition as in ASTM B337, B265, B348, B381)														
	USA	ASTM B363	WPT20	Smls Weld Fittings (Composition as in ASTM B337, B265, B348, B381)														
	USA	ASTM B265	Ti Grade 20	Sh Strp Plt Ann SHT	3.0-4.0	0.05 max	0.30 max	0.0200 max	0.03 max	0.12 max		7.5-8.5		0.4	Pd 0.04-0.08; Cr 5.5-6.5; Zr 3.5-4.5; OE 0.15 max; Mo 3.5-4.5	793	759	15
Beta C R58640	USA	MIL T-9046H	C(3Al8V6Cr4MoZr)	Sh Strp Plt ST (Sht,Strp)	3-4	0.05	0.3 max	0.02 max	0.03	0.12 max		7.5-8.5			Cr 5.5-6.5; Zr 3.5-4.5; Ti 71.5; Mo 3.5-4.5	862	827	8
Beta C R58640	USA	MIL T-9046H	C(3Al8V6Cr4MoZr)	Sh Strp Plt Ann (plt)	3-4	0.05	0.3 max	0.02 max	0.03	0.12 max		7.5-8.5			Cr 5.5-6.5; Zr 3.5-4.5; Ti 71.5; Mo 3.5-4.5	862	827	10
Beta C R58640	USA	MIL T-9046J	Code B-3	Sh Strp Plt SHT	3-4	0.05	0.3	0.02	0.03	0.12	3.8-5.25	7.5-8.5	0.01		Cr 5.5-6.5; Zr 3.5-4.5; Mo 3.5-4.5	890	820	8
Beta C R58640	USA	MIL T-9046J	Code B-3	Sh Strp Plt STA	3-4	0.05	0.3	0.02	0.03	0.12	3.8-5.25	7.5-8.5	0.01		Cr 5.5-6.5; Zr 3.5-4.5; Mo 3.5-4.5			
Beta C R58640	USA	MIL T-9047G	Ti-3Al-8V-6Cf-4Mo-4Zr	Bar Bil STA	3-4	0.05	0.3	0.02	0.03	0.12	3.8-5.25	7.5-8.5	0.01		Cr 5.5-6.5; Zr 3.5-4.5; Mo 3.5-4.5			

UNS numbers and US grades are provided as a means of cross referencing chemically similar alloys. Exchangability is only possible after independent examination of specifications. Tensile properties are minimum or typical. UTS and YS as Mpa, El as %. See Appendix for list of abbreviations used in Descriptions.

Grade UNS #	Country	Spec.	Designation	Comment	Al	C	Fe	H	N	O	S	V	W	OT x	Other	UTS	YS	ELG
Beta C R58640	USA	MIL T-9047G	Ti-3Al-8V-6Cr-4Mo-4Zr	Bar Bil SHT	3-4	0.05	0.3	0.02	0.03	0.12	3.8-5.25	7.5-8.5	0.01		Cr 5.5-6.5; Zr 3.5-4.5; Mo 3.5-4.5	890	820	8
Beta C R58640	USA	AMS 4957		Bar Wir CD	3-4	0.05	0.3	0.03	0.03	0.12	3.8-5.25	7.5-8.5	0.01		Cr 5.5-6.5; Zr 3.5-4.5; Mo 3.5-4.5			
Beta C R58640	USA	AMS 4958		Bar Rod STA	3-4	0.05	0.3	0.03	0.03	0.12	3.8-5.25	7.5-8.5	0.01	0.4	Cr 5.5-6.5; Zr 3.5-4.5; Mo 3.5-4.5	910	850	8
Beta C R58640	USA				3						3.8-5.25	8	0.01	0.4	Cr 6; Zr 4; Mo 4	880	815	5
Ti-13-11-3 R58010	Russia	GOST	IMP-10		3						3.8-5.25	13	0.01		Cr 11			
Ti-13-11-3 R58010	Spain	UNE 38-729	L-7701	Sh Strp Plt Wir Bar HT	2.5-3.5	0.05	0.35	0.02	0.05	0.18	3.8-5.25	13-15	0.01	0.4	Cr 10-12	895	825	10
Ti-13-11-3 R58010	Spain	UNE 38-729	L-7701	Sh Str Plt Wir Bar Ann	2.5-3.5	0.05	0.35	0.02	0.05	0.18	3.8-5.25	13-15	0.01		Cr 10-12			
Ti-13-11-3 R58010	USA	MIL T-9047E	13V-11Cr-3Al	Bar Ann or ST	2.5-3.5	0.05	0.3	0.02 max		0.2 max		13-15			Sn 3.75-5.25; Cr 10-12; Ti 69.4	862	827	10
Ti-13-11-3 R58010	USA	MIL T-9047E	13V-11Cr-3Al	Bar Ann (plt)	2.5-3.5	0.05	0.3	0.02 max		0.2 max		13-15			Sn 3.75-5.25; Cr 10-12; Ti 69.4	862	827	10
Ti-13-11-3 R58010	USA	MIL T-9046H	A(13V-11Cr-3Al)	Sh Strp Plt ST (Sht,Strp)	2.5-3.5	0.05	0.15-0.3	0.03 max	0.05	0.2 max		13-15			Cr 10-12; Ti 69.4	862	827	10
Ti-13-11-3 R58010	USA	MIL T-9046J	Code B-1	Sh Str Plt SHT	2.5-3.5	0.05	0.15-0.35	0.025	0.05	0.17	3.8-5.25	13-15	0.01	0.3	Cr 10-12	930	860	9
Ti 13 11 3 R58010	USA	MIL T 9046J	Code B-1	Sh Str Plt STA	2.5-3.5	0.05	0.15-0.35	0.025	0.05	0.17	3.8-5.25	13-15	0.01		Cr 10-12			
Ti-13-11-3 R58010	USA	MIL F-83142A	Comp 12	Frg HT	2.5-3.5	0.05	0.35	0.025	0.05	0.17	3.8-5.25	13-15	0.01	0.4	Cr 10-12	1000	930	8
Ti-13-11-3 R58010	USA	MIL F-83142A	Comp 12	Frg Ann	2.5-3.5	0.05	0.35	0.025	0.05	0.17	3.8-5.25	13-15	0.01		Cr 10-12	880	810	8
Ti-13-11-3 R58010	USA	AWS A5.16-70	ERTi-13V-11Cr-3Al	Wir Rod	2.5-3.5	0.05	0.25	0.008	0.03	0.12	3.8-5.25	13-15	0.01	0.4	Cr 10-12	895	825	10
Ti-13-11-3 R58010	USA	MIL T-9047G	Ti-13V-11Cr-3Al	Bar Bil STA	2.5-3.5	0.05	0.35	0.025	0.05	0.17	3.8-5.25	13-15	0.01	0.3	Cr 10-12	915	845	9
Ti-13-11-3 R58010	USA	MIL T-9047G	Ti-13V-11Cr-3Al	Bar Bil SHT	2.5-3.5	0.05	0.35	0.025	0.05	0.17	3.8-5.25	13-15	0.01	0.4	Cr 10-12	895	825	10
Ti-13-11-3 R58010	USA	AMS 4917D		Sh STr Plt SHT	2.5-3.5	0.05	0.35	0.025	0.05	0.17	3.8-5.25	13-15	0.01	0.4	Cr 10-12	895	825	10
Ti-13-11-3 R58010	USA	AMS 4917D		Sh Str Plt STA	2.5-3.5	0.05	0.35	0.025	0.05	0.17	3.8-5.25	13-15	0.01	0.4	Cr 10-12	1005	940	8
Ti-13-11-3 R58010	USA	AMS 4959B			2.5-3.5	0.05	0.35	0.03	0.05	0.17	3.8-5.25	13-15	0.01	0.4	Cr 10-12	895	825	6
Ti-13-11-3 R58010	USA				3						3.8-5.25	13	0.01	0.3	Cr 11	845	880	9

UNS numbers and US grades are provided as a means of cross referencing chemically similar alloys. Exchangability is only possible after independent examination of specifications. Tensile properties are minimum or typical . UTS and YS as Mpa, El as %. See Appendix for list of abbreviations used in Descriptions.

Grade UNS #	Country	Specification	Designation	Description	Al	Cd	Cu	Fe	Mg	Ni	Pb	Sn	Ti	Zn	Other	UTS	YS	EL
Z12001	France	NF A55-101	Z9	Ingot	0.01 max	0.0003 max	0.001 max	0.002 max				0.001 max		99.995 min	ST=Pb 0.0003			
Z12001	Germany	DIN 1706	2.2040/Zn99.99			0.003 max	0.001 max	0.002 max			0.003 max	0.001 max		99.995 min	0.004maxST,St=Pb+Cd			
Z12001	Germany	DIN 1706	2.2045/Zn99.995			0.003 max	0.001 max	0.002 max			0.003 max	0.001 max		99.995 min	0.004maxST,St=Pb+Cd			
Z12001	Europe	CEN EN 1179	Zn 99.995	Primary Ingot	0.001 max	0.003 max	0.001 max	0.002 max			0.003 max	0.001 max		99.995	Pb+Cd+Fe+Sn+Cu+Al 0.005 max			
Z12001	International	ISO 752	Zn99.995	Ingot	0.01 max	0.003 max	0.001 max	0.002 max			0.003 max	0.001 max		99.995 min				
Z12001	Pan America	COPANT 442	Zn99.995			0.003 max	0.001 max	0.002 max			0.003	0.001		99.995 min	0.04 Pb+Cd			
Z12001	South Africa	SABS 20	Zn99.995			0.002	0.001	0.001			0.002	0.001	0.001	99.995 min	In 0.005			
Z13001	Australia	AS 1242	Zn99.99	Ingot		0.003 max	0.002 max	0.003 max			0.003 max	0.001 max	0.001 max		In 0.0005 max			
Z13001	Canada	CSA HZ.2	9999	Slab											0.01max ST,ST=0.003Pb,0.005Fe,0.03 Cd max impurities			
Z13001	Europe	CEN EN 1179	Zn 99.99	Primary Ingot		0.005 max	0.002 max	0.003 max			0.005 max	0.001 max		99.99	Pb+Cd+Fe+Sn+Cu 0.01 max			
Z13001	International	ISO 752	Zn99.99	Ingot	0.01 max	0.003 max	0.002 max	0.003 max			0.003 max	0.001 max		99.99 min				
Z13001	Pan America	COPANT 442	Zn99.99			0.003 max	0.002	0.003			0.003	0.001		99.99 min	0.006 Pb+Cd			
Z13001	South Africa	SABS 20	Zn99.99		0.005	0.003	0.002	0.002			0.03	0.001	0.001	99.99 min	In 0.001			
Z13001	UK	BS 3436	Zn1	Ingot Obsolete Replaced by BS EN 1179		0.003 max		0.002 max			0.003 max	0.001 max	0.001 max	99.99 min	In 0.001 max			
Z13001	USA	ASTM B6	Special High Grade	Ingot	0.002 max	0.003 max	0.002 max	0.003 max			0.003 max	0.001 max		99.990 min	OT 0.010 max			
Z14001	Australia	AS 1242	Zn99.95	Ingot		0.02 max	0.002 max	0.01 max			0.03 max	0.001 max						
Z14001	France	NF A55-101	Z8	Ingot	0.01 max	0.02 max	0.002 max	0.02 max			0.03 max	0.001 max		99.95 min				
Z14001	Germany	DIN 1706	2.2035/Zn99.95			0.02 max	0.002 max	0.02 max			0.03 max	0.001 max		99.95 min	0.03max ST,St=Pb+Cd			
Z14001	Europe	CEN EN 1179	Zn 99.95	Primary Ingot	0.01 max	0.002 max	0.02 max				0.03 max	0.001 max		99.95	Pb+Cd+Fe+Sn+Cu 0.05 max			
Z14001	International	ISO 752	Zn99.95	Ingot	0.01 max	0.02 max	0.002 max	0.02 max			0.03 max	0.001 max		99.95 min				
Z14001	Pan America	COPANT 442	Zn99.95			0.02	0.002	0.02			0.03	0.001		99.95	OT 0.05 max			
Z14001	South Africa	SABS 20	Zn99.95		0.005	0.02	0.002	0.001			0.03	0.001	0.001	99.95	In 0.001			
Z14001	UK	BS 3436	Zn2	Ingot Obsolete Replaced by BS EN 1180		0.02 max		0.01 max			0.03 max	0.001 max		99.95 min	As 0.002 max			
Z15001	Canada	CSA HZ.2	9990	Slab		0.02		0.02							0.10maxST,ST=0.05Pb,			
Z15001	USA	ASTM B6	High Grade	Ingot	0.01 max	0.02 max		0.02 max			0.03 max			99.90 min	OT 0.10 max			
Z16001	France	NF A55-101	Z7	Ingot		0.15 max		0.05 max			0.45 max			99.5 min				
Z16001	Germany	DIN 1706	2.2095/Zn99.95			0.15 max		0.05 max			0.45 max			99.5 min				
Z16001	Europe	CEN EN 1179	Zn 9.5	Primary Ingot		0.01 max		0.05 max			0.45 max			99.5	Pb+Cd+Fe 0.5 max			
Z16001	International	ISO 752	Zn99.5	Ingot	0.01 max	0.15 max		0.05 max			0.45 max			99.5 min				
Z16001	Pan America	COPANT 442	Zn99.5			0.15		0.03			0.45	0.005		99.5	OT 0.5 max			
Z16001	South Africa	SABS 20	Zn99.5		0.05-0.005	0.05		0.03			0.35-0.45	0.001	0.001	99.5	In 0.001			

UNS numbers and US grades are provided as a means of cross referencing chemically similar alloys. Exchangability is only possible after independent examination of specifications. Tensile properties are minimum or typical. UTS and YS as Mpa, El as %. See Appendix for list of abbreviations used in Descriptions.

Worldwide Guide to Equivalent Nonferrous Metals and Alloys

Grade UNS #	Country	Specification	Designation	Description	Al	Cd	Cu	Fe	Mg	Ni	Pb	Sn	Ti	Zn	Other	UTS	YS	EL
Z16001	UK	BS 3436	Zn3	Obsolete Replaced by BS EN 1181		0.15 max		0.03 max			0.35 max	0.001 max		99.5 min				
Z17001	Canada	CSA HZ.2	9925	Slab		0.1		0.03							0.75max ST,ST=0.60Pb max impurities			
Z18001	Australia	AS 1242	Zn98.5	Ingot		0.2 max		0.05 max			1.40 max							
Z18001	Germany	DIN 1706	2.2085/Zn98.5			0.2 max		0.05 max			1.40 max			98.5 min				
Z18001	Europe	CEN EN 1179	Zn 98.5	Primary Ingot		0.01 max		0.05 max			1.4 max			98.5	Pb+Cd+Fe 1.5 max			
Z18001	International	ISO 752	Zn98.5	Ingot	0.02 max	0.2 max		0.05 max			1.40 max			98.5 min				
Z18001	Pan America	COPANT 442	Zn98.5			0.7		0.05			1.40			98.5	OT 1.50 max			
Z18001	South Africa	SABS 20	Zn98.5		0.005	0.15		0.04			0.95-1.35	0.02		98.5				
Z18001	UK	BS 3436	Zn4	Obsolete Replaced by BS EN 1182	0.01 max	0.15 max		0.04 max			1.35 max			98.5 min				
Z19001	Canada	CSA HZ.2	980	Slab				0.08							2.00 max ST, ST=1.25Pb, max impurities			
Z19001	USA	ASTM B6	Prime Western	Ingot	0.01 max	0.20 max	0.20 max	0.05 max			0.5-1.4			98.0 min	OT 2.0 max			
	Germany	DIN 1706	2.2075/Zn97.5			0.3 max		0.08			2.40 max			97.5 min				
Z13000	USA	ASTM B418	Type II	Cast Wrought Anode	0.005 max	0.003 max	0.002 max	0.0014 max			0.003 max							
Z32120	USA	ASTM B418	Type I	Cast Wrought Anode	0.1-0.5	0.025-0.07	0.005 max	0.005 max			0.006 max				OT 0.1 max			
Z32121	USA		Anodes Type III		0.1-0.5	0.025-0.15	0.005 max	0.005 max			0.006 max				Si 0.125 max			
Z21210	USA	ASTM B69	Zn-0.08Pb	Hot Rolled	0.001 max	0.005 max	0.001 max	0.012 max			0.10 max	0.001 max				134		65
Z21210	USA	ASTM B69	Zn-0.08Pb	Cold Rolled	0.001 max	0.005 max	0.001 max	0.012 max			0.10 max	0.001 max				145		50
Z21220	USA	ASTM B69	Zn-0.06Pb-0.06 Cd	Hot Rolled	0.001 max	0.05-0.08	0.005 max	0.012 max			0.05-0.10	0.001 max				150		52
Z21220	USA	ASTM B69	Zn-0.06Pb-0.06 Cd	Cold Rolled	0.001 max	0.05-0.08	0.005 max	0.012 max			0.05-0.10	0.001 max				150		40
Z21540	USA	ASTM B69	Zn-0.3Pb-0.3Cd	Hot Rolled	0.001 max	0.25-0.45	0.005 max	0.002 max			0.25-0.50	0.001 max				160		50
Z21540	USA	ASTM B69	Zn-0.3Pb-0.3Cd	Cold Rolled	0.001 max	0.25-0.45	0.005 max	0.002 max			0.25-0.50	0.001 max				170		45
	Japan	JIS H4321	Class 2 Obsolete	Plt, Sh		0.4 max	0.005 max	0.02 max			0.40 max			99 min				
	Japan	JIS H4321	Class 1 Obsolete	Plt, Sh		0.4 max		0.09 max		0.01	1.30 max			98.5 min				
	Japan	JIS H4321	Class 3 Obsolete	Plt, Sh		0.4 max		0.09 max		0.01	1.30 max			98.5 min				
	UK	BS 6561	Type A	Sh	0.01 max	0.005 max	0.08-0.18	0.002 max			0.003 max	0.001 max			Ti 0.03-0.15			
	UK	BS 6561	Type B	Sh	0.002 max	0.005 max	0.001 max	0.005 max			0.30-1.00	0.001 max						
Z40330	USA	ASTM B69	Zn-0.8Cu	Hot Rolled	0.005 max	0.02 max	0.70-0.90 max	0.01 max			0.02 max	0.02	0.02 max			170		50
Z44330	USA	ASTM B69	Zn-1Cu	Hot Rolled	0.001 max	0.005 max	0.85-1.25	0.012 max			0.10 max	0.001 max				170		50
Z44330	USA	ASTM B69	Zn-1Cu	Cold Rolled	0.001 max	0.005 max	0.85-1.25	0.012 max			0.10 max	0.001 max				210		40
Z45330	USA	ASTM B69	Zn-1Cu-0.010Mg	Hot Rolled	0.001 max	0.04 max	0.85-1.25	0.015 max	0.006-0.16		0.15 max	0.001 max				200		20
Z45330	USA	ASTM B69	Zn-1Cu-0.010Mg	Cold Rolled	0.001 max	0.04 max	0.85-1.25	0.015 max	0.006-0.16		0.15 max	0.001 max				248		25

UNS numbers and US grades are provided as a means of cross referencing chemically similar alloys. Exchangability is only possible after independent examination of specifications. Tensile properties are minimum or typical . UTS and YS as Mpa, El as %. See Appendix for list of abbreviations used in Descriptions.

Grade UNS #	Country	Specification	Designation	Description	Al	Cd	Cu	Fe	Mg	Ni	Pb	Sn	Ti	Zn	Other	UTS	YS	EL
Z41320	USA	ASTM B69	Zn-0.8Cu-0.15Ti	Hot Rolled	0.001 max	0.05 max	0.50-1.50	0.012 max			0.10 max	0.001 max	0.12-0.50				221		38
Z41320	USA	ASTM B69	Zn-0.8Cu-0.15Ti	Cold Rolled	0.001 max	0.05 max	0.50-1.50	0.012 max			0.10 max	0.001 max	0.12-0.50				200		60
Z81330	USA	ASTM B852	Z81330	Continuous Galvanizing	0.22-0.28	0.01 max	0.02 max	0.01 max			0.01-0.03					OT 0.01 max			
Z80410	USA	ASTM B852	Z80410	Continuous Galvanizing	0.32-0.38	0.01 max	0.02 max	0.01 max			0.01 max					OT 0.01 max			
Z81430	USA	ASTM B852	Z81430	Continuous Galvanizing	0.32-0.38	0.01 max	0.02 max	0.01 max			0.01-0.03					OT 0.01 max			
Z81450	USA	ASTM B852	Z81450	Continuous Galvanizing	0.32-0.38	0.01 max	0.02 max	0.01 max			0.075-0.125					OT 0.01 max			
Z80510	USA	ASTM B852	Z80510	Continuous Galvanizing	0.42-0.48	0.01 max	0.02 max	0.01 max			0.01 max					OT 0.01 max			
Z81530	USA	ASTM B852	Z81530	Continuous Galvanizing	0.42-0.48	0.01 max	0.02 max	0.01 max			0.01-0.03					OT 0.01 max			
Z81540	USA	ASTM B852	Z81540	Continuous Galvanizing	0.42-0.48	0.01 max	0.02 max	0.01 max			0.025-0.075					OT 0.01 max			
Z81550	USA	ASTM B852	Z81550	Continuous Galvanizing	0.42-0.48	0.01 max	0.02 max	0.01 max			0.075-0.125					OT 0.01 max			
Z33522	Australia	AS 1881	ZnAl4	Ingot	3.9-4.3	0.003 max	0.03 max	0.05 max	0.04-0.06		0.003 max	0.001 max	0.001 max		In 0.0005 max			
Z33522	Canada	CSA HZ.3	AG40	Ingot	3.9-4.3	0.003 max	0.10 max	0.08 max	0.04-0.06		0.003 max	0.001 max						
Z35520	Pan America	COPANT 443	ZnAl-4		3.9-4.3	0.003 max	0.03	0.01	0.03-0.06		0.003	0.001						
Z33522	USA	ASTM B240	AG40B	Ingot	3.9-4.3	0.0020 max	0.10 max	0.075 max	0.010-0.020	0.02 max	0.0020 max	0.0010 max			Cr 0.02 max; Mn 0.05 max; Si 0.035 max			
	Japan	JIS H5301	ZDC 1	Die, as cast	3.5-4.3	0.005 max	0.75-1.25	0.10 max	0.03-0.08		0.01	0.005				324		7
Z33523	USA	ASTM B86	AG40B	Die Cast	3.5-4.3	0.0020 max	0.25 max	0.075 max	0.005-0.020	0.005-0.020	0.0030 max	0.0010 max			Cr 0.02 max; Mn 0.6 max; Si 0.035 max	283		14
Z33521	USA	ASTM B240	AG40A	Ingot	3.9-4.3	0.003 max	0.10 max	0.075 max	0.025-0.05	0.02	0.004 max	0.002 max			Cr 0.02 max; Mn 0.05 max; Si 0.035 max			
Z33520	Australia	AS 1881	ZnAl4	As cast	3.5-4.3	0.005 max	0.25 max	0.10 max	0.03-0.06	0.02	0.005 max	0.002 max	0.001 max		In 0.0005 max; Mn 0.05 max; Si 0.03 max			
Z33520	Canada	CSA HZ.11	AG40	Die	3.5-4.3	0.005 max	0.10 max	0.10 max	0.03-0.06		0.005 max	0.002 max						
Z33520	Czech Republic	CSN 423558	ZnAl	Cast	3.9-4.3		0.10 max	0.075 max	0.02-0.05			0.001 max			Pb+Cd 0.009			
Z33520	Denmark	DS DS3013	7020-10		3.5-4.3	0.005 max	0.1	0.10	0.02-0.06		0.005	0.002						
Z33520	Denmark	DS DS3013	7020-11		3.5-4.3	0.005 max	0.1	0.10	0.02-0.06		0.005	0.002						
Z33520	Finland	SFS 3091	ZnAl4		3.5-4.3	0.005 max	0.10 max	0.10 max	0.02-0.06		0.005	0.02						
Z33520	France	NF A55-102	Z-A4G	As cast	3.9-4.3		0.10 max	0.10 max	0.03-0.06						0.008 Pb+Cd+Sn	240		2
Z33520	International	ISO R301	ZnAl4		3.9-4.3	0.003 max	0.03	0.05	0.03-0.06		0.003	0.001						
Z33520	Pan America	COPANT 442	Zn98					0.08			1.80			98	OT 2.00 max			
Z33520	Pan America	COPANT 443	ZnAl4	Ingot	3.9-4.3	0.003 max		0.01 max	0.03-0.06		0.003 max	0.001 max						
Z33520	South Africa	SABS 25	ZnAl4	Die-cast	3.90-4.30	0.003	0.03	0.05	0.04-0.06	0.001	0.003	0.001	0.001			286		15
Z33520	South Africa	SABS 26	ZBD1		3.80-4.30	0.005	0.10	0.10	0.03-0.06	0.006	0.005	0.002	0.001					
Z33520	Sweden	MNC 71E	7020(ZnAl4)	Die, Ingot	4				0.04							335		9
Z33520	UK	BS 1004	A		3.9-4.3				0.04-0.06						OT 0.0895 max			
Z33520	UK	BS 1004	Alloy A		3.8-4.3	0.005	0.1	0.10 max	0.03-0.06	0.01	0.005 max	0.002 max	0.001		In 0.0005; Mn 0.01			

UNS numbers and US grades are provided as a means of cross referencing chemically similar alloys. Exchangability is only possible after independent examination of specifications. Tensile properties are minimum or typical . UTS and YS as Mpa, El as %. See Appendix for list of abbreviations used in Descriptions.

Worldwide Guide to Equivalent Nonferrous Metals and Alloys

Grade UNS #	Country	Specification	Designation	Description	Al	Cd	Cu	Fe	Mg	Ni	Pb	Sn	Ti	Zn	Other	UTS	YS	EL
Z33520	USA	ASTM B86	AG40A	Die Cast	3.5-4.3	0.004 max	0.25 max	0.100 max	0.020-0.05	0.02 max	0.005 max		0.003 max		Cr 0.02x; Mn 0.06x;Si0.035x	283		10
Z35530	USA	ASTM B240	AC41A	Ingot	3.9-4.3	0.003 max	0.75-1.25	0.075 max	0.03-0.06	0.02 max	0.004 max	0.002 max			Cr 0.02x; Mn 0.05x; i0.035x			
Z33531	Australia	AS 1881	ZnAl4Cu1	Ingot	3.9-4.3	0.003 max	0.75-1.25	0.05 max	0.04-0.06		0.003 max	0.001 max	0.001 max		In 0.0005 max			
Z35531	Australia	AS 1881	ZnAl4Cu1	As cast	3.5-4.3	0.005 max	0.75-1.25	0.10 max	0.03-0.06	0.02	0.005 max	0.002 max	0.001 max		In 0.0005 max; Mn 0.05 max; Si 0.03 max			
Z35531	Canada	CSA HZ.11	AC41	Die	3.5-4.3	0.005 max	0.75-1.25	0.10 max	0.03-0.06		0.005 max	0.002 max						
Z35531	Canada	CSA HZ.3	AC41	Ingot	3.9-4.3	0.003 max	0.75-1.25	0.08	0.04-0.06		0.003 max	0.001 max						
Z33531	Czech Republic	CSN 423560	ZnAl4Cu1	Cast	3.9-4.3		0.75-1.25	0.075 max	0.02-0.05			0.001 max			Pb+Cd 0.009			
Z35531	Denmark	DS DS3013	7030-10		3.5-4.3	0.005	0.75-1.25	0.10	0.02-0.06		0.005	0.002						
Z35531	Denmark	DS DS3013	7030-11		3.5-4.3	0.005	0.75-1.25	0.10	0.02-0.06		0.005	0.002						
Z35531	Finland	SFS 3092	ZnAl4Cu1		3.5-4.3	0.005 max	0.75-1.25	0.10 max	0.02-0.06		0.005	0.002						
Z35531	France	NF A55-102	Z-A4U1G		3.9-4.3		0.75-1.25		0.03-0.06						0.008 Pb+Cd+Sn			
Z35531	International	ISO R301	ZnAl4Cu1		3.9-4.3	0.003 max	0.75-1.25	0.05	0.03-0.06		0.003	0.001						
Z35531	Japan	JIS H5301	ZDC 2	Die	3.5-4.3	0.004 max	0.25 max	0.10 max	0.02-0.06		0.005 max	0.003 max						
Z35531	Pan America	COPANT 443	ZnAl-4Cu1	Ingot	3.9-4.3	0.003 max	0.75-1.25	0.091 max	0.03-0.06		0.003 max	0.001 max						
Z35531	South Africa	SABS 25	ZnAl4Cu1	Die-cast	3.90-4.30	0.003	0.75-1.25	0.05	0.04-0.06	0.001	0.003	0.001	0.001					
Z35531	South Africa	SABS 26	ZBD2		3.80-4.30	0.005	0.75-1.25	0.10	0.03-0.06	0.01	0.005	0.002	0.001					
Z35531	Sweden	MNC 71E	7030(ZnAlCu1)	Die, Ingot	4		1.00		0.04									
Z35531	UK	BS 1004	Alloy B		3.8-4.3	0.005	0.75-1.25	0.10 max	0.03-0.06	0.01	0.005 max	0.002 max	0.001		In 0.0005; Mn 0.01			
Z35531	UK	BS 1004	B		3.8-4.3		0.75-1.25		0.03-0.06						OT 0.1195 max			
Z35531	USA	ANSI	AC41	Die, as cast	3.5-4.3	0.004 max	0.75-1.25	0.10	0.03-0.08		0.005	0.003				335		7
Z35531	USA	ASTM B86	AC41A	Die Cast	3.5-4.3	0.004 max	0.75-1.25	0.100 max	0.03-0.08	0.02 max	0.005 max		0.003 max		Cr 0.02 max; Mn 0.06 max; Si 0.035 max	329		7
Z35540	USA	ASTM B240	AC43A	Ingot	3.9-4.3	0.003 max	2.6-2.9	0.075 max	0.025-0.050	0.02 max	0.004 max	0.002 max			Cr 0.02 max; Mn 0.05 max; Si 0.035 max			
Z33541	Czech Republic	CSN 423562	ZnAlCu3	Cast	3.5-4.3		2.5-3.2	0.075 max	0.03-0.06			0.001 max			Pb+Cd 0.009			
Z35541	USA	ASTM B86	AC43A	Die Cast	3.5-4.3	0.004 max	2.5-3.0	0.100 max	0.020-0.050	0.02 max	0.005		0.003 max		Cr 0.02 max; Mn 0.06 max; Si 0.035 max	359		7
Z34510	USA	ASTM B792	Alloy A	Ingot for Slush Cast	4.5-5.00	0.005 max	0.2-0.3	0.100 max	0.010 max		0.007 max	0.005 max						
Z30500	USA	ASTM B792	Alloy B	Ingot for Slush Cast	5.25-5.75	0.005 max	0.1	0.100 max	0.010 max		0.007 max	0.005 max						
	Europe	CEN EN 1774	ZL 0010 (ZL16)	Ingot + Liquid	0.01-0.04	0.004 max	1.0-1.5	0.04 max	0.02		0.005 max	0.003 max	0.15-0.25		Cr 0.1-0.2; Si 0.04 max			
	Europe	CEN EN 1774	ZL 0400 (ZL3)	Ingot + Liquid	3.8-4.2	0.003 max	0.03 max	0.020 max	0.035-0.06	0.001	0.003 max	0.001 max			Si 0.02 max			
	Europe	CEN EN 1774	ZL 0410 (ZL5)	Ingot + Liquid	3.8-4.2	0.003 max	0.7-1.1	0.020 max	0.035-0.06	0.001	0.003 max	0.001 max			Si 0.02 max			
	Europe	CEN EN 1774	ZL 0430 (ZL2)	Ingot + Liquid	3.8-4.2	0.003 max	2.7-3.3	0.020 max	0.035-0.06	0.001	0.003 max	0.001 max			Si 0.02 max			
Z35542	USA			Forming Die	3.9-4.3	0.003 max	2.5-2.9	0.075 max	0.02-0.05		0.003 max	0.001 max						

UNS numbers and US grades are provided as a means of cross referencing chemically similar alloys. Exchangability is only possible after independent examination of specifications. Tensile properties are minimum or typical . UTS and YS as Mpa, El as %. See Appendix for list of abbreviations used in Descriptions.

11-4 Zinc

Grade UNS #	Country	Specification	Designation	Description	Al	Cd	Cu	Fe	Mg	Ni	Pb	Sn	Ti	Zn	Other	UTS	YS	EL
Z35542	USA	ASTM B793	Alloy B	Ingot for Dies/Molds	3.9-4.3	0.003 max	2.5-2.9	0.075 max	0.02-0.05			0.003 max	0.001 max						
Z35542	USA	MIL Z-7068	II		3.9-4.3	0.003 max	2.5-2.9	0.075	0.02-0.05			0.003 max	0.001 max						
Z35543	USA			Forming Die	3.5-4.5	0.005 max	2.5-3.5	0.100 max	0.02-0.10			0.007 max	0.005 max						
Z35543	USA	ASTM B793	Alloy A	Ingot for Dies/Molds	3.5-4.5	0.005 max	2.5-3.5	0.100 max	0.02-0.10			0.007 max	0.005 max						
	Europe	CEN EN 1774	ZL 0610 (ZL6)	Ingot + Liquid	5.6-6.0	0.003 max	1.2-1.6	0.020 max	0.005 max			0.003 max	0.001 max			Si 0.02 max			
Z38510	USA	ASTM B750	Zn-5Al-MM	Ingot for Hot-Dip Coating	4.2-6.2	0.005 max	0.1 max	0.075 max	0.05 max			0.005 max	0.002 max	0.02 max		Sb 0.002 max; Si 0.015 max; Zr 0.02 max; Ce+La 0.03-0.10; OE 0.02 max; OT 0.04 max			
	Europe	CEN EN 1774	ZL 0810 (ZL8)	Ingot + Liquid	8.2-8.8	0.005 max	0.9-1.3	0.035 max	0.02-0.03	0.001		0.005 max	0.002 max			Si 0.035 max			
Z35635	USA			Ingot	8.2-8.8	0.005 max	0.8-1.3	0.10 max	0.020-0.030			0.005 max	0.002 max						
Z35635	USA	ASTM B669	ZA-8	Ingot for Castings	8.2-8.8	0.005 max	0.8-1.3	0.065 max	0.020-0.030			0.005 max	0.002 max						
Z35636	USA	ASTM B791	ZA-8	Die Cast	8.0-8.8	0.006 max	0.8-1.3	0.075 max	0.015-0.030			0.006 max	0.003 max				372	290	6
Z35636	USA	ASTM B791	ZA-8	Sand Cast	8.0-8.8	0.006 max	0.8-1.3	0.075 max	0.015-0.030			0.006 max	0.003 max				248	200	1
Z35636	USA	ASTM B791	ZA-8	Permanent Mold Cast	8.0-8.8	0.006 max	0.8-1.3	0.075 max	0.015-0.030			0.006 max	0.003 max				221	207	I
Z55710	USA	ASTM B860	S-1 (90/10 Zn/Sb)	Hot Dip Galvanizing		0.003 max	0.003 max	0.03 max				0.015 max	0.01			As 0.015 max; Sb 9.5-10.5; OT 0.03 max			
Z30750	USA	ASTM B860	A-1 (90/10 Zn/Al)	Hot Dip Galvanizing	9.5-10.5	0.004 max	0.035 max	0.05 max				0.005 max	0.003 max			OT 0.01 max			
Z31710	USA	ASTM B860	A-2 (90/10 Zn/Al)	Hot Dip Galvanizing	9.5-10.5		0.5 max	0.15 max				0.4 max				OT 0.25 max			
Z35630	Australia	AS 1881	ZnAl11Cu1	Ingot	10.5-11.5	0.003 max	0.60-1.10	0.08 max	0.02-0.03			0.004 max	0.001 max	0.001 max		In 0.0005 max; Si 0.04 max			
	Europe	CEN EN 1774	ZL 1110 (ZL12)	Ingot + Liquid	10.8-11.5	0.005 max	0.5-1.2	0.05 max	0.02-0.03			0.005 max	0.002 max			Si 0.05 max			
Z35630	USA	ASTM B669	ZA-12	Ingot for Castings	10.8-11.5	0.005 max	0.5-1.2	0.065 max	0.020-0.030			0.005 max	0.002 max						
Z35631	Australia	AS 1881	ZnAl11Cu1	As cast	10-12	0.004 max	0.50-1.25	0.10 max	0.01-0.03	0.02		0.005 max	0.002 max	0.001 max		In 0.0005 max; Mn 0.05 max			
Z35631	USA	ASTM B791	ZA-12	Die Cast	10.5-11.5	0.006 max	0.5-1.2	0.075 max	0.015-0.030			0.006 max	0.003 max				400	317	4
Z35631	USA	ASTM B791	ZA-12	Sand Cast	10.5-11.5	0.006 max	0.5-1.2	0.075 max	0.015-0.030			0.006 max	0.003 max				276	207	1
Z35631	USA	ASTM B791	ZA-12	Permanent Mold Cast	10.5-11.5	0.006 max	0.5-1.2	0.075 max	0.015-0.030			0.006 max	0.003 max				310	207	1
Z33730	USA	ASTM B327	V12	Master Alloy for Die Casting	11.7-12.6	0.004 max	0.25 max	0.070 max	0.075-0.12			0.005 max	0.003 max						
Z30700	USA	ASTM B833	85/15 Zn/Al	Wire	14.0-16.0											OT 0.05 max			
Z35840	Australia	AS 1881	ZnAl27Cu2	Ingot	25.5-27.5	0.003 max	2.00-2.50	0.10 max	0.015-0.025			0.004 max	0.001 max	0.001 max		In 0.0005 max; Mn 0.03 max; Si 0.1 max			
Z35840	Australia	AS 1881	ZnAl27Cu2	As cast	25-28	0.004 max	1.80-2.60	0.12 max	0.01-0.03	0.02		0.005 max	0.002 max	0.001 max		In 0.0005 max; Mn 0.05 max			
	Europe	CEN EN 1774	ZL 2720 (ZL27)	Ingot + Liquid	25.5-28.0	0.005 max	2.0-2.5	0.07 max	0.012-0.02			0.005 max	0.002 max			Si 0.07 max			
Z35840	USA	ASTM B669	ZA-27	Ingot for Castings	25.5-28.0	0.005 max	2.0-2.5	0.072 max	0.012-0.020			0.005 max	0.002 max						
Z35841	USA	ASTM B791	ZA-27	Die Cast	25.0-28.0	0.006 max	2.0-2.5	0.075 max	0.010-0.020			0.006 max	0.003 max				421	365	1

UNS numbers and US grades are provided as a means of cross referencing chemically similar alloys. Exchangability is only possible after independent examination of specifications. Tensile properties are minimum or typical . UTS and YS as Mpa, El as %. See Appendix for list of abbreviations used in Descriptions.

Worldwide Guide to Equivalent Nonferrous Metals and Alloys

Grade UNS #	Country	Specification	Designation	Description	Al	Cd	Cu	Fe	Mg	Ni	Pb	Sn	Ti	Zn	Other	UTS	YS	EL
Z35841	USA	ASTM B791	ZA-27	Sand Cast	25.0-28.0	0.006 max	2.0-2.5	0.075 max	0.010-0.020		0.006 max	0.003 max				400	365	3
Z35841	USA	ASTM B791	ZA-27	Permanent Mold Cast	25.0-28.0	0.006 max	2.0-2.5	.075 max	0.010-0.020		0.006 max	0.003 max				421	365	1
	Germany	DIN 1707	2.2400/L-ZnSn20								1.00 max	18.00-20.00						
	Germany	DIN 8512	2.2360/L-ZnCd40		4 min	35-45								55-65				

Alloy	Grade UNS #	Country	Specification	Designation	Description	Ag	Au	C	Cd	Cu	Ni	Pb	Zn	OT max	Other	UTS	YS	EL
Ag	P07010	USA			Refined	99.99 min				0.010 max		0.001 max			Bi 0.0005 max; Fe 0.001 max; Pd 0.001 max; Se 0.0005 max; Te 0.0005 max			
Ag	P07016	USA				99.95 min				0.05 max				0.010	P 0.002 max			
Ag	P07015	USA			Refined	99.95 min				0.04 max		0.015 max			Bi 0.001 max; Fe 0.002 max			
Ag	P07017	USA			Braz	99.95 min		0.005 max	0.001 max	0.05 max		0.002 max	0.001 max		P 0.002 max			
Ag	P07017	USA	AWS A5.8-92	BVAg-0	Braze Weld Fill Grade 2 vacuum serv.	99.95 min		0.01 max	0.002 max	0.05 max		0.002 max	0.002 max		P 0.002 max			
Ag	P07020	USA			Braz	99.90 min				0.08 max		0.025 max			Ag+Cu 99.95 min; Bi 0.001 max; Fe 0.002 max			
Ag	P07017	USA	AWS A5.8-92	BVAg-0	Braze Weld Fill Grade 1 vacuum serv.	99.5 min		0.005 max	0.001 max	0.05 max		0.002 max	0.001 max		P 0.002 max			
Ag	P07932	USA			Sterling Silversmiths	92.50-93.50			0.05 max	6.50-7.50		0.03 max	0.06 max	0.06	Fe 0.05 max			
Ag	P07931	USA			Sterling	92.10-93.50			0.05 max	6.50-7.90		0.03 max	0.06 max	0.06	Fe 0.05 max			
Ag	P07925	USA	AWS A5.8-92	BAg-19	Braze Weld Fill	92.0-93.0								0.15	Li 0.15-0.30; bal Cu			
Ag	P07925	USA			Braz	92.0-93.0									Li 0.15-0.30; bal Cu			
Ag	P07900	USA			Elect Contact	89.6-91.0			0.05 max	9.0-10.4	0.01 max	0.03 max	0.06 max		Al 0.005 max; Fe 0.05 max; P 0.02 max			
Ag	P07850	USA			Braz	84.0-86.0									ba Mn			
Ag	P07850	USA	AWS A5.8-92	BAg-23	Braze Weld Fill	84.0-86.0								0.15	ba Mn			
Ag	P07723	USA	AWS A5.8-92	BAg-8a	Braze Weld Fill	71.0-73.0								0.15	Li 0.25-0.50; bal Cu			
Ag	P07720	USA	AWS A5.8-92	BAg-8	Braze Weld Fill	71.0-73.0								0.15	bal Cu			
Ag	P07720	USA			Braz	71.0-73.0									bal Cu			
Ag	P07723	USA			Braz	71.0-73.0									bal Cu			
Ag	P07727	USA			Braz Vacuum	71.0-73.0		0.005 max	0.001 max			0.002 max	0.001 max		P 0.002 max; bal Cu			
Ag	P07727	USA	AWS A5.8-92	BVAg-8	Braze Weld Fill Grade 1 vacuum serv.	71.0-73.0		0.005 max	0.001 max			0.002 max	0.001 max		P 0.002 max; bal Cu			
Ag	P07727	USA	AWS A5.8-92	BVAg-8	Braze Weld Fill Grade 2 vacuum serv.	71.0-73.0		0.01 max	0.002 max			0.002 max	0.002 max		P 0.02 max; bal Cu			
Ag	P07728	USA			Braz Vacuum	70.5-72.5		0.005 max	0.001 max		0.3-0.7	0.002 max	0.001 max		P 0.002 max; bal Cu			
Ag	P07728	USA	AWS A5.8-92	BVAg-8b	Braze Weld Fill Grade 1 vacuum serv.	70.5-72.5		0.005 max	0.001 max		0.3-0.7	0.002 max	0.001 max		P 0.002 max; bal Cu			
Ag	P07728	USA	AWS A5.8-92	BVAg-8b	Braze Weld Fill Grade 2 vacuum serv.	70.5-72.5		0.01 max	0.002 max		0.3-0.7	0.002 max	0.002 max		P 0.02 max; bal Cu			
Ag	P07700	USA	AWS A5.8-92	BAg-10	Braze Weld Fill	69.0-71.0				19.0-21.0			8.0-12.0	0.15				
Ag	P07700	USA			Braz	69.0-71.0				19.0-21.0			8.0-12.0					
Ag	P07687	USA			Braz Vacuum	67.0-69.0		0.005 max	0.001 max			0.002 max	0.001 max		P 0.002 max; Pd 4.5-5.5; bal Cu			
Ag	P07687	USA	AWS A5.8-92	BVAg-30	Braze Weld Fill Grade 1 vacuum serv.	67.0-69.0		0.005 max	0.001 max			0.002 max	0.001 max		P 0.002 max; Pd 4.5-5.5; bal Cu			

UNS numbers and US grades are provided as a means of cross referencing chemically similar alloys. Exchangability is only possible after independent examination of specifications. Tensile properties are minimum or typical . UTS and YS as Mpa, El as %. See Appendix for list of abbreviations used in Descriptions.

Worldwide Guide to Equivalent Nonferrous Metals and Alloys

Alloy	Grade UNS #	Country	Specification	Designation	Description	Ag	Au	C	Cd	Cu	Ni	Pb	Zn	OT max	Other	UTS	YS	EL
Ag	P07687	USA	AWS A5.8-92	BVAg-30	Braze Weld Fill Grade 2 vacuum serv.	67.0-69.0		0.01 max	0.002 max			0.002 max	0.002 max		P 0.02 max; Pd 4.5-5.5; bal Cu			
Ag		Czech Republic	CSN 055676	B-Ag66CuZn-790/735	Brazing Filler Metal	65.0-67.0				26.5-28.5		0.2 max			Bi 0.10 max; Fe 0.10 max; Sb 0.10 max; Sn 0.10 max; bal Zn			
Ag	P07650	USA	AWS A5.8-92	BAg-9	Braze Weld Fill	64.0-66.0				19.0-21.0			13.0-17.0	0.15				
Ag	P07650	USA			Braz	64.0-66.0				19.0-21.0			13.0-17.0					
Ag	P07630	USA	AWS A5.8-92	BAg-21	Braze Weld Fill	62.0-64.0				27.5-29.5	2.0-3.0			0.15	Sn 5.0-7.0			
Ag	P07630	USA			Braz	62.0-64.0				27.5-29.5	2.0-3.0				Sn 5.0-7.0			
Ag	P07627	USA			Braz Vacuum	60.5-62.5		0.005 max	0.001 max			0.002 max	0.001 max		In 14.0-15.0; P 0.002 max; bal Cu			
Ag	P07627	USA	AWS A5.8-92	BVAg-29	Braze Weld Fill Grade 1 vacuum serv.	60.5-62.5		0.005 max	0.001 max			0.002 max	0.001 max		In 14.0-15.0; P 0.002 max; bal Cu			
Ag	P07627	USA	AWS A5.8-92	BVAg-29	Braze Weld Fill Grade 2 vacuum serv.	60.5-62.5		0.01 max	0.002 max			0.002 max	0.002 max		In 14.0-15.0; P 0.02 max; bal Cu			
Ag	P07600	USA	AWS A5.8-92	BAg-18	Braze Weld Fill	59.0-61.0								0.15	Sn 9.5-10.5; bal Cu			
Ag	P07600	USA			Braz	59.0-61.0									bal Cu			
Ag		Czech Republic	CSN 055674	B-Ag60CuZn-760/700	Brazing Filler Metal	59.0-61.0				26.0-27.0		0.2 max			Bi 0.10 max; Fe 0.10 max; Sb 0.10 max; Sn 0.10 max; bal Zn			
Ag	P07607	USA			Braz Vacuum	59.0-61.0		0.005 max	0.001 max			0.002 max	0.001 max		P 0.002 max; Sn 9.5-10.5; bal Cu			
Ag	P07607	USA	AWS A5.8-92	BVAg-18	Braze Weld Fill Grade 1 vacuum serv.	59.0-61.0		0.005 max	0.001 max			0.002 max	0.001 max		P 0.002 max; Sn 9.5-10.5; bal Cu			
Ag	P07607	USA	AWS A5.8-92	BVAg-18	Braze Weld Fill Grade 2 vacuum serv.	59.0-61.0		0.01 max	0.002 max			0.002 max	0.002 max		P 0.02 max; Sn 9.5-10.5; bal Cu			
Ag	P07587	USA			Braz Vacuum	57.0-59.0		0.005 max	0.001 max	31.0-33.0		0.002 max	0.001 max		P 0.002 max; bal Pd			
Ag	P07587	USA	AWS A5.8-92	BVAg-31	Braze Wld Fill Gr 1 vac serv.	57.0-59.0		0.005 max	0.001 max	31.0-33.0		0.002 max	0.001 max		P 0.002 max; bal Pd			
Ag	P07587	USA	AWS A5.8-92	BVAg-31	Braze Wld Fill Gr 2 vac serv.	57.0-59.0		0.01 max	0.002 max	31.0-33.0		0.002 max	0.002 max		P 0.002 max; bal Pd			
Ag	P07560	USA	AWS A5.8-92	BAg-13a	Braze Weld Fill	55.0-57.0					1.5-2.5			0.15	bal Cu			
Ag	P07560	USA			Braz	55.0-57.0					1.5-2.5				bal Cu			
Ag	P07563	USA	AWS A5.8-92	BAg-7	Braze Weld Fill	55.0-57.0				21.0-23.0			15.0-19.0	0.15	Sn 4.5-5.5			
Ag	P07563	USA			Braz	55.0-57.0				21.0-23.0			15.0-19.0		Sn 4.5-5.5			
Ag	P07547	USA			Braz Vacuum	53.0-55.0		0.005 max	0.001 max	20.0-22.0		0.002 max	0.001 max		P 0.002 max; bal Pd			
Ag	P07547	USA	AWS A5.8-92	BVAg-32	Braze Wld Fill Gr 1 vac serv.	53.0-55.0		0.005 max	0.001 max	20.0-22.0		0.002 max	0.001 max		P 0.002 max; bal Pd			
Ag	P07547	USA	AWS A5.8-92	BVAg-32	Braze Wld Fill Gr 2 vac serv.	53.0-55.0		0.01 max	0.002 max	20.0-22.0		0.002 max	0.002 max		P 0.002 max; bal Pd			
Ag	P07540	USA	AWS A5.8-92	BAg-13	Braze Weld Fill	53.0-55.0					0.5-1.5		4.0-6.0	0.15	bal Cu			
Ag	P07540	USA			Braz	53.0-55.0					0.5-1.5		4.0-6.0		bal Cu			
Ag		Czech Republic	CSN 055672	B-Ag50CuZnCd-740/630	Brazing Filler Metal	49.0-51.0			19.0-21.0	19.0-21.0		0.2 max			Bi 0.10 max; Fe 0.10 max; Sb 0.10 max; Sn 0.10 max; bal Zn			

UNS numbers and US grades are provided as a means of cross referencing chemically similar alloys. Exchangability is only possible after independent examination of specifications. Tensile properties are minimum or typical . UTS and YS as Mpa, El as %. See Appendix for list of abbreviations used in Descriptions.

12-2 Miscellaneous

Alloy	Grade UNS #	Country	Specification	Designation	Description	Ag	Au	C	Cd	Cu	Ni	Pb	Zn	OT max	Other	UTS	YS	EL
Ag	P07507	USA			Braz Vacuum	49.0-51.0		0.005 max	0.001 max			0.002 max	0.001 max		P 0.002 max; bal Cu			
Ag	P07507	USA	AWS A5.8-92	BVAg-6b	Braze Weld Fill Grade 1 vacuum serv.	49.0-51.0		0.005 max	0.001 max			0.002 max	0.001 max		P 0.002 max; bal Cu			
Ag	P07507	USA	AWS A5.8-92	BVAg-6b	Braze Weld Fill Grade 2 vacuum serv.	49.0-51.0		0.01 max	0.002 max			0.002 max	0.002 max		P 0.02 max; bal Cu			
Ag	P07501	USA	AWS A5.8-92	BAg-3	Braze Weld Fill	49.0-51.0			15.0-17.0	14.5-16.5	2.5-3.5		13.5-17.5	0.15				
Ag	P07501	USA			Braz	49.0-51.0			15.0-17.0	14.5-16.5	2.5-3.5		13.5-17.5					
Ag	P07503	USA	AWS A5.8-92	BAg-6	Braze Weld Fill	49.0-51.0				33.0-35.0			14.0-18.0	0.15				
Ag	P07503	USA			Braz	49.0-51.0				33.0-35.0			14.0-18.0					
Ag	P07500	USA	AWS A5.8-92	BAg-1a	Braze Weld Fill	49.0-51.0			17.0-19.0	14.5-16.5			14.5-18.5	0.15				
Ag	P07500	USA			Braz	49.0-51.0			17.0-19.0	14.5-16.5			14.5-18.5					
Ag	P07502	USA			Braz	49.0-51.0			9.0-11.0	17.0-19.0			20.0-24.0					
Ag	P07505	USA	AWS A5.8-92	BAg-24	Braze Weld Fill	49.0-51.0				19.0-21.0	1.5-2.5		26.0-30.0	0.15				
Ag	P07505	USA			Braz	49.0-51.0				19.0-21.0	1.5-2.5		26.0-30.0					
Ag	P07490	USA	AWS A5.8-92	BAg-22	Braze Weld Fill	48.0-50.0				15.0-17.0	4.0-5.0		21.0-25.0	0.15	Mn 7.0-8.0			
Ag	P07490	USA			Braz	48.0-50.0				15.0-17.0	4.0-5.0		21.0-25.0		Mn 7.0-8.0			
Ag	P07453	Czech Republic	CSN 055670	B-Ag45CuZn-740/680	Brazing Filler Metal	44.0-46.0				29.0-31.0		0.2 max			Bi 0.10 max; Fe 0.10 max; Sb 0.10 max; Sn 0.10 max; bal Zn			
Ag	P07450	USA			Braz	44.0-46.0			23.0-25.0	23.0-25.0			14.0-17.0					
Ag	P07450	USA	AWS A5.8-92	BAg-1	Braze Weld Fill	44.0-46.0			23.0-25.0	14.0-16.0			14.0-18.0	0.15				
Ag	P07454	USA	AWS A5.8-92	BAg-36	Braze Weld Fill	44.0-46.0				26.0-28.0			23.0-27.0	0.15	Sn 2.5-3.5			
Ag	P07454	USA			Braz	44.0-46.0				26.0-28.0			23.0-27.0		Sn 2.5-3.5			
Ag	P07453	USA	AWS A5.8-92	BAg-5	Braze Weld Fill	44.0-46.0				29.0-31.0			23.0-27.0	0.15				
Ag	P07453	USA			Braz	44.0-46.0				29.0-31.0			23.0-27.0					
Ag		Czech Republic	CSN 055668	B-Ag40CuZnCd-630/595	Brazing Filler Metal	39.0-41.0			18.0-22.0	18.0-20.0		0.2 max			Bi 0.10 max; Fe 0.10 max; Sb 0.10 max; Sn 0.10 max; bal Zn			
Ag	P07401	USA	AWS A5.8-92	BAg-28	Braze Weld Fill	39.0-41.0				29.0-31.0			26.0-30.0	0.15	Sn 1.5-2.5			
Ag	P07401	USA			Braz	39.0-41.0				29.0-31.0			26.0-30.0		Sn 1.5-2.5			
Ag	P07400	USA	AWS A5.8-92	BAg-4	Braze Weld Fill	39.0-41.0				29.0-31.0	1.5-2.5		26.0-30.0	0.15				
Ag	P07400	USA			Braz	39.0-41.0				29.0-31.0	1.5-2.5		26.0-30.0					
Ag	P07402	USA			Braz	39.0-41.0				29.0-32.0			28.0-32.0	0.15				
Ag	P07380	USA	AWS A5.8-92	BAg-34	Braze Weld Fill	37.0-39.0				31.0-33.0			26.0-30.0	0.15	Sn 1.5-2.5			
Ag	P07380	USA			Braz	37.0-39.0				31.0-33.0			26.0-30.0		Sn 1.5-2.5			
Ag	P07350	USA	AWS A5.8-92	BAg-2	Braze Weld Fill	34.0-36.0			17.0-19.0	25.0-27.0			19.0-23.0	0.15				

UNS numbers and US grades are provided as a means of cross referencing chemically similar alloys. Exchangability is only possible after independent examination of specifications. Tensile properties are minimum or typical . UTS and YS as Mpa, El as %. See Appendix for list of abbreviations used in Descriptions.

Miscellaneous 12-3

Alloy	Grade UNS #	Country	Specification	Designation	Description	Ag	Au	C	Cd	Cu	Ni	Pb	Zn	OT max	Other	UTS	YS	EL
Ag	P07350	USA			Braz	34.0-36.0			17.0-19.0	25.0-27.0			19.0-23.0					
Ag	P07351	USA	AWS A5.8-92	BAg-35	Braze Weld Fill	34.0-36.0				31.0-33.0			31.0-35.0	0.15				
Ag	P07351	USA			Braz	34.0-36.0				31.0-33.0			31.0-35.0					
Ag	P07300	USA	AWS A5.8-92	BAg-2a	Braze Weld Fill	29.0-31.0			19.0-21.0	26.0-28.0			21.0-25.0	0.15				
Ag	P07300	USA			Braz	29.0-31.0			19.0-21.0	26.0-28.0			21.0-25.0					
Ag	P07301	USA	AWS A5.8-92	BAg-20	Braze Weld Fill	29.0-31.0				37.0-39.0			30.0-34.0	0.15				
Ag	P07301	USA			Braz	29.0-31.0				37.0-39.0			30.0-34.0					
Ag	P07251	USA	AWS A5.8-92	BAg-27	Braze Weld Fill	24.0-26.0			12.5-14.5	34.0-36.0			24.5-28.5	0.15				
Ag	P07251	USA			Braz	24.0-26.0			12.5-14.5	34.0-36.0			24.5-28.5					
Ag	P07252	USA	AWS A5.8-92	BAg-33	Braze Weld Fill	24.0-26.0			16.5-18.5	29.0-31.0			26.5-28.5	0.15				
Ag	P075252	USA			Braz	24.0-26.0			16.5-18.5	29.0-31.0			26.5-28.5					
Ag	P07253	USA	AWS A5.8-92	BAg-37	Braze Weld Fill	24.0-26.0				39.0-41.0			31.0-35.0	0.15	Sn 1.5-2.5			
Ag	P07253	USA			Braz	24.0-26.0				39.0-41.0			31.0-35.0		Sn 1.5-2.5			
Ag	P07250	USA	AWS A5.8-92	BAg-26	Braze Weld Fill	24.0-26.0				37.0-39.0	1.5-2.5		31.0-35.0	0.15	Mn 1.5-2.5			
Ag	P07250	USA			Braz	24.0-26.0				37.0-39.0	1.5-2.5		31.0-35.0		Mn 1.5-2.5			
Ag	P07200	USA			Braz	19.0-21.0				39.0-41.0			33.0-37.0		Mn 4.5-5.5			
Au	P00010	USA			Refined	0.001 max	99.995 min			0.001 max		0.001 max			Bi 0.001 max; Cr 0.003 max; Fe 0.001 max; Mg 0.001 max Mn 0.0003 max; Pd 0.001 max; Si 0.001 max; Sn 0.001 max			
Au	P00016	USA					99.99 min								Ag+Cu 0.009 max; OE 0.003 max			
Au	P00015	USA			Refined		99.99 min			0.005 max	0.0003 max	0.002 max			As 0.003 max; Bi 0.002 max; r 0.0003 max; Fe 0.002 max; Mg 0.003 max; Mn 0.0003 max; Pd 0.005 max; Si 0.005 max; Sn 0.001 max			
Au	P00020	USA			Refined	0.035 max	99.95 min			0.02 max		0.005 max			Fe 0.005 max; Ag+Cu 0.04 max; Pd 0.02 max			
Au	P00025	USA			Refined		99.5 min											
Au	P00100	USA					99.0 min								Ti 0.8-1.0			
Au	P00927	USA			Braz Vacuum		91.0-93.0	0.005 max	0.001 max			0.002 max			P 0.002 max; Zr 0.001 min; bal Pd			
Au	P00927	USA	AWS A5.8-92	BVAu-8	Braze Weld Fill Grade 1 vacuum serv.		91.0-93.0	0.005 max	0.001 max			0.002 max	0.001 max		P 0.002 max; bal Pd			
Au	P00927	USA	AWS A5.8-92	BVAu-8	Braze Weld Fill Grade 2 vacuum serv.		91.0-93.0	0.01 max	0.002 max			0.002 max	0.002 max		P 0.002 max; bal Pd			
Au	P00901	USA			Elect Contact		89.0-91.0			9.0-11.0								

UNS numbers and US grades are provided as a means of cross referencing chemically similar alloys. Exchangability is only possible after independent examination of specifications. Tensile properties are minimum or typical . UTS and YS as Mpa, El as %. See Appendix for list of abbreviations used in Descriptions.

12-4 Miscellaneous

Alloy	Grade UNS #	Country	Specification	Designation	Description	Ag	Au	C	Cd	Cu	Ni	Pb	Zn	OT max	Other	UTS	YS	EL
Au	P00900	USA				3.0-10.0	90.0 min											
Au	P00820	USA			Braz		81.5-82.5							0.15	bal Ni			
Au	P00820	USA	AWS A5.8-92	BAu-4	Braze Weld Fill		81.5-82.5							0.15	bal Ni			
Au	P00827	USA			Braz Vacum		81.5-82.5	0.005 max	0.001 max			0.002 max			P 0.002 max; Zr 0.001 min; bal Ni			
Au	P00827	USA	AWS A5.8-92	BVAu-4	Braze Weld Fill Grade 1 vacuum serv.		81.5-82.5	0.005 max	0.001 max			0.002 max	0.001 max		P 0.002 max; bal Ni			
Au	P00827	USA	AWS A5.8-92	BVAu-4	Braze Weld Fill Grade 2 vacuum serv.		81.5-82.5	0.01 max	0.002 max			0.002 max	0.002 max		P 0.002 max; bal Ni			
Au	P00800	USA			Braz		79.5-80.5								bal Cu			
Au	P00800	USA	AWS A5.8-92	BAu-2	Braze Weld Fill		79.5-80.5							0.15	bal Cu			
Au	P00807	USA			Braz Vacuum		79.5-80.5	0.005 max	0.001 max			0.002 max		0.15	P 0.002 max; Zr 0.001 min; bal Cu			
Au	P00807	USA	AWS A5.8-92	BVAu-2	Braze Weld Fill Grade 1 vacuum serv.		79.5-80.5	0.005 max	0.001 max			0.002 max	0.001 max		P 0.002 max; bal Cu			
Au	P008707	USA	AWS A5.8-92	BVAu-2	Braze Weld Fill Grade 2 vacuum serv.		79.5-80.5	0.01 max	0.002 max			0.002 max	0.002 max		P 0.002 max; bal Cu			
Au	P00750	USA			Elect Contact	21.4-22.6	74.2-75.8				2.6-3.4							
Au	P00260	USA			18 Karat Yellow Gold Jewelry	18.00-20.00	74.85-75.45						2.50-3.50		bal Cu			
Au	P00270	USA			18 Karat White Gold Jewelry		74.85-75.45				11.00-13.00		2.50-3.50		bal Cu			
Au	P00285	USA			18 Karat Red Gold Jewelry	4.5-5.50	74.70-75.30								bal Cu			
Au	P00250	USA			18 Karat Yellow Gold Jewelry	12.00-14.00	74.70-75.30								bal Cu			
Au	P00255	USA			18 Karat Yellow Gold Jewelry	14.00-16.00	74.70-75.30								bal Cu			
Au	P00280	USA			18 Karat Green gold Jewelry	21.50-23.50	74.70-75.30								bal Cu			
Au	P00275	USA			18 Karat White Gold Jewelry		74.70-75.30				16.80-18.80		4.97-5.97		bal Cu			
Au	P00710	USA			Elect Contact	4.0-5.0	70.5-72.5			13.5-15.5			0.7 max		Pt 8.0-9.0; Zr 1.3 min			
Au	P00700	USA	AWS A5.8-92	BAu-6	Braze Weld Fill		69.5-70.5				21.5-22.5			0.15	Pd 7.5-8.5			
Au	P00700	USA			Brax		69.5-70.5				21.5-22.5			0.15	Pd 7.5-8.5			
Au	P00692	USA			Elect Contact	24.5-25.5	68.5-69.5								Pt 5.5-6.5; S 0.01 max			
Au	P00580	USA			14 Karat Yellow Gold	3.0-4.0	58.03-58.63			30.74-31.74	0.10 max		5.68 max	0.15	Fe 0.05 max; Zr 7.18 min			
Au	P00230	USA			14 Karat Green Gold Jewelry	31.50-33.50	58.03-58.63						0.70 max		bal Cu			
Au	P00215	USA			14 Karat Yellow Gold Jewelry	23.78-25.78	58.03-58.63						0.14		bal Cu			
Au	P00180	USA			14 Karat Yellow Gold Jewelry	15.50-17.50	58.03-58.63						0.20		bal Cu			

UNS numbers and US grades are provided as a means of cross referencing chemically similar alloys. Exchangability is only possible after independent examination of specifications. Tensile properties are minimum or typical . UTS and YS as Mpa, El as %. See Appendix for list of abbreviations used in Descriptions.

Alloy	Grade UNS #	Country	Specification	Designation	Description	Ag	Au	C	Cd	Cu	Ni	Pb	Zn	OT max	Other	UTS	YS	EL
Au	P00235	USA			14 Karate Green Gold Jewelry	34.00-36.00	58.03-58.63						0.20		bal Cu			
Au	P00200	USA			14 Karat Yellow Gold Jewelry	20.20-22.20	58.03-58.63						0.30		bal Cu			
Au	P00220	USA			14 Karat Yellow Gold Jewelry	20.20-22.20	58.03-58.63						0.30		bal Cu			
Au	P00190	USA			14 Karat Yellow Gold Jewelry	9.00-11.00	58.03-58.63						1.50-2.50		bal Cu			
Au	P00205	USA			14 Karat Yellow Gold Jewelry	7.81-8.81	58.03-58.63						3.67-4.67		bal Cu			
Au	P00210	USA			14 Karat Yellow Gold Jewelry	3.50-4.50	58.03-58.63						5.93-6.93		bal Cu			
Au	P00225	USA			14 Karat Yellow Gold Jewelry	3.50-4.50	58.03-58.63						5.93-6.93		bal Cu			
Au	P00185	USA			14 Karat Yellow Gold Jewelry	6.96-7.96	58.03-58.63				1.00		6.13-7.13		bal Cu			
Au	P00150	USA			14 Karat White Gold Jewelry	1.50-2.50	58.184-58.484				7.50-8.50		7.02-8.02		bal Cu			
Au	P00170	USA			14 Karat Red Gold Jewelry	2.03-3.03	58.03-58.33								bal Cu			
Au	P00165	USA			14 Karat White Gold Jewelry		58.03-58.33				8.05-9.05		4.30-5.30		bal Cu			
Au	P00160	USA			14 Karat White Gold Jewelry		58.03-58.33				11.21-13.21		5.49-6.49		bal Cu			
Au	P00032	USA			14 Karat Yellow Gold Jewelry	3.50-4.50	58.03-58.33			30.08-32.08	0.60 max		6.13-6.93		Fe 0.55 max; Si 0.51 max			
Au	P00175	USA			14 Karat Yellow Gold Jewelry	4.50-5.50	58.03-58.33						6.13-7.13		Si 0.03; bal Cu			
Au	P00155	USA			14 Karat White Gold Jewelry		58.03-58.33				9.80-11.80		8.27-9.27		bal Cu			
Au	P00500	USA			Brax		49.5-50.5				24.5-25.5				Pd 24.5-25.5			
Au	P00507	USA			Braz Vaccuum		49.5-50.5	0.005 max	0.001 max		24.5-25.5	0.002 max	0.001 max		Co 0.06 max; P 0.002 max; bal Pd			
Au	P00507	USA	AWS A5.8-92	BVAu-7	Braze Wld Fill Gr 1 vac serv.		49.5-50.5	0.005 max	0.001 max		24.5-25.5	0.002 max	0.001 max		P 0.002 max; bal Pd			
Au	P99507	USA	AWS A5.8-92	BVAu-7	Braze Wld Fill Gr 2 vac serv.		49.5-50.5	0.01 max	0.002 max		24.5-25.5	0.002 max	0.002 max		P 0.002 max; bal Pd			
Au	P00125	USA			10 Karat White Gold Jewelry		41.517-41.817				0.65-1.65		9.35-11.35		bal Cu			
Au	P00105	USA			10 Karat Yellow Gold Jewelry	5.35-6.35	41.517-41.817				0.65-1.65		11.68-13.68		bal Cu			
Au	P00145	USA			10 Karat Red Gold Jewelry	3.32	41.40-41.70								bal Cu			
Au	P00140	USA			10 Karat Green Gold Jewelry	47.90-49.90	41.40-41.70						0.35		bal Cu			
Au	P00120	USA			10 Karat Yellow Gold Jewelry	6.10-7.10	41.40-41.70						3.20-4.20		bal Cu			

UNS numbers and US grades are provided as a means of cross referencing chemically similar alloys. Exchangability is only possible after independent examination of specifications. Tensile properties are minimum or typical . UTS and YS as Mpa, El as %. See Appendix for list of abbreviations used in Descriptions.

Alloy	Grade UNS #	Country	Specification	Designation	Description	Ag	Au	C	Cd	Cu	Ni	Pb	Zn	OT max	Other	UTS	YS	EL
Au	P00110	USA			10 Karat Yellow Gold Jewelry	10.66-12.66	41.40-41.70						5.33-6.33		Si 0.03; bal Cu			
Au	P00130	USA			10 Karat White Gold Jewelry		41.40-41.70				16.08-18.08		7.90-8.90		bal Cu			
Au	P00115	USA			10 Karat Yellow Gold Jewelry	5.00-6.00	41.40-41.70						8.50-9.50		bal Cu			
Au	P00135	USA			10 Karat White Gold Jewelry		41.36-41.66			1.47-2.47	14.05-16.05		11.62-12.62		bal Cu			
Au	P00375	USA	AWS A5.8-92	BAu-1	Braze Weld Fill		37.0-38.0							0.15	bal Cu			
Au	P00375	USA			Braz		37.0-38.0							0.15	bal Cu			
Au	P00350	USA	AWS A5.8-92	BAu-3	Braze Weld Fill		34.5-35.5				2.5-3.5			0.15	bal Cu			
Au	P00350	USA			Braz		34.5-35.5				2.5-3.5			0.15	bal Cu			
Au	P00300	USA	AWS A5.8-92	BAu-5	Braze Weld Fill		29.5-30.5				35.5-36.5			0.15	Pd 33.5-34.5			
Au	P00691	USA			Elect Contact	23.5-26.5	68.0-70.0								Pt 5.0-7.0			
Nb	R04200	USA			Reactor Grade			0.01 max			0.005 max				Fe 0.005 max; H 0.001 max; Hf 0.02 max; N 0.01 max; O 0.015 max; Ta 0.1 max; W 0.03 max; Zr 0.02 max			
Nb	R04200	USA	ASTM B391	Nb Type 1	Ingot Reactor Grade Unalloyed			0.01 max			0.005 max				Fe 0.005 max; Mo 0.010 max; H 0.0015 max; Hf 0.02 max; N 0.01 max; O 0.015 max; Si 0.005 max; Ta 0.1 x; W 0.03 x; Zr 0.02 max			
Nb	R04200	USA	ASTM B392	Nb Type 1	Bar Rod Wir Reactor Grade Unalloyed			0.01 max			0.005 max				Fe 0.005 max; H 0.0015 max; Hf 0.02 max; Mo 0.010 max; N 0.01 max; O 0.015 max; Si 0.005 max; Ta 0.1 x; W 0.03 x; Zr 0.02 max	125	85	25
Nb	R04200	USA	ASTM B393	Nb Type 1	Strp Sh Plt Reactor Grade Unalloyed			0.01 max			0.005 max				Fe 0.005 max; H 0.0015 max; Hf 0.02 max; Mo 0.010 max; N 0.01 max; O 0.015 max; Si 0.005 max; Ta 0.1 x; W 0.03 x; Zr 0.02 max	125	85	25
Nb	R04200	USA	ASTM B394	Nb Type 1	Smls Weld Tub Reactor Grade Unalloyed			0.01 max			0.005 max				Fe 0.005 max; H 0.0015 max; Hf 0.02 max; Mo 0.010 max; N 0.01 max; O 0.015 max; Si 0.005 max; Ta 0.1 x; W 0.03 x; Zr 0.02 max	125	85	25

UNS numbers and US grades are provided as a means of cross referencing chemically similar alloys. Exchangability is only possible after independent examination of specifications. Tensile properties are minimum or typical . UTS and YS as Mpa, El as %. See Appendix for list of abbreviations used in Descriptions.

Alloy	Grade UNS #	Country	Specification	Designation	Description	Ag	Au	C	Cd	Cu	Ni	Pb	Zn	OT max	Other	UTS	YS	EL
Nb	R04210	USA			Commercial Grade			0.01 max			0.005 max				Fe 0.01 max; H 0.001 max; Hf 0.02 max; N 0.01 max; O 0.025 max; Ta 0.2 max; W 0.05 max; Zr 0.02 max			
Nb	R04210	USA	ASTM B391	Nb Type 2	Ingot Commercial Grade Unalloyed			0.01 max			0.005 max				Fe 0.01 max; H 0.0015 max; Hf 0.02 max; Mo 0.010 max; N 0.01 max; O 0.025 max; Si 0.005 max; Ta 0.2 max; W 0.05 max; Zr 0.02 max			
Nb	R04210	USA	ASTM B392	Nb Type 2	Bar Rod Wir Commercial Grade Unalloyed			0.01 max			0.005 max				Fe 0.01 max; H 0.0015 max; Hf 0.02 max; Mo 0.010 max; N 0.01 max; O 0.025 max; Si 0.005 max; Ta 0.2 max; W 0.05 max; Zr 0.02 max	125	85	25
Nb	R04210	USA	ASTM B393	Nb Type 2	Strp Sh Plt Commercial Grade Unalloyed			0.01 max			0.005 max				Fe 0.01 max; H 0.0015 max; Hf 0.02 max; Mo 0.010 max; N 0.01 max; O 0.025 max; S i 0.005 max; Ta 0.2 x; W 0.05 x; Zr 0.02 max	125	85	25
Nb	R04210	USA	ASTM B394	Nb Type 2	Smls Weld Tub Commercial Grade Unalloyed			0.01 max			0.005 max				Fe 0.01 max; H 0.0015 max; Hf 0.02 max; Mo 0.010 max; N 0.01 max; O 0.025 max; Si 0.005 max; Ta 0.2 x; W 0.05 x; Zr 0.02 max	125	85	25
Nb	R04211	USA			Commercial Grade			0.005 max						0.15	Fe 0.010 max; H 0.002 max; N 0.010 max; O 0.030 max; Si 0.005 max; Ta 0.10 x; Zr 0.010x; OE 0.010 max			
Nb	Nb-1%Zr R04251	USA			Reactor Grade			0.01 max			0.005 max				Fe 0.005 max; H 0.001 max; Hf 0.02 max; N 0.01 max; O 0.015 max; Ta 0.1 x; W 0.03 x; Zr 0.8-1.2			
Nb	Nb-1%Zr R04251	USA	ASTM B391	Nb Type 3	Ingot Reactor Grade			0.01 max			0.005 max				Fe 0.005 max; H 0.0015 max; Hf 0.02 max; Mo 0.010 max; N 0.01 max; O 0.015 max; Si 0.005 max; Ta 0.1 x; W 0.03 x; Zr 0.8-1.2			

UNS numbers and US grades are provided as a means of cross referencing chemically similar alloys. Exchangability is only possible after independent examination of specifications. Tensile properties are minimum or typical . UTS and YS as Mpa, El as %. See Appendix for list of abbreviations used in Descriptions.

12-8 Miscellaneous

Alloy	Grade UNS #	Country	Specification	Designation	Description	Ag	Au	C	Cd	Cu	Ni	Pb	Zn	OT max	Other	UTS	YS	EL
Nb	Nb-1%Zr R04251	USA	ASTM B392	Nb Type 3	Bar Rod Wir Reactor Grade			0.01 max			0.005 max				Fe 0.005 max; H 0.0015 max; Hf 0.02 max; Mo 0.010 max; N 0.01 max O 0.015 max; Si 0.005 max; Ta 0.1 max; W 0.03 max; Zr 0.8-1.2	195	125	20
Nb	Nb-1%Zr R04251	USA	ASTM B393	Nb Type 3	Strp Sh Plt Reactor Grade			0.01 max			0.005 max				Fe 0.005 max; H 0.0015 max; Hf 0.02 max; Mo 0.010 max; N 0.01 max; O 0.015 max; Si 0.005 max; Ta 0.1 max; W 0.03 max; Zr 0.8-1.2	195	125	20
Nb	Nb-1%Zr R04251	USA	ASTM B394	Nb Type 3	Smls Weld Tub Reactor Grade			0.01 max			0.005 max				Fe 0.005 max; H 0.0015 max; Hf 0.02 max; Mo 0.010 max; N 0.01 max; O 0.015 max; Si 0.005 max; Ta 0.1 max; W 0.03 max; Zr 0.8-1.2	195	125	20
Nb	Nb-1%Zr R04261	USA			Commercial Grade			0.01 max			0.005 max				Fe 0.01 max; H 0.001 max; Hf 0.02 max; Si 0.005 max; N 0.01 max; O 0.025 max; Ta 0.2 max; W 0.05 max; Zr 0.8-1.2			
Nb	Nb-1%Zr R04261	USA	ASTM B391	Nb Type 4	Ingot Commercial Grade			0.01 max			0.005 max				Fe 0.01 max; H 0.0015 max; Hf 0.02 max; Mo 0.010 max; N 0.01 max; O 0.025 max; Si 0.005 max; Ta 0.2 max; W 0.05 max; Zr 0.8-1.2			
Nb	Nb-1%Zr R04261	USA	ASTM B392	Nb Type 4	Bar Rod Wir Commercial Grade			0.01 max			0.005 max				Fe 0.01 max; H 0.0015 max; Hf 0.02 max; Mo 0.010 max; N 0.01 max; O 0.025 max; Si 0.005 max; Ta 0.2 max; W 0.05 max; Zr 0.8-1.2	195	125	20
Nb	Nb-1%Zr R04261	USA	ASTM B393	Nb Type 4	Strp Sh Plt Commercial Grade			0.01 max			0.005 max				Fe 0.01 max; H 0.0015 max; Hf 0.02 max; Mo 0.010 max; N 0.01 max; O 0.025 max; Si 0.005 max; Ta 0.2 max; W 0.05 max; Zr 0.8-1.2	195	125	20

UNS numbers and US grades are provided as a means of cross referencing chemically similar alloys. Exchangability is only possible after independent examination of specifications. Tensile properties are minimum or typical . UTS and YS as Mpa, El as %. See Appendix for list of abbreviations used in Descriptions.

Worldwide Guide to Equivalent Nonferrous Metals and Alloys

Alloy	Grade UNS #	Country	Specification	Designation	Description	Ag	Au	C	Cd	Cu	Ni	Pb	Zn	OT max	Other	UTS	YS	EL
Nb	Nb-1%Zr R04261	USA	ASTM B394	Nb Type 4	Smls Weld Tub Commercial Grade			0.01 max			0.005 max				Fe 0.01 max; H 0.0015 max; Hf 0.02 max; Mo 0.010 max; N 0.01 max; O 0.025 max; Si 0.005 max; Ta 0.2 max; W 0.05 max; Zr 0.8-1.2	195	125	20
Nb	R04295	USA			Obsolete			0.015 max							H 0.0015 max; Hf 9-11; N 0.010 max; O 0.020 max; Ta 0.500 max; Ti 0.7-1.3; W 0.500 max; Zr 0.700 max			
Nb	R04271	USA			Obsolete			0.030 max							Fe 0.02 max; H 0.001 max; N 0.010 max; O 0.020 max; Ta 0.15 max; Ti 0.01 max; W 9.00-11.00; Zr 2.0-3.0			
Pd	P03995	USA			Refined	0.005 max	0.01 max			0.005 max	0.005 max	0.005 max			Al 0.005 max; Ca 0.005 max; Co 0.001 max; Fe 0.005 max; Mg 0.005 max; Mn 0.001 max; Si 0.005 max; Sb 0.002 max; Pt+Rh+Ru+Ir 0.03 max; Pd 99.95 min; Sn 0.005 max; Zr 0.0025 min			
Pd	P03980	USA			Refined										Ir 0.05 max; Pt 0.15 max; Rh 0.10 max; Ru 0.05 max; Pd 99.80 min			
Pd	P03657	USA			Braz Vacuum			0.005 max	0.001 max		0.06 max	0.002 max			P 0.002 max; Pd 64.0-66.0; Zr 0.001 min			
Pd	P03657	USA	AWS A5.8-92	BVPd-1	Braze Weld Fill Grade 2 vacuum serv.			0.01 max	0.002 max		0.06 max	0.002 max	0.002 max		P 0.002 max; Pd 64.0-66.0; bal Co			
Pd	P03657	USA	AWS A5.8-92	BVPd-1	Braze Weld Fill Grade 1 vacuum serv.			0.005 max	0.001 max		0.06 max	0.002 max	0.001 max		P 0.002 max; Pd 64.0-66.0; bal Co			
Pd	P03440	USA			Elect Contact	37.0-39.0				15.5-16.5	0.8-1.2				Pt 0.8-1.2; Pd 43.0-45.0			
Pd	P03300	USA				29.0-31.0	9.5-10.5			13.5-14.5				0.01	Pt 9.5-10.5; Pd 34.0-36.0			
Pd	P03350	USA			Elect Contact	29-31	9.5-10.5			13.5-14.5			0.6 max	0.1	Pt 9.5-10.5; Pd 34.0-36.0; Zr 1.2 min			
Ta	R05200	USA	ASTM B364	R05200	Ingot EB/VAR Cast Unalloyed			0.010 max			0.010 max				Fe 0.010 max; H 0.0015 max; N 0.010 max; Ti 0.010 max; Mo 0.020 max; Nb 0.10 max; O 0.015 max; Si 0.005; W 0.050 max			

UNS numbers and US grades are provided as a means of cross referencing chemically similar alloys. Exchangability is only possible after independent examination of specifications. Tensile properties are minimum or typical . UTS and YS as Mpa, El as %. See Appendix for list of abbreviations used in Descriptions.

Alloy	Grade UNS #	Country	Specification	Designation	Description	Ag	Au	C	Cd	Cu	Ni	Pb	Zn	OT max	Other	UTS	YS	EL
Ta	R05200	USA	ASTM B365	R05200	Rod Wir EB/VAR Cast Unalloyed			0.010 max			0.010 max				Fe 0.010 max; H 0.0015 max; N 0.010 max; Ti 0.010 max; Mo 0.020 max; Nb 0.10 max; O 0.015 max; Si 0.005; W 0.050 max	172	103	25
Ta	R05400	USA	ASTM B364	R05400	Ingot Sintered Unalloyed			0.010 max			0.010 max				Fe 0.010 max; H 0.0015 max; N 0.010 max; Ti 0.010 max; Mo 0.020 max; Nb 0.10 max; O 0.03 max; Si 0.005; W 0.050 max			
Ta	R05400	USA	ASTM B365	R05400	Rod Wir Sintered Unalloyed			0.010 max			0.010 max				Fe 0.010 max; H 0.0015 max; N 0.010 max; Ti 0.010 max; Mo 0.020 max; Nb 0.10 max; O 0.03 max; Si 0.005; W 0.050 max	172	103	25
Ta	Ta-2.5%W R05252	USA	ASTM B364	R05252	Ingot EB/VAR Cast			0.010 max			0.010 max				Fe 0.010 max; H 0.0015 max; N 0.010 max; Ti 0.010 max; Mo 0.020 max; Nb 0.50 max; O 0.015 max; Si 0.005; W 2.0-3.5			
Ta	Ta-2.5%W R05252	USA	ASTM B365	R05252	Rod Wir EB/VAR Cast			0.010 max			0.010 max				Fe 0.010 max; H 0.0015 max; N 0.010 max; Ti 0.010 max; Mo 0.020 max; Nb 0.50 max; O 0.015 max; Si 0.005; W 2.0-3.5	482	379	20
Ta	Ta-10%W R05255	USA	ASTM B364	R05255	Ingot EB/VAR Cast			0.010 max			0.010 max				Fe 0.010 max; H 0.0015 max; N 0.010 max; Ti 0.010 max; Mo 0.020 max; Nb 0.10 max; O 0.015 max; Si 0.005; W 9.0-11.0			
Ta	Ta-10%W R05255	USA	ASTM B365	R05255	Rod Wir EB/VAR Cast			0.010 max			0.010 max				Fe 0.010 max; H 0.0015 max; N 0.010 max; Ti 0.010 max; Mo 0.020 max; Nb 0.10 max; O 0.015 max; Si 0.005; W 9.0-11.0	276	193	20
Ta	Ta-40%Nb R05240	USA	ASTM B364	R05240	Ingot EB/VAR Cast			0.010 max			0.010 max				Fe 0.010 max; H 0.0015 max; N 0.010 max; Ti 0.010 max; Mo 0.020 max; O 0.020 max; Si 0.005; Nb 35.0-42.0; W 0.050 max			

UNS numbers and US grades are provided as a means of cross referencing chemically similar alloys. Exchangability is only possible after independent examination of specifications. Tensile properties are minimum or typical . UTS and YS as Mpa, El as %. See Appendix for list of abbreviations used in Descriptions.

Worldwide Guide to Equivalent Nonferrous Metals and Alloys

Alloy	Grade UNS #	Country	Specification	Designation	Description	Ag	Au	C	Cd	Cu	Ni	Pb	Zn	OT max	Other	UTS	YS	EL
Ta	Ta-40%Nb R05240	USA	ASTM B365	R05240	Rod Wir EB/VAR Cast			0.010 max			0.010 max				Fe 0.010 max; H 0.0015 max; N 0.010 max; Ti 0.010 max; Mo 0.020 max; O 0.020 max; Si 0.005; Nb 35.0-42.0; W 0.050 max	276	193	25
W	R07005	USA	ASTM F288											0.05	OE 0.01 max; W 99.95 min			
W	R07005	USA	ASTM F290											0.05	OE 0.01 max; W 99.95 min			
W	R07900	USA	AWS A5.12-92	EWP	El Arc Weld Cut									0.5	W 99.5 min			
W	R07900	USA	ASME SFA-5.12		Arc Weld									0.5	W 99.5 min			
W	R07920	USA	ASME SFA-5.12	EWZr-1	Arc Weld										ZrO 0.15-0.40; W 99.1 min			
W	R07920	USA	AWS A5.12-92	EWZr-1	El Arc Weld Cut									0.5	ZrO 0.15-0.40; W 99.1 min			
W	R07006	USA	AMS 7898					0.008 max			0.005 max				Al 0.005 max; Fe 0.005 max; H 0.001 max; Mo 0.020 max; N 0.002 max; O 0.005 max; Si 0.005 max			
W	R07911	USA	ASME SFA-5.12	EWTh-1	Arc Weld									0.5	ThO 0.8-1.2; W 98.3 min			
W	R07941	USA	AWS A5.12-92	EWLa-1	El Arc Weld Cut									0.5	ThO2 0.8-1.2; W 98.3 min			
W	R07941	USA	ASME SFA-5.12	EWLa-1	Arc Weld										La2O3 0.9-1.2; W 98.3 min			
W	R07911	USA	AWS A5.12-92	EWTh-1	El Arc Weld Cut									0.5	ThO2 0.8-1.2; W 98.5 min			
W	R07030	USA													W 96-98			
W	R07031	USA	ASTM F73											0.05	Re 2.5-3.5; OE 0.01 max			
W	R07912	USA	ASME SFA-5.12	EWTh-2	Arc Weld									0.5	ThO 1.7-2.2; W 97.3 min			
W	R07932	USA	AWS A5.12-92	EWCe-2	El Arc Weld Cut									0.5	CeO2 1.8-2.2; W 97.3 min			
W	R07932	USA	ASME SFA-5.12	EWCe-2	Arc Weld									0.5	CeO2 1.8-2.2; W 97.3 min			
W	R07912	USA	AWS A5.12-92	EWTh-2	El Arc Weld Cut									0.5	ZrO2 1.7-2.2; W 97.5 min			
W	R07050	USA													W 94-96			
W		USA	AWS A5.12-92	EWG	El Arc Weld Cut									0.5	W 94.5 min			
W	R07080	USA													W 91-94			
W	R07100	USA													W 89-91			
Zr	R60001	USA	ASTM B351	R60001	Bar Rod Wir HR CF Nuclear Applications Ann RT			0.027 max	0.00005 max	0.0050 max	0.0070 max				Al 0.0075 max; B 0.00005 max; Cr 0.020 max; Co 0.0020 max; Fe 0.150 max; H .0025 x; Hf0.010x; Mg 0.0020 max; Mn 0.0050 max; Mo 0.0050 max; N 0.0080 max; Sn 0.0050 max; Si 0.0120 max; U 0.00035 max; W 0.010 max	296	138	18

UNS numbers and US grades are provided as a means of cross referencing chemically similar alloys. Exchangability is only possible after independent examination of specifications. Tensile properties are minimum or typical . UTS and YS as Mpa, El as %. See Appendix for list of abbreviations used in Descriptions.

12-12 Miscellaneous

Alloy	Grade UNS #	Country	Specification	Designation	Description	Ag	Au	C	Cd	Cu	Ni	Pb	Zn	OT max	Other	UTS	YS	EL
Zr	R60001	USA	ASTM B352	R60001	Sh Strp Plt Nuclear ApplicationAnn RT longit			0.027 max	0.00005 max	0.0050 max	0.0070 max				Al 0.0075 max; B 0.00005 max; Cr 0.020 max; Co 0.0020 max; Fe 0.150 max; H 0.0025 max; Hf 0.010 max; Mg 0.0020 max; Mn 0.0050 max; Mo 0.0050 max; N 0.0080 max; Si 0.0120 max; Sn 0.0050 max; U 0.00035 max; W 0.010 max	296	138	18
Zr	R60001	USA	ASTM B353	R60001	Smls Weld Tub Nuclear Service			0.027 max	0.00005 max	0.0050 max	0.0070 max				Al 0.0075 max; B 0.00005 max; Cr 0.020 max; Co 0.0020 max; Fe 0.150 max; H 0.0025 max; Hf 0.010 max; Mg 0.0020 max; Mn 0.0050 max; N 0.0080 max; Mo 0.0050 max; Sn 0.0050 max; Si 0.0120 max: Ti 0.0050 max; U 0.00035 max; W 0.010 max			
Zr	R60701	USA			Obsolete			0.05 max							Zr+Hf 99.5 min; Fe+Cr 0.05 max; H 0.005 max; Hf 4.5; N 0.025 max			
Zr	R60702	USA	ASTM B495	R60702	Ingot			0.05 max							Zr+Hf 99.20 min; Fe+Cr 0.2 max; H 0.004 max; Hf 4.5 max; N 0.020 max; O 0.16 max			
Zr	R60702	USA	ASTM B493	R60702	Frg			0.05 max							Zr+Hf 99.2 min; Fe+Cr 0.2 max; H 0.005 max; Hf 4.5 max; N 0.025 max; O 0.16 max	379	207	16
Zr	R60702	USA	ASTM B523	R60702	Smls Weld Tub			0.05 max							Zr+Hf 99.2 min; Fe+Cr 0.2 max; H 0.005 max; Hf 4.5 max; N 0.025 max; O 0.16 max	379	207	16
Zr	R60702	USA	ASTM B550	R60702	Bar Wir			0.05 max							Zr+Hf 99.2 min; Fe+Cr 0.2 max; H 0.005 max; Hf 4.5 max; N 0.025 max; O 0.16 max	379	207	16
Zr	R60702	USA	ASTM B551	R60702	Strp Sh Plt			0.05 max							Zr+Hf 99.2 min; Fe+Cr 0.2 max; H 0.005 max; Hf 4.5 max; N 0.025 max; O 0.16 max	379	207	16

UNS numbers and US grades are provided as a means of cross referencing chemically similar alloys. Exchangability is only possible after independent examination of specifications. Tensile properties are minimum or typical . UTS and YS as Mpa, El as %. See Appendix for list of abbreviations used in Descriptions.

Alloy	Grade UNS #	Country	Specification	Designation	Description	Ag	Au	C	Cd	Cu	Ni	Pb	Zn	OT max	Other	UTS	YS	EL
Zr	R60702	USA	ASTM B658	R60702	Smls Weld Pip			0.05 max							Zr+Hf 99.2 min; Fe+Cr 0.2 max; H 0.005 max; Hf 4.5 max; N 0.025 max; O 0.16 max	379	207	16
Zr	R60702	USA	ASTM B494	R60702	Primary metal Sponge			0.05 max							Zr+Hf 99.2 min; Fe+Cr 0.2 max; H 0.005 max; Hf 4.5 max; N 0.025 max			
Zr	R60702	USA	AWS A5.24-90	ERZr2	Weld El Rod			0.05 max							Zr + Hf 99.01 min, Fe + Cr 0.2 max; H 0.005 max; Hf 4.5 max; N 0.025 max; O 0.16 max			
Zr	R60802	USA	ASTM B351	R60802	Bar Rod Wir HR CF Nuclear Applications Ann RT			0.027 max	0.00005 max	0.0050 max	0.03-0.08				Al 0.0075 max; B 0.00005 max; Co 0.002 max; Cr 0.05-0.15; Fe 0.07-0.20; H 0.0025 max; Hf 0.01 max; Mg 0.002 max; Mn 0.005 max; Mo 0.000 max; Mo 0.005 max; N 0.008 max; O 0.09-0.16; Si 0.012 max; Sn 1.20-1.70; U 0.00035 max; Fe+Cr+Ni 0.18-0.38; W 0.01 max	413	241	14
Zr	R60802	USA	ASTM B352	R60802	Sh Strp Plt Nuclear Application Ann RT longit			0.027 max	0.00005 max	0.0050 max	0.03-0.08				Al 0.0075 max; B 0.00005 max; Co 0.002 max; Cr 0.05-0.15; Fe 0.07-0.20; H 0.0025 max; Hf 0.01 max; Mg 0.002 max; Mn 0.005 max; Mo 0.005 max; Mo 0.005 max; N 0.008 max; O 0.09-0.16; Si 0.012 max; Sn 1.20-1.70; U 0.00035 max; Fe+Cr+Ni 0.18-0.38; W 0.01 max	400	241	25
Zr	R60802	USA	ASTM B353	R60802	Smls Weld Tub Nuclear Service			0.027 max	0.00005 max	0.0050 max	0.03-0.08				Al 0.0075 max; B 0.00005 max; Co 0.0020 max; Cr 0.05-0.15; Fe 0.04-0.20; H 0.0025 max; Hf 0.010 max; Mg 0.0020 max; Mn 0.0050 max; Mo 0.0050 max; N .0080x;Si0.0120 x; Sn 1.20-1.70; Ti 0.0050 max; U 0.00035 max; Fe+Cr+Ni 0.18-0.38; W 0.010 max			

UNS numbers and US grades are provided as a means of cross referencing chemically similar alloys. Exchangability is only possible after independent examination of specifications. Tensile properties are minimum or typical . UTS and YS as Mpa, El as %. See Appendix for list of abbreviations used in Descriptions.

Alloy	Grade UNS #	Country	Specification	Designation	Description	Ag	Au	C	Cd	Cu	Ni	Pb	Zn	OT max	Other	UTS	YS	EL
Zr	R60802	USA	ASTM B811	R60802	Smls Tub Nuclear Reactor Fuel Cladding Recryst Ann RT			0.027 max	0.00005 max	0.0050 max	0.03-0.08				Al 0.0075 max; B 0.00005 mx; Ca 0.003 mx; Nb 0.01 mx; Co 0.002 mx; Cr 0.05-0.15; Fe 0.07-0.2; Hf 0.01 mx; Mg 0.002 mx; Mn 0.005 mx; Mo 0.005 mx; N 0.008 mx; O 0.09-0.16; Si 0.012 mx; Sn 1.20-1.70; Ti 0.005 max; U 0.00035 max; Fe+Cr+Ni 0.18-0.38; W 0.01 mx	413	241	20
Zr	R60804	USA	ASTM B351	R60804	Bar Rod Wir HR CF Nuclear Applications Ann RT			0.027 max	0.00005 max	0.0050 max	0.0070 max				Al 0.0075 max; B 0.00005 max; Co 0.002 max; Cr 0.07-0.13; Fe 0.18-0.24; H 0.0025 max; Hf 0.01 max; Mg 0.002 max; Mn 0.005 max; Mo 0.005 max; O 0.09-0.16; Mo 0.0050 max; Si 0.012 max; Sn 1.20-1.70; U 0.00035 max; Fe+Cr 0.28-0.37; W 0.01 max	413	241	14
Zr	R60804	USA	ASTM B352	R60804	Sh Strp Plt Nuclear Application Ann RT longit			0.027 max	0.00005 max	0.0050 max	0.0070 max				Al 0.0075 max; B 0.00005 max; Co 0.002 max; Cr 0.07-0.13; Fe 0.18-0.24; O 0.09-0.16; H 0.0025 x; Hf 0.01 x; Mg 0.002 x; Mn 0.005 max; N 0.008 max; Mo 0.005 max; O 0.09-0.16; Si 0.0120 max; Sn 1.20-1.70; U 0.00035 max; Fe+Cr 0.28-0.37; W 0.01 max	400	241	25
Zr	R60804	USA	ASTM B353	R60804	Smls Weld Tub Nuclear Service			0.027 max	0.00005 max	0.0050 max	0.0070 max				Al 0.0075 max; B 0.00005 max; Co 0.0020 max; Cr 0.07-0.13; Fe 0.18-0.24; H .0025x; Hf .010 x; Mg 0.0020 max; Mn 0.0050 max; Mo 0.0050 max; N 0.0080 max; Si 0.0120 max; Sn 1.20-1.70; Ti 0.0050 max; U 0.00035 max; Fe+Cr 0.28-0.37; W 0.010 max			

UNS numbers and US grades are provided as a means of cross referencing chemically similar alloys. Exchangability is only possible after independent examination of specifications. Tensile properties are minimum or typical . UTS and YS as Mpa, El as %. See Appendix for list of abbreviations used in Descriptions.

Alloy	Grade UNS #	Country	Specification	Designation	Description	Ag	Au	C	Cd	Cu	Ni	Pb	Zn	OT max	Other	UTS	YS	EL
Zr	R60804	USA	ASTM B811	R60804	Smls Tub Nuclear Reactor Fuel Cladding Recryst Ann RT			0.027 max	0.00005 max	0.0050 max	0.0070 max				Al 0.0075 max; B 0.00005 mx; Ca 0.003 mx; Fe 0.18-0.24; H 0.0025 mx; Nb 0.01 mx; Co 0.002 mx; Cr 0.07-0.13; Hf 0.01 mx;Mg 0.002 mx;Mn 0.005 mx; Mo 0.005 mx; N 0.008 mx; O 0.09-0.16;Si 0.012 mx;Sn 1.2-1.7; Ti 0.005 mx; U 0.00035 mx; Fe+Cr 0.28-0.37; W 0.01 mx	413	241	20
Zr	R60703	USA	ASTM B494	R60703	Primary metal Sponge										Zr+Hf 98.0 min; Hf 4.5 max			
Zr	R60703	USA	ASTM B495	R60703	Ingot										Zr+Hf 98.0 min; Hf 4.5 max			
Zr	R60902	USA			Reactor Grade Obsolete			0.030 max	0.00005 max	0.3-0.7	0.0070 max	0.013 max			Al 0.0075 max; B 0.00005 mx; Co 0.002 mx; Cr .02 x; Fe 0.15 x; H 0.0025 mx; Hf 0.01 x; Mg .002 x; Mn .005 x; Mo 0.005 mx; N 0.0065 mx; Nb 2.4-2.8; O 0.08-0.12; Si 0.0012 mx; Sn 0.005 mx; Ta 0.02 mx; Ti 0.005 mx; U 0.00035 mx; V 0.005 mx; W 0.01 mx			
Zr	R60901	USA	ASTM B351	R60901	Bar Rod Wir HR CF Nuclear Applications Ann RT			0.027 max	0.00005 max	0.0050 max	0.0070 max				Al 0.0075 max; B 0.00005 max; Co 0.0020 max; Cr .020 x; Fe .150 x; H 0.0025 max; Hf 0.010 max; Mg 0.0020 max; Mn 0.0050 max; Mo 0.0050 max; N 0.0080 max; Nb 2.40-2.80; O 0.09-0.13; Si 0.0120 max; U 0.00035 max; W 0.010 max	448	310	15
Zr	R60901	USA	ASTM B352	R60901	Sh Strp Plt Nuclear Application Ann RT longit			0.027 max	0.00005 max	0.0050 max	0.0070 max				Al 0.0075 max; B 0.00005 max; Co 0.0020 max; Cr .020 x; Fe .150 x; H 0.0025 max; Hf 0.010 max; Mg .0020 x; Mn .0050 x; Mo 0.0050 x; N 0.0080 max; Nb 2.40-2.80; O 0.09-0.13; Si 0.0120 max; U 0.00035 max; W 0.010 max	448	338	20

UNS numbers and US grades are provided as a means of cross referencing chemically similar alloys. Exchangability is only possible after independent examination of specifications. Tensile properties are minimum or typical . UTS and YS as Mpa, El as %. See Appendix for list of abbreviations used in Descriptions.

12-16 Miscellaneous

Alloy	Grade UNS #	Country	Specification	Designation	Description	Ag	Au	C	Cd	Cu	Ni	Pb	Zn	OT max	Other	UTS	YS	EL
Zr	R60901	USA	ASTM B353	R60901	Smls Weld Tub Nuclear Service			0.027 max	0.00005 max	0.0050 max	0.0070 max				Al 0.0075 max; B 0.00005 max; Co 0.002 max; Cr 0.02 max; Fe 0.15 max; H 0.0025 max; Hf 0.01 max; Mg 0.002 max; Mo 0.005 max; Mn 0.005 max; Mo 0.005max; N 0.008 max; Nb 2.40-2.80; O 0.09-0.13; Si 0.012 max; Ti 0.005 max; U 0.00035 max; W 0.01 max			
Zr	R60704	USA	ASTM B493	R60704	Frg			0.05 max							Zr+Hf 97.5 min; Fe+Cr 0.2-0.4 ; H 0.005 max; Hf 4.5 max; N 0.025 max; O 0.18 max; Sn 1.0-2.0	413	241	14
Zr	R60704	USA	ASTM B523	R60704	Smls Weld Tub			0.05 max							Zr+Hf 97.5 min; Fe+Cr 0.2-0.4; H 0.005 max; Hf 4.5 max; N 0.025 max; O 0.18 max; Sn 1.0 2.0	413	241	14
Zr	R60704	USA	ASTM B550	R60704	Bar Wir			0.05 max							Zr+Hf 97.5 min; Fe+Cr 0.2-0.4; H 0.005 max; Hf 4.5 max; N 0.025 max; O 0.18 max; Sn 1.0-2.0	413	241	14
Zr	R60704	USA	ASTM B551	R60704	Strp Sh Plt			0.05 max							Zr+Hf 97.5 min; Fe+Cr 0.2-0.4; H 0.005 max; Hf 4.5 max; N 0.025 max; O 0.18 max; Sn 1.0-2.0	413	241	14
Zr	R60704	USA	ASTM B658	R60704	Smls Weld Pip			0.05 max							Zr+Hf 97.5 min; Fe+Cr 0.2-0.4; H 0.005 max; Hf 4.5 max; N 0.025 max; O 0.18 max; Sn 1.0-2.0	413	241	14
Zr	R60704	USA	ASTM B495	R60704	Ingot			0.050 max							Zr+Hf 97.5 min; Fe+Cr 0.20-0.40; H 0.005 max; Hf 4.5 max; N 0.025 max; O 0.18 max; Sn 1.00-2.00			

UNS numbers and US grades are provided as a means of cross referencing chemically similar alloys. Exchangability is only possible after independent examination of specifications. Tensile properties are minimum or typical . UTS and YS as Mpa, El as %. See Appendix for list of abbreviations used in Descriptions.

Alloy	Grade UNS #	Country	Specification	Designation	Description	Ag	Au	C	Cd	Cu	Ni	Pb	Zn	OT max	Other	UTS	YS	EL
Zr	R60704	USA	AWS A5.24-90	ERZr3	Weld El Rod			0.05 max							Zr + Hf 97.5 min, Fe+Cr .20-.40, H 0.005 max; Hf 4.5 max; N 0.025 max; O 0.16 max			
Zr	R60904	USA								0.0050 max	0.0035 max	0.0050 max			Al 0.0075 max; B 0.00005 max; Co 0.0020 max; Fe .065 x; H 0.002 x; Hf 0.005 max; Mg .002x; Mn .005x; Mo 0.005 max; N 0.0065 max; Nb 2.40-2.8; Si 0.012 max; Sn 0.1 max; Ta 0.01 max; Ti 0.005 max; U 0.00035 max; V .005 x; W .005 x			
Zr	R60705	USA	AWS A5.24-90	ERZr4	Weld El Rod			0.05 max							Zr+Hf 95.5 min, Fe+Cr .20 max.; H 0.005 max; Hf 4.5 max; N 0.025 max; Nb 2.0-3.0; O 0.16 max			
Zr	R60706	USA	ASTM B551	R60706	Strp Sh Plt			0.05 max							Zr+Hf 95.5 min; Fe+Cr 0.2 max; H 0.005 max; Hf 4.5 max; N 0.025 max; Nb 2.0-3.0; O 0.16 max	510	345	20
Zr	R60706	USA	ASTM B495	R60706	Ingot			0.050 max							Zr+Hf 95.5 min; Fe+Cr 0.2 max; H 0.005 max; Hf 4.5 max; N 0.025 max; Nb 2.0-3.0; O 0.16 max			
Zr	R60705	USA	ASTM B493	R60705	Frg			0.05 max							Zr+Hf 95.5 min; Fe+Cr 0.2 max; H 0.005 max; Hf 4.5 max; N 0.025 max; Nb 2.0-3.0; O 0.18 max	483	379	16
Zr	R60705	USA	ASTM B523	R60705	Smls Weld Tub			0.05 max							Zr+Hf 95.5 min; Fe+Cr 0.2 max; H 0.005 max; Hf 4.5 max; N 0.025 max; Nb 2.0-3.0; O 0.18 max	552	379	16
Zr	R60705	USA	ASTM B550	R60705	Bar Wir			0.05 max							Zr+Hf 95.5 min; Fe+Cr 0.2 max; H 0.005 max; Hf 4.5 max; N 0.025 max; Nb 2.0-3.0; O 0.18 max	552	379	16
Zr	R60705	USA	ASTM B551	R60705	Strp Sh Plt			0.05 max							Zr+Hf 95.5 min; Fe+Cr 0.2 max; H .005 x; Hf 4.5 x; N 0.025 max; Nb 2.0-3.0; O 0.18 max	510	345	20

UNS numbers and US grades are provided as a means of cross referencing chemically similar alloys. Exchangability is only possible after independent examination of specifications. Tensile properties are minimum or typical . UTS and YS as Mpa, El as %. See Appendix for list of abbreviations used in Descriptions.

12-18 Miscellaneous

Alloy	Grade UNS #	Country	Specification	Designation	Description	Ag	Au	C	Cd	Cu	Ni	Pb	Zn	OT max	Other	UTS	YS	EL
Zr	R60705	USA	ASTM B658	R60705	Smls Weld Pip			0.05 max							Zr+Hf 95.5 min; Fe+Cr 0.2 max; H 0.005 max; Hf 4.5 max; N 0.025 max; Nb 2.0-3.0; O 0.18 max	552	379	16
Zr	R60705	USA	ASTM B495	R60705	Ingot			0.050 max							Zr+Hf 95.5 min; Fe+Cr 0.2 max; H 0.005 max; Hf 4.5 max; N 0.025 max; Nb 2.0-3.0; O 0.18 max			
Zr	R60707	USA			Weld Fill Obsolete			0.05 max							Fe 0.20 max; Fe includes Cr; Zr+Hf 95.5 min; H 0.005 max; Hf 4.5 max; N 0.025 max; Nb 2.0-3.0			

UNS numbers and US grades are provided as a means of cross referencing chemically similar alloys. Exchangability is only possible after independent examination of specifications. Tensile properties are minimum or typical . UTS and YS as Mpa, El as %. See Appendix for list of abbreviations used in Descriptions.

Abbreviations

Aged	Aged		OE	Others each
AH	Age hardened		OT	Others total
Ann	Annealed		PH	Precipitation hardened
Apps	Applications		PHT	Precipitation heat treated
As Weld	As welded		Pip	Pipe
Auto	Automotive		Plt	Plate
Ave	Average		Powd	Powder
Bar	Bar		Press	Pressure
Bil	Billet		Prl	Pearlitic
Cast	Casting/Cast		Quen	Quenched and aged
CD	Cold drawn		QT	Quenched
Centrif	Centrifugal		Q/T	Quenched/tempered
CF	Cold finished/formed		Res	Resistant
Conds	Condenser		Rng	Ring
Cor	Corrosion		Rod	Rod
CR	Cold rolled		RT	Room temperature
CW	Cold worked		SA	Solution annealed
DA	Duplex annealed		Sand	Sand cast
Diam	Diameter		Sec	Section
El	Electrode or Elongation		Sh	Sheet
			Shp	Shape
ELI	Extra Low Interstitial		SHT	Solution heat treated
ERW	Electrical resistance welded		ST	Solution treated or Specials total
Ext	Extrusion		Sint	Sintered
Exch	Exchanger		SMAW	Shielded Metal Arc Welding
Fill	Filler		Smls	Seamless
Frg	Forging(s)		Sol	Solution
Hard	Hardness		SR	Stress relieved
HB	Brinell hardness		ST	Solution treated
HE	Hot extruded		STA	Solution treated and aged
Heat	Heat			
HF	Hot finished		Stab	Stabilized
HIP	Hot isostatic pressing		STOA	Solution treated and overaged
HR	Hot rolled			
HT	Heat treated		Strp	Strip
HW	Hot worked/wrought		Surg	Surgical
Imp	Implant		TA	Triplex annealed
Inv	Investment		Temp	Tempered
Liq	Liquid		Thk	Thickness
Mach	Machined		Tub	Tube
Mart	Martensitic		UTS	Ultimate tensile strength
Met	Metal		Ves	Vessel
Mult	Multiple		Weld	Welded/Welding
NHT	Not heat treated		Wir	Wire
Nom	Nominal		YS	Yield Strength
Norm	Normalized			

Africa

African Regional Organization for Standardization (ARSO)
PO Box 57363
Nairobi
Republic of Kenya
Tel: 254 2 330882
Fax: 254 2 218792

Founded in 1977, the African Regional Organization for Standardization (ARSO) has 24 members. It conducts quality control and metrology activities, certification marking operations, and laboratory tests.

Albania

Drejtoria e Standardizimit dhe Cilesise (DSC)
Rruga Mine Peza
Tirana
Albania
Tel: 355 42 2 62 55
Fax: 355 42 2 62 55

Algeria

Institut algerien de normalisation et de propriete industrielle (INAPI)
5, rue Abou Hamou Moussa
B.P. 403 - Centre de trl
Alger
Algeria
Tel: 213 2 63 96 42
Fax: 213 2 61 09 71

Argentina

Instituto Argentino de Racionalizacion de Materiales (IRAM)
Chile 1192
1098 Buenos Aires
Argentina
Tel: 54 1 383 37 51
Fax: 54 1 383 84 63
postmaster@iram.org.ar

IRAM is the Argentine standardization institute and the only organization in Argentina authorized by the government to deal, both in national and international ambits, with all the affairs related to standardization and quality control certification.

IRAM is a private society, and a member of ISO (International Organization for Standardization) and COPANT (Pan American Standards Commission).

Australia

Standards Australia (SAA)
P.O. Box 1055
Strathfield - NSW 2135
Australia
Tel: 61 2 746 47 00
    61 2 746 4748
Fax: 61 2 746 8450
    61 2 746 4765
intsect@saa.sa.telememo.au
http://www.standards.com.au/~sicsaa/

Standards Australia (SAA) was founded in 1922 and changed to its current name in 1988. It issues standards used primarily by firms doing business in Australia and the southwest Pacific area. There are currently around 6000 Australian standards, maintained by approximately 10,000 voluntary experts serving on a total of 1800 technical committees, and backed up by a full-time staff of over 300.

Australian standards appear as 1 to 4 digit numerical codes preceded by the upper case letters AS. Designations may also appear with the standard, and these should be separated by a space.

Example: AS 1446; AS 1565 80 A; AS 1867 1050

Austria

Osterreichisches Normungsinstitut (ON)
Heinestrasse 38
Postfach 130
A-1021 Wien
Austria
Tel: 43 1 213 00
Fax: 43 1 21 30 06 50
iro@tbxa.telecom.at

The Austrian Standards Institute (ON), founded in 1920, creates and publishes Austrian standards. The organization also recommends foreign standards for use in Austria.

Example: ONORM M3430; ONORM M3429; ONORM M3421 Al99.98

Belarus

Committee for Standardization, Metrology and Certification (BELST)
Starovilensky Trakt 93
Minsk 220053
Belarus
Tel: 375 172 37 52 13
Fax: 375 172 37 25 88
belst@mcsm.belpak.minsk.by

Belgium

Institut belge de normalisation (IBN)
Av. de la Brabanconne 29
B-1000 Bruxelles
Belgium
Tel: 32 2 738 01 11
Fax: 32 2 733 42 64

Created in 1946, the IBN consists of approximately 600 members, both individuals and businesses. The designations are prefixed with the letters NBN, Belgian designations are different for nonalloyed and alloyed steel. For nonalloyed steel (including carbon steel), the conventional designation usually consists of a letter, a number code, and a possible variable third part. There are three different criteria for classification: mechanical characteristics, technological characteristics, and chemical composition.

For alloyed steels, the designation system varies for heavily alloyed (above 5%) and slightly alloyed steel.

Example: NBN D 02-002

Brazil

Associacao Brasileira de Normas Tecnicas (ABNT)
Av. 13 de Maio, no 13, 27o andar
Caixa Postal 1680
20003-900 Rio de Janeiro-RJ
Brazil
Tol: 55 21 210 31 22
Fax: 55 21 532 21 43

The Brazilian Association of Technical Standards (ABNT) issues national standards. These designations now begin with upper case letters NBR and are followed by a 4-digit numerical code. Projects are coded as Committee: Subcommittee; Working Group- Sequential number.

Example: NBR 5000. Other examples are ABNT 1040; NB 82/79

Bulgaria

Committee for Standardization and Metrology at the Council of Ministers (BDS)
21, 6th September Str.
1000 Sofia
Bulgaria
Tel: 359 2 85 91
Fax: 359 2 80 14 02

Bulgarian standards are issued by the Committee for Standardization and Metrology. These designations begin with the upper case letters BDS and are followed by the standard's numerical code.

Example: BDS 7938; BDS 6751

Canada

Canadian Standards Association (CSA)
178 Rexdale Blvd.
Rexdale Ontario M9W 1R3
Canada
Tel: 416 747 4000
Fax: 416 747 4149

Founded in 1917, the Canadian Standards Association (CSA) has about 8,000 members both individual and corporate. The organization issues standards used primarily in Canada, but also in commerce between Canada and other countries.

All Canadian standards are preceded by the upper case letters CSA. The standard or designation then follows.

Examples: CSA GR20; CSA SG121; CSA S5; CSA GH.1.7.3; CSA HA.4.1100

Canada

Standards Council of Canada (SCC)
45 O'Connor Street, Suite 1200
Ottawa Ontario K1P 6N7
Canada
Tel: 1 613 238 32 22
Fax: 1 613 995 45 64
info@scc.ca

The Standards Council of Canada (SCC) is not responsible for writing the standards, but is the distributor of documents.

Chile

Instituto Nacional de Normalizacion (INN)
Matias Cousino 64 - 6o piso
Casilla 995 - Correo Central
Santiago
Chile
Tel: 56 2 696 81 44
Fax: 56 2 696 02 47

China

China State Bureau of Technical Supervision (CSBTS)
Dept. of Stand-4, Zhichun Road
Haidian District
P.O. Box 8010
Beijing 100088
Peoples Republic of China
Tel: 86 10 203 24 24
Fax: 86 10 203 10 10

Columbia

Instituto Columbiano de Normas Tecnicas (ICONTEC)
Carrera 37 52-95
Edificio ICONTEC
P.O. Box 14237
Santafe de Bogota
Columbia
Tel: 57 1 315 03 77
Fax: 57 1 222 14 35
sicontec@itecs5.telecom-co.net

Costa Rica

Instituto de Normas Tecnicas de Costa Rica (INTECO)
P.O. Box 6189-1000
San Jose
Costa Rica
Tel: 506 283 45 22
Fax: 506 283 48 31
inteco@sol.racsa.co.cr

Croatia

State Office for Standardization and Metrology (DZNM)
Ulica grada Vukovara 78
10000 Zagreb
Croatia
Tel: 385 1 53 99 34
Fax: 385 1 53 65 98

Cuba

Oficina Nacional de Normalizacion (NC)
Calle E. No. 261 entre 11 y 13
Vedado La Habana 10400
Cuba
Tel: 53 7 30 00 22
Fax: 53 7 33 80 48

Cyprus

Cyprus Organization for Standards and Control of Quality (CYS)
Ministry of Commerce, Industry and Tourism
Nicosia 1421
Cyprus
Tel: 357 2 37 50 53
Fax: 357 2 37 51 20

Czech Republic

Czech Office for Standards, Metrology and Testing (COSMT)
Biskupsky dvur 5
113 47 Praha 1
Czech Republic
Tel: 42 2 232 44 30
Fax: 42 2 232 43 73

The Czech Office for Standards, Metrology and Testing (COSMT) is a government agency concerned with standardization, metrology, testing, certification, and accreditation. The COSMT is a member of ISO and IEC, an affiliate of CEN and CENELEC.

Czech standards are arranged according to classes and subgroups by a six-digit reference number. All standards are preceded by CSN.

Example: CSN 01 0010

Denmark

Dansk Standard (DS)
Baunegaardsvej 73
Hellerup DK-2900
Denmark
Tel: 45 39 77 01 01
Fax: 45 39 77 02 02

The Danish Standards Association (DS) was founded in 1926 and is involved in the standardization of all fields except telecommunications.

DS is accredited to certify quality assurance systems according to ISO 9000 series. The organization is composed of 110 members.

Example: DS/EN 10025, DS/ISO 3798, DS/IEC 141-1, DS 13080-1

Ecuador

Instituto Ecuatoriano de Normalizacion (INEN)
Baquerizo Moreno 454 y
Av. 6 de Diciembre
Casilla, Quito 17-01-3999
Ecuador
Tel: 593 2 56 56 26
Fax: 593 2 56 78 15
inenl@inen.gov.ec

Egypt

Egyptian Organization for Standardization and Quality Control (EOS)
2 Latin America Street
Garden City Cairo
Egypt
Tel: 20 2 354 97 20
Fax: 20 2 355 78 41

The Egyptian Organization for Standardization was established in 1957 to be responsible for elaborating standard specifications for raw materials, products, technical operations, apparatuses, machines, measurement units, terminology, definitions, unified symbols classification. In 1979 the organization name was changed to include Quality Control.

Ethiopia

Ethiopian Authority for Standardization (ESA)
P O Box 2310
Addis Ababa
Ethiopia
Tel: 251 1 61 01 11
Fax: 251 1 61 31 77

Europe

Association Europeenne des Constructeurs de Materiel Aerospatial (AECMA)
88 Bd Malesherbes
F-75008 Paris
France
Tel: 33 1 4563 82 85
Fax: 33 1 4225 15 48

Within the European Association of Manufacturers of Aerospace Material (AECMA) standardization is carried out by the Standardization Committee (CN). The subcommittees responsible for this work are represented by the aerospace industry, the processing industries, public bodies and authorities, and commerce and science. The number of members is not limited. AECMA standards and designations begin with the prefix AECMA. A standard's numerical code is preceded with the lower and upper case letters prEN. Designations are alphanumeric.

Example: AECMA prEN2002-03; AECMA prEN2389; AECMA Co-P 92-HT; AECMA A1-P13 Pl-T3

Europe

European Committee for Standardization (CEN)
36, Rue de Strassart
B-1050 Brussels
Belgium
Tel: 32 2 519 68 11
Fax: 32 2 519 68 19

Founded in 1961, the European Committee for Standardization (CEN) is one of the three organizations responsible for voluntary standardization in the European Union. Development of standards related to electrotechnology and telecommunications is entrusted to CENELEC (The European Committee for Electronical Standardization) and ETSI (The European Telecommunications Standards Institute) respectively. CEN has an agreement for technical cooperation with the International Organization for Standardization (ISO). The adoption of ISO standards is optional for most of the countries who are members of ISO, in Europe the national standards bodies are obliged to adopt European Standards and withdraw conflicting national standards. However, no one is obliged to use European Standards. CEN's main sectors of activity are: information technology; biology and biotechnology; quality, certification, testing; transport and packaging; food; materials including ECISS; chemistry; and linked to European technical legislation: mechanical engineering; building and civil engineering; health technology; environment; health and safety at the workplace; gas and other energies; and consumer goods, sports, leisure.

Europe

Commission of the European Communities (CEC or CCE)
2100 M Street NW
7th Floor
Washington DC 20037
USA
Tel: 202 862 9500
Fax: 202 429 1766

Distribution of Euronorm standards is now handled by the national standards institutions of the Member States of the European Community. Standards issued are prefaced by the letters EURONORM. These are followed by a numerical code which simply numbers the standards chronologically followed by an indication of the year when issued or last updated.

Example: EURONORM 137-87

Finland

Finnish Electrotechnical Standards Assoc. (SESKO)
P O Box 134
SF-00211 Helsinki
Finland
Tel: 358 0 696 31
Fax: 358 0 677 059

The Finnish Electrotechnical Standards Association (SESKO) is a member of the Finnish Standards Association (SFS) which is a central coordinating body, also private by constitution.

SESKO is composed of member bodies representing professional associations and governmental institutions.

SESKO forms the Finnish National Committee of the International Electrotechnical Commission (IEC), the European committee for Electrotechnical Standardization (CENELEC), the CENELEC Electronic Components Committee (CECC) and the Nordic standardization co-operation (NOREK).

Finland

Finnish Standard Association SFS (SFS)
P O Box 116
Fin 00241 Helsinki
Finland
Tel: 358 0 149 9331
Fax: 358 0 146 4925
sfs@sfs.fi
http://www.sts.fi/

The Finnish Standards Association (SFS), founded in 1924, is an independent, non-profit organization. Its members include professional, commercial and industrial organizations, and the state of Finland represented by the ministries. The total number of SFS Standards amounts to over 6000. Finnish standards and designations were preceded by the letters SFS.

Most of the Finnish national material standards will disappear. As a member of the European Standards Committee (CEN) all European standards will be implemented.

France

Delegation Generale pour L'Armement (AIR)
Centre de Documentation de l'Armenment
26, Boulevard Victor
00460 - Armees
France
Tel: 33 1 4552 45 24
Fax: 33 1 4552 45 74

The French Ministry of Defense issues AIR standards. The prefix AIR in upper case letters appears with these designations.

Example: AIR 9165-001; AIR 9165-211

France

Association Francaise de Normalisation (AFNOR)
Tour Europe-Cedex 7
F-92080 Paris la Defense
France
Tel: 33 1 42 91 55 55
         33 1 42 91 58 07
Fax: 33 1 42 91 56 56

The Association Francaise de Normalisation (AFNOR) is a non-profit organization founded in 1926. Of its nearly 15,600 standards, more than 1,000 relate to metallurgy and are used widely in Europe, Africa, Asia, the Middle East, and the Caribbean. AFNOR standards usually begin with the letters NF.

AFNOR is the French member to ISO and CEN and is participating in the preparation of international standards.

Examples: NF A 35-550 for steel grade XC 38 (special unalloyed steel for heat treatment). NF A 35-5557 for steel grade 35 N CD 16 (a steel in which no alloying elements exceeds a proportion of 5% by weight). NF F 80-107 for steel grade Z 120 M 12 (a steel in which the carbon concentration lies between 1.05 and 1.35% and the manganese concentration lies between 11 and 14%). NF A 32-058 for cast steel grade Z 120 M 12 M.

Germany

DIN Deutsches Institut fur Normung e.V. (DIN)
Burggrafenstrasse 6
Postfach 1107
D-10787 Berlin
Germany
Tel: 49 30 2601 2344
Fax: 49 30 2601 1231
postmaster@din.de

The German Institute for Normalization (DIN) standards are developed by a non-profit organization consisting of about 26 Technical Groups.   Membership is voluntary and open to both German and foreign companies.

All German standards are preceded by the upper case letters DIN and followed by a numerical or alphanumerical code. An upper case letter sometimes precedes this code. German designations are reported in one of two methods. One method uses a descriptive code number with chemical symbols and numbers in the designation; the second, known as the Werkstoff number, uses numbers only with a decimal point after the first digit. (The latter method was devised to be more compatible with computerization.)

Examples: DIN E17440 X5CrNi1810 and DIN 17442 G-X20CrMo13 (standard and designation); DIN 17745 1.4120 (standard and Werkstoff number).

Ghana

Ghana Standards Board (GSB)
P O Box M 245
Accra
Ghana
Tel: 233 21 50 00 65
Fax: 233 21 50 00 92

Greece

Hellenic Organisation for Standardization (ELOT)
313 Acharnon Street
GR-111 45 Athens
Greece
Tel: 30 1 201 5025
Fax: 30 1 202 0776

Hellenic Organisation for Standardization (ELOT), founded in 1976, is a non-profit organization established under Greek law aiming at the promotion and application of Standardization, quality and related activities in Greece. ELOT is the body for the elaboration, publication and distribution of Hellenic Standards - the national member of ISO, TEC, CEN, CENELEC.

Hungary

Magyar Szabvanyugyi Hivatal (MSZH)
Postafiok 24
H1450 Budapest 9
Hungary
Tel: 36 1 218 30 11
Fax: 36 1 218 51 25

Hungarian standards are developed by the Hungarian Office for Standardization (MSZH) as founded in 1921. The agency is also a member of the ISO and affiliate in CEN and CENELEC.

Traditional steel grade designation have been adopted from other standards for Hungarian use. In some cases new designation methods have been developed, but for cast steel alloys Germany's DIN designation system has been adopted. Wrought nonferrous metals designations and compositions have been adopted for EN European Standards and ISO International Standards for Hungarian use.

Example: MSZ 1300; MSZ NI 499; MSZ KGST 483; MSZ ISO; MSZ EN

Iceland

Icelandic Council for Standardization (STRI)
Keldnaholt
IS-112 Reykjavik
Iceland
Tel: 354 587 70 00
Fax: 354 587 74 09
stri@iti.is

The Icelandic Council for Standardization is a non-governmental organization established in 1993. A predecessor of STRI, a special national standards board, was established in 1987. STRI is by law responsible for standardization in Iceland, including the adoption and publication of National Standards (identified by the prefix IST). STRI is also by law the national member of international and European standards organizations.

The Icelandic Council for Standardization is a member of CEN and therefore implements all EN standards for alloys. After implementation the standards get the Icelandic national prefix IST EN xxx.

India

Bureau of Indian Standards (BIS)
Manak Bhavan
9 Bahadur Shah Zafar Marg
New Delhi 110002
India
Tel: 91 11 323 79 91
Fax: 91 11323 40 62

The Bureau of Indian Standards (BIS) is responsible for issuing national standards. Indian standards begin with the prefix IS and are followed by a numerical code.

Example: IS:3930; IS:5517

Indonesia

Dewan Standardisasi Nasional-DSN (Standardization Council of Indonesia) (DSN)
Sasana Widya Sarwono Lantai 5
Jl. jend. Gatot Subroto No. 10
Jakarta 12710
Indonesia
Tel: 62 21 520 6574
Fax: 62 21 520 6574

The Standardization Council of Indonesia (DSN), established by the Presidential Decree No. 20/1984 and revised by the Presidential Decree No 7/1989, is a national body to coordinate, syncronize and develop standardization and metrology within a National Standardization System covering activities in standardization, testing, metrology, and quality assurance in Indonesia.

International

International Organization for Standardization (ISO)
1, rue de Varembe
Case postale 56
CH-1211 Geneve 20
Switzerland
Tel: 41 22 749 0111
Fax: 41 22 733 3430
central@isocs.iso.ch
http://www.iso.ch/

ISO (the International Organization for Standardization) is a worldwide federation of national standards bodies, usually comprised of over 100 members (one member from each country). The object of ISO is to promote the development of standardization and related activities in the world with a view to facilitating international exchange of goods and services, and to developing co-operation in the sphere of intellectual, scientific, technological and economic activity. The results of ISP technical work are published as International standards.

These standards and designations are prefixed with the letters ISO. Standards appear as a numerical code, and the designations as an alphanumeric code relating to the composition of the metal or alloy.

Example: ISO 3522; ISO AlMn1Cu; ISO AlZn6MgCu

Iran, Islamic Republic of

Institute of Standards and Industrial Research of Iran (ISIRI)
P. O. Box 31585-163
Karaj
Iran
Tel: 98 261 22 60 31
Fax: 98 261 22 50 15

Ireland

National Sstandards Authority of Ireland (NSAI)
Glasnevin
Dublin-9
Ireland
Tel: 353 1 837 01 01
Fax: 353 1 836 98 21

Israel

Standards Institution of Israel (SII)
42 Chaim Levanon Street
Tel Aviv 69977
Israel
Tel: 972 3 646 51 54
Fax: 972 3 641 96 83
standard@netvision.net.il

Italy

Ente Nazionale Italiano di Unificazione (UNI)
Via Battistotti Sassi 11 b
I-20133 Milano
Italy
Tel: 39 2 70 02 41
Fax: 39 2 70 10 61 06
webmaster@uni.unicei.it
http://www.unicei.it

The Italian National Standards Body (UNI) was founded in 1921, and is a member of both ISO (International Organisation of Standardisation) and CEN (European Committee for Standardisation).

Italian standards are preceded by the upper case letters UNI and followed by an alphanumeric code.

When UNI take over an international standards the upper case letter UNI is followed by ISO or EN and by an alphanumeric code.

Example: UNI 3159, UNI ISO 9000, UNI EN 29000

Jamaica

Jamaica Bureau of Standards (JBS)
6 Winchester Road
P. O. Box 113
Kingston 10
Jamacia
Tel: 1 809 926 31 40-6
Fax: 1 809 929 47 36

Japan

Japanese Industrial Standards Committee (JISC)
c/o Standards Department
Ministry of International Trade and Industry
1-3-1, Kasumigaseki, Chiyoda-ku
Tokyo 100
Japan
Tel: 81 3 35 01 92 95
Fax: 81 3 35 80 14 18
http://www.hike.te.chiba-u.ac.jp

The Japanese Industrial Standards Committee (JISC) issues standards that cover industrial or mineral products with the exception of those regulated by their own special standards organizations. The standards are divided into 17 divisions and are used both by commercial and government organizations involved in design engineering, quality assurance, research and development, construction, testing and maintenance.

JISC standards begin with the upper case letters JIS and are followed by an upper case letter with designates the standard's technology or division. This is then followed by a space and 4 digits.

Example: JIS G 3311; JIS S 20CK

Japan

Japanese Standards Association (JSA)
1-24-4, Akasaka
Minato-ku
Tokyo 107
Japan
Tel: 81 3 3583 8003
Fax: 81 3 3586 2029
http://www.jsa
jp/indexe.html

Kazakhstan

Committee for Standardization, Metrology and Certification (KAZMEMST)
pr. Altynsarina 83
480035 Almaty
Kazakhstan
Tel: 7 327 2 21 08 08
Fax: 7 327 2 28 68 22

Kenya

Kenya Bureau of Standards (KEBS)
Off Mombasa Road
Behind Belle Vue Cinema
P.O. Box 54974
Nairobi
Kenya
Tel: 254 2 50 22 10/19
Fax: 254 2 50 32 93
kebs@arso.gn.apc.org

Korea

Committee for Standardization of the Democratic Peoples's Republic of Korea (CSK)
Zung Gu Yok Seungli-Street
Pyongyang
Dem. P. Rep. of Korea
Tel: 85 02 57 15 76

Korea

Korean National Institute of Technology and Quality (KNITQ)
1599 Kwanyang-dong
Dongan-ku, Anyang-city
Kyonggi-do 430-060
Republic of Korea
Tel: 82 3 43 84 18 61
Fax: 82 3 43 84 60 77

Libyan Arab Jamahiriya

Libyan National Centre for Standardization and Metrology (LNCSM)
Industrial Research
Centre Building
P.O. Box 5178
Tripoli
Libyan Arab Jamahiriya
Tel: 218 21 499 49
Fax: 218 21 69 00 28

Malaysia

Standards and Industrial Research Institute of Malaysia (SIRIM)
Persiaran Dato Menteri, Section 2
P.O. Box 7035, 40911 Shah Alam
Selangor Darul Ehsan
Malaysia
Tel: 60 3 559 26 01
Fax: 60 3 550 80 95
http://www.sirim.my

The Standards and Industrial Institute of Malaysia (SIRIM) is a national multi-disciplinary research and development agency under the Ministry of Science, Technology and the Environment. It was established in 1975 and entrusted with the task of upgrading quality through standards and technical services, assistance and consultancy to industries.

Mauritius

Mauritius Standards Bureau (MSB)
Reduit
Mauritius
Tel: 230 464 7675
      230 454 1933
Fax: 230 464 1144

The Mauritius Standards Bureau (MSB) is a corporate body governed by an Act of parliament, responsible for the development, promotion and coordination of standardization, certification and related activities.

The Bureau has prepared Mauritian Standards (MS), but usually refers to international and foreign specifications for metals.

Mexico

Direccion General de Normas (DGN)
Calle Puente de Tecamachalco No 6
Lomas de Tecamachalco
Seccion Fuentes
Naucalpan de Juarez 53 950
Mexico
Tel: 52 5 729 93 00
Fax: 52 5 729 94 84

The General Directorate of Standards (DGN) issues national standards for the country. Mexican standards begin with the upper case letters NOM (Normas Oficiales Mexicanas). The code that follows consists of an upper case letter which denotes the standard's classification, followed by a hyphen and a number.

Example: NOM C-189

Mongolia

Mongolian National Institute for Standardization and Metrology (MNISM)
Ulaanbaatar-51
Mongolia
Tel: 976 1 35 83 49
Fax: 976 1 35 80 32

Morocco

Service de normalisation industrielle marocaine (SNIMA)
Ministere du commerce, de l'industrie et l'artisanat
Quartier administratif
Rabat Chellah
Morocco
Tel: 212 7 76 37 33
Fax: 212 7 76 62 96

Netherlands

Nederlands Normalisatie-instituut (NNI)
Kalfjeslaan 2
P.O. Box 5059
NL-2600 GB Delft
Netherlands
Tel: 31 15 2 69 03 90
Fax: 31 15 2 69 01 90
[name]@nni.nl

The Netherlands Normalization Institute (NNI) is composed of approximately 3000 individual firms and companies, and 200 various organizations. The association helps prepare Dutch standards and cooperates in the development of international standardization. It is affiliated with American National Standards Institute (ANSI), European Committee for Standardization (CEN), and International Organization for Standardization (ISO).

New Zealand

Standards New Zealand (SNZ)
Standards House
Private Bag 2439
Wellington 6020
New Zealand
Tel: 64 4 498 59 90
Fax: 64 4 498 59 94

Standards New Zealand (SNZ) is the national standards body for New Zealand, is a member of the International Standards Organization (ISO) and International Electromechanical Commission (IEC). It publishes New Zealand Standards (NZS) and Joint Australian/New Zealand Standards (AS/NZS).

Nigeria

Standards Organisation of Nigeria (SON)
Federal Secretariat
Phase 1, 9th Floor
Ikoyi Lagos
Nigeria
Tel: 234 1 68 26 15
Fax: 234 1 68 18 20

Norway

Norges Standardiseringsforbund (NSF)
P.O. Box 7020
Homansbyen
N-0306 Oslo
Norway
Tel: 47 22 46 60 94
Fax: 47 22 46 44 57
marked@norsk-standard.msmail.telemax.no

The Norwegian Standards Association (NSF) is the principal organization for standardization in Norway. The standardizing bodies affiliated with NSF are as follows: The Norwegian Council for Building Standardization - NBR, The Norwegian Electrotechnical Committee - NEK, Norwegian General Standardizating Body - Norsk Allmenstandardisering NAS, Norwegian Telecommunication Regulatory Authority - STF.

The Norwegian Standards Association (NSF) is the national member of ISO and CEN and the body responsible for the approval and publishing of all Norwegian Standards (Norsk Standard - NS)

Example: NS 824; NS 6097; NS 17570

Pakistan

Pakistan Standards Institution (PSI)
39 Garden Road, Saddar
Karachi 74400
Pakistan
Tel: 92 21 772 95 27
Fax: 92 21 772 81 24

Pan America

Pan American Standards Commission (COPANT)
Avenida Andres Bello, Torre Fondo Comun
Piso 11
Caracas 1050
Venezuela
Tel: 58 2 5742941
Fax: 58 2 5742941

The Pan American Standards Commission (COPANT) is comprised of national standards bodies of 18 countries (from the United States and many Latin American countries). COPANT is the Regional Standards Organization for America for the metals industry and other areas such as foods and agricultural products, plastics, quality assurance, etc. For its designations the acronym COPANT in upper case letters precedes the numeric code and the year of its aprobation.

Example: COPANT 1590-1992

Panama

Comision Panamena de Normas Industriales y Tecnicas (COPANIT)
Ministerio de Comercio e Industrias
Apartado Postal 9658
Panama, Zona 4
Panama
Tel: 507 2 27 47 49
Fax: 507 2 25 78 53

Philippines

Bureau of Product Standards (BPS)
Department of Trade and Industry
361 Sen. Gil J. Puyat Avenue
Makati
Metro Manila 1200
Philippines
Tel: 63 2 890 51 29
Fax: 63 2 890 49 26

Poland

Polish Committee for Standardization (PKN)
ul. Elektoralna 2
P.O. Box 411
PL-00-950 Warszawa
Poland
Tel: 48 22 620 54 34
Fax: 48 22 620 07 41

The Polish Committee for Standardization, formerly the Polish Committee for Standardization, Measures and Quality Control, was founded in 1972. It is a national agency and issues standards for that country. The standards are prefixed with the upper case letters PN. The designation or standards may appear in a number of ways.

Example: PN-79/H-88026.

Portugal

Instituto Portugues da Qualidade (IPQ)
Rua C a Avenida do Tres Vales
P-2825 Monte de Caparica
Portugal
Tel: 351 1 294 81 00
Fax: 351 1 294 81 01

Romania

Institutul Roman de Standardizare (IRS)
Str. Jean-Louis Calderon Nr. 13
70201Bucharest 2
Romania
Tel: 40 1 211 32 92
        40 1 615 58 70
Fax: 40 1 210 08 33
        40 1 312 47 44

The Romanian Standards Institute (IRS) was founded in 1948 and it has 140 staff members. They issue no standards of their own but use DIN, AA, STAS, NTR. IRS publishes "Standardizarea" Review and "Buletinul Standardizarii".

Russia

Committee of the Russian Federation for Standardization, Metrology and Certification (GOST R)
Leninsky Prospekt 9
Moskva 117049
Russian Federation
Tel: 7 095 236 40 44
Fax: 7 095 237 60 32

State standards for Russia number more than 23,000 and cover most areas of commerce, industry, agriculture and public health. The standards are defined within groups, i.e. mining minerals, petroleum products, metals and metallic products, etc. The standards are prefaced with the upper case letters GOST and are followed by a numerical code.

Example: GOST 13819; GOST 5.1491; GOST 22974.9

Saudi Arabia

Saudi Arabian Standards Organization (SASO)
Imam Saud Bin Abdul Aziz Bin Mohammed Road (West End)
P.O. Box 3437
Riyadh 11471
Saudia Arabia
Tel: 966 1 452 00 00
Fax: 966 1 452 00 86

The Kingdom of Saudi Arabia established a national standards organization by Royal Decree No. M/10 dated 03/03/1392 (April 16, 1972). SASO is the Saudi organization responsible for all of the activities related to standards and measurements, including the formulation, adoption, publication and distribution of national standards for all commodities and products as well as metrology, symbols, definitions of commodities and products, methods of sampling and testing and any other assignment approved by the Board of Directors. Participating in and cooperating with the Arab, regional and international standards organizations, including ISO and IEC.

Singapore

Singapore Productivity and Standards Board (PSB)
1 Science Park Drive
Singapore 118221
Singapore
Tel: 65 778 77 77
Fax: 65 776 12 80

Slovakia

Slovak Office of Standards, Metrology and Testing (UNMS)
Stefanovicova 3
814 39 Bratislava
Slovakia
Tel: 42 7 49 10 85
Fax: 42 7 49 10 50

Slovenia

Standards and Metrology Institute (SMIS)
Ministry of Science and Technology
Kotnikova 6
SI-61000 Ljubljana
Slovenia
Tel: 386 61 1312 322
Fax: 386 61 314 882
ic@usm.mzt.si
http://www.usm.mzt.si

The Standards and metrology Institute of the Republic of Slovenia operates under the aegis of the Ministry of Science and Technology. SMIS was established in June 1991, when Slovenia was declared an independent state. SMIS prepares, adopts and issues Slovenian standards, while also coordinating the tasks according to the rules of international standardization. Therefore, Slovenian standards are either international or European standards adopted according to the rules of ISO IEC Guide 21.

South Africa

South African Bureau of Standards (SABS)
1 Dr Lategan Rd, Groenkloof
Private Bag X191
Pretoria 0001
South Africia
Tel: 27 12 428 79 11
Fax: 27 12 344 15 68

The South African Bureau of Standards (SABS) was offically established by the South African government in 1945, although work in the area of standardization by other organizations began in the early 1900's. The number of the standard is preceded by the letters SABS and followed by the numeric or alphanumeric material type or grade designation.

Example: SABS 407 Type 1; SABS 1431 Grade 300WA; SABS 1465-2 Grade W4

Spain

Asociacion Espanola de Normalizacion y Certificacion (AENOR)
Fernandez de las Hoz, 52
E-28010 Madrid
Spain
Tel: 34 1 432 60 00
Fax: 34 1 310 49 76

The Spanish Association for Standardization and Certification (AENOR) is an independent organization of a private nature, set up to carry out Standardization and Certification activities, as a tool to improve the quality and competitiveness of products and services. AENOR is designated as a recognized body to develop Standardization and Certification (S+C) activities in Spain.

The designations begin with the letters UNE, representing the Spanish words une normal Espanola.

Sri Lanka

Sri Lanka Standards Institution (SLSI)
53 Dharmapala Mawatha
P.O. Box 17
Colombo 3
Sri Lanka
Tel: 94 1 32 60 51
Fax: 94 1 44 60 18

Sweden

SIS - Standardiseringen i Sverige (SIS)
St Eriksgatan 115
Box 6455
S-113 82 Stockholm
Sweden
Tel: 46 8 610 30 00
Fax: 46 8 30 77 57
info@sis.se

The Swedish Standards Institution (SIS) was founded in 1922 and has a membership of approximately 29 organizations. SIS is a member of the ISO and European Committee for Standardization. SIS is the central body with overall responsibility for standardization in Sweden. SIS is responsible for approval of swedish standards. SIS is also responsible for publishing, marketing and selling of swedish standards and for information, promotion and publicity of swedish standardization.

There are nine independent standardizing bodies affiliated to SIS. Each of them is responsible for drafting standards within its field of activity. Within the field of the metals industry there are three standardizing bodies.

All standards begin with the prefix SS or, if the standard was written prior to 1978, SIS.

Example: SS 11 21 19; SIS 14 01 00

Switzerland

Swiss Association for Standardization (SNV)
Muhlebachstrasse 54
CH-8008 Zurich
Switzerland
Tel: 41 1 254 54 54
Fax: 41 1 254 54 74

The Swiss Association for Standardization (SNV) was founded in 1919. It has about 560 members from associations, firms, and government bodies in Switzerland. It coordinates activities in industrial standardization. SNV is a member of the European Standards Organisation CEN and is working on replacing Swiss standards with those from CEN.

Syrian Arab Republic

Syrian Arab Organization for Standardization and Metrology (SASMO)
P.O. Box 11836
Damascus
Syrian Arab Republic
Tel: 963 11 445 05 38
Fax: 963 11 441 39 13

Tanzania

Tanzania Bureau of Standards (TBS)
P.O. Box 9524
Dar es Salaam
United Republic of Tanzania
Tel: 255 51 4 32 98
Fax: 255 51 4 32 98

Thailand

Thai Industrial Standards Institute (TISI)
Ministry of Industry
Rama VI Street
Bangkok 10400
Thailand
Tel: 66 2 245 78 02
Fax: 66 2 247 87 41

Trinidad

Trinidad and Tobago Bureau of Standards (TTBS)
P. O. Box 467
Port of Spain
Trinidad and Tobago
Tel: 1 809 662 88 27
Fax: 1 809 663 43 35
ttbs@opus-networx.com

Tunisia

Institut national de la normalisation et de la propriete industrielle (INNORPI)
B.P. 23
1012 Tunis-Belvedere
Tunisia
Tel: 216 1 78 59 22
Fax: 216 1 78 15 63

Turkey

Turk Standardlari Enstitusu (TSE)
Necatibey Cad.112
Bakanliklar 06100 Ankara
Turkey
Tel: 90 312 417 83 30
Fax: 90 312 425 43 99
tse-d@servis.net.tr

Founded in 1960, the Turkish Standards Institution (TSE) is a government agency dedicated to the preparation and publication of standards. It is also a member of the ISO. The prefix for Turkish standards are the letters TS. These are followed by a code number, or, in the case of a designation, an alphanumeric code.

Example: TS 2276; TS Mg-Al6Zn1

USA

Aeorspace Materials Specifications (AMS)
SAE International
400 Commonwealth Drive
Warrendale PA 15096-0001
USA
Tel: 412 776 4841
Fax: 412 776 0002/5760

Aerospace Materials Specifications (AMS) or Metric Aerospace Materials Specifications (MAM), are published by SAE, International. AMS and MAM designations pertain to materials intended for aerospace applications; the specifications typically include mechanical property requirements significantly more severe than those for non aerospace applications. Processing requirements are common in AMS steels. Thses specifications are generally used for procurement purposes.

Example: AMS 5356, AMS 5598B, MAM 5598

USA

Aluminum Association (AA)
900 19th St. NW
Ste. 300
Washington DC 20006
USA
Tel: 202 862 5104
Fax: 202 862 5164

The Aluminum Association, founded in 1933, has 75 members who are producers of aluminum and manufacturers of semi-fabricated aluminum products. The association is a primary source of statistics, standards and economic and technical information on aluminum and the aluminum industry in the United States.

Standards include a prefix which begins with AA.

USA

American National Standards Institute (ANSI)
11 West 42nd St.
13th Floor
New York N.Y. 10036
USA
Tel: 1 212 642 4900
Fax: 1 212 398 0023
info@ansi.org

The American National Standards Institute (ANSI) standards are used widely throughout industry. They cover a tremendous variety of items, from architectural products to consumer goods to nuclear safety standards. The Institute is the coordinator of the United States voluntary standards system and assists participants in the voluntary system to reach agreement on standards needs and priorities; arranging for competent organizations to undertake standards development work; providing fair and effective procedures for standards development; and resolving conflicts and preventing duplication of efforts.

ANSI standards shall be identified by a unique alphanumeric designation in accordance with the following: a) A designation assigned by the standards developer and adopted by ANSI for all new, revised, and reaffirmed standards. For example: ANSI/IEEE 123-1982; b) Standards developed by an Accredited Standards Committee (ASC) shall carry the committee designation. For example: ANSI X3.1-1982.

USA

American Petroleum Institute (API)
Publications and Distribution Section
1220 L Street N.W.
Washington DC 20005
USA
Tel: 202 682 8000
Fax: 202 962 4776

The American Petroleum Institute (API) fosters the development of standards, codes, and safe practices within the petroleum industries. These standards and codes are used by persons involved in the engineering, production, transportation, handling, and use of petroleum products. The API Standards appear with the letters APS before the specifications.

Example: API Spec 5AC; API Spec 5L

USA

American Society for Testing and Materials (ASTM)
100 Barr Harbor Drive
West Conshohocken PA 19428
USA
Tel: 610 832 9500
Fax: 610 832 9555
service@local.astm.org

The American Society for Testing and Materials (ASTM), founded in 1898, is a scientific and technical organization formed for the development of standards on characteristics and performance of materials, products, systems, and services. The organization issues the most widely used--in the United States--standard specifications for steel products, many of which are complete and generally adequate for procurement purposes. These frequently apply to specific products, which are generally oriented toward the performance of the fabricated end product.

ASTM is the world's largest source of voluntary "consensus" standards. That is, its documents represent a consensus drawn from producers, specifiers, fabricators, and users of metal products. In many cases, the dimensions, tolerances, limits and restrictions in the ASTM specifications are the same as corresponding items of the standard practices in the AISI Steel Products Manuals.

Many of the ASTM specifications have been adopted by the American Society of Mechanical Engineers (ASME) with little or no modification; ASME uses the prefix "S" along with the ASTM designation for these specifications. For example, ASME SA-213 and ASTM A213 are identical.

All ASTM standards begin with the prefix ASTM, followed by the actual standard code number.

Example: ASTM A311: ASTM A372 Class V Type B; ASTM A723 Grade 1 Class 1; ASTM A336 Grade F31

USA

American Society of Mechanical Engineers (ASME)
Codes and Standards Department
345 East 47th Street

New York NY 10017
USA
Tel: 212 605 333
Fax: 212 605 8750

The American Society of Mechanical Engineers (ASME) standards are used by personnel in research, testing, and design of power-producing machines such as internal combustion engines, steam and gas turbines, and jet and rocket engines. They are also used for the design and development of power-using machines such as refrigeration and air-conditioning equipment, elevators, machine tools, printing presses and, steel-rolling mills. ASME committees report to five different boards covering the following areas respectively: safety, nuclear, pressure technology, dimensional standards and performance test codes. The upper case letters ASME appear at the left of the specification followed by an alpha numeric code. When referenced by other ASME standards, standards dated prior to 1989 which carry ANSI in the designation, should be shown as the current ones (i.e. ASME followed by the alpha-numeric designation given by ANSI) followed by the title and the ANSI designation in parentheseis.

Example: ASME B16.14-1991; ASME SA194

USA

American Welding Society (AWS)
550 N. W. Lejeune Road
P.O. Box 351040
Miami FL 33135
USA
Tel: 305 443 9353
    800 443 9353
Fax: 305 443 7559

The American Welding Society (AWS) standards are used to support welding design, fabrication, testing, quality assurance and other related joining functions found in shipbuilding (design/construction), heavy construction, and a wide variety of other industries.

These standards always begin with the upper case letters AWS.

Example: AWS A5.24; AWS C5.7; AWS B4.0

USA

Copper Development Association (CDA)
260 Madison Ave.
New York NY 10016
USA
Tel: 212 251 7200
Fax: 212 251 7234

The Copper Development Association (CDA) was founded in 1963 to help expand the uses and applications of copper and copper products. The organization has also developed a standards system for wrought and cast coppers, and for copper alloy products. CDA's standards for copper alloys are identical to those numbers used by UNS designations.

All CDA standards begin with the upper case letter C, which is followed by 5 digits.

Example: CDA C52400

USA

Defense Printing Service
700 Robbins Avenue
Bldg. 4, Section D
Attn: Standardization Document Order Desk
Philadelphia PA 19111-5094
USA
Tel: 215 697 2179
      215 697 2667
Fax: 215 697 2978

The Department of Defense Single Stock Point (DODSSP) was created to centralize control and distribution, and provide access to extensive technical information within the collection of Military and Federal Specifications and standards and related documents produced or adopted by the DOD. The DODSSP mission was assumed by the Defense Printing Service in October 1990.

Military specifications (MIL) are issued by the United States Department of Defense (DOD) to define materials, products, or services used only or predominantly by military entities. Military standards provide procedures for design, manufacturing, and testing, rather than giving a particular material description.

All military specifications begin with the upper case letters MIL. The actual specification that follows begins with an upper case code letter that represents the first letter of the title for the item, followed immediately by a hyphen and then the serial number of digits.

Federal (QQ) specifications and standards are similar to the military, except they are issued by the General Services Administration (GSA) and are primarily for use by federal agencies. Their use, however, is now acceptable to the United States military establishment when there are no separate MIL specifications available.

Federal specifications begin with the upper case letters QQ followed by the code numbers and letters.

Example: MIL S-862 (military); FED QQ-S-763 (federal)

Both military and federal standards and specifications can be obtained through this address or by faxing the request on company letterhead.

USA

Electronic Industries Association/Institute for Interconnecting and Packaging Electronic Circuits (EIA)
The Institute for Interconnecting and Packaging Electronic Circuits (IPC)
7380 N. Lincoln Avenue
Lincolnwood IL 60646
USA
Tel: 708 677 2850
Fax: 708 677 9570

EIA and IPC Standards and Publications are designed to serve the public interest through eliminating misunderstandings between manufactures and purchasers, facilitating interchangeability and improvement of products and assisting the purchaser in selecting and obtaining with minimum delay the proper product for his particular need. Existence of such Standards and Publications shall not in any respect preclude any member or nonmenber of EIA or IPC from manufacturing or selling products not conforming to such Standards and Publications , nor shall the existence of such Standards and Publication preclude their voluntary use by those other than EIA or IPC members, whether the standard is to be used either domestically or internationally.

Recommended Standards and Publications are adopted by EIA and IPC without regard to whether their adoption may involve patents on articles, materials, or processes. By such action, EIA and IPC do not assume and liability to any patent owner, nor do they assume any obligation whatever to parties adopting the Recommended Standard or Publication. Users are also wholly responsible for protecting themsleves against all claims of liabilities for patent infrigement.

Example: EIA/IPC J-STD-006

USA

National Center for Standards and Certification Information
National Institute of Standards & Technology (NIST)
TRF Building, Room A163
Gaithersburg MD 20899
USA
Tel: 301 974 4040
Fax: 301 926 1559
http://www.nist.gov or {name}@nist.gov

The National Institute of Standards & Technology (NIST) is the centralized reference repository within the United States for the national standards of the world. Although the organization issues no standards or specifications, it attempts to maintain the most current copies of those standards issued by other organizations both inside and outside of the United States.

Two publications relating to standards: Directory of International and Regional Organizations Conducting Standards-Related Activities and Standards Activities of Organizations in the United States, are available from NIST.

USA

SAE, International (SAE)
400 Commonwealth Dr.
Warrendale PA 15096-0001
USA
Tel: 412 776 4841
Fax: 412 776 5760

The Society of Automotive Engineers (SAE) standards are used primarily by designers, manufacturers, and maintenance personnel in the automotive and aerospace industries. Thses standards are also a useful and effective series for the metals, plastics, rubber, chemical, and fastener industries in their standardization efforts. Automotive SAE standards begin with the upper case letters SAE. Immediately after this prefix the letter J appears, and it is followed by a numberical code. Prefixes for other standards vary, such as AS, AIR, and ARP with a numerical code.

Example: SAE J450; SAE J993b; ARP 111

Ukraine

State Committee of Ukraine for Standardization, Metrology and Certification (DSTU)
174 Gorky St.
GSP, Kiev-6,252650
Ukraine
Tel: 380 44 226 29 71
Fax: 380 44 226 29 70

United Kingdom

British Standards Institution (BSI)
2 Park Street
London W1A 2BS
England
Tel: 44 71 629 9000
Fax: 44 71 629 0506

Founded in 1901, the British Standards Institution (BSI) develops and publishes standards that are used extensively by exporters and importers. They are used both in government and industry by those who are involved in engineering, designing, production, testing and construction. The letters BS precede the standard's numerical code and may also include the alloy's designation.

Example: BS 3100; BS EN 10083-1

Uruguay

Instituto Uruguayo de Normas Tecnicas (UNIT)
San Jose 1031 P. 7
Galeria Elysee Montevideo
Uruguay
Tel: 598 2 91 20 48
Fax: 598 2 92 16 81

Uzbekistan

Uzbek State Centre for Standardization, Metrology and Certification (UZGOST)
Ulista Farobi, 333-A
700049 Tachkent
Uzbekistan
Tel: 7 371 246 17 10
Fax: 7 371 2 46 17 11

Venezuela

Comision Venezolana de Normas Industriales (COVENIN)
Avda. Andres Bello-Edf. Torre Fondo
Comun
Piso 12
Caracas 1050
Venezuela
Tel: 58 2 575 22 98
Fax: 58 2 574 13 12
covenin@dino.conicit.ve

COVENIN, created in 1958, is the Organization in charge of directing, planning and coordinating all the standardization activities in Venezuela. A series of measures of legal order culminated with the law on "Technical Standards and Quality Control" on December 31, 1979. This law represents the consolidation, from the legal view point, of the standardization and quality control process in Venezuela.

COVENIN coordinates the elaboration of the technical standards through committees and sub-committees dependent on COVENIN which include: Construction, Petroleum and Derivatives, Automotive, Ferrous Materials, Non-Ferrous Materials, Denistry, Electricity and Electronics, Chemistry, Metrology, Documentation, Containers, and Packing, Maintenance, Mechanics, Non-Destructive Testing, and Quality. COVENIN has approved over 2800 technical standards; 300 of them having been declared compulsory by the National Government.

Viet Nam

Directorate for Standards and Quality (TCVN)
70, Tran Hung Dao St.
Hanoi
Viet Nam
Tel: 84 4 26 62 20
Fax: 84 4 26 74 18

The Directorate for Standards and Quality is the The Governmental Body under the Ministry of Science, Technology and Enviroment having responsibility to advise the Government on issues in the fields of standardization, metrology and quality management in the country and representing Vietnam in internatinal and regional organizations in the fields concerned.

Yugoslavia (former)

> Savezni zavod za standardizaciju (SZS)
> Kneza Milosa 20
> Post Pregr. 933
> YU-11000 Beograd
> Yugoslavia
> Tel: 381 11 64 35 57
> Fax: 381 11 68 23 82
> etanasko@ubbg.etf.bg.ac.yu

The Yugoslavian Standardization Institute (SZS) was founded in 1946 and is concerned with the adoption and application of standards, technical norms for product quality and services, and regulations covered by legislation. Yugoslavian standards begin with the prefix JUS which is followed by an alphanumeric code. The first letter of the code denotes the section under which the standard is classified. Most standards relating to metallurgy are in section C.

Example: JUS C.AO.003; JUS C.K6.150; JUS C.T3.005

Zimbabwe

> Standards Association of Zimbabwe (SAZ)
> P.O. Box 2259
> Harare
> Zimbabwe
> Tel: 263 4 88 34 46
> Fax: 263 4 88 20 20

The Standards Association of Zimbabwe (SAZ) is a national organization whose membership is comprised of research and testing organizations, government agencies, consumers, and representatives of manufacturing. SAZ issues national standards governing products manufactured in Zimbabwe and conducts research and educational programs.

Trademarks:

EATONITE	Eaton Corporation	MONEL	Inco Alloys International
HASTELLOY	Haynes International	NICHROME	Driver Harris
HAYNES	Haynes International	NIMONIC	Inco Alloys International
INCO	Inco Alloys International	PROMET	American Crucible Products
INCONEL	Inco Alloys International	PYROMET	Carpenter Technology
MAR-M	Cannon-Muskegon	UDIMET	Special Metals

Designation Index

Worldwide Guide to Equivalent Nonferrous Metals and Alloys

Cross Reference to Standards Organizations in Section 14:

Prefix	Organization
AECMA	Europe, Association Europeenne des Constructeurs de Materiel Aerospatial
AIR	France, Delegation Generale pour L'Armement
AMS	USA, Aerospace Materials Specifications
ANSI	USA, American National Standards Institute
AS	Australia, Standards Australia
ASME	USA, American Society of Mechanical Engineers
ASTM	USA, American Society for Testing and Materials
AWS	USA, American Welding Society
BS	United Kingdom, British Standards Institution
CEN	Europe, European Committee for Standardization
COPANT	Pan America, Pan American Standards Commission
CSA	Canada, Canadian Standards Association
CSN	Czech Republic, Czech Office for Standards, Metrology and Testing
DGN	Mexico, Direccion General de Normas
DIN	Germany, Deutsches Institut fur Normung
DS	Denmark, Dansk Standard
DTD	United Kingdom, British Standards Institution
EIA/IPC	USA, Electronic Industries Association, Institute for Interconnecting and Packaging Electronic Circuits
GB	China, China State Bureau of Technical Supervision
GOST	Russia, Committee of the Russian Federation for Standardization, Metrology and Certification
IS	India, Bureau of Indian Standards
ISO	International, International Organization for Standardization
JIS	Japan, Japanese Industrial Standards Committee
JUS	Yugoslavia (former), Savezni zavod za standardizaciju
MIL	USA, Defense Printing Service
MNC	Sweden, Standardisering en I Sverige
NBN	Belgium, Institut Belge de Normalisation
NF	France, Association Francaise de Normalisation
NOM	Mexico, Direccion General de Normas
NS	Norway, Norges Standardisering sforbund
ONORM	Austria, Osterreichisches Normunginstitut
QQ	USA, Defense Printing Service
RWMA	Unknown, contact ANSI
SABS	South Africa, South African Bureau of Standards
SAE	USA, SAE International
SFS	Finland, Finnish Standard Association
SIS	Sweden, Standardisering en I Sverige
UNE	Spain, Asociacion Espanola de Normalizacion y Certificacion
VSM	Switzerland, Swiss Association for Standardization
WW	Unknown, contact ANSI

Prefix **Organization**

Specification Index